Demtröder

**Atoms, Molecules and Photons**

# GRADUATE TEXTS IN PHYSICS

Graduate Texts in Physics publishes core learning/teaching material for graduate- and advanced-level undergraduate courses on topics of current and emerging fields within physics, both pure and applied. These textbooks serve students at the MS- or PhD-level and their instructors as comprehensive sources of principles, definitions, derivations, experiments and applications (as relevant) for their mastery and teaching, respectively. International in scope and relevance, the textbooks correspond to course syllabi sufficiently to serve as required reading. Their didactic style, comprehensiveness and coverage of fundamental material also make them suitable as introductions or references for scientists entering, or requiring timely knowledge of, a research field.

Wolfgang Demtröder

# Atoms, Molecules and Photons

## An Introduction to Atomic-, Molecular- and Quantum-Physics

With 677 Figures and 42 Tables

Professor Dr. Wolfgang Demtröder
Universität Kaiserslautern
FB Physik
Erwin-Schrödinger-Str. 46
67663 Kaiserslautern
Germany
demtroed@physik.uni-kl.de



ISSN 1868-4513 ISSN 1868-4521 (electronic)
ISBN 978-3-642-10297-4 ISBN 978-3-642-10298-1 (eBook)
DOI 10.1007/978-3-642-10298-1
Springer Heidelberg Dordrecht London New York

Library of Congress Control Number: 2010935944



*Cover design*: Integra Software Services Pvt. Ltd., Pondicherry

Printed on acid-free paper

Springer is part of Springer Science+Business Media (www.springer.com)

# Preface to the Second Edition

The first edition of this textbook had found a friendly acceptance. This second edition does not change the concept of the representation which combines the experimental techniques for the investigation of atoms and molecules and their results that have lead to the development of quantum physics. Some new developments in laser physics and quantum optics have been inserted in Chapter 12 in order to give the reader some ideas about the frontiers in these fields regarding experimental techniques and physical insight.

This second edition represents a thoroughly revised version of the first edition, which, unfortunately, contained a lot of errors and misprints. I am grateful to many readers who have informed me about such errors and who offered corrections and improvements of the representation for a better understanding. I am particularly indebted to Dr. Nico Dam, Raboud University Nijmegen, Netherland and Prof. Zamik Rosenwaks, Ben Gurion University, Israel, who have sent me an extensive correction list.

The author hopes, that this new edition will be well accepted and that critical readers will send their comments or ideas about possible improvements.

I thank Dr. Th. Schneider, Springer Verlag for his continuous interest and encouragement.

Kaiserslautern,
October 2010

*Wolfgang Demtröder*

# Preface to the First Edition

The detailed understanding of matter, its phase transitions and its interaction with radiation could be only reached, after its microscopic structure determined by the kind of atoms or molecules as basic constituents of matter had been investigated. This knowledge allowed the controlled optimization of characteristic properties of matter. Atomic physics therefore represents not only an area of important fundamental research, but has furthermore many applications which have essentially formed our present technical world. The understanding of materials and their use in daily life, has major impact of our culture and our attitude towards nature and our environment.

This textbook is aimed as an introduction to the microscopic world of atoms, molecules and photons. It illustrates how our knowledge about the microscopic structure of matter and radiation came about and which crucial experiments forced an extension and refinement of existing classical theories, culminating in the development of quantum theory, which is now accepted as the basic theory of atomic and molecular physics.

The book therefore starts with a short historical review about the role of experiments for correcting erroneous ideas and proving the existence of atoms and molecules. The close interaction between experiments and theory has been one of the reasons for the rapid development of atomic physics in the $19^{\text{th}}$ and $20^{\text{th}}$ centuries. Examples are the kinetic theory of gases, which could be completely understood by the assumption of moving atoms in the gas, or the postulation of energy quanta in the radiation field, which could explain the discrepancy between measurements of the spectral energy distribution of thermal radiation fields and classical electrodynamics.

The new ideas of quantum physics and their corroboration by experiments are discussed in Chap. 3 while the fundamental equations of quantum mechanics and their applications to some simple examples are explained in Chap. 4.

A theory can be best understood by applications to a real situation. In Chap. 5 the quantum theory of the simplest real system, namely the hydrogen atom, is presented. Here it is again illustrated, that experiments enforced an extension of quantum mechanics to quantum electrodynamics in order to understand all experimental results. The description of larger atoms with many electrons is treated in Chap. 6, which also reduces the chemical properties of chemical elements to the structure of the electron shells and explains why all elements can be arranged in a periodic table.

The important subject of interaction of matter with radiation is discussed in Chap. 7. This prepares the ground for the explanation of lasers, treated in Chap. 8.

Molecules, consisting of two or more atoms, form the basis for the great variety of our world. They are discussed in Chaps. 9 and 10. In particular the question, why and how atoms can form stable molecules, and which kind of interaction occurs, is treated in more detail. In Chap. 11 the different experimental techniques for the investigation of atoms and molecules are presented, in order to give the reader a feeling for the inventive ideas and the necessary experimental skill for their realization. The last chapter presents a short overview on recent developments in atomic and molecular physics, which shall demonstrate that physics will be never a complete and finalized field. There is still much to explore and new ideas and scientific enthusiasm is needed, to push the border of our knowledge further ahead. Some examples in this chapter also illustrate possible important applications of new ideas such as the quantum computer or new techniques of frequency metrology used in the world wide global positioning system GPS.

Many people have helped to publish this book. First of all I would like to thank the team of LE-TeX, who have made the layout. In particular Uwe Matrisch, who has looked after the editing process and who has taken care of many handwritten remarks and corrections of the author with great patience. Dr. Schneider from Springer-Verlag has always supported this project, although it took longer as anticipated.

Many thanks go to all colleagues who have given their permission to reproduce figures or tables.

This book is an extended version of volume 3 of a German textbook consisting of 4 volumes. The authors hopes, that it will find a comparable good acceptance as the German version. He will be grateful for any reply of readers, giving corrections of possible errors or hints to improvements. Any of such replies will be answered as soon as possible. A textbook lives from the active collaboration of its readers and the author looks foreward to a lively correspondence with his readers. He hopes that this book can contribute to a better understanding of this fascinating field of atoms, molecules and photons.

Kaiserslautern,
August 2005

*Wolfgang Demtröder*

# Contents


## 1. Introduction

1.1 Contents and Importance of Atomic Physics . . . . . . . . . . . 1
1.2 Molecules: Building Blocks of Nature . . . . . . . . . . . . . . . 3
1.3 Survey on the Concept of this Textbook . . . . . . . . . . . . . 4

## 2. The Concept of the Atom

2.1 Historical Development . . . . . . . . . . . . . . . . . . . . . . . . 7
2.2 Experimental and Theoretical Proofs for the Existence of Atoms . . . . . . . . . . . . . . . . . . . . . . . . . . . . . . 9
2.2.1 Dalton's Law of Constant Proportions . . . . . . . . . 9
2.2.2 The Law of Gay-Lussac and the Definition of the Mole . . . . . . . . . . . . . . . . . . . . . . . . 11
2.2.3 Experimental Methods for the Determination of Avogadro's Constant . . . . . . . . . . . . . . . . . . 12
2.2.4 The Importance of Kinetic Gas Theory for the Concept of Atoms . . . . . . . . . . . . . . . . 17
2.3 Can One See Atoms? . . . . . . . . . . . . . . . . . . . . . . . . . 20
2.3.1 Brownian Motion . . . . . . . . . . . . . . . . . . . . . . . 20
2.3.2 Cloud Chamber . . . . . . . . . . . . . . . . . . . . . . . . 24
2.3.3 Microscopes with Atomic Resolution . . . . . . . . . . 24
2.4 The Size of Atoms . . . . . . . . . . . . . . . . . . . . . . . . . . . 29
2.4.1 The Size of Atoms in the Van der Waals Equation . . . . . . . . . . . . . . . . . . . . . . . . . . . . 29
2.4.2 Atomic Size Estimation from Transport Coefficients . . . . . . . . . . . . . . . . . . . . . . . . . 29
2.4.3 Atomic Volumes from X-Ray Diffraction . . . . . . . . 31
2.4.4 Comparison of the Different Methods . . . . . . . . . . 32
2.5 The Electric Structure of Atoms . . . . . . . . . . . . . . . . . . 33
2.5.1 Cathode Rays and Kanalstrahlen . . . . . . . . . . . . 34
2.5.2 Measurement of the Elementary Charge *e* . . . . . . . 35
2.5.3 How to Produce Free Electrons . . . . . . . . . . . . . 37
2.5.4 Generation of Free Ions . . . . . . . . . . . . . . . . . . 39
2.5.5 The Mass of the Electron . . . . . . . . . . . . . . . . . 41
2.5.6 How Neutral is the Atom? . . . . . . . . . . . . . . . . . 44
2.6 Electron and Ion Optics . . . . . . . . . . . . . . . . . . . . . . . 45
2.6.1 Refraction of Electron Beams . . . . . . . . . . . . . . 45
2.6.2 Electron Optics in Axially Symmetric Fields . . . . . . 47

2.6.3 Electrostatic Electron Lenses . . . . . . . . . . . . . . 49
2.6.4 Magnetic Lenses . . . . . . . . . . . . . . . . . . . . . . 50
2.6.5 Applications of Electron and Ion Optics . . . . . . . . 52
2.7 Atomic Masses and Mass Spectrometers . . . . . . . . . . . . . 53
2.7.1 J.J. Thomson's Parabola Spectrograph . . . . . . . . . 54
2.7.2 Velocity-Independent Focusing . . . . . . . . . . . . . 55
2.7.3 Focusing of Ions with Different Angles of Incidence . . . . . . . . . . . . . . . . . . . . . . . 57
2.7.4 Mass Spectrometer with Double Focusing . . . . . . . 57
2.7.5 Time-of-Flight Mass Spectrometer . . . . . . . . . . . 58
2.7.6 Quadrupole Mass Spectrometer . . . . . . . . . . . . . 61
2.7.7 Ion-Cyclotron-Resonance Spectrometer . . . . . . . . 63
2.7.8 Isotopes . . . . . . . . . . . . . . . . . . . . . . . . . . 64
2.8 The Structure of Atoms . . . . . . . . . . . . . . . . . . . . . 65
2.8.1 Integral and Differential Cross Sections . . . . . . . . . 65
2.8.2 Basic Concepts of Classical Scattering . . . . . . . . . 66
2.8.3 Determination of the Charge Distribution within the Atom from Scattering Experiments . . . . . 70
2.8.4 Thomson's Atomic Model . . . . . . . . . . . . . . . . . 71
2.8.5 The Rutherford Atomic Model . . . . . . . . . . . . . . 73
2.8.6 Rutherford's Scattering Formula . . . . . . . . . . . . 74
Summary . . . . . . . . . . . . . . . . . . . . . . . . . . . . . . . . 77
Problems . . . . . . . . . . . . . . . . . . . . . . . . . . . . . . . . 79

**3. Development of Quantum Physics**

3.1 Experimental Hints to the Particle Character of Electromagnetic Radiation . . . . . . . . . . . . . . . . . . . 81
3.1.1 Blackbody Radiation . . . . . . . . . . . . . . . . . . . . 82
3.1.2 Cavity Modes . . . . . . . . . . . . . . . . . . . . . . . . 84
3.1.3 Planck's Radiation Law . . . . . . . . . . . . . . . . . . 86
3.1.4 Wien's Law . . . . . . . . . . . . . . . . . . . . . . . . . 88
3.1.5 Stefan–Boltzmann's Radiation Law . . . . . . . . . . . 88
3.1.6 Photoelectric Effect . . . . . . . . . . . . . . . . . . . . 89
3.1.7 Compton Effect . . . . . . . . . . . . . . . . . . . . . . . 91
3.1.8 Properties of Photons . . . . . . . . . . . . . . . . . . . 93
3.1.9 Photons in Gravitational Fields . . . . . . . . . . . . . 94
3.1.10 Wave and Particle Aspects of Light . . . . . . . . . . 95
3.2 Wave Properties of Particles . . . . . . . . . . . . . . . . . . . 97
3.2.1 De Broglie Wavelength and Electron Diffraction . . . . 97
3.2.2 Diffraction and Interference of Atoms . . . . . . . . . . 98
3.2.3 Bragg Reflection and the Neutron Spectrometer . . . . . 100
3.2.4 Neutron and Atom Interferometry . . . . . . . . . . . . 100
3.2.5 Application of Particle Waves . . . . . . . . . . . . . . 101
3.3 Matter Waves and Wave Functions . . . . . . . . . . . . . . . . 102
3.3.1 Wave Packets . . . . . . . . . . . . . . . . . . . . . . . . 103
3.3.2 The Statistical Interpretation of Wave Functions . . . . 105
3.3.3 Heisenberg's Uncertainty Principle . . . . . . . . . . . 106

3.3.4 Dispersion of the Wave Packet . . . . . . . . . . . . . . 109
3.3.5 Uncertainty Relation for Energy and Time . . . . . . . 110
3.4 The Quantum Structure of Atoms . . . . . . . . . . . . . . . . . 111
3.4.1 Atomic Spectra . . . . . . . . . . . . . . . . . . . . . . . . 112
3.4.2 Bohr's Atomic Model . . . . . . . . . . . . . . . . . . . . 113
3.4.3 The Stability of Atoms . . . . . . . . . . . . . . . . . . . 117
3.4.4 Franck–Hertz Experiment . . . . . . . . . . . . . . . . . 118
3.5 What are the Differences Between Classical and Quantum Physics? . . . . . . . . . . . . . . . . . . . . . 120
3.5.1 Classical Particle Paths Versus Probability Densities in Quantum Physics . . . . . . . . . . . . . . 120
3.5.2 Interference Phenomena with Light Waves and Matter Waves . . . . . . . . . . . . . . . . . . . . 121
3.5.3 The Effect of the Measuring Process . . . . . . . . . . . 123
3.5.4 The Importance of Quantum Physics for our Concept of Nature . . . . . . . . . . . . . . . . 124
Summary . . . . . . . . . . . . . . . . . . . . . . . . . . . . . . . . 125
Problems . . . . . . . . . . . . . . . . . . . . . . . . . . . . . . . . 127

**4. Basic Concepts of Quantum Mechanics**

4.1 The Schrödinger Equation . . . . . . . . . . . . . . . . . . . . . 129
4.2 Some Examples . . . . . . . . . . . . . . . . . . . . . . . . . . . 131
4.2.1 The Free Particle . . . . . . . . . . . . . . . . . . . . . . . 131
4.2.2 Potential Barrier . . . . . . . . . . . . . . . . . . . . . . . 132
4.2.3 Tunnel Effect . . . . . . . . . . . . . . . . . . . . . . . . . 135
4.2.4 Particle in a Potential Box . . . . . . . . . . . . . . . . . 138
4.2.5 Harmonic Oscillator . . . . . . . . . . . . . . . . . . . . . 141
4.3 Two-and Three-Dimensional Problems . . . . . . . . . . . . . 144
4.3.1 Particle in a Two-dimensional Box . . . . . . . . . . . . 144
4.3.2 Particle in a Spherically Symmetric Potential . . . . . . 145
4.4 Expectation Values and Operators . . . . . . . . . . . . . . . . 149
4.4.1 Operators and Eigenvalues . . . . . . . . . . . . . . . . . 150
4.4.2 Angular Momentum in Quantum Mechanics . . . . . . 152
Summary . . . . . . . . . . . . . . . . . . . . . . . . . . . . . . . . 155
Problems . . . . . . . . . . . . . . . . . . . . . . . . . . . . . . . . 157

**5. The Hydrogen Atom**

5.1 Schrödinger Equation for One-electron Systems . . . . . . . . . 159
5.1.1 Separation of the Center of Mass and Relative Motion . . . . . . . . . . . . . . . . . . . . 159
5.1.2 Solution of the Radial Equation . . . . . . . . . . . . . . 161
5.1.3 Quantum Numbers and Wave Functions of the H Atom . . . . . . . . . . . . . . . . . . . . . . . 163
5.1.4 Spatial Distributions and Expectation Values of the Electron in Different Quantum States . . . . . . . 166

5.2 The Normal Zeeman Effect . . . . . 168
5.3 Comparison of Schrödinger Theory with Experimental Results . . 170
5.4 Relativistic Correction of Energy Terms . . . . . 172
5.5 The Electron Spin . . . . . 174
5.5.1 The Stern–Gerlach Experiment . . . . . 175
5.5.2 Experimental Confirmation of Electron Spin . . . . . 176
5.5.3 Einstein–de Haas Effect . . . . . 177
5.5.4 Spin-Orbit Coupling and Fine Structure . . . . . 178
5.5.5 Anomalous Zeeman Effect . . . . . 181
5.6 Hyperfine Structure . . . . . 184
5.6.1 Basic Considerations . . . . . 184
5.6.2 Fermi-contact Interaction . . . . . 186
5.6.3 Magnetic Dipole-Dipole Interaction . . . . . 187
5.6.4 Zeeman Effect of Hyperfine Structure Levels . . . . . 187
5.7 Complete Description of the Hydrogen Atom . . . . . 188
5.7.1 Total Wave Function and Quantum Numbers . . . . . 188
5.7.2 Term Assignment and Level Scheme . . . . . 188
5.7.3 Lamb Shift . . . . . 191
5.8 Correspondence Principle . . . . . 194
5.9 The Electron Model and its Problems . . . . . 195
Summary . . . . . 198
Problems . . . . . 200

**6. Atoms with More Than One Electron**

6.1 The Helium Atom . . . . . 201
6.1.1 Approximation Models . . . . . 202
6.1.2 Symmetry of the Wave Function . . . . . 203
6.1.3 Consideration of the Electron Spin . . . . . 204
6.1.4 The Pauli Principle . . . . . 205
6.1.5 Energy Levels of the Helium Atom . . . . . 206
6.1.6 Helium Spectrum . . . . . 208
6.2 Building-up Principle of the Electron Shell for Larger Atoms . . 209
6.2.1 The Model of Electron Shells . . . . . 209
6.2.2 Successive Building-up of Electron Shells for Atoms with Increasing Nuclear Charge . . . . . 210
6.2.3 Atomic Volumes and Ionization Energies . . . . . 212
6.2.4 The Periodic System of the Elements . . . . . 216
6.3 Alkali Atoms . . . . . 218
6.4 Theoretical Models for Multielectron Atoms . . . . . 221
6.4.1 The Model of Independent Electrons . . . . . 221
6.4.2 The Hartree Method . . . . . 222
6.4.3 The Hartree–Fock Method . . . . . 224
6.4.4 Configuration Interaction . . . . . 224
6.5 Electron Configurations and Couplings of Angular Momenta . . . 224
6.5.1 Coupling Schemes for Electronic Angular Momenta . . . . . 224
6.5.2 Electron Configuration and Atomic States . . . . . 229

6.6 Excited Atomic States . . . . . 231
6.6.1 Single Electron Excitation . . . . . 232
6.6.2 Simultaneous Excitation of Two Electrons . . . . . 232
6.6.3 Inner-Shell Excitation and the Auger Process . . . . . 233
6.6.4 Rydberg States . . . . . 234
6.6.5 Planetary Atoms . . . . . 236
6.7 Exotic Atoms . . . . . 237
6.7.1 Muonic Atoms . . . . . 238
6.7.2 Pionic and Kaonic Atoms . . . . . 239
6.7.3 Anti-hydrogen Atoms and Other Anti-atoms . . . . . 240
6.7.4 Positronium and Muonium . . . . . 241
Summary . . . . . 243
Problems . . . . . 245

**7. Emission and Absorption of Electromagnetic Radiation by Atoms**

7.1 Transition Probabilities . . . . . 248
7.1.1 Induced and Spontaneous Transitions, Einstein Coefficients . . . . . 248
7.1.2 Transition Probabilities, Einstein Coefficients and Matrix Elements . . . . . 250
7.1.3 Transition Probabilities for Absorption and Induced Emission . . . . . 253
7.2 Selection Rules . . . . . 253
7.2.1 Selection Rules for Spontaneous Emission . . . . . 253
7.2.2 Selection Rules for the Magnetic Quantum Number . . 254
7.2.3 Parity Selection Rules . . . . . 255
7.2.4 Selection Rules for Induced Absorption and Emission . 256
7.2.5 Selection Rules for the Spin Quantum Number . . . . . 256
7.2.6 Higher Order Multipole Transitions . . . . . 257
7.2.7 Magnetic Dipole Transitions . . . . . 259
7.2.8 Two-Photon-Transitions . . . . . 259
7.3 Lifetimes of Excited States . . . . . 260
7.4 Line Profiles of Spectral Lines . . . . . 261
7.4.1 Natural Linewidth . . . . . 262
7.4.2 Doppler Broadening . . . . . 264
7.4.3 Collision Broadening . . . . . 267
7.5 X-Rays . . . . . 270
7.5.1 Bremsstrahlung . . . . . 271
7.5.2 Characteristic X-Ray-Radiation . . . . . 272
7.5.3 Scattering and Absorption of X-Rays . . . . . 273
7.5.4 X-ray Fluorescence . . . . . 278
7.5.5 Measurements of X-Ray Wavelengths . . . . . 278
7.6 Continuous Absorption and Emission Spectra . . . . . 280
7.6.1 Photoionization . . . . . 281
7.6.2 Recombination Radiation . . . . . 284
Summary . . . . . 286
Problems . . . . . 287

## 8. Lasers


8.1 Physical Principles . . . 289
8.1.1 Threshold Condition . . . 290
8.1.2 Generation of Population Inversion . . . 292
8.1.3 The Frequency Spectrum of Induced Emission . . . 295
8.2 Optical Resonators . . . 295
8.2.1 The Quality Factor of Resonators . . . 295
8.2.2 Open Optical Resonators . . . 296
8.2.3 Modes of Open Resonators . . . 297
8.2.4 Diffraction Losses of Open Resonators . . . 300
8.2.5 The Frequency Spectrum of Optical Resonators . . . 301
8.3 Single Mode Lasers . . . 301
8.4 Different Types of Lasers . . . 304
8.4.1 Solid-state Lasers . . . 305
8.4.2 Semiconductor Lasers . . . 307
8.4.3 Dye Lasers . . . 308
8.4.4 Gas Lasers . . . 310
8.5 Nonlinear Optics . . . 313
8.5.1 Optical Frequency Doubling . . . 314
8.5.2 Phase Matching . . . 314
8.5.3 Optical Frequency Mixing . . . 316
8.6 Generation of Short Laser Pulses . . . 316
8.6.1 Q-Switched Lasers . . . 316
8.6.2 Mode-Locking of Lasers . . . 318
8.6.3 Optical Pulse Compression . . . 321
8.6.4 Measurements of Ultrashort Optical Pulses . . . 322
Summary . . . 324
Problems . . . 324


## 9. Diatomic Molecules


9.1 The $H_2^+$ Molecular Ion . . . 327
9.1.1 The Exact Solution for the Rigid $H_2^+$ Molecule . . . 328
9.1.2 Molecular Orbitals and LCAO Approximations . . . 331
9.1.3 Improvements to the LCAO ansatz . . . 334
9.2 The $H_2$ Molecule . . . 335
9.2.1 Molecular Orbital Approximation . . . 336
9.2.2 The Heitler–London Method . . . 337
9.2.3 Comparison Between the Two Approximations . . . 338
9.2.4 Improvements to the Approximations . . . 339
9.3 Electronic States of Diatomic Molecules . . . 340
9.3.1 The Energetic Order of Electronic States . . . 340
9.3.2 Symmetry Properties of Electronic States . . . 341
9.3.3 Electronic Angular Momenta . . . 341
9.3.4 Electron Spins, Multiplicity and Fine Structure Splittings . . . 343

9.3.5 Electron Configurations and Molecular Ground States . . . 344
9.3.6 Excited Molecular States . . . 346
9.3.7 Excimers . . . 347
9.3.8 Correlation Diagrams . . . 348
9.4 The Physical Reasons for Molecular Binding . . . 349
9.4.1 The Chemical Bond . . . 349
9.4.2 Multipole Interaction . . . 350
9.4.3 Induced Dipole Moments and van der Waals Potential . . . 352
9.4.4 General Expansion of the Interaction Potential . . . 355
9.4.5 The Morse Potential . . . 355
9.4.6 Different Binding Types . . . 356
9.5 Rotation and Vibration of Diatomic Molecules . . . 357
9.5.1 The Born-Oppenheimer Approximation . . . 357
9.5.2 The Rigid Rotor . . . 359
9.5.3 Centrifugal Distortion . . . 361
9.5.4 The Influence of the Electron Motion . . . 361
9.5.5 Vibrations of Diatomic Molecules . . . 363
9.5.6 Interaction Between Rotation and Vibration . . . 364
9.5.7 The Dunham Expansion . . . 366
9.5.8 Rotational Barrier . . . 366
9.6 Spectra of Diatomic Molecules . . . 367
9.6.1 Transition Matrix Elements . . . 367
9.6.2 Vibrational-Rotational Transitions . . . 369
9.6.3 The Structure of Electronic Transitions . . . 372
9.6.4 Continuous Spectra . . . 377
Summary . . . 380
Problems . . . 381

**10. Polyatomic Molecules**

10.1 Electronic States of Polyatomic Molecules . . . 383
10.1.1 The $H_2O$ Molecule . . . 383
10.1.2 Hybridization . . . 384
10.1.3 The $CO_2$ Molecule . . . 388
10.1.4 Walsh Diagrams . . . 389
10.2 Molecules with more than Three Atoms . . . 390
10.2.1 The $NH_3$ Molecule . . . 390
10.2.2 Formaldehyde and Other $H_2AB$ Molecules . . . 392
10.2.3 Aromatic Molecules and $\pi$-Electron Systems . . . 392
10.3 Rotation of Polyatomic Molecules . . . 394
10.3.1 Rotation of Symmetric Top Molecules . . . 397
10.3.2 Asymmetric Rotor Molecules . . . 399
10.4 Vibrations of Polyatomic Molecules . . . 399
10.4.1 Normal Vibrations . . . 399
10.4.2 Quantitative Treatment . . . 399
10.4.3 Couplings Between Vibrations and Rotations . . . 402

10.5 Spectra of Polyatomic Molecules . . . 403
10.5.1 Vibrational Transitions within the Same Electronic State . . . 404
10.5.2 Rotational Structure of Vibrational Bands . . . 406
10.5.3 Electronic Transitions . . . 407
10.6 Clusters . . . 408
10.6.1 Production of Clusters . . . 410
10.6.2 Physical Properties of Clusters . . . 410
10.7 Chemical Reactions . . . 412
10.7.1 First Order Reactions . . . 412
10.7.2 Second Order Reactions . . . 413
10.7.3 Exothermic and Endothermic Reactions . . . 414
10.7.4 Determination of Absolute Reaction Rates . . . 415
10.8 Molecular Dynamics and Wave Packets . . . 416
Summary . . . 418
Problems . . . 420

**11. Experimental Techniques in Atomic and Molecular Physics**

11.1 Basic Principles of Spectroscopic Techniques . . . 422
11.2 Spectroscopic Instruments . . . 423
11.2.1 Spectrometers . . . 423
11.2.2 Interferometers . . . 429
11.2.3 Detectors . . . 433
11.3 Microwave Spectroscopy . . . 437
11.4 Infrared Spectroscopy . . . 440
11.4.1 Infrared Spectrometers . . . 440
11.4.2 Fourier Transform Spectroscopy . . . 440
11.5 Laser Spectroscopy . . . 444
11.5.1 Laser-Absorption Spectroscopy . . . 444
11.5.2 Optoacoustic Spectroscopy . . . 445
11.5.3 Optogalvanic Spectroscopy . . . 447
11.5.4 Cavity-Ringdown Spectroscopy . . . 448
11.5.5 Laser-Induced Fluorescence Spectroscopy . . . 450
11.5.6 Ionization Spectroscopy . . . 452
11.5.7 Laser Spectroscopy in Molecular Beams . . . 453
11.5.8 Nonlinear Laser Spectroscopy . . . 455
11.5.9 Saturation Spectroscopy . . . 456
11.5.10 Doppler-Free Two-Photon Spectroscopy . . . 459
11.6 Raman Spectroscopy . . . 460
11.6.1 Basic Principles . . . 460
11.6.2 Coherent Anti-Stokes Raman Spectroscopy . . . 462
11.7 Spectroscopy with Synchrotron Radiation . . . 463
11.8 Electron Spectroscopy . . . 465
11.8.1 Experiments on Electron Scattering . . . 465
11.8.2 Photoelectron Spectroscopy . . . 467
11.8.3 ZEKE Spectroscopy . . . 469

11.9 Measurements of Magnetic and Electric Moments in Atoms and Molecules . . . . . 470
11.9.1 The Rabi-Method of Radio-Frequency Spectroscopy . . 471
11.9.2 Stark-Spectroscopy . . . . . 473
11.10 Investigations of Atomic and Molecular Collisions . . . . . 474
11.10.1 Elastic Scattering . . . . . 475
11.10.2 Inelastic Scattering . . . . . 478
11.10.3 Reactive Scattering . . . . . 479
11.11 Time-Resolved Measurements of Atoms and Molecules . . . . . 480
11.11.1 Lifetime Measurements . . . . . 480
11.11.2 Fast Relaxation Processes in Atoms and Molecules . . . 484
Summary . . . . . 485
Problems . . . . . 486

**12. Modern Developments in Atomic and Molecular Physics**

12.1 Optical Cooling and Trapping of Atoms . . . . . 487
12.1.1 Photon Recoil . . . . . 487
12.1.2 Optical Cooling of Atoms . . . . . 489
12.1.3 Optical Trapping of Atoms . . . . . 491
12.1.4 Bose–Einstein Condensation . . . . . 493
12.1.5 Molecular Spectroscopy in a MOT . . . . . 495
12.2 Time-resolved Spectroscopy in the Femtosecond Range . . . . . 497
12.2.1 Time-resolved Molecular Vibrations . . . . . 497
12.2.2 Femtosecond Transition State Dynamics . . . . . 498
12.2.3 Coherent Control . . . . . 499
12.3 Optical Metrology with New Techniques . . . . . 501
12.3.1 Frequency Comb . . . . . 501
12.3.2 Atomic Clocks with Trapped Ions . . . . . 503
12.4 Squeezing . . . . . 504
12.5 New Trends in Quantum Optics . . . . . 510
12.5.1 Which Way Experiments . . . . . 510
12.5.2 The Einstein–Podolski–Rosen Paradox . . . . . 512
12.5.3 Schrödinger's Cat . . . . . 513
12.5.4 Entanglement and Quantum Bits . . . . . 513
12.5.5 Quantum Gates . . . . . 515
Summary . . . . . 517
Problems . . . . . 518

**Chronological Table for the Development of Atomic and Molecular Physics** . . . . . 519

**Solutions to the Exercises** . . . . . 523

**References** . . . . . 571

**Subject Index** . . . . . 581

# 1. Introduction

This book deals with the microscopic building blocks of matter: *atoms and molecules*. These are the smallest particles responsible for the characteristic properties of gases, liquids and solids. Although with modern techniques they can be split into still smaller particles, such as electrons, protons and neutrons, these latter "elementary particles" do not bear the characteristic features of the specific macroscopic body formed by atoms or molecules. We will discuss in detail in this textbook how the diversity of macroscopic bodies and their properties are related to their composition of atoms and molecules. We will, however, restrict the treatment to free atoms and molecules because a detailed discussion of the microscopic structure of solids would increase the size of this book beyond reason.

A very important issue of atomic physics is the interaction of atoms and molecules with electromagnetic radiation, which can be absorbed or emitted by these particles. Photons, or "energy quanta," are the constituents of electromagnetic radiation and are created or annihilated by matter. They therefore form an essential part of the microscopic world.

"Classical physics" was already a well-established closed theory at the end of the 19th century and could explain nearly all aspects of fields such as mechanics, electrodynamics and optics. Only the theory of relativity and the physics of nonlinear phenomena, leading to the discovery of chaos, were later developed.

On the other side, most of the discoveries about atoms and molecules were made during the 20th century and even the last decade brought us still many surprises in atomic and molecular physics. The reasons for this relatively late development of atomic physics are manifold. First of all, the objects in this field are very small and cannot be viewed by the naked eye. Many sophisticated experimental techniques had to be invented first in order to gain reliable information on these microparticles. Furthermore it turned out that classical theories were not adequate to describe atoms and molecules and their interactions. After a new theory called "quantum theory" was developed in the first three decades of the 20th century, a rapid progress in atomic and molecular physics took place, and our knowledge on this field increased explosively. Nevertheless there are still a large number of open questions and poorly understood phenomena that await their solutions by future generations of researchers.

## 1.1 Contents and Importance of Atomic Physics

Atomic physics deals with the structure of atoms, their mutual interaction and their dynamics, i.e., their time-dependent properties. The goal of experimental and theoretical efforts in this field is the full understanding of macroscopic properties of matter on the basis of its microscopic composition of the constituent atoms and a quantitative description of the relations between microscopic and macroscopic features. We will later see that this goal has, besides its essential contribution to fundamental physics and a new concept of nature, an enormous influence on technical applications.

At the beginning of the 20th century, when atomic physics started to develop as an original field, it was regarded as pure fundamental science, with no practical application. *Lord Ernest Rutherford* (1871–1937), one of the pioneers of early atomic physics, wrote as early as 1927, after the discovery of possible transformations of atoms through impact by energetic particles, "Anyone who expects a source of power from transformation of atoms is talking moonshine." This point of view has radically changed. Besides the quite intensive

W. Demtröder, *Atoms, Molecules and Photons*, 2nd ed., Graduate Texts in Physics,
DOI 10.1007/978-3-642-10298-1_1, © Springer-Verlag Berlin Heidelberg 2010

fundamental research in atomic physics, the number of scientific and technical applications has increased enormously.

The methods developed in atomic physics are meanwhile used routinely in chemistry, biology, medicine and industry. In particular the instruments invented during research work in atomic physics, such as the X-ray tube, the electron microscope, the oscilloscope, spectrometers, tomographers, lasers etc., are now indispensable tools in other scientific fields or for the solution of technical problems.

The importance of atomic physics is therefore not restricted to physics. Atomic physics, together with molecular physics, forms the foundations of chemistry. It explains the chemical properties of atoms and the order of elements in the periodic table, the binding of molecules and the molecular structure. Chemical reactions are reduced to collisions between atoms and molecules. Because of its importance, a new branch of chemistry called "quantum chemistry" has been established, which deals with the theoretical foundation of chemistry based on quantum theory. The famous natural philosopher *Georg Christoph Lichtenberg* (1742–1799) wrote, "Someone who only knows chemistry does not really understand it either."

The complex reactions in the earth's atmosphere are started by the interaction of sunlight with atoms and molecules leading to energy deposition in molecules, their ionization and dissociation into fragments. Collisions between these particles can further increase the number of possible chemical reactions. The reaction probability depends not only on the temperature but also on the internal energy and structure of the collision partners. A more detailed understanding of these processes and the influence of man-made pollutant substances on such processes is of crucial importance for the survival of mankind [1.1–5].

During recent years the molecular basis of biological processes has been widely investigated. New experimental techniques of atomic physics have been applied to the studies of living cells and the reactions proceeding inside a cell. It is now possible to follow the paths of single molecules intruding a cell using spectroscopic methods of high spatial and spectral resolution [1.6–8].

Also in medicine, many diagnostic tools are borrowed from atomic physics and even therapeutic methods, such as specific laser treatment of cancer or irradiation with particle beams, are based on investigations in atomic physics.

The development of star models in astrophysics has gained important stimulation from laboratory experiments on absorption and emission of radiation by atoms or ions, on recombination processes between free electrons and ions or on lifetimes of excited atoms and on collision processes between electrons, ions and neutral atoms and molecules. Besides high-energy physics, atomic physics has considerably contributed to a better understanding of the formation of stars, on radiation transport and on the structure of star atmospheres [1.9–10].

Atomic physics has also played an essential role for the optimization of modern technical developments. One famous example is the rapidly increasing manifold of lasers and their various applications [1.11]. Modern illumination techniques with energy saving lamps, discharge tubes or light emitting diodes are essentially applied atomic physics [1.12–13]. New procedures for the nondestructive inspection of materials or for the enhancement of catalytic reactions on surfaces are based on results of research in atomic physics. For many technical developments in the production of semiconductor chips, such as the controlled diffusion of impurity atoms into the semiconductor or the interaction of gases and vapors with solid surfaces, which are processes studied in atomic physics, play an essential role [1.14, 15]. Without exaggeration, one may therefore say that atomic physics has an important share in the development of modern technology and this will certainly increase even more in the future.

For metrology the measuring techniques developed in atomic physics have increased the achievable accuracy by several orders of magnitude [1.16]. With laser spectroscopic methods, for example, the absolute values of fundamental physical constants, such as the Rydberg constant, the fine structure constant or the ratio $m_e/m_p$ of electron mass to proton mass, could be measured with such high precision that the question of whether these "constants" are really constant or change slightly with time over millions of years can now be attacked experimentally with measurement times of a few years.

The central importance of atomic physics for many other fields is schematically illustrated by the block diagram in Fig. 1.1.

Besides its influence on the technological development, atomic physics and quantum theory have

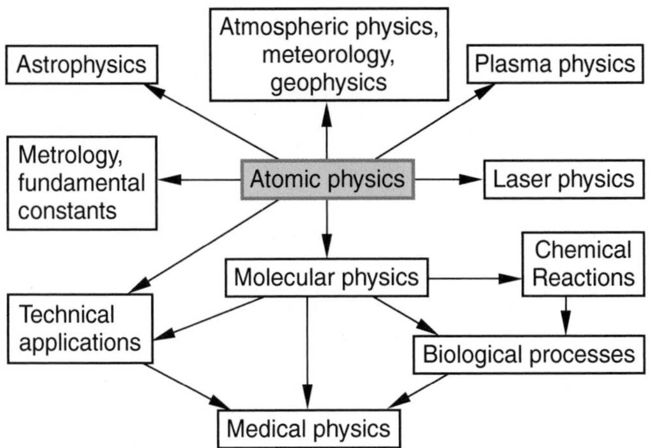


**Fig. 1.1.** The central role of atomic physics

essentially contributed to a modern view of nature that replaces the former mechanistic concept of our world [1.17]. The belief of a strict separation between matter and energy had to be modified by the recognition that both manifestations of nature are interchangeable and the anticipation of a strict causality for all processes in our surrounding has now been limited by the uncertainty relations in quantum mechanics. Maxwell's daemon of classical physics, who could exactly predict the future outcome of events as long as he knew the initial conditions sufficiently accurately, has to be replaced by probability statements, since the exact knowledge of all initial conditions is not possible. The deterministic view of nature, where all future events were already determined by the present conditions had to undergo a critical revision. This change in the concept of nature has considerably influenced philosophy and epistemology, i.e., the theory of knowledge, and has induced hot discussions about the question of whether objective cognition is possible independent of the thinking subject [1.18].

These few examples should have illustrated the importance of atomic physics for our modern world and why it is therefore worthwhile to study this fascinating field in more detail.

## 1.2 Molecules: Building Blocks of Nature

In nature we find 92 different elements that correspond to stable atoms. These atoms can form larger entities, called molecules. The smallest molecules consist of two atoms, such as $H_2$, $N_2$, $O_2$, NaCl, etc., while large molecules (for instance proteins or DNA) are composed of many thousands of atoms (Fig. 1.2).

The large variety and the manifold of species in nature is due to the huge number of possible combinations of these 92 stable atoms to form molecules. The chemical and therefore the biological properties of these molecules depend on:

- The specific kind of atoms they are composed of.
- The spatial structure of the molecules, i.e., the way in which the atoms are arranged within the molecule.
- The binding energy of atoms or atomic groups in the molecule.
- The stability, depending on the heights of the energy barrier, that has to be overcome to change the geometrical structure of the molecule.

Only recently has it become possible to calculate the structure and the binding energies of small- and medium-sized molecules by ab initio methods using fast computers. In many cases, however, experimental

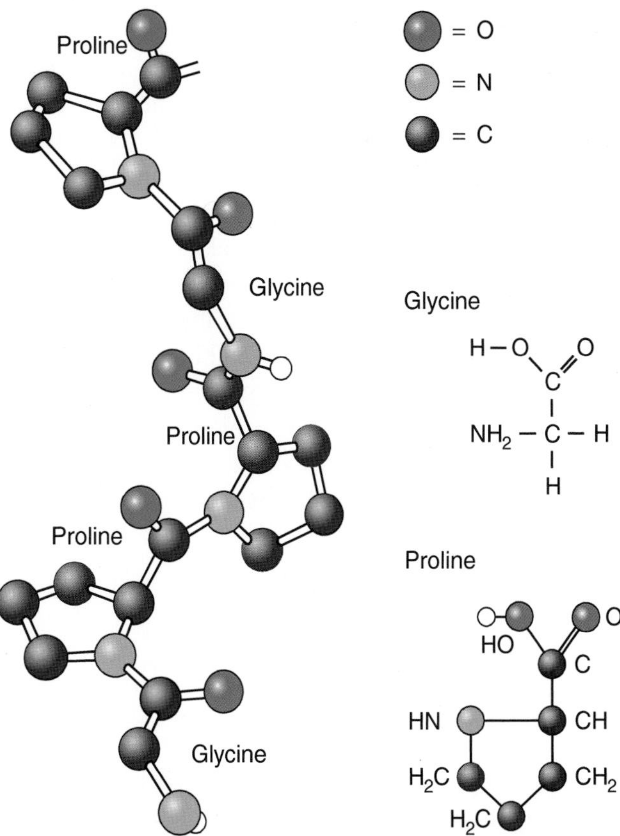


**Fig. 1.2.** Section of a left-handed coiled strand of the collagen triple helix, where three such strands coil right-handed

methods are still indispensable because sufficiently accurate calculations surpass the capacity of even large computers.

The goal of such investigations is a better knowledge of molecular structure and the potential surfaces that determine this structure and the relevant binding energies. In recent years the dynamics of excited molecules, i.e., the way the energy, pumped into a molecule (for example by absorption of light), is distributed within the molecule over the course of time, has attracted more and more interest from researchers. With a time resolution of a few femtoseconds ($1\,\text{fs} = 10^{-15}\,\text{s}$) obtained with ultrashort laser pulses, it is now possible to observe the motions of atoms in molecules in real-time and to gain much information on molecular dynamics, such as dissociation or isomerization. This allows one to follow more closely the atomic processes in chemical reactions. In special cases it is even possible to control such reactions, i.e., to enhance wanted reaction channels and to suppress unwanted ones. This opens the way for controlled synthesis of larger molecules from smaller constituents.

Many biological processes, such as energy production in living cells, photosynthesis, ion migration through cell walls, signal transport in nerves or the time sequence of the visual process from the illuminated retina in the eye to the recognition of the light image in the brain, can now be studied in more detail due to advanced experimental techniques developed in atomic physics [1.19].

The experimental and theoretical molecular physics therefore gains increasing attention for many fields in modern chemistry and biology. In many laboratories, researchers are working on the ambitious goal of unraveling the structure and the arrangement of different amino acid molecules in large biomolecules, to understand their role in genes and to clarify the genetic code and its relevance for the characteristic features of life [1.20–22].

## 1.3 Survey on the Concept of this Textbook

The goal of this textbook is to facilitate the understanding of the structure and dynamics of atoms and molecules by starting from basic concepts and experimental facts in atomic and molecular physics. It is also interesting to learn a little bit about the way our present knowledge has developed. Therefore, a short historical review is first provided about the successive improvement of the atomic concept, which has led to more and more refined atomic models. In particular, the experimental key investigations resulting either in the confirmation, modification or even change of existing theories are discussed in order to give a better appreciation for the skill and imagination of earlier researchers.

The most important theoretical approach for the description of the microworld is certainly the development of quantum physics during the first three decades of the 20th century. We will discuss in Chap. 3 the basic experimental results that forced a correction of classical physics. Then the basic features of quantum physics, particle-wave duality, the uncertainty relation and its experimental verification are presented and the probability concept for describing processes in the microworld is explained.

In Chap. 4 we then introduce the formal representation of quantum mechanics, in particular the Schrödinger equation and its application to some simple problems, in order to illustrate differences to and similarities with classical physics.

In Chap. 5 the simplest of all atoms, the hydrogen atom is treated with the tools acquired in the foregoing chapters. Here we can learn about many features that are also relevant for other atoms but can be calculated more accurately for the H atom because it is the only system for which the Schrödinger equation can be solved exactly. Even here, new characteristic features such as the spin of the electron, resulting in the fine structure of the measured spectra could not immediately be explained and demanded the broadening of the quantum theory and the development of a new branch of quantum physics, called quantum electrodynamics.

Chapter 6 deals with atoms consisting of more than one electron, where new phenomena occur, which are related to the Coulomb repulsion between the electrons and to the fact that electrons cannot be distinguished from each other. The treatment of many-electron systems is illustrated by the example of the two-electron helium atom and is then extended to larger atoms.

The absorption and emission of light by atoms is a source of detailed information on the structure of atoms, on the possible atomic energy levels and on dynamical processes in excited atoms. This also includes X-rays, which are discussed in Chap. 7. After treating the interaction of electromagnetic radiation with atoms, we have laid the fundaments for the understanding of lasers. Their basic principle and their various technical realizations are presented in Chap. 8.

In Chap. 9 we start the discussion of the basic physics of molecules. The simplest stable molecules, the $H_2^+$ ion (two protons and one electron) and the $H_2$ molecule (two protons and two electrons) serve as examples to explain the nomenclature and the principles of theoretical approximations for the description of diatomic molecules. Both examples illustrate the origin of the chemical binding of atoms forming a stable molecule. While for small atomic distances in a diatomic molecule the quantitative treatment of chemical binding demands quantum theory, at large distances the binding energy is small and can be treated by classical methods, which will be also discussed in this chapter.

The most important source of information on molecular structure is provided by molecular absorption and emission spectra, which are discussed in more detail in Chap. 10. We start with diatomic molecules and treat polyatomic molecules in Chap. 11.

The chapter following is devoted to experimental techniques in atomic and molecular physics. Here we will illustrate how all knowledge of atomic and molecular structure discussed in the foregoing chapters has been achieved by experimental results and how experiment and theory supplement each other to efficiently achieve optimum progress in our understanding of the microscopic structure of matter. The last chapter of this text books treats some modern developments of modern atomic physics, such as optical cooling of atoms, Bose-Einstein condensation of free atoms, time-resolved molecular dynamics on the femtosecond time scale and some fundamental experiments in quantum optics.

For a more detailed study of the subjects presented in this textbook the reader is referred to the literature given in the corresponding sections. Besides modern treatments, sometimes the original historical papers on new discoveries are also cited. This provides the reader direct access to the way new ideas came about and to the original interpretations of experimental results, which, although often ingenious, did not always agree with our present point of view, since our ancestors did not have all of facts now available to us. Nevertheless the study of this original papers will give us some admiration of the inspiration and brilliance of the early researchers

# 2. The Concept of the Atom

Our present knowledge about the size and internal structure of atoms is the result of a long development of ideas and concepts that were initially based both on philosophical speculations and on experimental hints, but were often not free of errors. Only during the 19th century did the increasing number of detailed and carefully planned experiments, as well as theoretical models that successfully explained macroscopic phenomena by the microscopic atomic structure of matter, could collect sufficient evidence for the real existence of atoms and therefore convinced more and more scientists. However, even around the year 1900, some well-reputed chemists, such as *Wilhelm Ostwald* (1853–1932), and physicists, e.g., *Ernst Mach* (1838–1916), still doubted the real existence of atoms. They regarded the atomic model as only a working hypothesis that could better explain many macroscopic phenomena, but should not be taken as reality.

In this chapter we will therefore discuss, after a short historical survey, the most important experimental proofs for the real existence of atoms. Furthermore, some measurements are explained that allow the quantitative determination of all atomic characteristics, such as their size, mass, charge distribution and internal structure. These experiments prove without doubt that atoms do exist, even though nobody has ever seen them directly because of their small size.

**Fig. 2.1.** *Democritus* ($\sim$ 460–370 BC) (from K. Faßmann: Die Großen, BD I/2, Kindler-Verlag, Munich)

## 2.1 Historical Development

Historically, the first concept of the atomic structure of matter was developed by the Greek philosopher *Leucippus* (around 440 B.C.) and his disciple *Democritus* (460–370 B.C.) (Fig. 2.1), who both taught that all natural bodies consist of "infinitely small" particles that completely fill the volume of the bodies and are not further divisible. They called these particles "atoms" (from the Greek word atomos = indivisible). Outside the atoms there is only the empty space (a vacuum). Different atoms differ in size and shape and the characteristic properties of matter are, according to this model, due to different arrangements of equal or of differing atoms. All observable changes in the macroscopic world are caused by corresponding changes in atomic composition. Atom movements and collisions between atoms create and modify matter.

We meet here for the first time the idea that the properties of macroscopic bodies can be explained by the characteristics of their constituents. This hypothesis, which comes close to our modern concept of atomic physics, had been an extension and refinement

W. Demtröder, *Atoms, Molecules and Photons*, 2nd ed., Graduate Texts in Physics,
DOI 10.1007/978-3-642-10298-1_2, © Springer-Verlag Berlin Heidelberg 2010

of former ideas by *Empedocles* (490–430 B.C.), who believed that everything is composed of the four elemental constituents: fire, water, air and soil.

The concept of *Democritus* represents in a way a symbiosis of the different doctrines of pre-Socratic philosophers. First, the static hypothesis of *Parmenides* (around 480 B.C.) about the never-changing eternal existence of the world and secondly the dynamical doctrine of *Heraclitus* (around 480 B.C.), which stresses as the most important point the evolution instead of the static nature of things, since everything changes with time (nobody can submerge twice into the same river as the same man, because the river, as well as the man, is changing in time).

According to *Democritus*, atoms represent static nature while their movements and their changing composition explain the diversity of matter and its time evolution.

The famous Greek philosopher *Plato* (427–347 B.C.) pushed the abstraction of the concept further. He used the hypothesis of the four "elements" fire, water, air, and soil but attributed to these elements four regular three-dimensional geometric structures, which are formed by symmetric triangles or squares (Fig. 2.2). Fire is related to the tetrahedron (four equilateral triangles), air to the octahedron (eight equilateral triangles), water to the icosahedron (20 equilateral triangles), and the soil, particularly important to mankind, to the cube (six squares or 12 isosceles triangles). Plato's ideas therefore reduced the atoms to mathematical structures that are not necessarily based on the real existence of matter. These "mathematical atoms" can change their characteristics by changing the arrangement of the elemental triangles. This is, according to Plato, equivalent to the observable evolution of matter.

*Aristoteles* (384–322 B.C.), a student of *Plato*, did not accept this concept of atoms since it contradicted his idea of a continuous space filled with matter. He also did not believe in the existence of empty space between the atoms. His influence was so great that *Democritus'* hypothesis was almost abandoned and nearly forgotten until it was revived and modified later by *Epicurus* (341–271 B.C.), who attributed atoms not only size but also a mass to explain why bodies fell down.

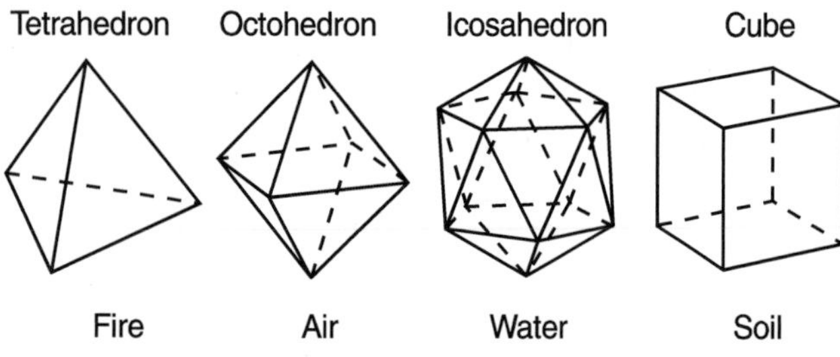

**Fig. 2.2.** The platonic bodies

After *Epicurus* the atomic theory was forgotten for many centuries. This was due to the influence of the Christian church, which did not accept the materialistic view that everything, even human beings, should be composed of atoms, because this seemed to be in contradiction to the belief in God as the creator of bodies and soul. There had occasionally been attempts to revive the atomic idea, partly induced by Arabic scientists, but they did not succeed against church suppression. One example was the Prior *Nikolaus of Autrecourt* in France, who was forced in 1348 to "withdraw" his newly developed atomic concept.

The large shortcoming of all these philosophical hypotheses was the lack of experimental guidance and proof. They were more speculative.

The real breakthrough of modern atomic physics was achieved by chemists in the 18th century. They found for many chemical reactions, by accurately weighing the masses of reactants and reaction products, that their results could be best explained by the hypothesis that all reactants consist of atoms or molecules that can recombine into other molecules (see below).

Besides this increasing amount of experimental evidence for the existence of atoms, the atomic hypothesis won a powerful ally from theoretical physics when James Prescott Joule (1818–1889), *Rudolf Julius Clausius* (1822–1888), *James Clark Maxwell* (1831–1879), and *Ludwig Boltzmann* (1884–1906) developed the kinetic theory of gases, which could derive all macroscopic quantities of gases, such as pressure, temperature, specific heat, viscosity, etc., from the assumption that the gas consists of atoms that collide with each other and with the walls of the container. The temperature is a measure of the average kinetic energy of the atoms and the pressure represents the mean momentum the atoms transfer to the wall per second per unit wall area. The first ideas about the kinetic theory of gases and its relation to heat were published by John Herapath (1790–1868) in a book, which was known to Joule.

Quantitative information about the size of atoms and their internal structure, i.e., mass and charge distribution inside the atoms was only obtained in the 20th century. The complete theoretical description was possible after the development of quantum theory around 1930 (see Chaps. 3 and 4).

In the Chronological Table at the end of this book one finds a compilation of historical landmarks in the development of atomic physics. For more detailed information on the history of atomic and molecular physics the reader is referred to the literature [2.1–6].

## 2.2 Experimental and Theoretical Proofs for the Existence of Atoms

Before we discuss the different experimental techniques developed for the proof of atoms, a general remark may first be useful. The objects of atomic physics are not directly visible since they are much smaller than the wavelength of visible light, unlike bodies in the macroscopic world. Therefore, indirect method for their investigation are required. The results of such experiments need careful interpretation in order to allow correct conclusions about the investigated objects. This interpretation is based on assumptions that are derived from other experiments or from theoretical models.

Since it is not always clear whether these assumptions are valid, the gain of information in atomic physics is generally an iterative process. Based on the results of a specific experiment, a model of the investigated object is developed. This model often allows predictions about the results of other experiments. These new experiments either confirm the model or they lead to its refinement or even modification.

> In this way, through collaboration between experimentalists and theoreticians, a successively refined and correct model can be established that reflects the reality as accurately as possible.

This means that it allows correct predictions for all future experimental results. This will be illustrated by the successive development of more and more refined models of the atom, which will be discussed in the following sections and in Chap. 3.

### 2.2.1 Dalton's Law of Constant Proportions

The first basic experimental investigations that have lead to a more concrete atomic model, beyond the more speculative hypothesis of the Greek philosophers, were performed by chemists. They determined the mass ratios of the reactants and reaction products for chemical reactions. The basic ideas had already been prepared by investigations of *Daniel Bernoulli* (1700–1782), who explained the experimental results of the Boyle–Marriotte Law:

$$p \cdot V = \text{const} \quad \text{at constant temperature}$$

by the movements of tiny particles in a gas with volume $V$ which exert the pressure $p$ onto the walls around $V$ through collisions with the wall. These ideas laid the foundations of the kinetic gas theory, which was later more rigorously developed by *Clausius*, *Maxwell*, and *Boltzmann*.

Following the more qualitative findings of *Joseph Louis Proust* (1754–1826) on mass ratios of reactants and reaction products in chemical reactions, the English chemist *John Dalton* (1766–1844) (Fig. 2.3) recognized, after many experiments of quantitative analyses and syntheses for various chemical compounds, that the mass ratios of reactants forming a chemical compound, are always the same for the same reaction, but may differ for different reactions.

**Fig. 2.3.** *John Dalton* (1766–1844)

## EXAMPLES

1. 100 g of water are always formed out of 11.1 g of hydrogen and 88.9 g of oxygen. The mass ratio of the reactants is then 1 : 8.
2. 100 g of copper oxide CuO contains 79.90 g Cu and 20.10 g oxygen with a mass ratio of about 4 : 1.
3. Some reactants can combine in different mass ratios to form different products. For example, there are five different manganese oxides where 100 g of manganese combines either with 29.13 g, 43.69 g, 58.26 g, 87.38 g or 101.95 g of oxygen. The different amounts of oxygen represent the mass ratios 2 : 3 : 4 : 6 : 7.

From these experimental results *Dalton* developed his atomic hypothesis in 1803, which stated that the essential feature of any chemical reaction is the recombination or separation of atoms. He published his ideas in the paper "A New System of Chemical Philosophy" [2.7], which contains the three important postulates:

- All chemical elements consist of very small particles (atoms), which can not be further divided by chemical techniques.
- All atoms of the same chemical element have equal size, mass and quality, but they differ from the atoms of other elements. This means that the properties of a chemical element are determined by those of its atoms.
- When a chemical element $A$ reacts with an element $B$ to form a compound $AB_n$ ($n = 1, 2, \ldots$) each atom of $A$ recombines with one or several atoms of $B$ and therefore the number ratio $N_B/N_A$ is always a small integer.

*Dalton's* atomic hypothesis can immediately explain the experimental results given in the above examples:

1. Two hydrogen atoms H recombine with one oxygen atom O to form the molecule $H_2O$ (Fig. 2.4). The observed mass ratio 11.1/88.9 is determined by the masses of the atoms H and O. From the mass ratio $m(\mathrm{H})/m(\mathrm{O}) = 1/16$ (see Sects. 2.2.2 and 2.7), the measured mass ratio of the reactants follows as

$$m(\mathrm{H_2})/m(\mathrm{O}) = 2/16 = 11.1/88.9\,.$$

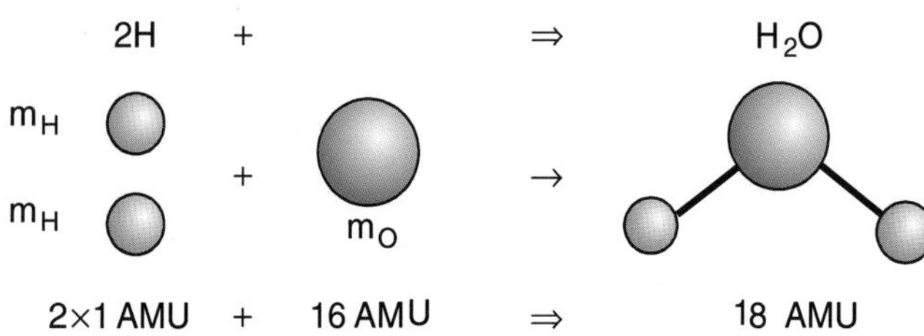


**Fig. 2.4.** Reaction of hydrogen and oxygen to form water molecules as an example of Dalton's atomic hypothesis

2. For the reaction $\mathrm{Cu} + \mathrm{O} \rightarrow \mathrm{CuO}$ the mass ratio of the reactants corresponds to the relative masses $m(\mathrm{Cu})/m(\mathrm{O}) = 64/16 = 4 : 1$.
3. The different manganese oxides are MnO, $Mn_2O_3$, $MnO_2$, $MnO_3$, and $Mn_2O_7$. Therefore, the number of O atoms that combine with two Mn atoms have the ratios 2 : 3 : 4 : 6 : 7 for the different compounds, which is exactly what had been found experimentally.

Since *Dalton's* laws only deal with mass ratios and not with absolute atomic masses, the reference mass can be chosen arbitrarily. *Dalton* related all atomic masses to that of the H atom as the lightest element. He named these relative masses ***atomic weights***.

**Note:**

"Atomic weights" are not real weights but dimensionless quantities since they represent the ratio $m(\mathrm{X})/m(\mathrm{H})$ of the atomic masses of an atom X to the hydrogen atom H.

*Jörg Jakob Berzelius* (1779–1848) started to accurately determine the atomic weights of most elements in 1814. Nowadays this historic definition of atomic weight is no longer used. Instead of the H atom the $^{12}$C atom is defined as reference. The atomic weight has been replaced by the ***atomic mass unit (AMU)***

$$1\,\mathrm{AMU} = (1/12)\,m(^{12}\mathrm{C}) = 1.66053886 \times 10^{-27}\,\mathrm{kg}.$$

All relative atomic masses are given in these units.

## EXAMPLES

The mass of a Na atom is $m(\mathrm{Na}) = 23\,\mathrm{AMU}$, that of Uranium 238 is $m(\mathrm{U}) = 238\,\mathrm{AMU}$ and that of the nitrogen molecule $N_2$ is $2 \times 14 = 28$ AMU.

### 2.2.2 The Law of Gay-Lussac and the Definition of the Mole

*Joseph Louis Gay-Lussac* (1778–1850) and *Alexander von Humboldt* (1769–1859) (Fig. 2.5) discovered in 1805 that the volume ratio of oxygen gas and hydrogen gas at equal pressures was always 1 : 2 when the two gases recombined completely to form water vapor. Further detailed experiments with other gases lead to the following conclusion:

> When two or more different gases completely recombine to form a gaseous chemical compound, the ratio of the volumes of reactands and reaction products at equal pressure and temperature is always given by the ratio of small integer numbers.

**Fig. 2.5.** *Alexander von Humboldt* (1769–1859) (with kind permission from the Alexander von Humboldt foundation, Bonn)

**EXAMPLES**

1. $2\,dm^3$ hydrogen gas $H_2$ and $1\,dm^3$ oxygen gas $O_2$ recombine to form $2\,dm^3$ water vapor $H_2O$ (not $3\,dm^3$ $H_2O$ as might be naively expected!).
2. $1\,dm^3$ $H_2$ and $1\,dm^3$ $Cl_2$ form $2\,dm^3$ HCl gas.

*Amadeo Avogadro* (1776–1856) (Fig. 2.6) explained these results by introducing the definition of molecules:

> A molecule is the smallest particle of a substance that determines the properties of this substance. It is composed of two or more atoms.

Referring to the experimental results of *Gay-Lussac*, *Avogadro* concluded:

> At equal pressures and temperatures, the same volume of different gases always contains the same number of molecules.

With this hypothesis the two preceding examples are described by the reaction equations:

$$2\,H_2 + O_2 \rightarrow 2\,H_2O\,,$$
$$H_2 + Cl_2 \rightarrow 2\,HCl\,.$$

The total mass $M$ of a gas with volume $V$ containing $N$ molecules with mass $m$ is then:

$$M = N \cdot m\,. \tag{2.1}$$

**Fig. 2.6.** *Amadeo Avogadro* (1776–1856) with kind permission from the Deutsche Museum, Munich

The mass ratio $M_1/M_2$ of equal volumes of different gases at equal pressure and temperature therefore equals the mass ratios $m_1/m_2$ of the corresponding molecules, since the number $N$ of molecules is the same for both gases.

It is convenient to introduce a specific reference quantity of molecules, called one ***mole*** [1 mol]. The volume occupied by one mole of a gas is called the mole volume $V_M$. The definition of a mole is as follows:

> 1 mol is the quantity of a substance that contains the same number of particles (atoms or molecules) as 0.012 kg of carbon $^{12}$C.

This definition is equivalent to: 1 mol of atoms or molecules with atomic mass number X AMU has a mass of X grams.

EXAMPLES

1. 1 mol helium He $\hat{=}$ 4 g helium
2. 1 mol oxygen $O_2 \hat{=} 2 \cdot 16\,\text{g} = 32\,\text{g}$ oxygen
3. 1 mol water $H_2O \hat{=} (2 \cdot 1 + 16)\,\text{g} = 18\,\text{g}$ water
4. 1 mol iron oxide $Fe_2O_3 \hat{=} (2 \cdot 56 + 3 \cdot 16)\,\text{g} = 160\,\text{g}$ iron oxide

The number $N_A$ of atoms or molecules contained in 1 mol is the ***Avogadro constant***. Its experimental value is

$$N_A = 6.0221415(10) \times 10^{23}\,\text{mol}^{-1}\,.$$

From the hypothesis of Avogadro the statement follows:

> Under standard conditions ($p = 1000\,\text{hPa}$, $T = 0\,°\text{C}$) 1 mol of an arbitrary gas always occupies the same volume $V_M$, called the mole volume:
>
> $$V_M = 22.710981(39)\,\text{dm}^3\,\text{mol}^{-1}\,.$$

**Note:**

The value of $V_M$ depends on the definition of the standard conditions. There are two sets of these conditions in use: The old one uses

a) $T = 273{,}15\,\text{K}$ ($0^0$C) and $p = 1\,\text{atm}$ (1013.25hPa) $\rightarrow$ $V_M = 22{,}413996\,\text{dm}^3\text{mol}^{-1}$

The new one, adapted in the CODATA list and NIST reference tables uses:

b) $T = 273{,}15\,\text{K}$ ($0^0$C) and $p = 1\,\text{bar}$ (1000hPa) $\rightarrow$ $V_M = 22.710981\,\text{dm}^3\text{mol}^{-1}$

### 2.2.3 Experimental Methods for the Determination of Avogadro's Constant

Since the Avogadro constant $N_A$ is a fundamental quantity that enters many basic physical equations, several experimental methods have been developed for the accurate measurement of $N_A$ [2.8]. We will only present some of them here.

#### a) Determination of $N_A$ from the general equation of gases

From the kinetic theory of gases the general equation

$$p \cdot V = N \cdot k \cdot T \tag{2.2}$$

can be derived for the volume $V$ of an ideal gas under the pressure $p$ at a temperature $T$, which contains $N$ molecules. Here $k$ is the Boltzmann constant. For 1 mol of a gas with volume $V_M$, $N$ becomes $N_A$ and (2.2) converts to

$$p \cdot V_M = N_A \cdot k \cdot T = R \cdot T\,. \tag{2.3}$$

The gas constant

$$R = N_A \cdot k \tag{2.4}$$

is the product of Avogadro's and Boltzmann's constants. It can be determined from (2.3) when $p$, $V_M$ and $T$ are measured. If the Boltzmann constant $k$ and the gas constant $R$ can be measured independently, the Avogadro constant $N_A$ can be determined from (2.4).

#### α) Measurements of the gas constant R

The gas constant $R$ can be obtained from measurements of the specific heat. The internal energy of 1 mol is

$$U = f \cdot \tfrac{1}{2}kT \cdot N_A = \tfrac{1}{2}f \cdot R \cdot T\,, \tag{2.5}$$

where $f$ is the number of degrees of freedom of the atoms or molecules of the substance. For example

$f = 3$ for atoms, $f = 3 + 2 = 5$ for diatomic molecules at low temperatures where the vibrations are not excited and $f = 7$ at higher temperatures.

The molar specific heat $C_\mathrm{v}$ for a constant mole volume of a gas is

$$C_\mathrm{v} = \left(\frac{\partial U}{\partial T}\right)_\mathrm{v} = \frac{1}{2} f \cdot R \,. \tag{2.6}$$

This is the energy that increases the temperature of 1 mol of a gas by 1 K and can therefore be readily measured, giving the value of $R$, if the number of degrees of freedom $f$ is known.

Another way to measure the gas constant $R$ is based on the difference

$$R = C_\mathrm{p} - C_\mathrm{v} \tag{2.7}$$

of the molar specific heats $C_\mathrm{p}$ at constant pressure and $C_\mathrm{v}$ at constant volume.

The most accurate determination of $R$ uses the measurement of the velocity of sound waves $v_\mathrm{s}$ in an acoustic resonator (Fig. 2.7). A spherical volume is filled with argon at a pressure $p$ and temperature $T$. A small loudspeaker S produces sound waves that lead to resonant standing waves if the sound frequency matches one of the radial eigenfrequencies $f_{0,n} \propto v_\mathrm{s}/\lambda_{0,n}$ with $\lambda_{0,n} = r_0/n$ of the spherical acoustic resonator with radius $r_0$. These resonantly enhanced sound waves are detected by a microphone Mi. The frequencies $f_{0,n}$ of different resonances are measured. As is outlined in the solution of Problem 2.6, the gas constant is related to the measurable acoustic eigenfrequencies $f_{0,n}$, the sound velocity $v_\mathrm{s}$, the molar specific heats $C_\mathrm{p}$ and $C_\mathrm{v}$, the temperature $T$ and the volume $V$ by

$$R = \frac{M \cdot v_\mathrm{s}^2}{T \cdot \kappa} = \frac{M}{T} \frac{f_{0,n}^2 \cdot r_0^2}{\kappa \cdot n^2} \,, \tag{2.8}$$

where $M$ is the molar mass and $\kappa = C_\mathrm{p}/C_\mathrm{v}$ [2.9]. Since the acoustic losses of the spherical resonator are low, the resonances are very sharp and the resonance frequencies $\mathrm{f}_{0,\mathrm{n}}$ can be determined with high accuracy. Tuning the loudspeaker-frequency through several resonances from $\mathrm{n} = 1$ to higher values the squares $\mathrm{f}_{0,\mathrm{n}}{}^2$ of the resonance frequencies can be plotted against $\mathrm{n}^2$ which gives a straight line with the slope $\mathrm{R\kappa T/(Mr_0{}^2)}$. Since M and $\kappa$ are known and T and $\mathrm{r}_0$ can be accurately measured, the gas constant R can be determined [2.9].

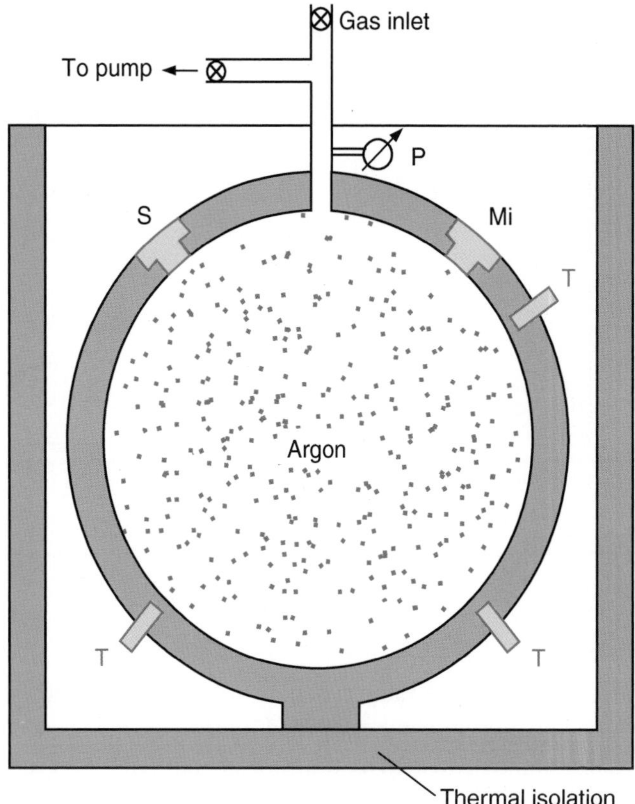


**Fig. 2.7.** Determination of the gas constant $R$ from measurements of the velocity of sound in argon (Mi := microphone, T = thermometer, S = loudspeaker as sound source, P = pressure meter)

### β) Measurement of the Boltzmann Constant

The Boltzmann constant $k$ was first determined in 1906 by *Jean Baptiste Perrin* (1870–1942). He observed the vertical density distribution $n(z)$ of small latex particles in a liquid within a glass cylinder (Fig. 2.8). At equilibrium the Boltzmann distribution

$$n(z) = n(0) \cdot \mathrm{e}^{-m^* gz/kT} \tag{2.9}$$

is obtained, where $m^* g = (m - \varrho_\mathrm{L} \cdot V_\mathrm{p})g$ is the effective weight of a particle with volume $V_\mathrm{p}$, (i.e., its real weight minus its buoyancy in the liquid with density $\varrho_\mathrm{L}$). This gives the gradient

$$\frac{\mathrm{d}n}{\mathrm{d}z} = -n(z) \cdot \frac{m^* \cdot g}{k \cdot T} \,, \tag{2.10}$$

The mass $m$ of the particles can be determined by measuring their size (volume) under a microscope and their density with standard techniques.

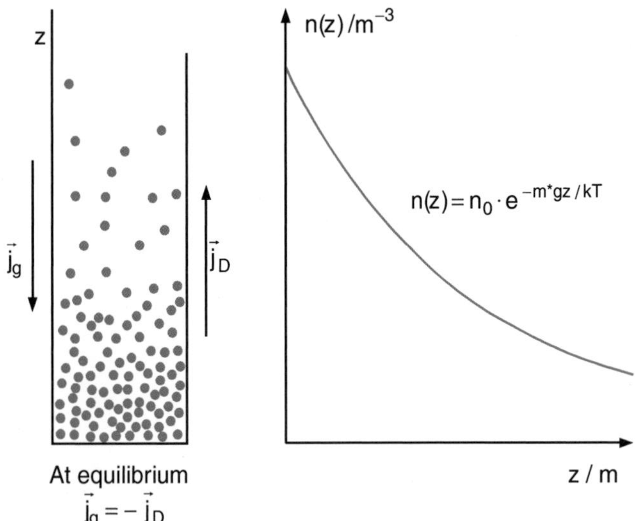


**Fig. 2.8.** Stationary distribution $n(z)$ of small particles in a liquid

Counting the number of $n(z)$ yields $\mathrm{d}n/\mathrm{d}z$ and therefore the Boltzmann constant from (2.6). The rather tedious counting can be avoided by the following consideration. Due to gravity the particles sink down. If the gravity force

$$\boldsymbol{F}_{\mathrm{g}} = (m - \varrho_{\mathrm{L}} \cdot V_{\mathrm{p}})\boldsymbol{g} \tag{2.11a}$$

is just compensated by the friction force

$$\boldsymbol{F}_{\mathrm{f}} = -6\pi\eta r \boldsymbol{v} \,, \tag{2.11b}$$

which spherical particles of radius $r$ experience when they fall with the velocity $\boldsymbol{v}$ in a medium with viscosity $\eta$, the net force is zero. The constant sink velocity is then

$$v_{\mathrm{g}} = \frac{(m - \varrho_{\mathrm{L}} \cdot V_{\mathrm{p}}) \cdot g}{6\pi\,\eta r} \quad \text{where} \quad V_{\mathrm{p}} = \tfrac{4}{3}\pi r^3 \,. \tag{2.12}$$

The downward flux of particles $j_{\mathrm{g}} = v_{\mathrm{g}} \cdot n$ creates a concentration gradient $\mathrm{d}n/\mathrm{d}z$, which leads to an upward diffusion flux

$$j_{\mathrm{diff}} = -D \cdot \frac{\mathrm{d}n}{\mathrm{d}z} = D \cdot n \cdot \frac{(m - \varrho_{\mathrm{L}} \cdot V_{\mathrm{p}})g}{k \cdot T} \,, \tag{2.13}$$

where $D$ is the diffusion coefficient.

Finally, stationary conditions are reached when both fluxes just cancel. This means

$$\boldsymbol{j}_{\mathrm{diff}} + \boldsymbol{j}_{\mathrm{g}} = 0 \Rightarrow k = \frac{6\pi\,\eta r \cdot D}{T} \,. \tag{2.14}$$

Therefore, the Boltzmann constant $k$ can be determined from the measurements of viscosity $\eta$, diffusion coefficient $D$, temperature $T$, and the radius $r$ of the spherical particles.

The most accurate method to measure $k$ will be discussed in Sect. 2.3.1.

### b) Direct Determination of Avogadro's Constant

From measurements of the absolute mass $m$ of atoms X (see Sect. 2.7) and the molar mass $M_{\mathrm{X}}$ (i.e., the mass of a gas of atoms X within the molar volume $V = 22.4\,\mathrm{dm}^3$ under normal conditions $p$ and $T$) the Avogadro constant

$$N_{\mathrm{A}} = M_{\mathrm{X}}/m_{\mathrm{X}}$$

can be directly determined.

The molar mass $M_{\mathrm{X}}$ can be also obtained for nongaseous substances from the definition

$$M_{\mathrm{X}} = 0.012\,m_{\mathrm{X}}/m(^{12}\mathrm{C})\,\mathrm{kg}$$

when the absolute mass of the carbon atoms $m(^{12}\mathrm{C})$ is measured (see Sect. 2.7).

### c) Determination of Avogadro's Constant from Electrolysis

Another method for the determination of $N_{\mathrm{A}}$ is based on Faraday's law for electrolytic processes. It states that the electric charge

$$\boxed{F = N_{\mathrm{A}} \cdot e = 96{,}485.3383(83)\,\mathrm{C/mol}} \tag{2.15}$$

is transported to the electrode in an electrolytic cell, when 1 mol of singly charged ions with mass $m_{\mathrm{X}}$ and elementary charge $e$ has been deposited at the electrode. Therefore, weighing the mass increase $\Delta m$ of the electrode after a charge $Q$ has been transferred, yields:

$$\Delta m = \frac{Q}{e} m_{\mathrm{X}} = \frac{Q}{e}\frac{M_{\mathrm{X}}}{N_{\mathrm{A}}}$$
$$\Rightarrow N_{\mathrm{A}} = \frac{Q}{e}\frac{M_{\mathrm{X}}}{\Delta m} \tag{2.16a}$$

where $M_{\mathrm{X}}$ is the molar mass of the ions with mass $m_{\mathrm{X}}$

EXAMPLE

In the electrolytic process

$$AgNO_3 \leftrightarrow Ag^+ + NO_3^-$$

of silver nitrate the transport of charge $F$ means a simultaneous deposition of the molar mass $M = N_A \cdot m(\text{Ag})$ at the negative electrode, which can be measured by weighing the cathode before and after the charge transport. With the atomic mass number of silver AM(Ag) = 107.89 AMU the Avogadro number

$$N_A = \frac{107.89\,\text{AMU}}{\Delta m} \cdot \frac{Q}{e} \tag{2.16b}$$

is obtained from the measured mass increase $\Delta m$ of the electrode and the transported charge $Q = (\Delta m/M) N_A \cdot e$.

### d) Determination of $N_A$ from X-Ray Diffraction

The most accurate method for the determination of $N_A$ is based on X-ray diffraction or X-ray interferometry, which are used to measure the distances between atoms in a regular crystal [2.10]. This yields the total number of atoms per volume if the crystal structure is known.

Let us consider a cubic crystal, where the atoms sit at the corners of small cubes with sidelength $a$ (Fig. 2.9). When a plane wave with wavelength $\lambda$ is incident on the crystal under an angle $\vartheta$ against a crystal plane (Fig. 2.10) the partial waves scattered by the different atoms of adjacent planes with distance $d$ interfere with each other. In the direction $-\vartheta$, which corresponds to the direction of specular reflection, their path difference is $\Delta s = 2d \cdot \sin\vartheta$. If $\Delta s$ equals an integer $m$ of the wavelength $\lambda$, the interference is constructive and the amplitude of the different partial waves add up. This is expressed by the Bragg condition

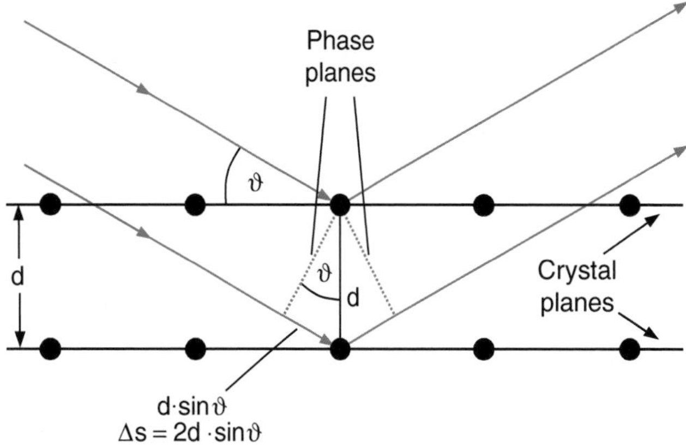


Fig. 2.10. Bragg-reflection of X-rays by two crystal planes

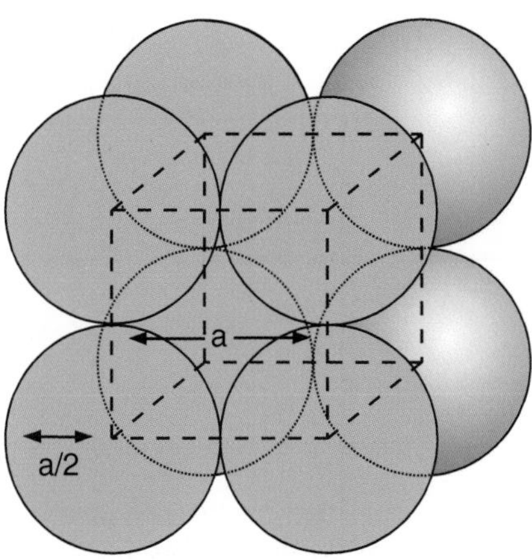


Fig. 2.9. Elementary cell of a cubic crystal

$$2d \cdot \sin\vartheta = m \cdot \lambda\,. \tag{2.17}$$

At a given wavelength $\lambda$ one obtains maxima of intensity $I(\vartheta)$ of the scattered radiation only for those inclination angles $\vartheta$, for which (2.17) is fulfilled.

One sees from (2.17) that for $m > 0$ the wavelength

$$\lambda = \frac{2d}{m}\sin\vartheta < 2d$$

has to be smaller than twice the distance d between adjacent crystal planes. For visible light $\lambda \gg d$, but for X-rays of sufficient energy $\lambda < 2d$ can be achieved (see Sect. 7.6).

**Note:**

In (2.17) $\vartheta$ is the angle of the incident radiation against the crystal planes *not* against the normal to the planes, different from the conventional definition in optics.

The distances $d_k$ between neighboring parallel planes depend on the orientation angle $\alpha$ of these planes against the surface planes at the cube. For a cubic crystal we conclude from Fig. 2.11:

$$\begin{aligned} d_k &= a \cdot \sin\alpha_k && \text{for} \quad \alpha_k \neq 0 \\ d_k &= a && \text{for} \quad \alpha_k = 0\,, \end{aligned} \tag{2.18}$$

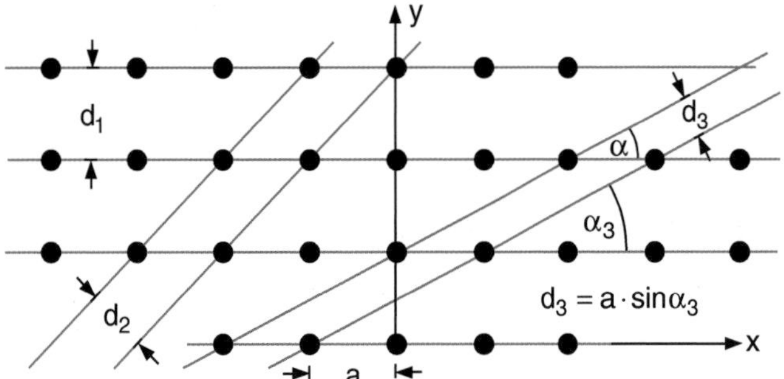


**Fig. 2.11.** Examples of crystal planes perpendicular to the drawing plane with different separations $d_i$

where the lattice constant $a$ gives the distance between neighboring atoms. If the crystal is turned against the direction of the incident beam, one obtains for different angles $\vartheta_m$, maxima of the diffracted radiation, when

$$d_k \cdot \sin\vartheta = a \cdot \sin\alpha_k \cdot \sin\vartheta_m = m \cdot \lambda \quad \text{for} \quad m = 1, 2, 3, \dots .$$

If the wavelength $\lambda$ is known (see Sect. 7.6) the distances $d_k$ between adjacent planes and therefore the lattice constant $a$ can be determined from the measured angles $\vartheta_\mathrm{m}$.

A macroscopic crystal cube with sidelength $D$ contains $N = (D/a)^3$ atoms if one atom sits at every "lattice point". The Avogadro number is then

$$N_\mathrm{A} = N \cdot \frac{M_\mathrm{m}}{M_\mathrm{c}} = \frac{D^3}{a^3} \cdot \frac{M_\mathrm{m}}{M_\mathrm{c}} , \tag{2.19}$$

where $M_\mathrm{c}$ is the mass of the crystal, measured by weighing, and $M_\mathrm{m}$ is its molar mass.

**Note:**

There are considerations to replace the reference for a mass unit (until present it has been a 1 kg cylinder of platinum-iridium in Paris) by a perfectly polished and more readily measurable spherical crystal of silicon, which forms a cubic-face-centered crystal [2.11]. In such a crystal each elementary cubic cell with sidelength $a$ contains 4 Si-atoms (One at a corner and three at the centers of 3 side faces). A sphere with radius $R_\mathrm{s} \gg a$ then contains

$$N = 4 \cdot \frac{4}{3}\pi \frac{R_\mathrm{s}}{a^3} = \frac{4V}{a^3} \tag{2.20}$$

Si atoms. With the density $\varrho$ (kg/m$^3$) we obtain from (2.11a,b) the Avogadro number

$$N_\mathrm{A} = N \cdot \frac{M_\mathrm{m}}{M_\mathrm{c}} = \frac{4M_\mathrm{m}}{\varrho \cdot a^3} = \frac{16\pi}{3}\left(\frac{R_\mathrm{s}}{a}\right)^3 \tag{2.21}$$

from measurements of the lattice constant $a$ (by X-ray diffraction) and the density $\varrho$ (or the mass $M$ and radius $R_\mathrm{s}$ of the sphere), since the molar mass $M_\mathrm{m}$ is known.

The experimental results for the distance $a$ obtained from such measurements range from 0.1–0.5 nm, depending on the specific crystal. Since the diameters $d_\mathrm{a}$ of the atoms cannot be larger than the lattice constant $a$ they have to be smaller than 0.1–0.5 nm (see Sect. 2.4). This gives an upper limit for the size of the atoms, forming the crystal.

The most accurate technique for measuring atomic distances in crystals is X-ray interferometry which combines X-ray diffraction with interferometric techniques. Its basic principle is illustrated in Fig. 2.12. The interferometer consists of three parallel slices $S_i$, all milled out of one large single crystal of Si. In the first slice $S_1$ the incident X-ray beam is split into a transmitted beam 1 and a diffracted beam 2. Diffraction at the crystal planes parallel to the $xy$ plane (in Fig. 2.12 schematically drawn as horizontal lines) occurs if the Bragg condition (2.10) is fulfilled. Both beams again suffer diffraction in the second parallel slice $S_2$ where

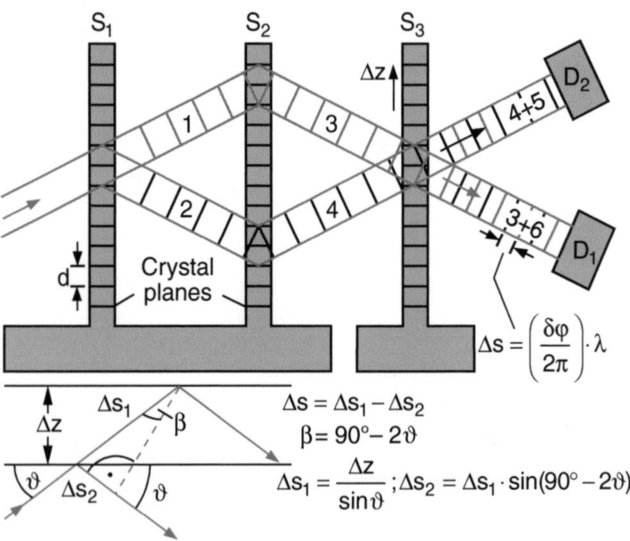


**Fig. 2.12.** An X-ray interferometer that has been milled out of a single crystal

(besides the transmitted beams, not shown in Fig. 2.12) the diffracted beams 3 and 4 are generated, which overlap in the third slice $S_3$ where they can again be transmitted or diffracted.

The transmitted part of beam 4 now interferes with the diffracted part 5 of beam 3 and the detector $D_2$ monitors the total intensity, which depends on the phase difference between the partial waves 4 and 5. Detector $D_1$ measures the interference intensity of the superimposed transmitted beam 3 and the diffracted beam 6 of beam 4.

When the slice $S_3$, which can be moved against the others, is shifted into the $z$-direction by an amount $\Delta z$ the path difference $\Delta s$ between the interfering beams is changed by

$$\delta s = \frac{\Delta z}{\sin\theta}[1 - \sin(90° - 2\vartheta)] = 2\Delta z \cdot \sin\vartheta \,. \tag{2.22}$$

The arrangement is similar to that of a Mach–Zehnder interferometer in optics. However, since the wavelength $\lambda$ of X-rays is about $10^4$ times smaller than that of visible light, the accuracy of the device must be correspondingly higher. If the slice $S_3$ is shifted continuously, the detectors monitor maxima or minima of the inferference intensity every time the path difference $\delta s$ becomes an integer multiple of $\lambda$.

The maxima are counted and its total number $N$ at a total shift $\Delta z$ is

$$N = \frac{2\Delta z \cdot \sin\vartheta}{\lambda} \,. \tag{2.23}$$

The total shift $\Delta z$ is measured with a laser interferometer to within an uncertainty of $\Delta z/z = 10^{-6}$–$10^{-7}$ [2.12, 13].

EXAMPLE

$d = 0.2\,\text{nm}$, $\Delta z = 1\,\text{mm}$, $\vartheta = 30° \to N = 5 \times 10^6$, which allows an accuracy with a relative uncertainty of $2 \times 10^{-7}$.

Instead of X-rays also neutrons can be used [2.14].

Table 2.1 compiles the different methods for the determination of the gas constant $R$, the Boltzmann constant $k$, the Faraday constant $F$, the elementary charge $e$ and Avogadro's number $N_a$. The values of these constants, which are regarded today as the most reliable ones, are given on the inside cover of this book, according to the recommendation of the International Union of Pure and Applied Physics IUPAP (CODATA 2006).

**Table 2.1.** Different methods for the determination of Avogadro's number

| method | fundamental constant | Avogadro's number |
|---|---|---|
| general gas equation | universal gas constant $R$ | $N_A = R/k$ |
| barometric pressure formula (*Perrin*) | Boltzmann's constant $k$ | |
| diffusion (*Einstein*) | | |
| torsionsal oscillations (*Kappler*) | | |
| electrolysis | Faraday's constant $F$ | $N_A = F/e$ |
| Millikan's oil-drop experiment | elementary charge $e$ | |
| X-ray diffraction and interferometry | distance $d$ between crystal planes in a cubic crystal | $N_A = (V/a^3)\frac{M_m}{M_c}$ for cubic primitive crystal; $N_A = 4M_m/\varrho a^3$ for cubic face centered crystal |
| Measurement of atom number $N$ in a single crystal with mass $M_c$ and molar mass $M_m$. | $N_A = N \cdot \frac{M_m}{M_c}$ | |

## 2.2.4 The Importance of Kinetic Gas Theory for the Concept of Atoms

The first ideas of a possible relation between the internal energy $U$ of a gas and the kinetic energies of its molecules were put forward in *1821 by John Herapath. Later in* 1848 *James Prescott Joule* (1818–1889) (Fig. 2.13) who had read the publication of Herapath found by accurate measurements that mechanical energy could be converted into heat energy. Initiated by suggestions from *August Karl Krönig* (1822–1879), *Clausius* and *Maxwell* put these ideas on a more quantitative basis. They derived independently the general equation of gases $pV_m = RT$ from the kinetic energies of the gas molecules. We will here only give a simplified version of the gas kinetic model, which assumes that the gas inside a container with

**Fig. 2.13.** James Prescott Joule

volume $V$ consists of atoms or molecules that can be treated as small rigid balls with radius $r_0$. They can undergo elastic collisions with each other and with the wall. For these collisions energy and momentum are conserved. Collisions with the wall can only occur, if the balls approach the wall within a distance $r_0$. Collisions with each other can only happen when the distance between the ball centers becomes $d = 2r_0$. For larger distances the interaction between the balls is zero. The interaction energy between two hard spheres is therefore (Fig. 2.14):

$$\begin{aligned} E_{\text{pot}}(r) &= 0 \quad \text{for} \quad r \geq 2r_0 \,, \\ E_{\text{pot}}(r) &= \infty \quad \text{for} \quad r < 2r_0 \,. \end{aligned} \tag{2.24}$$

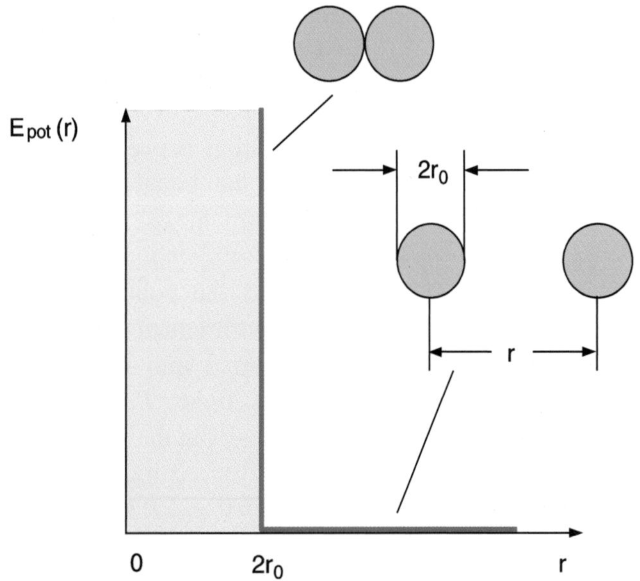


**Fig. 2.14.** Interaction potential between two hard spheres with radius $r_0$

If the density of such a model gas is sufficiently small ($r_0$ should be small compared to the mean distance $\langle d \rangle$ between the particles) it is called an ***ideal gas***. For an ideal gas the eigenvolume $V_e = (4N/3)\pi r_0^3$ of the $N$ particles is small compared to the total volume $V$ of the gas. The particles can therefore be approximately treated as point-like particles.

### EXAMPLE

At a pressure of 1 bar and room temperature $T = 300$ K, 1 cm$^3$ of a gas contains about $3 \times 10^{19}$ atoms or molecules. Their mean distance is $\langle d \rangle = 3$ nm. For helium atoms $r_0 = 0.5$ nm. This gives $r_0 / \langle d \rangle = 0.017 \ll 1$ and $V_e/V = 5 \times 10^{-6}$. Helium under these conditions can therefore be regarded as an ideal gas.

The gas exerts a pressure $p$ onto the wall of the container, which is caused by momentum transfer during collisions of the gas molecules with the wall (Fig. 2.15). Since the pressure is equal to the force per surface $A$ and the force equals the time derivative of the momentum transfer, we obtain the relation for the pressure $p$:

$$p = \frac{\mathrm{d}}{\mathrm{d}t}\left(\frac{\text{momentum transfer to } A}{\text{area } A}\right) . \tag{2.25}$$

If, for example, $N_x$ atoms with velocity $v_x$ hit a wall in the $yz$-plane per second, the momentum transfer per second for completely elastic collisions is $2N_x m v_x$ and the pressure exerted onto the wall is

$$p = 2N_x m v_x / A \,. \tag{2.26}$$

The number density $n$ of $N$ atoms in the volume $V$ is $n = N/V$. Let us first only consider that fraction $n_x$ in a cuboid with volume $V$ that has velocities $v_x$ in the

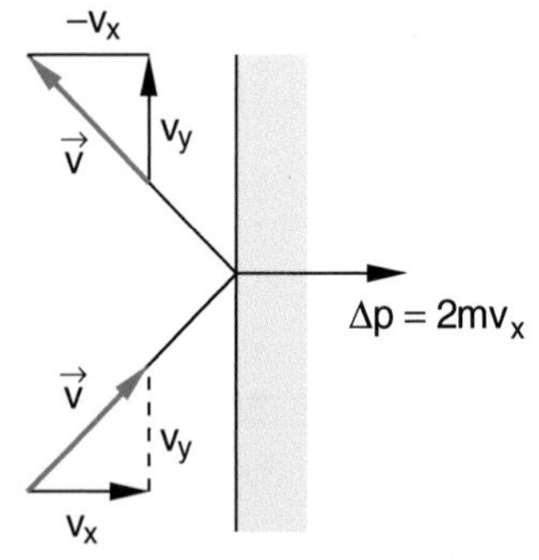


**Fig. 2.15.** Momentum transfer at a particle collision with the wall

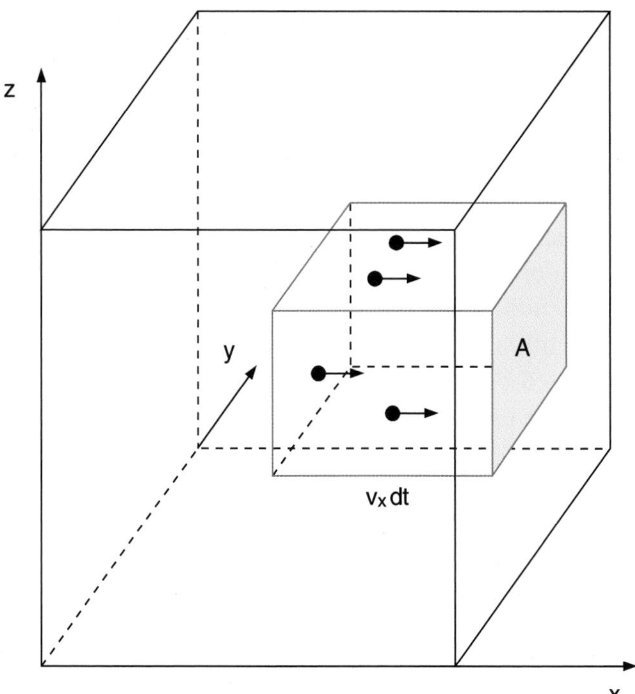


**Fig. 2.16.** Illustration of (2.27)

$x$-direction (Fig. 2.16). Within the time interval $\Delta t$ the number of atoms $Z$ hitting the area $A$ of a wall in the $yz$-plane is

$$Z = n_x v_x A \Delta t \ . \tag{2.27}$$

These are just the atoms inside the blue volume in Fig. 2.16 with length $v_x \Delta t$ and cross section $A$. Each atom transfers the momentum $2mv_x$. Therefore the force acting on the surface element $A$ is $F = 2Zmv_x/\Delta t = 2n_x m v_x^2 A$ and the pressure $p$ acting on the wall is

$$p = 2mn_x v_x^2 \ . \tag{2.28}$$

If an atom moves with the velocity $\boldsymbol{v} = \{v_x, v_y, v_z\}$ at an arbitrary angle against the wall, the momentum transfer to the wall is only caused by its component $v_x$, because the tangential components parallel to the wall do not transfer any momentum (Fig. 2.15).

There is another point we have to consider. Not all atoms have the same velocity. At thermal equilibrium the velocities of a resting gas are isotropic, i.e., each direction is equally probable. Since the pressure of the gas is isotropic the momentum transfer must be the same in all directions. Therefore we obtain for the mean square values

$$\left\langle v_x^2 \right\rangle = \frac{1}{N} \int N_x(v_x) v_x^2 \, \mathrm{d}v_x = \left\langle v_y^2 \right\rangle = \left\langle v_z^2 \right\rangle , \tag{2.29}$$

where $N$ is the total number of atoms or molecules in the gas and $N_x(v_x)$ is the number of molecules with velocity components within the interval $v_x$ to $v_x + \mathrm{d}v_x$. Since on the time average the number of atoms moving into the $+x$-direction equals that into the $-x$-direction, the pressure on a wall in the $yz$-plane is

$$p = \frac{1}{2} n \, 2m \left\langle v_x^2 \right\rangle = nm \left\langle v_x^2 \right\rangle , \tag{2.30}$$

where $n$ is the total number density. From the relation $v^2 = v_x^2 + v_y^2 + v_z^2$ we obtain with (2.29)

$$\left\langle v_x^2 \right\rangle = \left\langle v_y^2 \right\rangle = \left\langle v_z^2 \right\rangle = \frac{1}{3} \left\langle v^2 \right\rangle . \tag{2.31}$$

with the mean kinetic energy of each molecule $\langle E_{\mathrm{kin}} \rangle = (m/2)v^2$ this gives the pressure

$$p = \frac{2}{3} n \cdot \langle E_{\mathrm{kin}} \rangle \ . \tag{2.32}$$

Using (2.30) this gives with $n \cdot V = N$

$$pV = \frac{2}{3} N \frac{1}{2} m \left\langle v^2 \right\rangle , \tag{2.33}$$

Many experiments have proved that the product $pV$ at a constant number $N$ of molecules in the volume $V$ solely depends on the temperature $T$. This means that the mean kinetic energy of the molecules is a function of $T$. One defines the absolute temperature $T$ by the relation

$$\boxed{\frac{m}{2} \left\langle v^2 \right\rangle \underset{\mathrm{def}}{=} \frac{3}{2} kT ,} \tag{2.34}$$

where $k = 1.38054 \times 10^{-23}\,\mathrm{J/K}$ is the ***Boltzmann constant***.

With this definition (2.33) transfers into

$$\boxed{pV = NkT ,} \tag{2.35}$$

which represents a generalization of Boyle–Mariotte's law $pV = \mathrm{const}$ at constant temperature $T$.

This illustrates that the kinetic gas theory can "explain" the macroscopic law (2.35) for ideal gases by a microscopic atomic model where the mean kinetic energy of the atoms are responsible for the pressure and the temperature of the gas.

## 2.3 Can One See Atoms?

The spatial resolution of an optical microscope is limited by the wavelength $\lambda$ of the light used to illuminate an object. With some tricks one may achieve a resolution of $\Delta x \geq \lambda/2$. Using the special technique of near-field microscopy, structures on surfaces can be resolved with about 30 nm resolution. Since we have seen in the preceding paragraph, that the size of atoms is around 0.2–0.5 nm, we cannot expect to see atoms directly through a microscope with visible light ($\lambda \approx 500$ nm).

However, several techniques have been developed that allow an indirect observation of atoms and give detailed information on atomic sizes, structure and dynamics.

First of all, one can mark the location of atoms with size $d \ll \lambda$ through the light scattered by the atoms. When an atom travels through an intense beam of visible laser light, it can absorb and reemit many photons during its flight time through the beam (Fig. 2.17). One can then "see" the atom as a light spot, i.e., as a structureless point and no information about its size or structure can be obtained. One can only say: "It's there."

There are several other methods that give similar information. With computer graphics one can produce nice pictures of such "atom images" on the screen, which may be impressive because they appear to give a magnified picture of the microworld of atoms and molecules. However, one should always keep in mind that such pictures are produced due to the interaction of light or particles with atoms. Only if this interaction is fully understood can the interpretation of the images give a true model of atoms or molecules. This will be illustrated by the different techniques discussed in the following sections.

### 2.3.1 Brownian Motion

The biologist and medical doctor *Robert Brown* (1773–1858) discovered in 1827 that small particles suspended in liquids performed small irregular movements, which can be viewed under a microscope. Although he first thought that these movements were caused by small living bacteria, he soon found out that the movement could also be observed for inorganic particles that are definitely not alive.

The observation can be explained if one assumes that the particles are permanently hit by fast moving atoms or molecules coming from statistically distributed directions (Fig. 2.18).

The visualization of Brownian motion is very impressive. It is possible to demonstrate it to a large auditorium when using cigarette smoke particles in

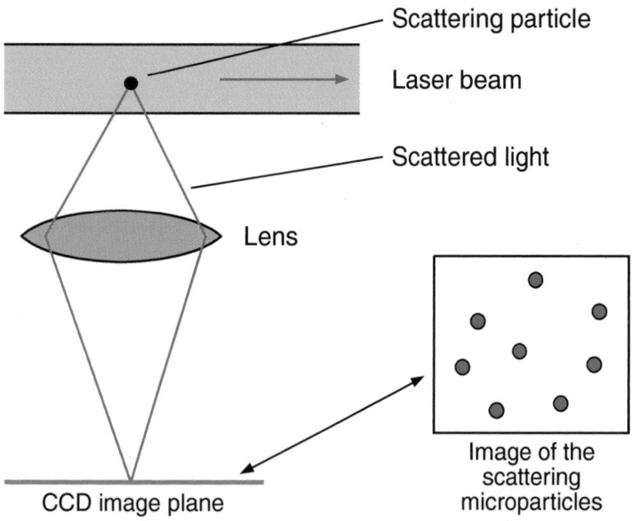

**Fig. 2.17.** Scattering of visible light by single atoms. Each image point corresponds to one atom

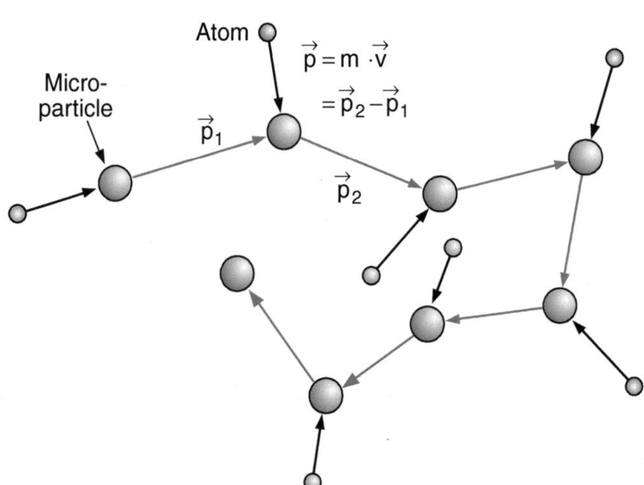


**Fig. 2.18.** Schematic illustration of Brownian motion

air, illuminated by a laser beam and viewed through a microscope with a video camera.

Also here, the atoms are not directly seen but their impact on the smoke particle can be measured and, provided the mass of the smoke particle is known, the atomic momentum transferred to the particle, can be determined.

There is a nice demonstration that simulates Brownian motion. A larger disk on an air table is hit by many small discs, which simulate the air molecules. If the large disc carries a small light bulb, its statistical path over the course of time can be photographed and the path lengths between two successive collisions (the free path) can be measured (Fig. 2.19).

The basic theory of Brownian motion was developed independently in 1905 by *Albert Einstein* (1879–1955) and *Marian Smoluchowski* (1872–1917). It is closely related to diffusion [2.15]. We will only briefly outline the basic ideas here.

Assume particles in a gas show a small gradient $\mathrm{d}n/\mathrm{d}x$ of their number density $n$, described by the linear relation (Fig. 2.20)

$$n(x) = n(0) - Gx \,. \tag{2.36}$$

Under the influence of mutual collisions the particles perform statistical movements with a probability distribution $f(\xi)$ where $\xi$ is the length of such a displacement in the $x$-direction between two collisions. The number density of particles with movement $\xi$, is then:

$$n(\xi)\,\mathrm{d}\xi = nf(\xi)\,\mathrm{d}\xi \quad \text{with} \quad n = \int n(\xi)\,\mathrm{d}\xi \,, \tag{2.37}$$

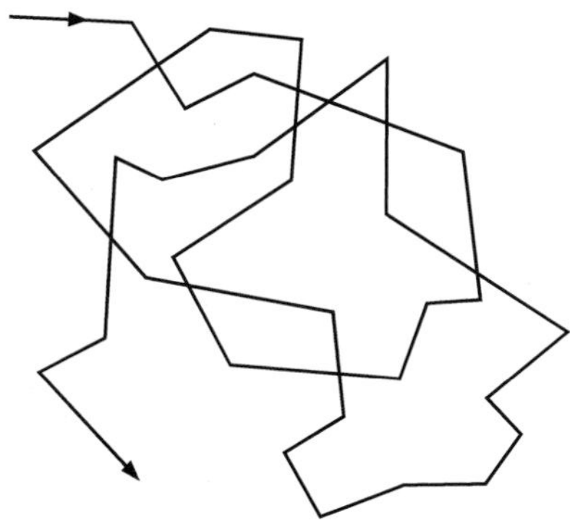

**Fig. 2.19.** Irregular path of a puck on an air table, which is hit statistically by smaller pucks (lecture demonstration of Brownian motion)

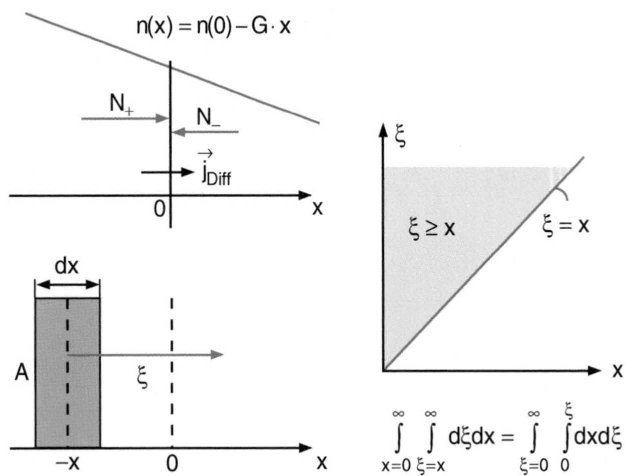


**Fig. 2.20.** Illustrating drawings for the derivation of (2.45)

where the distribution function $f(\xi)$ is defined as

$$f(\xi)\,\mathrm{d}\xi = \frac{1}{n}n(\xi)\,\mathrm{d}\xi \,.$$

For a positive gradient $G$ in (2.36) the number $N_+$ of particles moving through a unit area in the plane $x = 0$ into the $+x$-direction is larger than the corresponding number $N_-$ in $-x$-direction. Therefore, the net particle diffusion flux through a unit area in the plane $x = 0$ is (Fig. 2.20)

$$\boldsymbol{j}_{\text{diff}} = \frac{N_+ - N_-}{\Delta t}\hat{\boldsymbol{e}}_x \,. \tag{2.38}$$

Out of all $n(x)\,\mathrm{d}x$ particles within the volume $\mathrm{d}V = A\,\mathrm{d}x$ centered around the plane $x = -x_i$ with unit area $A$, only those particles with an elongation $\xi > x_1$ can pass through the plane $x = 0$. Their number is

$$\mathrm{d}N_+ = \left[\int_{\xi=-x}^{\infty} n(x)f(\xi)\,\mathrm{d}\xi\right]\mathrm{d}x \,. \tag{2.39}$$

Integration over all volume elements along the negative $x$-axis yields with (2.36)

$$N_+ = \int_{x=-\infty}^{0}\left(\int_{\xi=-x}^{\infty}(n(0) - Gx)f(\xi)\,\mathrm{d}\xi\right)\mathrm{d}x \,. \tag{2.40a}$$

Renaming the variable $x = -x'$ gives

$$N_+ = \int\limits_{x'=0}^{\infty} \left( \int\limits_{\xi=x'}^{\infty} (n(0) + Gx') f(\xi)\, d\xi \right) dx' . \tag{2.40b}$$

In a similar way we obtain for the rate $N_-$ of particles moving from right to left in Fig. 2.20

$$N_- = \int\limits_{x=0}^{\infty} \left( \int\limits_{\xi=-x}^{-\infty} (n(0) - Gx) f(\xi)\, d\xi \right) dx , \tag{2.41a}$$

which can be transformed by the substitution $\xi \to -\xi'$ (note that the distribution function $f(\xi)$ is symmetric and therefore $f(-\xi) = f(\xi)$) into

$$N_- = \int\limits_{x=0}^{\infty} \left( \int\limits_{\xi'=x}^{\infty} (n(0) - Gx) f(\xi')\, d\xi' \right) dx . \tag{2.41b}$$

Since the name of a variable is irrelevant, we can rename $x' \to x$ in (2.40b) and $\xi' \to \xi$ in (2.41b). Subtracting (2.41b) from (2.40b) we obtain the difference

$$\begin{aligned} N_+ - N_- &= 2G \int\limits_{x=0}^{\infty} \left( \int\limits_{\xi=x}^{\infty} f(\xi)\, d\xi \right) x\, dx \\ &= 2G \int\limits_{\xi=0}^{\infty} \left( \int\limits_{x=0}^{\xi} x\, dx \right) f(\xi)\, d\xi , \end{aligned} \tag{2.42a}$$

where the interchange of the integration limits does not change the double integral since both cover the blue area in Fig. 2.20.

Integration over $x$ gives

$$\Delta N = G \int\limits_{0}^{\infty} \xi^2 f(\xi)\, d\xi . \tag{2.42b}$$

Since the distribution function $f(\xi)$ is symmetric ($f(\xi) = f(-\xi)$) we can write (2.42b) as

$$\Delta N = \frac{1}{2} G \int\limits_{\xi=-\infty}^{+\infty} \xi^2 f(\xi)\, d\xi = \frac{1}{2} G \left\langle \xi^2 \right\rangle \tag{2.42c}$$

because the average $\left\langle \xi^2 \right\rangle$ is defined as

$$\left\langle \xi^2 \right\rangle = \int\limits_{-\infty}^{+\infty} \xi^2 f(\xi)\, d\xi . \tag{2.43}$$

Inserting (2.42c) into (2.38) yields the relation

$$j_{\text{diff}} = \frac{1}{2} \frac{\left\langle \xi^2 \right\rangle}{\Delta t} G \tag{2.44a}$$

between particle diffusion flux density and density gradient $G$.

According to (2.13) we can also write $j_{\text{diff}}$ as

$$j_{\text{diff}} = -D \frac{\partial n}{\partial x} = -D \cdot G . \tag{2.44b}$$

The comparison of (2.44a) with (2.44b) gives the diffusion coefficient

$$\boxed{D = \frac{1}{2} \frac{\left\langle \xi^2 \right\rangle}{\Delta t}} \tag{2.45a}$$

expressed by the average squared elongation of the particles on their statistical path (Fig. 2.19).

The diffusion process is due to the kinetic energy of the particles and their collisions in a medium with a density gradient. The influence of collisions can be macroscopically described by the viscosity $\eta$. For spherical particles with radius $r$, where the friction force equals $6\pi\eta r v$ (Stokes' law), we obtain according to (2.14)

$$D = \frac{kT}{6\pi\eta r} \tag{2.45b}$$

and therefore the mean square deviation of a particle from its position at time $t = 0$ in a gas with viscosity $\eta$ is

$$\left\langle \xi^2 \right\rangle = \frac{kT}{3\pi\eta r} \Delta t . \tag{2.46}$$

It depends on the temperature $T$, the viscosity $\eta$ and the radius $r$ of the particle and increases linearly with time $\Delta t$.

> The quantity $\sqrt{\langle \xi^2 \rangle}$, which is a measure of the mean deviation of a particle from its original location at time $t = 0$ increases during the time interval $\Delta t$ only with the square root $\sqrt{\Delta t}$ (Fig. 2.21).

If the mean quadratic deviation $\langle \xi^2 \rangle$ can be measured, the Boltzmann constant $k$ can be determined from (2.46).

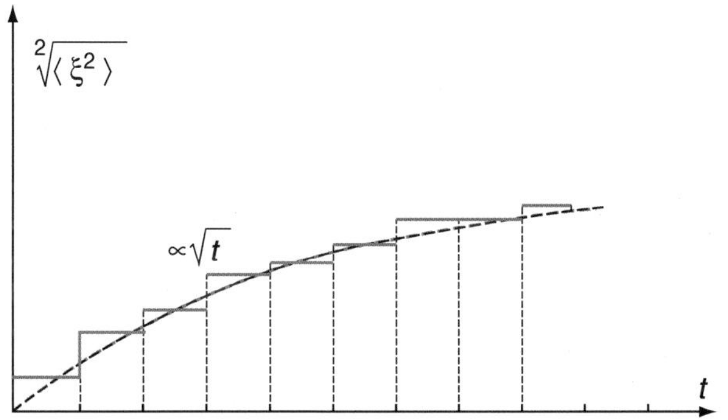


**Fig. 2.21.** Square root of mean square deviation $\langle \xi^2 \rangle$ of a diffusing particle as a function of time.

*Eugen Kappler* (1905–1977) demonstrated an elegant experimental technique in 1939 [2.16]. A modern version of it is shown in Fig. 2.22a. A small mirror is suspended on a thin torsional wire. The air molecules impinge on the mirror surface and cause, by their momentum transfer, small statistical angular deviations $\Delta\varphi$ of the mirror from its equilibrium position at $\varphi = 0$, which can be monitored by the reflection of a laser beam, detected with a position-sensitive CCD detector.

The system has only one degree of freedom; it can only perform torsional vibrations around the axis defined by the torsional wire. With the deviation angle $\varphi$ from the equilibrium position $\varphi = 0$, the mean potential and kinetic energies are:

$$\langle E_{\text{pot}} \rangle = \frac{1}{2} D_{\text{r}} \langle \varphi^2 \rangle = \frac{1}{2} kT \,, \tag{2.47a}$$

$$\langle E_{\text{kin}} \rangle = \frac{1}{2} I \langle \dot{\varphi}^2 \rangle = \frac{1}{2} kT \,, \tag{2.47b}$$

where $D_{\text{r}}$ is the restoring torque of the torsional wire and $I$ the moment of inertia of the system.

The statistical deviations of $\varphi$ from the equilibrium position $\varphi = 0$ of the mirror (Fig. 2.22b) follow a Gaussian probability distribution (Fig. 2.22c)

$$P(\varphi) = P(0) \mathrm{e}^{-\varphi^2 / \langle \varphi^2 \rangle} \,. \tag{2.48a}$$

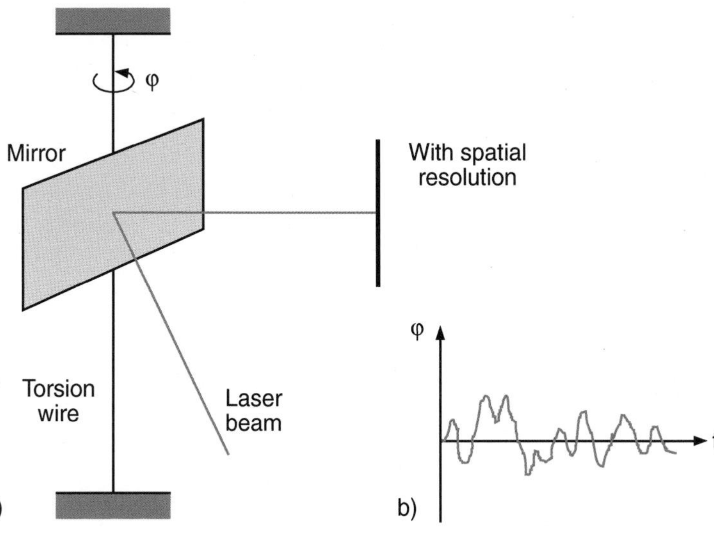


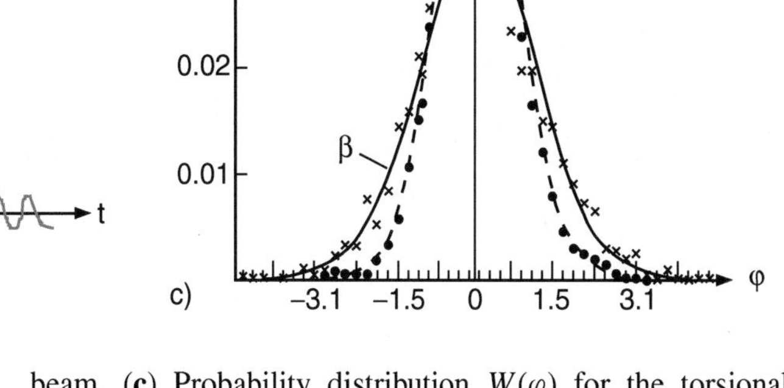


**Fig. 2.22a–c.** Determination of the Boltzmann constant $k$ from the Brownian torsional motion of a mirror. (**a**) Experimental setup. (**b**) Statistical path of the reflected light beam. (**c**) Probability distribution $W(\varphi)$ for the torsional elongation $\varphi$ for an averaging time of 0.55 s ($\alpha$) and 0.27 s ($\beta$)

The measured full half-width

$$\Delta\varphi = 2\sqrt{\langle\varphi^2\rangle \ln 2} \tag{2.48b}$$

of this distribution yields the mean square deviation $\langle\varphi^2\rangle$ and from (2.47) the Boltzmann constant

$$k = (D_r/T)\langle\varphi^2\rangle = (D_r/T)\frac{(\Delta\varphi)^2}{4\ln 2} . \tag{2.49}$$

### 2.3.2 Cloud Chamber

*Charles T. Wilson* (1869–1959) developed his cloud chamber in 1911, which allowed him to view the spur of single fast atoms, ions or electrons entering the chamber. The basic principle is as follows: Incident particles with sufficient kinetic energy can ionize the atoms or molecules in the cloud chamber, which is filled with supersaturated water vapor (nowadays alcohol vapor is generally used). The ions, formed along the spur of the incident particle, attract the polar water molecules and act as condensation nuclei for the formation of small water droplets. When the chamber is illuminated by visible light, the droplets cause Mie-scattering which makes the particle track visible as a thin bright spur of tiny water droplets (Fig. 2.23).

Although the observation of the different tracks in the cloud chamber is impressive, it does not allow a direct view of the incident particles themselves but only locates their paths through the chamber. In former times the cloud chamber was used for the observation of nuclear reactions (see the example in Fig. 2.23). Nowadays a new device, the bubble chamber, has replaced it and the cloud chamber is mainly used for demonstration purposes.

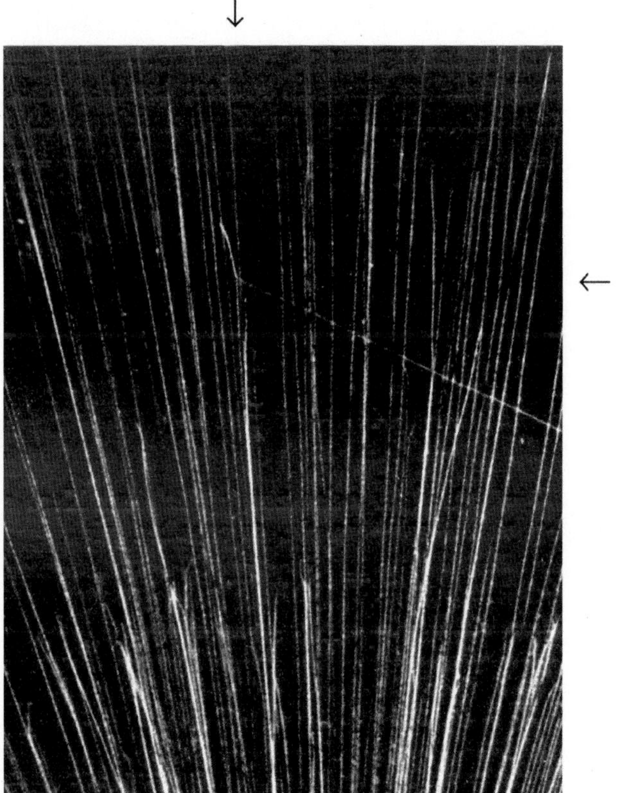

**Fig. 2.23.** Cloud chamber tracks of $\alpha$ particles (= He nuclei), which are emitted from a source below the lower edge of the photograph. One $\alpha$ particle collides with a (not visible) nitrogen nucleus at the crossing point of the two arrows, forming an $^{17}_{8}$O nucleus and a proton. The O nucleus flies towards 11 o'clock, the proton towards 4 o'clock (from W. Finkelnburg: Einführung in die Atomphysik, Springer, Berlin Heidelberg New York, 1976)

### 2.3.3 Microscopes with Atomic Resolution

During the last four decades of the 20th century, new devices have been developed that allow a spatial resolution within the subnanometer range, and are therefore capable of making single atoms "visible." Since their basic understanding demands the knowledge of atomic physics and solid state physics, they can only be explained here in a more qualitative way while for their quantitative description the reader is referred to the literature [2.17, 18].

#### a) Field Emission Microscope

The oldest of these devices is the field emission electron microscope (Fig. 2.24) developed by *Ernst Müller* in 1937 [2.19]. A very sharp tip at the end of a thin tungsten wire serves as a cathode in the middle of an evacuated glass bulb. The anode has the form of a sphere and is covered on the inside with a fluorescent layer (like a television screen). When a voltage $V$ of several kilovolts is applied between cathode and anode, the electric field strength at the cathode surface is

$$\boldsymbol{E} = \frac{V}{r} \cdot \hat{\boldsymbol{e}}_r ,$$

where $r$ is the radius of the nearly spherical tip of the tungsten wire (Fig. 2.24b). With special etching

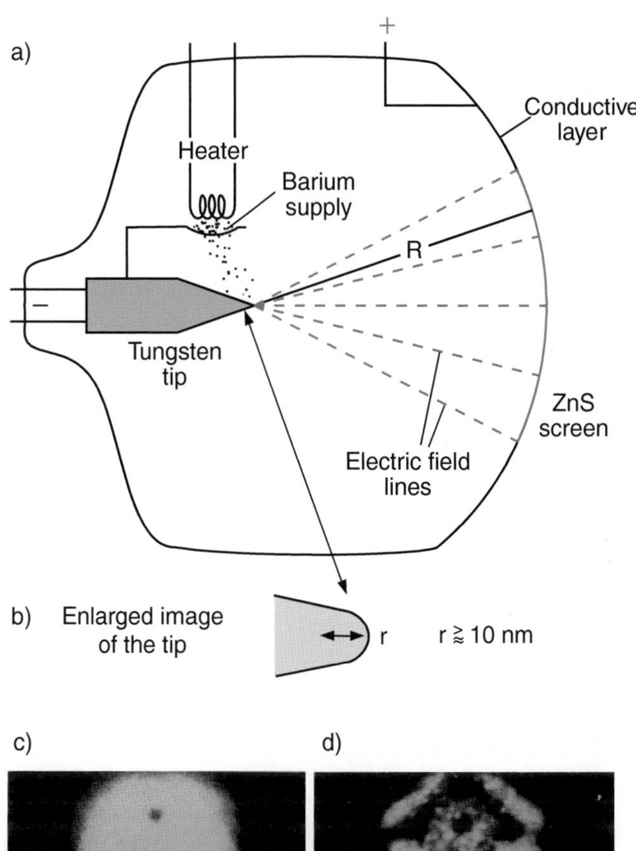


**Fig. 2.24.** (**a**) Basic concept of the field emission microscope. (**b**) Enlarged view of the tungsten tip. (**c**) Image of the tungsten surface around the tip, $10^7$-fold enlarged on the screen of the field emission microscope. (**d**) Visualization of Ba atoms on the tungsten tip

techniques it is possible to fabricate tips with $r < 10\,\mathrm{nm}$! This means that for a moderate voltage $V = 1\,\mathrm{kV}$ the electric field at the surface of the tungsten tip is already $E \geq 10^{11}\,\mathrm{V/m}$. Such high electric fields exceed the internal atomic fields (see Sect. 3.5) and are sufficiently large to release electrons from the metal surface (field emission, see Sect. 2.5.3). These electrons are accelerated by the electric field, follow the electric field lines, and impinge on the fluorescent screen at the anode where every electron causes a small light flash, similar to the situation at the screen of an oscilloscope. Most of the electrons are emitted from places at the cathode surface where the work function (i.e., the necessary energy to release an electron) is minimum. These spots are imaged by the electrons on the spherical anode (radius $R$) with a magnification factor $M = R/r$. With $R = 10\,\mathrm{cm}$ and $r = 10\,\mathrm{nm}$ a magnification of $M = 10^7$ is achieved (Fig. 2.24).

Even with this device, only the locations of electron emission are measured but no direct information on the structure of atoms is obtained. If other atoms with a small work functions are brought to the cathode surface (for example by evaporating barium atoms from an oven near the cathode) then the electron emission mainly comes from these atoms. One can now see these atoms and their thermal motions on the cathode surface with $10^7$ fold magnification (Fig. 2.24d).

### b) Transmission Electron Microscope

The electron microscope, first invented by Ernst Ruska in 1932 has meanwhile been improved so much that it reaches a spatial resolution of 0.1 nm [2.20–22]. The electrons are emitted from a heated cathode wire with a sharp kink (hair needle cathode) and are accelerated by a high voltage (up to 500 kV). With specially formed electric or magnetic fields, serving as electron optics (see Sect. 2.6) the electrons are imaged onto the sample, which is prepared as a thin foil (Fig. 2.25). While transmitting through the sample, the electrons are deflected by elastic collisions or loose energy by inelastic collisions. The transmitted electrons are imaged again onto a fluorescent screen where a magnified image of the absorption or scattering centers in the sample is produced (Fig. 2.26), which can be viewed either through an optical microscope or with a CCD camera and an electronic image converting system.

The spatial resolution of the electron microscope increases with decreasing size of the electron source. A nearly point-like source can be realized with field emission from a sharp edged tungsten tip (Fig. 2.27) like that in the field emission microscope. The emitted electrons can than be imaged by the electron optics to form a nearly parallel beam that traverses the sample. Each point of the sample is then imaged with a large magnification onto the screen.

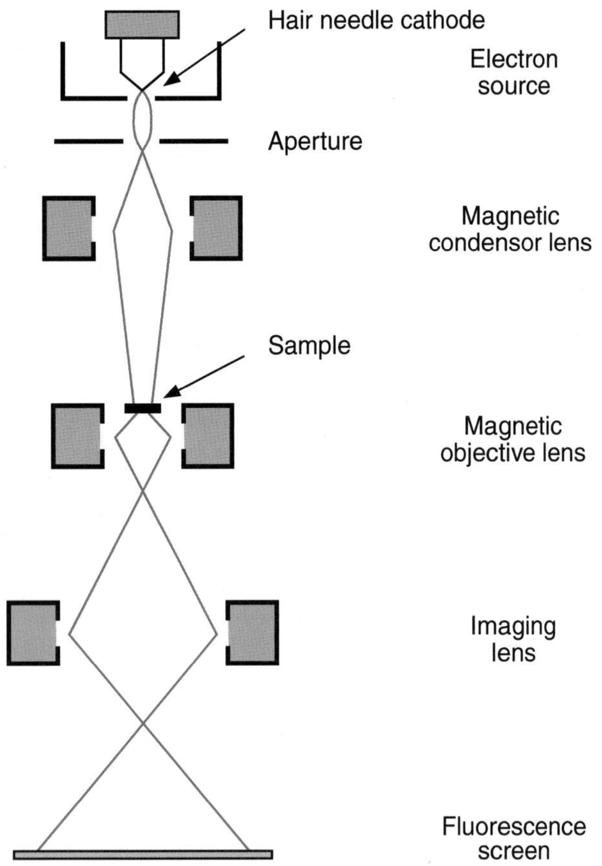


**Fig. 2.25.** Principle setup of the transmission electron microscope

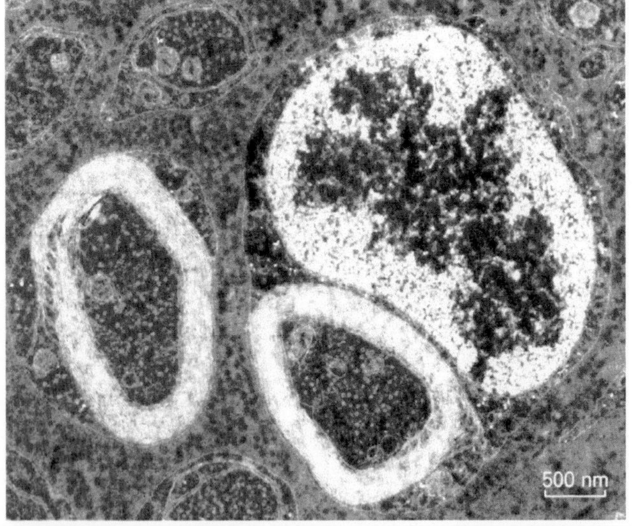


**Fig. 2.26.** Image of nerve cells in a thin undyed frozen slice taken with a transmission electron microscope (with kind permission of Zeiss, Oberkochen)

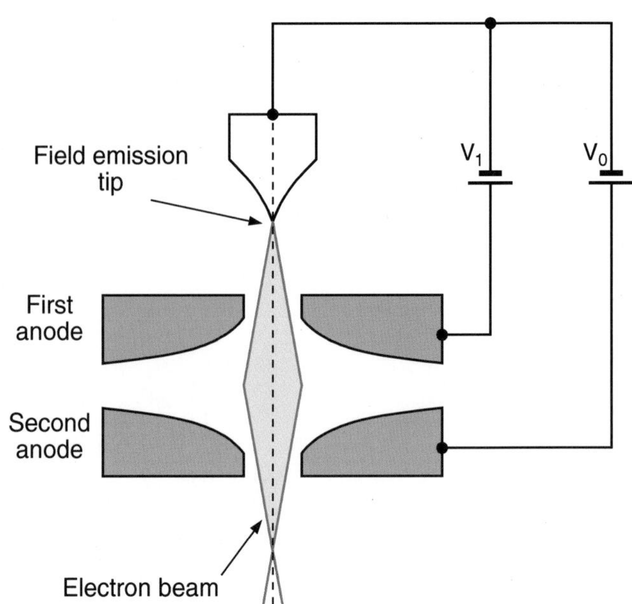


**Fig. 2.27.** Field-emission electron source where the electrons are emitted from a point-like tungsten tip and imaged by electrostatic lenses

The drawbacks of the transmission electron microscope are the following:

- Due to strong absorption of electrons by solid materials, the penetration depth is very small. One therefore has to prepare the sample as a thin sheet.
- The electron beam has to be intense in order to obtain sufficient image quality with a high contrast. This means a larger current density $j$ and total electron current $I = Aj$, where $A$ is the illuminated area.
- The unavoidable absorption heats the sample up, which may change its characteristics or may even destroy parts of the sample. This is particularly critical for biological samples.

Most of these drawbacks can be avoided with the scanning electron microscope.

### c) Scanning Electron Microscope

In the scanning electron microscope (Fig. 2.28) the electron beam is focused onto the surface of the sample (which now is not necessarily a thin sheet), where it produces light emission by excitation of the sample molecules and secondary electrons by impact ionization. The electron beam is scanned over the surface of

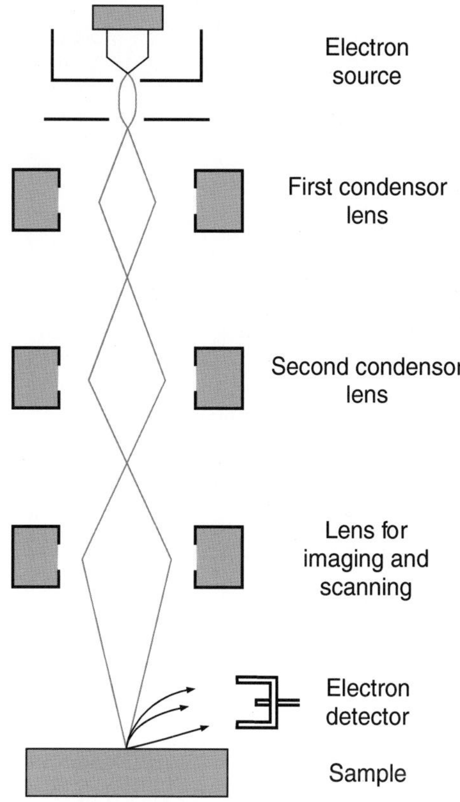


**Fig. 2.28.** Scanning electron microscope

the sample by an appropriate deflection program for the electron optics. This is quite similar to the situation in a TV tube.

The fluorescence light can be viewed through an optical microscope or the secondary electrons, emitted from the surface element $\mathrm{d}x\,\mathrm{d}y$ of the sample, are extracted by an electric extraction field and imaged onto a detector where a signal $S(x, y, t)$ is produced that depends on the intensity of the secondary electrons emitted from the small focal area $\mathrm{d}x\,\mathrm{d}y$ around the point $(x, y)$, which in turn depends on the characteristic properties of the sample at that location [2.23, 24].

### d) Scanning Tunneling Microscope

The highest spatial resolution of structures on electrical conducting solid surfaces has so far been achieved with the scanning tunneling microscope, invented at the research laboratories of IBM in Rüchlikon, Switzerland [2.25, 26] in 1984 by *Gerd Binning* (*1947) and *Heinrich Rohrer* (*1933), who were awarded the Nobel Prize in 1986 for this invention.

Similar to the electron field microscope a tungsten needle with a very sharply etched tip is used, which is however, not fixed but is scanned in a controllable way at a very small distance (a few tenths of a nanometer) over the surface (Fig. 2.29).

If a small voltage of a few volts is applied between the tip (cathode) and the surface (anode) the electrons can jump from the needle into the surface by a process called tunneling (see Sect. 4.2.3). The electric current depends exponentially on the distance between tip and surface. When the tip is scanned over the surface by piezo elements (these are ceramic cylinders that change their length when an electric voltage is applied to them), any deviation of the surface in the $z$-direction from the exact $xy$-plane results in a change of the tunnel current.

Generally the tunnel current is kept constant by a controlled movement of the tip in vertical direction, which always keeps it at the same distance $\Delta z$ from the real surface and therefore reflects the topography $z(x, y)$ of the surface. The control current of the piezo element for the movement in z-direction is then taken as the signal (Fig. 2.30), which is transferred to a computer where a magnified picture of the surface can be seen (Fig. 2.31). If single atoms or molecules are adsorbed at the surface, they can be viewed by this technique, because the distance to the needle is changed at the location of the atom.

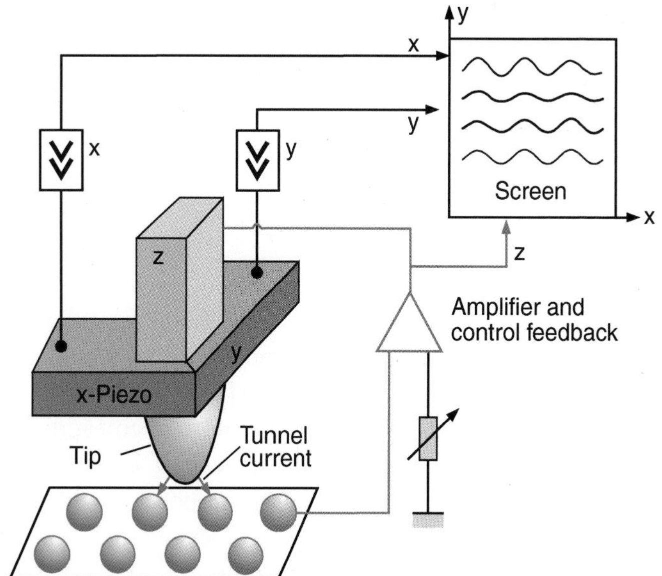


**Fig. 2.29.** Scanning tunneling microscope

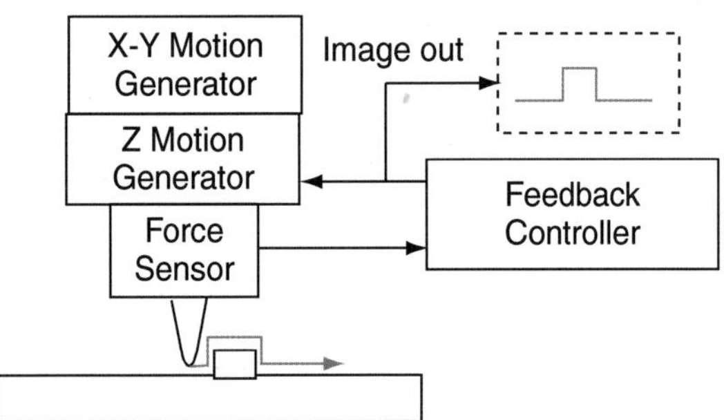


**Fig. 2.30.** Electronic Control and imaging of a tunnelling microscope

**Fig. 2.31.** Arsenic atoms at the surface of a galliumarsenate single crystal, visualized by a scanning tunneling microscope (image size 17nm × 17nm). Non-periodic structures and steps on the surface can be resolved on an atomic scale (with kind permission from A.J. Heinrich, W. Wenderath and R.G. Ulbricht, University of Göttingen)

An impressive demonstration of the capabilities of the tunneling microscope was published by scientists at the IBM research center in San Jose [2.29]. A clean nickel surface at low temperatures was covered by a few Xenon atoms. With the tunneling microscope these atoms were picked up and transported to selected places on the surface, where they were released in such a way that the atoms formed the letters IBM (Fig. 2.32). A similar experiment was performed by Karl-Heinz Rieder and his group at the Free University of Berlin [2.30], who wrote the letters "FU" on a copper surface using CO molecules (Fig. 2.33).

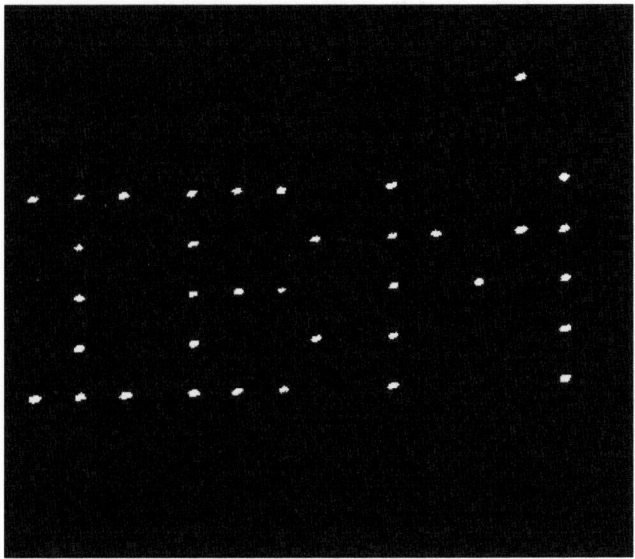

**Fig. 2.32.** Manipulation of single Xe atoms on a Ni(110) surface with an atomic force microscope (with kind permission of Dr. Eigler)

**Fig. 2.33.** Arrangement of single CO molecules on a copper surface to form the letters FU (with kind permission of Prof. Rieder, FU Berlin)

This is probably the most impressive way to make single atoms "visible", since here not only the location but also the size of the atoms or molecules can be measured.

### e) Atomic Force Microscope

The tunneling microscope is restricted to the investigation of conductive surfaces because for nonconductive surfaces the tunnel current would result in surface charges that alter the potential and therefore the voltage between needle and surface. This limitation can be overcome with the atomic force microscope, which uses a similar design as the tunneling microscope [2.31]. However, here it is not the tunnel current that is measured but the force between the atoms of the tip and those at the surface. These very small forces are measured by monitoring, through the reflection of a laser beam, the small upwards or downwards shift of the carefully balanced lever carrying a small and sharp-edged tip [2.33].

For a survey on these different modern techniques the reader is referred to the literature [2.25–35].

## 2.4 The Size of Atoms

There are many experimental methods that allow the estimation of atomic size. We will provide a brief survey here.

In a liquid (for instance in liquid helium or argon) the atoms can be assumed to be densely packed (otherwise the liquid could be more readily compressed). If one mole of a liquid with density $\rho_l$ occupies the volume $V_M$ and has a mass $M_M$ the volume of a single atom is

$$V_a \lesssim V_M/N_A = M_M/(\rho_l N_A) \ , \tag{2.50a}$$

where $N_A$ is the Avogadro number. Assuming a spherical size of the atom, its radius $r_0$ is

$$r_0 = (3V_a/4\pi)^{1/3} \ . \tag{2.50b}$$

After discussing two further methods we will see in Sect. 2.8 that atoms cannot be regarded as rigid balls with a well-defined radius. The electron cloud around the atomic nucleus can be described by a charge distribution that gradually decreases with increasing radius and differs from the mass distribution within the atom. The definition of atomic size and atomic radius is therefore dependent on the interaction between the atom and the probe used to measure these quantities. Different methods will therefore yield slightly different atomic sizes.

### 2.4.1 The Size of Atoms in the Van der Waals Equation

While an ideal gas (point-like particles without interactions) obeys the general equation

$$pV_M = RT \tag{2.51a}$$

between pressure $p$, mole volume $V_M$ and temperature $T$, a real gas with atoms of volume $V_a$ that interact with each other, is described by the van der Waals equation

$$\left(p + \frac{a}{V_M^2}\right)(V_M - b) = RT \ . \tag{2.51b}$$

The constant $b = 4N_A V_a$ equals four times the "eigenvolume" of all $N_A$ atoms in the mole volume $V_M$ (see Problem 2.10), while $a/V_M^2$ gives the "eigenpressure" of the interacting atoms. The constant $a$ depends on the strength of the interaction between the atoms.

Measuring the relation between $p$ and $V_M$ at different temperatures $T$ allows the separate determination of the two constants a and b and therefore yields the "covolume $b$" and the volume

$$V_a = \frac{b}{4N_A} \tag{2.51c}$$

of a single atom.

### 2.4.2 Atomic Size Estimation from Transport Coefficients

When the characteristic quantities of a gas such as mass density, energy density or momentum are not constant over the volume of the gas, the gradients of these quantities cause transport phenomena that finally lead to equilibrium at a homogeneous distribution if the gradient is not maintained by external influences.

For density gradients, diffusion takes place where mass is transported, for temperature gradients, heat conduction occurs where energy is transported and for velocity gradients, the momentum of the molecules is transferred.

All these transport phenomena are realized on a microscopic scale by collisions between atoms or molecules and therefore the mean free path length $\Lambda$ (i.e., the mean distance an atom travels between two collisions) plays an important role for the quantitative description of all these phenomena.

In a gas at thermal equilibrium with atom number density $n$ the mean free path length is given by

$$\Lambda = (1/\mathrm{n} \cdot \sigma)\sqrt{<\mathrm{v}^2> / <\mathrm{v_r}^2>} \tag{2.52a}$$

where $<\mathrm{v}^2>$ is the mean square velocity of the gas particles, $<\mathrm{v_r}^2>$ is the mean square relative velocity and $\sigma = (r_1 + r_2)^2$ is the collision cross section. The cross section $\sigma = \pi(r_A + r_B)^2$ for collisions between atoms A and B is defined as the circular disc with radius $(r_A + r_B)$ around the center of atom A with radius $r_A$ through which atom B with radius $r_B$ has to pass in order to touch atom A and suffer a collision (Fig. 2.34).

The relative velocity of two particles with velocities $\mathrm{v_i}$ and $\mathrm{v_k}$ is $\mathrm{v_r} = \mathrm{v_i} - \mathrm{v_k}$. Therefore $<\mathrm{v_r}^2>= <(\mathrm{v_i} - \mathrm{v_k})^2>=<\mathrm{v_i}^2> + <\mathrm{v_k}^2> -2 <\mathrm{v_i} \cdot \mathrm{v_k}>$. Since the directions of the velocities $\mathrm{v_i}$ and $\mathrm{v_k}$ are randomly orientated, the mean scalar product is zero. If all particles of the gas have the same mass, $\mathrm{v_i}^2 = \mathrm{v_k}^2 = \mathrm{v}^2$. Therefore $<\mathrm{v_r}^2>= 2 <\mathrm{v}^2>$. **This means the the mean square relative velocity of a gas of equal atoms is twice the mean square of the absolute velocity**. The free mean pathlength $\Lambda$ then becomes

$$\Lambda = \frac{1}{n \cdot \sigma \cdot \sqrt{2}} \tag{2.52b}$$

replacing the number density $n$ by the gas pressure $p = n \cdot k \cdot T$ gives

$$\Lambda = \frac{kT}{p \cdot \sigma \cdot \sqrt{2}} \tag{2.52c}$$

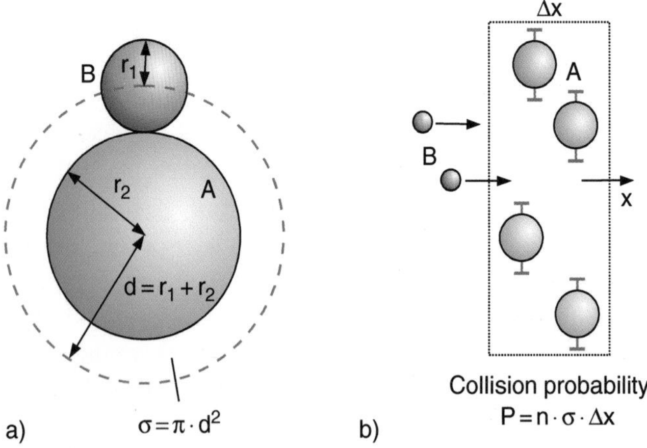


**Fig. 2.34.** Determination of atomic size from the collision cross section $\sigma = \pi d^2$

Measuring the mean free pathlength $\Lambda$ gives information on the collision cross section and therefore on the size of the colliding atoms.

The above mentioned transport phenomena are directly related to $\Lambda$. They can therefore be used to measure the size of atoms or molecules.

- ***Diffusion***:
  If a density gradient $\mathrm{d}n/\mathrm{d}z$ exists in a gas, there will be a net mass transport $\mathrm{d}M/\mathrm{d}t$ per second through the area $A$ perpendicular to the $z$-direction. The mass flux density is then

$$j_{ZM} = \frac{1}{A}\frac{\mathrm{d}M}{\mathrm{d}t} = -Dm\frac{\mathrm{d}n}{\mathrm{d}z}\,. \tag{2.53}$$

  The diffusion coefficient $D$ for atoms with mass $m$ and number density $n$ can be calculated as

$$D = \frac{1}{3}\bar{v}\Lambda = \frac{2}{3p\sigma}\frac{(kT)^{3/2}}{(\pi m)^{1/2}} \tag{2.54}$$

  because the mean velocity of the atoms is $\bar{v} = (8kT/\pi m)^{1/2}$.

- ***Heat conduction***
  In a gas with a temperature gradient $\mathrm{d}T/\mathrm{d}z$ the heat energy transported per second through the area $A$ is given by

$$\frac{\mathrm{d}Q}{\mathrm{d}t} = -\lambda A\frac{\mathrm{d}T}{\mathrm{d}z}\,, \tag{2.55}$$

  where $\lambda$ is the coefficient of heat conduction. It is related to the specific heat $c_v$ of the gas at constant volume by

$$\lambda = \frac{1}{3}nmc_v\bar{v}\Lambda = \frac{2c_v}{3\sigma}\sqrt{\frac{kTm}{\pi}}\,. \tag{2.56}$$

  Measuring the coefficient $\lambda$ therefore yields the collision cross section $\sigma$ and with it the atomic radius.

- ***Viscosity of a Gas***
  If a velocity gradient $\mathrm{d}v_y/\mathrm{d}x$ exists in a gas flowing in the $y$-direction with the velocity $\mathrm{v_y}$, part of their momentum per unit volume

$$p = n \cdot m \cdot <v_y>$$

  is transferred between adjacent layers $x = a$ and $x = a + dx$ (Fig. 2.35). This momentum transfer

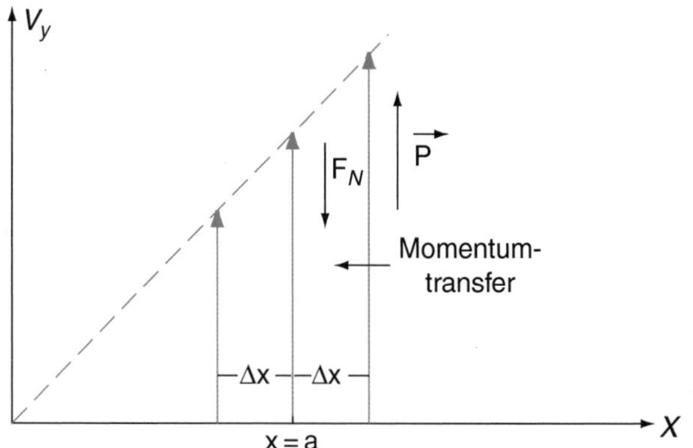


**Fig. 2.35.** Momentum transfer in a gas flow with velocity gradients. At $x = 0$ is a wall which forces, due to friction, $v_y = 0$.

depends on the velocity difference between adjacent layers $x = a$ and $x = a + \Delta x$ and on the frictional force between the molecules and therefore on the viscosity $\eta$. The frictional force exerted per area A in the plane x = const. can be written as

$$F_r = \eta \cdot A \cdot dv_y/dx \tag{2.57a}$$

This force is due to collisions between the atoms. The momentum transferred per sec through the area A between adjacent layers $x = a$ and $x = a+\Delta x$ of the gas is:

$$dp/dt = n \cdot m \cdot A \cdot < v_y > (dv_y/dx) \cdot \Delta x \text{ with } \Delta x = \Lambda \tag{2.57b}$$

because the mean distance the atoms travel between two collisions equals the mean free path $\Lambda$. The mean thickness $\Delta x$ of a layer therefore has to be equal to the mean free path $\Lambda$. Since the force $F_r$ equals the time derivative d*p*/d*t* of the transferred momentum, we obtain from the comparison of (2.57a and b) for the viscosity $\eta$ the relation

$$\eta = n \cdot m \cdot < v_y > \cdot \Lambda. \tag{2.58a}$$

where $< v_y >$ is the thermal average about the velocity of the flowing gas. Since there is a gradient of $v_y$ we have to take the average over the different velocities in the area $y =$ const, through which the gas flows. This gives a factor $^1/_2$ and we obtain for the viscosity

$$\eta = {}^1/_2 n \cdot m \cdot < v > \cdot \Lambda = m < v > /(2\sigma\sqrt{2}) \tag{2.58b}$$

The viscosity $\eta$ can be measured with different arrangements. One possible example is the measurement of the force which is necessary to maintain a constant mass flux d*M*/d*t* of a gas with density $\rho$ through a circular pipe with radius r and length L. This force is $\pi r^2 \cdot \Delta p$, where $\Delta p$ is the pressure difference between the two ends of the pipe. The mass flux $dM/dt = \rho \cdot dV/dt$ is related to the viscosity by

$$\rho \cdot dV/dt) = \pi r^4 \Delta p/(8\eta L) \tag{2.59}$$

In summary:

> Measurements of diffusion coefficient $D$ or heat conduction coefficient $\lambda$ or viscosity $\eta$ yield the corresponding collision cross sections and therefore the size of atoms. Since atoms are not really hard spheres their mutual interactions do not abruptly drop at distances $r_1 + r_2$ but fall off only gradually. Therefore, the different methods give slightly different values of the atomic size.

### 2.4.3 Atomic Volumes from X-Ray Diffraction

In Sect. 2.2.3 we have seen that the diffraction of X-rays by periodic crystals is one of the most accurate methods for the determination of the distances between adjacent lattice planes. From such distances the volume $V_E$ of the elementary lattice cell (often called a primitive cell) (Fig. 2.36) can be obtained. In order to derive the volume $V_a$ of the atoms of the crystal, one has to know which fraction $f$ of the elementary cell volume is actually filled by atoms. If there

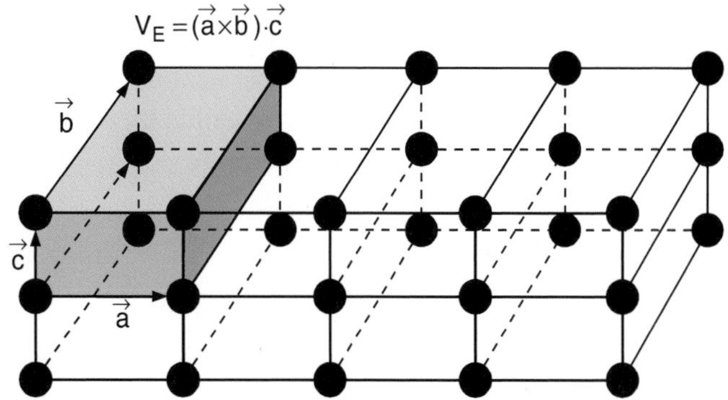


**Fig. 2.36.** Elementary cell of a regular crystal

are $N_E$ atoms per elementary cell, we get for the atomic volume

$$V_a = f V_E / N_E . \qquad (2.60)$$

The following three examples illustrate different values of $f$ for some simple lattice structures, assuming the atoms to be described by hard spheres with radius $r_0$.

## EXAMPLES

1. ***Primitive cubic crystal***
   It consists of atoms placed at the eight corners of the cubic elementary cell, which touch each other (Fig. 2.37a). The figure shows that only 1/8 of the volume of each atom is inside the elementary cell. This means that $N_E = 8 \times 1/8 = 1$. The side length of the cubic primitive cell is $a = 2r_0$, and the filling factor is

$$f = \frac{(4/3)\pi r_0^3}{(2r_0)^3} = 0.52 . \qquad (2.61a)$$

2. ***Body-centered Cubic Crystal***
   Here an additional atom is sitting at the center of the primitive cubic cell, which touches the neighboring atoms at the corners of the cube along the spatial diagonal, so that $4r_0 = a\sqrt{3}$. With $N_E = 2$ we obtain

$$f = \frac{2(4/3)\pi r_0^3}{\left[\left(4/\sqrt{3}\right) r_0\right]^3} = 0.68 . \qquad (2.61b)$$

3. ***Face-centered Cubic Crystal***
   In addition to the atoms at the eight corners of the elementary cell one atom sits at the center of each of the six faces (Fig. 2.37b).
   These atoms touch each other along the side face diagonal, which has the length $a\sqrt{2} = 4r_0$. From the figure it is clear, that only one half of each atom at the side faces belongs to the elementary cell. The number of atoms per elementary cell is then: $N_E = 8 \times 1/8 + 6 \times 1/2 = 4$ and therefore

$$f = \frac{4(4/3)\pi r_0^3}{\left[\left(4/\sqrt{2}\right) r_0\right]^3} = 0.74 . \qquad (2.61c)$$

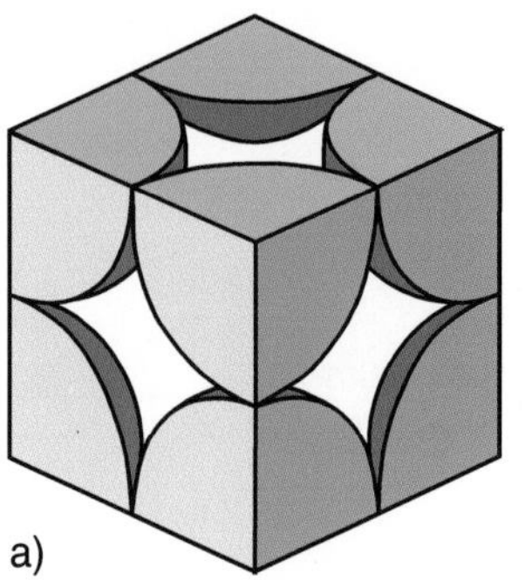

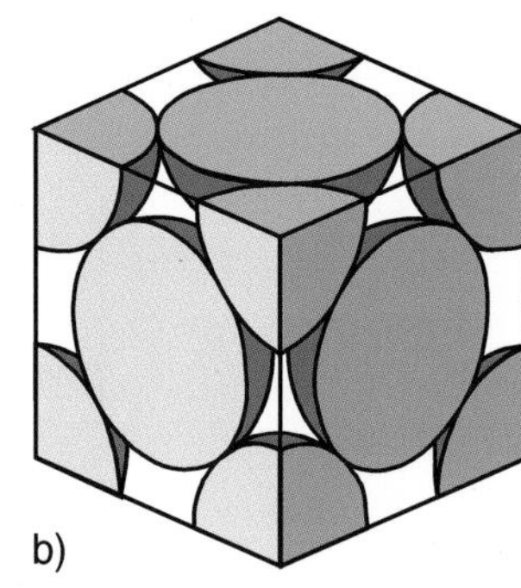


**Fig. 2.37a,b.** Illustration of the determination of the volume filling factor (**a**) for a primitive cubic crystal (**b**) for a cubic face-centered crystal

> This shows that the face centered cubic crystal has the highest packing density.

The atomic volume $V_a$ can now be obtained from (2.61), where $V_E$ is determined by X-ray diffraction and $N_E$ from the crystal structure.

### 2.4.4 Comparison of the Different Methods

The different methods all give the same order of magnitude for the atomic size although their values for the atomic radii differ slightly, as can be seen from Table 2.2. These differences have to do with the above-mentioned difficulty in defining an exact atomic radius as can be done for a rigid sphere. The real atoms experience long-range attractive forces and short-range repulsive forces when interacting with other atoms or molecules. The interaction potential between two

**Table 2.2.** Atomic radii in units of $10^{-10}$ m = 1 Å for a hard sphere model as determined from a) the van der Waals equation, b) the collision cross section obtained from measured transport coefficients, c) X-ray diffraction in noble gas crystals at low temperatures

| Atom | a) | b) | c) |
|---|---|---|---|
| He | 1.33 | 0.91 | 1.76 |
| Ne | 1.19 | 1.13 | 1.59 |
| Ar | 1.48 | 1.49 | 1.91 |
| Kr | 1.59 | 1.61 | 2.01 |
| Xe | 1.73 | 1.77 | 2.20 |
| Hg | 2.1 | 1.4 | — |

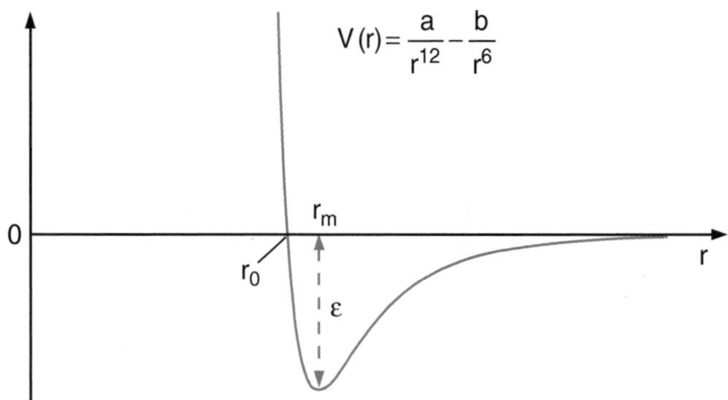


**Fig. 2.38.** Lenard–Jones-potential of the interaction between two neutral atoms

atoms A and B can be fairly well described by the empirical Lenard–Jones potential (Fig. 2.38)

$$E_{\text{pot}}(r) = \frac{a}{r^{12}} - \frac{b}{r^6} , \tag{2.62}$$

where the constants $a$ and $b$ depend on the kind of interacting atoms. One possible definition for the atomic radius is the value

$$r_{\text{m}} = \left(\frac{2a}{b}\right)^{1/6} , \tag{2.63a}$$

where the potential energy has its minimum value

$$E_{\text{pot}}(r_{\text{m}}) = -b^2/4a = -\varepsilon , \tag{2.63b}$$

which equals the potential depth $-\varepsilon$.

Another possible definition is

$$r_0 = \left(\frac{a}{b}\right)^{1/6} , \tag{2.63c}$$

where $E_{pot}(r_0) = 0$. For $R < r_0$ the step increase of the repulsive part of the potential comes close to that of a rigid sphere.

In summary:

> The atomic radii all lie within the range of $(0.5–5) \times 10^{-10}\,\text{m} = 0.5–5\,\text{Å}$. Their exact values depend on the atomic model. Different methods for their measurement give slightly different values because they probe different parts of the interaction potential.

## 2.5 The Electric Structure of Atoms

Various experimental investigations had already shown at the end of the 19th century that matter consists of electrically charged particles. The essential evidence came from:

- Investigations of electrolytic conductivity in polar liquids, which proved that molecules could dissociate into positively and negatively charged constituents that drift in opposite directions when in an external electric field. They were called "ions" (from the Greek word "ιον" for "the moving"). *Michael Faraday* found that the charge transported to the electrodes was proportional to the mass transport (see Sect. 2.2.3.e).
- Experiments on gas discharges, where the observed light emission could be drastically influenced by electric or magnetic fields. This proves that electrically charged particles are moving within the discharge region.
- Observations of the influence of magnetic fields on the electric current in metals and semiconductors (Hall effect, Barlow's wheel).
- The discovery that particles emitted from radioactive substances show different deflections in magnetic fields. They should therefore consist of positively charged heavy particles (called $\alpha$ particles) and negatively charged light particles (called $\beta$-particles, which are identical with electrons) (Fig. 2.39).

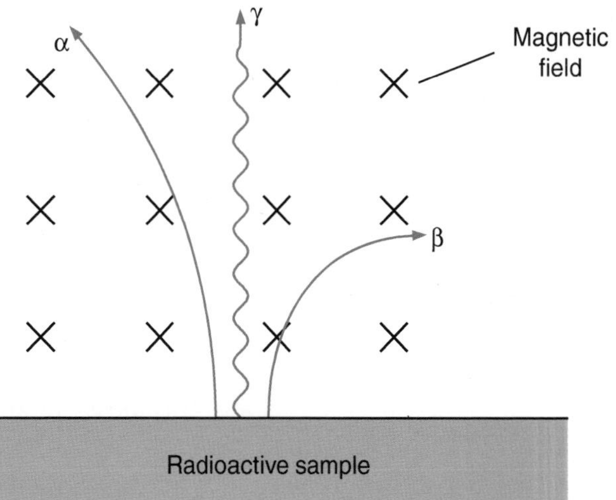


**Fig. 2.39.** Different deflections of $\alpha$ and $\beta$ particles in a magnetic field

These experimental findings together with the assumption that matter is composed of atoms led to the following hypothesis:

> Atoms are built up of charged particles. They can therefore not be "indivisible," but have a substructure, which, however, was unknown at this time. The electrically charged positive and negative constituents have different masses.

This raises the questions:

- What properties do these constituents have?
- What force keeps them together to form stable atoms?
- What is the charge distribution inside the atom?
- How can the microscopic properties of matter be explained by this model?

We can answer some of these questions immediately; the others are discussed at the end of this chapter.

Since atoms are neutral, the amount of positive and negative charge in an atom must be equal in order to compensate each other. Many classical experiments have shown that the electric Coulomb force is larger than the gravitational force by about 20 orders of magnitude. The latter is therefore completely negligible for the stability of atoms. Electric forces are responsible for the interaction between the constituents of an atom. Why the attractive Coulomb force between the positive and negative atomic constituents does not lead to the collapse of atoms has only been later answered by quantum mechanics (see Sect. 3.4.3).

### 2.5.1 Cathode Rays and Kanalstrahlen

Investigations of gas discharges by *J. Plucker* (1801–1868), *Johann Wilhelm Hittorf* (1824–1914), *Joseph John Thomson* (1856–1940), *Phillip Lenard* (1862–1947) (Nobel Prize 1905), and many others have all contributed much to our understanding of the electric structure of atoms. It is worthwhile to note that the essential experimental progress was only possible after the improvement of vacuum technology (the invention of the mercury diffusion pump, for example, allowed one to generate vacua down to $10^{-6}$ hPa).

In a gas discharge tube at low pressures, Hittorf observed particle rays emitted from the cathode that followed (without external fields) straight lines, which he could prove by the shadow that was produced on a fluorescent screen when obstacles were put in the path of the cathode rays. From the fact that these particle rays could be deflected by magnetic fields, Hittorf correctly concluded that they must be charged particles and from the direction of the deflection it became clear that they were negatively charged (Fig. 2.40). The first quantitative, although not very accurate, determination of the magnitude of their charge was obtained in 1895 by *J.B. Perrin* (and with an improved apparatus in 1897 by *Thomson*, who collimated the particles through a slit in the anode, deflected them after the anode by 90° through a magnetic field and detected them by an electrometer (Fig. 2.41a)).

With the design of Fig. 2.41b, where the cathode rays are better collimated by two slits $B_1$ and $B_2$, thus producing a small spot on the fluorescent screen, Thomson could measure the ratio $e/m$ of charge $e$ to mass $m$ of the particles by applying electric and magnetic fields for beam deflection (see Sect. 2.6). This was the first example of a cathode ray oscilloscope. Thomson could also show that the ratio $e/m$ was independent of the cathode material, but was about $10^4$ times larger than that for the "Kanalstrahlen" discovered in 1886 by *Eugen Goldstein* (1850–1930) in a discharge tube, which fly through a hole in the cathode in the opposite direction of the cathode rays (Fig. 2.42). *Wilhelm Wien* (1864–1928) measured in 1897 the value of $e/m$ for the particles in the Kanalstrahlen and he proved that they are positively charged atoms of the gas inside the discharge tube [2.36].

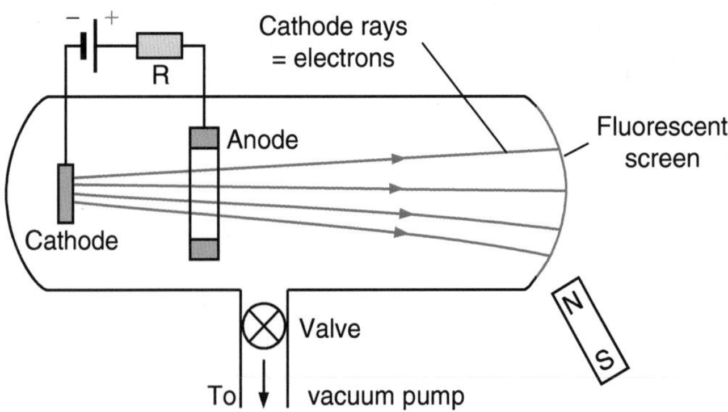


**Fig. 2.40.** Schematic drawing of the experimental setup for observing cathode rays. The deflection of the rays by an external magnet can be observed on the screen

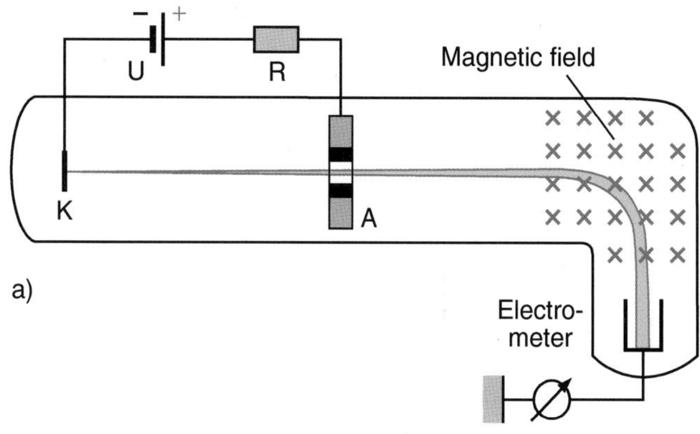


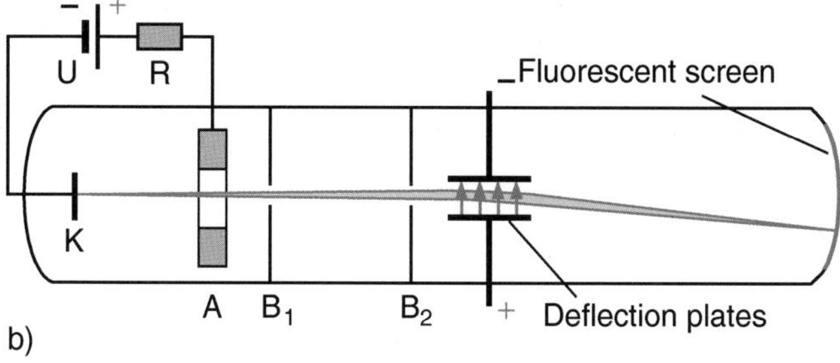


**Fig. 2.41a,b.** Experimental arrangement of *Thomson* for the determination of the ratio $e/m$ of cathode rays through their deflection (**a**) in a magnetic field and (**b**) in an electric field

The negative light particles of the cathode rays were named *electrons* after a proposal by *J. Stoney* and *G. Fitzgerald* in 1897. The positively charged heavy particles were named ions according to the existing name for charged atoms or molecules in the electrolysis.

This short survey on the history of gas discharges shows that all of these discoveries were made within the short time span of a few years. They gave the following picture of the charged constituents of atoms:

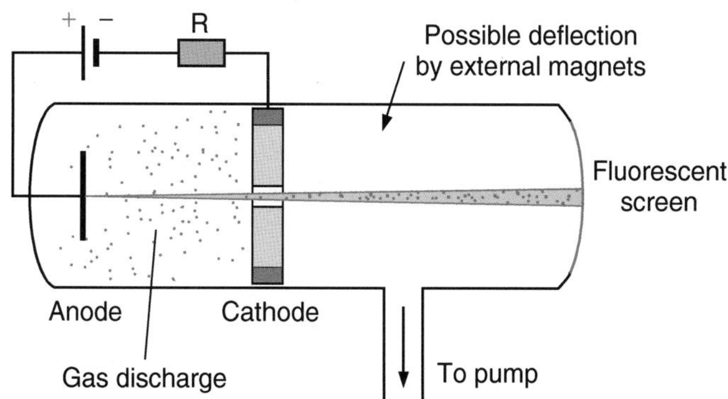


**Fig. 2.42.** Apparatus for demonstrating "channel-rays" (positive charged ions) in a discharge with a hole (channel) in the massive cathode

> Atoms consist of negatively charged electrons and positively charged particles that just compensate the negative charge to make the whole atom neutral.

Still nothing was known about the spatial distribution of negative and positive charges within the atom and about the sizes of the charged particles.

### 2.5.2 Measurement of the Elementary Charge *e*

The first measurement of the absolute value of the charge of a positive ion was made in 1899 by *Thomson* and his student *Charles Wilson*, who had developed his cloud chamber (see Sect. 2.3). The ions generated in the cloud chamber by external radiation act as nucleation centers for water droplets by attraction of water molecules in supersaturated water vapor. These small droplets, which are made visible by illumination, sink slowly due to gravity. They reach a constant terminal sink velocity in air with viscosity $\eta$ when the gravitational force $F_g = m^*g$ ($m^* = m - \rho_{air}V$ is the apparent mass, taking into account the buoyancy) just equals the opposite frictional force $F_f = -6\pi\eta r v$ for spherical particles. This gives for the apparent mass $m^*$ the equation

$$gm^* = g\left(\rho_{water} - \rho_{air}\right)\frac{4}{3}\pi r^3 = 6\pi\eta r v\,. \tag{2.64}$$

From the measured values of the sink velocity

$$v = \frac{m^* g}{6\pi\eta r}\,, \tag{2.64a}$$

which depends on the viscosity $\eta$ of the gas, the radius $r$ of the droplets can be determined and from the total mass, condensed at the lower plate of the cloud chamber per unit time and the transported charge $Q$, the number of particles $N$ can be obtained. This gives the average charge $q = Q/N$ of each particle. Assuming that each particle carries only one elementary charge, Thomson estimated this charge to be around $10^{-19}$ C, close to the correct value of $1.6 \times 10^{-19}$ C.

A much more accurate value was obtained 1910 by *Robert Andrew Millikan* (1868–1953) in his famous oil droplet experiment [2.37]. With an atomizer, he blew a fine spray of minute oil droplets into the air between the two horizontal plates of a condenser (Fig. 2.43).

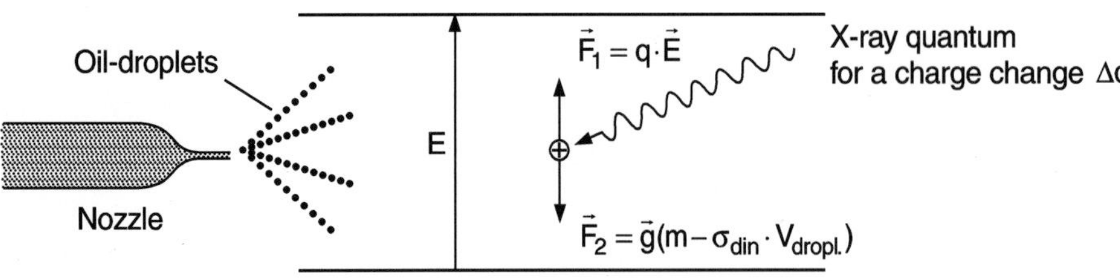


**Fig. 2.43.** Principle scheme of Millikan's oil-droplet experiment for measuring the unit of charge

These droplets could be viewed through a microscope by illuminating them with an arc lamp. They were charged by irradiation with X-rays, which strip off a small number $n$ of electrons ($n = 0, 1, 2, 3, \ldots$) from the droplets, leaving them positively charged. Without an electric field, the droplets sink down with constant terminal velocity $v_0$, if the gravitational force $m^* g$ is just compensated for by the frictional force. From (2.64) we then obtain

$$v_0 = \frac{g\,(\rho_{oil} - \rho_{air})\,\frac{4}{3}\pi r^3}{6\pi\eta r}$$
$$\Rightarrow r = \sqrt{\frac{9\eta v_0}{(\rho_{oil} - \rho_{air})\,2g}} \,. \tag{2.65a}$$

From the measured sink velocity $v_0$, therefore, the radius $r$ of the droplets is obtained if the viscosity $\eta$ of air is known.

If an electric field $E_0$ is now applied, an additional force $F_a = q\boldsymbol{E}_0$ acts on the droplet, where $q = -ne$ and $-e$ is the charge of one electron. With the right field polarity the droplet can be kept at a constant height if the electric force just compensates for the gravitation:

$$n \cdot e\boldsymbol{E}_0 = -\boldsymbol{g}\,(\rho_{oil} - \rho_{air})\,\frac{4}{3}\pi r^3 \,. \tag{2.65b}$$

This gives the charge of the droplet

$$-q = ne = -\frac{g}{E_0}\frac{4}{3}\pi r^3\,(\rho_{oil} - \rho_{air}) \,, \tag{2.65c}$$

where $r$ is taken from (2.65a).

In a variation of the experiment a larger electric field $E_1$ is applied, which moves the droplets upwards. One droplet now experiences the total force

$$\boldsymbol{F} = q\boldsymbol{E}_1 - g\frac{4}{3}\pi r^3\,(\rho_{oil} - \rho_{air}) - 6\pi\eta r v \,, \tag{2.65d}$$

which becomes zero for the terminal velocity

$$v_1 = \frac{qE_1 - g\frac{4}{3}\pi r^3\,(\rho_{oil} - \rho_{air})}{6\pi\eta r} \,. \tag{2.65e}$$

Subtracting (2.65a) from (2.65e) gives

$$v_1 - v_0 = \frac{qE_1}{6\pi\eta r}$$

from which the charge

$$q = -ne = 6\pi\eta r(v_1 - v_0)$$
$$= \frac{36\pi}{E_1}(v_1 - v_0)\left(\frac{\eta}{2}\right)^{3/2}\sqrt{\frac{v_0}{g\,(\rho_{oil} - \rho_{air})}} \tag{2.65f}$$

can be obtained. When the droplet changes its charge, its terminal velocity $v_1$ will change. The smallest change is observed for $\Delta n = 1$. This gives the elementary charge $e$.

Millikan had already noted that the results he obtained scattered for droplets with different radii. The reason for this is that Stokes' law for the viscosity force on moving spheres is only valid if the radius $r$ of the droplet is large compared with the mean free path $\Lambda$ of the air molecules. This was not strictly true for the oil droplets, and particularly not for the smaller ones.

## EXAMPLE

At a pressure of 1 bar the mean free path length is $\Lambda \approx 5 \times 10^{-6}\,\mathrm{m} = 5\,\mu\mathrm{m}$. For droplets with $r < 10\,\mu\mathrm{m}$ a correction to (2.65a) has to be made.

The presently accepted value for $e$ is $e = 1.602176487(40) \times 10^{-19}\,\mathrm{C}$, where the number in parentheses gives the uncertainty of the two last digits.

### 2.5.3 How to Produce Free Electrons

Free electrons can be produced in many different ways. We will discuss the most important methods.

#### a) Thermal Emission from Solid Surfaces

When a metal is heated to a high temperature $T$ a fraction of the free conduction electrons can get a sufficiently large kinetic energy to overcome the attraction by the solid and can leave the metal (thermal emission) (Fig. 2.44a). If these electrons are collected onto an anode by an accelerating electric field one observes an electric current $I$ with a current density $j_s$ [A/m$^2$] per unit area of the emitting surfaces, which follows the Richardson equation

$$j_s = AT^2 e^{-W_a/kT} , \tag{2.66}$$

where $W_a$ is the work function of the metal, i.e., the necessary minimum energy of the electrons to leave the metal. The constant $A$ depends on the material and the condition of the surface (Table 2.3). For a regular crystal, $A$ also depends on the direction of the surface normal against the crystal axes. In order to achieve high current fluxes, materials with low values of $W_a$ and high melting temperatures are needed. A commonly used composite material is tungsten doped with barium or cesium. At high temperatures the barium or cesium atoms diffuse to the surface where they contribute to a low work function $W_a$. The measured anode current increases with the applied voltage until it reaches a saturation value, where all electrons, emitted from the cathode, are collected by the anode (Fig. 2.44a).

**Table 2.3.** Work functions $W_a = eU_a$ and electric current density emission coefficients $A$ for some commonly used cathode materials

| Material | $W_a$/eV | $A/(\mathrm{Am^{-2}K^{-2}})$ |
|---|---|---|
| barium | 2.1 | $6 \times 10^4$ |
| tungsten-barium | 1.66 | $\sim 10^4$ |
| tungsten-cesium | 1.4 | $\sim 3 \times 10^4$ |
| thorium | 3.35 | $6 \times 10^5$ |
| tantalum | 4.19 | $5.5 \times 10^5$ |
| tungsten | 4.54 | $(1.5–15) \times 10^5$ |
| nickel | 4.91 | $(3–130) \times 10^5$ |
| thoriumoxyd | 2.6 | $(3–8) \times 10^4$ |

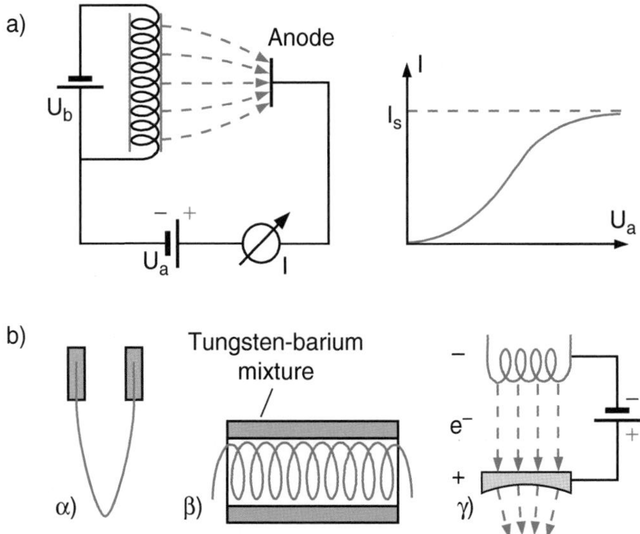


**Fig. 2.44a,b.** Thermal electron emission devices. (**a**) Arrangement for measuring the emission current and its saturation value. (**b**) Technical realization of different thermal emission cathodes. ($\alpha$) Hair-needle cathode, ($\beta$) cathode hollow cylinder heated by a current through a tungsten helix inside the cylinder. ($\gamma$) Focusing cathode, heated by electron bombardment on the backside

Thermal emission represents the most important technique for the production of free electrons in oscilloscope tubes, TV tubes, broadcasting electron valves, and for all instruments where high current densities are required, such as electron beam welding or ion sources for accelerators. In Fig. 2.44b different types of cathodes are shown.

#### b) Field Emission

When a voltage $U$ is applied between an anode and a cathode formed as a sharp tip with radius $r$ on a tungsten wire, the electric field $(\boldsymbol{E}) = U/r$ may become as large as $10^{11}$ V/m (see Sect. 2.3). Such high fields change the potential at the surface of the tip and can extract electrons. Field emission is technically used in cases where a nearly point-like electron source is required, such as in the field emission electron microscope (Sect. 2.2.3) and in some types of high resolution electron microscopes (Fig. 2.27).

#### c) Photoeffect at Metal Surfaces

When the surface of a metal is illuminated by UV light, electrons with kinetic energy

$$E_{kin} = h\nu - W_a \tag{2.67}$$

are emitted from the surface where $\nu$ is the optical frequency of the illuminating light and $h$ is a constant (Planck's constant, see Sect. 3.2), $W_a$ is the work function of the metal (i.e., the minimum energy required to release an electron (see Table 2.3)).

### d) Secondary Electron Emission

When the surface of a metal is irradiated with fast electrons or ions instead of photons, so-called "secondary" electrons are emitted (Fig. 2.45). The average number of emitted electrons per single incident particle is called the coefficient of secondary emission. It depends on the material, the angle $\alpha$ of incidence and on the kind of incident particles and their energy. Some values for $\eta$ are given in Table 2.4.

Secondary emission plays an important role in many optical and spectroscopic devices. One example is the photomultiplier (Fig. 2.46), where the incident light releases electrons from the cathode (photoeffect), which are then accelerated by an electric field and impinge on a specially formed electrode. There they release secondary electrons, which are again accelerated onto a second electrode, and so on. Finally an electron avalanche arrives at the anode with a charge $q = Me$, where the magnification factor $M = \eta^m$ depends on the secondary emission coefficient and on the number $m$ of electrodes. Typical values of $M$ are: $M = 10^5$–$10^7$. The electric charge pulse $q(t)$ produces a voltage $V_a(t) = q(t)/C_a$ at the exit capacitance $C_a$, which is discharged through a resistor $R$ between anode and ground potential.

**Table 2.4.** Maximum values $\eta_{max}$ of secondary emission coefficients $\eta$ for the optimum energy $W_{max}$ of incident electrons

| Material | $\eta_{max}$ | $W_{max}/eV$ |
|---|---|---|
| Ag | 1.5 | 800 |
| Al | 1.0 | 300 |
| C (diamond) | 2.8 | 750 |
| Na | 0.8 | 300 |
| W | 1.4 | 650 |
| KBr | 14 | 1800 |
| LiF | 8.5 | 700 |
| NaI | 19 | 1300 |
| MgO-crytal | 20–25 | 1500 |
| MsO-layer | 5–15 | 500–1500 |
| GaP + Cs | 120 | 2500 |

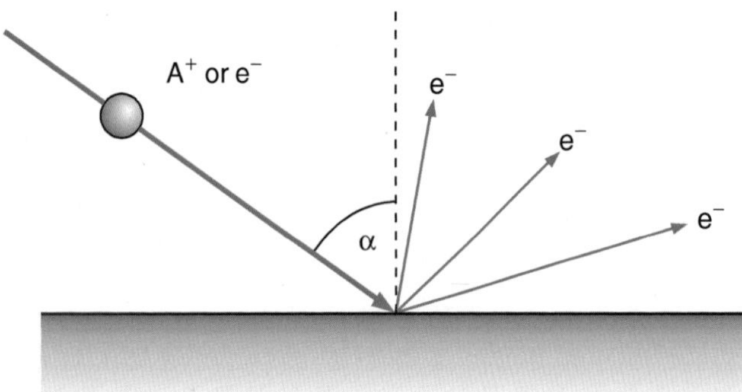


**Fig. 2.45.** Schematic illustration of secondary electron emission under ion bombardment of a metal surface

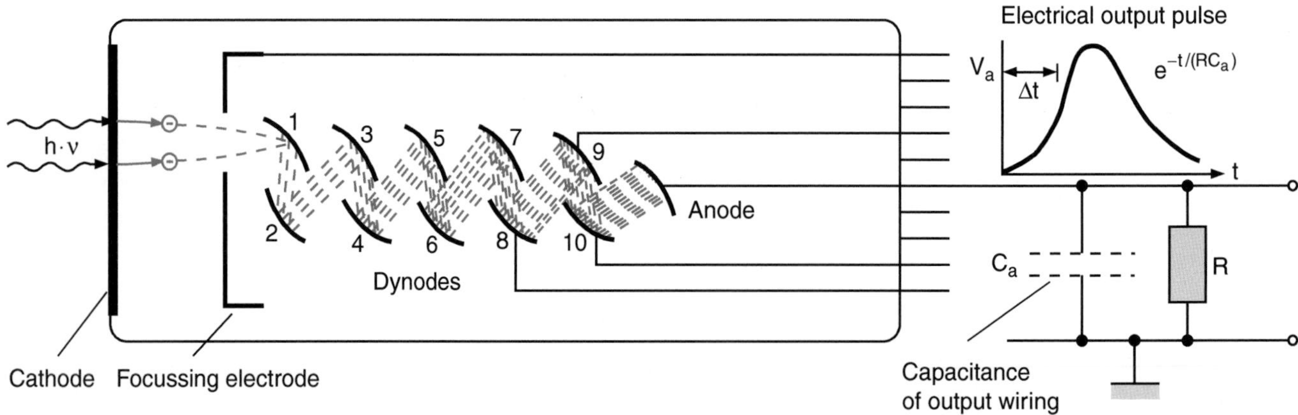


**Fig. 2.46.** Principle of a photomultiplier. The rise time of the output pulse reflects the time spread of the electrons in the tube, the decay time solely depends on the product $RC_a$

EXAMPLE

$N_{ph} = 1$, $\eta = 4$, $m = 10$, $e = 1.6 \times 10^{-19}\,\mathrm{C}$, $C_a = 100\,\mathrm{pF} \Rightarrow V_a = \frac{4^{10} \times 1.6 \times 10^{-19}}{10^{-10}} = 17\,\mathrm{mV}$.

If the photocathode is replaced by a metal electrode with a high secondary emission coefficient $\eta$, single particles (electrons or ions) that are impinging on the cathode can be detected (electron- or ion-multiplier).

Other examples of secondary electron emission applications are image intensifier or scanning surface electron microscopes (Sect. 2.3.3).

### 2.5.4 Generation of Free Ions

While the techniques described above produced free electrons emitted from solid surfaces, in the following processes pairs of ions and electrons, generally in the gas phase, are being formed.

#### a) Electron Impact Ionization

The most important mechanism for the production of free ion-electron pairs is the electron impact ionization, where an electron with sufficient kinetic energy $E_{kin}$ hits an atom A and releases another electron

$$e^-(E_{kin}) + A \rightarrow A^+ + e^-(E_1) + e^-(E_2)\,. \tag{2.68a}$$

The final energies $E_1$, $E_2$ of the electrons after the reaction have to obey energy conservation

$$E_1 + E_2 = E_{kin} - E_{ion}\,. \tag{2.68b}$$

Here a neutral atom looses one electron, which had a binding energy $E_{ion}$ (also called the "ionization energy") and is split into a positively charged ion $A^+$ and an electron $e^-$ (Fig. 2.47). The probability of this process depends on the kinetic energy $E_{kin}$ of the incident electron, the atomic particle A and the ionization energy $E_{ion}$ of the released electron. It is generally described by the ionization cross section $\sigma_{ion}(E_{kin})$, which gives the circular area around atom A through which the primary electron has to pass in order to ionize A. Figure 2.48 shows ionization cross sections $\sigma_{ion}(E_{kin})$ for some atoms A as a function of the electron impact energy $E_{kin}$.

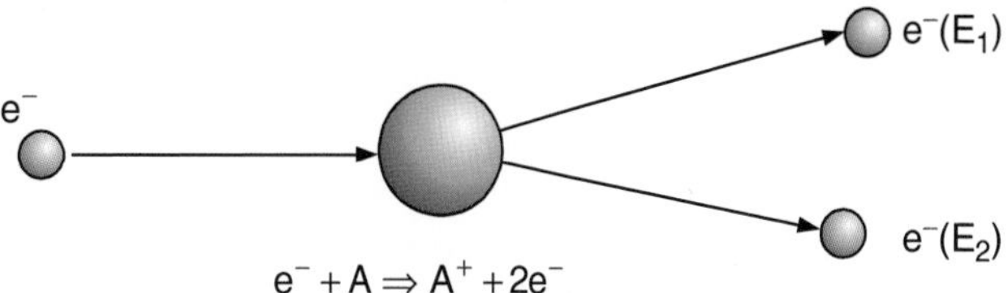


**Fig. 2.47.** Ionization of atoms by electron impact

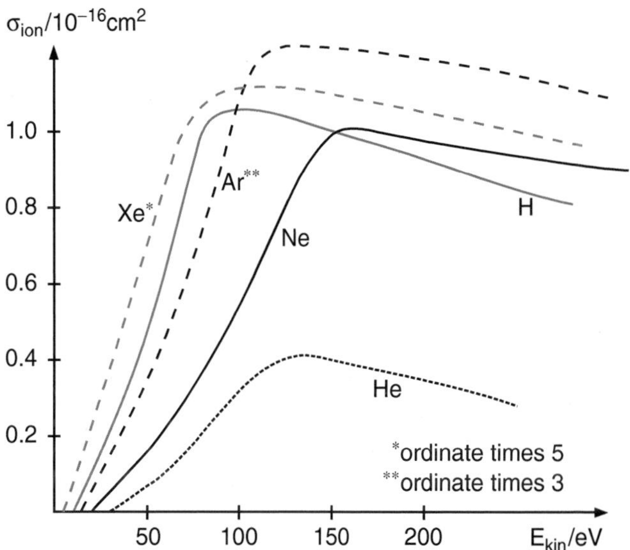


**Fig. 2.48.** Electron impact ionization cross sections $\sigma(E_{kin})$ for some atoms. The values for Ar and Xe are really three times and five times larger than shown

> Electron impact ionization in gas discharges represents the major contribution to the generation of charge carriers.

When ions $B^+$ instead of electrons collide with atoms A to produce ionization

$$B^+(E_{kin}) + A \rightarrow A^+ + B^+ + e^- \tag{2.68c}$$

the relative kinetic energy $E_{kin}$ of the collision partners must be much higher than the ionization energy $E_{ion}$ of the electron, because only a small fraction ($\approx m_e/m_B$) of the kinetic energy of the heavy particle B can be transferred during the collision to the electron of atom A which has to be removed from the atom.

### b) Photoionization of Atoms

When atoms are irradiated with light of sufficiently short wavelengths (generally ultraviolet light), the absorbed light can excite an atomic electron to an energy above the ionization limit (photoionization, see Sect. 7.6.1). This process represents the leading mechanism for the production of ions in the upper atmosphere (ionosphere). Here the UV light of the sun can ionize nearly all atoms and molecules in this region. Since the photoionization cross section is generally small, efficient photoionization of atoms demands high light intensities in the UV. This process therefore gained increasing importance after the introduction of lasers, which can deliver intensities many orders of magnitude larger than conventional light sources (see Chap. 8).

### c) Charge Exchange Collisions

When ions $A^+$ pass through a gas or metal vapor of neutral atoms B, an electron can be transferred from the atom B to the ion $A^+$ during the close passage of $A^+$ with B

$$A^+ + B \rightarrow A + B^+ \tag{2.68d}$$

if the ionization energy of B is smaller than that of A.

When slow electrons pass through a gas of neutral atoms A, they may be captured by the atoms to form negative ions

$$e^- + A \rightarrow A^- \tag{2.68e}$$

if the relative kinetic energy $E_{kin}$ can be transferred to a third partner. This process plays an important role in the earth's atmosphere and also in the sun's photosphere where the process

$$H + e^- \rightarrow H^- + h\nu \tag{2.68f}$$

is mainly responsible for the emission of the sun's continuous visible radiation. The inverse process

$$H^- + H + E_{kin}(H^-, H) \rightarrow H + H + e^- \tag{2.68g}$$

resupplies the neutral H atoms lost by the process (2.68f).

### d) Thermal Ionization

At very high temperatures the kinetic energy of atoms may become sufficiently large to allow ionizing atomic collisions.

$$\begin{aligned} A + B &\rightarrow A + B^+ + e^- \\ &\rightarrow A^+ + B + e^- \\ &\rightarrow A^+ + B^+ + 2e^- \end{aligned} \tag{2.69}$$

depending on the kinetic energy of the relative motion of A and B. Such processes occur, for instance, in the hot atmospheres of stars. The state of matter at these high temperatures consisting of a mixture of neutral atoms, ions and electrons, is called a ***plasma***.

The different mechanisms for the production of ions are summarized in Fig. 2.49.

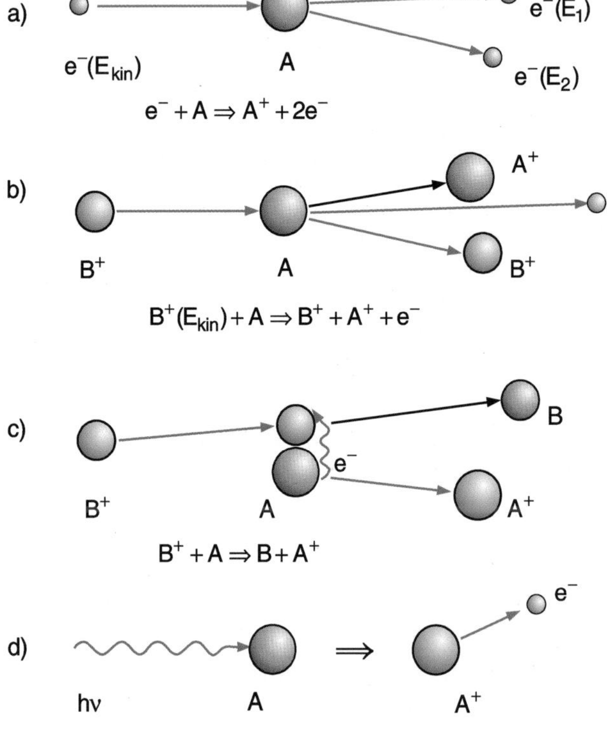


**Fig. 2.49a–d.** Survey of the various processes for the production of ions. (**a**) Electron impact ionization. (**b**) Ion impact ionization. (**c**) Charge exchange for grazing collisions. (**d**) Photoionization

### e) Technical Ion Sources

For the practical realization of atomic or molecular ionization, special devices for the ion production have been developed which are called ion sources.

The most widely spread source is the electron impact ion source (Fig. 2.50). Electrons are emitted by a hot cathode and are accelerated by a cylindrical anode in the form of a grid around the cathode. They pass

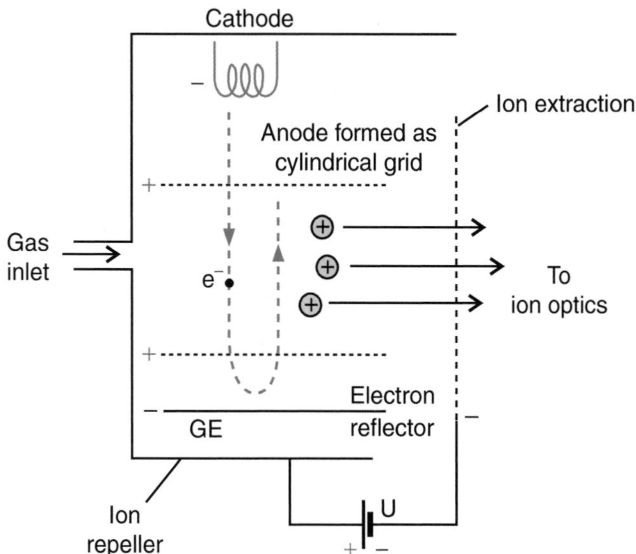

**Fig. 2.50.** Electron impact ion source

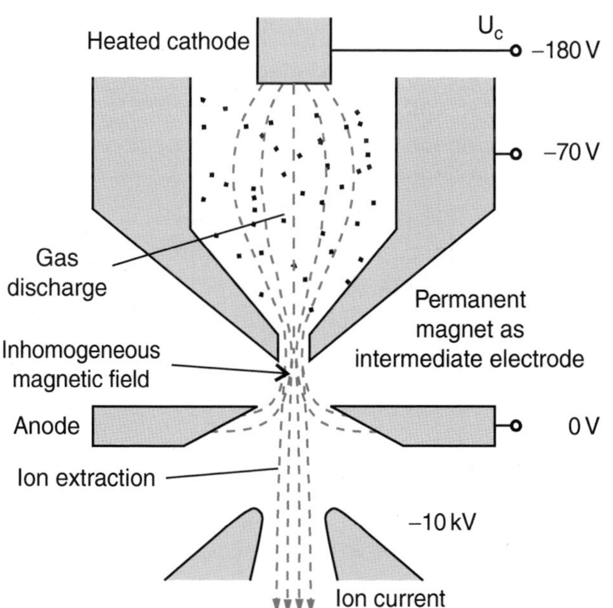

**Fig. 2.51.** Duo plasmatron ion source

the ionization volume where the atoms or molecules are inserted. The electrons can be reflected by the electrode GE allowing them to pass through the ionization volume a second time.

The ions formed by this electron impact are extracted by properly designed electrodes at a negative potential. They are imaged by special ion optics (see Sect. 2.6) and form a nearly parallel beam of ions, which can then be mass selected by electric or magnetic fields (see Sect. 2.7).

The electron impact ion sources are operated at low pressures ($10^{-3}$–$10^{-5}$ mbar). This implies that the achievable ion currents are relatively small. Higher ion currents can be realized with plasma ion sources, where a gas discharge at high pressures is maintained. One example is the duo-plasmatron source (Fig. 2.51) where a low voltage gas discharge is initiated between the heated cathode and the anode. The ions are extracted by a high voltage (several kV) through the small hole in an auxiliary electrode that compresses the plasma and therefore increases its spatial density. A magnetic field keeps the plasma away from the walls and further increases the ion density. Even substances with low vapor pressure can be vaporized (for example by electron- or ion impact) and can then be ionized inside the discharge.

A more detailed discussion of different techniques for the production of ions can be found in [2.38–40].

### 2.5.5 The Mass of the Electron

All methods for the determination of the electron mass use the deflection of electrons in electric or magnetic fields, where the Lorentz force

$$\boldsymbol{F} = q(\boldsymbol{E} + \boldsymbol{v} \times \boldsymbol{B}) \tag{2.70a}$$

acts on a particle with charge $q$, which moves with a velocity $\boldsymbol{v}$ across the fields (Fig. 2.52). Inserting Newton's equation $\boldsymbol{F} = m\ddot{\boldsymbol{r}}$ into (2.70a) we obtain the three coupled differential equations

$$\begin{aligned}
\ddot{x} &= \frac{q}{m}\left(E_x + v_y B_z - v_z B_y\right) ,\\
\ddot{y} &= \frac{q}{m}\left(E_y + v_z B_x - v_x B_z\right) ,\\
\ddot{z} &= \frac{q}{m}\left(E_z + v_x B_y - v_y B_x\right) .
\end{aligned} \tag{2.70b}$$

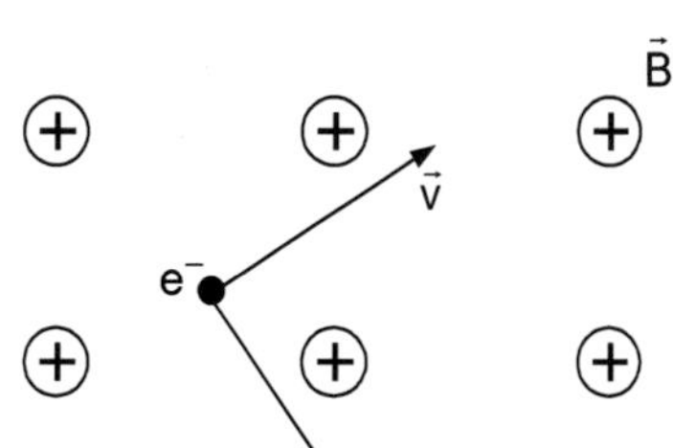

**Fig. 2.52.** Lorentz-force $\boldsymbol{F}$ acting on an electron $e^-$ that moves with velocity $\boldsymbol{v}$ in a homogeneous magnetic field $\boldsymbol{B}$, pointing perpendicularly into the drawing plane

These equations show that it is not the mass $m$ directly, but only the ratio $q/m$ that can be obtained from measuring the path of a charged particle in these fields. One therefore needs an additional measurement (for instance the Millikan experiment) in order to determine the charge $q$ seperately. The mass $m$ can then be obtained from one of the following experiments, which illustrate (2.70) by several examples.

### a) Fadenstrahlrohr

Electrons emitted from a hot cathode in a glass bulb are accelerated in the $y$-direction and enter a magnetic field that points into the $z$-direction (Fig. 2.53). Since here $\boldsymbol{v} = \{0, v_y, 0\}$ and $B = \{0, 0, B_z\}$, (2.70b) reduces with $q = -e$ to

$$\ddot{x} = -\frac{e}{m} v_y B_z \,. \tag{2.70c}$$

This shows that the electrons are bent into the $x$-direction and acquire a velocity component $v_x$ but remain within the plane $z =$ const. The Lorentz force is always perpendicular to their velocity $\boldsymbol{v} = \{v_x, v_y\}$ and therefore does not change the magnitude of the velocity. The path of the electrons is therefore a circle with a radius $R$ (Fig. 2.54) defined by the compensation of centrifugal and Lorentz force

$$\frac{mv^2}{R} = evB_z \,. \tag{2.71a}$$

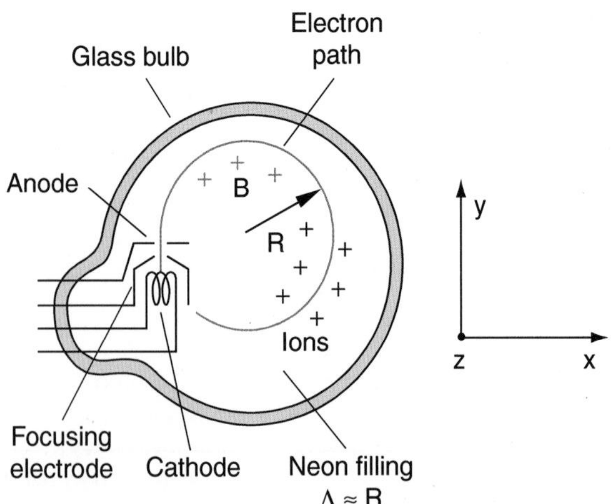


**Fig. 2.53.** Experimental device ("Fadenstrahlrohr") for measuring the ratio $e/m$

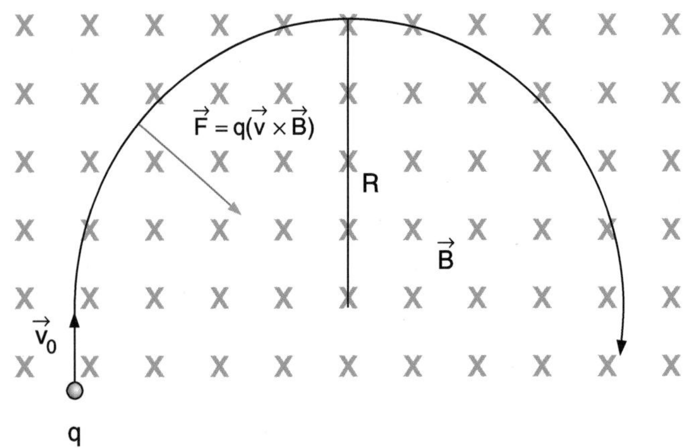


**Fig. 2.54.** Circular path of an electron beam in a homogeneous magnetic field perpendicular to the initial velocity $v_0$ of the electrons

This gives the radius

$$R = \frac{mv}{eB} = \frac{1}{B}\sqrt{2Vm/e} \,, \tag{2.71b}$$

because the velocity $v$ of the electrons is determined by the acceleration voltage $V$ according to $(m/2)v^2 = eV$.

The path of the electrons can be made visible, if the glass bulb is filled with a gas at low pressure so that the mean free path of the electrons is comparable to the circumference of the circle. Through collisions with the electrons, the atoms are excited and emit light (see Sect. 3.4). This visible circular path of the electrons allows the measurement of its radius $R$ and of the ratio

$$\frac{e}{m} = \frac{2V}{R^2B^2} \,. \tag{2.72}$$

If the electrons enter the homogeneous magnetic field under the angle $\alpha$ against the field direction, the electron velocity $v = \{v_x, 0, v_z\}$ can be composed of the two components $v_x$ and $v_z$ (Fig. 2.55). The $v_x$ component is perpendicular to the field and leads to a circular motion with radius $R = mv_x/(eB)$ and a circulation time $T = 2\pi R/v_x = 2\pi m/(eB)$ independent of the velocity component $v_x$! The $v_z$ component is parallel to the field lines and therefore the Lorentz force is zero. This component leads to a straight line. The superposition of the two movements results in a helical path around the field lines with a pitch $\Delta z = Tv_z = 2\pi R v_z/v_x = 2\pi R/\tan\alpha$. For $\alpha = 90°$ the electron path lies in the $x$-$y$ plane and becomes a circle.

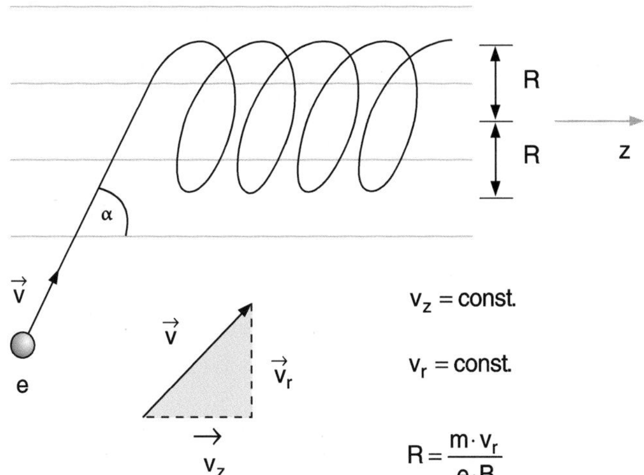


**Fig. 2.55.** Helical path of electrons that enter a homogeneous magnetic field under an angle $\alpha \neq 90°$ against the field lines

A more accurate measurement of $e/m$ is possible with the Wien filter of Fig. 2.56, where an electron beam is accelerated through a voltage $V$ and enters with the velocity $v$ a region where a homogeneous electric field $\boldsymbol{E} = \{-E_x, 0, 0\}$ and a magnetic field $\boldsymbol{B} = \{0, -B_y, 0\}$ are superimposed in such a way that the two forces are antiparallel. The electron beam is collimated by the aperture $S_1$ to assure that the velocity components $v_x, v_y$ become negligibly small. With the proper selection of field strengths one can achieve that the total force $\boldsymbol{F} = -e(\boldsymbol{v} \times \boldsymbol{B}) - e\boldsymbol{E}$ acting on the electrons becomes zero. This gives

$$v_z \approx v = \sqrt{2Ve/m} = E/B \Rightarrow \frac{e}{m} = \frac{E^2}{2VB^2} . \tag{2.73}$$

Only electrons with a velocity $v$ that give the exact compensation of the electric and magnetic forces can pass through the second aperture $S_2$ and reach the detector.

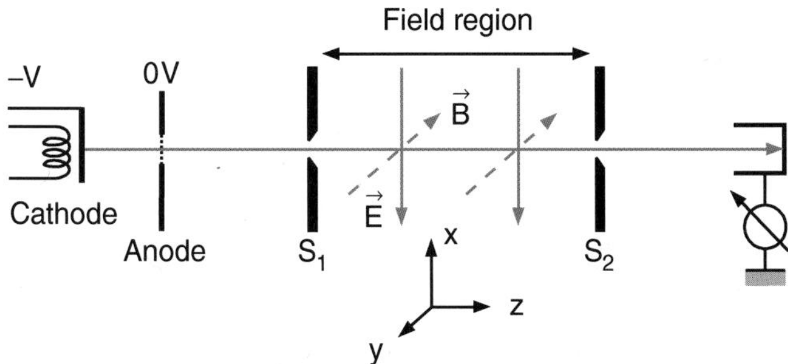


**Fig. 2.56.** Wien filter

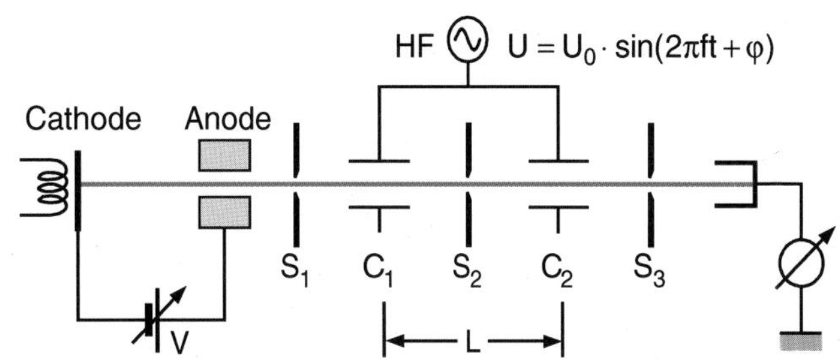


**Fig. 2.57.** Precision methods for the measurement of $e/m$ with two radio frequency deflection plates separated by the distance $L$

Besides its application to the determination of the ratio $e/m$, the Wien filter can be used as a narrow velocity filter for electron or ion beams.

Instead of the Wien filter two condensers $C_1$ and $C_2$ can be used (Fig. 2.57). When an AC voltage with frequency $f$ is applied to the condensors the electric field is time dependent. Electrons can only pass through the aperture $S_2$, if they transverse the first condenser $C_1$ at a time where the voltage is zero and they pass through the next aperture $S_3$ if they also traverse the second condenser at zero voltage. With the separation $L$ between the two condensers their flight time $T = L/v = n/2f$ must be an integer multiple $n$ of the half period $\Delta T/2 = 1/2f$ of the AC voltage. This imposes the condition

$$v_z = \sqrt{\frac{2Ve}{m}} = 2Lf/n . \tag{2.74}$$

Varying the frequency $f$ or the acceleration voltage $V$ yields maxima of the signal for the different values of the integers $n = 1, 2, 3, \ldots$. This allows the determination of $n$ and thus of $e/m$.

The accuracy of $e/m$ measurements has steadily increased over the course of time. However, systematic errors have often mislead experimenters and the results of different laboratories often differed by more than the error limits stated by the different scientists (Fig. 2.58). The main uncertainty for the determination of the electron mass $m$ from the measured ratio $e/m$ stems from the error limits in the measurement of the charge $e$. The best value accepted today is

$$m_e = (9.10938215 \pm 0.0000005) \times 10^{-31}\,\text{kg} .$$

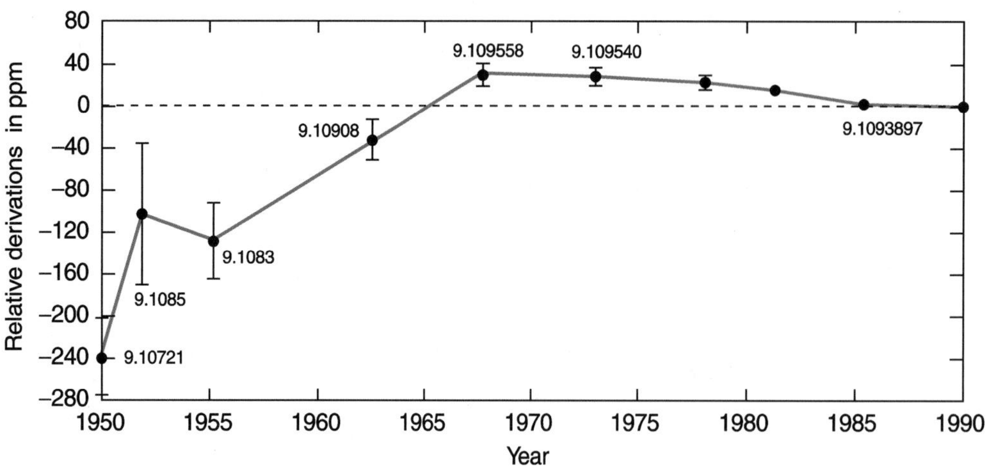


**Fig. 2.58.** Historical experimental results for the electron mass in units of $10^{-31}$kg. The ordinate gives the relative deviations $\Delta m/m$ from the value accepted today. Note that the error bars, stated by the authors, are mostly much smaller than the real deviations

### 2.5.6 How Neutral is the Atom?

The experiments discussed so far have shown that atoms can be split into negatively charged electrons and positive ions. Millikan had measured the charge of positive ions that missed one or several electrons.

The question now is how exactly the negative charge of the atomic electrons is compensated by the positive charge in a neutral atom. We will see later that this positive charge is provided by the protons in the atomic nucleus. Our question can therefore also be stated as:

> Is there any difference in the absolute values of electron and proton charges?

This question is of fundamental importance, because a tiny difference would cause huge macroscopic effects. For example, if there was a difference

$$\Delta q = |\mathrm{e}^+| - |\mathrm{e}^-| \geq 2 \times 10^{-18}\, e$$

the expansion of the universe could be explained by electrostatic repulsion forces [2.41].

In order to give upper limits for such a possible difference $\Delta q$, precision experiments have been designed. We will only discuss a few of them.

#### a) Gas Effusion

Through the orifice of a large, electrically isolated metal container containing $N$ atoms or molecules $(\mathrm{d}N/\mathrm{d}t)\Delta t$ atoms effuse during the time interval $\Delta t$ into vacuum and are condensed at a cold wall

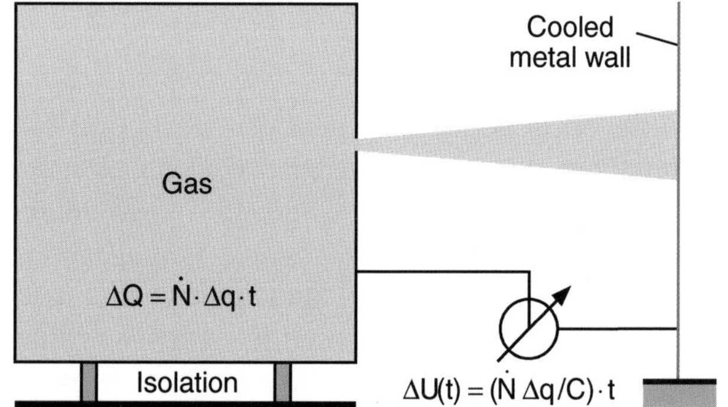


**Fig. 2.59.** Experimental arrangement for obtaining an upper limit for a possible difference $\Delta q = |\mathrm{e}^+| - |\mathrm{e}^-|$

(Fig. 2.59). If there was a net charge $\Delta q$ per atom, the isolated container would loose the charge $\Delta Q = \Delta q(\mathrm{d}N/\mathrm{d}t)\Delta t$. With a capacitance of the container $C$, this would lead to a voltage $V = \Delta Q/C$ of the container against the ground, which could be measured with an electrometer. The experiment showed that no measurable voltage appeared.

**EXAMPLE**

$\mathrm{d}N/\mathrm{d}t = 10^{20}\,\mathrm{s}^{-1}$, $\Delta t = 100\,\mathrm{s}$, $C = 10^{-9} F \Rightarrow V = 10^{22}(\Delta q/C)\,\mathrm{Volt} \Rightarrow \Delta q = 10^{-22} \cdot 10^{-9}$ As. Since the voltage can be measured with an uncertainty of $10^{-9}$ V, the upper limit of a possible charge $\Delta q$ per atom must be $\Delta q \leq 10^{-40}\,\mathrm{As} \approx 10^{-21}\, e$.

#### b) Deflection of an Atomic Beam

Out of a cold reservoir, atoms effuse and are collimated by two slits $S_1$ and $S_2$ to form a well collimated

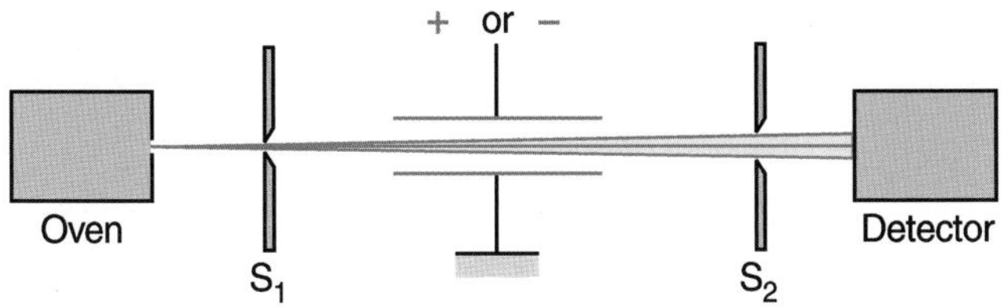


**Fig. 2.60.** Measuring the neutrality of atoms in an atomic beam experiment

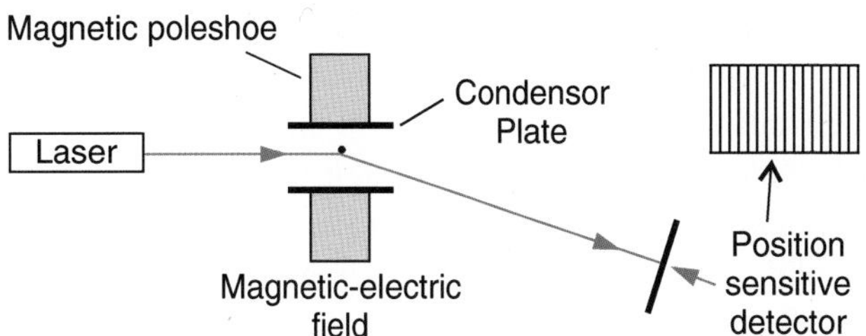


**Fig. 2.61.** Checking the neutrality of atoms by measuring the position of a small ferromagnetic ball in a magnetic + electric field, monitored by reflection of a laser beam

horizontal beam, which is slightly bent due to gravitation. Between the two apertures, a condenser is placed where a positive or negative voltage is alternatively applied (Fig. 2.60). If there was any charge $\Delta q$, the beam would be deflected and the signal monitored by the detector behind $A_2$ would change. No such change was observed.

In summary, all of these experiments showed that a possible difference $\Delta q = (e^+) - (e^-)$ between the amounts of the charges of proton and electron has to be smaller than $\Delta q < 10^{-21}\,e \approx 1.6 \times 10^{-40}\,$C.

EXAMPLE

The orifice of the reservoir had a diameter of 0.04 mm and the width of the two slits is also 0.04 mm, the distance between orifice and slit $S_1$ was 400 cm, the length of the condenser 200 cm. With a voltage of 10 kV and a plate separation of 2 mm the electric field is $E = 5 \times 10^6\,$V/m and the force acting on charged particles is $F = \Delta q\, E$. For Cs atoms this experiment gave with a mean velocity of $v = 300\,$m/s an upper limit of $\Delta q \le 10^{-32}\,$As $\approx 10^{-13} \cdot e$ (see Problem 2.12).

### c) Floating Ball

This experiment has similarities with Millikan's oil droplet experiment. Here a small ferromagnetic ball with diameter 0.1 mm is kept floating in a vacuum by an electromagnet, where the magnetic force just cancels the gravity force. The ball position can be monitored by a laser beam reflected at grazing incidence by the ball (Fig. 2.61). The surfaces of the two magnetic poles carry the plates of a condenser, which produces an electric field $E$. If the ball with $N$ atoms each with $Z$ electrons carried a net charge

$$\Delta Q = Q^+ - Q^- = NZ\left(|e^+| - |e^-|\right) \qquad (2.75)$$

the ball would experience a force $F = \Delta Q\, E$ that would shift the ball out of its equilibrium position. This could be measured sensitively through the corresponding deviation of the reflected laser beam [2.42].

## 2.6 Electron and Ion Optics

Charged particles can be deflected and imaged by properly designed electric or magnetic fields that act on these particles in a similar way as lenses, mirrors and prisms act on light in optics. Such field arrangements are therefore named electron or ion optics. They have allowed the development of electron microscopes (Figs. 2.25–2.28) and mass spectrometers (see Sect. 2.7). Since electron microscopes reach a spatial resolution down to 0.1 nm they have become indispensable for the investigations of small structures in biological cells, material and surface phenomena and molecular structure. Mass spectrometers play an important role in analyzing the atomic composition of compounds, in measuring isotopic abundances, determining age in geology and archeology. They are used to measure the absolute masses of atoms and molecules and to define the absolute value of the atomic mass unit AMU (see Sect. 2.2.1).

It is therefore worthwhile to study the basic principles of electron and ion optics.

### 2.6.1 Refraction of Electron Beams

An electric field exerts a force on a particle with charge $q$

$$F = qE = -q\,\mathbf{grad}\,\phi_{el}\,, \qquad (2.76)$$

which is always perpendicular to the equipotential surfaces $\phi_{el} = \text{const.}$

Assume a particle with mass $m$ and charge $q$ moves with constant velocity $\boldsymbol{v}_1$ through a field-free region and enters at an angle $\alpha$ against the surface normal of the planar boundary surface of a homogeneous electric field (Fig. 2.62).

Such an arrangement can be, for instance, realized by four planar parallel metal grids kept at constant potentials. Between grids 2 and 3 in Fig. 2.62 a homogeneous electric field $E = (\phi_1 - \phi_2)/d$ is maintained, while in the regions above and below the electric field is $E = 0$. Energy conservation demands

$$\frac{m}{2}v_2^2 = \frac{m}{2}v_1^2 + qV \tag{2.77}$$

while passing through the field, the $x$-component of the velocity remains unchanged. From Fig. 2.62 we realize the relations: $\sin\alpha = v_{1x}/v_1$, $\sin\beta = v_{2x}/v_2$ and because of $v_{1x} = v_{2x}$ we obtain the law of refraction of a parallel electron beam

$$\frac{\sin\alpha}{\sin\beta} = \frac{v_2}{v_1}\,, \tag{2.78}$$

which corresponds to Snellius' law of refraction $\sin\alpha/\sin\beta = n_2/n_1$ in optics, if we replace the ratio $v_2/v_1$ of the electron velocity by the ratio $n_2/n_1$ of the refractive indices.

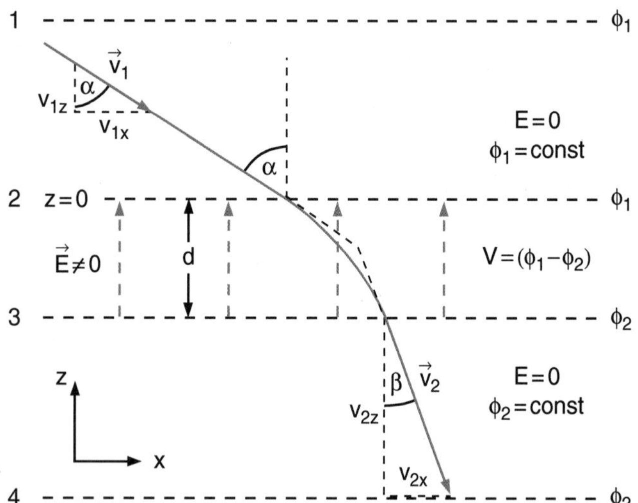


**Fig. 2.62.** Deflection of an electron beam at plane boundaries between regions of different electric fields

The electrons have acquired their initial velocity $v_1$ by a potential difference $V_0 = \phi_0 - \phi_1$ where $(m/2)v_1^2 = qV_0$. Inserting this into (2.77) gives

$$\frac{m}{2}v_2^2 = q(V_0 + V) \Rightarrow \frac{v_2}{v_1} = \sqrt{\frac{V_0 + V}{V_0}}\,. \tag{2.79a}$$

Within the homogeneous electron field $E$ the component $v_x$ remains constant while $v_z$ is increasing. The path of the electrons then follows a parabola

$$z(x) = -\frac{1}{2}\frac{qE}{m}\frac{x^2}{v_{1x}^2} - \frac{v_{1z}}{v_{1x}}x\,. \tag{2.79b}$$

If the distance $d$ between the grids at $z = 0$ and $z = d$ is sufficiently small, we can approximate the path of the electron by two straight lines in the following way.

We extrapolate the straight path of the incident electron for $z < 0$ until the mid plane at $z = d/2$ and that of the outgoing electron for $z > d$ back to $z = d/2$ (dashed straight lines in Fig. 2.62). Then the analogy to the optical refraction at the boundary plane between two media with refractive indices $n_1$, $n_2$ becomes more suggestive. According to (2.74) and (2.79) we can attribute refractive indices to the electron optics arrangement by the relation

$$\frac{n_2}{n_1} = \frac{\sin\alpha}{\sin\beta} = \sqrt{1 + V/V_0}\,. \tag{2.80}$$

This shows that the ratio of the refractive indices is solely determined by the potential difference $V = \phi_1 - \phi_2$ and the initial energy $(m/2)v_1^2 = eV_0$ of the incident electron.

The direction of a parallel electron beam can be also changed by traversing a perpendicular electric field (Fig. 2.63). When electrons enter the homogeneous field $E_z$ of a condenser at $z = 0$ with the velocity $\boldsymbol{v} = \{v_x, 0, 0\}$ the force $F = qE$ gives a parabolic path

$$z = \frac{1}{2}\frac{q}{m}E\frac{x^2}{v^2} \Rightarrow \left(\frac{\mathrm{d}z}{\mathrm{d}x}\right)_{x=L} = \frac{qE}{m}\frac{L}{v^2} = \tan\delta\,. \tag{2.81}$$

In both cases (Figs. 2.62 and 2.63) the electron beam is refracted analogously to a light beam in a prism.

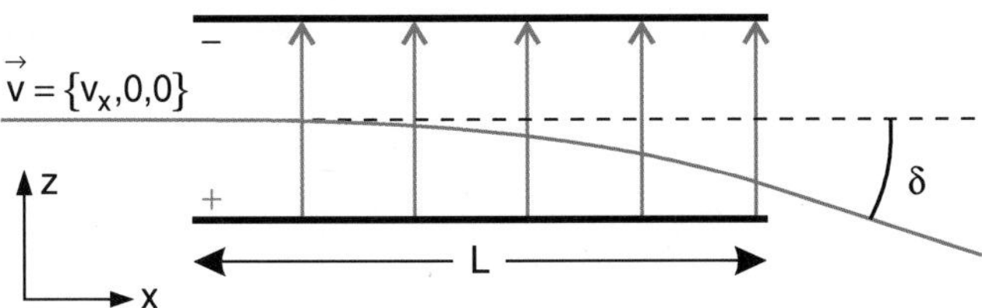


Fig. 2.63. Deflection of an electron beam in the homogeneous electric field between two charged parallel metal plates

## 2.6.2 Electron Optics in Axially Symmetric Fields

We will now look at the imaging of electron beams in inhomogeneous electric fields with axial symmetry.

From (2.76) one obtains with $\boldsymbol{E} = -\ \mathbf{grad}\ \phi$ and without magnetic field ($B = 0$) the basic equations:

$$m\frac{d^2x}{dt^2} = e\frac{\partial\phi}{\partial x}, \quad m\frac{d^2y}{dt^2} = e\frac{\partial\phi}{\partial y}, \quad m\frac{d^2z}{dt^2} = e\frac{\partial\phi}{\partial z}, \tag{2.82}$$

which can be solved, at least numerically, if the potential $\phi$ is determined by the arrangement of charged metal surfaces. If there are no space charges, the Laplace equation becomes:

$$\frac{\partial^2\phi}{\partial x^2} + \frac{\partial^2\phi}{\partial y^2} + \frac{\partial^2\phi}{\partial z^2} = 0\,. \tag{2.83}$$

There is no analytical solution for the general case, but there are, of course, always numerical ones.

Most of the electro-optic lenses are realized by axial symmetric electric or magnetic fields, where the former can be obtained by charged circular apertures or tubes, which can be set at a freely selectable potential.

One example is given by Fig. 2.64 where two cylindrical tubes are connected with a voltage source that sets the potentials $\phi_1$ and $\phi_2$ for the two tubes. The equipotential surfaces are indicated by the dotted lines. Depending on the values of $\phi_1$ and $\phi_2$ a collecting or a diverging electron lens can be realized. The appropriate coordinates here are cylindrical coordinates $r$, $\varphi$, $z$. Since the system is axially symmetric, the potential cannot depend on the angle $\varphi$, but only on $r$ and $z$. While for $r = 0$ the derivative $\partial\phi/\partial r = 0$, this is not the case for $r \neq 0$. Therefore, a force acts on those electrons that move away from the symmetry axis. Electrons coming from the left side in Fig. 2.64 are attracted towards the axis (note that the force is always

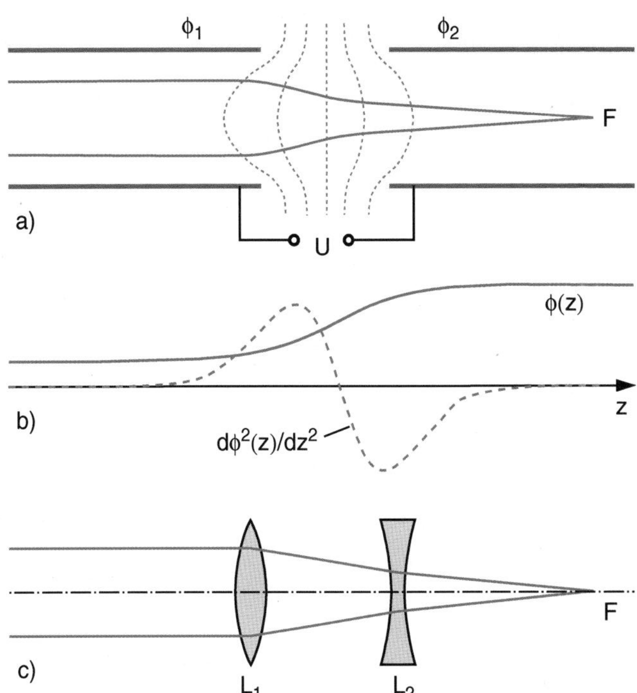


Fig. 2.64a–c. Electron lens consisting of two cylindrical tubes at different potentials. (a) Schematic representation. (b) Potential $\phi(z)$ and its second derivative $d^2\phi/dz^2$. (c) Optical analogy

perpendicular to the dotted equipotential lines), while they are driven away from the axis in the right tube. For $\phi_2 > \phi_1$ the electrons are accelerated when moving from the first to the second tube. Their velocity is therefore larger at the right side as at the left one. This means that the defocusing effect on the right side is smaller than the focusing effect on the left side. The system acts as a converging lens. For $\phi_2 < \phi_1$ the situation is reversed and the system represents a diverging lens.

Let us prove more quantitatively the focusing effect of an electric field with axial symmetry. We use as an example the potential

$$\phi(r,z) = a\left(z^2 - \frac{1}{2}r^2\right) \tag{2.84}$$

shown in Fig. 2.65, which is produced by two pairs of axially symmetric hyperbolic electrodes, where the $z$-axis is the symmetry axis. The electrodes are kept at constant potentials $\phi = 0$ and $\phi = \phi_0$, respectively. The general equipotential surfaces inside the system are obtained by inserting $\phi = C$ into (2.84), which yields

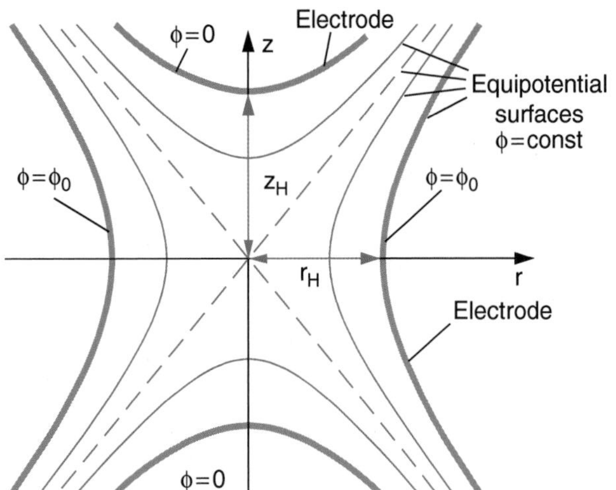


**Fig. 2.65.** Hyperbolic electrostatic field with axial symmetry around the $z$-axis

$$\frac{z^2}{C/a} - \frac{r^2}{2C/a} = 1 \,. \tag{2.85}$$

This proves that the equipotential surfaces $\phi = C$ are hyperboloids around the symmetry axis $r = 0$ ($z$-axis). Their minimum distance from the origin $(r, z) = (0, 0)$ is $z_\mathrm{H} = \sqrt{C/a}$ for $r = 0$ and $r_\mathrm{H} = \sqrt{2C/a}$ for $z = 0$. The potential at the $z$-axis ($r = 0$) is:

$$\phi(r = 0, z) = az^2 \Rightarrow \left(\frac{\partial^2 \phi}{\partial z^2}\right)_{r=0} = 2a \,. \tag{2.86a}$$

Inserting this into (2.84) gives

$$\phi(r, z) = \phi(0, z) - \frac{1}{4}\phi''(0, z)r^2 \,. \tag{2.86b}$$

It is therefore always possible to determine the hyperbolic potential $\phi(r, z)$ at an arbitrary point $(r, \varphi, z)$ when its value $\phi(0, z)$ and its second derivative $\phi''(r = 0) = (\partial^2\phi/\partial z^2)_0$ at the symmetry axis are known!

The radial components $E_r$ of the electric field can be obtained from (2.84) as

$$E_r = -\frac{\partial \phi}{\partial r} = ar \,. \tag{2.87}$$

Electrons away from the symmetry axis always experience a linear force

$$\boldsymbol{F}_r = -e\boldsymbol{E}_r = -a \cdot e \cdot \boldsymbol{r} \,, \tag{2.88}$$

which drives them back to the axis. The $r$-component of their movement therefore represents a harmonic oscillation. Furthermore, all electrons that start from a single point ($z = z_1, r = 0$) on the symmetry axis are again focused into a point ($z = z_2, r = 0$) on the axis, as long as their velocity components $v_z$ are equal, even if the other components $v_r$, $v_\varphi$ differ for the different electrons (Fig. 2.66). These properties of hyperbolic potentials are used to realize ion traps.

The general axial symmetric potential will not necessarily have hyperbolic equipotential surfaces. In order to obtain its value $\phi(r, z)$ at points not too far from the axis $r = 0$ from its values on the axis, we use the Taylor expansion

$$\phi(r, z) = \phi(0, z) + b_2(z)r^2 + b_4(z)r^4 + \ldots \,, \tag{2.89}$$

where only terms with even powers of $r$ can be nonzero because of the axial symmetry ($\phi(-r) = \phi(r)$). For small deviations from the axis ($b_n r^n \ll \phi(0, z)$) we can neglect all higher order terms with $n > 2$. Inserting (2.89) into the Laplace equation (2.83), which reads in cylindrical coordinates as

$$\frac{1}{r}\frac{\partial \phi}{\partial r} + \frac{\partial^2 \phi}{\partial r^2}\frac{\partial^2 \phi}{\partial z^2} = 0 \,, \tag{2.90}$$

(because $\partial\phi/\partial\varphi = 0$), we obtain

$$\phi(r, z) = \phi(0, z) - \frac{1}{4}\left(\frac{\partial^2 \phi}{\partial z^2}\right)_{0,z} r^2 \,. \tag{2.91}$$

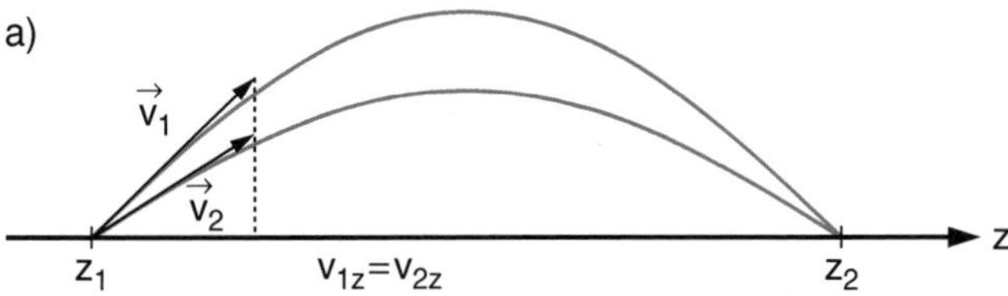


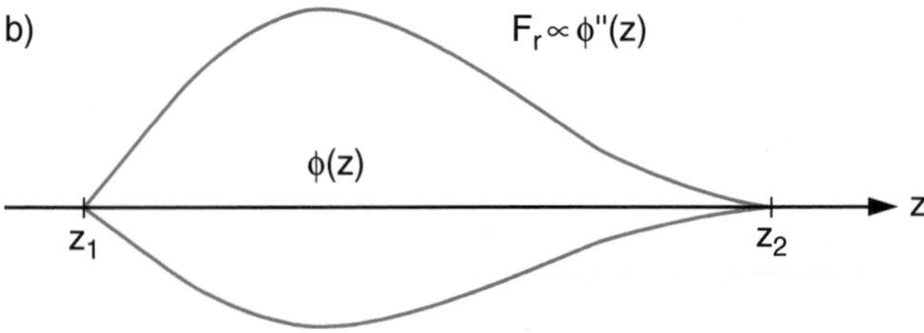


**Fig. 2.66a,b.** Focusing in a cylindrical symmetric electric field. (**a**) Hyperbolic field. (**b**) Arbitrary axially symmetric field

The comparison with (2.86) shows that within this paraxial approximation ($b_2(z)r^2 \ll \varphi(0,z)$) the same relation exists for an arbitrary axially symmetric potential (2.89) as for a hyperbolic potential.

> This implies the remarkable fact that for all axial symmetric potentials $\phi(r,z)$ the potential in an arbitrary point not too far from the axis can be calculated from its values $\phi(r=0)$ and $\phi''$ $(r=0)$ at the axis.

The electron path in such fields can now be obtained from the general equations (2.82), which reduce here to

$$m\frac{\mathrm{d}^2r}{\mathrm{d}t^2} = e\frac{\partial\phi}{\partial r}\ ; m\frac{\mathrm{d}^2z}{\mathrm{d}t^2} = e\frac{\partial\phi}{\partial z}\ . \tag{2.92}$$

With the paraxial approximation $(\mathrm{d}^2\phi/\mathrm{d}r^2)r^2 \ll (\mathrm{d}\phi/\mathrm{d}r)r$ we obtain from (2.90)

$$\frac{\partial\phi}{\partial r} = -\frac{1}{2}\frac{\partial^2\phi}{\partial z^2}r\ ; \left(\frac{\partial\phi}{\partial z}\right)_{r,z} = \left(\frac{\mathrm{d}\phi}{\mathrm{d}z}\right)_{r=0}\ . \tag{2.93}$$

The equations of motion are then

$$\begin{aligned} F_r &= m\frac{\mathrm{d}^2r}{\mathrm{d}t^2} = -\frac{e}{2}\left(\frac{\mathrm{d}^2\phi}{\mathrm{d}z^2}\right)_{r=0}\cdot r = -a(z)r \\ F_z &= m\frac{\mathrm{d}^2z}{\mathrm{d}t^2} = e\left(\frac{\mathrm{d}\phi}{\mathrm{d}z}\right)_{r=0}\ . \end{aligned} \tag{2.94}$$

Close to the axis $r=0$, $v_r \ll v_z$ and therefore $v = \sqrt{v_r^2+v_z^2} \approx v_z$.

The radial component of the electron motion can now differ from a harmonic oscillator, because the restoring force $F_r = a(z)r$ may change with $z$.

### 2.6.3 Electrostatic Electron Lenses

In Fig. 2.67 the path of an electron is shown schematically for a "thin lens," where the extension $d$ of the electric field is small compared to the focal length $f$. Assume the electric field is restricted to the space between the planes $z=z_1$ and $z=z_2$. In the left side field-free region the path is then a straight line, where for paraxial rays, the angle $\alpha_1$ is small. Within the field region $z_1 \le z \le z_2$ the path is bent, but the distance $r$ from the axis does not vary much within the small interval $d$, which implies that $r_a \approx r_m \approx r_b$. From Fig. 2.67

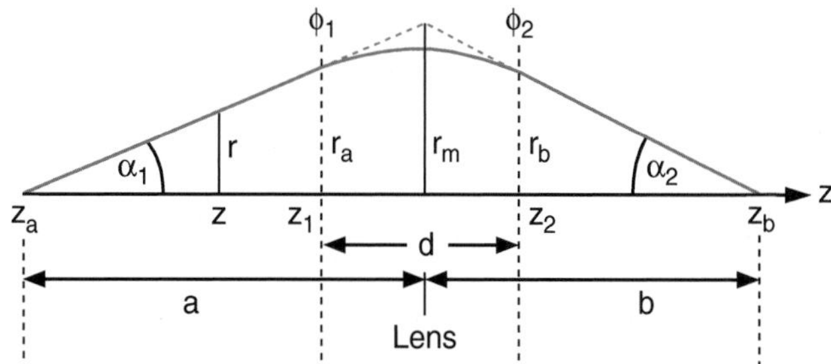


**Fig. 2.67.** Illustration of the lens equation for an electron lens

we can then derive the relations

$$\left(\frac{\mathrm{d}r}{\mathrm{d}z}\right)_{z\le z_1} = \tan\alpha_1 \approx \frac{r_m}{a}\ . \tag{2.95}$$

In the field-free region to the right we obtain

$$\left(\frac{\mathrm{d}r}{\mathrm{d}z}\right)_{z\ge z_1} = -\tan\alpha_2 = -\frac{r_m}{b}\ . \tag{2.96}$$

Adding (2.96) and (2.97) yields the lens equation for thin electron lenses

$$\boxed{\frac{1}{f} = \frac{1}{a} + \frac{1}{b} = \frac{1}{r_m}\left[\left(\frac{\mathrm{d}r}{\mathrm{d}z}\right)_{z=z_1} - \left(\frac{\mathrm{d}r}{\mathrm{d}z}\right)_{z=z_2}\right],} \tag{2.97}$$

which exactly corresponds to the lens equation in geometrical optics when the difference in the brackets in (2.97) is defined as the ratio $r_m/f$. The focal length $f$ of the electron lens can be derived from (2.91) as

$$f = \frac{4\sqrt{\phi_0}}{\int_{z_1}^{z_2}\frac{1}{\sqrt{\phi}}\left(\frac{\mathrm{d}^2\phi}{\mathrm{d}z^2}\right)_{0,z}\mathrm{d}z}\ . \tag{2.98}$$

It depends on the potential $\phi(z)$ and its second derivative on the axis $r=0$ and also on the initial energy $(m/2)v_0^2 = e\phi_0$ of the incoming electrons.

Figure 2.68 shows two possible experimental realizations of such electrostatic lenses using plane grids and apertures. The voltage $V = \phi_1 - \phi_2$ is applied between the grid and a circular aperture at a distance $d$ from the grid. The equipotential surfaces are axially symmetric around the symmetry axis $r=0$ ($z$-axis). Since the electric field $\boldsymbol{E} = -\ \mathbf{grad}\ \phi$ is perpendicular to these surfaces, the forces ($\boldsymbol{F} = -e\boldsymbol{E}$) acting on the electrons is always perpendicular to the equipotential surfaces. If the circular aperture is, for instance,

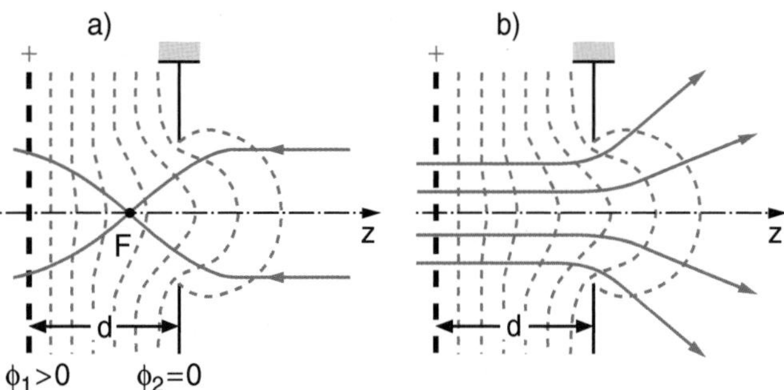


**Fig. 2.68.** Electron lens consisting of an aperture and a grid at different potentials. A parallel beam of electrons coming from the right is focussed into the point F while the beam diverges if it comes from the left

grounded ($\phi_2 = 0$) and the grid is set at a positive potential ($\phi_1 > 0$), a parallel electron beam entering from the right side into the system, is focused into the focal point $F$ (Fig. 2.68a). The focal length $f$ depends on the voltage $V = \phi_1 - \phi_2$ and the kinetic energy of the incoming electrons. While a parallel electron beam entering from the left diverges (Fig. 2.68b).

The symmetrical arrangement of three apertures in Fig. 2.69, with potentials $\phi_1 = \phi_3 = 0$ and $\phi_2 \neq 0$ represents a combination of collimating and diverging

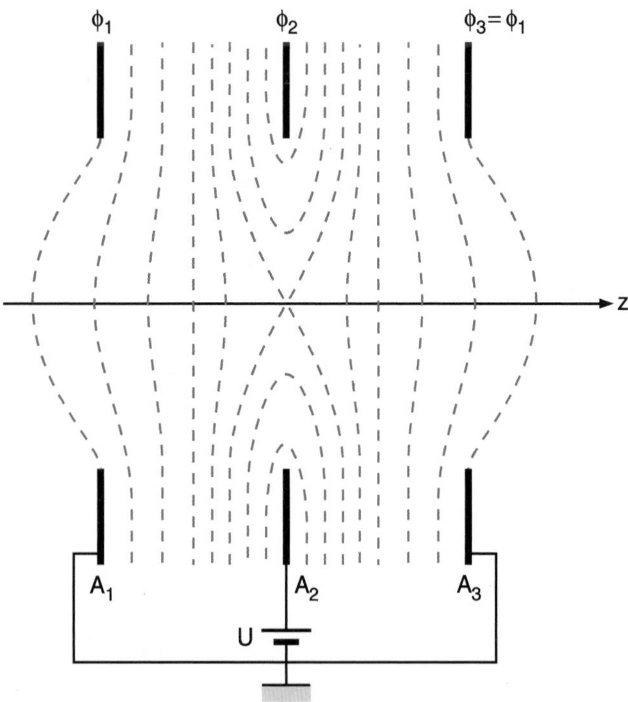


**Fig. 2.69.** Einzel lens realized by a symmetric arrangement of three apertures at different potentials

lenses. Depending on the polarity of the applied voltage between $A_2$ and $A_1$ or $A_2$ and $A_3$ either the collimating or the diverging influence dominates. For example, with a positive voltage applied to $A_2$ ($\phi_2 > 0$) the electrons entering from the left are accelerated between $A_1$ and $A_2$ but decelerated between $A_2$ and $A_3$. The diverging effect then dominates and the whole system acts as diverging lens, while for $\phi_2 < 0$ a collecting lens is realized.

Electrostatic cylindrical lenses, which only focus in one direction, can be, for instance, realized by a cylindrical condenser (Fig. 2.70), where the two cylindrical surfaces are set at voltages $+V/2$ and $-V/2$, respectively. Electrons passing through an entrance slit $S_0$ and entering the field region with the energy $eV_0$, are imaged onto a line $S_1$ in the focal plane of the electron lens, while the image line $S_2$ of electrons with a different energy $e(V_0 + \Delta V)$ is spatially shifted from $S_1$ to $S_2$. The cylindrical condenser can therefore be used as energy analyzer for electrons with an energy distribution $N(E)$ (see Problem 2.15). The analogue in optics is a cylindrical lens plus prism. The focal length of the cylindrical condenser depends on the angle $\varphi$ and the voltage $\pm V/2$ at the condenser surfaces has to be adapted to the kinetic energy $(m/2)v^2 = eV_0$ of the electrons. In the solution to Problem 2.15 it is shown that the optimum path of the electrons through the condensor is obtained for $V = 2V_0 \ln(R_2/R_1)$, where the $R_i$ are the radii of the condenser plates.

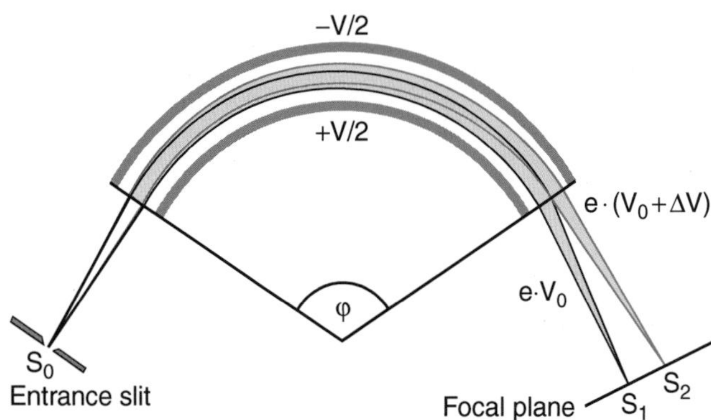


**Fig. 2.70.** Electrical cylindrical capacitor acting as cylindrical electron lens

### 2.6.4 Magnetic Lenses

Electrons entering a homogeneous magnetic field under the angle $\alpha$ against the field lines (Fig. 2.71a)

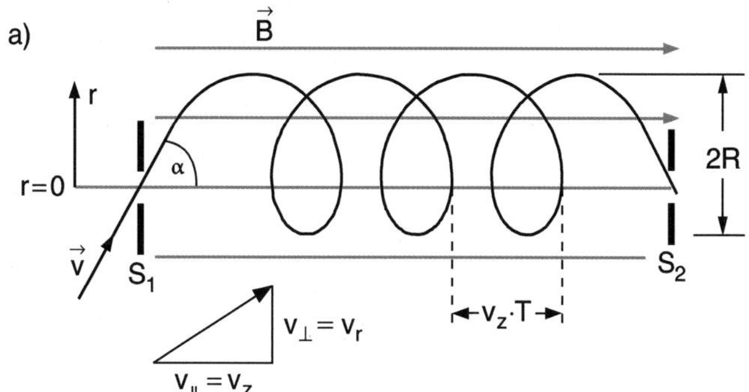


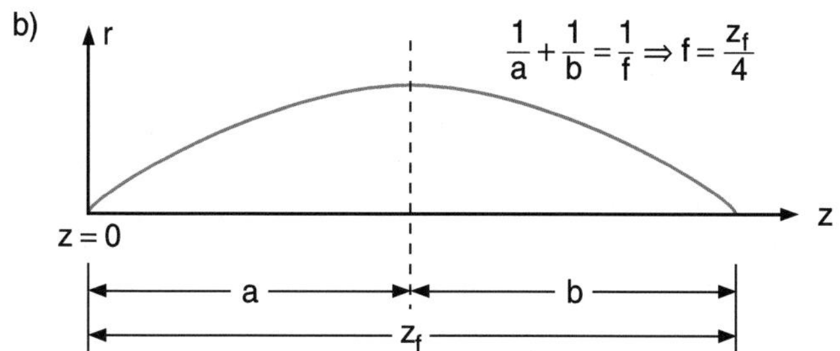


**Fig. 2.71a,b.** Homogeneous longitudinal magnetic field as an electron lens. (**a**) Illustration of the helical path. (**b**) Definition of the focal length

experience the Lorentz force $\boldsymbol{F} = -e(\boldsymbol{v} \times \boldsymbol{B})$, which deflects their path. For $B = \{0, 0, B_z\}$ we obtain $F_z = 0$. We separate the velocity $\boldsymbol{v} = \{v_x, v_y, v_z\}$ into a component $v_\parallel = v_z$ parallel to the field lines and a perpendicular component $v_\perp = \left(v_x^2 + v_y^2\right)^{1/2}$. The parallel component $v_\parallel$ is always perpendicular to the force $\boldsymbol{F}$. This means that the amount of $v_\perp$ remains constant, but the direction of $\boldsymbol{v}$ changes. For $v_z = 0$, the electron path would be a circle with radius $R$, given by the balance between the Lorentz and centrifugal forces:

$$\frac{mv_\perp^2}{R} = ev_\perp B \Rightarrow R = \frac{m}{e}\frac{v_\perp}{B}\,. \tag{2.99a}$$

The time $T$ for transversing one cycle

$$T = \frac{2\pi R}{v_\perp} = \frac{2\pi}{e}\frac{m}{B} \tag{2.99b}$$

is independent of $v_\perp$ and $R$!

For $v_z \neq 0$ the electron path is a helix with a pitch $\Delta z = T \cdot v_z$. An electron starting from the axis $r = 0$ at $z = 0$ and $t = 0$ is imaged onto a point $z = v_z \cdot T$ onto the axis.

Any homogeneous magnetic field (which can be, for instance, obtained by a current $I$ flowing through a cylindrical coil) represents an electron lens, which focuses all electrons starting from a point $z = 0$ on the $z$-axis into the focal point

$$z_f = v_z T = \frac{2\pi m}{eB} v_z \tag{2.99c}$$

independent on their angle $\alpha$ against the axis, if only their velocity component $v_z$ is the same.

For practical realizations, generally $v_\perp \ll v_\parallel$. Therefore, electrons accelerated by a voltage $V$ all have approximately the same velocity $v_z \approx v = \sqrt{2eV/m}$. We can then define a focal length $f$ of the magnetic lens according to Fig. 2.71b given by

$$f = \frac{1}{4} z_f = \frac{\pi}{2B}\sqrt{\frac{2Vm}{e}}\,. \tag{2.100}$$

Besides such longitudinal magnetic fields transverse fields can also be used as magnetic lenses for imaging of electron or ion beams. A transverse sector field with sector angle $2\varphi$, shown in Fig. 2.72 represents a cylindrical lens, which focuses all electrons or ions starting from the entrance slit $S_1$, into a line $S_2$ parallel to $S_1$, where both $S_1$ and $S_2$ are located in a plane passing through the point $M$, which is the center of the circular electron or ion path in the magnetic field. This can be seen as follows.

We divide the sector field in Fig. 2.72 into two half parts and regard in Fig. 2.73 only the right half. Ions in a parallel ion beam entering the field from the left, traverse the fields on circular paths with a radius $R = mv/(qB)$ where $q$ is the charge of the ions (or the electrons). The center of the circular path $\overline{SA_0}$ for ions in the middle of the parallel beam with width $b$ is denoted by $M_0$, the center $M_1$ for ions entering on the path 1 is shifted upwards by $b/2$. The ions on the center path $\overline{SA_0}$ have been deflected by the angle $\varphi$ when leaving the field, while those on path $S_1$ by $(\varphi + \alpha)$.

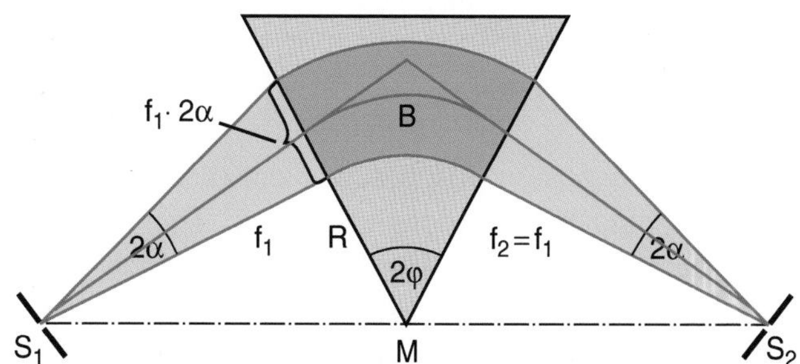


**Fig. 2.72.** Magnetic sector field as cylindrical lens for ions or electrons

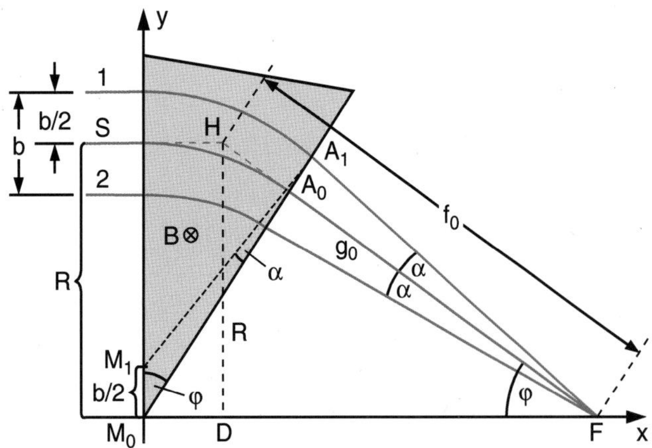


**Fig. 2.73.** Focusing of a parallel incident ion beam by one half of the magnetic sector field in Fig. 2.72

The straight paths after leaving the field are always perpendicular to the radius $M_0A_0$ or $M_1A_1$, respectively. This implies that the straight lines are inclined against each other by an angle $\alpha$. They intersect at the point F.

The distance $g_0 = \overline{A_0F}$ is then:

$$g_0 = \frac{\overline{A_0A_1}}{\tan\alpha} = \frac{\overline{M_0A_1} - \overline{M_0A_0}}{\tan\alpha} . \tag{2.101}$$

For the triangle $\Delta M_1A_1M_0$ we obtain with $\overline{M_0A_0} = R$ the relation

$$\overline{M_0A_1} = \frac{\sin(\varphi+\alpha)}{\sin\varphi}R . \tag{2.102}$$

This yields

$$\overline{A_0A_1} = \overline{M_0A_1} - R = R\left(\frac{\sin(\varphi+\alpha)}{\sin\alpha} - 1\right) \approx R(\cos\alpha - 1 + \cot\varphi\sin\alpha) . \tag{2.103}$$

For small angles $\alpha$ the approximation $\cos\alpha \approx 1$ and $\sin\alpha \approx \tan\alpha$ holds. This gives

$$g_0 = \frac{\overline{A_0A_1}}{\sin\alpha} \approx R\cot\varphi . \tag{2.104}$$

The distance $g_0$ is nearly independent of the width $b$ of the ingoing beam as long as $b \ll R$. Then all ions in the beam are focused into the point $F$, which is the focal point. From (2.104) and the rectangular triangle $\Delta M_0A_0F$ (with the 90° angle at $A_0$) it follows that the angle $\angle M_0FA_0$ equals $\varphi$, which implies that the line $\overline{FM_0}$ is parallel to the direction of the incoming beam.

Similar to the treatment of thick lenses in geometrical optics we can define a principal plane $DH$ at $x = D$ where the extrapolated lines of the incoming center ray $S$ and the outgoing straight line $A_0F$ intersect. The focal length $f_0$ of the magnetic sector field is then defined as the distance $f_0 = \overline{HF}$. With the relation $\overline{HD} = R$ and $\sin\varphi = \overline{HD}/\overline{HF}$ we obtain for the focal length of one half sector field

$$f_0 = \frac{R}{\sin\varphi} . \tag{2.105}$$

The focal length of a magnetic sector field with a full sector angle $2\varphi$ is the ratio of radius of curvature $R$ and $\sin\varphi$. For $2\varphi = 180° \Rightarrow f_0 = R$

When we now reflect the sector in Fig. 2.73 at the plane $x = 0$ and add this mirror image to the sector in Fig. 2.73 we obtain the arrangement of Fig. 2.72. Ions that start as a divergent beam from the slit $S_1$ form a parallel beam at the symmetry plane through $M$ in Fig. 2.72 and are therefore, according to the foregoing discussion, imaged onto $S_2$. From (2.104) and the rectangular triangle $\overline{DHF}$ in Fig. 2.73 it follows that $\sphericalangle A_0FM_0 = \varphi$ and that $S_1$, $M$ and $S_2$ must lie on the same straight line.

In the direction perpendicular to the drawing plane of Fig. 2.72 (i.e., parallel to the magnetic field) no focusing occurs. This means that for particle paths parallel to the plane $z = 0$ ($z$ gives the direction of the magnetic field) every point in the entrance slit is imaged onto its corresponding image point in $S_2$, quite analogous to the situation for cylindrical lenses in optics.

For more details on electron- and ion optics see the text books [2.43–45].

### 2.6.5 Applications of Electron and Ion Optics

The applications of electron optics can be illustrated by the example of a modern transmission electron microscope, shown in Fig. 2.74. The incident parallel electron beam is transmitted through a thin slice of the sample. The electrons suffer elastic and inelastic collision with the atoms and molecules of the sample. Since the energy loss by inelastic collisions depends on

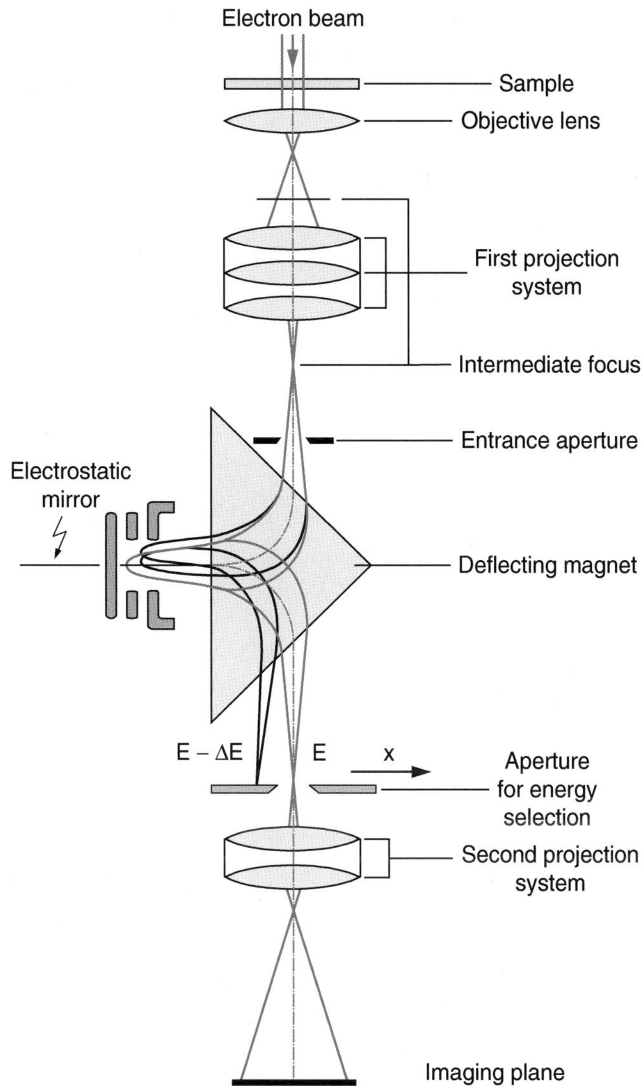


**Fig. 2.74.** Modern electron microscope. The lenses symbolize electron lenses not optical lenses (with kind permission from Zeiss, Oberkochen)

the kind of atoms it can be used to analyze the atomic composition of the sample.

In order to measure this energy loss, the transmitted electrons are focused by an electron lens system into a first focal plane. The focal point serves as point like source for the following magnetic sector field, where the electrons are deflected to the left and reflected by an electrostatic mirror, realized by an electro-optic system at negative potential. The reflected electrons pass through the magnetic field again and are deflected according to their reversed velocity. Electrons with differing energies are imaged at different positions $x$ in the focal plane. A movable aperture selects only those electrons with the wanted energy. One may select the elastically scattered electrons or those that suffered inelastic collisions with a selectable energy loss in the sample. This selection enhances the contrast of the final magnified image and allows one to filter selected parts of the sample, e.g., spots containing heavy atoms that cause larger energy losses [2.46–48].

Further examples of applications of ion optics are the various kinds of mass spectrometers that will be discussed in the following section.

## 2.7 Atomic Masses and Mass Spectrometers

After having discussed in the previous sections experimental methods for the determination of atomic sizes and electrical properties of atoms we will now deal with the measurement of atomic masses [2.49–52].

The most simple method for measuring atomic masses is based on the knowledge of Avogadro's constant $N_A$. When the mass $M$ of a mole of atoms $x$ in a gas is measured (it has a volume of $22.4\,\text{dm}^3$ at $p = 1033$ hPa and $T = 0\,°\text{C}$), the mass $m_x$ of an atom is

$$m_x = M/N_A\,.$$

If the relative atomic mass

$$A = 12\frac{m_x}{m\left(^{12}\text{C}\right)}$$

in AMU is known (see Sect. 2.2.1) then the absolute mass $m_x$ is directly obtained from $M = A \cdot 10^{-3}$ kg as

$$m_x = \frac{A \cdot 10^{-3}}{N_A}\,\text{kg}$$

without further measurements.

The mass $m_x = M/N$ of atoms in a regular crystal with mass $M$ is determined from the total number $N$ of atoms in the crystal that can be obtained from the geometry of the crystal and the distances between crystal planes measured by X-ray diffraction (see Sect. 2.4.4).

> The most accurate method for the determination of atomic masses is, however, based on the deflection of ions in electric or magnetic fields.

From the measured mass $m(A^+)$ of an atomic ion, the mass of the neutral atom is

$$m(\mathrm{A}) = m(\mathrm{A}^+) + m(\mathrm{e}^-) - \frac{1}{c^2} E_\mathrm{B} , \tag{2.106}$$

where the last term (which is generally neglected) represents the mass equivalent of the binding energy $E_\mathrm{B}$ of the electron in the atom A.

It is quite instructive to briefly follow the historical development and gradual improvement of mass spectrometers, in order to appreciate the work of our scientific ancestors and to understand why modern devices are more accurate.

### 2.7.1 J.J. Thomson's Parabola Spectrograph

Ions with charge $q$ are produced in a gas discharge, are accelerated by a voltage $V$ and pass in the $z$-direction through a homogeneous magnetic field $\boldsymbol{B} = \{B_x, 0, 0\}$, which is superimposed by a homogeneous electric field $\boldsymbol{E} = \{E_x, 0, 0\}$ (Fig. 2.75). The equations of motion are then with $B_x = B$; $E_x = E$

$$\frac{\mathrm{d}^2x}{\mathrm{d}t^2} = \frac{q}{m} E ; \quad \frac{\mathrm{d}^2y}{\mathrm{d}t^2} = \frac{q}{m} vB ; \quad \frac{\mathrm{d}^2z}{\mathrm{d}t^2} = 0 . \tag{2.107a}$$

The time $t$ can be eliminated by using the relations

$$\frac{\mathrm{d}x}{\mathrm{d}t} = \frac{\mathrm{d}x}{\mathrm{d}z}\frac{\mathrm{d}z}{\mathrm{d}t} = \frac{\mathrm{d}x}{\mathrm{d}z} v_z \approx v \frac{\mathrm{d}x}{\mathrm{d}z} \tag{2.107b}$$

because the velocity increase of $v_x$ in the electric field is very small compared to the initial velocity $v_z$. This gives:

$$\frac{\mathrm{d}^2x}{\mathrm{d}z^2} = \frac{q}{mv^2} E \tag{2.107c}$$

$$\frac{\mathrm{d}^2y}{\mathrm{d}z^2} = \frac{q}{mv} B . \tag{2.107d}$$

Integration of (2.107c) over $z$ for values $-L/2 \le z \le +L/2$ within the field region yields

$$\frac{\mathrm{d}x}{\mathrm{d}z} = \int_{-L/2}^{z} \frac{qE}{mv^2} \mathrm{d}z' = \frac{qE}{mv^2} \left( \frac{L}{2} + z \right)$$

$$\Rightarrow x(z) = \frac{qE}{2mv^2} \left( \frac{L}{2} + z \right)^2 . \tag{2.108}$$

Integration of (2.107d) gives

$$y(z) = \frac{qB}{2mv} \left( \frac{L}{2} + z \right)^2 . \tag{2.109}$$

For $z > \frac{L}{2}$ the two fields are zero, i.e., $E = 0$ and $B = 0$. The total force on the ion is then zero and its path follows a straight line with an inclination against the $x$-direction, resp the $y$-direction given by

$$\left( \frac{\mathrm{d}x}{\mathrm{d}z} \right)_{L/2} = \frac{qE}{mv^2} L ; \quad \left( \frac{\mathrm{d}y}{\mathrm{d}z} \right)_{L/2} = \frac{q \cdot B}{mv} L \tag{2.110}$$

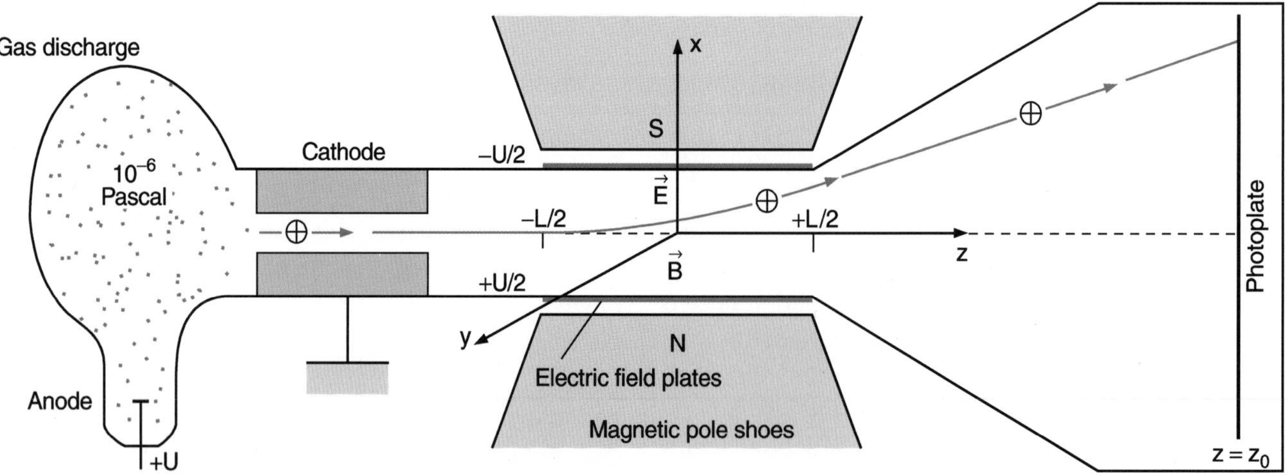


**Fig. 2.75.** *Thomson's* parabola spectrograph

The ions therefore impinge on a photoplate at the plane $z = z_0$ at an $x$-coordinate

$$\begin{aligned} x(z_0) &= \frac{qEL^2}{2mv^2} + \frac{qEL}{mv^2}\left(z_0 - \frac{L}{2}\right) \\ &= \frac{qEL}{mv^2} z_0 \end{aligned} \tag{2.111a}$$

while the $y$-coordinate is

$$\begin{aligned} y(z_0) &= \frac{qBL^2}{2mv} + \frac{qBL}{mv}\left(z_0 - \frac{L}{2}\right) \\ &= \frac{qBL}{mv} z_0 \,. \end{aligned} \tag{2.111b}$$

For a given initial velocity $v = (2qV/m)^{1/2} = v_z$ every ion hits a point $\{x(z_0), y(z_0)\}$ at the photo plate, which depends on the ratio $q/m$. In the original mass spectrograph by Thomson, the ions were produced in a gas discharge and had a broad velocity distribution. They therefore arrive at different points $(x(z_0), y(z_0))$, depending on their velocity. In order to find a relation between $x(z_0)$ and $y(z_0)$ one has to eliminate $v$. Solving (2.111b) for $v$ and inserting this expression into (2.111a) gives the relation

$$x(z_0) = \frac{m}{q}\frac{E}{B^2 L z_0} y^2 = a(m) y^2 \,. \tag{2.112}$$

This represents a separate parabola $x = ay^2$ for every value of $m/q$ (Fig. 2.76a). Measuring this parabola allows the determination of the factor $a$ and therefore the ratio $q/m$. Figure 2.76b shows for illustration that such parabolas are obtained for different isotopes of neon ions from a gas discharge, where water vapor and benzene vapor had also been added for calibration purposes since they deliver $H^+, C^+, CH^+, O^+, OH^+$ and $H_2O^+$ ions.

According to (2.111), the velocity spread $\Delta v$ corresponds to a length $\Delta s_p$ of the parabola given by

$$\Delta s_p \simeq \sqrt{\Delta x^2 + \Delta y^2} = \frac{qL}{mv^2} z_0 \sqrt{B^2 + \left(\frac{2E}{v}\right)^2}\, \Delta v \,. \tag{2.113}$$

This spread decreases the sensitivity of photographic recording.

**Note:**

The equations (2.111a) and (2.111b) show that the deflection $x$ of a charged particle by an electric field is inversely proportional to its kinetic energy, while in the magnetic field it is inversely proportional to its momentum $mv$. The deflection in electric fields is therefore a measure of the kinetic energy that in a magnetic field is a measure of the momentum of the particle.

### 2.7.2 Velocity-Independent Focusing

In Thomson's mass spectrograph, ions of equal masses but different velocities were spread out along a parabola. This diminishes the intensity of the ions impinging onto a specific point $(x, y)$ on the photoplate. One would obtain a much higher signal if all ions of equal mass could be focused onto the same spot (for

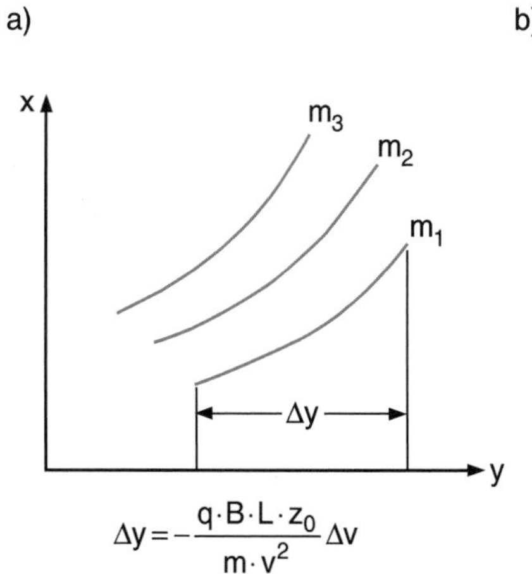


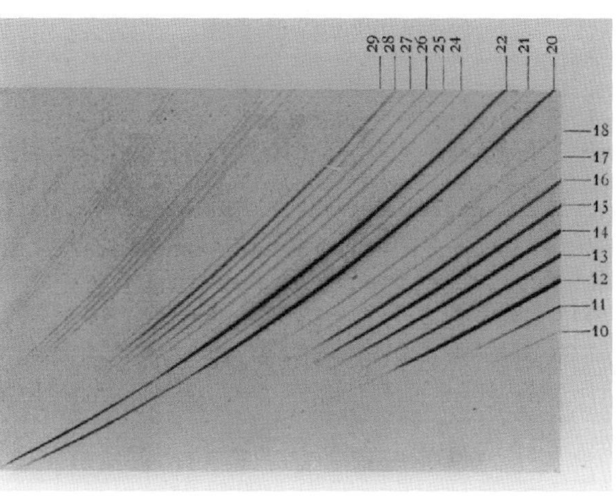


**Fig. 2.76a,b.** The ions arrive at the photoplate along a parabolic curve, for each mass $m_i$ on a separate parabola. (**a**) Schematic drawing. (**b**) Real measurement of the different neon isotopes in an isotopically enriched neon discharge with admixtures of water and benzene $C_6H_6$ (From J. Mattauch [2.53])

instance the entrance aperture of an ion detector) independent of their velocity. This can be achieved by the mass spectrograph constructed by *Francis William Aston* (1877–1945). Here the electric and magnetic field regions are spatially separated (Fig. 2.77) and the direction of the fields is chosen in such a way that the deflections of the ions are in opposite directions. The incident ion beam is collimated by the two slits $S_1$ and $S_2$ parallel to the $y$-direction and enters the electric field as a parallel beam in the $z$-direction. The deflection angle $\alpha$ of the beam by the electric field is, according to (2.110)

$$\tan\alpha = \left(\frac{\mathrm{d}x}{\mathrm{d}z}\right)_{L_{1/2}} = \frac{qEL_1}{mv^2} \tag{2.114a}$$

while the deflection angle $\beta$ caused by the magnetic field is

$$\tan\beta = \frac{qBL_2}{mv} . \tag{2.114b}$$

For small deflection angles ($\tan\alpha \ll 1$, $\tan\beta \ll 1$) we can approximate $\tan\alpha \approx \alpha$, $\tan\beta \approx \beta$ and derive from (2.114a), (2.114b) the relations:

$$\frac{\mathrm{d}\alpha}{\mathrm{d}v} = -\frac{2qEL_1}{mv^3} = -\frac{2\alpha}{v} \quad \text{and} \tag{2.114c}$$

$$\frac{\mathrm{d}\beta}{\mathrm{d}v} = -\frac{qBL_2}{mv^2} = -\frac{\beta}{v} . \tag{2.114d}$$

The lateral deflection by the electric field is

$$D_E = -(a+b)\cdot\tan\alpha \approx -(a+b)\cdot\alpha$$

and by the magnetic field

$$D_M = b\cdot\tan(\beta-\alpha) + b\cdot\tan\beta \approx b\cdot\beta$$

The total lateral deflection $D$ of the ions against the $z$-axis is then approximately (see Fig. 2.77)

$$D \approx -(a+b)\alpha + b\beta . \tag{2.115}$$

For $\mathrm{d}D/\mathrm{d}v = 0$ the deflection becomes independent of the ion velocity $v$. This yields

$$-(a+b)\frac{\mathrm{d}\alpha}{\mathrm{d}v} + b\frac{\mathrm{d}\beta}{\mathrm{d}v} = +\frac{2(a+b)\alpha}{v} - \frac{b\beta}{v} = 0$$

$$\Rightarrow \beta b = 2(a+b)\alpha \Rightarrow D = +(a+b)\alpha . \tag{2.116}$$

This must be valid for all masses $m$ i.e. for all angles $\alpha = qE\cdot L_1/mv^2$, which means for all points of the photoplate surface. This can be fullfilled, if the photoplate is arranged in a plane that is inclined by the angle $\alpha$ against the $z$-direction and that intersects the $z$-axis in the middle of the electric field (see Fig. 2.77). The slit $S_3$ defines a range $\Delta\alpha$ of deflection angles $\alpha$ around a medium value $\alpha_m$, which defines the position of the photoplate. This means, that not all ions with vastly differing velocities are transmitted but only those within a selectable range $\Delta v$. Velocity-independent focusing therefore means:

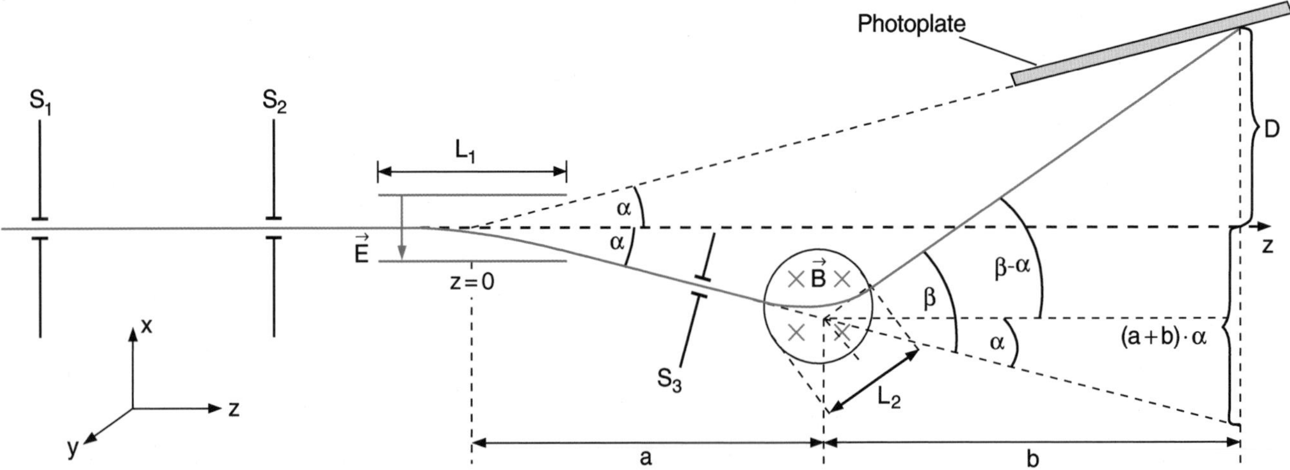


**Fig. 2.77.** Aston's mass spectrograph with velocity independent focusing

All ions with velocities within the interval $v_m - \Delta v/2$ up to $v_m + \Delta v/2$ are imaged onto a small spot with diameter $\Delta s$ on the photoplate. The value of $\Delta s$ is much smaller than in Thomson's parabola spectrograph, which implies that one really wins in intensity (number of ions impinging per time onto a given spatial interval $\Delta s$).

### 2.7.3 Focusing of Ions with Different Angles of Incidence

Up to here we have assumed that the incident ions form a parallel beam in the $z$-direction. Although this can be approximately realized by apertures the loss in intensity quadratically increases with increasing collimation. It is therefore highly desirable if the ions with different directions of their initial velocities could be refocused and therefore contribute to the signal without loss of mass resolution.

This goal was first realized by *Arthur Jeffrey Dempster* (1886–1950) who built a magnetic sector field with $2\varphi = 180°$ where the ion paths are half-circles with radius $R = mv/(qB)$. As is shown in Fig. 2.78a all ions passing through the entrance slit A within the angular range from $-\alpha$ to $+\alpha$ against the $y$-axis are imaged onto the exit plane $y = 0$ within a small interval $\Delta x \approx R\alpha^2$. This can be understood from the relations obvious from Fig. 2.78a:

$$\overline{AC} = 2R\,; \quad \overline{AB} = 2R\cos\alpha \approx 2R\left(1 - \alpha^2/2\right)$$
$$\Rightarrow \Delta s = \overline{AC} - \overline{AB} \approx R\alpha^2\,.$$

**EXAMPLE**

$R = 10\,\text{cm}$, $\alpha = 3° = 0.05\,\text{rad} \Rightarrow \Delta s = 2.5 \times 10^{-2}\,\text{cm}$. Placing a 0.25-mm wide slit in the exit plane allows all incident ions with equal ratios $q/m$, but velocity directions ranging from $-3°$ to $+3°$, to be transmitted through the slit.

As has been shown in Sect. 2.6.4, magnetic sector fields with arbitrary sector angles $\varphi_m$ act as cylindrical lenses with a focal length

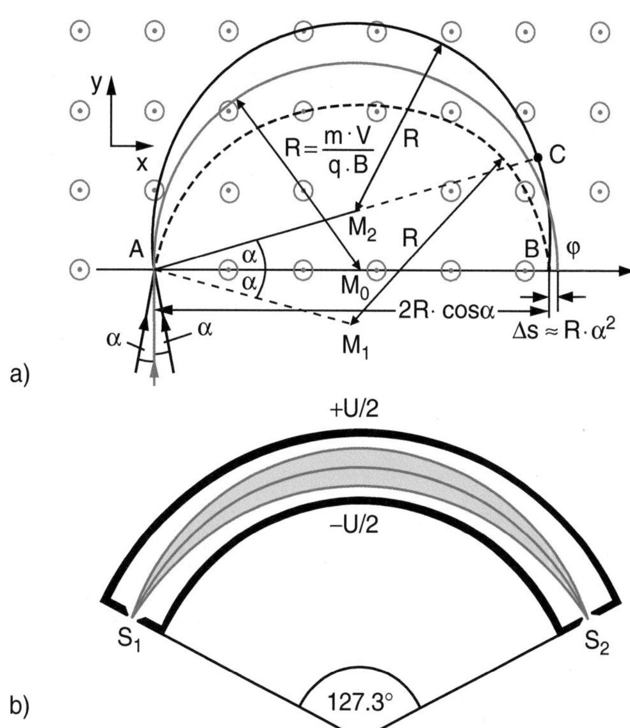


**Fig. 2.78a,b.** Comparison of directional focusing (**a**) in a 180° magnetic sector field, (**b**) in an electrostatic 127.3° cylindrical field

$$f = \frac{R_0}{\sin(\varphi_m/2)} = \frac{m}{q}\frac{v}{B}\frac{1}{\sin(\varphi_m/2)}\,.$$
$$= \frac{mv}{qB} \text{ for } \varphi_m = \pi\,.$$

The comparison of the 180° magnetic sector field with the cylindric electric condenser reveals (see Problem 2.15) that both fields act the same way as cylindrical lenses, if the sector angle $\varphi_m$ is replaced by $\varphi_{el} = \varphi_m/\sqrt{2}$. The magnetic 180° sector field therefore corresponds to a 127.3° electric cylinder condenser (Fig. 2.78b).

### 2.7.4 Mass Spectrometer with Double Focusing

When using a combination of electric and magnetic sector fields with proper sector angles, simultaneous velocity and directional focusing can be achieved, as shown in the example in Fig. 2.79. The incident ion beam passing the entrance slit has an angular divergence of $\pm\Delta\alpha$. Behind the cylindrical condenser the slit $S_2$ selects ions with charge $q$ of within the energy

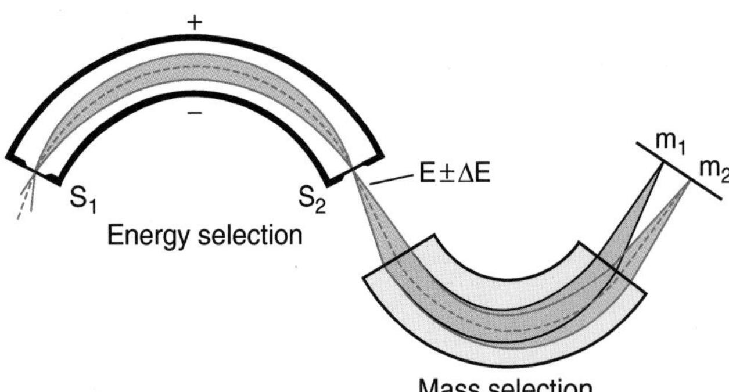


**Fig. 2.79.** Example of a double-focusing mass spectrograph

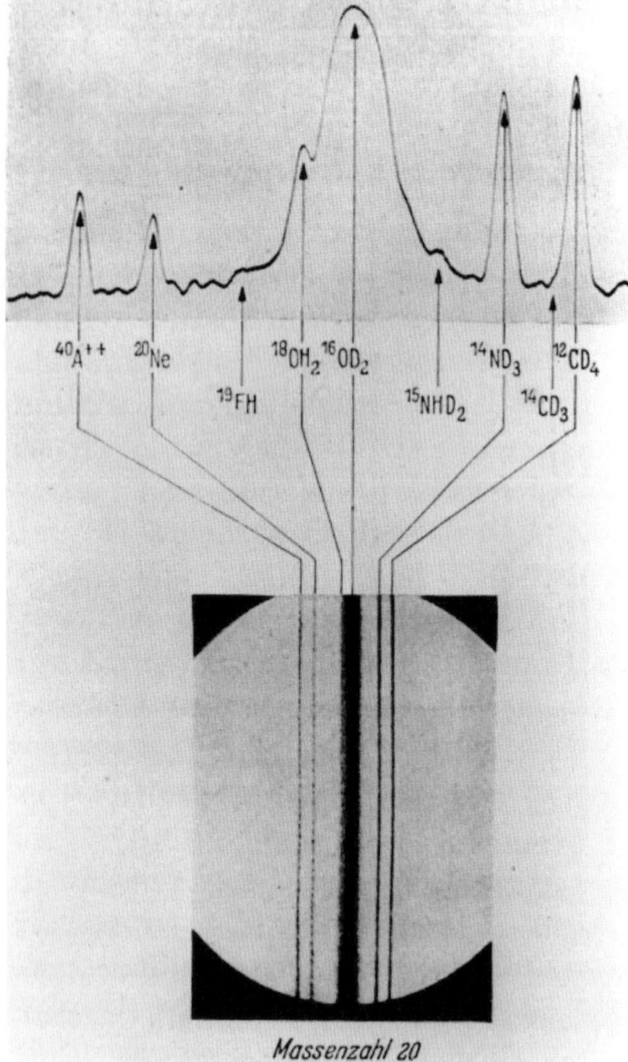


**Fig. 2.80.** Selection of the high resolution mass spectrum of ions within the mass range around 20 AMU, obtained from a gas discharge of argon and neon, mixed with methane $CH_4$, ammonia $NH_3$, water vapor $H_2O$ and their isotopomers (From Mattauch [2.53])

range $E \pm \Delta E$. Since all ions have been accelerated by the same voltage $V$ before reaching $S_1$ they all should have the same energy $q \cdot V$. Their energy spread comes about because their thermal velocity distribution in the ion source has to be added to their acceleration energy. This spreads the ions with slightly different velocities over a spatial interval in the plane of slit $S_2$. The slit width of $S_2$ determines the energy interval $\Delta E$ of those ions entering the magnetic sector field, where the mass separation takes place, because the magnetic field separates ions with different momenta $mv = \sqrt{2mqE}$, see (2.73). Both fields focus divergent ion beams if the sector angles $\varphi_{\mathrm{el}}$ and $\varphi_{\mathrm{m}}$ have been chosen correctly as has been shown in the previous sections. For $\varphi_{\mathrm{el}} = \pi/\sqrt{2}$ all ions with energies within the interval $E + \Delta E$, which pass through $S_1$, are imaged onto $S_2$. For the magnetic sector angle $\varphi_{\mathrm{m}} = 60°$ $S_2$ is then imaged with a focal length $f_{\mathrm{m}} = R/\sin 30° = 2R = 2mv/(qB)$ onto the photoplate or a slit $S_3$ in front of the detector. The imaging of ions with different velocities is achieved in first order in the same way as in Aston's spectrometer, because the deflections in both fields are in opposite directions.

For an illustration of the achievable resolution $m/\Delta m = 6 \times 10^3$, Fig. 2.80 shows a section of a mass spectrum of different masses around AMU = 20, measured with such a double-focusing spectrograph. Their masses differ only by small fractions of 1 AMU, due to different mass defects of the individual nuclei.

### 2.7.5 Time-of-Flight Mass Spectrometer

The basic principle of a time-of-flight (TOF) mass spectrometer is simple (Fig. 2.81). At time $t = 0$ ions are produced in a small volume (for example the crossing volume of a laser beam and a molecular beam). They are accelerated by a voltage $V$ to a velocity $v = (2qV/m)^{1/2}$ where $m$ is their mass and $q$ their charge. The ions drift through a field-free region of length $L$ before they are detected. Measuring the time delay between ion production at $t = 0$ and mass-dependent arrival time

$$t_m = \frac{L}{v} = \frac{L}{\sqrt{2qV/m}} \tag{2.117a}$$

at the detector gives the mass $m$ of the ion:

$$m = \frac{2qV}{L^2} t_m^2 \,. \tag{2.117b}$$

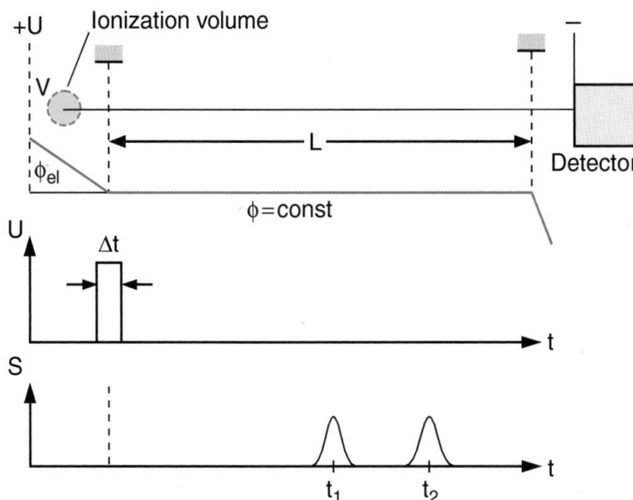


**Fig. 2.81.** Principal scheme of a time-of-flight mass spectrometer

**EXAMPLE**

$L = 1\,\text{m}$, $V = 1\,\text{kV}$, $m = 100\,\text{AMU} = 1.6\times 10^{-25}\,\text{kg}$, $q = e = 1.6\times 10^{-19}\,\text{C} \Rightarrow t_m = 52\,\mu\text{s}$.

The accuracy of mass determination depends on how accurate the distance $L$, the time of flight $T_m$, and the acceleration voltage can be measured. The mass resolution $m/\Delta m$ depends on the shortest time interval $\Delta t$ that can still be resolved. From (2.117b) we obtain

$$\frac{\Delta m}{m} = 2\frac{\Delta t}{t_m}\,. \tag{2.117c}$$

Since the time resolution $\Delta t$ cannot be better than the initial time spread, the duration $\Delta t_0$ of the initial ion pulse should be as short as possible. If the ions are, for instance, formed through photo-ionization by a short laser pulse of $\Delta t_0 = 10^{-8}\,\text{s}$, the initial pulse width is short compared to the flight time and can therefore often be neglected.

**EXAMPLE**

$\Delta t_0 = 10^{-8}\,\text{s}$, $t_m = 50\,\mu\text{s} \Rightarrow \Delta m/m = 4\times 10^{-4}$. This means that two masses $m_1 = 2500\,\text{AMU}$ and $m_2 = 2501\,\text{AMU}$ can be still separated.

Another problem arises because the ions are not all produced at the same location, but within a finite volume. The electric field used to extract the ions varies over this volume and therefore the kinetic energy, the ions acquire by the extraction voltage, varies accordingly. Therefore, the velocities of ions, produced at different locations, differ, which smears out the arrival times and limits the mass resolution.

In order to improve this, a modification of the extraction field was introduced by McLaren et al. [2.54], where the acceleration occurs in two steps. This can be achieved by three plane grids kept at the potentials $\phi_1, \phi_2$, and $\phi_3 = 0$ with distances $d_1$ and $d_2$ (Fig. 2.82) and homogeneous fields $E_1 = (\phi_2 - \phi_1)/d_1$, $E_2 = (\phi_3 - \phi_2)/d_2$ between the grids.

Assume the $x$-axis to be the symmetry axis of the spectrometer. An ion produced at the distance $x_1$ from grid 2 has a flight time $t_1$ until it reaches the grid 2. We obtain for ions starting at $x = d_1 - x_1$ with a velocity $v = 0$

$$x_1 = \frac{1}{2}\frac{qE_1}{m}t_1^2 \Rightarrow t_1 = \sqrt{\frac{2mx_1}{qE_1}}\,. \tag{2.118a}$$

At grid 2 ($x = d_1$) it has acquired the velocity

$$v_1 = \left(\frac{\mathrm{d}x}{\mathrm{d}t}\right)_{t_1} = \frac{qE_1}{m}t_1\,. \tag{2.118b}$$

At grid 3 with potential $\phi_3$ the velocity has increased to

$$v_2 = v_1 + \frac{qE_2}{m}t_2, \tag{2.118c}$$

where $t_2$ is the flight time from $G_2$ to $G_3$. Integration yields the relation between the distance $d_2$ and the flight times $t_1, t_2$ and the electric fields $E_1, E_2$

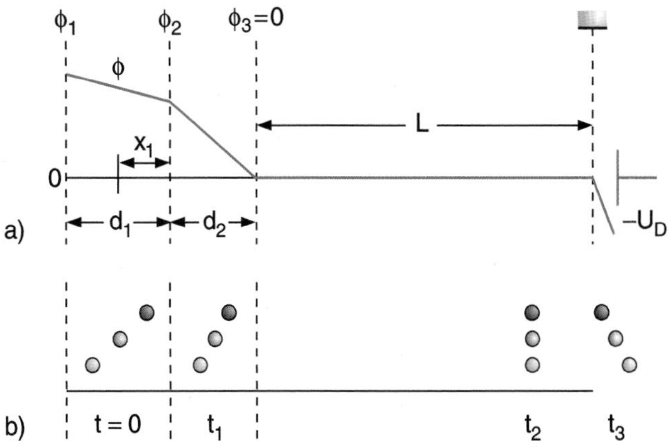


**Fig. 2.82.** Arrangement of grids and potentials in a McLaren time-of-flight spectrometer with improved mass resolution

$$\begin{aligned} d_2 &= \int\limits_{t_1}^{t_1+t_2} v\,\mathrm{d}t = \int\limits_{t_1}^{t_1+t_2} \left(v_1 + \frac{qE_2}{m}t\right)\mathrm{d}t \\ &= v_1 t_2 + \frac{1}{2}\frac{qE_2}{m}\left(2t_1t_2 + t_2^2\right) \\ &= \frac{q}{m}\left[(E_1+E_2)\,t_1t_2 + \frac{1}{2}E_2t_2^2\right] . \end{aligned} \tag{2.119}$$

The solution of the quadratic equation in $t_2$ gives $t_2 = t_2(t_1, E_1, E_2)$.

The drift time across the field-free region is then $t_3 = L/v_2$ and the total time of flight

$$t = t_1 + t_2 + t_3 .$$

Inserting $t_2$ and $v_2$ gives the total times $t$ as a function of $t_1, E_1, E_2$.
If $t$ should be independent of the location $x$ of ion production, one has to demand $\mathrm{d}t/\mathrm{d}x_1 = 0$. Inserting $t_1$, $t_2$, and $t_3$ into (2.119) and taking the derivative $\mathrm{d}t/\mathrm{d}x_1$ yields the optimum flight path length $L_{\mathrm{opt}}$ in the field-free region from $G_3$ to the detector

$$L_{\mathrm{opt}} = d_1 k^{3/2}\left(1 - \frac{d_2}{d_1}\frac{1}{k+\sqrt{k}}\right) \tag{2.120}$$

with

$$k = 1 + \frac{2d_2}{d_1}\frac{E_2}{E_1} = 1 + 2\frac{V_2}{V_1} ,$$

where $V_1 = \phi_2 - \phi_1$ and $V_2 = \phi_3 - \phi_2$ are the potential differences between the grids in Fig. 2.82. It is therefore possible to construct the spectrometer in such a way that the distance from ion source to detector equals the optimum flight distance $L_{\mathrm{opt}}$ where the total flight time is the same for all ions independent of the location where they are produced. In Fig. 2.82b the total flight times of these ions produced at different distances $x_1$ from grid $\phi_2$ are illustrated. Ions with larger values of $x_1$ have larger flight paths to the detector but also gain higher kinetic energies because they are produced at a higher potential $\phi(x)$. Therefore they pass the slower ions produced at smaller $x$ after a total flight time $t_2$ at a total distance $x(t_2) = L_{\mathrm{opt}} + d_2 + x_1$ from their production point. That is where the detector has to be placed.

The advantages of time-of-flight spectrometers are the following:

- It is possible to measure all mass components of a mixture of different species
- Even atoms or molecules with very large masses (for example biological molecules with $m \le 10^5$ AMU) can be detected, although they have a correspondingly long flight time and the ion detector needs a high acceleration voltage to achieve a sufficient secondary electron emission coefficient.
- The TOF design is simple and easy to construct. It is much cheaper than other types of mass spectrometer.

The mass resolution of TOF can be considerably increased if the ions are reflected at the end of the drift distance $L$ by an electrostatic reflector, which consists of grids at a positive voltage producing an electric field that repels the ions (Fig. 2.83). Ions entering the field under an incident angle $\alpha$ against the field direction are reflected into an angle $2\alpha$ against their incident direction, where they reach the detector after a further drift distance $L$. The faster ions penetrate deeper into the reflecting field and therefore travel a larger distance, just compensating for an earlier arrival time at the reflecting field. This device, called a "***reflectron***" [2.55], achieves the same total travel time for all ions within a velocity interval $\Delta v$ (see also Problem 2.16e). Time-of-flight spectrometers are particularly useful in combination with photo-ionisation by short-pulse lasers, because here start time and ionization volume are precisely defined [2.56].

For illustration, Fig. 2.84 shows the TOF mass spectrum of $Na_n$ clusters [2.57]. These are loosely bound compounds of $n$ sodium atoms. Such clusters are of current interest, since they allow detailed studies of the transition regime between free molecules and liquid droplets or solid micro particles.

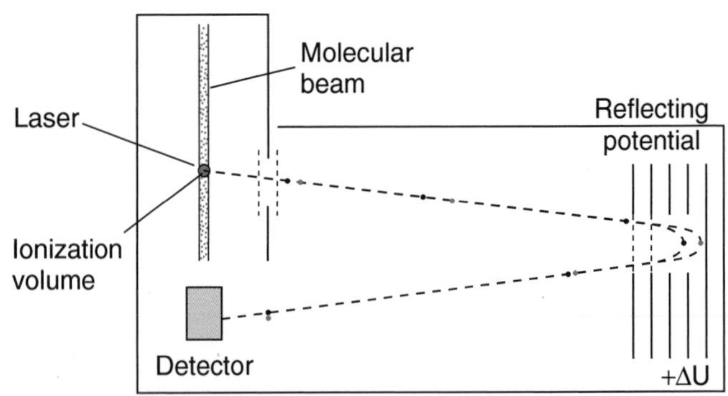


**Fig. 2.83.** Two ions (*black* and *blue dots*) with the same mass but different energies travel along different paths in the reflector field and arrive at the detector at the same time

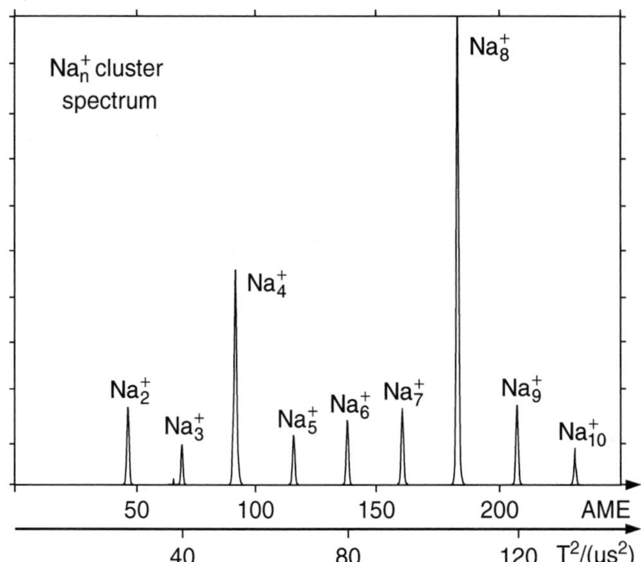


**Fig. 2.84.** Time-of-flight mass spectrum of sodium cluster ions $Na_n^+$ [2.57]

### 2.7.6 Quadrupole Mass Spectrometer

In Sect. 2.6.2 it was shown that an axially symmetric hyperbolic electrostatic field causes focusing or defocusing of charged particles, depending on the polarity of the applied dc voltage. The quadrupole mass spectrometer developed in 1953 by *Wolfgang Paul* (1913–1994), (Nobel Prize 1992) and *H. Steinwedel* [2.58] uses a hyperbolic potential

$$\phi(x,z) = \frac{\phi_0}{2r_0^2}\left(x^2 - z^2\right) , \tag{2.121}$$

which is *not* axially symmetric. It is, however, invariant against translation in the $y$-direction. It is formed by four metal electrodes with hyperbolic inner surfaces, where two opposite electrodes are electrically connected and are kept at the potential $\pm\phi_0/2$ (Fig. 2.85b). Note the difference between the potential diagram of Fig. 2.85a and that of Fig. 2.65. While the latter has axial symmetry around the $z$-axis and is generated by electrodes with cylindrical symmetry, the potential in Fig. 2.85a has no axial symmetry, although the diagram looks similar. The reason for this difference is due to the rods, forming the electrodes, which extend linearly into the $y$-direction and Fig. 2.85a just shows the potential in an arbitrary plane $y = y_0$, while all electrodes in Fig. 2.61 have axial symmetry around the $z$-axis.

The ions are formed in an ion source and are accelerated by a voltage $U_0$ before they enter the mass spectrometer and fly into the $y$-direction. With a time-independent constant voltage $U = \phi_0$ between neighboring electrodes the field component $E_x = -\phi_0 x/r_0^2$ causes a force $F_x = +qE_x$ that drives the ions back to the center $x = 0$. The ions therefore perform harmonic oscillations in the $xy$-plane. Because of the opposite polarity of the field component $E_z = +\phi_0 z/r_0^2$ the force component $E_z = qE_z$ is directed away from the center $z = 0$ and the ions are driven away from the central axis $z = 0$ along their flight in $y$-direction. Their movement in the $yz$-plane is therefore instable and such a dc device would not be useful. However, using a trick

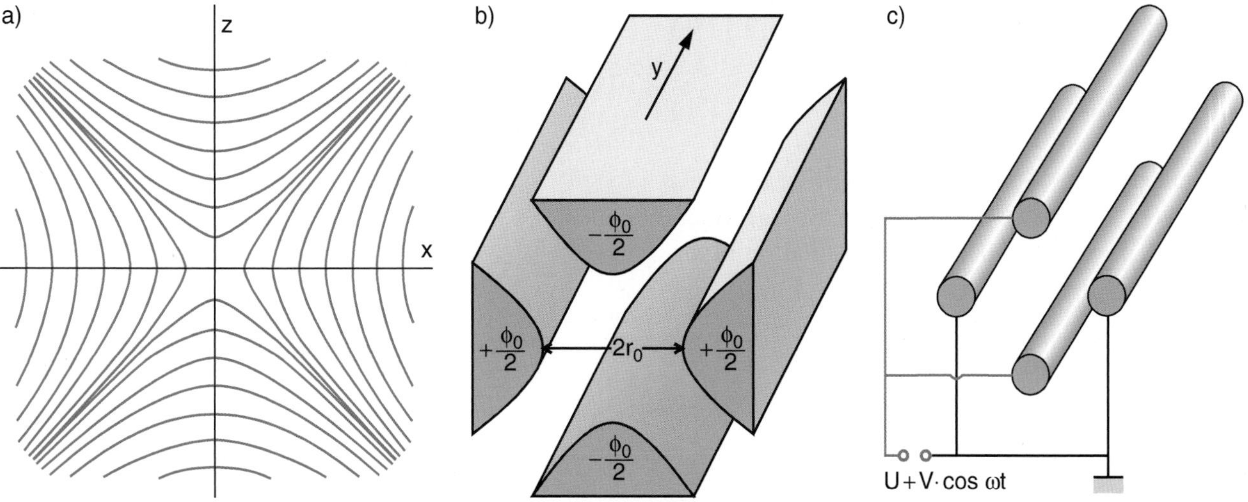


**Fig. 2.85a–c.** Quadrupole mass spectrometer. (**a**) Equipotential lines. (**b**) Hyperbolic electrodes. (**c**) The experimental approximation of (**b**) uses cylindrical rods

the ions can be stabilized in both directions, if in addition to the dc voltage $U$ an ac voltage $V \cos\omega t$ is applied to the electrodes. The potential $\phi$ then becomes

$$\phi_0(t) = U + V\cos\omega t\ . \tag{2.122}$$

The polarity of the electrodes changes periodically. This means that within a half-period of the ac field the ions are stabilized in the $x$-direction and destabilized in the $z$-direction while in the next half period the situation is just reversed. Before the ions during their destabilization period can fly too far away from the axis they are stabilized again and are brought back. It can be shown mathematically, that on the time average this device leads to a stabilization in both directions for ions of a selected mass, but to a destabilization for ions of different masses. The mass selection is determined by the frequency $\omega$ and the ratio $U/V$ of dc and ac amplitude.

The equations of motion for the ions are

$$\ddot{x} + \frac{q}{mr_0^2}(U + V\cos\omega t)x = 0\ , \tag{2.123a}$$

$$\ddot{z} - \frac{q}{mr_0^2}(U + V\cos\omega t)z = 0\ . \tag{2.123b}$$

Introducing the dimensionless parameters

$$a = \frac{4qU}{mr_0^2\omega^2}\ ,\quad b = \frac{2qV}{mr_0^2\omega^2}\ ,\quad \tau = \frac{1}{2}\omega t \tag{2.124}$$

transforms these equations into the (well-known to mathematicians) Mathieu's differential equations

$$\frac{d^2x}{d\tau^2} + (a + 2b\cos 2\tau)x = 0\ , \tag{2.125a}$$

$$\frac{d^2z}{d\tau^2} - (a + 2b\cos 2\tau)z = 0\ . \tag{2.125b}$$

The parameter $a$ represents twice the ratio of the ion's potential energy $qV$ in the dc field to the average kinetic energy $(m/2)v^2 = mr_0^2\omega^2/2$ of its oscillation in the ac field, while $b$ gives the average ratio of $E_{pot}$ to $E_{kin}$ in the ac field.

Mathieu's equations have stable and unstable solutions depending on the values of the parameters $a$ and $b$.

The stable solutions describe oscillations of the ions with limited amplitude. These ions pass through the quadruple spectrometer in the $y$-directions without hitting the electrodes.

The unstable solutions describe ions with oscillation amplitudes in the $x$- or $z$-direction, which exponentially increase while the ion is moving into the $y$-directions. The ion hits the electrodes before it can reach the detector.

The stable regions can be represented in the $a$–$b$-diagram of Fig. 2.86. Note, that the conditions for stability solely depend on the parameters $a$ and $b$ and not on the initial conditions of the ions. Choosing the optimum combination of $a$ and $b$ allows one to transmit

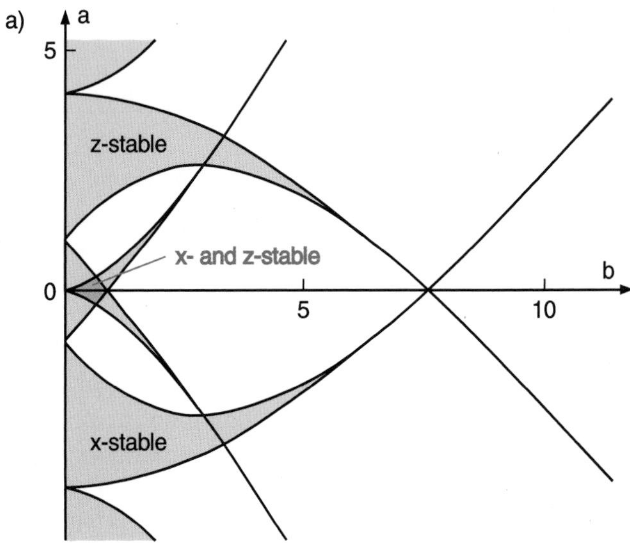


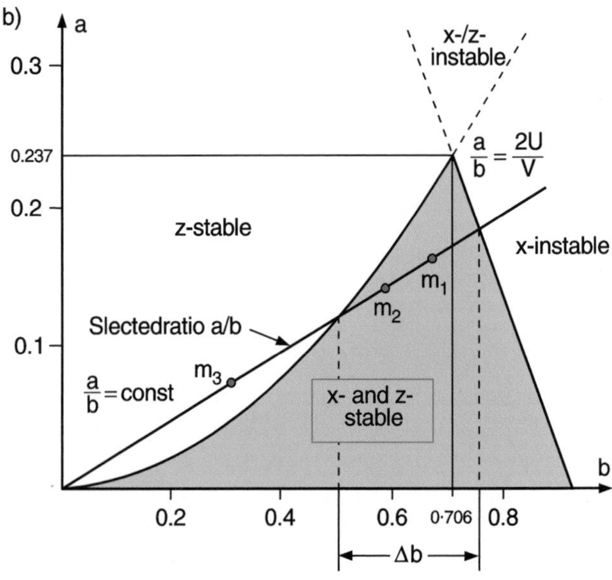


**Fig. 2.86.** (**a**) Different stabiltity ranges (*grey*) of the quadrupole mass filter. (**b**) Enlarged section of the blue stability range in (**a**). The straight line with $a/b = \text{const}$ gives the location for the masses $m_i$. Only masses inside the colored region are transmitted

the wanted mass $m$ and to suppress all other masses. This is illustrated by Fig. 2.86b, where the first stability region for both the $x$- and $z$-direction limited by $a < 0.237$ and $b < 0.9$ is plotted on an expanded scale. For given values of $U$ and $V$ the different masses all lie on the straight line $a/b = 2U/V = \text{const}$, as can be seen from (2.124). The position of a mass $m_i = 4qU/\left(ar_0^2\omega^2\right)$ depends for a mass spectrometer with fixed values of $r_0$ and $\omega_0$ on the parameter $a$. Only those ions that have masses within the stable region reach the detector. For our example in Fig. 2.86b these are the masses $m_1$ and $m_2$. The closer the straight lines approaches the peak of the stability region, the smaller is the mass range $\Delta m$ transmitted to the detector. Selecting the ratio $a/b$ therefore allows one to set the mass range of transmitted ions, which determines the mass resolution of the spectrometer. The mass resolution $m/\Delta m$ of the quadruple mass spectrometer can therefore be easily adjusted (within certain limits) by choosing the appropriate ratio $U/V$ of dc voltage $U$ and ac amplitude $V$ [2.59].

### 2.7.7 Ion-Cyclotron-Resonance Spectrometer

This type of mass spectrometer was developed in 1965 and since then has been greatly improved. Today it represents the device with the highest accuracy in absolute mass measurements and the highest mass resolution ($m/\Delta m \geq 10^8$!).

Its basic design [2.60] is illustrated in Fig. 2.87. It consists of an axially symmetric hyperbolic electric field (like that in Fig. 2.65) with the $z$-axis as symmetry axis, superimposed by a homogeneous magnetic field $B$ in the $z$-direction. The ions, produced in an ion source are injected into the device and then the electric field is switched on. The electric field stabilizes the ion in $z$-direction and the magnetic field causes them to move on circles around the magnetic field lines, thus stabilizing them in the radial directions ($x$- and $y$-directions). In order to avoid collisions of the ions with the background gas atoms the vacuum has to be very good ($p < 10^{-16}$ Pa). Without the electric field the ions with an initial velocity $\boldsymbol{v} = \{v_x, v_y, 0\}$ would move on circles with radius $R = mv/(qB)$. The angular velocity (see Sect. 2.7.4)

$$\omega_c = \frac{qB}{m} \tag{2.126}$$

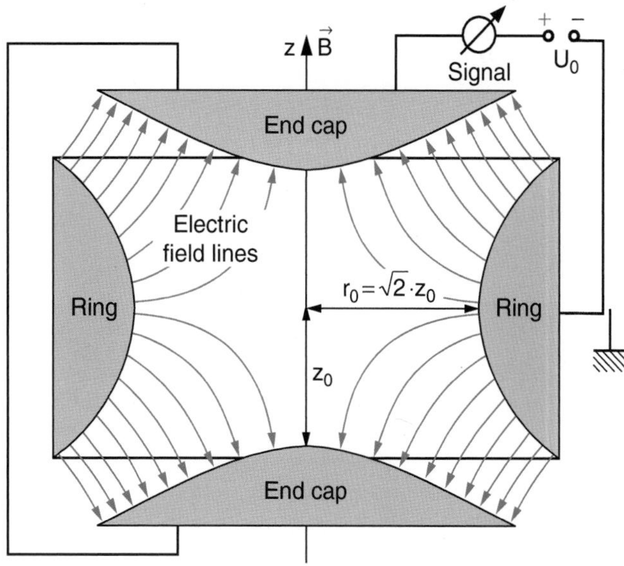


**Fig. 2.87.** Cyclotron-resonance mass spectrometer (penning trap)

is the cyclotron frequency. It is independent of the radius $R$.

The electric field is formed by hyperbolic electrodes, consisting of two hyperbolic caps and one ring symmetric around the $z$-axis. A positive voltage at the caps stabilizes the ions in the $z$-direction. The axially symmetric electric field has the components (see Sect. 2.6.2).

$$E_r = +\frac{U_0}{2z_0^2}, \quad E_z = -\frac{U_0}{z_0^2}. \tag{2.127}$$

Without a magnetic field the ions would perform harmonic oscillations in the $\pm z$-direction, due to the linear restoring force $qE_z$, but they were not stabilized in the radial direction. The superposition of the homogeneous magnetic field $B_z$ stabilizes the ions in all directions but their movement becomes more complicated (Fig. 2.88). It can be composed of the cyclotron movement (circles around an axis in the $z$-direction), a second component, where the center of these circles performs oscillations in the $\pm z$-direction (axial oscillations) and a third component, where the center of the circles undergoes a slow drift on a large circle around the $z$-axis (magnetron movement).

The angular frequency of the periodic ion movement

$$\omega_\pm = \frac{\omega_c}{2} \pm \sqrt{\left(\frac{\omega_c}{2}\right)^2 - \frac{\omega_{el}^2}{2}} \tag{2.128}$$

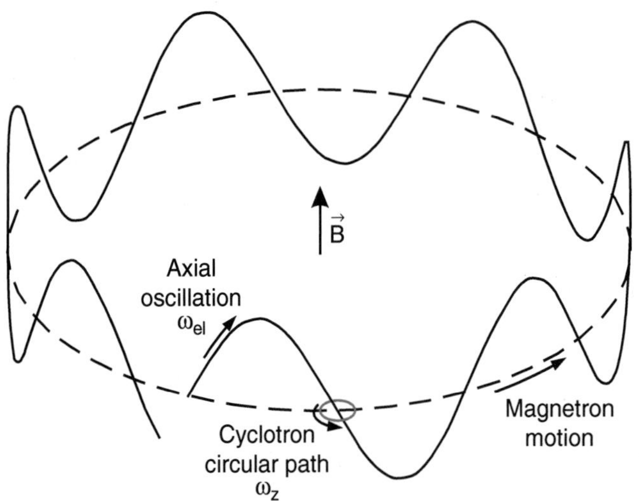


**Fig. 2.88.** Composition of the ion-motion of magnetron motion around the field direction, axial oscillations and circular cyclotron motion

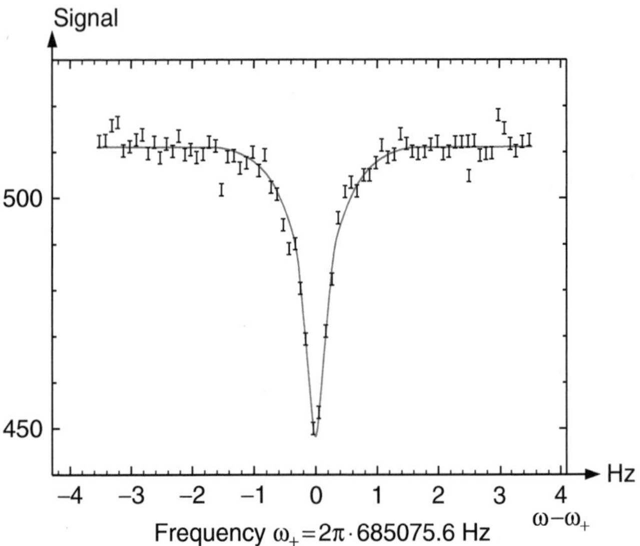


**Fig. 2.89.** Illustration of mass resolution of the cyclotron resonance spectrometer by showing the width of the resonance frequency $\omega_+$ for the $Cs^+$ ion [2.61]

is determined by the cyclotron frequency $\omega_c$ (2.93) and the frequency $\omega_{el}$ of the harmonic oscillation due to the electric field. The periodic ion movement induces an ac voltage $U(t)$ at the caps, which can be used to monitor the frequency of this movement.

The Fourier transform

$$U(\omega - \omega_\pm) = \int U(t) e^{i(\omega-\omega_\pm)t} \, dt \qquad (2.129)$$

of the measured voltage exhibits sharp peaks at $\omega = \omega_+$ and $\omega = \omega_-$, which allows the accurate determination of the cyclotron frequency $\omega_c$ and therefore, using (2.126), the ion mass as well, if the magnetic field $B$ is known. The magnetic field can be calibrated using $^{12}C^+$ ions because their mass represents the unit of the atomic mass scale (see Sect. 2.2.1).

As an illustration of the achievable accuracy, Fig. 2.89 shows the resonance peak around the frequency $\omega_+$ of $^{133}Cs^+$ ions, which has a line width of only 0.3 Hz (!) at a cyclotron frequency $\omega_c = 2\pi \cdot 685{,}075.6$ Hz. The central frequency can be determined within $\pm 0.05$ Hz, which implies, according to (2.93), an accuracy of $m/\Delta m \geq 10^8$ [2.61].

### 2.7.8 Isotopes

Measurements of atomic weights with chemical methods (Sect. 2.1) brought the result that most of the natural chemical elements have atomic mass numbers $x$ AMU, where $x$ generally is close to an integer. For some elements, however, large deviations from an integer were found. The explanation of these findings became possible through accurate measurements of the atomic masses with mass spectrometers. These measurements showed that most chemical elements in nature consist of a mixture of components with slightly different masses, differing from each other by one or a few atomic mass units. These components of a chemical element have exactly the same chemical properties, their only difference lies in their masses. They were called *isotopes*.

**EXAMPLES**

1. The natural isotopic abundance of oxygen is 99.75% $^{16}O$ with 16 AMU and 0.2% of $^{18}O$ with 18 AMU. The average mass number (weighted without isotope separation) is therefore $0.9975 \times 16 + 0.002 \times 18 = 16.005$ AMU.
2. Natural chlorine consists of 77.5% $^{35}Cl$ and 24.5% $^{37}Cl$, which gives an average mass number of $0.755 \times 35 + 0.245 \times 37 = 35.49$ AMU.

The atomic mass number of each isotope is written as an upper left index before the chemical symbol, whereas the number of its electrons, which determines its chemical properties, is written as a lower left index. Then $^{37}_{17}Cl$ is a chlorine isotope with 17 electrons and a mass of 37 AMU.

The real explanation for isotopes was only possible after the discovery of the neutron in 1932. It then became clear, that the atomic nucleus (see Sect. 2.8) consists of positively charged particles, called protons, and neutral particles, called neutrons. The total charge of all protons cancels that of all electrons in the atom. *Isotopes only differ in their number of neutrons.*

Figure 2.90 shows the abundances of the molybdenum isotopes measured with a high-resolution double-focusing mass spectrometer.

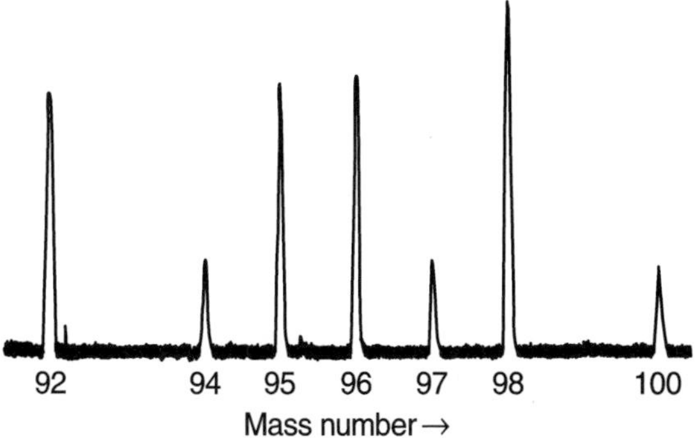


**Fig. 2.90.** Relative abundances of molybdenum isotopes, measured with the double-focusing mass spectrograph of Mattauch [2.53]

## 2.8 The Structure of Atoms

The experiments discussed so far, have given us information on the size and masses of atoms and also on the fact that neutral atoms carry negative and positive charges. How these charges are distributed over the volume of an atom was only discovered in 1911 by the scattering experiments of Rutherford and his group.

Such scattering experiments can also give information on the interaction potential between two colliding atoms and its dependence on the distance between the collision partner. We will therefore discuss in this section the scattering of particles by each other and the atomic models resulting from such experiments.

### 2.8.1 Integral and Differential Cross Sections

When a parallel beam of particles A with a particle flux density $\dot{N} = \mathrm{d}N/\mathrm{d}t$ pass per second and unit area in the $x$-direction through a layer with thickness $d_x$, which contains particles B with a particle density $n_\mathrm{B}$ (Fig. 2.91a) a fraction of the incident particles A will

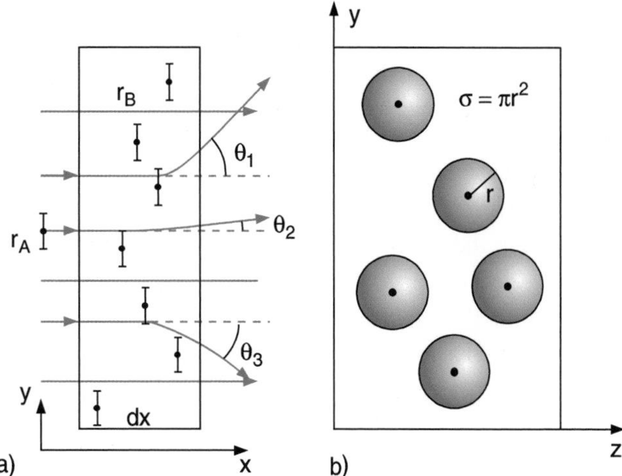


**Fig. 2.91.** (**a**) Scattering of atoms A by atoms B with number density $n_\mathrm{B}$ in a layer with thickness d$x$. (**b**) Illustration of the collision cross section by circles with radius $r = r_\mathrm{A} + r_\mathrm{B}$

be scattered out of the original $x$-direction due to their collisions with particles B. The deflection angle at such a collision depends on many parameters: the distance between A and B, the interaction potential, the masses $m_\mathrm{A}$ and $m_\mathrm{B}$, and the relative velocity $\boldsymbol{v}_\mathrm{A} - \boldsymbol{v}_\mathrm{B}$.

If the number $n_\mathrm{B}\,\mathrm{d}x$ of scattering particles B along the path d$x$ is sufficiently small, each particle A will pass, during its way through d$x$, at most one atom B closely enough to be scattered by a measurable angle. This means multiple scattering can be neglected.

> We define the integral scattering cross section $\sigma_\mathrm{int}$ for scattering of particles A by particles B as that area $\sigma_\mathrm{int} = \pi r^2$ around B, through which A has to pass in order to be deflected by an angle $\Theta$ larger than the minimum detectable deflection angle $\Theta_\mathrm{min}$ (Fig. 2.91b and 2.88a).

Due to these deflections, the particle flux density $\dot{N}$ decreases over the distance d$x$ by

$$\mathrm{d}\dot{N} = -\dot{N}\sigma_\mathrm{int}n_\mathrm{B}\,\mathrm{d}x\,. \tag{2.130}$$

Dividing by $\dot{N}$ and integrating over d$x$ gives the particle flux after having passed a distance $x$ through the scattering region

$$\dot{N}(x) = \dot{N}(0)\,\mathrm{e}^{-n_\mathrm{B}\sigma_\mathrm{int}x}\,. \tag{2.131}$$

The integral cross section is related to the mean free path length $\Lambda$ (for particles B at rest) by (see Problem 2.17)

$$\Lambda = \frac{1}{n\sigma} . \tag{2.132}$$

A possible experimental realization for measuring integral scattering cross sections is shown in Fig. 2.92a. The incident beam of particles A is collimated by two slits $S_1$, $S_2$ and passes either through a thin foil of atoms B (in the case of fast particles A which can penetrate the foil) or through a gaseous sample restricted to a volume $V$ with thickness $\mathrm{d}x$. Such a volume is realized either by a differentially pumped cell with holes for the entrances and exit of the beam A or by a second collimated beam of particle B, which crosses the beam A perpendicularly (Fig. 2.92b). In the case of a differentially pumped cell, the particles B, which effuse through the holes of the cell have to be pumped away in order to maintain sufficiently low pressures outside the cell so that collisions of atoms A only occur inside the defined volume of the cell but not outside.

The detector is located behind a third slit $S_3$, that transmits only those particles A that have not been deflected by collisions.

More information is obtained by measuring that fraction of the incident particles A, that is scattered into a defined solid angle $\mathrm{d}\Omega$, and which is determined by the differential cross section.

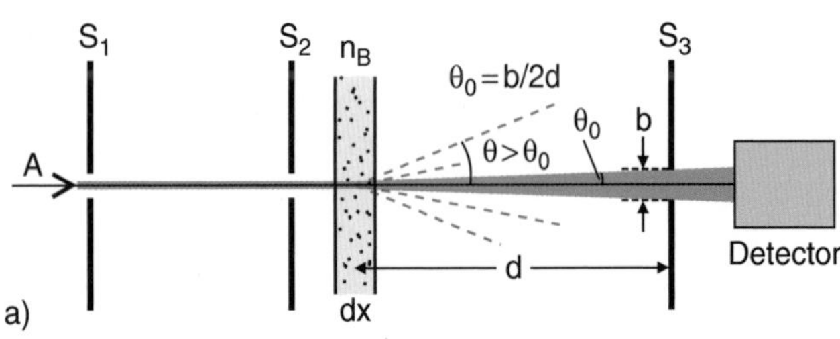


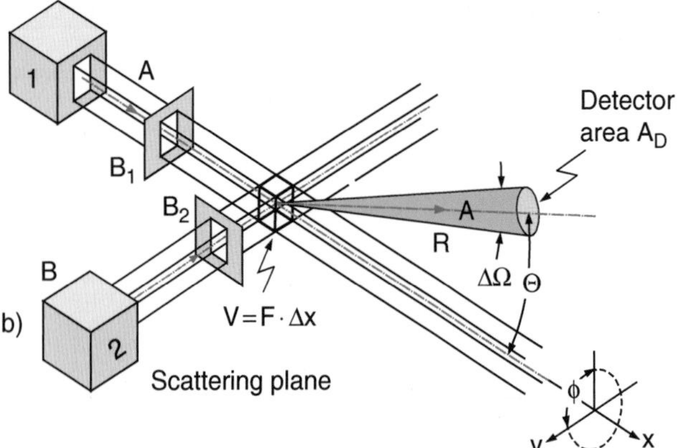


**Fig. 2.92.** (**a**) Measurement of the integral cross section $\sigma$. (**b**) Measurement of the differential cross section $\mathrm{d}\sigma/\mathrm{d}\Omega$

While for the determination of the integral cross section, the decrease of the intensity of the incident particles A (that is, the unscattered particles) is measured, the differential cross section $\mathrm{d}\sigma(\Theta)/\mathrm{d}\Omega$ is a measure for those particles that have been deflected by a certain angle $\Theta$ into the solid angle $\mathrm{d}\Omega$.

We will now derive an expression for the differential cross section.

Assume $\dot{N}A$ incident particles pass per second through the area $A$ in the scattering volume $V = A \cdot \Delta x$, and $\Delta\dot{N}(\Theta, \Delta\Omega)$ is the rate of particles scattered by a deflection angle $\Theta$ and detected by the detector with an acceptance solid angle $\Delta\Omega$. Then

$$\frac{\Delta\dot{N}}{\dot{N}A} = \frac{n_\mathrm{B}}{A} V \frac{\mathrm{d}\sigma}{\mathrm{d}\Omega}\Delta\Omega = n_\mathrm{B}\Delta x \frac{\mathrm{d}\sigma}{\mathrm{d}\Omega}\Delta\Omega \tag{2.133}$$

is the fraction of incident particles that is scattered into the solid angle $\Delta\Omega$ accepted by the detector. It is determined by the particle density $n_\mathrm{B}$ of scatterers B, the length $\Delta x$ which the incident beam of particles A traverse through the scattering volume $V$ and the differential scattering cross section $\mathrm{d}\sigma/\mathrm{d}\Omega$ which depends on the interaction potential between particles A and B.

For measuring $\mathrm{d}\sigma/\mathrm{d}\Omega$ the setup of Fig. 2.92b can be used. Two beams, collimated by the apertures $S_1$ and $S_2$ cross each other in the scattering volume $V = A\Delta x$. The particles A scattered by an angle $\Theta$ into the solid angle of $\Delta\Omega$ are monitored by the detector with sensitive area $A_\mathrm{D} = R^2\Delta\Omega$ in a distance $R$ from the scattering volume $V$ where $R \gg \Delta x$.

We will now look into the relation between $E_\mathrm{pot}(r)$ and $\mathrm{d}\sigma/\mathrm{d}\Omega$ in more detail.

## 2.8.2 Basic Concepts of Classical Scattering

As is generally shown in classical mechanics, the movements of two particles with masses $m_1$, $m_2$, velocities $\boldsymbol{v}_1, \boldsymbol{v}_2$ and a mutual interaction potential $E_\mathrm{pot}(|\boldsymbol{r}_1 - \boldsymbol{r}_2|)$ can be represented in the center of mass coordinate frame as the movement of a single particle with reduced mass

$$\mu = \frac{m_1 m_2}{m_1 + m_2}$$

and relative velocity $\boldsymbol{v} = \boldsymbol{v}_1 - \boldsymbol{v}_2$ in a potential $E_{\text{pot}}(r)$, where $r = |\boldsymbol{r}_1 - \boldsymbol{r}_2|$ is the distance between the two particles. The description of the scattering of the two particles by each other in this center of mass frame is named "potential scattering," because it demands, besides the reduced mass $\mu$ and the initial conditions $(\boldsymbol{r}_0, \boldsymbol{v}_0)$ solely the knowledge of the interaction potential $E_{\text{pot}}(r)$.

We will here restrict the discussion to the most simple case of spherically symmetric potentials $E_{\text{pot}}(r)$, which is adequate for many real collision events. In such a potential, the angular momentum $\boldsymbol{L}$ of the particle remains constant (see Problem 2.20). This implies that the path of the particle is planar. It always stays within the so-called "scattering plane." Therefore, polar coordinates $(r, \varphi)$ are best suited for the description of the particles time-dependent position. The deflection angle of our particles, measured in the center of mass frame is named $\vartheta$, while it is described by $\Theta$ in the laboratory frame (Fig. 2.93).

The deflection of the incident particle A depends on its impact parameter $b$. This is the smallest distance of A to the target particle B, if there is no deflection, i.e., if A passes along a straight line (Fig. 2.94a). For the potential scattering (i.e., the description of the scattering process in the center of mass frame) the particle B is fixed at the origin of our coordinate frame and it can be regarded as a point-like particle with infinite mass.

When the initial velocity of particle A is $|\boldsymbol{v}(-\infty)| = v_0$, energy conservation demands

$$\frac{1}{2}\mu v^2 + E_{\text{pot}}(r) = \frac{1}{2}\mu v_0^2 = \text{const}\,, \qquad (2.134)$$

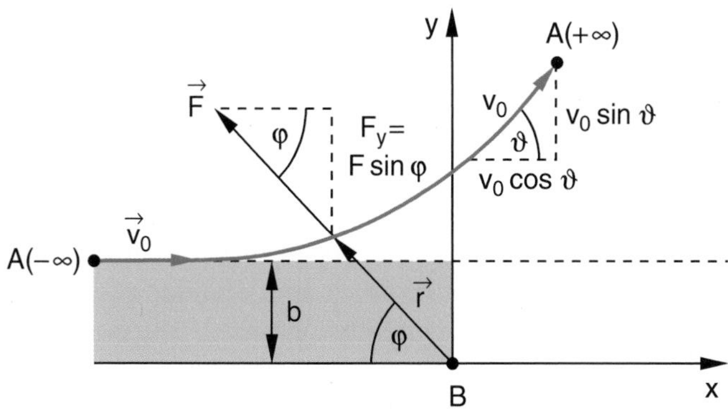


**Fig. 2.93.** Scattering of a charged particle A in a potential $V(r) \propto 1/r$, where $r$ is the distance between A and B

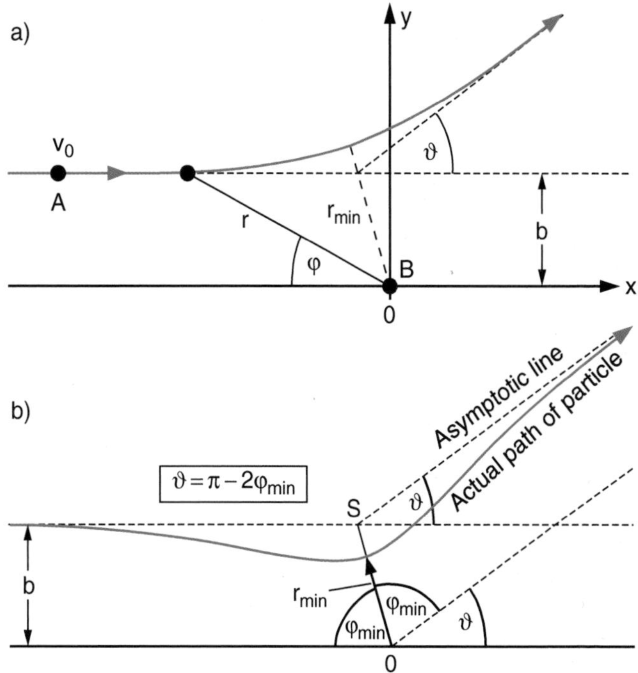


**Fig. 2.94.** (**a**) Scattering of a particle A with reduced mass $\mu = m_A m_B/(m_A + m_B)$ in a potential $V(r)$ with the origin in B. (**b**) Relation between scattering angle $\vartheta$ in the center of mass system and the polar angle $\varphi_{\min}$ at the closest approach between A and B (point S)

because $E_{\text{pot}}(r = \pm\infty) = 0$. The angular momentum $\boldsymbol{L}$, with respect to the scattering center at $\boldsymbol{r} = 0$ is

$$\begin{aligned} \boldsymbol{L} &= \mu(\boldsymbol{r} \times \boldsymbol{v}) = \mu\left(\boldsymbol{r} \times \left[\frac{\mathrm{d}r}{\mathrm{d}t}\hat{e}_r + r\frac{\mathrm{d}\varphi}{\mathrm{d}t}\hat{e}_t\right]\right) \\ &= \mu r\dot{\varphi}\left(\boldsymbol{r} \times \hat{e}_t\right)\,, \end{aligned} \qquad (2.135)$$

because $\hat{e}_r$ is parallel to $\boldsymbol{r}$. The unit vector $\hat{e}_t$ points along the tangent to the path of A. For $\boldsymbol{L}$ we obtain:

$$|\boldsymbol{L}| = L = \mu r^2\dot{\varphi} = \mu v_0 b\,, \qquad (2.136)$$

because $L(x =! -\infty) = \mu \cdot v_0 \cdot r \cdot \sin\varphi = \mu \cdot v_0 \cdot b$. The kinetic energy in the center of mass frame is

$$\begin{aligned} E_{\text{kin}} &= \frac{1}{2}\mu v^2 = \frac{1}{2}\mu\left(\dot{r}^2 + r^2\dot{\varphi}^2\right) \\ &= \frac{1}{2}\mu\dot{r}^2 + \frac{L^2}{2\mu r^2}\,. \end{aligned} \qquad (2.137)$$

The total energy $E = T + E_{\text{pot}}$ in the center of mass frame can then be written as

$$E_{\text{total}} = E_0 = \frac{1}{2}\mu\dot{r}^2 + \frac{L^2}{2\mu r^2} + E_{\text{pot}}(r) = \text{const}\,. \qquad (2.138)$$

Solving (2.138) and (2.136) for $\dot{r}$ and $\dot{\varphi}$ yields

$$\dot{r} = \left[\frac{2}{\mu}\left(E_0 - E_{\text{pot}}(r) - \frac{L^2}{2\mu r^2}\right)\right]^{1/2} \tag{2.139a}$$

$$\dot{\varphi} = \frac{L}{\mu r^2} \,. \tag{2.139b}$$

In a real experiment the path $(r(t), \varphi(t))$ of a single particle cannot be followed. However, the measured deflection angle $\vartheta$ allows to determine the asymptotic values of the path for $r \to \infty$. Since for a spherically symmetric potential this path must be mirror-symmetric to the line OS through the point S of closest approach in Fig. 2.94b. (This means that the scattering process is invariant against time-reversal.) We can relate the asymptotic scattering angle $\vartheta$ to the polar angle $\varphi_{\min} = \varphi(r_{\min})$ by

$$\vartheta = \pi - 2\varphi_{\min} \,.$$

This yields the relation

$$\varphi_{\min} = \int\limits_{\varphi=0}^{\varphi_{\min}} \mathrm{d}\varphi = \int\limits_{r=-\infty}^{r_{\min}} \frac{\mathrm{d}\varphi}{\mathrm{d}t}\frac{\mathrm{d}t}{\mathrm{d}r}\,\mathrm{d}r = \int\limits_{r=-\infty}^{r_{\min}} (\dot{\varphi}/\dot{r})\,\mathrm{d}r = \int\limits_{r_{\min}}^{+\infty} \frac{\dot{\varphi}}{\dot{r}}\,\mathrm{d}r \,.$$

With (2.139a) and (2.139b) the scattering angle in the CM-frame becomes:

$$\vartheta(E_0, L) = \pi - 2\int\limits_{r_{\min}}^{+\infty} \frac{(L/(\mu r^2))\,\mathrm{d}r}{\left[\frac{2}{\mu}\left(E_0 - E_{\text{pot}}(r) - \frac{L^2}{2\mu r^2}\right)\right]^{1/2}} \,. \tag{2.140}$$

With the total energy $E_0 = \frac{1}{2}\mu v_0^2$ the amount of the angular momentum

$$L = \mu r v \sin\varphi = \mu b v_0 \Rightarrow L^2 = \mu^2 b^2 v_0^2 = 2\mu b^2 E_0 \tag{2.141}$$

is uniquely defined by the initial energy $E_0$ and the impact parameter $b$ of the incident particle B. Inserting these relations into (2.140) we obtain

$$\vartheta(E_0, b) = \pi - 2b \int\limits_{r_{\min}}^{+\infty} \frac{\mathrm{d}r}{r^2\left[1 - \frac{b^2}{r^2} - \frac{E_{\text{pot}}(r)}{E_0}\right]^{1/2}} \,. \tag{2.142}$$

This shows that the deflection angle $\vartheta$ is determined by the interaction potential $E_{\text{pot}}(r)$, the impact parameter $b$ and the initial energy $E_0$.

The lower integration limit $r_{\min}$ is fixed by the condition $\dot{r}(r_{\min}) = 0$. This gives with (2.139) and (2.141)

$$r_{\min} = \frac{b}{\left[1 - \frac{E_{\text{pot}}(r_{\min})}{E_0}\right]^{1/2}} \,. \tag{2.143}$$

**Note:**

- For $r = r_{\min}$ the integrand in (2.140) becomes infinite. Whether the integral remains finite depends on the exponent $n$ in the power dependence of the interaction potential $(E_{\text{pot}}(r^n))$.
- For $b = 0$ is $L = 0 \Rightarrow \vartheta = \pi$. Particles with $b = 0$ suffer central collisions with B. They are reflected back into the incident direction.
- If $\vartheta_{\min}$ is the smallest still detectable deflection angle then all particles with $\vartheta < \vartheta_{\min}$ are regarded as not scattered. These are all particles with $b > b_{max}(\vartheta_{\min})$. The integral scattering cross section is then $\sigma_{\text{int}} = \pi b_{\max}^2$. This shows that with such a definition the cross section, which should be solely dependent on the particle characteristics, becomes dependent on the design of the apparatus. This contradiction is removed by the quantum-mechanical treatment of collisions.
- For monotonic potentials $E_{\text{pot}}(r)$ (for example pure repulsive potentials $E_{\text{pot}} \propto r^{-1}$) there is, for a given energy $E_0$, a well-defined unique deflection angle $\vartheta$ for each value $b$ of the impact parameter (Fig. 2.95a). This is no longer true for non-monotonic potentials (Fig. 2.95b), where, for example, two different impact parameters $b_1$ and $b_2$ may lead to the same deflection angle $\vartheta$. Plotting the curves $\vartheta(b)$ at a given initial energy $E_0$ yields deflection curves such as those shown in Fig. 2.95. Their form depends on $E_0$ and $E_{\text{pot}}(r)$.

We should keep in mind that the only quantity obtained from a scattering experiment is the differential or integral scattering cross section. The impact parameter $b$ itself cannot be directly measured! The measured scattering cross section yields, however, the wanted information on the deflection curve $\vartheta(b)$ from which the interaction potential can be derived. This can be seen as follows.

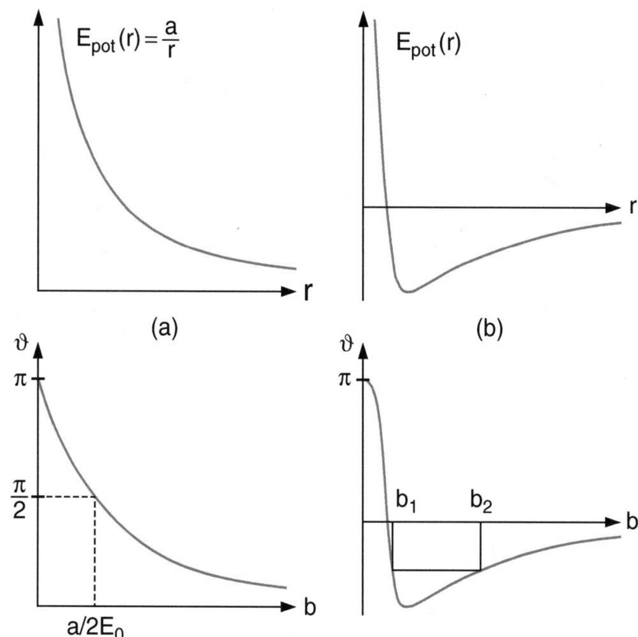


**Fig. 2.95a,b.** Qualitative relation interaction potential and deflection function $\vartheta(b)$. (**a**) Monotonic potential. (**b**) Potential with a minimum

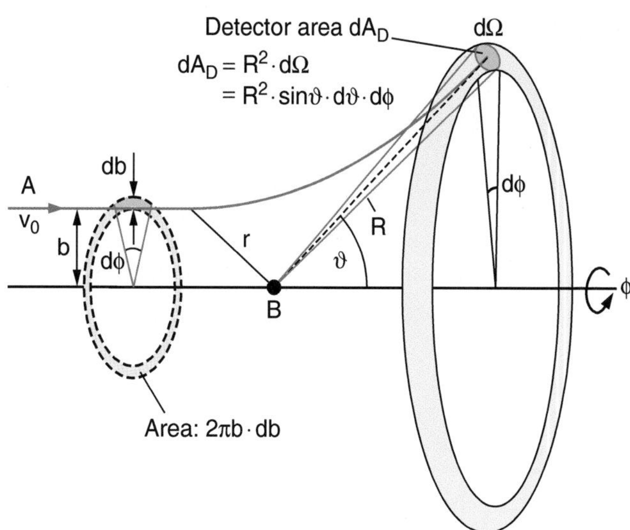


**Fig. 2.96.** Relation between impact parameter $b$, scattering angle $\vartheta$ and differential cross section $\mathrm{d}\sigma/\mathrm{d}\Omega$

Let us assume a parallel beam of incident particles A with particle flux density $\dot{N}_A = n_A v_A$ that passes through a layer of particles B in rest with density $n_B$. All particles A passing through an annular ring with radius $b$ and width d$b$ around an atom B are deflected by the angle $\vartheta \pm \mathrm{d}\vartheta/2$, assuming a spherically symmetric interaction potential (Fig. 2.96). Through this annular ring $\mathrm{d}\dot{N}_A = \dot{N}_A\,\mathrm{d}A = n_A v_A 2\pi b\,\mathrm{d}b$ particles A pass per second. One particle B therefore scatters the fraction

$$\frac{\mathrm{d}\dot{N}_A\left(\vartheta \pm \frac{1}{2}\mathrm{d}\vartheta\right)}{\dot{N}_A} = 2\pi b\,\mathrm{d}b = 2\pi b\frac{\mathrm{d}b}{\mathrm{d}\vartheta}\,\mathrm{d}\vartheta \quad (2.144)$$

of all particles A, incident per second and unit area onto the target, into the range of deflection angles $\vartheta \pm \mathrm{d}\vartheta/2$. The detector with area $A_D = R^2\,\mathrm{d}\Omega = R^2 \sin\vartheta\,\mathrm{d}\vartheta\,\mathrm{d}\phi$ in a distance $R$ from the scattering center B, receives the fraction

$$\frac{\mathrm{d}\dot{N}_A(\vartheta,\phi)}{\dot{N}_A}\frac{\mathrm{d}\phi}{2\pi} = b\frac{\mathrm{d}b}{\mathrm{d}\vartheta}\,\mathrm{d}\vartheta\,\mathrm{d}\phi\,, \quad (2.145)$$

which passes through the segment $b\,\mathrm{d}b\,\mathrm{d}\phi$ of the annular ring in Fig. 2.96.

The fraction of all incident particles A, scattered by *all* atoms B with density $n_B$ in the volume $V = A\Delta x$ is then:

$$\frac{\mathrm{d}\dot{N}_A(\vartheta,\mathrm{d}\Omega)}{\dot{N}_A} = n_B A\Delta x b\frac{\mathrm{d}b}{\mathrm{d}\vartheta}\,\mathrm{d}\vartheta\,\mathrm{d}\phi\,. \quad (2.146)$$

The comparison with (2.133) gives, with $\mathrm{d}\Omega = \sin\vartheta$ $\mathrm{d}\vartheta\,\mathrm{d}\phi$, the differential scattering cross section

$$\boxed{\frac{\mathrm{d}\sigma}{\mathrm{d}\Omega} = b\frac{\mathrm{d}b}{\mathrm{d}\vartheta}\frac{1}{\sin\vartheta}\,.} \quad (2.147)$$

From (2.142) we obtain $\mathrm{d}b/\mathrm{d}\vartheta = (\mathrm{d}\vartheta/\mathrm{d}b)^{-1}$ as a function of the interaction potential $E_{pot}(r)$. This shows:

> The differential cross section gives information on the interaction potential $E_{pot}(r)$ between the colliding particles A and B at a distance $r$.

We can therefore also write (2.146) as

$$\frac{\mathrm{d}\dot{N}_A(\mathrm{d}\Omega)}{\dot{N}_A} = n_B A\Delta x\frac{\mathrm{d}\sigma}{\mathrm{d}\Omega}\,\mathrm{d}\Omega\,. \quad (2.148)$$

Measuring the relative fraction $\mathrm{d}\dot{N}_A/\dot{N}_A$ yields the differential cross section and with it the interaction potential.

The integral scattering cross section is obtained by integration over d$\Omega$, where the integration limits are

$\vartheta(b=0)=\pi$ and $\vartheta(b_{max})=\vartheta_{min}$:

$$\sigma_{int}=\int_{\Omega}\frac{d\sigma}{d\Omega}\,d\Omega=\int_{\vartheta=\pi}^{\vartheta_{min}}\int_{\phi=0}^{2\pi}\frac{d\sigma}{d\Omega}\sin\vartheta\,d\vartheta\,d\phi\,, \tag{2.149a}$$

where $\vartheta_{min}$ is the smallest detectable deflection angle. The integration over $\phi$ gives $2\pi$. With (2.147) we get:

$$\sigma_{int}=2\pi\int_{\vartheta=\pi}^{\vartheta_{min}}\frac{b}{\sin\vartheta}\left|\frac{db}{d\vartheta}\right|\sin\vartheta\,d\vartheta$$

$$=2\pi\int_{b=0}^{b_{max}}b\,db=\pi b_{max}^2\,. \tag{2.149b}$$

## EXAMPLE

Collisions of hard spheres A and B with equal diameters $D$. The potential energy in this case is:

$$E_{pot}(r)=\begin{Bmatrix}\infty & \text{for} & r\le D\\ 0 & \text{for} & r>D\end{Bmatrix}\,.$$

From Fig. 2.97a it is seen that at the closest approach $\sin\varphi_m=b/D$, which implies that a collision can only take place for $b\le D$. For the scattering angle $\vartheta$ we find $\vartheta/2=\pi/2-\varphi_m$.

The impact parameters for $b\le D$ are therefore

$$b=D\sin\varphi_m=D\cos(\vartheta/2)\,.$$

Then the derivative $db/d\vartheta$ becomes

$$\left|\frac{db}{d\vartheta}\right|=\frac{D}{2}\sin\vartheta/2$$

and the differential scattering cross section is:

$$\frac{d\sigma}{d\Omega}=\frac{b}{\sin\vartheta}\frac{db}{d\vartheta}=\frac{D\cos(\vartheta/2)D\sin(\vartheta/2)}{2\sin\vartheta}=\frac{D^2}{4}$$

$$\Rightarrow\sigma_{int}=\int\frac{d\sigma}{d\Omega}\,d\Omega=4\pi\frac{D^2}{4}=\pi D^2\,.$$

The deflection function $\vartheta(b)$ for hard spheres (Fig. 2.97c) is

$$\vartheta(b)=\pi-2\varphi_m=\pi-2\arcsin(b/2D)\,.$$

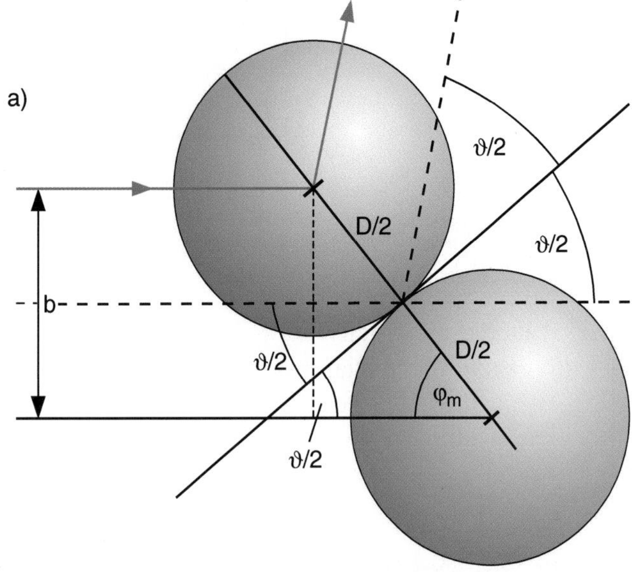


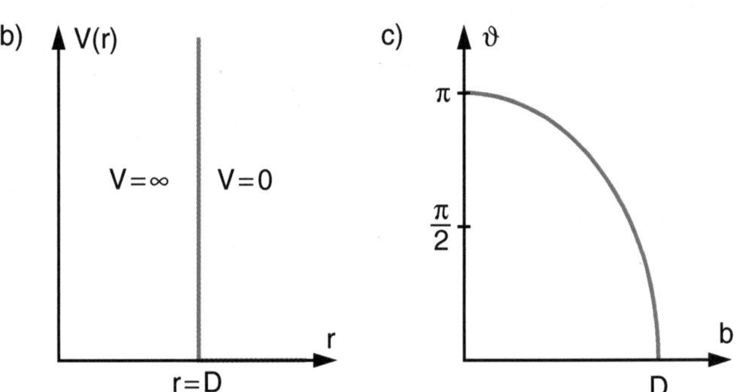


**Fig. 2.97a–c.** Collision of hard spheres with diameter $D$. (**a**) Scattering angle for impact parameters $b<D$. (**b**) Potential $V(r)$. (**c**) Deflection function $\vartheta(b)$

### 2.8.3 Determination of the Charge Distribution within the Atom from Scattering Experiments

In order to find the charge distribution in atoms the best choice is to use charged incident particles A with charge $q_1$ as probes. The charge $q_2=\rho_{el}\,dV$ of atoms B inside the volume element $dV$ contributes a force

$$dF_c(r)=\frac{1}{4\pi\varepsilon_0}\frac{q_1q_2}{r^2}\hat{r}\,. \tag{2.150}$$

The total force that is responsible for the deflection of the charged particles A is obtained by integration of all volume elements of atom B.

At the beginning of the 20th century charged projectiles were available in the form of $\alpha$ particles from the radioactive decay of some radioactive substances with charge $q_1 = +2e$, mass $m = m_{He} = 4\,\text{AMU}$ and kinetic energies $E_{kin} = 1$–$9\,\text{MeV}$. Also, electrons with lower energies $E_{kin} < 10\,\text{keV}$ could be produced from cathode rays in gas discharges.

When the heavy $\alpha$ particles pass through an atom, the light electrons of this atom, because of their small mass, contribute little to the deflection of the $\alpha$ particles which is mainly caused by the positive charges with larger masses. The measured angular distribution $N(\vartheta)$ of the scattered $\alpha$ particles therefore mainly gives information about the spatial distribution of the positive charges while the presence of the atomic electrons only gives a small correction.

### 2.8.4 Thomson's Atomic Model

The results of his experiments and those of others brought *Thomson* to the conclusion that each neutral atom consists of $Z$ electrons with the total charge $q^- = -Ze$ and constituents with a total positive charge $q^+ = +Ze$. Since the atom is neutral, Thomson proposed for the spatial distribution of the charges his "raisin cake model," where all charges were equally distributed over the volume of the atom with radius $R$ because this would result in a distribution of minimum energy, if only electric forces were present (Fig. 2.98).

How can this model be tested experimentally?

The electric field of a homogeneously charged sphere with radius $R$ and charge $Ze$ at a distance $r \leq R$ from the center is given as

$$\boldsymbol{E} = \frac{Q}{4\pi\varepsilon_0 r^2} = \frac{Zer}{4\pi\varepsilon_0 R^3}\hat{r}\,, \tag{2.151}$$

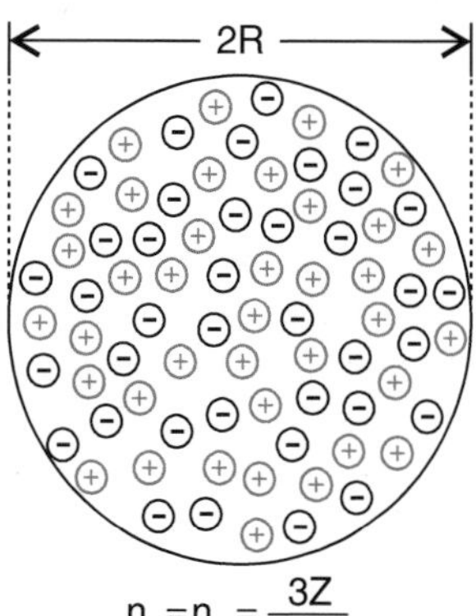


**Fig. 2.98.** Thomson's "raisin cake" model for the distribution of positive and negative charges in the atom

with $\hat{r} = \vec{r}/|\vec{r}|$, because the charge inside the radius $r$ is $Q = Z \cdot e \cdot r^3/R^3$. If we first neglect the negative charges, an electron would experience the force

$$\boldsymbol{F} = -e\boldsymbol{E} = -k\boldsymbol{r} \quad \text{with} \quad k = \frac{Ze^2}{4\pi\varepsilon_0 R^3}\,. \tag{2.152a}$$

Any radial displacement of an electron from its equilibrium position would then lead to a harmonic oscillation (since the restoring force is linearly dependent on the displacement) with a frequency

$$\omega = \sqrt{k/m}\,. \tag{2.152b}$$

If we now consider the other $Z - 1$ electrons, we assume, according to the raisin cake model, a uniform density

$$n_e = \frac{Z}{\frac{4}{3}\pi R^3} \tag{2.152c}$$

of the electrons that equals the density of the positive charges. This cloud of electrons can oscillate against the cloud of positive charges with the so-called "plasma-frequency," which can be derived as

$$\omega_p = \sqrt{\frac{n_e e^2}{\varepsilon_0 m_e}} = \sqrt{\frac{3Ze^2}{4\pi\varepsilon_0 m_e R^3}}\,, \tag{2.152d}$$

which differs from the simple model of a single electron (2.152b) only by a factor of $\sqrt{3}$. When illuminating these atoms with light the atoms would preferentially absorb at their resonance frequency $\omega_p$ and its higher harmonics $\omega_n = n\omega_p$.

Atoms excited by light or electron impact should emit light preferentially at these frequencies.

However, the observed frequencies of light absorbed or emitted by atoms do not at all agree with those estimated by Thomson's model.

The strongest argument against the raisin cake model is supported by scattering experiments, first performed by Sir *E. Rutherford* and his coworkers, using $\alpha$ particles emitted by radioactive atoms. These experiments give a different angular distribution of scattered charged particles than expected from Thomson's model. This will now be outlined in more detail.

In Fig. 2.99 we consider the deflection of an $\alpha$ particle with charge $q = +2e$ and mass $m_\alpha \approx 7350\,m_e$ by a spherical homogeneous distribution of the positive charge $Q = +Ze$. Because of their small mass $m_e$

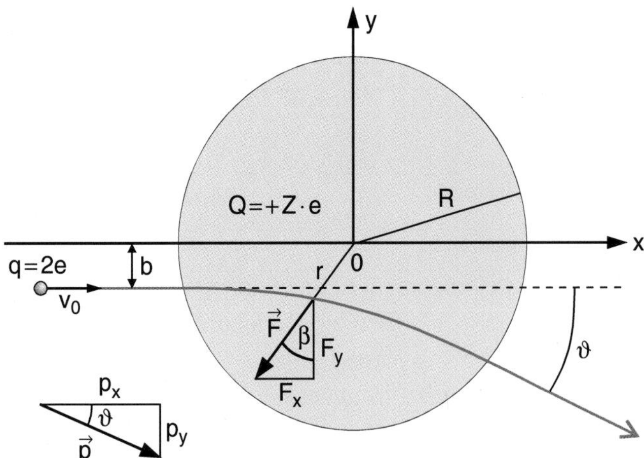


**Fig. 2.99.** Scattering of a particle with charge $q$ by a homogeneous spherical charge distribution with total charge $Q$

the electrons of the atom do not significantly contribute to the deflection of heavy $\alpha$ particles. They are nevertheless important because they bring about that the total atom is neutral for distances $b > R$. A charged particle passing the atom with radius $R$ at an impact parameter $b > R$ is therefore not deflected by much. In order to measure the charge distribution inside the atom we therefore need to include only impact parameters $b \le R$. The following estimation gives an upper limit for the maximum possible deflection angle $\vartheta_{\max}$ for a homogeneous distribution of the positive charges. The existence of the negatively charged light electrons will decrease this angle only slightly.

A projectile particle with momentum $mv_0$ in the $x$-direction is deflected by the angle $\vartheta$ while passing through the atom. The deviation is due to the repulsive fore component

$$F_y = F(r)\cos\beta\,, \tag{2.153a}$$

which acts at any point of the path within the atom and causes a change

$$\Delta p_y = \int F_y\,\mathrm{d}t \tag{2.153b}$$

of the momentum (Fig. 2.99). The force $\boldsymbol{F} = q\boldsymbol{E}$ at a distance $r$ from the center is determined by the electric field $\boldsymbol{E}$ (2.151). We will later see that the deflection is very small. We can therefore neglect the curvature of the path and approximate the path by a slightly inclined straight line with a length $d = 2\sqrt{R^2 - b^2}$ and $\cos\beta \approx b/r$. With this approximation, we obtain during the time-of-flight

$$T = \frac{2}{v_0}\sqrt{R^2 - b^2}$$

the momentum change

$$\Delta p_y = \frac{2Ze^2 b}{4\pi\varepsilon_0 R^3} \approx \frac{4Zkb}{v_0}\sqrt{R^2 - b^2}$$

$$\text{with}\quad k = \frac{e^2}{4\pi\varepsilon_0 R^3}\,. \tag{2.154}$$

Since $\Delta p_y \ll p_x$ we may regard $p_x \approx p = \text{const}$. This yields

$$\frac{\Delta p_y}{p_x} \approx \frac{\Delta p_y}{p} = \tan\vartheta = \frac{4Zkb}{mv_0^2}\sqrt{R^2 - b^2}\,. \tag{2.155}$$

The deflection angle $\vartheta$ depends on the impact parameter $b$. Its maximum value is obtained when the derivative $\mathrm{d}\vartheta/\mathrm{d}b$ is zero. With $\tan\vartheta \approx \vartheta$ we obtain

$$\frac{\mathrm{d}\vartheta}{\mathrm{d}b} = \frac{4Zk}{mv_0^2}\left[\sqrt{R^2 - b^2} - \frac{b^2}{\sqrt{R^2 - b^2}}\right] = 0\,,$$

which yields

$$b(\vartheta_{\max}) = R/\sqrt{2} \quad\text{and}\quad \vartheta_{\max} = \frac{2ZkR^2}{mv_0^2}\,. \tag{2.156}$$

We can define with (2.155) an average deflection angle averaged over all impact parameters $b \le R$. This gives

$$\begin{aligned}\overline{\vartheta} &= \int_{b=0}^{R} \vartheta\,\frac{2\pi b}{\pi R^2}\,\mathrm{d}b = \frac{8Zk}{mv_0^2 R^2}\int_{b=0}^{R}\sqrt{R^2 - b^2}\,b^2\,\mathrm{d}b \\ &= \frac{\pi}{2}\frac{ZkR^2}{mv_0^2} = \frac{\pi}{4}\vartheta_{\max} = \frac{Ze^2}{8\varepsilon_0 R m v_0^2}\,.\end{aligned} \tag{2.157}$$

The average deflection angle $\overline{\vartheta}$ equals approximately the ratio of potential energy $E_{\text{pot}} = 2Ze^2/(4\pi\varepsilon_0 R)$ at the distance $R$ from the center and kinetic energy $(m/2)v_0^2$.

For typical radii $R \approx 0.2\,\text{nm}$ of gold atoms the mean deflection angle $\overline{\vartheta}$ for $\alpha$ particles with $E_{\text{kin}} \approx 5\,\text{MeV}$, scattered by gold atoms ($Z = 79$) should be according to Thomson's model (2.157) with $mv_0^2 = 10\,\text{MeV} = 1.9\times 10^{-12}\,\text{Nm}$

$$\Rightarrow \overline{\vartheta} = 7.6\times 10^{-5}\,\text{rad} \mathrel{\hat{=}} 4.6\times 10^{-3\circ} = 0.27'\,. \tag{2.158}$$

This is an extremely small deflection angle, which is not easy to measure. However, until now we have only considered the deflection of $\alpha$ particles by a single atom.

In the experiment performed by *Rutherford* and his coworkers *Geiger* and *Marsden*, the $\alpha$ particles pass through a thin gold foil and are therefore scattered by many gold atoms. For an atomic diameter of 0.4 nm and a 20-$\mu$m thick foil, the $\alpha$ particles have to pass through $5 \times 10^4$ atomic layers. The impact parameters $b_i$, related to the centers of the different atoms are more or less statistically distributed (see Fig. 2.100a). Therefore, the average deflection angles $\overline{\vartheta}$ caused by the different atoms, will also be statistically distributed. The statistical average $\langle\overline{\vartheta}\rangle$ of the total deflection angle after $n$ scattering events is (see books on probability theory)

$$\langle\overline{\vartheta}\rangle = \sqrt{n} \cdot \overline{\vartheta} \, .$$

The situation is completely analogous to the random walk problem (see Feynman, vol. I) where a drunken sailor throws a coin and goes for every step forward one step to the left or to the right according to whether the coin shows its foreside or its backside. The probability $P$ that the sailor deviates after $n$ steps by a distance $\Delta y$ from the straight line $y = 0$ is given by the Gaussian distribution

$$P(y) = C\mathrm{e}^{-(y^2/n\Delta y^2)} \, .$$

By a similar consideration, one obtains

$$N(\vartheta) = N_0 \mathrm{e}^{-\vartheta^2/(n\overline{\vartheta^2})} \, . \tag{2.159}$$

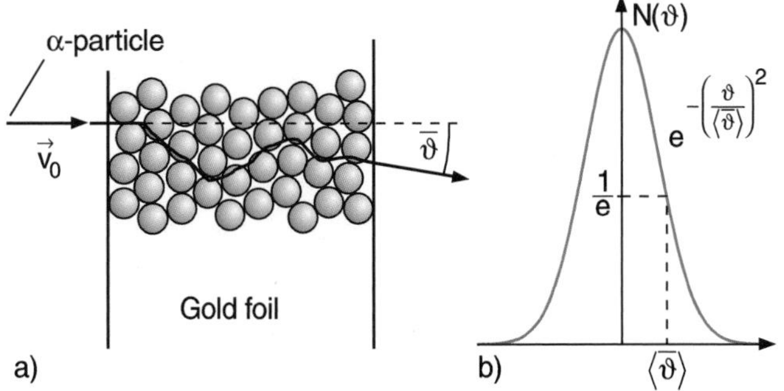


**Fig. 2.100.** (**a**) Multiple scattering of an $\alpha$ particle by gold atoms in a foil according to Thomson's model. (**b**) Expected angular distribution of the scattered particles

for the distribution $N(\vartheta)$ of particles deflected by an angle $\vartheta$ after the foil.

### EXAMPLE

For $n = 5 \times 10^4$, $\overline{\vartheta} = 7.6 \times 10^{-5}$ rad $\Rightarrow \langle\overline{\vartheta}\rangle = 1.7 \times 10^{-2}$ rad $\approx 1°$. The Gaussian distribution with a maximum at $\vartheta = 0$ has a full halfwidth of $(\Delta\vartheta)_{1/2} = 3.4 \times 10^{-2}$ rad $= 2°$.

This is in sharp contrast to the experimental results discussed in the next section.

## 2.8.5 The Rutherford Atomic Model

In order to test Thomson's model, Geiger and Marsden performed extensive scattering measurements [2.62] with the experimental setup illustrated in Fig. 2.101.

The $\alpha$ particles were emitted by radon gas in the tube T, and were collimated by the narrow channel D. The nearly parallel beam of $\alpha$ particles then passes through a thin gold foil F and the scattered $\alpha$ particles

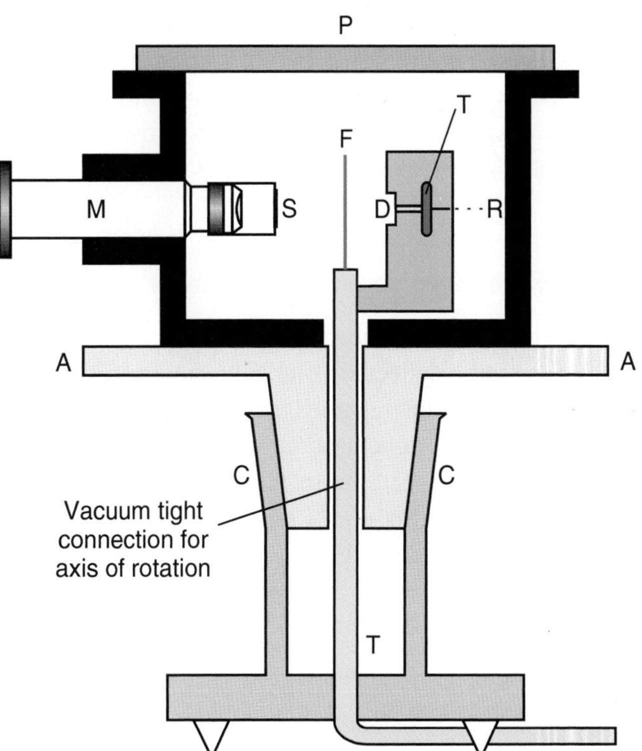


**Fig. 2.101.** Experimental setup for Rutherford's scattering experiment

produced faint light flashes on a phosphorous screen S, which were observed through a microscope. The detector (screen and microscope) could be turned against the direction of the incident beam. This allowed the detection of $\alpha$ particles scattered by an arbitrary angle $\vartheta$ into a deflection range $\vartheta \pm 1/2\Delta\vartheta$, where $\Delta\vartheta$ is the angular resolution of the detector.

The experiments clearly showed that even particles with large deflection angles up to $\vartheta = 150°$ (limited by the experimental set-up) could also be observed, in sharp contrast to Thomson's atomic model. Rutherford, who was very much surprised by these unexpected results said, "*This is as improbable as a bullet being reflected back when shooting it into a cotton-wool ball.*" [2.62].

After many discussions, long thoughts and the examination of several models proposed in the literature, Rutherford recognized that the positive charge must be contained in a very small volume around the center of the atom. This volume, which should carry nearly the total mass of the atom, although it covers only a tiny fraction of the atomic value, was given the name "atomic nucleus" by Rutherford.

According to this model the $\alpha$ particles are deflected only by the nucleus because the masses of the electrons are small compared with that of the $\alpha$ particles ($m_e/m_\alpha \approx 1.4 \times 10^{-4}$).

Based on these considerations Rutherford derived his famous scattering formula, which shows excellent agreement with experimental findings.

### 2.8.6 Rutherford's Scattering Formula

When the $\alpha$ particles, according to Rutherford's model, are essentially deflected solely by the atomic nucleus, which may be regarded as a point-like particle, the theoretical treatment of the scattering is reduced in the center-of-mass frame to the scattering of a particle with reduced mass $\mu = m_\alpha m_N/(m_\alpha + m_N) \approx m_\alpha$ in a Coulomb potential (see Sect. 2.8.1). Following the argument in Sect. 2.8.1, we obtain for the angular momentum $L = |\boldsymbol{L}|$ from (2.136)

$$L = \mu r^2 \dot{\varphi} = \mu v_0 b \tag{2.160}$$

and for the component $F_y$ of the Coulomb-force, responsible for the deflection of the $\alpha$ particles,

$$F_y = \mu \frac{\mathrm{d}v_y}{\mathrm{d}t} = \frac{a \sin\varphi}{r^2}$$

$$\text{with} \quad a = \frac{qQ}{4\pi\varepsilon_0}\,; \quad q = 2e\,, \quad Q = Ze\,. \tag{2.161}$$

This gives

$$\frac{\mathrm{d}v_y}{\mathrm{d}t} = \frac{a \sin\varphi}{\mu r^2} = \frac{a \sin\varphi}{L}\frac{\mathrm{d}\varphi}{\mathrm{d}t}\,. \tag{2.162a}$$

The $\alpha$ particle comes from the point A ($r = -\infty$) in Fig. 2.102 and finally reaches, after the scattering event, the point B ($r = +\infty$). The angle $\varphi$ changes during this scattering for a particle with scattering angle $\vartheta$ from $\varphi = 0$ to $\varphi = \varphi_{\max} = \pi - \vartheta$. Integration of (2.162a) gives

$$\int\limits_0^{v_0 \sin\vartheta} \mathrm{d}v_y = \frac{a}{\mu v_0 b} \int\limits_0^{\pi-\vartheta} \sin\varphi\, \mathrm{d}\varphi\,. \tag{2.162b}$$

The solution of the integrals on both sides of (2.162b) is

$$v_0 \sin\vartheta = \frac{a}{\mu v_0 b}(1 + \cos\vartheta)\,. \tag{2.162c}$$

Because $(1 + \cos\vartheta)/\sin\vartheta = \mathrm{cotan}\,(\vartheta/2)$, the relation between the deflection angle $\vartheta$ and impact parameter $b$ of a particle in a potential with potential energy $E_{\text{pot}} = a/r$ becomes

$$\mathrm{cotan}\,(\vartheta/2) = \frac{\mu v_0^2 b}{a}\,. \tag{2.163a}$$

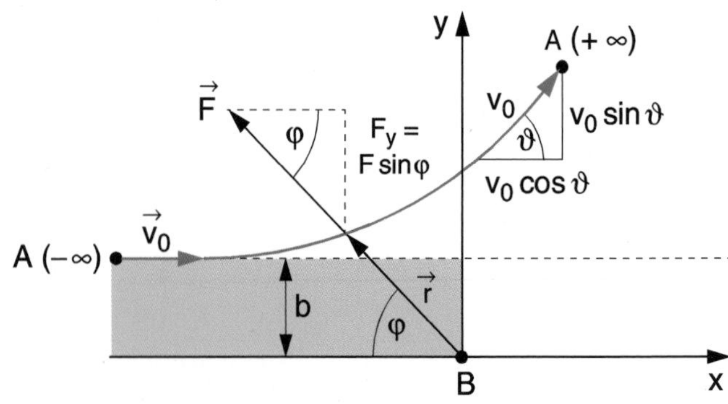


**Fig. 2.102.** Scattering in a Coulomb potential

The ratio $a/b$ represents the potential energy at the distance $r = b$. Inserting this into (2.163a) gives:

$$\cot(\vartheta/2) = \frac{2E_{\text{kin}}(r=-\infty)}{E_{\text{pot}}(r=b)} = \frac{4\pi\varepsilon_0}{qQ}\mu v_0^2 b\,. \quad (2.163b)$$

> The scattering angle $\vartheta$ in the CM-system is for a Coulomb potential determined by the ratio
>
> $$\mu v_0^2/(a/b) = 2E_{\text{kin}}(r=-\infty)/E_{\text{pot}}(b)$$
>
> of twice the initial kinetic energy to the potential energy at the distance $r = b$.

## EXAMPLE

$b = 2\times10^{-12}$ m ($\approx 1/100$ atomic diameter), $\mu v_0^2 = 10\,\text{MeV} = 1.6\times10^{-12}$ J; $q = 3.2\times10^{-19}$ C, $Q = 1.26\times10^{-17}$ C, $\mu = 3.92$ AMU $\Rightarrow \vartheta = 1.3°$. For $b = 2\times10^{-13}$ m (1/1000 atomic diameter) $\Rightarrow \vartheta = 13.2°$, and for $b = 2\times10^{-14}$ m $\Rightarrow \vartheta = 51°$.

This example illustrates that for $\vartheta > 1°$ the scattering cross section $\sigma = \pi b^2 = 10^{-4}\pi R_{\text{A}}^2$ becomes very small compared to the atomic cross section $\pi R_{\text{A}}^2$. This illustrates that in spite of the large number of gold atoms in the foil each $\alpha$ particle is scattered by $\vartheta > 1°$ once at most when passing through the foil.

In order to obtain the differential scattering cross section, we have to calculate that fraction of all incident $\alpha$ particles which is scattered into the angular range $\vartheta \pm \frac{1}{2}\Delta\vartheta$ and can reach the detector with area (Fig. 2.103)

$$\Delta A_{\text{D}} = (R\sin\vartheta)R\Delta\vartheta\Delta\phi = R^2\Delta\Omega\,. \quad (2.164)$$

In (2.147) the differential cross section was derived as

$$\frac{d\sigma}{d\Omega} = b\frac{db}{d\vartheta}\frac{1}{\sin\vartheta}\,. \quad (2.165a)$$

From (2.163a) we obtain

$$\frac{db}{d\vartheta} = \frac{1}{2}\frac{qQ}{4\pi\varepsilon_0\mu v_0^2}\frac{1}{\sin^2(\vartheta/2)}\,. \quad (2.165b)$$

Inserting this into (2.165a) and using the relation $\sin\vartheta = 2\sin(\vartheta/2)\cos(\vartheta/2)$ and (2.163b) for the

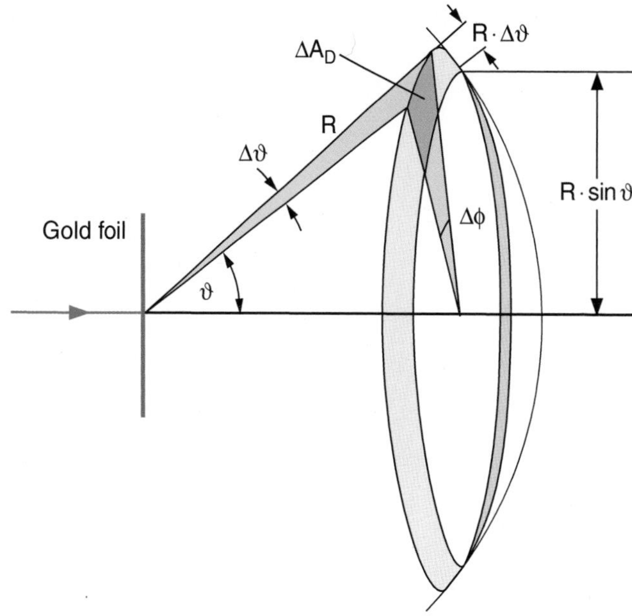


Fig. 2.103. Definition of solid angle $\Delta\Omega$ and detector area $\Delta A_{\text{D}} = R^2\Delta\Omega = R^2\sin\vartheta\,\Delta\vartheta\,\Delta\varphi$

impact parameter $b$ we finally get the differential cross section

$$\boxed{\frac{d\sigma}{d\Omega} = \frac{1}{4}\left(\frac{qQ}{4\pi\varepsilon_0\mu v_0^2}\right)^2\frac{1}{\sin^4(\vartheta/2)}} \quad (2.166)$$

for the scattering of particles with initial kinetic energy $E_{\text{kin}} = \frac{1}{2}\mu v_0^2$ and charge $q$ in a Coulomb potential produced by a point-like charge $Q$. This yields the fraction

$$\frac{\Delta\dot{N}}{\dot{N}A} = \frac{n_{\text{gold}}\Delta V}{4R^2A}\left(\frac{qQ}{8\pi\varepsilon_0 E_{\text{kin}}}\right)^2\frac{\Delta A_{\text{D}}}{\sin^4(\vartheta/2)} \quad (2.167)$$

of incident particles $\dot{N}A$ in a parallel beam with cross section $A$, that are scattered by $n\Delta V$ gold atoms within the volume $\Delta V = A\cdot\Delta x$ into the solid angle $\Delta\Omega$ around $\vartheta$ and reach the detector with area $\Delta A_{\text{D}} = \Delta\Omega/R^2$ at a distance $R \gg R_{\text{A}}$ from the scattering center.

**Note:**

$\dot{N}A$ is the total number of particles that passes per second through the sample area $A$.

**Table 2.5.** Measured counting rates for different scattering angles $\vartheta$ [2.63]

| angle $\vartheta$ | counting rate $\Delta\dot{N}$ | $\Delta\dot{N}\cdot\sin^4\vartheta/2$ |
|---|---|---|
| 15° | 132,000 | 38.3 |
| 30° | 7 800 | 35.0 |
| 37.5° | 3 300 | 35.3 |
| 45° | 1 435 | 30.8 |
| 60° | 477 | 29.8 |
| 75° | 211 | 29.1 |
| 105° | 70 | 27.7 |
| 120° | 52 | 29.1 |
| 135° | 43 | 31.2 |
| 150° | 33 | 28.7 |

The measured angular distribution (Table 2.5 and Fig. 2.104) agrees very well with (2.167). Note, that the product $\Delta\dot{N}\cdot\sin^4\vartheta/2$ in Table 2.5 is fairly constant for $\vartheta \geq 45°$. The slightly larger values for small angles $\vartheta$ are due to the fact, that here $\Delta\vartheta$ is no longer very small compared to $\vartheta$. Here $\sin^4(\vartheta/2)\cdot\Delta\vartheta$ has to be replaced by

$$\int\limits_{\vartheta-\Delta\vartheta/2}^{\vartheta+\Delta\vartheta/2} \sin^4\vartheta/2\,\mathrm{d}\vartheta = \frac{3}{8}\Delta\vartheta + \frac{1}{8}\cos 2\vartheta\sin\Delta\vartheta - \cos\vartheta\sin(\Delta\vartheta/2)\,.$$

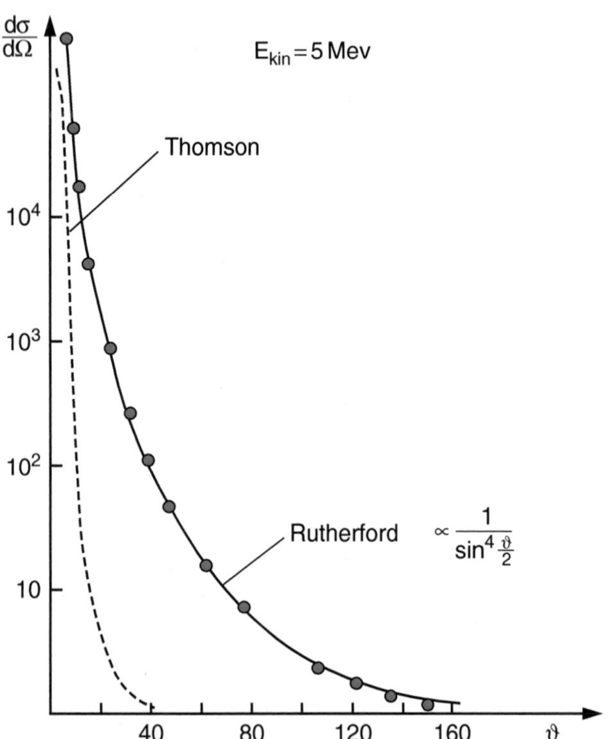


**Fig. 2.104.** Comparison of experimental results (*points*) with the predictions by Thomson (*dashed curve*) and Rutherford (*solid curve*)

The experiments showed that for large angles $\vartheta > \vartheta_c$ which correspond to small impact parameters deviations from the expected values were observed (Fig. 2.105). Rutherford had already recognized that these deviations are due to the finite size $R_N$ of the nucleus. If the impact parameter $b$ becomes smaller than $R_N$, the $\alpha$ particle penetrates into the nucleus. For $b < R_N$ the deflection should be described by (2.157). However, for $b < R_N$ a new short range force, called the nuclear force, becomes important and the Coulomb law (2.150) is no longer valid. The attractive nuclear force is much stronger than the repulsive Coulomb force and changes the deflection of the $\alpha$ particles.

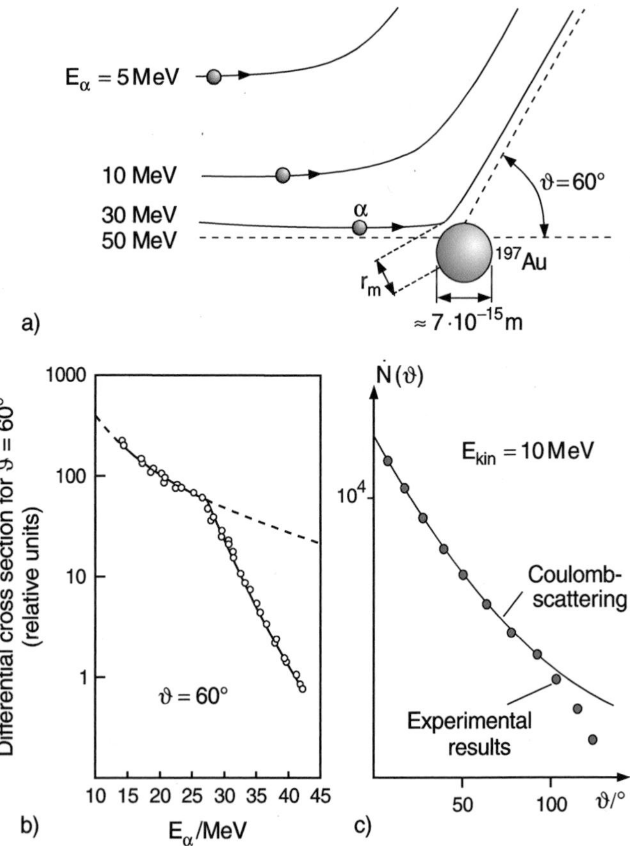


**Fig. 2.105.** (**a**) Paths of particles with different initial energies, all scattered by $\vartheta = 60°$. (**b**) Deviation from Coulomb potential for paths with $E_{\text{kin}} > 25\,\text{MeV}$ (i.e., $b < b_c$) for $\vartheta = 60°$. (**c**) Deviation at fixed initial energy $E_{\text{kin}} = 10\,\text{MeV}$ for scattering angle $\vartheta > 100°$

Thus the impact parameter at which the measured distribution deviates from the predicted one gives a measure for the size of the atomic nucleus. One obtains values of

$$R_N \approx r_0 A^{1/2} ,$$

where $A$ is the atomic mass in AMU and $r_0 = 1.3 \times 10^{-15}$ m.

The volume of the nucleus therefore only accounts for the fraction $(R_N/R_A)^3 \approx 10^{-13}$–$10^{-15}$ of the atomic volume $V_A$!

While the atomic volume of the gold atom is about $V_A = 10^{-29}\,\mathrm{m}^3$ that of its nucleus is only $V_N \approx 10^{-42}\,\mathrm{m}^3$.

## SUMMARY

- The initially diffuse and sometimes incorrect picture of atoms has been concretized by more and more refined experiments during the past 200 years. This has lead to a quantitative atomic model that describes most observations correctly.
- In a first crude model atoms are described by spherical charge and mass distributions with mean radii of 0.05–0.5 nm, which can be determined by scattering experiments or by X-ray diffraction in crystals.
- 1 mol is that quantity of matter that contains as many atoms or molecules as 0.012 kg $^{12}$C, or that contains as many grams as the mass number of its atoms or molecules (in atomic units AMU) indicates.
- The Avogadro constant $N_A = 6.022 \times 10^{23}$ /mol gives the number of atoms or molecules in 1 mol of the substance.
- Each neutral atom consists of $Z$ electrons with mass $m_e = (1/1836)$ AMU and charge $-e = -1.6 \times 10^{-19}$ C and a much more massive nucleus with mass $A$ (in AMU) and charge $+Ze$.
- Free electrons can be produced by thermal emission from metal surfaces, by field emissions from sharp metal peaks in high electric fields, by electron impact on atoms or molecules and by photoionization, following light absorption by atoms or molecules, and finally by the photoeffect, where light incident on metal surfaces can result in the emission of electrons.
- Neutral atoms can be ionized through electron impact, photon absorption, by collisions with fast ions or by charge exchange in collisions with other atoms. Atoms that have lost $n$ electrons are called $n$-fold ionized.
- Negative ions can be formed by recombination of free electrons with neutral atoms. They have a surplus of electrons and can be readily ionized to become neutral atoms again.
- Charged particles can be deflected in electric or magnetic fields. Specially formed fields can act as electron- or ion-optics. The total length of electrostatic or magnetic lenses can be continuously varied with the field strength.
- The ratio $q/m$ of charge $q$ to mass $m$ of electrons and ions can be measured with instruments based on the deflection of charged particles in electric or magnetic fields. The mass can be determined separately only if the charge $q$ is known. Mass spectrometers are devices that separate ions of different masses and can measure the absolute values of masses after calibration with carbon atoms representing the mass unit.
- In time-of-flight spectrometers the mass dependent flight time through a field-free region of ions accelerated by a known voltage $U$ is used for mass determination.
- The elementary charge unit $q = e$ can be measured with Millikan's oil droplet experiment
- Scattering of $\alpha$ particles ($He^{++}$) by gold atoms, or more recent modern variants of Rutherford's initial experiments, using high energy electrons or protons, support the Rutherford atomic model, which proposes the following structure of atoms: By far, the major part of the atomic mass is united in a very small volume, called the atomic nucleus with a typical radius of $(1–5) \times 10^{-15}$ m, which is about five orders of magnitude smaller than ▶

the atomic radius $R_A \approx 10^{-10}$ m. The volume of the nucleus is therefore only about $10^{-13}$–$10^{-15}$ of the atomic volume. The rest of the volume contains the $Z$ electrons, but is nearly completely empty of mass although filled with the electric field of the charges.

- The positive charge $+Ze$ of the atomic nucleus is compensated by the negative charges $-Ze$ of the $Z$ electrons to form a neutral atom. Experiments show that possible differences $\Delta q$ between the absolute values of positive and negative charges are smaller than $\Delta q/q \leq 10^{-21}$.

## PROBLEMS

1. In 1 $m^3$ of air there are $2.6 \times 10^{25}$ molecules under normal conditions ($p = 101{,}325$ Pa = 1 atm and $T = 273.2$ K = 0 °C). How large is
a) the mean distance between two molecules?
b) the spatial filling factor $\eta = V_{mol}/1\,m^3$, when the molecules are described by spheres with radius $R = 0.1$ nm?
c) the mean free path length $\Lambda$?
2. The main constituents of air are: 78% $N_2$, 21% $O_2$ and 1% Ar. Using these numbers calculate the mass density $\rho$ of air under normal conditions.
3. How many atoms are in
a) 1 g of $^{12}C$?
b) 1 $cm^3$ of helium at a pressure $p = 10^5\,Pa$ and $T = 273$ K?
c) 1 kg of nitrogen gas ($N_2$)?
d) In a steel bottle with 10 $dm^3$ volume of $H_2$ gas at $p = 10^6$ Pa?
4. In interstellar space the mean density of H atoms is about 1 atom/$cm^3$ and the mean temperature is about 10 K. What is the pressure under these conditions? Why can such low pressures not be obtained on earth under laboratory conditions?
5. Imagine that an international commission has defined a new temperature scale, where the absolute zero is defined as 0° N and the freezing point of water as 100° N. What is the value of the boiling point of water on this new scale? What would be the value of the Boltzmann constant $k$ in J/° N?
6. Prove the relation $v_s = v_{ph} = (\kappa RT/M)^{1/2}$ between sound velocity $v_s$ molar mass $M$ and temperature $T$ given in Sect. 2.2.3b? How large are the frequencies of radial acoustic resonances in a spherical resonator with radius $r_0$?
7. In his experiments about the number density distribution of latex particles in water, Perrin found $49 \cdot \Delta h$ particles per $cm^3$ in a slice $\Delta h$ at a height $h$ and 14 particles at $h + 60\,\mu$m. The mass density of the particles was $\rho_T = 1.194\,kg/dm^3$ and their radius $r = 2.12 \times 10^{-7}$ m. What was the mass of the particles, the Avogadro constant and their mole mass?
8. a) What is the incidence angle $\alpha$ for X-rays with $\lambda = 0.5$ nm falling onto a grating with 1200 grooves per mm when the first diffraction maximum should be observed under the angle $\beta_1 = 87°$. Where does the the second diffraction order appear? How large must $\alpha$ be to give $\beta_1 - \beta_2 \geq 0.75°$?
b) X-rays with $\lambda = 0.2$ nm are diffracted at the (100) plane (parallel to two of the endfaces of the cube) of a cubic NaCl crystal. The first diffraction order appears at $\beta = 21°$. What is the lattice constant of the NaCl crystal? How large is the Avogadro constant calculated from this experimental result when the mass density is $\rho_{NaCl} = 2.1\,kg/dm^3$?
c) What are the radius and the volume of Ar atoms in a cold Ar crystal (face centered cubic lattice = highest density package of spheres), when the specular reflection maximum is at $\vartheta = 43°$ for an X-ray beam with wavelength $\lambda = 0.45$ nm, incident on the crystal under an angle $\vartheta$ against the plane (100) parallel to one side of the cubic crystal?
9. The general equation for a real gas can be written either in the form of the Van der Waals equation (2.51b) or as a Taylor series in powers $(1/V_M)^n$ of the inverse mole volume $V_M$:

$$pV_M = RT\left(1 + B(T)/V_M + C(T)/V_M^2 + \ldots\right).$$

Compare the virial coefficients $B(T), C(T), \ldots$ with the constants $a$ and $b$ in the van der Waals equation and discuss their physical meaning.
10. Derive the equation (2.52).
11. How accurate can the ratio $e/m$ for electrons be determined under the following conditions?
a) In a longitudinal magnetic field, when the electrons pass through apertures with 1 mm diameter located before their entrance into the field and in the focal plane. The current reaching the detector can be measured with a relative accuracy of $10^{-3}$, the magnetic field $B$ and the acceleration voltage $U$ with $10^{-4}$ and the distance $L = 100$ mm between entrance and exit aperture with $2 \times 10^{-3}$?
b) With a Wien filter, where entrance and exit slits with width $b = 0.1$ mm are separated by $d = 10$ cm and the acceleration voltage is $U = 1$ kV. The accuracies for all necessary measurements are the same as above.

12. A beam of Cs-atoms with velocity $v = 300\,\mathrm{m/s}$ is emitted into vacuum through an orifice with diameter $d_0 = 40\,\mu\mathrm{m}$. The beam is collimated by a slit $S_1$ with width $b_1 = 40\,\mu\mathrm{m}$, placed 200 cm downstreams of the orifice.
a) Calculate the vertical deviation of the beam from a horizontal line at a distance $d_2 = 200\,\mathrm{cm}$ away from $S_1$, caused by gravity. b) What is the deflection of the atoms with assumed charge $\Delta q$ after passing an electric field $E = 5 \times 10^6\,\mathrm{V/m}$ with length $L = 200\,\mathrm{cm}$, placed between $S_1$ and $S_2$? c) How large is the relative change $\Delta N/N$ of particles passing through the second slit $S_2$ 200 cm away from $S_1$, when the electric field is switched from $+E$ to $-E$? d) Estimate the accuracy limit for the determination of $\Delta q = |\mathrm{e}^+| - |\mathrm{e}^-|$, when a change $\Delta \dot{N} = 10^{-4}\,\dot{N}$ with $\dot{N} = 10^9\,\mathrm{s}^{-1}$ can still be monitored?
13. $\mathrm{Ar}^+$ ions with a kinetic energy of $10^3\,\mathrm{eV}$ pass through a magnetic 60° sector field. What is the magnitude of the magnetic field for achieving a focal length of $f = 80\,\mathrm{cm}$?
14. The electric potential along the axis of an electron lens with cylindrical symmetry shall be described by

$$\begin{aligned} \phi &= \phi_0 + az^2 \quad \text{for} \quad 0 \le z \le z_0 \\ \phi &= \phi_0 \quad \text{for} \quad z \le 0 \\ \phi &= \phi_0 + az_0^2 \quad \text{for} \quad z \ge z_0 \end{aligned}$$

How large is the focal length of this lens for electrons entering the field with a velocity $v_0 = (2e\Phi_0/m)^{1/2}$ for a potential $\phi_0 = 10^4\,\mathrm{V/m}$ and $a = 10^3\,\mathrm{V/m^3}$?
15. a) Derive the relation $U = 2V_0 \ln(R_2/R_1)$ between the voltage $U$ between the two cylindrical plates of a condensor and the optimum acceleration voltage $V_0$ for electrons passing through the condensor on the central curved path?
b) For which angle $\varphi$ in Fig. 2.70 acts the condensor as cylindrical lens? What is its focal length?
16. Within a thin layer of thickness $b = 2\,\mathrm{mm}$ between two grids, separated by $d = 30\,\mathrm{mm}$ with a voltage of 300 V between the grids, ions with mass $m$ are produced and sent through a time-of-flight mass spectrometer.
a) What is the time of flight and its spread in a 1-m long field-free drift tube? Is it possible to separate two masses with $m_1 = 110\,\mathrm{AMU}$ and $m_2 = 100\,\mathrm{AMU}$?
b) Show that the reflectron has a larger mass resolution than the linear time-of-flight spectrometer. What is the reason for this? Which parameters influence the mass resolution?
c) Compare the mass resolution of a time of flight spectrometer with that of a magnetic 180° mass spectrometer, where ions within the velocity interval $v_0 \pm \Delta v/2$ enter the field as a parallel beam through an aperture that is 1ămm in width. The exit slit is also 1ămm wide. What is the beam divergence at the exit plane?
17. Prove, that the angular momentum of incident particles, defined with respect to the centre point is conserved during a collision, if the interaction potential is spherical symmetric.
18. $\alpha$ particles with $E_{\mathrm{kin}} = 5\,\mathrm{MeV}$ are scattered by gold atoms in a thin foil ($d = 5 \times 10^{-6}\,\mathrm{m}$, $\rho = 19.3\,\mathrm{g/cm^3}$, $M = 197\,\mathrm{g/mole}$).
a) What is the impact parameter $b$ for particles with a scattering angle $\vartheta = 90°$?
b) What is the value for $r_{\min}$ for backward scattering $\vartheta = 180°$?
c) What fraction of all incident particles is scattered into angles $\vartheta \ge 90°$?
d) What fraction is scattered into the range $45° \le \vartheta \le 90°$?
19. Compare the relative numbers of particles scattered into the range $\vartheta = (1 \pm 0.5)°$ and $\vartheta = (5 \pm 0.5)°$ for both the Thomson and the Rutherford model for the gold foil of Problem 2.18 when the angular resolution is 1°.
20. Protons are shot into a copper foil with $d = 12\,\mu\mathrm{m}$.
a) What is the proton energy if the central collisions $r_{\min}$ become equal to the nuclear radius $r_{\mathrm{N}} = 5 \times 10^{-15}\,\mathrm{m}$ of the copper atoms?
b) For $r_{\min} < r_{\mathrm{N}}$, deviations from the Rutherford scattering formula are expected. For which scattering angles $\vartheta$ does this happen at a proton energy of 9.5 MeV?

# 3. Development of Quantum Physics

At the beginning of the 20th century several experimental findings could not be explained by the existing theories of the time, which we will name "classical physics". These experiments indicated that the conception of classical physics had to be modified. Examples are the measured spectral distribution of radiation from black bodies, which was in disagreement with theoretical predictions, the photo effect, the explanation of the Compton effect and a satisfactory answer to the question of why atoms in their lowest energetic state are stable.

It turned out that the particle model of classical mechanics had to be reviewed. This model attributed to each particle a well-defined path in space that could be predicted for all times provided the initial conditions (location and velocity at time $t = 0$) and the force field acting on the particle, were known.

Also the classical description of electromagnetic fields and waves by the Maxwellian equations seemed to need a critical revision when applied to the microscale of atoms and molecules.

This chapter will present the most important experimental proofs for the necessary modification and extension of classical physics, which have led to the development of quantum physics. The basic ideas of quantum physics will be presented here, while a brief description of the mathematical framework of quantum theory along with some examples are given in the next chapter.

## 3.1 Experimental Hints to the Particle Character of Electromagnetic Radiation

During the 18th century a long-lasting quarrel was fought among scientists about the correct description of light. *Newton* and his school postulated that light should consist of small particles [3.1]. Their model could explain the straight paths of light rays and also the refraction of light at the boundary between two media with different refractive indices. *Huygens* and other scientists, on the other side, believed that light was a wave phenomenon. Their experiments on diffraction and interference seemed to prove their theory [3.2].

The wave model was generally accepted when *Heinrich Hertz* discovered the electromagnetic waves and when it was recognized that visible light was just a special case of electromagnetic waves restricted to the wavelength region between $\lambda = 0.4\,\mu\text{m}$–$0.7\,\mu\text{m}$, which could be described by the Maxwellian equations like all electromagnetic waves.

The following sections shall illustrate that both models, the particle model and the wave description of light, were partly correct but that both models must be combined for a complete characterization of light. The important point, shown by the quantum physical model is that particle and wave descriptions do not contradict but rather supplement each other. Depending on which property of light is described, the wave model or the particle model is a more proper description. Let us first summarize the classical model of electromagnetic waves.

In classical physics a plane electromagnetic wave

$$\boldsymbol{E} = \boldsymbol{A}\cos(\boldsymbol{k}\cdot\boldsymbol{r} - \omega t)$$

is described by its amplitude $\boldsymbol{A} = |\boldsymbol{A}|\hat{e}_\text{p}$, its frequency $\omega$, and its wave vector $\boldsymbol{k}$, which points into the propagation direction and has the amount $|\boldsymbol{k}| = 2\pi/\lambda$, determined by the wavelength $\lambda = c/\nu = 2\pi c/\omega$. In case of a polarized wave the direction of the electric field vector is given by the polarization unit vector $\hat{e}_\text{p}$. The energy density of this electromagnetic wave in a vacuum

W. Demtröder, *Atoms, Molecules and Photons*, 2nd ed., Graduate Texts in Physics,
DOI 10.1007/978-3-642-10298-1_3, 

$$w_{\mathrm{em}} = \varepsilon_0 |E|^2 = \frac{1}{2}\varepsilon_0 \left(E^2 + c^2 B^2\right) \quad \left[\mathrm{J/m^3}\right] \tag{3.1}$$

is determined by its electric field amplitude $E = |\boldsymbol{E}|$ and can be also described by the magnetic field amplitude $B$.

The intensity (the incident power per unit area) of an electromagnetic wave is

$$I = c\varepsilon_0 E^2 = c w_{\mathrm{em}} \tag{3.2a}$$

and the power, transmitted through the area $A$ with the normal unit vector $\hat{e}_{\mathrm{n}}$

$$\frac{\mathrm{d}W}{\mathrm{d}t} = IA\frac{\boldsymbol{k}}{|k|}\hat{e}_{\mathrm{n}} \tag{3.2b}$$

is determined by the relative orientation of wave vector $\boldsymbol{k}$ and normal vector $\boldsymbol{e}_{\mathrm{n}}$ of the area $A$.

Note that in this classical description both power density and intensity depend on the square of the electric field amplitude $E$. They are continuous functions of $E$ and of the space coordinates inside the radiation field.

The classical electromagnetic wave also has a momentum density (momentum per unit volume),

$$\pi_{\mathrm{em}} = \varepsilon_0(\boldsymbol{E} \times \boldsymbol{B}) = \frac{1}{c^2}\boldsymbol{S} \tag{3.2c}$$

described by the pointing vector $\boldsymbol{S} = \varepsilon_0 c^2(\boldsymbol{E} \times \boldsymbol{B})$, where the amount of $\boldsymbol{S}$

$$|\boldsymbol{S}| = \varepsilon_0 c E^2 = I \tag{3.2d}$$

equals the intensity of the wave.

An important quantity is the spectral intensity $I_\nu$ [W m$^{-2}$ s] with

$$I = \int\limits_{\nu=0}^{\infty} I_\nu(\nu)\mathrm{d}\nu\,, \tag{3.2e}$$

where $I_\nu(\nu)\mathrm{d}\nu$ gives the incident power density within the spectral interval $\mathrm{d}\nu$.

All these results can be derived from Maxwell's equations and the continuity equation, which describes all phenomena observed until the end of the 19th century very well.

The first hints that corrections were necessary came from experiments measuring the spectral distribution of the radiation emitted by a hot blackbody, which will be discussed in the following section.

### 3.1.1 Blackbody Radiation

Material that absorbs all incident radiation (its absorption is $A^* = 1$) is called a blackbody. It can be approximately realized by a closed cavity with absorbing walls and a small hole in one of the walls (Fig. 3.1). If the area $\Delta A$ of this hole is very small compared to the area $A$ of the inner walls, radiation passing from outside through the hole into the cavity has a negligible chance to leave the cavity again, i.e., it is completely absorbed. This means that the absorption of the hole is $A^* \approx 1$.

When the walls of the cavity are heated to a temperature $T > T_{\mathrm{s}}$, where $T_{\mathrm{s}}$ is the temperature of the surrounding, the hole acts as radiation source with an intensity that is larger than that of any other body at the same temperature. This can be demonstrated by a simple experiment. Into one side of a solid graphite cube the letter H is mill-cut (Fig. 3.2). At low temperatures the letter appears much darker than its surroundings, but at higher temperatures (about 1000 K) it appears

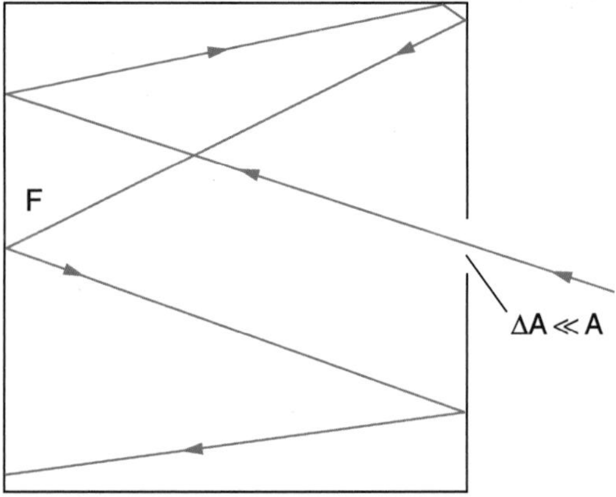


**Fig. 3.1.** A closed cavity absorbs nearly all radiation entering the cavity through a small hole

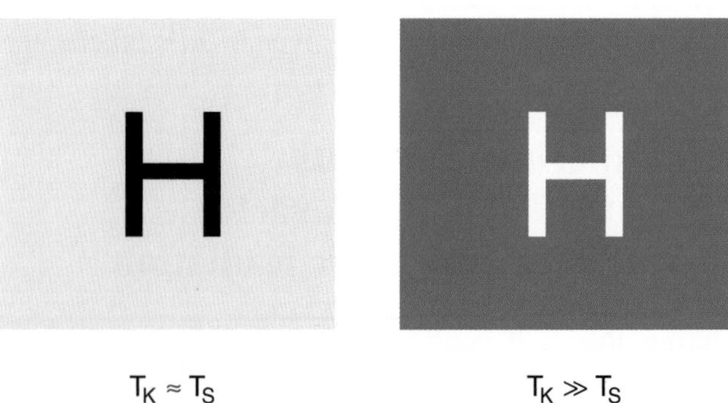


**Fig. 3.2.** The letter H mill-cut into a graphite cube appears completely black at temperature $T_{\mathrm{k}} < T_{\mathrm{s}}$ = temperature of the surrounding, but appears bright for $T_{\mathrm{k}} \gg T_{\mathrm{s}}$

bright yellow on a dark red surrounding. This means that at low temperatures it absorbs nearly all incident radiation while at higher temperatures it emits more than the surface of the cube at the same temperature.

Inside the closed cavity of Fig. 3.1 a stationary radiation field exists that depends solely on the temperature of the cavity walls and not on the dimensions $d$ of the cavity as long as $d \gg \lambda$, where $\lambda$ is the wavelength of the enclosed radiation. The application of basic laws of thermodynamics lead to the following considerations.

- For a stationary state of the cavity radiation the radiation power emitted by the walls must equal that absorbed by them for all frequencies $\nu$ of the radiation (otherwise the radiation field would change in time). This means

$$\frac{\mathrm{d}W_\mathrm{a}(\nu)}{\mathrm{d}t} = \frac{\mathrm{d}W_\mathrm{e}(\nu)}{\mathrm{d}t} . \tag{3.3}$$

  For such a stationary state we define the temperature $T$ of the radiation field by the temperature of the cavity walls. The radiation field has the following characteristics.
- The cavity radiation field is isotropic, which means that the spectral radiation density $S_\nu^*$ (this is the radiation power per frequency interval $\mathrm{d}\nu = 1\,\mathrm{s}^{-1}$ radiated into the solid angle $\mathrm{d}\Omega = 1$ Sterad) is independent of the direction at every point of the cavity. If this wasn't the case, one could insert a black disc with surface area $\mathrm{d}A$ into the radiation field and orient it in such a way that its surface normal would point into the direction of maximum radiation density (Fig. 3.3). The disc would then absorb more energy than it emits and would be heated above the temperature $T$ of its surrounding. This, however, contradicts the second law of thermodynamics.
- The cavity radiation is homogeneous, i.e., its energy density is independent of a special point inside the cavity. Otherwise a similar argument would hold as in the previous paragraph and a perpetual motion machine of the second kind could be constructed.

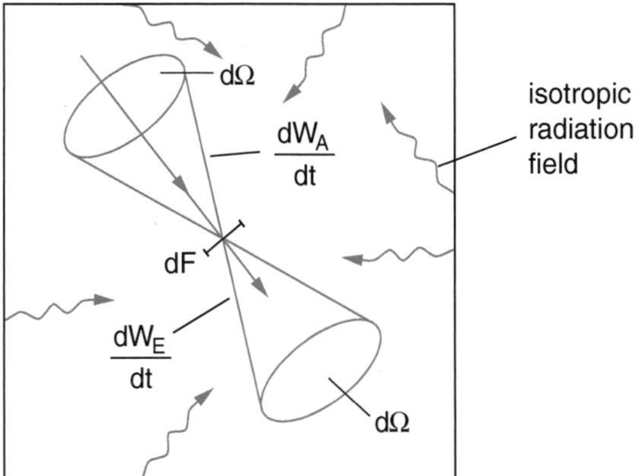


**Fig. 3.3.** A body in the cavity is at thermal equilibrium with the thermal isotropic radiation inside the cavity

When a body is placed inside the cavity radiation field, its surface element absorbs the power

$$\frac{\mathrm{d}W_\mathrm{a}(\nu)}{\mathrm{d}t} = A_\nu^* S_\nu^* \,\mathrm{d}A\,\mathrm{d}\Omega\,\mathrm{d}\nu \tag{3.4a}$$

from the radiation with spectral radiation power density $S_\nu^*$ within the frequency interval $\mathrm{d}\nu$ incident on $\mathrm{d}A$ within the solid angle $\mathrm{d}\Omega$. The constant $A_\nu^*$ is the spectral absorbance of the body. The surface element $\mathrm{d}A$ emits, on the other hand, the power

$$\frac{\mathrm{d}W_\mathrm{e}(\nu)}{\mathrm{d}t} = E_\nu^* \,\mathrm{d}A\,\mathrm{d}\Omega\,\mathrm{d}\nu \tag{3.4b}$$

into the solid angle $\mathrm{d}\Omega$ within the frequency interval $\mathrm{d}\nu$.

For thermal equilibrium conditions both quantities must be equal. Since the cavity radiation is isotropic and homogeneous, this must be valid for every direction $(\theta, \varphi)$ and for every location inside the cavity. Therefore we obtain Kirchhoff's law:

$$\boxed{\frac{E_\nu^*}{A_\nu^*} = S_\nu^*(T) .} \tag{3.5}$$

> For all bodies in thermal equilibrium with the cavity radiation the ratio of spectral emittance $E_\nu^*$ and absorbance $A_\nu^*$ equals the spectral radiation density $S_\nu^*$ of the cavity radiation, which itself depends on the temperature $T$.

For a black body is $A_\nu^* \equiv 1$ for all frequencies $\nu$. We therefore conclude from (3.5):

> The spectral emittance $E_\nu^*$ of a blackbody equals the spectral radiation density $S_\nu^*$ of the cavity radiation.

Our next task is now to determine the spectral dependence of $S_\nu^*$, which equals the spectral intensity distribution $E_\nu^*(\nu)$ of the blackbody radiation.

### 3.1.2 Cavity Modes

Let us assume a radiation field inside a cubic box with side lengths $L \gg \lambda$ that is in thermal equilibrium with the walls of the cavity at temperature $T$. The field can be described as the superposition of waves with wave vectors $\boldsymbol{k} = \{k_x, k_y, k_z\}$. A stationary field distribution, which means a standing wave field, can be only realized if certain boundary conditions are fulfilled. These conditions demand that the field amplitude be zero at the walls of the box (Fig. 3.4a,b). This restricts the possible values of the components of the $\boldsymbol{k}$-vector to

$$k_x = \frac{\pi}{L} n_1\,, \quad k_y = \frac{\pi}{L} n_2\,, \quad k_z = \frac{\pi}{L} n_3$$

$$\Rightarrow k = |\boldsymbol{k}| = \frac{\pi}{L}\sqrt{n_1^2 + n_2^2 + n_3^2}\,, \tag{3.6}$$

where the $n_i$ are arbitrary integers. For the wavelengths $\lambda = 2\pi/k$ and the angular frequency $\omega = kc$ we then obtain

$$\lambda = \frac{2L}{\sqrt{n_1^2 + n_2^2 + n_3^2}}\,, \tag{3.7a}$$

$$\omega = ck = \frac{\pi c}{L}\sqrt{n_1^2 + n_2^2 + n_3^2}\,. \tag{3.7b}$$

Every stationary field distribution with a specified triple $(n_1, n_2, n_3)$ is called a ***mode*** of the cavity radiation field.

The question now is how many modes with frequencies $\omega < \omega_m$ can exist, where $\omega_m$ is a number determined by the specific problem.

In a coordinate system in $k$-vector space with coordinates $(\pi/L)(n_1, n_2, n_3)$ each triple of integers $(n_1, n_2, n_3)$ represents a point in a three-dimensional lattice with the lattice constant $\pi/L$. In this system (3.7) describes all possible lattice points within the positive octand of a sphere with radius $\omega/c$ (see Fig. 3.4c). If this radius is large compared to the lattice constant $\pi/L$, (which means $L \gg \lambda_m$) the number $N = (L/\pi)^3 \cdot V$ of lattice points $(n_1, n_2, n_3)$ with $\omega \le \omega_m$ is approximately given by the volume

$$V = \frac{1}{8}\frac{4}{3}\pi\left(\frac{\omega_m}{c}\right)^3$$

of the octant of the sphere in Fig. 3.4c. This also gives the number of modes of a stationary radiation field.

Each mode can be composed of two standing waves with independent polarization directions of the electric field vector

$$\boldsymbol{E} = a_1\hat{e}_1 + a_2\hat{e}_2\,, \quad \hat{e}_1 \cdot \hat{e}_2 = \delta_{12}\,, \quad \hat{e}_1, \hat{e}_2 \perp \boldsymbol{k}\,.$$

For the total number of possible standing waves with frequencies $\omega < \omega_m$ inside a cubic cavity with length $L$ including the polarization we then obtain

$$N(\omega \le \omega_m) = 2\left(\frac{L}{\pi}\right)^3 \cdot V = 2\frac{1}{8}\frac{4\pi}{3}\left(\frac{L\omega_m}{\pi c}\right)^3 = \frac{1}{3}\frac{L^3\omega_m^3}{\pi^2 c^3}\,. \tag{3.8}$$

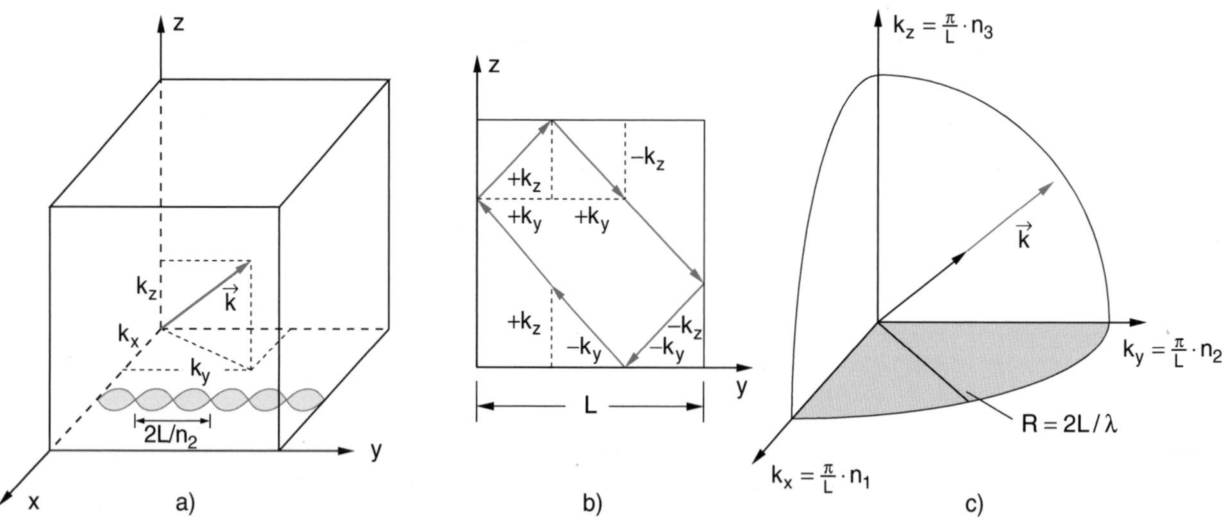


**Fig. 3.4a–c.** Modes of a stationary EM field in a cavity. (**a**) Standing waves in a cubic cavity (**b**) Superposition of possible $\boldsymbol{k}$ vectors to form standing waves, illustrated in a two-dimensional coordinate system (**c**) Illustration of the calculation of the maximum number of modes in momentum space

The mode density (i.e., the number of modes within the unit volume)

$$n(\omega \le \omega_m) = \frac{1}{L^3} N(\omega < \omega_m) = \frac{1}{3} \frac{\omega_m^3}{\pi^2 c^3} \qquad (3.9a)$$

becomes independent of the size of the cavity as long as $L \gg \lambda$.

The spectral mode density (i.e., the number density of modes within the spectral interval $d\omega = 1\,s^{-1}$ can be obtained by differentiating (3.8)). This gives

$$n_\omega = \frac{d}{d\omega}(n(\omega)) = \frac{\omega^2}{\pi^2 c^3} . \qquad (3.9b)$$

Spectroscopists prefer the frequency $\nu$ instead of the angular frequency $\omega = 2\pi\nu$. This converts (3.9b) because $d\omega = 2\pi\, d\nu$ into

$$n_\nu(\nu) = \frac{8\pi\nu^2}{c^3} . \qquad (3.10a)$$

In Fig. 3.5 this quantity is plotted against the frequency $\nu$ on a double logarithmic scale.

The number of modes per unit volume within the frequency interval between $\nu$ and $\nu + d\nu$ is

$$\boxed{n_\nu(\nu)d\nu = \frac{8\pi\nu^2}{c^3} d\nu .} \qquad (3.10b)$$

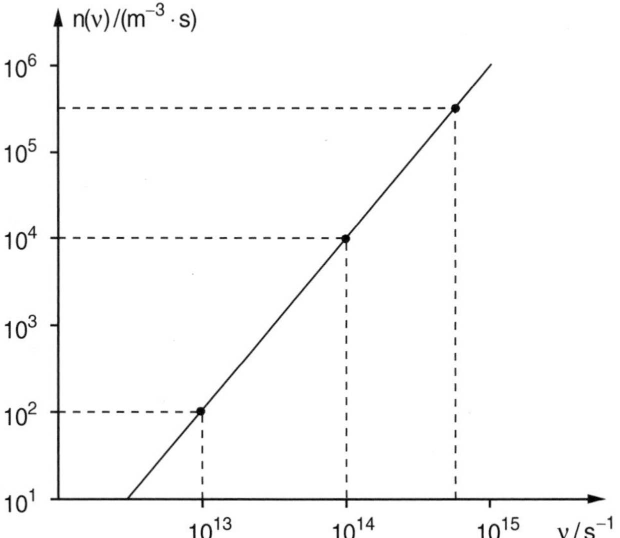


**Fig. 3.5.** Spectral mode density $n(\nu)$ as a function of frequency $\nu$, represented on a double-logarithmic scale

**EXAMPLE**

In the visible range ($\nu = 6 \times 10^{14}\,s^{-1} \hat{=} \lambda = 500\,nm$), we obtain from (3.10): $n(\nu) = 3 \times 10^5\,m^{-3}\,Hz^{-1}$. Within the frequency interval $d\nu = 10^9\,s^{-1}$ (this corresponds to the frequency width of a Doppler broadened spectral line (see Sect. 7.3)), there are $3 \times 10^{14}$ modes/m$^3$.

---

The spectral energy density $w_\nu(\nu)$ of the cavity radiation field is then

$$w_\nu(\nu)d\nu = n(\nu)\bar{w}_\nu(T)d\nu , \qquad (3.11)$$

where $\bar{w}_\nu(T)$ is the average energy per mode, which depends on the temperature $T$.

For the determination of $\bar{w}_\nu(T)$ *Rayleigh and Jeans* used a classical model. They assumed that each mode of the field could be treated like a harmonic oscillator with the mean energy $w_\nu(T) = kT$ (remember the proof in thermodynamics that the mean energy of a system with $f$ degrees of freedom is $\frac{f}{2}kT$). The one-dimensional harmonic oscillator has potential and kinetic energies with equal mean values. Therefore its total mean energy is $2\frac{1}{2}kT$. This yields for the spectral energy density of the cavity field

$$w_\nu(\nu) = d\nu = \frac{8\pi\nu^2}{c^3} kT d\nu \qquad (3.12)$$

(the Rayleigh–Jeans radiation law).

Through a small hole in a cavity wall the radiation density $S^*(\nu)d\nu = (c/4\pi)w_\nu d\nu d\Omega$ is emitted into the solid angle $d\Omega$. Using (3.11) this gives

$$S_\nu^*(\nu)d\nu = \frac{2\nu^2}{c^2} kT d\nu . \qquad (3.13)$$

The experimental check shows, that for small frequencies (in the infrared spectral region) (3.13) agrees quite well with the experimental results. However, for larger frequencies (in the ultraviolet region) drastic discrepancies were found. While the Rayleigh–Jeans law predicts

$$\lim_{\nu \to \infty} S_\nu^*(\nu) \to \infty$$

the experiments proved that with increasing $\nu$ the radiation density $S_\nu^*(\nu)$ increases at first, passes through a maximum and then decreases again. This discrepancy, which could not be explained at that time, was called the *ultraviolet catastrophe*.

### 3.1.3 Planck's Radiation Law

*What is wrong with the Rayleigh–Jeans model?*

This question was finally answered by *Max Planck*, who in 1904 developed a new theory he called quantum hypothesis [3.3, 4]. It is based on the following assumptions.

Planck also described the cavity modes of the radiation field as oscillators. However, he postulated, that these oscillators could not increase or decrease their energies $w_\nu$ by arbitrary small amounts (as this would be for $w_\nu = kT$), but only in integer multiples of a minimum energy quantum $h\nu$. The constant

$$h = 6.6260755 \times 10^{-34}\,\text{Js}$$

is named *Planck's constant.*

This energy quantum $h\nu$ with the minimum possible energy $w_\nu > 0$ is called a ***photon***. The energy stored in a mode containing $n$ photons is then

$$w_\nu = nh\nu \,.$$

At thermal equilibrium the energy distribution $w_\nu(T)$ of a system is governed by the Boltzmann factor $\exp[-w/kT]$. Therefore the probability $p(w_\nu)$ that a mode contains $n$ photons, i.e., has the energy $w_\nu = nh\nu$, is given by

$$p(w_\nu) = \frac{\mathrm{e}^{-nh\nu/(kT)}}{\sum_{n=0}^{\infty} \mathrm{e}^{-nh\nu/(kT)}} \,, \tag{3.14a}$$

where the denominator represents the partition function which normalizes the probablity $p(w_\nu)$ in such a way that

$$\sum_{n=0}^{\infty} p(nh\nu) = 1$$

as can be immediately seen from (3.14a). This means, that the total probability, that a mode contains an energy between 0 and $\infty$ must, of course, be 100%.

The mean energy per mode is then (see Problem 3.2)

$$\begin{aligned}\bar{w}_\nu &= \sum_{n=0}^{\infty} nh\nu p(nh\nu) \\ &= \frac{\sum nh\nu \mathrm{e}^{-nh\nu/kT}}{\sum \mathrm{e}^{-nh\nu/kT}} = \frac{h\nu}{\mathrm{e}^{h\nu/kT} - 1} \,.\end{aligned} \tag{3.14b}$$

The spectral energy density $w_\nu(\nu)$ of the blackbody radiation is then

$$w_\nu(\nu, T) = n(\nu)\bar{w}_\nu(\nu, T) \,. \tag{3.14c}$$

Inserting (3.10b) and (3.14b) yields the famous Planck's radiation law

$$w_\nu(\nu)\mathrm{d}\nu = \frac{8\pi h\nu^3}{c^3} \frac{\mathrm{d}\nu}{\mathrm{e}^{h\nu/kT} - 1} \tag{3.15}$$

for the spectral energy density $w_\nu$ [J m$^{-3}$ s], i.e., the energy per unit volume and unit frequency interval $\mathrm{d}\nu = 1\,\text{s}^{-1}$.

The spectral radiation energy, emitted within the frequency interval $\mathrm{d}\nu$ by a surface element $\mathrm{d}A$ of a blackbody into the solid angle $\mathrm{d}\Omega$ is then

$$\begin{aligned}S_\nu^* \mathrm{d}\nu\, \mathrm{d}\Omega\, \mathrm{d}A &= \frac{c}{4\pi} w_\nu \mathrm{d}\nu\, \mathrm{d}\Omega \\ &= \frac{2h\nu^3}{c^2} \frac{\mathrm{d}\nu\, \mathrm{d}\Omega}{\mathrm{e}^{h\nu/kT - 1}} \mathrm{d}A \,,\end{aligned} \tag{3.16}$$

which is in complete agreement with experimental results!

For $h\nu \ll kT$ the denominator in (3.16) can be approximated by $h\nu/kT$ because $\mathrm{e}^x \approx 1 + x$ for $x \ll 1$. Then we obtain

$$S_\nu^*(\nu) = \frac{2\nu^2}{c^2} kT \Rightarrow w(\nu) = \frac{8\pi\nu^2}{c^3} kT \,. \tag{3.17}$$

This is identical to the Rayleigh–Jeans formula (3.13), which turns out to be the asymptotic case of the general Planck distribution for the long wavelength region $(h\nu/kT \ll 1)$.

Using the relation $\lambda = c/\nu$, Planck's law (3.15) can also be written in terms of the wavelength $\lambda$. Note that $\mathrm{d}\lambda/\mathrm{d}\nu = -c/\nu^2 \Rightarrow \mathrm{d}\lambda = -(c/\nu^2)\mathrm{d}\nu$. Defining $w_\lambda(\lambda)\mathrm{d}\lambda$ as the spectral energy per wavelength interval $\mathrm{d}\lambda$ gives

$$w_\lambda(\lambda)\mathrm{d}\lambda = \frac{8\pi hc}{\lambda^5} \frac{\mathrm{d}\lambda}{\mathrm{e}^{hc/(\lambda kT)} - 1} \,. \tag{3.18}$$

From $\mathrm{d}w_\lambda/\mathrm{d}\lambda = 0$ one obtains (see Problem 3.3) for the wavelength $\lambda_\mathrm{m}$, at which $w_\lambda(\lambda)$ has its maximum

$$\lambda_\mathrm{m} = \frac{2.88 \times 10^{-3}[\text{m}]}{T[\text{K}]}$$

$$\Rightarrow \lambda_\mathrm{m} T = 2.88 \times 10^{-3}[\text{mK}] = \text{const} \,. \tag{3.18a}$$

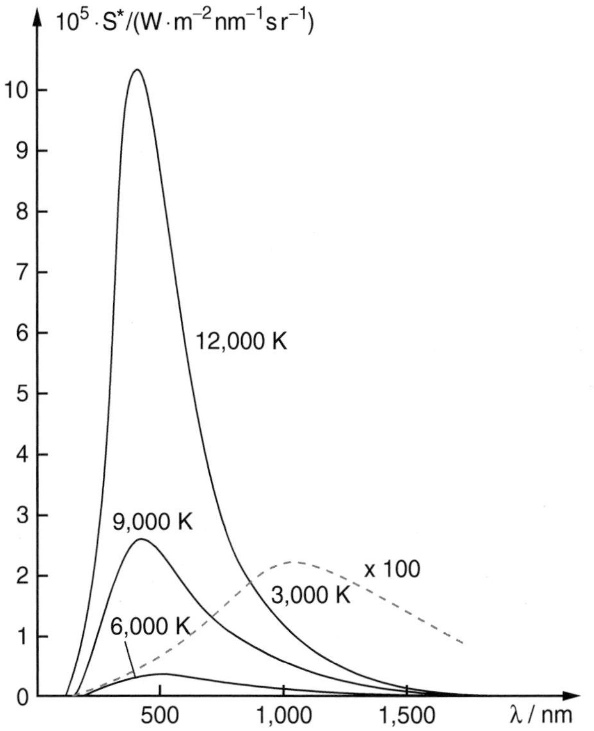


**Fig. 3.6.** Spectral distribution of the radiation density $S^*(\lambda)$ of a blackbody within the spectral interval $\Delta\lambda = 1$ nm. The ordinate for the curve for 3000 K is increased 100-fold

To illustrate, Fig. 3.6 shows the wavelength-dependent distribution of the spectral radiation density of a blackbody at four different temperatures and Fig. 3.7 demonstrates the difference between the Rayleigh–Jeans and Planck distribution.

## EXAMPLE

The sun can be regarded, to a good approximation, as a blackbody with a surface temperature of 5800 K. Its spectral radiation density emitted at $\lambda = 500\,\mathrm{nm}$ ($\nu = 6 \times 10^{14}\,\mathrm{s^{-1}}$) into the solid angle $\mathrm{d}\Omega = 1$ Sterad within a wavelength interval $\mathrm{d}\lambda = 1\,\mathrm{nm}$ ($\Delta\nu = 1.2 \times 10^{12}\,\mathrm{s^{-1}}$) is, according to (3.16),

$$S^*_\nu \Delta\nu = 4.5 \times 10^4 \frac{\mathrm{W}}{\mathrm{m^2\,Sterad}} .$$

Integration over all wavelengths gives the total radiation density

$$S^* = 1 \times 10^7\,\mathrm{W/(m^2\,Sterad)} .$$

Integration over the suns surface yields the total power

$$P_\odot = 4\pi R_\odot^2 10^7\,\mathrm{W} = 3.82 \times 10^{26}\,\mathrm{W}$$

radiated by the sun into all directions.

The earth, seen from the center of the sun, at a distance $r = 1.5 \times 10^{11}$ m covers a solid angle

$$\Delta\Omega = \frac{\pi R_E^2}{4\pi r^2} = \frac{R_\mathrm{E}^2/4}{(1.5 \times 10^{11})^2} = 2.5 \times 10^{-7}\,\mathrm{Sterad} .$$

It therefore receives from the sun the total radiation power

$$\mathrm{d}W/\mathrm{d}t = P_\odot \Delta\Omega/4\pi = 7.6 \times 10^{18}\,\mathrm{W} .$$

About 37% of this irradiation falls into the visible range between $\nu_1 = 4 \times 10^{14}$ Hz ($\lambda = 750\,\mathrm{nm}$) and $\nu_2 = 7 \times 10^{14}$ Hz ($\lambda = 430\,\mathrm{nm}$).

The intensity $I = P_\mathrm{E}/A$ received per $\mathrm{m^2}$ of the earth surface at vertical incidence is the solar constant

$$\mathrm{SC} = 1.36 \times 10^3\,\mathrm{W/m^2} .$$

About $37\% \mathrel{\hat{=}} 500\,\mathrm{W/m^2}$ are in the visible range.

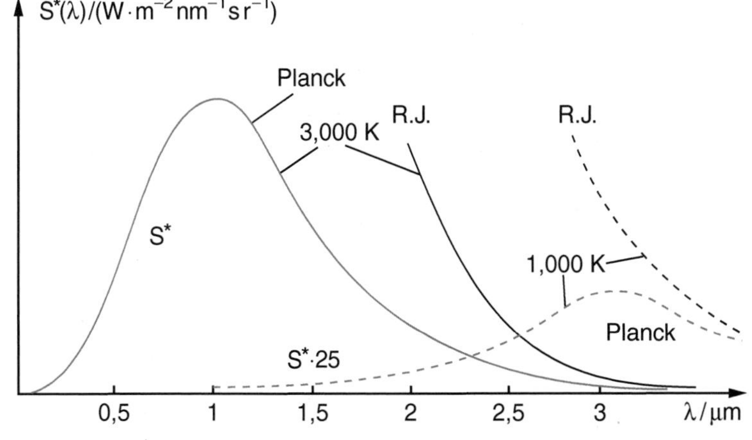


**Fig. 3.7.** Comparison between Planck's spectral distribution and the Rayleigh–Jeans spectral distribution of the blackbody radiation at two different temperatures. Note, that $S^*(T = 1000\,\mathrm{K})$ has been enlarged by a factor of 25

Less than half of this radiation power heats the surface of the earth. Part of it is reflected by the atmosphere and the earth's surface and part of it is absorbed in the atmosphere.

### 3.1.4 Wien's Law

The maximum of the intensity distribution $S^*_\nu(\nu)$ (3.16) is obtained by setting the derivative $\mathrm{d}S^*_\nu/\mathrm{d}\nu = 0$. This is more tedious than looking for $\mathrm{d}(\ln S^*_\nu)\mathrm{d}\nu = 0$, which, of course, gives the same frequency $\nu_\mathrm{m}$ for the intensity maximum.

The result is (see Problem 3.3)

$$\nu_\mathrm{m} = \frac{2.82}{h}kT \Rightarrow \frac{\nu_\mathrm{m}}{T} = 5.87 \times 10^{10}\,\mathrm{s}^{-1}\,\mathrm{K}^{-1}\,. \tag{3.19}$$

The frequency $\nu_\mathrm{m}$ of the maximum intensity therefore increases linearly with the temperature $T$, which means that the ratio $\nu_\mathrm{m}/T$ is a constant.

**Note:**

$\lambda_\mathrm{m}$ does not equal $\nu_\mathrm{m}/c$, because $w_\lambda(\lambda)$ is defined for a spectral interval $\mathrm{d}\lambda = 1\,\mathrm{nm}$, while $w_\nu(\nu)$ is defined for $\mathrm{d}\nu = 1\,\mathrm{s}^{-1}$. Because of the nonlinear relation $\mathrm{d}\lambda = -(c/\nu^2)\mathrm{d}\nu$ the variation of $w_\lambda(\lambda)$ with decreasing $\lambda$ differs from that of $w_\nu(\nu)$ with increasing $\nu$.

The product $\lambda_\mathrm{m}T = 2.88 \times 10^{-3}\,\mathrm{m\,K}$ is a constant (see Fig. 3.8). Wien's law can therefore be written as

$$\boxed{\lambda_\mathrm{m}T = \text{const} \quad \text{or} \quad \nu_\mathrm{m}/T = \text{const}\,,} \tag{3.20}$$

where the two constants are different.

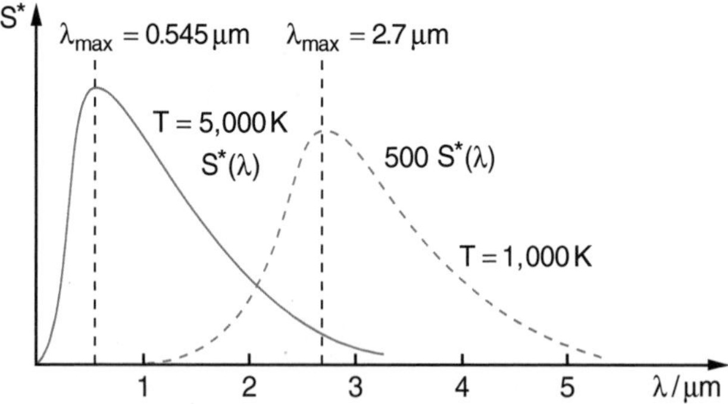


**Fig. 3.8.** Illustration of Wien's law. The ordinate of the 1000 K distribution has been expanded by a factor 500

The experimental results are in perfect agreement with Wien's law, which is derived from Planck's radiation law.

### 3.1.5 Stefan–Boltzmann's Radiation Law

The total energy density of the blackbody radiation, integrated over all frequencies $\nu$ is

$$W(T) = \int\limits_{\nu=0}^{\infty} w_\nu(\nu, T)\mathrm{d}\nu = \frac{8\pi h}{c^3}\int \frac{\nu^3\mathrm{d}\nu}{\mathrm{e}^{h\nu/kT-1}}\,. \tag{3.21}$$

We use the abbreviation $x = h\nu/kT$ and expand

$$\frac{1}{\mathrm{e}^x - 1} = \frac{\mathrm{e}^{-x}}{1-\mathrm{e}^{-x}} = \sum_{n=1}^{\infty} \mathrm{e}^{-nx}\,.$$

Inserting this expansion into (3.21) yields, with

$$\nu = (kT/h)x \Rightarrow \mathrm{d}\nu = \frac{kT}{h}\mathrm{d}x\,,$$

$$w(T) = \frac{8\pi h}{c^3}\left(\frac{kT}{h}\right)^4 \sum_{n=1}^{\infty}\int\limits_0^{\infty} x^3\mathrm{e}^{-nx}\mathrm{d}x\,. \tag{3.22a}$$

The integration of each member of the sum gives the final result for the energy density

$$w(T) = aT^4 \quad \text{with} \quad a = \frac{4\pi^5k^4}{15h^3c^3}\,. \tag{3.22b}$$

The radiation power $S^*$ emitted from a surface element $A = 1\,\mathrm{m}^2$ into the solid angle $\mathrm{d}\Omega = 1$ Sterad is, with $S^* = (c/4\pi)W$,

$$\boxed{S^*(T) = \frac{\pi^4k^4}{15h^3c^2}T^4\,.} \tag{3.23a}$$

Into the solid angle $\mathrm{d}\Omega = 2\pi$ the radiation power of $1\,\mathrm{m}^2$ surface of a blackbody is then

$$\boxed{\frac{\mathrm{d}W}{\mathrm{d}t} = 2\pi S^*(T) = \sigma T^4} \tag{3.23b}$$

$$\text{with} \quad \sigma = \frac{c}{2}a = \frac{2\pi^5k^4}{15h^3c^2} = 5.67 \times 10^{-8}\,\mathrm{W\,m^{-2}\,K^{-4}}\,.$$

The constant $\sigma$ is named the *Stefan–Boltzmann constant*.

Planck's law, Wien's law and the Stefan–Boltzmann law are all in complete agreement with experimental findings. This strongly corroborates Planck's hypothesis of the quantized radiation field, which postulates that the energy density $w_\nu(\nu, T)$ is not a continuous function of $T$ but a discontinuous step function with smallest steps $h \cdot \nu$. In most cases these steps are so small that they are not directly noticed. We will, however, soon discuss experiments where they can be directly detected.

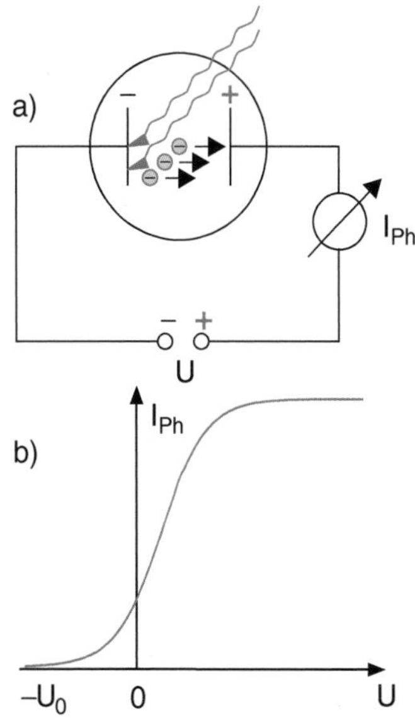


**Fig. 3.10.** (**a**) Phototube for measuring the photo-induced electric current $I_{ph}$ as a function of the applied voltage; (**b**) Photo-current $I_{ph}(U)$

### 3.1.6 Photoelectric Effect

When a negatively charged isolated metal plate is irradiated by ultraviolet light (Fig. 3.9) the electric charge on the plate decreases, as was found in 1887 by *Heinrich Hertz* (1857–1894) and later in 1895 through more detailed experiments by *Wilhelm Hallwachs* (1859–1922). This means that electrons must have left the plate.

This light-induced electron emission can be quantitatively measured with the device shown in Fig. 3.10. The irradiated plate within an evacuated glass bulb serves as cathode, which is opposed by a similar plate with positive voltage that forms the anode of an electric diode. The photocurrent $I_{ph}(U)$ is measured as a function of the voltage $U$ between cathode and anode. The measurements show that $I_{ph}(U)$ starts already at slightly negative voltages $U_0$ (i.e., the anode has a negative voltage $U_0$ against the cathode), rises with increasing voltage until it reaches a saturation value that depends on the radiative power incident on the cathode (Fig. 3.10b). When electrons can reach the anode with a negative bias voltage $-U \geq -U_0$ they must have been emitted from the cathode with a kinetic energy

$$E_{kin} = \frac{m_e}{2} v^2 \geq eU \; .$$

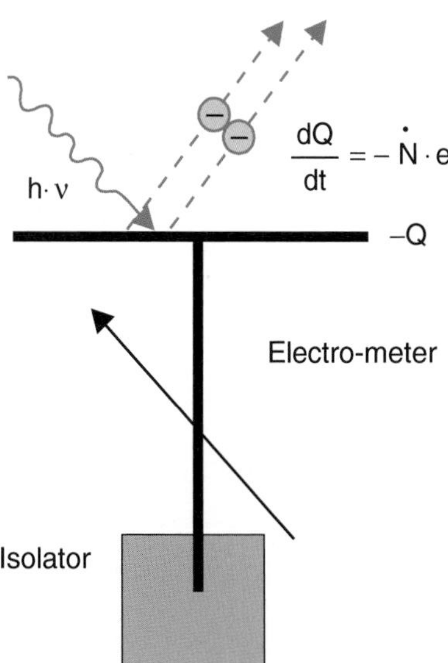


**Fig. 3.9.** Experimental arrangement of Hallwachs for the demonstration of the photoelectric effect

In 1902 Lennard obtained the following results after careful measurements:

- The kinetic energy $(m/2)v^2$ of the photoelectrons is dependent solely on the wavelength $\lambda$ of the incident light, not on its intensity!
- The number of ejected photoelectrons is proportional to the light intensity.
- There is no measurable time delay between irradiation and electron ejection.

Einstein was able to explain Lennard's experimental results in 1905 using the model of light quanta (photons). Each absorbed photon transfers its energy $h\nu$ completely to an electron inside the metal, which is bound to the metal by attractive forces and needs a minimum energy $W_a$ (work-function) to leave the metal.

The maximum kinetic energy of the photo electron is then

$$E_{\mathrm{kin}}^{\max} = h\nu - W_\mathrm{a} \,. \tag{3.24}$$

In the experiment this maximum kinetic energy

$$E_{\mathrm{kin}}^{\max} = eU_0 \tag{3.25}$$

can be determined from the bias voltage $-U_0$ at which the photocurrent starts. The relation (3.24) can then be written as

$$eU_0 = h\nu - W_\mathrm{a} \,. \tag{3.26}$$

Plotting $eU_0$ against the photon energy $h\nu = hc/\lambda$, gives the straight line shown in Fig. 3.11. From its slope $\tan\alpha = eU_0/(h\nu)$, Planck's constant $h$ can be determined. The intersection with the vertical axis $h\nu$ yields the work function $W_\mathrm{a}$ of the cathode material.

In the classical model of an electromagnetic wave the radiation power $P_\mathrm{L} = I_\mathrm{L} A$ incident onto the area $A$ of the metal should be equally shared by all $N\Delta V$ electrons with a number density $N$ within the irradiated volume $V = A\Delta z$ ($\Delta z \approx \lambda$ is the penetration depth of the light wave with wavelength $\lambda$ into the metal).

Each electron would then receive, on average, the energy

$$\bar{w} = \frac{P_\mathrm{L}\Delta t}{NA\lambda} \,. \tag{3.27}$$

Since $\bar{w}$ has to be larger than the workfunction $W_\mathrm{a}$, we conclude from (3.27) that the irradiation time must be at least $\Delta t \geq W_\mathrm{a} N A\lambda / P_\mathrm{L}$.

## EXAMPLE

A zinc plate (electron density $N = 10^{23}\,/\mathrm{cm}^3$, workfunction $W_\mathrm{a} \approx 4\,\mathrm{eV}$) is irradiated at a distance of 1 m from the light source (Hg arc lamp) that emits (through a spectral fitter) 1 W radiation power at $\lambda = 250\,\mathrm{nm}$. The power incident onto 1 cm² of the zinc plate is then the intensity

$$I_\mathrm{c} = \frac{1\,\mathrm{W}}{4\pi R^2} = 8 \times 10^{-6}\,\mathrm{W/cm^2} \,.$$

This power is shared by $N\Delta V = \lambda \cdot 10^{23} = 2.5 \times 10^{19}$ electrons. The average radiation power received by one electron is $P_\mathrm{el} = 3 \times 10^{-24}\,\mathrm{W} = 2 \times 10^{-5}\,\mathrm{eV/s}$. This means it would take $\Delta t = W_\mathrm{a}/P_\mathrm{el} = 2 \times 10^5\,\mathrm{s}$ before an electron had accumulated sufficient energy to escape from the metal. This strongly contradicts the experimental results.

Einstein's model could explain the experimental results completely. In 1921 he received the Nobel Prize in Physics for his photoeffect theory (not for his theory of relativity!).

There are numerous descriptions of detailed experiments in the literature that unambiguously corroborate Einstein's theory of the photo effect [3.5]. One example is the experiment performed by Joffe and Dobronrawov in 1925 [3.6]. They measured the changes $\Delta Q$ of the charge $Q$ of a small bismuth particle balanced between the plates of a capacitor (Fig. 3.12) during illumination of the particle with low intensity X-rays. Every change $\Delta Q$, due to the loss of an electron results in a perturbation of the balance between gravitational and electric force and leads to a movement of the particle, which is observed through a microscope.

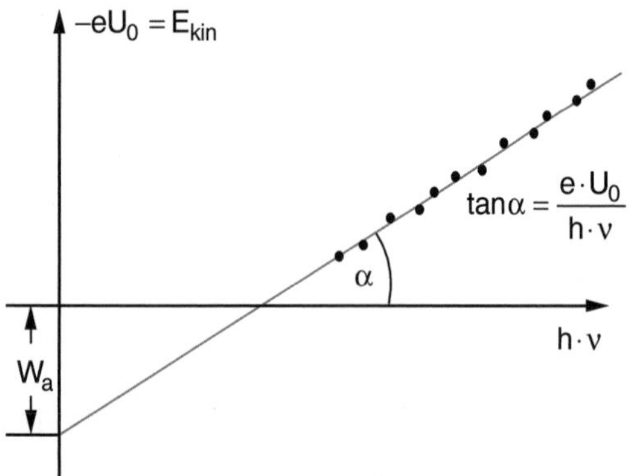


**Fig. 3.11.** Plot of the maximum bias voltage $U_0$ as a function of the frequency $\nu$ of the incident light

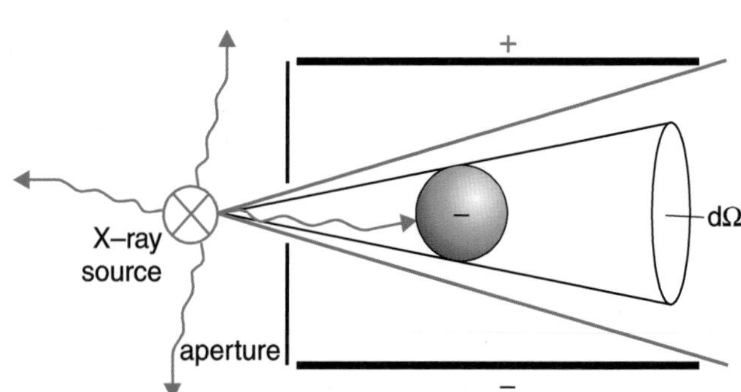


**Fig. 3.12.** Experiment of Joffé and Dobronrawov for confirming the photon model of Planck

The X-ray source emitted $\dot{N} = 10^3$ X-ray quanta with $h\nu = 10^4$ eV per second into all directions, which corresponds to a radiation power of $10^{-12}$ W. On the average a charge change $\Delta Q$ of the particle was observed every 30 min. The quantum hypothesis explains this experimental finding as follows.

The number of X-ray photons $h\nu$, hitting the bismuth particle within the time interval $\Delta t$ is $Z = \dot{N}\Delta t \Delta\Omega/4\pi$, where $\Delta\Omega$ is the solid angle accepted by the particle. With $\Delta\Omega = 6 \times 10^{-6}$ Sterad and $\Delta t = 1800\,s$ the estimated number becomes $Z = 1$ in accordance with the experiment.

The classical model describes X-rays as a spherical wave, propagating from a point-like source in all directions. Like in the quantum model, the fraction $\mathrm{d}P$ of the radiation power $P$ absorbed by the particle is $\mathrm{d}P = P\mathrm{d}\Omega/4\pi$. However, the absorbed power $\mathrm{d}P$ should be shared by all $N \approx 10^{12}$ electrons of the bismuth particle. In order to emit an electron after 30 min, all electrons must simultaneously transfer their energy share to the same electron, which is highly improbable. This example shows that the deficiency of the wave model stems from the fact that it does not concentrate the wave energy onto a single electron but spreads it out to a larger volume determined by the irradiated area.

**Note:**

However, the discrepancy between the wave model of light and the experimental results can be removed if not only the radiation but also the electrons are described as waves (see Sect. 3.2).

### 3.1.7 Compton Effect

The particle character of light quanta became particularly clear by the results of an experiment performed in 1922 by *Arthur Holly Compton* (1892–1962). When irradiating material by X-rays with wavelength $\lambda_0$, Compton found that the scattered radiation did not only contain the wavelengths $\lambda_0$ expected for elastic scattering, but also larger wavelength $\lambda_s > \lambda_0$ (Fig. 3.13). The intensity distribution $I(\lambda_s)$ of this scattered radiation is strongly dependent on the scattering angle $\varphi$, but only slightly on the kind of material used.

The quantum model describes the Compton effect as an elastic collision between a photon with energy $h\nu$ and momentum $\boldsymbol{p} = h\boldsymbol{k}$ and a loosely bound electron of the scattering material (Fig. 3.13b). If the binding energy $E_b \ll h\nu$ is small compared to the photon energy $h\nu$ we can neglect it and regard the electron as a free particle. In order to simplify the following calculation we further assume that the electron is at rest before the collision, although this assumption is not essential for the result.

For the elastic collision

$$h\nu_0 + \mathrm{e}^-\,(E_{\mathrm{kin}} = 0) \rightarrow h\nu_s + \mathrm{e}^-\left(E'_{\mathrm{kin}} > 0\right) \tag{3.28}$$

total energy and total momentum have to be conserved. Since the photon travels with the velocity of light and the ejected electron might also have a large velocity, we must use the relativistic description. We choose our coordinate system in such a way that the incident photon travels into the $x$ direction, and the $xy$-plane is the scattering plane (Fig. 3.13c). The energy conservation then reads with $\beta = v/c$ as:

$$\begin{aligned} h\nu_0 &= h\nu_s + E_{\mathrm{kin}}(\mathrm{e}^-) \\ \text{with}\quad E_{\mathrm{kin}} &= \frac{m_0c^2}{\sqrt{1-\beta^2}} - m_0c^2 \end{aligned} \tag{3.29}$$

and momentum conservation demands

$$\boldsymbol{p}_0 = \hbar\boldsymbol{k}_0 = \hbar\boldsymbol{k}_s + \boldsymbol{p}_e \quad \text{with} \quad \boldsymbol{p}_e = \frac{m_0\boldsymbol{v}}{\sqrt{1-\beta^2}}\,. \tag{3.30}$$

Solving (3.30) for $\boldsymbol{p}_e$ yields for $|\boldsymbol{p}_e|^2$:

$$\begin{aligned} \frac{m_0^2v^2}{1-\beta^2} &= \hbar^2\left(\vec{k}_0 - \vec{k}_s\right)^2 \\ &= \frac{h^2}{c^2}\left(\nu_0^2 + \nu_s^2 - 2\nu_0\nu_s\cos\varphi\right)\,, \end{aligned} \tag{3.31}$$

where $\varphi$ is the angle between $\boldsymbol{k}_0$ and $\boldsymbol{k}_s$ and the relation $|\boldsymbol{p}| = \hbar k = h/\lambda = h\nu/c$ has been used. Squaring (3.29) yields

$$\left(h\nu_0 - h\nu_s + m_0c^2\right)^2 = \frac{m_0^2c^4}{1-\beta^2}$$

which gives with $c^2 = v^2/\beta^2$, after rearranging the terms,

$$\frac{m_0^2v^2}{1-\beta^2} = \frac{h^2}{c^2}(\nu_0 - \nu_s)^2 + 2h(\nu_0 - \nu_s)\,m_0\,. \tag{3.32}$$

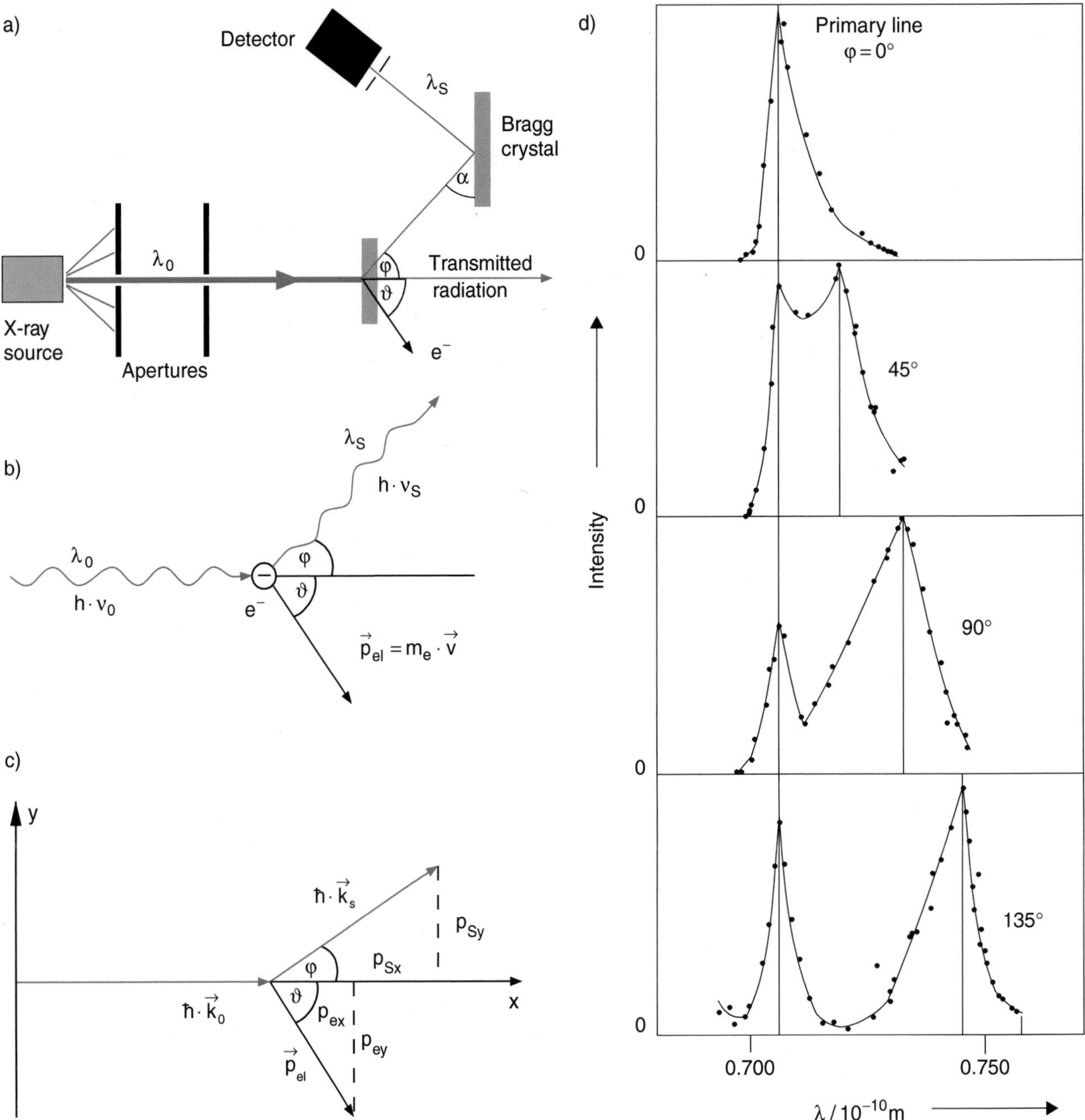


**Fig. 3.13a–d.** The Compton effect. (**a**) Experimental setup (**b**) Schematic diagram (**c**) Momentum diagram (**d**) Wavelength distribution $I_S(\lambda)$ of scattered radiation intensity for different scattering angles $\varphi$

A comparison between (3.31) and (3.32) yields

$$\nu_0 - \nu_s = \frac{h}{m_0}\frac{\nu_0 \nu_s}{c^2}(1 - \cos\varphi) .$$

Inserting $\lambda = c/\nu$ and $(1 - \cos\varphi) = 2\sin^2(\varphi/2)$ we obtain Compton's scattering formula

$$\boxed{\lambda_s = \lambda_0 + 2\lambda_c \sin^2(\varphi/2) ,} \tag{3.33}$$

where the constant

$$\lambda_c = \frac{h}{m_0 c} = 2.4262 \times 10^{-12}\,\mathrm{m} \tag{3.34a}$$

is the ***Compton wavelength of the electron***, which represents the wavelength change $\Delta\lambda = \lambda_s - \lambda_0$ of the X-rays for a scattering angle $\varphi = 90°$. Multiplying (3.34a) by $c$ yields

$$\frac{\lambda_c}{\lambda_0} = \frac{h\nu_0}{m_0 c^2} . \tag{3.34b}$$

This shows that the ratio of Compton wavelength $\lambda_c$ and incident wavelength $\lambda_0$ equals the ratio of incident photon energy $h\nu_0$ to the rest energy $m_0c^2$ of the electron.

The experimental results are in excellent agreement with this theoretical result and again corroborate the particle model of electromagnetic radiation.

From measurements of $\lambda_0$, $\lambda_s$ and $\varphi$ the Compton wavelength $\lambda_c$ and Planck's constant $h$ can be determined, since the electron mass $m_0$ is known (see Sect. 2.6).

### 3.1.8 Properties of Photons

The experiments described in the previous sections have demonstrated the particle character of electromagnetic radiation. Each electromagnetic field with frequency $\nu$ consists of energy quanta $h\nu$, called ***photons***. A field with energy density $w_{em}$ contains

$$n = \frac{w_{em}}{h\nu} \tag{3.35}$$

photons per m$^3$. In an electromagnetic wave with field amplitude $E$ and intensity $I = c\varepsilon_0 E^2$

$$\dot{N} = \frac{I}{h\nu} = nc \tag{3.36}$$

photons pass per second through an illuminated unit area $A = 1\,\text{m}^2$, perpendicular to the wave vector $\boldsymbol{k}$. The Compton effect shows that the momentum

$$\boldsymbol{p} = \hbar\boldsymbol{k} \quad \text{with} \quad |\boldsymbol{p}| = \hbar\frac{2\pi}{\lambda} = h\nu/c \tag{3.37}$$

can be attributed to a photon $h\nu$. The total momentum per unit volume inside a plane electromagnetic wave with energy density $w_{em}$ is then

$$\boldsymbol{\pi}_{em} = n\hbar\boldsymbol{k} \quad \text{and} \quad |\boldsymbol{\pi}_{em}| = nh\nu/c = w_{em}/c . \tag{3.38a}$$

When a photon is absorbed by an atom (see Sect. 7.3) the angular momentum of the atom changes by $1\hbar = h/2\pi$. Since the total angular momentum of the system

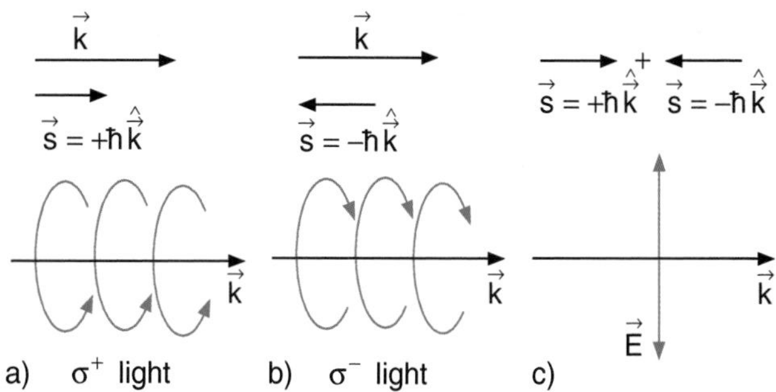


**Fig. 3.14.** Photon model of polarized radiation. (**a**) $\sigma^+$ = left circular (**b**) $\sigma^-$ = right circular (**c**) $\pi$ = linear polarization

"atom plus photon" must be conserved, the photon has to carry the angular momentum $\pm 1\hbar$, which is also called the spin of the photon

$$\boldsymbol{s}_{\text{photon}} = \pm\hbar\frac{\boldsymbol{k}}{k} \tag{3.38b}$$

and is independent of its energy $h\nu$. If left circularly-polarized light ($\sigma^+$-polarization) propagating in the $z$ direction is absorbed by atoms, the $z$ component of their angular momentum $J_z$ is changed by $\Delta J_z = +\hbar$, if $\sigma^-$-light (right circular polarization) is absorbed $J_z$ changes by $\Delta J_z = -\hbar$. We therefore must conclude that $\sigma^+$ light consists of photons with spins $\boldsymbol{s}_{ph} = +\hbar\boldsymbol{k}/|\boldsymbol{k}|$ pointing into the propagation direction, while $\sigma^-$ light represents photons with $\boldsymbol{s}_{ph} = -\hbar\boldsymbol{k}/|\boldsymbol{k}|$ (Fig. 3.14).

**Note:**

In older literature $\sigma^+$ light is called left circularly-polarized, because the polarization vector rotates counterclockwise for an observer looking against the direction of light propagation. Similary $\sigma^-$ light was named "right circular polarized".

Since linearly polarized light ($\pi$-polarization) can be regarded as a superposition of $\sigma^+ + \sigma^-$ components, it should consist of equal numbers of photons with $\boldsymbol{s}_{ph} = +\hbar\boldsymbol{k}/|\boldsymbol{k}|$ and $\boldsymbol{s}_{ph} = -\hbar\boldsymbol{k}/|\boldsymbol{k}|$. Indeed, the absorption of $\pi$-polarized light does not change the angular momentum component $J_z$ of the atoms.

According to the relation $E = mc^2$ between mass $m$ and total energy $E$ of a particle we can formally attribute the mass

$$m = \frac{E}{c^2} = \frac{h\nu}{c^2} \tag{3.39}$$

to a photon with energy $h\nu$. From the relativistic energy relation

$$E = \sqrt{p^2c^2 + m_0^2c^4}$$

between energy $E$ and momentum $\boldsymbol{p}$ of a particle with rest mass $m_0$ it follows for the photon with $E = h\nu$ and $p = E/c$ that its rest mass $m_0$ must be zero. This result can be also obtained from the relativistic relation

$$m = \frac{m_0}{\sqrt{1 - v^2/c^2}} \Rightarrow m_0 = m\sqrt{1 - v^2/c^2}\ .$$

Since $v_{\text{photon}} = c$, we obtain for a finite mass $m$ the result $m_0 = 0$.

**Note:**

Photons at rest do not exist! It is therefore somehow artificial to speak of the photon rest mass. It is more logical to use the relation $m = h \cdot \nu/c^2$ for the definition of the photon mass.

### 3.1.9 Photons in Gravitational Fields

When we attribute the mass $m = h\nu/c^2$ to a photon, it must experience a gravitational force. If a photon is send from a point $\boldsymbol{r}_1$ with gravitation potential $\phi_G(\boldsymbol{r}_1)$ to a point $\boldsymbol{r}_2$ with $\phi_G(\boldsymbol{r}_2)$ the potential energy changes by

$$\Delta E_{\text{pot}} = m\Delta\phi_G = \frac{h\nu}{c^2}\left(\phi_G(\boldsymbol{r}_1) - \phi_G(\boldsymbol{r}_2)\right)\ . \tag{3.40}$$

Because of energy conservation the photon energy $h\nu$ must change by this amount. The frequency $\nu_1$ of a photon is therefore changing to

$$\nu_2 = \nu_1\left(1 - \frac{\Delta\phi_G}{c^2}\right) \Rightarrow \frac{\Delta\nu}{\nu} = \frac{\Delta\phi_G}{c^2} \tag{3.41a}$$

when it travels from the point $\boldsymbol{r}_1$ to the point $\boldsymbol{r}_2$ in a gravitational field.

EXAMPLES

1. A light source at the earth's surface at $H_0 = 0$ emits light into the vertical direction (Fig. 3.15). A detector at the height $H$ measures the frequency $\nu_2 = \nu\left(1 - \frac{gH}{c^2}\right) \Rightarrow \frac{\Delta\nu}{\nu} = \frac{gH}{c^2}$. With $H = 20\,\text{m}$, $g = 9.81\,\text{m/s}^2$ we obtain $\Delta\nu/\nu = 2.5 \times 10^{-15}$. This gravitational redshift was indeed measured by Pound and Rebka [3.7] using the Mößbauer-effect as a sensitive frequency detector.

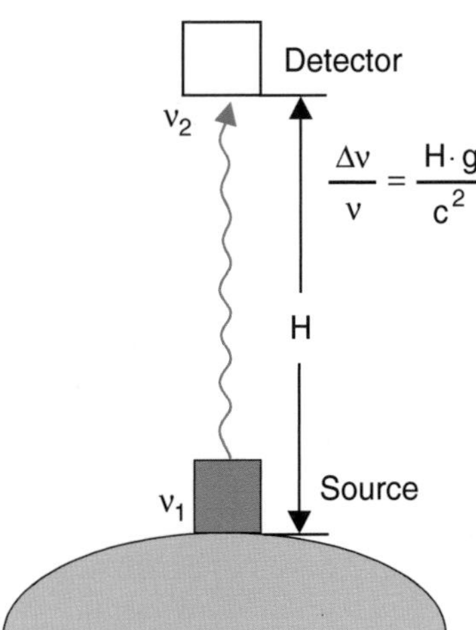


**Fig. 3.15.** Demonstration experiment by Pound and Rebka for the proof of photon redshifts in the gravitational field of the earth

2. Light with frequency $\nu_1$ emitted from the surface of the sun, is received on earth with the smaller frequency

$$\nu_2 = \nu_1\left(1 - G\frac{M_\odot}{R_\odot c^2}\right)\ , \tag{3.41b}$$

where $M_\odot$ is the mass of the sun, $R_\odot$ its radius and $G$ the gravitational constant. The frequency increase of photons entering the gravitational field of the earth is very small compared to this decrease and has been neglected. Inserting numerical values yields $\Delta\nu/\nu = 5 \times 10^{-7}$. This frequency shift can be readily measured with modern interferometers [3.8].

3. For $M = R \cdot c^2/G$ we obtain $\nu_2 = 0$. This represents a "Black Hole" where no light from locations $R < R_s = M \cdot G/c^2$ can escape. The radius $R_s$ is called Schwarzschild Radius.

Light suffers a red shift when propagating away from an attracting mass. This red shift corresponds to an increase $\Delta W_{\text{pot}}$ in potential energy of a particle with mass $m = h\nu/c^2$. If $\Delta W_{\text{pot}} > h\nu_0$ the light cannot leave the attracting mass. This situation is found for black holes.

### 3.1.10 Wave and Particle Aspects of Light

The discussion in the last section has shown that the particle properties of photons, such as mass $m = h\nu/c^2$, energy $h\nu$ and momentum $\boldsymbol{p} = \hbar\boldsymbol{k}$; $|\boldsymbol{p}| = \hbar/\lambda$ can be only defined by using the wave characteristics frequency $\nu$ and wavelength $\lambda = c/v$. This itself illustrates the tight connection between the particle model and the wave model of electromagnetic fields. As an example we will consider the relation between the intensity $I$ of an electromagnetic wave and the corresponding photon flux density. When $\dot{N} = \mathrm{d}N/\mathrm{d}t$ photons $h\nu$ pass per second with velocity $c$ through a unit area perpendicular to the propagation direction, the intensity $I$ (energy per m$^2$ per s) is

$$I = \dot{N}h\nu . \tag{3.42a}$$

In the wave model the intensity is

$$I = \varepsilon_0 c E^2 . \tag{3.42b}$$

The comparison of (3.42a) and (3.42b) gives the relation between the electric field amplitude $E$ of the electromagnetic wave and the photon flux $\dot{N}$

$$E = \sqrt{\frac{\dot{N}h\nu}{\varepsilon_0 \cdot c}} \tag{3.42c}$$

which shows that the field amplitude is proportional to the square root of the photon flux $\dot{N}$.

The photon structure of light can be demonstrated by various experiments besides those already discussed in the previous sections. A famous example is the experiment by Taylor (Fig. 3.16) where many equal detectors $\mathrm{D}_i$ are arranged on a circle with radius $R$ around the light source S [3.12]. The light, isotropically emitted by the source into all directions, is described in the classical wave model as a spherical wave with field amplitude

$$E = \frac{A}{r}\mathrm{e}^{\mathrm{i}(kr-\omega t)} . \tag{3.43a}$$

All detectors with sensitive area $S$ at a distance $R$ from the centrum $r = 0$ receive, per s, the same radiation power

$$\frac{\mathrm{d}W}{\mathrm{d}t} = c\varepsilon_0 \frac{A^2}{R^2} S . \tag{3.43b}$$

This is indeed observed for sufficiently large light intensities.

If, however, the emitted power of the light source is diminished so much that $\mathrm{d}W/\mathrm{d}t \ll h\nu/\tau$, where $\tau$ is the time interval that can just be resolved by the detectors then at most one photon can reach a detector per resolved time interval. Not all detectors give simultaneous signals but only one or none of them during the time $\tau$. The output signals of the detectors $\mathrm{D}_i$ are statistically distributed in time. Averaged over a time interval $\Delta t \gg \tau$ all detectors again show the same total number of counts.

This means that in this experiment the quantum structure of light becomes apparent at low light intensities. The radiation energy of the source is in this case *not* simultaneously emitted into all directions, but at one time a photon flies only in one direction and at another time into another direction.

The total number of photons received by each detector over a time interval $\Delta t$ shows a Poisson distribution

$$N = \bar{N}\mathrm{e}^{-(N-\bar{N})/\bar{N}} \tag{3.44}$$

with a standard deviation $\sigma = \sqrt{\bar{N}}$. The probability $P$ that any one of the detectors has counted $N = \bar{N} \pm 3\sqrt{\bar{N}}$ photons is $P = 0.997$.

This illustrates that the classical wave description of light represents the asymptotic case of large photon fluxes. The relative fluctuations of the photon flux

$$\frac{\Delta\dot{N}}{\dot{N}} \propto \frac{1}{\sqrt{\dot{N}}} \tag{3.45}$$

decreases with increasing photon flux $\dot{N}$ and the quantum structure of electromagnetic waves (i.e., the steps

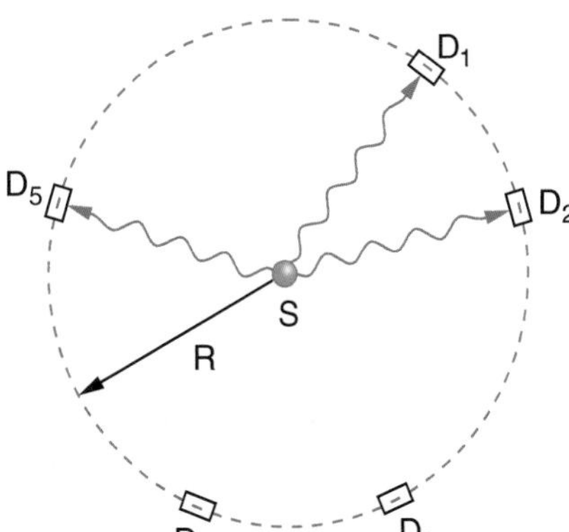


**Fig. 3.16.** Taylor's experiment to prove the photon structure of a light wave

$h\nu$ in the energy density) becomes significant only for small photon flux densities $\dot{N}$.

An essential characteristic of a classical particle is its localization within a small spatial volume, which is the volume of the particle, in contrast to a wave that is spread out over a larger volume. How this apparent contradiction of particle and wave description can be solved will be discussed in Sect. 3.3.

The question of whether light can be regarded as a wave or a stream of particles was controversially and vehemently discussed for a long time between *Isaac Newton* (1642–1727), who represented the particle model, and *Christian Huygens* (1629–1695) who believed in the wave model [3.2]. Both scientists met in London in 1689 and discussed their controversial views but could not come to an agreement.

Experiments by Huygens, *Thomas Young* (1773–1829), and many other researchers, looking at diffraction and interference of light decided the quarrel in favor of the wave model, since at that time it was generally accepted that particles will not show any interference effects. It is very instructive to discuss the diffraction and the interference of light by a double slit (Young's double slit) experiments (Fig. 3.17). The detector can be a photoplate in the observation plane or, in a modern version, a two-dimensional CCD array, where one can decide which diode of the array has received a photon. The experiment shows that at sufficiently low intensities one first sees a statistical spatial distribution of photons without recognizable interference fringes (Fig. 3.17a). Increasing the illumination time at a constant source intensity, the interference pattern emerges more and more clearly (Fig. 3.17b). As long as the difference $N_{\max} - N_{\min}$ of the photon numbers in the maxima and minima of the interference pattern is less than $\sqrt{N}$, the pattern can hardly be seen. However, if by accumulation over a time interval $\Delta t$ the number $\dot{N}\Delta t$ becomes sufficiently large, interference clearly appears, even if the intensity is so small that only one photon passes the slits at a time. From a classical point of view this cannot be understood. How should a photon know through which of the two slits the foregoing photon passed? This apparent paradox has been solved by the quantum theory, which will be discussed in Chap. 4.

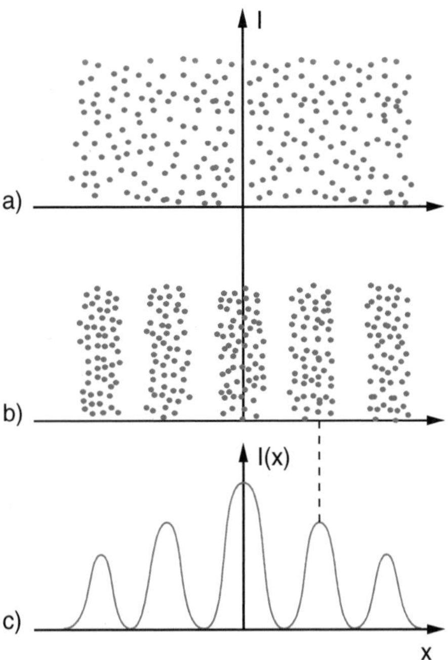


**Fig. 3.17. a–c** Observation of an interference pattern when light passes through a double slit. (**a**) Very small light intensity where $\Delta N > N_{\max} - N_{\min}$ (**b**) Medium intensity $\Delta N \approx N_{\max} - N_{\min}$ (**c**) High intensity $\Delta N \ll N_{\max} - N_{\min}$

> The above examples show that according to our present concepts, light has wave as well as particle character. The important point is that both models do not contradict but supplement each other, as will become clearer in the next sections.

At the end of this section Table 3.1 compiles both the wave and particle descriptions of light.

**Table 3.1.** Characteristic properties of the photon and the electromagnetic field

| Energy | Momentum | Spin | Mass equivalent |
|---|---|---|---|
| $E = h\nu$ | $p = \hbar k$ | $s = \pm\hbar\hat{k}$ | $m = E/c^2$ $= h/(c\cdot\lambda)$ |
| $E = \hbar\omega$ | $\lvert p\rvert = h/\lambda$ $= E/c$ | $\lvert s\rvert = \hbar$ | $m_0 = 0$ |

| Spectral energy density | Intensity | Momentum density |
|---|---|---|
| $w_0 = n\cdot h\nu$ $= \varepsilon_0\lvert E\rvert^2$ | $I = n\cdot c\cdot h\nu$ $= c\varepsilon_0\lvert E\rvert^2$ | $\pi_{St} = \left(\frac{1}{c^2}\right)\cdot S$ $= n\hbar k$ |

## 3.2 Wave Properties of Particles

We will now show that objects that had always been regarded unambiguously as particles, such as electrons, nuclei, atoms or molecules also show wave properties, since diffraction and interference phenomena have been meanwhile observed for these particles (Fig. 3.18).

*Louis de Broglie* (1892–1987) (Fig. 3.19) was the first to propose (in 1924) a wave description of both light and particles, although at this early time, no wave properties of particles had been observed experimentally [3.9].

For this idea, which he later outlined in more detail, de Broglie was awarded the Nobel Prize in 1929.

### 3.2.1 De Broglie Wavelength and Electron Diffraction

If the relation $\boldsymbol{p} = \hbar \boldsymbol{k}$ found for photons is also applied to particles with mass $m$, moving with the velocity $\boldsymbol{v}$, we can attribute to them a wavelength $\lambda = 2\pi/k$, which can be written with $k = p/\hbar$ and $p = m \cdot v$ as

$$\lambda = \frac{h}{p} = \frac{h}{mv} = \frac{h}{\sqrt{2mE_{\mathrm{kin}}}} \,. \tag{3.46}$$

> The de Broglie wavelength is inversely proportional to the momentum of a particle.

When electrons are accelerated by a voltage $U$ reaching the velocity $v \ll c$, we obtain with $E_{\mathrm{kin}} = eU$ the de Broglie wavelength

$$\lambda = \frac{h}{\sqrt{2meU}} \,. \tag{3.47a}$$

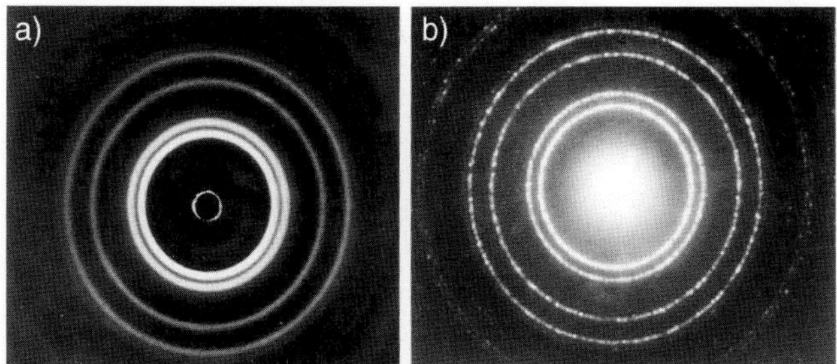


**Fig. 3.18a,b.** Comparison of (**a**) electron diffraction and (**b**) X-ray diffraction by a thin-foil of crystalline aluminium

**Fig. 3.19.** *Loius de Broglie* (1892–1987). From: E. Bagge: Die Nobelpreisträger der Physik (Heinz-Moos-Verlag, München 1964)

**EXAMPLE**

$U = 100\,\mathrm{V}$, $m_{\mathrm{e}} = 1.9 \times 10^{-31}\,\mathrm{kg}$, $h = 6.6 \times 10^{34}\,\mathrm{Js} \Rightarrow \lambda = 1.2 \times 10^{-10}\,\mathrm{m} = 0.12\,\mathrm{nm}$.

*Clinton Joseph Davisson* (1881–1938) (Nobel Prize 1937) and *Lester Halbat Germer* (1896–1971) could indeed demonstrate in 1926, two years after de Broglie's proposal, that electrons with the kinetic energy $E_{\mathrm{kin}} = e \cdot U$ in a collimated beam from an electron gun at the voltage $-U$ produced interference patterns, when they were reflected by a single crystal of nickel (Fig. 3.20). There are two ways to detect these interferences:

a) When the electrons, accelerated by a constant voltage $U$, are reflected by the Nickel-crystal, a diffraction pattern $I(\varphi)$ is measured, when the detector has a variable direction $\varphi$ against the incident beam.
b) At fixed angles $\varphi$ and $\vartheta$ the voltage $U$ is varied and thus the de-Broglie wavelength (Fig. 3.20b). Maxima of the reflected intensity are observed for the Bragg-condition (see Sect. 2.2.3) $m \cdot \lambda = 2d \cdot \sin\vartheta$; $m = 1, 2, 3, \ldots$, where $d$ is the distance

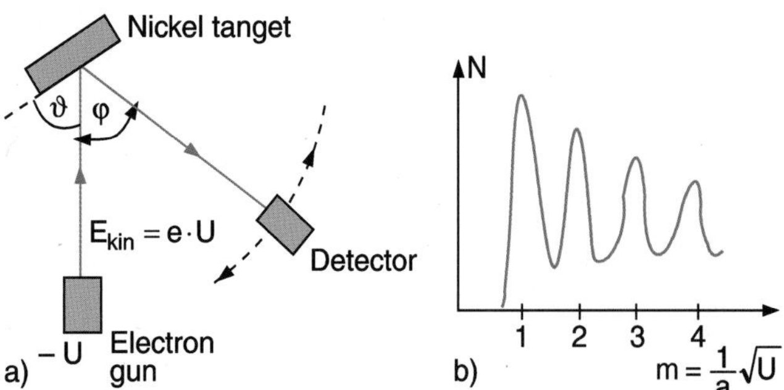


**Fig. 3.20.** (**a**) Schematic drawing of the experimental setup of Davisson and Germer. (**b**) Reflected intensity as a function of the acceleration voltage $U$, according to (3.47b)

between the crystal planes. With $\lambda = h/\sqrt{2m_e eU}$ the maxima appear at voltages $U$ with

$$\sqrt{U} = \frac{m \cdot h}{2d \cdot \sin\vartheta \cdot \sqrt{2e \cdot m_e}} = m \cdot a\,. \quad (3.47b)$$

With high energy electrons, transmitted through a thin foil of crystaline aluminium, circular diffraction patterns can be detected on a photoplate behind the foil which were quite similar to those patterns produced by X-rays (Fig. 3.18b), which had already been recognized as short wavelength electromagnetic waves (see Chap. 7).

This proved that electrons, which had always been clearly regarded as particles, also show wave properties in this experiment, in accordance with de Broglie's hypothesis (3.41). In Fig. 3.21 the diffraction patterns of light and of electrons produced by the sharp edge of a MgO crystal are compared. It illustrates that the same Fresnel diffraction pattern is produced in both cases if the product $\lambda r_0$ of wavelength $\lambda$, and distance $r_0$ between the edge and the photoplate is the same.

a)

b)

**Fig. 3.21a,b.** Comparison of (**a**) light diffraction and (**b**) electron diffraction with $E_{kin} = 38$ keV at the edge of a MgO single crystal. For (**b**) the distance $r_0$ of the photoplate from the edge was chosen such that the product $r_0\lambda$ was equal for (**a**) and (**b**) (From: H. Raether: Elektroneninterferenzen in: Handbuch der Physik Bd. **32**, 443 (1957))

### 3.2.2 Diffraction and Interference of Atoms

More recent experiments have shown that particle diffraction and interference is not restricted to electrons but can also be observed for atoms [3.10]. In Fig. 3.22, helium atoms in an atomic beam pass in the $x$ direction through a narrow slit of width $b = 12\,\mu$m. The atomic waves diffracted by $S_0$ then reach two slits $S_1$ and $S_2$, with $b = 1\,\mu$m each, separated by 8 μm, which are located 64 cm downstream of $S_0$.

In the observation plane $x = x_0$ an interference pattern $I(y)$ is produced that can be monitored with a detector, movable in the $y$ direction. The observed

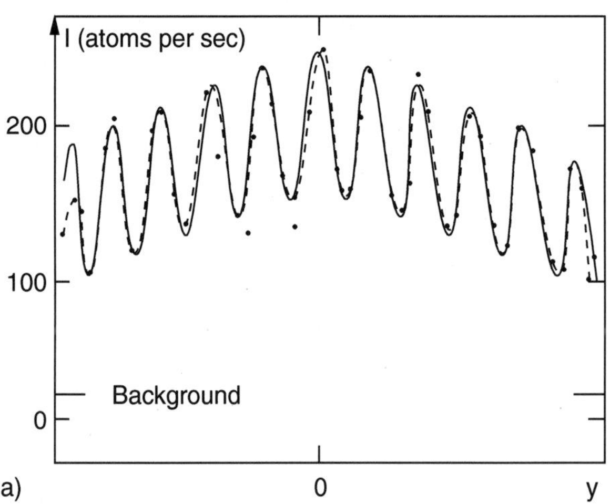


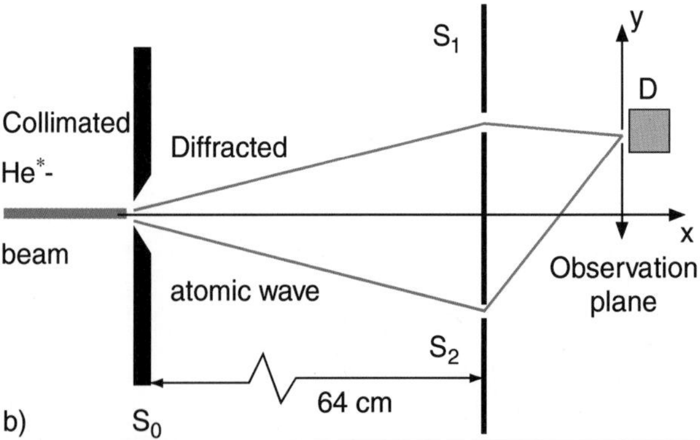


**Fig. 3.22a,b.** Diffraction of a collimated beam of He atoms by a slit Sp and observation of the interference pattern behind a double slit. (**a**) Observed interference pattern (**b**) Experimental setup [3.10]

interference pattern looks like that of light in Young's double slit experiment.

**Note:**

He atoms in their ground state cannot be detected very efficiently. Therefore energetically excited He atoms He* are used, which are in a "metastable state" (see Sect. 6.1) with a lifetime longer than the transit time from the He* source to the detector. These excited He* atoms release electrons when impinging on a metal plate. The resulting ions can be effectively monitored.

In a second experiment, shown in Fig. 3.23, a collimated beam of metastable He atoms passes through a standing light wave with optical wavelength $\lambda_L$,

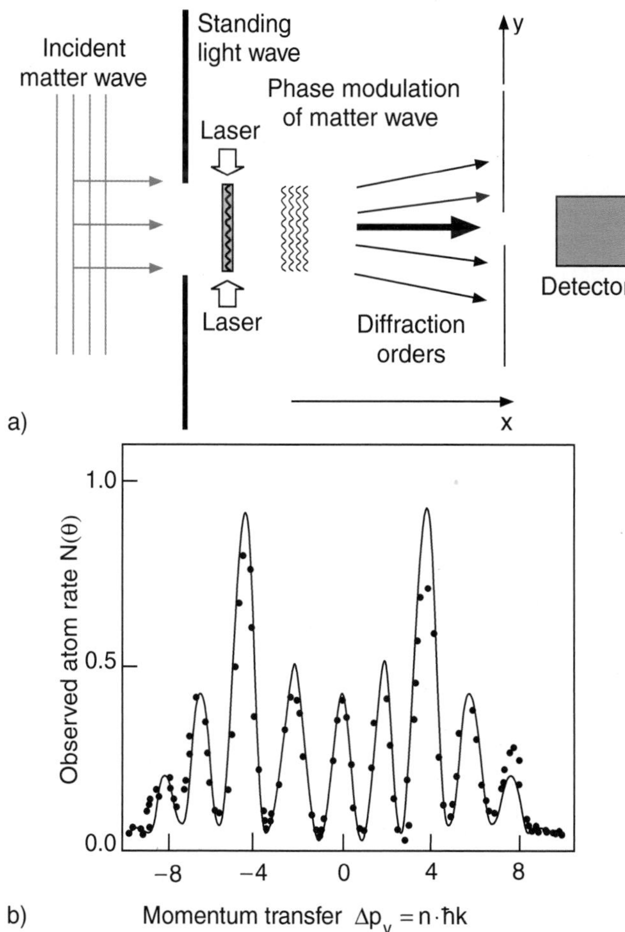


**Fig. 3.23a,b.** Diffraction and interference phenomena observed when a beam of metastable He atoms passes through a standing light wave. (**a**) Experimental setup (**b**) Observed interference pattern $N(\theta)$ in the plane $x = \text{const}$, $y = L \sin\theta$ [3.10]

produced by two laser beams traveling in opposite $\pm y$ directions. In the nodes of the standing wave the light amplitude is zero and the atoms can pass undeflected. In the maxima, the light intensity is high and the atoms can absorb photons if the laser frequency is tuned to an absorbing transition of the atoms. These photons transfer their momentum $\hbar \boldsymbol{k}$ in the $\pm y$ direction to the atoms, resulting in a deflection of atoms out of the $x$ direction. With a detector behind the light wave the atomic interference pattern shown in Fig. 3.23b is observed, where the number $N(\Theta)$ of He* atoms is measured when the detector is moved along the $\pm y$ direction [3.10]. The result can be explained by two different models, which do not contradict each other.

In the wave model the standing light wave acts onto the atoms like a phase grating, quite analogous to phase gratings used in optics, because the periodic variation of the electric field $E(y)$ of the standing light wave acts on the atoms like a periodic variation of the index of refraction. The reason for this is that the potential energy of the atoms depends on the electric field amplitude and since the total energy of the atoms is constant, their kinetic energy changes accordingly. Diffraction of the atomic de Broglie wave by this phase grating with a grating constant $d = \lambda_L/2$ results in intensity maxima if the phase difference between neighboring diffracted partial waves $\Delta\varphi = (2\pi/\lambda_{dB})\Delta s$, becomes $\Delta\varphi = n2\pi$ $(n = 1, 2, 3, \ldots)$. With the path difference $\Delta s = d \sin\Theta = \lambda_L \sin\Theta$ we obtain for the $n$th diffraction maximum at angle $\Theta_n \ll 1$

$$n\lambda_{dB} = \Delta s = \lambda_L \sin\Theta_n \Rightarrow \Theta_n \approx \frac{n\lambda_{dB}}{\lambda_L} . \quad (3.48)$$

The result of Fig. 3.23b can be also explained with the particle model if photons and atoms are both treated as particles. The absorption of $n$ photons in the $\pm$ direction causes a recoil of the atoms with a transferred momentum

$$\Delta p_y = \pm n\hbar k = \pm nh/\lambda_L . \quad (3.49a)$$

The atoms therefore change their flight direction by angles $\Theta_n$, determined by

$$\sin\Theta_n = \frac{\Delta p_y}{p_x} \approx \frac{\Delta p_y}{p} = \frac{n\lambda_{dB}}{\lambda_L} , \quad (3.49b)$$

which gives the same result as (3.48) and illustrates that the particle model as well as the wave model correctly describe the same experimental result [3.11].

### 3.2.3 Bragg Reflection and the Neutron Spectrometer

When a collimated beam of particles with momentum $p = mv$ and de Broglie wavelength $\lambda_{dB} = h/p$ impinges on a crystal under the angle $\vartheta$ against parallel crystal planes, the partial waves reflected by the different planes with distance $d$ can interfere constructively if the path difference $\Delta s = 2d \sin\alpha$ becomes an integer multiple of the wavelength $\lambda_{dB}$ (see Fig. 2.10). This leads to the Bragg condition:

$$\boxed{2d \sin\vartheta = n\lambda_{dB}} \quad (n = 1, 2, 3, \ldots) . \tag{3.49c}$$

Measuring the different angles $\vartheta_n$ of incidence, where maxima occur in the reflected beam, yields the de Broglie wavelength $\lambda_{dB}$, if the distance $d$ between adjacent crystal planes is known.

This is illustrated in Fig. 3.24 by the example of a neutron spectrometer [3.12]. The neutrons, produced inside the core of a nuclear reactor are slowed down by inelastic collisions with H-atoms in paraffin and leave the reactor through collimating apertures as a collimated beam with thermal velocity distribution. They are directed towards a crystal that can be turned around a vertical axis to choose a wanted angle $\vartheta_1$ against the direction of the incident neutron beam. They have a velocity

$$v = \frac{h}{2md \sin\vartheta_1} . \tag{3.49d}$$

The crystal acts as velocity filter and reflects only neutrons with the wavelength $\lambda_{dB} = 2d \sin\vartheta_1$ into the direction $\vartheta_2$. It therefore acts like a grating monochromator in optical spectroscopy. When these "monochromized neutrons" impinge onto a second crystal with unknown distance $d$ between its crystal planes, $d$ can be determined if the crystal is turned around a vertical axis and the intencity maxima are counted by a borotrifluoride detector.

Instead of using the Bragg reflection by a crystal with known distance $d$ between its planes, a time-of-flight technique can be used to select neutrons with a given velocity (Fig. 3.24b). A rotating absorbing disc with a narrow slit transmits neutrons only during the short time interval $\Delta t$ at time $t = 0$. At a distance $L$ from the first disc a second disc with a narrow slit rotates synchronously with the rotation period $T$. Only neutrons with velocities $v = nL/T$ $(n = 1, 2, 3 \ldots)$ can pass through both slits. These neutrons with known velocities and therefore known de Broglie wavelength can now be directed towards a Bragg crystal with unknown distances $d_x$ of its planes. Measuring the angles $\vartheta_n$, where maxima of the reflected beam are observed, yields the wanted distances $d_x$.

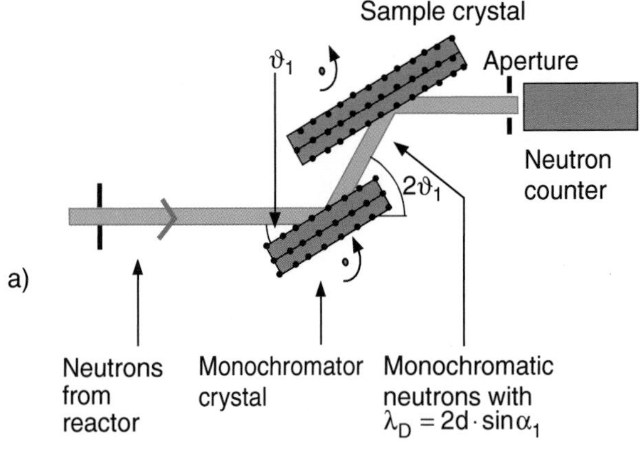


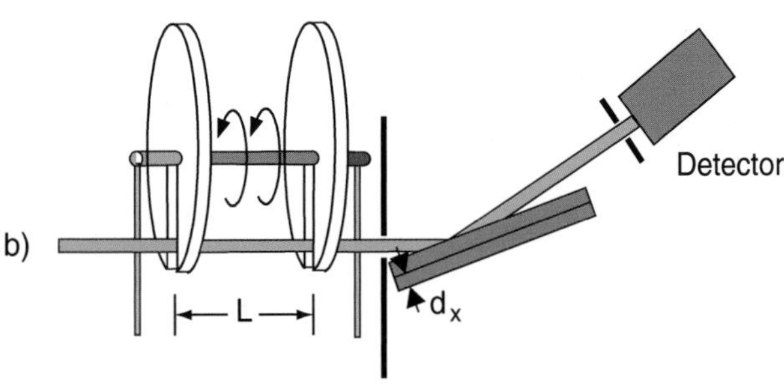


**Fig. 3.24a,b.** Neutron spectrometer. (**a**) Neutrons from a nuclear reactor are monochromatized by Brass reflection (**b**) Velocity selection with two rotating discs and time-resolved detection

### 3.2.4 Neutron and Atom Interferometry

The wave properties of neutrons allow the construction of a neutron interferometer, analogous to an X-ray interferometer (see Sect. 2.2.3). Such a particle interferometer can be used to measure many quantities important for basic physics and for many applications.

The device is shown in Fig. 3.25 (compare with Fig. 2.12!). Three thin slices of a silicon crystal are cut out of a single crystal. This guarantees that the three

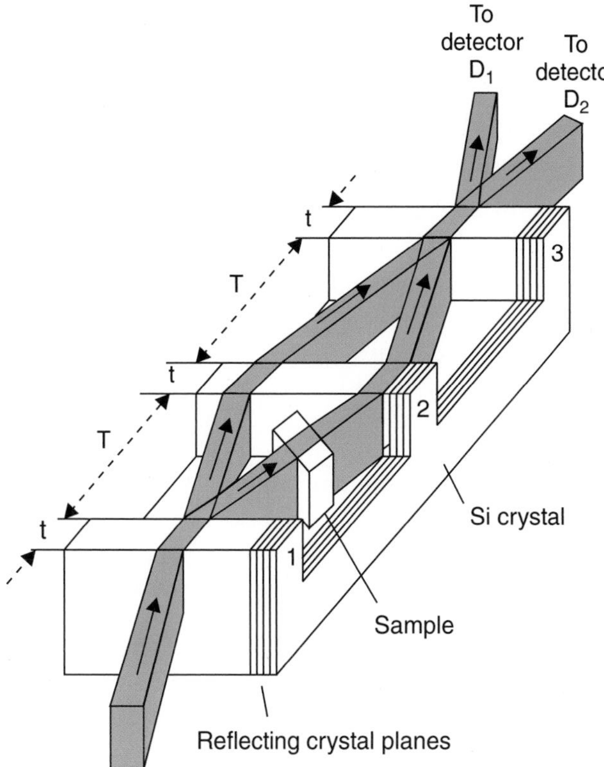


**Fig. 3.25.** Neutron interferometer [3.14]

slices have the same crystal orientation. The intensities $I_1$ and $I_2$, measured by the detectors $D_1$ and $D_2$ depend on the phase differences of the interfering partial neutron beams. The sum of the intensities $I_1 + I_2$ is, however, independent of phase shifts.

If a phase shifting sample is now brought into one of the partial beams, the signals of $D_1$ and $D_2$ change. Due to interactions of the neutrons with the nuclei of the sample, their potential is altered inside the sample. This causes a phaseshift of their de Broglie wave.

The phase shift can also be caused by the gravitational field of the earth, if the two partial beams pass at different heights $H_i$ through the interferometer, because then the potential energy of the neutrons $mgH_i$ is different and therefore their kinetic energy changes by

$$\Delta E_{\text{kin}} = -m \cdot g \cdot \Delta H .$$

Because $E_{\text{kin}} = h^2 / \left(2m\lambda_{\text{dB}}^2\right)$ we obtain

$$\Delta\lambda_{\text{dB}} = \frac{2mg \cdot \Delta H}{E_{\text{kin}}} .$$

**Note:**

Phase shifts caused by $\Delta H = 1\,\text{cm}$ can already be detected [3.14].

With such an interferometer, phase shifts caused by electric or magnetic fields or by materials transparent for neutrons can be measured. They give information on the interaction of neutrons with the fields or with the atoms of the sample.

Also for atoms interferometers can be realized [3.13]. Replacing the diffracting Bragg crystal slices by standing optical waves, which deflects the atoms by photon recoil allows the realization of an atomic interferometer (see Sect. 3.2.3), which can measure the phase shift of atomic de Broglie waves, when interacting with external fields or with matter.

### 3.2.5 Application of Particle Waves

The de Broglie wavelength $\lambda_{\text{dB}} = h/(mv)$ can be adopted to the special problem by selecting the appropriate particle velocity $v$. For example, the measurement of the distances $d$ between adjacent crystal planes demands $\lambda_{\text{dB}} < d$. In Table 3.2 some numerical values of $\lambda_{\text{dB}}$ for electrons, neutrons and He atoms at three different kinetic energies are compiled, in order to give a feeling of the order of magnitude for $\lambda_{\text{dB}}$.

**EXAMPLE**

Helium atoms at room temperature ($T \approx 300\,\text{K}$) have a mean velocity $\bar{v} \approx 1{,}300\,\text{m/s}$ and a mean kinetic energy $E_{\text{kin}} \approx 0.03\,\text{eV}$. Their de Broglie wavelength is then $\lambda_{\text{dB}} = 8.3 \times 10^{-11}\,\text{m}$. This is about half of the typical atomic distances in a crystal.

With such thermal He atoms, the surface structure of solids can be probed by measuring the diffraction

**Table 3.2.** De Broglie wavelengths in units of $10^{-10}\,\text{m} = 1\,\text{Å}$ for electrons, neutrons and He-atoms with different kinetic energies

| $E_{\text{kin}}$/eV | Electrons | Neutrons | He Atoms |
|---|---|---|---|
| 0.03 | 70.9 | 1.65 | 0.83 |
| 1 | 12.3 | 0.28 | 0.143 |
| $10^4$ | 0.123 | 0.003 | 0.001 |

pattern of the reflected atoms obtained when a parallel beam of He atoms falls under an angle $\vartheta$ onto the surface. Since the atoms, contrary to neutrons, do not penetrate into the solid, only the surface layer of the crystal atoms contributes to the diffraction. The diffraction pattern therefore gives information on the structure of this surface layer and will change when other atoms are adsorbed at the surface.

When we regard the electrons in an electron microscope as de Broglie waves, their wavelength at a kinetic energy of $E_{kin} = 10^5$ eV is $\lambda_{dB} \approx 4 \times 10^{-12}$ m, i.e., by five orders of magnitude smaller than the wavelength of visible light. Therefore, the principle lower limit $\Delta x$ for the spatial resolution ($\Delta x \geq \lambda/2$) is much smaller than in a light microscope. However, although the resolution in a real electron microscope is not limited by diffraction and reaches a spatial resolution of $\Delta x < 0.1$ nm, it is still limited by imaging errors in the electron-optics.

## 3.3 Matter Waves and Wave Functions

In the wave model of a particle with mass $m$, moving with a velocity $v$ in the $x$ direction, we write the wave function quite analogously to that of light waves as

$$\psi(x,t) = C\mathrm{e}^{\mathrm{i}(kx-\omega t)} = C\mathrm{e}^{\mathrm{i}/\hbar(px-Et)}, \qquad (3.50a)$$

where we have used the relations $E = E_{kin} = \hbar\omega$ between the energy $E$ of a free particle ($E_{pot} = 0$) and the angular frequency $\omega$ of its matter wave, and

$$\boldsymbol{p} = \hbar \boldsymbol{k} \qquad (3.50b)$$

between its momentum $\boldsymbol{p}$ and the wave vector $\boldsymbol{k}$ with $|\boldsymbol{k}| = 2\pi/\lambda_{dB}$.

Note, however, that there exists an important difference between electromagnetic waves and matter waves. The phase velocity of electromagnetic waves in vacuum, obtained from the condition

$$\frac{\mathrm{d}}{\mathrm{d}t}(kx - \omega t) = 0 \Rightarrow v_{ph} = \frac{\mathrm{d}x}{\mathrm{d}t} = \frac{\omega}{k} = \mathrm{const}$$
$$\Rightarrow \frac{\mathrm{d}v_{ph}}{\mathrm{d}\omega} = 0 \qquad (3.51)$$

is independent of $\omega$, since $k = 2\pi/\lambda = \omega/c \Rightarrow v_{ph} = c$.

This means that electromagnetic waves propagating in a vacuum do not show dispersion.

For matter waves this does not hold! We obtain for a free particle with $E = E_{kin} = p^2/2m$

$$\omega = \frac{E}{\hbar} = \frac{\hbar k^2}{2m} \Rightarrow v_{ph} = \frac{\omega}{k} = \frac{\hbar k}{2m} = \frac{p}{2m}$$
$$\Rightarrow \frac{\mathrm{d}v_{ph}}{\mathrm{d}\omega} = \frac{1}{k} \neq 0\,. \qquad (3.52)$$

The phase velocity depends on the momentum $p$ of the particle. This can be seen, when we calculate $\omega(k)$. We have to use the relativistic relation

$$E = (p^2c^2 + m^2c^4)^{1/2} = mc^2 + p^2/(2m) + \ldots..$$

between energy E and momentum p. With $E = h\omega$ we obtain:

$$\omega(k) = E/\hbar = mc^2/\hbar + \hbar k^2/2m + \ldots..$$

The phase velocity $v_{ph} = \omega/k$ becomes therefore:

$$v_{ph} = h\omega/\hbar k = mc^2/(\hbar k) + \hbar k/2m + \ldots..$$

This shows that the phase velocity does depend on k, which means that matter waves show dispersion.

The absolute value of $v_{ph}$ is

$$v_{ph} = h\omega/\hbar k = E/p.$$

with $E = mc^2/(1 - \beta^2)$ and $p = mv_T/(1 - \beta^2)$ we obtain

$$v_{ph} = E/p = c^2/v_T \rightarrow v_{ph} > c!!!$$

which implies that the phase velocity of matter waves is larger than the velocity of light.

Question: Why is this not in contradiction to the theory of special relativity?

Therefore the matter wave (3.52) and its phase velocity $v_{ph}$ is not directly useful for the description of the moving particle. Furthermore, the particle is restricted to a certain volume $\mathrm{d}V = \mathrm{d}x\mathrm{d}y\mathrm{d}z$ around a point $(x(t), y(t), z(t))$, which moves with the particle velocity $v_p$ in space, while the matter wave (3.50a) extends over the total space. This deficiency can be removed by introducing the model of wave packets.

### 3.3.1 Wave Packets

Wave packets represent a superposition of monochromatic waves with frequencies $\omega_j$ in an interval $\Delta\omega$ (Fig. 3.26). The introduction of wave packets (sometimes also called wave groups) localizes the matter wave within a certain space interval $\Delta x$ or for three dimensional problems within the volume $\mathrm{d}V$. This can be seen as follows.

When many monochromatic plane waves with amplitudes $C_j$, frequencies $\omega_j$ within a restricted interval $\Delta\omega$ and parallel wave vectors $\boldsymbol{k}$ within the interval $\Delta k$ (Fig. 3.27a), propagating in the $x$ direction, are superimposed, their sum

$$\psi(x,t) = \sum_j C_j \mathrm{e}^{\mathrm{i}(k_j x - \omega_j t)} \tag{3.53}$$

shows maximum amplitudes $\psi(x_\mathrm{m})$ at locations $x_\mathrm{m}$ (Fig. 3.27b), which propagate into the $x$ direction with the group velocity $v_\mathrm{g} = \mathrm{d}\omega/\mathrm{d}k$.

With an infinite number of such waves with frequencies $\omega$ filling the interval

$$\omega_0 - \Delta\omega/2 \le \omega \le \omega_0 + \Delta\omega/2$$

and with wave numbers $k_0 - \Delta k/2 \le k \le k_0 + \Delta k/2$ the sum (3.53) transforms into the integral

$$\psi(x,t) = \int\limits_{k_0-\Delta k/2}^{k_0+\Delta k/2} C(k)\mathrm{e}^{\mathrm{i}(kx-\omega t)}\mathrm{d}k\ . \tag{3.54}$$

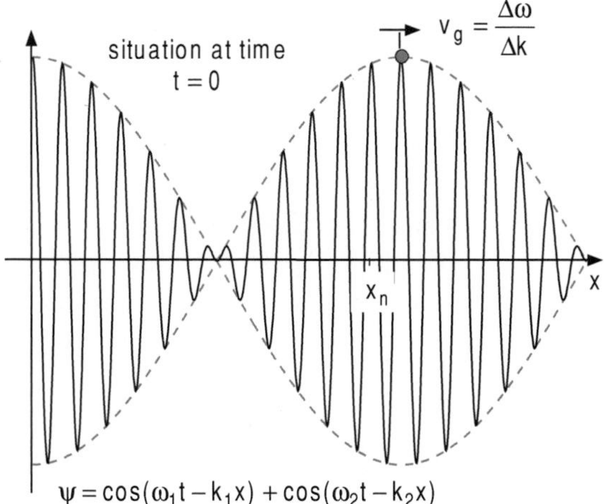


**Fig. 3.26.** Superposition of two monochromatic waves with slightly different frequencies $\omega_1$ and $\omega_2$ and equal amplitudes results resolutions in a beat pattern

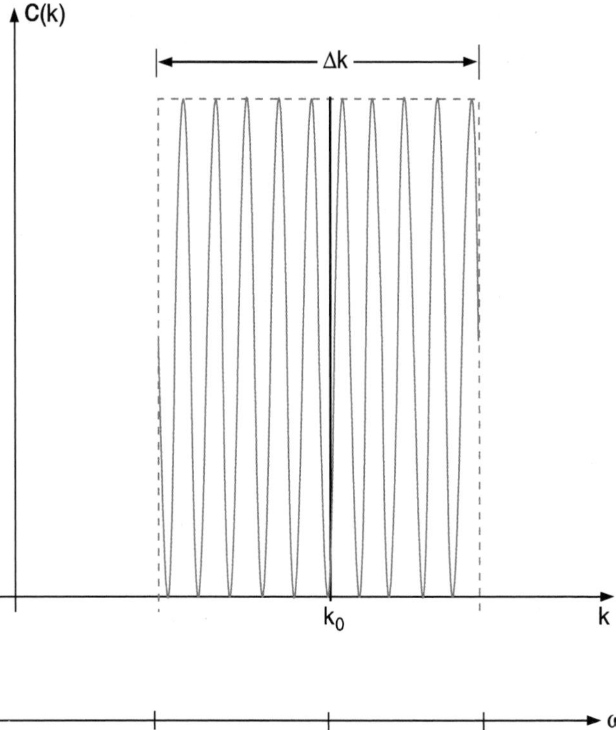


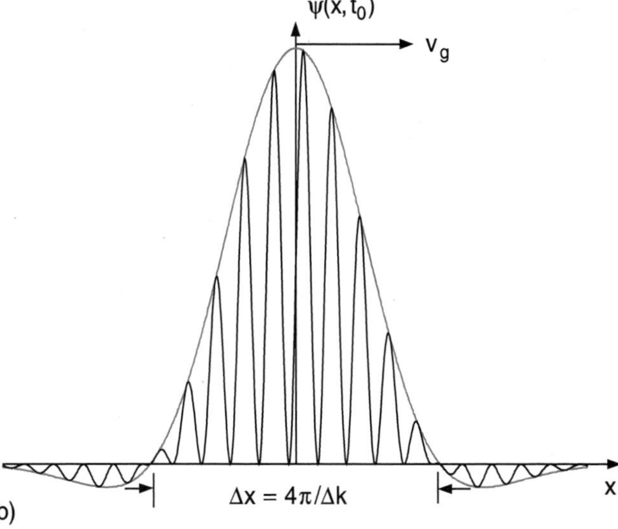


**Fig. 3.27a,b.** Wave packet as superposition of waves with frequencies $\omega$ within the range $k_0 \pm \Delta k/2$ or $\omega_0 \pm \Delta\omega/2$. (**a**) Equal amplitudes $C(k) = C(k_0)$ of all partial waves. (**b**) Resulting wave packet $\psi(x, t_0)$ at a given time $t = t_0$

For sufficiently small intervals $\Delta k \ll k_0$ we can expand $\omega(k)$ into the Taylor series

$$\omega(k) = \omega_0 + \left(\frac{\mathrm{d}\omega}{\mathrm{d}k}\right)_{k_0} (k - k_0) + \dots\ , \tag{3.55}$$

where we neglect the higher order terms. If the amplitude $C(k)$ does not change much over the small interval $\Delta k$, we can replace $C(k)$ by its constant value $C(k_0)$. Inserting (3.55) into (3.54) we obtain with the abbreviations

$$\kappa = k_0 - k \quad \text{and} \quad u = (\mathrm{d}\omega/\mathrm{d}k)_{k_0} t - x$$

the matter wave function

$$\psi(x,t) = C(k_0)\mathrm{e}^{\mathrm{i}(k_0 x - \omega_0 t)} \int\limits_{-\Delta k/2}^{+\Delta k/2} \mathrm{e}^{\mathrm{i}u\kappa}\,\mathrm{d}\kappa \,. \tag{3.56a}$$

The integration is simple and yields

$$\psi(x,t) = A(x,k,t)\mathrm{e}^{\mathrm{i}(k_0 x - \omega_0 t)}$$
$$\text{with} \quad A(x,k,t) = 2C(k_0)\frac{\sin(u\Delta k/2)}{u} \,. \tag{3.56b}$$

This function describes a plane matter wave with an amplitude $A(x,t)$ that has a maximum for $u = 0$, corresponding to a position

$$x_\mathrm{m} = \left(\frac{\mathrm{d}\omega}{\mathrm{d}k}\right)_{k_0} t \,. \tag{3.56c}$$

The maximum of the wave packet propagates with the group velocity $v_\mathrm{g} = (\mathrm{d}\omega/\mathrm{d}k)_{k_0}$ into the $x$ direction. We call $\psi(x,t)$ a ***wave packet*** (Fig. 3.27b). The form of the wave packet (maximum amplitude and width $\Delta x$, i.e., the distance between the first side minima of both sides of the central maximum depends on the chosen interval $\Delta k$ and the amplitude distribution $C(k)$ in (3.25). In Figs. 3.27b and 3.28b the wavepackets (3.54) with constant amplitudes $C(k) = C(k_0)$ and with a Gaussian distribution of $C(k)$ are compared.

With the relations

$$\omega = \frac{E}{\hbar} = \frac{p^2}{2m\hbar} = \frac{\hbar k^2}{2m}$$
$$\Rightarrow v_\mathrm{g} = \left(\frac{\mathrm{d}\omega}{\mathrm{d}k}\right)_{k_0} = \frac{\hbar k_0}{m} = \frac{p}{m} = v_\mathrm{p} \tag{3.57}$$

it follows that the group velocity of the wave packet equals the particle velocity $v_\mathrm{p}$.

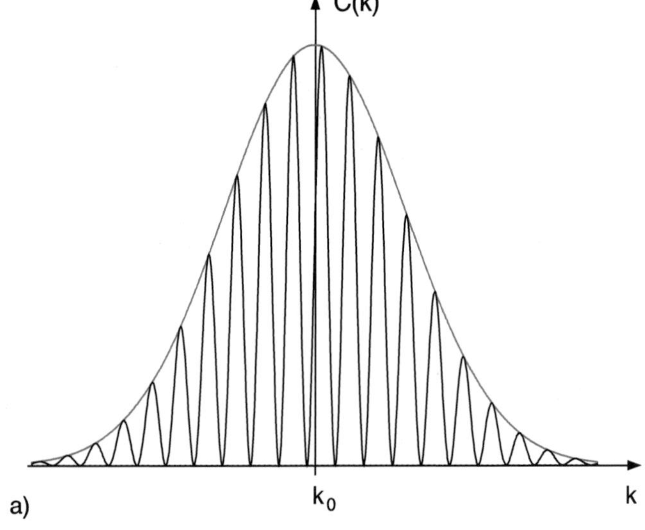


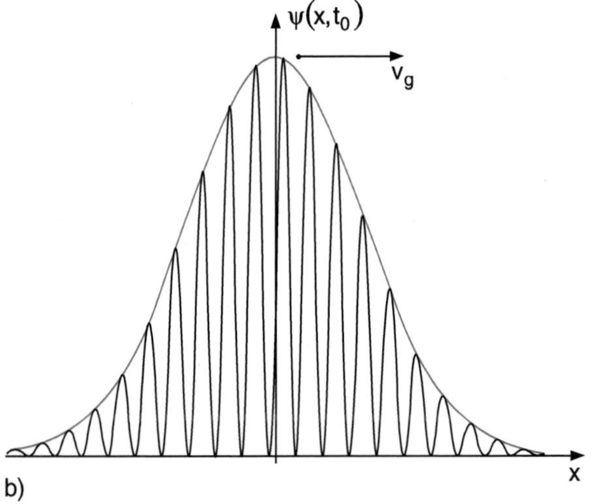


**Fig. 3.28a,b.** Wave packet $\psi(x,t_0)$ composed of partial waves with Gaussian amplitude distribution (**a**). $C(k) = C(k_0)\exp\left[-(k-k_0)^2/k_0^2\right]$; (**b**) $\psi(x)$ at given time $t = t_0$

> Such a wave packet represents a better description of a particle than the spatially extended plane wave, because its center wave vector $\boldsymbol{k}_0$ and its group velocity $v_g$ correspond to particle properties: $\vec{\boldsymbol{k}}_0 = \vec{\boldsymbol{p}}/\hbar$ and $v_\mathrm{g} = v_\mathrm{p}$

- The group velocity $v_\mathrm{g}$ equals the particle velocity $v_\mathrm{p}$
- The wave vector $\boldsymbol{k}_0$ of the group center determines the particle momentum $\boldsymbol{p}_\mathrm{p} = \hbar\boldsymbol{k}_0$
- The wave packet is localized in space. The amplitude $A(x,t)$ of the wavefunction (3.56b) has non vanishing values only within a limited space

interval $\Delta x$. From (3.56) we obtain at time $t = 0$ the width $\Delta x$ of the central maximum between the two minima

$$\Delta x = \frac{4\pi}{\Delta k} \geq \frac{2\pi}{k_0} = \lambda_{\mathrm{dB}} .$$

This shows that the width $\Delta x$ of the wave packet has a minimum value equal to the de Broglie wavelength $\lambda_{\mathrm{dB}}$.

We summarize:

> Particles can be described by wave packets. The particle velocity equals the group velocity of the wave packet and the particle can be localized not better than within its de Broglie wavelength.

Note that this spatial restriction corresponds to the spatial resolution in optics, when illuminating objects with light. Here, however, the spatial resolution is limited by the wavelength $\lambda$ of the light.

**Remark**

The additional small side maxima in Fig. 3.27b that appear when a constant amplitude $C(k) = C(k_0)$ is used, disappear when $C(k)$ is represented by a Gaussian distribution

$$C(k) = C(k_0)\exp\left[-(k-k_0)^2/2\Delta k^2\right]$$

(see Fig. 3.28).

In spite of this close correspondence between the wavepacket and its particle, the wave packet amplitude (3.56a) can not be the real representation of the particle for the following reasons:

- The wave function $\psi(x,t)$ in (3.56a) can assume complex and negative values, which do not correspond to real and measurable particle properties
- The width $\Delta x(t)$ of the wave packet increases with time because of the dispersion of the matter waves of which it is composed (see next section). It changes its form while propagating in space, in contrast to a real particle, which keeps its form.
- Any elementary particle, such as the electron, is indivisible according to our present knowledge. A wave, however, can be split, for instance by a beam splitter, into two components, which then move into different directions.

**Fig. 3.29.** *Max Born* (1882–1970) From: E. Bagge: Die Nobelpreisträger der Physik (Heinz Moos-Verlag, München 1964)

These difficulties induced *Max Born* (1882–1970, Fig. 3.29) to propose a statistical interpretation of the physical meaning of matter waves [3.15].

### 3.3.2 The Statistical Interpretation of Wave Functions

Since a particle incident on a boundary plane between two media has to be either reflected or transmitted, the splitting of the corresponding matter wave into a reflected and transmitted partial wave could be related to the probability of reflection or transmission of the particle. However, the probability is per definition a positive real number between zero and one. Therefore, the complex wave amplitude can not be directly interpreted as a probability. Max Born proposed the following definition.

The probability $P(x,t)\mathrm{d}x$ of finding a particle at time $t$ in the space interval from $x - \Delta x/2$ to $x + \Delta x/2$ is proportional to the absolute square $|\psi(x,t)|^2$ of the wave function $\psi(x,t)$ representing the particle

$$P(x,t)\Delta x \propto |\psi(x,t)|^2 \Delta x . \tag{3.58}$$

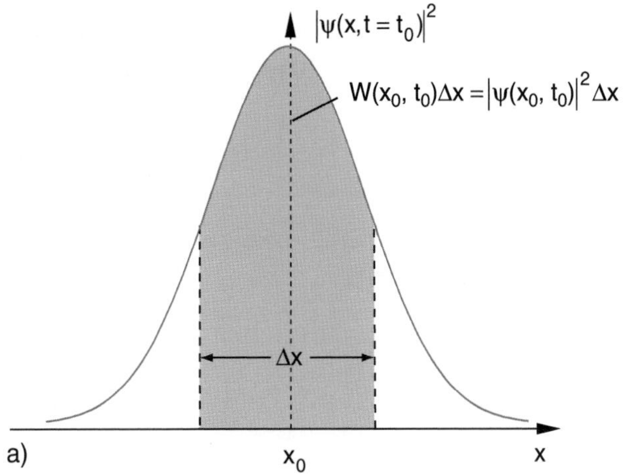


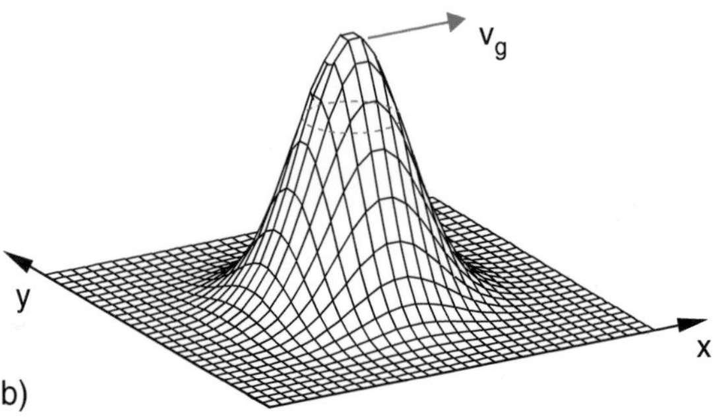


**Fig. 3.30.** (**a**) Absolute square of a one-dimensional wave packet, representing the probability density to find a particle at $x = x_0$ at the time $t = t_0$. (**b**) Two-dimensional wave packet

$|\psi(x,t)|^2$ is called the ***probability density*** at the location $x$ at time $t$ (Fig. 3.30a).

A particle moving along the $x$-axis can be found with certainty within the interval $-\infty \leq x \leq +\infty$. Therefore the condition

$$\int_{-\infty}^{+\infty} |\psi(x,t)|^2 \mathrm{d}x = 1 \tag{3.59}$$

can be used for the normalization of the wave function. With this normalization the proportionality factor in (3.58) becomes unity and we obtain

$$P(x,t)\Delta x = |\psi(x,t)|^2 \Delta x \ . \tag{3.60}$$

To a particle moving freely in three-dimensional space, a three-dimensional wave packet $\psi(x,y,z,t)$ can be attributed (Fig. 3.30b) and the same arguments as above result in the normalization

$$\iiint |\psi(x,y,z,t)|^2 \mathrm{d}x\mathrm{d}y\mathrm{d}z \equiv 1 \ .$$

In summary:

> Each "particle" can be represented by a three-dimensional wave packet, described by the wave function $\psi(x,y,z,t)$. The real positive quantity
>
> $$P(x,y,z,t)\mathrm{d}x\mathrm{d}y\mathrm{d}z = |\psi(x,y,z,t)|^2 \mathrm{d}x\mathrm{d}y\mathrm{d}z$$
>
> gives the probability of finding the particle at the time $t$ within the volume $\mathrm{d}V = \mathrm{d}x\mathrm{d}y\mathrm{d}z$ around the point $(x,y,z)$. The probability is maximum at the center of the wave packet, which moves with the group velocity $v_\mathrm{g}$ through space which is identical with the particle velocity $v_\mathrm{p}$.

Note, however, that the probability is larger than zero not only at the point $(x,y,z)$, but also within a finite volume $\mathrm{d}V$ around $(x,y,z)$. This means that it is not possible to localize the particle exactly at the point $(x,y,z)$. The localization of the particle is only possible within an uncertainty that is related to the spatial width of the corresponding wave packet. We will now discuss this aspect in more detail.

### 3.3.3 Heisenberg's Uncertainty Principle

We choose a wave packet as a superposition of plane waves with amplitudes $C(k)$, following the Gaussian distribution

$$C(k) = C_0 \mathrm{e}^{-[(a/2)^2(k-k_0)^2]} \tag{3.61}$$

around the central wavenumber $k_0$. The one-dimensional wave packet then becomes

$$\psi(x,t) = C_0 \int_{-\infty}^{+\infty} \mathrm{e}^{-[(a/2)(k-k_0)]^2} \mathrm{e}^{\mathrm{i}(kx-\omega t)} \mathrm{d}k \ . \tag{3.62}$$

with $C_0 = \sqrt{a}/(2\pi)^{3/4}$. The integration over $k$ is analytically possible and yields at $t = 0$

$$\psi(x,0) = \left(\frac{2}{\pi a^2}\right)^{1/4} \mathrm{e}^{-x^2/a^2} \mathrm{e}^{\mathrm{i}k_0 x} \ , \tag{3.63a}$$

where we have chosen $C_0 = \sqrt{a}/(2\pi)^{3/4}$ in order to obtain the normalized wave function with the probability density

$$|\psi(x,0)|^2 = \sqrt{\frac{2}{\pi a^2}} \mathrm{e}^{-2x^2/a^2} \ , \tag{3.63b}$$

which satisfies the condition

$$\int_{-\infty}^{+\infty} |\psi(x,0)|^2 \mathrm{d}x = 1$$

as can be readily verified.

The wave packet (3.63) has its maximum amplitude at $x = 0$. At the points $x_{1,2} = \pm a/2$ the probability density has dropped to $1/\sqrt{\mathrm{e}}$ of the maximum value. The interval $\Delta x = x_1 - x_2 = a$ is defined as the full width of the wave packet (3.63). The full width $\Delta k = k_1 - k_2$ of the amplitude distribution $C(k)$ between the values $k_1$ and $k_2$ where $C(k_{1,2}) = C_0/\sqrt{\mathrm{e}}$ is, according to (3.62) $\Delta k = 1/a$.

This yields the important result (Fig. 3.31):

$$\Delta x \cdot \Delta k = 1 \qquad (3.64a)$$

The product of spatial width $\Delta x$ and wavenumber width $\Delta k$ is for a gaussian wave packet $\Delta x \cdot \Delta k = 1$.

This result might already be familiar to you from optics. In every spectrometer the smallest, still resolvable frequency interval $\Delta\omega_{\min} = 1/\Delta t_{\max}$ is limited by the maximum traversal time difference between interfering light waves.

With $\Delta\omega = c\Delta k$ and $\Delta x = c\Delta t_{\max}$ this gives again $\Delta k_{\min} \cdot \Delta x_{\max} = 1$.

Therefore the result $\Delta k \cdot \Delta x \geq 1$ is not specific for quantum physics, but is typical for any wave model. Its significance for the quantum mechanical description of particles stems from the interpretation of $|\psi|^2$ as probability density of the location of the particle.

With the de Broglie relation $p_x = \hbar k_x$ for the momentum $p_x$ of a particle moving in $x$ direction, one obtains from (3.64a) the relation

$$\Delta x \cdot \Delta p_x = \hbar\ . \qquad (3.64b)$$

It can be proved [3.16] that a Gaussian wave packet has the minimum product $\Delta x \cdot \Delta p_x$. For all other amplitude distributions $C(k)$, the product becomes larger. We, therefore arrive at the uncertainty principle, first formulated by *Werner Karl Heisenberg* (1901–1975, Fig. 3.32)

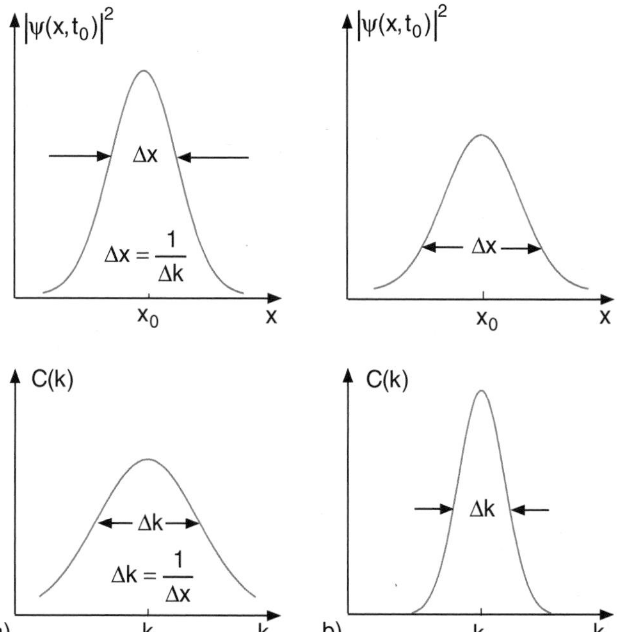


**Fig. 3.31a,b.** Illustration of the uncertainty relation between localization and momentum uncertainty. (**a**) Small spatial uncertainty (**b**) Large spatial uncertainty

**Fig. 3.32.** *Werner Heisenberg* (1901–1975) From: E. Bagge: Die Nobelpreisträger der Physik (Heinz-Moos-Verlag, München 1964)

$$\Delta x \cdot \Delta p_x \geq \hbar \,. \tag{3.64c}$$

The product of the uncertainty $\Delta x$ in the localization of a particle, defined as the spatial width of its wave packet, and the uncertainty $\Delta p_x$ of the particle's momentum $p_x$, defined as the width of the momentum distribution $C(k)$ with $k = p/\hbar$, cannot be smaller than $\hbar$.

This illustrates that Planck's constant $h$ (or the reduced constant $\hbar = h/2\pi$) does not only describe the quantization of photon energies but also governs the uncertainty relation. This is due to the de Broglie relation $\lambda_{dB} = h/p$ between the momentum $p$ of a particle and the wavelength $\lambda_{dB}$ of its matter wave.

**Remark**

Often, one defines the width $\Delta x$ of a Gaussian distribution as the interval between the points $x_1$ and $x_2$ where the function $|\psi(x)|^2$ in (3.63b) drops down to $1/e$ (instead of $1\sqrt{e}$) of its maximum value and the width $\Delta k$ between the values $k_1$ and $k_2$ where $C(k)$ drops to $1/e$ of $C(k_0)$. This definition then results in $\Delta k = \sqrt{2}/a$ and $\Delta x = \sqrt{2} \cdot a$ and instead of (3.64a) and (3.64c)one obtains the uncertainty product

$$\Delta x \cdot \Delta k = 2 \quad \text{or} \quad \Delta x \cdot \Delta p_x \geq 2\hbar \tag{3.65}$$

If the width of the wave packet with constant amplitudes $C_0$ is chosen as the distance between the first zero points on both sides of the central maximum (Fig. 3.27a) we obtain from (3.56b) instead of (3.64b), the relation

$$\Delta x \cdot \Delta p_x \geq 2h = 4\pi\hbar \,, \tag{3.66}$$

which means that for this definition of $\Delta x$, $\Delta p_x$ the uncertainty product is $4\pi$ times larger than in (3.64b).

This should remind you that the numerical value of the lower limit for the product $\Delta x \cdot \Delta p_x$ depends on the definition of the uncertainties $\Delta x$, $\Delta p_x$ and on the form of the wave packet.

For the other directions $y$ and $z$ of a three-dimensional Gaussian wave packet one obtains in an analogous way:

$$\Delta y \cdot \Delta p_y \geq \hbar \,, \quad \Delta z \cdot \Delta p_z \geq \hbar \,. \tag{3.67}$$

We will now illustrate the uncertainty principle using some examples.

### a) Diffraction of Electrons Passing Through a Slit

We regard a parallel beam of electrons with momentum $\boldsymbol{p} = \{0, p_y, 0\}$ and a large beam diameter incident onto a narrow slit with width $\Delta x = b$ (Fig. 3.33). Before the slit, the $x$ component of the momentum of the electrons is $p_x = 0$, but we cannot know the $x$ coordinates of an individual electron, that is, $\Delta x = \infty$. Of all electrons only those with $x$ coordinates within the interval $x = 0 \pm b/2$ can pass through the slit. For these electrons we know their $x$ coordinates within the uncertainty $b$, i.e., $\Delta x = b$.

According to the uncertainty principle (3.66), their momentum components $p_x$ must have the uncertainty $\Delta p_x \geq h/b$. Since $\Delta p_x = p_y \cdot \sin\theta \approx p \cdot \sin\theta$, this means that the electrons can be found within the angular interval $-\Theta \leq \varphi \leq \Theta$ where

$$\sin\Theta = \pm\frac{\Delta p_x}{p} = \pm\frac{h}{bp} \,. \tag{3.68a}$$

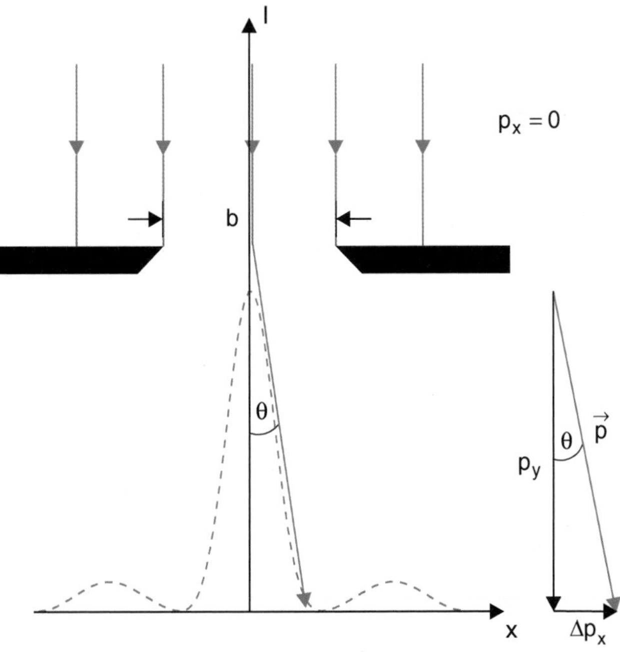


**Fig. 3.33.** Diffraction of electrons by a slit as explained by the uncertainty relation

When we describe the electrons by their matter wave with the de Broglie wavelength $\lambda_{dB} = h/p$, the wave is diffracted by the slit, as in wave optics, and the central diffraction maximum has an angular width $\Delta\varphi = 2\Theta$ between the two first minima. The wave model gives

$$\sin\Theta = \frac{\lambda}{b} = \frac{h}{bp}\,, \tag{3.68b}$$

which turns out to be identical to (3.68a).

> This illustrates that the uncertainty principle just takes into account the description of particles by matter waves. If this is accepted, the relations (3.64) and (3.68) follow from classical diffraction theory.

#### b) Spatial Resolution Limit of the Microscope

Assume one wants to measure the location $x$ of a particle at rest by illuminating it with light of wavelength $\lambda$ (Fig. 3.34).

The photon must be scattered by the particle into the solid angle with apex angle $2\alpha$ in order to reach the collimating lens of the microscope with diameter $d$, where $\sin\alpha \approx \tan\alpha = d/2y$. The uncertainty $\Delta p_x$ of the photon momentum $p_x$ is then

$$\Delta p_x = p_x \sin\alpha \approx \frac{h}{\lambda_{dB}}\frac{d}{2y} = \frac{h}{\lambda}\cdot\frac{d}{2y}\,, \tag{3.69}$$

because the de Broglie wavelength of a photon

$$\lambda_{dB} = \frac{h}{p} = \frac{h\cdot c}{h\cdot\nu} = \lambda$$

equals the wavelength $\lambda$ of the light wave.

Conservation of momentum requires that the scattering particle must suffer a recoil $p_x$ with the uncertainty $\Delta p_x$.

Parallel light entering from above the collimating lens (Fig. 3.34b) produces in the focal plane, at a distance $y$ from the lens, a diffraction pattern where the central maximum has the diameter

$$D = 1.2\cdot 2y\sin\Theta \approx 2y\frac{\lambda}{d} \tag{3.70}$$

(see textbooks on optics).

This shows that the location $x$ of the scattering particle can not be defined more accurately than within

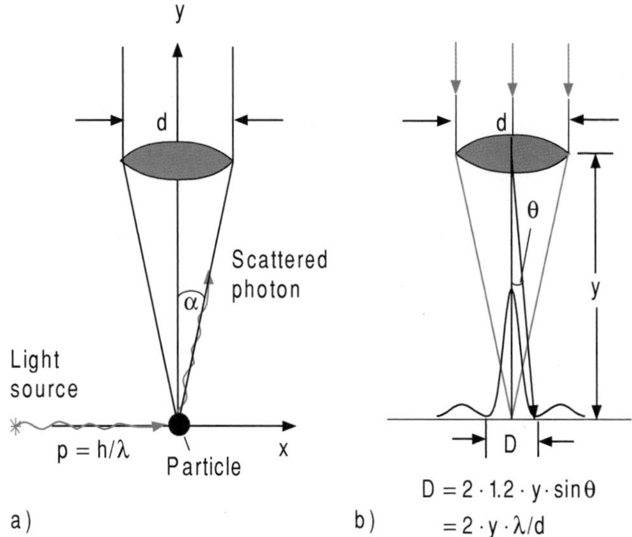


**Fig. 3.34.** Explanation of spatial resolution limit of a microscope by the uncertainty relation

the interval $\Delta x \approx D$. From (3.69) and (3.70) we again obtain the relation

$$\Delta p_x \cdot \Delta x \geq \frac{h}{\lambda}\frac{d}{2y}2y\frac{\lambda}{d} = h\,. \tag{3.71}$$

When using light with shorter wavelength $\lambda$, the uncertainty $\Delta x$ becomes smaller, but the uncertainty $\Delta p_x$ of the particles momentum becomes accordingly larger.

> This example illustrates that the measuring process itself (here the illumination of the particle) changes the state of the measured object (here its momentum and location).

### 3.3.4 Dispersion of the Wave Packet

According to (3.57) the group velocity $v_g$ of the wave packet is related to the momentum $p$ and the velocity $v_p$ of the corresponding particle by

$$v_g = \frac{p}{m} = v_p\,. \tag{3.72}$$

The initial momentum $p$ can be determined only within the uncertainty interval $\Delta p$. This implies a corresponding uncertainty

$$\Delta v_g = \frac{1}{m}\Delta p \geq \frac{1}{m}\frac{\hbar}{\Delta x_0}\,, \tag{3.73}$$

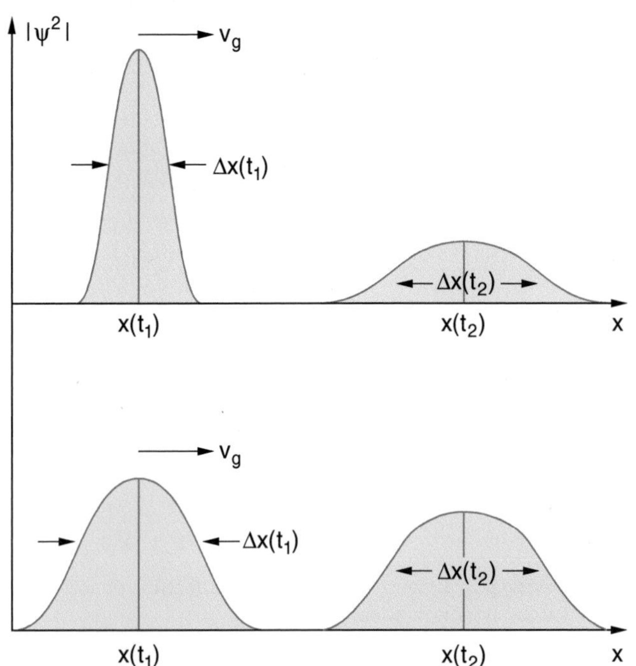


**Fig. 3.35.** Spread of a wave packet in time for two different initial uncertainties $\Delta x(t_1)$

where $\Delta x_0$ is the initial width of the wave packet, i.e., the uncertainty of the determination of the starting point $x_0$. Because of the uncertainty $\Delta v_g$ of the velocity the uncertainty $\Delta x = \Delta v_g t + \Delta x_0$ increases with time. We obtain

$$\Delta x(t) = \Delta v_g t + \Delta x_0 \geq \frac{\hbar}{m \Delta x_0} t + \Delta x_0 .$$

The width $\Delta x(t)$ of the wave packet therefore increases with time while the area $\int |\psi(x,t)|^2 \mathrm{d}x$ remains constant because of the normalization

$$\int_{-\infty}^{+\infty} |\psi(x,t)|^2 \mathrm{d}x = 1 .$$

The rate with which $\Delta x(t)$ increases becomes larger when $\Delta x_0$ is smaller, because the uncertainty $\Delta v_g$ increases as $\hbar/(m\Delta x_0)$ (Fig. 3.35). The localization of a particle moving with velocity $v_g \pm \Delta v_g$ becomes more and more uncertain after its initial parameters $(x_0 \pm \Delta x_0,\ v_g \pm \Delta v_g)_{t=0}$ had been determined at $t = 0$.

### 3.3.5 Uncertainty Relation for Energy and Time

In Sect. 3.3.3 we have discussed the spatial width $\Delta x$ of a wave packet which is composed of an infinite number of monochromatic waves with wavenumbers $k$ within the interval $k - \Delta k/2 \leq k \leq k + \Delta k/2$. We will now investigate how accurate the energy $\hbar\omega_0$ at the center frequency $\omega_0$ of a wave packet can be measured, when the measuring time is $\Delta t$. We consider a wave packet composed of monochromatic waves $A_i \exp[\mathrm{i}(\omega_i t - k_i x)]$, where the frequencies $\omega_i$ are spread over the

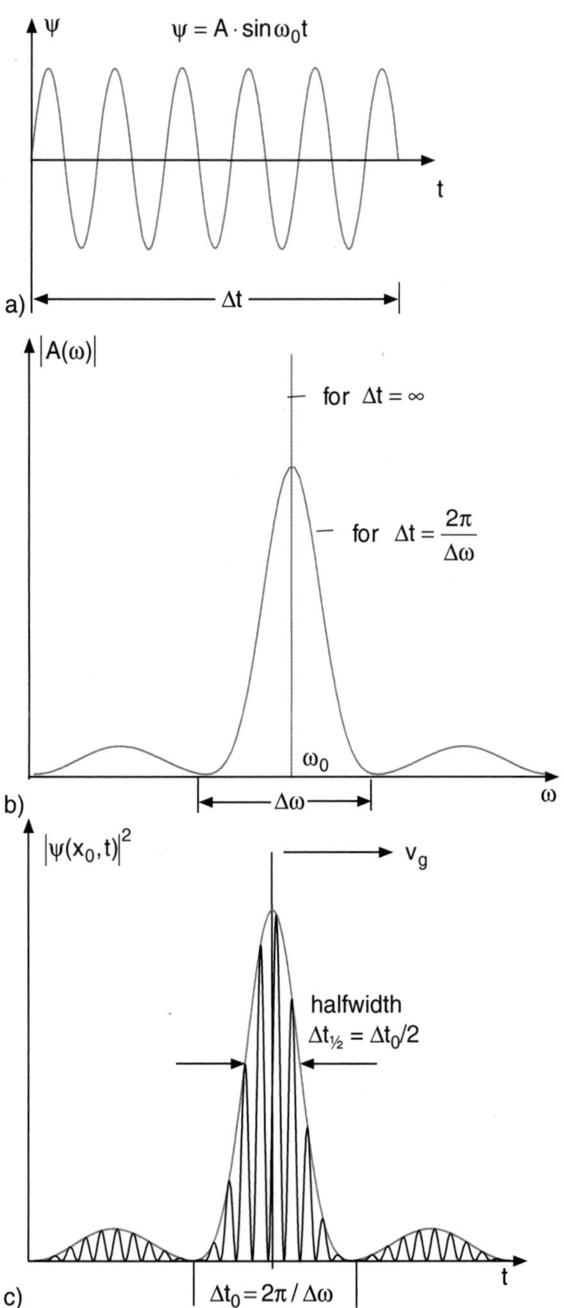


**Fig. 3.36.** Illustration of the uncertainty relation $\Delta\omega \cdot \Delta t \geq 2\pi$. (**a**) Sine wave measured during the time interval $\Delta t$ only. (**b**) Amplitude distribution $A(\omega)$ of this sine wave. (**c**) Wave packet with frequency uncertainty $\Delta\omega$, passing through the detection point $x = x_0$

interval $\omega_0 - \Delta\omega/2 \le \omega_i \le \omega_0 + \Delta\omega/2$ (Fig. 3.36). We now integrate over the frequency interval $\Delta\omega$ instead over the wavenumber interval $\Delta k$. We therefore write the wave packet as the superposition

$$\psi(x,t) = \int_{\omega_0-\Delta\omega/2}^{\omega_0+\Delta\omega/2} A(\omega)\mathrm{e}^{\mathrm{i}(kx-\omega t)}\mathrm{d}\omega \,. \tag{3.74}$$

The procedure is quite analogous to that in Sect. 3.3.3. Inserting the Taylor expansion

$$k = k_0 + \left(\frac{\mathrm{d}k}{\mathrm{d}\omega}\right)_{\omega_0}(\omega - \omega_0) + \dots$$

into (3.74) we obtain for constant amplitudes $C(\omega) \equiv C_0$ with the abbreviations

$$u = t - (\mathrm{d}k/\mathrm{d}\omega)_{\omega_0}x \,; \quad \Delta\omega = \omega_0 - \omega$$

the solution:

$$\psi(x,t) = 2A_0\frac{\sin(u\Delta\omega)}{u}\mathrm{e}^{\mathrm{i}(k_0x-\omega_0t)} \,. \tag{3.75}$$

At a fixed position $x_0$ the maximum of the wave packet at $u = 0$ appears at the time $t_0 = (\mathrm{d}k/\mathrm{d}\omega)x_0$. The two minima at both sides of the central maximum pass the position $x_0$ at the times

$$t_{1,2} = \left(\frac{\mathrm{d}k}{\mathrm{d}\omega}\right)_{\omega_0} x_0 \pm \frac{\pi}{\Delta\omega} \,. \tag{3.76}$$

It therefore takes the time $\Delta t = 2\pi/\Delta\omega$ for the central maximum to pass through the point $x_0$, where the measurement is performed.

On the other hand, when a wave packet is only observed during the time interval $\Delta t$, its central frequency $\omega_0$ can only be measured within an uncertainty $\Delta\omega = 2\pi/\Delta t$. This can be proved as shown below.

When a monochromatic wave

$$\psi(x,t) = A_0\mathrm{e}^{\mathrm{i}(\omega_0t-k_0x)}$$

is measured only during the time interval $\Delta t$ at $x = 0$ the Fourier transform of the wave train gives the amplitude distribution

$$A(\omega) = \int_{-\Delta t/2}^{+\Delta t/2} A_0\mathrm{e}^{\mathrm{i}(\omega_0t-k_0x)}\mathrm{d}t = \frac{A_0\sin\left(\frac{(\omega-\omega_0)\Delta t}{2}\right)}{\frac{1}{2}(\omega-\omega_0)} \,, \tag{3.77}$$

which determines the frequency spectrum of the wave train. The central maximum of this distribution has a width $\Delta\omega = 2\pi/\Delta t$, defined by half the distance $\omega_1 - \omega_2$ between the zero points of (3.77) at $\omega_1 = \omega_0 + 2\pi/\Delta t$ and $\omega_2 = \omega_0 - 2\pi/\Delta t$.

Since the energy $E = \hbar\omega$ is related to the frequency $\omega$ we obtain the uncertainty relation

$$\Delta E \cdot \Delta t \ge 2\pi\hbar = h \tag{3.78}$$

between energy $E$ and observation time $\Delta t$.

When a particle is observed only during the limited time interval $\Delta t$, its energy $E$ can be determined only within the uncertainty limits $\Delta E \ge h/\Delta t$.

**Remark**

1. If a Gaussian amplitude distribution $A(\omega)$ is assumed instead of the constant amplitude $A(\omega) \equiv A_0$, the smallest uncertainty $\Delta E \cdot \Delta t \ge \hbar$ is obtained analogously to (3.64).
2. Equation (3.78) has been derived from classical physics, using the Fourier theorem. The quantum mechanical aspect comes only from the relation $E = h\nu = \hbar\omega$.

## 3.4 The Quantum Structure of Atoms

Rutherford's scattering experiments discussed in Chap. 2 have proved that atoms consist of a nucleus with positive charge $Q = +Ze$, a small radius $R_\mathrm{K} \le 10^{-14}$ m and a mass that nearly equals that of the total atom, and of $Z$ electrons with very small masses, occupying, however, a volume that is about $10^{12}$–$10^{15}$ times larger than that of the nucleus.

Up to now we have not discussed any possible structure of the spatial and energy distribution of the atomic electrons. In particular it must be cleared whether the electrons move around within the atom or whether they are at rest and form a static charge distribution. Both assumptions lead to certain contradictions: Because of the attractive Coulomb force between the positively charged nucleus and the negatively charged electrons a static configuration would not be stable. A dynamical model in which the electrons move on closed paths within the atom and must therefore be continuously accelerated, cannot explain why these accelerated charges do not radiate, as postulated by electrodynamics. They would then loose energy and would also become unstable.

The experiments presented in the following sections have essentially contributed to clarifying these questions.

### 3.4.1 Atomic Spectra

In 1859 *Gustav Kirchhoff* (1824–1887) and *Robert Bunsen* (1811–1899) had already found, through joint research, that atoms only absorb or emit light at certain discrete wavelengths $\lambda_i$. These specific wavelengths that are characteristic of each chemical element, are called the absorption or emission spectra of the atom. These spectra are like a fingerprint of the atom, since every atomic species can be unambiguously recognized by its spectrum.

A possible experimental arrangement for measuring emission spectra is shown in Fig. 3.37a.

The light emitted by atoms in the light source (which might be, for instance, a gas discharge or a high current arc between two electrodes of the material to be investigated) is collected by the lens $L_1$ and imaged onto the entrance slit $S_1$ of the spectrograph, which is placed in the focal plane of lens $L_2$. The parallel light bundle passes through the dispersing prism and the lens $L_3$ images the entrance slit onto the observation plane, where the position of the slit image $S_2(\lambda)$ depends on the wavelength $\lambda$. If the light source emits light with discrete wavelengths $\lambda_K$, the photoplate in the observation plane shows after being developed dark lines at all those positions $x_K$ that correspond to slit images $S_2(\lambda_K)$. Such a spectrum as that in Fig. 3.37b) is therefore called a line spectrum. Many light sources emit continuous spectra, i.e., their emitted intensity $I(\lambda)$ is a continuous function of the wavelength $\lambda$. Examples are the radiation of the sun's photosphere, the emission of a blackbody (see Sect. 3.2) and generally the emission of hot solid bodies.

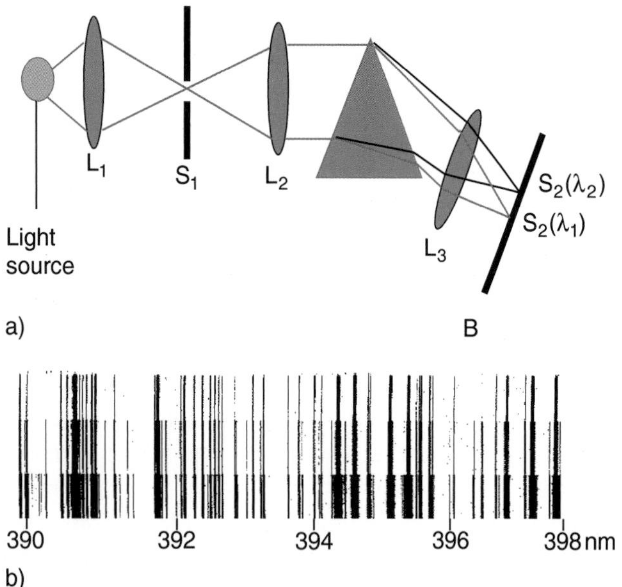


**Fig. 3.37.** (**a**) Prism spectrograph for measuring the emission spectrum of a light source. (**b**) Emission spectrum of an iron arc in the spectral interval 390–398 nm, taken for three different exposure times

Absorption spectra can be measured with the setup shown in Fig. 3.38. The radiation from a continuous light source is collimated by the lens $L_1$ and the parallel light bundle is sent through the absorbing gaseous sample. The lens $L_2$ focuses the transmitted radiation onto the entrance slit of a spectrograph. At those wavelengths $\lambda_K$, where the sample atoms absorb light, the transmitted intensity is smaller than at nonabsorbing wavelengths. The negative of the illuminated photoplate therefore shows bright lines on a dark background (Fig. 3.38b).

Such experiments brought about the following results:

- Each wavelength observed in an absorption spectrum also appears in the emission spectrum of the same kind of atoms if the atoms have been excited into the emitting state by absorption of light or by collisional excitation.
- The absorption and emission spectra are characteristic for specific atoms. They allow the unambiguous determination of the chemical element corresponding to these spectra. The spectral analysis therefore yields the composition of chemical elements in a sample. This is particularly important in astrophysics where the spectrum of the starlight gives information on the number and the composition of chemical elements in the atmosphere of the star.
- The spectral lines are not completely narrow, even if the spectral resolution of the spectrograph is extremely high. This means that the atoms do not emit strictly monochromatic radiation but show an intensity distribution $I(\lambda_K)$ around each wavelength $\lambda_K$

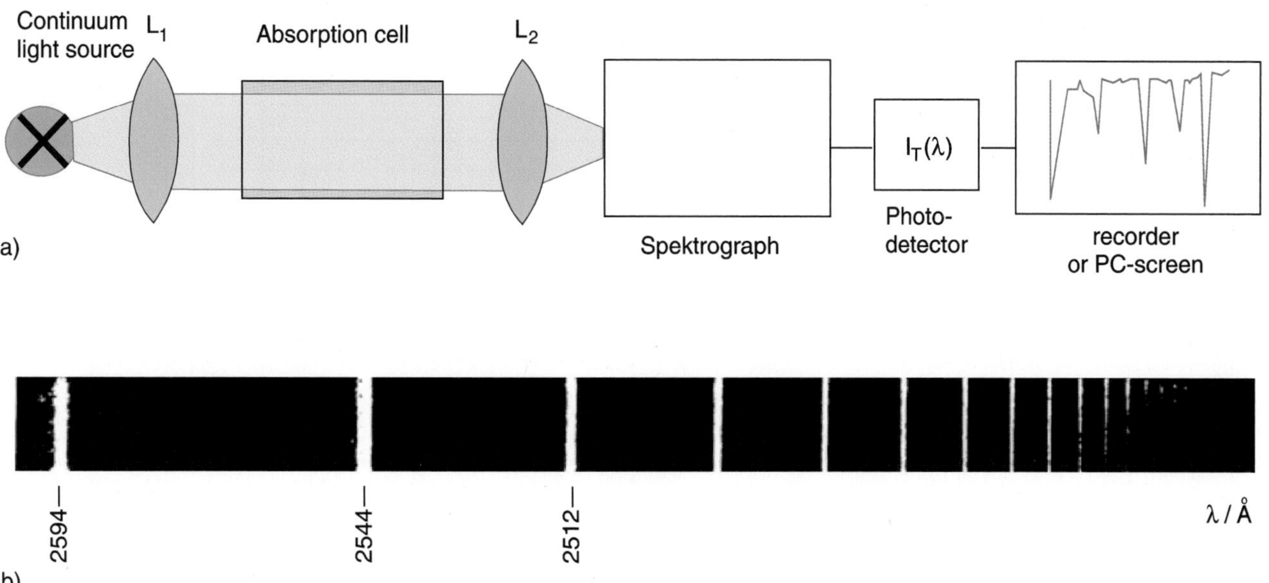


**Fig. 3.38a,b.** Measurement of absorption spectra. (**a**) Experimental arrangement (**b**) Absorption spectrum of sodium vapor

with a finite halfwidth $\Delta\lambda$. The reasons for these halfwidths will be discussed in Sect. 7.5.

The most simple of all atoms is the H atom, consisting of only one proton and one electron. Its emission spectrum (Fig. 3.39) was measured in 1885 by *Johann Jakob Balmer* (1825–1898). He could fit the wavenumbers $\bar{\nu}_K = 1/\lambda_K$ of its emission lines by the simple formula

$$\bar{\nu}_K = Ry\left(\frac{1}{n_1^2} - \frac{1}{n_2^2}\right), \tag{3.79}$$

where the integer numbers $n_1$, $n_2$ take the values $n_1 = 2$ and $n_2 = 3, 4, 5, \ldots$. The constant $Ry = 109{,}678\,\mathrm{cm}^{-1}$ is the Rydberg constant, which is historically given by spectroscopists in units of inverse centimeters $\mathrm{cm}^{-1}$, since all wavenumbers $\bar{\nu}_K = 1/\lambda_K$ are measured in these units.

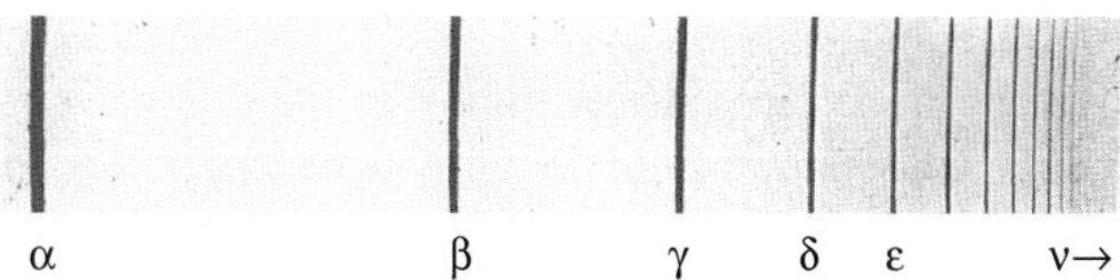


**Fig. 3.39.** Balmer series of the hydrogen atom emitted from a hydrogen gas discharge

## EXAMPLE

A spectral line with a wavelength $\lambda_K = 500\,\mathrm{nm} = 5 \times 10^{-5}\,\mathrm{cm}$ has a wavenumber $\bar{\nu}_K = 2 \times 10^4\,\mathrm{cm}^{-1}$.

Later on *Theodore Lyman* (1874–1954) and *Friedrich Paschen* (1865–1947) found further series in the emission and absorption spectrum of the H atom, which could all be described by the Balmer formula (3.79), but with $n_1 = 1$ (Lyman series) or $n_1 = 3$ (Paschen series) (Fig. 3.40).

How can we understand these experimental results?

### 3.4.2 Bohr's Atomic Model

Many theoriticians tried to develop models that could explain the experimental findings. However, most of these models could describe some results but not all of them in a consistent way without any contradictions. After many efforts *Nils Bohr* (1885–1962) (Fig. 3.41) starting from Rutherford's atomic model finally developed in 1913 the famous planetary model of the atoms [3.4, 17], which we will now discuss for atomic systems with only one electron (H atom, $He^+$ ion, $Li^{++}$ ion, etc.).

In Bohr's atomic model the electron (mass $m_e$, charge $-e$) and the nucleus (mass $m_N$, charge $+Ze$) both move on circles with radius $r_e$ or $r_N$, respectively,

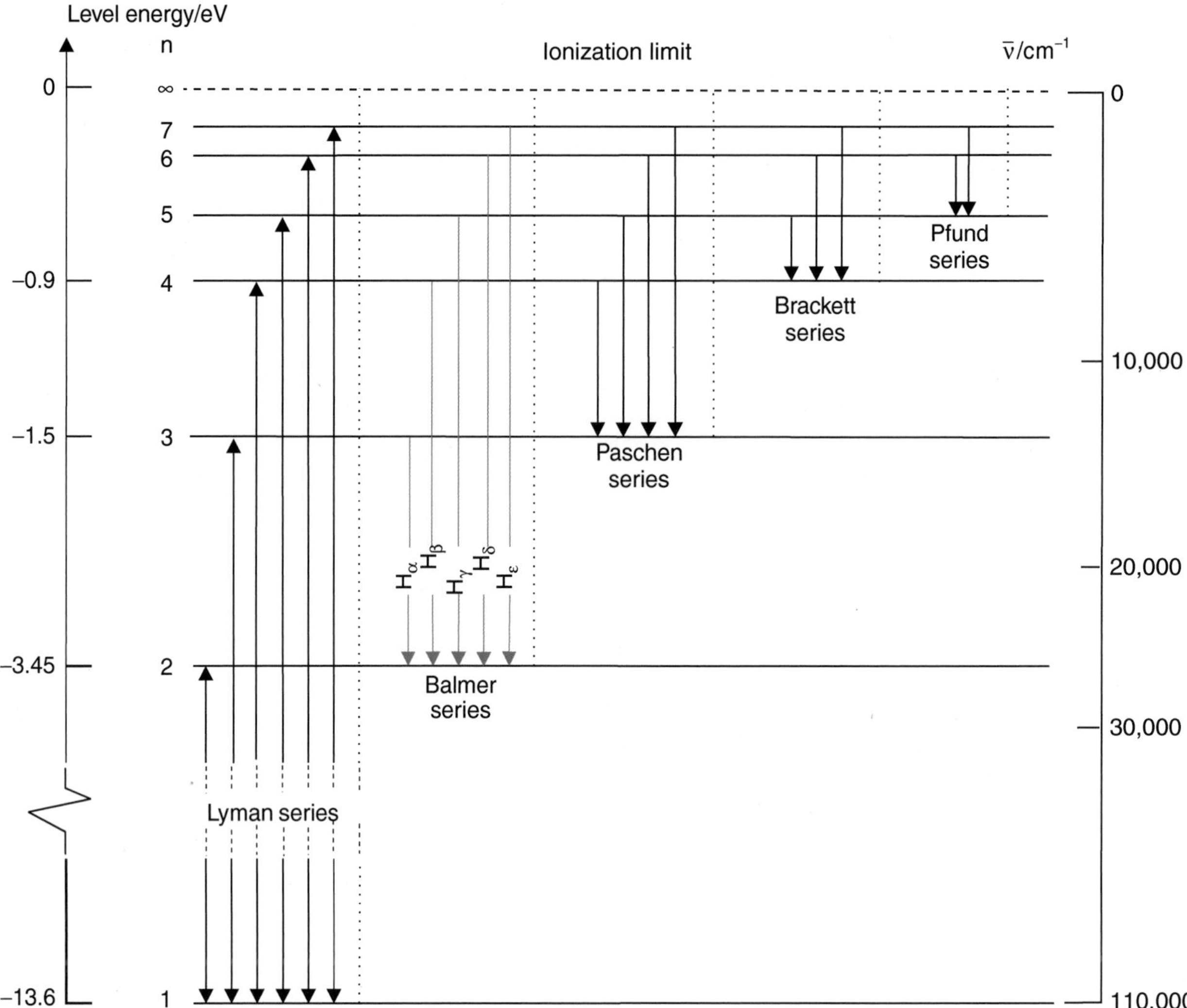


**Fig. 3.40.** Simplified level scheme of the hydrogen atom and the different absorption or emission series

around their center of mass. This movement of two bodies can be described in the center of mass system by the movement of a single particle with reduced mass $\mu = (m_e m_N)/(m_e + m_N) \approx m_e$ in the Coulomb potential $E_{pot}(r)$ around the center $r = 0$, where $r$ is the distance between electron and nucleus. The balance between Coulomb force and centripetal force yields the equation

$$\frac{\mu v^2}{r} = \frac{1}{4\pi\varepsilon_0}\frac{Ze^2}{r^2}, \tag{3.80}$$

which determines the radius

$$r = \frac{Ze^2}{4\pi\varepsilon_0 \mu v^2} \tag{3.81}$$

of the circular path of the electron. As long as there are no further restrictions for the kinetic energy $(\mu/2)v^2$ any radius $r$ is possible, according to (3.81).

If, however, the electron is described by its matter wave with wavelength $\lambda_{dB} = h/(\mu v)$ a stationary state of the atom must be described by a standing wave along the circle (Fig. 3.42) since the electron should not leave the atom. This gives the quantum condition:

$$2\pi r = n\lambda_{dB} \quad (n = 1, 2, 3, \ldots), \tag{3.82}$$

which restricts the possible radii $r$ to the discrete values (3.82). With the de Broglie wavelength $\lambda_{dB} = h/(\mu v)$ the relation

$$v = \frac{h}{\lambda \cdot \mu} = n\frac{h}{2\pi\mu r} \tag{3.83}$$

**Fig. 3.41.** *Niels Bohr* (1885–1962) From E. Bagge: Die Nobelpreisträger (Heinz-Moos-Verlag, München 1964)

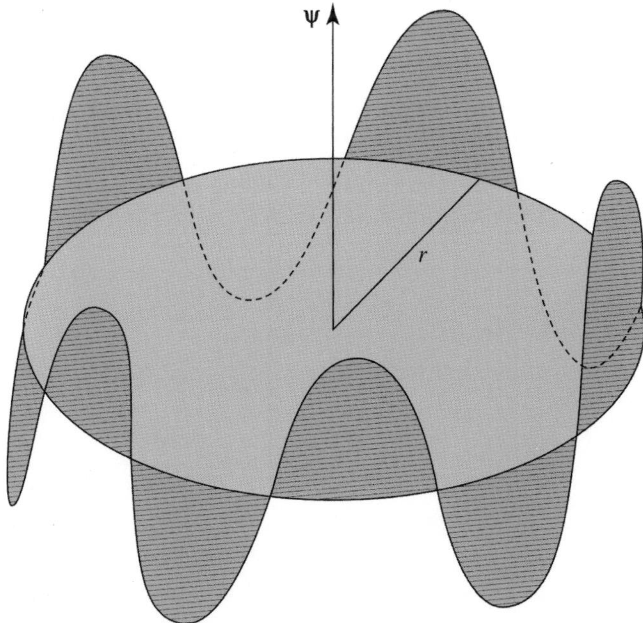


**Fig. 3.42.** Standing deBroglie wave $\psi(\varphi =$ with n = 5 and a wavelength $\lambda_D = 2\pi r/5$ for illustration of the quantum condition in Bohr's atomic model

between velocity and radius is obtained. Inserting this into (3.81) yields the possible radii for the electron circles:

$$r_n = \frac{n^2 h^2 \varepsilon_0}{\pi \mu Z e^2} = \frac{n^2}{Z} a_0 \,, \tag{3.84}$$

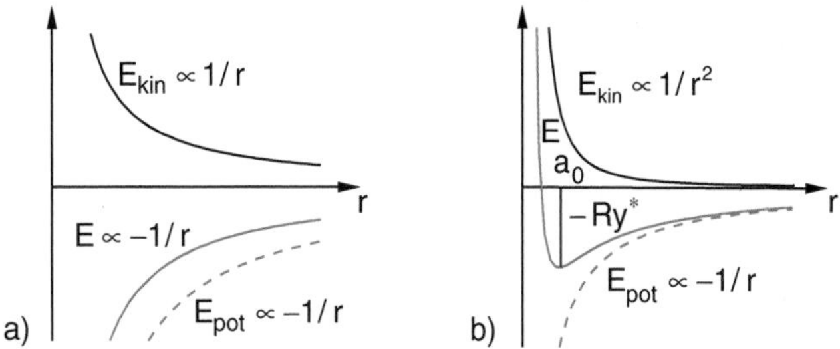


**Fig. 3.43a,b.** Radial dependence of kinetic, potential, and total energy of the electron in the Coulomb field of the nucleus. (**a**) Classical model (**b**) Quantum mechanical model

where

$$a_0 = \frac{\varepsilon_0 h^2}{\pi \mu e^2} = 5.2917 \times 10^{-11}\,\text{m} \approx 0.5\,\text{Å}$$

is the smallest radius of the electron ($n = 1$) in the hydrogen atom ($Z = 1$), which is named the *Bohr radius*.

The kinetic energy $E_{kin}$ of the atom in the center of mass system is obtained from (3.80) as

$$E_{kin} = \frac{\mu}{2} v^2 = \frac{1}{2} \frac{Ze^2}{4\pi \varepsilon_0 r} = -\frac{1}{2} E_{pot} \tag{3.85}$$

and equals $-1/2$ times its potential energy. The total energy (Fig. 3.43)

$$E = E_{kin} + E_{pot} = +\frac{1}{2} E_{pot} = -\frac{1}{2} \frac{Ze^2}{4\pi \varepsilon_0 r} \tag{3.86}$$

is negative and approaches zero for $r \to \infty$. Inserting (3.84) for $r$ yields for the possible energy values $E_n$ of an electron moving in the Coulomb potential of the nucleus:

$$E_n = -\frac{\mu e^4 Z^2}{8 \varepsilon_0^2 h^2 n^2} = -Ry^* \frac{Z^2}{n^2} \tag{3.87}$$

with the Rydberg constant

$$Ry^* = hcRy = \frac{\mu e^4}{8 \varepsilon_0^2 h^2} \tag{3.88}$$

expressed in energy units Joule.

This illustrates that the total energy of the atom in the center of mass system (which nearly equals the energy of the electron) can only have discrete values for stationary energy states, which are described by the *quantum number* $n = 1, 2, 3 \ldots$ (Fig. 3.40). Such a stationary energy state of the atom is called a quantum state. In Bohr's model, the quantum number $n$ equals the number of periods of the standing de Broglie wave along the circular path of the electron.

**Note:**

1. The exact value of the Rydberg constant $Ry$ depends, according to (3.88), on the reduced mass $\mu$ of the electron nucleus system. It differs, therefore, slightly for different masses of the nucleus. In order to have a unique definition, the Rydberg constant $Ry_\infty$ for infinite nuclear mass $m_\mathrm{N} = \infty \Rightarrow \mu = m_\mathrm{e}$ is defined. Its numerical value is $Ry_\infty = 109{,}737.31534\,\mathrm{cm}^{-1}$.

   The Rydberg constant for finite nuclear mass $m_\mathrm{N}$ is then:

   $$Ry = Ry_\infty \mu / m_\mathrm{e} \,.$$

2. Bohr's atomic model is a "semiclassical model", which treats the movement of the electron as that of a point mass on a classical path but adds an additional quantum condition (which is in fact a boundary condition for the de Broglie wavelength of the moving electron).
3. This quantum condition can also be formulated using the angular momentum $\boldsymbol{L}$ of the electron. Multiplying (3.83) by $\mu r$ yields

   $$\mu r v = |\boldsymbol{L}| = n\hbar \,,$$

   where $\hbar = h/2\pi$. This means:

> The angular momentum of the electron on its path around the nucleus is quantized. The absolute value $n\hbar$ is an integer multiple of Planck's constant $\hbar$.

The two conditions:

a) The angular momentum of the atom in the center of mass system is $|\boldsymbol{L}| = n\hbar$
b) The circumference of the circular path of the electron $2\pi r = n\lambda_\mathrm{dB}$ must be an integer multiple of the de Broglie wavelength

are identical. They are both due to the boundary condition for the standing de Broglie wave.

In order to explain the line spectra observed in absorption or emission, the following hypothesis is added to Bohr's model.

By absorption of a photon $h\nu$ the atom can be excited from a lower energy State $E_i$ into a higher state $E_k$, if the energy conservation

$$h\nu_{ik} = E_k - E_i \tag{3.89}$$

is fulfilled. Inserting the relation (3.87) for the energies $E_k$, $E_i$ yields the frequencies

$$\nu_{ik} = \frac{Ry^*}{h} Z^2 \left( \frac{1}{n_i^2} - \frac{1}{n_k^2} \right) \tag{3.90}$$

of the absorbed light. With the wave numbers $\bar{\nu} = \nu/c$ and $Ry^* = hc \cdot Ry$ we obtain for the hydrogen atom ($Z = 1$) exactly Balmer's formula (3.79) for his observed spectra.

When emitting a photon $h\nu$, the atom undergoes a transition from a higher energy state $E_k$ to a lower state $E_i$, where again energy conservation (3.89) has to be fulfilled.

We will summarize the preceding results of Bohr's model of hydrogenic atoms with a single electron.

> - The electron moves on circles around the nucleus with quantized radii
>
>   $$r_n = \frac{n^2}{Z^2} a_0 = \frac{n^2 h^2 \varepsilon_0}{\pi \mu Z e^2}$$
>
>   that increase quadratically with the integer quantum number $n$.
> - The possible values $r_n$ are inversely proportional to the nuclear charge $Ze$. For the $\mathrm{He}^+$ ion with $Z = 2$ they are only half as large as in the hydrogen atom.
> - In each quantum state the atom has a well-defined total energy
>
>   $$E_n = -Ry^* \frac{Z^2}{n^2} \,, \quad E_\mathrm{pot} = +2E_n \,,$$
>   $$E_\mathrm{kin} = -E_n \,.$$
>
>   The energy $E_\infty = 0$ for $n = \infty$ and $r_n \to \infty$ is chosen as zero. In its lowest possible state the energy is $E_1 = -Ry^* Z^2$. Therefore the

positive energy $-E_1$ is necessary to ionize the atom in its ground state (i.e., to bring the electron from $r = r_1$ to $r = \infty$). It is called the *ionization energy*. For the H atom the ionization energy is $E_{\text{ion}} = 13.6\,\text{eV}$.

- By absorption of a photon $h\nu = E_k - E_i$ the atom can be excited from its lower energy state $E_i$ into the higher state $E_k$. Emission of a photon by an excited atom causes a transition from $E_k$ to $E_i$.

**Note:**

The first excited state ($n = 2$) of the H atom already needs an excitation energy of about 10.2 eV, which is 3/4 of the ionization energy.

Although Bohr's semiclassical atomic model explains the observed spectra very well, and also brings some esthetical satisfaction, because of its resemblance to the planetary system, it leaves several questions open. One essential point is that, according to classical electrodynamics, every accelerated charge should emit radiation. The electron on its circular path is such an accelerated charge. It should, therefore, loose energy by emitting radiation and should spiral down into the nucleus. Therefore, the Bohr model cannot explain the existence of stable atoms.

### 3.4.3 The Stability of Atoms

The stability of atoms is consistently explained by quantum theory. We will here give a conspicuous argument based on the uncertainty relation. It should be only regarded as a simple estimation that is not restricted to circular paths of the electron. If $a$ is the mean radius of the atom, we can give the distance $r$ of the electron from the nucleus with an uncertainty $\Delta r \le a$, since we know that the electron has to be found somewhere within the atom. According to the uncertainty relation the uncertainty $\Delta p_r$ of the radial component of the electron momentum $p$ must be larger than $\hbar/a$. Therefore we conclude for the uncertainty $\Delta p \ge \Delta p_r \ge \hbar/a$ (otherwise we could determine $p$ within narrower limits than its component $p_r$). We find the relation $p > \Delta p \ge \hbar/a$. The mean kinetic energy of the electron is:

$$E_{\text{kin}} = \frac{p^2}{2m_e} \ge \frac{(\Delta p)^2}{2m_e} \ge \frac{\hbar^2}{2m_e a^2} . \tag{3.91}$$

Its potential energy at a distance $a$ from the nucleus is

$$E_{\text{pot}} = -\frac{e^2}{4\pi\varepsilon_0 a} \tag{3.92}$$

and its total energy $E = E_{\text{kin}} + E_{\text{pot}}$ at the distance $a$ is then:

$$E \ge \frac{\hbar^2}{2ma^2} - \frac{e^2}{4\pi\varepsilon_0 a} . \tag{3.93}$$

The largest probability of finding the electron is at a distance $a_{\text{min}}$ where the total energy is minimum, i.e., where $dE/\text{d}a = 0$. This gives

$$a_{\text{min}} = \frac{4\pi\varepsilon_0\hbar^2}{me^2} = \frac{\varepsilon_0 h^2}{\pi\mu e^2} = a_0 \tag{3.94}$$

which is identical to the Bohr radius $a_0$.

Therefore, a stable state exists with the minimum energy limit

$$E_{\text{min}} = -\frac{me^4}{2(4\pi\varepsilon_0\hbar^2)^2} = -\frac{me^4}{8\varepsilon_0^2 h^2} = -Ry^* , \tag{3.95}$$

which is consistent with the energy of the lowest state with $n = 1$ in Bohr's model.

Although the quantum mechanical results for the energy confirms Bohr's result, the explanation of the stability is different.

According to the uncertainty principle the atom in its lowest state cannot radiate because it has minimum energy. In order to emit a photon, it would have to make a transition to a higher energy state, which contradicts energy conservation. The reason for this energy minimum is the sharp increase of the kinetic energy of the electron with decreasing distance $a$, due to the uncertainty of its momentum (Fig. 3.43). In higher energy states the atom can radiate, in accordance with the experimental results.

In Bohr's model the stability is explained by the assumption of standing waves for the electron, where the Poynting vector is zero. However, this does not explain why higher energy states, which are also represented by standing waves, do radiate.

### 3.4.4 Franck–Hertz Experiment

*James Franck* and *Gustav Hertz* [3.18] gave in 1914 an impressive experimental proof for the energy quantization of atoms based on the following experimental arrangement (Fig. 3.44a).

Electrons, emitted from a hot cathode in a bulb, filled with mercury vapor at low pressures, are accelerated to the energy $E_{\text{kin}} = eU$ by the grid G at the voltage $U$ against the cathode. The electrons can only reach the collecting anode A that is kept at a lower voltage $U_A = U - \Delta U$, if their energy after having passed the grid G is at least $e\Delta U$.

When measuring the electron current $I_A(U)$ as a function of the acceleration voltage $U$, one obtains a curve like that in Fig. 3.44b. The current increases with $U$ between $U = 0$ and $U = 4.9\,\text{eV}$ and follows a typical diode characteristics. Above $U_1 = 4.9\,\text{V}$ the current decreases sharply, goes through a minimum, rises again until it reaches a second maximum at about $U_e = 9.8\,\text{eV}$.

How can this be explained?

The electrons suffer elastic and inelastic collisions with the Hg atoms. In inelastic collisions,

$$e^- (E_{\text{kin}}) + \text{Hg} \to \text{Hg}^* (E_a) + e^- (E_{\text{kin}} - E_a)$$

the electrons excite the Hg atoms and transfer the amount $\Delta E_{\text{kin}} = E_{\text{kin}} - E_a$ of their kinetic energy to the excitation energy $E_a$ of the atom. Because of this loss of energy the electrons cannot overcome the bias voltage $-\Delta U$ and therefore cannot reach the detector.

During elastic collisions the electron can at most transfer the fraction $4m_e/m_{\text{Hg}} \approx 10^{-5}$ of its kinetic energy. At sufficiently low pressures each electron suffers only a few elastic collisions and the total energy loss due to elastic collisions is then completely negligible. However, elastic collisions may result in large angular changes of the electron's flight direction and the electrons may therefore hit the walls of the tube before they reach the anode. Without inelastic collisions the electron current would follow the dashed curve in Fig. 3.44b, which resembles the electron current in a diode tube. The further maxima and minima in the actually measured current $I(U_A)$ are due to the fact that at sufficiently large voltages $U$ the electron can regain, after $n$ inelastic collisions, the minimum required kinetic energy $e\Delta U$ during its flight path to the grid G for

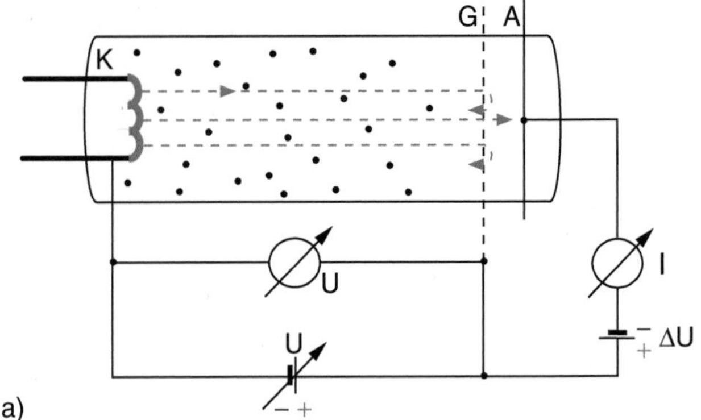


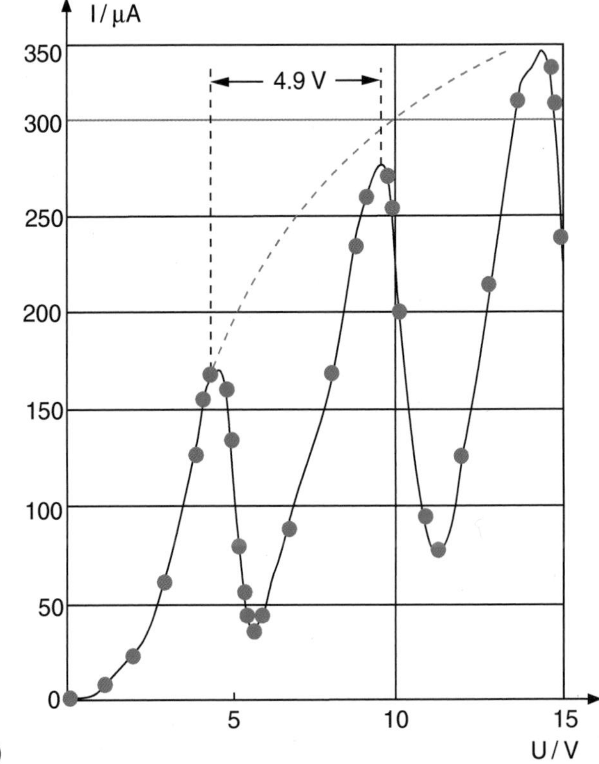


**Fig. 3.44a,b.** Franck–Hertz experiment. (**a**) Experimental setup (**b**) Electron current as function of the acceleration voltage $U$ in a tube with mercury vapor

overcoming the bias voltage $\Delta U$ but has not enough energy for the $(n + 1)$th inelastic collision [3.19].

The separation between subsequent maxima corresponds to the excitation energy $E_a = 4.9\,\text{eV}$ of Hg atoms. The exact form of the curve $I(U)$ in Fig. 3.44 is determined by

- The energy dependence of the excitation probability
- The energy distribution of the electrons emitted from the hot cathode.

With the improved experimental setup of Fig. 3.45a the energy resolution could be substantially improved. Here, two grids are used and the acceleration of the electrons is essentially restricted to the short flight path between K and $G_1$, while the small adjustable voltage $U_2$ between $G_2$ and $G_1$ does not change the electron energy much. The excitation probability is then nearly the same for all points between $G_1$ and $G_2$. With such an improved apparatus the finer details of the excitation

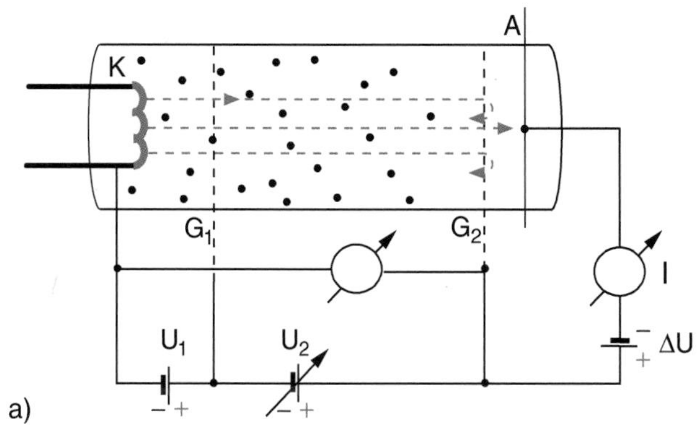


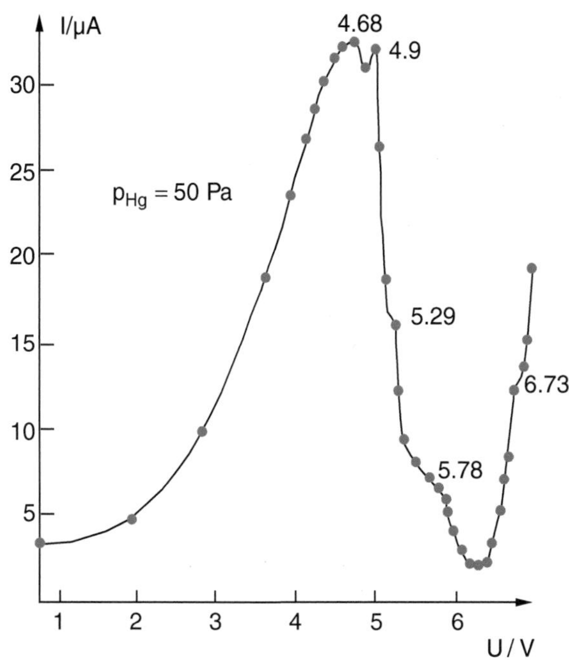


**Fig. 3.45.** (**a**) Improved experimental arrangement for the Franck–Hertz experiment with higher energy resolution. (**b**) Electron current $I(U)$ measured with the apparatus shown in (**a**), where the excitation of many higher levels in the Hg atom can be seen. The structured maximum corresponds to the first maximum in Fig. 3.44b

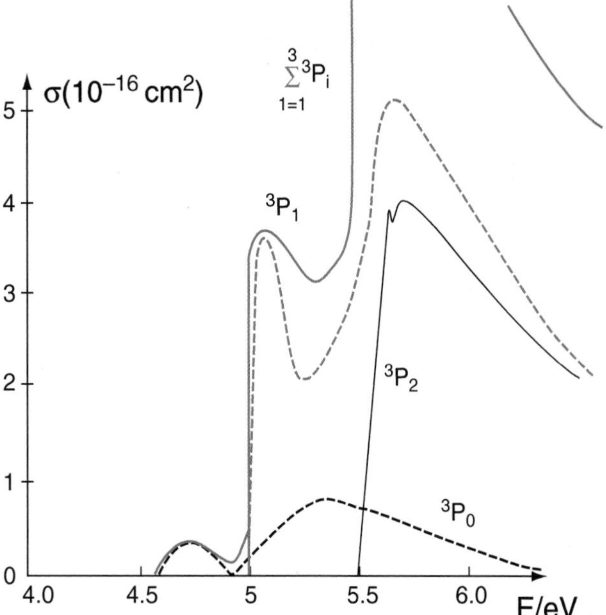


**Fig. 3.46.** Excitation cross section for electron excitation of the singlet → triplet transition $6^1S_0 \rightarrow 6^3P_{0,1,2}$ in the Hg-atom [3.19]

function could be resolved, which correspond to different excited states of the Hg atoms (Fig. 3.46).

The excited Hg* atoms release their excitation energy by emission of light

$$\mathrm{Hg}^* \rightarrow \mathrm{Hg} + h\nu \, .$$

Measuring this fluorescence light through a monochromator shows that the emitted spectral lines have wavelengths $\lambda_k$, which exactly correspond to the measured absorption lines of Hg vapor. Time-resolved measurements of this fluorescence prove that the excited atomic levels $E_i$ are not stable. They decay within a very short time (typically $\approx 10^{-8}$ s) into lower states $E_k$, where

$$\Delta E = E_k - E_i = h\nu_{ik} = hc/\lambda_{ik} \, .$$

Only the lowest atomic states (called the ground states) are stable. Their lifetimes are infinitely long (if not excited by collisions or absorption of photons). The wavelength $\lambda_{ik}$ of the emission spectrum measured through a spectrograph allow a much higher accuracy in the determination of energy levels than those obtained from the electron impact measurements in Figs. 3.45 and 3.46.

The experimental results of the electron impact excitation prove that atoms can acquire energy only in discrete energy quanta $\Delta E$. Their magnitude depend on the specific atom and its level structure.

## 3.5 What are the Differences Between Classical and Quantum Physics?

In the quantum physical description of microparticles, such as atoms, molecules, electrons and photons, there is no distinct separation between particle model and wave model. The matter wave function is characterized by the particle momentum $p$ and energy $E$ as well as by the de Broglie wavelength $\lambda$ or the frequency $\nu = E/h$. The examples given in the previous sections have illustrated the particle nature of light and the wave properties of particles. In this section we will discuss the particle-wave duality of microparticles and make clear, by some more instructive examples, that this duality does not give contradictory but rather complementary descriptions of nature.

### 3.5.1 Classical Particle Paths Versus Probability Densities in Quantum Physics

The classical path of a particle can be exactly predicted, at least in principle, for all times, if the initial conditions (e.g., $\boldsymbol{r}(t=0)$ and $\boldsymbol{v}(t=0)$) and the forces acting on the particle are known. For the model of point-like massive particles the equation of motion (Newton's equation $\boldsymbol{F} = m\boldsymbol{a}$) can be solved either analytically or numerically with computers within any wanted accuracy.

For linear equations of motion, small inaccuracies of the initial conditions results in only small uncertainties of the further path $\boldsymbol{r}(t)$ of the particles.

However, many phenomena in nature have to be described by nonlinear equations of motion (e.g., the motion of a particle in a turbulent flow). Here, tiny changes in the initial conditions may already change the future development of the particles motion drastically. For such "chaotic" movements the exact calculation of the motion $\boldsymbol{r}(t)$ is in principle not possible, even in "classical physics."

Quantum physics brings, through the uncertainty relations, an additional principal limit to the calculation of the time development of a physical system.

- The initial conditions $\boldsymbol{r}(0)$ and $\boldsymbol{p}(0)$ for location and momentum of a particle can not be both given simultaneously exactly but only within uncertainty limits. The product $\Delta x_i \cdot \Delta p_i$ $(i = x, y, z)$ of the uncertainty $\Delta x_i$, $\Delta p_i$ cannot be smaller than Planck's constant $\hbar$ (Fig. 3.47). Instead of the classical well-defined path represented by the solid curve in Fig. 3.47 the location $x(t)$ can be only determined within a certain area $\Delta x \cdot \Delta t$, schematically shown by the coloured area in Fig. 3.47, which becomes larger over the course of time.
- The determination of the exact paths $\boldsymbol{r}(t)$ of single particles is replaced in quantum physics by probability statements. It is only possible to determine the probability $P(x, p, t)$ to find a microparticle with momentum $p(t)$ at the location $x(t)$ at time $t$.
- Measurements of $x$ and $p$ changes the state of the micro-particle (see Sect. 3.3.3).
- The probability of finding a particle at time $t$ at the location $x$ is related to its wave function $\psi(x,t)$. Averaging over a large number of identical measurements gives the mean probability $|\psi(x,t)|^2\mathrm{d}x$ to find the particle at time $t$ within the spatial interval $\mathrm{d}x$ around $x$. In the classical wave description this corresponds to the normalized intensity of the wave at $(x, t)$.

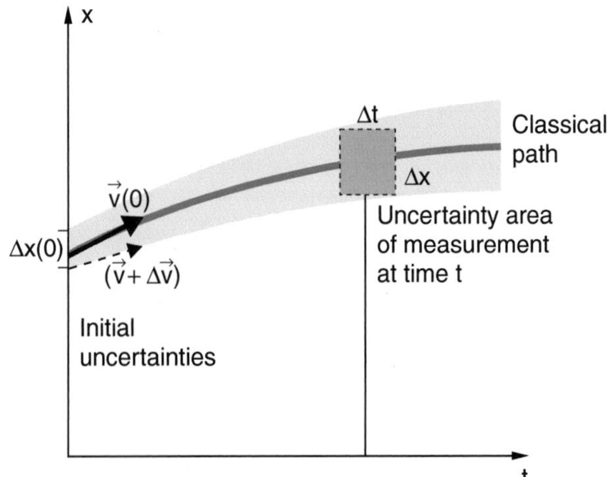


**Fig. 3.47.** Uncertainty limits of the path of a microparticle, determined by the uncertainty $\Delta x$, $\Delta p_x$ of the initial conditions and by the principal uncertainty of the measurement of $\Delta x$ at the particle location at time $t$ and $\Delta t$ of the time $t$

- The spatial uncertainty $\Delta x$ of a particle corresponds to its de Broglie wavelength $\lambda_{dB} = h/p = h/mv$. While this uncertainty plays an essential role for microparticles, it is generally completely negligible for macro particles because of their large mass. Only under special conditions (electrons in metals or neutrons in neutron stars) the uncertainty relation remains essential for macroscopic bodies. Of course, it is responsible for the stability of atoms (see Sect. 3.4.3) and therefore of matter in general.

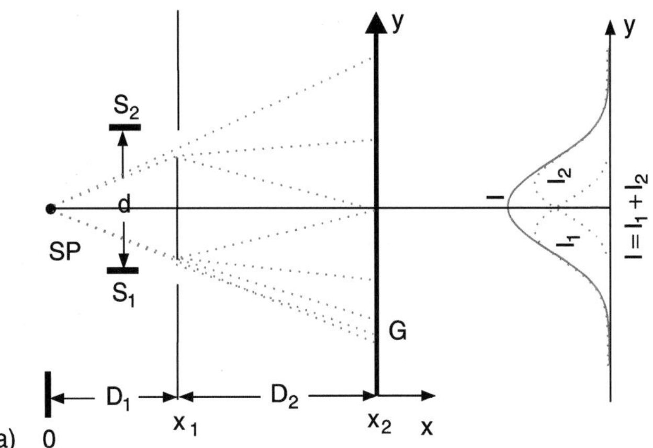


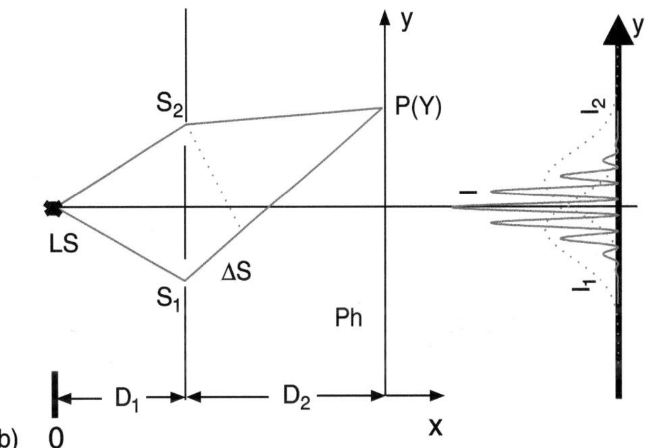


**Fig. 3.48. a,b** Young's double slit experiment (**a**) with macroscopic dye particles and (**b**) with photons

### 3.5.2 Interference Phenomena with Light Waves and Matter Waves

The observation of interference phenomena had been always regarded as convincing proof for the wave model of light. In this section we will illustrate the physical essence of the quantum description of particles by matter waves, discussing several modifications of Young's double slit experiment, performed with:

- Macroscopic particles
- Light
- Electrons

#### a) Macroscopic Particles

A spray gun SP in Fig. 3.48a produces a divergent beam of small dye particles ($\varnothing \approx 1\,\mu$m), which hits a screen with the two narrow slits $S_1$ and $S_2$ (width $b$, separation $d >> b$.) at the position $x = x_1$. At the position $x_2$ behind the screen the transmitted particles hit a glass plate G, where they stick to the surface. The intensity $I(y)$ of the transmitted particles can be measured by the density of the deposited dye layer.

When we close the slit $S_2$, we obtain the density-distribution $I_1(y)$ (dashed curve) when closing $S_1$, a slightly shifted distribution $I_2(y)$ is measured. When both slits are open, the measured intensity distribution $I(y) = I_1(y) + I_2(y)$ equals the sum of the two distributions of each slit, as could have been expected.

> With macroscopic particles no interference phenomena are observed!

#### b) Light

When we replace the spray gun by a light source LS and the glass plate by a photoplate Ph we observe for a proper choice of the slit dimensions ($b \approx 2\lambda$, $d > b$) similar intensity distributions as in Sect. a, if only one of the slits is open. They correspond to the central diffraction maximum for the diffraction by a single slit. However, if both slits are open, the observed intensity does not equal the sum $I_1 + I_2$, but an interference pattern appears that can be described by

$$\begin{aligned} I(y) &= |A_1(y) + A_2(y)|^2 \\ &= I_1 + I_2 + 2A_1A_2\cos(\Delta\varphi(y))\,, \end{aligned} \tag{3.96}$$

where $A_i$ is the amplitude of the partial wave transmitted through the slit $S_i$ and $\Delta\varphi = (2\pi/\lambda)\Delta s$ is the phase difference between the two interfering partial waves at the point $P(y)$ on the photoplate, resulting from the path difference $\Delta s(y) = \overline{S_1P(y)} - \overline{S_2P(y)}$.

Now we perform an important modification of the experiment.

We gradually decrease the intensity $I_0$ of the light source until at most one photon is traversing the system during the passage time $\Delta t = D/c$ with $D = D_1 + D_2$ from the source to the detector. There is then always at most one photon on the way. Its arrival time and location in the detector plane can be measured when replacing the photoplate by an array of many small photo detectors (CCD-array). The photon can only pass through *one* of the two slits. If it passed simultaneously through both slits, each slit would have transmitted half a photon $\frac{1}{2}\hbar\omega$, in contradiction to Planck's hypothesis that $\hbar\omega$ is the smallest energy unit of the electromagnetic field.

If the experiment is continued over a sufficiently long time period in order to collect a sufficiently large signal, one again observes an interference pattern. This interference could *not* have been produced by the interaction between different photons, passing simultaneously through the two slits.

If two shutters are placed in front of the two slits, which open alternately at a defined time, so that the observer knows which slit is open, the interference pattern disappears! The intensity in the observation plane then becomes

$$I = I_1 + I_2 \,.$$

This illustrates that the interference pattern only appears when we do not know through which of the two slits a photon has passed, although we know that it must have passed through one of the two slits. We can only state that the probability is $P = 0.5$ for passing through one of the two slits.

The quantum physical description is unambiguous. If $\psi_1$ is the wave function of the photon passing through slit $S_1$ and $\psi_2$ for passing through $S_2$ the wave function must be $\psi = \psi_1 + \psi_2$ for the case that both slits are open and we do not know through which slit the photon has passed. The probability of detecting a photon in the observation plane is therefore:

$$\begin{aligned} |\psi(x = x_1 + x_2, y)|^2 &= |\psi_1 + \psi_2|^2 \\ &= |\psi_1|^2 + |\psi_2|^2 + \psi_1^*\psi_2 + \psi_1\psi_2^* \,. \end{aligned} \tag{3.97a}$$

The last two terms represent the interference. If we assume (following Huygen's principle) that each point of the slit acts as a source of an outgoing spherical wave

$$\psi_i = \frac{A_i}{r} \mathrm{e}^{\mathrm{i}(k_i r - \omega t)}$$

the interference term becomes for $A_1 = A_2 = A$:

$$\psi_1^*\psi_2 + \psi_2^*\psi_1 = \frac{A^2}{r^2} \cos\left[k(s_1 - s_2)\right] \,, \tag{3.97b}$$

where $s_i$ is the distance between slit $S_i$ and the point of observation $P(y)$.

If the slit $S_1$ is closed, $\psi_1$ becomes zero and the interference disappears.

### c) Electrons

Instead of the light source in Sect. b we now use an electron gun, which emits a divergent electron beam covering both slits. A spatially resolving electron detector (CCD-array) in the plane $x = x_1 + x_2$ measures the transmitted electrons. The observed intensity pattern is quite similar to that in Sect. b if all quantities $(b, d, x_1 + x_2)$ are scaled by the ratio $\lambda/\lambda_{\mathrm{dB}}$ of light wavelength $\lambda$ and the de Broglie wavelength $\lambda_{\mathrm{dB}}$. The experiment shows that electrons also produce interference patterns due to their wave properties if the ratios $\lambda_{\mathrm{dB}}/b$ and $b/d$ are chosen properly.

We will now discuss another, very instructive "Gedanken-experiment" as a modification of the previous experiment. When a light source LS is placed

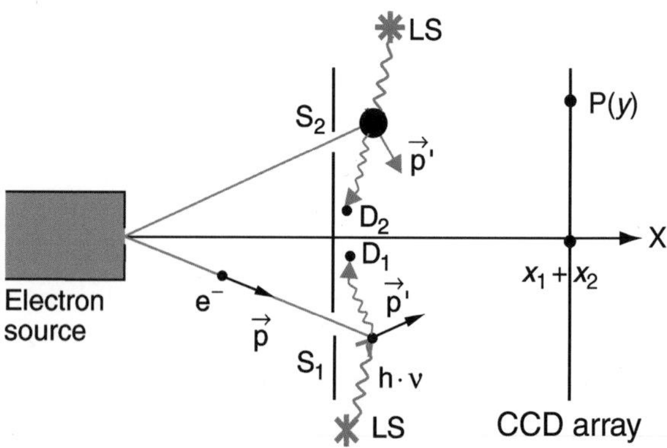


**Fig. 3.49.** Double slit experiment with electrons. The light from a source LS scattered by an electron after having passed slit $S_1$ allows one, in principle, to know through which of the two slits the electron has passed

behind the plane of the two slits, it can illuminate electrons passing through one of the slits (Fig. 3.49). The light scattered from the electron could be detected by one of the two detectors $D_i$ $(i = 1, 2)$ and gives information about the location of the electron just behind the slit. In this way it should be possible to decide through which of the two slits the electron has passed.

If the detector $D_i$ and simultaneously a pixel of the CCD-array in the plane $x = x_1 + x_2$ deliver a signal pulse we know that an electron has passed through slit $S_1$ and has impinged onto the point $P(y)$ in the plane $x = x_1 + x_2$. Will there be still an interference pattern? Quantum mechanics says "no," in accordance with the experimental results. How can this be understood?

The collision between photon and electron changes the electron's momentum and the direction of its velocity. As was shown in Sect. 3.4 for the example of the microscope, the momentum change is $\Delta p_y \leq h/d$, if $d$ is the spatial resolution, i.e., the uncertainty, with which the location of the scatterer can be determined. The electron then arrives at the detection plane at another point $(x = x_1 + x_2, y + \Delta y)$. The statistical distribution with a width $\Delta y = (\Delta p_y / p_y)x_2$ becomes larger than the distance $\Delta = (\lambda/d)x_2 = h/(p_y d)x_2$. This means, the interference is completely washed out if the uncertainty of the electron's location in the plane of the slits $S_1$, $S_2$ becomes smaller than the slit distance $d$. The measuring process change the conditions for observation and destroys partly or completely the interference.

New experiments reveal, however, that the recoil during the measuring processes of the electron's location is not the main cause of the disappearance of the interference pattern, but the knowledge about the way of the electron (see Sect. 12.5). This has been proved by measuring the interference pattern in a specific designed atom interferometer. The path of an atom in an atom-interferometer can be inferred from preparing specific atomic states by microwave absorption. Because the microwave frequency $\nu$ is much smaller than that of visible light, its momentum $p = h\nu/c$ transferred to an atom is completely negligible. Nevertheless, any interference phenomena at the exit of an atom interferometer disappeared when knowledge was gained about which way an atom had gone [12.4]. We can therefore summarize:

> Interference phenomena for light or particle waves are due to the principal lack of knowledge about the exact path of particles through the interference device.

The probability description of Born's statistical interpretation of matter waves can be illustrated by the following example. When in the arrangement of Fig. 3.48 only a single electron is detected at $(x = D, y)$, we do not know, whether it has passed through slit $S_1$ or $S_2$. However, we can give the probability amplitude $\psi_1(y)$ that the electron has passed through $S_1$ and is detected at $(x = D, y)$ or $\psi_2(y)$ for its passage through $S_2$. For a coherent superposition the amplitudes have to be added. The total probability, to find the electron at the position $y$ in the detector plane, independent of its way through $S_1$ or $S_2$ is then

$$\begin{aligned} |\psi(y)|^2 &= |\psi_1(y - y_1) + \psi_2(y - y_2)|^2 \\ &= |\psi_1|^2 + |\psi_2|^2 + \psi_1^* \psi_2 + \psi_1 \psi_2^* \,. \end{aligned} \quad (3.98)$$

Since the probability to find the electron somewhere in the detector plane is

$$\int |\psi(y)|^2 \mathrm{d}y = 1$$

the detection probability for a detector that covers the interval from $y$ to $y + \mathrm{d}y$ is

$$P(y)\mathrm{d}y = |\psi(y)|^2 \mathrm{d}y \,.$$

When $N$ electrons have passed per second through either of the two slits, the counting rate of the detector is

$$Z(y)\mathrm{d}y = N|\psi(y)|^2 \mathrm{d}y \,. \quad (3.99)$$

Inserting (3.98) gives the spatial interference structure of the counting rate $Z(y)$.

### 3.5.3 The Effect of the Measuring Process

The examples in the preceding sections have illustrated that in the quantum mechanical description of a measurement, the measuring process itself affects the result of the measurement. When we measure the location of a particle, we change by this measurement its momentum $p$. This change $\Delta p$ is the larger the more accurate we measure its position. If this measurement of position is performed with light of wavelength $\lambda$ (remember that the position $x$ cannot be measured more accurately than by $\Delta x \approx \lambda$). The photon transfers the momentum $\Delta p = h/\lambda$ to the particle and the random direction of the recoil momentum results in an uncertainty $\Delta p$ of the particles momentum.

The influence of the measuring process on the state of the measured object can impose a limit to the accuracy of the measurement not only for microparticles but even for macroscopic bodies.

One example is the measurement of gravitational waves by a large metal cylinder with about $10^4$ kg mass suspended by elastic strings [3.20].

The gravitational waves, which are produced, for example, by an exploding star (supernova) cause a periodic contraction and expansion $\Delta L$ of the length $L$ of the cylinder. For a supernovae at a distance of $10^5$ light years the amplitude $\Delta L$ is estimated as $10^{-21}$ m. For its measurements, two points $(x_1, y_1)$ and $(x_2, y_2 = y_1)$ at the ends of the cylinder have to be marked. The measurement of their position with an uncertainty $\Delta x = 10^{-21}$ m will result in the momentum transfer $\Delta p \geq \hbar/(2\Delta x)$. If the cylinder was at rest before the measurement, it will afterwards move with a velocity $v = \Delta p/m \geq \hbar/(2m\Delta x)$. For a frequency $\nu_G = 10^3\,\mathrm{s}^{-1}$ of the gravitational wave the time between two contractions is $\tau \approx 10^{-3}$ s. Within this time the cylinder moves by $\Delta x_m = v\tau \approx \hbar\tau/(2m\Delta x)$. Inserting the values $m = 10^4$ kg, $\Delta x = 10^{-21}$ m, $\tau = 10^{-3}$ s, gives

$$\Delta x_m \geq 5 \times 10^{-21}\,\mathrm{m} = 5\Delta x\ .$$

The uncertainty of the local position of the massive cylinder, caused by measuring its position, is therefore larger than the shift expected due to the gravitational wave. A way out of this dilemma is a larger mass $m$ and averaging over many measurements.

> We note that the measuring process itself changes the state of the measured object.

**Remark**

Meanwhile, several sophisticated experimental arrangements have been set up that allow "quantum non-demolishing" experiments, where information about the measured system is obtained without altering the state of the measured object. For more details see Sect. 12.5 and the literature [3.21–25].

### 3.5.4 The Importance of Quantum Physics for our Concept of Nature

Quantum theory can answer all open questions discussed in the preceding sections (for example concerning the stability of atoms, the diffraction of electrons, the ultraviolet disaster and the photoeffect). Its extension to quantum electrodynamics (QED) is in complete agreement with all known experimental results within the range of its validity, which means that all phenomena associated with atoms, molecules and solids are satisfactorily described by this QED-theory. Its limitations only become visible when investigating the nuclear structure and the properties of elementary particles. Its disadvantage is its lack of vividness. There is, for instance, no vivid picture of the photon because of its dual character as particle or as wave. Also the quantum mechanical postulate that a lack of knowledge about the way in which a particle has reached the detector, causes interference phenomena, contradicts our usual picture of classical physics.

The concept of probability and the uncertainty relations have essential philosophical consequences. The future destiny of a microparticle is no longer completely determined by its past. First of all, we only know its initial state (location and momentum) within limits set by the uncertainty relations. Furthermore, the final state of the system shows (even for accurate initial conditions) a probability distribution around a value predicted by classical physics.

The possibility of exact predictions, postulated in classical physics for exactly known initial conditions, is restricted in quantum physics in a two-fold way. The initial conditions are not exactly known and the measuring process itself affects the state of the system.

This is further illustrated by the following example in Fig. 3.50. A light beam is split by the 50% beam splitter BS into two equally intense partial beams, which are detected by the detectors $D_1$ and $D_2$. Each of the two detectors counts the statistically impinging photons and both measure on the average an equal mean photon rate. Since a photon is indivisible, the beam splitter BS can only either transmit or reflect the photon. Which of the photons is reflected and which is transmitted, is *not* predictable. One can only say that the probability for each of the two processes is 0.5.

Now a second beam splitter $BS_1$ is inserted (Fig. 3.50b) that splits the incoming light beam into two

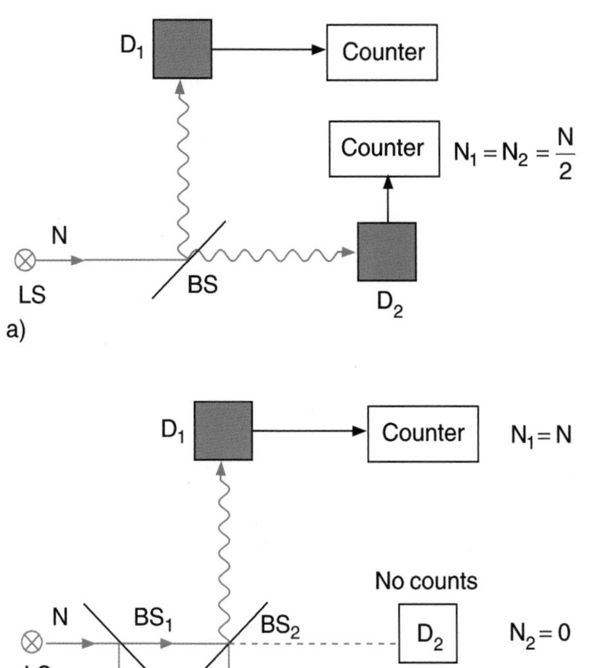


**Fig. 3.50.** (**a**) The role of the beam splitter without observing an interference pattern. (**b**) An additional beam splitter to form an interferometer

partial beams that are again superimposed at the beam splitter $BS_2$. From interference experiments in classical optics we know that the intensity received by the two detectors $D_1$, $D_2$ depends on the phase difference between the two interfering beams at S. At a proper phase difference, destructive interference occurs for the beam directed towards $D_2$ and constructive interference for the beam towards $D_1$. This means that $D_2$ receives no light, but $D_1$ all of the light. This is observed even at very low light intensities, where only one photon is simultaneously on its way from the source to the detector.

How can the photons "know" that they now all should travel towards $D_1$ and not towards $D_2$?

This example illustrates that the experimentator can influence the future fate of photons by choosing the corresponding experimental arrangement. Changing the distance $\overline{BS_1 - M_1 - M_2 - BS_2}$ affects the splitting ratio of $BS_2$. This demonstrates that it is not meaningful to attribute a definite path to the photons, but rather only a detection probability for $D_1$ and $D_2$, which depends on the experimental arrangement.

During recent years several modifications of such "which way" experiments have been demonstrated. Their results are all in complete agreement with quantum theory but cannot be explained by classical concepts.

For a more detailed discussion the reader is referred to the recommendable book of Paul [3.22] and to several other books or review articles about this interesting field [3.23, 24, 25].

## SUMMARY

- Many experimental results prove the particle character of electromagnetic waves. Examples are the spectral distribution of blackbody radiation, the photoelectric effect, the Compton effect or measurements of the photon structure of the light emitted by a weak light source.
- The derivation of Planck's radiation formula based on the photon model gives results that are in complete agreement with experiments.
- The energy quanta $h\nu$ of the electromagnetic field are called photons. One can formally define the photon mass as $m = h\nu/c^2$. Photons are deflected by gravitational fields, like other particles with mass $m$. There are no photons at rest. Nevertheless one defines a rest mass $m_0 = 0$, in order to describe photons by the same relativistic equations for energy and momentum used for other particles with $m_0 \neq 0$.
- To define the characteristic properties of a photon, such as momentum $\boldsymbol{p}_{\text{phot}} = \hbar \boldsymbol{k} = (h/\lambda)\hat{\boldsymbol{k}}$, energy $E = h\nu = h\omega$ and mass $m = h\nu/c^2$, one needs the wave properties frequency $\nu$, wavelength $\lambda$ and wave vector $\boldsymbol{k}$. This shows the duality in the description of light.
- The wave character of particles is proved experimentally by diffraction and interference phenomena. Examples are the electron diffraction, the Bragg reflection of neutrons by single crystals, neutron interferometry and numerous experiments in atom optics, demonstrating diffraction and interference of matter waves.

▶

- Matter waves show dispersion, even in a vacuum. Their phase velocity depends on the frequency $\omega$.
- Particles can be described by wave packets. The particle velocity equals the group velocity of the wave packet.
- The absolute square $|\psi(x,t)|^2$ of the matter wave function represents a probability density. This means that $|\psi(x,t)|^2 \mathrm{d}x$ gives the probability to find the particle at time $t$ in the interval $\mathrm{d}x$ around $x$.
- Position $x$ and momentum $p_x$ of a particle cannot be simultaneously accurately measured. Heisenberg's uncertainty relation $\Delta x \cdot \Delta p_x \geq \hbar$ puts a principle lower limit to the uncertainties $\Delta x$ of the particle position and $\Delta p_x$ to its momentum, if both are measured simultaneously. The uncertainty limits depend on the form of the wave packed. The lowest limit is set for a gaussion profile of the wave packet.
- Analogous to the situation in classical optics the uncertainty $\Delta x$ for the location of a particle cannot be smaller than the wavelength $\lambda = h/p$ of its matter wave.
- A similar uncertainty relation $\Delta E \cdot \Delta t \geq \hbar$ is valid for the measurement of the particle energy $E$ and the time duration $\Delta t$ of this measurement. The energy of an excited atomic state with mean lifetime $\tau$ can only be measured with an uncertainty $\Delta E = \hbar/\tau$.
- Bohr's classical atomic model, where the electrons orbit around the nucleus on circles like planets around the sun, needs an additional quantum condition in order to explain the discrete energy levels of the atom. The condition for the radius $r$ of the orbit can be formulated as $2\pi r = n\lambda$, where $\lambda = h/p$ is the de Broglie wavelength of the electron's matter wave, or as the quantization $|l| = rp = nh$ of the angular momentum $l$ of the electron. Both conditions are equivalent and lead to a quantization of the energy levels.
- The allowed energy levels of atoms or ions with only one electron and nuclear charge $Ze$ are

$$E_n = -Ry^* Z^2/n^2 \quad (n = 1, 2, 3, \dots) ,$$

  where $Ry^* = \mu e^4 / \left(8\pi \varepsilon_0^2 h^2\right)$ is the Rydberg constant for the system consisting of the nucleus and the electron with reduced mass $\mu$ in the center of mass system.
- Bohr's model predicts many features of atomic spectra correctly, but has to be modified on some essential points.
- The uncertainty relation explains the stability of atomic ground states.
- All excited atomic states $E_k$ are unstable. They decay under emission of a photon $h\nu = E_k - E_i$ into lower states $E_i$.
- The quantization of atomic energy levels is corroborated by the results of the Franck–Hertz experiment and by the observation of line spectra in absorption and emission of atoms.
- The quantum mechanical description replaces the exactly determined path of a microparticle by a probability distribution $|\psi(x,t)|^2$ of a wave packet. This distribution spreads in time. The uncertainty $\Delta x$ becomes the larger the more accurate the initial location $x(t_0)$ had been known.
- The interference phenomena observed for matter waves in interference experiments, where the matter wave is split and later recombined, are due to the imperfect knowledge of which path the particle has taken. The final state of the particle therefore has to be described by a linear combination of two or more wave functions. If the path of the particle is defined by additional experiments the interference pattern disappears, because now the linear combination can be reduced to one of the terms in the linear combination.

## PROBLEMS

1. What is the velocity and the kinetic energy of a neutron with the de Broglie wavelength $\lambda_{dB} = 10^{-10}$ m? Is this still a thermal neutron?
2. Calculate the mean energy per mode of the cavity radiation field and prove (3.14b).
3. Derive Wien's law in the form of (3.18) and (3.18a).
4. a) Show that energy conservation and momentum conservation cannot be simultaneously fulfilled when a free electron with velocity $v_1$ absorbs a photon $h\nu$, increasing its velocity to $v_2 > v_1$. Why are both quantities conserved in the Compton effect? b) What is the momentum of a photon $h\nu = 0.1$ eV (infrared), $h\nu = 2$ eV (visible), and $h\nu = 2$ MeV ($\gamma$-Quant)?
5. A slit with width $b$ is illuminated by a parallel beam of electrons with kinetic energy $E_{kin}$. For which slit width $b$ is the width $B$ of the central diffraction maximum observed on a screen at a distance $D$ from the slit?
6. What are radius and electron velocity $v$ on the first Bohr orbit with $n = 1$

   a) in the H atom with $Z = 1$?

   b) in the gold atom with $Z = 79$?

   c) How large is the relativistic mass increase for the two cases? How much do the energy values change if this mass increase is taken into account?
7. Free neutrons have a mean lifetime of about 900 s. After what distance from the neutron source has the number of neutrons with a de Broglie wavelength $\lambda_{dB} = 1$ nm in a collimated neutron beam in vacuum decayed to one half of its initial value?
8. Calculate the wavelength of the Lyman-$\alpha$ line

   a) for tritium atoms ($Z = 1$, $A = 3$)

   b) for positronium $e^+e^-$.
9. An atom with one electron has the energy levels $E_n = -a/n^2$. Its spectrum has two neighboring lines with $\lambda_1 = 97.5$ nm and $\lambda_2 = 102.8$ nm. What is the value of the constant $a$ and which atomic element belongs to this spectrum?
10. The Balmer series of the hydrogen atom should be measured with a grating spectrograph with a spectral resolution $\lambda/\Delta\lambda = 5 \times 10^5$. Up to which principal quantum number $n$ are two neighboring lines still resolvable?
11. What are the numerical values for the first Bohr radius of the electron and its kinetic energy in the $He^+$ ion? Which value of the minimum energy is obtained from the uncertainty relation?

# 4. Basic Concepts of Quantum Mechanics

In Chap. 3 we saw that because of the uncertainty relation, the location $x$ and the momentum $p_x$ of an atom cannot be simultaneously measured exactly. The classical well-defined path $\boldsymbol{r}(t)$ of a particle moving with velocity $\boldsymbol{v}(t)$ has to be replaced by the probability

$$P(x, y, z, t)\,\mathrm{d}V = |\psi(x, y, z, t)|^2\,\mathrm{d}V \tag{4.1}$$

of finding a particle at a given time $t$ in the volume $\mathrm{d}V$ around the point $\boldsymbol{r} = \{x, y, z\}$. The probability density is determined by the absolute square of the wave function $\psi$.

In this chapter we will demonstrate, using several simple examples, how the wave function $\psi$ can be obtained. These examples also illustrate the physical ideas behind the mathematical framework of quantum mechanics and their differences from classical concepts. They furthermore show under which conditions the quantum mechanical results converge to the classical ones and they demonstrate that classical mechanics represent the correct theory for all situations where the de Broglie wavelength $\lambda_{\mathrm{dB}}$ approaches zero (this is nearly always the case in daily life phenomena).

We will also see that most of the quantum mechanical phenomena are well known in classical optics. This means that the essential new concept of quantum mechanics is the description of classical particles by matter waves, where the de Broglie wavelength $\lambda_{\mathrm{dB}} = h/(mv)$ gives the minimum spatial uncertainty range when the momentum $p = mv$ is known within $\Delta p = \pm\hbar/\lambda_{\mathrm{dB}} = \pm mv/2\pi$. The deterministic classical model of the time-dependent location $(\boldsymbol{r}(t), \boldsymbol{p}(t))$ of a particle in phase space is replaced by a statistical description, where only probabilities can be given for the result of a measurement. There is a principal uncertainty for the simultaneous determination of location and momentum of a particle, which cannot be overcome even with sophisticated measuring techniques.

## 4.1 The Schrödinger Equation

In this section we will discuss the basic equation of quantum mechanics that was introduced in 1926 by *Erwin Schrödinger* (1887–1961) (Fig. 4.1). The solutions of this equation are the wave functions $\psi(x, y, z, t)$

**Fig. 4.1.** *Erwin Schrödinger* (1887–1961). From E. Bagge: Die Nobelpreisträger der Physik (Heinz-Moos-Verlag, München 1964)

W. Demtröder, *Atoms, Molecules and Photons*, 2nd ed., Graduate Texts in Physics,
DOI 10.1007/978-3-642-10298-1_4, 

we are looking for. They can be obtained, in analytical form, only for a few simple problems. For real and therefore more complex situations the solutions have to be calculated numerically. However, with present day computing power, this generally poses no major problems.

Although the Schrödinger equation cannot be derived in a direct mathematical way, it is possible to show how Schrödinger arrived at this equation from classical wave concepts if the description of particles by matter waves is accepted.

We regard a freeparticle with mass $m$ moving with uniform velocity $v$ in the $x$ direction. With the relations $\lambda = h/p \Rightarrow \boldsymbol{p} = \hbar \boldsymbol{k}$ and $E = \hbar\omega = E_{\text{kin}}$ (because $E_{\text{pot}} = 0$) we obtain, analogously to an optical wave, the matter wave function

$$\psi(x,t) = A\mathrm{e}^{\mathrm{i}(kx-\omega t)} = A\mathrm{e}^{(\mathrm{i}/\hbar)(px-E_{\text{kin}}t)}\,, \tag{4.2}$$

where $E_{\text{kin}} = p^2/2m$ is the kinetic energy of the particle. With the derivatives

$$\frac{\partial^2\psi}{\partial x^2} = -\frac{p^2}{\hbar^2}\psi \quad \text{and} \quad \frac{\partial^2\psi}{\partial t^2} = -\omega^2\psi = -\frac{E_{\text{kin}}^2}{\hbar^2}\psi$$

we obtain the wave equation

$$\frac{\partial^2\psi}{\partial x^2} = \frac{1}{u^2}\frac{\partial^2\psi}{\partial t^2} \tag{4.3}$$

for waves that propagate with the phase velocity $u = \omega/k = E/p$.

For stationary problems, where $p$ and $E$ do not explicitly depend on time, the wave function (4.2) can be split into a product of a factor $\psi(x) = A\mathrm{e}^{\mathrm{i}kx}$, which only depends on $x$ (not on $t$) and a time-dependent phase factor $\mathrm{e}^{-\mathrm{i}\omega t}$. We can then write

$$\psi(x,t) = \psi(x)\mathrm{e}^{-\mathrm{i}\omega t}\,. \tag{4.4}$$

Inserting (4.4) into the wave equation (4.3) yields, for the space function $\psi(x)$ with $k^2 = p^2/\hbar^2 = 2mE_{\text{kin}}/\hbar^2$, the equation:

$$\frac{\partial^2\psi}{\partial x^2} = -k^2\psi = -\frac{2m}{\hbar^2}E_{\text{kin}}\psi\,. \tag{4.5}$$

For the case of a particle moving in the $x$ direction in an external field with potential energy $E_{\text{pot}}$, the particle has the total energy $E = E_{\text{kin}} + E_{\text{pot}}$. Replacing $E_{\text{kin}}$ in (4.5) by $E - E_{\text{pot}}$ we obtain the one-dimensional Schrödinger equation for stationary situations:

$$\boxed{-\frac{\hbar^2}{2m}\frac{\partial^2\psi}{\partial x^2} + E_{\text{pot}}\psi = E\psi\,.} \tag{4.6a}$$

For the more general case of a particle moving in three-dimensional space the wave equation (4.3) becomes

$$\Delta\psi = \frac{1}{u^2}\frac{\partial^2\psi}{\partial t^2}$$

with the Laplace operator

$$\Delta = \frac{\partial^2}{\partial x^2} + \frac{\partial^2}{\partial y^2} + \frac{\partial^2}{\partial z^2}\,.$$

For stationary conditions, the wave function $\psi(x,y,z,t)$ can be again split into a space dependent factor and a phase factor $\Psi(x,y,z,t) = \psi(x,y,z)\mathrm{e}^{-\mathrm{i}\omega t}$, where $\psi(x,y,z)$ is the solution of the three dimensional stationary Schrödinger equation

$$\boxed{-\frac{\hbar^2}{2m}\Delta\psi + E_{\text{pot}}\psi = E\psi\,.} \tag{4.6b}$$

For time-dependent problems we obtain the Schrödinger equation as follows: Taking the partial derivative $\partial\psi/\partial t$ of (4.2) gives

$$\frac{\partial\psi}{\partial t} = -\frac{\mathrm{i}}{\hbar}E_{\text{kin}}\psi\,. \tag{4.6c}$$

For a free particle with $E_{\text{pot}} = 0$ is $E_{\text{kin}} = E = \text{const.}$ Inserting (4.6c) into (4.5) gives

$$\mathrm{i}\hbar\frac{\partial\psi(x,t)}{\partial t} = -\frac{\hbar^2}{2m}\frac{\partial^2\psi(x,t)}{\partial x^2}\,, \tag{4.7a}$$

which can be generalized for the three-dimensional case to

$$\mathrm{i}\hbar\frac{\partial\psi(\boldsymbol{r},t)}{\partial t} = -\frac{\hbar^2}{2m}\Delta\psi(\boldsymbol{r},t)\,. \tag{4.7b}$$

If the particle is moving in an external potential with potential energy $E_{\text{pot}}(r)$, the general three-dimensional time-dependent Schrödinger equation is:

$$\boxed{\mathrm{i}\hbar\frac{\partial\psi(\boldsymbol{r},t)}{\partial t} = -\frac{\hbar^2}{2m}[\Delta\psi(\boldsymbol{r},t) + E_{\text{pot}}(\boldsymbol{r})]\cdot\psi(r,t)\,.} \tag{4.8}$$

**Remarks**

1. The "derivation" of the stationary Schrödinger equation (4.6b) is based on the de Broglie relation $\boldsymbol{p} = h\boldsymbol{k}$ between momentum $\boldsymbol{p}$ of a particle and wave vector $\boldsymbol{k}$ of the matter wave, which has been proved by many experiments but cannot be derived in a mathematical sense.

2. The stationary Schrödinger equation (4.6) represents the energy conservation $E_{\rm kin}\psi + E_{\rm pot}\psi = E\psi$ in quantum mechanics. Just as the classical law of energy conservation (first law of thermodynamics) it cannot be derived from first principles but is solely based on experimental experiences.
3. The Schrödinger equations (4.6) and (4.7) are linear equations. This means that with solutions $\psi_1$ and $\psi_2$ any linear combination $c_1\psi_1 + c_2\psi$ is also a solution (superposition principle).
4. Since (4.7) is a complex equation, the wave functions $\psi$ may also be complex. However, the absolute square $|\psi|^2$ is always real, as it should be, because it describes a physical quantity, namely the spatial probability density of a particle.
5. There is an important difference compared to the linear dispersion relation $\omega(k) = kc$ of electromagnetic waves. For the matter wave of a particle with energy $E$ and momentum $\boldsymbol{p}$, where $E = p^2/(2m)$, the dispersion relation becomes a quadratic relation $\omega(k) = (\hbar/2m)k^2$, where $E = \hbar\omega$ and $\boldsymbol{p} = \hbar\boldsymbol{k}$. Therefore the group velocity $\mathrm{d}\omega/\mathrm{d}k = (\hbar/m)k$ of the matter wave depends on $k$ (see Sect. 3.3.4).

For nonstationary problems (this means that energy $E = E(t)$ and momentum $p = p(t)$ may be time dependent) $\partial^2\psi/\partial t^2$ can no longer be written as $-\omega^2\psi$ and (4.7b) cannot be derived from the wave function (4.2).

Schrödinger postulated that the generalized equation (4.8) for a particle moving in a time-dependent external field with potential energy $E_{pot}(\boldsymbol{r}, t)$ should still be valid for this more general case.

> Although there is no mathematical proof for this postulate, numerous experiments have confirmed that the time-dependent Schrödinger equation (4.8) is a correct description for all nonrelativistic phenomena observed so far. It is the master equation of nonrelativistic quantum mechanics and all phenomena in this field can be (at least numerically) calculated by this equation.

For stationary problems the wave function $\psi(r, t)$ can be separated into

$$\psi(r, t) = \psi(\boldsymbol{r})\mathrm{e}^{-\mathrm{i}(E/\hbar)t} .$$

Inserting this product into (4.8) again yields the stationary Schrödinger equation (4.6a) for the spatial part $\psi(\boldsymbol{r})$.

**Note:**

The solutions of the time-independent Schrödinger-equation describe stationary problems. If moving particles or time-dependent processes are described, these solutions represent time averages over time-dependent processes. In order to gain more insight into the dynamics of such processes one can either use the time-dependent Schrödinger equation (which often demands extensive computer programs), or semi-classical models which often can give more insight into the physics of the problem. In this textbook we will use the second choice in order to give the reader a feeling of the physical essence of a problem.

## 4.2 Some Examples

In this section we will demonstrate the application of the Schrödinger equation (4.6) to some simple one-dimensional problems. These examples shall illustrate the wave description of particles and its physical consequences.

### 4.2.1 The Free Particle

A particle is called a free particle when it moves or rests in a constant homogeneous potential $E_{\rm pot} = \text{const.}$ Because $\boldsymbol{F} = -\,\mathbf{grad}\,E_{\rm pot} = 0$ there is no force acting on the particle. We can choose this constant potential energy as $E_{\rm pot} = 0$ without restricting the generality of the problem. From (4.6) we obtain the Schrödinger equation of the free particle:

$$-\frac{\hbar^2}{2m}\frac{\mathrm{d}^2\psi(x)}{\mathrm{d}x^2} = E\psi(x) . \tag{4.9a}$$

The total energy $E = E_{\rm kin} + E_{\rm pot}$ is now, because $E_{\rm pot} = 0$,

$$E = \frac{p^2}{2m} = \frac{\hbar^2k^2}{2m} .$$

And (4.9a) reduces to

$$\frac{\mathrm{d}^2\psi}{\mathrm{d}x^2} = -k^2\psi . \tag{4.9b}$$

With the general solution

$$\psi(x) = A\mathrm{e}^{\mathrm{i}kx} + B\mathrm{e}^{-\mathrm{i}kx}$$

the time-dependent wave function

$$\psi(x, t) = \psi(x)\mathrm{e}^{-\mathrm{i}\omega t} = A\mathrm{e}^{\mathrm{i}(kx-\omega t)} + B\mathrm{e}^{-\mathrm{i}(kx+\omega t)} \tag{4.9c}$$

represents the superposition of a plane wave traveling into the $+x$ direction and a wave into the $-x$ direction.

The coefficients $A$ and $B$ are the amplitudes of these waves, which can be determined from the boundary conditions. If, for example, the wave function shall describe electrons that are emitted from a cathode and fly into $+x$ direction to the detector, the coefficient $B$ must be zero. From the experimental setup we know that the electrons can only be found within the distance $L$ between cathode and detector (Fig. 4.2). This implies that the amplitude $A$ can be nonzero only for $0 < x < L$.

With the normalization condition

$$\int_0^L |\psi(x)|^2 \,\mathrm{d}x = 1 \quad \Rightarrow \quad A^2 L = 1 \tag{4.10}$$

we obtain for the amplitude $A = 1/\sqrt{L}$.

**Note:**

In order to define the location of a particle at time $t$ more precisely, we have to use wave packets instead of the plane waves (4.21) for its description:

$$\psi(x,t) = \int_{k_0-\Delta k/2}^{k_0+\Delta k/2} A(k)\mathrm{e}^{\mathrm{i}(kx-\omega t)}\,\mathrm{d}k \tag{4.11b}$$

(see Sect. 3.3.1).

The local uncertainty $\Delta x \geq \hbar/(\Delta p_x) = 1/(\Delta k)$ at time $t = 0$ depends on the momentum uncertainty $\Delta p_x = \hbar \Delta k$ (Fig. 3.31). The larger the uncertainty $\Delta k$ the smaller the uncertainty $\Delta x$ $(t = 0)$, however, the faster the wave packet spreads for $t > 0$.

This can be nicely illustrated by our example in Fig. 4.2. If we apply a short voltage pulse between the cathode K and the anode A at $t = 0$, electrons can start moving from K through the grid A to the detector D only during the pulse. Due to the thermal velocity distribution of the electrons emitted from the hot cathode their velocities are spread over the interval $\Delta v \propto \Delta k$. Electrons with different values of $v$ emitted at $t = 0$ are "smeared out" over the interval $\Delta x$ at later times $t$. The uncertainty $\Delta x$ increases with time according to

$$\frac{\mathrm{d}}{\mathrm{d}t}(\Delta x(t)) = \Delta v(t) = \frac{\hbar}{m}\Delta k(t) \propto \frac{\Delta p(t=0)}{m}\,.$$

The spatial spread of the wave packet is proportional to its initial spread of momentum.

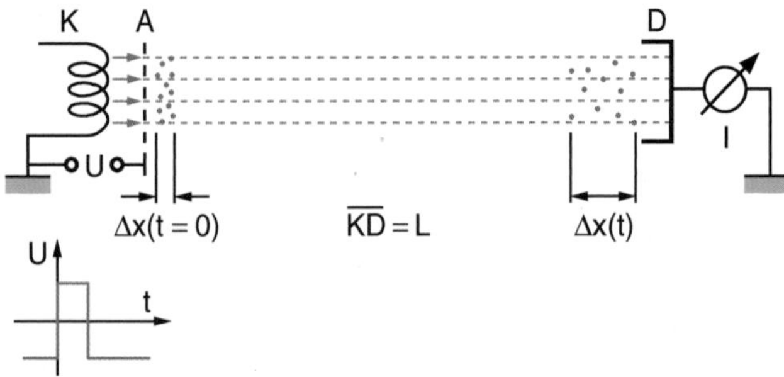


**Fig. 4.2.** Illustration of the spread of the wave packet by observing a bunch of electrons emitted at $t = 0$ from the cathode with velocity spread $\Delta v(t = 0)$

**Note:**

The following examples are mainly concerned with moving particles which have to be correctly described by wave packets. However, the essential features can be already seen when the particles are represented by the spatial part of (4.4), i.e. by stationary wave functions. The treatment by wave packets gives much more complex results because time-dependent interference effects complicate the spatial structure of the wave packet.

### 4.2.2 Potential Barrier

We divide the total $x$ range into two parts (Fig. 4.3). For $x < 0$ the potential energy is zero, while for $x \geq 0$ $E_{\mathrm{pot}} = E_0 \neq 0$. At $x = 0$ we have a step $\Delta E_{\mathrm{pot}} = E_0$ of the potential energy. From the left side of Fig. 4.3 free particles with energy $E$ fly into the $+x$ direction. We describe them, as in (4.10), by the wave function

$$\psi_{\mathrm{I}}(x) = A\mathrm{e}^{\mathrm{i}kx} + B\mathrm{e}^{-\mathrm{i}kx}\,, \tag{4.11}$$

where $B$ is the amplitude of the wave functions for particles reflected at the potential barrier and traveling in the $-x$ direction. Note, that only the spatial part of the wave function is written in (4.11), the time dependent part is here omitted because we are dealing here

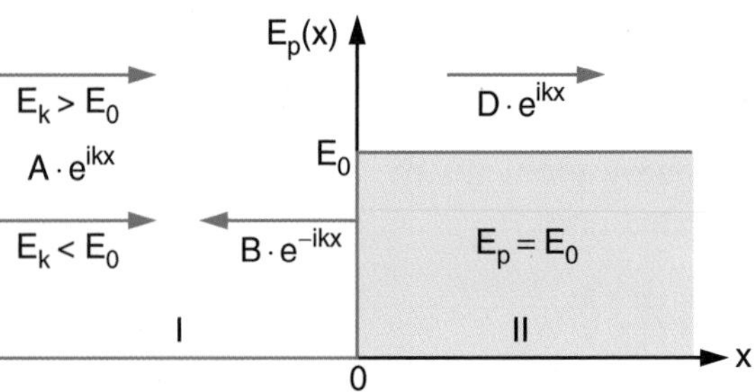


**Fig. 4.3.** One-dimensional potential barrier

with stationary problems. The complete wave function is represented by (4.9c).

For $x \geq 0$ the Schrödinger equation is

$$\frac{d^2\psi}{dx^2} + \frac{2m}{\hbar^2}(E - E_0)\psi = 0 . \tag{4.12a}$$

With the abbreviation $\alpha = \sqrt{2m(E_0 - E)}/\hbar$ this reduces to

$$\frac{d^2\psi}{dx^2} - \alpha^2\psi = 0 . \tag{4.12b}$$

The solutions of (4.12b) are

$$\psi_{II} = (x \geq 0) = Ce^{\alpha x} + De^{-\alpha x} . \tag{4.13}$$

Since the total wave function

$$\psi(x) = \begin{cases} \psi_I & \text{for} \quad x < 0 \\ \psi_{II} & \text{for} \quad x \geq 0 \end{cases}$$

has to be uniquely defined within the whole range $-\infty \leq +\infty$, the derivative $d\psi/dx$ must be a continuous function and has to be finite within this range. Otherwise, the second derivative in (4.12b) would not be defined and (4.12) could not be applied.

This gives the following boundary conditions for $x = 0$:

$$\psi_I(x = 0) = \psi_{II}(x = 0)$$

$$\Rightarrow \quad A + B = C + D \tag{4.14a}$$

$$\left.\frac{d\psi_I}{dx}\right|_{x=0} = \left.\frac{d\psi_{II}}{dx}\right|_{x=0}$$

$$\Rightarrow \quad ik(A - B) = +\alpha(C - D) . \tag{4.14b}$$

We distinguish the two situations where the energy $E_{kin} = E$ of the incident particles is smaller ($E < E_0$) or larger ($E > E_0$) than the potential step (Fig. 4.3).

### a) $E < E_0$:

In this case $\alpha = \sqrt{2m(E_0 - E)}/\hbar > 0$ is real and the coefficient $C$ in (4.13) has to be zero. Otherwise $\psi_{II}(x)$ becomes infinite for $x \to \infty$ and could no longer be normalized. With (4.14) we then obtain:

$$B = \frac{ik + \alpha}{ik - \alpha}A \quad \text{and} \quad D = \frac{2ik}{ik - \alpha}A . \tag{4.15}$$

The wave function in the region $x < 0$ is then

$$\psi_I(x) = A\left[e^{ikx} + \frac{ik + \alpha}{ik - \alpha}e^{-ikx}\right] . \tag{4.15a}$$

Its real part is illustrated in Fig. 4.4.

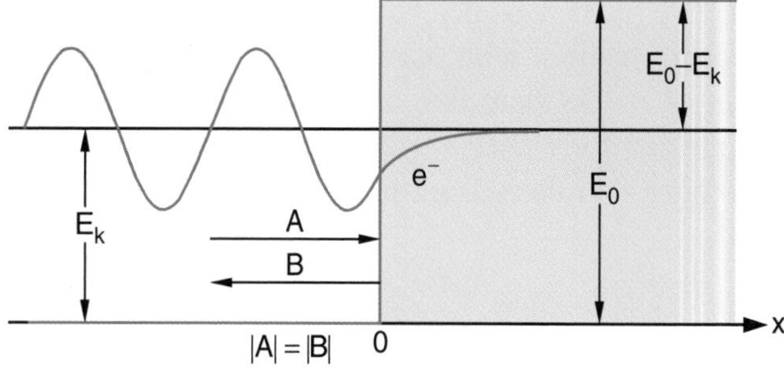


**Fig. 4.4.** Complete reflection of the incident wave for $E_{kin} < E_0$, in spite of its penetration into the potential wall

The fraction $R$ of the reflected particles is

$$R = \frac{|Be^{-ikx}|^2}{|Ae^{ikx}|^2} = \frac{|B|^2}{|A|^2} = \left|\frac{ik + \alpha}{ik - \alpha}\right|^2 = 1 . \tag{4.16}$$

This means that all particles are reflected.

This would also be expected from classical mechanics. Since the particles do not have sufficient energy, they cannot reach the region $x > 0$.

However, there is an important difference in the wave description:

> The particles are not reflected exactly at $x = 0$ but penetrate into the region $x > 0$ before they return to the $-x$ direction.

This can be seen from closer inspection of $\psi_{II}$. The probability of finding a particle at $x \geq 0$ is

$$P(x) = |\psi_{II}(x)|^2 = \left|De^{-\alpha x}\right|^2 = \frac{4k^2}{\alpha^2 + k^2}|A|^2 e^{-2\alpha x}$$

$$= \frac{4k^2}{k_0^2}|A|^2 e^{-2\alpha x} , \tag{4.17}$$

where $k_0^2 = 2mE_0/\hbar^2$. After a penetration depth $x = 1/(2\alpha)$, the probability has dropped to $1/e$ of its value $\left(4k^2/k_0^2\right) A^2$ at $x = 0$.

> Particles with energy $E$, described by matter waves can penetrate into regions with potential energy $E_0 > E$, where they would not be allowed in classical particle mechanics.

However, if we accept the wave description of particles the result, which is surprising at a first glance, follows quite naturally and completely analogously to the well-known phenomena of total reflection in wave optics. If a light wave traveling in a medium with refractive index $n$ is incident on the boundary plane to a second medium with refractive index $n' < n$, under an angle $\theta$ less than the critical angle $\theta_c$ it is totally reflected for $n \sin\theta > n'$. The wave, however, still penetrates into the second medium with a penetration depth

$$x_p = \frac{1}{k\gamma} \quad \text{with} \quad \gamma = \sqrt{n^2 \sin^2\theta - n'^2}\,.$$

This shows that quite analogous formulas are obtained for both cases if $2\alpha$ is replaced by $k\gamma$.

### b) $E > E_0$

Now the kinetic energy $E_{\text{kin}} = E$ of the incident particles is larger than the potential barrier $E_0$. In the classical particle model, *all* particles would travel into the region $x \geq 0$, while their kinetic energy $E_{\text{kin}}(x \geq 0) = E - E_0$ would become smaller (Fig. 4.5).

What does the wave model say?

The quantity $\alpha$ in (4.12b) is now purely imaginary and we replace it by the real quantity

$$k' = \mathrm{i}\alpha = \sqrt{2m(E - E_0)}/\hbar\,. \tag{4.18}$$

The solutions for $\psi_{\text{I}}(x)$ are again (4.11) and for $\psi_{\text{II}}(x)$ we obtain instead of (4.13) for $x \geq 0$

$$\psi_{\text{II}}(x) = C\mathrm{e}^{-\mathrm{i}k'x} + D\mathrm{e}^{\mathrm{i}k'x}\,. \tag{4.19a}$$

Since for $x > 0$ no particles travel in the $-x$ direction, the amplitude $C$ has to be zero and we obtain

$$\psi_{\text{II}}(x) = D\mathrm{e}^{-\mathrm{i}k'x}\,. \tag{4.19b}$$

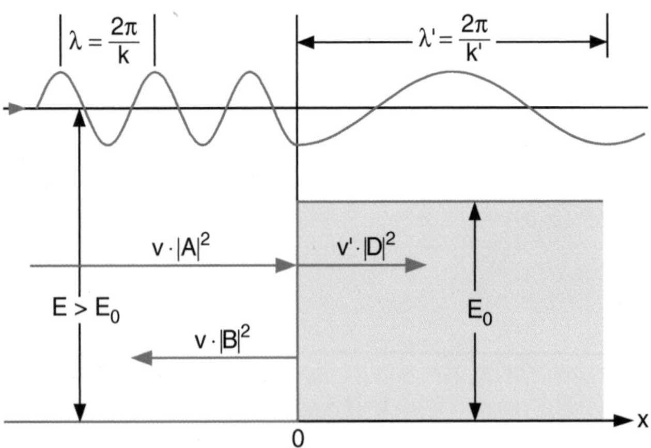


**Fig. 4.5.** Transmission and reflection for $E_{\text{kin}} > E_0$

From the boundary conditions (4.14) we deduce that

$$B = \frac{k - k'}{k + k'}A \quad \text{and} \quad B = \frac{2k}{k + k'}A\,. \tag{4.20}$$

The wave functions $\psi_{\text{I}}$ and $\psi_{\text{II}}$ then become:

$$\psi_{\text{I}}(x) = A\left(\mathrm{e}^{\mathrm{i}kx} + \frac{k - k'}{k + k'}\mathrm{e}^{-\mathrm{i}kx}\right) \quad x < 0 \tag{4.21a}$$

and

$$\psi_{\text{II}}(x) = A\frac{2k}{k + k'}\mathrm{e}^{\mathrm{i}k'x} \quad x \geq 0 \tag{4.21b}$$

The reflection coefficient

$$R = \frac{|B|^2}{|A|^2} = \left|\frac{k - k'}{k + k'}\right|^2 \tag{4.22}$$

gives the fraction of reflected particles, while the transmission coefficient

$$T = \frac{v'|D|^2}{v|A|^2} = \frac{4kk'}{|k + k'|^2} \tag{4.23}$$

gives the fraction of the fluxes of transmitted particles (Fig. 4.5). Here the different velocities $v$ and $v'$ in both regions have to be taken into account. Since $D^2$ gives the spatial density of transmitted particles, the flux is $v'D^2$.

From (4.22) and (4.23) it follows that $R + T = 1$ (conservation of the particle number!)

The reflection coefficient $R$ strongly depends on the ratio $E_0/E$ (Fig. 4.6), since $k' = \sqrt{2mE_0(E/E_0 - 1)}/\hbar$.

For $E = E_0 \Rightarrow k' = 0$ and $T = 0$; $R = 1$. All particles are reflected.

### Remarks

1. Instead of the positive potential step a negative step with $E_0 < 0$ can also occur. This can be realized when in Fig. 4.3 the particles enter from the right side, i.e., all $k$-vectors are inverted.
2. Since in optics the wave number $k = 2\pi/\lambda$ is related to the refractive index $n$ by $k = nk_0$, (4.22) can be directly converted to

   $$R = \left|\frac{n_1 - n_2}{n_1 + n_2}\right|^2,$$

   which gives the optical reflection coefficient for a boundary plane between two media with refractive indices $n_1$ and $n_2$. This again shows the close relationship between wave optics and matter wave description.

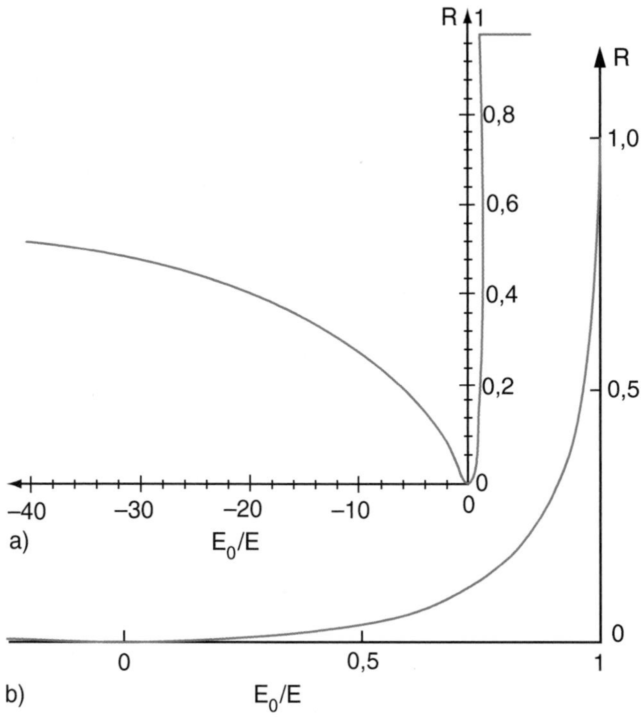


**Fig. 4.6.** (**a**) Reflectivity $R$ of a potential barrier plotted as a function of the ratio $E_0/E$ of the heights $E_0$ of the barrier to the energy $E$ of the incident particle. Negative values represent negative potentials $E_0 < 0$. (**b**) An enlarged section for $E \geq E_0 > 0$

3. The reflection and transmission at a negative potential step are analogous to the transition from an optical dense medium with refractive index $n_1$ into a less dense medium with $n_1 < n_2$.

### 4.2.3 Tunnel Effect

We will now consider the case of a potential barrier with finite width $\Delta x = a$ (Fig. 4.7). We therefore divide the whole $x$ range into three sections:

$$\begin{aligned} &\mathrm{I}: \quad x < 0\,, \quad E_{\text{pot}} = 0 \\ &\mathrm{II}: \quad 0 \leq x \leq a\,, \quad E_{\text{pot}} = E_0 \\ &\mathrm{III}: \quad x > a\,, \quad E_{\text{pot}} = 0\,. \end{aligned}$$

The wave functions are similar to the discussion in the previous section:

$$\psi_{\mathrm{I}}(x) = A\mathrm{e}^{\mathrm{i}kx} + B\mathrm{e}^{-\mathrm{i}kx}\,, \tag{4.24a}$$

$$\psi_{\mathrm{II}} = C\mathrm{e}^{\alpha x} + D\mathrm{e}^{-\alpha x}\,, \quad \text{and} \tag{4.24b}$$

$$\psi_{\mathrm{III}} = A'\mathrm{e}^{\mathrm{i}kx} + B'\mathrm{e}^{-\mathrm{i}kx}\,. \tag{4.24c}$$

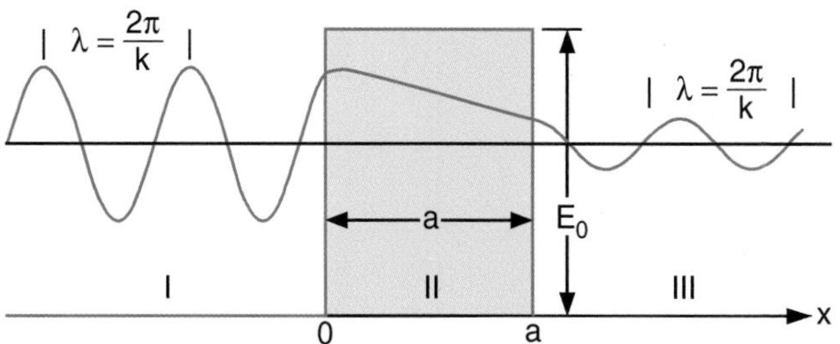


**Fig. 4.7.** The tunnel effect. Penetration of a matter wave through a rectangular potential barrier

Since there is no wave in region III that travels in the $-x$ direction, the coefficient $B'$ has to be zero. The boundary conditions:

$$\psi_{\mathrm{I}}(0) = \psi_{\mathrm{II}}(0)\,, \quad \psi_{\mathrm{II}}(a) = \psi_{\mathrm{III}}(a)\,, \tag{4.25a}$$

$$\left.\frac{\mathrm{d}\psi_{\mathrm{I}}}{\mathrm{d}x}\right|_{x=0} = \left.\frac{\mathrm{d}\psi_{\mathrm{II}}}{\mathrm{d}x}\right|_{x=0}\,, \quad \text{and}$$

$$\left.\frac{\mathrm{d}\psi_{\mathrm{II}}}{\mathrm{d}x}\right|_{x=a} = \left.\frac{\mathrm{d}\psi_{\mathrm{III}}}{\mathrm{d}x}\right|_{x=a} \tag{4.25b}$$

yield relationships between the coefficients $A, B, C, D, A'$. The velocity $v$ is in region III the same as in region I. This gives for the transmission, after some algebra

$$\begin{aligned} T &= \frac{v|A'|^2}{v|A|^2} = \frac{|A'|^2}{|A|^2} \\ &= \frac{1 - E/E_0}{(1 - E/E_0) + (E_0/4E)\sinh^2(a\alpha)} \end{aligned} \tag{4.26a}$$

with $\alpha = \sqrt{2m(E_0 - E)}/\hbar$ (see Problem 4.4).

For large barrier widths ($a\alpha \gg 1$) we can approximate

$$\sinh(a\alpha) = \frac{1}{2}\left(\mathrm{e}^{a\alpha} - \mathrm{e}^{-a\alpha}\right) \approx \frac{1}{2}\mathrm{e}^{a\alpha}$$

and obtain with $E_0^2\mathrm{e}^{2a\alpha}/(16E(E_0 - E)) \gg 1$ from (4.26)

$$T \approx \frac{16E}{E_0^2}(E_0 - E)\mathrm{e}^{-2a\alpha}\,. \tag{4.26b}$$

The transmission of the matter wave (and therefore of the particle represented by this matter wave as well) through the potential barrier, depends on the barrier heights $E_0$, on the difference $(E_0 - E)$ and exponentially on the product $a\alpha$ of barrier widths $a$ and coefficient $\alpha = \sqrt{2m(E_0 - E)}/\hbar$.

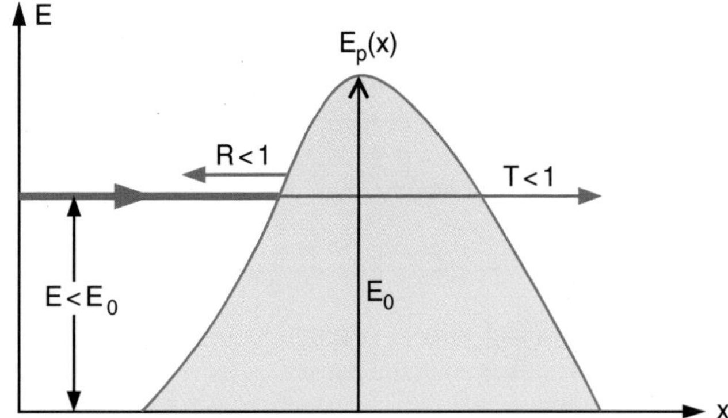


**Fig. 4.8.** Illustration of the tunnel effect for a potential barrier of general shape

In the classical particle model, particles with $E < E_0$ could never overcome the barrier. The transmission of matter waves through a potential barrier is called the *tunnel effect*, because the particles seem to penetrate the barrier at energy $E$ on a horizontal path, as if this were a tunnel. The potential barrier $E_p(x)$ can have an arbitrary form, as indicated in Fig. 4.8. In our examples we use rectangular barriers because the calculation of transmission coefficients is much simpler for this case.

## EXAMPLE

$$E = E_0/2\,,$$

$$a = \lambda_{dB}/2\pi = \hbar/p = \frac{\hbar}{\sqrt{2m(E_0 - E)}} \Rightarrow a\alpha = 1\,.$$

With sinh(1) = 1.543 we obtain from (4.26a):

$$T = \frac{0.5}{0.5 + 0.5 \cdot 1.543^2} = 0.3\,.$$

This means that 30% of all incident particles can transmit through the barrier.

If $a = \lambda_{dB} \Rightarrow a\alpha = 2\pi$. With $\sinh^2(2\pi) = 7 \times 10^4$ the transmission becomes $T = 1.5 \times 10^{-5}$.

In Fig. 4.9 the reflection coefficient $R$ and the transmission $T = 1 - R$ are plotted as a function of the ratio $E/E_0$ for the special case of a barrier with width $a = 0.48\lambda_{dB}(E_0)$ that is about half of the de Broglie wavelength $\lambda_{dB}$ of the incident particles. The diagram shows that even for $E > E_0$ the transmission is smaller than one, which means that not all of the incident particles pass the barrier, although they would do so in the classical particle model.

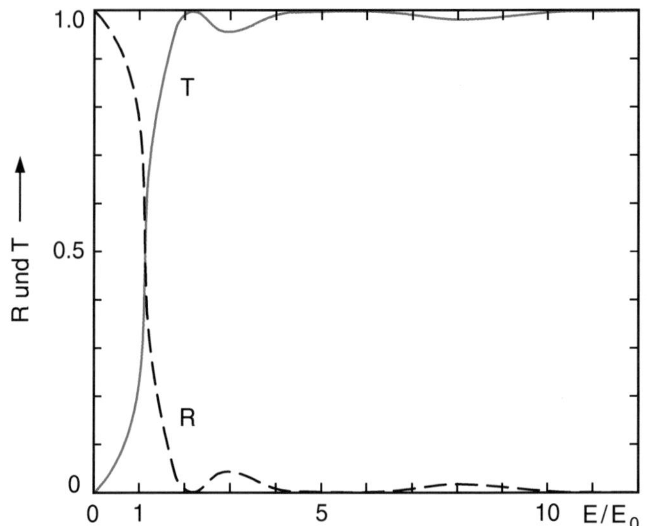


**Fig. 4.9.** Transmission $T$ and reflectivity $R$ of a rectangular barrier of width $a = 3\hbar/\sqrt{2mE_0} = (3/2\pi)\lambda_{dB} \approx 0.48\lambda_{dB}$ as a function of the energy of the incident particle

For $E > E_0$ the quantity $\alpha$ in (4.26) becomes imaginary. We, therefore, introduce the real quantity $k' = i\alpha = \sqrt{2m(E - E_0)}/\hbar$. This transfers (4.26) into

$$T = \frac{E/E_0 - 1}{(E/E_0 - 1) + (E_0/4E)\sin^2(k'a)}\,. \tag{4.26c}$$

This formula is completely analogous to the Airy-formula in optics, which describes the transmission of an electromagnetic wave through a glass plate. The undulations of $T$ in Fig. 4.9 are due to interferences between the waves reflected at the front side and the backside of the barrier. For $k'a = m\pi \Rightarrow a = n\lambda/2$, destructive interference occurs, which results in $R = 0$ and $T = 1$ for $E = E_0$ because in this case the de Broglie wavelength $\lambda'_{dB} = 2\pi/k'$ becomes infinite, the parameter $\alpha$ therefore becomes zero and the transmission is $T = 1$ independent of the width a of the barrier (Fig. 4.10). This can be verified using the rule of de-l'Hôspital, i.e., by differentiating the nominator and denominator in (4.26c) with respect to $E$ and then inserting $E = 0$.

In classical wave optics the tunnel effect is a well-known phenomenon. When a light wave is totally reflected at the boundary plane between glass and air (Fig. 4.11) the reflection coefficient is $R = 1$ and the detector $D_1$ receives the full incident intensity. If, however, a second glass plate is brought within a distance $d$ parallel to the first interface, part of the light is transmitted to the detector $D_2$. The transmission $T$ can

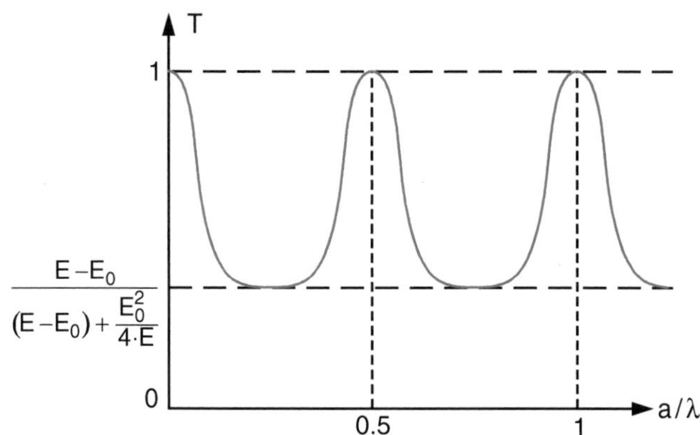


**Fig. 4.10.** Transmission $T$ as a function of the ratio $a/\lambda_{dB}$ of barrier width $a$ and de Broglie wavelength $\lambda'_{dB}$ for $E > E_0$

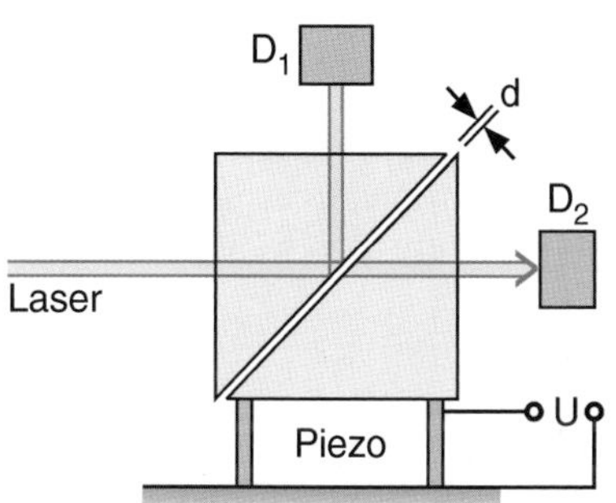


**Fig. 4.11.** Measurement of frustrated total reflection in optics, by changing the separation $d$ of the two prism surfaces

be varied between 0 and 1 depending on the distance $d$. The reason for this is that part of the light wave penetrates through the first interface and this evanescent wave travels parallel to the interface plane. Its amplitude decreases but the transmission through the interface remains zero unless a second air-glass-interface is brought within $d \approx \lambda$ to the first one.

> The tunnel effect is, therefore, *not* a specific quantum mechanical phenomenon (as is often stated in text books), but occurs for all waves (optical, acoustic or matter waves). The crucial quantum mechanical aspect is the description of classical particles by waves.

We will illustrate the tunnel effect, discussed so far by a model barrier, for some real situations in atomic and nuclear physics.

### a) Field Ionisation of Atoms

The electron of the hydrogen atom is bound by the Coulomb force

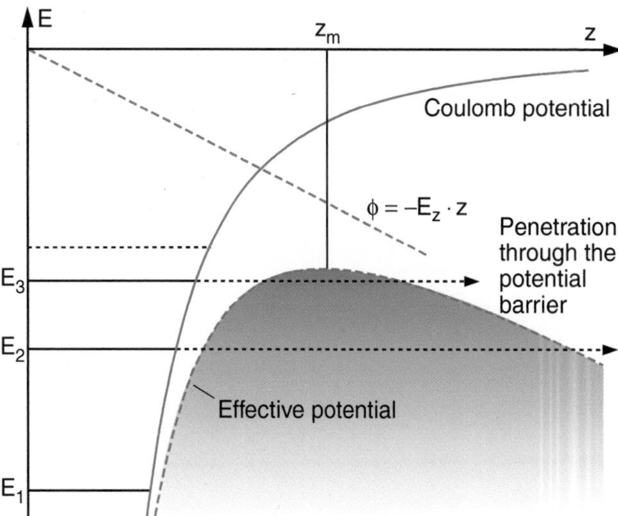


**Fig. 4.12.** Field ionization of high lying atomic energy levels by an external electric field through the tunnel effect

$$\boldsymbol{F}_\mathrm{C} = -\frac{1}{4\pi\varepsilon_0}\frac{e^2}{r^2}\hat{r} \tag{4.27}$$

between electron and proton. If the atom is placed in an external electric field, $\boldsymbol{E} = \{0, 0, E_z\}$ in $z$ direction, an additional force $\boldsymbol{F} = -e\boldsymbol{E} = -eE_z\hat{e}_z$ is acting on the electrons (Fig. 4.12). With $z = r\cos\vartheta$ we obtain for $\vartheta = 0$ the $z$ dependence of the potential

$$\phi(z) = -\frac{e}{4\pi\varepsilon_0 z} - E_z z \tag{4.28}$$

which has a maximum for $[\mathrm{d}\phi/\mathrm{d}z]_{z_\mathrm{m}} = 0$. This yields

$$z_\mathrm{m} = \sqrt{\frac{e}{4\pi\varepsilon_0 E_z}}\,. \tag{4.29}$$

The potential for $z_\mathrm{m}$ becomes

$$\phi(z_\mathrm{m}) = -2\sqrt{\frac{e\cdot E_z}{4\pi\varepsilon_0}}$$

The ionization energy is lowered by this amount. Due to the tunnel effect the electron can escape even for $E < e\phi(z_\mathrm{m})$.

The energy level $E_3$ in Fig. 4.12 would be stable in a classical particle model but can decay by tunnel effect in the quantum mechanical model, which describes the electron by its matter wave.

### b) Nuclear α-decay

The radioactive $\alpha$-decay rate of instable radioactive nuclei is determined by the tunnel effect. The $\alpha$-particle (2 protons + 2 neutrons) is kept within the nucleus by

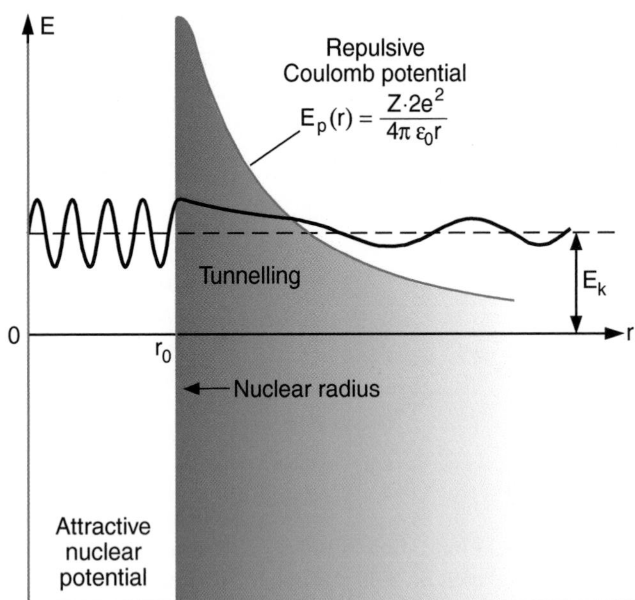


**Fig. 4.13.** Emission of $\alpha$ particles out of the nucleus of heavy atoms through the tunnel effect

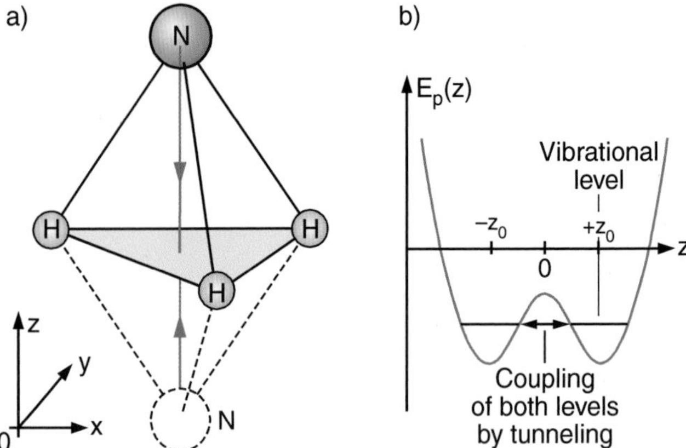


**Fig. 4.14.** Tunneling of the N atom through the plane of the three H atoms in the $NH_3$ molecule, resulting in the inversion splitting of vibrational levels

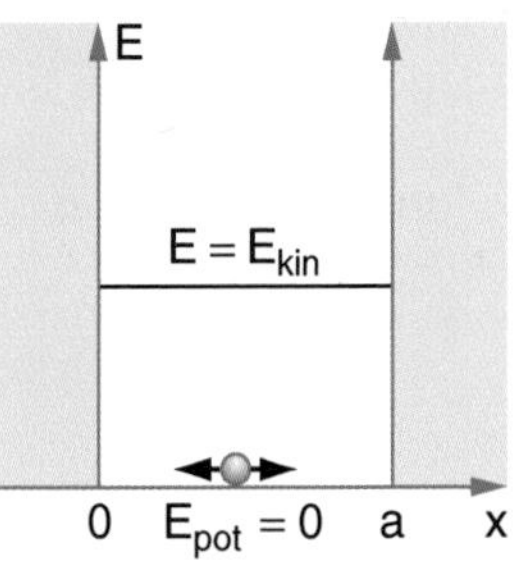


**Fig. 4.15.** Particle in a one-dimensional potential well with infinitely high walls

the strong attractive nuclear force, which can be modeled by the potential well in Fig. 4.13. Superimposed on the attractive potential is the repulsive Coulomb-potential due to the electrostatic repulsion between the positively charged $\alpha$-particle and the positive charge of the nucleus. If the energy of the $\alpha$-particle is above the total potential energy at $r \to \infty$ ($E = E_{pot} + E_{kin} > E_{pot}(\infty)$) the $\alpha$-particle can tunnel through the potential barrier and can leave the nucleus. The energy difference $\Delta E = E(r_0) - E_{pot}(\infty)$ appears as kinetic energy of the emitted $\alpha$-particle.

### c) Inversion Tunneling in the $NH_3$ Molecule

A famous example of the tunnel effect in molecular physics is the $NH_3$ molecule (Fig. 4.14). The potential energy $E_p(z)$ of the N atom has two minima at $z = \pm z_0$ above and below the plane of the three H atoms, separated by a maximum at $z = 0$. Even if the energy of the N atom vibrating against the three H atoms is less than the potential energy $E_p(z = 0)$, it can tunnel through the potential barrier and has, therefore, an equal probability of being at $z = z_0$ or $z = -z_0$.

## 4.2.4 Particle in a Potential Box

We will now discuss the case that a particle is restricted to a finite interval $0 \le x \le a$ (Fig. 4.15). This can be realized, if the potential energy is chosen as

$$E_{pot}(x) = \begin{cases} 0 & \text{for} \quad 0 \le x \le a \\ \infty & \text{elsewhere} \, . \end{cases} \tag{4.30}$$

In order to calculate the possible energies of the particle with mass $m$, we have to solve the one-dimensional Schrödinger equation with the given boundary conditions.

Within the interval $0 \le x \le a$ where $E_{pot} = 0$ the Schrödinger equation is

$$\frac{d^2\psi}{dx^2} + k^2\psi = 0 \quad \text{with} \quad k^2 = \frac{2mE}{\hbar^2} \tag{4.31}$$

with the boundary condition $\psi(x < 0) = \psi(x > a) = 0$, since the particle cannot reach these $x$ ranges.

We use, as we did in Sect. 4.2.2, the trial wave functions

$$\psi(x) = A e^{ikx} + B e^{-ikx} \tag{4.32}$$

as possible solutions. The boundary condition $\psi(x = 0)$ yields

$$A + B = 0 \tag{4.33a}$$

while $\psi(x = a) = 0$ gives

$$A\mathrm{e}^{\mathrm{i}ka} + B\mathrm{e}^{-\mathrm{i}ka} = 0 . \qquad (4.33b)$$

Because of (4.33a) this reduces to

$$A\left(\mathrm{e}^{\mathrm{i}ka} - \mathrm{e}^{-\mathrm{i}ka}\right) = 0 . \qquad (4.33c)$$

When inserting (4.32) and (4.33a) into (4.31) we obtain the possible wave functions

$$\psi(x) = A\left(\mathrm{e}^{\mathrm{i}kx} - \mathrm{e}^{-\mathrm{i}kx}\right) = 2\mathrm{i}A\sin(kx) . \qquad (4.34a)$$

The second boundary condition (4.33b) demands

$$2\mathrm{i}A\sin(ka) = 0 \Rightarrow ka = n\pi \quad (n = 1, 2, 3, \ldots) . \qquad (4.34b)$$

The wave functions are then

$$\psi(x) = 2\mathrm{i}A\sin\left(\frac{n\pi}{a}x\right) = C\sin\left(\frac{n\pi}{a}x\right) \qquad (4.35)$$

with $C = 2\mathrm{i}A$. They describe standing waves with amplitudes $C$. The boundary conditions allow only certain wavelengths

$$\lambda_n = \frac{2a}{n} \qquad (4.36a)$$

or wave numbers

$$k_n = \frac{2\pi}{\lambda_n} = n\frac{\pi}{a} . \qquad (4.36b)$$

These standing waves are completely analogous to vibrations of a string fixed at both ends (Fig. 4.16). The probability $|\psi(x)|^2\,\mathrm{d}x$ of finding the particle in the interval $x - \mathrm{d}x/2 \le x + \mathrm{d}x/2$ is shown in Fig. 4.16b.

The possible energies of the particle in the one-dimensional box follow from (4.36b) as:

$$E_n = \frac{p^2}{2m} = \frac{\hbar^2 k_n^2}{2m} = \frac{\hbar^2}{2m}\frac{\pi^2}{a^2}n^2 \quad (n = 1, 2, 3, \ldots). \qquad (4.37a)$$

The energies are quantized. Not every energy is possible but only discrete values $E_n$, which increase as $n^2$ and are proportional to $1/a^2$. This means that the larger the width $a$ of the potential well is, the smaller the eigenenergies (Fig. 4.17a,b).

The minimum energy is *not zero* but

$$E_1 = \frac{\hbar^2}{2m}\frac{\pi^2}{a^2} . \qquad (4.38)$$

This zero-point energy is caused by restricting the location $x$ of the particle to $\Delta x = a$. The larger $\Delta x$ the smaller is the zero-point energy.

This is directly related to Heisenberg's uncertainty relation that demands

$$\Delta p \Delta x \ge h/2 .$$

With $\Delta x = a$ this gives

$$p \ge \Delta p \ge \frac{h}{2a} \Rightarrow E_{\min} = \frac{p_{\min}^2}{2m} \ge \frac{h^2}{8ma^2} = \frac{\hbar^2\pi^2}{2ma^2} . \qquad (4.39)$$

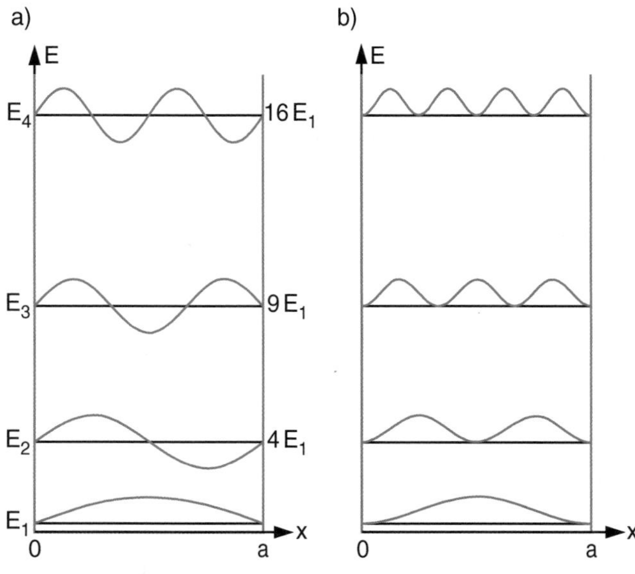


**Fig. 4.16a,b.** Energy levels and eigenfunctions of a particle in a one-dimensonal box with infinitely high walls. (**a**) Eigenfunctions $\psi(x)$ as standing waves. (**b**) Spatial probability distribution $|\psi(x)|^2$

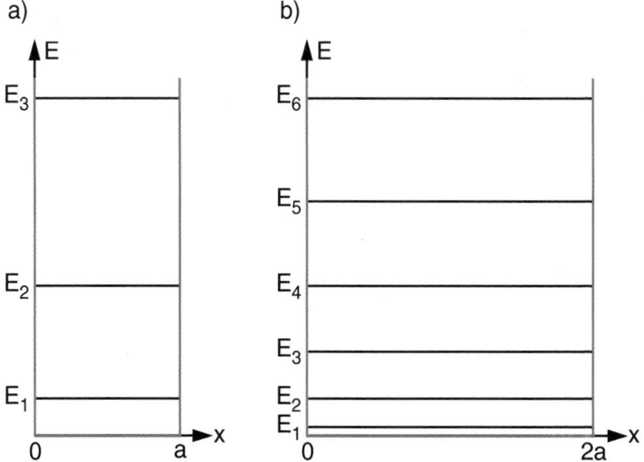


**Fig. 4.17a,b.** Comparison of energy levels in a one-dimensional potential box with different widths $\Delta x$. (**a**) $\Delta x = a$, (**b**) $\Delta x = 2a$

For the minimum wave number we obtain:

$$k_{\min} = \frac{1}{\hbar} p_{\min} = \frac{\pi}{a} \Rightarrow \lambda_{\max} = \frac{2\pi}{k_{\min}} = 2a . \tag{4.40}$$

## EXAMPLE

For an electron in a potential well with $a = 1\,\text{nm}$ the energy values are

$$E_n = \frac{\hbar^2\pi^2}{2m_e a^2} n^2 = 0.368\,\text{eV} \cdot n^2 .$$

The excitation of the electron from its lowest state $E_1$ to the first excited state $E_2$ demands the energy $\Delta E = E_2 - E_1 = 3 \cdot 0.368\,\text{eV} = 1.1\,\text{eV}$.

### Remarks

1. When the walls of the potential box have a finite height $E_0$ the particle can partly penetrate into these walls, as was shown in Sect. 4.2.2. Its wave function decays inside the walls exponentially (Fig. 4.18). The wave functions slightly change because the boundary conditions $\psi(0) = 0$ and $\psi(a) = 0$ are no longer valid. The energies $E_n$ decrease. As is outlined in Problem 4.4 the energies $E_n$ can be numerically calculated from the transcendent equations

$$k_n \tan(k_n a/2) = \alpha \tag{4.41a}$$

or

$$-k_n \text{cotan}(k_n a/2) = \alpha \tag{4.41b}$$

depending on the boundary conditions, with

$$\alpha = \frac{1}{\hbar}\sqrt{2m(E_0 - E)}$$

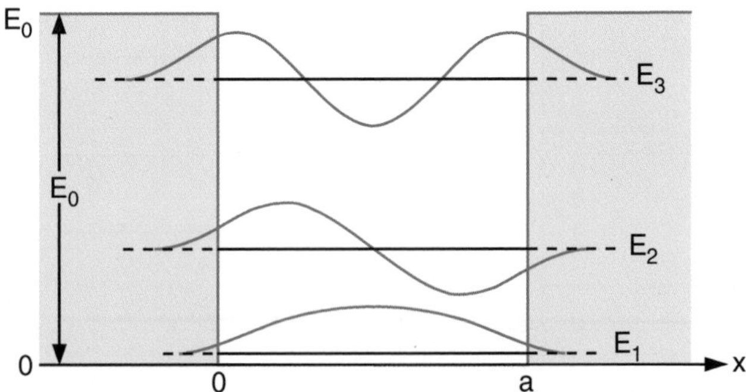


**Fig. 4.18.** Energy levels and eigenfunctions in a one-dimensional potential box with finite heights of the walls

and

$$k_n = \frac{1}{\hbar}\sqrt{2mE_n} . \tag{4.41c}$$

2. Energies $E > E_0$ are not quantized, since the particle is now no longer restricted to the interval $0 \le x \le a$. A particle with $E > E_0$ traveling into the $x$ direction traverses the potential well, however, not with the transmission probability $T = 1$. Part of its matter wave is reflected, as has already been discussed in Sect. 4.2.2.
3. Up to now we have described the particle by its matter wave with a well- defined k-value. In Sect. 3.3 we have, however, learned that the adequate description is a wave packet that contains many different $k$-values within the interval $k_0 - \Delta k/2 \le k + \Delta k/2$. While the solutions (4.35) describe the stationary time-averaged situation of a particle in a one-dimensional box, the time-dependent description of a particle moving back and forth in the box has to be performed with wave packets. Here the interference pattern is much more complex, because many different wavelengths of the wave packet contribute to it. An impression of the time-dependent description gives the colored picture 1 in the Appendix.
4. If the energy scale in Fig. 4.18 is shifted in such a way that

$$E_{\text{pot}} = \begin{cases} -E_0 & \text{for } 0 \le x \le a \\ 0 & \text{elsewhere} \end{cases}$$

all formulas in Sect. 4.2.3 can still be used if $E_0$ is replaced by $-E_0$. Since $E > -E_0$, the transmission $T$ is now obtained from (4.26c), but with a different value of $k'$

$$k' = \sqrt{2m(E + E_0)}/\hbar .$$

At energies $E_n = \hbar^2 k_n^2/2m$ with $k_n = n\pi/a$ the transmission shows maxima. This phenomenon, caused by interferences between waves reflected at the front and backside of the potential can be also observed for more general (i.e., non-square well) potentials. For instance, the scattering cross section of particles scattered by other atoms in a gas shows minima for energies of the incident particle when destructive interference occurs for the deflected de Broglie waves (Ramsauer effect).

In Summary

> If a particle is spatially restricted to the interval $\Delta x \le a$, it can only have discrete energy values
>
> $$E_n = \frac{\hbar^2}{2m}\frac{\pi^2}{a^2}n^2 = \frac{h^2}{8ma^2}.$$
>
> Its minimum energy is not zero but
>
> $$E_1 = \frac{\hbar^2}{2m}\frac{\pi^2}{a^2} \quad \text{(zero-point energy).}$$

### 4.2.5 Harmonic Oscillator

A very famous example of one-dimensional problems is the harmonic oscillator, which plays an important role in many branches of physics, such as molecular physics, solid-state physics and nuclear physics. If a particle moves in a parabolic potential with potential energy $E_{\text{pot}} = \frac{1}{2}Dx^2$ the restoring force is $\boldsymbol{F} = -\,\mathbf{grad}\,E_{\text{pot}} = -Dx$. The classical description uses a point mass $m$, that is attached to a spring with restoring force $\boldsymbol{F} = -D\boldsymbol{x}$ (Hook's law) (Fig. 4.19). If the point mass is removed from its equilibrium position at $x = 0$ and then released it performs harmonic oscillations $x = A\sin(\omega t)$ or $x = A\cos(\omega t)$ depending on the initial conditions. The frequency of these oscillations is

$$\omega = \sqrt{D/m} \Rightarrow D = m\omega^2 . \tag{4.42}$$

It depends on the mass $m$ and the restoring force constant $D$ of the spring.

The quantum mechanical treatment starts from the Schrödinger equation (4.6a), which now becomes:

$$-\frac{\hbar^2}{2m}\frac{\mathrm{d}^2\psi}{\mathrm{d}x^2} + \frac{1}{2}Dx^2\psi = E\psi . \tag{4.43}$$

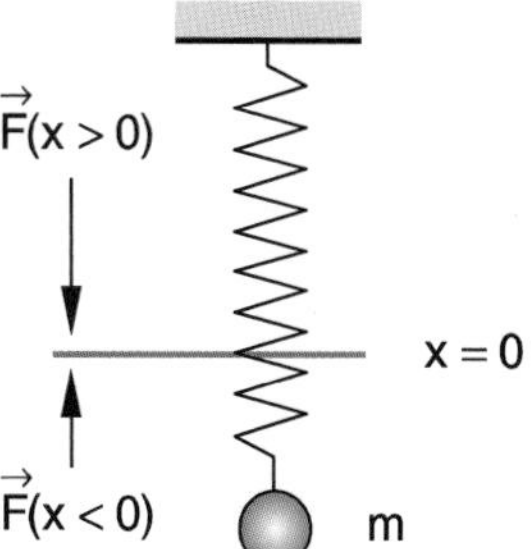


**Fig. 4.19.** Classical model of the harmonic oscillator

Inserting (4.42) gives

$$-\frac{\hbar^2}{2m}\frac{\mathrm{d}^2\psi}{\mathrm{d}x^2} + \frac{1}{2}\omega^2 m x^2\psi = E\psi . \tag{4.44a}$$

Using the transformation of the variable $x$ to

$$\xi = x\sqrt{m\omega/\hbar} \tag{4.45}$$

and the abbreviation

$$C = \frac{2E}{\hbar\omega}$$

we obtain from (4.44a) the Schrödinger equation in the form

$$\frac{\mathrm{d}^2\psi}{\mathrm{d}\xi^2} + (C - \xi^2)\psi = 0 . \tag{4.44b}$$

For the special case $C = 1$ $\left(E = \frac{1}{2}\hbar\omega\right)$ the solution of (4.44b) is:

$$\psi_0(\xi) = A\mathrm{e}^{-\xi^2/2} \tag{4.46a}$$

as can be readily verified by inserting (4.46a) into (4.44b). We now make the more general ansatz:

$$\psi(\xi) = H(\xi)\mathrm{e}^{-\xi^2/2} . \tag{4.46b}$$

Inserting this into (4.44b) yields the equation

$$\frac{\mathrm{d}^2H}{\mathrm{d}\xi^2} - 2\xi\frac{\mathrm{d}H}{\mathrm{d}\xi} + (C - 1)H = 0 \tag{4.47}$$

for the function $H(\xi)$. This is a Hermitian differential equation, well known to mathematicians. Its solutions are the Hermitian polynomials $H_\upsilon(\xi)$ of order $\upsilon$, which are defined by the generation equation

$$H_\upsilon(\xi) = (-1)^\upsilon \mathrm{e}^{\xi^2}\frac{\mathrm{d}^\upsilon}{\mathrm{d}\xi^\upsilon}\left(\mathrm{e}^{-\xi^2}\right) \tag{4.48}$$

with $\upsilon = 0, 1, 2, \ldots$, as can be verified by inserting (4.48) into (4.47). The Hermitian polynomials can be found in mathematical tables. The eigenfunctions $\psi(\xi)$ in (4.46b) for the first four polynomials $H_\upsilon$ with $\upsilon = 0, 1, 2$ and 3 are listed in Table 4.1, where the $N_i$ are normalization factors, which ensure that

$$\int\limits_{\xi=-\infty}^{\infty} |\psi(\xi)|^2\,\mathrm{d}\xi = \int\limits_{x=-\infty}^{\infty} |\psi(x)|^2\,\mathrm{d}x = 1 . \tag{4.49}$$

**Table 4.1.** Eigenfunctions of the harmonic oscillator for the four lowest vibrational levels

| $v$ | $E(v)$ | $\psi_v(\xi)$ |
|---|---|---|
| 0 | $\frac{1}{2}\hbar\omega$ | $N_0 \cdot e^{-\xi^2/2}$ |
| 1 | $\frac{3}{2}\hbar\omega$ | $N_1 \cdot 2\xi \cdot e^{-\xi^2/2}$ |
| 2 | $\frac{5}{2}\hbar\omega$ | $N_2 \cdot (4\xi^2 - 2) \cdot e^{-\xi^2/2}$ |
| 3 | $\frac{7}{2}\hbar\omega$ | $N_3 \cdot (8\xi^3 - 12\xi) \cdot e^{-\xi^2/2}$ |

The Hermitian polynomials can be represented by the power series

$$H(\xi) = \sum_{i=0}^{v} a_i \xi^i \,, \tag{4.50}$$

which must be finite. Otherwise $H_v(\xi)$ would become infinite for $\xi > 1$ and the normalization (4.49) would not be possible for the wave functions

$$\psi(x) = \tilde{H}(x) e^{-(mE/\hbar^2)x^2/2} \,. \tag{4.51}$$

with $\tilde{H}(x) = N_v \cdot H(x)$ where $N_v = \sqrt[4]{m \cdot E_v/\hbar^2}$ is a normalization factor.

Inserting (4.50) into (4.47) and comparing the coefficients of equal powers $\xi^i$ yields the recursion formula

$$(i+2)(i+1)a_{i+2} = [2i - (C-1)]\, a_i \,. \tag{4.52}$$

Assume that $\xi^v$ is the highest power of $\xi^i$, then $a_{v+2}$ has to be zero. From (4.52) we can, therefore, conclude

$$(2v - C + 1) = 0 \Rightarrow v = \frac{1}{2}(C-1) \,. \tag{4.53}$$

Since $C = 2E/\hbar\omega$ we obtain the relation between the "quantum numbers" $v$ and the energy values

$$E(v) = \left(v + \frac{1}{2}\right)\hbar\omega \quad (v = 0, 1, 2, \ldots) \,. \tag{4.54}$$

The possible energy values of the harmonic oscillator are equidistant with the separation $\Delta E = \hbar\omega$. The lowest energy for $v = 0$

$$E_0 = \frac{1}{2}\hbar\omega \tag{4.55}$$

is not zero but equals half of the level separation $\Delta E = \hbar\omega$. $E_0$ is called zero-point energy.

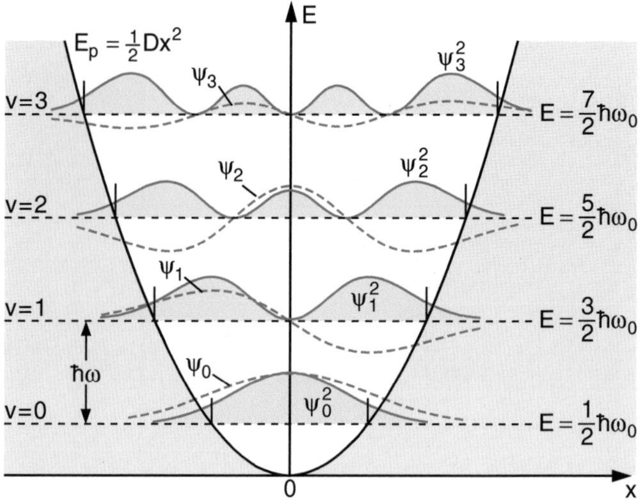


**Fig. 4.20.** Equidistant energy levels wave functions $\psi_n(x)$ and spatial probabilities $|\psi(x)|^2$ for a particle in a parabolic potential (harmonic oscillator)

The normalized wave functions of the harmonic oscillator are

$$\psi(x) = \tilde{H}(x) e^{-(mE_v/\hbar^2)x^2} \,. \tag{4.56}$$

They are depicted in Fig. 4.20 for $v = 0, 1, 2, 3$ together with their absolute squares $|\psi|^2\, dx$, which give the probability to find the particle in the interval $dx$ around the position $x$.

In classical physics, the probability of finding the particle during the time interval $dt$ within the spatial interval $dx$ during the harmonic oscillation with period $T = 2\pi/\omega$ is

$$P(t)\, dt = \frac{dt}{T} = \frac{dx}{v(x)T} \,, \tag{4.57}$$

where $dt = dx/v(x)$ is the time interval the particle needs to traverse the spatial interval $dx$. Since the velocity $v(x) = dx/dt$ becomes zero at the two turning points of the oscillation, the probability $P(x)$ has maxima at these points (Fig. 4.21).

Also the quantum mechanical description yields, for $v > 0$, maxima around the turning points, but has an oscillatory probability $P(x)\, dx$ of finding the particle in the interval $dx$ around $x$. For large quantum numbers $v$ the classical description (dashed curve in Fig. 4.21, shown for $v = 10$.) represents the average of the quantum mechanical probability $|\psi|^2\, dx$.

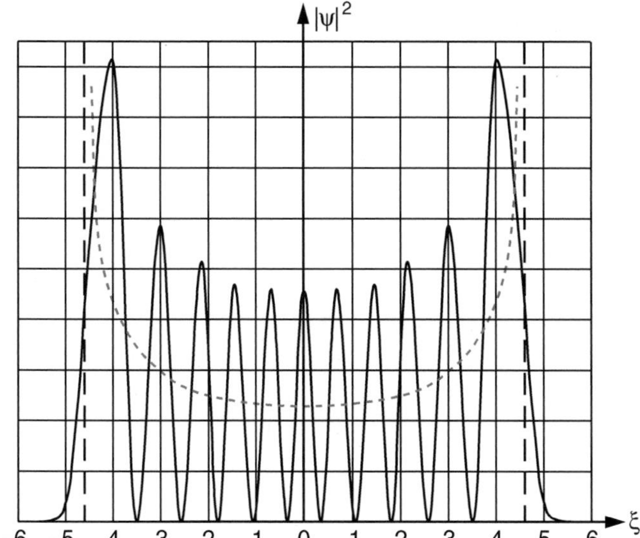


**Fig. 4.21.** Comparison of classical (*dashed blue line*) and quantum mechanical probabilities for a particle with high vibrational quantum number (here $v = 10$) of the harmonic oscillator. The abscissa is plotted in units of $\xi = x\sqrt{m\omega/\hbar}$

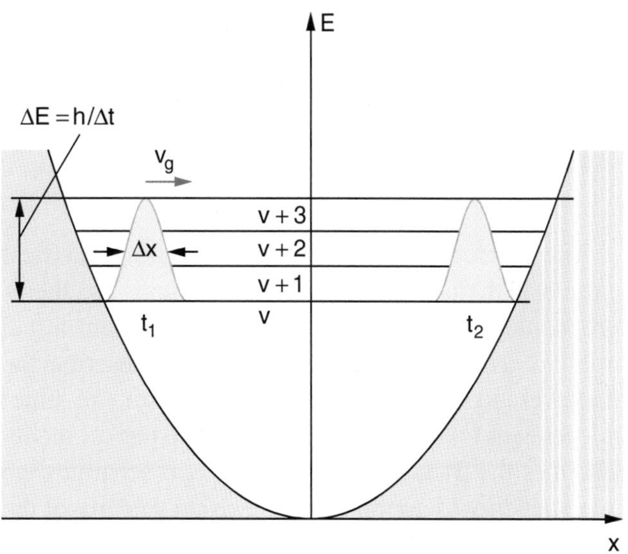


**Fig. 4.22.** Two momentary pictures of a wave packet oscillating back and forth in the harmonic potential. The wave packet is composed here of four vibrational wave functions of levels $v$ to $v + 3$

For the lowest state $v = 0$, both descriptions differ considerably. The quantum mechanical probability has a Gaussian distribution with its maximum at $x = 0$, whereas the classical description assumes a particle at rest and the probability would therefore be a delta function at $x = 0$. The classical model gives the lowest energy as $E_0 = 0$, while the quantum mechanical model yields the zero-point energy $E_0 = \hbar\omega/2$, which is justified by the uncertainty relation.

All experiments performed so far (see Chap. 11) have proven that the quantum mechanical model of the harmonic oscillator is the correct description.

### Remark

The quantities $|\psi|^2\,\mathrm{d}x$ in (4.39) give the probabilities of finding the particle within the interval $\mathrm{d}x$ only for the stationary case. This case is relevant if the energies $E_n$ are measured with a low time resolution, which means that the measuring interval $\Delta t$ is large compared to the vibrational period $T$ of the oscillator. For measurements with a very high time resolution, the dynamics of the vibrating particle can be resolved. In this case the stationary wave description is no longer adequate, but wave packets must be used in order to localize the particle better during its vibration (Fig. 4.22). The wave packets are composed of the superposition of several vibrational wave functions within the energy interval $\Delta E = \hbar/\Delta t$, where $\Delta t$ is the time resolution of the experiment. The wave packets move back and forth between the potential walls. This wave packet description comes closer to the classical model (see Chap. 11).

### EXAMPLE

The vibration period of a vibrating molecule depends on the mass and the restoring force (i.e., the potential coefficient $D$ in (4.30)). For $Na_2$ molecules $T = 2 \times 10^{-13}$ s, for $N_2$ molecules $T = 1.4 \times 10^{-14}$ s and for the light hydrogen molecule $H_2$ the vibrational period is only $T = 7.6 \times 10^{-15}$ s.

In order to measure the dynamics of a vibrating molecule a time resolution in the femtosecond range ($1\,\text{fs} = 10^{-15}$ s) is necessary.

If the wave packet should have a spatial resolution $\Delta x$ corresponding to a resolved time interval $\Delta t$ that is small compared with the vibrational period $T$, then its energy resolution $\Delta E > h/\Delta t$ becomes worse. For instance, if $\Delta t = 0.2\,T$ than $\Delta E$ is about five times as large as the energy separation $\Delta E_{\text{vib}} = E_{n+1} - E_n$. This means that five vibrational levels are simultaneously involved in the oscillation. The superposition of the wave functions (including their time factor $\exp(i\omega_n t)$) of these five levels determines the form of the wave packet, which oscillates, like the classical particle, between the turning points within the harmonic potential (Fig. 4.22).

The discrepancy between the oscillating classical particle and the stationary wave functions is, therefore, somewhat artificial and does not represent a real problem. The stationary wave functions describe the time-averaged motion of the particle with the time-independent total energy, while the wave packet model describes the dynamics of the particle, it separates the kinetic energy (determined by the group velocity of the wave packet) and the potential energy. Both energies change during the oscillation, while the total energy remains constant.

## 4.3 Two-and Three-Dimensional Problems

For the solution of two- or three-dimensional problems we have to start from the three dimensional Schrödinger equation (4.6b)

$$-\frac{\hbar^2}{2m}\Delta\psi + E_{\text{pot}}\psi = E\psi \tag{4.58}$$

with the Laplace operator in Cartesian coordinates

$$\Delta\psi = \frac{\partial^2\psi}{\partial x^2} + \frac{\partial^2\psi}{\partial y^2} + \frac{\partial^2\psi}{\partial z^2} \,. \tag{4.59}$$

The solutions $\psi = \psi(x,y,z)$ can be obtained analytically only for potentials with high symmetry. For all other cases the Schrödinger equation has to be solved numerically. We will start with the more simple case of a two-dimensional potential box.

### 4.3.1 Particle in a Two-dimensional Box

If the potential energy is

$$E_{\text{pot}}(x,y) = \begin{cases} 0 & \text{for } 0 \le x \le a\,; \quad 0 \le y \le b \\ \infty & \text{elsewhere} \end{cases} \tag{4.60}$$

(Fig. 4.23a) we can use the separable ansatz

$$\psi(x,y) = f(x)g(y) \tag{4.61}$$

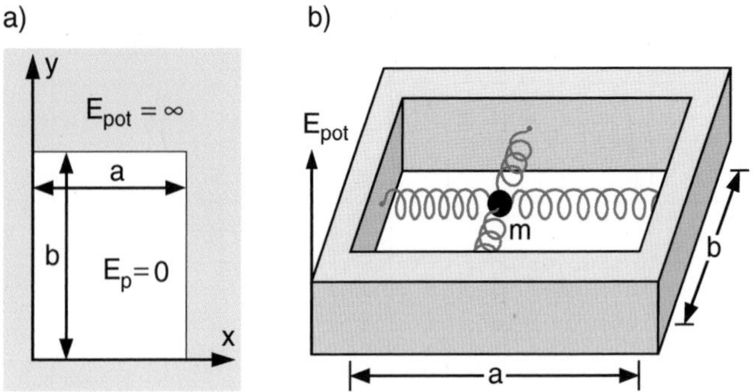


**Fig. 4.23.** (**a**) Rectangular two-dimensional potential box. (**b**) Classical analogy of a mass bound by two springs to the walls of a box as example for a two-dimensional harmonic oscillator

for the solution of (4.58). Inserting (4.61) into (4.58) we obtain, with the boundary conditions

$$\psi(x=0, y=0) = \psi(x=a, y=b) = 0\,, \tag{4.62}$$

analogously to (4.35), the solutions are

$$\begin{aligned} f(x) &= A\sin\left(\frac{n_x\pi x}{a}\right) \\ g(y) &= B\sin\left(\frac{n_y\pi y}{b}\right) \end{aligned} \tag{4.63}$$

with the integer numbers $n_x$, $n_y$. Our wave function then becomes

$$\psi(x,y) = C\sin\left(\frac{n_x\pi x}{a}\right)\sin\left(\frac{n_y\pi y}{b}\right) \tag{4.64}$$

with $C = AB$. The normalization

$$\int_{x=0}^{a}\int_{y=0}^{b} |\psi(x,y)|^2\,\mathrm{d}x\,\mathrm{d}y = 1 \tag{4.65}$$

yields the condition $C = 2/\sqrt{ab}$.

Inserting this normalized wave function into the Schrödinger equation (4.58) gives the possible energies

$$E(n_x,n_y) = \frac{h^2}{8m}\left(\frac{n_x^2}{a^2} + \frac{n_y^2}{b^2}\right) = E_{1x}n_x^2 + E_{2y}n_y^2 \tag{4.66a}$$

with

$$E_{1x} = \frac{h^2}{8ma^2}\,; \quad E_{1y} = \frac{h^2}{8mb^2}\,. \tag{4.66b}$$

This shows, that every combination $(n_x, n_y)$ yields a possible energy value. The number of allowed energy levels within a given energy interval is, therefore, much larger than in the one-dimensional case (Fig. 4.24).

It may happen, that two different combinations $(n_x, n_y)$ result in the same energy. For example for the quadratic potential box with $a = b$ the two configurations with $(n_x = 7, n_y = 1)$ and $(n_x = n_y = 5)$ both have the same energy $E = 50E_1$, although their wave functions differ completely (see Fig. 4.25). Such states with different wave functions, but with the same energy are called ***degenerate***.

The absolute squares $|\psi(x, y)|^2$ of the wave functions represent the probability to find the particle at the position $(x, y)$ in the energy state $E(n_x, n_y)$. For degenerate states different spatial distributions of this probability give the same energy value.

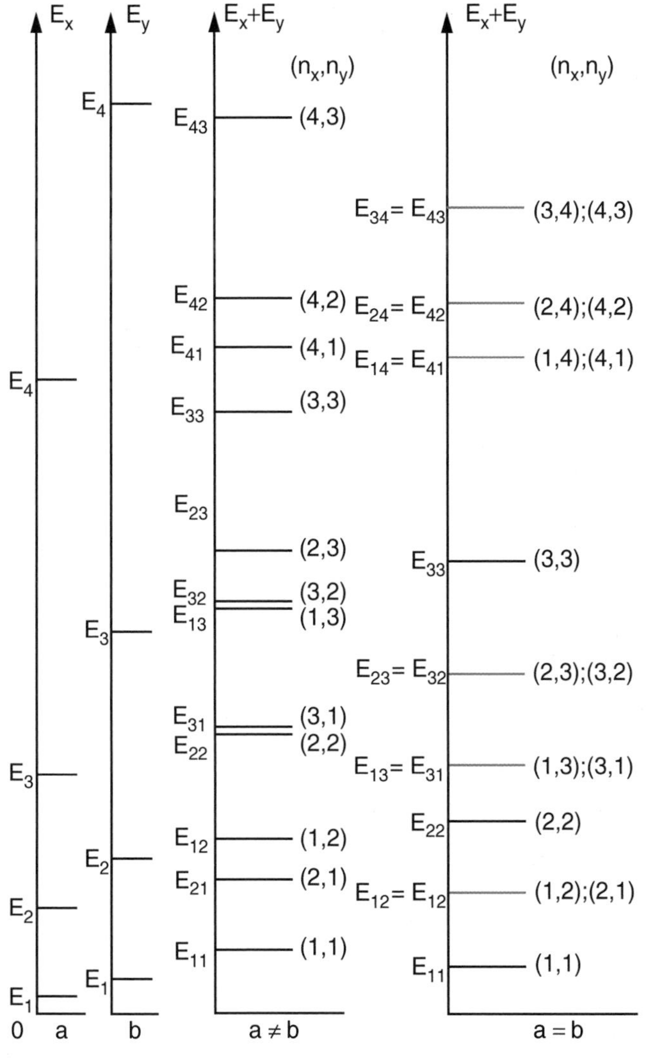


**Fig. 4.24.** Energy eigenvalues $E_{mx,my}$ of a particle in a two-dimensional rectangular potential box with unequal rectangle sides $a > b$ and for a quadratic box with $a = b$

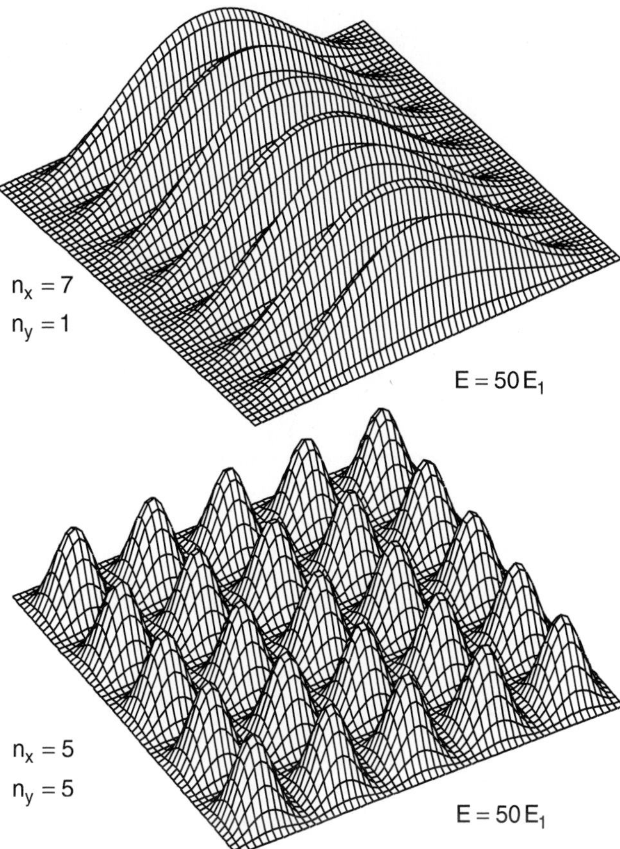


**Fig. 4.25.** Spatial probability $|\psi(x, y)|^2$ of a particle in a quadratic potential box for two degenerated energy states $(n_x = 7, n_y = 1$ and $n_x = n_y = 5)$

A similar treatment gives for a two-dimensional parabolic potential the energy values of a two-dimensional harmonic oscillator, illustrated by the classical analogon in Fig. 4.23b.

### 4.3.2 Particle in a Spherically Symmetric Potential

For a potential with spherical symmetry the solutions of the Schrödinger equation are easier to obtain when we use spherical coordinates $(r, \vartheta, \varphi)$ instead of the Cartesian coordinates $(x, y, z)$. The relations between the two sets of coordinates (Fig. 4.26) is given by

$$\left.\begin{aligned} x &= r \sin\vartheta \cos\varphi \\ y &= r \sin\vartheta \sin\varphi \\ z &= r \cos\vartheta \end{aligned}\right\} \Rightarrow \begin{aligned} r &= \sqrt{x^2 + y^2 + z^2} \\ \vartheta &= \arccos\left(\frac{z}{\sqrt{x^2+y^2+z^2}}\right) \\ \varphi &= \arctan(y/x) \end{aligned} \tag{4.67}$$

The differentiation in the Laplace operator (4.59) using the relations (4.67) gives $\Delta(r, \vartheta, \varphi)$ as

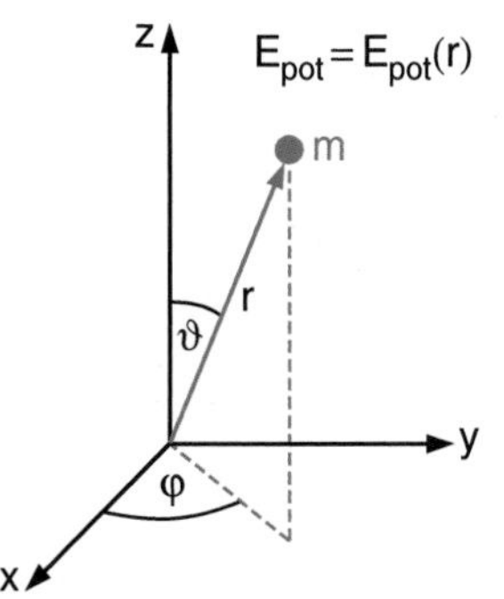


**Fig. 4.26.** Spherical coordinates for the description of a particle in a spherical potential

$$\Delta(r,\vartheta,\varphi)=\frac{1}{r^2}\frac{\partial}{\partial r}\left(r^2\frac{\partial}{\partial r}\right)+\frac{1}{r^2\sin\vartheta}\frac{\partial}{\partial\vartheta}\left(\sin\vartheta\frac{\partial}{\partial\vartheta}\right)+\frac{1}{r^2\sin^2\vartheta}\frac{\partial^2}{\partial\varphi^2}\,. \tag{4.68}$$

Then the Schrödinger equation becomes, in spherical coordinates,

$$\frac{1}{r^2}\frac{\partial}{\partial r}\left(r^2\frac{\partial\psi}{\partial r}\right)+\frac{1}{r^2\sin\vartheta}\frac{\partial}{\partial\vartheta}\left(\sin\vartheta\frac{\partial\psi}{\partial\vartheta}\right)+\frac{1}{r^2\sin^2\vartheta}\frac{\partial^2\psi}{\partial\varphi^2}+\frac{2m}{\hbar^2}\left(E-E_{\text{pot}}(r)\right)\psi=0\,. \tag{4.69}$$

For its solutions we try the product ansatz

$$\psi(r,\vartheta,\varphi)=R(r)\Theta(\vartheta)\Phi(\varphi)\,. \tag{4.70}$$

Inserting this into (4.69) gives, after multiplication of both sides with $r^2\sin^2\vartheta/\psi$,

$$\frac{\sin^2\vartheta}{R(r)}\frac{\mathrm{d}}{\mathrm{d}r}\left(r^2\frac{\mathrm{d}R}{\mathrm{d}r}\right)+\frac{\sin\vartheta}{\Theta(\vartheta)}\frac{\mathrm{d}}{\mathrm{d}\vartheta}\left(\sin\vartheta\frac{\mathrm{d}\Theta}{\mathrm{d}\vartheta}\right)+\frac{2m}{\hbar^2}\left(E-E_{\text{pot}}(r)\right)r^2\sin^2\vartheta=-\frac{1}{\Phi(\varphi)}\frac{\mathrm{d}^2\Phi}{\mathrm{d}\varphi^2}\,. \tag{4.71}$$

Now we draw an important conclusion:

The left side of (4.71) solely depends on $r$ and $\vartheta$, the right side solely on $\varphi$. Since the equation has to be valid for all values of $r$, $\vartheta$, and $\varphi$ it follows that both sides have to be equal to a constant $C_1$ (prove this statement!). For the right side of the equation this gives

$$\frac{\mathrm{d}^2\Phi}{\mathrm{d}\varphi^2}=-C_1\Phi \tag{4.72}$$

with has a solution

$$\Phi=A\mathrm{e}^{\pm\sqrt{C_1}\varphi}\,. \tag{4.73}$$

The function $\Phi(\varphi)$ has to be uniquely defined for all possible values of $\varphi$. This results in the condition

$$\begin{aligned}&\Phi(\varphi)=\Phi(\varphi+n2\pi)\\&\Rightarrow \mathrm{e}^{\pm\sqrt{C_1}2n\pi}=1\Rightarrow\sqrt{C_1}=m\\&\text{with}\quad m=\text{integer}\,,\end{aligned} \tag{4.74}$$

which means that $m$ must be a positive or negative integer. The solutions $\Phi(\varphi)$ are then:

$$\Phi_m(\varphi)=A\mathrm{e}^{\mathrm{i}m\varphi}\,. \tag{4.75a}$$

We will normalize them in such a way that

$$\int_0^{2\pi}\Phi_m^*(\varphi)\Phi_m(\varphi)\,\mathrm{d}\varphi=1\Rightarrow A=\frac{1}{\sqrt{2\pi}}\,.$$

This gives the normalized functions

$$\boxed{\Phi_m(\varphi)=\frac{1}{\sqrt{2\pi}}\mathrm{e}^{\mathrm{i}m\varphi}.} \tag{4.75b}$$

They are orthogonal because

$$\int_0^{2\pi}\Phi_m^*\Phi_n\,\mathrm{d}\varphi=\delta_{m,n}\,.$$

Now we will determine the solutions $\Theta(\vartheta)$. We divide both sides of (4.71) that equal the constant $C_1=m^2$ by $\sin^2\vartheta$ and rearrange the different terms in such a way that only terms that solely depend on $\vartheta$ remain on the right side, while on the left side only terms depending on $r$ remain. This yields:

$$\frac{1}{R}\frac{\mathrm{d}}{\mathrm{d}r}\left(r^2\frac{\mathrm{d}R}{\mathrm{d}r}\right)+\frac{2m}{\hbar^2}r^2\left(E-E_{\text{pot}}(r)\right)=-\frac{1}{\Theta\sin\vartheta}\frac{\mathrm{d}}{\mathrm{d}\vartheta}\left(\sin\vartheta\frac{\mathrm{d}\Theta}{\mathrm{d}\vartheta}\right)+\frac{m^2}{\sin^2\vartheta}=C_2\,, \tag{4.76}$$

where the same conclusion as for (4.71) is again drawn that both sides have to be equal to a constant $C_2$.

### Remark

Unfortunately the symbol $m$ in eqation (4.76) is used for two different quantities. On the left side it represents the mass of the particle and on the right side the integer $m=\sqrt{C_1}$. Since in most of the literature, the symbol $m$ is used for the magnetic quantum number $m=\sqrt{C_1}$ (see later) we did not want to depart from this common nomenclature.

From (4.76) we obtain for the function $\Theta(\vartheta)$ the equation

$$\frac{1}{\Theta \sin\vartheta}\frac{\mathrm{d}}{\mathrm{d}\vartheta}\left(\sin\vartheta\frac{\mathrm{d}\Theta}{\mathrm{d}\vartheta}\right)-\frac{m^2}{\sin^2\vartheta}=-C_2\,. \quad (4.77)$$

**a) $m$=0**

For the special case $m=0$ (4.77) reduces with the abbreviation $\xi=\cos\vartheta$, to the Legendre differential equation

$$\frac{\mathrm{d}}{\mathrm{d}\xi}\left[\left(1-\xi^2\right)\frac{\mathrm{d}\Theta}{\mathrm{d}\xi}\right]+C_2\Theta=0\,. \quad (4.78)$$

The solutions can be written as the power series

$$\Theta=a_0+a_1\xi+a_2\xi^2+\ldots\,. \quad (4.79)$$

Since the function $\Theta(\vartheta)$ must also be finite for $\xi=\pm1$, i.e., for $\vartheta=0°$ and $\vartheta=180°$, the power series can only have a limited number of terms.

Inserting (4.79) into (4.78) yields, when comparing the coefficients of equal powers of $\xi$ the recursion formula,

$$a_{k+2}=a_k\frac{k(k+1)-C_2}{(k+2)(k+1)}\,. \quad (4.80)$$

If the power series ends with the highest term $a_l\xi^l$ the conditions $a_l\neq 0$ but $a_{l+2}=0$ must be fulfilled. This gives

$$C_2=l(l+1)\quad l\in\mathbb{N}\,. \quad (4.81)$$

The real solutions

$$\Theta_l(\xi)=\mathrm{const}P_l(\cos\vartheta) \quad (4.82)$$

of Legendre's equation (4.78) are named ***Legendre's polynomials***. The boundary conditions demand:

$$\Theta^2(\vartheta)=\Theta^2(\vartheta+\pi)\Rightarrow\Theta(\vartheta)=\pm\Theta(\vartheta+\pi)\,.$$

Each of the functions represented by the power series (4.79) can therefore contain either even powers of $\xi$ only, or odd powers only.

**b) $m\neq0$**

For $m\neq 0$ the solutions of equation (4.78) are the associated Legendre functions $P_l^m(\cos\vartheta)$. As is shown in mathematical textbooks, they can be obtained from the Legendre's polynomials

$$\Theta_l(\xi)=P_l(\cos\vartheta)\quad\text{with}\quad\xi=\cos\vartheta$$

by solving the equation

$$P_l^m(\cos\vartheta)=\mathrm{const}(1-\xi^2)^{|m/2|}\frac{\mathrm{d}^{|m|}}{\mathrm{d}\xi^{|m|}}P_l(\xi)\,. \quad (4.83)$$

Since $P_l(\xi)$ is represented by a power series up to the highest power $\xi^l$, we see from (4.83) that the condition $|m|\leq l$ has to be fulfilled, otherwise the $m$th derivative would not be defined. The integer numbers $m$ can be positive as well as negative. Therefore the range of possible $m$-values is restricted to

$$-l\leq m\leq +l\,. \quad (4.84)$$

The functions $P_l^m$ are defined by (4.83) only apart from a constant factor, which is determined by the normalization condition

$$\int\limits_{\vartheta=0}^{\pi}\left|P_l^m(\cos\vartheta)\right|^2\sin\vartheta\,\mathrm{d}\vartheta=1\,. \quad (4.85)$$

The product functions

$$Y_l^m(\vartheta,\varphi)=P_l^m(\cos\vartheta)\Phi_m(\varphi) \quad (4.86)$$

are the ***spherical surface harmonic functions*** (Table 4.2). Their normalization

**Table 4.2.** Spherical surface harmonics

| $l$ | $m$ | $Y_l^m$ |
|---|---|---|
| 0 | 0 | $\frac{1}{2\sqrt{\pi}}$ |
| 1 | $\pm1$ | $\mp\frac{1}{2}\sqrt{\frac{3}{2\pi}}\sin\vartheta\,\mathrm{e}^{\pm\imath\varphi}$ |
| | 0 | $\frac{1}{2}\sqrt{\frac{3}{\pi}}\cos\vartheta$ |
| 2 | $\pm2$ | $\frac{1}{4}\sqrt{\frac{15}{2\pi}}\sin^2\vartheta\,e^{\pm2\imath\varphi}$ |
| | $\pm1$ | $\mp\frac{1}{2}\sqrt{\frac{15}{2\pi}}\cos\vartheta\,\sin\vartheta\,\mathrm{e}^{\pm\imath\varphi}$ |
| | 0 | $\frac{1}{4}\sqrt{\frac{5}{\pi}}(2\cos^2\vartheta-\sin^2\vartheta)$ |
| 3 | $\pm3$ | $\mp\frac{1}{8}\sqrt{\frac{35}{\pi}}\sin^3\vartheta\,\mathrm{e}^{\pm3\imath\varphi}$ |
| | $\pm2$ | $\frac{1}{4}\sqrt{\frac{105}{2\pi}}\cos\vartheta\,\sin^2\vartheta\,\mathrm{e}^{\pm2\imath\varphi}$ |
| | $\pm1$ | $\mp\frac{1}{8}\sqrt{\frac{21}{\pi}}\sin\vartheta\,(5\cos^2\vartheta-1)\,\mathrm{e}^{\pm\imath\varphi}$ |
| | 0 | $\frac{1}{4}\sqrt{\frac{7}{\pi}}(5\cos^3\vartheta-3\cos\vartheta)$ |

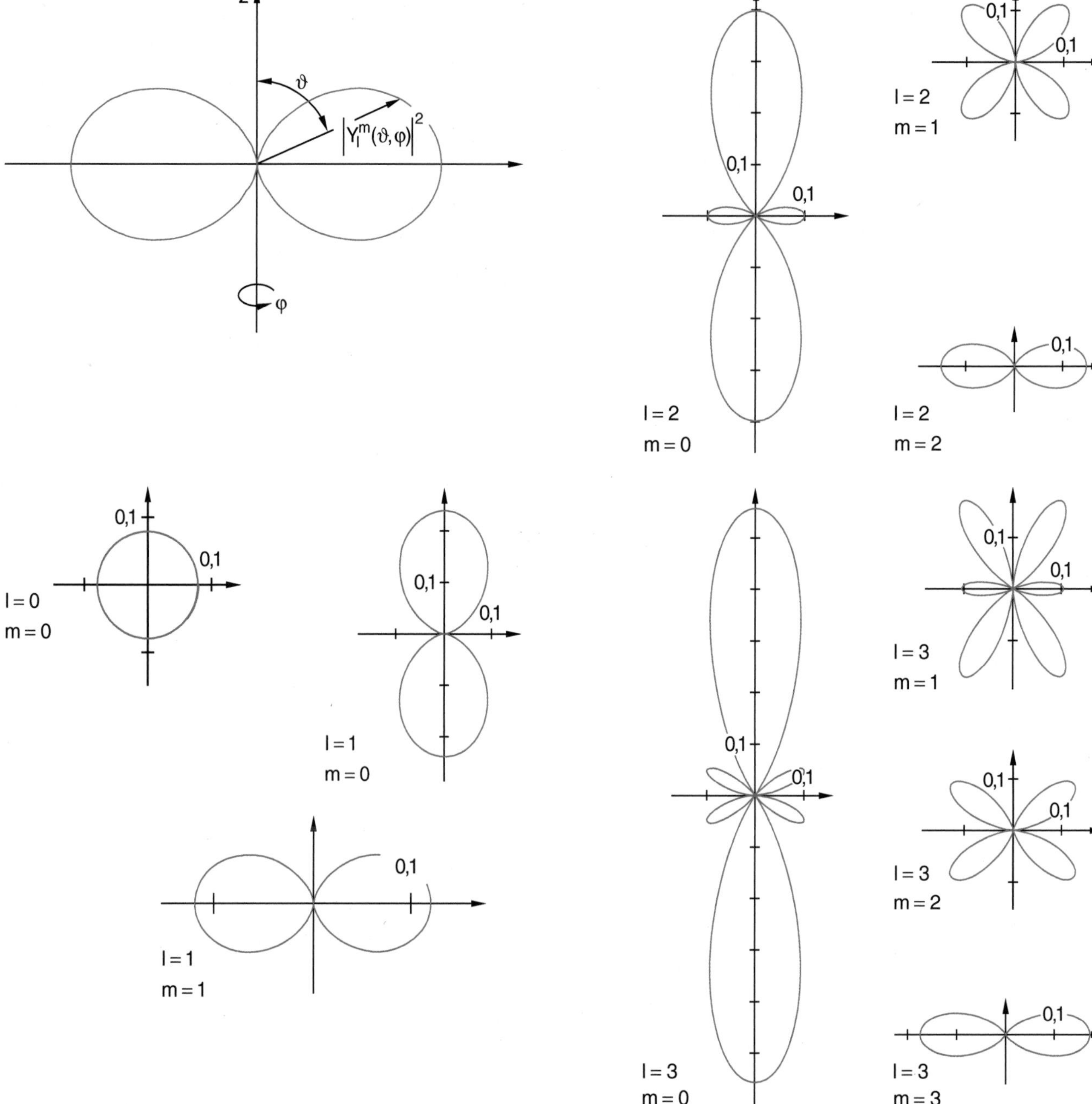


**Fig. 4.27.** Polar diagrams of the absolute squares $|Y_l^m(\vartheta,\varphi)|^2$ of the spherical harmonics. The length $|\boldsymbol{r}|$ of the vector $\boldsymbol{r}$ gives the values $|Y_l^m(\cos\vartheta)|^2$ for the different values of $\vartheta$. All diagrams are symmetric with respect to rotations around the $z$-axis, which has been chosen here as the vertical axis

$$\int_{\vartheta=0}^{\pi}\int_{\varphi=0}^{2\pi} |Y_l^m(\vartheta,\varphi)|^2 \sin\vartheta \, \mathrm{d}\vartheta \, \mathrm{d}\varphi = 1 \tag{4.87}$$

is automatically fulfilled because both factors are normalized separately. The absolute squares $|Y_l^m(\vartheta,\varphi)|^2$ are illustrated in Fig. 4.27 for some values of $l$ and $m$.

The quantity $|Y_l^m|^2 \sin\vartheta \, \mathrm{d}\vartheta \, \mathrm{d}\varphi$ gives the probability of finding the particle within the cone $\vartheta \pm \mathrm{d}\vartheta/2; \varphi \pm \mathrm{d}\varphi/2$.

**Remark**

The angular part $Y_l^m(\vartheta,\varphi)$ of the wave function is independent of the $r$-dependence of the potential

$E_{\text{pot}}(r)$. The only demand for the factorization (4.70) is the spherical symmetry of the potential. This means:

> All spherical symmetric potentials have the same angular part $Y_l^m(\vartheta,\varphi)$ of the eigenfunctions $\psi(r,\vartheta,\varphi)$.

The radial function $R(r)$ in (4.70) can be determined from the left side of (4.76). Based on the results (4.55) and (4.81) we obtain after multiplication with $R(r)$:

$$\frac{\mathrm{d}}{\mathrm{d}r}\left(r^2\frac{\mathrm{d}R}{\mathrm{d}r}\right)+\left[\frac{2m}{\hbar^2}r^2\left(E-E_{\text{pot}}(r)\right)-l(l+1)\right]R(r)=0\,. \tag{4.88}$$

The solutions of (4.88) do depend on the radial dependence of the potential $E_{\text{pot}}(r)$ and on the total energy $E$. We will discuss these solutions for the case of a Coulomb potential in the next chapter.

From (4.84) and (4.86) we can conclude that for a given allowed energy $E$ and a fixed quantum number $l$ there are $(2l+1)$ different spherical harmonics $Y_l^m$, because the energy does ***not*** depend on the quantum number $m$ which can be any integer number within the interval

$$-l\le m\le +l\,. \tag{4.88x}$$

**Remark**

The separation (4.70) is possible only for spherical symmetric potentials that do not depend on the angles $\vartheta$ and $\varphi$. The spherical harmonics can, nevertheless, depend on $\vartheta$ and $\varphi$. This means that even for a spherically symmetric potential the probability of finding a particle generally depends on the angles and is therefore not spherically symmetric! The spherical harmonics are solutions for every potential with spherical symmetry, independent of the radial form of this potential. In acoustics the functions $Y_l^m$ describe the possible deformations of the surface of a sphere caused by resonant standing acoustic waves within the sphere. They are therefore also called spherical surface functions.

## 4.4 Expectation Values and Operators

For the statistical description of properties of a many particle system one has to define average values, which depend on the distribution of these properties over the particles of the system. For example, the mean velocity $\bar{v}$ in a system of particles with velocity distribution $f(v)$ is defined by

$$\bar{v}=\int\limits_{v=0}^{\infty} vf(v)\,\mathrm{d}v\,. \tag{4.88y}$$

The function $f(v)\,\mathrm{d}v$ gives the probability of finding a particle within the interval $v$ to $v+\mathrm{d}v$. The mean square velocity

$$\overline{v^2}=\int\limits_{v=0}^{\infty} v^2 f(v)\,\mathrm{d}v \tag{4.88z}$$

gives the average value of $v^2$.

In quantum mechanics, the probability of finding a particle within the interval from $x$ to $x+\mathrm{d}x$ is given by $|\psi(x)|^2\,\mathrm{d}x$. The average value

$$\langle x\rangle=\int\limits_{-\infty}^{+\infty} x|\psi(x)|^2\,\mathrm{d}x \tag{4.89}$$

is called the expectation value for the location $x$ of the particle. The exact location in classical mechanics is replaced by a probability statement. The above definition of the expectation value has the following meaning.

When we perform a series of measurements of the location $x$ of a particle, the result will be a distribution of slightly different values $x$ around the mean value

$$\langle x\rangle=\int\psi^*(x)x\psi(x)\,\mathrm{d}x\,.$$

This distribution is not caused by errors or inaccuracies of measurement, but by the fact that due to the uncertainty relation $\Delta x\ge h/\Delta p$, it is in principle not possible to measure the location $x$ more accurately.

In an analogous way, we can define the expectation value of the three dimensional radius vector

$$\begin{aligned}\langle \boldsymbol{r}\rangle&=\int\psi^*(\boldsymbol{r})\boldsymbol{r}\psi(\boldsymbol{r})\,\mathrm{d}\tau\\&=\int\limits_x\int\limits_y\int\limits_z\psi^*(x,y,z)\boldsymbol{r}\psi(x,y,z)\,\mathrm{d}x\,\mathrm{d}y\,\mathrm{d}z\,.\end{aligned} \tag{4.90}$$

When a particle with charge $q$ is moving in an external electric field with potential $\phi(r)$, its mean potential energy is

$$\langle E_{\text{pot}}\rangle=q\int\psi^*(r)\phi(r)\psi(r)\,\mathrm{d}\tau\,, \tag{4.91}$$

where the volume element $\mathrm{d}\tau = \mathrm{d}x\mathrm{d}y\mathrm{d}z$ and the integral indicates the integration over all three dimensions.

> The expectation value of a measurable quantity of a particle is equal to its mean value obtained from the wave function $\psi$ of the particle replacing the classical distribution function.

### 4.4.1 Operators and Eigenvalues

The general expectation value $\langle A \rangle$ of a measurable quantity $A$ (called observable), is defined as

$$\langle A \rangle = \int \psi^* \hat{A} \psi \, \mathrm{d}\tau \tag{4.92}$$

where $\hat{A}$ is the operator related to the observable $A$. The operator $\hat{A}$ performs a definite operation on the wave function $\psi$. For example the operator $\hat{r}$ corresponding to the spatial coordinate $\boldsymbol{r}$ just multiplies the wave function $\psi$ with the radius vector $\boldsymbol{r}$.

The expectation value of the kinetic energy can be obtained from the Schrödinger equation, which represents the quantum mechanical analogy to the classical law of energy conservation $E_{\text{kin}} + E_{\text{pot}} = E$. From (4.6b) we see that

$$\langle E_{\text{kin}} \rangle = -\frac{\hbar^2}{2m} \int \psi^* \Delta \psi \, \mathrm{d}\tau \, . \tag{4.93}$$

The operator of the kinetic energy of a particle

$$\hat{E}_{\text{kin}} = -\frac{\hbar^2}{2m} \Delta \tag{4.94}$$

causes the second derivative of the particle's wave function. Performing the first derivative of the wave function

$$\psi = A\mathrm{e}^{(\mathrm{i}/\hbar)(\boldsymbol{pr} - Et)} \tag{4.95}$$

with respect to $x$, $y$ and $z$ we obtain

$$-\mathrm{i}\hbar \frac{\partial}{\partial x} \psi = p_x \psi \, , \quad -\mathrm{i}\hbar \frac{\partial}{\partial y} \psi = p_y \psi \, , \quad \text{and}$$

$$-\mathrm{i}\hbar \frac{\partial}{\partial z} \psi = p_z \psi \, . \tag{4.96}$$

This means that the operator of the particle's momentum can be expressed by

$$\hat{p} = -\mathrm{i}\hbar \nabla \, . \tag{4.97}$$

If $A$ is a measurable physical quantity that is measured $N$ times, the measured values $A_n$ will show deviations $\Delta A_n = A_n - \langle A \rangle$ from the mean value

$$\langle A \rangle = \frac{1}{N} \sum_{n=1}^{N} A_n$$

that are caused by the uncertainty $\Delta A$ of the observable $A$. There may be additional deviations caused by statistical or systematic errors of the measurement.

The deviations $\Delta A$ become zero, if the operator $A$ applied to the function $\psi$ reproduces this function apart from a constant factor $A$, i.e., if

$$\hat{A}\psi = A\psi \, . \tag{4.98}$$

The function $\psi$ is called an ***eigenfunction*** of the operator $\hat{A}$ and the constant $A$ is the ***eigenvalue*** of the operator $\hat{A}$. In this case we obtain for the expectation value

$$\langle A \rangle = A \int \psi^* \psi \, \mathrm{d}\tau = A \, , \tag{4.99}$$

which means that the expectation value of an operator $\hat{A}$ formed with an eigenfunction $\psi$ of this operator equals the eigenvalue $A$ that is well-defined and has no uncertainty.

This can be seen as follows. For eigenvalues we have the relations

$$\begin{aligned} \left\langle A^2 \right\rangle - \langle A \rangle^2 &= \int \psi^* \hat{A}^2 \psi \, \mathrm{d}\tau - \left( \int \psi^* \hat{A} \psi \, \mathrm{d}\tau \right)^2 \\ &= \int \psi^* \hat{A}\hat{A} \psi \, \mathrm{d}\tau - A^2 \left( \int \psi^* \psi \, \mathrm{d}\tau \right)^2 \\ &= A^2 \int \psi^* \psi \, \mathrm{d}\tau - A^2 (\int \psi^* \psi \, \mathrm{d}\tau)^2 \\ &= 0 \quad \text{because} \quad \int \psi^* \psi \, \mathrm{d}\tau = 1 \, . \end{aligned} \tag{4.100}$$

The mean quadratic deviation $\left\langle (\Delta A)^2 \right\rangle$ becomes zero, because

$$\begin{aligned} \left\langle (\Delta A)^2 \right\rangle &= \left\langle (A - \langle A \rangle)^2 \right\rangle = \left\langle A^2 \right\rangle + \langle A \rangle^2 - 2 \langle A \langle A \rangle \rangle \\ &\Rightarrow \left\langle (\Delta A)^2 \right\rangle = 2 \left\langle A^2 \right\rangle - 2 \langle A \rangle^2 = 0 \, . \end{aligned} \tag{4.101}$$

Here we have used the relation $\langle A \langle A \rangle \rangle = \langle A \rangle^2$.

This results means:

> If the wave function $\psi$ is eigenfunction of an operator $\hat{A}$, then the mean quadratic deviation of an observable $A$ is zero. The system is in a state where the quantity $A$ is constant in time and therefore the same value of $A$ is obtained (apart from measuring errors) when several measurements are performed over the course of time.

Since $A$ is a measurable quantity, it must be real. We therefore only allow those operators for physical quantities that have real eigenvalues and not complex ones. This demand is fulfilled for all Hermitian operators.

Definition:

> An operator is called Hermitian, if
>
> $$\int \psi^* \hat{A}\psi \, \mathrm{d}\tau = \int \psi \left(\hat{A}\psi\right)^* \mathrm{d}\tau \,, \tag{4.102}$$
>
> where $\psi^*$ is the conjugate complex of the function $\psi$.

If the operators $\hat{A}$ and $\hat{B}$ of two observables $A$ and $B$ have the same eigenfunctions $\psi$, both quantities $A$ and $B$ of a particle described by the wave function $\psi$, can be measured simultaneously.

This can be seen as follows. From the relations

$$\hat{A}\psi = A\psi \quad \text{and} \quad \hat{B}\psi = B\psi$$

it follows that

$$\hat{B}\hat{A}\psi = \hat{B}(A\psi) = A\left(\hat{B}\psi\right) = AB\psi$$

and

$$\hat{A}\hat{B}\psi = \hat{A}(B\psi) = B\left(\hat{A}\psi\right) = BA\psi \,.$$

Since $A$ and $B$ are real numbers which are commutative, i.e., $AB = BA$, we obtain

$$\left(\hat{A}\hat{B} - \hat{B}\hat{A}\right)\psi = 0 \Rightarrow \hat{A}\hat{B}\psi = \hat{B}\hat{A}\psi \,. \tag{4.103}$$

Two operators that fulfill the condition (4.103) are called commutable.

> If two operators are commutable their eigenvalues can be measured simultaneously with no uncertainty (apart from measuring errors).

This will be illustrated by some examples.

We have seen from the discussion of the stationary Schrödinger equation that the operator of the classical kinetic energy $E_{\text{kin}} = p^2/2m$ of a particle is given in quantum mechanics by the Laplace operator:

$$\hat{E}_{\text{kin}} = -\frac{\hbar^2}{2m}\Delta \,, \tag{4.104}$$

which performs the second derivative operation on the wave function.

The operator of the total energy $E = E_{\text{kin}} + E_{\text{pot}}$ is the Hamiltonian operator

$$\hat{H} = \hat{E}_{\text{kin}} + \hat{E}_{\text{pot}} = -\frac{\hbar^2}{2m}\Delta + E_{\text{pot}}(\boldsymbol{r}) \,, \tag{4.105}$$

which is analogous to the classical Hamiltonian function and describes the total energy of stationary systems.

> The total energy of a stationary state is the eigenvalue of the Hamiltonian operator. The Schrödinger equation (4.8) can be written as
>
> $$H\psi = E\psi \,.$$

From (4.104) the operator of the momentum of a particle follows as

$$\hat{p} = -\mathrm{i}\hbar\nabla \tag{4.106}$$

because the relation

$$\hat{E}_{\text{kin}} = \frac{\hat{p}^2}{2m} = -\frac{\hbar^2}{2m}\nabla^2 = -\frac{\hbar^2}{2m}\Delta$$

can then be fulfilled.

The expectation value of the momentum is

$$\langle \boldsymbol{p} \rangle = -\mathrm{i}\hbar \int \psi^* \nabla\psi \, \mathrm{d}\tau \,. \tag{4.107}$$

The $x$-components $\hat{p}_x$ of the momentum operator $\hat{p}$ and $x$ of the operator $\hat{x}$ are not commutable. This can be readily seen as follows:

$$\begin{aligned}
&\hat{x} \cdot \hat{p}_x(f(x)) - \hat{p}_x \cdot \hat{x}(f(x)) \\
&= -\hat{x}\mathrm{i}\hbar\frac{\partial f}{\partial x} + \mathrm{i}\hbar\frac{\partial}{\partial x}(xf(x)) \\
&= -\mathrm{i}\hbar\left(x\frac{\partial f}{\partial x} - f(x) - x\frac{\partial f}{\partial x}\right). \\
&= +\mathrm{i}\hbar f(x) \neq 0
\end{aligned}$$

If $f(x)$ is an eigenfunction of $\hat{x}$ it cannot be an eigenfunction of $\hat{p}_x$. This means, that the observables $x$ and $p_x$ cannot be exactly measured simultaneously.

We can now make the general statement:

> To every measurable physical quantity we can relate a Hermitian operator, where the quantity is the eigenvalue of this operator.

In Table 4.3 some physical quantities are listed together with their corresponding operators.

**Table 4.3.** Some measurable quantities and their operators

| Physical quantity | Operator |
|---|---|
| position vector $\boldsymbol{r}$ | $\boldsymbol{r}$ |
| potential energy $E_{\text{pot}}$ | $\hat{E}_{\text{pot}} = q \cdot \hat{V}(r)$ |
| kinetic energy $E_{\text{kin}}$ | $\dfrac{-\hbar^2}{2m}\Delta$ |
| total energy $E = E_{\text{pot}} + E_{\text{kin}}$ | $\hat{H} = \hat{E}_{\text{pot}} - \dfrac{\hbar^2}{2m}\Delta$ |
| momentum $\boldsymbol{p}$ | $\hat{\boldsymbol{p}} = -\imath\hbar\nabla$ |
| angular momentum $\boldsymbol{L}$ | $\hat{\boldsymbol{L}} = -\imath\hbar(\boldsymbol{r} \times \nabla)$ |
| $z$-component of $\boldsymbol{L}$ | $\hat{L}_z = -\imath\hbar\dfrac{\partial}{\partial\varphi}$ |

**Remark**

The representations of the vector operator $\hat{\boldsymbol{r}}$ as the vector $\boldsymbol{r}$ and of the momentum operator $\hat{p}$ as $-\mathrm{i}\hbar\nabla$ are useful if the wave functions are functions of the coordinate $\boldsymbol{r}$. There are physical situations (for example in solid-state physics) where problems can be solved easier when the wave functions $\psi(\boldsymbol{r})$ are Fourier transformed into functions $\Phi(\boldsymbol{p})$ of the momentum $\boldsymbol{p}$, which are defined in momentum space. One example is the eigenfunction

$$\Phi(\boldsymbol{p}) = \mathrm{e}^{\mathrm{i}(\boldsymbol{p}/\hbar)\boldsymbol{r}}$$

of the momentum operator $\hat{p}$.

One obtains the location $\boldsymbol{p}_p$ in momentum space, if the operator $\hat{\boldsymbol{p}}_p = -\mathrm{i}\hbar\nabla_p$ is applied to the function $\Phi(\boldsymbol{p})$. This yields

$$\hat{\boldsymbol{p}}_p\Phi = p_p\Phi\,.$$

### 4.4.2 Angular Momentum in Quantum Mechanics

The classical definition of the angular momentum $\boldsymbol{L}$ of a particle with mass $m$ and velocity $\boldsymbol{v}$ with respect to the origin $\boldsymbol{r} = 0$ is

$$\boldsymbol{L} = \boldsymbol{r} \times \boldsymbol{p} = m(\boldsymbol{r} \times \boldsymbol{v})\,. \tag{4.108a}$$

With the definition of the momentum operator $\boldsymbol{p} = -\mathrm{i}h\nabla$ we obtain the operator of the ***angular momentum***

$$\hat{L} = -\mathrm{i}\hbar(\boldsymbol{r} \times \nabla)\,. \tag{4.108b}$$

In Cartesian coordinates the components of $\hat{L}$ are

$$\hat{L}_x = -\mathrm{i}\hbar\left(y\frac{\partial}{\partial z} - z\frac{\partial}{\partial y}\right), \quad \hat{L}_y = -\mathrm{i}\hbar\left(z\frac{\partial}{\partial x} - x\frac{\partial}{\partial z}\right), \quad \text{and} \quad \hat{L}_z = -\mathrm{i}\hbar\left(x\frac{\partial}{\partial y} - y\frac{\partial}{\partial x}\right). \tag{4.108c}$$

In order to obtain $L$ in spherical coordinates, we use the transformations

$$\frac{\partial}{\partial x} = \frac{\partial r}{\partial x}\frac{\partial}{\partial r} + \frac{\partial\vartheta}{\partial x}\frac{\partial}{\partial\vartheta} + \frac{\partial\varphi}{\partial x}\frac{\partial}{\partial\varphi} \tag{4.109}$$

with similar expressions for $y$ and $z$. This gives the components (see Problem 4.7)

$$\hat{L}_x = \mathrm{i}\hbar\left(\sin\varphi\frac{\partial}{\partial\vartheta} + \text{cotan}\vartheta\,\cos\varphi\frac{\partial}{\partial\varphi}\right), \quad \hat{L}_y = \mathrm{i}\hbar\left(-\cos\varphi\frac{\partial}{\partial\vartheta} + \text{cotan}\vartheta\,\sin\varphi\frac{\partial}{\partial\varphi}\right), \quad \text{and} \quad \hat{L}_z = -\mathrm{i}\hbar\frac{\partial}{\partial\varphi}\,. \tag{4.110}$$

From (4.110) we get the operator of the square $L^2$ as

$$\hat{L}^2 = \hat{L}_x^2 + \hat{L}_y^2 + \hat{L}_z^2 = -\hbar^2\left[\frac{1}{\sin\vartheta}\frac{\partial}{\partial\vartheta}\left(\sin\vartheta\frac{\partial}{\partial\vartheta}\right) + \frac{1}{\sin^2\varphi}\frac{\partial^2}{\partial\varphi^2}\right]. \tag{4.111}$$

A comparison with (4.71) reveals that $\hat{L}^2$ is proportional to the angular part of the Laplace operator $\Delta$.

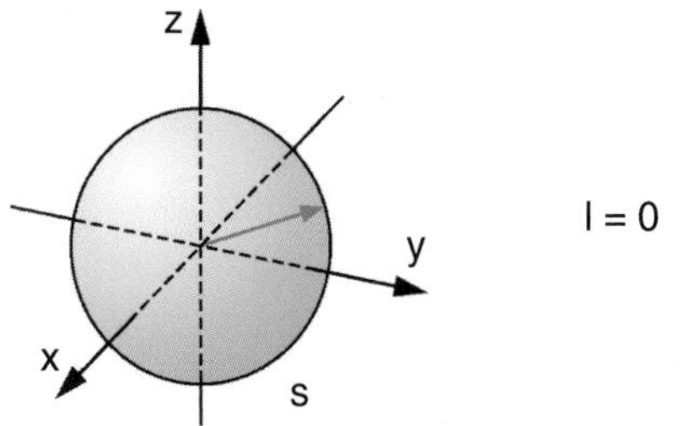


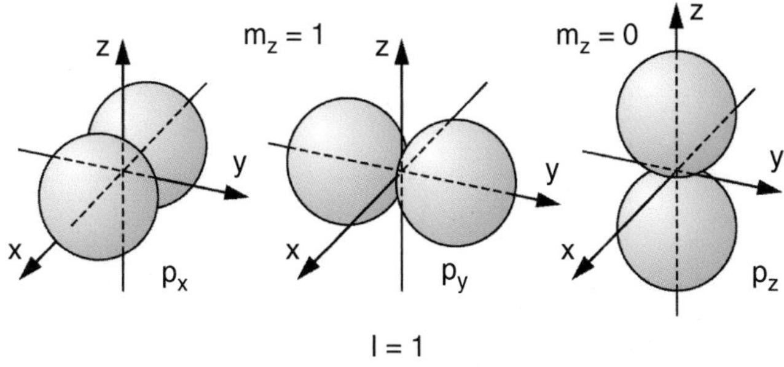


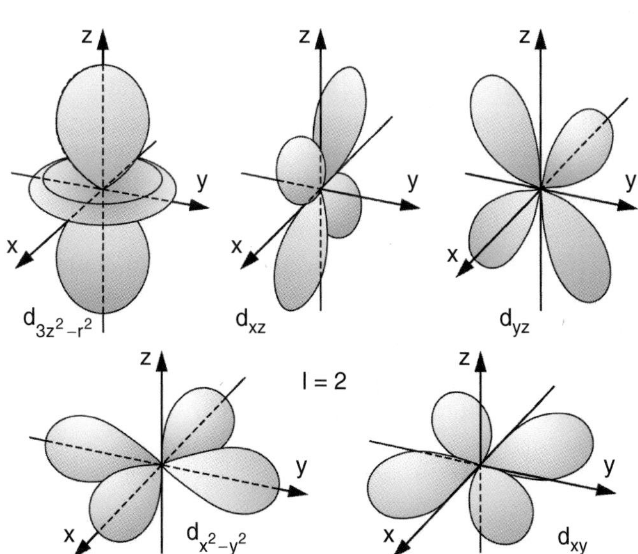


**Fig. 4.28.** Absolute squares $\left|Y_l^m(\vartheta,\varphi)\right|^2$, which are proportional to the spatial probability density of a particle in a spherical symmetric potential, are plotted here in Cartesian coordinates

This implies that the spherical harmonics (Fig. 4.28) are eigenfunctions of the operator $\hat{L}^2$.

We will prove this by applying $\hat{L}^2$ to the wave function $\psi(r,\vartheta,\varphi) = R(r)Y_l^m(\vartheta,\varphi)$. With (4.111) we obtain from (4.77), (4.81)

$$\begin{aligned}\hat{L}^2\psi &= \hat{L}^2(R(r)Y_l^m(\vartheta,\varphi)) = R(r)\hat{L}^2Y_l^m(\vartheta,\varphi)\\ &= R(r)l(l+1)\hbar^2Y_l^m(\vartheta,\varphi) = l(l+1)\hbar^2\psi\ .\end{aligned} \tag{4.112a}$$

The expectation value of the square of the angular momentum $\boldsymbol{L}$ is therefore

$$\left\langle \boldsymbol{L}^2\right\rangle = \int \psi^*\hat{L}^2\psi\,\mathrm{d}\tau = l(l+1)\hbar^2 \tag{4.112b}$$

because the functions $\psi$ are normalized.

The integer $l$ is therefore named the ***angular momentum quantum number***. For the absolute value of the angular momentum we obtain from $\boldsymbol{L}^2 = l(l+1)\hbar^2$:

$$|\boldsymbol{L}| = \sqrt{l(l+1)}\hbar\ . \tag{4.112c}$$

For the $z$-component $L_z$ we get, using (4.110),

$$\begin{aligned}\hat{L}_z\psi &= -\mathrm{i}\hbar\frac{\partial}{\partial\varphi}(R(r)\Theta(\vartheta)\Phi(\varphi))\\ &= -\mathrm{i}\hbar R(r)\Theta(\vartheta)\frac{\partial}{\partial\varphi}\mathrm{e}^{\mathrm{i}m\varphi} = m\hbar\psi\ .\end{aligned}$$

The eigenvalues of $L_z$ are therefore

$$\langle L_z\rangle = m\hbar\ . \tag{4.113}$$

where $m$ is the projection quantum number of the angular momentum. It is introduced in Sect. 4.3.2.

The two operators $\hat{L}_z$ and $\hat{L}$ have the same eigenfunctions. They are, therefore, commutable and the quantities $L_z$ and $L^2$ can be simultaneously measured.

If the operators $\hat{L}_x$ or $\hat{L}_y$ are applied to the function $\psi$, one can prove that $\psi$ is not an eigenfunction of $\hat{L}_x$, i.e.,

$$\hat{L}_x\psi \neq m_x\psi\ . \tag{4.114}$$

**Table 4.4.** Labeling of levels with angular momentum quantum number $l$ and the degree of degeneracy

| $l$ | $m$ | Name | Degree of degeneracy |
|---|---|---|---|
| 0 | 0 | $s$ | 1 |
| 1 | $-1, 0, +1$ | $p$ | 3 |
| 2 | $-2$ to $+2$ | $d$ | 5 |
| 3 | $-3$ to $+3$ | $f$ | 7 |
| 4 | $-4$ to $+4$ | $g$ | 9 |
| 5 | $-5$ to $+5$ | $h$ | 11 |

**Table 4.5.** Expression of the angular functions $Y(x,y,z)$ in Fig. 4.28 as functions of $\vartheta$ and $\varphi$

| $l$ | $\lvert m_l \rvert$ | Angular function |
|---|---|---|
| 0 | 0 | $s = 1/\sqrt{4\pi}$ |
| 1 | 0 | $p_z = \sqrt{3/4\pi}\cos\vartheta$ |
| | 1 | $p_x = \sqrt{3/4\pi}\sin\vartheta\cos\varphi$<br>$p_y = \sqrt{3/4\pi}\sin\vartheta\sin\varphi$ |
| 2 | 0 | $d_{3z^2-r^2} = \sqrt{5/16\pi}\,(3\cos^2\vartheta - 1)$ |
| | 1 | $d_{xz} = \sqrt{15/4\pi}\sin\vartheta\cos\vartheta\cos\varphi$<br>$d_{yz} = \sqrt{15/4\pi}\sin\vartheta\cos\varphi\sin\varphi$ |
| | 2 | $d_{x^2-y^2} = \sqrt{15/4\pi}\sin^2\vartheta\cos 2\varphi$<br>$d_{xy} = \sqrt{15/4\pi}\sin^2\vartheta\sin 2\varphi$ |

The same is true for $L_y$. However, for the operator $L_x^2 + L_y^2 = L^2 - L_z^2$ the function $\psi$ in (4.70) is an eigenfunction. The eigenvalues are

$$\left(\hat{L}_x^2 + \hat{L}_y^2\right)\psi = \left[l(l+1) - m^2\right]\hbar^2\psi\,. \qquad (4.115)$$

Historically the eigenfunctions with $l = 0$ are called $s$-functions, with $l = 1$ $p$-functions, with $l = 2$ $d$-functions (see Table 4.4). The quantum number $m$ gives the orientation of the symmetry axis of these functions against the $z$ direction. If an external magnetic field $B = \{0, 0, B_z\}$ is applied, the quantum number $m$ gives the projection of the angular momentum $L$ onto the field direction. It is therefore called the magnetic quantum number. Since $m$ can take all integer values in the interval $-l \le m \le +l$ there are $2l + 1$ different functions describing states with the same energy in a spherical potential without magnetic field. Sometimes it is useful to represent the functions $Y_l^m(\vartheta,\varphi)$ in Cartesian coordinates, for instance to illustrate the orientation of chemical bonds in molecules.

With

$$\sin\vartheta\, e^{\pm i\varphi} = \frac{1}{r}(x \pm iy)$$

the representation of the $Y_l^m(\vartheta,\varphi)$ in Table 4.2 transforms into the Cartesian representation $Y(x,y,z)$ in Table 4.5, illustrated in Fig. 4.28.

We will demonstrate this transformation for the example of $p$-functions with $l = 1$:

$$\begin{aligned}
\psi(p_x) &= f_p(r)\frac{x}{r} = f_p(r)\sin\vartheta\cos\varphi \\
&= f_p(r)\sqrt{\frac{2\pi}{3}}\left(Y_1^{+1} + Y_1^{-1}\right) \\
\psi(p_y) &= f_p(r)\frac{y}{r} = f_p(r)\sin\vartheta\sin\varphi \\
&= f_p(r)\sqrt{\frac{2\pi}{3}}\left(Y_1^{-1} - Y_1^{+1}\right) \\
\psi(p_z) &= f_p(r)\frac{z}{r} = Y_1^0 = \sqrt{3/4\pi}\cos\vartheta\,. \qquad (4.116)
\end{aligned}$$

While in classical mechanics the angular momentum of a particle moving in a spherically symmetric potential is constant in time (because there is no torque on the particle), which implies that the amount $|L|$ and all three components are constant and well defined, the quantum mechanical description gives a different answer. The amount $|L|$ and only one of the components have a time-independent constant value, the two other components cannot be measured simultaneously. Generally the $z$-axis is chosen as the preferential direction (***quantization axis***), which means that the$z$-component $L_z$ is chosen as the component that can be measured simultaneously with $L^2$. With this convention we get:

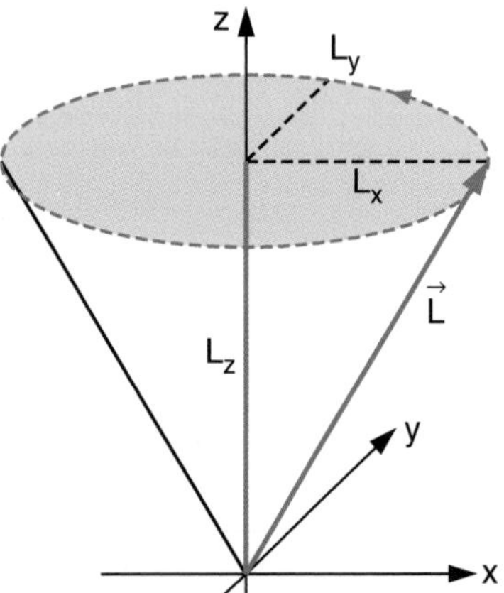


**Fig. 4.29.** The vector $\boldsymbol{L}$ of the orbital angular momentum has a definite length $|\boldsymbol{L}|$ and projection $L_z$, but no defined orientation in space

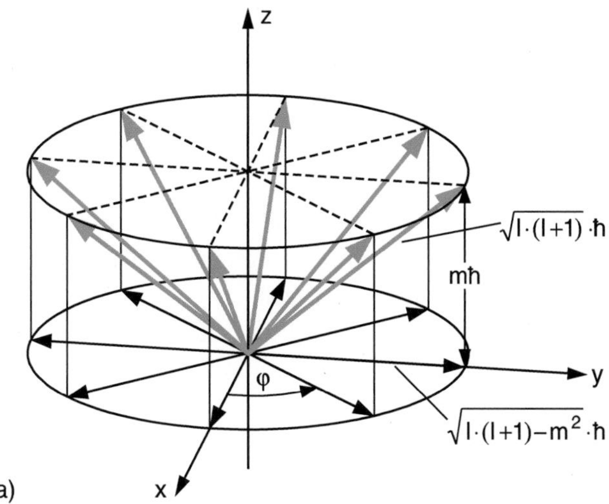


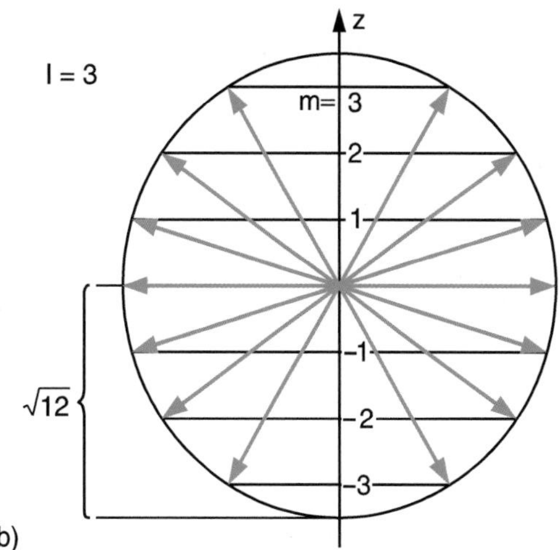


**Fig. 4.30.** (**a**) Possible orientations of the angular momentum $\boldsymbol{l}$ with a fixed projection $\langle l_z \rangle = m\hbar$ and length $|\boldsymbol{l}| = \sqrt{l(l+1)}$, where the vector $\boldsymbol{l}$ can be everywhere on the surface of the cone with angle $\vartheta$ and $\cos\vartheta = m\hbar/|\boldsymbol{l}|$. (**b**) Possible projections of the angular momentum $\boldsymbol{l}$ with $l = 3$ and $-3 \le m \le +3$

$$\hat{L}^2\psi = l(l+1)\hbar^2\psi \; ; \quad \hat{L}_z\psi = m\hbar\psi \; . \tag{4.117}$$

In Fig. 4.29 this is illustrated by a vector model. The length of the angular momentum $L$ (blue vector) is well defined as is its projection onto the $z$-axis. The two other components $L_x$ and $L_y$ are not defined. This means in our vector model that the endpoint of the vector $L$ can have any location on the dashed blue circle (Fig. 4.30).

In summary

> The operators $L^2$ and $L_z$ have common eigenfunctions. Their eigenvalues are simultaneously measurable. $L_x$ and $L_y$ have no common eigenfunctions with $L^2$. Only the sum $L_x^2 + L_y^2 = L^2 - L_z^2$ can be measured simultaneously with $L^2$ and $L_z^2$.

## SUMMARY

- The essential part for the quantum mechanical description of particles is the wave function $\psi(\boldsymbol{r}, t)$, which gives the probability amplitude for finding the particle at the position $\boldsymbol{r}$ at time $t$.
- The wave functions $\psi(\boldsymbol{r}, t)$ are solutions of the time-dependent Schrödinger equation, which allows one to calculate (at least numerically) the behavior of a particle with mass $m$ in an arbitrary potential $\Phi(r, t)$ as a function of position $r$ and time $t$.
- For stationary problems the time-independent Schrödinger equation

$$-\frac{\hbar^2}{2m}\Delta\psi(\boldsymbol{r}) + E_{\text{pot}}(\boldsymbol{r})\psi(\boldsymbol{r}) = E\psi(\boldsymbol{r})$$

is used. The solutions are stationary wave functions $\psi(r)$, that depend only on the position $r$ but not on time. They might be complex functions. Their real absolute square $|\psi(r)|^2\,\mathrm{d}\tau$ gives the probability to find the particle in the volume element $\mathrm{d}\tau = \mathrm{d}x \cdot \mathrm{d}y \cdot \mathrm{d}z$ around the position $\boldsymbol{r}$.
- For time-dependent problems, the solutions of the time-dependent Schrödinger equation

$$\frac{\partial\psi(\boldsymbol{r}, t)}{\partial t} = -\frac{\mathrm{i}}{\hbar}\hat{H}\psi(\boldsymbol{r}, t)$$

with the Hamilton operator

$$\hat{H} = -\frac{\hbar^2}{2m}\Delta + E_{\text{pot}}(\boldsymbol{r}, t)$$

▶

are wave functions $\psi(r,t)$. The absolute square $|\psi(r,t)|^2$ describes the movement of the particle in space. If $E_{pot}$ does not depend on time, the wave functions

$$\psi(r,t) = \psi(r)e^{i(E/\hbar)t}$$

can be split into a product of a spatial part $\psi(r)$, which is a solution of the stationary Schrödinger equation, and a phase factor with an exponent that depends on time and on the energy of the particle.

- A particle with energy $E$ can tunnel through a potential barrier of height $E_0$, even if $E < E_0$ (tunnel effect). The tunnel probability depends on the difference $E_0 - E$ and on the width $\Delta x$ of the potential barrier. The tunnel effect is a wave phenomenon and is also observed in classical wave optics.
- When a particle is restricted to a spatial interval $\Delta x$ (one-dimensional potential box) it has discrete energy values

$$E_n = \frac{\hbar^2}{2m}\frac{\pi^2}{a^2}n^2 ,$$

which increase with the square of the integer $n$, called a quantum number. The lowest energy is realized for $n = 1$ and is larger than zero. This zero-point energy is a consequence of the uncertainty relation $\Delta p_x \cdot \Delta x \geq \hbar$.

- In a parabolic potential $E_{pot} = ax^2$ the energy eigenvalues

$$E_n = (n+1)\hbar\omega$$

of the harmonic oscillator are equidistant.

- For a spherical symmetric potential the three-dimensional wave function

$$\psi(r,\vartheta,\varphi) = R(r)\Theta(\vartheta)\Phi(\varphi)$$

can be separated into a product of three one-dimensional wave functions.

- The angular part $\Theta(\vartheta)\Phi(\varphi)$ is the same for all spherically symmetric potentials independent of the radial form of the potential. It is described by the spherical surface harmonics $Y_l^m(\vartheta,\varphi)$, while the radial part $R(r)$ reflects the radial dependence of the potential.
- The functions $Y_l^m(\vartheta,\varphi)$ are eigenfunctions of the square $\hat{L}^2$ of the angular momentum operator and of the operator $\hat{L}_z$ of its $z$-component.
- The expectation values of a measurable physical quantity $A$ with the operator $\hat{A}$ is given by

$$\langle A \rangle = \int \psi^* \hat{A} \psi \, d\tau .$$

- If the functions $\psi$ are eigenfunctions of the operator $\hat{A}$, the expectation value $\langle A \rangle$ of the measurable quantity $A$ is equal to the exactly measurable eigenvalue $A$.
- Two operators $\hat{A}$ and $\hat{B}$ are commutable, if the relation $\hat{A}\hat{B}\psi = \hat{B}\hat{A}\psi$ holds.
- Two commutable operators have common eigenfunctions with eigenvalues that are simultaneously measurable.

## PROBLEMS

1. Show that for time-independent potentials the solutions of the time-dependent Schrödinger equation can always be written as the product $\psi(r,t) = f(r) \cdot g(t)$ of two functions. What is the form of $g(t)$ for a constant total energy of the particle?
2. What is the reflectivity $R$ for a proton with $E_{\text{kin}} = 0.4\,\text{meV}$ that hits a rectangular potential barrier with $E_{\text{pot}} = 0.5\,\text{meV}$ and a width $\Delta x = 1\,\text{nm}$?
3. What is the reflectivity of a potential well ($E_{\text{pot}} = -E_0 = -1\,\text{eV}$, $\Delta x = 5\,\text{nm}$) as a function of the energy $E$ of an incident particle? Prove that relations analogous to (4.26) are also valid for a potential well with $E_{\text{pot}} = -E_0$. What is the exact form of these relations for this case?
4. Derive equations (4.26a) and (4.26c) for the tunnel transmission.
5. How many energy levels of a particle with mass $m$ fit into a rectangular potential well with width $a = 0.7\,\text{nm}$ up to the energy $E = 10\,\text{eV}$, if the bottom of the well is at $E = 0$ and the walls are infinitely high (a) for a proton and (b) for an electron? (c) How do the values change when the walls have finite heights $E_{\text{max}} = 10\,\text{eV}$?
6. How large is the spatial uncertainty $\Delta x$ of a particle in a parabolic potential $E_{\text{pot}} = (1/2)Dx^2$, if its zero-point energy is $E(v = 0) = (1/2)\hbar(D/m)^{1/2}$? How large is $\Delta x$ for $v = 20$?
7. Derive from (4.108c) the relation (4.110) for $L$ and (4.111) for $L^2$.
8. A particle with kinetic energy $E$ is confined in a potential well with width $\Delta x = a$ and depth $E = -E_0$. What is the penetration depth into the walls of the well where the probability $|\psi(x)|^2$ has decreased to $1/e$ of its maximum value?
9. An electron with energy $E = E_0/2$ or $E = E_0/3$ hits a rectangular potential barrier with heights $E_0$ and width $a = \hbar/(2mE)^{1/2} = \lambda_{\text{dB}}$. How large is its transmission probability? For which value of the ratio $E/E_0$ is $T$ a maximum?
10. Electrons with $E_{\text{kin}} = 0.8\,\text{eV}$ and $1.2\,\text{eV}$, respectively, hit a potential barrier with $E_0 = 1\,\text{eV}$ and $\Delta x = a = 1\,\text{nm}$. Calculate, for both cases, the

    a) transmission $T$ and the reflection $R$. Prove that always $R + T = 1$.
11. An electron is confined in a two-dimensional quadratic potential well with $a = 1\,\text{nm}$ and $E^0 = -1\,\text{eV}$. How many bound states exist in this well? (Use the formulas for a well with infinitely high walls).

# 5. The Hydrogen Atom

Based on the discussions in Chap. 4 we will now apply the quantum mechanical treatment to the simplest atom, the H atom, which consists of one proton and one electron moving in the spherical symmetric Coulomb potential of the proton. These one-electron systems, such as the hydrogen atom and the ions $He^+$, $Li^{++}$, $Be^{+++}$, etc., are the only real systems for which the Schrödinger equation can be exactly (i.e., analytically) solved. For all other atoms or molecules approximations have to be made. Either the Schrödinger equation for these systems can be solved numerically (which offers a mathematical solution within the accuracy of the computer program, but generally gives little insight into the physical nature of the approximation), or the real atoms are described by approximate models that can be calculated analytically. In any case, for all multielectron systems, one has to live with approximations, either in the numerical solution of the exact atomic model or for the exact solution of the approximate model.

A closer inspection of the spectrum of the hydrogen atom and other atoms reveals, however, that at higher spectral resolution the lines show a substructure that cannot be described by the Schrödinger theory, but is due to new effects, such as fine structure, hyperfine structure or the anomalous Zeeman effect. Therefore, even for the simple hydrogen atom the quantum mechanical model of an electron in the Coulomb field of the proton has to be modified by introducing new properties of electron and proton, such as electron spin or proton spin and their mutual interactions. These effects, which are small compared to the Coulomb energy, are included in a relativistic theory, based on the ***Dirac equation***, which is called ***quantum electrodynamics***. The Schrödinger equation can be regarded as the fundamental equation of nonrelativistic quantum theory.

The treatment of the hydrogen atom illustrates in a very clear way the basic ideas of quantum mechanics, it explains the physical interpretation of quantum numbers, the description of the Zeeman effect and the fine structure by the model of angular momentum vector couplings and gives a better understanding of the more complex many-electron systems that will be discussed in the next chapter.

## 5.1 Schrödinger Equation for One-electron Systems

The Schrödinger equation for a system consisting of one electron (mass $m_1$, charge $q = -e$ and radius vector $\boldsymbol{r}_1$) and a nucleus (mass $m_2 \gg m_1$, charge $q = +Ze$ and radius vector $\boldsymbol{r}_2$) is:

$$-\frac{\hbar^2}{2m_1}\Delta_1\psi - \frac{\hbar^2}{2m_2}\Delta_2\psi - \frac{Ze^2}{4\pi\varepsilon_0 r}\psi = E\psi(\boldsymbol{r}_1, \boldsymbol{r}_2)\,, \tag{5.1}$$

where $\Delta_i$ is the Laplace operator with respect to $\boldsymbol{r}_i$. The first term describes the kinetic energy of the electron, the second describes that of the nucleus and the third one the potential energy of the Coulomb interaction between the two particles, where $r = |\boldsymbol{r}_1 - \boldsymbol{r}_2|$ is the distance between the two particles. The wave function $\Psi(\boldsymbol{r}_1, \boldsymbol{r}_2)$ depends on the location of the electron and nucleus, which means it depends on six coordinates.

### 5.1.1 Separation of the Center of Mass and Relative Motion

In classical mechanics it is shown that the movement of a closed system of particles can always be separated into the motion of the center of mass and the relative motion of the particles in the center- of-mass system. This is also possible in quantum mechanics as can be seen by the following derivation.

W. Demtröder, *Atoms, Molecules and Photons*, 2nd ed., Graduate Texts in Physics,
DOI 10.1007/978-3-642-10298-1_5, © Springer-Verlag Berlin Heidelberg 2010

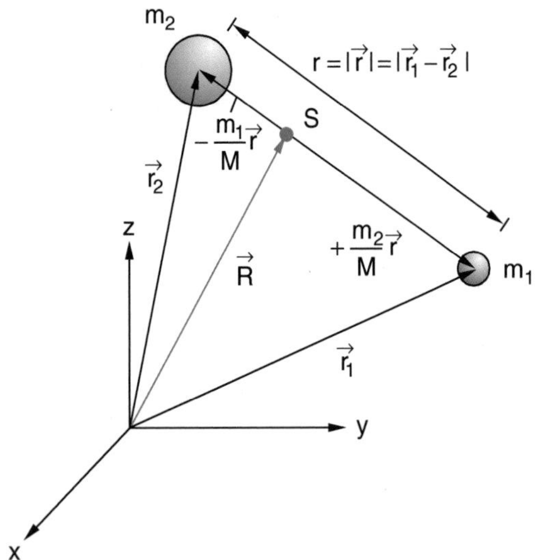


**Fig. 5.1.** Transformation of laboratory frame into the center-of-mass coordinate system

We regard a system of two particles with coordinates $\boldsymbol{r}_1 = \{x_1, y_1, z_1\}$ and $\boldsymbol{r}_2 = \{x_2, y_2, z_2\}$ written in lower case letters and the coordinates of the center of mass (written in capital letters) (Fig. 5.1)

$$\boldsymbol{R} = \frac{m_1\boldsymbol{r}_1 + m_2\boldsymbol{r}_2}{M} \quad \text{with} \quad M = m_1 + m_2 \,,$$
$$\boldsymbol{R} = \{X, Y, Z\} \quad \vec{r}_i = \{x_i, y_i, z_i\}$$

With the relative distance $r = \{x, y, z\} = |\boldsymbol{r}_1 - \boldsymbol{r}_2| = |\{x_1 - x_2, y_1 - y_2, z_1 - z_2\}|$ we obtain from Fig. 5.1 the relations:

$$\boldsymbol{r}_1 = \boldsymbol{R} + \frac{m_2}{M}\boldsymbol{r} \,, \quad \boldsymbol{r}_2 = \boldsymbol{R} - \frac{m_1}{M}\boldsymbol{r} \,. \tag{5.2}$$

In order to properly write the Schrödinger equation (5.1) in the coordinates $\boldsymbol{r}$ and $\boldsymbol{R}$ we have to consider that the differentiation of the function $\Psi(\boldsymbol{r}, \boldsymbol{R})$ with respect to the variable $x_i$ $(i = 1, 2)$ follows the chain rule:

$$\begin{aligned}
\frac{\partial\psi}{\partial x_1} &= \frac{\partial\psi}{\partial X}\frac{\partial X}{\partial x_1} + \frac{\partial\psi}{\partial x}\frac{\partial x}{\partial x_1} = \frac{m_1}{M}\frac{\partial\psi}{\partial X} + \frac{\partial\psi}{\partial x} \\
\frac{\partial^2}{\partial x_1^2} &= \frac{\partial}{\partial X}\left(\frac{m_1}{M}\frac{\partial\psi}{\partial X} + \frac{\partial\psi}{\partial x}\right)\frac{\partial X}{\partial x_1} \\
&\quad + \frac{\partial}{\partial x}\left(\frac{m_1}{M}\frac{\partial\psi}{\partial X} + \frac{\partial\psi}{\partial x}\right)\frac{\partial x}{\partial x_1} \\
&= \frac{m_1^2}{M^2}\frac{\partial^2\psi}{\partial X^2} + \frac{2m_1}{M}\frac{\partial^2\psi}{\partial X\partial x} + \frac{\partial^2\psi}{\partial x^2} \,.
\end{aligned} \tag{5.3}$$

Analogous expressions are obtained for $x_2$. When (5.1) is written in the new coordinates $(X, Y, Z)$ and $(x, y, z)$, the mixed terms in (5.3) cancel and the Schrödinger equation becomes

$$\left[-\frac{\hbar^2}{2M}\left(\frac{\partial^2}{\partial X^2} + \frac{\partial^2}{\partial Y^2} + \frac{\partial^2}{\partial Z^2}\right) + \frac{\hbar^2}{2\mu}\left(\frac{\partial^2}{\partial x^2} + \frac{\partial^2}{\partial y^2} + \frac{\partial^2}{\partial z^2}\right)\right]\psi + E_{\text{pot}}(r)\psi = E\psi \,, \tag{5.4}$$

where $\mu = (m_1 m_2)/(m_1 + m_2)$ is the reduced mass of the system.

For the solution of (5.4) we try the ansatz:

$\psi(\boldsymbol{R}, \boldsymbol{r}) = f(\boldsymbol{r})g(\boldsymbol{R})$.

Inserting this into (5.4) yields after division by $\psi$

$$-\frac{\hbar^2}{2M}\frac{\Delta_R g}{g} = \frac{\hbar^2}{2\mu}\frac{\Delta_r f}{f} - E_{\text{pot}}(r) + E \,, \tag{5.5}$$

where $\Delta_R$ is the Laplace operator for $(X, Y, Z)$ and $\Delta_r$ that for $(x, y, z)$. We can abreviate (5.5) by $T_1 = T_2 + T_3 + E$.

The term $T_1$ on the left side of (5.5) depends solely on the center-of-mass coordinates $X, Y, Z$. The two other terms $T_2$ and $T_3$, on the right side, depend solely on the coordinates $x, y, z$. The total energy $E$ of the system is constant.

Now we draw the same conclusion as in Sect. 4.3.2: Since equation (5.5) has to be valid for arbitrary values of the coordinates $X, Y, Z$ and $x, y, z$, both sides of the equation have to be constant. This means $T_1$ as well as $T_2 + T_3$ have to be constant, otherwise (5.5) can not be fulfilled for arbitrary choices of the coordinates. This gives the two conditions:

$$\begin{aligned}
&\frac{\hbar^2}{2M}\frac{\Delta_R g}{g} = \text{const} = -E_g \\
&\frac{\hbar^2}{2\mu}\frac{\Delta_r f}{f} - E_{\text{pot}}(r) = \text{const} = -E_f \,,
\end{aligned} \tag{5.6}$$

with $E_g + E_f = E$. We then obtain the two separate equations

$$-\frac{\hbar^2}{2M}\Delta_R g(R) = E_g g(R) \tag{5.7a}$$

$$-\frac{\hbar^2}{2\mu}\Delta_r f(r) + E_{\text{pot}}(r) f(r) = E_f f(r) \,. \tag{5.7b}$$

The first equation describes the kinetic energy $E_g = E_{\text{kin}}^{(\text{CM})}$ of the center-of-mass motion, which means the movement of the whole atom. Its solution is, as outlined in Chap. 4, the spatial part of the plane wave

$$g(X,Y,Z) = A\mathrm{e}^{\mathrm{i}(kR-(E_g/\hbar)t)} .$$

With the de Broglie wavelength

$$\lambda_{\mathrm{CM}} = \frac{2\pi}{k} = \frac{h}{\sqrt{2ME_g}} ,$$

which depends on the translational energy $E_g$ of the center-of-mass motion. The correct description of this motion has to use wave packets (see Sect. 3.3.1–3.3.4).

The relative motion of electron and nucleus is described by (5.7b). Renaming $f(r)$ as $\psi(r)$, $E_f$ as $E$ and $\Delta_r$ as $\Delta$, we obtain the Schrödinger equation

$$-\frac{\hbar^2}{2\mu}\Delta\psi + E_{\mathrm{pot}}(r)\psi = E\psi , \tag{5.8}$$

which is identical to the Schrödinger equation (4.40) for a particle in a spherically symmetric potential if the mass $m$ of the particle is replaced by the reduced mass $\mu = m_1 \cdot m_2/(m_1 + m_2)$.

The Schrödinger equation of a one-electron atom can be separated into a term describing the center of mass and a second term that describes the position $r$ of a particle with reduced mass $\mu$ relative to that of the nucleus at $r = 0$ in a potential with potential energy $E_{\mathrm{pot}}$.

In Sect. 4.3.2 we have already discussed the separation of this equation in spherical coordinates $(r,\vartheta,\varphi)$. It was shown there that the wave function

$$\psi(r,\vartheta,\varphi) = R(r)Y_l^m(\vartheta,\varphi)$$

can be separated for arbitrary spherical potentials into a radial function $R(r)$ that depends on the $r$-dependence of the potential and the angular part, which equals the spherical functions $Y_l^m$ independent of the radial coordinate $r$.

In order to obtain the wave functions for the hydrogen atom we have to look for the radial wave function for the Coulomb potential. Inserting this function into the Schrödinger equation yields the energy eigenvalues $E$.

**Note:**

The function $R(r)$ has nothing to do with the coordinate $R$ of the center of mass!

### 5.1.2 Solution of the Radial Equation

With the product-ansatz

$$\psi(r,\vartheta,\varphi) = R(r)Y_l^m(\vartheta,\varphi)$$

for the wave function $\Psi(r,\vartheta,\varphi)$ in Sect. 4.3.2 we had already obtained (4.65) for the radial part $R(r)$, which converts for $m \to \mu$ and $C_2 = l(l+1)$ into

$$\frac{1}{r^2}\frac{\mathrm{d}}{\mathrm{d}r}\left(r^2\frac{\mathrm{d}R}{\mathrm{d}r}\right) + \frac{2\mu}{\hbar^2}\left(E - E_{\mathrm{pot}}(r)\right)R(r) = \frac{l(l+1)}{r^2}R(r) . \tag{5.9}$$

The integer $l$ describes, according to (4.89), the integer quantum number of the orbital angular momentum of the particle with respect to the origin $r = 0$ in our relative coordinate system, where the nucleus is at rest at $r = 0$.

Differentiation of the first term and introducing the Coulomb-potential for $E_{\mathrm{pot}}(r)$ yields

$$\frac{\mathrm{d}^2R}{\mathrm{d}r^2} + \frac{2}{r}\frac{\mathrm{d}R}{\mathrm{d}r} + \left[\frac{2\mu}{\hbar^2}\left(E + \frac{Ze^2}{4\pi\varepsilon_0 r}\right) - \frac{l(l+1)}{r^2}\right]R = 0 . \tag{5.10}$$

In the limit $r \to \infty$ all terms with $1/r$ and $1/r^2$ in the bracket approach zero and (5.10) becomes for this limiting case:

$$\frac{\mathrm{d}^2R(r)}{\mathrm{d}r^2} + \frac{2}{r}\frac{dR}{dr} = -\frac{2\mu}{\hbar^2}ER(r) . \tag{5.11}$$

The solutions of this equation describe the asymptotic behavior of the radial wave function $R(r)$. The probability of finding the electron in a spherical shell with volume $4\pi r^2\,\mathrm{d}r$ around the nucleus between the radii $r$ and $r + \mathrm{d}r$ is given by $4\pi|R|^2r^2\mathrm{d}r$. The absolute square of the function $r \cdot R$ therefore gives the probability of finding the electron within the unit volume of the spherical shell. We therefore introduce the function $W(r) = r \cdot R(r)$ into (5.11). This yields, with $k = \sqrt{2\mu E}/\hbar$,

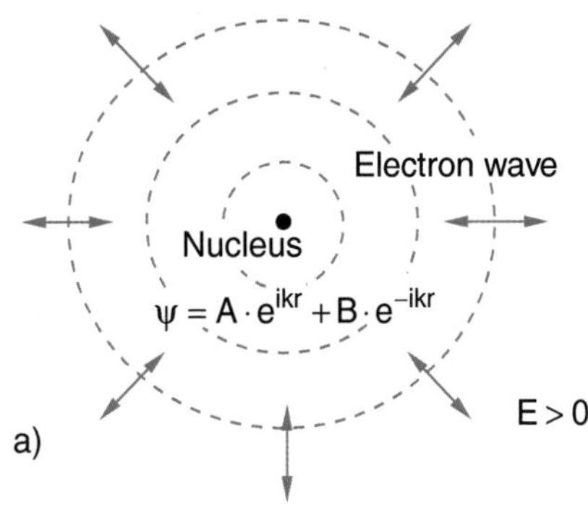


**Fig. 5.2.** (**a**) Ingoing and outgoing spherical waves as solutions to the Schrödinger equation for an electron with $E > 0$ in a spherical potential. (**b**) Experimentally decreasing wave amplitude for $E < 0$

$$\frac{d^2 W}{dr^2} = -k^2 W \tag{5.12a}$$

with the asymptotic solution

$$W(r \to \infty) = A\mathrm{e}^{ikr} + B\mathrm{e}^{-ikr} . \tag{5.12b}$$

This gives for $R(r) = W(r)/r$

$$R(r) = \frac{A}{r}\mathrm{e}^{ikr} + \frac{B}{r}\mathrm{e}^{-ikr} . \tag{5.12c}$$

For $E > 0$ $k$ is real and the first term in (5.12b) represents the spatial part of an outgoing spherical wave

$$\psi(r,t) = \frac{A}{r}\mathrm{e}^{\mathrm{i}(kr-\omega t)} , \tag{5.12d}$$

which describes an electron that can, with a positive total energy, leave the atom and can reach $r \to \infty$ (Fig. 5.2a). The second term corresponds to an incoming spherical wave that represents an electron coming from $R = \infty$ and approaching the nucleus (this is called a collision process).

For $E < 0$ we substitute $\kappa = \sqrt{-2\mu E}/\hbar = ik$ and obtain the real asymptotic solutions

$$R(r \to \infty) = A\mathrm{e}^{-\kappa r} + B\mathrm{e}^{+\kappa r} . \tag{5.12e}$$

Since $R(r)$ must be finite for all values of $r$ (otherwise the function $R(r)$ could not be normalized) it follows that $B = 0$. We then obtain the asymptotic solution

$$R(r \to \infty) = A\mathrm{e}^{-\kappa r} .$$

This is an exponentially decreasing function which has decayed to $1/\mathrm{e}$ for $r = 1/\kappa$.

For the general solution, valid for all values of $r$, we try the ansatz

$$R(r) = u(r)\mathrm{e}^{-\kappa r} . \tag{5.12f}$$

Inserting this into (5.10) we obtain for $u(r)$ the equation

$$\frac{\mathrm{d}^2 u}{\mathrm{d}r^2} + 2\left(\frac{1}{r} - \kappa\right)\frac{\mathrm{d}u}{\mathrm{d}r} + \left[\frac{2a - 2\kappa}{r} - \frac{l(l+1)}{r^2}\right]u = 0 . \tag{5.13}$$

Where the abbreviation $a$ is

$$a = \frac{\mu Z e^2}{4\pi\varepsilon_0 \hbar^2}.$$

The reciprocal value $r_1 = 1/a = 4\pi\varepsilon_0\hbar^2/(\mu Z e^2)$ gives, according to (3.85), the Bohr radius of the lowest energy level.

We write $u(r)$ as the power series

$$u(r) = \sum_j b_j r^j . \tag{5.14}$$

Inserting this into (5.13) the comparison of the coefficients of equal powers in $r$ yields the recursion formula

$$b_j = 2b_{j-1}\frac{\kappa j - a}{j(j+1) - l(l+1)} . \tag{5.15}$$

Since $R(r)$ must be finite for all values of $r$, the power series can only have a finite number of summands. If the last nonvanishing coefficient in the power series (5.14) is $b_{n-1}$, then $b_j$ becomes zero for $j = n$. This immediately gives, from (5.15), the condition that only the coefficients $b_j$ with $j < n$ contribute to the series (5.14).We therefore have the condition

$$j < n . \tag{5.16}$$

Since for $j = n \Rightarrow b_j = 0$ we obtain from (5.15)

$$a = n\kappa . \tag{5.17}$$

With $\kappa = +\sqrt{-2\mu E}/\hbar$ this yields the condition for the energy values

$$E_n = -\frac{a^2\hbar^2}{2\mu n^2} = -\frac{\mu Z^2 e^4}{8\varepsilon_0^2 h^2 \cdot n^2} = -Ry^* \frac{Z^2}{n^2} \tag{5.18}$$

with the Rydberg constant

$$Ry^* = \frac{\mu e^4}{8\varepsilon_0^2 h^2} . \tag{5.18a}$$

**Note** that (5.18) is identical to eq. (3.88) of Bohr's model.

> The quantum mechanical calculation of one-electron systems gives the same energy values as Bohr's atomic model.

**Note:**

1. From the derivation of (5.18) it can be recognized that the discrete eigenvalues $E_n$ of possible energies stem from the restraint $\psi(r \to \infty) \to 0$, which implies that the electron is confined within a finite spatial volume (see also Sect. 4.2.4).
2. Besides the condition (5.18) for the energies there is also a restraint for the angular momentum quantum numbers $l$ following from (5.15). According to (4.59) $l$ must be an integer. For the values $j < n$ that are allowed according to (5.16) the denominator in (5.15) would become zero for $l = j$, which would result in an infinite coefficient $b_j$. We therefore have to demand that in (5.15) all terms with $l \geq j$ must be zero in order to keep the function $u(r)$ finite.
   We have then the condition
   $$l < j < n,$$
   which gives the restraint for the angular momentum quantum number $l$
   $$\boxed{l \leq n - 1 .} \tag{5.19a}$$

The projection of the angular momentum $\vec{l}$ onto the quantization axis ($z$-direction) is $l_z = m \cdot \hbar$ with $m = -l, -l+1, \ldots.. l-1, l$. There are $2l + 1$ possible $m$-values.

With the recursion formula (5.15) the functions $u(r)$ and with (5.12f) also the radial wave functions $R(r)$ can be calculated successively. Table 5.1 lists, for the lowest values of $n$ and $l$, the radial functions $R(r)$. They depend on $n$ because of the condition (5.16) and on $l$ because of (5.15).

**Table 5.1.** Normalized radial wave functions $R(r)$ (Laguerre-Polynomials) of an electron in the Coulomb potential of the nucleus with charge $Ze$ ($N = (Z/na_0)^{3/2}$; $x = Zr/na_0$; $a_0 = 4\pi\varepsilon_0\hbar^2/(Z\mu e^2)$)

| $n$ | $l$ | $R_{n,l}(r)$ |
|---|---|---|
| 1 | 0 | $2Ne^{-x}$ |
| 2 | 0 | $2Ne^{-x}(1-x)$ |
| 2 | 1 | $\frac{2}{\sqrt{3}} Ne^{-x} x$ |
| 3 | 0 | $2Ne^{-x}\left(1 - 2x + \frac{2x^2}{3}\right)$ |
| 3 | 1 | $\frac{2}{3}\sqrt{2}Ne^{-x}x(2-x)$ |
| 3 | 2 | $\frac{4}{3\sqrt{10}} Ne^{-x}x^2$ |
| 4 | 0 | $2Ne^{-x}\left(1 - 3x + 2x^2 - \frac{x^3}{3}\right)$ |
| 4 | 1 | $2\sqrt{\frac{5}{3}}Ne^{-x}x\left(1 - x + \frac{x^2}{5}\right)$ |
| 4 | 2 | $2\sqrt{\frac{1}{5}}Ne^{-x}x^2\left(1 - \frac{x}{3}\right)$ |
| 4 | 3 | $\frac{2}{3\sqrt{35}}Ne^{-x}x^3$ |

The energies $E_n$ can be calculated from (5.18) without the knowledge of the functions $R(r)$. These functions give, however, the radial electron distribution and therefore the electric structure of the atom around the nucleus. This will be discussed in more detail in the following section.

### 5.1.3 Quantum Numbers and Wave Functions of the H Atom

In the preceding section we have seen, that the energy $E(n, l)$ of an atomic state depends on certain quantum numbers: The principal quantum number $n$ and the quantum number $l$ of the orbital angular momentum.

The stationary wavefunctions, which describe the time-averaged spatial distribution of the electron, are also characterized by these quantum numbers and in addition by the quantum number $m$ of the $z$-component $l_z = m\hbar$ of the orbital angular momentum (see Sect. 4.4.2).

The normalized wave functions

$$\psi(r, \vartheta, \varphi) = R_{n,l}(r) Y_l^m(\vartheta, \varphi)$$

discussed in Sects. 4.3.2 and 5.1.2 are also called atomic orbitals, because in the old Bohr-Sommerfeld theory the electron was assumed to move on certain

**Table 5.2.** Normalized total wave functions of an electron in the Coulomb potential $E_{\text{pot}} = -Ze^2/(4\pi\varepsilon_0 r)$

| $n$ | $l$ | $m$ | Eigenfunction $\psi_{n,l,m}(r,\vartheta,\varphi)$ |
|---|---|---|---|
| 1 | 0 | 0 | $\frac{1}{\sqrt{\pi}}\left(\frac{Z}{a_0}\right)^{\frac{3}{2}} \mathrm{e}^{\frac{-Zr}{a_0}}$ |
| 2 | 0 | 0 | $\frac{1}{4\sqrt{2\pi}}\left(\frac{Z}{a_0}\right)^{\frac{3}{2}} \left(2-\frac{Zr}{a_0}\right) \mathrm{e}^{\frac{-Zr}{2a_0}}$ |
| 2 | 1 | 0 | $\frac{1}{4\sqrt{2\pi}}\left(\frac{Z}{a_0}\right)^{\frac{3}{2}} \frac{Zr}{a_0} \mathrm{e}^{\frac{-Zr}{2a_0}} \cos\vartheta$ |
| 2 | 1 | ±1 | $\frac{1}{8\sqrt{\pi}}\left(\frac{Z}{a_0}\right)^{\frac{3}{2}} \frac{Zr}{a_0} \mathrm{e}^{\frac{-Zr}{2a_0}} \sin\vartheta\, \mathrm{e}^{\pm \imath\varphi}$ |
| 3 | 0 | 0 | $\frac{1}{81\sqrt{3\pi}}\left(\frac{Z}{a_0}\right)^{\frac{3}{2}} \left(27-18\frac{Zr}{a_0}+2\frac{Z^2r^2}{a_0^2}\right) \mathrm{e}^{\frac{-Zr}{3a_0}}$ |
| 3 | 1 | 0 | $\frac{\sqrt{2}}{81\sqrt{\pi}}\left(\frac{Z}{a_0}\right)^{\frac{3}{2}} \left(6-\frac{Zr}{a_0}\right) \frac{Zr}{a_0} \mathrm{e}^{\frac{-Zr}{3a_0}} \cos\vartheta$ |
| 3 | 1 | ±1 | $\frac{1}{81}\cdot\sqrt{\frac{2}{\pi}}\left(\frac{Z}{a_0}\right)^{\frac{3}{2}} \left(6-\frac{Zr}{a_0}\right) \frac{Zr}{a_0} \mathrm{e}^{\frac{-Zr}{3a_0}} \sin\vartheta\, \mathrm{e}^{\pm \imath\varphi}$ |
| 3 | 2 | 0 | $\frac{1}{81\sqrt{6\pi}}\left(\frac{Z}{a_0}\right)^{\frac{3}{2}} \frac{Z^2r^2}{a_0^2} \mathrm{e}^{\frac{-Zr}{3a_0}} (3\cos^2\vartheta-1)$ |
| 3 | 2 | ±1 | $\frac{1}{81\sqrt{\pi}}\left(\frac{Z}{a_0}\right)^{\frac{3}{2}} \frac{Z^2r^2}{a_0^2} \mathrm{e}^{\frac{-Zr}{3a_0}} \sin\vartheta\cos\vartheta\, \mathrm{e}^{\pm \imath\varphi}$ |
| 3 | 2 | ±2 | $\frac{1}{162\sqrt{\pi}}\left(\frac{Z}{a_0}\right)^{\frac{3}{2}} \frac{Z^2r^2}{a_0^2} \mathrm{e}^{\frac{-Zr}{3a_0}} \sin^2\vartheta\, \mathrm{e}^{\pm 2\imath\varphi}$ |

**Table 5.3.** Labeling of atomic states $(l, m)$ using Latin and Greek letters

| $l$ | state label | $\lvert m\rvert$ | state label |
|---|---|---|---|
| 0 | $s$ | 0 | $\sigma$ |
| 1 | $p$ | 1 | $\pi$ |
| 2 | $d$ | 2 | $\delta$ |
| 3 | $f$ | 3 | $\varphi$ |
| 4 | $g$ | 4 | $\gamma$ |

*Note*: Often the nomenclature $\sigma, \pi, \delta \ldots$ is reserved for molecules (see Table 9.1). Since for atoms as well as for diatomic molecules this greek nomenclature gives the values of the projection $m\hbar$, it is justified to use it for atoms as well

orbitals around the nucleus. This expression is, however, misleading, because we know from the discussion in Sect. 3.5 that we can not attribute to the electron a definite path, but only a probability of finding it within a volume $\mathrm{d}V$, given by $|\psi|^2 \mathrm{d}V$.

The normalized total wave functions $\psi$ for the lowest energy states of the hydrogen atom are compiled in Table 5.2. They depend on the Quantum numbers $n, l$ and $m$. This also means that the probability of finding the electron at the position $(r, \vartheta, \varphi)$, i.e., the spatial electron distribution depends on these quantum numbers.

> Each atomic state, described by its energy and its spatial electron distribution is defined uniquely by the three quantum numbers $n, l$ and $m$.

In Fig. 4.28 the spatial density distributions $\left|Y_l^m(\vartheta,\varphi)\right|^2$ are illustrated for $l = 0, 1, 2$. The different $(l, m)$ states are labeled according to an international convention with lower case Latin letters, as compiled in Table 5.3. For example, a state with quantum numbers $n = 2, l = 1$ is a $2p$ state, one with $n = 4, l = 3$ is a $4f$ state. The magnetic quantum number $m$ is indicated by a coordinate index, e.g. the $2p(m = 0)$ state is designated as $2p_z$, the $2p(m = \pm 1)$ by $2p_x$ on $2p_y$, where the index gives the direction of the lobe of the p-function (Fig. 4.27).

In Fig. 5.3 the radial wave functions of some atomic states are illustrated. Together with the angular part $Y_l^m(\vartheta, \varphi)$, drawn in Fig. 4.24, the total wave functions can be visualized, as shown for two examples in Fig. 5.4, which represents the three-dimensional electron distribution in the Coulomb potential for the $2s$ and the $2p\sigma$ state.

Since, according to (5.18), the energy $E_n$ of an atomic state depends in this model solely on $n$ and not on $l$ or $m$, all states with possible combinations of $l$ and $m$ for the same $n$ have the same energy. For each quantum number $l$ there are $2l + 1$ possible $m$ values, because $-l \le m \le +l$. The total number of different

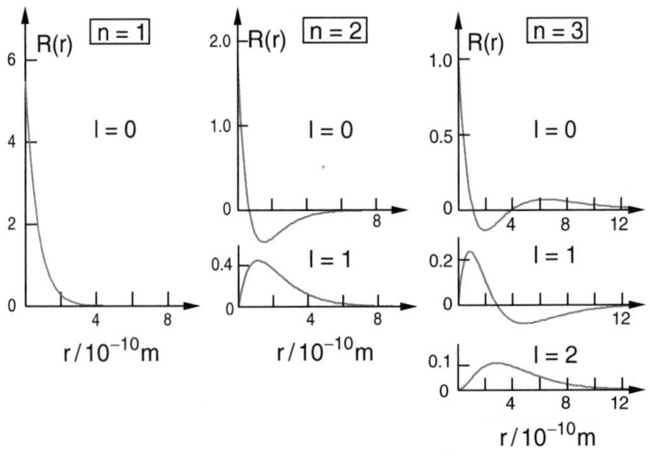


**Fig. 5.3.** The radial wave function $R_{n,l}(r)$ for the principle quantum numbers $n = 1, 2, 3$. The ordinate is scaled in units of $10^8 \mathrm{m}^{-3/2}$

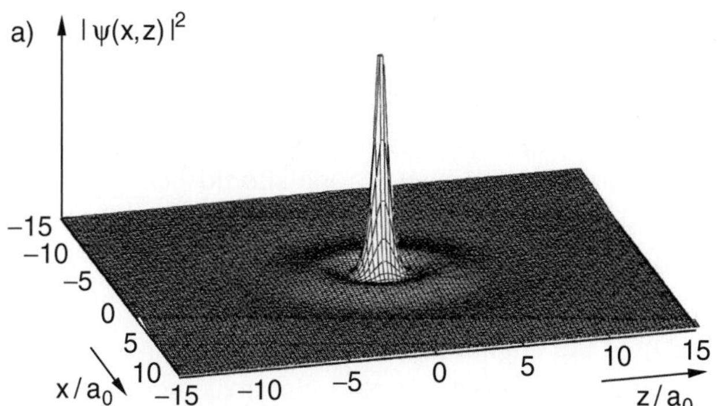


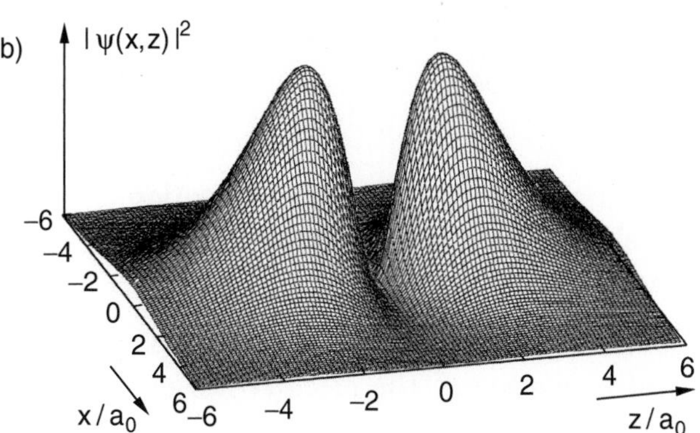


**Fig. 5.4a,b.** Illustration of the three-dimensional electron charge distribution (**a**) for the spherical symmetric function of the $2s$ state and (**b**) for the $2p$ $(m = 0)$ state. In both cases the two-dimensional projection of $|\psi|^2$ onto the $xz$-plane is shown (calculated by H. von Busch, Kaiserslautern, Germany)

atomic states with the same energy is then, because of $l < n$,

$$k = \sum_{l=0}^{n-1}(2l + 1) = n^2 . \tag{5.19b}$$

Different states with the same energy are called energetically degenerate. The number of degenerate states is called the degeneracy order. The states in the Coulomb potential (e.g., for the H atom or the $He^+$ ion) are $n^2$-fold degenerate, which means that $n^2$ states with different quantum numbers $l$ and $m$ and therefore different wave functions, but the same quantum number $n$ all have the same energy (Fig. 5.5).

## EXAMPLES

The state with the lowest energy (ground state) with $n = 1, l = 0, m = 0$ is a nondegenerate $1s$-state.

States with $n = 2$ may have angular momentum quantum numbers $l = 0, m = 0$ $(2s)$ or $l = 1$ and $m = 0 (2p_z)$ and $m = \pm 1$ $(2p_{x,y})$. Such states are therefore fourfold degenerate.

**Note**, that several effects (such as electron spin, nuclear spins, external fields or the relativistic mass dependence), which are not included in the Schrödinger theory, may lift the degeneracy and split degenerate

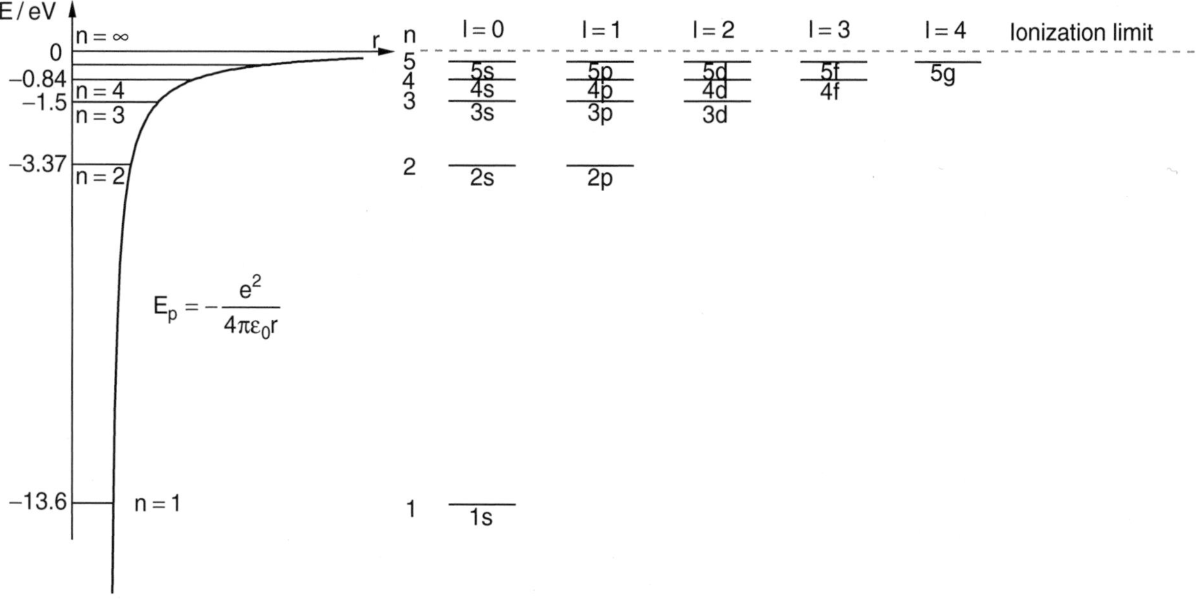


**Fig. 5.5.** Level scheme of the H atom, drawn on a correct scale according the Schrödinger equation

levels into components with different energies (see Sects. 5.3–5.6).

### 5.1.4 Spatial Distributions and Expectation Values of the Electron in Different Quantum States

The spatial distribution of the electron in $s$ states is spherically symmetric. The electron has the angular momentum

$$|l| = \sqrt{l(l+1)}\hbar = 0$$

in contrast to the Bohr-model, where the electron moves on a circular path around the nucleus with an angular momentum $|l| = h$. We can see from Table 5.2, that the spatial probability density $|\psi|(r,\vartheta,\varphi)|^2$ in the 1$s$ state has its maximum at $r = 0$, i.e, at the location of the nucleus.

When we want to calculate the probability $P(r)\mathrm{d}r$ to find the electron within a spherical shell in a distance between $r$ and $r + \mathrm{d}r$ from the nucleus, independent of the angles $\vartheta$ and $\varphi$, we have to solve the integral

$$P(r)\,\mathrm{d}r = \int\limits_{\vartheta=0}^{\pi}\int\limits_{\varphi=0}^{2\pi} |\psi(r,\vartheta,\varphi)|^2 r^2\,\mathrm{d}r\,\sin\vartheta\,\mathrm{d}\vartheta\,\mathrm{d}\varphi\,. \tag{5.20}$$

Inserting the wave function $\psi$ for $n = 1$, $l = 0$ and $m = 0$ (i.e., for the ground state of the hydrogen atom), we obtain

$$P(r)\,\mathrm{d}r = \frac{4Z^3}{a_0^3} r^2 \mathrm{e}^{-2Zr/a_0}\mathrm{d}r\,. \tag{5.21}$$

Comparing this with the wave function for the 1$s$ state we get the result:

> The probability to find the electron within the distance $r$ to $r + \mathrm{d}r$ from the nucleus is in the 1$s$ state given by
>
> $$P(r)\,\mathrm{d}r = 4\pi r^2|\psi(r,\vartheta,\varphi)|^2\,\mathrm{d}r\,.$$

**Note:**

Spherical symmetric electron distributions are obtained for all $s$ states with arbitrary quantum number $n$.

The function $P(r)$ for $n = 1$ is maximum for $r_\mathrm{m} = a_0/Z$, as can be seen immediately by differentiation of (5.21). For $Z = 1$ one obtains the Bohr-radius $r_\mathrm{m} = a_0$ (the maximum probability of finding the electron is at the Bohr-radius!). However, one should keep in mind that the angular momentum obtained from the quantum mechanical treatment is $l = 0$, while the Bohr model gives $l = 1$. All experiments performed so far have confirmed the quantum mechanical result.

If one would like to use a classical model for the movement of the electron in the 1$s$ state, one has to replace the circular path of the Bohr model by periodic linear motions of the electron through the nucleus. The direction of this oscillation is, however, randomly distributed, causing an average electron distribution that is spherically symmetric (Fig. 5.6). *Arnold Sommerfeld* (1868–1951) showed that the electron motion can be described to proceed on very eccentric elliptical orbits passing close to the nucleus, which causes a fast precession of the large axis and brings about that the outer turning points are uniformly distributed on a spherical surface.

The expectation value $\langle r\rangle$ for the mean distance between electron and nucleus is given by

$$\langle r\rangle = \int\limits_{r=0}^{\infty}\int\limits_{\vartheta=0}^{\pi}\int\limits_{\varphi=0}^{2\pi} r|\psi(r,\vartheta,\varphi)|^2 r^2\sin\vartheta\,\mathrm{d}\vartheta\,\mathrm{d}\varphi\,\mathrm{d}r\,. \tag{5.22a}$$

For the 1$s$ state this yields, after inserting the 1$s$ wave function from Table 5.2,

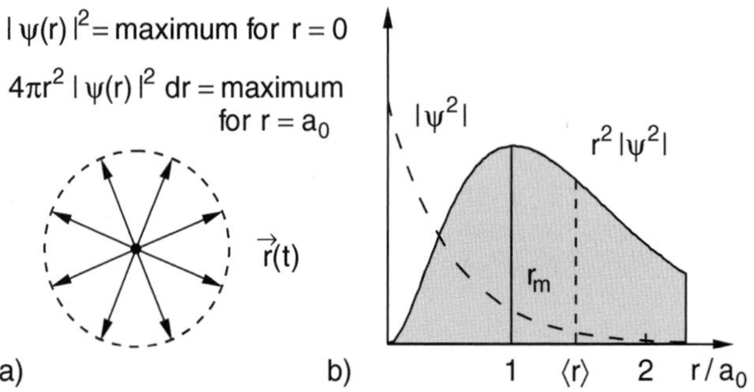


**Fig. 5.6.** (**a**) Classical model of electron paths as oscillations on straight lines through the nucleus of the 1$s$ state. The orientation of the lines is statistically distributed to give an isotropic average. (**b**) Comparison between the probability density $|\psi(1s)|^2$ and the probability $4\pi r^2|\psi(1s)|^2\,dr$ of finding the electron within the spherical shell $4\pi r^2\,\mathrm{d}r$

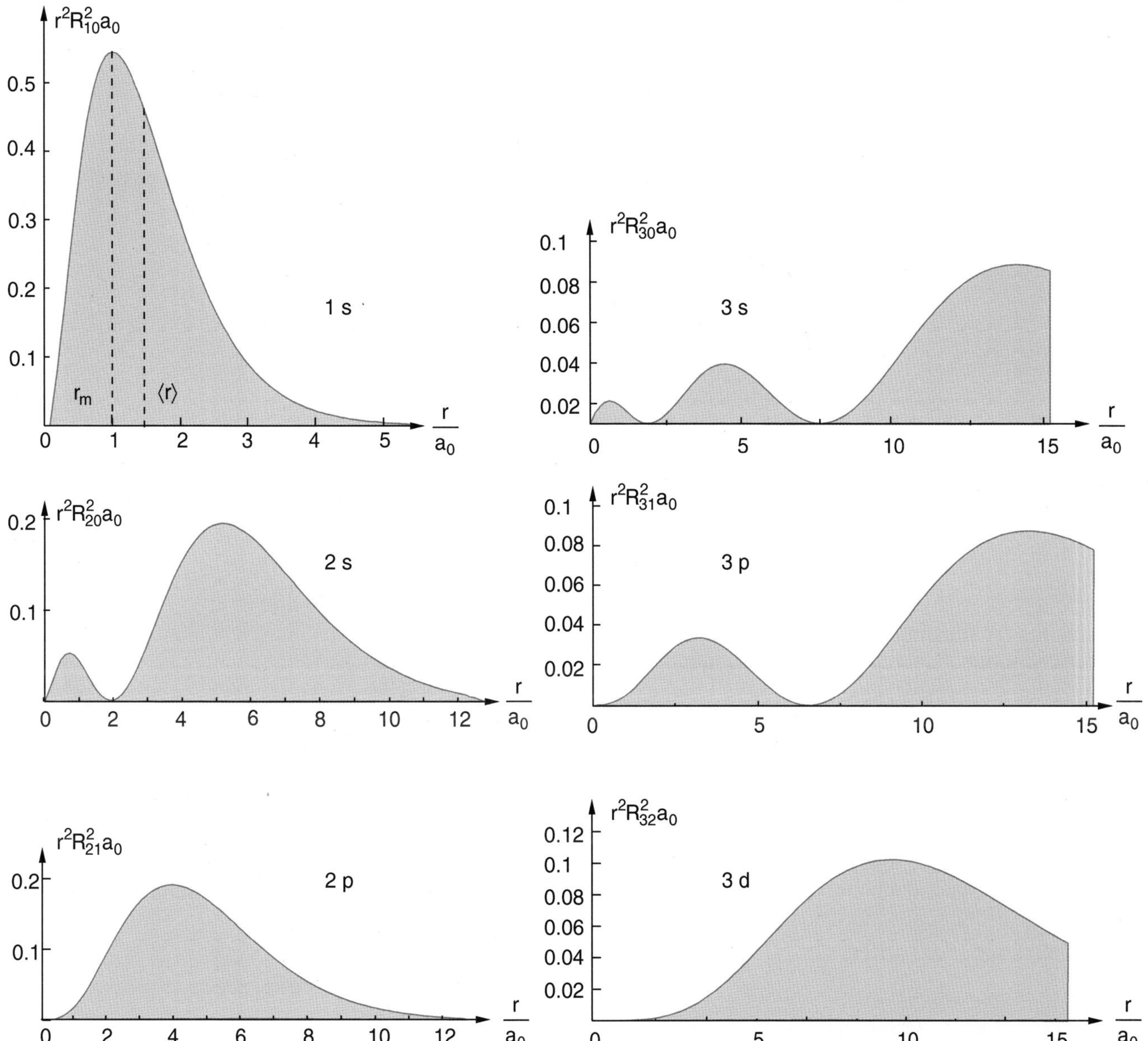


**Fig. 5.7.** Radial charge distribution of the electron in different states of the H atom. Note the different ordinate and abzissa scales

$$\langle r\rangle = \int_{r=0}^{\infty} \frac{r}{\pi a_0^3} 4\pi r^2 \mathrm{e}^{-2r/a_0}\,\mathrm{d}r = \frac{3}{2}a_0\,, \qquad (5.22b)$$

which differs from the Bohr radius $a_0$!

In Fig. 5.7 the functions $r^2 a_0 |R_{nl}(r)|^2$ are plotted for some states against the abscissa $r/a_0$, i.e., in units of the Bohr radius. They are normalized in such a way that the shaded area under the curve becomes

$$\int_{r=0}^{\infty} r^2 |R_{n,l}(r)|^2 a_0 \frac{\mathrm{d}r}{a_0} = 1\,.$$

The plotted curves are directly proportional to the probability $4\pi r^2 |R_{nl}(r)|^2 \mathrm{d}r$ of finding the electron within the spherical shell between $r$ and $r + \mathrm{d}r$.

The probability $P(r < a_0)$ of finding the electron within the Bohr radius $a_0$ is, for $s$-functions ($l = 0$),

$$P_{n,l}(r \le a_0) = 4\pi \int_{r=0}^{a_0} r^2 |\psi_{n,0}(r)|^2\,\mathrm{d}r\,, \qquad (5.23)$$

and can be calculated with the functions in Tables 5.1 and 4.2.

For $n = 1$ and $l = 0$, for example, we obtain:

$$P_{1,0}(r \le a_0) = \frac{4}{a_0^3} \int_{r=0}^{a_0} r^2 \mathrm{e}^{-2r/a_0} \mathrm{d}r = 0.32 .$$

For $n = 2$ and $l = 0$ one can verify the result:

$$P_{2,0}(r \le a_0) = \frac{1}{8a_0^3} \int_{r=0}^{a_0} \left(4r^2 - \frac{4r^3}{a_0} + \frac{r^4}{a_0^2}\right) \mathrm{e}^{-r/a_0}\, \mathrm{d}r = 0.034 .$$

While for $n = 2$ and $l = 1$ one obtains the smaller probability

$$P_{2,1}(r \le a_0) = \frac{1}{24a_0^5} \int_{r=0}^{a_0} r^4 \mathrm{e}^{-r/a_0}\, \mathrm{d}r = 0.0037 .$$

These results are illustrated by the curves in Fig. 5.7.

In the simplified classical model this means that the orbits with $l = 0$ correspond to very eccentric elliptical paths where the electron is often close to the nucleus, while orbits with maximum possible $l$ are close to circular orbits. With increasing principal quantum number $n$ the maximum possible values of $l < n$ become larger and the quantum mechanical spatial probabilities approach more and more closely the classical circular orbits.

An interesting result arises:

> Summing the spatial probability $|\psi(r, \vartheta, \varphi)|^2$ at a given $n$ over all allowed values of $l$ and $m$ gives the total probability in the state $n$, which is always spherically symmetric! Therefore the sum over the electron distributions in all possible states $(l, m)$ for a given value of $n$ is called an ***electron shell***.

## 5.2 The Normal Zeeman Effect

We will now discuss the behavior of the H atom in an external magnetic field. In the beginning we will use a semiclassical model (called the vector model), where the electron motion is described by a classical circular orbit, while the angular momentum is given by the quantum mechanical expression

$$|\boldsymbol{l}| = \sqrt{l(l+1)}\hbar .$$

An electron with charge $-e$ moving with the velocity $v$ and the circular frequency $\nu = v/(2\pi r)$ on a circle with radius $r$ represents an electric current

$$I = -e\nu = -\frac{ev}{2\pi r} , \tag{5.24}$$

which causes a magnetic moment

$$\boldsymbol{\mu} = I\boldsymbol{A} = I\pi r^2 \hat{\boldsymbol{n}} , \tag{5.25}$$

where $\boldsymbol{A} = \pi r^2 \hat{\boldsymbol{n}}$ is the area vector perpendicular to the plane of the motion (Fig. 5.8).

The angular momentum of the circulating electron is

$$\boldsymbol{l} = \boldsymbol{r} \times \boldsymbol{p} = m_\mathrm{e} r v \hat{\boldsymbol{n}} . \tag{5.26}$$

The comparison of (5.25) and (5.26) gives the relation

$$\boxed{\boldsymbol{\mu} = -\frac{e}{2m_\mathrm{e}} \boldsymbol{l}} \tag{5.27}$$

between magnetic moment $\boldsymbol{\mu}$ and angular momentum $\boldsymbol{l}$ of the electron. Since $\mu$ is proportional to $l$, the orbital magnetic moment is often labeled $\boldsymbol{\mu}_l$.

In an external magnetic field the potential energy of a magnetic dipole with magnetic moment $\boldsymbol{\mu}$ is

$$E_\mathrm{pot} = -\boldsymbol{\mu} \cdot \boldsymbol{B} . \tag{5.28}$$

Using the relation (5.27) this can be expressed by the angular momentum $l$ as

$$E_\mathrm{pot} = +\frac{e}{2m_\mathrm{e}} \boldsymbol{l} \cdot \boldsymbol{B} . \tag{5.29}$$

When the magnetic field points into the $z$-direction ($\boldsymbol{B} = \{0, 0, B_z = B\}$), we obtain from (5.29), because of $l_z = m\hbar$,

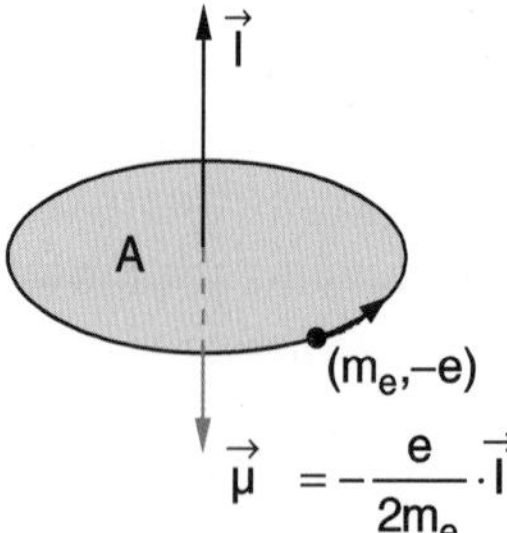


**Fig. 5.8.** Classical model of orbital angular momentum $\boldsymbol{l}$ and magnetic moment $\boldsymbol{\mu}$

$$E_{\text{pot}} = \frac{e\hbar}{2m_{\text{e}}} mB \,, \tag{5.30}$$

where $m$ (which had been introduced before as the projection of $l$ onto the $z$-axis) is called the magnetic quantum number, that can take the values $-l \le m \le +l$.

The constant factor in (5.30)

$$\mu_{\text{B}} = \frac{e\hbar}{2m_{\text{e}}} = 9.274015 \times 10^{-24}\ \text{J/T} \tag{5.31}$$

is called the ***Bohr magneton***.

We can now write the additional energy caused by the magnetic field as

$$\Delta E_{\text{m}} = \mu_{\text{B}} m B \,, \tag{5.32}$$

which gives for the energies of the hydrogen atomic states in an external magnetic field:

$$\boxed{E_{n,l,m} = E_{\text{Coul}}(n) + \mu_{\text{B}} m B \,.} \tag{5.33}$$

The $2l+1$ $m$-sublevels that are degenerate without magnetic field split into $2l+1$ equidistant Zeeman components with energies between $E_n^{\min} = E_{\text{Coul}}(n) - \mu_{\text{B}} l B$ and $E_n^{\max} = E_{\text{Coul}}(n) + \mu_B l B$, with an energetic distance (Fig. 5.9)

$$\Delta E = E_{n,l,m} - E_{n,l,m-1} = \mu_{\text{B}} B \,, \tag{5.34}$$

between adjacent levels which is determined by the product of Bohr magneton $\mu_{\text{B}}$ and magnetic field strength $B$.

> The splitting of the $2l+1$ degenerate $m$ components in an external magnetic field B due to the orbital magnetic moment related to the angular momentum $|l| = \sqrt{l(l+1)}\hbar$ is called the ***normal Zeeman effect***.

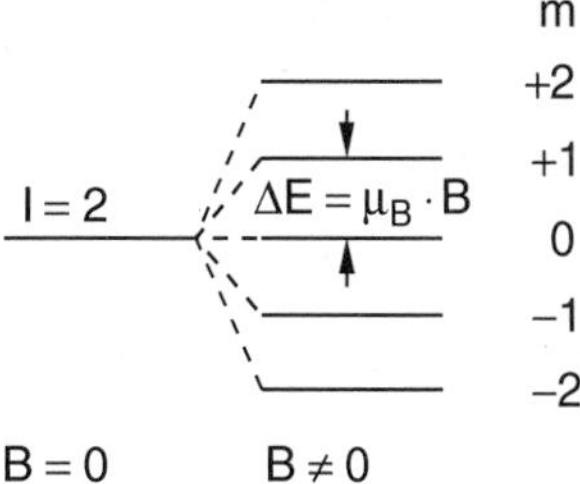


**Fig. 5.9.** Zeeman splitting of a level with $l = 2$ in a homogeneous magnetic field (normal Zeeman effect)

Using the Bohr magneton (5.31) we can write the orbital magnetic moment of the electron as

$$\boxed{\boldsymbol{\mu}_l = -(\mu_B/\hbar)\boldsymbol{l} \,.} \tag{5.35a}$$

Since the external magnetic field with cylindrical symmetry breaks the spherical symmetry of the Coulomb potential the orbital angular momentum $l$ of the electron is no longer constant, because the torque

$$\boldsymbol{D} = \boldsymbol{\mu}_l \times \boldsymbol{B} \tag{5.35b}$$

acts on the electron. In the case of a magnetic field $\boldsymbol{B} = \{0, 0, B_z = B\}$ in the $z$ direction the $z$ component of $l$ stays constant. The vector $\boldsymbol{l}$ precesses around the $z$-axis on a cone with the apex angle $2\alpha$ (Fig. 5.10), where

$$\cos\alpha = \frac{l_z}{|\boldsymbol{l}|} = \frac{m}{\sqrt{l(l+1)}} \,. \tag{5.35c}$$

The component $l_z$ has the values

$$l_z = m\hbar \quad \text{with} \quad -l \le m \le +l \,. \tag{5.35d}$$

Also, the absolute value of $l$

$$|\boldsymbol{l}| = \sqrt{l(l+1)}\hbar \tag{5.36}$$

is well defined, while the two other components $l_x$ and $l_y$ are not defined (see Sect. 4.4.2). Their quantum mechanical expectation value is zero, as is the classical time averaged value. Since the magnetic moment $\vec{\mu}_l$ is anti-parallel to $\vec{l}$, it also precesses around the magnetic field axis with a well defined component $\mu_z$.

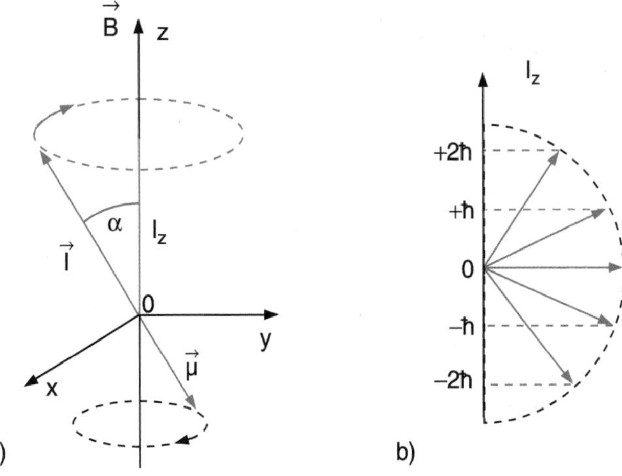


**Fig. 5.10a,b.** Vector model of the normal Zeeman effect. (**a**) Classical model of orbital angular momentum precessing around the field axis. (**b**) Possible orientations of $\boldsymbol{l}$ and projections $m\hbar$ in the quantum mechanical description

For the absorption or emission of light by atoms in a magnetic field, our model makes the following predictions.

When a circularly polarized $\sigma^+$-light wave propagates into the $z$ direction all photons have the spin $+\hbar\hat{\mathbf{k}}$. If they are absorbed by atoms in the magnetic field $B = \{0, 0, B_z\}$ they transfer their spin $+\hbar\hat{\mathbf{k}}$ to the atoms (because of conservation of angular momentum) and therefore cause transitions with $\Delta l_z = +\hbar$, where the quantum number $m$ changes by $+1$.

For $\sigma^-$, polarization of the light wave transitions with $\Delta m = -1$ and $\Delta l_z = -\hbar$ are induced.

A similar consideration is valid for the emission of light by atoms in a magnetic field. For the light emitted into the direction of the magnetic field (i.e., in the $z$ direction) the two circularly polarized $\sigma^+$ and $\sigma^-$ components are observed, while for light emitted into the direction perpendicular to the field three linearly polarized components are observed. One component with the $E$-vector parallel to $B$, which is not shifted against the field-free transition, and two components with $E \perp B$, which are shifted to opposite sides of the unshifted line (Fig. 5.11).

According to (5.34), the Zeeman splitting $\Delta E = \mu_B B$ is independent of the quantum numbers $n$ and $l$. This implies that all atomic states should have the same separation of the Zeeman components. Therefore every spectral line corresponding to a transition $(n_1 l_1) \to (n_2, l_2)$ should always split in a magnetic field into three Zeeman components (Fig. 5.12) with $\sigma^+$, $\sigma^-$ and $\pi$-polarization and a frequency separation of

$$\Delta\nu = \mu_B B / h \, .$$

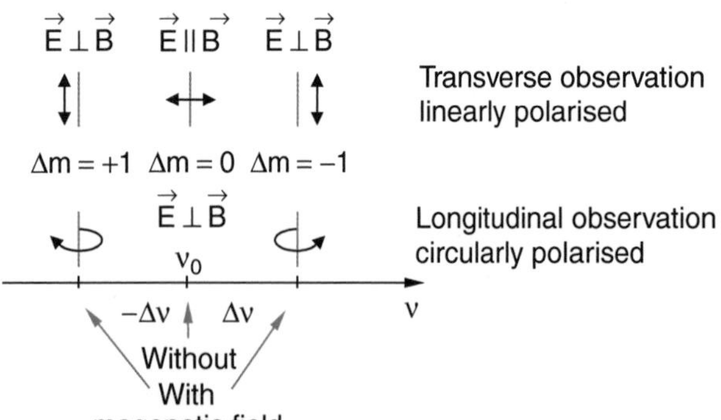


Fig. 5.11. Normal Zeeman effect. Zeeman splitting and polarizations of a spectral line with frequency $\nu_0$ observed in emission. The splitting is $\Delta\nu = \mu_B B / h$

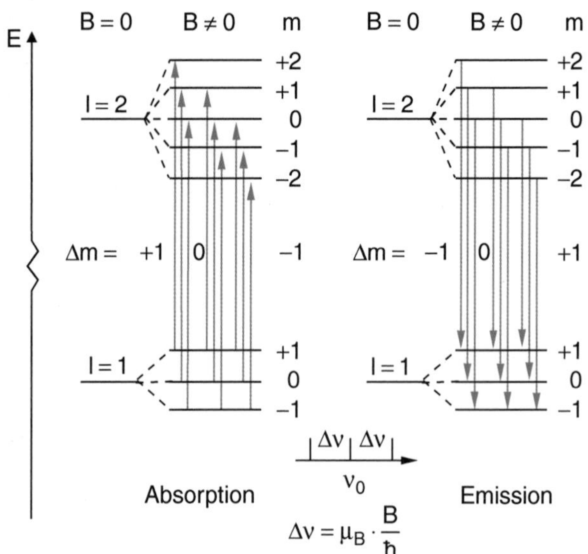


Fig. 5.12. Level scheme and transitions $\Delta m = 0, \pm 1$ between Zeeman level in absorption and emission for the normal Zeeman effect

Next we will discuss how this prediction and all the other conclusions drawn from the Schrödinger model of the atoms match the experimental results?

## 5.3 Comparison of Schrödinger Theory with Experimental Results

Although the hydrogen atom is, from a theoretical point of view, the simplest atomic system, and can be calculated analytically (at least within the Schrödinger model), its experimental investigation is not as simple. There are several reasons for this:

1. First of all hydrogen atoms cannot be bought in a bottle. They have to be produced by dissociation of $H_2$ molecules. This can be achieved in several ways.
   The easiest method is dissociation by electron impact in gas discharges, where electrons with sufficient kinetic energy collide with hydrogen molecules $H_2$ to form H atoms, according to the scheme

   $$H_2 + e^- \to H^* + H + e^- \, .$$

   The excited atoms $H^*$ in the state $|n_k\rangle$ release their excitation energy partly or completely by emitting photons with energy $h\nu = E(n_k) - E(n_i)$ with $E_i < E_k$:

$$\mathrm{H}^*(n_k) \xrightarrow{h \cdot \nu} \mathrm{H}(n_i)\,.$$

Another technique is the thermal dissociation of $H_2$ molecules at high temperatures ($T = 1500 - 2000K$) in the presence of catalysts (e.g., tungsten surfaces).

Nowadays the preferred method is the dissociation by microwave discharges, which has proved to be the most efficient way of forming H atoms.

2. The absorption spectrum of ground state hydrogen atoms lies in the vacuum ultraviolet (VUV) spectral region. It therefore can be measured only in the vacuum, i.e, in evacuated spectrographs (Fig. 5.13b). In most experiments the emission rather than the absorption is measured. The hydrogen discharge is placed in front of the entrance slit of the evacuated spectrograph (Fig. 5.13) and the dispersed emission spectrum of the hydrogen atoms is detected on a photoplate. Since most materials absorb in the VUV, no lenses are used and the curved grating (Rowland arrangement) images the entrance slit onto the photoplate. For wavelengths below 120 nm no entrance window can be used and the air coming from the outside through the open entrance slit has to be pumped away by differential pumping in order to maintain the vacuum inside the spectrograph.

Since the emission of excited H atoms has a spectrum covering the whole range from the infrared to the VUV region, that part of the emission spectrum with wavelengths above 200nm can be measured with spectrographs in air. Here, the intensity can be enlarged by imaging the discharge source S by a lens onto the entrance slit and two curved mirrors $M_1$ and $M_2$ image the entrance slit $S_1$ onto the CCD camera at the exit (Fig. 5.13a). As has already been discussed in Sect. 3.4.1 the lines in the spectrum of the H Atom can be arranged in series (Fig. 3.40) with wavenumbers that can be fit by the simple relations

$$\bar{\nu}_{ik} = Ry\left(\frac{1}{n_i^2} - \frac{1}{n_k^2}\right)$$
$$\text{with} \quad Ry = Ry_\infty \frac{m_\mathrm{p}}{m_\mathrm{e} + m_\mathrm{p}}$$
$$= 109{,}677.583\,\mathrm{cm}^{-1} \qquad (5.37)$$

in accordance with the formula (3.79) and (5.18).

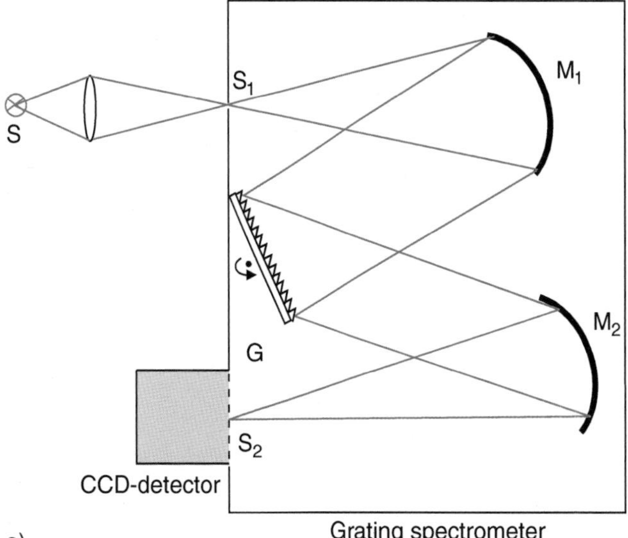


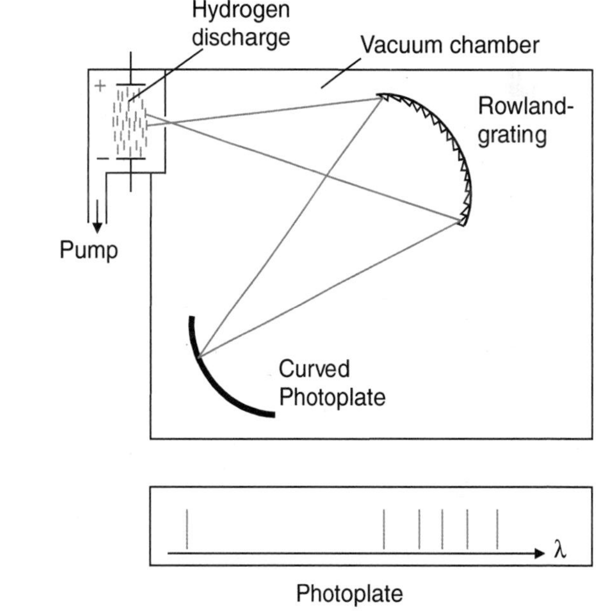


**Fig. 5.13.** (**a**) Experimental setup for measuring the emission spectrum of atomic hydrogen for lines with $\lambda > 200$ nm. (**b**) Vacuum UV spectrograph for measuring the Lyman-series with $\lambda < 200$ nm

Accurate measurements with higher spectral resolution showed, however, significant deviations of the measured line positions and line structures from the predictions of the Schrödinger theory discussed so far.

a) The wavenumbers $\nu_{ik}$ of the different transitions between levels $|i\rangle$ and $|k\rangle$ depend not only on the principal quantum number $n$ but also slightly on the

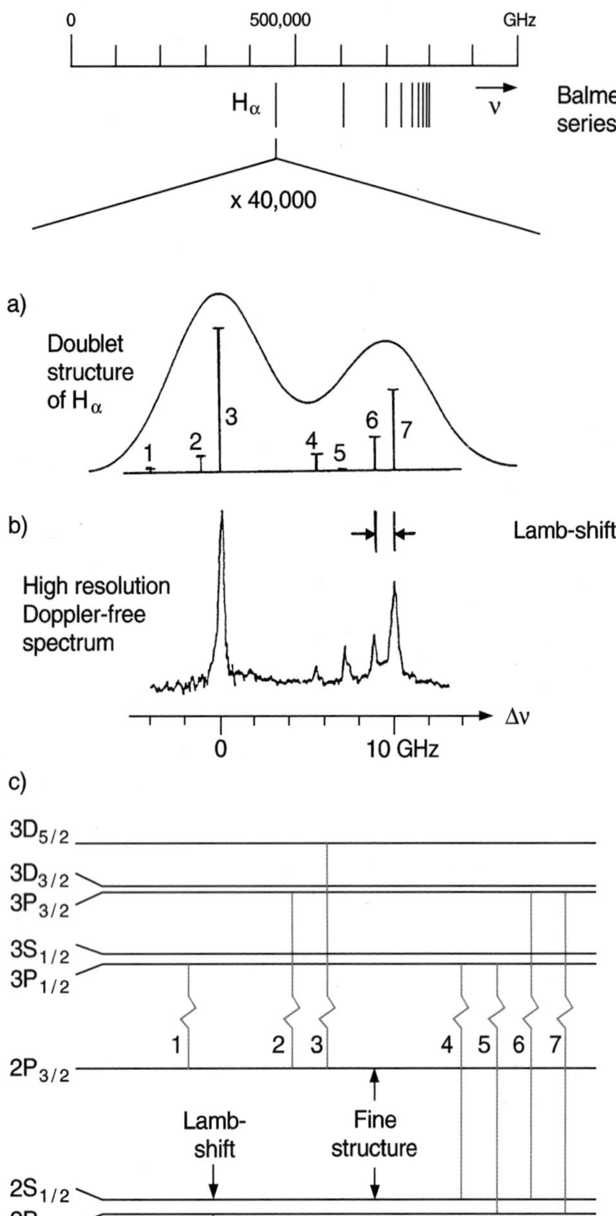


**Fig. 5.14a–c.** Balmer series of the hydrogen atom. (**a**) Fine structure of $H_\alpha$ measured with conventional Doppler-limited spectroscopy. (**b**) High-resolution Doppler-free spectrum of $H_\alpha$ showing the Lamb shift of the $2^2S_{1/2}$ level. (**c**) Level scheme

angular momentum quantum number $l$. The absolute wavenumbers for the H atom deviate from the predictions by up to $0.2\,\mathrm{cm}^{-1}$.

b) All spectral lines starting from $s$ levels with $l = 0$ consist of two narrowly spaced components (doublets). Those starting from levels with $l > 0$ contain even more components (Fig. 5.14).

**Table 5.4.** Comparison of the measured wavenumbers $\bar{\nu}_{\mathrm{vac}}$ of the Balmer series and the calculated values $\bar{\nu}_{Ry}$ obtained from the Rydberg formula

| | $n$ | $\lambda_{\mathrm{air}}/\mathrm{Å}$ | $\nu_{\mathrm{vac}}/\mathrm{cm}^{-1}$ | $\nu_{Ry}/\mathrm{cm}^{-1}$ |
|---|---|---|---|---|
| $H_\alpha$ | 3 | 6,562.79 | 15,233.21 | 15,233.00 |
| $H_\beta$ | 4 | 4,861.33 | 20,564.77 | 20,564.55 |
| $H_\gamma$ | 5 | 4,340.46 | 23,032.54 | 23,032.29 |
| $H_\delta$ | 6 | 4,101.73 | 24,373.07 | 24,372.80 |
| $H_\varepsilon$ | 7 | 3,970.07 | 25,181.33 | 25,181.08 |
| $H_\zeta$ | 8 | 3,889.06 | 25,705.84 | 25,705.68 |
| $H_\eta$ | 9 | 3,835.40 | 26,065.53 | 26,065.35 |
| $H_\vartheta$ | 10 | 3,797.91 | 26,322.80 | 26,322.62 |
| $H_\iota$ | 11 | 3,770.63 | 26,513.21 | 26,512.97 |
| $H_\kappa$ | 12 | 3,750.15 | 26,658.01 | 26,657.75 |
| $H_\lambda$ | 13 | 3,734.37 | 26,770.65 | 26,770.42 |
| $H_\mu$ | 14 | 3,721.95 | 26,860.01 | 26,859.82 |
| $H_\nu$ | 15 | 3,711.98 | 26,932.14 | 26,931.94 |

c) The experimentally observed splittings of the Zeeman components agrees only for a few atomic species with the prediction of the normal Zeeman pattern. For most atoms it is more complicated. For the H atom, for instance, it looks completely different from the regular triplet pattern in Fig. 5.12.

d) The ground state of the H atom ($n = 1$, $l = 0$) shows a very narrow splitting into two components (hyperfine structure), which differs for the two isotopes $^1$H and $^2\mathrm{H} = {}^2\mathrm{D}$.

In order to explain these deviations, the Schrödinger theory has to be extended and new atomic parameters have to be included. This will be discussed in the next sections.

## 5.4 Relativistic Correction of Energy Terms

The last section has shown, that the coarse structure of the hydrogen spectrum is well described by the Schrödinger-theory for an electron in the Coulomb-field of the atomic nucleus. The small discrepancies found by comparison of experimental results with theoretical predictions are due to relativistic effects. There are essentially three contributions to these effects:

1. The relativistic dependence of the electron mass on its velocity in the Coulomb-field which results in a small decrease of its kinetic energy

2. A closer inspection of the relativistic treatment shows that the electron charge is "smeared out" over a volume $\lambda_c{}^3 = (\hbar/m_e c)^3$ where $\lambda_c$ is the Compton wavelength of the electron. This leads to a change of the potential energy of the electron (Darwin-term)
3. The interaction between the magnetic moment of the electron due to its orbital angular momentum and the magnetic moment due to the electron spin results in a shift and a splitting of the energy levels (fine-structure).

We will now briefly discuss these three contributions and their magnitude.

1) **Relativistic mass increase**

Instead of the nonrelativistic energy relation

$$E = p^2/2m + E_{\text{pot}} \tag{5.38}$$

anticipated by the Schrödinger theory, we have to use the relativistic energy relation

$$E = c\sqrt{m_0^2c^2 + p^2} - m_0c^2 + E_{\text{pot}} \,. \tag{5.39}$$

For the electron in the hydrogen atom, the velocity $v$ of the electron is still small compared to the velocity of light $c$, which means that $E_{\text{kin}} \ll m_0c^2$ or $p^2/m_0^2c^2 \ll 1$. We can therefore expand the square root in (5.39) into the power series

$$\sqrt{1+\frac{p^2}{m_0^2c^2}} = 1 + \frac{1}{2}\frac{p^2}{m_0^2c^2} - \frac{1}{8}\frac{p^4}{m_0^4c^4} + \ldots\,,$$

which gives for the energy expression (5.38)

$$E = \left(\frac{p^2}{2m_0} + E_{\text{pot}}\right) - \frac{p^4}{8m_0^3c^2} + \ldots = E_{\text{nr}} - \Delta E_{\text{r}} \,. \tag{5.40}$$

For $E_{\text{kin}} \ll m_0c^2$ we can neglect the higher order terms. In this approximation the last term in (5.40) represents the relativistic correction $\Delta E_{\text{r}}$ to the non-relativistic energy (5.38).

We can obtain the quantum mechanical expectation value of this correction by substituting $p \to -i\hbar\nabla$, which leads to the expression

$$\Delta E_{\text{r}} = \frac{\hbar^4}{8m_0^3c^2}\int \psi^*_{n,l,m}\nabla^4\psi_{n,l,m}\,\mathrm{d}\tau \,. \tag{5.41}$$

Inserting for $\psi$ the wave functions of the hydrogen atom $\psi_{n,l,m}$ gives (see Problem 5.10)

$$\Delta E_{\text{r}} = -E_{\text{nr}}\frac{Z^2\alpha^2}{n}\left(\frac{3}{4n} - \frac{1}{l+1/2}\right) \,. \tag{5.42}$$

The constant

$$\alpha = \frac{e^2}{4\pi\varepsilon_0\hbar c} = 7.297353 \times 10^{-3} = \frac{1}{137} \tag{5.43}$$

is called ***Sommerfeld's fine structure constant***.

The total energy of an eigen-state for the H atom is then

$$E_{n,l} = -Ry\frac{Z^2}{n^2}\left[1 - \frac{\alpha^2Z^2}{n}\left(\frac{3}{4n} - \frac{1}{l+1/2}\right)\right]\,, \tag{5.44}$$

which now depends not only on $n$ but also on $l$! The relativistic correction is maximum for $n = 1$ and $l = 0$.

## EXAMPLES

1. For $n = 1$, $l = 0$ the magnitude of the relativistic correction is
$$\Delta E_{\text{r}} = -Z^2E_1 5\alpha^2/4 \,.$$
2. For $Z = 1$ this becomes
$$\Delta E_{\text{r}} = -9 \times 10^{-4}\,\text{eV} \,.$$
3. For $n = 2, l = 0$ we obtain
$$\Delta E_{\text{r}}(n = 2, l = 0) = -\frac{13}{16}Z^2E_2\alpha^2 = 1.5 \times 10^{-4}\,\text{eV. for } Z = 1$$
4. For $n = 2, l = 1$ the correction is only
$$\Delta E_{\text{r}}(n = 2, l = 1) = -\frac{7}{48}E_2Z^2\alpha^2 = Z^2 \cdot 2.6 \times 10^{-5}\,\text{eV} \,.$$

This illustrates that

a) The relativistic energy shift is maximum for the ground state of atoms ($n = 1, l = 0$).

b) The correction depends on both quantum numbers $n$ and $l$. The $n$-fold degeneracy of states $(n, l)$, deduced from the Schrödinger theory is lifted by the relativistic correction.
c) At a given value of $n$, the electron comes closest to the nucleus (and therefore acquires the largest velocity) for small values of $l$ (the Sommerfeld orbits are then ellipses with large eccentricity). The relativistic mass increase is then maximum, which decreases the energy term value. For the maximum allowed $l = n - 1$ the orbit is circular and the velocity of the electron has a constant medium value. The relativistic mass correction is then minimum.

**Note:**

As the numerical examples show, the relativistic mass correction only amounts to less than $10^{-4}$ of the Coulomb energy.

2) **Darwin-Term**

The second relativistic contribution comes from the fact, that, even in a model which describes the electron as a point charge, the momentary position of the electron cannot be defined more precisely than within the volume $\lambda_c^3 = (\hbar/m_e c)^3$ given by the Compton wavelength $\lambda_c$ of the electron. The potential energy of the electron in the Coulomb field of the nucleus is then the weighted average of all values of the electric field within the volume $\lambda_c^3$ around the point $\boldsymbol{r}$. This means: the potential energy is no longer $E_{pot}(r)$ but is determined by the integral

$$E_{pot}(\boldsymbol{r}) = \int f(\boldsymbol{\rho}) E_{pot}(\boldsymbol{r} + \boldsymbol{\rho}) \, d^3\rho \tag{5.45}$$

over the volume $\lambda_c^3$ around the point $\boldsymbol{r}$. Expanding $E_{pot}(r + \rho)$ into a Taylor series of powers of $\rho$ around the point $\rho = 0$ gives

$$E_{pot}(r + \rho) = E_{pot}(r) + (dE_{pot}/d\rho)_{\rho=0} \cdot \rho + 1/2(d^2E_{pot}/d\rho^2) \cdot \rho^2 + \ldots\ldots \tag{5.46}$$

the first term gives the unperturbed potential used in the Schrödinger theory. The second term vanishes because of the spherical symmetry of the Coulomb field. The third term, inserted into (5.45) gives the relativistic **Darwin correction** which is of the order of magnitude $(\hbar/m_e c)^2 \Delta E_{pot}(r)$, where $\Delta$ is the Laplace operator.

For the Coulomb potential $E_{pot}(r) = -(Ze^2/4\pi\varepsilon_0 r)$ we obtain

$$\Delta E_{pot} = (Ze^2/\varepsilon_0)\delta(r)$$

(where $\delta(\boldsymbol{r})$ is the delta function $\delta(r) = 1$ for $r = \boldsymbol{0}$ and $\delta(\boldsymbol{r} \neq 0) = 0$) and the relativistic correction becomes

$$W_{rel} = +(Ze^2\hbar^2/\varepsilon_0 m_e^2 c^2) \cdot \delta(\boldsymbol{r}) \tag{5.47}$$

The quantum mechanical expectation value, obtained with the hydrogen wavefunction $\psi$ gives the **Darwin term**

$$W_D = +(Ze^2\hbar^2/\varepsilon_0 m_e^2 c^2)\, |\psi(0)|^2 \tag{5.48}$$

where $\psi(0)$ is the wavefunction at the origin $r = 0$. Since only the $s$-functions have nonvanishing values at $r = 0$ we can insert the hydrogen $1s$-function

$$\psi_{1s} = 1/\sqrt{\pi}(Z/a_0)^{3/2} e^{-Zr/a0}$$

and obtain with the definition of $a_0$ and the fine-structure constant $\alpha$

$$W_D = 4Z^4 m_e c^2 \alpha^4 \, . \tag{5.49}$$

With the nonrelativistic energy

$$W_0 = E_{nr} = (1/2) m_e c^2 (Z^2/n^2)\alpha^2 \tag{5.50}$$

the ratio $W_D/W_0$ of Darwin term to the nonrelativistic energy becomes

$$W_D/W_0 = 2Z^2\alpha^2 \approx 8(Z/137)^2 \, . \tag{5.51}$$

For $Z = 1$ the Darwin correction is about $10^{-4} E_{nr}$.

We will now discuss the third relativistic correction, namely the interaction between the orbital magnetic moment and the spin moment of the electron, causing the fine structure of spectral lines.

## 5.5 The Electron Spin

Several experimental results such as the Stern-Gerlach experiment, the fine-structure of spectral lines or the anomalous Zeeman effect indicated that the electron must have, besides its charge $-e$ and mass $m_e$ an additional characteristic property, which was called electron spin and which must cause an additional magnetic moment $\mu_s$ in addition to the orbital magnetic moment $\mu_l$. This had been already postulated by Fermi, before it could be experimentally confirmed.

### 5.5.1 The Stern–Gerlach Experiment

The space quantization of the angular momentum introduced by *Arnold Sommerfeld* in 1916 was considered by many physicists as a purely theoretical model without any real correspondence in nature. It was therefore in doubt if the explanation of the Zeeman effect by the corresponding space quantization of the magnetic moment could ever be proved experimentally. Nevertheless this issue was intensely discussed among experimental physicists. *Otto Stern* (1888–1969), who had a lot of experience with atomic beams, proposed to test the idea of space quantization by measuring the deflection of atoms in a collimated atomic beam passing through a transverse inhomogeneous magnetic field. He found in *Walther Gerlach* (1889–1979) an enthusiastic and experimentally skillful assistant whom he convinced to try this experiment. They started in 1919 and after many unsuccessful efforts and following improvements to their beam apparatus, the two researchers were finally rewarded with an unexpected result in 1921 [5.1].

They chose as test objects silver atoms because these atoms could be detected on a glass plate, where they condensate and form a thin layer with an optical transparency depending on the thickness of the layer and therefore the number of incident atoms.

The silver atoms were evaporated in a hot furnace and emerged through a narrow hole A in the furnace into the vacuum chamber (Fig. 5.15). They were then collimated by the slit S before they entered, traveling in the $x$ direction, the inhomogeneous magnetic field $B_z$, which pointed in the $z$ direction. In the plane $x = x_0$ the atoms were condensed on a cold glass plate. With a densitometer the density $N(z)$ of silver atoms on the glass plate could be measured.

Without a magnetic field, the symmetric blue density profile corresponding to the central curve in Fig. 5.15c was obtained. It represents the density profile of the non-deflected atomic beam due to the spread of the transverse velocity components of the silver atoms behind the collimating slit S.

In the inhomogeneous magnetic field the force

$$\boldsymbol{F} = -\boldsymbol{\mu} \cdot \mathbf{grad}\, \boldsymbol{B}$$

on the atoms depends on the spatial orientation of the magnetic moment $\boldsymbol{\mu}$ relative to $\boldsymbol{B}$. The intensity distribution $I(z)$ is therefore expected to split into as many peaks as the possible values of the scalar product $\boldsymbol{\mu} \cdot \boldsymbol{B}$. At that time the quantum number $l$ of the orbital angular momentum of silver was assumed to be $l = 1$. Therefore three possible values of $\mu B$ were expected, corresponding to the three magnetic quantum numbers

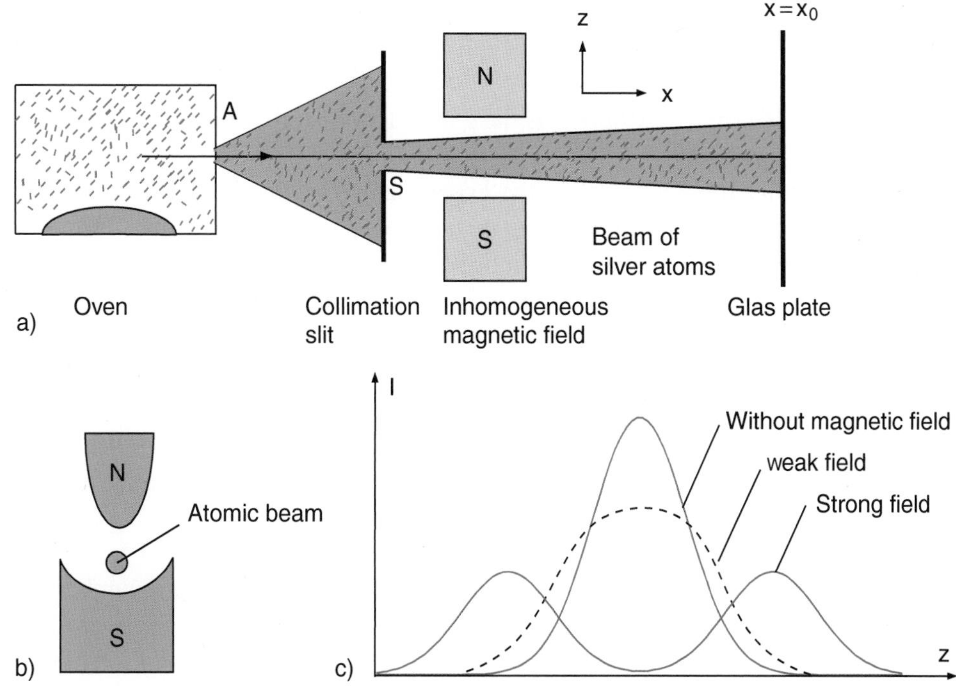


**Fig. 5.15.** (**a**) Stern–Gerlach apparatus for measuring the space quantization of angular momentum $l_z = m\hbar$. (**b**) Cut in the $yz$-plane of the inhomogeneous magnetic field. (**c**) Observed density pattern $\infty N(z)$ of silver atoms in the detection plane $x = x_0$

$m_l = 0, \pm 1$. The experimental result clearly showed only two peaks and a minimum intensity in between the peaks where atoms with $m = 0$ should arrive. Bohr, who was asked for advice, explained this result by a model that assumed that atoms with their magnetic moment perpendicular to the direction of the magnetic field ($m = 0$) were unstable and would flip into one of the other quantum states with $m = \pm 1$ [5.1].

Stern and Gerlach had proved with their pioneering experiment that space quantization is a real effect and does not only exist in the brain of theoreticians. O. Stern later on received the Nobel Price 1944 for his contribution to the development of the molecular beam technique and the discovery of the magnetic moment of the proton.

Although their experiment was a very ingenious demonstration of space quantization, it turned out that the interpretation of their results was not correct, because the splitting was not due to the orbital angular momentum, but to a new quantity called the electron spin, which was postulated as a new characteristic property of the electron after further convincing experimental discoveries.

### 5.5.2 Experimental Confirmation of Electron Spin

When spectroscopic measurements showed that the ground state of silver atoms is in fact an $s$ state with $l = 0$, Bohr's explanation of the results of the Stern–Gerlach experiment could no longer be regarded as correct.

*Samuel A. Goudsmit* (1902–1978) and *George E. Uhlenbeck* (1900–1988) proposed a new model where the electron possesses an intrinsic angular momentum, called the electron spin in addition to a possible orbital angular momentum. This model attributes a new property to the electron, which is then characterized by its mass $m$, its charge $q = -e$, its orbital angular momentum $\boldsymbol{l}$ and its spin $\boldsymbol{s}$ which is connected with a magnetic moment $\boldsymbol{\mu}_s$ that can interact with magnetic fields. Many further experimental findings such as the fine structure in atomic spectra and deviations from the normal Zeeman splittings (called the anomalous Zeeman effect) corroborated this hypothesis.

This spin can be mathematically treated like an angular momentum and is therefore often regarded as a mechanical "eigen-angular momentum" of the electron. The astronomical analogy is the earth revolving around the sun in one year but simultaneously turning around its own axis within one day. The total angular momentum of the earth is the orbital angular momentum plus the "spin" of the earth. Similarly, the electron of the hydrogen atom has for $l > 0$ an orbital angular momentum $\boldsymbol{l} = \boldsymbol{r} \times \boldsymbol{p}$ and in addition its spin $\boldsymbol{s}$. However, as will be discussed in Sect. 5.8 this interpretation of the electron spin as a mechanical angular momentum runs into serious difficulties. Nevertheless the spin can be treated as a vector obeying the same mathematical rules as the orbital angular momentum.

In analogy to the orbital angular momentum the absolute value of the spin $\boldsymbol{s}$ is written as

$$\boxed{|\boldsymbol{s}| = \sqrt{s(s+1)}\hbar} \ , \tag{5.52}$$

where $s$ is the spin quantum number, which had already been introduced by Pauli as an additional quantum number in order to explain the different components in the fine structure of observed spectra (see below). Pauli, however, regarded this as a pure mathematical quantity and resisted giving it a physical meaning as a real angular momentum.

The magnetic spin moment $\boldsymbol{\mu}_s$ is related to the spin by

$$\boldsymbol{\mu}_s = \gamma_s \boldsymbol{s} \ , \tag{5.53}$$

where $\gamma$ is the gyromagnetic ratio of magnetic moment to spin.

From the experimental result of the Stern–Gerlach experiment, that the beam of silver atoms was split in the inhomogeneous magnetic field into two components, it can be concluded that the angular momentum should have two possible orientations. Since spectroscopic investigations of the silver atom proved that the orbital angular momentum in the ground state is zero, the splitting must be due to electron spin. The magnetic spin moment $\mu_s$ must have two orientations and therefore the electron spin must also have two components $s_z = m_s h$. If the electron spin should be treated as an angular momentum then the quantum number $m_s$ must obey the relation $-s \le m_s \le +s$. Since $m_s$ can only change by an integer value, the explanation of two

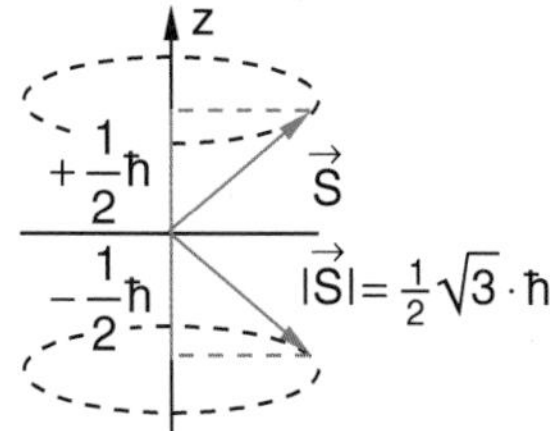


**Fig. 5.16.** Space quantization of electron spin

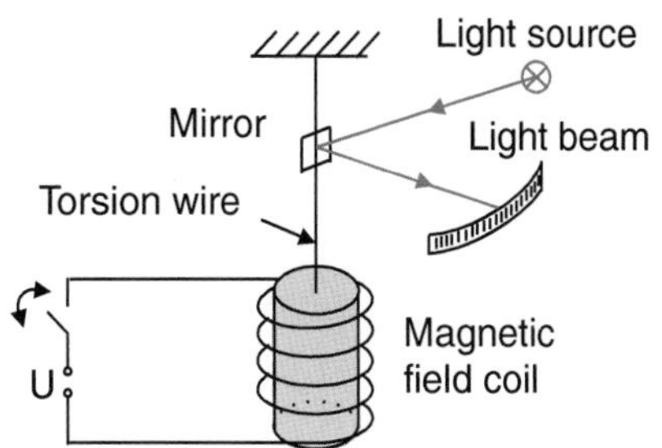


**Fig. 5.17.** Einstein–de Haas experiment

components in the Stern–Gerlach experiment gives the condition (Fig. 5.16):

$$s = 1/2 \Rightarrow m_s = \pm 1/2 \,. \tag{5.54}$$

*Note*. that for a one-electron atom the quantum number s is always $+1/2$, but $m_s$ can be $\pm 1/2$.

The absolute value of the electron spin is then

$$|s| = \sqrt{s(s+1)}\hbar = \frac{1}{2}\sqrt{3}\hbar \,. \tag{5.54a}$$

And the two components in $\pm z$ direction are

$$s_z = \pm\frac{1}{2}\hbar \,. \tag{5.54b}$$

We will at first introduce further experiments leading to the introduction of the electron spin and will then discuss some basic theoretical considerations.

### 5.5.3 Einstein–de Haas Effect

Einstein proposed the following experiment in order to gain insight into the causes of magnetic properties of solids. It was actually performed some years later by the Dutch physicist *Wander Johannes de Haas* (1878–1960).

An iron cylinder with mass $m$ and radius $R$ hangs on a thin wire (Fig. 5.17) in a vertical magnetic field $B = \{0, 0, B_z\}$, produced by an electric current through a coil. The magnetic field was chosen sufficiently strong to saturate the magnetization $\boldsymbol{M} = N\boldsymbol{\mu}$ of the cylinder with $N$ free electrons, each possessing the magnetic moment $\mu$. In cases of saturation, all magnetic moments point in one direction opposite to that of the magnetic field. Since the free electrons in the conduction band of iron have no orbital angular momentum and therefore also no orbital magnetic moment (there is no force center to cause a circular motion of the electrons) the magnetic moment must have another reason. We will see that it is caused by the spins of the free electrons.

**Note:**

In iron, the magnetic moment is not exclusively caused by the free electrons in the conduction band. A minor contribution also comes from the electrons bound in the iron atoms. They can have orbital angular momentum and spin. However, this contribution is small and we will neglect it for the following discussion. (For a more detailed discussion see textbooks on solid state physics).

If the magnetic field is reversed, all magnetic moments flip into the opposite direction. This causes a change in magnetization

$$\Delta \boldsymbol{M} = 2\boldsymbol{M} = 2N\boldsymbol{\mu} \,. \tag{5.55}$$

From the measurement of $\Delta \boldsymbol{M}$ the product $N\boldsymbol{\mu}$ can be determined.

The reversal of the magnetic moments also results in a flip of the corresponding angular momenta $\boldsymbol{s}$ of all electrons. The resulting change

$$\Delta \boldsymbol{S} = 2Ns_z = -\Delta \boldsymbol{L} = -I\omega \tag{5.56}$$

has to be compensated by the opposite change $\Delta \boldsymbol{L}$ of the mechanical angular momentum $\vec{L}$ of the cylinder with the moment of inertia $I = (1/2)MR^2$. With a reversal of the magnetic field, the cylinder that was initially at rest acquires the angular momentum $-\boldsymbol{L} = -(1/2)MR^2\omega$ and a rotational energy

$$E_{\text{rot}} = L^2/2I = L^2/(MR^2) \,, \tag{5.57a}$$

which results in a torsion of the suspension wire with a restoring force $F_r = -D_r\varphi$. The maximum torsion angle $\varphi_{\max}$ is reached when the potential energy

$$E_{\text{pot}} = \frac{1}{2} D_{\text{r}} \varphi_{\max}^2 = E_{\text{rot}} = L^2/2I \tag{5.57b}$$

of the twisted suspension wire equals the kinetic rotational energy. The measurement of $\varphi_{\max}$ therefore allows the determination of $L$. The experiment gave the surprising result

$$\Delta L = N\hbar = 2N\hbar/2 \,. \tag{5.57c}$$

This means that the $z$ component of the angular momentum of each electron must have the amount $s_z = \hbar/2$. From the measured magnetization change $\Delta M$ it was, however, clear that the magnetic moment related to this angular momentum must be $\mu = \mu_{\text{B}}$, i.e, the same as for the orbital angular momentum with $l = 1\hbar$.

The gyromagnetic ratio

$$\Delta M/\Delta S = \frac{\Delta \mu_s}{\Delta s_z} = \frac{|\boldsymbol{\mu}_s|}{|\boldsymbol{s}|} = \frac{\mu_B}{{}^1\!/_2\hbar} = \gamma_s \tag{5.58}$$

of magnetic moment $|\boldsymbol{\mu}_s|$ and angular momentum $|\boldsymbol{s}|$ is therefore twice as large as for the orbital angular momentum, where it is

$$|\boldsymbol{\mu}_l|/|\boldsymbol{l}| = \mu_{\text{B}}/\hbar = \gamma_l \,. \tag{5.59}$$

This means that $\gamma_s$ is twice as large as $\gamma_l$!

For the electron spin, the ratio of magnetic moment to mechanical angular momentum is twice as large as for the orbital angular momentum of the electron.

The magnetic spin moment is written analogously to the orbital moment $\boldsymbol{\mu}_l = -(\mu_{\text{B}}/\hbar)\boldsymbol{l}$ as

$$\boldsymbol{\mu}_s = -g_s(\mu_{\text{B}}/\hbar)\boldsymbol{s} \,. \tag{5.60a}$$

The factor $g_s \approx 2$ is called the ***Landë factor***.

The absolute value of the magnetic spin moment is

$$|\boldsymbol{\mu}_s| = g_s \mu_{\text{B}} \sqrt{s(s+1)} \,. \tag{5.60b}$$

**Remark**

Here the electron spin has been introduced phenomenologically. The exact value of the Landë factor $g_s = 2.0023$ can only be explained by a theory that already includes the electron spin in the basic equations. Such a theory has been developed by *Paul Dirac* (1902–1984) who replaced the Schrödinger equation with the Dirac equation. Its representation exceeds, however, the level of this textbook.

### 5.5.4 Spin-Orbit Coupling and Fine Structure

We will now discuss why the energy levels of the H atom with $l > 0$, split into two components, which could not be explained by the Schrödinger theory. Since this splitting is very small and can be only resolved with high resolution spectrographs, where the hydrogen lines appear as a fine substructure, it was named fine structure.

We start with a semiclassical model, treating the angular momenta as vectors with quantized absolute values and quantized $z$ components. In Sect. 5.2 it was shown that an electron with charge $-e$, moving with the orbital angular momentum $l$ on a circle around the nucleus, produces a magnetic moment

$$\boldsymbol{\mu}_l = -\frac{e}{2m_{\text{e}}}\boldsymbol{l} = -(\mu_{\text{B}}/\hbar) \cdot \boldsymbol{l} \,.$$

That is proportional to $l$.

In a coordinate system where the electron rests at the origin, the nucleus with positive charge $Z \cdot e$ moves with the frequency $\nu$ on a circle around the electron. This causes a circular current $Ze\nu$ that produces a magnetic field $B$ at the location of the electron (Fig. 5.18). According to Biot–Savart's law (see textbooks on magnetic fields) this magnetic field is

$$\begin{aligned} \boldsymbol{B}_l &= \frac{\mu_0 Ze}{4\pi r^3}(\boldsymbol{v} \times (-\boldsymbol{r})) = -\frac{\mu_0 Ze}{4\pi r^3}(\boldsymbol{v} \times \boldsymbol{r}) \\ &= +\frac{\mu_0 Ze}{4\pi r^3 m_{\text{e}}}\boldsymbol{l} \end{aligned} \tag{5.61}$$

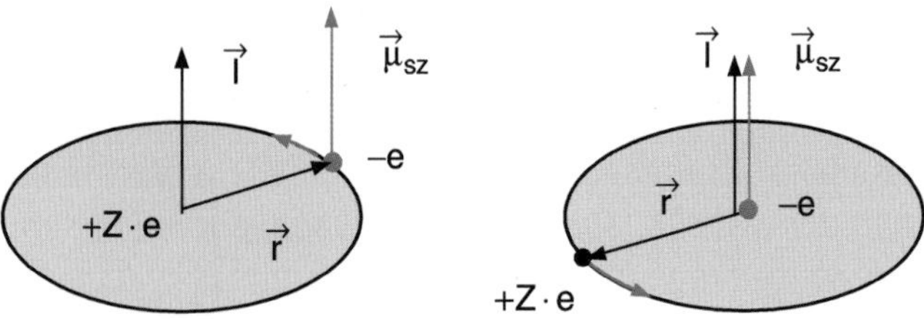


**Fig. 5.18a,b.** Vector model of spin-orbit interaction. (**a**) vector model with the electron circling around the at. (**b**) transformation to a coordinate system, where the electron rests at $r = 0$

because the angular momentum of the electron in a coordinate system where the electron moves around the proton at rest is $\vec{l} = m_e(\vec{v} \times \vec{r})$ and the transformation to the rest frame of the electron changes the sign of the vector product.

The magnetic spin moment of the electron has two spatial orientations in this field according to the two spin directions $s_z = \pm\hbar/2$. This causes an additional energy (in addition to the Coulomb energy)

$$\Delta E = -\boldsymbol{\mu}_s \cdot \boldsymbol{B}_l = g_s \mu_B \frac{\mu_0 Z e}{4\pi r^3 m_e \hbar}(\boldsymbol{s} \cdot \boldsymbol{l})$$

$$\approx \frac{\mu_0 Z e^2}{4\pi m_e^2 r^3}(\boldsymbol{s} \cdot \boldsymbol{l}) . \qquad (5.62)$$

because $g_s \approx 2$ and $\mu_B = e\hbar/2m_e$. In this coordinate system the electron is not in an inertial frame of reference. Transforming the coordinate system back to the rest-frame of the nucleus by a Lorentz transformation gives a factor 1/2 (Thomas factor [5.2]), which is due to the fact that the electron spin in the rest-frame of the nucleus precesses when moving around the nucleus (Thomas precession).

The energy levels $E_n$ of (5.18), which had been obtained without taking into account the electron spin, now split, due to the spin-orbit coupling, into the fine structure components with energies

$$E_{n,l,s} = E_n - \boldsymbol{\mu}_s \cdot \boldsymbol{B}_l = E_n + \frac{\mu_0 Z e^2}{8\pi m_e^2 r^3}(\boldsymbol{s} \cdot \boldsymbol{l}) . \qquad (5.63)$$

The scalar product $(\boldsymbol{s} \cdot \boldsymbol{l})$ may be positive or negative depending on the orientation of the spin relative to the orbital angular momentum.

When we introduce the total angular momentum

$$\boldsymbol{j} = \boldsymbol{l} + \boldsymbol{s} \quad \text{with} \quad |\boldsymbol{j}| = \sqrt{j(j+1)}\hbar \qquad (5.64a)$$

as a vector sum of orbital angular momentum $\boldsymbol{l}$ and electron spin $\boldsymbol{s}$ (Fig. 5.19), we can square this sum and obtain

$$\boldsymbol{j}^2 = \boldsymbol{l}^2 + \boldsymbol{s}^2 + 2\boldsymbol{l} \cdot \boldsymbol{s} . \qquad (5.64b)$$

This gives for the scalar product of the vectors $\vec{l}$ and $\vec{s}$

$$\boldsymbol{l} \cdot \boldsymbol{s} = \frac{1}{2}\hbar^2[j(j+1) - l(l+1) - s(s+1)] . \qquad (5.65)$$

In operator notation this is written as

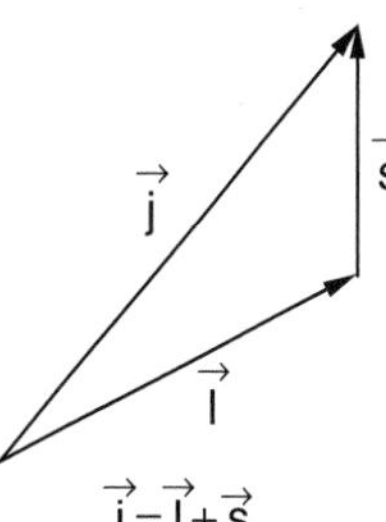


**Fig. 5.19.** Vector coupling of orbital angular momentum $\boldsymbol{l}$ and electron spin $\boldsymbol{s}$ to form the total angular momentum $\boldsymbol{j}$ of the electron

$$\hat{l} \cdot \hat{s}\psi_{n,l,s,j,m_l} = \frac{\hbar^2}{2}[j \cdot (j+1) - l(l+1) - s(s+1)]\psi_{n,l,s,j,m_l} .$$

With this relation we can write (5.63) as

$$E_{n,l,j} = E_n + \frac{a}{2}[j(j+1) - l(l+1) - s(s+1)] . \qquad (5.66a)$$

With the spin-orbit coupling constant

$$\boxed{a = \frac{\mu_0 Z e^2 \hbar^2}{8\pi m_e^2 r^3}} . \qquad (5.66b)$$

For $s = \frac{1}{2}$ the energy levels split, depending on the orientation of the spin, into the two components with $j = l + 1/2$ and $j = l - 1/2$ (see Fig. 5.20).

*Note*: Finestructure splittings are observed only for levels with $l \geq 1$, i.e. for $p, d, f, \ldots$ levels, ***not*** for $s$-levels with $l = 0$.

> The fine structure may be regarded as Zeeman splitting due to the interaction of the magnetic spin moment with the internal magnetic field generated by the orbital motion of the electron.

In the quantum mechanical model the distance $r$ of the electron from the nucleus cannot be given exactly. Only the time-averaged value of $r$ related to the probability of finding the electron at the location $r$ is a measurable quantity

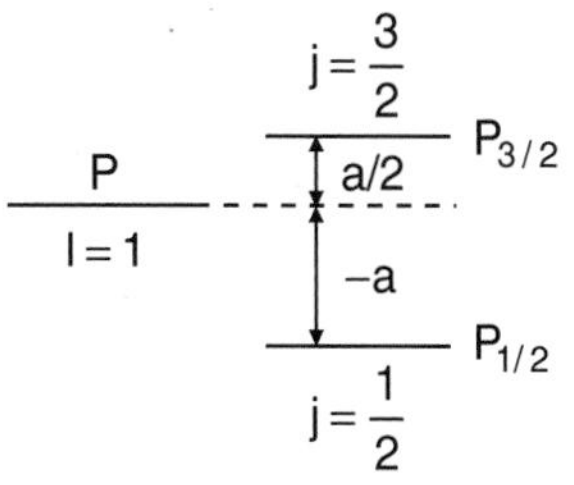

$j = \frac{3}{2}$
P
$l = 1$
a/2
$P_{3/2}$
−a
$P_{1/2}$
$j = \frac{1}{2}$

**Fig. 5.20.** Energy level scheme of fine structure splitting of a $^2P(l = 1)$ state

$$\langle r \rangle = \int \psi^*_{n,l,m} r \psi_{n,l,m} \, \mathrm{d}\tau \Rightarrow$$
$$\left\langle \frac{1}{r^3} \right\rangle = \int \psi_{n,l,m} \frac{1}{r^3} \psi_{n,l,m} \, \mathrm{d}\tau \, . \tag{5.67a}$$

The quantum mechanical average $\langle a \rangle$ of the spin-orbit coupling constant is then

$$\langle a \rangle = \frac{\mu_0 Z e^2 \hbar^2}{8\pi m_\mathrm{e}^2} \int \psi^*_{n,l,m} \frac{1}{r^3} \psi_{n,l,m} \, \mathrm{d}\tau \, . \tag{5.67b}$$

Inserting the hydrogen wave functions $\psi_{n,l,m}(r, \vartheta, \varphi)$, the integral can be solved and one obtains

$$\langle a \rangle = -E_n \frac{Z^2 \alpha^2}{nl(l+1/2)(l+1)} \, , \tag{5.68}$$

where the constant

$$\alpha = \frac{\mu_0 c e^2}{4\pi\hbar} = \frac{e^2}{4\pi\varepsilon_0 \hbar c} \approx \frac{1}{137} \tag{5.69}$$

is **Sommerfeld's fine structure constant**, which was already introduced in Sect. 5.4 for the relativistic correction of the level energies.

The energy separation of the two fine structure components $(n, l, j = l + 1/2)$ and $(n, l, j = l - 1/2)$ is then, according to (5.66) and (5.68),

$$\begin{aligned} \Delta E_{n,l,s} &= \langle a \rangle \left( l + \frac{1}{2} \right) = -E_n \frac{Z^2 \alpha^2}{nl(l+1)} \\ &\approx -5.3 \times 10^{-5} E_n \frac{Z^2}{nl(l+1)} \, . \end{aligned} \tag{5.70}$$

This shows that the splitting is very small compared to the energy $E_{n,l}$ of the levels $(n, l)$ and justifies the name "fine structure." *Note*, that $E_n < 0$!

As can be seen from (5.70), the fine structure splitting decreases with increasing quantum numbers $n$ and $l$, but it increases proportionally to the product $Z^2 E_n$. Since the energies $E_n$ of the levels with principal quantum number $n$ follow the relation $E_n \propto Z^2/n^2$, we can write the fine structure splitting as

$$\Delta E_{n,l,s} \propto \frac{Z^4}{n^3 l(l+1)} \, . \tag{5.71}$$

## EXAMPLE

For the $2p$ level of the H atom, we have $Z = 1$, $n = 2$, $l = 1$ and $E_n = -3.4\,\mathrm{eV}$. From (5.70) we therefore obtain for the fine structure splitting $\Delta E_{l,s} = 4.6 \times 10^{-5}\,\mathrm{eV} \Rightarrow \Delta E_{l,s}/hc = \Delta\bar{\nu} = 0.37\,\mathrm{cm}^{-1}$.

If both effects, the relativistic increase of the electron mass and the spin-orbit coupling, are taken into account we have to add (5.42) and (5.66a) and obtain for the energy of a fine structure component $(n, l, j)$ (see Problem 5.11)

$$E_{n,j} = E_n \left[ 1 + \frac{Z^2 \alpha^2}{n} \left( \frac{1}{j + 1/2} - \frac{3}{4n} \right) \right] , \tag{5.72}$$

which turns out to be independent of $l$ (Fig. 5.21).

*Note*: Since $E_n < 0$, levels with larger $j$-values have a higher energy (see Fig. 5.22).

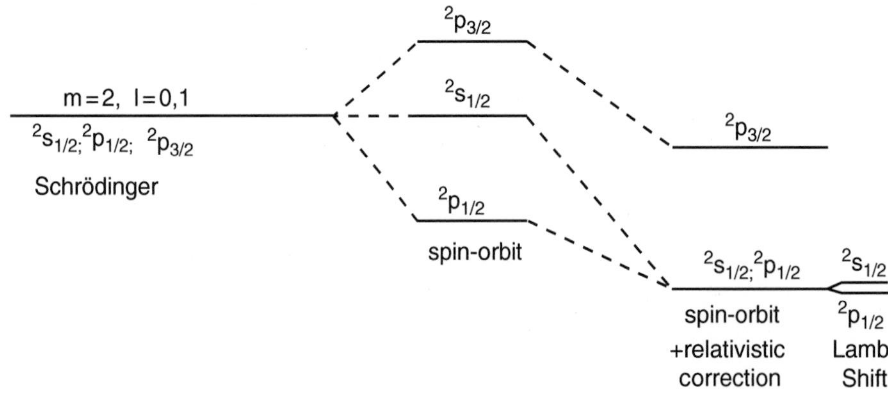


**Fig. 5.21.** Energies of the levels $n = 1, l = 0, 1, s = 1/2$ of the H-atom within the Schrödinger theory and including the different relativistic effects

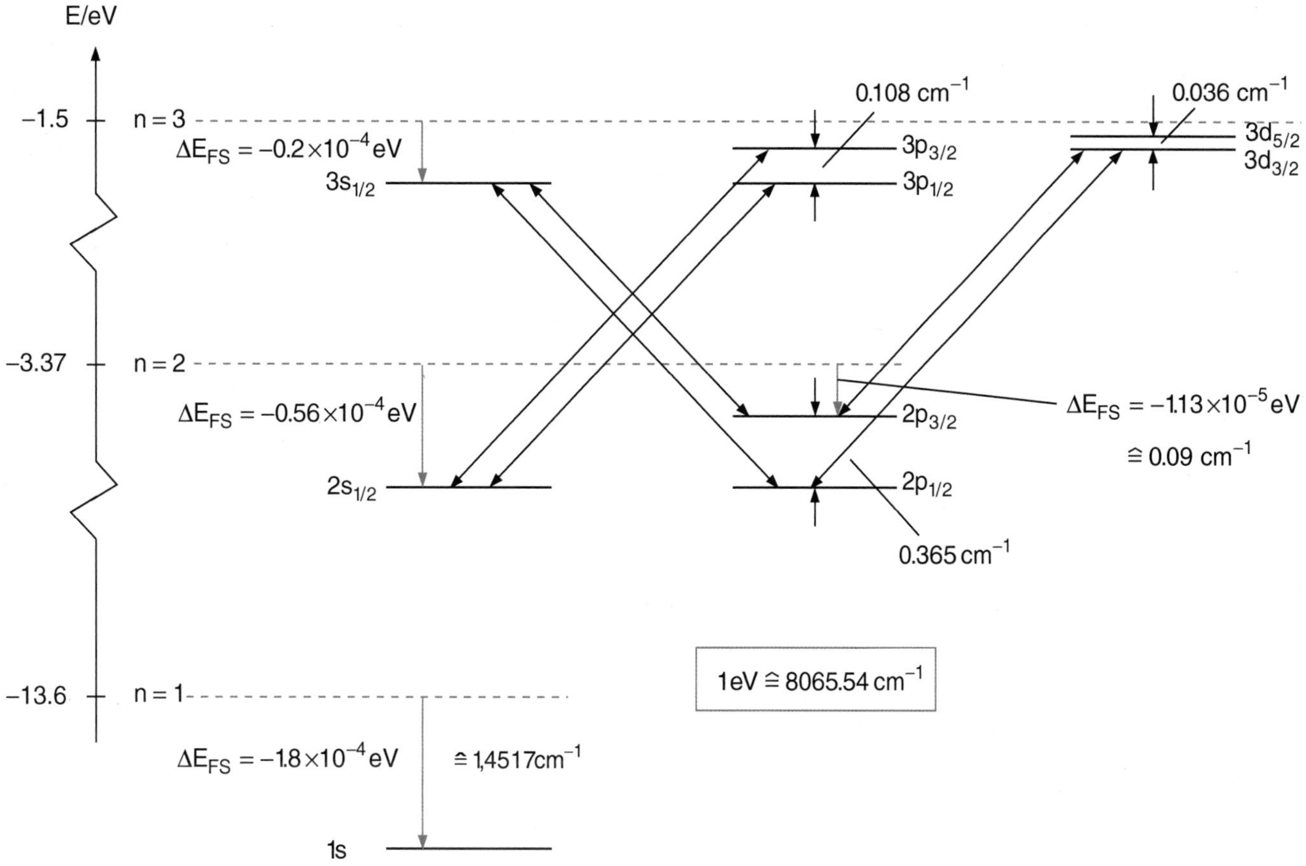


**Fig. 5.22.** Energy level scheme of the hydrogen atom, taking into account the relativistic mass increase and the spin-orbit coupling. The dashed lines mark the energies obtained from the Schrödinger equation

In the Coulomb field with $E_{\text{pot}} \propto 1/r$ the energy of a fine structure component $(n, l, j)$ does not depend on the quantum number $l$. All levels with equal quantum numbers $n$ and $j$ have the same energy (Fig. 5.22).

EXAMPLE

The two levels $2s_{1/2}$ and $2p_{1/2}$ or $3p_{3/2}$ and $3d_{3/2}$ have the same energy (Fig. 5.22).

*Note*: This degeneracy applies only to one-electron systems such as the hydrogen atom or the ions $\text{He}^+$, $\text{Li}^{++}$, etc., where the electron moves in a Coulomb potential. This is because the assumption $E_{\text{pot}} \propto 1/r$ enters into the relativistic mass correction, as well as in (5.67), for the calculation of the fine structure constant. For atoms with more than one electron there is no longer a Coulomb potential because of the mutual interaction between the electrons. Here levels with different values of the quantum number $l$ have different energies even for equal quantum numbers $j$.

### 5.5.5 Anomalous Zeeman Effect

When the electron spin $\boldsymbol{s}$ and the magnetic spin moment $\boldsymbol{\mu}_s$ are taken into account, the total magnetic moment depends on the coupling of the two vectors $\boldsymbol{\mu}_l + \boldsymbol{\mu}_s$. The Zeeman splittings of levels in a magnetic field now become more complicated than those for the normal Zeeman effect shown in Figs. 5.11 and 5.12, which are only observed, if the total spin of the atomic electrons is $\boldsymbol{S} = \sum \boldsymbol{s}_i = \boldsymbol{0}$. This is, for instance, the case for the helium atom with two electrons with antiparallel spins in its ground state (see Sect. 6.1).

*Without a magnetic field*, the total angular momentum $\boldsymbol{j} = \boldsymbol{l} + \boldsymbol{s}$ of the electron in a Coulomb field (central force field) is constant, which means that its

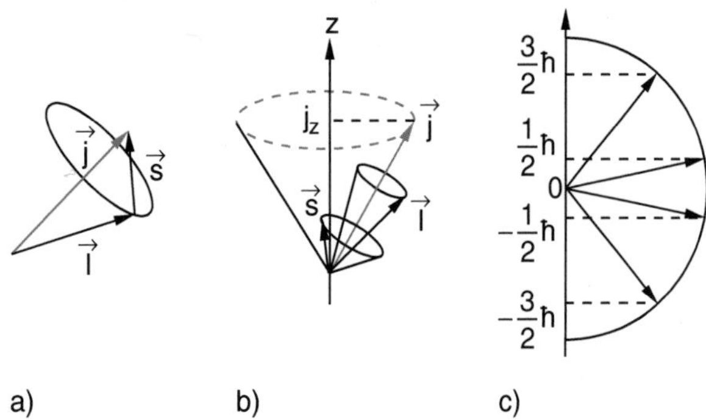


**Fig. 5.23.** (**a**) Coupling of $\boldsymbol{l}$ and $\boldsymbol{s}$ and their precession around the space-fixed vector $\boldsymbol{j} = \boldsymbol{l} + \boldsymbol{s}$ without external field. (**b**) Precession of $\boldsymbol{j}$ in an external magnetic field $B_z$. (**c**) Possible orientations of $\boldsymbol{j}$ with components $j_z = m\hbar$

orientation in space and its absolute value are constant, independent of time (Fig. 5.23a). *In an external homogeneous magnetic field* $\boldsymbol{B} = \{0, 0, B_z\}$, the magnetic moment $\boldsymbol{\mu}_j$, and therefore $\boldsymbol{j}$, precess around the field axis with constant components $\mu_z$ and $j_z$ (Fig. 5.23b).

If the external magnetic field is weaker than the magnetic field generated by the orbital movement of the electron, the Zeeman splitting is smaller than the fine structure splittings. In other words, the coupling between orbital angular momentum and spin is stronger than the coupling of $\boldsymbol{\mu}_l$ and $\boldsymbol{\mu}_s$ to the external field. The spin-orbit coupling is still valid and the absolute value $|\boldsymbol{j}|$ of the total angular momentum

$$\boldsymbol{j} = \boldsymbol{l} + \boldsymbol{s} \quad \text{with} \quad |\boldsymbol{j}| = \sqrt{j(j+1)}\hbar \tag{5.72a}$$

is conserved in the external magnetic field. Its direction is, however, no longer space-fixed because the magnetic moment $\boldsymbol{\mu}_j = \boldsymbol{\mu}_l + \boldsymbol{\mu}_s$, which is related to $\boldsymbol{j}$, experiences a torque

$$\boldsymbol{D} = \boldsymbol{\mu}_j \times \boldsymbol{B} \,. \tag{5.72b}$$

For one-electron systems the component $j_z$ can take the values $j_z = m_j\hbar$ with the half-integer values $-j_z \le m_j \le +j_z$ (Fig. 5.23c).

The magnetic moments of orbital motion and spin of the electron are:

$$\boldsymbol{\mu}_l = -(\mu_B/\hbar)\boldsymbol{l} \quad \text{and} \quad \boldsymbol{\mu}_s = -g_s(\mu_B/\hbar)\boldsymbol{s} \,. \tag{5.72c}$$

There is an important point to mention:

> Because of the Landë factor $g_s \approx 2$ for the spin moment $\boldsymbol{\mu}_s$, the total magnetic moment
>
> $$\boldsymbol{\mu}_j = \boldsymbol{\mu}_l + \boldsymbol{\mu}_s = -(\mu_B/\hbar)(\boldsymbol{l} + g_s\boldsymbol{s}) \tag{5.73}$$
>
> is for $l > 0$ no longer parallel to the total angular momentum $\boldsymbol{j} = \boldsymbol{l} + \boldsymbol{s}$!

*Without an external field* the absolute value and the direction of $\boldsymbol{j}$ is constant in time. Since the vector $\boldsymbol{s}$ precesses around the axis of the internal magnetic field produced by the orbital movement of the electron and $\boldsymbol{\mu}_j$ is not parallel to $\boldsymbol{j}$, $\boldsymbol{\mu}_j$ has to precess around the direction of the space-fixed vector $\boldsymbol{j}$. The time average $\langle\boldsymbol{\mu}_j\rangle$ of $\boldsymbol{\mu}_j$ is then the projection of $\boldsymbol{\mu}_j$ onto $\boldsymbol{j}$ (Fig. 5.24a). This gives

$$\langle\boldsymbol{\mu}_j\rangle = \frac{\boldsymbol{\mu}_j \cdot \boldsymbol{j}}{|\boldsymbol{j}|^2}\vec{j} = -\frac{e}{2m_e}\left(\frac{\boldsymbol{l} \cdot \boldsymbol{j}}{|\boldsymbol{j}|^2} + g_s\frac{\boldsymbol{s} \cdot \boldsymbol{j}}{|\boldsymbol{j}|^2}\right)\vec{j} \,. \tag{5.74}$$

From $\boldsymbol{j} = \boldsymbol{l} + \boldsymbol{s}$ follows

$$\begin{aligned}\boldsymbol{l} \cdot \boldsymbol{j} &= \frac{1}{2}[\boldsymbol{j}^2 + \boldsymbol{l}^2 - \boldsymbol{s}^2] \\ &= \frac{1}{2}[j(j+1) + l(l+1) - s(s+1)]\hbar^2\end{aligned} \tag{5.74a}$$

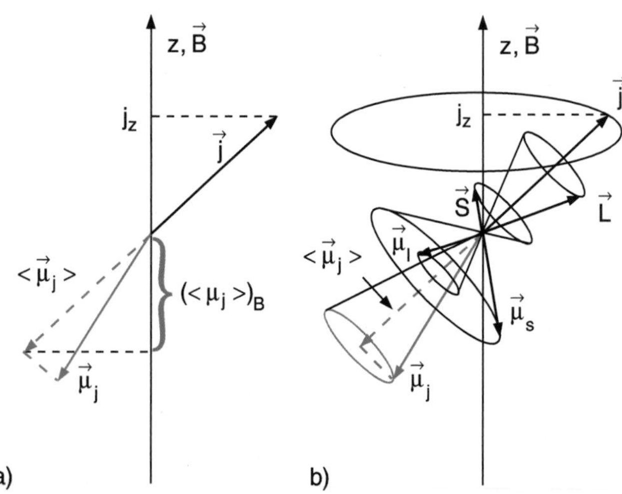


**Fig. 5.24.** (**a**) Projection of $\boldsymbol{\mu}_j$ onto the direction of $\boldsymbol{j}$. (**b**) Precession of the angular momentum $\boldsymbol{j}$ and the average magnetic moment $\langle\boldsymbol{\mu}_j\rangle$ around the $z$-axis in an external magnetic field $B = \{0, 0, B_z\}$

and similar from $\boldsymbol{l} = \boldsymbol{j} - \boldsymbol{s}$:

$$\boldsymbol{s} \cdot \boldsymbol{j} = \frac{1}{2}[j(j+1) + s(s+1) - l(l+1)]\hbar^2 \,.$$

We can therefore write (5.74) with $g_s \approx 2$ as

$$\begin{aligned} \langle \mu_j \rangle &= -\frac{3j(j+1) + s(s+1) - l(l+1)}{2j(j+1)} \mu_B \\ &= -g_j \, \mu_B \,. \end{aligned} \tag{5.75}$$

The Landë factor $g_j$ is defined here as

$$\boxed{g_j = 1 + \frac{j(j+1) + s(s+1) - l(l+1)}{2j(j+1)} \,.} \tag{5.76}$$

For $s = 0$ (pure orbital magnetism) it follows $j = l$ and we obtain $g_j = 1$. For $l = 0$ (pure spin magnetism) is $j = s$ and therefore $g_j \approx 2$. If orbital angular momentum and spin both contribute to the magnetic moment, the value of the Lande factor $g$ can be calculated from eq. (5.76).

In an external magnetic field $\boldsymbol{B} = \{0, 0, B_z\}$, the spatial orientation of the total angular momentum is no longer constant. The vector $\boldsymbol{j}$ precesses around the field direction. The projection of $j$ can take the values

$$j_z = m_j h \quad \text{with} \quad -j \le m_j \le +j \,.$$

The precession of $\boldsymbol{\mu}_j$ around $\boldsymbol{j}$ is faster than that of $\boldsymbol{j}$ around $\boldsymbol{B}$ as long as the finestructure splitting is larger than the Zeeman splitting. Therefore the $z$ component $\langle \mu_j \rangle_z$ of the average magnetic moment $\langle \mu_j \rangle$ is

$$\langle \mu_j \rangle_z = -m_j g_j \mu_B \tag{5.76a}$$

and the additional energy of the Zeeman component $m_j$ is

$$\Delta E_{m_j} = -\langle \mu_j \rangle_z B = m_j g_j \mu_B B \,. \tag{5.77a}$$

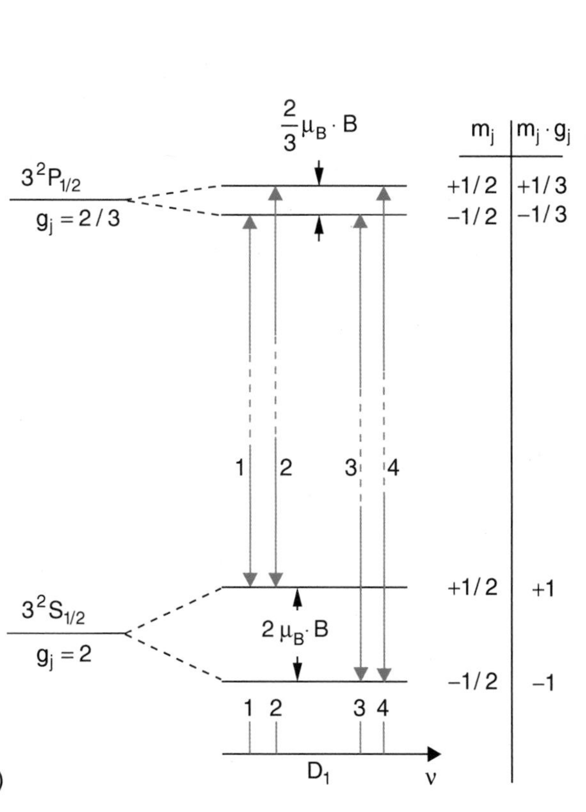


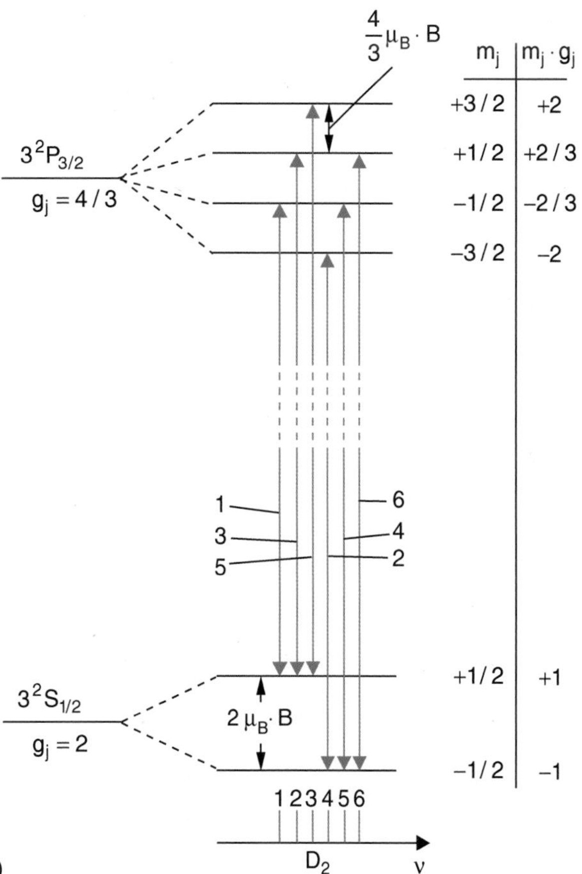


**Fig. 5.25a,b.** Anomalous Zeeman effect of the transitions (**a**) $^2P_{1/2} \leftarrow {}^2S_{1/2}$ and (**b**) $^2P_{3/2} \leftarrow {}^2S_{1/2}$ neglecting hyperfine structure

The energy of a specific Zeeman component $m_j$ of a fine structure level $|n,l,j>$ is them obtained from (5.76) and (5.77) as

$$E_{n,j,m_j} = E_n \left[1 + \frac{Z\alpha^2}{n}\left(\frac{1}{j+1/2} - \frac{3}{4n}\right)\right] + m_j g_j \mu_B B \tag{5.77b}$$

The energy separation between two adjacent Zeeman components comes out as

$$\Delta E_{m_j} - \Delta E_{m_{j-1}} = g_j \mu_B B \,. \tag{5.78}$$

Since the Landë factor $g_j$ depends on the quantum numbers $l$ and $j$,according to (5.76), the Zeeman splitting for the anomalous Zeeman effect differs for the different levels $(n,l,j)$, contrary to the situation for the normal Zeeman effect. Therefore, the Zeeman pattern of spectral lines is more complicated here. There are generally more than three Zeeman components.

The following examples shall illustrate the situation for the anomalous Zeeman effect. In Fig. 5.25 the Zeeman pattern of the two D-lines in the sodium spectrum are shown, corresponding to the transitions Na $3^2S_{1/2} \to 3^2P_{1/2}$ ($D_1$ line) and Na $3^2S_{1/2} \to 3^2P_{3/2}$ ($D_2$ line). For the H atom a completely similar pattern is obtained. Only the spin-orbit coupling constant $a$ is smaller and therefore the fine structure splitting smaller. The Lande factors of the different levels are

$$g_j\left({}^2S_{1/2}\right) = 2\,, \quad g_j\left({}^2P_{1/2}\right) = 2/3\,,$$
$$g_j\left({}^2P_{3/2}\right) = 4/3\,.$$

The spectrum shows four Zeeman components for the transition $3^2S_{1/2} \to 3^2P_{1/2}$ and six components for the $3^2S_{1/2} \to 3^2P_{3/2}$, which are not equidistant.

As for the normal Zeeman effect, transitions with $\Delta m_j = \pm 1$ are circularly polarized and those with $\Delta m_j = 0$ are linearly polarized with the electric field vector $E$ in the direction of the external magnetic field.

## 5.6 Hyperfine Structure

In the previous sections we have described the atomic nucleus as a point-like charge $Ze$ that interacts with the electron merely through the electric Coulomb-potential

$$\phi(r) = -\frac{Ze}{4\pi\varepsilon_0 r}\,.$$

With this potential the Schrödinger equation allowed the calculation of the term values of all levels in the H atom and the wavenumbers of all transitions between these levels. The fine structure of the spectral lines was explained by the magnetic interaction between the magnetic moments of the orbital angular momentum and the electron spin. This magnetic interaction was just added to the Coulomb interaction. It cannot be calculated from the Schrödinger equation, which does not include the electron spin.

### 5.6.1 Basic Considerations

If the hydrogen spectrum is observed with very high spectral resolution, one finds that even the fine structure components are split into two subcomponents. The separation of these sub-components is, for the H atom, smaller than the Doppler width of the spectral lines and therefore these components cannot be recognized with Doppler-limited resolution. This very small splitting, which for many atoms can only be resolved with special Doppler-free spectroscopic techniques (see Chap. 12), is called ***hyperfine structure***. It is explained as follows.

Atomic nuclei have a small but finite volume and possess, besides their electric charge $Ze$ another characteristic quantity $\vec{I}$, called the nuclear spin analogue to the electron spin $\vec{s}$. Its absolute value

$$|\boldsymbol{I}| = \sqrt{I(I+1)}\hbar \tag{5.79}$$

is described by the nuclear spin quantum number $I$. The projection of $\boldsymbol{I}$ onto the $z$-axis is

$$I_z = m_I \hbar \quad \text{with} \quad -I \le m_I \le +I\,. \tag{5.80}$$

The nuclear spin $\vec{I}$ can be treated as an angular momentum in complete analogy to the electron spin.

A magnetic moment is connected with the nuclear spin

$$\mu_N = \gamma_K I\,. \tag{5.81}$$

The unit of the nuclear magnetic moment is the ***nuclear magneton***

$$\mu_K = \frac{e}{2m_p}\hbar = \frac{m_e}{m_p}\mu_B = \frac{\mu_B}{1836} = 5.05 \times 10^{-27}\,\mathrm{J\,T^{-1}} \tag{5.82}$$

analogue to the Bohr magneton $\mu_B$. However, the nuclear magneton is smaller by a factor $m_e/m_p \approx 1/1836$. The magnetic moment of the proton is $\mu_I(\mathrm{p}) = 2.79\mu_K$ and is determined by the movements and charges of the three quarks (u,u,d) inside the proton.

The magnetic moment of any atomic nucleus can be written in units of the nuclear magneton as

$$\mu_N = \gamma_K I = g_N \frac{\mu_K}{\hbar} I\,, \tag{5.83}$$

where the dimensionless factor $g_N = \gamma_K \hbar/\mu_K$ is called the ***nuclear g-factor***.

The nuclear magnetic moment gives two contributions to the shift and splitting of energy levels of the atomic electrons:

a) The interaction of the nuclear magnetic moment $\mu_N$ with the magnetic field produced by the electrons at the nucleus (Zeeman effect of $\mu_N$ in the internal magnetic field produced by the electrons).
b) The interaction of the electronic magnetic moment $\boldsymbol{\mu}_j$ with the nuclear moment $\boldsymbol{\mu}_N$ (magnetic dipole-dipole interaction).

The potential energy of the nuclear magnetic moment $\boldsymbol{\mu}_N$ in the magnetic field produced by the electron at the location of the nucleus is (Fig. 5.26)

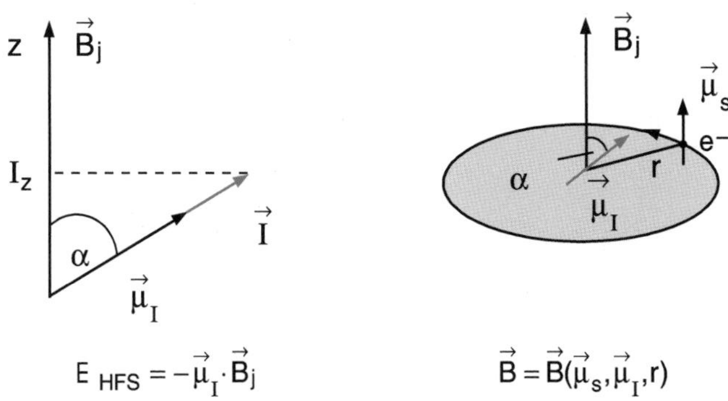


**Fig. 5.26.** Interaction between nuclear magnetic moment $\boldsymbol{\mu}_I$ and the magnetic field $\boldsymbol{B}_j$ produced at the location of the nucleus by the orbital motion of the electron and the magnetic moment $\boldsymbol{\mu}_s$ due to the electron spin $\boldsymbol{s}$

$$E_{\mathrm{pot}}(I, j) = -\boldsymbol{\mu}_N \cdot \boldsymbol{B}_{\mathrm{int}} = -|\boldsymbol{\mu}_N| B_j \cos(\angle \boldsymbol{j}, \boldsymbol{I})\,. \tag{5.84}$$

Introducing the total angular momentum $\boldsymbol{F} = \boldsymbol{j} + \boldsymbol{I}$ of the atom as the vector sum of the total electronic angular momentum $\boldsymbol{j} = \boldsymbol{l} + \boldsymbol{s}$ and the nuclear spin $I$ (Fig. 5.27), we obtain, because of $\boldsymbol{j} \cdot \boldsymbol{I} = 1/2(\boldsymbol{F}^2 - \boldsymbol{j}^2 - \boldsymbol{I}^2) = |\boldsymbol{j}||\boldsymbol{I}| \cos(\angle \boldsymbol{j}, \boldsymbol{I})$,

$$\cos \angle(\boldsymbol{j}, \boldsymbol{I}) = \frac{\boldsymbol{j} \cdot \boldsymbol{I}}{|\boldsymbol{j}||\boldsymbol{I}|} = \frac{1}{2} \frac{F(F+1) - j(j+1) - I(I+1)}{\sqrt{j(j+1)I(I+1)}}\,. \tag{5.84a}$$

The hyperfine energy of the H atom is then

$$\Delta E_{\mathrm{HFS}} = \frac{A}{2}[F(F+1) - j(j+1) - I(I+1)]\,, \tag{5.85}$$

where the hyperfine constant

$$A = \frac{g_N \mu_K B_j}{\sqrt{j(j+1)}} \tag{5.86}$$

depends on the internal magnetic field produced by the electron, and is therefore dependent on the electronic angular momentum $j$.

Each energy level $E_{n,l,j}$ splits into hyperfine components, due to the interaction between nuclear magnetic moment and electronic magnetic moments. The energy of these components is

$$E_{\mathrm{HFS}} = E_{n,l,j} + \frac{1}{2}A[F(F+1) - j(j+1) - I(I+1)]\,. \tag{5.87}$$

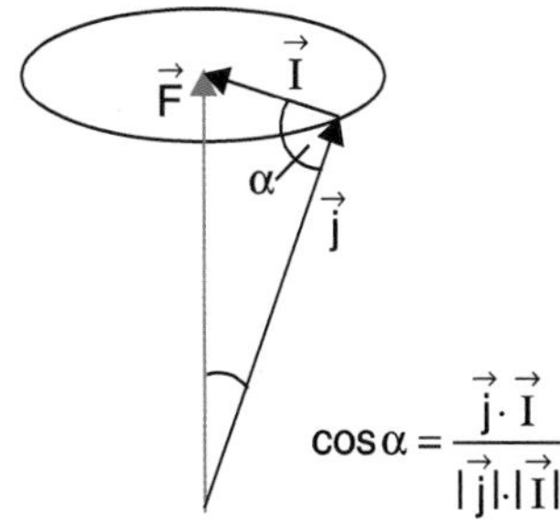


**Fig. 5.27.** Coupling of total electronic angular momentum $\boldsymbol{j} = \boldsymbol{l} + \boldsymbol{s}$ and nuclear spin $\boldsymbol{I}$ to the total atomic angular momentum $\boldsymbol{F} = \boldsymbol{j} + \boldsymbol{I}$

For the H atom, with a proton as the nucleus, the experiments give the values

$$I = 1/2, \quad g_I = +5.58 \Rightarrow (\mu_N)_z = \pm 2.79 \mu_K .$$

For the ground state $^2S_{1/2}$ is $j = 1/2, I = 1/2 \Rightarrow F = 0$ or $F = 1$. This gives the two hyperfine components (Fig. 5.28).

$$E_{HFS}(F=0) = E_{1,0,1/2} - \frac{3}{4}A$$
$$E_{HFS}(F=1) = E_{1,0,1/2} + \frac{1}{4}A , \tag{5.88}$$

with the separation $\Delta E = A = 5.8 \cdot 10^{-6}$ev.

$\Rightarrow \Delta\bar{\nu} = \Delta E/hc = 0.047\,\text{cm}^{-1} \Rightarrow \Delta\nu = 1.42\,\text{GHz}, \ \Delta\lambda = 21\,\text{cm}$.

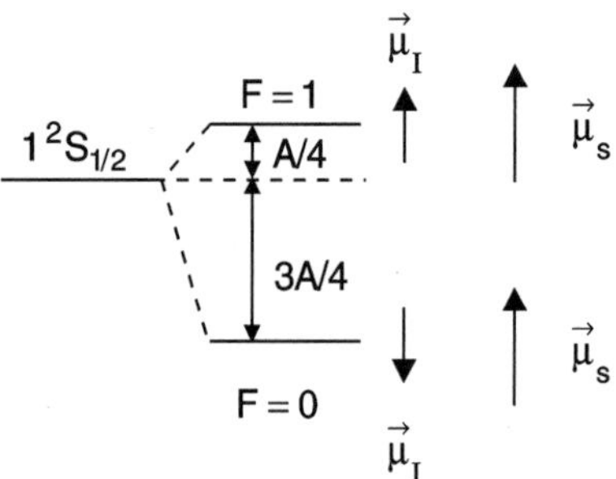


**Fig. 5.28.** Hyperfine structure of the $1^2S_{1/2}$ state of the H atom. The hyperfine coupling constant is $A = 0.047\text{cm}^{-1}$

absorption $1S \to 2S$ (Fig. 5.29), where two photons are simultaneously absorbed out of two antiparallel laser beams. The splitting of the two lines in Fig. 5.29 reflects the difference $\Delta\nu = (\Delta E(1S) - \Delta E(2S))/h$ of the hyperfine splittings of the lower and the upper state of the transition with $\Delta F = F(1S) - F(2S) = 0$. The splitting $\Delta E(2S)$ is small compared to that of the ground state.

The total angular momentum $F$ has to be conserved for the two photon transition because the two absorbed photons have opposite spins. They therefore do not transfer angular momentum to the atom.

The hyperfine splitting of the ground state $1S$ can be directly measured by a magnetic dipole transition (see Sect. 7.2.6) between the two HFS components. This transition lies in the microwave range with a wavelength of $\lambda = 21$ cm. It plays an important role in radio astronomy, because H atoms are the most abundant species in the universe and H atoms in interstellar clouds can be excited by star radiation into the upper HFS level and can emit this transition as radio signals received on earth by large parabolic radio antennas. The measurements of the signal amplitude with spatial resolution gives information about the density distribution, velocities and temperatures of H atoms in the universe.

## 5.6.2 Fermi-contact Interaction

The internal magnetic field at the location $r = 0$ of the nucleus depends on $j$ and on the spatial probability distribution $|\psi_{n,l}|^2$ of the electron. The hyperfine splitting is particularly large for $1S$ states where $\psi$ has a maximum for $r = 0$. The magnetic interaction of the $1s$ electron with the nucleus is called ***Fermi-contact interaction***, because there is close contact between the electron and the nucleus. A more detailed calculation shows that for $S$ states the hyperfine constant is given by

$$A = \frac{2}{3}\mu_0 g_I \mu_B g_N \mu_K |\psi_n(r=0)|^2 . \tag{5.89}$$

This is the dominant contribution for the HFS of the H atom. The absolute value of the hyperfine splitting in the $^2S_{1/2}$ ground state of the H atom is $\Delta\bar{\nu} = 0.0474\,\text{cm}^{-1}$. In the optical spectral region it can only be resolved with special Doppler-free techniques. One example is the Doppler-free two-photon

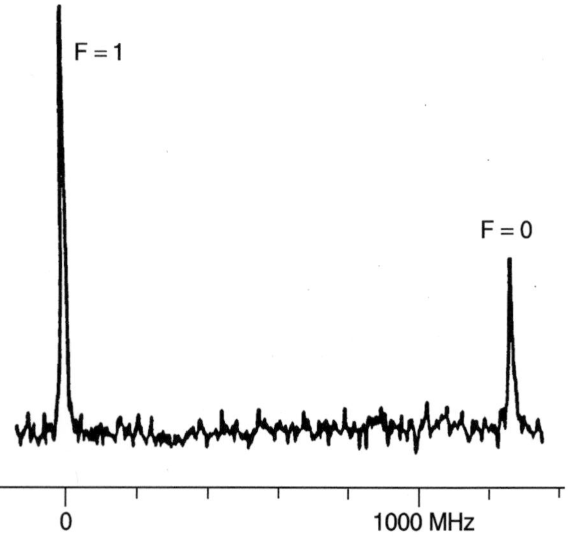


**Fig. 5.29.** The two resolved hyperfine components $F = 0, 1$ of the two-photon transition $2S \leftarrow 1S$ in the H atom (with kind permission of Th.W. Hänsch from G.F. Bassani, M. Inguscio, T.W. Hänsch (eds) The Hydrogen Atom (Springer, Berlin Heidelberg New York, 1989))

### 5.6.3 Magnetic Dipole-Dipole Interaction

The second contribution to the hyperfine splitting is the dipole-dipole interaction between the magnetic dipoles of the electron and the nucleus. This contribution is zero for $S$ states with a spherically symmetric charge distribution, because the average of the electronic magnetic moment is zero. It therefore plays a role only for states with $l \geq 1$ (Fig. 5.30), where the first contribution is small because the electron density at $r = 0$ is zero (Fig. 5.3).

For larger atoms there are also electrostatic contributions to the hyperfine structure if the nucleus has an electric quadrupole moment. For the H atom, however, this electrostatic contribution is absent because the proton has no electric quadrupole moment.

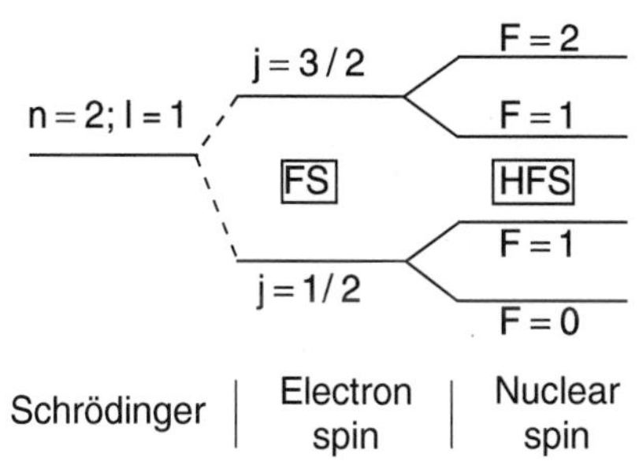


**Fig. 5.30.** Fine and hyperfine splittings of the $2P$ state of the H atom with $n = 2$, $l = 1$, $j = 1/2, 3/2$, $F = 0, 1, 2$ (not to scale)

### 5.6.4 Zeeman Effect of Hyperfine Structure Levels

In a weak external magnetic field $B$ the hyperfine component with $F = 1$ splits into three Zeeman sublevels with $m_F = 0, \pm 1$, while the component with $F = 0$ does not split. This is, however, only observed for weak fields as long as the interaction energy $\Delta E_{\text{HFS}}$ between nuclear magnetic moment and electron moments is larger than the Zeeman coupling energy $\mu_s B$ between the electronic spin moment and the external magnetic field. This gives rise to the anomalous Zeeman effect of the hyperfine levels.

For stronger fields, when $\boldsymbol{\mu}_s \cdot \boldsymbol{B}$ becomes larger than $\Delta E_{\text{HFS}}$ the electron spin $\boldsymbol{s}$ and the nuclear spin $\boldsymbol{I}$ become uncoupled and the energy $E_F$ of the levels is governed by the interaction energy $\boldsymbol{\mu}_s \cdot \boldsymbol{B}$ between electronic magnetic moment and external field. In this case there are only two Zeeman components with $s_z = \pm 1/2$. Each of these components shows a hyperfine splitting into two HFS components with $m_I = \pm 1/2$ (Fig. 5.31).

This uncoupling of angular momentum by the magnetic field is called the ***Paschen–Back effect*** (Fig. 5.32). It appears for hyperfine structure at rather small magnetic fields. It is also observed for the fine structure levels, but only at higher fields, because the

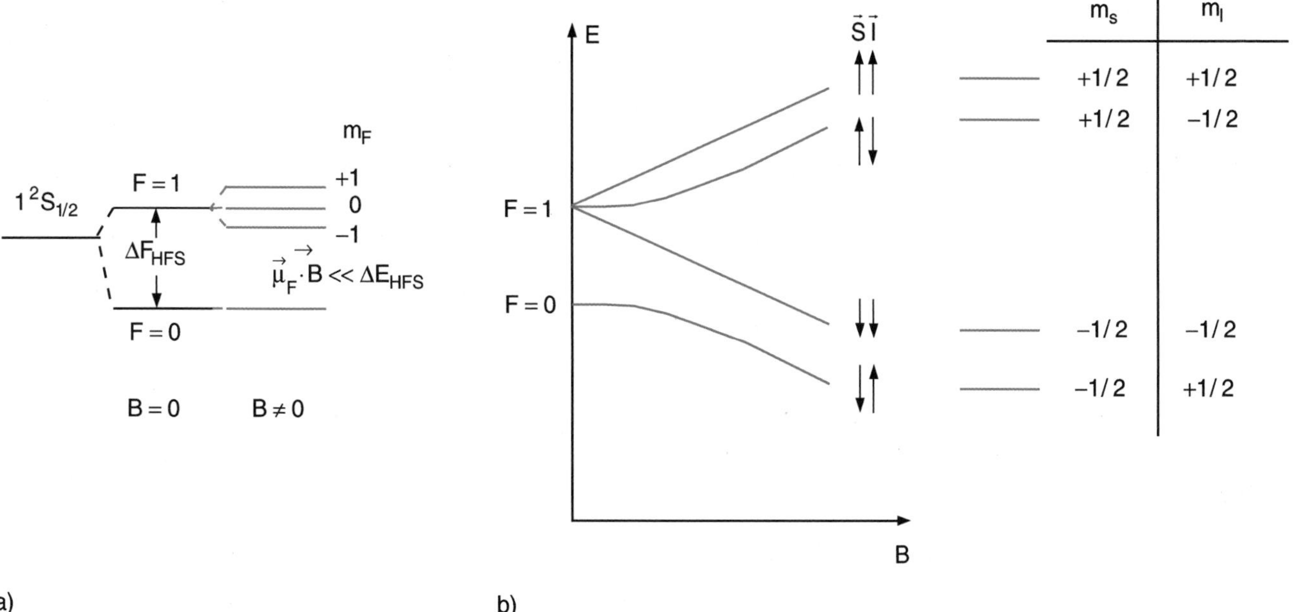


**Fig. 5.31a,b.** Zeeman effect of the $1^2S_{1/2}$ ground state of the hydrogen atom. (**a**) Weak magnetic field. (**b**) Energy dependence $E_{F,I,S}(B)$ of hyperfine components

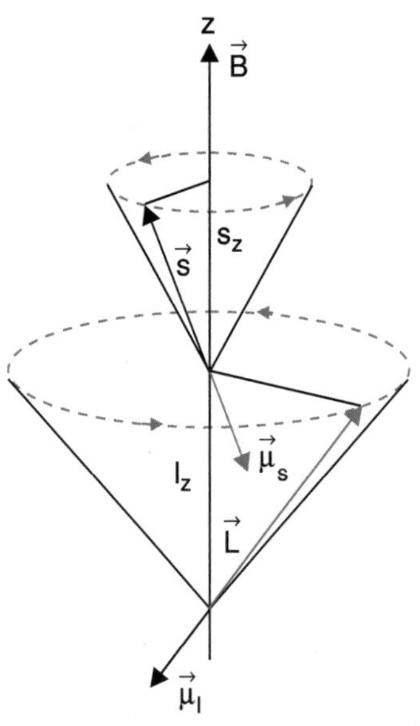


**Fig. 5.32.** Vector model of the Paschen–Back effect

interaction energy between $\boldsymbol{\mu}_s$ and the internal magnetic field produced by $\boldsymbol{l}$ is generally much larger than the Zeeman energy $\boldsymbol{\mu}_j \cdot \boldsymbol{B}_{\text{ext}}$, while the Zeeman energy $\boldsymbol{\mu}_{\text{N}} \cdot B_{\text{ext}}$ can exceed the hyperfine energy $\Delta E_{\text{HFS}}$.

## 5.7 Complete Description of the Hydrogen Atom

The preceding sections have shown that all the effects discussed so far make the spectrum of the simplest atom more complicated than was assumed in Bohr's model of the H atom. In this section we will summarize all phenomena discussed in this chapter and some new ones for a complete description of the hydrogen spectrum.

### 5.7.1 Total Wave Function and Quantum Numbers

The solutions of the Schrödinger equation for the H atom gave (without taking into account the electron spin) $n^2$ different wave functions for each value of the principal quantum number $n$. They represent $n^2$ different atomic states with the same energy (they are $n^2$-fold degenerate), but with different spatial distributions of the electron density. Each of these $n^2$ wave functions

$$\psi_{n,l,m_l}(x,y,z) = R_{n,l}(r) Y_l^m(\vartheta,\varphi) \tag{5.90a}$$

is unambiguously defined by the quantum numbers $n, l,$ and $m_l$.

The introduction of the electron spin with its two possible orientations $s_z = \pm 1/2\hbar$ against the $z$-axis (which is chosen as preferential axis by general agreement and is called the ***quantization axis***) adds a new quantum number $m_s = \pm 1/2$, which defines the projection of the electron spin onto the quantization axis. Each of the spatial electron distributions $\psi_{n,l,m}(x,y,z)$ can be realized with two spin orientations. This is described by multiplying the spatial wave function $\psi_{n,l,m_l}(x,y,z)$ with a spin function $\chi_{m_s}(s_z)$ that defines the projection $s_z = m_s\hbar$ of the electron spin $s$ onto the quantization axis. We label the spin function as $\chi^+$ for $m_s = +1/2$ and as $\chi^-$ for $m_s = -1/2$. The total wave function, including the electron spin, is then

$$\psi_{n,l,m_l,m_s}(x,y,z,s_z) = \psi_{n,l,m_l}(x,y,z)\chi_{m_s} \,. \tag{5.90b}$$

> Each electronic state of a one-electron atom is unambiguously defined by the four quantum numbers $n, l, m_l$ and $m_s$. It is described by a single wave function (5.90).

### 5.7.2 Term Assignment and Level Scheme

For the complete assignment of an atomic state by its quantum numbers $(n, l, m_l, s, m_s)$ the short hand notation

$$n\,^{2s+1}X_j \tag{5.91a}$$

is used. The letter $n$ denotes the principal quantum number. The capital letter $X$ stands for $S$ $(l = 0)$, $P$ $(l = 1)$, $D$ $(l = 2)$, $F$ $(l = 3)$, .... The upper left index $2s + 1$ is the multiplicity, which gives the number of fine structure components for $s > 0$, provided the orbital angular momentum quantum number is $l > 0$. For systems with only one electron outside closed shells is $s = 1/2$ and the multiplicity is $2s + 1 = 2$. Atoms with a single electron always have doublet states, which split into two fine structure components for $l > 0$. The lower right index in (5.91a) gives the quantum number $j$ of the total electronic angular momentum $\boldsymbol{j} = \boldsymbol{l} + \boldsymbol{s}$ (Fig. 5.19).

The hyperfine components are labeled by the quantum number $F$ of the total angular momentum $\boldsymbol{F} = \boldsymbol{j} + \boldsymbol{I}$, including the nuclear spin $\boldsymbol{I}$ (Fig. 5.27).

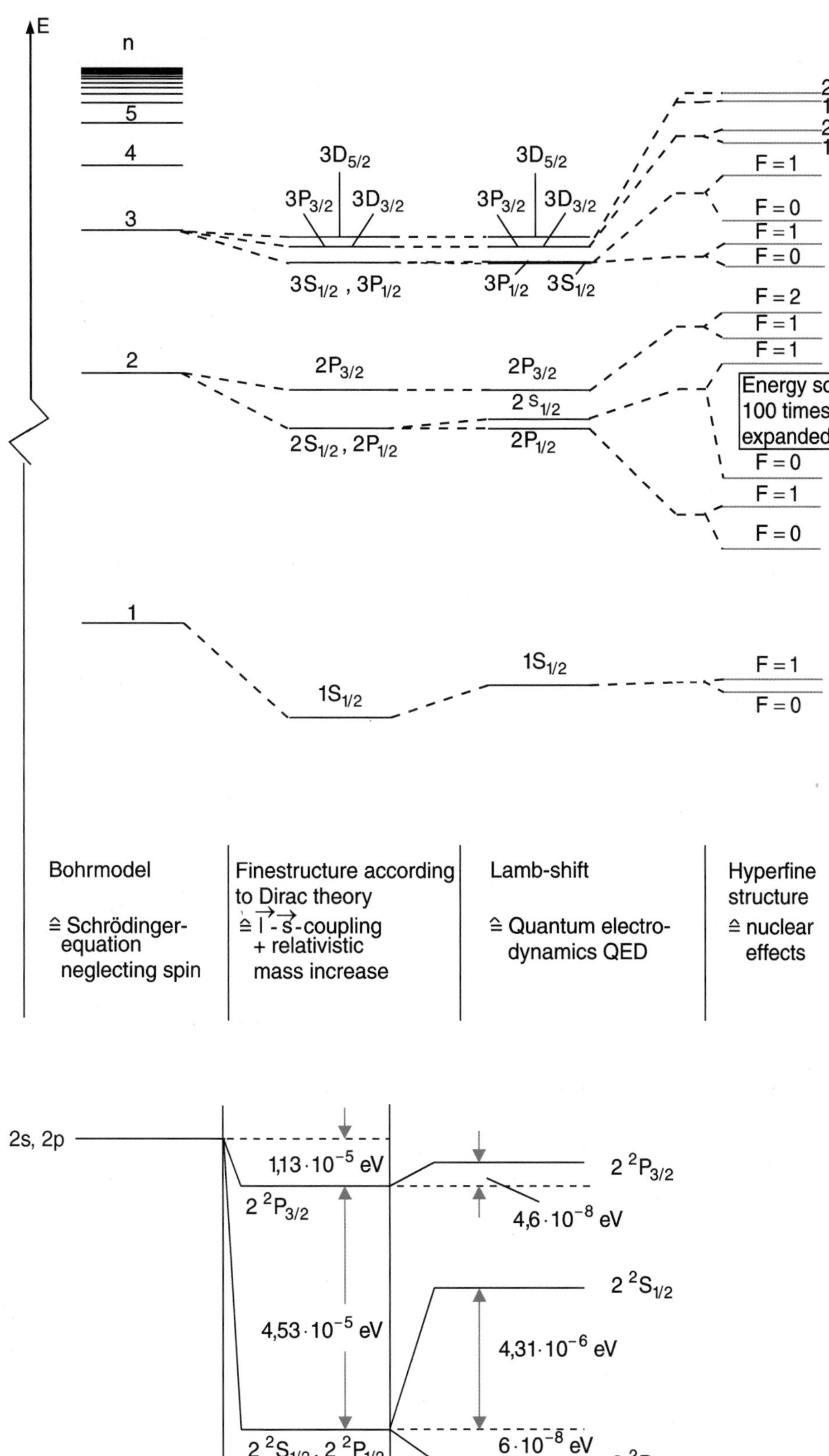


**Fig. 5.33.** Complete level scheme of the H atom including all interactions known so far. *Note*: The fine, HFS structure, and the Lamb shift are not drawn to scale. They are exaggerated in order to illustrate the splittings and shifts

**Fig. 5.34.** Fine structure and Lamb shift of the $n = 2$ level of the H atom. Note the different energy scale compared to Fig. 5.33.

## EXAMPLE

The first excited state $2\,^2P$ of the H atom that can be reached by one-photon excitation from the $1\,^2S_{1/2}$ ground state is defined by the quantum numbers $n = 2$, $s = 1/2$, $l = 1$, and $j = 1/2$ or $j = 3/2$. The two fine structure components are therefore labeled as $2\,^2P_{1/2}$ and $2\,^2P_{3/2}$. Both of them split into two hyperfine levels $2\,^2P_{1/2}$ ($F = 0$ or 1) and $2\,^2P_{3/2}$ ($F = 1$ or 2).

---

Without nuclear spin interaction and without Lamb shift (see Sect. 5.7.3) all levels of the H atom with equal quantum numbers $(n, j)$ have the same energy, because the energy shift due to the relativistic increase of the electron mass $m_e$ and that due to spin-orbit coupling just cancel. This $(2j + 1)$ fold degeneracy is lifted by the hyperfine interaction, because the magnitude of this interaction depends on the spatial distribution of the electron density and is therefore different for different values of the quantum number $l$. Levels with equal quantum numbers $n$ and $j$ but different values of $l$ do experience different hyperfine shifts and splittings.

In an external magnetic field, each atomic state $(n, l, s, j)$ splits without hyperfine interaction into $2j + 1$ Zeeman components. The energy separation of these components depends on the Lande factor $g_j$ (5.76), which might be different for the different levels. Generally the Zeeman splittings of different states are therefore different (anomalous Zeeman effect). For states with total electron spin $S = 0$ (which can be only realized for atoms with an even number of electrons) the normal Zeeman effect applies and the total Zeeman splittings are equal for all states with the same orbital angular momentum quantum number $l$, while the energy difference between neighbouring Zeeman levels is independent of $l$.

If the Zeeman splittings are small compared to the hyperfine splittings ($\boldsymbol{\mu}_j \cdot \boldsymbol{B} \ll A\boldsymbol{j} \cdot \boldsymbol{I}$), the external magnetic field can not break the coupling between electronic and nuclear magnetic moments. The total angular momentum including nuclear spin is then the vector sum $\boldsymbol{F} = \boldsymbol{j} + \boldsymbol{I}$, which has $2F + 1$ possible orientations against the external magnetic field with different energies. Therefore the hyperfine levels with the quantum number $F$ split into $2F + 1$ Zeeman components.

For the ground state $1^2S_{1/2}$ of the H atom no Zeeman splitting is observed for the HFS component $F = 0$, while the other HFS component with $F = 1$ splits into three Zeeman sublevels (Fig. 5.31).

For higher magnetic fields ($\boldsymbol{\mu}_j \cdot \boldsymbol{B} > \Delta E_{\text{HFS}}$) the coupling between $\boldsymbol{j}$ and $\boldsymbol{I}$ breaks down. The quantum number $F$ is no longer defined and the Zeeman shift of the levels depends on $\boldsymbol{\mu}_j \cdot \boldsymbol{B}$. For still higher magnetic fields ($\boldsymbol{\mu}_j \cdot \boldsymbol{B} > \Delta E_{\text{FS}}$) even the coupling between $\boldsymbol{s}$ and $\boldsymbol{l}$ breaks down. In this case there is no longer a defined total electronic angular momentum $j$ but $l$ and $s$ precess separately around the field axis (Paschen–Back effect Fig. 5.32).

The complete level scheme of the H atom is shown in Fig. 5.33 where, on the left side, the energy levels without effects of the electron spin are plotted, which are the energies obtained from Bohr's atomic model and also from the Schrödinger equation. The level energies plotted in the second column take into account the relativistic mass increase and the fine structure due to spin-orbit coupling. The next column adds the Lamb shift (see next section) and the last column includes the hyperfine interaction.

*Note* that the energy scales for fine and hyperfine interactions are widely spread, in order to show these small splittings in the same diagram. The absolute values for the splitting of the $2P_{1/2}$ level are illustrated in Fig. 5.34.

### Remark

In this chapter the electron spin was introduced in a phenomenological way, based on the results of experiments, such as the Einstein–de Haas effect, the fine structure in the atomic spectra and the Stern–Gerlach experiment. Mathematically, the total wave function was written as the product of spatial wave function (solution of the Schrödinger equation) and spin function. This heuristic introduction of the electron spin is able to explain all experimental results discussed so far, although it does not meet the requirements of a strict mathematical derivation.

A complete theory, including the electron spin ab initio, was developed by *Paul A.M. Dirac* (1902–1984) who used as a master equation, instead of the Schrödinger equation, an equation that includes all relativistic effects (Dirac equation). This equation can be solved analytically for all one-electron systems as long as they can be reduced to real one body systems (for instance the treatment of the H atom can be reduced to a one-body system, where one particle with the reduced mass $\mu$ moves in the spherical symmetric Coulomb potential). This treatment is no longer possible for the two particle system $e^+e^-$ (positronium consisting of positron and electron) because the interaction between the two spins of $e^+$ and $e^-$ represents a strong perturbation of the Coulomb potential (see Sect. 6.7.4).

### 5.7.3 Lamb Shift

An atom can absorb or emit electromagnetic radiation. The correct description has to take into account the interaction of this atom with the radiation field. This interaction is not only present during the absorption or emission of photons, but also for so-called "virtual interactions," where the atomic electron in the Coulomb field of the nucleus can absorb and then emit a photon of energy $\hbar\omega$ during a time interval $\Delta t < \hbar/\Delta E = 1/\omega$. The uncertainty relation $\Delta E \Delta t \geq \hbar$ allows such processes without violating the energy conservation law.

This interaction leads to a small shift of the energy levels (Lamb-shift), which depends on the spatial probability distribution of the electron in the Coulomb field of the nucleus and therefore on the quantum numbers $n$ and $l$.

The Lamb shift can be understood at least qualitatively by an illustrative simple model. Because of the photon recoil, the statistical virtual absorption and emission of photons results in a shaky movement of the electron in the Coulomb field of the nucleus (Fig. 5.35), where its distance from the nucleus $r$ varies in a random way by $\delta r$. Its average potential energy is then

$$\langle E_{\text{pot}} \rangle = -\frac{Ze^2}{4\pi\varepsilon_0} \left\langle \frac{1}{r + \delta r} \right\rangle . \tag{5.91b}$$

For a random distribution of $\delta r$ is $\langle \delta r \rangle = 0$ but $\langle (r + \delta r)^{-1} \rangle \neq \langle r^{-1} \rangle$. Therefore an energy shift occurs. Its quantitative calculation is not possible within the framework of the Schrödinger theory but can be performed in an extended theory called ***quantum electrodynamics***, which contains the complete description of atoms and their electron shells including the interaction with the radiation field [5.3].

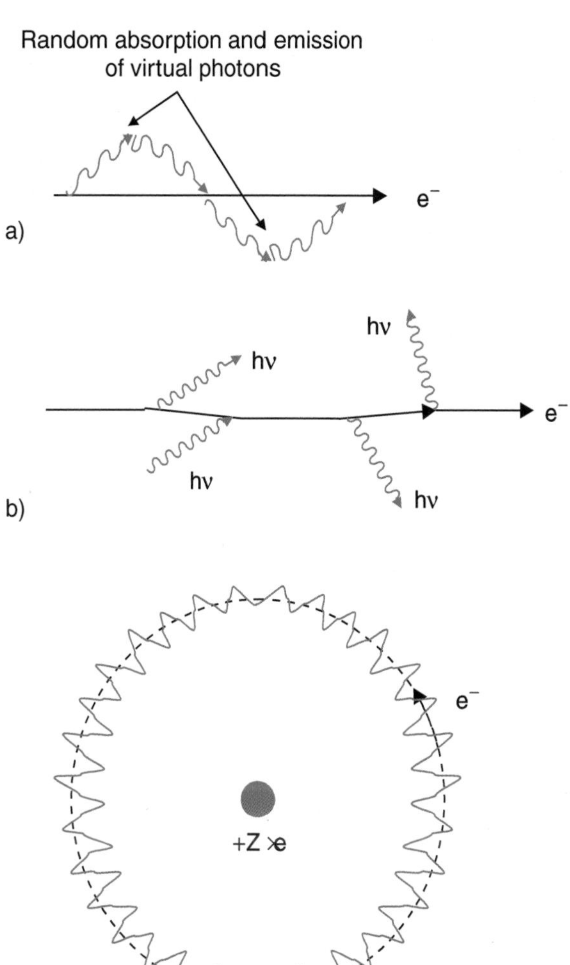


**Fig. 5.35a–c.** Illustration of the random shaky motion of the electron due to absorption and emission of virtual photons. Motion of a free electron in a radiation field without taking into account the photon recoil (**a**), with recoil (**b**), and shaky motion of an electron in a Coulomb field on the first Bohr orbit including the photon recoils (**c**)

The effects of these interactions are generally very small. Therefore, in most cases the Schrödinger theory, including the electron spin, is sufficiently accurate to match the experimental results. Only in special cases, and in particular for high precision measurements, does the Lamb shift have to be taken into account.

The complete term diagram of the levels with $n = 2$ in the H atom is drawn in Fig. 5.34. The Lamb shift $\Delta E_{\text{L}}$ is maximum for the $S$ states, because the wave function has a maximum at the position of the nucleus and the effect of the random variations $\delta r$ are largest for small $r$ values.

The numerical values for the Lamb shifts are

$$\Delta E_{\mathrm{La}}(1^2S_{1/2}) = +3.55 \times 10^{-5}\,\mathrm{eV}$$
$$\Rightarrow \Delta\nu_{\mathrm{La}} = +8.176\,\mathrm{GHz}$$
$$\Delta E_{\mathrm{La}}(2^2S_{1/2}) = +4.31 \times 10^{-6}\,\mathrm{eV}$$
$$\Rightarrow \Delta\nu_{\mathrm{La}} = +1.056\,\mathrm{GHz}$$
$$\Delta E_{\mathrm{La}}(2^2P_{1/2}) = -5.98 \times 10^{-8}\,\mathrm{eV}$$
$$\Rightarrow \Delta\nu_{\mathrm{La}} = -14\,\mathrm{MHz}\,.$$

The first measurement of the Lamb shift was performed in 1947 by *Willis Lamb* (*1912) [5.4] and *Robert Retherford* (*1912) using the experimental setup shown in Fig. 5.36.

In a heated tungsten oven, hydrogen is thermally dissociated. (In modern devices, a higher degree of dissociation is achieved with a microwave discharge.) The H atoms emerging from a hole in the oven into the vacuum are collimated by the aperture B into a nearly parallel atomic beam. The atoms are excited into the metastable $2^2S_{1/2}$ state by collisions with electrons crossing the atomic beam. The lifetime of the $2S$ state is about $1s$ and therefore longer than the flight time of the atoms through the apparatus. After a pathlength $L$, the metastable atoms impinge onto a tungsten target, where they transfer their excitation energy, which is higher than the energy necessary to release electrons from the conduction band, which are collected by a detector. The rate of emitted electrons represents a small electric current that can be measured.

During their flight to the detector the electrons pass a radio frequency field with a tunable frequency. If the frequency matches the energy separation $\Delta E = E(2^2S_{1/2}) - E(^2P_{1/2}) = 4.37 \times 10^{-6}\,\mathrm{eV}$ ($\Rightarrow \nu_{\mathrm{res}} = 1.05 \times 10^9\,\mathrm{Hz}$ or $\lambda = 30\,\mathrm{cm}$) between the $2^2S_{1/2}$ state and the $2^2P_{1/2}$ state, transitions $2^2S_{1/2} \to 2^2P_{1/2}$ are induced. The lifetime of the $2^2P_{1/2}$ state is only $\tau \approx 2 \times 10^{-9}\,\mathrm{s}$, because it decays spontaneously into the $1S$ state by emitting Lyman-$\alpha$ radiation. Therefore atoms in the $2P$ state cannot reach the detector. Hydrogen atoms in the $1S$ ground state cannot release electrons from the tungsten target. Therefore the measured electron current decreases and $I(\nu_{\mathrm{rf}})$ shows a sharp dip at the resonance radio frequency.

An alternative way for detecting the transitions between the $2^2S_{1/2}$ and the $2^2P_{1/2}$ states is the measurement of the Lyman $\alpha$-fluorescence emitted from the $^2P_{1/2}$ state. It can be detected with a solar blind photomultiplier viewing the rf field region.

The numerical value $\nu_{\mathrm{res}} = 1.05 \times 10^9\mathrm{Hz} = 1.05\,\mathrm{GHz}$ obtained from these experiments is in good agreement with theory. However, recent, much more accurate measurements, show that for a reliable comparison with theory the charge distribution in the proton, which affects the Lamb shift, must be known more accurately than is presently possible from high energy scattering experiments.

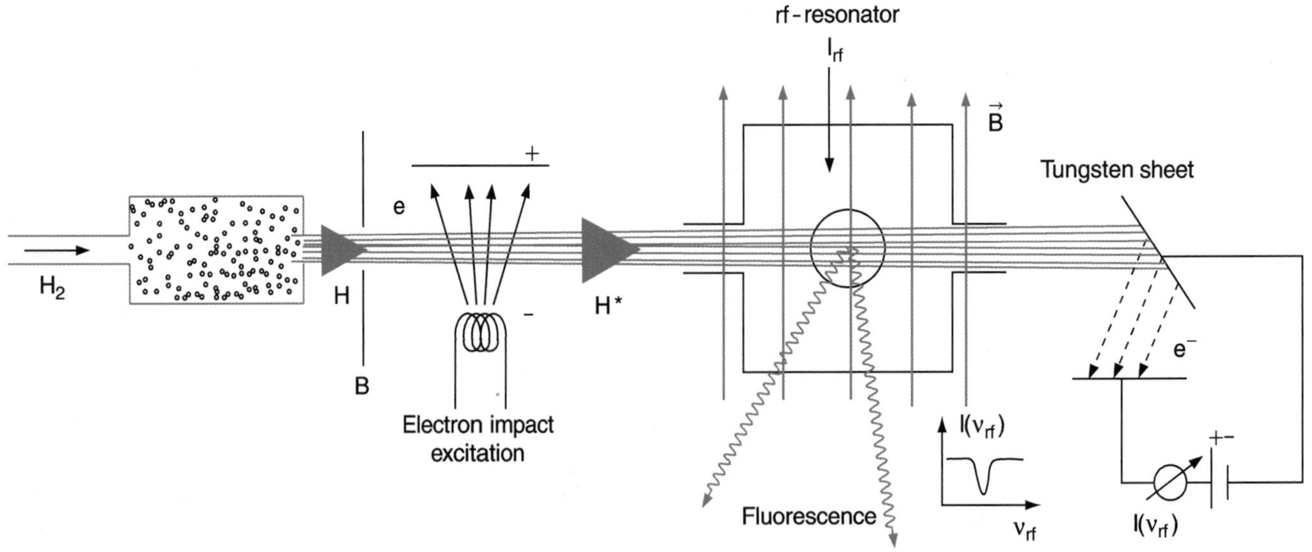


**Fig. 5.36.** Lamb–Retherford experiment

**Note:**

In real experiments [5.4] very small electric stray fields, which are difficult to eliminate completely, already cause Stark shifts that are different for the $2S$ and the $2P$ levels. These shifts not only add to the Lamb shift but can also mix the $2S$ and $2P$ levels, causing Lyman-$\alpha$ emission without applying the rf field. This effect can be avoided by applying a static magnetic field $B$, which causes a Zeeman splitting and an increase of the energy separation between the $2^2S_{1/2}$ and $2^2P_{1/2}$ levels (Fig. 5.37). Instead of tuning the rf field, the magnetic field is now varied at a fixed radio frequency until the resonance is reached for transitions between the Zeeman levels. This has the additional advantage that the radiofrequency can always stay in resonance with the rf resonator and therefore the rf field amplitude in the interaction zone is always constant at its maximum value.

If the experiment is repeated at different radio frequencies, the resonance will occur at different magnetic fields $B$. Plotting the measured values of $\nu_{rf}$ as a function of $B$ (Fig. 5.37b) allows the extrapolation towards $B = 0$, which yields the field-free Lamb shift.

While the Lamb–Retherford experiment measured only the Lamb shift of the $2S$ state, a modern version of Lamb shift measurement can also determine the much larger Lamb shift of the $1^2S_{1/2}$ ground state [5.5]. It is based on the precise comparison of the frequencies of two different optical transitions in the H atom (Fig. 5.38):

Firstly the two-photon transition $1^2S_{1/2} \to 2^2S_{1/2}$, which is only possible if two photons are simultaneously absorbed (see Sects. 7.2.6 and 10.5). And secondly, the one-photon transition $2^2S_{1/2} \to 4^2P_{1/2}$.

According to the Schrödinger theory (and also the Dirac theory) the relation

$$\nu_1^0\left(1^2S_{1/2} \to 2^2S_{1/2}\right) = 4\nu_2^0\left(2^2S_{1/2} \to 4^2P_{1/2}\right) \tag{5.91c}$$

holds. Taking into account the Lamb shift (which is negligible for the $4^2P_{1/2}$ level) we obtain for the actual frequencies

$$\begin{aligned} \nu_1 &= \nu_1^0 - \Delta E_{La}(1S) + \Delta E_{La}(2S), \\ \nu_2 &= \nu_2^0 - \Delta E_{La}(2S)\,. \end{aligned} \tag{5.91d}$$

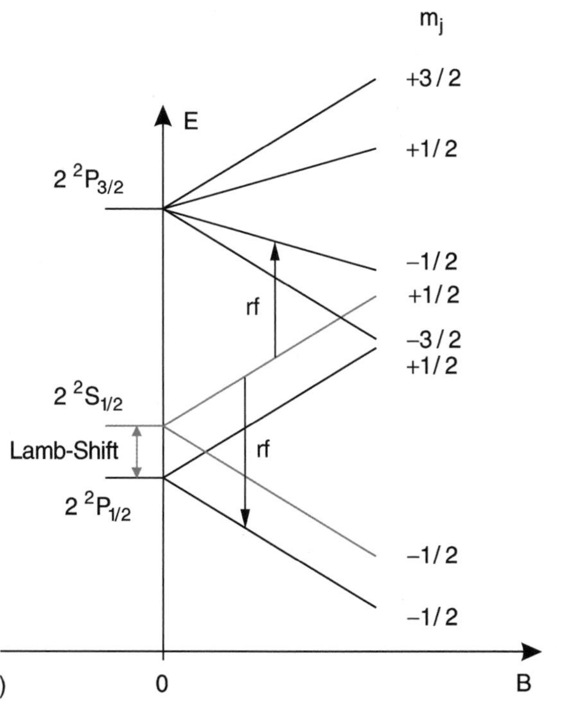


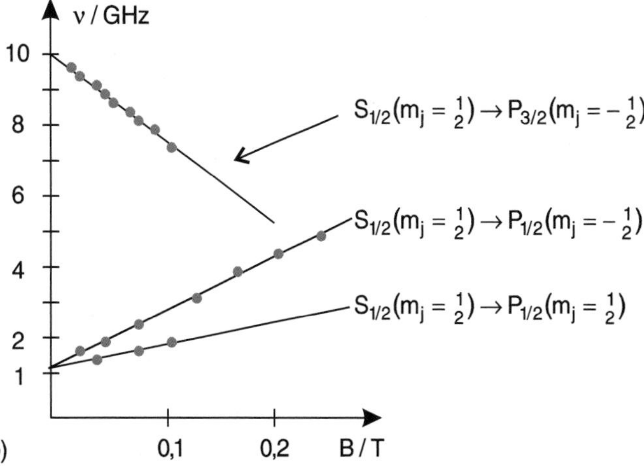


**Fig. 5.37.** (**a**) Zeeman splittings of the $2^2P_{1/2}$, $2^2S_{1/2}$ and $2^2P_{3/2}$ levels for measuring the Lamb shift. (**b**) Frequencies of the rf transitions as a function of the magnetic field strength

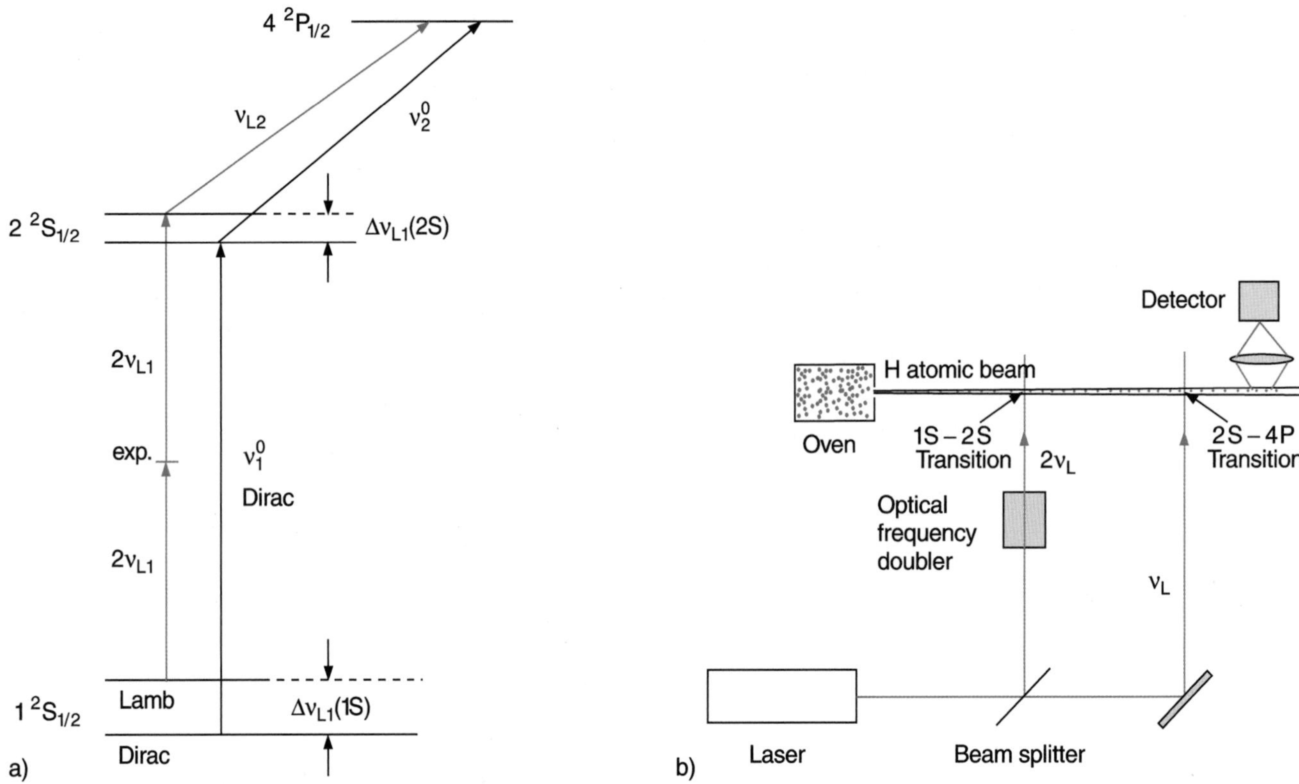


**Fig. 5.38a,b.** Optical measurement of the Lamb shift. (**a**) Level scheme. (**b**) Experimental arrangement

The difference

$$\begin{aligned}\Delta\nu &= \nu_1 - 4\nu_2\\ &= \nu_1^0 - 4\nu_2^0 - (\Delta E_{\mathrm{La}}(1S) - 5\Delta E_{\mathrm{La}}(2S))/h\\ &= (-\Delta E_{\mathrm{La}}(1S) + 5\Delta E_{\mathrm{La}}(2S))/h \qquad (5.92\mathrm{a})\end{aligned}$$

is measured. Since the Lamb shift of the $2S$ state is known from the Lamb–Retherford experiment, the shift of the $1S$ state can be determined from (5.92a).

The two-photon transition $1S \to 2S$ is excited by two photons from the frequency-doubled output of a dye laser, tuned to the optical frequency $\nu_L = (1/4)\nu(1S - 2S) = (1/4)\nu_1$. The Lamb shift of the $1S$ ground state is then

$$\Delta E_{\mathrm{La}}(1S) = 5\Delta E_{\mathrm{La}}(2S) - h\cdot(\nu_1 - 4\nu_2)\,. \qquad (5.92\mathrm{b})$$

The very precisely measured frequencies $\nu_1$ and $\nu_2$ [5.6] furthermore yields the present most accurate value of the Rydberg constant

$$Ry_\infty = 10{,}973{,}731.568639(91)\,\mathrm{m}^{-1}\,.$$

## 5.8 Correspondence Principle

For many qualitative results, estimates are sufficient and can save much of the time necessary for more detailed calculations. Here, a correspondence principle, formulated by *Niels Bohr*, is very useful. It illustrates the relation between classical and quantum physical quantities [5.7]. Its statements are as follows.

The predictions of quantum mechanics have to converge against classical results for the limit of large quantum numbers.

Selection rules for transitions between atomic states are valid for all quantum numbers. This means that rules obtained from classical considerations for large quantum numbers must also be valid for quantum mechanical selection rules for small quantum numbers.

This correspondence principle allows a quantitative relation between classical and quantum physics and gives the validity area for a classical description and its correspondence to a quantum mechanical model at the borderline of the classical realm. We will illustrate this using some examples.

**EXAMPLES**

1. According to classical electrodynamics, the frequency of an electromagnetic wave emitted by an electron on an orbit around the nucleus equals the revolution frequency of the electron. In Bohr's atomic model, this frequency is on the $n$th orbital

$$\nu_{\text{cla}} = \frac{v}{2\pi r} = \frac{mZ^2e^4}{4\varepsilon_0^2 n^3 h^3} . \tag{5.92c}$$

The quantum theory demands that $h\nu = \Delta E = E_i - E_k$. This gives

$$\begin{aligned}\nu_{\text{QM}} &= \frac{m_e Z^2 e^4}{8\varepsilon_0^2 h^3}\left(\frac{1}{n_i^2} - \frac{1}{n_k^2}\right) \\ &= \frac{m_e Z^2 e^4}{8\varepsilon_0 h^3}\frac{(n_k+n_i)(n_k-n_i)}{{n_k}^2 \cdot {n_i}^2} .\end{aligned} \tag{5.93}$$

For large quantum numbers $n$ and small quantum jumps $\Delta n = n_k - n_i \ll n_i$ we can approximate $(n_k + n_i)(n_h - n_i) \approx 2n\Delta n$ and we obtain:

$$\nu_{\text{QM}} \approx \frac{mZ^2e^4}{4\varepsilon_0^2 n^3 h^3}\Delta n . \tag{5.94}$$

For $\Delta n = 1$ the quantum model gives the classical fundamental frequency (5.92b) and for $\Delta n = 2, 3, \ldots$ the corresponding harmonics (Table 5.5).

2. The angular momentum of the electron is, according to Bohr's model,

$$|\boldsymbol{l}| = n\hbar \quad \text{with} \quad n = 1, 2, 3, \ldots , \tag{5.95a}$$

while the Schrödinger theory yields

$$|\boldsymbol{l}| = \sqrt{l(l+1)}\hbar . \tag{5.95b}$$

For small values of $l$ the differences between the two models are significant, because the lowest state is described by $l = 1$ in the Bohr model while the quantum theory demands $l = 0$.
For large values of $l$ and $n$ both models converge against $l \approx [l(l+1)]^{1/2} \leq [(n-1)n]^{1/2} \approx n$ (because $l \leq n - 1$).

3. For the limiting case of small frequencies (large wavelengths) Planck's radiation law converges against the Rayleigh–Jeans law (see Sect. 3.1). The mean energy of the black body radiation at the frequency $\nu$ is $\langle E\rangle = \langle n\rangle h\nu$, where $\langle n\rangle$ is the mean population density of photons $h\nu$ in a mode of the radiation field. From Planck's formula we can see that for $\nu \to 0$, the energy converges as $E \to kT$. This gives

$$\langle n\rangle h\nu \to kT \Rightarrow \langle n\rangle \to kT/(h\nu) . \tag{5.95c}$$

For $h\nu \ll kT$ the mean photon density $\langle n\rangle$ becomes very large, and $\langle n\rangle h\nu \gg h\nu$. The quantum structure of the photon field becomes less prominent, because the energy $E = \langle n\rangle h\nu$ is now a nearly continuous function of $n$ and the classical model does not differ much from the quantum mechanical one.

4. For the harmonic oscillator, the probability $|\psi_n(R)|^2$ of finding the system in the $n$th vibrational level at a distance $R$ is for small quantum numbers $n$ very different for the classical and the quantum mechanical models. However, for large values of $n$ the classical probability $P_{\text{cl}}(R)$ approaches more and more the average of $|\psi(R)|^2$ (see Fig. 4.21).

**Table 5.5.** Comparison of quantum mechanical and classical transition frequencies $\Delta n = 1$ for the H atom

| $n$ | $\nu_{\text{QM}}$ | $\nu_{\text{cla}}$ | Difference (%) |
|---|---|---|---|
| 5 | $5.26 \cdot 10^{13}$ | $7.38 \cdot 10^{13}$ | 29 |
| 10 | $6.57 \cdot 10^{12}$ | $7.72 \cdot 10^{12}$ | 14 |
| 100 | $6.578 \cdot 10^{9}$ | $6.677 \cdot 10^{9}$ | 1.5 |
| 1000 | $6.5779 \cdot 10^{6}$ | $6.5878 \cdot 10^{6}$ | 0.15 |
| 10,000 | $6.5779 \cdot 10^{3}$ | $6.5789 \cdot 10^{3}$ | 0.015 |

The correspondence principle is particularly useful for the discussion of selection rules for radiative transitions between atomic or molecular levels (see Chap. 7).

## 5.9 The Electron Model and its Problems

We have learned so far that the electron has a rest mass $m_e = 9.1 \times 10^{-31}$ kg, a negative electric charge $e = -1.6 \times 10^{-19}$ Coulomb, a spin $\boldsymbol{s}$ with the absolute value

$$|\boldsymbol{s}| = \frac{1}{2}\sqrt{3}\hbar , \tag{5.95d}$$

which can be mathematically treated like an angular momentum, and a magnetic moment

$$|\boldsymbol{\mu}_s| = g_s\mu_B \approx 2\mu_B ,$$

which is related to the spin by

$$\boldsymbol{\mu}_s = -\gamma_s \cdot \boldsymbol{s} \quad \text{with} \quad \gamma_s = e/m_e .$$

Up to now we have neither discussed the size of the electron, nor the spatial mass and charge distribution.

In a simple classical model, one assumes that the electron can be described by a charged sphere where the mass is uniformly distributed over the volume of this sphere and, because of the electric repulsion between charges of equal sign, the charge is uniformly distributed over its surface (Fig. 5.39). The radius $r_e$ of this sphere (the ***classical electron radius***) can then be calculated as follows.

The capacity of the charged surface is

$$C = 4\pi\varepsilon_0 r_e . \tag{5.95e}$$

In order to bring a total charge $Q = -e$ onto this capacitor, one needs the energy

$$W = \frac{1}{2}Q^2/C = \frac{1}{2}e^2/C = e^2/(8\pi\varepsilon_0 r_e) = E_{pot} . \tag{5.95f}$$

This potential energy corresponds to the energy $W = \frac{1}{2}\varepsilon_0|E|^2$ of the static electric field produced by the charged electron. If this energy equals the mass energy $m_e c^2$ of the electron, the classical electron radius becomes

$$r_e = \frac{e^2}{8\pi\varepsilon_0 m_e c^2} = 1.4 \times 10^{-15}\,\text{m} . \tag{5.96a}$$

If the charge is not only on the surface of the sphere but is uniformly distributed over the volume, an analogue consideration yields twice the energy, i.e., $W = e^2/(4\pi\varepsilon_0 r_e)$ and a radius

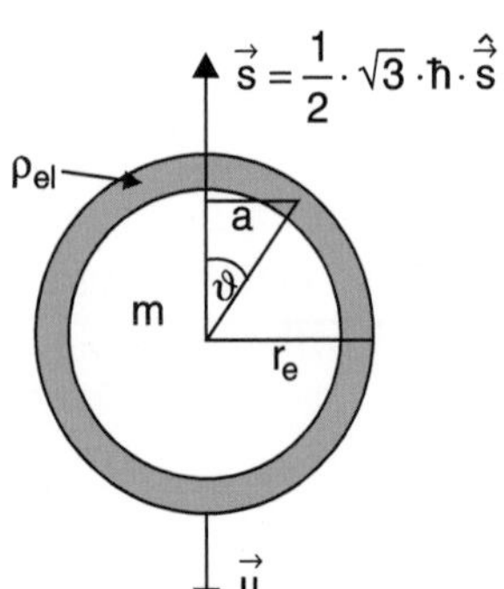


**Fig. 5.39.** Classical model of the electron as a sphere with mass $m$, uniform surface charge $-e$, spin $s$ and magnetic moment $\boldsymbol{\mu}_s$

$$r_e = e^2/\left(4\pi\varepsilon_0 m_e c^2\right) = 2.8 \times 10^{-15}\,\text{m} . \tag{5.96b}$$

In this model the magnetic moment $\mu_s$ of the electron is produced by the rotating charge. The elementary calculation gives the relation

$$\mu_s = \frac{1}{3}\omega \cdot e \cdot r_e^2 \text{ with } \mu_s = |\vec{\mu}_s| \tag{5.97}$$

between $\mu_s$ and the angular rotation frequency $\omega$. Inserting the absolute value $\mu_s = 2\mu_B = 1.85 \times 10^{-23}\,\text{Am}^2$, obtained from the Einstein–de Haas experiment and the classical electron radius $r_e = 1.4 \times 10^{-15}$ m, yields the angular frequency

$$\omega = \frac{3\mu_s}{e \cdot r_e^2} = 1.7 \times 10^{26}\,\text{s}^{-1} . \tag{5.98}$$

This would result in a velocity at the equator of the sphere of

$$v = \omega r_e = 2.3 \times 10^{11}\,\text{m/s} \gg c = 3 \times 10^8\,\text{m/s!!} \tag{5.99}$$

This is clearly a problematic result.

A similar result is obtained, in contradiction to special relativity, when the electron spin is interpreted as mechanical angular momentum of a sphere with the classical electron radius.

The moment of inertia of the sphere is $I = \frac{2}{5}m_e r_e^2$ and the angular momentum

$$|s| = \frac{1}{2}\sqrt{3}\cdot\hbar = I \cdot \omega = \frac{2}{5}m_e r_e^2 \cdot \omega . \tag{5.100}$$

This gives an angular velocity

$$\omega = \frac{5\cdot\sqrt{3}\cdot\hbar}{4 m_e r_e^2} \tag{5.101}$$

and a velocity of a point at the equator of

$$v = \frac{5\cdot\sqrt{3}\cdot\hbar}{4 m_e r_e} . \tag{5.102}$$

Inserting the numerical values yields

$$v = 9 \times 10^8\,\text{m/s} > c = 3 \times 10^8\,\text{m/s} .$$

From high energy scattering experiments it can be concluded that the charge $e$ of the electron is localized within a smaller volume with $r < 10^{-16}$ m. The resultant smaller value of $r$ would, however, increase the discrepancies of this mechanical model even more, because a smaller $r_e$ in the denominator of the expression (5.102) would further increase the equator velocity $v \propto 1/r_e$.

> Apparently the mechanical model of the electron as a charged sphere and the interpretation of its spin as mechanical angular momentum must be wrong. Up to now there does not exist a convincing realistic model of the electron.

The high energy experiments and precision measurements of the magnetic spin moment indicate that the electron can be treated as a point-like charge. Its mass $m_e = E/c^2$ can be interpreted as the energy $E$ of the electric field produced by its charge $-e$. The spin is an additional characteristic of the electron. Although it follows the same mathematical relations as other angular momenta, such as the commutation relations, and it has the properties of a vector, it apparently cannot be regarded as a mechanical angular momentum in the classical sense.

The charge distribution

$$\begin{aligned} \mathrm{d}q(r,\vartheta,\varphi) &= \varrho_{\mathrm{el}}(r,\vartheta,\varphi)\,\mathrm{d}\tau \\ &= -e|\psi(r,\vartheta,\varphi)|^2 \cdot r^2 \sin\vartheta\,\mathrm{d}r\,\mathrm{d}\vartheta\,\mathrm{d}\varphi \end{aligned}$$

of the electron in the atomic electron shell gives the probability to find the (probable point-like electron) in the volume element $\mathrm{d}\tau$ around the location $(r,\vartheta,\varphi)$.

These considerations illustrate a general problem in the realm of microparticles. Is the distinction between particles with mass $m$ and field energy $E = mc^2$ still meaningful? What are the lower limits of volumes $\Delta V = \Delta x \cdot \Delta y \cdot \Delta z$ in space, where our geometrical concept of space is still valid? Do we have to go to a higher dimensional space when we want to describe elementary particles?

There have been several attempts to answer these questions, but a definite indisputable model has not yet been developed. There are, however, mathematical theories which are consistent with all experimental results, although they do not provide a clear and vivid picture of the electron.

The Dirac theory starts from a relativistic equation (the Dirac equation) that describes all properties of the electron correctly (except its self-interaction with its radiation field resulting in the Lamb shift). Analogous to the situation for the Schrödinger equation the Dirac equation cannot be derived in a mathematical way from first physical principles. The complete theory that includes all aspects of atomic and molecular physics is ***quantum electrodynamics (QED)*** [5.8–10]. Its introduction is, however, beyond the scope of this book.

## SUMMARY

- The three-dimensional Schrödinger equation for the hydrogen atom can be separated in the center-of-mass system into three one-dimensional equations. This is possible because of the spherically symmetric potential. The solutions of the Schrödinger equation are wave functions $\psi(r,\vartheta,\varphi) = R(r)\Theta(\vartheta)\Phi(\varphi)$, which can be written as the product of three functions of only one variable. While the radial part $R(r)$ depends on the special $r$-dependence of the potential, the angular part $Y_l^m(\vartheta,\varphi) = \Theta(\vartheta)\Phi(\varphi)$ represents spherical surface harmonics $Y_l^m$ for all spherical potentials. These functions depend only on the quantum numbers $l$ of the orbital angular momentum $\boldsymbol{l}$ and $m_l$ of its projection $l_z$.
- The constraints of normalization and unambiguity for the wave function lead to the quantization of bound energy states with $E < 0$ (only discrete energy levels exist) while for states with $E > 0$ all energies are allowed (continuous states). One can also say that if the wave function is restricted to a finite volume in space, the energies are quantized. If the particle can move all over the space, a continuous energy spectrum appears.
- Each wave function $\psi = \psi_{n,l,m}(r,\vartheta,\varphi)$ of the H atom is unambiguously defined by the three quantum numbers $n$ (principal quantum number), $l$ (quantum number of orbital angular momentum $\boldsymbol{l}$) and $m_l$ (projection quantum number of $l_z$).
- The absolute square $|\psi(r,\vartheta,\varphi)|^2$ of the wave function describes the probability density function. This means, that $|\psi|^2\,\mathrm{d}V$ gives the probability to find the particle within the volume $\mathrm{d}V$.
- The energy eigenvalues $E_n$ are obtained by inserting the wave functions $\psi_{n,l,m}$ into the Schrödinger equation.
- Within the Schrödinger model the energies $E_n$ of the discrete states of the hydrogen atom depend solely on $n$, not on $l$ and $m$. All states with equal $n$ but different values of $l$ or $m$ have the same energy (they are degenerate). For each possible value of $E_n$ there are $k = \sum_{l=0}^{n-1}(2l+1) = n^2$ different wave functions $\psi_{n,l,m}(r,\vartheta,\varphi)$ that describe $n^2$ different spatial charge distributions of the electron. The energy states of the hydrogen atom are therefore $n^2$-fold degenerate.
- The normal Zeeman effect results from the interaction of the magnetic moment $\mu_l$ (due to the orbital motion of the electron) with an external magnetic field. This interaction splits the energy states $E_{n,l}$ into $(2l+1)$ equidistant Zeeman components with energies shifted by $\Delta E = \mu_\mathrm{B} m_l B$ against the field-free energies, where $\mu_\mathrm{B}$ is the Bohr magneton.
- Several experimental results (anomalous Zeeman effect, Stern–Gerlach experiment, Einstein–de Haas experiment) force an extension of the Schrö dinger theory. This was achieved by the introduction of the electron spin $\vec{s}$ with an additional spin magnetic moment $\boldsymbol{\mu}_s = -g_s(\mu_\mathrm{B}/\hbar)\boldsymbol{s}$ with the Lande factor $g_s \approx 2$. The total angular momentum of the electron is the vector sum $\boldsymbol{j} = \boldsymbol{l} + \boldsymbol{s}$. The total wave function is now written as a product of the spatial part and a spin function.
- The fine structure, observed in the atomic spectra, can be interpreted as Zeeman splitting, caused by the interaction of the spin magnetic moment $\mu_s$ with the internal magnetic field, produced by the orbital motion of the electron. The energies of the fine structure components are

$$E_{n,l,j} = E_n + \frac{a}{2}[j(j+1) - l(l+1) - s(s+1)]\,,$$

where

$$a = \frac{\mu_0 Z e^2 \hbar^2}{8\pi m_\mathrm{e}^2 r^3}$$

is the spin-orbit coupling constant.
- In the Coulomb potential all energy terms with equal quantum number $j$ are degenerate when the Lamb-shift is neglected. This is due to the cancellation of the energy shift due to the relativistic increase of the electron mass and the shift caused by the spin-orbit interaction. This degeneracy is lifted in non-Coulombic potentials, even if they are spherically symmetric, because here the two shifts are different. It is furthermore lifted by the Lamb-shift.
- The anomalous Zeeman effect is observed for all states with total spin $S \neq 0$: and total orbital angular momentum $L \neq 0$. The energy shift of the Zeeman components is $\Delta E = -\boldsymbol{\mu}_j \cdot \boldsymbol{B}$,

with $\boldsymbol{\mu}_j = \boldsymbol{\mu}_l + \boldsymbol{\mu}_s$. Each term $E_{n,j}$ splits into $(2j+1)$ Zeeman components, which are generally not equidistant as for the normal Zeeman effect.

- Atoms with a nuclear spin $I$ and a corresponding (very small) nuclear magnetic moment $\boldsymbol{\mu}_\mathrm{N}$ show an additional small energy shift $\Delta E = -\boldsymbol{\mu}_\mathrm{N} \cdot \boldsymbol{B}$ of the atomic states, caused by the interaction of the nuclear magnetic moment with the internal magnetic field produced by the electrons at the position of the nucleus (hyperfine structure). The energy levels split into $(2F+1)$ hyperfine-components, where $F$ is the quantum number of the total angular momentum $\boldsymbol{F} = \boldsymbol{J} + \boldsymbol{I} = \boldsymbol{L} + \boldsymbol{S} + \boldsymbol{I}$, including the nuclear spin $\boldsymbol{I}$.
- If the interaction of the electron with the radiation field produced by virtual emission and absorption of photons is taken into account, the energy levels experience a small additional shift, called the Lamb shift. The shift is maximum for the $1S$ state, smaller for the $2S$ state and negligible for the $P$ or $D$ states. The Lamb shift can only be calculated within the framework of quantum electrodynamics.
- The Schrödinger theory describes the hydrogen atom correctly if relativistic effects (mass increase and electron spin) are neglected. The Dirac theory includes these effects, but does not take into account the Lamb shift. A complete description of all effects observed so far, is possible within the quantum electrodynamic theory.
- Up to now no concrete model of the electron exists that consistently describes all characteristics such as mass, size, charge, spin and magnetic moment.

## PROBLEMS

1. Calculate the expectation values $\langle r\rangle$ and $\langle 1/r\rangle$ for the two states 1$s$ and 2$s$ in the hydrogen atom.
2. Which spectral lines in the emission spectrum of hydrogen atoms can be observed if the atoms are excited by electrons with kinetic energy $E_{\mathrm{kin}} = 13.3\,\mathrm{eV}$?
3. By what factor does the radius of the Bohr orbit increases if the H atom in its ground state is excited by (a) 12.09 eV and (b) 13.387 eV?
4. Show that within the Bohr model the ratio $\mu_l/l$ of orbital magnetic moment and angular momentum is independent on the principal quantum number $n$.
5. By how much does the mass of the hydrogen atom differ in the state with $n = 2$ from that in the state $n = 1$ (a) because of the relativistic increase of the electron mass and (b) because of the higher potential energy? Assume circular motion of the electron.
6. In the classical model, the electron is described as a rigid sphere with radius $r$, mass $m$, charge $-e$ and uniform charge distribution. (a) What is the velocity of a point on the equator of this sphere when the angular momentum is $1/2\sqrt{3}h$? (b) What would the rotational energy of this sphere be? Compare the result with the mass energy $m_{\mathrm{e}}c^2$. Use both numerical values $r_{\mathrm{e}} = 1.4\times 10^{-15}\,\mathrm{m}$ (obtained from the classical model of the electron) and $r_{\mathrm{e}} = 10^{-18}\,\mathrm{m}$ (obtained from scattering experiments).
7. Assume you want to measure the Zeeman splitting of the Balmer $\alpha$-line on the transition $2^2S_{1/2} \to 3^2P_{1/2}$ in a magnetic field of $B = 1\,\mathrm{T}$. (a) What should the minimum spectral resolution of a grating spectrograph be in order to resolve all components? What is the minimum number of grooves that must be illuminated if you observe in the second diffraction order? (b) What is the minimum magnetic field $B$ needed to resolve the Zeeman components with a Fabry–Perot interferometer (plate separation $d = 1\,\mathrm{cm}$, reflectivity of each plate $R = 95\%$)?
8. How large is the internal magnetic field produced by the 1$s$ electron in the H atom at the location of the proton that causes the splitting of the two hyperfine components observed in the transition with $\lambda = 21\,\mathrm{cm}$ between the two components?
9. Compare the frequencies of the absorption lines $1S \to 2P$ for the three isotopes $^1$H, $^2$D, and $^3$T of the hydrogen atom (a) by taking into account the different reduced masses and (b) by calculating the hyperfine shifts and splittings with the nuclear spin quantum numbers $I(\mathrm{H}) = 1/2$, $I(\mathrm{D}) = 1$ and $I(\mathrm{T}) = 3/2$ and the nuclear magnetic moments $\mu_{\mathrm{N}}(\mathrm{H}) = 2.79\mu_{\mathrm{K}}$, $\mu_{\mathrm{N}}(\mathrm{D}) = 0.857\mu_{\mathrm{K}}$; $\mu_{\mathrm{N}}(\mathrm{T}) = 2.98\mu_{\mathrm{K}}$.
10. Derive the expression (5.49) for the Darwin term.
11. Derive eq. (5.72) by adding (5.42) and (5.66a) using (5.68).

# 6. Atoms with More Than One Electron

In atoms with more than one electron additional problems arise that are caused by mutual electrostatic and magnetic interactions between the electrons. In addition, we are now confronted with new symmetry principles that are valid if two electrons are exchanged. These stem from the fact that electrons cannot be distinguished from each other.

We will first study these phenomena for the helium atom, which represents the simplest system with two electrons. This will help us understand the building up principle for the structure of electron shells for larger atoms. We will see that the electron configurations for all atoms can be obtained from the minimum energy principle, the correct coupling of the different angular momenta of the electrons and the observation of certain symmetry rules. This results in the determination of all possible energy states of the atoms and the characterization of these states by quantum numbers.

## 6.1 The Helium Atom

The helium atom consists of a nucleus with charge $+Ze = +2e$ and mass $m_K \approx 4m_H$ and of two electrons each with charge $q = -e$. The spatial distribution of the two electrons depends on their wave function $\psi(\boldsymbol{r}_1, \boldsymbol{r}_2)$, which is a function of the spatial coordinates $\boldsymbol{r}_1 = (x_1, y_1, z_1)$ and $\boldsymbol{r}_2 = (x_2, y_2, z_2)$ of the two electrons. Their distances from the nucleus are $r_1 = |\boldsymbol{r}_1|$ and $r_2 = |\boldsymbol{r}_2|$ and their mutal distance is $r_{12} = |\boldsymbol{r}_1 - \boldsymbol{r}_2|$.

The potential energy of the electrons is then:

$$E_{\text{pot}} = -\frac{e^2}{4\pi\varepsilon_0}\left(\frac{Z}{r_1} + \frac{Z}{r_2} - \frac{1}{r_{12}}\right) . \qquad (6.1a)$$

The operator of the kinetic energy in the center of mass system is

$$\hat{E}_{\text{kin}} = -\frac{\hbar^2}{2\mu}\left(\Delta_1(\boldsymbol{r}_1) + \Delta_2(\boldsymbol{r}_2)\right)$$

$$\text{with} \quad \mu = \frac{m_e m_K}{m_e + m_K} \qquad (6.1b)$$

where the operator $\Delta_i$ acts on the coordinate $\boldsymbol{r}_i$. Since $m_K \approx 7300\, m_e$, we can use the approximation $\mu \approx m_e = m$. The Schrödinger equation is then:

$$-\frac{\hbar^2}{2m}\Delta_1\psi(\boldsymbol{r}_1, \boldsymbol{r}_2) - \frac{\hbar^2}{2m}\Delta_2\psi(\boldsymbol{r}_1, \boldsymbol{r}_2) + E_{\text{pot}}\psi(\boldsymbol{r}_1, \boldsymbol{r}_2) = E\psi(\boldsymbol{r}_1, \boldsymbol{r}_2) . \qquad (6.2)$$

The last term on the left side is the potential energy, which is no longer spherically symmetric as in the hydrogen atom, but depends on the angle $\alpha$ between the radius vectors $\boldsymbol{r}_i$ to the electrons, because of their mutual repulsion. From Fig. 6.1 we can derive the relation

$$r_{12}^2 = |\boldsymbol{r}_1 - \boldsymbol{r}_2|^2 = r_1^2 + r_2^2 - 2r_1 r_2 \cos\alpha .$$

We therefore *can not* separate the total wave function into a radial part and an angular part, as we could in the case of one-electron systems. This implies that

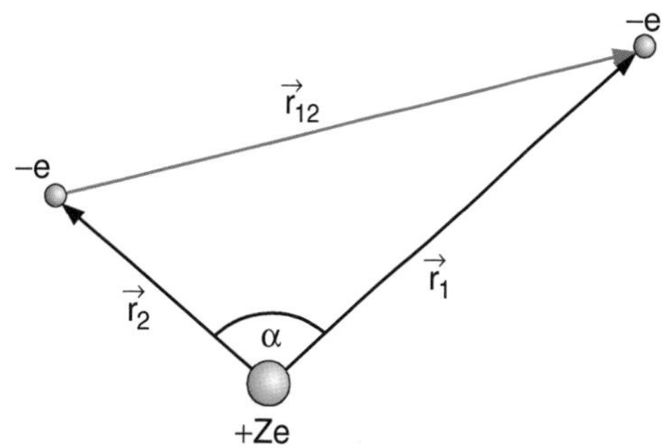


**Fig. 6.1.** The helium atom

W. Demtröder, *Atoms, Molecules and Photons*, 2nd ed., Graduate Texts in Physics,
DOI 10.1007/978-3-642-10298-1_6, 

the Schrödinger equation (6.2) is no longer solvable analytically and we have to use approximations.

### 6.1.1 Approximation Models

Because of the mutual repulsion, the charge distribution of the electrons will be such, that the total energy becomes a minimum. This means: The sum of kinetic energy and potential energy, due to electron-electron repulsion and attraction between the electrons and the nucleus becomes a minimum. On the time average is $\langle r_{12} \rangle > \langle r_1 \rangle = \langle r_2 \rangle$.

In a first crude approximation we can therefore neglect the last term in (6.1a). Then we can separate the wave function into the product

$$\psi(\boldsymbol{r}_1, \boldsymbol{r}_2) = \psi_1(\boldsymbol{r}_1)\psi_2(\boldsymbol{r}_2) \,. \tag{6.3}$$

Inserting this into the Schrödinger equation (6.2) yields two separate equations for the two electrons

$$\begin{aligned} &-\frac{\hbar^2}{2m}\Delta_1\psi_1(\boldsymbol{r}_1) - \frac{e^2}{4\pi\varepsilon_0}\frac{Z}{r_1}\psi_1(\boldsymbol{r}_1) = E_1\psi_1(\boldsymbol{r}_1) \\ &-\frac{\hbar^2}{2m}\Delta_2\psi_2(\boldsymbol{r}_2) - \frac{e^2}{4\pi\varepsilon_0}\frac{Z}{r_2}\psi_2(\boldsymbol{r}_2) = E_2\psi_2(\boldsymbol{r}_2) \end{aligned} \tag{6.4}$$

with $E_1 + E_2 = E$. Each of these equations is identical to the Schrödinger eqation (5.8) for the one electron system and can be solved accordingly.

With $Z = 2$ we obtain in this approximation for the energy of the two electrons in the lowest state with $n = 1$:

$$\begin{aligned} E_{\mathrm{He}}(1s) &= -2Z^2E_{\mathrm{H}} = -2 \cdot 4 \cdot 13.6\,\mathrm{eV} \\ &= -108.8\,\mathrm{eV} \,. \end{aligned}$$

The experimental value for this energy that is necessary to remove both electrons from the atom (this means to convert the He atom into the doubly charged ion $\mathrm{He}^{++}$) is, however, only $E_{\exp} = 78.98\,\mathrm{eV}$.

The neglection of the electron repulsion therefore introduces an absolute error of 30 eV, i.e., a relative error of about 40%.

A much better approximation is obtained by a model that assumes that each of the two electrons moves in the Coulomb potential of the nucleus, shielded by the charge distribution of the other electron (which is assumed to have a spherically symmetric time average). The resulting potential for each electron is then a spherically symmetric Coulomb potential generated by

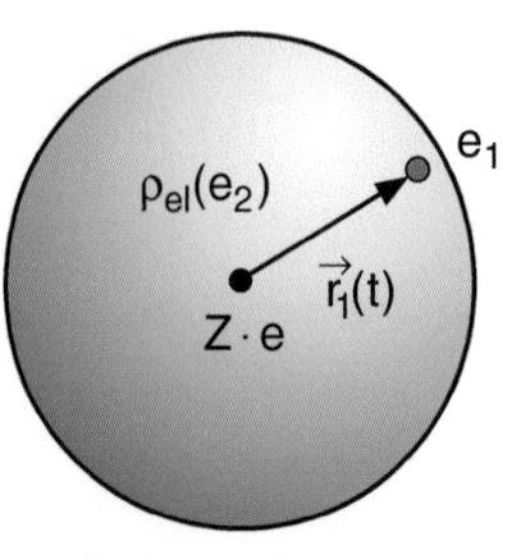


**Fig. 6.2.** Partial shielding of the nuclear charge $+Ze$ by the negative charge distribution $\rho_{\mathrm{el}}(e_2) = -e|\psi_{1s}(\boldsymbol{r}_2)|^2$ of a $1s$ electron

the effective charge $Q_{\mathrm{eff}} = (Z - S)e$ (Fig. 6.2). The quantity $S$ $(0 \le S \le 1)$ is called the shielding constant.

For total shielding $S = 1$ and one would need the energy $E_{\mathrm{H}}$ to remove the first electron from the atom. The remaining ion $\mathrm{He}^+$ now has the nuclear charge $+2e$ and the binding energy of the second electron is therefore $-Z^2E_{\mathrm{H}} = -4E_{\mathrm{H}}$. The total ionization energy of the He atom is then

$$E_{\mathrm{He}}(1s) = -E_{\mathrm{H}} - 4E_{\mathrm{H}} = -5E_{\mathrm{H}} = -68\,\mathrm{eV} \,, \tag{6.5}$$

which comes much closer to the experimental value $E_{\mathrm{He}} = -78.983\,\mathrm{eV}$. For a shielding constant $S = 0.656$ the experimental value is exactly reproduced. In our model the correct energy is therefore obtained for an effective nuclear charge of $Z_{\mathrm{eff}}e = +1.344\,e$. This implies that about 33% of the real nuclear charge $+2e$ is shielded for one electron by the other electron in the $1s$ state.

**Note:**

The shielding for an electron in higher energy states (for instance the $2s$ or $2p$ state) by an $1s$ electron can be much larger, because the spatial charge distribution for the higher state has only small values within the $1s$ distribution of the shielding electron.

The spatial charge distribution of the shielding electron in the $1s$ state is given by

$$\varrho_{\mathrm{el}} = -e\psi_2^*(1s)\psi_2(1s) \,. \tag{6.6}$$

The potential energy of the other electron is then

$$E_{\mathrm{pot}}(\boldsymbol{r}_1) = -\frac{e^2}{4\pi\varepsilon_0}\left(\frac{Z}{r_1} - \int\limits_{\vartheta}\int\limits_{\varphi}\int\limits_{r_2}\frac{\psi_2^*\psi_2}{r_{12}}\,\mathrm{d}\tau_2\right) \,. \tag{6.7}$$

In a first approximation we can assume that the charge distribution of the shielding electron is not changed much by the presence of the second electron. This means that we can take the unperturbed hydrogenic wave functions for its spherically symmetric spatial distribution. This yields for the potential energy

$$E_{pot}(\boldsymbol{r}_1) = -\frac{e^2}{4\pi\varepsilon_0}\left(\frac{Z}{r_1} - \frac{1}{\pi}\left(\frac{Z}{a_0}\right)^3 \int\limits_{\tau_2} \frac{e^{-2Zr_2/a_0}}{r_{12}}\, d\tau_2\right), \tag{6.8}$$

where $d\tau_2 = r_2{}^2 \sin\eta_2\, d\eta_2 d\varphi_2$

Equation (6.8) has the solution (see problem 6.1)

$$E_{pot}(\boldsymbol{r}_1) = -\frac{e^2}{4\pi\varepsilon_0}\left[\frac{Z-1}{r_1} + \left(\frac{Z}{a_0} + \frac{1}{r_1}\right) e^{-2Zr_1/a_0}\right]. \tag{6.9a}$$

For $Z = 2$ this gives:

$$E_{pot}(\boldsymbol{r}_1) = -\frac{e^2}{4\pi\varepsilon_0 r_1}\left[1 + \left(\frac{2r_1}{a_0} + 1\right) e^{-4r_1/a_0}\right] \tag{6.9b}$$

Inserting this expression into the Schrödinger equation gives a much better value for the energy of the helium ground state than by inserting the shielding factor $S = 1$. In addition the shielding is now dependent on the distance $r_1$ of the electron $e_1$ from the nucleus. This is reasonable because the more the electron $e_1$ penetrates into the charge distribution of $e_2$, the lower the shielding by $e_2$ becomes.

### 6.1.2 Symmetry of the Wave Function

We will label the two electrons $e_1$ and $e_2$. The two factors $\psi_1(n_1, l_1, m_{l_1})$ and $\psi_2(n_2, l_2, m_{l_2})$ of the separated wave function (6.3) depend on the three quantum numbers $(n.l, m_l)$ of the two electrons, which we will abbreviate with

$$a = (n_1, l_1, m_{l_1}) \quad \text{and} \quad b = (n_2, l_2, m_{l_2}).$$

The probability $P(a,b) = |\Psi_{ab}(r_1, r_2)|^2$, that the atomic state $(a,b)$ is realized, (this means that $e_1$ is in state $a$ and $e_2$ in state $b$) can be expressed in the approximate model of independent electrons (the interaction term $e^2/r_{12}$ is neglected) by the absolute square of the product function (6.3)

$$\psi_{ab}^{\mathrm{I}} = \psi_1(a)\psi_2(b). \tag{6.10a}$$

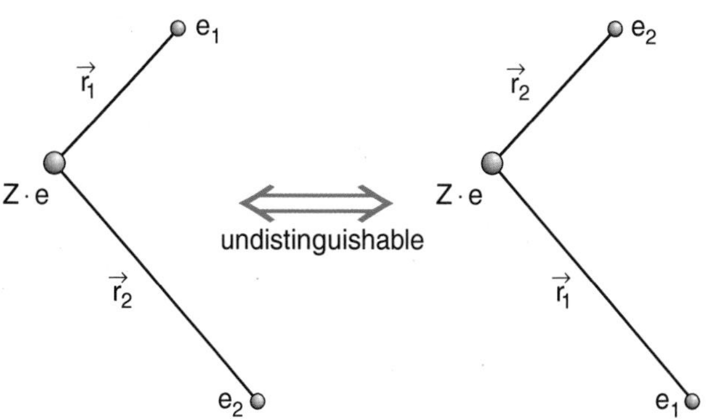


**Fig. 6.3.** The two-electron configurations are undistinguishable

If we exchange the two electrons (Fig. 6.3), ($e_1$ is now in state $b$ and $e_2$ in state $a$) our product function becomes

$$\psi_{ab}^{\mathrm{II}} = \psi_2(a)\psi_1(b). \tag{6.10b}$$

However, the two electrons are indistinguishable! This means that the charge distribution of the total atom should not be changed under exchange of the two electrons. We can write this as

$$\left|\psi_{ab}^{\mathrm{I}}\right|^2 = \left|\psi_{ab}^{\mathrm{II}}\right|^2 \Rightarrow \psi_{ab}^{\mathrm{I}} = e^{i\varphi}\psi_{ab}^{\mathrm{II}}. \tag{6.11}$$

Applying the permutation of the two electrons twice brings the state back into its original configuration. This demands $\varphi = 0$ or $\varphi = \pi \Rightarrow$

$$\psi_{ab}^{\mathrm{I}} = \pm\psi_{ab}^{\mathrm{II}}. \tag{6.12}$$

Neither of the two functions $\psi^{\mathrm{I}}$ nor $\psi^{\mathrm{II}}$ fullfill this condition. They therefore cannot represent the correct eigenfunctions for the description of our atomic state. We can, however, form a symmetric and an antisymmetric linear combination of these product functions, which obey the condition (6.12):

$$\psi_{\mathrm{atom}}^{\mathrm{s}} = \psi_1(a)\psi_2(b) + \psi_2(a)\psi_1(b) \tag{6.13a}$$

$$\psi_{\mathrm{atom}}^{\mathrm{a}} = \psi_1(a)\psi_2(b) - \psi_2(a)\psi_1(b). \tag{6.13b}$$

The symmetric function $\psi_{\mathrm{atom}}^{\mathrm{s}}$ reproduces itself under exchange of the two electrons, while $\psi_{\mathrm{atom}}^{\mathrm{a}}$ only changes its sign.

**Note:**

$\psi^{\mathrm{s}}$ and $\psi^{\mathrm{a}}$ represent the probability amplitudes for the configuration that one electron is in state $a$ and the other in state $b$. However, we do not know which of the two electrons is in $a$ and which is in $b$. This problem is quite analogous to the situation in Young's double slit experiment (see Sect. 3.5.2), where the probability of finding a photon on the screen behind the double slit is given by the absolute square of the sum of two probability amplitudes. In a similar way the probability for the realization of an atomic state where one electron is in state $a$ and the other in state $b$ is given either by $|\psi^{s}_{atom}|^2$ or by $|\psi^{a}_{atom}|^2$. Which of the two functions gives the correct description of the atomic state depends on the total electron spin of this state, as will be explained below.

If both electrons are in the same state ($a = b$), we obtain from (6.13b)

$$\psi^{a}_{atom} = \psi_1(a)\psi_2(a) - \psi_2(a)\psi_1(a) \equiv 0 . \quad (6.14)$$

This means: The probability to find two electrons with parallel spins in the same state $(n, l, m_l)$ is zero.

> Two electrons with the same quantum numbers $(n, l, m_l)$ are described by the symmetric spatial wave function $\psi^{s}_{atom}$.

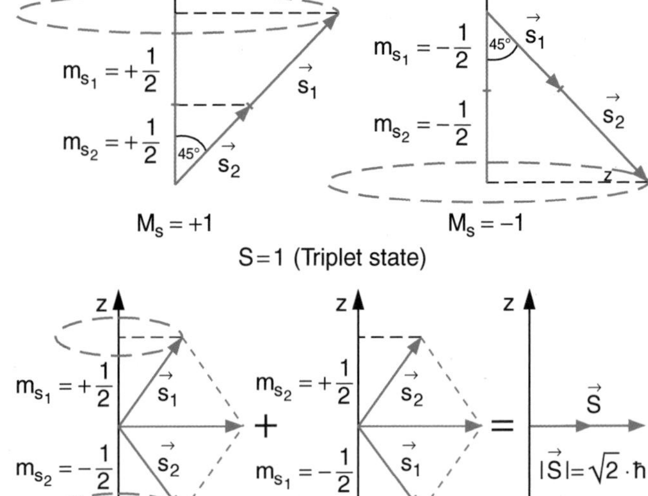


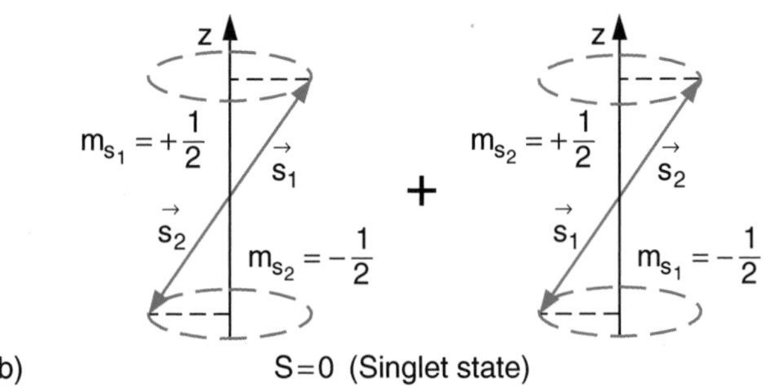


**Fig. 6.4a,b.** Vector model of (**a**) the three triplet sublevels with $S = 1$, $M_S = 0, \pm 1$ and (**b**) of the singlet level with $S = 0$

### 6.1.3 Consideration of the Electron Spin

Based on the experimental facts (fine structure and the anomalous Zeeman effect) described in the preceding chapter we know that each electron has a spin $\boldsymbol{s}$ with a value $|\boldsymbol{s}| = \sqrt{s(s+1)}\hbar$, where the spin quantum number $s$ takes the value $s = +1/2$ and a component $s_z = m_s\hbar$ where the spin projection quantum number $m_s$ can only have the values $m_s = +1/2$ or $m_s = -1/2$. We will describe these two possible spin orientations by spin functions $\chi^+(m_s = +1/2)$ and $\chi^-(m_s = -1/2)$. The correct mathematical description of these functions, which are represented by vectors with two components (spinors), is not important for the following considerations.

The spin state of the atom where both electrons have parallel spins must be described by symmetric spin functions. There are 3 possible ways to form symmetric spin functions:

$$\begin{aligned} \chi^{s}_1(1,2) &= c_1\chi^+(1)\chi^+(2) \quad \text{and} \\ \chi^{s}_2(1,2) &= c_2\chi^-(1)\chi^-(2) , \\ \chi^{s}_3(1,2) &= c_3\left[\chi^+(1)\chi^-(2) + \chi^-(1)\chi^+(2)\right] . \end{aligned} \quad (6.15)$$

which remain unchanged, when the two electrons are exchanged.

If we normalize the spin functions ($|\chi^*\chi|^2 = 1$), the coefficients in (6.15) become $c_1 = c_2 = 1$; $c_3 = 1/\sqrt{2}$. This gives the three normalized symmetric spin functions (Fig. 6.4a)

$$\begin{aligned} \chi^{s}_1(1,2) &= \chi^+(1)\chi^+(2) ; \quad M_s = m_{s_1} + m_{s_2} = +1 \\ \chi^{s}_2(1,2) &= \chi^-(1)\chi^-(2) ; \quad M_s = m_{s_1} + m_{s_2} = -1 \\ \chi^{s}_3(1,2) &= \frac{1}{\sqrt{2}}\left[\chi^+(1)\chi^-(2) + \chi^+(2)\chi^-(1)\right] M_s = 0 , \end{aligned} \quad (6.16)$$

which describe atomic states with total electron spin $\boldsymbol{S} = \boldsymbol{s}_1 + \boldsymbol{s}_2$, its amount $|\boldsymbol{S}| = \sqrt{S(S+1)}\hbar$, the total spin quantum number $S = 1$, and the total spin projection quantum number $M_s = m_{s_1} + m_{s_2} = 0, \pm 1$.

The total electron spin $S$ with $S = 1$ has three possible projections onto the quantization axis with quantum numbers $M_S = 0, \pm 1$. If the electron spin interacts with other angular momenta or with external fields, the corresponding atomic state splits into three components. We therefore name such states ***triplet states***.

The antisymmetric wave function

$$\chi^{\mathrm{a}} = \chi^{+}(1)\chi^{-}(2) - \chi^{+}(2)\chi^{-}(1) \qquad (6.17)$$

represents an atomic state with total electron spin-quantum number $S = 0$ and therefore $M_S = 0$, which we call a ***singlet state*** (Fig. 6.4b).

The total wave function of an atomic state can now be written as the product

$$\psi_{\text{total}} = \psi_{ab}(r_1, \vartheta_1, \varphi_1, r_2, \vartheta_2, \varphi_2) \cdot \chi(S, M_S) \qquad (6.18)$$

of the spatial wave function $\psi(r, \vartheta, \varphi)$ that is determined by the two sets of quantum numbers $a = (n_1, l_1, m_{l_1})$ and $b = (n_2, l_2, m_{l_2})$, and the spin wave function $\chi(S, M_S)$, which depends on the quantum number $S$ of the total electron spin $\boldsymbol{S} = \boldsymbol{s}_1 + \boldsymbol{s}_2$ and the quantum number $M_S = m_{s_1} + m_{s_2}$ of the projection $M_z = M_S \hbar$.

**Note:**

This separation into the product (6.18) is only possible if the interaction between the spin and the orbital angular momentum can be neglected (see Sect. 6.5).

### 6.1.4 The Pauli Principle

The observation and the analysis of the helium spectrum (see next section) and of many other atoms with more than one electron brought the following surprising result.

The only atomic states that are observed in nature are described by total wave functions (spatial function times spin function) that are antisymmetric against a permutation of two electrons.

Based on these experimental results and on fundamental symmetry arguments *Wolfgang Pauli* (1900–1958) (Fig. 6.5) postulated the general symmetry rule in 1925 (called the Pauli principle):

**Fig. 6.5.** *Wolfgang Pauli* (1900–1958) From: E. Bagge: Die Nobelpreisträger der Physik, Heinz Moos-Verlag, München, 1964

The total wave function of a system with more than one electron is always antisymmetric with repect to an exchange of two electrons.

Up until now, no exception to this rule has been found!

A more involved theoretical treatment of systems with identical particles shows that this Pauli principle is valid for all particles with spin quantum number $s = (n + 1/2)$ $(n = 0, 1, 2, \ldots)$. Such particles are called ***Fermions***, whereas particles with integer spin quantum number $s$ are called ***Bosons***.

The Pauli principle is therefore also valid for protons and neutrons (both have a nuclear spin quantum number of 1/2).

If two electrons of an atom are both in the same state described by the spatial wave function $\psi_{n,l,m}$, (this means that they have the same quantum numbers $n$, $l$, and $m_l$), their antisymmetric spatial wave functions becomes, according to (6.14), zero. This implies that such a state has to be described by a spatial wave function that is symmetric against exchange of two electrons. Since the Pauli principle demands that the total wave function has to be antisymmetric, it follows that the spin wave function must be antisymmetric and can be described by (6.17). The two spin projections $m_{s1}\hbar$ and $m_{s2}\hbar$ must differ in their sign.

In other words, two electrons with the same quantum numbers $n, l, m_l$ must have different spin projection quantum numbers $m_{s_1} \neq m_{s_2}$! When we describe an atomic state by the four quantum numbers $(n, l, m_l, m_s)$ for each electron we can formulate the Pauli principle as:

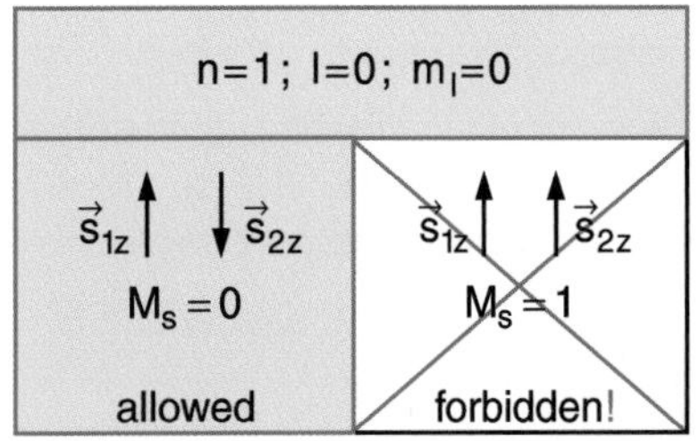


**Fig. 6.6.** Ground state $^1 1 S_0$ of the helium atom with $n=1$, $l=0$, $m_l=0$, $M_s=0$

> An atomic state characterized by the set of four quantum numbers $(n, l, m_l, m_s)$ can be occupied by at most one electron.

or:

> An atomic state with the three quantum numbers $(n, l, m_l)$, can be occupied by at most two electrons with opposite spin orientations $m_s = +1/2$ and $m_s = -1/2$ (Fig. 6.6).

### 6.1.5 Energy Levels of the Helium Atom

The lowest energy level of the He atom (ground state) is obtained if both electrons have the lowest possible principal quantum number $n = 1$. The other quantum numbers must then be $l = 0$ and $m_l = 0$. The two electrons now have identical quantum numbers $(n, l, m_l)$ of the spatial wave function and therefore their spin quantum numbers $m_{s_1} = +1/2 \neq m_{s_2} = -1/2$ must be different. Since the spatial wave function is symmetric, the spin function must be antisymmetric. For the total spin we get $S = s_1 + s_2 = 0$. Both spins are antiparallel and $M_S = m_{s_1} + m_{s_2} = 0$. The helium ground state is a singlet state. The state does not split in an external magnetic field but has only one Zeeman component because its total angular momentum and therefore also its magnetic moment are zero.

The number $2S + 1$ of possible orientations $M_S$ of the total spin $S$ is called the ***multiplicity*** of the atomic state. The multiplicity of an atomic state is written as an upper index in front of the symbol for the total orbital angular momentum. Each atomic state is characterized by the symbol $n^{S+1}L_J$, where $n$ is the principal quantum number, $L$ the quantum number of the total orbital angular momentum $\vec{L} = \vec{l}_1 + \vec{l}_2$ (apart from spin) and $J$ the quantum number of the total angular momentum $\vec{J} = \vec{L} + \vec{S}$ including the spin.

The helium ground state is then labeled as the $1^1S_0$ state ($n = 1$, $2S + 1 = 1$, $L = 0$ and $J = 0$).

The helium atom can be excited into higher electronic states by absorption of photons or by electron impact or by collisions with other particles, if their energy is sufficiently high. If one electron, say $e_1$, is

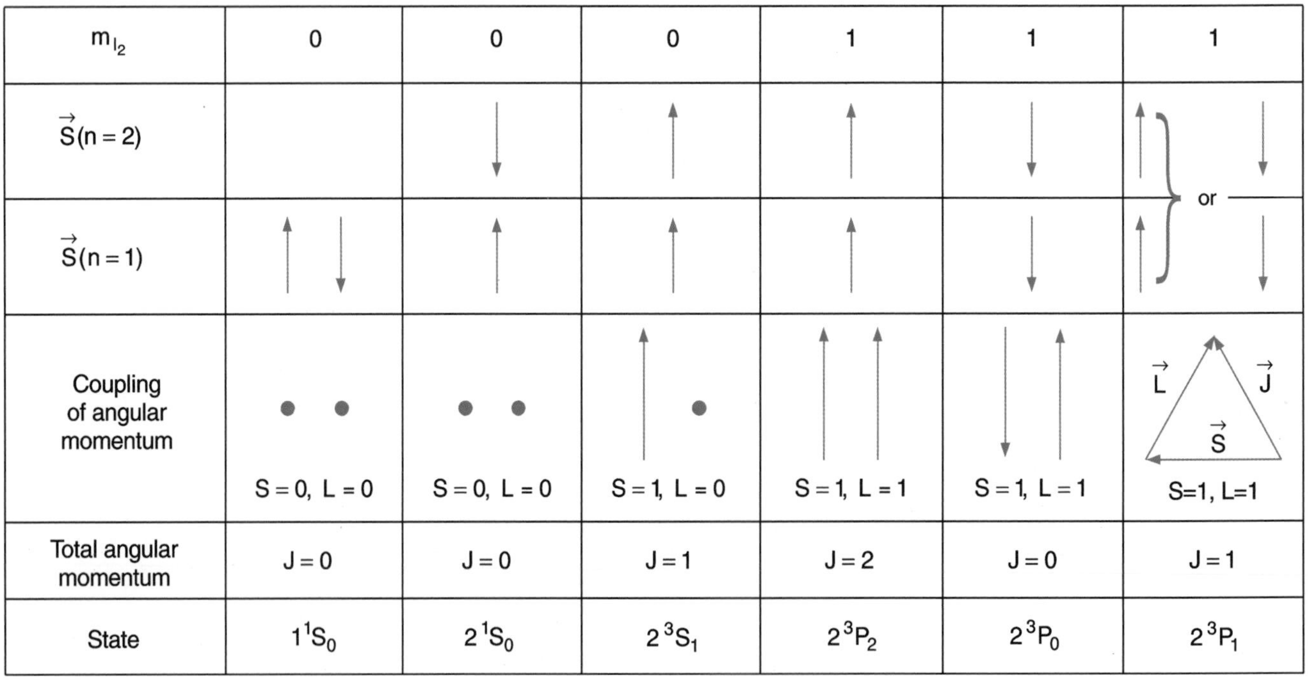

| $m_{l_2}$ | 0 | 0 | 0 | 1 | 1 | 1 |
|---|---|---|---|---|---|---|
| $\vec{S}(n=2)$ | | ↓ | ↑ | ↑ | ↓ | ↑ or ↓ |
| $\vec{S}(n=1)$ | ↑↓ | ↑ | ↑ | ↑ | ↓ | ↑ or ↓ |
| Coupling of angular momentum | S = 0, L = 0 | S = 0, L = 0 | S = 1, L = 0 | S = 1, L = 1 | S = 1, L = 1 | $\vec{L}$ $\vec{J}$ $\vec{S}$ S=1, L=1 |
| Total angular momentum | J = 0 | J = 0 | J = 1 | J = 2 | J = 0 | J = 1 |
| State | $1^1S_0$ | $2\,^1S_0$ | $2\,^3S_1$ | $2\,^3P_2$ | $2\,^3P_0$ | $2\,^3P_1$ |

**Fig. 6.7.** Symbolic representation of the quantum numbers $n$, $L$ and $S$ for the ground state and some excited states of the helium atom. The electron $e_1$ is always in the $1S$ ground state

excited into a state with $n = 2$ and the other electron $e_2$ stays in the lower state with $n = 1$, the quantum number $l_1$ can take the values $l_1 = 0$ or $l_1 = 1$. Since the principle quantum numbers $n_1 = 2$ and $n_2 = 1$ differ, all other quantum numbers can be the same for the two electrons or they can differ (see Fig. 6.7). Therefore the following excited states of the He atom can be realized for $n_1 = 2$ and $(n_2 = 1, l_2 = 0, m_{l_2} = 0, m_{s_2} = +1/2)$:

$$2^1S_0\left(l_1 = 0, m_{l_1} = 0, m_{s_1} = -\frac{1}{2}, J = 0\right)$$
$$2^1P_1\left(l_1 = 1, m_{l_1} = 0, \pm 1, m_{s_1} = -\frac{1}{2}, J = 1\right)$$
$$2^3S_1\left(l_1 = 0, m_{l_1} = 0, m_{s_1} = +\frac{1}{2}, J = 1\right)$$
$$2^3P_0\left(l_1 = 1, m_{l_1} = -1, m_{s_1} = +\frac{1}{2}, J = 0\right)$$
$$2^3P_1\left(l_1 = 1, m_{l_1} = 0, m_{s_1} = +\frac{1}{2}, J = 1\right)$$
$$2^3P_2\left(l_1 = 1, m_{l_1} = +1, m_{s_1} = +\frac{1}{2}, J = 2\right).$$

While the ground state $^1S_0$ of the helium atom must be a singlet state according to the Pauli principle, the excited states can be either singlet or triplet states.

Because of spin-orbit coupling (see Sects. 5.6.2 and 6.5) all triplet states with the spin quantum number $S = 1$ and the orbital quantum number $L \geq 1$ split into three fine structure components that differ in the quantum number $J$ of the total angular momentum $\boldsymbol{J} = \boldsymbol{l}_1 + \boldsymbol{l}_2 + \boldsymbol{s}_1 + \boldsymbol{s}_2$ (Fig. 6.8).

The magnitude of the splitting and the energetic order of the fine structure components depend on the kind and strength of the coupling between the different angular momenta (see Sect. 6.5).

The level system of the helium atom therefore consists of a singlet system (single components with $S = 0 \Rightarrow J = L$) and a triplet system with $S = 1$ (Fig. 6.9).

The energy of the singlet levels is quite different from that of the triplet levels with the same quantum numbers $(n, l, m_l)$. The reason for this difference is *not* the magnetic interaction of the spin-orbit coupling (which only causes small fine structure splittings), but a consequence of the Pauli principle. For example, the energetic difference between the $2^1S_0$ level and the

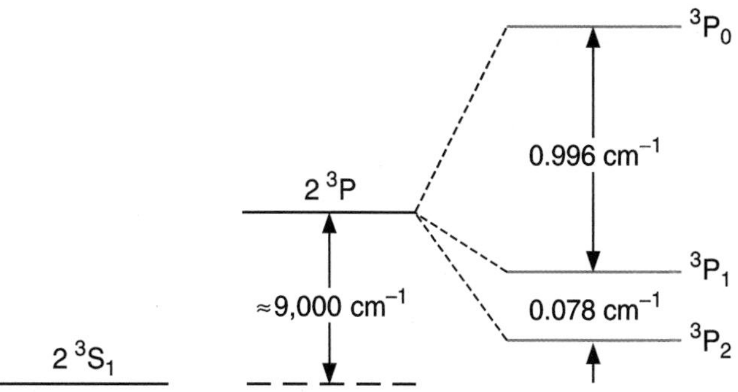


**Fig. 6.8.** Fine structure of the $2^3P$ state compared with the unsplit $2^3S_1$ state of the helium atom

$2^3S_0$ level is $\Delta E = E(^1S_0) - E(^3S_0) = 0.78\,\text{eV}$! The $2^3S$ level is described by the antisymmetric spatial wave function, where the mutual time-averaged distance $\langle r_{12} \rangle$ between the two electrons is larger than in the $2^1S$ state, where the electrons can come much closer together since the spatial wave function is, even for $r_{12} = 0$, not equal to zero. The time-averaged electrostatic repulsion between the two electrons

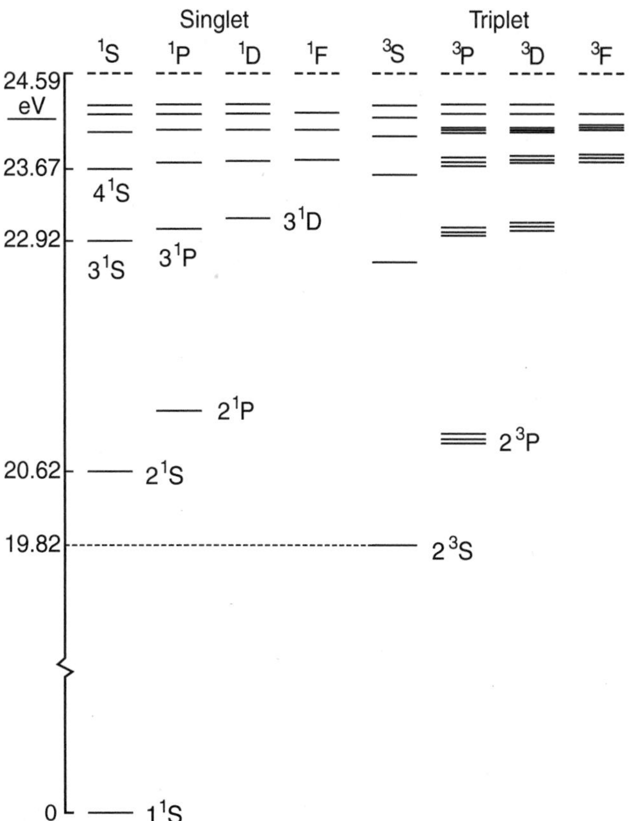


**Fig. 6.9.** Level scheme of singlet and triplet states of the helium atom from $L = 0$ up to $L = 3$. The ground state $1^1S_0$ is chosen to have the energy $E = 0$

$$\langle E_{\text{pot}}(r_{12})\rangle = \left\langle \frac{e^2}{4\pi\varepsilon_0 r_{12}} \right\rangle = \frac{e^2}{4\pi\varepsilon_0} \int \psi^* \frac{1}{r_{12}} \psi \, d\tau$$

is therefore larger in the $2^1S$ state than in the $2^3S$ state. This pushes the energy of the $2^1S$ state above that of the $2^3S$ state. A second reason is, that the shielding of the nuclear Coulomb field by the electrons is smaller in the triplet state, because the two electrons are farther apart. This increases the attraction between nucleus and electrons and decreases the energy.

**Note:**

While this explanation is correct for the $S$ states the situation is more complex for the $P$ states. Detailed calculations [6.1] have shown that for the $^3P$ state the average distance between the two electrons is smaller than for the $^1P$ state, which implies that the positive Coulomb repulsion energy is larger in the $^3P$ state than in the $^1P$ state. The reason for the lower total energy of the $^3P$ state is that for both electrons the average distance from the nucleus is smaller, which gives a larger negative potential energy of the electrons in the $^3P$ state. This second effect overcompensates the first effect.

### 6.1.6 Helium Spectrum

The spectrum of the helium atom consists of all allowed transitions between two arbitrary energy levels $E_i$, $E_k$ (see Chap. 7). For all excited states where only one of the two electrons is excited only this electron is involved in such a transition, the other stays in the ground state. The energy of the absorbed or emitted photons is

$$h\nu_{ik} = E_i - E_k \Rightarrow \lambda_{ik} = \frac{hc}{E_i - E_k} \tag{6.19}$$

(Fig. 6.10). However, not every transition obeying the energy relation (6.19) is actually observed in the spectrum, because certain selection rules exist (see Chap. 7) for possible transitions. For the Helium atom with the total orbital angular momentum $\boldsymbol{L} = \boldsymbol{l}_1 + \boldsymbol{l}_2$, the total spin $\boldsymbol{S} = \boldsymbol{s}_1 + \boldsymbol{s}_2$ and the total electronic angular momentum $\boldsymbol{J} = \boldsymbol{L} + \boldsymbol{S}$ these selection rules can be expressed by the corresponding quantum numbers L, S and J:

Only those transitions are allowed for which the selection rules

$$\Delta L = \pm 1;\ \Delta M_L = 0, \pm 1;\ \Delta S = 0;\ \Delta J = 0 \text{ or } \pm 1 \text{ except for } J = 0 \leftrightarrow J = 0 \tag{6.20}$$

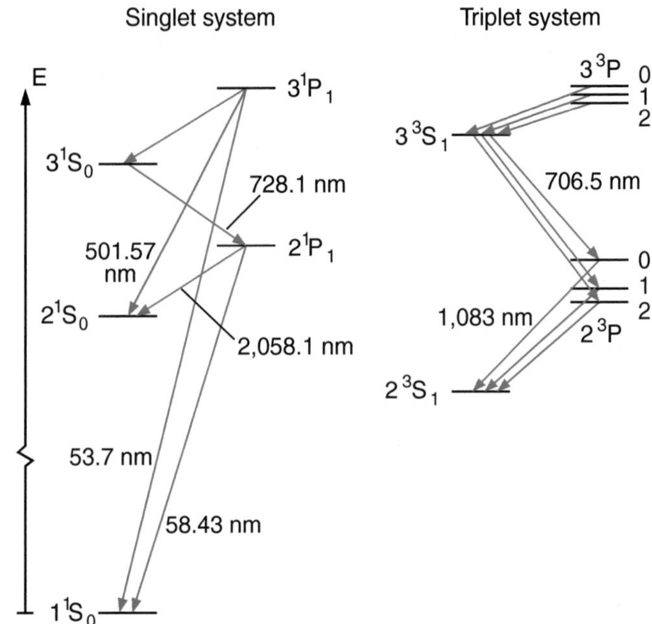


**Fig. 6.10.** Possible transitions within the singlet and the triplet system

are fulfilled. Transitions between two levels with J = 0 are forbidden.

These selection rules just reflect the conservation of angular momenta of the system atom + absorbed or emitted photon.

According to these selection rules, transitions between the singlet system ($S = 0$) and the triplet system ($S = 1$) are forbidden.

For transitions between triplet levels with $L \geq 1$, more than three components are often observed, as can be seen from Fig. 6.11, which shows the six possible transitions between the fine structure components of $3^3D$ and $2^3P$ levels.

Since the spectrum of the singlet system looks quite different from that of the triplet system regarding the

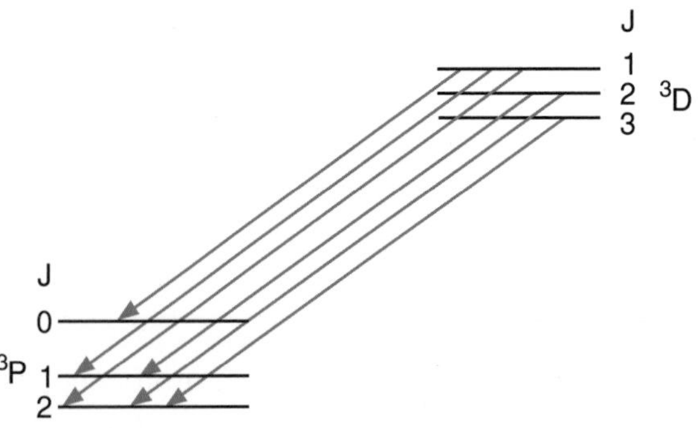


**Fig. 6.11.** All allowed transitions between the fine structure levels of the $^3D$ and the $^3P$ state

line positions and the fine structure (Fig. 6.10) the two spectra were initially regarded as originating from different kinds of atoms. Because the chemical analysis had unambiguously identified both systems as belonging to helium, it was believed that two kinds of helium might exist, which were named ***para-helium*** and ***ortho-helium***.

Today we know that there exists only one kind of helium and that the difference in the spectra stems from the different total electron spin $\boldsymbol{S} = \boldsymbol{s}_1 + \boldsymbol{s}_2$. For para-helium the total spin quantum number is $S = 0$ and for ortho-helium it is $S = 1$.

## 6.2 Building-up Principle of the Electron Shell for Larger Atoms

Since the Pauli principle does not allow more than two electrons in the $1s$ state with $n = 1$, the additional electrons in atoms with more than two electrons have to occupy higher energy states with $n \geq 2$ even in the lowest energy state (ground state) of these atoms.

The population of electrons in atoms with energy levels $(n, l, m_l, m_s)$ occurs in such a way that

1. The Pauli principle is obeyed and
2. The total energy of all electrons is minimum for the atomic ground state.

It is remarkable that the structure of the electron shells of all existing atomic elements can be explained by these two principles. In particular, the arrangement of the elements in the periodic table postulated by *D. Mendelejew* 1869 and indepently by *L. Meyer* 1870 by comparing the chemical properties of the elements, follows quite naturally from these principles in a very satisfactory way. It explains the periodic table using the structure of the atomic electron shells, governed by these two principles.

> Without the Pauli principle the electron shells of all atoms would collapse into the $1s$ shell with the lowest energy. One can therefore say: The Pauli principle guarantees the stability of atoms and the great variety of chemical properties of the different elements.

We will explain these general remarks by some specific examples.

### 6.2.1 The Model of Electron Shells

The radial distributions of atomic electrons, according to Sect. 5.1.4, is given by

$$P(r) = r^2 |R_{n,l}(r)|^2 ,$$

where $R_{n,l}(r)$ is the radial part of the wave function for an electron with principal quantum number $n$ and orbital angular momentum quantum number $l$.

We discussed in Sect. 4.3 that for each value of $l$ there are $(2l + 1)$ degenerate wave functions $Y_l^m$, with different quantum numbers $m_l$, describing different angular distributions. For each value of the principal quantum $n$ there are $n$ possible values $l = 0, 1, 2, \ldots, n-1$ of the angular momentum quantum number. Therefore, there are

$$\sum_{l=0}^{n-1} (2l + 1) = n^2 \tag{6.21}$$

different states described by the wave functions $\psi_{n,l,m}(r, \vartheta, \varphi)$ that can be occupied by at most $2n^2$ electrons with pairs of opposite spins, according to the Pauli principle.

The time-averaged total charge distribution of all $2n^2$ electrons with the same principal quantum number $n$

$$e|\psi_n|^2 = e \sum_l \sum_{m_l} |\psi_{n,l,m_l}|^2 = C \cdot e \sum_l |R_{n,l}(r)|^2 \tag{6.22}$$

is obtained by summation over the squares of all possible wave functions with $l < n$ and $-l \leq m_l \leq +l$, where $C$ is a normalization factor. This gives a spherically symmetric charge distribution, as can be seen by summing over all squared spherical harmonics $Y_l^m$ for a given value of $n$. This charge distribution has maxima at certain values of the distance $r$ from the nucleus, which solely depend on the principal quantum number $n$. The main part of the electron charge is contained within the spherical shell between the radii $r - \Delta r/2$ and $r + \Delta r/2$ (Fig. 6.12). Such a spherically symmetric charge distribution is called an ***electron shell***. The different shells are labeled as follows:

$$\begin{aligned} n &= 1 : \text{K-shell} , & n &= 4 : \text{N-shell} \\ n &= 2 : \text{L-shell} , & n &= 5 : \text{O-shell} \\ n &= 3 : \text{M-shell} , & n &= 6 : \text{P-shell} \end{aligned}$$

Each of these electron shells has, including the electron spin, $2n^2$ states $(n, l, m_l, m_s)$, where each of these states

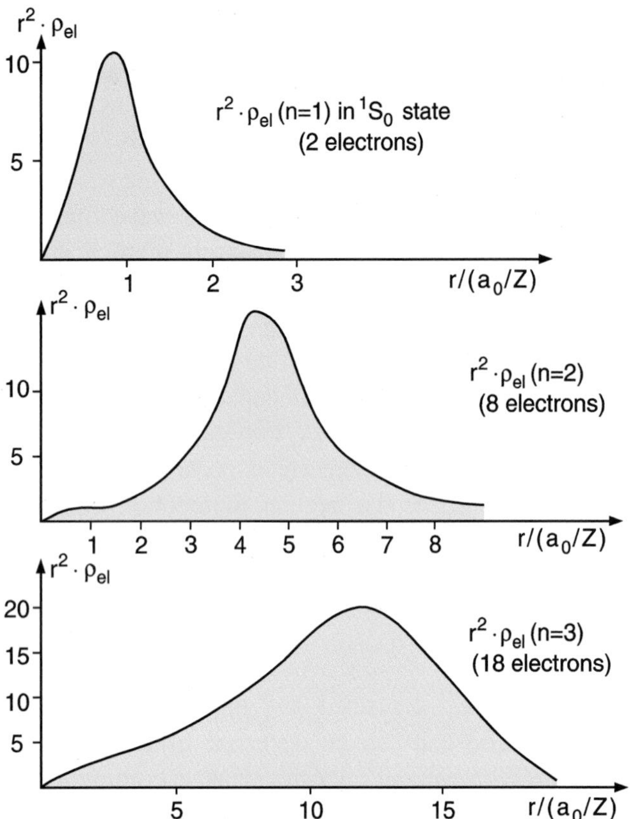


**Fig. 6.12.** Radial electron density distribution for fully occupied shells with $n = 1, 2$ and 3

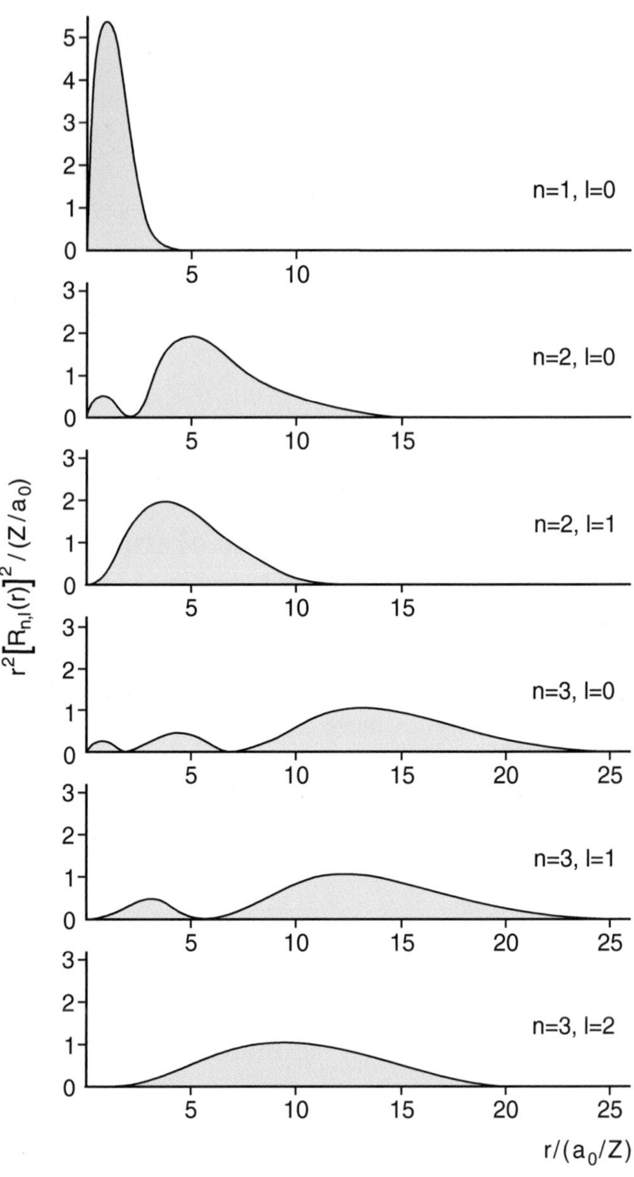


**Fig. 6.13.** Radial dependence of the probability density for an electron between the spherical shells for $r$ and $r + dr$ for different quantum numbers $n, l$

can be occupied by at most one electron. Some of these states can be degenerate (for instance all $2l + 1$ levels of a given $l$-value are degenerate without an external magnetic field).

> According to the Pauli principle each electron shell can be occupied by at most $2n^2$ electrons.

Since the radial wave function for non-Coulomb potentials also depends on the angular momentum quantum number $l$ (Fig. 6.13), one calls the arrangement of all electrons with given values of $n$ and $l$ a ***subshell*** (Table 6.1).

> For each value of $n$ there are $n$ different values of $l$ and therefore $n$ subshells.

### 6.2.2 Successive Building-up of Electron Shells for Atoms with Increasing Nuclear Charge

The successive building-up of the electron shell with increasing total number $Z$ of atomic electrons according to the Pauli principle is illustrated in Fig. 6.14 for the ground states of atoms with the ten smallest values of $Z$ from hydrogen ($Z = 1$) to neon ($Z = 10$). The two possible spin states $m_s = \pm 1/2$ are symbolized by upwards or downwards arrows. Fully occupied states are marked as dark blue, states with only one electron as light blue and unoccupied states as white.

For lithium, with $Z = 3$, the third electron cannot occupy the K-shell ($1s$), because there are already two electrons. It has to be in the next highest energy L-subshell $2s$ with $n = 2$ and $l = 0$. The electron configuration of the Li atom is then $(1s)^2(2s)$, where

**Table 6.1.** Maximum number of electrons in the different atomic electron shells and subshells

| Shell | K | L | M | N | O |
|---|---|---|---|---|---|
| Maximum number of electrons in shell X | 2 | 8 | 18 | 32 | 50 |
| Subshells<br>Number of electrons | $1s$<br>2 | $2s$ $2p$<br>2 6 | $3s$ $3p$ $3d$<br>2 6 10 | $4s$ $4p$ $4d$ $4f$<br>2 6 10 14 | $+5g$<br>18 |
| Total number of electrons up to the filled shell X | 2 | 10 | 28 | 60 | 110 |

the exponent gives the number of electrons in the corresponding subshell. The quantum numbers of the third electron are $n = 2$; $l = 0$, $m_l = 0$; $m_s = \pm 1/2$ and the Li ground state is labeled as $2^2S_{1/2}$ (see Sect. 6.1.4).

The fourth electron in the beryllium atom can still occupy the $2s$ state ($n = 2$; $l = 0$; $m_l = 0$; $m_s = -1/2$) if the spin quantum number $m_s$ differs from that of the third electron. The ground state of the Be atom is therefore $2^1S_0$.

For the fifth electron in the boron atom the state $2s$ is already occupied and it has to go into the $2p$ state with $n = 2$ and $l = 1$. The ground state of B is then $2^2P_{1/2}$.

The next two electrons for the elements carbon C and nitrogen N still fit into the subshell $2p$ with $l = 1$ and $m_l = 0, \pm 1$. It turns out that the lowest energy is realized, if the three electrons have parallel spins. The reason for this rule is that the spatial wave functions for electrons with parallel spins are antisymmetric (see Sect. 6.1.4) and describe an electron distribution where the electrons are farther apart than for symmetric wave functions. Therefore their mutual Coulomb repulsion is smaller and the energy is lower. This is summarized in Hund's rule:

For every atomic ground state, the total electron spin has the maximum value tolerated by the Pauli principle.

The quantum numbers ($L$, $S$, and $J$) of the atomic ground states are determined by the total orbital angular momentum $\boldsymbol{L} = \sum \boldsymbol{l}_i$, the total spin $\boldsymbol{S} = \sum \boldsymbol{s}_i$ and their coupling to $\boldsymbol{J} = \boldsymbol{L} + \boldsymbol{S}$. The ground state of C is then $2^3P_0$ because $|\vec{L}| = |\vec{l}_1 + \vec{l}_2| = 1\hbar$, and of N it is $2^4S_{3/2}$ because $\sum \vec{l}_i = \vec{0}$. For the next three atoms O, F and Ne the three additional electrons still fit into the $2p$ shell, but according to the Pauli principle, their spins must be opposite to that of the three electrons, already occupying this subshell. The total spin quantum number therefore decreases from oxygen ($S = 1$) to fluorine ($S = 1/2$) to neon ($S = 0$). For neon the L-shell with $n = 2$ is fully occupied. The total orbital angular momentum is $L = \sum l_i = 0$ and the total spin $S = \sum s_i = 0$. The time-averaged electron charge distribution for neon is spherically symmetric. The spectroscopic labels of the ground states of the first ten elements are given in Fig. 6.14.

With sodium ($Z = 11$) the building-up of the M-shell with $n = 3$ starts, until eight electrons fill this

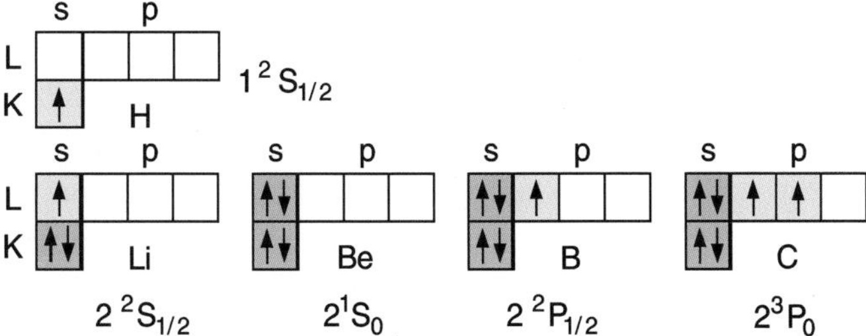


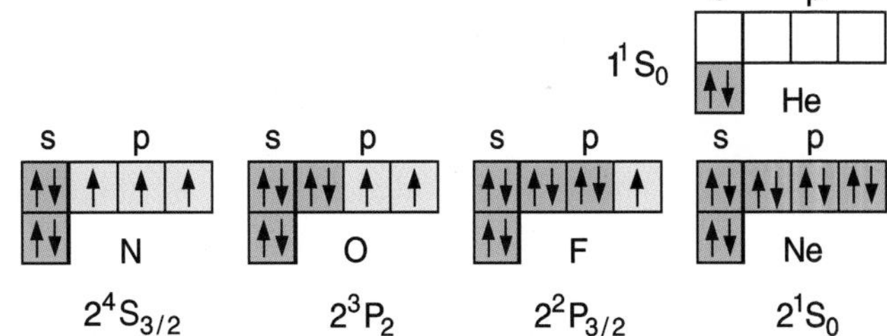


**Fig. 6.14.** Building up principle of the electron configurations for the ground states of the first ten elements in the periodic table. The arrows indicate the electron spins with $m_s = \pm 1/2$. In states occupied by only one electron the spin projection quantum number can be $m_s = \pm 1/2$

**Table 6.2.** Electron configuration in the ground states of the chemical elements

| Shell | | | K | L | | M | | | O |
|---|---|---|---|---|---|---|---|---|---|
| Z | | Element | 1s | 2s | 2p | 3s | 3p | 3d | 4s |
| 1 | H | Hydrogen | 1 | | | | | | |
| 2 | He | Helium | 2 | | | | | | |
| 3 | Li | Lithium | 2 | 1 | | | | | |
| 4 | Be | Beryllium | 2 | 2 | | | | | |
| 5 | B | Boron | 2 | 2 | 1 | | | | |
| 6 | C | Carbon | 2 | 2 | 2 | | | | |
| 7 | N | Nitrogen | 2 | 2 | 3 | | | | |
| 8 | O | Oxygen | 2 | 2 | 4 | | | | |
| 9 | F | Fluorine | 2 | 2 | 5 | | | | |
| 10 | Ne | Neon | 2 | 2 | 6 | | | | |
| 11 | Na | Sodium | 2 | 2 | 6 | 1 | | | |
| 12 | Mg | Magnesium | 2 | 2 | 6 | 2 | | | |
| 13 | Al | Aluminum | 2 | 2 | 6 | 2 | 1 | | |
| 14 | Si | Silicon | 2 | 2 | 6 | 2 | 2 | | |
| 15 | P | Phosphorus | 2 | 2 | 6 | 2 | 3 | | |
| 16 | S | Sulfur | 2 | 2 | 6 | 2 | 4 | | |
| 17 | Cl | Chlorine | 2 | 2 | 6 | 2 | 5 | | |
| 18 | Ar | Argon | 2 | 2 | 6 | 2 | 6 | | |
| 19 | K | Potassium | 2 | 2 | 6 | 2 | 6 | | 1 |
| 20 | Ca | Calcium | 2 | 2 | 6 | 2 | 6 | | 2 |
| 21 | Sc | Scandium | 2 | 2 | 6 | 2 | 6 | 1 | 2 |
| 22 | Ti | Titanium | 2 | 2 | 6 | 2 | 6 | 2 | 2 |
| 23 | V | Vanadium | 2 | 2 | 6 | 2 | 6 | 3 | 2 |
| 24 | Cr | Chromium | 2 | 2 | 6 | 2 | 6 | 5 | 1 |
| 25 | Mn | Manganese | 2 | 2 | 6 | 2 | 6 | 5 | 2 |
| 26 | Fe | Iron | 2 | 2 | 6 | 2 | 6 | 6 | 2 |
| 27 | Co | Cobalt | 2 | 2 | 6 | 2 | 6 | 7 | 2 |

| Shell | | | K | L | | M | | | N | | | O | |
|---|---|---|---|---|---|---|---|---|---|---|---|---|---|
| Z | | Element | 1s | 2s | 2p | 3s | 3p | 3d | 4s | 4p | 4d | 5s | 5p |
| 28 | Ni | Nickel | 2 | 2 | 6 | 2 | 6 | 8 | 2 | | | | |
| 29 | Cu | Copper | 2 | 2 | 6 | 2 | 6 | 10 | 1 | | | | |
| 30 | Zn | Zink | 2 | 2 | 6 | 2 | 6 | 10 | 2 | | | | |
| 31 | Ga | Gallium | 2 | 2 | 6 | 2 | 6 | 10 | 2 | 1 | | | |
| 32 | Ge | Germanium | 2 | 2 | 6 | 2 | 6 | 10 | 2 | 2 | | | |
| 33 | As | Arsenic | 2 | 2 | 6 | 2 | 6 | 10 | 2 | 3 | | | |
| 34 | Se | Selenium | 2 | 2 | 6 | 2 | 6 | 10 | 2 | 4 | | | |
| 35 | Br | Bromium | 2 | 2 | 6 | 2 | 6 | 10 | 2 | 5 | | | |
| 36 | Kr | Krypton | 2 | 2 | 6 | 2 | 6 | 10 | 2 | 6 | | | |
| 37 | Rb | Rubidium | 2 | 2 | 6 | 2 | 6 | 10 | 2 | 6 | | 1 | |
| 38 | Sr | Strontium | 2 | 2 | 6 | 2 | 6 | 10 | 2 | 6 | | 2 | |
| 39 | Y | Yttrium | 2 | 2 | 6 | 2 | 6 | 10 | 2 | 6 | 1 | 2 | |
| 40 | Zr | Zirconium | 2 | 2 | 6 | 2 | 6 | 10 | 2 | 6 | 2 | 2 | |
| 41 | Nb | Niobium | 2 | 2 | 6 | 2 | 6 | 10 | 2 | 6 | 4 | 1 | |
| 42 | Mo | Molybdenum | 2 | 2 | 6 | 2 | 6 | 10 | 2 | 6 | 5 | 1 | |
| 43 | Tc | Technetium | 2 | 2 | 6 | 2 | 6 | 10 | 2 | 6 | 6 | 1 | |
| 44 | Ru | Ruthenium | 2 | 2 | 6 | 2 | 6 | 10 | 2 | 6 | 7 | 1 | |
| 45 | Rh | Rhodium | 2 | 2 | 6 | 2 | 6 | 10 | 2 | 6 | 8 | 1 | |
| 46 | Pd | Palladium | 2 | 2 | 6 | 2 | 6 | 10 | 2 | 6 | 10 | | |
| 47 | Ag | Silver | 2 | 2 | 6 | 2 | 6 | 10 | 2 | 6 | 10 | 1 | |
| 48 | Cd | Cadmium | 2 | 2 | 6 | 2 | 6 | 10 | 2 | 6 | 10 | 2 | |
| 49 | In | Indium | 2 | 2 | 6 | 2 | 6 | 10 | 2 | 6 | 10 | 2 | 1 |
| 50 | Sn | Tin | 2 | 2 | 6 | 2 | 6 | 10 | 2 | 6 | 10 | 2 | 2 |
| 51 | Sb | Antimony | 2 | 2 | 6 | 2 | 6 | 10 | 2 | 6 | 10 | 2 | 3 |
| 52 | Te | Tellurium | 2 | 2 | 6 | 2 | 6 | 10 | 2 | 6 | 10 | 2 | 4 |
| 53 | I | Iodine | 2 | 2 | 6 | 2 | 6 | 10 | 2 | 6 | 10 | 2 | 5 |
| 54 | Xe | Xenon | 2 | 2 | 6 | 2 | 6 | 10 | 2 | 6 | 10 | 2 | 6 |

M-shell, which is still not fully occupied for argon with $Z = 18$ because the $d$ subshell is not yet occupied (Table 6.2). The analysis of the atomic spectra proves that with potassium ($Z = 19$) in the first row of the third period in the periodic table the building up of the $4s$ shell starts, which is fully occupied for calcium ($Z = 20$) before the $3d$ shell is filled. The reason for this apparent deviation from the regular scheme stems from the fact that the $3d$ electrons are, on average, farther away from the atomic nucleus than the $4s$ electrons. Therefore their energy is higher and the principle of energy minimization favors the $4s$ electrons. Indeed extensive computer calculations of the total energy $E_{\text{total}}$ prove that $E_{\text{total}}$ of the electron shell of K and Ca is smaller if the $4s$ shell is filled instead of the $3d$ shell.

In Fig. 6.17 the successive building-up of the different electron shells is illustrated by an arrow diagram (without taking into account the peculiarity for Cu, where electrons are rearranged within a subshell).

This model of atomic electron shells can explain all peculiarities in the periodic table such as the group of rare earth elements, sitting all in the same row of the table. Here inner shells $4f$ and $5d$ are successively filled without changing the occupation of the outer $6s$ shell, which determines the chemical character of an element (see below). Also for all actinides from radium ($Z = 88$) to Rutherfordium ($Z = 104$) the occupation of the outer shell $7s$ is the same while the inner shells $5f$ and $6d$ are successively filled up.

### 6.2.3 Atomic Volumes and Ionization Energies

The shell structure of the atomic electron distribution is substantiated by many experimental results. We will only present some of them here.

The experimental techniques discussed in Sect. 2.4 allow the determination of atomic sizes and volumes. The dependence of these volumes on the number $Z$ of atomic electrons exhibit a typical periodicity (Fig. 6.15) corresponding to that of the periodic table. Each time a new electron shell starts to be occupied (for the elements Li, Na, K, Rb and Cs), the atomic volumes jump upwards. The atomic shell model explains this readily, because the new shell with a higher principal quantum number $n$ has a larger mean radius $\langle r \rangle$

**Table 6.2.** Electron configuration in the ground states of the chemical elements (continued)

| Shell | | | N | O | | | | P | Shell | | | N | O | | | | P | | | Q |
|---|---|---|---|---|---|---|---|---|---|---|---|---|---|---|---|---|---|---|---|---|
| Z | | Element | 4f | 5s | 5p | 5d | 5f | 6s | Z | | Element | 4f | 5s | 5p | 5d | 5f | 6s | 6p | 6d | 7s |
| 55 | Cs | Cesium | | 2 | 6 | | | 1 | 80 | Hg | Mercury | 14 | 2 | 6 | 10 | | 2 | | | |
| 56 | Ba | Barium | | 2 | 6 | | | 2 | 81 | Tl | Thallium | 14 | 2 | 6 | 10 | | 2 | 1 | | |
| 57 | La | Lanthanium | | 2 | 6 | 1 | | 2 | 82 | Pb | Lead | 14 | 2 | 6 | 10 | | 2 | 2 | | |
| 58 | Ce | Cerium | 2 | 2 | 6 | | | 2 | 83 | Bi | Bismuth | 14 | 2 | 6 | 10 | | 2 | 3 | | |
| 59 | Pr | Praseodymium | 3 | 2 | 6 | | | 2 | 84 | Po | Polonium | 14 | 2 | 6 | 10 | | 2 | 4 | | |
| 60 | Nd | Neodymium | 4 | 2 | 6 | | | 2 | 85 | At | Astatine | 14 | 2 | 6 | 10 | | 2 | 5 | | |
| 61 | Pm | Promethium | 5 | 2 | 6 | | | 2 | 86 | Rn | Radon | 14 | 2 | 6 | 10 | | 2 | 6 | | |
| 62 | Sm | Samarium | 6 | 2 | 6 | | | 2 | 87 | Fr | Francium | 14 | 2 | 6 | 10 | | 2 | 6 | | 1 |
| 63 | Eu | Europium | 7 | 2 | 6 | | | 2 | 88 | Ra | Radium | 14 | 2 | 6 | 10 | | 2 | 6 | | 2 |
| 64 | Gd | Gadolinium | 7 | 2 | 6 | 1 | | 2 | 89 | Ac | Actinium | 14 | 2 | 6 | 10 | | 2 | 6 | 1 | 2 |
| 65 | Tb | Terbium | 9 | 2 | 6 | | | 2 | 90 | Th | Thorium | 14 | 2 | 6 | 10 | | 2 | 6 | 2 | 2 |
| 66 | Dy | Dysprosium | 10 | 2 | 6 | | | 2 | 91 | Pa | Protactinium | 14 | 2 | 6 | 10 | 2 | 2 | 6 | 1 | 2 |
| 67 | Ho | Holmium | 11 | 2 | 6 | | | 2 | 92 | U | Uranium | 14 | 2 | 6 | 10 | 3 | 2 | 6 | 1 | 2 |
| 68 | Er | Erbium | 12 | 2 | 6 | | | 2 | 93 | Np | Neptunium | 14 | 2 | 6 | 10 | 5 | 2 | 6 | | 2 |
| 69 | Tm | Thulium | 13 | 2 | 6 | | | 2 | 94 | Pu | Plutonium | 14 | 2 | 6 | 10 | 6 | 2 | 6 | | 2 |
| 70 | Yb | Ytterbium | 14 | 2 | 6 | | | 2 | 95 | Am | Americium | 14 | 2 | 6 | 10 | 7 | 2 | 6 | | 2 |
| 71 | Lu | Lutetium | 14 | 2 | 6 | 1 | | 2 | 96 | Cm | Curium | 14 | 2 | 6 | 10 | 7 | 2 | 6 | 1 | 2 |
| 72 | Hf | Hafnium | 14 | 2 | 6 | 2 | | 2 | 97 | Bk | Berkelium | 14 | 2 | 6 | 10 | 8 | 2 | 6 | 1 | 2 |
| 73 | Ta | Tantalium | 14 | 2 | 6 | 3 | | 2 | 98 | Cf | Californium | 14 | 2 | 6 | 10 | 10 | 2 | 6 | | 2 |
| 74 | W | Tungsten | 14 | 2 | 6 | 4 | | 2 | 99 | Es | Einsteinium | 14 | 2 | 6 | 10 | 11 | 2 | 6 | | 2 |
| 75 | Re | Rhenium | 14 | 2 | 6 | 5 | | 2 | 100 | Fm | Fermium | 14 | 2 | 6 | 10 | 12 | 2 | 6 | | 2 |
| 76 | Os | Osmium | 14 | 2 | 6 | 6 | | 2 | 101 | Md | Mendelevium | 14 | 2 | 6 | 10 | 13 | 2 | 6 | | 2 |
| 77 | Ir | Iridium | 14 | 2 | 6 | 7 | | 2 | 102 | No | Nobelium | 14 | 2 | 6 | 10 | 14 | 2 | 6 | | 2 |
| 78 | Pt | Platinum | 14 | 2 | 6 | 9 | | 1 | 103 | Lr | Lawrencium | 14 | 2 | 6 | 10 | 14 | 2 | 6 | 1 | 2 |
| 79 | Au | Gold | 14 | 2 | 6 | 10 | | 1 | 104 | Rf | Rutherfordium | 14 | 2 | 6 | 10 | 14 | 2 | 6 | 2 | 2 |

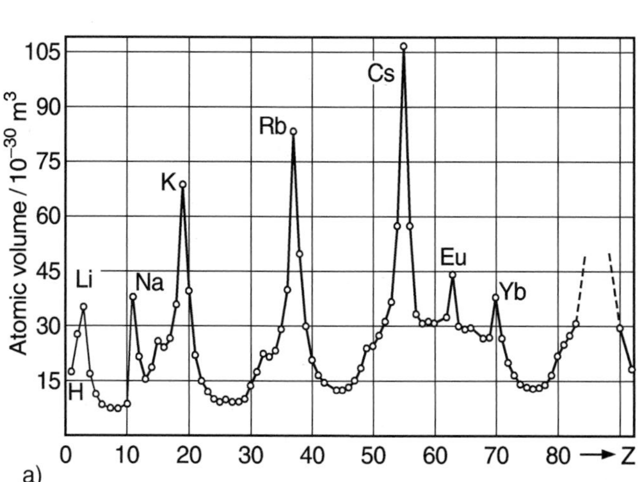


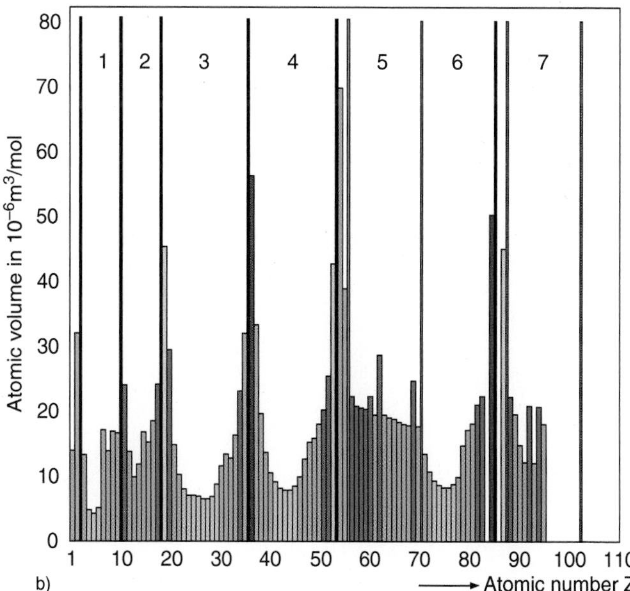


**Fig. 6.15a,b.** Variation of atomic volume $V = \frac{4}{3}\pi <r>^3$ where $<r>=$ mean atomic radius with the number $Z$ of electrons. Atomic volumes from the ratio $V_{mol} = M_{mol}/S$ of molar mass $M_{mol}$ and density $S$. (adapted from MIT course 3.091). The black bars indicate the end of a row in the periodic table

(see Fig. 5.7) than the shells with lower $n$ values. If the atomic volume

$$V_a = V_M/N_a = M_M/(\rho N_a)$$

is determined from the molar mass $M_M$, the density $\rho$ and the Avogadro number $N_a$, slight deviations from the values in Fig. 6.15a are obtained (Fig. 6.15b). The reason is that the density not only depends on the atomic masses but also on the interatomic forces, which can be attractive or repulsive. One example is the He-atom, which has a negligible interaction with neighbouring atoms and therefore a smaller density resulting in a larger molar volume.

Also, the ionization energies $E_{ion}$ show this periodicity (Fig. 6.16). The energy necessary to remove the outer electron (which is the most weakly bound electron) from its state $(n.l.m_l)$ to infinity is

$$W_{ion} = \frac{1}{2}\int_{r_n}^{\infty} \frac{Z_{eff}e^2}{4\pi\varepsilon_0 r^2}\,dr = \frac{Z_{eff}e^2}{8\pi\varepsilon_0 r_n} = Ry^*\frac{Z_{eff}^2}{n^2} \quad (6.23)$$

which depends on the average distance $\langle r\rangle = r_n$ of the electron from the nucleus with the effective charge $eZ_{eff} = e(Z - S)$, partly shielded by the inner electrons, with the shielding parameter $S$.

The noble gases, with their closed, fully occupied shells have the smallest value of $\langle r\rangle$, which means the largest effective charge $eZ_{eff}$ of all elements in the same row of the periodic table and therefore the highest ionization energy. They form the sharp peaks in the curve in Fig. 6.16 while the alkali atoms, where the electron in the outer shell, occupied by only one electron, is more shielded by the lower closed shells, represent the minima in the ionization curve $E_{ion}(Z)$. In Table 6.3 all measured ionization energies and the effective charge numbers $Z_{eff}$ are listed for the first 36 elements in the periodic table.

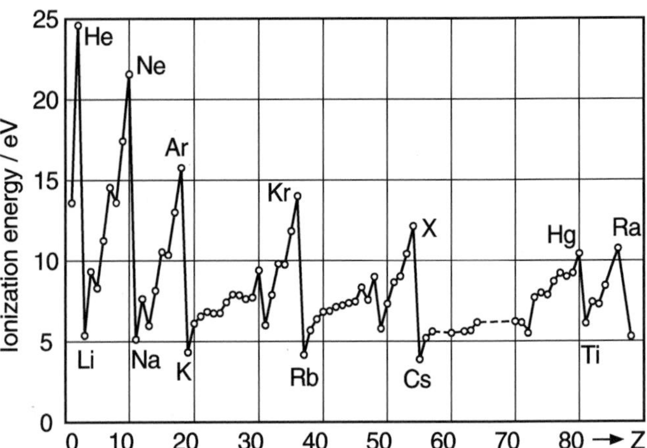


**Fig. 6.16.** Variation of ionization energies with the nuclear charge number $Z$ = total number of electrons

In Fig. 6.17 the successive filling of the different shells with electrons is illustrated. There are some anomalies in this process. For example: After the $3p$

**Table 6.3.** Nuclear charge number $Z$, principle quantum number $n$, ionization energy $E_{ion}$, effective nuclear charge number $Z_{eff}$ and shielding constant $S$ for the leucht-electron in the ground states of the first 36 chemical elements

| Element | $Z$ | $n$ | $E_{ion}$/eV | $Z_{eff}$ | $S = Z - Z_{eff}$ | Elementt | $Z$ | $n$ | $E_{ion}$/eV | $Z_{eff}$ | $S = Z - Z_{eff}$ |
|---|---|---|---|---|---|---|---|---|---|---|---|
| H | 1 | 1 | 13.595 | 1.00 | | K | 19 | 4 | 4.339 | 2.26 | 16.74 |
| He | 2 | 1 | 24.580 | 1.36 | 0.64 | Ca | 20 | 4 | 6.111 | 2.68 | 17.32 |
| Li | 3 | 2 | 5.390 | 1.25 | 1.75 | Sc | 21 | 4 | 6.56 | 2.78 | 18.22 |
| Be | 4 | 2 | 9.320 | 1.66 | 2.34 | Ti | 22 | 4 | 6.83 | 2.84 | 19.16 |
| B | 5 | 2 | 8.296 | 1.56 | 3.44 | V | 23 | 4 | 6.738 | 2.82 | 20.18 |
| C | 6 | 2 | 11.264 | 1.82 | 4.18 | Cr | 24 | 4 | 6.76 | 2.82 | 21.18 |
| N | 7 | 2 | 14.54 | 2.07 | 4.93 | Mn | 25 | 4 | 7.432 | 2.96 | 22.04 |
| O | 8 | 2 | 13.614 | 2.00 | 6.00 | Fe | 26 | 4 | 7.896 | 3.05 | 22.95 |
| F | 9 | 2 | 17.42 | 2.26 | 6.74 | Co | 27 | 4 | 7.86 | 3.04 | 23.96 |
| Ne | 10 | 2 | 21.559 | 2.52 | 7.48 | Ni | 28 | 4 | 7.633 | 3.00 | 25.00 |
| Na | 11 | 3 | 5.138 | 1.84 | 9.16 | Cu | 29 | 4 | 7.723 | 3.01 | 25.99 |
| Mg | 12 | 3 | 7.644 | 2.25 | 9.75 | Zn | 30 | 4 | 9.391 | 3.32 | 26.68 |
| Al | 13 | 3 | 5.984 | 1.99 | 11.01 | Ga | 31 | 4 | 5.97 | 2.66 | 28.34 |
| Si | 14 | 3 | 8.149 | 2.32 | 11.68 | Ge | 32 | 4 | 8.13 | 2.09 | 28.91 |
| P | 15 | 3 | 10.55 | 1.64 | 12.36 | As | 33 | 4 | 9.81 | 3.40 | 29.60 |
| S | 16 | 3 | 10.357 | 2.62 | 13.38 | Se | 34 | 4 | 9.75 | 3.38 | 30.62 |
| Cl | 17 | 3 | 13.01 | 2.93 | 14.07 | Br | 35 | 4 | 11.84 | 3.73 | 31.27 |
| Ar | 18 | 3 | 15.755 | 3.23 | 14.77 | Kr | 36 | 4 | 13.996 | 4.06 | 31.94 |

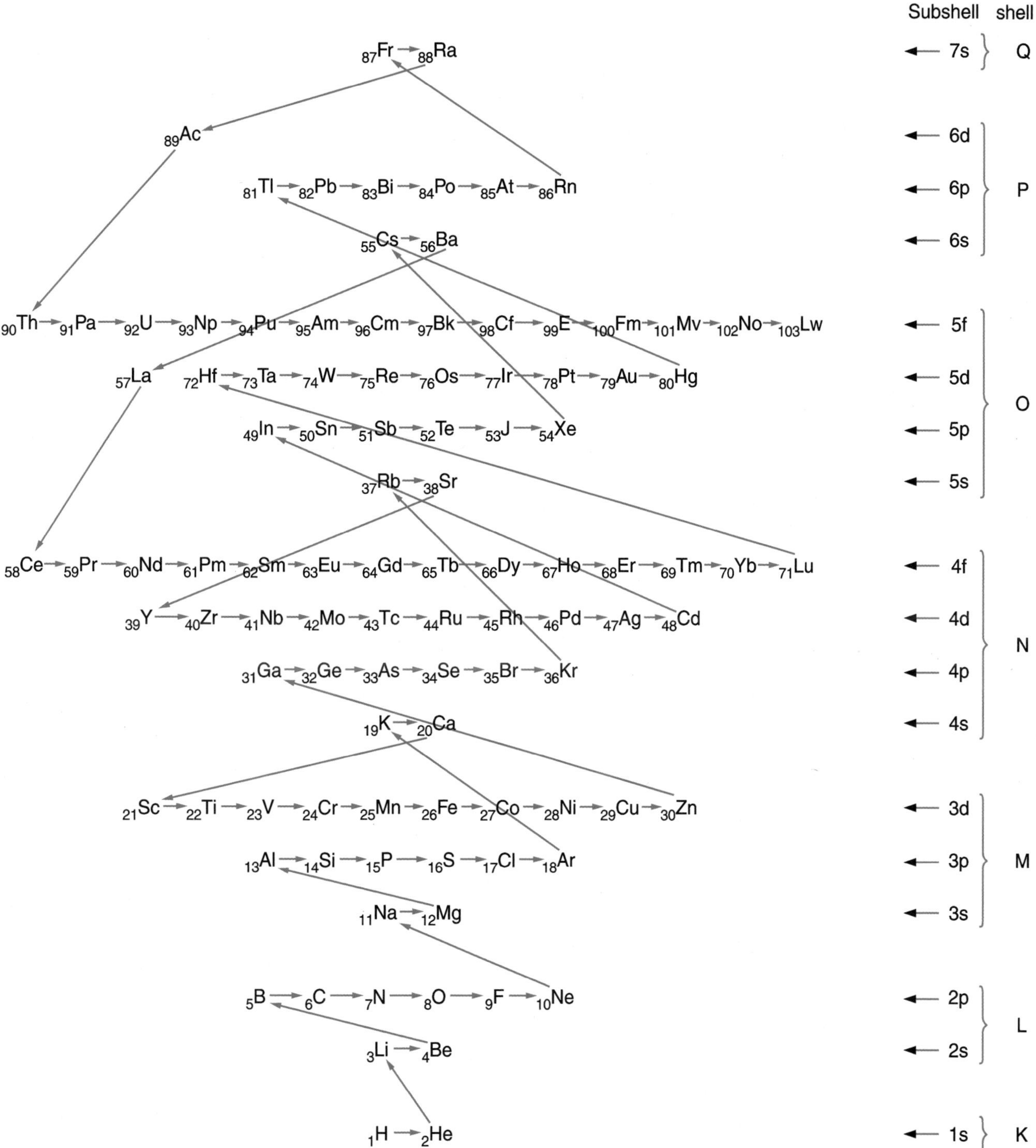


**Fig. 6.17.** Building up the electron shells of all chemical elements

subshell of the $M$-shell has been completely filled, the next two electrons are not put into the empty $3d$ subshell but into the $4s$ subshell of the $N$-shell. Only after the $4s$ shell is filled with 2 electrons, the next electrons are filling the $3d$ shell. The reason for this anomaly is the energetic order of the different subshell. Quantum mechanical calculations clearly show, that the electrons in the $4s$ shell have lower energies than two electrons in the $3d$ shell. This is due to the fact, that the $3d$ wavefunction has its maximum at a much larger radius r than the 4*s* wavefunction (see Fig. 5.7 and Table 5.2). The mean distance of the electrons from the nucleus is therefore larger in the 3*d* state than in the 4*s* state.

A similar situation occurs for the transition from the full 4*p* shell to the 5*s* shell before the 4*d* shell is filled. Before the 4*f* shell is filled the 5*p* and the 6*s*-shells are occupied because their energies are lower than that of the 4*f* shell. The successive filling of the subshells can be better memorized with the help of Fig. 6.18.

Since the mean radius $r_n$ in the Bohr's atomic model is given by (3.84), the ionization energy for an electron in an unshielded Coulomb potential of the nuclear charge $Ze$ is

$$W_0 = \frac{Ze^2}{4\pi\varepsilon_0 r_n} = Ry^* \frac{Z^2}{n^2} . \tag{6.24}$$

A comparison of (6.24) with (6.23) allows the determination of the effective charge $Z_{\text{eff}}e$ and the shielding constant $S = Z - Z_{\text{eff}}$:

$$\begin{aligned} Z_{\text{eff}} &= n\sqrt{\frac{W_{\text{ion}}}{Ry^*}} \\ \Rightarrow S &= \frac{n}{\sqrt{Ry^*}}\left(\sqrt{W_0} - \sqrt{W_{\text{ion}}}\right) \end{aligned} \tag{6.25}$$

from measured ionization energies $W_{\text{ion}}$.

More detailed information on the quantitative characteristics of the different elements can be found in [6.2, 3].

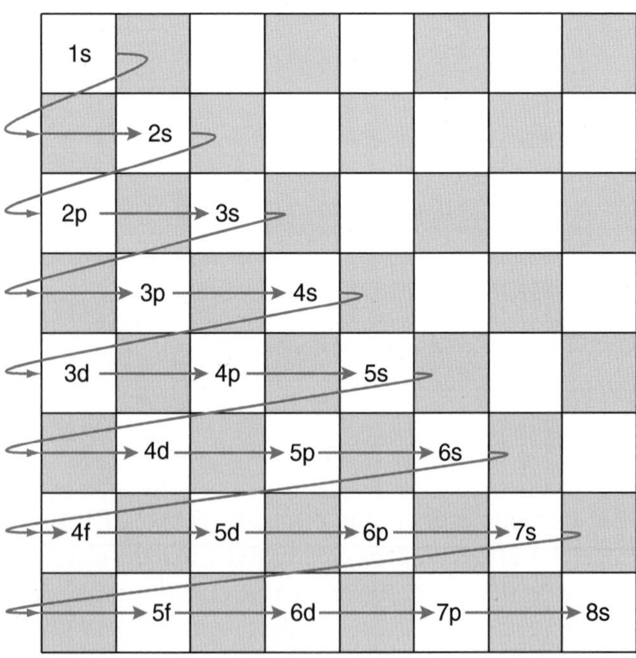


**Fig. 6.18.** Memory aid for the successive filling of the subshells

### 6.2.4 The Periodic System of the Elements

*Dijmitrij Iwanowitsch Mendelejew* (1834–1907) and *Julius Lothor Meyer* (1830–1895) had the idea, independent from each other, to arrange all chemical elements with increasing atomic numbers $A$ in a table with several rows in such a way, that elements with similar chemical properties are all placed in the same column of the table (Fig. 6.19). Later on, it turned out that the number $Z$ of atomic electrons, rather than the atomic mass number $N$ is the correct ordering parameter. This gives seven rows (periods) and eight columns (groups of elements), where all elements with similar chemical properties are included in one of these groups. The alkali elements form the first group, the alkaline earth elements form the second, the halogens the seventh and the noble gases are in the eighth column.

In the sixth period, the third row comprises all rare-earth elements (lanthanides) from La to Lu and in the seventh period, the third row includes all actinides and trans-uranium elements from Th to the artificially produced heavy element Lawrencium Lr with nuclear charge $Z = 103$.

The explanation of this arrangement of all elements is now completely understood based on the structure of the atomic electron shells, discussed in the foregoing section.

> The ordering parameter in the periodic system of the elements is the nuclear charge number $Z$ that equals the number of atomic electrons.

The atomic mass numbers $A$ are nearly integers for all elements with a single isotope. Isotopes of the same chemical element have the same nuclear charge number $Z$ (i.e., the same number of protons in the nucleus) but differ in the number $N$ of neutrons. Their atomic mass numbers

$$A = \left[Z(m_{\text{p}} + m_{\text{e}}) + N \cdot m_{\text{n}}\right] / (M_{\text{C}}/12)$$

therefore differ.

The small difference in $A$ from an integer for elements with only a single isotope is caused by the mass defect $\Delta M = \Delta E_{\text{NB}}/c^2$ due to the nuclear binding energy. If several isotopes of an elements exists, the atomic mass number can be far from an integer. It depends on the relative abundances of these isotopes

| Period | Group I | II | III | IV | V | VI | VII | VIII |
|---|---|---|---|---|---|---|---|---|
| | 1 H 1.00797 | | | | | | | 2 He 4.0026 |
| | 3 Li 6.939 | 4 Be 9.022 | 5 B 10.81 | 6 C 12.01115 | 7 N 14.0067 | 8 O 15.9994 | 9 F 18.9984 | 10 Ne 20.183 |
| | 11 Na 22.9696 | 12 Mg 24.312 | 13 Al 26.9815 | 14 Si 28.086 | 15 P 30.9738 | 16 S 32.064 | 17 Cl 35.453 | 18 Ar 39.948 |
| | 19 K 39.102 | 20 Ca 40.08 | 21 Sc 44.956 | 22 Ti 47.90 | 23 V 50.942 | 24 Cr 51.996 | 25 Mn 54.938 | 26 Fe 55.847 27 Co 58.9332 28 Ni 58.71 |
| | 29 Cu 63.54 | 30 Zn 65.37 | 31 Ga 69.72 | 32 Ge 72.59 | 33 As 74.9216 | 34 Se 78.96 | 35 Br 79.909 | 36 Kr 83.80 |
| | 37 Rb 85.47 | 38 Sr 87.62 | 39 Y 88.905 | 40 Zr 91.22 | 41 Nb 92.906 | 42 Mo 95.94 | 43 Tc 99 | 44 Ru 101.07 45 Rh 102.905 46 Pd 106.4 |
| | 47 Ag 107.87 | 48 Cd 112.40 | 49 In 114.82 | 50 Sn 118.69 | 51 Sb 121.75 | 52 Te 127.60 | 53 J 126.9044 | 54 Xe 131.3 |
| | 55 Cs 132.905 | 56 Ba 137.34 | 57 La 138.91 | 72 Hf 178.49 | 73 Ta 180.948 | 74 W 183.85 | 75 Re 186.2 | 76 Os 190.2 77 Ir 192.2 78 Pt 195.09 |
| | 79 Au 196.967 | 80 Hg 200.59 | 81 Tl 204.37 | 82 Pb 207.19 | 83 Bi 208.98 | 84 Po 210 | 85 At 210 | 86 Rn 222 |
| | 87 Fr 223 | 88 Ra 226.05 | 89 Ac 227 | 104 Rf 258? | 105 Db 260? | 106 Sg | 107 Bh | 108 Hs 109 Mt 110 Ds |

| 58 Ce | 59 Pr | 60 Nd | 61 Pm | 62 Sm | 63 Eu | 64 Gd | 65 Tb | 66 Dy | 67 Ho | 68 Er | 69 Tm | 70 Yb | 71 Lu |
|---|---|---|---|---|---|---|---|---|---|---|---|---|---|
| 140.12 | 140.907 | 144.24 | 145 | 150.35 | 151.96 | 157.25 | 158.924 | 162.50 | 164.93 | 167.26 | 168.934 | 173.04 | 174.97 |

| 90 Th | 91 Pa | 92 U | 93 Np | 94 Pu | 95 Am | 96 Cm | 97 Bk | 98 Cf | 99 Es | 100Fm | 101 Md | 102 No | 103 Lr |
|---|---|---|---|---|---|---|---|---|---|---|---|---|---|
| 232.038 | 231 | 238.03 | 237 | 244 | 243 | 247 | 247 | 251 | 254 | 257 | 256 | 256 | 258? |

**Fig. 6.19.** The periodic table of the elements. The mean mass numbers are averaged over all natural isotopes of an element. The lanthanides (upper blue row) are all in column III of the sixth row, the actinides (lower blue row) are in column III of the seventh row

and is the weighted average of the mass numbers of the different isotopes:

$$\bar{A} = \sum \eta_i A_i \quad \text{with} \quad \eta_i = \frac{N_i}{\sum N_i} , \tag{6.26}$$

where $N_i$ is the number of atoms per mol of the isotope with mass $M_i$ and $N = \sum N_i$ is the total number per mol of atoms of this element.

The chemical properties of the elements are mainly determined by the outer electrons with the smallest binding energy, which are therefore called the valence electrons. The reason for this is as follows.

In chemical reactions where atoms collide with each other and form molecules

$$\mathrm{A} + \mathrm{B} + \mathrm{M} \rightarrow \mathrm{AB} + \mathrm{M} \tag{6.27a}$$

the atomic electrons are rearranged during such a reaction. The third collision partner M (which could be the wall of the container) is necessary to take away the excess kinetic energy and to allow the binding of AB.

For instance, the electron from the Na atom is transferred to the Cl atom. The energy that is necessary for such a rearrangement is provided by the kinetic energy of the reactants and the binding energy of the reaction product. These energies are, however, small and amount to only a few eV. Therefore the binding energy of the electrons can not be larger and inner shell electrons can not participate in such reactions.

If atoms collide with molecules, the molecule can be dissociated and new reaction products are produced, such as

$$\mathrm{A} + \mathrm{BC} \rightarrow \mathrm{AB} + \mathrm{C} . \tag{6.27b}$$

Such reactions also lead to a rearrangement of atomic electrons or to a transfer of electrons from one atom to the other. For such rearrangements the electron cannot be too tightly bound, otherwise the electron could not leave "its atom". Therefore only valence electrons can participate.

Since the ordering parameter in the periodic table is the number $Z$ of electrons, and after a shell is fully occupied a new period starts, elements in the same column have the same number of electrons in the outer shell. The binding energies of electrons in the outer shell of atoms in the same column in Fig. 6.19 are nearly the same, because their effective charge number $Z_{\mathrm{eff}}$ is nearly equal. These elements should therefore show a similar chemical behavior.

**EXAMPLES**

1. The alkali atoms Li, Na, K, Rb, Cs and Fr all have only one valence electron in the outer shell. They are all monovalent and have a similar chemical behavior.
2. All noble gases He, Ne, Ar, Kr, Xe and Rn have a fully occupied outer shell. These are therefore placed in the last column of the periodic table. In order to excite an electron, it has to be lifted into a higher nonoccupied shell. This demands a large energy (for He, e.g., about 20 eV). Such a large energy is not available for most chemical reactions. Therefore noble gases are chemically inactive and do not react with other elements under normal conditions. If, however, one of the electrons is excited into a higher shell by other means (for instance by electron impact in a gas discharge), then the noble gas atom can react, because now much less energy is required to transfer the electron from the excited state to the other reaction partners.
3. The halogens F, Cl, Br and I all have one empty place in their nearly filled outer shell. They all behave chemically similar and react with alkali atoms readily, because the energy gain achieved by bringing the electron from the alkali atom into this hole is larger than the binding energy of the electron in the alkali atom. The alkali-halogen molecule forms an ionic bond $Na^+Cl^-$, where the electron has a larger probability of being in the electron shell of the halogen atom than in the alkali atom.
4. All lanthanides from Lanthanum to Lutetium have the same number of electrons in the outer P-subshell. They only differ in the number of electrons in inner, incompletely filled shells (see Table 6.2).

Other physical properties of the elements, such as the electrical conductivity, also depend on the structure of their electron shell and can be satisfactorily explained by our atomic model.

## 6.3 Alkali Atoms

The alkali atoms are the most similar to the hydrogen atom. They are therefore called "hydrogen-like". They have, besides $n_0$ closed shells with principal quantum numbers $n \leq n_0 = 1, 2, 3, \ldots$, a single electron in the outer shell with $n = n_0 + 1$.

Since the time-averaged electron distribution in the closed shell is spherically symmetric with quantum numbers $L = 0$ and $S = 0$ for the total orbital angular momentum and the total electron spin, the outer electron moves in a spherically symmetric potential, which, however, differs from the Coulomb-potential of the hydrogen atom. It consists of the Coulomb potential produced by the nearly point-like nucleus with charge $Ze$ and the spatially extended spherical distributions of the other electrons with charge $-(Z-1)e$ in the filled shells.

The outer electron can be excited into higher states by absorption of visible light and can emit visible light afterwards. It is therefore called a *"leucht-electron"* from the German verb leuchten, which means "to shine".

When $r_c$ is the mean radius of the highest closed electron shell, the potential $\Phi(r)$ of the leucht-electron can be approximated by a Coulomb potential for all values $r > r_c$. Since the nuclear charge $Ze$ is very effectively shielded by the $Z - 1$ electrons in closed shells, the effective charge number is $Z_{\mathrm{eff}} \approx 1$.

For $r < r_c$ this is no longer true, because here the outer electron submerges into the closed shells and the screening of the nuclear charge becomes less effective. Here the potential depends on the radial distribution in the electron shell. The effective radial dependence of the potential $\Phi(r)$ changes from a Coulomb potential

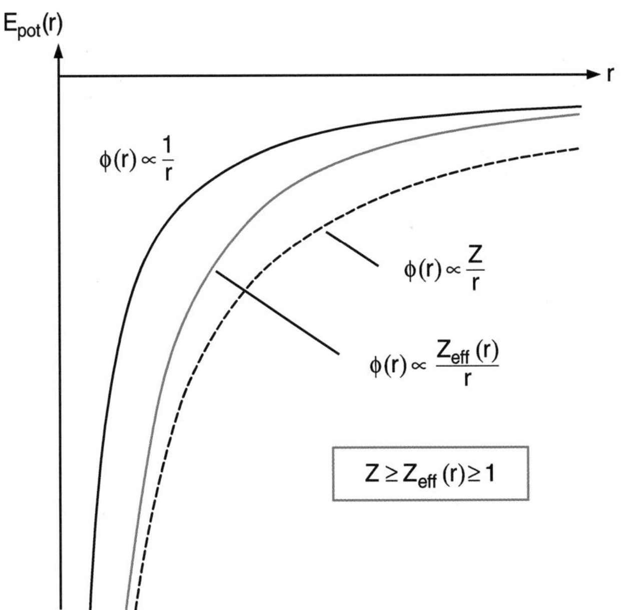


**Fig. 6.20.** Radial dependence of the effective potential energy for the outer electron in an alkali atom

with central charge $Ze$ at small values of $r$ to one with a completely screened charge $(Z-(Z-1))e = e$ at very large distances $r$ (Fig. 6.20):

$$\lim_{r\to 0}\phi(r) = \frac{-Ze}{4\pi\varepsilon_0 r}\,; \quad \lim_{r\to\infty}\phi(r) = \frac{-e}{4\pi\varepsilon_0 r}\,. \tag{6.28}$$

We will illustrate this by the simplest case of the Li atom, where the outer electron in the $2s$ state moves in the potential of the nucleus with charge $Q = +3e$ and the two screening electrons in the $1s$ state. If $r_i$ is the distance of the $2s$ electron from the nucleus and $r_{ij}$ to the $j$th electron in the 1s state (Fig. 6.21), the potential for the $2s$ electron is given by

$$\phi(r_i) = \frac{-Ze}{4\pi\varepsilon_0 r_i} + \frac{e}{4\pi\varepsilon_0}\left[\int \frac{|\psi_{1s}(r_1)|^2}{r_{i1}}\mathrm{d}\tau_1 + \int \frac{|\psi_{1s}(r_2)|^2}{r_{i2}}\mathrm{d}\tau_2\right], \tag{6.28x}$$

where $\psi(r_j)$ is the wave function of the $j$th electron and the integration is performed over the coordinates of the $j$th electron. In a crude approximation the influence of the interaction between the two $1s$ electrons on their wave function can be neglected and the wave functions $\psi_j$ can be written as hydrogenic $1s$ wave functions, listed in Table 5.2. Inserting these wave functions into (6.28x) yields the potential for the $2s$ electron (see Problem 6.2)

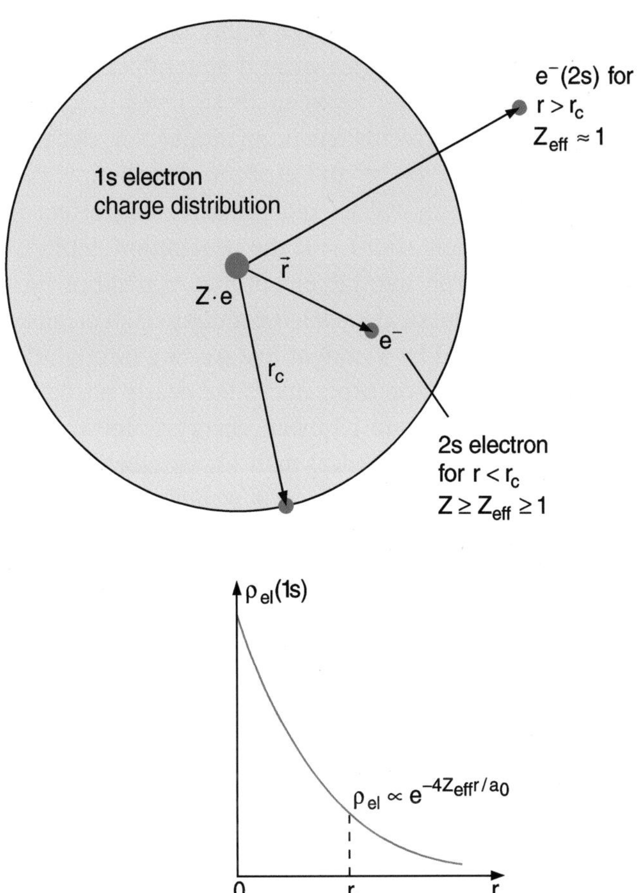


**Fig. 6.21.** Illustration of shielding of the nuclear charge $Ze$ for the $2s$ electron by the charge distribution of the two $1s$-electrons

$$\phi(r) = \frac{-Ze}{4\pi\varepsilon_0 r}\left[1 - 2\mathrm{e}^{-2Z_{\mathrm{eff}}r/a_0}\left(\frac{r\cdot Z_{\mathrm{eff}}}{a_0}+1\right)\right]. \tag{6.28y}$$

For $r \to 0$ the potential equals the Coulomb potential for $Z = 3$, while for $r \to \infty$ the potential becomes a Coulomb potential with $Z_{\mathrm{eff}} = 1$, which means that the two $1s$ electrons have the screening factor $S = 2$. For the potential energy $E_{\mathrm{pot}}(r) = -e\cdot\phi_{\mathrm{eff}}(r)$ we therefore obtain:

$$-\frac{Ze^2}{4\pi\varepsilon_0 r} < E_{\mathrm{pot}}(r) < -\frac{e^2}{4\pi\varepsilon_0 r}\,. \tag{6.29}$$

For the hydrogen atom, levels with the same principal quantum number $n$ but different quantum numbers $l$ are degenerate. This $n$-fold degeneracy is due to the Coulomb potential and no longer holds for another potential, even if it is spherically symmetric. Therefore this $l$-degeneracy is lifted for the alkali atoms, where levels with different values of $l$ within the same $n$ shell do have different energies, even if spin effects are neglected. The splitting between these levels becomes larger for small principal quantum numbers $n$, because the mean radius $\langle r\rangle$ of the electron distribution differs for different $l$-values and the screening effect is therefore different. For $l = 0$ the penetration depth of the outer electron into the electron core is more pronounced (in a classical model, the motion of an electron with $l = 0$ would be a straight line passing through the nucleus), the electron experiences the nearly unshielded nuclear charge and its mean energy is lower than that for $l > 0$. The classical path for an electron with $l = n - 1$ would be a circle with a radius $\langle r\rangle$ and the penetration into the electron core is minimum, the electron experiences an optimum shielded nuclear charge and its energy is higher than that for all lower $l$-values with a given principal quantum number $n$ (Fig. 6.21).

This implies that the energies follow the sequence

$$E(n,l=0) < E(n,l=1) < E(n,l=2)\ldots \quad \text{(Fig. 6.22)}\,.$$

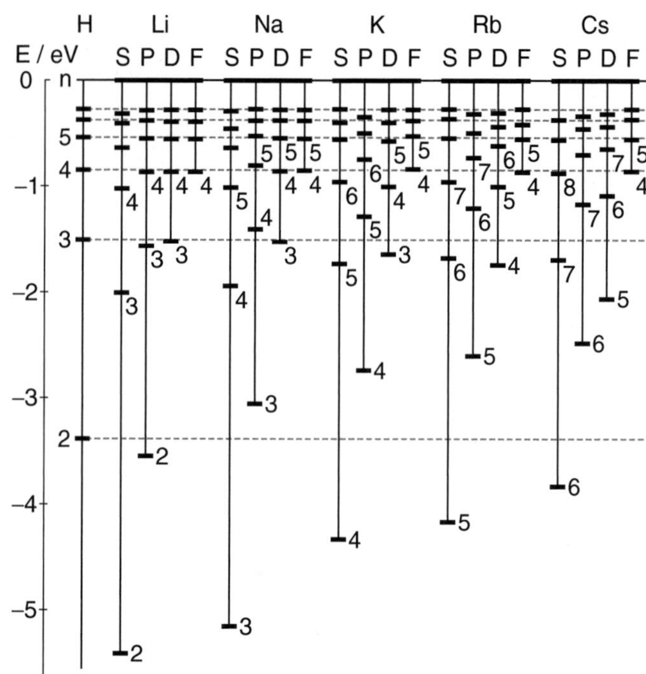


**Fig. 6.22.** Simplified level scheme $E(n,l)$ of the alkali atoms compared with the hydrogen atom (dashed red lines)

Using different nomenclature we can write

$$E_n(s) < E_n(p) < E_n(d)\ldots\,. \tag{6.30}$$

The zero reference point $E = 0$ is either chosen as the ionization limit $E(n \to \infty) = 0$, in which case all energies of bound states become negative (left scale in Fig. 6.22), or the energy of the ground state is chosen to be $E_g = 0$. In this case all energies of excited states become positive and the ionization energy $E_{\text{ion}} = -E_B$ becomes equal to the negative binding energy of the electron in the ground state. In spectroscopic nomenclature, generally the second possibility is chosen. Instead of energies $E$ the term values $T = E/hc$ are given. This allows one to express the wavenumber $\bar{\nu} = 1/\lambda$ [cm$^{-1}$] of a transition between level $|i\rangle$ and $|k\rangle$ as the difference of the two term values:

$$\bar{\nu}_{ik} = T_i - T_k\,.$$

For illustration, the level scheme of the sodium atom and the possible transitions are given in Fig. 6.23. The left scale gives the energies in electron volts eV with $E(n = \infty) = 0$, while the right scale gives the corresponding term values in cm$^{-1}$ with $E_g = 0$. The conversion factor is

$$1\,\text{eV} = 8065.541 cm^{-1}\,. \tag{6.31}$$

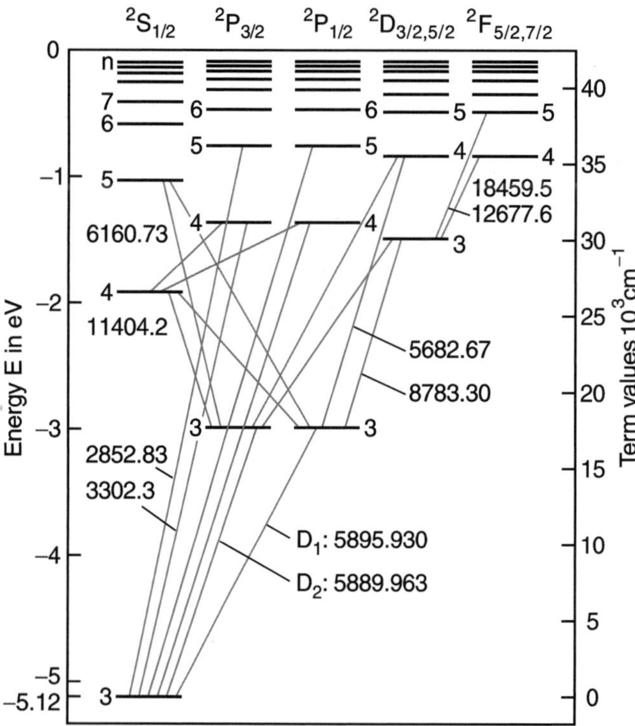


**Fig. 6.23.** Level scheme and transition of the sodium atom

For large principal quantum numbers $n$ (which means large mean distances $\langle r\rangle$ of the electron from the nucleus), where the potential approaches the Coulomb potential of the H-atom (Fig. 6.20), the energy levels of the alkali atoms can be described by the modified Rydberg formula (see Sect. 6.6.4)

$$E_{n,l} = -\frac{Ry^*}{n_{\text{eff}}^2} = -\frac{Ry^*}{(n-\delta_{n,l})^2}\,, \qquad (6.32)$$

where the integer principal quantum number $n$ is replaced by an effective quantum number

$$n_{\text{eff}} = n - \delta_{nl}\,.$$

The quantity $\delta_{nl}$, that depends on $n$ and $l$ is called the ***quantum defect***, which expresses the changes of the energy values $E_{nl}$ against that of the hydrogen atom ($\delta = 0$) by a dimensionless number.

To summarize we can say that the shifts of the alkali energy levels $E_{nl}$ against those of the hydrogen atom are caused by the following effects.

- The deviation of the effective potential from the Coulomb potential, which causes energy shifts $\Delta E_{nl}$ that depend on $n$ and $l$, because of the $n$- and $l$-dependent penetration depths of the outer electron into the core of the other electrons.
- The outer electron interacts with the other electrons in the core and polarizes the electron shell. This leads to a deviation from the spherical charge distribution even for closed shells. The magnitude of this polarization depends on the angular momentum $\boldsymbol{l}$ of the outer electron.
- When the outer electron penetrates into the core, it can collide with the other electrons. This may result in an exchange of the outer with an inner electron, which causes an additional energy shift.

All these effects are included in the quantum defects $\delta_{nl}$. In Table 6.4 the measured quantum defects for different levels of the sodium atom are compiled.

**Note:**

The numbers in Table 6.4 illustrate that the quantum defects depend only slightly on the principal quantum number $n$. Therefore the energies shifts $\Delta E$ between

$$E_n = -Ry^*/(n-\delta_{nl})^2$$

and the energy levels $E_n = -Ry^*/n^2$ in the H atom decrease with increasing values of $n$.

**Table 6.4.** Measured quantum defects $\delta_{nl}$ for different Rydberg states of the sodium atom

| Term | $n=3$ | $n=4$ | $n=5$ | $n=6$ | $n=7$ | $n=8$ |
|---|---|---|---|---|---|---|
| $s$: $l=0$ | 1.373 | 1.357 | 1.352 | 1.349 | 1.348 | 1.351 |
| $p$: $l=1$ | 0.883 | 0.867 | 0.862 | 0.859 | 0.858 | 0.857 |
| $d$: $l=2$ | 0.010 | 0.011 | 0.013 | 0.011 | 0.009 | 0.013 |
| $f$: $l=3$ | – | 0.000 | −0.001 | −0.008 | −0.012 | −0.015 |

**EXAMPLE**

The shift of the level $E_{n,l=0}$ is:

$\Delta E = Ry^*[1/n^2 - 1/(n-\delta)^2]$. For $n = 3$ this amounts to $\Delta E = Ry^*[(1/9) - 1/1.627^2] = 0.27R^*$, while for $n = 20$ the shift is only $\Delta E = Ry^*[1/400 - 1/336.7] = 0.0005Ry^*$.

## 6.4 Theoretical Models for Multielectron Atoms

In Section 6.1 we already saw that even for the simplest case of a multielectron atom, namely the two-electron He atom, an exact theoretical treatment is not possible. The reason for these difficulties are the interactions between the electrons, which because they have a non-spherical symmetry, prevent a separation of the Schrödinger equation as in one-electron atoms. Either numerical methods have to be used or approximate models that might be calculated analytically. If one starts from a crude, but easier-to-calculate model and then improves this model in successive steps, the physical insight into the effect of the improvements and their physical significance is much better than for a numerical treatment of the accurate atomic model without approximations. Furthermore this successive way of improving the atomic model is not restricted to a specific atom but can be applied to all multielectron atoms.

In the following we will discuss some commonly used approximation models for the description of larger atoms.

### 6.4.1 The Model of Independent Electrons

If we consider an arbitrary electron $e_i$ from a many-electron atom, its electrostatic interaction with the other electrons $e_j$ results in a potential energy

$$E_{\text{pot}}(r_i) = \frac{e^2}{4\pi\varepsilon_0}\sum_{j\neq i}\frac{1}{|\vec{r}_i - \boldsymbol{r}_j|}\,. \tag{6.33}$$

In the model of independent electrons, this interaction is not explicitly introduced but it is taken into account implicitly by using an effective potential $\Phi_{\text{eff}}(r)$ that depends on the nuclear charge $Ze$ and the time averaged spherical charge distribution of all other electrons. Any arbitrarily chosen electron $e_i$ moves in this potential that is independent of the momentary location of the other electrons. This model therefore reduces the problem to a one-electron model, which can be solved more easily with numerical techniques. Contrary to the situation in the H atom, the radial dependence of this potential is different from that of the Coulomb potential, although both potentials have spherical symmetry. When we insert this effective potential into the Schrödinger equation, we get the wave functions and the energy eigenvalues for any of these arbitrarily chosen electrons.

These one-electron wave functions can be separated into an angular part and a radial part. While the angular part is the same as that of the hydrogen wave functions (because the potential has spherical symmetry), its radial part is different (because the potential is not Coulombic).

The energy states $E_i\ (n_i, l_i, m_{l_i}, m_{s_i})$ in the one-particle model are defined by the four quantum numbers $n_i$, $l_i$, $m_{l_i}$ and $m_{s_i}$ of the $i$-th electron. The Pauli principle demands that each of these states can only be occupied by at most one electron. Starting from the lowest energy state and filling all electrons of the atom successively with increasing energy into the different states gives the structure and electronic energy of the atom.

The question now is how to obtain the effective potential. If the one-electron wave functions $\psi_j$ of all electrons $e_j\ (i \neq j)$ were known, the potential for the $i$th electron could be calculated according to

$$\phi_{\text{eff}}(r_i) = \frac{-e}{4\pi\varepsilon_0}\left[\frac{Z}{r_i} - \sum_{j\neq i}\int\frac{1}{r_{ij}}|\psi_j(\mathbf{r}_j)|^2\,\mathrm{d}\tau_j\right]\,, \tag{6.34}$$

where the first term gives the attractive interaction due to the nuclear charge $Ze$ and the second term the repulsive interaction with the electronic charge, which is written as the sum over the charge distributions $-e|\psi_j|^2$ of all electrons $e_j$ with $j \neq i$, shielding the nuclear charge. Since the integration extends over all angular coordinates, the mutual distance $r_{ij}$ is averaged and the second term becomes a function that depends solely on $r$. The total effective potential is then spherically symmetric.

We can therefore reduce the problem of finding the optimum effective potential to that of finding the optimum one-electron wave functions.

The procedure for calculating wave functions as solutions of the Schrödinger equation is called ***ab-initio method***, because it starts from the basic equation, where the approximation is solely due to the approximate potential inserted into the Schrödinger equation.

### 6.4.2 The Hartree Method

The optimum wave functions can be obtained by an iterative procedure, first proposed by *Douglas Rayner Hartree* (1897–1958). It is illustrated by the flow diagram in Fig. 6.24.

We start with a guessed spherically symmetric potential $\phi^{(0)}(r)$, that approximates, in a crude way, the screening of the nuclear charge by the electrons. A possible ansatz for such a zeroth-order potential could be, for the lithium atom,

$$\phi^{(0)}(r) = \frac{-e}{4\pi\varepsilon_0}\left(\frac{Z}{r} - a\mathrm{e}^{-br}\right)\,, \tag{6.35}$$

which describes the screening of the nuclear charge by the two $1s$ electrons and provides the effective potential for the $2s$ electron. The two parameters $a$ and $b$ can be adjusted to optimize the effective potential.

Inserting this potential into the one-electron Schrödinger equation for the $i$th electron, the one-electron wave function $\varphi_i^{(0)}$ and the energy eigenvalue $E_i$ can be calculated. This is now done for all $N$ electrons of the atom.

The different energy states are then filled with electrons, starting from the lowest state and obeying the Pauli principle, until each of all $N$ electrons is assigned to a specific state.

Now these wave functions are used to calculate, according to (6.34), the improved potential for the $i$th electron

$$\phi_i^{(1)}(r_i) = \frac{-e}{4\pi\varepsilon_0}\left[\frac{Z}{r_i} - \sum_{j\neq i}\int \frac{\left|\varphi_j^{(0)}(r_j)\right|^2}{r_{ij}}\mathrm{d}\tau_j\right], \tag{6.36}$$

which is produced by the nuclear charge and all other electrons with time-averaged charge distributions. This potential is inserted into the Schrödinger equation, giving improved wave functions $\varphi_i^{(1)}(r)$ and energy eigenvalues $E_i^{(1)}$ for the $i$th electron, which are compared with the wave functions $\varphi_i^{(0)}$ and energies $E_i^{(0)}$.

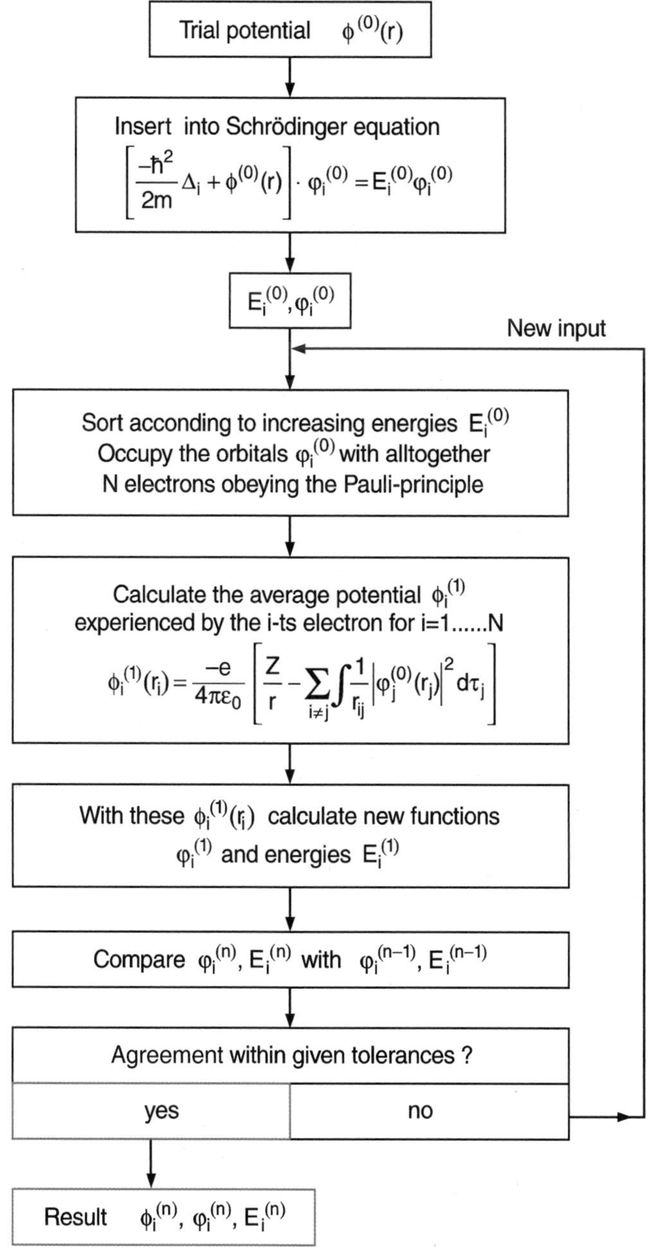


**Fig. 6.24.** Flow charge diagram of the Hartree procedure

If they do not agree with each other within given limits, the procedure is continued until a minimum value is approached for the lowest energy, because it can be shown that the energies, obtained with the "true" wave functions, are always lower than those obtained with approximate functions. After a certain number of iterations, depending on the quality of the zeroth order potential of the wave functions, the effective potential and the energy values converge and further iterations do not change them noticeably. The procedure is therefore called "***self-consistent field approximation***" (SCF).

The important point of the Hartree method is, that the total wave function $\psi(r_1, r_2, \ldots, r_N)$ is reduced to a product

$$\psi(\boldsymbol{r}_1, \boldsymbol{r}_2, \ldots, \boldsymbol{r}_N) = \varphi_1(\boldsymbol{r}_1)\varphi_2(\boldsymbol{r}_2)\ldots\varphi_N(\boldsymbol{r}_N) \tag{6.37}$$

of one-electron wave functions [6.4].

**Note:**

This approximation is much better than the product (6.3) discussed for the He atom. There the interaction between the electrons had been completely neglected. Here it is implicitly introduced in a global way by choosing the best effective potential for obtaining the one electron wave functions.

In Section 6.1 we have already noted that the total wave function must be antisymmetric with respect to the exchange of two arbitrary electrons. This can be achieved if linear combinations of product functions, such as (6.37), are formed that meet this symmetry condition. Such an antisymmetric linear combination can be written as the determinant

$$\psi(\boldsymbol{r}_1, \boldsymbol{r}_2, \ldots, \boldsymbol{r}_N) = C\begin{vmatrix} \varphi_1(\boldsymbol{r}_1)\varphi_1(\boldsymbol{r}_2)\ldots\varphi_1(\boldsymbol{r}_N) \\ \varphi_2(\boldsymbol{r}_1)\varphi_2(\boldsymbol{r}_2)\ldots\varphi_2(\boldsymbol{r}_N) \\ \vdots \\ \varphi_N(\boldsymbol{r}_1)\varphi_N(\boldsymbol{r}_2)\ldots\varphi_N(\boldsymbol{r}_N) \end{vmatrix} \tag{6.38}$$

which automatically fulfills this demand. When two electrons are exchanged, two columns of the determinant are interchanged, which inverts the sign of the determinant. This representation of an antisymmetric multielectron wave function is called a ***Slater determinant***. It describes the wave function of a level in an atom with many electrons as an antisymmetric linear combination of products of one-electron wave functions. More details can be found in [6.4].

### 6.4.3 The Hartree–Fock Method

So far we have neglected the electron spin. In Sect. 6.1.3 we saw that the total wave function can be written as the product

$$\psi = \varphi(\boldsymbol{r})\chi(\boldsymbol{s}) \tag{6.38a}$$

of the spatial part and spin function. This is valid if the interaction energy between the magnetic moments of spins and orbital momenta (causing the fine structure splitting) is small compared to the electrostatic energy, which is generally the case. Instead of (6.38) we then obtain the determinant

$$\psi(\boldsymbol{r},\boldsymbol{s}) = C\begin{vmatrix} \varphi_1(1)\chi_1(1)\varphi_1(2)\chi_1(2)\ldots\varphi_1(N)\chi_1(N) \\ \vdots \\ \varphi_N(1)\chi_N(1)\varphi_N(2)\chi_N(2)\ldots\varphi_N(N)\chi_N(N) \end{vmatrix} \tag{6.38b}$$

which is again antisymmetric with respect to an exchange of two arbitrarily chosen electrons. When using these Hartree–Fock functions for computing the energy eigenvalues, the necessary computer time is much longer than for the Hartree functions, but the results are much more accurate and nowadays such functions are nearly exclusively used for ab initio calculations.

### 6.4.4 Configuration Interaction

The best and most often used method for including the interaction between the electrons (electron correlation) is the configuration interaction (CI) technique. Combined with the Hartree–Fock method it provides the most accurate wave functions and energy values for multielectron atoms.

The wave function of an atomic state is written as the linear combination

$$\Psi(\boldsymbol{r},\boldsymbol{s}) = \sum_k c_k\psi_k \tag{6.38c}$$

of Slater determinants (6.38b), where each of the Slater determinants gives the distribution of the atomic electrons over the different one-electron levels, called a configuration. If the energy

$$E = \langle\Psi|H|\Psi\rangle$$

of an atomic state is calculated with the functions (6.38c), integrals of the form $\langle\psi_i|H|\psi_k\rangle$ also contribute to the total energy. These integrals describe the contributions of interactions between the different configurations $\psi_i$, represented by the one-electron functions (6.38b) to the energy.

In the sum (6.38c) only functions $\psi_k$ with the same symmetry and the same spin are included. Otherwise the integrals $\langle\psi_i|H|\psi_k\rangle$ vanish.

## 6.5 Electron Configurations and Couplings of Angular Momenta

Besides electrostatic interactions, the magnetic interactions between the magnetic moments of electrons must also be taken into account. These much smaller interactions cause a splitting of the energy states into fine structure components. While for one-electron atoms there are only two fine structure components for all levels with $l \geq 1$, corresponding to the two different orientations of the electron spin $\boldsymbol{s}$ with respect to the orbital angular momentum $\boldsymbol{l}$ (see Sect. 5.6.3), there might be more than two components in multielectron atoms. The manifold of fine structure components of a given state $(n, l)$ is called a ***multiplet***.

The different electrons characterized by their one-electron wave function, are labeled according to their principal quantum number $n$ and the quantum number $l$ of their orbital angular momentum. The electron configuration describes these quantum numbers for all electrons of the atom. For instance, the electron configuration $1s^2 2s2p$ represents a four-electron atom with two electrons in the $1s$ state with $n = 1$ and $l = 0$, one in the $2s$ state with $n = 2$ and $l = 0$ and one in the $2p$ state with $n = 2$ and $l = 1$. The configuration $2s^2 2p^3$ of a seven-electron atom has two electrons in the filled $1s$ state (which are not included in the labeling, because it is self-evident that the $1s$ shell has to be occupied in the ground state of atoms with more than one electron), two electrons in the $2s$ state and three electrons in the $2p$ state. The total quantum numbers of the atomic state depend on the quantum numbers of the individual electrons and on the couplings of their angular momenta.

### 6.5.1 Coupling Schemes for Electronic Angular Momenta

The way the orbital angular momenta $l_i$ and the spins $s_i$ of the individual electrons are coupled to form the total angular momentum $J$ of the atom, depends on the energetic order of the different interactions. We will discuss two limiting cases.

### a) $L$-$S$ Coupling

If the interaction energies

$$W_{l_i l_j} = a_{ij} \boldsymbol{l}_i \cdot \boldsymbol{l}_j \tag{6.39a}$$

between the orbital magnetic moments of electrons $e_i$ and $e_j$ and

$$W_{s_i s_j} = b_{ij} \boldsymbol{s}_i \boldsymbol{s}_j \tag{6.39b}$$

between their spin moments are large compared to the interaction energy

$$W_{l_i \bar{s}_i} = c_{ii} \boldsymbol{l}_i \boldsymbol{s}_i \tag{6.39c}$$

between orbital magnetic moment $\boldsymbol{\mu}_{li} = \mu_B \boldsymbol{l}_i$ and spin moment $\boldsymbol{\mu}_s = g_s \mu_B \boldsymbol{s}_i$ of the same electron, then the orbital angular momenta $l_i$ of the different electrons couple to a total orbital momentum

$$\boldsymbol{L} = \sum \boldsymbol{l}_i \quad \text{with} \quad |\boldsymbol{L}| = \sqrt{L(L+1)}\hbar \tag{6.40a}$$

and the individual spins $s_i$ to a total spin

$$\boldsymbol{S} = \sum_i \boldsymbol{s}_i \quad \text{with} \quad |\boldsymbol{S}| = \sqrt{S(S+1)}\hbar \tag{6.40b}$$

of the atomic state. The total angular momentum of the electron shell is then

$$\boldsymbol{J} = \boldsymbol{L} + \boldsymbol{S} \quad \text{with} \quad |\boldsymbol{J}| = \sqrt{J(J+1)}\hbar\ . \tag{6.40c}$$

This limiting coupling case is named $\boldsymbol{L} - \boldsymbol{S}$ coupling (Fig. 6.25). The electron configuration with total orbital angular momentum $\boldsymbol{L}$ and total spin $\boldsymbol{S}$ results (depending on the coupling of $\boldsymbol{L} + \boldsymbol{S} = \boldsymbol{J}$) in different fine structure components of a multiplet, which only differ in their quantum number $J$. The number of possible fine structure components equals the smaller of the two numbers $(2S+1)$ or $(2L+1)$, because this gives the number of possible relative orientations between the two vectors $\boldsymbol{S}$ and $\boldsymbol{L}$, and therefore the number of different couplings $\boldsymbol{L} + \boldsymbol{S} = \boldsymbol{J}$.

The energy of a fine structure component is

$$E_J = E(n, L, S) + C \cdot \boldsymbol{L} \cdot \boldsymbol{S}\ , \tag{6.41}$$

where the last term gives the coupling energy of the interaction between total orbital angular momentum $\boldsymbol{L}$ and total spin $\boldsymbol{S}$. The coupling constant $C$ is given in units of $[1\,\mathrm{kg}^{-1}\mathrm{m}^{-2}]$.

Because of the vector relation

$$\boldsymbol{J}^2 = (\boldsymbol{L} + \boldsymbol{S})^2 = \boldsymbol{L}^2 + \boldsymbol{S}^2 + 2\boldsymbol{L} \cdot \boldsymbol{S} \tag{6.42}$$

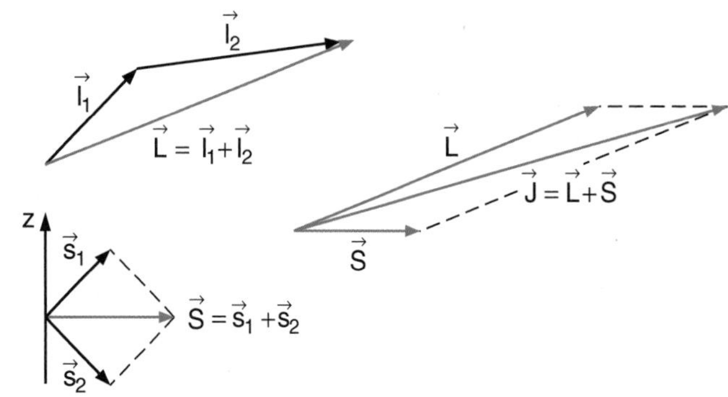


**Fig. 6.25.** Vector model of $L$-$S$ coupling

we obtain for the fine structure coupling energies

$$C \cdot \boldsymbol{L} \cdot \boldsymbol{S} = \frac{1}{2} C[J(J+1) - L(L+1) - S(S+1)]\hbar^2\ . \tag{6.43}$$

The labeling of a fine structure component is $n^{2S+1}L_J$.

**EXAMPLES**

$3^3P_1(n = 3, S = 1, L = 1, J = 1)$ ;
$4^2D_{3/2}(n = 4, S = 1/2, L = 2, J = 3/2)$ ,

The following nomenclature is used in accordance with the labeling of levels in one-electron atoms:

$L = 0$ : S-terms ; $L = 1$ : P-terms ;
$L = 2$ : D-terms , . . . .

**Note:**

Unfortunately the letter S is used in the literature for two different things, namely, for the total electron spin and for levels with $L = 0$.

**EXAMPLES**

1. The electron configuration with $L = 2$ and $S = 1$ results in three fine structure components with quantum numbers $J = 1, 2, 3$ (Fig. 6.26a). The corresponding vector couplings are shown in

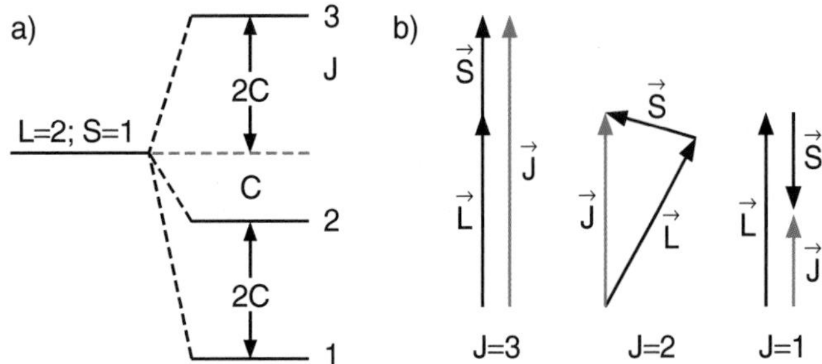


**Fig. 6.26a,b.** $L$-$S$ coupling for the case $L = 2$, $S = 1$. (**a**) Level scheme (**b**) Coupling possibilities

Fig. 6.26b. The energies of the fine structure components are calculated according to (6.41)–(6.43)

$$\begin{aligned} &E_J(n, L, S, J) \\ &= E(n, L, S) + C/2[J(J+1) - 6 - 2]\hbar^2 \\ &= E(n, L, S) + 2C\hbar^2 \quad \text{for} \quad J = 3 \\ &= E(n, L, S) - 1C\hbar^2 \quad \text{for} \quad J = 2 \\ &= E(n, L, S) - 3C\hbar^2 \quad \text{for} \quad J = 1 \end{aligned}$$

The fine structure components are ***not*** equally spaced!

2. The number of fine structure components is $2S + 1$ if $L > S$ or $2L + 1$ if $S > L$.
3. The configuration with $L = 1$ and $S = 3/2$ has three possible fine structure components with $J = 5/2$, $3/2$ and $1/2$. The components are labeled $^4P_{5/2}$, $^4P_{3/2}$, and $^4P_{1/2}$.

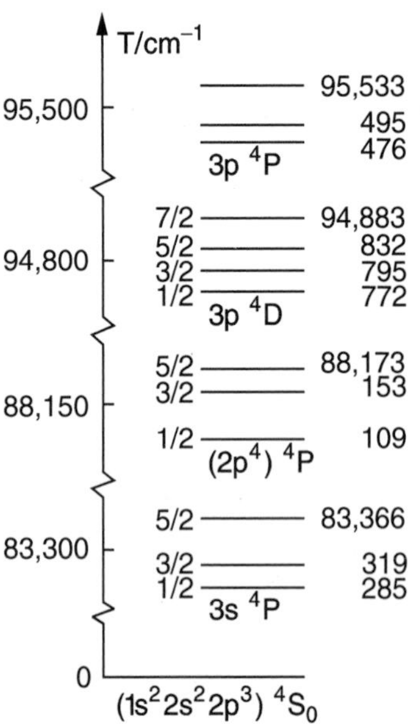


**Fig. 6.27.** Level scheme of the lowest quartet states of the nitrogen atom

For $L$-$S$ coupling, the fine structure splitting $\Delta E_{\mathrm{FS}} = E_J - E_{J'}$ is small compared to the energetic separation of levels with different values of $L$ or $S$. In the spectrum of an atom, following $L$-$S$ coupling, one recognizes a distinct multiplet structure of narrow fine structure components (Fig. 6.27).

- The fine structure constant $C$ is largest for the lowest atomic levels (small values of $n$) with $L \neq 0$ and $S \neq 0$. The magnitude of the multiplet splitting decreases with increasing principal quantum number $n$ (Fig. 6.28).
- $L$-$S$ coupling is valid mainly for light atoms with small $Z$ values. Quantum mechanical calculations show that the fine structure constant $C$ is proportional to

$$C \propto Z^4/n^3 , \tag{6.43y}$$

while the energy separation between levels with different values of $L$ only increases with $Z^2/n^3$.
- For large $Z$ values, the fine structure splittings become comparable with the separation of levels with different $L$ values and the validity of the $L$-$S$ coupling scheme breaks down.

Note that the weighted average of the energies of all fine structure components

$$\bar{E} = \frac{1}{k}\sum_{j=1}^{k}(2j+1)E_j = E(n, L, S) , \tag{6.43x}$$

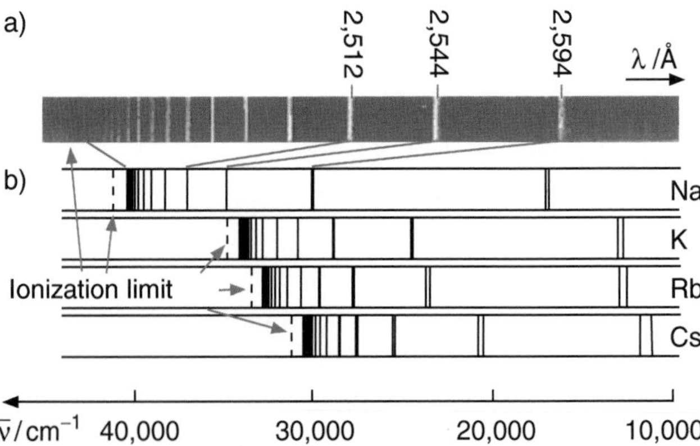


**Fig. 6.28a–c.** Absorption spectra of the alkali atoms. (**a**) Enlarged section of the Na spectrum in (**b**). (**c**) Spectra and ionization limits of K, Rb, and Cs

where each component is weighted by its statistical weight factor $2J+1$ (according to the number of possible spatial orientations of $\boldsymbol{J}$) coincides with the energy $E(n, L, S)$ of the unsplit level ((2.45) without the coupling term). This weighted average is indicated in Fig. 6.26a by a dotted line.

The small fine structure splitting is, for instance, visible in the spectrum of the smaller alkali atoms (Fig. 6.28), where transitions from the ground state $n\,^2S$, which has no fine structure splitting, to higher levels $(n+x)\,^2P$ are observed. The splittings of the lines therefore directly give the splittings of the upper levels. The figure illustrates that the splitting increases with $Z$ although the principle quantum number $n$ of the ground states also increase ($n=3$ for Na, $n=4$ for K, $n=5$ for Rb and $n=6$ for Cs). Table 6.5 shows the fs-splittings of elements in different $Z$ but equal $n$-values.

In Fig. 6.29 the different interaction terms in the case of $L$-$S$ coupling are schematically illustrated for the example of two interacting electrons with $l_1 = 1$ and $l_2 = 2$. The Pauli principle demands for singlet terms ($S=0$) another spatial distribution of the two electrons (symmetric spatial wave function) than for triplet terms ($S=1$) with an antisymmetric spatial wave function. In the latter case the two electrons are farther apart than in the first case and the electrostatic repulsion energy is smaller. The triplet states therefore have a lower energy than the singlet states. It should be stressed again that the splitting between singlet and triplet levels with equal $L$ values is not due to a magnetic interaction, but due to the electrostatic interactions between the electrons, which is different for singlet and triplet levels.

Since the potential for electrons in multielectron atoms is no longer a Coulomb potential, the $L$ degeneracy discussed for the H atom is lifted and levels with different $L$ values have different energies. The vector sum $\boldsymbol{L} = \boldsymbol{l}_1 + \boldsymbol{l}_2$ of the two orbital angular

**Table 6.5.** Fine structure splittings of the levels ($n=2, l=1$) for some light atoms in cm$^{-1}$

| Element | $Z$ | State | fs-splitting / cm$^{-1}$ |
|---|---|---|---|
| He | 2 | $2p\ ^3P^0$ | 1 |
| Be | 3 | $2p^2\,^3P^0$ | 3 |
| C | 6 | $2p^2\,^3P$ | 42 |
| O | 8 | $2p^4\,^3P$ | 226 |
| F | 9 | $2p^5\,^3P_0$ | 404 |

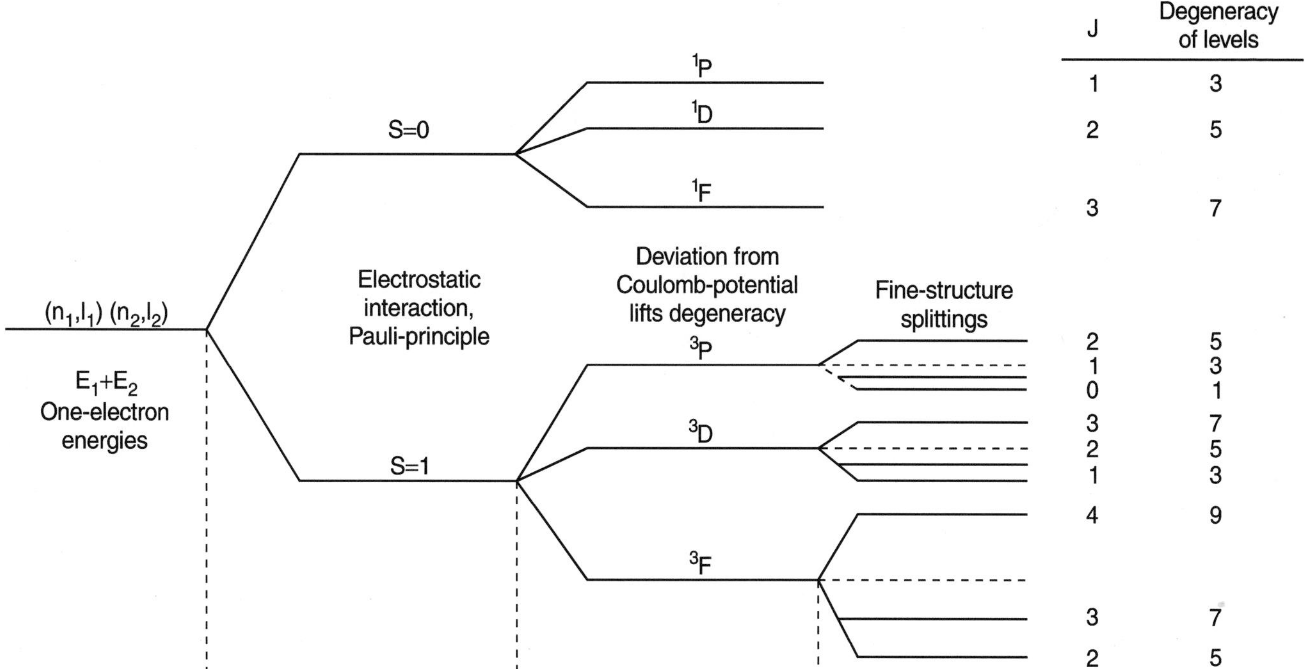


**Fig. 6.29.** Energetic sequences of the different interactions with the corresponding level splittings in the case of $L$-$S$ coupling for the example of the $(n_1p)^1(n_2d)^1$ configuration

momenta $\boldsymbol{l}_i$ results in possible quantum numbers $L = 1, 2, 3$. The singlet term with $S = 0$ splits into the levels $^1P(L = 1)$, the $^1D$ level ($L = 2$) and the $^1F$ levels with $L = 3$ and the triplet states in the corresponding triplet components.

The singlet levels show no fine structure (because the total spin is zero), while the triplet states split into three fine structure components. The splitting is largest for the largest $L$ value (see (6.41)).

The different fine structure components are labeled by the quantum number $J$, which is given in Fig. 6.29, together with the degeneracy $2J + 1$.

### b) $j$-$j$ Coupling

If the interaction energy

$$W_{l_i s_i} = c_{ii} \boldsymbol{l}_i \boldsymbol{s}_i \tag{6.44}$$

between the magnetic moment of an electron due to its orbital angular momentum and its spin moment becomes larger than the magnetic interactions

$$W_{l_i l_j} = a_{ij} \boldsymbol{l}_i \boldsymbol{l}_j \quad \text{or} \quad W_{s_i s_j} = b_{ij} \boldsymbol{s}_i \boldsymbol{s}_j$$

between different electrons, the order of couplings changes. Now $\boldsymbol{l}_i$ and $\boldsymbol{s}_i$ initially couple to form the resultant angular momentum

$$\boldsymbol{j}_i = \boldsymbol{l}_i + \boldsymbol{s}_i \tag{6.45a}$$

of the electron $e_i$, and the vectors $\boldsymbol{j}_i$ of the different electrons couple to the total angular momentum $J$ of the atomic state

$$\boldsymbol{J} = \sum \boldsymbol{j}_i \,. \tag{6.45b}$$

This limiting coupling case, which is mainly observed for heavy atoms with large $Z$ values, is called ***j*-*j*** coupling. The vector coupling diagram is shown in Fig. 6.30.

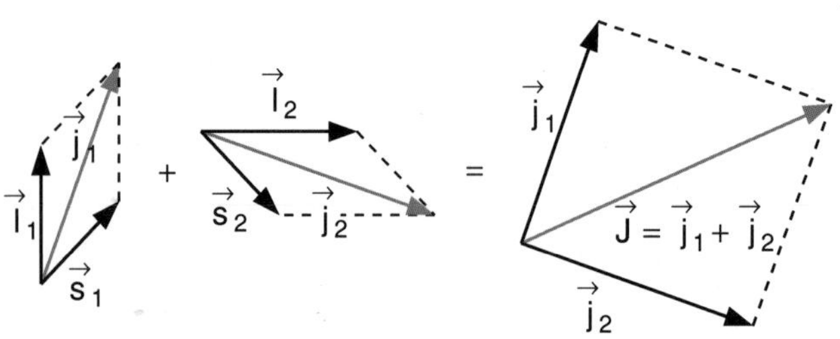


**Fig. 6.30.** Vector model of $j$-$j$ coupling

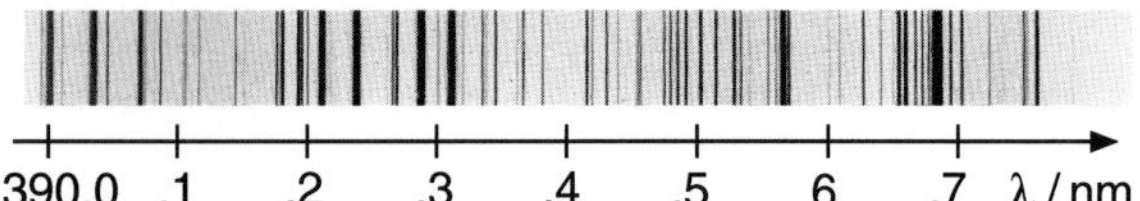


**Fig. 6.31.** Section of the emission spectrum of iron atoms in the near UV

**Note:**

In the limiting case of $j$-$j$ coupling the total orbital angular momentum $L$ and the total spin $S$ are no longer defined, although the individual vectors $\boldsymbol{l}_i$ and $\boldsymbol{s}_i$ are known. There are no longer S, P, D... levels and also no distinction between singlet, doublet or triplet levels can be made. The only well-defined "good" quantum number is $J$ for the total angular momentum $\boldsymbol{J}$ with $|\boldsymbol{J}| = (J(J+1))^{1/2}\hbar$. Levels with equal quantum numbers $l_i$ for the individual electrons but different spins $s_i$ no longer form narrowly spaced fine structure components of multiplets but are energetically mixed with levels of different $l_i$.

The spectra of such atoms with large $Z$ numbers are therefore confusing and not easy to assign. The spectrum is very crowded as can be seen in the example for the iron spectrum, from which a small section is shown in Fig. 6.31.

EXAMPLE

The tin atom Sn ($Z = 50$), has the ground state configuration $5s^2 5p^2$. When one of the two $p$ electrons is excited into the $6s$ level, the electron configuration $(5s^2, 5p, 6s)$ is obtained with $L = 1$, $S = 1$, and $J = 0, 1, 2$.

If the $6s$ electron is completely removed, the $\mathrm{Sn}^+$ ion with the configuration $(5s^2, 5p)$ results in the $5p^2P$ state, which shows a fine structure splitting of the same magnitude as the $(5s^2 5p, 6s)$ configuration of the Sn atom. This demonstrates that the main part of the fine structure splitting is caused by the interaction of the $p$ electron with the other electrons, because the $p$ electron submerges into the electron shell of the Sn atom. Only the minor part is due to spin-orbit coupling between the $6s$ and the $5p$ electron. This is a further indication for $j$-$j$-coupling [6.3].

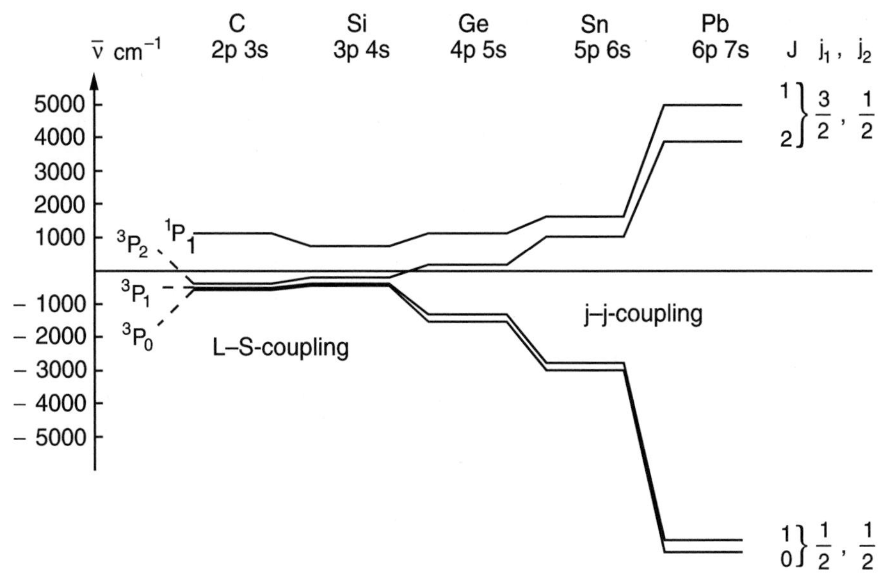


**Fig. 6.32.** Transition range between $L$-$S$ coupling and $j$-$j$ coupling for equivalent states of atoms in the fourth column of the periodic table

For most atoms, intermediate coupling cases apply, which are between pure $L$-$S$ coupling and $j$-$j$ coupling. In Fig. 6.32 the transition from $L$-$S$ coupling for the carbon atom ($Z = 6$), over the intermediate coupling for Germanium ($Z = 32$) to the $j$-$j$ coupling for lead Pb ($Z = 82$) is illustrated.

The total number of possible levels for a given electron configuration $(l_i, s_i)$ is the same for the two limiting cases. It is therefore possible to draw for the transition from $L$-$S$ coupling to $j$-$j$ coupling in such a diagram unambiguously connecting lines for levels with a given $J$-value. Such a diagram is called correlation diagram (Fig. 6.33).

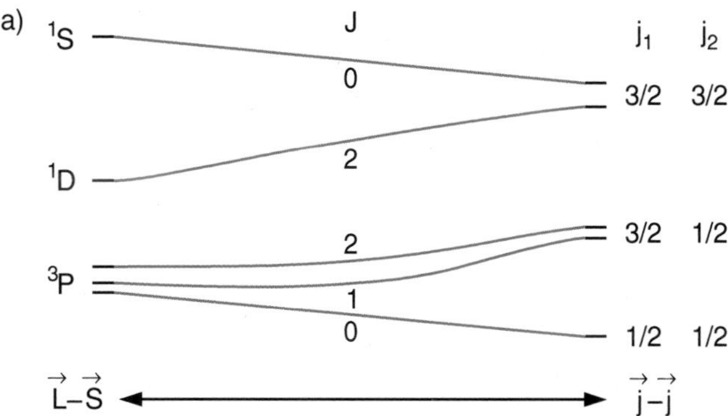


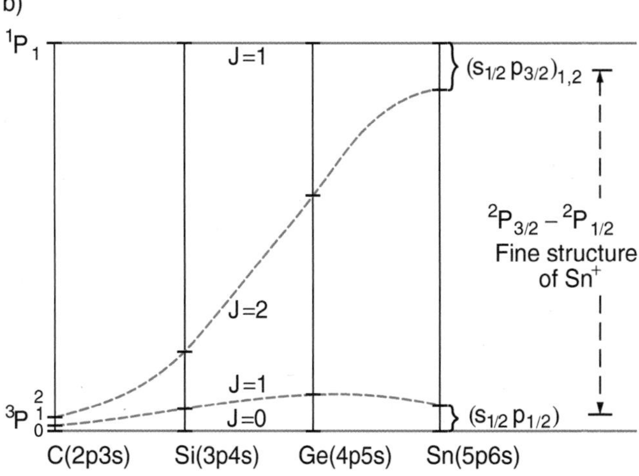


**Fig. 6.33a,b.** Correlation diagram for the transition from $L$-$S$ coupling for light atoms to $j$-$j$ coupling for heavy atoms. (**a**) For the $p^2$-configuration. (**b**) For the $(np, (n+1)s)$ configuration of some atoms

### 6.5.2 Electron Configuration and Atomic States

In this section we will discuss how the different atomic states and their spectroscopic assignment can be deduced from the electron configurations.

From the building-up principle, discussed in Sect. 6.2 it follows that in the case of $L$-$S$ coupling the total orbital angular momentum $L = \sum l_i$ for a filled electron shell must be zero, because for each quantum number $l$, all levels with projection quantum numbers $m_l$ $(-l \le m_l \le +l)$ are occupied. This means that all possible orientations of the orbital angular momentum are realized and therefore the vector sum $L = \sum l_i$ must be zero.

Since this shell is filled with pairs of electrons with antiparallel spins, according to the Pauli principle, the total spin $S = \sum s_i$ must also be zero. This is illustrated in Fig. 6.34 for the neon atom.

> All noble gases in their ground states have the quantum numbers $L = S = J = 0$, their ground state is ${}^1S_0$.
> For all other atoms the values of $L$ and $S$ can be determined by counting only electrons in unfilled shells. All filled shells with a given principal quantum number $n$ do not affect the angular momenta of the unfilled shells.

We will illustrate this for the example of the carbon atom with six electrons.

The electron configuration in the ground state is $1s^2 2s^2 2p^2$. The $1S$ shell and the $2S$ subshell are filled. Their quantum numbers are $S = L = J = 0$. We only need to consider the two $2p$ electrons with $l = 1$ and $s = 1/2$. Depending on the relative orientation of the two orbital angular momenta $\boldsymbol{l}_i$ and their spins $\boldsymbol{s}_i$ different atomic states can be realized, which differ in their quantum numbers $L$, $S$ and $J$. This is illustrated in Fig. 6.35a, where the possible orientations of the two vectors $\boldsymbol{l}_1$ and $\boldsymbol{l}_2$ of the two $p$ electrons are indicated.

Ne $1s^2 2s^2 2p^6$

K shell: n=1 2 electrons

$$\left.\begin{array}{l}\vec{l}_1 = \vec{l}_2 = 0 \Rightarrow \vec{L} = 0\\ \vec{s}_1 + \vec{s}_2 = 0 \Rightarrow \vec{S} = 0\end{array}\right\} \Rightarrow \vec{J} = 0$$

L shell: n=2 8 electrons

$$\vec{l}_1 = \vec{l}_2 = 0 \Rightarrow \vec{s}_1 = -\vec{s}_2$$

a) $\Rightarrow \vec{L} = 0$ and $\vec{S} = 0$

b) $\vec{l}_3 + \vec{l}_4 + \vec{l}_5 = 0, \quad \vec{S}_3 + \vec{S}_4 + \vec{S}_5 = \vec{S}_a \neq 0$

c) $\vec{l}_6 + \vec{l}_7 + \vec{l}_8 = 0, \quad \vec{S}_6 + \vec{S}_7 + \vec{S}_8 = \vec{S}_b = -\vec{S}_a$

$\Rightarrow \vec{L} = 0; \vec{S} = \vec{S}_a + \vec{S}_b = 0$

$\vec{l}_3$ $\vec{l}_4$ $\vec{l}_5$ $\vec{s}_5$ $\vec{s}_4$ $\vec{s}_3$ $\vec{l}_6$ $\vec{l}_7$ $\vec{l}_8$ $\vec{s}_6$ $\vec{s}_7$ $\vec{s}_8$

**Fig. 6.34.** Illustration of the vector sums $\sum \boldsymbol{l}_o = \boldsymbol{L} = \boldsymbol{0}$ for orbital angular momentum and $\sum s_i = S = 0$ for the electron spins for closed shell atoms such as neon

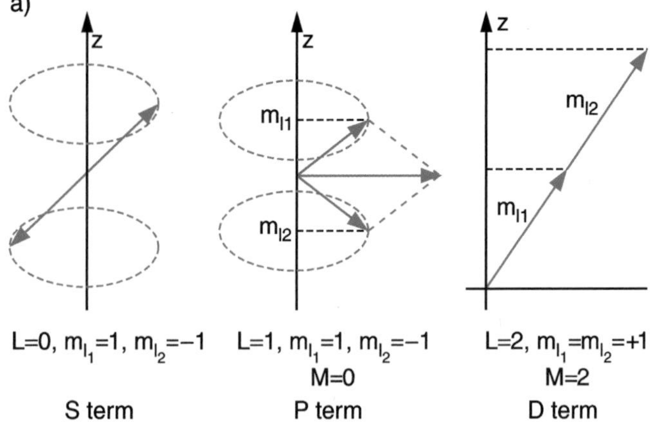


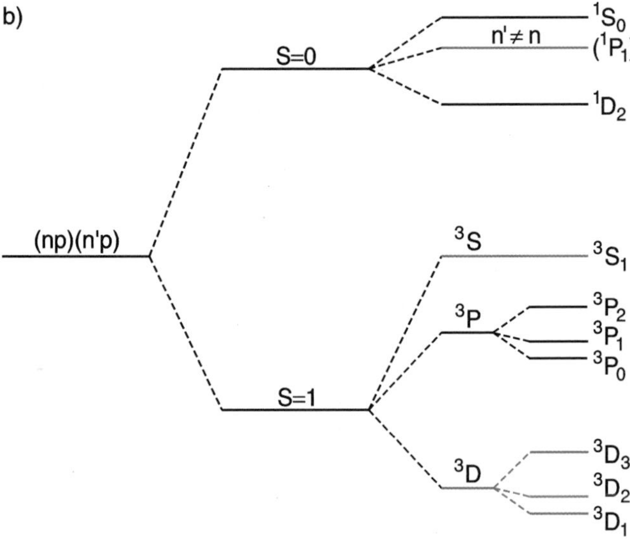


**Fig. 6.35.** (**a**) Vector model of the $p^2$-configuration. (**b**) Term diagram. The blue levels are only possible for $n_1 \neq n_2$

The level diagram of Fig. 6.35b shows the different energies of the resulting states, that are allowed by the Pauli principle. If the two electrons have the same principal quantum number ($n = n'$), only the black energy terms are allowed. For these states the sum $L + S$ is always an even integer. For electrons with $n_1 \neq n_2$, five additional terms can be obtained, which are marked in blue in Fig. 6.35b. The triplet states are lower than the singlet states. Within the triplet states the $D$ states are lower than the $P$ states and for a given value of $L$ the levels with the smallest $J$ value have the lowest energy. Table 6.6 lists the possible atomic states, resulting from different two-electron configurations.

The possible states resulting from the $p^2$ configuration are illustrated in another way by the Slater diagram of Fig. 6.36. Here the left part shows all $M_L$ and $M_S$

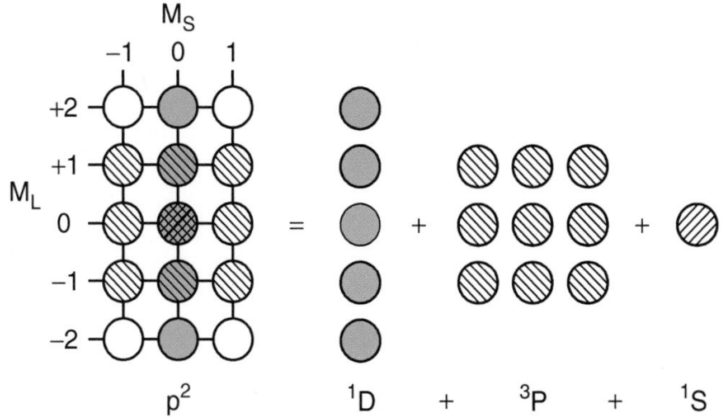


**Fig. 6.36.** Slater diagram of all levels $(M_L, M_S)$ for a $p^2$-configuration of equivalent electrons $(n_1 = n_2)$. The white circles are not observed because of the Pauli principle

values that can be realized. The white circles indicate states that are forbidden by the Pauli principle. The right part of Fig. 6.36 shows how the different states contribute to all possible $M_L$ and $M_S$ values on the left side. This diagram shows that for equivalent $2p$ electrons with $n_1 = n_2$ fifteen combinations of $M_L$ and $M_S$ are possible. Five of these combinations result in the $^1D$ state, nine in the $^3P$ state and one in the $^1S$ state. The combination $M_L = M_S = 0$ comes from three different couplings of the individual angular momenta of the two $p$ electrons, and $M_L = \pm 1$, $M_S = 0$ from two, as explained by Table 6.7. Table 6.7 shows in detail the different quantum numbers $L$, $S$, $m_{l_1}$, $m_{l_2}$, $m_{s_1}$, $m_{s_2}$, $M_S M_J$ and the resulting atomic states. In Table 6.8 the possible multiplicities due to the coupling of the electron spins are compiled.

**Table 6.6.** Possible total angular momenta and resulting atomic levels for different two-electron configurations

| Electron configuration | Quantum numbers of angular momenta | | | Level assignment |
|---|---|---|---|---|
| | $L$ | $S$ | $J$ | |
| $s$ | 0 | $\frac{1}{2}$ | $\frac{1}{2}$ | $^2S_{1/2}$ |
| $s^2$ | 0 | 0 | 0 | $^1S_0$ |
| | 0 | 1 | 1 | $^3S_1$ for $n_1 \neq n_2$ |
| $sp$ | 1 | 0 | 1 | $^1P_1$ |
| | 1 | 1 | 0, 1, 2 | $^3P_0, {}^3P_1, {}^3P_2$ |
| $p^2$ | 0 | 0 | 0 | $^1S_0$ |
| | 1 | 1 | 0, 1, 2 | $^3P_0, {}^3P_1, {}^3P_2$ |
| | 2 | 0 | 2 | $^1D_2$ |
| | 0 | 1 | 1 | $^3S_1$ } only for $n_1 \neq n_2$ |
| | 1 | 0 | 1 | $^1P_1$ } only for $n_1 \neq n_2$ |
| | 2 | 1 | 1, 2, 3 | $^3D_{1,2,3}$ } only for $n_1 \neq n_2$ |

**Table 6.7.** Possible quantum numbers for levels resulting from a $np^2$ electron configuration with $n_1 = n_2 = n$

| $L$ | $S$ | $m_{l_1}$ | $m_{l_2}$ | $m_{s_1}$ | $m_{s_2}$ | $M_S$ | $M_J$ | Term |
|---|---|---|---|---|---|---|---|---|
| 0 | 0 | 0 | 0 | $+\frac{1}{2}$ | $-\frac{1}{2}$ | 0 | 0 | $^1S_0$ |
| 1 | 0 | 0 | −1 | $+\frac{1}{2}$ | $-\frac{1}{2}$ | 0 | 0 | $^3P_0$ |
| | | 1 | −1 | $+\frac{1}{2}$ | $+\frac{1}{2}$ | +1 | +1 | |
| 1 | 1 | 1 | 0 | $-\frac{1}{2}$ | $-\frac{1}{2}$ | −1 | 0 | $^3P_1$ |
| | | 1 | −1 | $-\frac{1}{2}$ | $-\frac{1}{2}$ | −1 | −1 | |
| | | 1 | 0 | $+\frac{1}{2}$ | $+\frac{1}{2}$ | +1 | +2 | |
| | | 1 | −1 | $+\frac{1}{2}$ | $+\frac{1}{2}$ | +1 | +1 | |
| 1 | 1 | 0 | 0 | $+\frac{1}{2}$ | $-\frac{1}{2}$ | 0 | 0 | $^3P_2$ |
| | | 1 | −1 | $-\frac{1}{2}$ | $-\frac{1}{2}$ | 0 | −1 | |
| | | 0 | −1 | $-\frac{1}{2}$ | $-\frac{1}{2}$ | −1 | −2 | |
| | | +1 | +1 | $+\frac{1}{2}$ | $-\frac{1}{2}$ | 0 | +2 | |
| | | +1 | 0 | $+\frac{1}{2}$ | $-\frac{1}{2}$ | 0 | +1 | |
| 2 | 0 | 1 | −1 | $+\frac{1}{2}$ | $-\frac{1}{2}$ | 0 | 0 | $^1D_2$ |
| | | 0 | −1 | $+\frac{1}{2}$ | $-\frac{1}{2}$ | 0 | −1 | |
| | | −1 | −1 | $+\frac{1}{2}$ | $-\frac{1}{2}$ | 0 | −2 | |

The total spin $S$ and with it the multiplicity $2S + 1$ of an atomic state depends on the number of electrons in not completely filled shells (Table 6.8).

## 6.6 Excited Atomic States

In Sect. 6.1.5 we illustrated for the case of the He atom that for excited atomic states, the number of possible ways to couple the different angular momenta becomes much larger than for the ground state. The reason for this larger manifold of possible states is that now the excited electron has a different principal quantum number and therefore the Pauli principle imposes fewer restrictions.

In this section we will present the different possibilities of populating excited atomic states.

Such excitations can be experimentally realized by the absorption of photons, by collisions with electrons, or with high energy ions.

One speaks of *one-electron excitation* if in the independent electron model (Sect. 6.4.1) only the quantum

**Table 6.8.** Total electron spin and multiplicity of atomic states resulting from different numbers of valence electrons

| Electrons | Spin quantum number $m_s$ | Multiplicity |
|---|---|---|
| 1 | $m_s = \frac{1}{2}$ | Doublet |
| 2 | $m_{s_1} = +\frac{1}{2}, m_{s_2} = +\frac{1}{2} \Rightarrow S = 1$ | Triplet |
| | $m_{s_1} = +\frac{1}{2}, m_{s_2} = -\frac{1}{2} \Rightarrow S = 0$ | Singlet |
| 3 | $+\frac{1}{2}, +\frac{1}{2}, -\frac{1}{2} \Rightarrow S = \frac{1}{2}$ | Doublet |
| | $+\frac{1}{2}, +\frac{1}{2}, +\frac{1}{2} \Rightarrow S = \frac{3}{2}$ | Quartet |
| 4 | $+\frac{1}{2}, +\frac{1}{2}, -\frac{1}{2}, -\frac{1}{2} \Rightarrow S = 0$ | Singlet |
| | $+\frac{1}{2}, +\frac{1}{2}, +\frac{1}{2}, -\frac{1}{2} \Rightarrow S = 1$ | Triplet |
| | $+\frac{1}{2}, +\frac{1}{2}, +\frac{1}{2}, +\frac{1}{2} \Rightarrow S = 2$ | Quintet |

numbers of one electron are changed while they remain the same for all other electrons. One should, however, keep in mind that because of the electron correlation the excitation of one electron does affect the energies of the others, because the electrostatic interaction is changed and also, to a smaller extent, the magnetic interactions might change. The energy transferred to the atom by photon absorption or electron impact is, even for one-electron excitation not completely transferred to the excited electron but also changes the energy of the residual electron shell. This is particularly true for the excitation of an electron from inner shells, because here the average distance between the electrons is smaller and the correlation larger.

### 6.6.1 Single Electron Excitation

The smallest energy is required, when an electron in the outer valence shell is excited into higher states. The excitation energy of these valence electrons ranges from 1–10 eV. Exceptions are the noble gases, where the outer shell is completely filled, the binding energy is larger and the excitation energy of an electron in this shell is higher. For example, for helium the first excited state lies about 20 eV above the ground state.

The excited state $E_k$ is not stable. It can decay spontaneously into lower states $E_i$ by emitting a photon with energy $h\nu = E_k - E_i$. The mean lifetime $\tau_k$ of the excited state depends on the total probability of radiative transitions into lower states (see Sect. 7.3). For some states this transition probability is very small and the lifetime correspondingly long. Such states are called ***metastable states***.

EXAMPLES

1. Lifetimes of some excited states:

$$\begin{aligned} &\mathrm{H}(2^2P_{1/2}): \quad \tau = 1.5 \times 10^{-9}\,\mathrm{s} \\ &\mathrm{He}(2^1P_1): \quad \tau = 0.5 \times 10^{-9}\,\mathrm{s} \\ &\mathrm{Na}(3^2P_{1/2}): \quad \tau = 16 \times 10^{-9}\,\mathrm{s} \end{aligned}$$

2. Lifetimes of metastable states:

$$\begin{aligned} &\mathrm{H}(2^2S_{1/2}): \quad \tau = 8\,\mathrm{s} \\ &\mathrm{He}(2^1S_0): \quad \tau = 19.6\,\mathrm{ms} \\ &\mathrm{He}(2^3S_1): \quad \tau = 7870s \end{aligned}$$

### 6.6.2 Simultaneous Excitation of Two Electrons

Under special conditions two electrons can be excited simultaneously. Assume, for instance, that two electrons from the $2s$ state of Be (Fig. 6.14) are excited into higher states. The two excited electrons can populate the $2p$ state, or one electron is excited into the $2p$ state and the other into the $3p$ state, etc. The total excitation energy is then

$$E = E_1 + E_2 + \Delta E\,, \tag{6.46}$$

where $E_i$ is the excitation energy for the single excitation of electron $e_i$ and $\Delta E$ is the change in the interaction energy between the two electrons and with the core, caused by the excitation of the two electrons.

The doubly excited atomic state can decay either by emission of two photons, or the energy of one excited electron can be transferred to the other excited electron, due to the Coulomb interaction between the two electrons. This causes a further excitation of one electron into still higher states and a de-excitation of the other. This process becomes less likely if the two electrons are excited into states with very different principal quantum numbers, because then the mean distance between the two electrons becomes larger and their mutual interaction weaker. This increases the lifetime of such doubly excited states if their energy is still below the ionization energy.

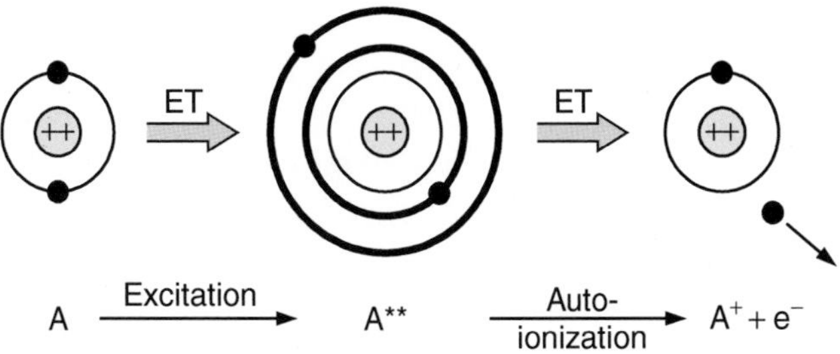

Fig. 6.37. Illustration of autoionization of a doubly excited atomic state (ET = energy transfer)

Since the total energy must be conserved, this process of energy transfer from one electron to the other is only possible, if an excited atomic state exists, that matches the energy $E^*$ of this highly excited electron. This is very unlikely for the discrete atomic energy levels below the ionization limit, i.e, for $E^* < E_{\text{ion}}$, but is always possible for the continuous energy spectrum $E^* > E_{\text{ion}}$. In this case the transfer of the excitation energy of one of the two excited electrons to the other allows this electron to leave the atom and the atom becomes ionized (Fig. 6.37). This process, which is illustrated in Fig. 6.38, is called ***autoionization***. The term diagram is shown in Fig. 6.38b for the case of the doubly excited Li atom. One electron is excited from the $2s$ into the $2p$ level, the other from the $1s$ into the $3p$ level. The total energy of the doubly excited state

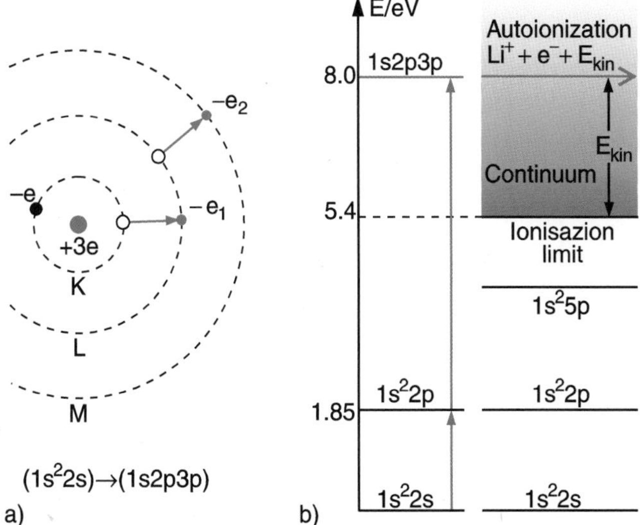

Fig. 6.38a,b. Simultaneous excitation of two electrons in the Li atom. (a) Bohr model. (b) Level scheme with autoionization

$1s2p3p$ lies above the ionization energy

$$E_{\text{ion}} = \lim_{n\to\infty} (1s^2 np) \tag{6.46a}$$

of singly excited states at about 5.4 eV. If the $2p$ electron transfers its energy to the $3p$ electron, the energy of the latter becomes higher than $E_{\text{ion}}$ and the electron can leave the atom.

### 6.6.3 Inner-Shell Excitation and the Auger Process

When an electron from an inner shell is excited into higher unoccupied states, larger energies are required than for the excitation of a valence electron, because electrons in inner shell are much less screened from the nuclear charge and their binding energy is accordingly larger. Inner shell excitation therefore needs either UV or even X-ray photons or collisions between inner shell electrons and incident electrons having sufficient energy.

The hole created in the inner shell by the excitation of an electron can be refilled when an electron from a higher state $E_i$ falls into this hole. The energy $\Delta E$ released by this process is generally emitted as UV or X-ray photons $h\nu = \Delta E$ (Fig. 6.39). This is the source for the discrete X-ray emission in X-ray tubes (see Sect. 7.6.2).

The energy $\Delta E = E_i - E_k$, available during the transition of an electron $E_1$ from level $|i\rangle$ to $|k\rangle$, can also be transferred directly onto another electron $e_2$ of the same atom, due to the interaction between the electrons. If the binding energy $E_{\text{B}}$ of $e_2$ is smaller than $\Delta E$, it can leave the atom, which means that

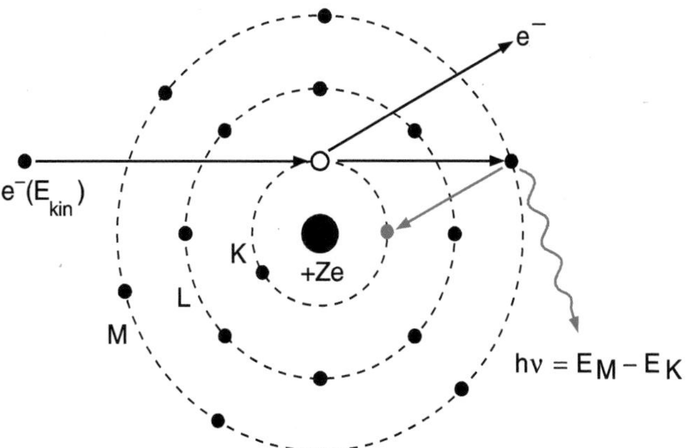

Fig. 6.39. Inner shell excitation with subsequent emission of characteristic X-rays

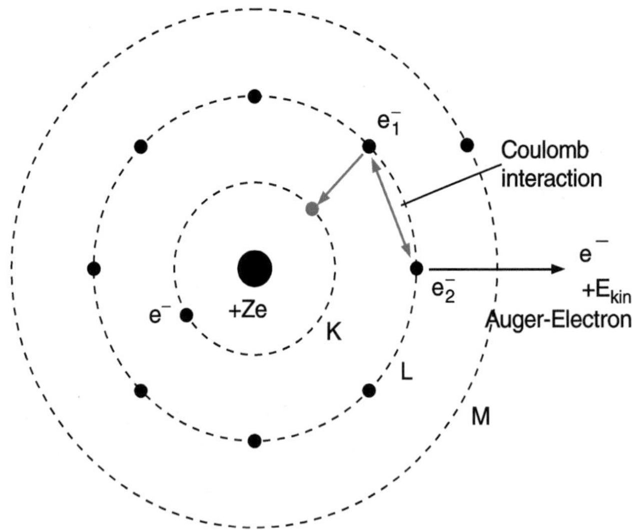


**Fig. 6.40.** The Auger effect. The electron $e_1^-$ in the $L$ shell falls down into the vacancy in the $K$ shell and transfers its energy to the electron $e_2^-$, which leaves the atom

autoionization takes place (Fig. 6.40). This special kind of autoionization is called the ***Auger effect***.

The kinetic energy of the ejected Auger electron is

$$E_{\text{kin}} = E_i - E_k - E_{\text{B}} \ .$$

Measuring this energy allows the determination of the atomic state out of which it was ionized.

The emission of X-rays and the Auger process are competing processes. The fraction of all inner shell excitations that lead to X-ray fluorescence on the transition $E_i \to E_k$ is called the fluorescence yield. It depends on the level $E_k$ and on the nuclear charge $Ze$ of the atom. For $Z < 30$ the Auger process is dominant, for $Z > 60$ the fluorescence yield reaches 90%, if the excitation starts from the $K$ shell.

### 6.6.4 Rydberg States

In Sect. 4.3 it was shown that the mean radius of the electron in the H atom

$$\langle r \rangle = a_0 n^2 \quad \text{with} \quad a_0 = 5 \times 10^{-11}\,\text{m}$$

is proportional to the square of the principal quantum number $n$. An electron with $n = 100$, therefore, has a mean radius of its orbit $\langle r \rangle = 5 \times 10^{-7}\,\text{m} = 0.5\,\mu\text{m}$! When its orbital angular momentum has the maximum value $|\boldsymbol{l}| = (n-1)h$ the electron orbit approaches a circular path with radius $r = \langle r \rangle$, since for large values of $n$ the quantum mechanical description converges towards the classical one (see Sect. 5.8).

For $l \ll n$ the classical orbits are elliptical. The inner turning point (perihelion) of the ellipse comes close to the nucleus and therefore experiences a revolution of the major axis (Fig. 6.41). The maximum probability of finding the electron is around the black dashed circle in Fig. 6.41, because there are the outer turning points (aphelions) of the elliptical orbits where the electron has its smallest velocity.

According to the Rydberg formula (6.32) in Sect. 6.3,

$$E_n = -Ry^*/(n-\delta)^2 \ , \tag{6.46b}$$

the binding energy of a Rydberg electron becomes very small for large $n$ values. It can therefore be readily ionized by small perturbations, such as collisions with atoms or electrons or by external electric fields.

The field ionization of Rydberg atoms is used for a very sensitive detection of these atoms.

The level scheme is shown in Fig. 6.42. For sufficiently large values of $n$, the potential is close to a Coulomb potential with $Z = 1$ because the nuclear charge is nearly completely shielded by the electron core. The homogeneous external field in the $x$ direction has the potential energy $E_{\text{pot}} = -e\boldsymbol{E}_0 x$. With

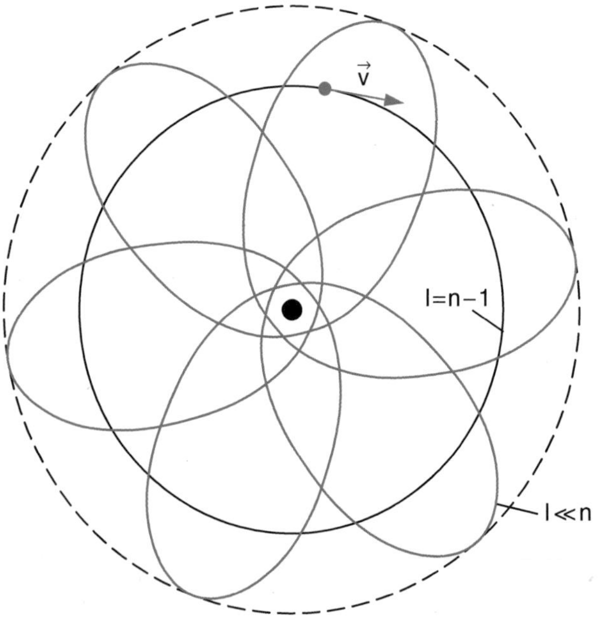


**Fig. 6.41.** Classical paths of Rydberg electrons with $l \ll n$ (*blue curves*) and $l = l_{\max} = n - 1$ (*black curve*)

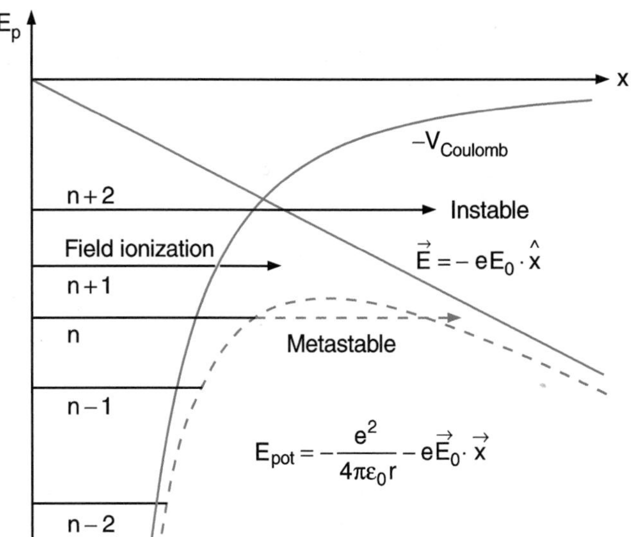


**Fig. 6.42.** Field ionization of a Rydberg level

$x = r\cos\vartheta$, the sum of the two potentials

$$\begin{aligned} E_{pot} &= -\frac{e}{4\pi\varepsilon_0 r} - e\boldsymbol{E}_0 x \\ &= -e\left(\frac{1}{4\pi\varepsilon_0 r} + E_0 r\cos\vartheta\right) \end{aligned} \tag{6.46c}$$

has a maximum for $\mathrm{d}E_{pot}/\mathrm{d}r = 0$. This yields a radius $r_c$ of maximum energy

$$r_c = \sqrt{\frac{1}{4\pi\varepsilon_0 E_0\cos\vartheta}}$$

and the potential energy

$$E_{pot}(r_c) = -e\sqrt{\frac{E_0\cos\vartheta}{\pi\varepsilon_0}}$$

The ionization energy is therefore lowered for $\vartheta = 0$ by

$$\Delta E_{pot} = -e\sqrt{\frac{E_0}{\pi\varepsilon_0}}\,. \tag{6.46d}$$

## EXAMPLE

With the external electric field $E = 5\times10^3\,\mathrm{V/m}$, the maximum of the total potential energy is $E_{pot}(r_c) = -0.858\times10^{-21}\,\mathrm{J} = -5.4\,\mathrm{meV}$. All Rydberg levels with $n > 50$ are above this energy and are therefore ionized. In fact, due to the tunnel effect even levels with $n < 50$ can be ionized, although their ionization probability decreases exponentially with decreasing $n$.

For the mean kinetic energy of a Rydberg electron in a Coulomb potential, the virial theorem states that

$$\langle E_{kin}\rangle = -1/2\langle E_{pot}\rangle\,.$$

From energy conservation $E_n = E_{kin} + E_{pot}$ ist follows then that

$$E_{kin} = -E_n = Ry^*/n^2\,.$$

With increasing $n$ the velocity of the electron decreases as $1/n$.

The absorption or emission frequencies for transitions between neighboring Rydberg levels are

$$\nu = (E_n - E_{n-1})/h \approx \frac{2Ry^*}{n(n-1)}\,.$$

For $n \geq 60$ they are in the microwave range, for $n \geq 300$ in the rf range. Table 6.9 summarizes some properties of Rydberg states.

## EXAMPLE

For the Rydberg level of the hydrogen atom with $n = 100$ the kinetic energy is $E_{kin} = 2\times10^{-22}\,\mathrm{J} = 0.6\times10^{-3}\,\mathrm{eV}$. The velocity of the Rydberg electron on a circular path with $r = n^2 a_0$ is then $v = 2.2\times10^4\,\mathrm{m/s}$, its revolution period is $T_n = 2\pi/v = 1.4\times10^{-10}\,\mathrm{s}$ and its revolution frequency $\nu = 7\times10^9\,\mathrm{s^{-1}}$. Compared to the period $T_1 = 1.4\times10^{-16}\,\mathrm{s}$ for the lowest orbital with $n = 1$ the Rydberg electron moves very slowly.

For heavy atoms with $Z \gg 1$ the difference becomes even larger, because the velocity on the inner orbit where the nuclear charge $Ze$ is barely shielded, scales with $Z$.

The frequency $\nu(n \to n+1)$ of transitions between neighboring Rydberg levels $n = 100$ and $n+1$ is $\nu = 6.5\,\mathrm{GHz}$.

If in atoms with more than one electron, one of the electrons is excited into such a high lying Rydberg state, it mainly moves outside the charge distribution of the other electrons. The potential for the Rydberg electron is therefore nearly the Coulomb potential of an ion with the effective charge $Q_{eff} = +e$, because the nuclear charge is nearly cancelled by the $Z-1$

**Table 6.9.** Characteristic data of atomic Rydberg states ($a_0 = 5.29 \times 10^{-11}$ m, $Ry = 1.09737 \times 10^7 \mathrm{m}^{-1}$)

| Physical quantity | $n$-dependence | $H(n=2)$ | $H(n=50)$ |
|---|---|---|---|
| Binding energy | $-Ry^* n^{-2}$ | 3.4 eV | $0.0054\,\mathrm{eV} \mathrel{\hat{=}} 43.5\,\mathrm{cm}^{-1}$ |
| Energy $E(n+1) - E(n)$ difference | $\Delta E_n = Ry\left(\frac{1}{n^2} - \frac{1}{(n+1)^2}\right)$ | $\frac{5}{36} R \sim 2\,\mathrm{eV}$ | $0.2\,\mathrm{meV} \mathrel{\hat{=}} 2\,\mathrm{cm}^{-1}$ |
| Mean Bohr radius | $a_0 n^2$ | $4a_0$ | $2500 a_0 = 132\,\mathrm{nm}$ |
| Geometric cross section | $\pi a_0^2 n^4$ | $16\pi a_0^2$ | $6\pi \times 10^6\, a_0^2 = 5 \times 10^{-14}\,\mathrm{m}^2$ |
| Revolution period | $T_n \propto n^3$ | $10^{-15}$ s | $2 \times 10^{-11}$ s |
| Radiative lifetime | $\propto n^3$ | $5 \times 10^{-9}$ s | $1.5 \times 10^{-4}$ s |
| Critical electric field for ionization | $E_c = \pi\varepsilon_0 Ry^{*2} e^{-3} n^{-4}$ | $5 \times 10^9$ V/m | $5 \times 10^3$ V/m |

other electrons. The screening of the nuclear Coulomb potential depends on the penetration depth of the Rydberg electron into the atomic core, which in turn depends on the angular momentum of the Rydberg electron. For the maximum angular momentum quantum number $l = n - 1$ the classical orbit is circular and the penetration depth is minimum while for small values of $l \ll n$, the classical path is an ellipse with large eccentricity and the Rydberg electron submerges deep into the core where it experiences the deviation from the Coulomb potential.

In the quantum mechanical language the deviation of the potential for a Rydberg electron from the pure Coulomb potential of a one-electron atom depends on the spatial overlap of the wave functions of the Rydberg electron with those of the core electrons. These wave functions depend on the quantum numbers $n$ and $l$ (see Fig. 5.7).

The energies of Rydberg levels in multielectron atoms is therefore shifted against the energy

$$E_n = -Ry^*/n^2 \tag{6.47}$$

of hydrogen like atoms. This energy shift can be expressed by the so-called quantum defect $\delta_{nl}$. The Rydberg formula (6.47) is then generalized to

$$E_{n,l} = -\frac{Ry^*}{(n - \delta_{n,l})^2} = -\frac{Ry^*}{n_{\mathrm{eff}}^2}\,, \tag{6.48}$$

where the quantum defect $\delta_{nl}$ depends on $n$ and $l$ (Table 6.4). Often, the effective principal quantum number $n_{\mathrm{eff}} = n - \delta_{nl}$ is introduced, which deviates from an integer. This allows one to use a similar Rydberg formula for all atoms [6.5]. From the equation

$$E_{n,l} - E_n = Ry^*\left(\frac{1}{(n-\delta)^2} - \frac{1}{n^2}\right) \approx \frac{2\delta}{n^3}$$

the quantum defect can be calculated. For $n \gg \delta$ the difference between observed level energies $E_{n,l}$ and those, calculated for a Coulomb potential decreases as $1/n^3$.

### 6.6.5 Planetary Atoms

When two valence electrons are excited into different high lying Rydberg states $(n, l)$ and $(n', l')$ (Fig. 6.43) the total excitation energy is way above the ionization limit (Fig. 6.44) and autoionization can occur (see Sect. 6.6.2). Since the radii of Rydberg orbits scale with $n^2$ the mean distance between the two excited electrons is large for $n \neq n'$ and their mutual interaction becomes small for large values of $n$. This decreases the probability of autoionization and the lifetime of such doubly excited Rydberg states may be much longer than those for doubly excited lower states. This allows the observation and spectroscopic characterization of such unusual atoms called planetary atoms, because the two electrons circle around the inner atomic core like planets around the sun [6.6].

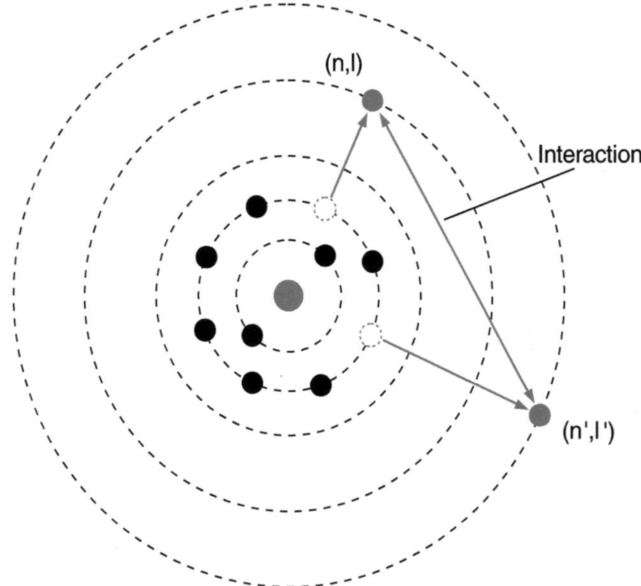


**Fig. 6.43.** Classical model of a planetary atom with two excited electrons

If the two electrons come close to each other, the mutual interaction increases and one electron can transfer its excitation energy onto the other electron, which can then leave the core (autoionization, Fig. 6.44). The

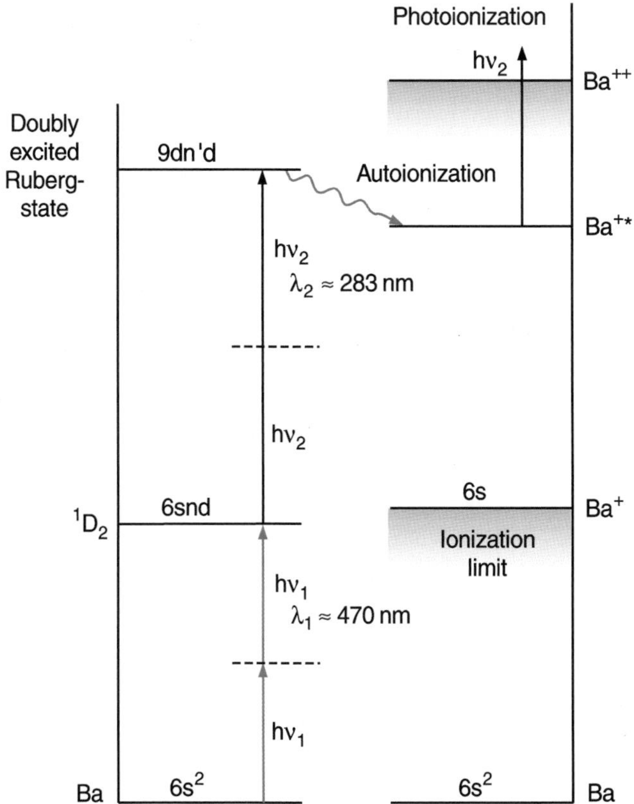


**Fig. 6.44.** Level scheme, excitation and autoionization of a planetary atom for the example of the Ba atom

ionic state resulting from the autoionization can be detected by measuring the kinetic energy of the ejected electron or by photoionization of the excited ion state into a doubly ionized atom $A^{++}$, which can be monitored by mass spectrometry. Measurements of the decay times and their dependence on the quantum numbers $(n,l,s)$ and $(n',l',s')$ gives much information on the correlation energy between two electrons in defined states, which are nearly unaffected by the core electrons.

One example is the investigation of planetary barium atoms (Fig. 6.44). By simultaneous absorption of two photons $h\nu_1$ from a pulsed laser L1 one electron is excited from the ground state $6s^2$ into the $^1D_2$ state with the electron configuration $6snd$. Further excitation by absorption of two other photons $h\nu_2$ from a second laser L2 brings the other 6s electron from the $6snd$ configuration into the $9dn'd$ configuration which correspond to a doubly excited high-lying Rydberg state with an energy above the ionization limit. This state can therefore decay into an excited state of the $Ba^+$ ion, which can be detected by further ionization into $Ba^{++}$ by absorption of another photon $h\nu_2$ from the second laser. The $Ba^{++}$ ions are detected with time resolved techniques. This gives information on the lifetime of the doubly excited state [6.7].

## 6.7 Exotic Atoms

Up to now we have assumed that the spatial extension of the atomic nucleus can be neglected and the nucleus can be treated as a point-like charge. This assumption is justified as long as the mean nuclear radius ($r_N \approx 10^{-15}$ m) is very small compared to the mean distance $\langle r \rangle$ between electron and nucleus, which can be estimated by the Bohr radius ($r_1 \approx 10^{-10}$ m) of the electrons for the lowest electron state with $n = 1$.

In Sects. 5.1 and 5.7 we have already discussed that for $1S$ states with $l = 0$ the electron wave function has its maximum at $r = 0$ at the position of the nucleus. The energy of the $1S$ states should therefore be affected by the spatial distribution of the nuclear charge, in particular for atoms with a high nuclear charge $Ze$. Measurements with high spectral resolution can indeed detect energy shifts caused by the deviation of the nucleus from a point charge, which are part of the hyperfine shifts (Sect. 5.6).

Much larger shifts are observed in exotic atoms where one atomic electron is replaced by a heavier particle with negative charge and mass $m_x \gg m_e$, such as a muon $\mu^-$, a $\tau$ lepton $\tau^-$, a $\pi^-$ meson or an antiproton $p^-$. The Bohr-radii (see (3.85))

$$r_n = \frac{4\pi\varepsilon_0\hbar^2 n^2}{Ze^2\mu} \tag{6.49}$$

of these particles, which scale inversely proportional to the reduced mass

$$\mu = \frac{m_x M_N}{m_x + M_N}$$

are much smaller than for the corresponding orbits of an electron in the Coulomb field of the nucleus with charge $Ze$. The influence of the spatial distribution of the nuclear charge on the energy levels of such exotic atoms is therefore much more pronounced. Measurements of these energy shifts give detailed information on the spatial charge distribution and the mass distribution within the nucleus and their dependence on the nuclear spin.

Unfortunately the elementary particles $\mu^-$, $\pi^-$ or $\tau^-$ are not stable. They decay within $10^{-6}$ s to $10^{-8}$ s into other particles. Therefore the exotic atoms only exist for a short time. This makes their spectroscopic characterization difficult. Nevertheless it has been possible in recent years to produce sufficient numbers of exotic atoms and to perform accurate spectroscopic measurements of their energy states and transition probabilities [6.8, 9, 10]. This will be illustrated by some examples.

### 6.7.1 Muonic Atoms

A muonic atom consists of the atomic nucleus, a negatively charged muon $\mu^-$ and the electron shell with $(Z - p)$ electrons. When the muon is captured by the neutral atom with $Z$ electrons, the released energy (kinetic energy and binding energy of the muon) can be transferred to the electron shell and $p$ electrons $(p = 1, 2, 3, \ldots)$ can leave the atom due to the Auger effect (see Sect. 6.6.3). Because of the large muon mass $m_\mu = 206.76\, m_e$ the lowest possible Bohr orbit $(n = 1)$ of the muon is for a nuclear charge $Ze$ with $Z = 30$ according to (6.49) only

$$r_1(\mu^-) = 7.7 \times 10^{-15}\,\text{m}\,,$$

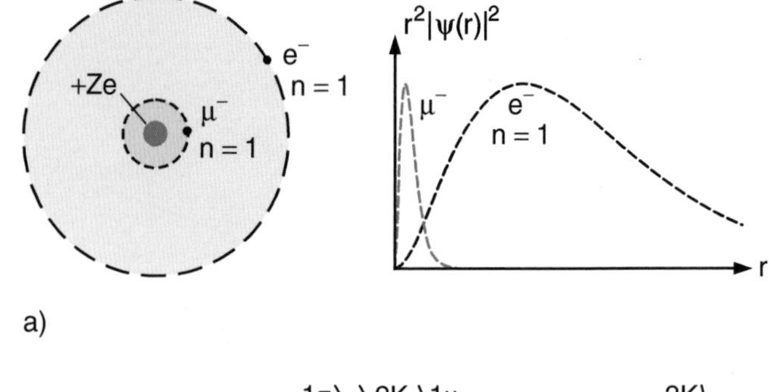


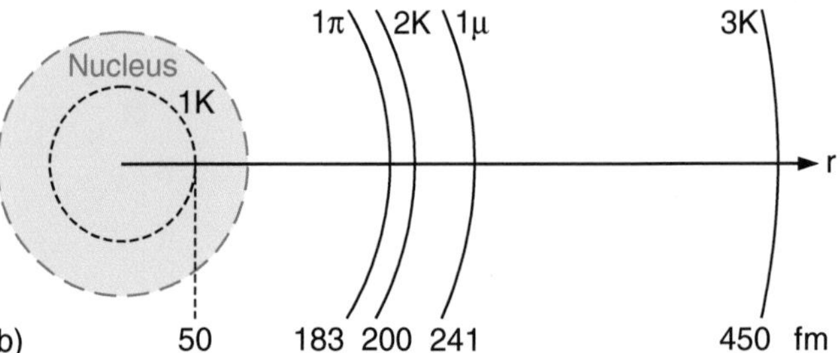


**Fig. 6.45a,b.** Muonic atom. (**a**) Comparison of radial charge density for the muon $\mu^-$ and the electron $e^-$ in the $1s$ state. (**b**) Bohr-radius of levels with principle quantum number $n$ for exotic atoms, where the electron is replaced by different muons or mesons

which is of the same order of magnitude as the nuclear radius (Fig. 6.45). This means that the muon experiences the unshielded nuclear Coulomb field and the energies $E_n$ of the muonic atom levels are very much influenced by the spatial distribution of the nuclear charge, while the other electrons have a much smaller effect on the muon, because their average distance from the muon is much larger.

Measuring the wavelength of the radiation emitted when the muon jumps from level $E_i$ into the lower level $E_k$ allows the determination of the energy differences $\Delta E_{ik} = E_i - E_k$ and therefore the deviations of the level energies from those in a pure Coulomb potential. These deviations are caused by the spatial charge distribution within the nucleus. The potential experienced by the muon can be expanded into a power series of $r^{-n}$, where the different terms represent the monopole potential, the quadrupole, octopole, etc., potential. A model calculation yields that nuclear charge distribution which fits best the measured term energies [6.9]. For muonic lead atoms $(Z = 82)$ the photon energies are in the MeV range.

Since the mean lifetime of $\mu^-$ is $2.2\,\mu$s muonic atoms are unstable even in their ground state. For light atoms $(Z < 10)$ the $\mu^-$ decays according to the scheme

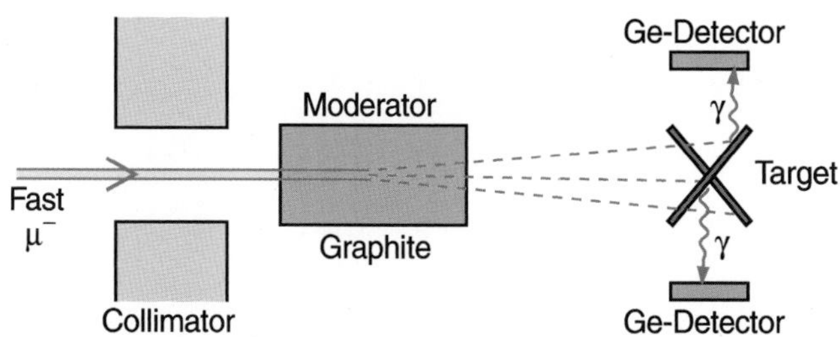


**Fig. 6.46.** Generation of exotic atoms and their detection by X-ray spectroscopy

$$\mu^- \to e^- + \overline{\nu}_e + \nu_\mu \tag{6.50}$$

into an electron, an electron antineutrino and a muon neutrino. For heavy atoms ($Z > 10$) the lowest $\mu^-$ orbit is already within the nucleus. In this case the muon induces the nuclear reaction

$$\mu^- + p \to n + \nu_\mu \,, \tag{6.51}$$

where a proton in the nucleus is converted into a neutron. The probability of this reaction is high for the lowest muon level and the mean lifetime of the muon is therefore much smaller than for lighter atoms where the muon orbit is outside the nucleus.

A possible experimental arrangement for the spectroscopy of muonic atoms is shown in Fig. 6.46. Fast protons from a proton synchrotron collide with a target producing an intense beam of $\pi^-$ mesons, which decay during their flight within $2.2 \times 10^{-8}$ s into $\mu^- + \overline{\nu}_\mu$ forming a fast beam of $\mu^-$. The fast muons are slowed down in a graphite block and are completely stopped in two crossed thin sheets of a specific material. Here they are captured by the atoms in the sheets forming myonic atoms or ions in high lying levels from where they can cascade down into lower levels. The X-ray radiation, emitted during this cascading is measured with a germanium semiconductor detector with high energy resolution.

Meanwhile even the fine structure of energy levels in muonic atoms and Zeeman splittings have been resolved. The fine structure splittings are here much larger than in normal atoms and amount to several eV. Transitions between fine structure components can therefore be induced with visible lasers. These measurements yield very accurate absolute values of mass and magnetic moment of the $\mu^-$ muon [6.11, 12].

### 6.7.2 Pionic and Kaonic Atoms

Instead of the myon, a negative $\pi^-$ meson can also be captured by a neutral atom. The energy released by this capture process is sufficient to eject one or several electrons from the atomic electron shell (Fig. 6.47). For a $\pi^-$ meson in atomic orbits with $n < 17$ the Bohr radius is already sufficiently small to make the interaction of the $\pi^-$ with the electrons of the atom negligibly small.

The nucleons (protons and neutrons) in the atomic nucleus interact with the $\pi^-$ meson not only through Coulomb forces but also through the short range, but much stronger, nuclear force. A comparison of the energy levels in the myonic and the pionic atoms gives information about the nuclear forces and their radial dependence (because the lepton $\mu^-$ does not feel the strong nuclear force contrary to the $\pi^-$ meson).

Exotic atoms with heavier negative mesons ($K^-$, $\eta^-$) allows probing of charge and mass distribution at even smaller distances from the center of the nucleus. They can give information on deviations of these distributions from a spherical symmetry. Since the lifetime of the $K^-$ mesons is only 12 ns, measurements of the

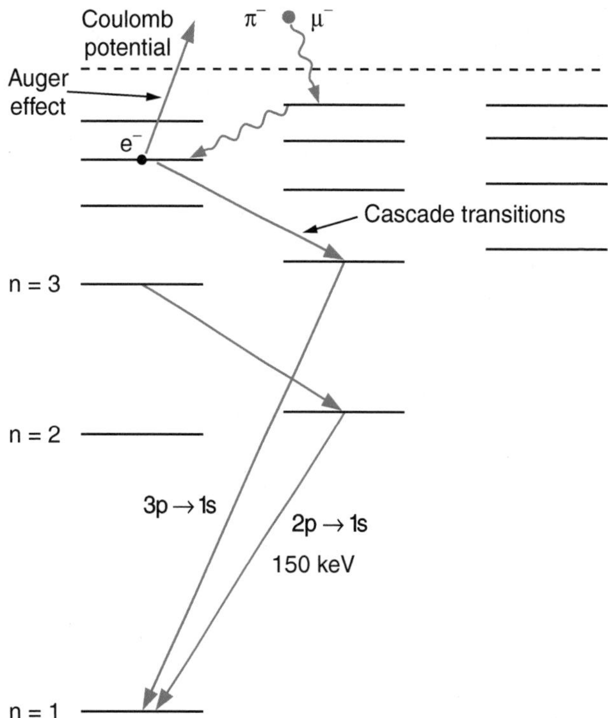


**Fig. 6.47.** Capture of a $\mu^-$ myon or a $\pi^-$ meson with following cascading transitions into the final ground state

**Table 6.10.** Characteristic features of exotic atoms

| Particle | $e^-$ | $\mu^-$ | $\pi^-$ | $K^-$ |
|---|---|---|---|---|
| $m/m_e$ | 1 | 207 | 273 | 967 |
| Bohr radius $r_1$ in fm | $\frac{5.3}{Z}\cdot 10^4$ | $\frac{256}{Z}$ | $\frac{194}{Z}$ | $\frac{54.8}{Z}$ |
| Term energy for $n=1$, $Z=1$ | $-13.6\,\text{eV}$ | $-2.79\,\text{keV}$ | $-3.69\,\text{keV}$ | $-13.1\,\text{keV}$ |
| $\Delta E(n=2\to 1)$ for $Z=20$ | 4.1 keV | 837 keV | 1.1 MeV | 3.9 MeV |
| Mean lifetime of free particle $\tau$ / s | $\infty$ | $2.2\cdot 10^{-6}$ | $2.6\cdot 10^{-8}$ | $1.2\cdot 10^{-8}$ |
| Fine structure splitting $2^2P$ for $Z=20$, $n=2$ | 6.6 eV | 1.3 keV | 1.8 keV | 6.4 keV |

spectra of these exotic atoms becomes more and more difficult [6.13].

Instead of leptons or mesons, an electron in the atomic shell can also be replaced by negatively charged hadrons such as the antiproton $p^-$ or the $\Sigma^-$ particle which have a larger mass and therefore even smaller Bohr radii [6.14].

In Table 6.10 some characteristic properties of different exotic atoms are compiled.

### 6.7.3 Anti-hydrogen Atoms and Other Anti-atoms

If the proton and electron in the hydrogen atom are both replaced by their anti-particles, the anti-proton $p^-$ and the anti-electron = positron $e^+$, the exotic bound system $(p^-e^+)$ of the two anti-particles can be formed, which is called **anti-hydrogen**. Its production is by no means trivial but recently the first anti-hydrogen atoms have been observed [6.15]. Their detailed spectroscopy and the comparison of the energy levels with those of the H-atom provide a stringent test of possible differences between the absolute values of positive and negative charges, of the masses and the magnetic moments of elementary particles and anti-particles.

In particle accelerators a large number of anti-protons $p^-$ can be produced by high energy collisions of protons $p^+$ with protons. Positrons can be obtained from radioactive $\beta^+$-emitters and are subsequently accelerated. Both anti-particles are stored and accumulated in storage rings where they circulate with high energies. However, only a tiny fraction of these high energy anti-particles can be slowed down to thermal energies, where they can be captured in special magnetic traps in order to enhance the formation of anti-hydrogen. This is a difficult task, since on one hand a large number of slow antiparticles is needed in order to produce a sufficient number of anti-hydrogen atoms. On the other hand collisions between antiprotons $p^-$ and protons $p^+$ from the residual background gas in the trap or between positrons $e^+$ and electrons $e^-$ will immediately annihilate these particles by the reactions

$$p^+ + p^- \to 2\gamma\ ;\quad e^+ + e^- \to 2\gamma\ .$$

After an extremely good vacuum could be achieved in the trap anti-protons could be captured and stored for several weeks in specially designed magnetic traps (Fig. 6.48). This long storage time has allowed the researchers to obtain precise spectroscopic data about the charge and the magnetic moment of the anti-proton [6.16]. The result of these measurements was that the relative mass difference $\Delta m/m = [m(p^+) - m(p^-)]/m(p^+)$ and the relative charge difference $|\Delta q/q|$ are both smaller than $10^{-8}$.

Recently a research collaboration at the European high energy center CERN also reported that anti-hydrogen atoms had been observed, which could be stored in a sophisticated trap design that can store both particles simultaneously for a short time. In order to perform precision measurements, a larger number and longer storage time of both anti-particles is needed. Experiments for improving the situation are underway and first results are expected in the near future.

Slowing down anti-protons $p^-$ in a hydrogen target at low temperatures, protonium $(p^+p^-)$, a bound system of a proton and an anti-proton has been observed.

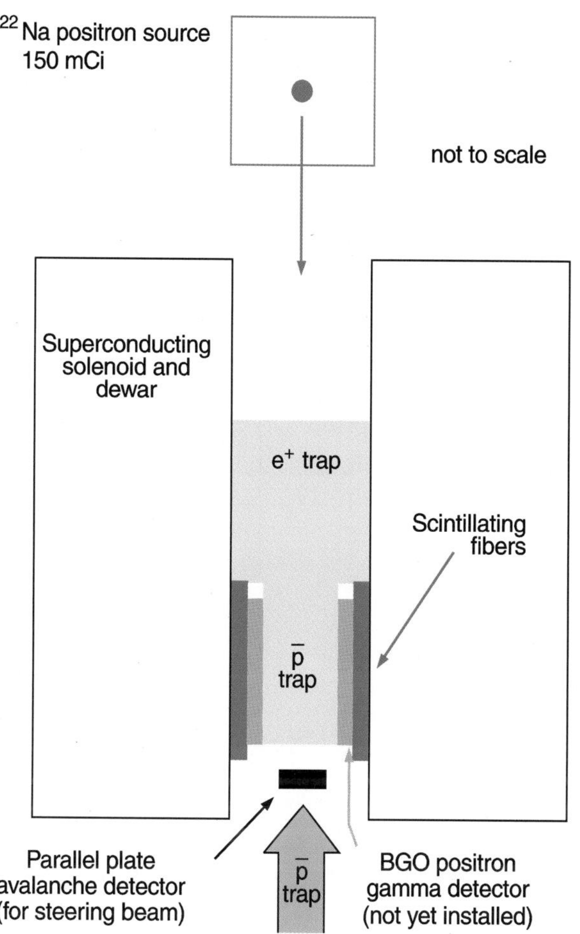


**Fig. 6.48.** Magnetic trap for anti-protons $\bar{p}$ and positrons $e^+$ to form anti-hydrogen $(\bar{p}e^+)$ (www.Atrap Collaboration, CERN)

Its reduced mass is

$$\mu = 1/2m_\mathrm{p} = 469\,\mathrm{MeV}/c^2 .$$

The radius of the first Bohr orbit is $57 \times 10^{-15}$ m, photon energies for transitions between different energy levels are in the range of keV. For instance the photons of the Balmer $\alpha$-line emitted on the transition $3p \to 2s$ have the energy $h\nu = 1.7\,\mathrm{keV}$.

The capture of anti-protons $p^-$ by heavy atoms has a higher probability than capturing by light atoms. Recently, the Lyman spectrum of the exotic atom of anti-protonic argon was observed. This system consists of an argon nucleus with 18 protons and 22 neutrons, but instead of 18 electrons in the atomic shell one electron is replaced by an anti-proton. The antiproton has a much smaller Bohr radius than the electrons and the Coulomb-interaction between the electrons and the anti-proton is therefore small. Transitions between excited energy levels and the lowest level of the anti-proton produce the Lyman series with energies of the emitted photons in the range between 20–200 keV. These energies can be measured with high accuracy using germanium semiconductor detectors [6.15]. They give valuable information about the interaction of the anti-proton with the protons and neutrons in the atomic nucleus.

### 6.7.4 Positronium and Muonium

Positronium is a hydrogen-like system consisting of an electron $e^-$ and a positron $e^+$. Its investigation gives very interesting information about a pure leptonic system of two light particles with equal masses, which have opposite charges and magnetic moments. Since the reduced mass $\mu = 1/2m_\mathrm{e}$ is only about half of that in the hydrogen atom the radii of the Bohr orbits are twice as large. Both particles circulate around the center of mass, which is located in the middle between the two particles. The sum of kinetic and potential energy is about one-half of that in the hydrogen atom. Accurate measurements of the spectral lines emitted from excited states of the system allow one to prove whether the electron is indeed a point-like charge [6.16].

The positronium can be produced by recombination of slow positrons and electrons. Fast positrons from a radioactive $^{58}$Co source ($\beta^+$-emitter) are monochromatized by Bragg reflection at the (110)-surface of a tungsten single crystal and are then slowed down by an electric bias field. The slow positrons can be stored in a magnetic bottle (Fig. 6.49a), where positive voltages at both ends of the bottle prevent the positrons from escaping the bottle. By high negative voltage pulses they can be extracted and impinge as a positron pulse onto an aluminum foil. Here they can capture electrons to form positronium. By heating the foil, the positronium can evaporate from the foil and pass through a pulsed laser beam, where it can be optically excited into various levels if the laser wavelength $\lambda$ is tuned to the corresponding transition wavelength $\lambda_{ik}$.

Positronium is one of the few systems where the lifetime of the ground state is smaller than that of excited states, because in the $1S$ state the wave functions of the two particles overlap and therefore the particles can come into contact and annihilate by the process

$$e^+ + e^- \to 2\gamma .$$

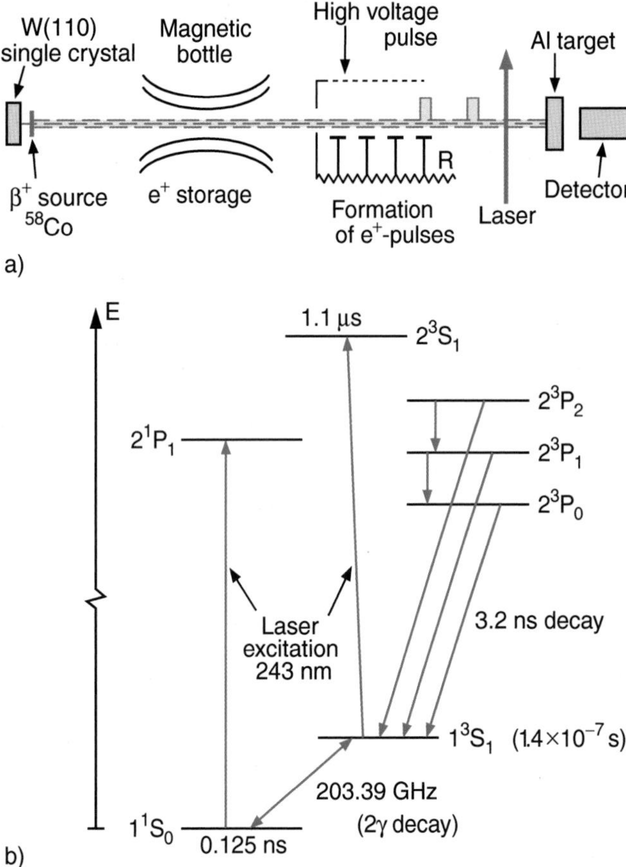


**Fig. 6.49(a).** Generation of a positronium $e^+e^-$. (**b**) Level scheme of $e^+e^-$

Since the center of mass of the positronium is nearly at rest, the two $\gamma$-quanta with the energy $h\nu = 0.5\,\mathrm{MeV}$ are emitted into opposite directions and are detected by a germanium detector.

Since the magnetic moments of electron and positron have the same magnitude but opposite sign, the magnetic interaction between the two particles is much larger than the hyperfine interaction in the hydrogen atom, where the small nuclear magnetic moment only causes a small splitting of the energy levels. The magnetic interaction can therefore no longer be treated as a small perturbation and the theoretical treatment of the positronium can not be based on the Schrödinger equation, but has to use the framework of quantum electrodynamics.

Meanwhile a series of accurate spectroscopic measurements of positronium have been performed that allow a precise test of predictions of quantum electrodynamics [6.16]. Here, nuclear size effects, which are important for precise interpretations of very high resolution spectroscopy of the hydrogen atom (see Sect. 5.7) are completely absent and the measurements give information on a pure leptonic system, where, besides the Coulomb interaction, only magnetic and weak forces play a role.

Another leptonic system is muonium ($\mu^+e^-$), which consists of a positive muon $\mu^+$ and a negative electron $e^-$. The $\mu^+$ leptons are produced by irradiating beryllium with 500 MeV protons (Fig. 6.50). They are then slowed down in a target of $SiO_2$-powder, where they can capture an electron. The neutral system $\mu^+e^-$ can diffuse out of the target into the interaction region with a laser beam where it is excited in a similar way as the positronium [6.17].

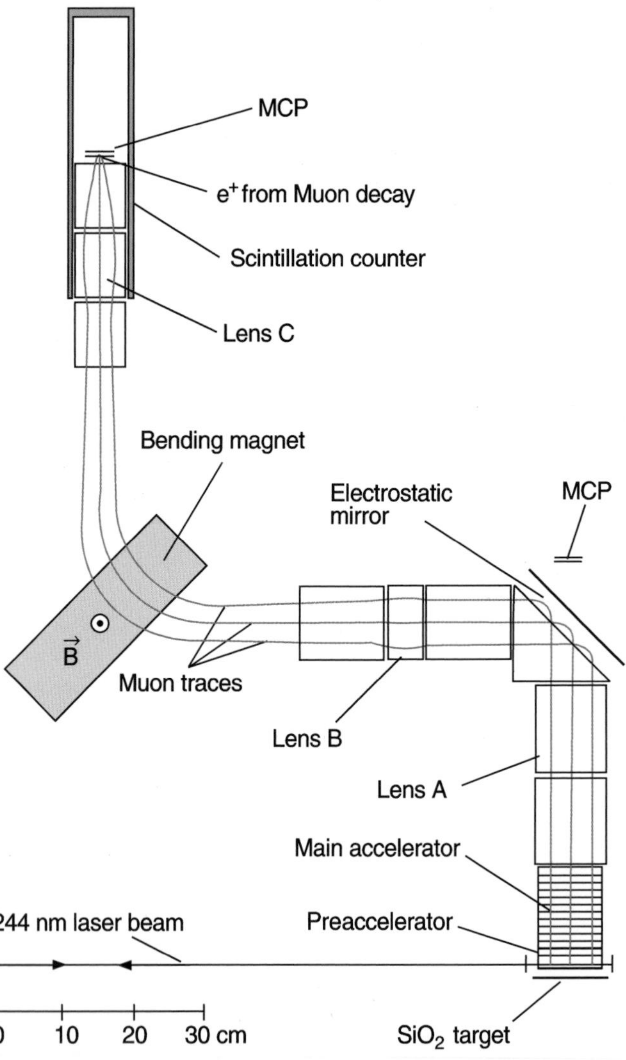


**Fig. 6.50.** Two-photon spectroscopy of the $1S$-$2S$-transition in muonium $\mu^+e^-$. After photoionization the ionized $\mu^+$ is accelerated, energy selected and detected by the decay products $e^+$ from $\mu^+ \to e^+ + \nu$ [6.18]

## SUMMARY

- The theoretical treatment of atoms with more than one electron has to take into account the interaction between the different electrons. The potential is no longer spherically symmetric and the atomic wave function cannot be separated into functions of only one variable.
- The total wave function has to be antisymmetric with respect to the exchange of two arbitrary atomic electrons (Pauli principle). If the total wavefunction is written as a product of spatial wavefunction and spin-function symmetric spatial functions can be only combined with antisymmetric spin functions and vice versa.
- Another formulation of the Pauli-principle is: An atomic state $(n, l, m_l, m_s)$, characterized by the four quantum numbers $n$ (principal quantum number), $l$ (orbital angular momentum quantum number), $m_l$ (projection quantum number of orbital angular momentum) and $m_s$ (electron spin projection quantum number) can only be occupied by at most one electron.
- The Pauli principle and the principle of minimum energy govern the building-up of the electron shells of all atoms. The shell structure of atomic electrons resulting from these principles, explains the arrangement of all chemical elements in the periodic system of elements.
- The dependence of atomic volumes and ionization energies on the number $Z$ of atomic electrons reflect the shell structure of the electron arrangement in atoms. Alkali atoms have the smallest ionization energy and the largest atomic volume of all elements in the same row of the periodic table. Noble gas atoms have the highest ionization energy and the smallest atomic volume in their row.
- Alkali atoms are hydrogen-like. They have only one electron in the highest occupied shell. The potential for this electron is spherically symmetric but deviates from a Coulomb potential. The Rydberg term energies can be described by a Rydberg formula in the same way as for the H atom if the integer principal quantum number $n$ is replaced by a non-integer effective principal quantum number $n_{\text{eff}} = n - \delta_{n,l}$, where the quantum defect $\delta_{n,l}$ depends on the quantum numbers $n$ and $l$.
- The spatial charge distribution and the term energies of multi-electron atoms with $Z$ electrons can be approximately calculated with the Hartree method, which is an iterative optimization procedure based on the assumption that each electron moves in an effective spherical symmetric potential produced by the atomic nucleus and the average charge distribution of all other electrons.
- The total wave function $\psi(r_1, r_2, r_3, \ldots, r_Z)$ of these multielectron atoms can be written as the linear combination of products of one-electron functions $\phi(r_i)$ (Slater determinant), which is antisymmetric with respect to an exchange of two arbitrary electrons, thus obeying the Pauli principle.
- The vector coupling of the angular momenta of the different electrons depends on the energetic order of the different interactions. For light atoms (small $Z$) the interaction $a_{ik}\boldsymbol{l}_i\boldsymbol{l}_k$ between the orbital angular momenta $\boldsymbol{l}_i, \boldsymbol{l}_k$ of the different electrons and $b_{ik}\boldsymbol{s}_i\boldsymbol{s}_k$ between their spins $\boldsymbol{s}_i$, $\boldsymbol{s}_k$ is stronger than the interaction $c_{ii}\boldsymbol{l}_i\boldsymbol{s}_i$ between $\boldsymbol{l}_i$ and $\boldsymbol{s}_i$. The different $\boldsymbol{l}_i$ couple to

$$\boldsymbol{L} = \sum \boldsymbol{l}_i \,.$$

And the $s_i$ to

$$\boldsymbol{S} = \sum \boldsymbol{s}_i \,.$$

The total electronic angular momentum is

$$\boldsymbol{L} + \boldsymbol{S} = \boldsymbol{J}$$

($\boldsymbol{L}$-$\boldsymbol{S}$ coupling is dominant). The spectra of atoms obeying $L$-$S$ coupling show narrow fine structure multiplets.
- For closed shells is $L = S = J = 0$.
- For heavy atoms (large $Z$) the $j$-$j$ coupling is dominant. The interaction energy $c_{ii}l_i s_i$ is larger than the interaction $a_{ik}l_i l_k$ between the different $l_i$, $l_k$. Here the coupling scheme is:

$$l_i + s_i = j_i \quad \text{and} \quad J = \sum j_i \,. \tag{6.52}$$

- The quantum numbers $L$ and $S$ are no longer "good" quantum numbers, i.e., they are not well-defined. The different components of a spin-orbit multiplet do not form a fine structure pattern in the spectrum, but are so widely separated that components from different multiplets may overlap.

▶

- For medium size atoms (medium values of $Z$) an intermediate coupling scheme between $L$-$S$ coupling and $j$-$j$ coupling is observed.
- In excited states of multielectron atoms each electron configuration $(n, l)$ and $(n', l')$ of two electrons can give rise to many atomic states with different energies, due to the different coupling possibilities of the angular momenta of the two electrons.
- Rydberg states of atoms are excited states, where one electron is brought into a state with large principal quantum number $n$. This electron has a large Bohr radius $r_n = a_0 n^2$ and moves in a nearly spherical potential formed by the atomic nucleus and the residual electron core. The ionization energy of these Rydberg states is $E_{\text{ion}} = Ry^* Z_{\text{eff}}^2 / n^2$.
- In planetary atoms, two electrons are excited into different Rydberg states. They decay preferentially by auto-ionization.
- Exotic atoms are formed by replacing one of the atomic electrons by a heavier negative elementary particle. For the myonic atoms, this is a $\mu^-$ myon, for pionic atoms a $\pi^-$ meson. Such exotic atoms are not stable, because the elementary particles have a short lifetime. Their Bohr radii $r_n \propto 1/m$ are much smaller than those of ordinary atoms. The spectroscopy of these exotic atoms probes the charge and mass distribution in the atomic nucleus.
- Positronium is a system consisting of an electron $e^-$ and a positron $e^+$. Its lifetime ranges between 1 ns and 1 $\mu$s, depending on the energy state of the system. The ground state has the smallest lifetime.
- Antimatter is formed by atoms consisting of antiprotons and anti-electrons = positrons. The anti-hydrogen atom has been experimentally formed by recombination of antiprotons and positrons in a special trap.

## PROBLEMS

1. When the first electron in the He atom is described by the $1s$ wave function, what is the potential for the second electron? (In this model the interaction between the two electrons is only indirectly taken into account by the time-averaged charge distribution of the first electron.)
2. Derive (6.28y) for the potential of the $2s$ electron in the Li-atom, assuming that the two $1s$-electrons can be described by hydrogenic $1s$-wavefunctions for $Z_{\text{eff}} = 2$.
3. A system of Na atoms with number density $n$ is cooled down to the temperature $T$. What is the critical temperature, where the de Broglie wavelength of the Na atoms becomes equal to their mean distance. Numerical example: $n = 10^{10}\,/\text{cm}^3$.
4. In a classical model of the He atom, the two electrons move around the nucleus on a circle with radius $r = 0.025$ nm. What is the minimum potential energy where the two electrons are at opposite locations on the circle, and what is the kinetic energy of the two electrons? Compare this with the measured energy of the $1s^2$ ground state of the He atom and discuss the difference.
5. How large would be the energy difference between the $1s2s$ and the $1s3s$ states of the helium atom for the potential of Problem 6.1. Compare this with the measured energy difference $\Delta E$, obtained from Fig. 6.9 (or more accurately from *CH. Moores* Tables of Atomic Energy Levels, NBS monograph).
6. Give a simple vivid explanation for Hund's rule that the lowest level of a multielectron atom is realized by the maximum electron spin compatible with the Pauli principle.
7. What is the relation between the shielding constant $S$ and the quantum defect $\delta_{n,l}$ of a Rydberg state $(n, l)$ with large principal quantum number $n$ and maximum angular momentum quantum number $l = n - 1$ in an alkali atom?
8. What is the photon energy for a transition $n = 2 \rightarrow n = 1$ in a myonic atom with a mass of 140 AMU and a nuclear charge number $Z = 60$? For which principal quantum number $n$ has the Bohr radius $r_n$ of the myon the same value as the lowest radius $r_1$ for an electron in this atom?
9. Why is the energy of the $3P$ term in the Na atom higher than that of the $3S$ term?
10. The negative $H^-$ ion is a two-electron system like the He atom. How large is the binding energy of the second electron according to a similar calculation as in Problem 6.1?
11. The energy of the ground state $2S$ in the Li atom is $E = -5.39$ eV, that of the Rydberg state with $n = 20$ is $E = -0.034$ eV. How large is the effective charge $Z_{\text{eff}}e$, the mean Bohr radius $r_n$ and the quantum defect $\delta = n - n_{\text{eff}}$ of the valence-electron in the two states?
12. The absolute value of the binding energy of the alkali atoms decreases with increasing atomic size as $E_B(\text{Li}) = -5.395$ eV; $E_B(\text{Na}) = -5.142$ eV; $E_B(\text{K}) = -4.34$ eV; $E_B(\text{Rb}) = -4.17$ eV; $E_B(\text{Cs}) = -3.90$ eV. Give a qualitative explanation for this sequence. How would you determine these values of $E_B$ experimentally and how can one calculate them?
13. How large is the maximum of the potential barrier and at which value of $x$ is it located, if a hydrogen atom is placed in a homogeneous electric field $E = -E_0 x$ with $E_0 = 3 \times 10^4$ V/m. Determine the critical principal quantum number $n_c$ where field ionization (without a tunnel effect) starts. Calculate the field-ionization probability for the levels with $n_c - 1$, possible through the tunnel effect.

# 7. Emission and Absorption of Electromagnetic Radiation by Atoms

We have so far discussed primarily stationary atomic states that are described by a stationary wave function $\Psi_{n,l,m_l,m_s}$ or by the corresponding quantum numbers $n, l, m_l, m_s$, which give all angular momenta $\boldsymbol{l}, \boldsymbol{s}$ and $\boldsymbol{j} = \boldsymbol{l} + \boldsymbol{s}$ of single electron atoms and the energies $E_i$ of the states $|i\rangle$, where the index $i$ stands for all four quantum numbers. The spatial parts of the wave functions are obtained by solving the stationary Schrödinger equation and the spin is described by the corresponding spin function.

For atoms with more than one electron the couplings of the different angular momenta depends on the coupling strength between them and only those states are realized that obey the Pauli principle, which demands that the total wave function (including the spin part) has to be antisymmetric with regard to the exchange of two electrons. The wave functions that fulfill this demand can be written as linear combination of Slater determinants (see Sect. 6.4).

When discussing Bohr's atomic model we mentioned that atoms can undergo transitions between different states with energies $E_i$ and $E_k$, when a photon with energy

$$\hbar\omega = E_i - E_k \tag{7.1}$$

is emitted or absorbed.

Experiments show, however, that the absorption or emission spectrum of an atom does not contain all possible frequencies $\omega$ according to (7.1). There must be certain "selection rules" that select from all possible combinations $E_i$ and $E_k$ only those between which a radiative transition can take place. Furthermore, the intensity of the spectral lines can vary by many orders of magnitude, which means that the probability of a transition generally depends strongly on the specific combination of the two atomic states in (7.1).

Besides the energy conservation expressed by (7.1) also the total angular momentum of the system (atom + photon) has to be conserved. The transition probability therefore depends on the polarization of the emitted or absorbed electromagnetic radiation.

In this chapter we will discuss how such transition probabilities can be calculated from the wave functions of the states involved in the transition (Sect. 7.1). Also, experimental methods for measuring such transition probabilities are presented.

The selection rules are then discussed in Sect. 7.2 and the problem of measuring and calculating lifetimes of emitting atomic states is covered in Sect. 7.3.

For transitions of an electron in an outer shell of the atom the energy difference in (7.1) amounts to a few electron volts. The transition frequency then falls into the spectral region between the near infrared to the near ultraviolet (1 eV corresponds to a wavelength $\lambda$ of $1.234\,\mu$m = near infrared, while 3 eV corresponds to $\lambda = 411$ nm = blue spectral region). Since these transitions mostly give rise to emission of visible light, the electron in the outer atomic shell is often referred to the German word "Leucht-Elektron", which means "light-emitting electron."

If an electron in an inner atomic shell is excited into higher unoccupied states, its excitation energy can be several keV. When it returns to its initial state, radiation with short wavelengths (X-rays) are emitted. The emission or absorption spectrum corresponding to transitions of inner shell electrons therefore fall between the far ultraviolet and X-ray region with wavelengths between 0.01–50 nm (a wavelength of 0.01 nm corresponds to an energy of about 120 keV). The generation, absorption and detection of X-rays is discussed in Sect. 7.4.

When measuring the frequency dependence of the emitted or absorbed radiation intensity $I(\omega)$, it turns

W. Demtröder, *Atoms, Molecules and Photons*, 2nd ed., Graduate Texts in Physics,
DOI 10.1007/978-3-642-10298-1_7, 

out that discrete spectra are found where the intensity peaks around certain frequencies $\omega_{ik}$, which obeys the energy relation (7.1), but also continuous spectra where the intensity $I(\omega)$ is a smooth function of $\omega$. However, even for discrete spectra, the intensity $I(\omega)$ of a spectral line is not a delta function, but has an intensity profile with a halfwidth $\Delta\omega$, which depends on the lifetime of the states involved, on the temperature of the atomic sample and on its pressure. This subject of line profiles and line broadening will be treated in Sect. 7.5.

## 7.1 Transition Probabilities

In this section we will discuss what a transition probability means and how it depends on the wave functions of the atomic states involved in a specific transition between these states. Some techniques for measuring transition probabilities are presented.

### 7.1.1 Induced and Spontaneous Transitions, Einstein Coefficients

If an atom in the state $|k\rangle$ with energy $E_k$ is brought into an electromagnetic radiation field with spectral energy density $w_\nu(\nu)$ (this is the field energy per unit volume and unit frequency interval $\Delta\nu = 1\,\mathrm{s}^{-1}$) it can absorb a photon $h\nu$, which brings the atom into a state with higher energy $E_i = E_k + h\nu$.

The probability per second for such an absorbing transition

$$\frac{\mathrm{d}\mathcal{P}_{ki}^{\mathrm{abs}}}{\mathrm{d}t} = B_{ki} w_\nu(\nu) \tag{7.2}$$

is proportional to the spectral energy density $w_\nu(\nu) = n(\nu)h\nu$ of the radiation field (where $n(\nu)$ is the number of photons $h\nu$ per unit volume within the frequency interval $\Delta\nu = 1\,\mathrm{s}^{-1}$). The proportionality factor $B_{ki}$ is the ***Einstein coefficient for absorption***. Each absorption takes one photon from a specific mode of the radiation field (see Sect. 3.1) and therefore decreases the number of photons in this mode by one.

The radiation field can also induce atoms in an excited state with energy $E_i$ to emit a photon with energy $h\nu = E_i - E_k$ into a specific mode of the radiation field and to go into the lower state $E_k$. This process is called ***induced (or stimulated) emission***. It increases by one the number of photons in this mode from which the inducing photon came. Since the two photons are in the same mode, they have identical propagation directions. The energy of the atom is reduced by $\Delta E$ and that of the mode of the radiation field is increased by the same amount $\Delta E = h\nu$.

The probability per second for the induced emission is analogous to (7.2) given by

$$\frac{\mathrm{d}\mathcal{P}_{ik}^{\mathrm{ind.\,em}}}{\mathrm{d}t} = B_{ik} w_\nu(\nu)\,. \tag{7.3}$$

The factor $B_{ik}$ is the ***Einstein coefficient for induced emission***.

An excited atom can also give away its excitation energy spontaneously without an external radiation field. This process is called ***spontaneous emission***. Different from the induced emission, the spontaneous photon can be emitted into an arbitrary direction, i.e., into any one of the modes of the radiation field. The probability per second for such a spontaneous emission is

$$\frac{\mathrm{d}\mathcal{P}_{ik}^{\mathrm{sp.\,em.}}}{\mathrm{d}t} = A_{ik}\,. \tag{7.4}$$

The factor $A_{ik}$ is the ***Einstein coefficient for spontaneous emission***. It is solely dependent on the wave functions of the states $|i\rangle$ and $|k\rangle$ but independent of the radiation field. In Fig. 7.1 all three processes are depicted schematically.

We will now look for relations between the three Einstein coefficients.

Consider $N_i$ atoms in state $E_i$ and $N_k$ atoms in state $E_k$ within the unit volume inside a radiation field with spectral energy density $w(\nu)$. Under stationary conditions the number densities $N_i$ and $N_k$ do not vary with time, i.e., they are constant. This means that the absorption rate must be equal to the total emission rate:

$$B_{ki} w_\nu(\nu) N_k = \left(B_{ik} w_\nu(\nu) + A_{ik}\right) N_i\,. \tag{7.5}$$

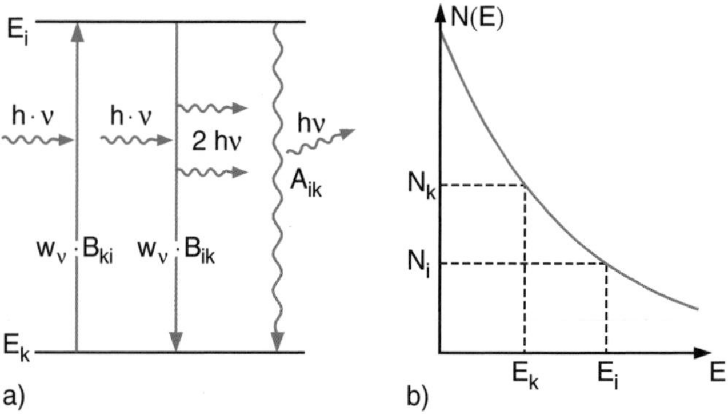


**Fig. 7.1.** (**a**) Absorption, induced emission and spontaneous emission in a two-level system. (**a**) Thermal population distribution $N(E)$

At thermal equilibrium, the ratio $N_i/N_k$ follows the Boltzmann distribution (Fig. 7.1b)

$$\frac{N_i}{N_k} = \frac{g_i}{g_k}\mathrm{e}^{-(E_i-E_k)/kT} = \frac{g_i}{g_k}\mathrm{e}^{-h\nu/kT} , \qquad (7.6)$$

where $g = 2J+1$ is the statistical weight (i.e., the number of possible realizations) of a state with energy $E$ and total angular momentum quantum number $J$. This state has $2J+1$ possible orientations of the angular momentum vector $\boldsymbol{J}$, which all have (without an external magnetic field) the same energy; they are energetically degenerate.

Inserting (7.6) into (7.5) and solving for $w(\nu)$ yields

$$w_\nu(\nu) = \frac{A_{ik}/B_{ik}}{(g_k/g_i)(B_{ki}/B_{ik})\,\mathrm{e}^{h\nu/kT} - 1} . \qquad (7.7)$$

The spectral energy density of the thermal radiation field is, on the other side, given by Planck's formula (see Sect. 3.1)

$$w_\nu = \frac{8\pi h\nu^3}{c^3}\frac{1}{\mathrm{e}^{h\nu/kT} - 1} . \qquad (7.8)$$

Since the two equations (7.7) and (7.8) describe the same radiation field for all frequencies $\nu$ and at arbitrary temperatures $T$, we get for the nominator in (7.7):

$$A_{ik}/B_{ik} = \frac{8\pi h\nu^3}{c^3} \qquad (7.9a)$$

and for the constant factor in the denominator

$$g_k B_{ki}/(g_i B_{ik}) = 1 . \qquad (7.9b)$$

This yields the relations between the Einstein coefficients:

$$B_{ik} = \frac{g_k}{g_i}B_{ki} \qquad (7.10a)$$

$$A_{ik} = \frac{8\pi h\nu^3}{c^3}B_{ik} . \qquad (7.10b)$$

These important relations give us a deeper insight into the three processes of induced emission and absorption and spontaneous emission:

If both states have equal statistical weights ($g_i = g_k$) the Einstein coefficients for induced absorption and emission are equal.

Since $8\pi\nu^2/c^3$ gives the number of modes within the unit frequency interval $\Delta\nu = 1\,\mathrm{s}^{-1}$ (see Sect. 3.1) the ratio $A_{ik}/(8\pi\nu^2/c^3)$ gives the probability per second that a spontaneous photon is emitted by an atom into one mode of the radiation field. On the other hand, the product $B_{ik}h\nu$ gives the probability per second that induced emission is induced by one photon, i.e., that an induced photon is emitted into a mode of the radiation field, which had contained one inducing photon. Rearranging (7.10b) into

$$\frac{A_{ik}}{8\pi\nu^2/c^3} = B_{ik}h\nu \qquad (7.10c)$$

shows that the spontaneous emission probability into one mode equals the induced emission probability if this mode contains just one photon (after the induced emission it contains two photons). In other words:

Spontaneous and induced emission rates are equal in a radiation field that contains, on average, one photon per mode.

If the radiation field contains, on average, $n$ photons per mode then the ratio of induced to spontaneous emission rates is

$$\frac{P_{ik}^{\text{ind. em}}}{P_{ik}^{\text{sp. em}}} = \frac{B_{ik}nh\nu}{A_{ik}c^3/(8\pi\nu^2)} = n . \qquad (7.10d)$$

The ratio of induced to spontaneous emission rates into one mode of the radiation field equals the number of photons in this mode.

Figure 7.2 illustrates the average number of photons per mode of a thermal radiation field as a function of frequency $\nu$ for different temperatures $T$. It demonstrates that in thermal radiation fields at temperatures $T < 10^3$ K the average photon number $n$ in the visible spectral range is small compared to one. In this case the spontaneous emission in the visible range exceeds by far the induced emission.

In order to enhance the induced emission beyond the spontaneous emission, non-thermal radiation fields have to be realized, where the photon number $n$ is not equally distributed among all modes but is concentrated into one or a few modes. In these modes, $n \gg 1$ and then the induced emission rate becomes much larger

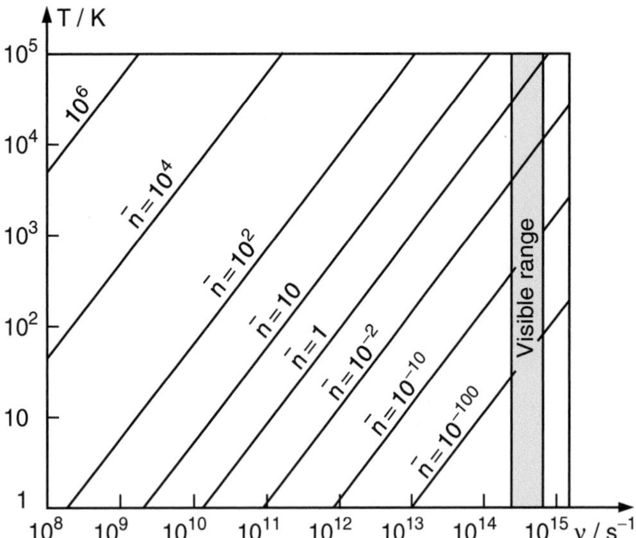


**Fig. 7.2.** Mean photon number $\overline{n}$ per mode of the thermal radiation field as a function of frequency $\nu$ and temperature $T$

than the spontaneous one. This situation is realized in lasers, which will be discussed in Chap. 8.

**EXAMPLES**

1. At a distance of 10 cm away from the filament of a 100 W light bulb, the average photon number per mode at $\lambda = 500\,\text{nm}$ is about $10^{-8}$. For atoms in this radiation field the spontaneous emission rate exceeds by far the induced rate.
2. In the brightest spot of a high-pressure mercury lamp in the maximum of the intense mercury emission line at $\lambda = 253.7\,\text{nm}$ the photon number per mode is about $10^{-2}$. Even here the induced emission plays a minor role compared with the spontaneous emission.
3. Within the optical resonator of a helium-neon laser (output power 1 mW through the output mirror with transmittance of 1%), which oscillates in a single mode, the photon number in this mode is about $10^7$! Here, the spontaneous emission into this mode is completely negligible. Note, however, that the total spontaneous emission within the Doppler-width of the neon transition at $\lambda = 632.18\,\text{nm}$, which is distributed over $3 \times 10^8$ modes of the active volume of $1\,\text{cm}^3$ and is emitted into all directions, exceeds 1 W and is therefore stronger than the induced emission.

**Note:**

When using the angular frequency $\omega = 2\pi\nu$ instead of $\nu$, the unit frequency interval $\mathrm{d}\omega = 2\pi\,\mathrm{d}\nu$ is larger by a factor of $2\pi$. Since $w(\nu)\,\mathrm{d}\nu$ must be equal to $w(\omega)\,\mathrm{d}\omega$, the spectral energy density

$$w_\omega(\omega) = \frac{\hbar\omega^3}{\pi^2c^3}\frac{1}{\mathrm{e}^{\hbar\omega/kT}-1} = w_\nu(\nu)/2\pi$$

of the radiation field is then smaller by this factor. The ratio of the Einstein coefficients

$$A_{ik}/B_{ik}^{(\omega)} = \frac{\hbar\omega^3}{\pi^2c^3}$$

is then also smaller by the factor $2\pi$. However, the ratio

$$A_{ik}/\left(B_{ik}^{(\omega)}w_\omega(\omega)\right) = A_{ik}/\left(B_{ik}^{(\nu)}w_\nu(\nu)\right)$$

of spontaneous to induced emission rates remains the same.

### 7.1.2 Transition Probabilities, Einstein Coefficients and Matrix Elements

The relation between transition probabilities and the quantum mechanical description by matrix elements can be illustrated in a simple way by a comparison with classical oscillators emitting electromagnetic radiation.

A classical oscillating electric dipole (Hertzian dipole) with electric dipole moment

$$\boldsymbol{p} = q\boldsymbol{r} = \boldsymbol{p}_0 \sin\omega t$$

emits the average power, integrated over all directions $\vartheta$ against the dipole axis (Fig. 7.3a) [5.2]

$$\overline{P} = \frac{2}{3}\frac{\overline{p^2}\omega^4}{4\pi\varepsilon_0c^3} \quad \text{with} \quad \overline{p^2} = \frac{1}{2}p_0^2\,. \tag{7.11}$$

The emitted radiation power is therefore proportional to the average of the squared dipole moment. The quantum mechanical pendant can be rigorously derived by time-dependent perturbation theory, which is, however, beyond the scope of this book. We will therefore use a more intuitive way.

During the absorption or emission of a photon the atom undergoes a transition between two levels $|i\rangle$ and $|k\rangle$, i.e it changes its eigenstate in time. Therefore it cannot be described by the stationary Schrödinger equation (4.6), but we have to use the time-dependent Schrödinger equation (4.8).

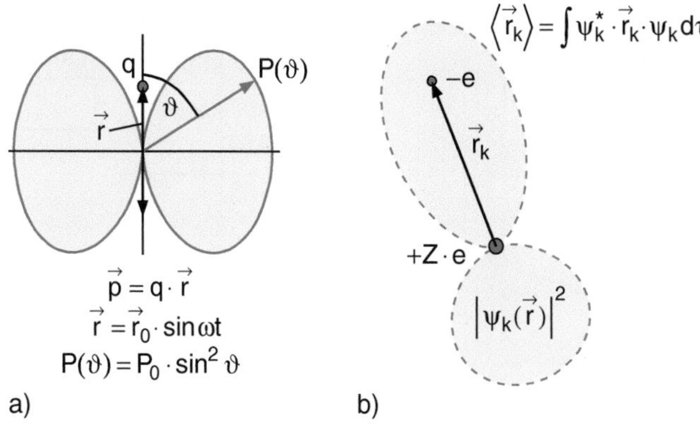


**Fig. 7.3.** (**a**) Spatial radiation characteristics of a classical oscillating electric dipole. (**b**) The expectation value $\langle p_k \rangle = -e \langle \boldsymbol{r}_k \rangle$ of the quantum mechanical dipole moment in level $|k\rangle$, determined by its wave function $\psi_k$

With the time-dependent wavefunction

$$\Psi(r,t) = \psi(r) \cdot e^{-i(Ek/\hbar)\cdot t} = \psi(r) \cdot e^{-i\omega k \cdot t} \tag{7.12a}$$

for the state $|k\rangle$ with energy $E_k$, every linear combination

$$\Psi = \Sigma c_k \psi_{k'} e^{-i(Ek/\hbar)\cdot t} \tag{7.12b}$$

is also a solution of eq. (4.8). If we replace the classical average dipole moment by the quantum mechanical expression

$$-e \cdot \int \Psi^* \cdot r \cdot \Psi d\tau = -e \Sigma\Sigma c_i^* c_k \cdot e^{-i\omega k \cdot t} \int \psi_i^* \mathbf{r} \psi_k d\tau \tag{7.13}$$

only those terms in (7.13b) with $i \neq k$ do not vanish. The reason for this is that $|\psi_i|^2$ is an even function but r an odd function of the coordinates x,y,z. Since the integral extends over the whole coordinate space it vanishes for an odd integrand.

In the classical model a periodically oscillating dipole moment emits radiation.

In analogy to the classical expression for the average dipole moment we define the integral

$$M_{ik} = -e \int \psi_i^* \mathbf{r} \psi_k d\tau \tag{7.14}$$

as transition dipole moment for a transition between the atomic states $|i\rangle$ and $|k\rangle$ where the two indices $i = (n_i, l_i, m_{li}, m_{si})$ and $k = (n_k, l_k, m_{lk}, m_{sk})$ are abbreviations for the four quantum numbers of each state. The vector $\boldsymbol{r}$ is the radius vector of the electron from the origin at the atomic nucleus (Fig. 7.3b).

The integration extends over the three spatial coordinates of the electron. The volume element is $d\tau = dx\,dy\,dz$ in Cartesian coordinates or $d\tau = r^2\,dr\,\sin\vartheta\,d\vartheta\,d\varphi$ in spherical coordinates.

If the statistical weights $g_i$ and $g_k$ of levels $|i\rangle$ and $|k\rangle$ are equal it follows that $M_{ik} = M_{ki}$.

We can define the average squared transition dipole moment by

$$\langle |M_{ik}|^2 \rangle = \frac{1}{2}(|M_{ik}| + |M_{ki}|)^2 = 2|M_{ik}|^2 .$$

Replacing the average classical dipole moment $\langle p^2 \rangle$ in (7.11) by $\langle |M_{ik}|^2 \rangle$ we obtain the average radiation power, emitted by an atom in level $\langle i|$ on the transition $\langle i| \rightarrow \langle k|$ as

$$\langle P_{ik} \rangle = \frac{4}{3} \frac{\omega_{ik}^4}{4\pi\varepsilon_0 c^3} |M_{ik}|^2 , \tag{7.15}$$

which is equivalent to the classical expression (7.11) for the radiation power of the Hertzian dipole, if the average $\overline{p^2}$ is replaced by $2|M_{ik}|^2$.

$N_i$ atoms in level $\langle i|$ emit the average radiation power $\langle P \rangle = N_i \langle P_{ik} \rangle$ on the transition $\langle i| \rightarrow \langle k|$ with frequency $\omega_{ik}$.

Using the Einstein coefficient $A_{ik}$ for spontaneous emission, which gives the probability per second that one atom emits a photon on the transition $\langle i| \rightarrow \langle k|$ the average power emitted by $N_i$ atoms (Fig. 7.4) is

$$\langle P \rangle = N_i A_{ik} h\nu_{ik} = N_i A_{ik} \hbar\omega_{ik} . \tag{7.16}$$

The comparison of (7.15) with (7.16) yields the relation

$$\boxed{A_{ik} = \frac{2}{3} \frac{\omega_{ik}^3}{\varepsilon_0 h c^3} |M_{ik}|^2} \tag{7.17}$$

between the Einstein coefficient $A_{ik}$ and the transition moment $M_{ik}$. If the wave functions $\psi_i$, $\psi_k$ of the two states involved in the transition, are known, the spontaneous transition probability $A_{ik}$ can be calculated

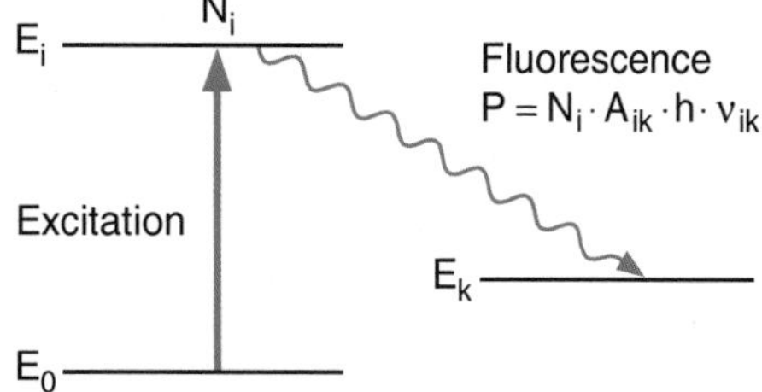


**Fig. 7.4.** Mean radiation power $\langle p_i k \rangle$ emitted by $N_i$ excited atoms as fluorescence on the transition $|i\rangle \rightarrow |k\rangle$

from (7.17) and therefore the total radiation power emitted by $N_i$ atoms in level $\langle i|$ on the transition $\langle i| \rightarrow \langle k|$ can also be calculated.

The expectation values $M_{ik}$ for all possible transitions between arbitrary levels $i, k = 1, 2, \ldots, n$ can be arranged in an $n \times n$ matrix. The $M_{ik}$ are therefore called ***Matrix elements***. If some of the matrix elements are zero, the corresponding transition does not occur. One says that this transition is "not allowed" but "forbidden." The absolute square $|M_{ik}|^2$ of the matrix element is directly proportional to the probability of the transition $\langle i| \rightarrow \langle k|$, i.e., of the intensity of the corresponding line in the atomic spectrum.

**Note:**

Equation (7.17), called the dipole approximation (see appendix) is only valid, when the wavelength $\lambda$ of the radiation is large compared to the dimensions of the dipole. This is completely analogous to the classical case of the Hertzian dipole.

For visible light this is readily fulfilled since $\lambda \approx 500\,\text{nm}$ is very large compared to the average size $r \approx 0.5\,\text{nm}$ of the emitting atomic dipole. This means that $r/\lambda \approx 10^{-3}$. However, the dipole approximation is no longer valid for X-rays when the wavelength becomes smaller than 1 nm.

The experimental arrangement for measuring the emitted radiation power is depicted in Fig. 7.5. The radiation, emitted from the atoms is collected by a lens and imaged onto the entrance slit of a spectrograph, which has the transmission $T(\omega)$. A detector behind the spectrograph receives the signal

$$S(\omega) = N_i\,\langle P_{ik}\rangle\,\varepsilon\,\mathrm{d}\Omega\,T(\omega)\eta(\omega)\,, \tag{7.18}$$

where $N_i$ is the number of emitting atoms in level $\langle i|$, $\langle P_{ik}\rangle$ is the average power emitted by a single atom into the solid angle $4\pi$, $\mathrm{d}\Omega$ is the solid angle accepted by the spectrograph, $\varepsilon$ is the fractional area of the image of the light source that passes through the entrance slit of the spectrograph, $T(\omega)$ is the transmission of the spectrograph and $\eta$ is the spectral efficiency of the detector.

Generally, the image of the light source is larger than the width of the entrance slit, which implies that $\varepsilon < 1$. Here, an optical fiber bundle can be used to increase the total light collection efficiency (Fig. 7.5b).

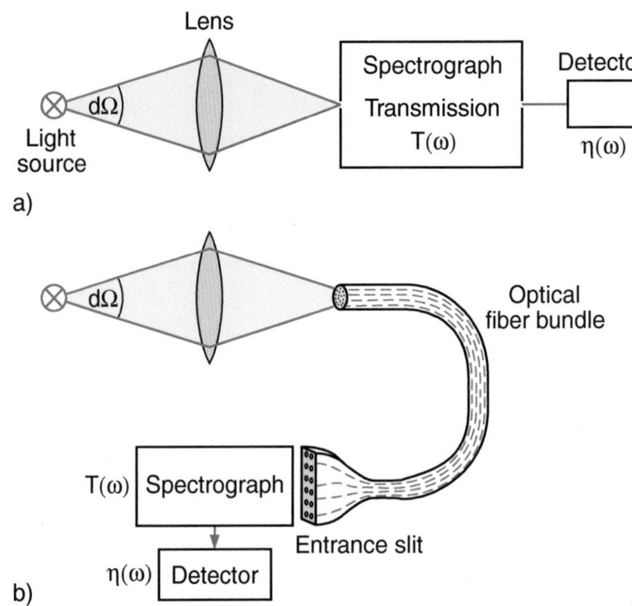


**Fig. 7.5a,b.** Experimental setup for measuring the radiation power emitted by the source $S$ on the transition $|i\rangle \rightarrow |k\rangle$. (**a**) Conventional arrangement. (**b**) Use of an optical fiber bundle for increasing the collection efficiency

The light source is imaged onto the circular entrance cross section of the fiber bundle. The exit cross section of the bundle can be formed into a narrow rectangular area, that matches the entrance slit of the spectrograph and makes the fractional area $\varepsilon$ close to $\varepsilon = 1$.

The ratio $S_{ik}/S_{nm}$ of the measured signals for two spectral lines at frequencies $\omega_{ik}$ and $\omega_{nm}$ is then, according to (7.18) and (7.16),

$$\frac{S_{ik}}{S_{nm}} = \frac{N_i A_{ik}\omega_{ik}}{N_m A_{nm}\omega_{nm}}\,\frac{T(\omega_{ik})\eta(\omega_{ik})}{T(\omega_{nm})\eta(\omega_{nm})}\,, \tag{7.19}$$

where $N_i$ and $N_n$ are the numbers of emitting atoms in levels $\langle i|$ and $\langle n|$ (Fig. 7.6), respectively.

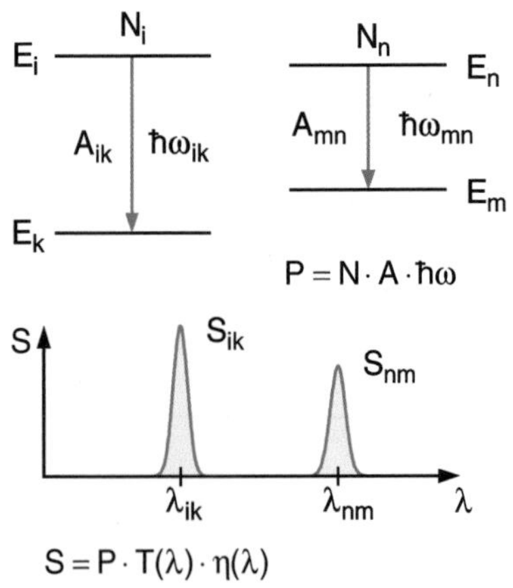


**Fig. 7.6.** Measurement of the relative line strengths of two spectral lines

### 7.1.3 Transition Probabilities for Absorption and Induced Emission

While the transition probabilities of spontaneous emission are independent of an external radiation field and solely depend on the wave functions of the atomic states, the induced processes do depend on the spectral energy density $w_\omega(\omega)$ of the inducing radiation field, as was already discussed in Sect. 7.1.1.

We describe the electromagnetic wave incident on the atom with its nucleus at $\boldsymbol{r} = 0$ by

$$\boldsymbol{E} = \boldsymbol{E}_0 \mathrm{e}^{\mathrm{i}(\boldsymbol{k}\cdot\boldsymbol{r}-\omega t)}\,, \tag{7.20}$$

where $\boldsymbol{k}$ is the wave vector of the electromagnetic wave with the wavenumber $k = |\boldsymbol{k}| = 2\pi/\lambda$. The quantum mechanical treatment gives the probability per sec

$$\frac{\mathrm{d}\mathcal{P}_{ni}}{\mathrm{d}t} = \frac{\pi e^2}{h^2} \left| \int \psi_n^* \boldsymbol{E}_0 \cdot \boldsymbol{r} \mathrm{e}^{\mathrm{i}\boldsymbol{k}\cdot\boldsymbol{r}} \psi_i \,\mathrm{d}\tau \right|^2 \tag{7.21}$$

for absorbing a photon $h\omega$, which induces the atomic transition $\langle n| \to \langle i|$. We have here selected a transition from a level $\langle n|$ instead of $\langle k|$ in order to avoid confusion with the wavenumber $k$. For $\boldsymbol{k}\cdot\boldsymbol{r} \ll 1$ (this means that the wavelength $\lambda$ is much larger than the size of the atom) we obtain the dipole approximation with $\mathrm{e}^{ikr} \approx 1$

$$\frac{\mathrm{d}\mathcal{P}_{ni}}{\mathrm{d}t} = \frac{\pi e^2}{h^2} E_0^2 \left| \int \psi_n^* \hat{\varepsilon} \cdot \boldsymbol{r} \psi_i \,\mathrm{d}\tau \right|^2 , \tag{7.22}$$

where $\boldsymbol{\varepsilon} = \boldsymbol{E}_0/|\boldsymbol{E}_0|$ is the unit vector in the direction of the electric field $\boldsymbol{E}$ of the wave.

Equation (7.22) shows that the transition probability $\mathcal{P}_{ni}$ depends on the scalar product $\boldsymbol{E}_0 \cdot \boldsymbol{r}$, i.e, on the relative orientation of electric field vector $\boldsymbol{E}$ of the light wave and atomic dipole moment $\boldsymbol{p} = e\boldsymbol{r}$.

When the radiation field is isotropic (for example, the thermal radiation field discussed in Sect. 3.1) the scalar product can be averaged over all directions. This gives

$$\langle |\hat{e}\cdot\vec{r}|^2 \rangle = \frac{\int_\varphi \int_\vartheta r^2 \cos^2\vartheta \sin\vartheta \,\mathrm{d}\vartheta \,\mathrm{d}\varphi}{\int \mathrm{d}\Omega}$$

$$\int \mathrm{d}\Omega = \int \sin\vartheta \,\mathrm{d}\vartheta \,\mathrm{d}\varphi$$

$$= \frac{1}{3} r^2$$

$$\left\langle |\boldsymbol{\varepsilon}\cdot\boldsymbol{r}|^2 \right\rangle = \frac{1}{3} |\boldsymbol{r}|^2 . \tag{7.23}$$

Using the relation $w = \frac{1}{2}\varepsilon_0 |E_0|^2$ between the spectral energy density $w$ and the electric field $E$ of the radiation field, we can write (7.22) for isotropic radiation fields as

$$\frac{\mathrm{d}\mathcal{P}_{ki}}{\mathrm{d}t} = \frac{2\pi e^2}{3\varepsilon_0 h^2} \left| \int \psi_k \boldsymbol{r} \psi_i \,\mathrm{d}\tau \right|^2 w_\nu(\nu)\,, \tag{7.24}$$

where we have replaced the index $n$ by $k$ and the total energy density $w$ is related to the spectral energy density by $w = \int w_\nu(\nu)\,\mathrm{d}\nu$.

Comparing this result with (7.2) we obtain for the Einstein coefficient $B_{ki}$ for absorption

$$\boxed{B_{ki}^{(\nu)} = \frac{2}{3} \frac{\pi^2 e^2}{\varepsilon_0 h^2} \left| \int \psi_k^* \boldsymbol{r} \psi_i \,\mathrm{d}\tau \right|^2 .} \tag{7.25}$$

$$B_{ki}^{(\nu)} = \frac{2\pi^2}{3\varepsilon_0 h^2} |M_{ki}|^2 \,;\; B_{ki}^{(\omega)} = \frac{\pi}{3\varepsilon_0 \hbar^2} |M_{ki}|^2$$

A comparison of (7.25) and (7.17) again yields the relation (7.9b) between the Einstein coefficients $A_{ik}$ and $B_{ik}$.

## 7.2 Selection Rules

Not every transition possible according to the energy conservation rule (7.1) is actually observed in atomic spectra. The reason for this is that besides energy conservation, the conservation of angular momentum and certain symmetry rules must also be obeyed. This is all included in the transition matrix elements.

### 7.2.1 Selection Rules for Spontaneous Emission

From (7.16) it follows that for the spontanously emitted radiation only those transitions $\langle i| \to \langle k|$ are allowed for which the transition dipole matrix element

$$\boldsymbol{M}_{ik} = e \int \psi_i^* \boldsymbol{r} \psi_k \,\mathrm{d}\tau \tag{7.26}$$

is not zero. This means that at least one of the components

$$(M_{ik})_x = e \int \psi_i^* x \psi_k \,\mathrm{d}\tau$$

$$(M_{ik})_y = e \int \psi_i^* y \psi_k \,\mathrm{d}\tau \tag{7.27}$$

$$(M_{ik})_z = e \int \psi_i^* z \psi_k \,\mathrm{d}\tau$$

must be different from zero.

We will illustrate this for the hydrogen atom. In order to make the calculation not too complicated we will disregard the electron spin and only deal with the spatial part of the wave function, since in the matrix elements discussed so far we have only used the spatial part of the wave functions and the integration extends only over the spatial coordinates.

The hydrogenic wave functions are, according to Sect. 5.1.3:

$$\psi_{n,l,m_l} = \frac{1}{\sqrt{2\pi}} R_{n,l}(r) \Theta_m^l(\vartheta) \mathrm{e}^{\mathrm{i}m_l\varphi} . \tag{7.28}$$

For linearly polarized light with the electric field vector $E = \{0, 0, E_0\}$ only the $z$-component of $M_{ik}$ contributes to the spontaneous emission. We choose the $z$-axis as quantization axis. With $z = r\cos\vartheta$ the $z$-component becomes

$$(M_{ik})_z = \frac{1}{2\pi} \int\limits_{r=0}^{\infty} R_i R_k r^3 \,\mathrm{d}r \times \int\limits_{\vartheta=0}^{\pi} \Theta_{m_k}^{l_k} \Theta_{m_i}^{l_i} \sin\vartheta \cos\vartheta \,\mathrm{d}\vartheta \times \int\limits_{\varphi=0}^{2\pi} \mathrm{e}^{\mathrm{i}(m_k - m_i)\varphi} \,\mathrm{d}\varphi . \tag{7.29}$$

Only those transitions $\langle i| \to \langle k|$ appear in the spectrum, for which all three factors are nonzero.

For circularly polarized light emitted into the $z$-direction the $x$- and $y$-components of $\boldsymbol{M_{ik}}$ can contribute to the transition probability. The electric field vector for circularly polarized $\sigma^+$-light can be written as $E^+ = E_x + \mathrm{i}E_y$ and for $\sigma^-$-light is $E^- = E_x - \mathrm{i}E_y$. Therefore only the $x$- and $y$-components of the matrix element (7.26) contribute to the transitions. Forming the linear combinations $(M_{ik})_x \pm \mathrm{i}(M_{ik})_y$ of the matrix elements we obtain from (7.27) with $x = r\sin\vartheta \cos\varphi$ and $y = r\sin\vartheta \sin\varphi$

$$(M_{ik})_x + \mathrm{i}\,(M_{ik})_y = \frac{1}{2\pi} \int\limits_{r=0}^{\infty} R_i R_k r^3 \,\mathrm{d}r \times \int\limits_{\vartheta=0}^{\pi} \Theta_{m_i}^{l_i} \Theta_{m_k}^{l_k} \sin^2\vartheta \,\mathrm{d}\vartheta \times \int\limits_{\varphi=0}^{2\pi} \mathrm{e}^{\mathrm{i}(m_k - m_i + 1)\varphi} \,\mathrm{d}\varphi \tag{7.30a}$$

$$(M_{ik})_x - \mathrm{i}\,(M_{ik})_y = \frac{1}{2\pi} \int\limits_{r=0}^{\infty} R_i R_k r^3 \,\mathrm{d}r \times \int\limits_{\vartheta=0}^{\pi} \Theta_{m_i}^{l_i} \Theta_{m_k}^{l_k} \sin^2\vartheta \,\mathrm{d}\vartheta \times \int\limits_{\varphi=0}^{2\pi} \mathrm{e}^{\mathrm{i}(m_k - m_i - 1)\varphi} \,\mathrm{d}\varphi . \tag{7.30b}$$

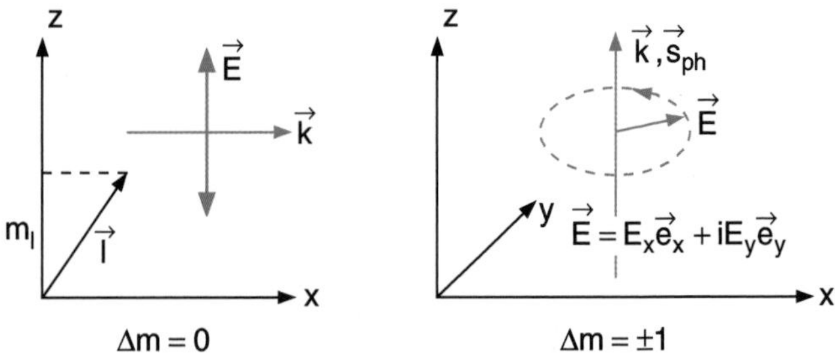


**Fig. 7.7.** Transitions with $\Delta m = 0$ (emission of linearly polarized light) and $\Delta m = \pm 1$ (circular polarization). Quantization axis is the $z$-axis

### 7.2.2 Selection Rules for the Magnetic Quantum Number

The last factor in the matrix elements (7.29) for linearly polarized light is always zero, except for $m_i = m_k$. This gives the selection rule

$$(M_{ik})_z \neq 0 \quad \text{only for} \quad \Delta m = m_i - m_k = 0 . \tag{7.31}$$

The last integrals in (7.30a) and (7.30b) for circular polarized light show that $(M_{ik})_x + \mathrm{i}(M_{ik})_y \neq 0$ only for $m_k = m_i - 1$ and $(M_{ik})_x - \mathrm{i}(M_{ik})_y \neq 0$ only for $m_k = m_i + 1$ (Fig. 7.7).

> In conclusion, we obtain for transitions $E_i \to E_k$ the selection rules for the change $\Delta m = m_i - m_k$ of the magnetic quantum number $m$
>
> $\Delta m = \pm 1$ for circularly polarized light (7.32a)
>
> $\Delta m = 0$ for linearly polarized light . (7.32b)

This selection rule also follows from the conservation of angular momentum for the system atom +

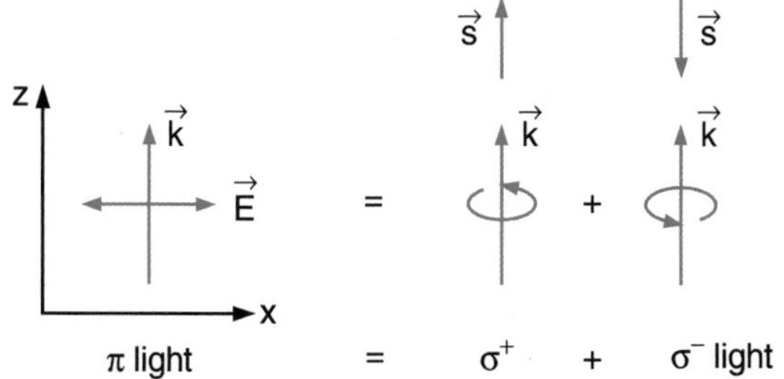


**Fig. 7.8.** Linearly polarized light with average photon spin $s = \mathbf{0}$ zero as the superposition of $\sigma^+$ and $\sigma^-$ light

photon. For $\sigma^+$ light, the photon spin is $s_z = +1\hbar$, pointing in the $+z$ direction. When the photon is emitted into the $z$-direction the atom has to decrease the $z$ component of its angular momentum by the same amount (Fig. 7.7). For $\sigma^-$ light, the photon spin is $s_z = -1\hbar$, giving rise to atomic transitions with $\Delta m = -1$. Since linearly polarized light can be regarded as superposition of $\sigma^+$ and $\sigma^-$ light, the expectation value for the photon spin is the quantum mechanical average of $-\hbar$ (for $\sigma^-$-light) and $+\hbar$ (for $\sigma^+$-light) and is therefore zero (Fig. 7.8). This implies that the z-component of the atomic angular momentum does not change during a $\pi$-transition.

We can this also explain by the following argument: The electric field vector $E$ points into a direction perpendicular to the z-direction, which is our quantization axis. Since $E_z = 0$ the $z$-component of the transition dipole moment vanishes.

For the emission of light $E_i \to E_k + h\nu$, the angular momentum $m_i\hbar$ in the initial state must be equal to the sum of angular momentum $m_k\hbar$ in the final state and the photon spin (Table 7.1).

When the atom is placed in an external static magnetic field $\boldsymbol{B} = \{0, 0, B_z\}$, which causes the degenerate magnetic sublevels to split into Zeeman components, one observes for the light emitted into the field direction, two circularly polarized components. In the direction perpendicular to the magnetic field direction, one observes three linearly polarized components,

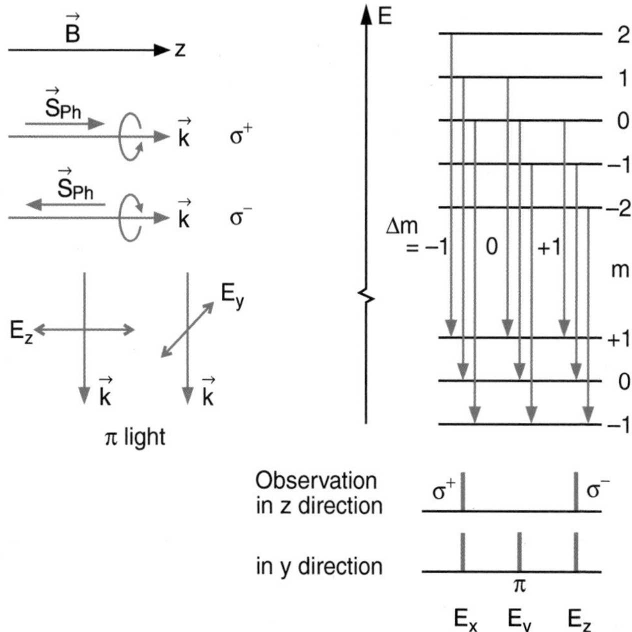


**Fig. 7.9.** Possible transitions $\Delta m = \pm 1$ of Zeeman components for the normal Zeeman effect and the corresponding polarization of the radiation for the observation parallel and perpendicular to the magnetic field direction

one polarized in the $z$ direction, which is due to the component $(M_{ik})_z$ with $\Delta m = 0$, and two due to the sum $(M_x + \mathrm{i}M_y) + (M_x - \mathrm{i}M_y) = 2M_x$ with $\Delta m = \pm 1$ (Fig. 7.9) (see also Sect. 5.2).

**Table 7.1.** Change $\Delta m$ of the magnetic quantum number $m$ under absorption or emission of photons

| Photon | Absorption | Emission |
|---|---|---|
| $\sigma^+$: $\boldsymbol{s}_{\text{phot}} \uparrow\uparrow \boldsymbol{k}$ | $\Delta m = +1$ | $\Delta m = -1$ |
| $\sigma^-$: $\boldsymbol{s}_{\text{phot}} \downarrow\uparrow \boldsymbol{k}$ | $\Delta m = -1$ | $\Delta m = +1$ |
| $\pi$: $\langle \boldsymbol{s}_{\text{phot}} \rangle = 0$ | $\Delta m = 0$ | $\Delta m = 0$ |

### 7.2.3 Parity Selection Rules

Even when the selection rules for the magnetic quantum number are fulfilled (which implies that the third integrals in (7.29) or (7.30) are nonzero, the second integrals can still vanish, bringing the transition probability to zero.

The somewhat lengthy calculation (see [7.1]) shows that the integrals

$$\int \Theta_{m_k}^{l_k} \Theta_{m_i}^{l_i} \sin\vartheta \,\cos\vartheta \,\mathrm{d}\vartheta$$

$$\text{and} \quad \int \Theta_{m_k}^{l_k} \Theta_{m_i}^{l_i} \sin^2\vartheta \,\mathrm{d}\vartheta$$

in (7.29) and (7.21) are nonzero only if $l_k - l_i = \pm 1$.

Only those transition are allowed for which the quantum number $l$ of the electronic orbital angular momentum $\boldsymbol{l}$ obeys the selection rule

$$\Delta l = l_i - l_k = \pm 1 . \tag{7.33}$$

Also, this selection rule is a consequence of the conservation of angular momentum. The absorbed or emitted photon has the spin $s = \pm 1\hbar$. Since the total angular momentum of the system atom plus photon has to be constant when a photon is absorbed or emitted, the atom must change its angular momentum during the transition $\langle i| \rightarrow \langle k|$ by $\pm 1\hbar$.

The selection rule (7.33) can be also derived from symmetry arguments. We consider the matrix element

$$\boldsymbol{M}_{ik} = \int\limits_x \int\limits_y \int\limits_z \psi_i^*(x,y,z)\boldsymbol{r}\psi_k(x,y,z)\,\mathrm{d}x\,\mathrm{d}y\,\mathrm{d}z$$

in Cartesian coordinates. Since the integration extends from $x, y, z = -\infty$ to $+\infty$ the integrand has to be an even function of the coordinates $x, y, z$, otherwise the integral vanishes. Since $\boldsymbol{r} = \{x, y, z\}$ is an odd function, the product $\psi_i^* \psi_k$ must also be an odd function in order to make the integrand an even function.

The symmetry of a function with regard to a reflection of all coordinates at the origin is called its parity. The function $f(x, y, z)$ has even (or positive) parity if $f(x, y, z) = +f(-x, -y, -z)$, it has odd (or negative) parity, if $f(x, y, z) = -f(-x, -y, -z)$.

The transition moment $M_{ik}$ can be only nonzero, if the wave functions $\psi_i$ and $\psi_k$ of the two states of the transition $\langle i| \rightarrow \langle k|$ have opposite parities.

The hydrogenic wave functions in Table 5.2 have the parity $(-1)^l$. This implies, that the quantum number $l$ of the angular momentum has to change by an odd number for an allowed dipole transition. Since the angular momentum of the photon is $\pm 1\hbar$, this odd number has to be $\pm 1$ for electric dipole transitions. This again gives the selection rule (7.33).

### 7.2.4 Selection Rules for Induced Absorption and Emission

According to equ. (7.22) the polarization of the inducing electromagnetic wave, expressed by the polarization unit vector $\hat{\boldsymbol{\varepsilon}} = \mathbf{E_0}/|\mathrm{E}_0|$ is important for the specific selection rules. The scalar product $\mathbf{e} \cdot \mathbf{r}$ can be written as the sum

$$\varepsilon \cdot \mathbf{r} = \varepsilon_x x + \varepsilon_y y + \varepsilon_z z$$

Only those transitions where at least one the three components is nonzero can contribute to the transition probability. Consider for example a linearly polarized light wave travelling into the z-direction with the electric field vector $\mathbf{E} = \{\mathrm{E}_0, 0, 0\}$ pointing into the x-direction. Because $\mathrm{e_y} = \mathrm{e_z} = 0$ the y- and z-components of the transition dipole moment are zero.

For circular polarized $\sigma^+$ light, travelling into the z-direction we can write

$$\mathbf{E} = \mathbf{E}_0(\varepsilon_x + i\varepsilon_y).$$

this shows that the z component of the transition dipole moment does not contribute to the absorption. Whether the other components are nonzero, depends on the wavefunctions $\psi_\mathrm{i}$ and $\psi_\mathrm{k}$ of the levels $|\mathrm{i}\rangle$ and $|\mathrm{k}\rangle$. Here the same selection rules are valid as discussed in Section 7.2.2 and 7.2.3.

### 7.2.5 Selection Rules for the Spin Quantum Number

Up to now we have neglected the electron spin. For atoms with only one electron the absolute value of the spin is always $|\boldsymbol{s}| = \sqrt{3/4}\hbar$, which does not change under electric dipole transitions.

The same consideration applies for one-electron transitions in multi-electron atoms with $\boldsymbol{S} = \sum \boldsymbol{s}_i$, where only one electron is involved in the transition. This gives the selection rule

$$\Delta S = S_i - S_k = \left(S_i^* + \frac{1}{2}\right) - \left(S_k^* + \frac{1}{2}\right) = 0\,, \tag{7.34}$$

where $S^*$ is the total spin quantum number of all other electrons, not involved in the transition. Since their spin quantum number $S^*$ does not change, we obtain $S_i^* = S_k^*$ and therefore $\Delta S = 0$.

For atoms with two electrons where each of the two could interact with the light wave the wave functions depend on the spatial coordinates $(\boldsymbol{r}_1, \boldsymbol{r}_2)$ of both electrons. The transition dipole matrix element now becomes

$$M_{ik} = e \int \psi_i^*(\boldsymbol{r}_1, \boldsymbol{r}_2)(\boldsymbol{r}_1 + \boldsymbol{r}_2)\psi_k(\boldsymbol{r}_1, \boldsymbol{r}_2)\,\mathrm{d}\tau_1\,\mathrm{d}\tau_2\,, \tag{7.35}$$

where the integration extends over all six coordinates of the two electrons. Because the two electrons are indistinguishable, $M_{ik}$ should not change under exchange of the two electrons.

For a singlet state, the spatial part $\psi(\boldsymbol{r}_1, \boldsymbol{r}_2)$ of the wave function is symmetric under electron exchange (see Sect. 6.1), for a triplet state it is antisymmetric. The matrix element (7.35) is only independent of an electron exchange, if both wave functions $\psi_i$ and $\psi_k$ are either symmetric or both are antisymmetric with regard to an electron exchange. This means, that both states have to be either singlet or triplet states.

> Transitions between singlet and triplet states are forbidden. The selection rule is
>
> $$\Delta S = 0\,. \tag{7.36}$$

**Note:**

This selection rule is not as strict as the parity selection rule. It only holds if the spin-orbit coupling is small, which means that the spin quantum number is well defined. One says in this case that it is a "good quantum number." The total wave function can then be separated into a product of spatial part and spin function.

In heavier atoms the spin-orbit coupling increases strongly with the nuclear charge $Ze$ (see Sect. 6.4). The separation into a spatial and a spin part is no longer possible and S is no longer a good quantum number. In such cases, one observes transitions between different multiplet systems with $\Delta S = \pm 1$, called "intercombination lines." Their intensity is still much weaker than that of the allowed transitions with $\Delta S = 0$. One example is the intercombination line of Hg atoms at $\lambda = 253.7\,\mathrm{nm}$, which is emitted on the transition $6^3P \to 6^1S$.

Although the absolute value $|\boldsymbol{S}|$ of the total spin of the electron shell does not change for an allowed electric dipole transition, the orientation of the spin $\boldsymbol{S}$ relative to the electronic orbital angular momentum can change.

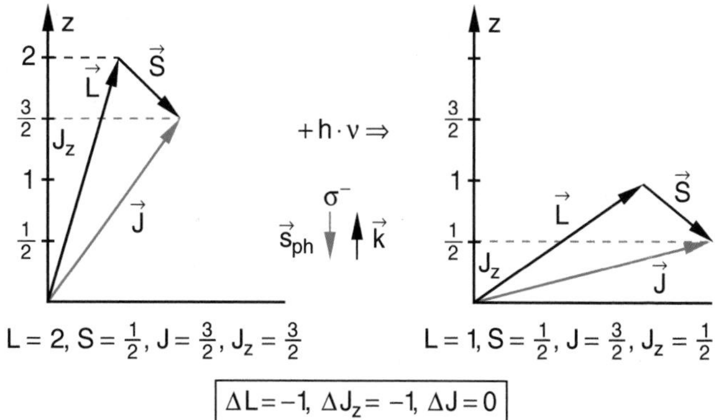


**Fig. 7.10.** Example for the conservation of electron spin $S$ for transitions in atoms with $L$-$S$-coupling, illustrated by the transition ${}^2D_{1/2} \overset{h\nu}{\leftrightarrow} {}^2P_{1/2}$

For the quantum number $J$ of the total angular momentum $\boldsymbol{J} = \boldsymbol{L} + \boldsymbol{S}$ we obtain the general selection rule

$$\Delta J = 0, \pm 1\,, \quad \text{but} \quad J = 0 \nrightarrow J = 0\,. \tag{7.37}$$

For $\Delta S = 0$ the necessary change $\Delta L = \pm 1$ can be compensated by the opposite change $\Delta M_S = \mp 1$ in order to obtain $\Delta J = 0$ (Fig. 7.10).

There is, for instance the allowed transition ${}^2P_{3/2}(l=1, m_s=+1/2) \to {}^2D_{3/2}(l=2, m_s=-1/2)$ in alkali atoms.

In Table 7.2 all selection rules are compiled.

### 7.2.6 Higher Order Multipole Transitions

Besides the electric dipole transitions with transition probability (7.17) or (7.21) given by the absolute square of the transition dipole matrix element (7.26) there are also electric quadrupole transitions and magnetic dipole transitions that have transition probabilities which are smaller by several orders of magnitude. They become important for cases where the electric dipole transitions are forbidden [7.2].

Electric quadrupole radiation is emitted by a quadrupole moment that changes in time. Analogously to the derivation of the radiation emitted by an oscillating electric dipole moment in Sect. 7.1.2, the amplitude of the quadrupole radiation is proportional

**Table 7.2.** Selection rules for electric dipole transitions

| Selection rule | Remark |
|---|---|
| $\Delta l = \pm 1$ for one-electron systems | Strictly valid |
| $\Delta L = \pm 1$ for multi electron systems with $\boldsymbol{L}$-$\boldsymbol{S}$-coupling | Gerade levels are solely combined with ungerade levels |
| $\Delta M = 0, \pm 1$ | $\Delta M = 0$: linear polarized light<br>$\Delta M = \pm 1$: $\sigma^+$ or $\sigma^-$ circularly polarized light |
| $\Delta S = 0$ | Valid for light atoms. Exceptions for heavy atoms with large spin-orbit coupling (weak Intercombination lines) |
| $\Delta J = 0, \pm 1$ | $J = 0 \to J = 0$ is forbidden |

to the second time derivative of the oscillating electric quadrupole moment.

With the spatial extension $a_c$ of the electric charge distribution and the wavelength $\lambda$ of the emitted radiation the amplitude ratio of quadrupole to dipole radiation is on the order $a_c/\lambda$ and the intensity ratio therefore about $(a_c/\lambda)^2$.

### EXAMPLE

$a_c = 10^{-10}$ m, $\lambda = 500$ nm $= 5 \times 10^{-7}$ m $\Rightarrow I_Q/I_D = 4 \times 10^{-8}$.

This illustrates that in the visible spectral range, electric quadrupole transitions can be only detected on transitions for which electric dipole transitions are forbidden. Otherwise they would be completely masked by the much stronger dipole transitions.

The electric quadrupole moment $QM$ of a distribution of charges $q_i(x, y, z)$ can be written in the form of a matrix ($QM$ is in fact a second rank tensor)

$$QM = \begin{pmatrix} QM_{xx} & QM_{xy} & QM_{xz} \\ QM_{yx} & QM_{yy} & QM_{yz} \\ QM_{zx} & QM_{zy} & QM_{zz} \end{pmatrix} . \tag{7.38a}$$

The components are

$$\begin{aligned} QM_{xx} &= \sum q_i \left(3x_i^2 - r_i^2\right); \\ QM_{yy} &= \sum q_i \left(3y_i^2 - r_i^2\right); \\ QM_{zz} &= \sum q_i \left(3z_i^2 - r_i^2\right); \\ QM_{xy} &= QM_{yx} = 3\sum q_i x_i y_i; \\ QM_{xz} &= QM_{zx} = 3\sum q_i x_i z_i; \\ QM_{yz} &= QM_{zy} = 3\sum q_i y_i z_i, \\ \text{where} \quad r_i &= \left(x_i^2 + y_i^2 + z_i^2\right)^{1/2} . \end{aligned} \tag{7.38b}$$

All components depend on the product of two coordinates or on the square of coordinates. Under reflection of all coordinates at the origin, the components are therefore not changed. This means that the quadrupole moment has positive parity, contrary to the electric dipole moment, which has negative parity. The two wave functions of the two states of a quadrupole transition therefore must have the same parity. Since the parity of the wave function is determined by $(-1)^l$, we obtain the following result.

The selection rule for the orbital angular momentum quantum number $l$ on quadrupole transitions is

$$\Delta l = 0, \pm 2 . \tag{7.39}$$

The same result holds for the quantum number $L$ of multielectron atoms.

For the quantum number $J$ of the total angular momentum $\boldsymbol{J} = \boldsymbol{L} + \boldsymbol{S}$ the selection rules are $\Delta J = 0, \pm 1, \pm 2$, where the transition $J = 0 \to 0$ is again forbidden.

The different selection rules for electric dipole and electric quadrupole transitions are illustrated in Fig. 7.11.

### EXAMPLES

The transitions ${}^2S_{1/2} \to {}^2D_{5/2}$; ${}^2S_{1/2} \to {}^2D_{3/2}$ and $2^3P_0 \to 3^3P_2$ are both allowed quadrupole transitions.

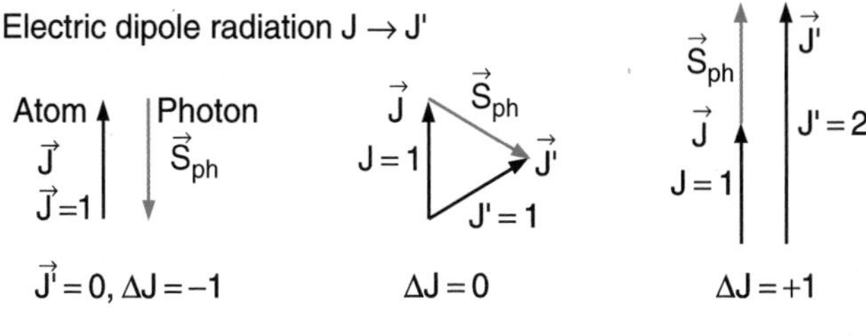


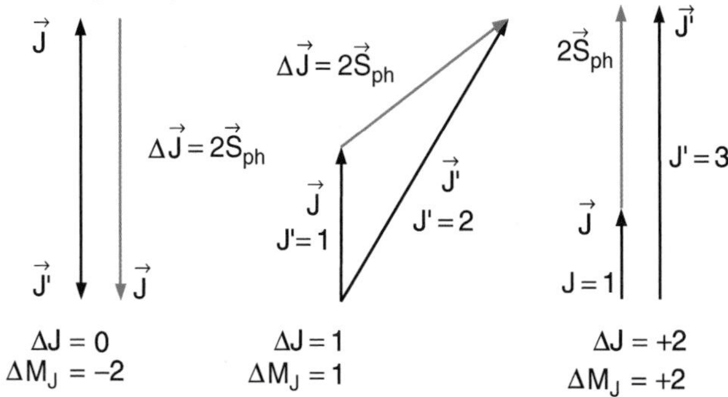


**Fig. 7.11.** Possible changes $\Delta J = J' - J$ of the quantum number $J$ of total angular momentum $\vec{J}$ for electric dipole transitions $\Delta\vec{J} = \vec{S}_{ph}$ and electric quadrupole transitions $\Delta\vec{J} = n \cdot \vec{S}_{ph}$ $(n = 0, 1, 2)$

### 7.2.7 Magnetic Dipole Transitions

Magnetic dipole transitions appear when the amount or the direction of the atomic magnetic dipole moment changes for a transition. Examples are transitions with $\Delta m = \pm 1$ between the Zeeman components of an atomic level, or between the fine structure components of an atomic state $(n, l, m_l, m_s)$, e.g., for the transition $3P_{3/2} \to 3P_{1/2}$ of the Na $(3P)$ state.

The square of the magnetic dipole matrix element is about 2–3 orders of magnitude smaller than that of an electric dipole transition. In addition, the following fact further drastically diminishes the intensity of magnetic dipole transitions. Most of these transitions occur between levels with a small energy separation $\Delta E$. The frequency $\nu = \Delta E/h$ is therefore smaller by several orders of magnitude compared with optical transitions. Since the spontaneous transition probability scales with $\nu_{ik}^3$ the emitted or absorbed line intensity is indeed very small compared with lines in the optical range, unless the latter belong to forbidden transitions.

### 7.2.8 Two-Photon-Transitions

By "simultaneous" absorption of two photons two-photon transitions with $\Delta L = 0, \pm 2$ become possible. However, they demand high light intensities, which are only achievable with lasers, because the two photons both have to be present within the absorption volume of the atom within the time interval $\Delta t$ of the transition.

The transition probability of a two-photon transition between the initial level $\langle i|$ and the final level $\langle f|$

$$\frac{\mathrm{d}\mathcal{P}_{if}}{\mathrm{d}t}(2\omega) \propto \left| \sum_k \frac{\boldsymbol{M}_{ik}\hat{e}_1 \boldsymbol{M}_{kf}\hat{e}_2}{\omega_{ki} - \omega_1 - \boldsymbol{v}\boldsymbol{k}_1} + \sum_k \frac{\boldsymbol{M}_{ik}\hat{e}_2 \boldsymbol{M}_{kf}\hat{e}_1}{\omega_{ki} - \omega_2 - \boldsymbol{v}\boldsymbol{k}_2} \right|^2 \tag{7.40}$$

is given by products of one-photon transition probabilities for transitions between the initial level and an intermediate level $\langle v|$ at the energy $E_i + h\nu$ for the first photon $\hbar\omega_1$ and $\langle v| \to \langle f|$ for the second photon $\hbar\omega_2$. The energy of the level $\langle v|$ may not coincide with a real atomic level, which means that the first absorbed photon excites the atom far off-resonance with a real level $\langle k|$. In order to simplify the diagrams, the level $\langle v|$ is often called a "virtual level" (Fig. 7.12). The sums extend over all real atomic levels $\langle k|$ that are connected by allowed one-photon transitions with the initial level $\langle i|$. The two terms in (7.40) take account of the fact that either the first photon with $\omega_1$ can be absorbed on the first step $i \to k$ or the second photon with $\omega_2$. Since both possibilities are not distinguishable the total transition amplitude has to be the sum of the two individual amplitudes (see Sect. 3.5). The denominator describes the detuning of the photon frequency $\omega_i$

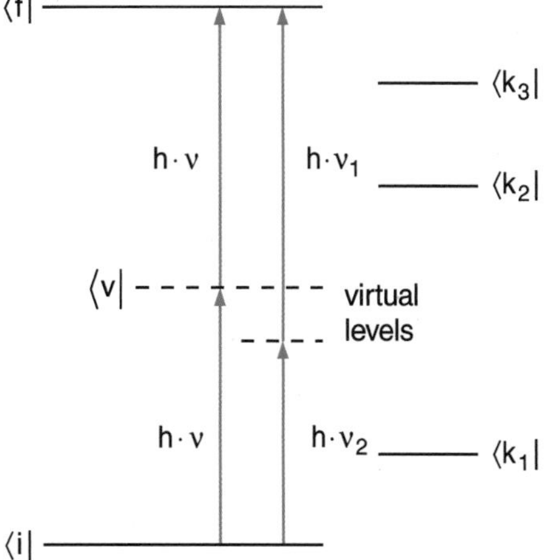


**Fig. 7.12.** Two-photon transition with two equal or two different photons

from the frequency $\omega_{ik}$ of the one photon transition to the real level $E_k$ including the Doppler-shift.

Examples for observed two-photon transitions are the $1^1S_0 \to 2^1S_0$ transition in the H atom with $\Delta L = 0$ or the $3^2S_{1/2} \to 4^2D_{3/2,5/2}$ transition in the Na atom with $\Delta L = 2$ [7.3, 4].

## 7.3 Lifetimes of Excited States

If an atom is excited (for instance by absorption of a photon, or by collisions with electrons) into a state with energy $E_i$ above that of the ground state, it can spontaneously relax back into a lower state with energy $E_j$ by emitting a photon $h\nu = E_i - E_j$. This spontaneous emission is called **fluorescence**. This lower state $E_j$ may be still above the ground state $E_k$. In this case it can further relax into the ground state by photon emission or by a collision-induced transition.

The probability per second for the fluorescence transition per atom is (see Sect. 7.2) given by the Einstein coefficient $A_{ij}$. If $N_i$ atoms are in the energy state $E_i$, the emission rate on the transition $\langle i| \to \langle j|$ is

$$\mathrm{d}N_i = -A_{ij} N_i \,\mathrm{d}t \,. \tag{7.41a}$$

If the state $E_i$ can relax into several lower states $E_j < E_i$ (Fig. 7.13), we obtain for the total rate

$$\mathrm{d}N_i = -A_i N_i \,\mathrm{d}t \quad \text{with} \quad A_i = \sum_j A_{ij} \,. \tag{7.41b}$$

Integration of (7.41b) gives the time dependent population density

$$\boxed{N_i(t) = N_i(0)\mathrm{e}^{-A_i t} \,.} \tag{7.41c}$$

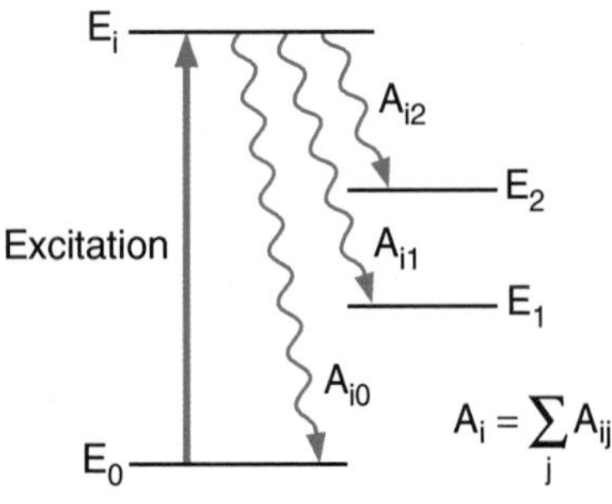


**Fig. 7.13.** Spontaneous radiative decay (fluorescence) from the excited level $E_i$ into several lower levels $E_j$

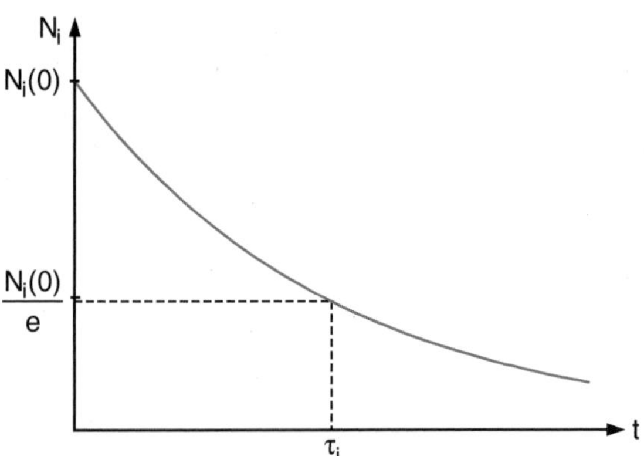


**Fig. 7.14.** Experimental decay curve of the population $N_i$ of an excited level $E_i$ with mean lifetime $\tau_i$

The population of the excited state $E_i$ decreases exponentially from its initial value $N_i(0)$ at time zero to $N_i(\infty) = 0$ (Fig. 7.14).

The constant $\tau_i = 1/A_i$ is the mean lifetime of the state $E_i$. This can be seen as follows. The mean lifetime is defined as

$$\begin{aligned} \langle t_i \rangle &= \frac{1}{N_0} \int_{N_0}^{0} t \cdot \mathrm{d}N_i(t) \\ &= -\int_0^\infty t A_i \mathrm{e}^{-A_i t} \,\mathrm{d}t = \frac{1}{A_i} = \tau_i \,, \end{aligned} \tag{7.42}$$

where $(1/N_0)\mathrm{d}N_i(t)\,\mathrm{d}t$ is the probability of a decay of an atom within the time interval between $t$ and $t + \mathrm{d}t$.

After the mean lifetime $\langle t_i \rangle = \tau_i$ the initial population $N_i(t = 0)$ has decreased to $N_i(0)/\mathrm{e}$.

Measurements of the mean lifetime of a level with energy $E_i$ allows the determination of the total transition probability $A_i = \sum A_{in} = 1/\tau_i$. The specific transition probabilities $A_{in}$ can be inferred from measurements of the relative line intensities of the individual transitions $E_i \to E_n$ using the relation

$$A_{in} = A_i \frac{I_{in}/(h\nu_{in})}{\sum_n I_{in}/(h\nu_{in})} \,. \tag{7.43}$$

This yields, with (7.17), the transition matrix elements $M_{ik}$ [7.5].

If other relaxation processes (for example collisions) add with the probability $R_i$ per second to the deactivation of level $E_i$ (Fig. 7.15), (7.41b) must be modified to

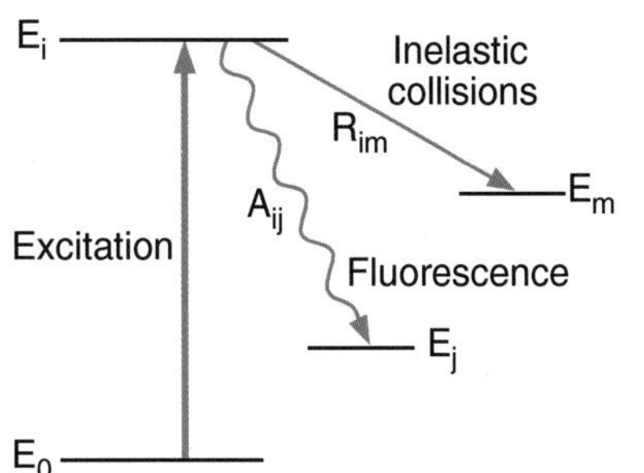


**Fig. 7.15.** Inelastic collisions can contribute to the depopulation of level $E_i$

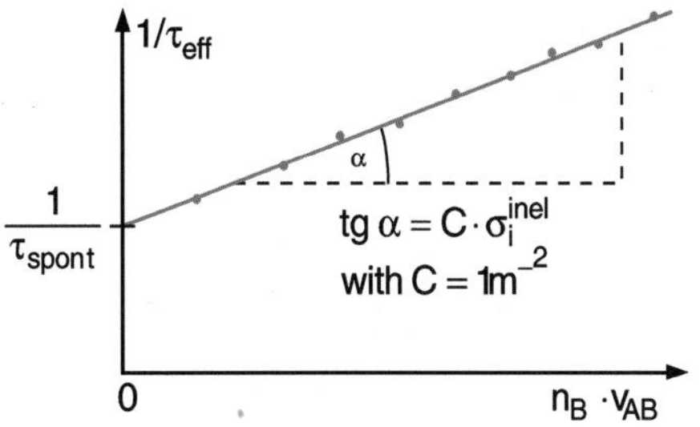


**Fig. 7.16.** Inverse effective lifetime $1/\tau_{\text{eff}}$ as a function of the density $n_{\text{B}}$ of collision partners B (Stern–Volmer plot)

$$\text{d}N_i = -(A_i + R_i)N_i\,\text{d}t\,. \tag{7.44}$$

We then obtain the time dependent population density

$$N_i(t) = N_i(0)\text{e}^{-(A_i+R_i)t} \tag{7.45}$$

and the effective lifetime $\tau_{\text{eff}}$ becomes

$$\tau_i^{\text{eff}} = \frac{1}{A_i + R_i}\,. \tag{7.46}$$

If level $E_i$ of atom A is depopulated by inelastic collisions with other atoms B, the collision-induced rate becomes

$$R_i = n_{\text{B}}\overline{v}_{\text{AB}}\sigma_i^{\text{inel}}\,, \tag{7.47}$$

where $n_{\text{B}} = N_{\text{B}}/V$ is the number density of atoms B and

$$\overline{v}_{\text{AB}} = \sqrt{\frac{8kT}{\pi\mu}} \quad \text{with} \quad \mu = \frac{M_{\text{A}}M_{\text{B}}}{M_{\text{A}} + M_{\text{B}}} \tag{7.48}$$

is the mean relative velocity of the two collision partners with reduced mass $\mu$ in a gas cell at temperature $T$.

Plotting the inverse effective lifetime

$$\frac{1}{\tau_i^{\text{eff}}} = \frac{1}{\tau_i^{\text{spont}}} + n_{\text{B}}\overline{v}_{\text{AB}}\sigma_i^{\text{inel}} \tag{7.49}$$

as a function of the product $n_{\text{B}}v_{\text{AB}}$ (Stern–Vollmer plot (Fig. 7.16)) one obtains a straight line with the slope equal to the inelastic collision cross section $\sigma_i^{\text{inel}}$.

The intersection of this straight line with the axis $n_{\text{B}} = 0$ yields the inverse spontaneous lifetime $1/\tau_i^{\text{spont}}$.

From the general equation of state

$$pV = NkT$$

the number density $n_{\text{B}} = N/V = p/kT$ can be expressed by the pressure $p$ and the temperature $T$, which can be determined experimentally much easier than the number density $n_{\text{B}}$. This yields the relation

$$\boxed{\frac{1}{\tau_i^{\text{eff}}} = \frac{1}{\tau_i^{\text{spont}}} + \sigma_i^{\text{inel}}\sqrt{\frac{8}{\pi\mu kT}}\,p} \tag{7.50}$$

between effective lifetime $\tau_{\text{eff}}$ of an excited state and the pressure $p$ of the collision partners B [7.6, 7].

## 7.4 Line Profiles of Spectral Lines

The absorption or emission of radiation on an atomic transition

$$\Delta E = E_i - E_k = h\nu_{ik}$$

does not result in a strictly monochromatic spectral line, but rather in a frequency distribution around the central frequency $\nu_0$. This gives a line profile $I(\nu - \nu_0)$ with a full-width at half-maximum $\delta\nu$, which not only depends on the spectral resolution of the measuring apparatus but also on basic physical properties, such as the lifetimes of the atomic states involved in the transition, the velocity distribution of the moving atoms and the pressure of the gaseous sample.

The linewidth in frequency units is defined as the interval $\delta\nu = \nu_2 - \nu_1$ between the frequencies $\nu_1 < \nu_0$ and $\nu_2 > \nu_0$ on both sides of the central frequency $\nu_0$, where the intensity $I(\nu_1) = I(\nu_2) = \frac{1}{2}I(\nu_0)$ of the spectral line profile has dropped to half its maximum value at the line center (**full width half maximum FWHM**). Often the linewidth is given in units of the circular frequency $\omega = 2\pi\nu \Rightarrow \delta\omega = 2\pi\delta\nu$ or in units of the wavelength $\lambda$. Because $\lambda = c/\nu \Rightarrow$

$$\delta\lambda = -(c/\nu^2)\delta\nu = -(\lambda/\nu)\delta\nu\,. \tag{7.51}$$

The relative half-widths are equal in all these notations, because from (7.51) it follows that

$$\left|\frac{\delta\lambda}{\lambda}\right| = \left|\frac{\delta\nu}{\nu}\right| = \left|\frac{\delta\omega}{\omega}\right| . \tag{7.52}$$

The spectral interval inside the full half-width is called the ***line kernel***, the ranges outside the ***line wings*** (Fig. 7.17).

There are several reasons for the finite linewidths of spectral lines:

- The energy levels of the atoms have an energy uncertainty $\delta E = h/\tau$ related to their finite lifetime $\tau$. The frequency width of a spectral line corresponding to a transition between levels $E_i$ and $E_k$ is

  $$\delta\nu_{ik} = (\delta E_i + \delta E_k)/h$$

  (natural linewidth, see Sect. 7.4.1).
- Atoms in a gas move with a mean velocity $v$, depending on their mass and on the temperature of the gas. This leads to a Doppler shift of their emitted or absorbed radiation. The statistical distribution of the velocities of many atoms in a gas results in statistically distributed Doppler shifts, which gives rise to a Gaussian line profile (Doppler broadening, see Sect. 7.4.2).
- Every atom in a gas interacts with other neighboring atoms. This results in a shift of the atomic energy levels. Since the interaction decreases with increasing distance between the interacting atoms the level shifts and with it the line-shifts increase with the density or pressure of the gas, resulting in a pressure broadening and shift of the spectral lines (Sect. 7.4.3).

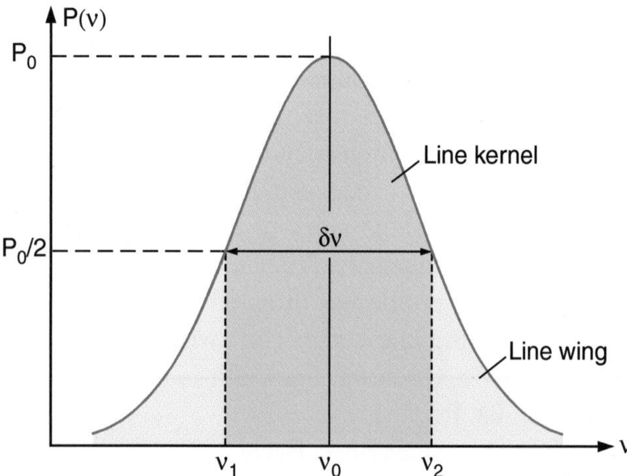

**Fig. 7.17.** Line profile of a spectral line

We will now discuss these effects in more detail.

### 7.4.1 Natural Linewidth

An excited atom can deliver its excitation energy in the form of spontaneous emission. We will describe the excited electron by the classical model of a damped harmonic oscillator with mass $m$, spring constant $D$ and eigenfrequency $\omega_0 = \sqrt{D/m}$ (Hertzian dipole). The time dependent amplitude of its oscillation can be obtained from the equation of motion

$$\ddot{x} + \gamma\dot{x} + \omega_0^2 x = 0 , \tag{7.53}$$

where $\gamma$ is the damping constant. The real solution of (7.53) is:

$$x(t) = x_0 \mathrm{e}^{-(\gamma/2)t}\left[\cos\omega t + (\gamma/2\omega)\sin\omega t\right] . \tag{7.54a}$$

Note that the frequency $\omega = \sqrt{\omega_0^2 - (\gamma/2)^2}$ of the damped oscillator is slightly lower than $\omega_0$ of the undamped oscillator. However, we will see that for excited atoms the damping constant $\gamma$ is in most cases much smaller than $\omega_0$. We can therefore neglect the second term and approximate $\omega$ by $\omega_0$. The solution of (7.53) then becomes

$$x(t) \approx x_0 \mathrm{e}^{-(\gamma/2)t}\cos\omega_0 t . \tag{7.54b}$$

Because of the time dependent amplitude $x(t)$ the frequency of the emitted radiation will no longer be strictly monochromatic, as would be the case for an infinitely long undamped oscillation. The Fourier transformation of $x(t)$ yields the frequency distribution of the amplitudes

$$\begin{aligned} A(\omega) &= \frac{1}{\sqrt{2\pi}}\int\limits_{-\infty}^{+\infty} x(t)\mathrm{e}^{-\mathrm{i}\omega t}\,\mathrm{d}t \\ &= \frac{1}{\sqrt{2\pi}}\int\limits_{0}^{+\infty} x_0\mathrm{e}^{-(\gamma/2)t}\cos\omega_0 t\,\mathrm{e}^{-\mathrm{i}\omega t}\,\mathrm{d}t , \end{aligned} \tag{7.55a}$$

where we have assumed that the excitation of the atom occurs at $t = 0$, which means that the oscillation starts at $t = 0$, giving $x(t < 0) = 0$ (Fig. 7.18a).

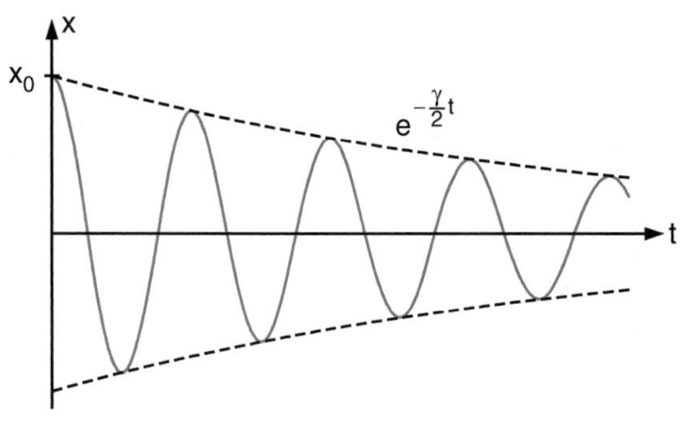


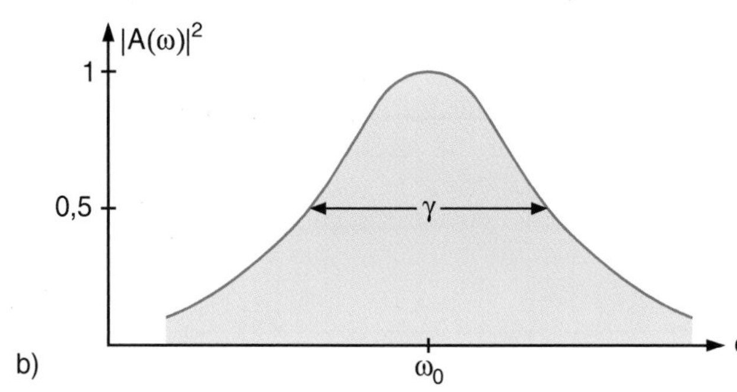


**Fig. 7.18.** (**a**) Damped oscillation. (**b**) Lorentzian line profile $|A(\omega)|^2$ as Fourier transform of a damped oscillation

The elementary integration of (7.55a) yields the complex amplitudes

$$A(\omega) = \frac{x_0}{\sqrt{8\pi}}\left[\frac{1}{\mathrm{i}(\omega_0 - \omega) + \gamma/2} + \frac{1}{\mathrm{i}(\omega_0 + \omega) + \gamma/2}\right] . \tag{7.55b}$$

In the vicinity of the resonance frequency $\omega_0$ is $|\omega - \omega_0| \ll \omega_0$. We can therefore neglect the second term in (7.55b).

The amplitude $A(\omega)$ is proportional to the Fourier component $E(\omega)$ of the electric field of the emitted radiation. Therefore the spectral radiation power density $P_\omega(\omega)$ is

$$P_\omega(\omega) \propto A(\omega)A^*(\omega) .$$

In the vicinity of the central eigenfrequency $\omega_0$ we then obtain the spectral profile of the radiation power density

$$P_\omega(\omega) = \frac{C}{(\omega - \omega_0)^2 + (\gamma/2)^2} . \tag{7.56}$$

The constant $C$ is chosen in such a way that the total power becomes

$$\int_0^\infty P_\omega(\omega)\,\mathrm{d}\omega = P_0 . \tag{7.57a}$$

The substitution $\omega' = \omega - \omega_0$ transforms the integral into

$$\int_0^\infty P_\omega(\omega)\,\mathrm{d}\omega \approx \int_{-\infty}^{+\infty} P_\omega(\omega')\,\mathrm{d}\omega' = P_0 . \tag{7.57b}$$

The integration of this integral yields $C = P_0\gamma/2\pi$.

The normalized line profile

$$\boxed{P_\omega(\omega) = P_0 \frac{\gamma/2\pi}{(\omega - \omega_0)^2 + (\gamma/2)^2}} \tag{7.58a}$$

is called the ***Lorentzian profile*** (Fig. 7.18b). The full-width at half-maximum (FWHM) is derived from (7.58) as

$$\delta\omega_n = \gamma \Rightarrow \delta\nu_n = \gamma/2\pi . \tag{7.59}$$

This FWHM is called the ***natural linewidth***, because it is caused by the spontaneous emission of the atom, without any external influences.

**Note:**

Sometimes one finds in the literature another normalization where the constant $C$ is chosen in such a way that $P_0 = P_\omega(\omega_0)$ becomes the spectral power density at the line center. This gives $C = (\gamma/2)^2 P_0$ and the line profile becomes

$$P_\omega(\omega) = \frac{P_\omega(\omega_0)}{1 + \left(\frac{\omega - \omega_0}{\gamma/2}\right)^2} . \tag{7.58b}$$

We can gain more physical insight into the energetic conditions during the emission of radiation, when we multiply both sides of (7.53) by $m\dot{x}$, which gives

$$m\dot{x}\ddot{x} + m\omega_0^2 x\dot{x} = -\gamma m\dot{x}^2 . \tag{7.60}$$

This can be written as

$$\frac{\mathrm{d}}{\mathrm{d}t}\left[\frac{m}{2}\dot{x}^2 + \frac{m}{2}\omega_0^2 x^2\right] = \frac{\mathrm{d}W}{\mathrm{d}t} = -\gamma m\dot{x}^2 . \tag{7.61}$$

The expression in the brackets represents the total energy $W = E_{\text{kin}} + E_{\text{pot}}$ as the sum of kinetic and potential energy. Inserting $x(t)$ from (7.54b) into the left side of (7.61) gives for $\gamma \ll \omega_0$ the radiation power

$$P = \frac{\mathrm{d}W}{\mathrm{d}t} = -\gamma m x_0^2 \omega_0^2 \mathrm{e}^{-\gamma t} \sin^2 \omega_0 t \,. \tag{7.62}$$

The average over one oscillation period becomes, with $\langle \sin^2 \omega_0 t \rangle = 1/2$,

$$\overline{P} = \frac{\mathrm{d}\overline{W}}{\mathrm{d}t} = -\frac{1}{2}\gamma m x_0^2 \omega_0^2 \mathrm{e}^{-\gamma t} \,. \tag{7.63}$$

Since the decrease of the oscillator energy equals the radiation power, we can see from (7.63) that the radiation power decreases exponentially and after the mean lifetime $\tau = 1/\gamma$ it has only 1/e of its initial power $P(t = 0) = P_0$.

In Sect. 7.3 we have seen that the mean lifetime $\tau_i = 1/A_i$ of an excited atomic state is related to the Einstein coefficient $A_i$ for spontaneous emission. This means that when we replace the classical damping constant $\gamma$ by the Einstein coefficient $A_i$ we can directly adopt the formulas for a classical damped harmonic oscillator to describe the time dependence of spontaneous emission of excited atoms. We then obtain for the natural linewidth of a transition from level $|i\rangle$ to the ground state

$$\boxed{\delta\omega_n = A_i = \frac{1}{\tau_i} \Rightarrow \delta\nu_n = \frac{A_i}{2\pi} = \frac{1}{2\pi\tau_i} \,.} \tag{7.64}$$

These equations can be also derived from Heisenberg's uncertainty relations. If an atomic state has a mean lifetime $\tau$, its energy can be determined only within an uncertainty $\Delta E = \hbar/\tau$. The frequency uncertainty of the emission from this state into the ground state (with $\tau_0 = \infty$) is then

$$\Delta\nu = \Delta E/h = \frac{1}{2\pi\tau} \Rightarrow \Delta\nu = \delta\nu_n \,.$$

For a transition between two excited states $E_k$ and $E_i$ with lifetimes $\tau_k$ and $\tau_i$ both energy uncertainties contribute to the linewidth (Fig. 7.19), which yields

$$\Delta E = \Delta E_i + \Delta E_k \Rightarrow \delta\nu_n = \frac{1}{2\pi}\left(\frac{1}{\tau_i} + \frac{1}{\tau_k}\right) \,. \tag{7.65}$$

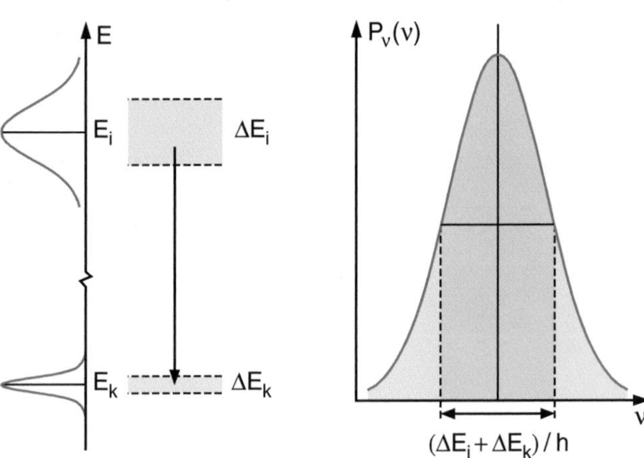


**Fig. 7.19.** Natural linewidth caused by the energy uncertainties $\Delta E_i, \Delta E_k$ of the atomic levels connected by the transition $h\nu_{ik} = E_i \leftrightarrow E_k$

## EXAMPLES

1. The natural linewidth of the Na D-line, emitted on the transition from the excited level $3P_{1/2}$ ($\tau = 16\,\text{ns}$) into the ground state $3S_{1/2}$ ($\tau = \infty$) is
$$\delta\nu_n = \frac{10^9}{16 \cdot 2\pi} \approx 10^7 \mathrm{s}^{-1} = 10\,\text{MHz} \,.$$
Since the frequency at the line center is $\nu_0 = 5 \times 10^{14}\,\mathrm{s}^{-1}$ the damping constant $\gamma = 6.25 \times 10^7$ is very small compared to $\omega_0 = 2\pi\nu_0$. Only after $8 \times 10^6$ oscillation periods the amplitude has decreased to 1/e of its initial value. This illustrates that the approximation $\gamma \ll \omega_0$, made above, is indeed well justified.
2. For metastable excited states, the lifetimes may become very long, because there are no allowed spontaneous transitions to lower states. Examples are the $2^2S_{1/2}$ states of the H atom, or the $2^3S_1$ state of the He atom. The excited $2^2S_{1/2}$ state of the hydrogen atom can only undergo a two-photon transition into the ground state. Its lifetime is about 0.14 s (!) and its natural linewidth is accordingly $\delta\nu_n = 1.1\,\mathrm{s}^{-1}$. Such a small linewidth can be measured only with special techniques (see Chap. 11).

### 7.4.2 Doppler Broadening

If an excited atom moves with the velocity $\boldsymbol{v} = \{v_x, v_y, v_z\}$, the center frequency of the radiation,

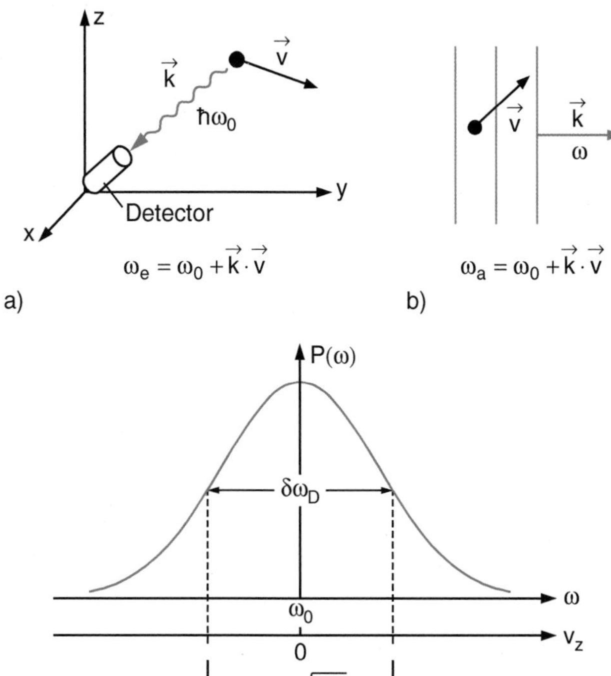


**Fig. 7.20a–c.** Doppler-broadening of spectral lines. (**a**) Doppler-shift emitted. (**b**) Absorbed photons. (**c**) Gaussian line profile caused by the thermal velocity distribution of absorbing or emitting atoms

emitted into the direction of the wave vector $\boldsymbol{k}$ of the wave appears for an observer at rest Doppler-shifted to

$$\omega_e = \omega_0 + \boldsymbol{k}\cdot\boldsymbol{v} \quad \text{with} \quad |\boldsymbol{k}| = 2\pi/\lambda \tag{7.66}$$

(Fig. 7.20a). Also, the absorption frequency $\omega_a$ of an atom, moving with the velocity $\boldsymbol{v}$ changes. If a plane wave with wave vector $\boldsymbol{k}$ and frequency $\omega$ hits the moving atom, the frequency $\omega$ appears in the moving frame of the atom shifted to $\omega' = \omega - \boldsymbol{k}\cdot\boldsymbol{v}$. Since the absorption frequency of the atom in its rest frame is $\omega_0$, the wave can only be absorbed if $\omega' = \omega_0$. The frequency of the incident wave therefore has to be

$$\omega = \omega_a = \omega_0 + \boldsymbol{k}\cdot\boldsymbol{v} \tag{7.67}$$

in order to be absorbed (i.e., coincide with the center frequency $\omega_0$ of the atom in its rest frame).

If the wave travels in the $z$ direction its wave vector is $\boldsymbol{k} = \{0, 0, v_z\} \Rightarrow \boldsymbol{k}\cdot\boldsymbol{v} = k_z v_z$ and the absorption frequency becomes

$$\omega_a = \omega_0 + k_z v_z = \omega_0(1 + v_z/c) \tag{7.68}$$

(Fig. 7.20b).

At thermal equilibrium, the velocities of the atom follow a Maxwell–Boltzmann distribution. The number density $n_i(v_z)\,\mathrm{d}v_z$ of atoms in the absorbing level $|i\rangle$ with velocity components $v_z$ within the interval from $v_z$ to $v_z + \mathrm{d}v_z$ is given by

$$n_i(v_z)\,\mathrm{d}v_z = \frac{N_i}{v_w\sqrt{\pi}} \mathrm{e}^{-(v_z/v_w)^2}\,\mathrm{d}v_z\,, \tag{7.69}$$

where $v_w = (2k_B T/m)^{1/2}$ is the most probable velocity component, $k_B$ is the Boltzmann constant and

$$N_i = \int\limits_{-\infty}^{+\infty} n_i(v_z)\,\mathrm{d}v_z$$

is the total number of all atoms in level $E_i$ per unit volume.

If $v_z$ and $\mathrm{d}v_z$ in (7.69) are expressed by the frequency $\omega$ and the frequency shift $\mathrm{d}\omega$ according to the relations (7.68) we obtain $v_z = (c/\omega_0)(\omega - \omega_0)$ and $\mathrm{d}v_z = (c/\omega_0)\,\mathrm{d}\omega$. Inserting this into (7.69) gives the number of atoms

$$n_i(\omega)\,\mathrm{d}\omega = \frac{cN_i}{\omega_0 v_w\sqrt{\pi}} \mathrm{e}^{-[c(\omega-\omega_0)/(\omega_0 v_w)]^2}\,\mathrm{d}\omega \tag{7.70}$$

that absorb or emit radiation within the frequency interval between $\omega$ and $\omega + \mathrm{d}\omega$. Since the absorbed or emitted spectral radiation power density is proportional to $n(\omega)\,\mathrm{d}\omega$, the intensity profile of a Doppler-broadened absorption or emission line becomes

$$P_\omega(\omega) = P_\omega(\omega_0)\mathrm{e}^{-[c(\omega-\omega_0)/(\omega_0 v_w)]^2}\,. \tag{7.71}$$

This a Gaussian function that is symmetric to the center frequency $\omega_0$ (Fig. 7.20c). The full half-width is

$$\delta\omega_D = |\omega_1 - \omega_2|$$

$$\text{with} \quad P(\omega_1) = P(\omega_2) = \frac{1}{2}P(\omega_0)\,.$$

Using the relation $v_w = (2k_B T/m)^{1/2}$ converts this into

$$\delta\omega_D = 2\sqrt{\ln 2}\,\omega_0 v_w/c = (\omega_0/c)\sqrt{8k_B T \ln 2/m}\,. \tag{7.72a}$$

Because $(4\ln 2)^{-1/2} \approx 0.6$ we can write (7.71) as

$$\boxed{P(\omega) = P(\omega_0)\mathrm{e}^{-[(\omega-\omega_0)/0.6\delta\omega_D]^2}\,.} \tag{7.71b}$$

The ***Doppler-width*** $\delta\omega_D$ increases proportionally with the frequency $\omega_0$, with $T^{1/2}$ and decreases with increasing mass $m$ of the atom as $m^{-1/2}$.

With the Avogadro number $N_A$, the molar mass $M = N_A m$ and the gas constant $R = N_A k_B$ we can transform (7.72a) into the expression

$$\begin{aligned}\delta\nu_D &= \frac{2\nu_0}{c}\sqrt{(2RT/M)\ln 2}\\ &= 7.16\times 10^{-7}\nu_0\sqrt{T/M}\,\mathrm{s}^{-1}\end{aligned}\,, \tag{7.72b}$$

which is more handy for fast calculations, where $T$ is given in K and $M$ in g/mol.

## EXAMPLES

1. The Lyman $\alpha$-line of the transition $2P \to 1S$ in the hydrogen atom ($M = 1$ g/mol): $\lambda = 121.6$ nm $\Rightarrow$ $\nu_0 = 2.47\times 10^{15}\,\mathrm{s}^{-1}$. At a temperature of $T =$ 1000 K in a hydrogen discharge the Doppler-width becomes $\delta\nu_D = 5.6\times 10^{10}\,\mathrm{s}^{-1} = 56$ GHz, $\delta\lambda_D =$ $2.8\times 10^{-3}$ nm.
2. The Na D-line of the transition $3P_{1/2} \to 3S_{1/2}$ in the sodium atom ($M = 23$ g/mol) has a wavelength $\lambda = 589.1$ nm and a frequency $\nu_0 = 5.1\times 10^{14}\,\mathrm{s}^{-1}$. At a temperature $T = 500$ K the Doppler-width is $\delta\nu_D = 1.7\times 10^9\,\mathrm{s}^{-1} = 1.7$ GHz $\Rightarrow \delta\lambda =$ $2\times 10^{-3}$ nm.
3. The infrared transition in the $CO_2$ molecule ($M = 44$ g/mol) at $\lambda \approx 10\,\mu$m ($\nu_0 = 3\times 10^{13}\,\mathrm{s}^{-1}$) has, at room temperature $T = 300$ K, a Doppler width $\delta\nu_D = 5.6\times 10^6\,\mathrm{s}^{-1} = 56$ MHz $\Rightarrow \delta\lambda =$ $1.3\times 10^{-3}$ nm.

These examples show that in the visible range the Doppler width exceeds the natural linewidth by about two orders of magnitude.
This is also true for vibrational transitions in the infrared, where the lifetimes of the vibrational levels are very long and the natural linewidths accordingly small.

This implies that generally the natural linewidth is completely masked by the much larger Doppler-width. Without special experimental tricks (see Sect. 11.5.8) it is therefore not possible to measure the natural linewidth directly. It is, however, possible to determine it from measured spontaneous lifetimes of excited states (see Sect. 7.3).

**Note:**

The Gaussian line profile of Doppler broadening decreases exponentially with increasing distance $|\omega-\omega_0|$ from the line center $\omega_0$, (see (7.73)), while the Lorentzian line profile decreases only as $(\omega-\omega_0)^{-2}$ (7.58). Therefore, it is possible to extract from the extreme line wings, information on the natural line profile in spite of the much larger Doppler-width (Fig. 7.21).

More detailed consideration shows that a Doppler-broadened spectral line cannot be strictly represented by a pure Gaussian profile as has been assumed in the foregoing discussion, since not all atoms with a definite velocity component $v_z$ emit or absorb radiation at the same frequency $\omega' = \omega_0(1 + v_z/c)$. Because of the finite lifetime of the atomic energy levels, the frequency response of atoms with a velocity component $v_z$ is represented by the Lorentzian profile (see Sect. 7.4.1)

$$L(\omega-\omega') = \frac{\gamma/2\pi}{(\omega-\omega')^2 + (\gamma/2)^2} \tag{7.73}$$

with a central frequency $\omega' = \omega_0(1 + v_z/c)$ (Fig. 7.22).

Let $n(\omega')\,\mathrm{d}\omega' = n(v_z)\,\mathrm{d}v_z$ be the number of atoms per unit volume with velocity components within the interval $v_z$ to $v_z + \mathrm{d}v_z$. The spectral intensity

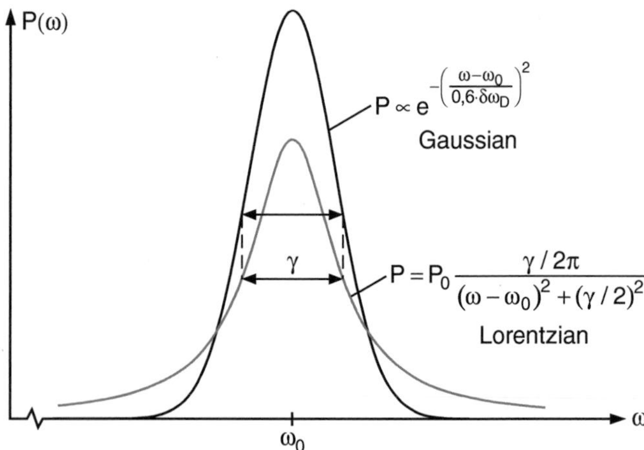


**Fig. 7.21.** Comparison between Lorentzian and Gaussian line profiles with equal half-widths

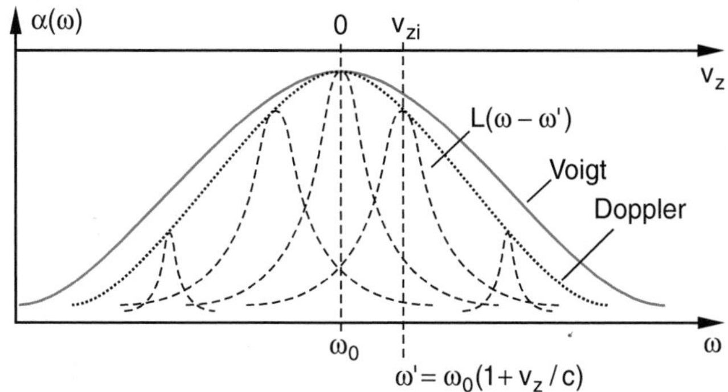


**Fig. 7.22.** Voigt profile as a convolution of Lorentzian line shapes $L(\omega_0 - \omega')$ with $\omega' = \omega_0(1 + v_z/c)$

distribution $I(\omega)$ of the total absorption or emission of all atoms on the transition $E_i \to E_k$ is then

$$I(\omega) = I_0 \int_0^\infty n(\omega')L(\omega - \omega')\,\mathrm{d}\omega'\,. \tag{7.74a}$$

Inserting (7.74a) for $L(\omega - \omega')\,\mathrm{d}\omega'$ and (7.70) for $n(\omega')$ we obtain

$$I(\omega) = C \int_0^\infty \frac{\mathrm{e}^{-[(c/v_\mathrm{p})(\omega_0-\omega')/\omega_0]^2}}{(\omega - \omega')^2 + (\gamma/2)^2}\,\mathrm{d}\omega' \tag{7.74b}$$

with

$$C = \frac{\gamma N_i c}{2 v_\mathrm{p} \pi^{3/2} \omega_0}\,.$$

This intensity profile, which represents a convolution of Lorentzian and Gaussian profiles (Fig. 7.22) is called a ***Voigt profile***. Such profiles play an important role for the spectroscopy of stellar atmospheres, where accurate measurements of line wings allow one to separate the contributions of Doppler-broadening and natural linewidths or collisional broadening. From such measurements, the temperatures and the pressures of the atmospheric layers of stars or planets from which the radiation is emitted or where radiation from the inner part of the star is absorbed can be deduced.

### 7.4.3 Collision Broadening

If an atom A, with energy levels $E_i$ and $E_k$, approaches another atom or molecule B, the interaction between A and B shifts the energies of both levels. This energy shift depends on the structure of the electron shell of both partners A and B, on the specific energy levels and on the mutual distance $R$(A,B), which we define as the distance between the centers of mass of A and B. The energy shift generally differs for the different levels. It may be positive (for repulsive potentials between A and B) or negative (for attractive potentials). Plotting the energies $E_i(R)$ and $E_k(R)$ of atom A as a function of the distance $R$ we obtain potential curves, such as those schematically drawn in Fig. 7.23. The approach of two particles to a distance $R_\mathrm{c}$, where the interaction energy becomes noticeable (this means that the potential curves deviate noticeably from $E(R = \infty)$), is called a two-body collision and the system AB($R$) is called a collision pair. If the densities of A and B are not too high, the probability that three collision partners approach each other simultaneously within $R < R_\mathrm{c}$ (three body collision) is very small and we can neglect it.

The distance $R_\mathrm{c}$ where the interaction becomes noticeable is the collision radius. If the relative velocity of A and B is $v$, the duration of the collision (also called the collision time) can be defined as $\tau_\mathrm{coll} = R_\mathrm{c}/v$.

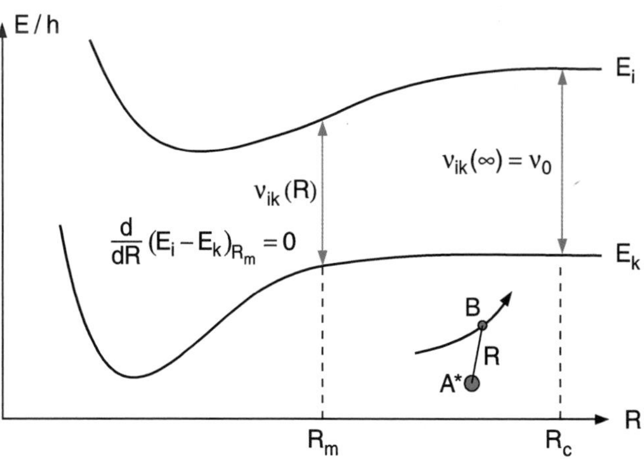


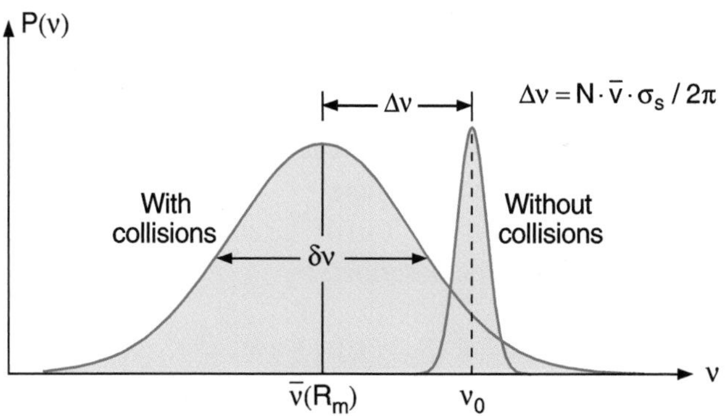


**Fig. 7.23a,b.** Collision broadening. (**a**) Potential curves of the collision pair AB and A*B. (**b**) Shift and broadening of a spectral line by collisions

EXAMPLE

At thermal velocities of $v = 5 \times 10^2$ m/s and a typical collision radius $R_c = 1$ nm the collision time becomes $\tau_{coll} = 2 \times 10^{-12}$ s. This illustrates that the collision times are very short.

Note, that one has to distinguish between the collision time and the mean time $\langle\tau\rangle = n^{-1/3}/v$ between two collisions, which is proportional to the average distance $d = n^{-1/3}$ between two atoms and therefore depends on the density $n$ of the collision partners. At sufficiently low pressures, $\langle\tau\rangle$ is much longer than $\tau_{coll}$.

EXAMPLE

At a pressure of 1 mbar the density is about $3 \times 10^{22}$ m$^{-3}$. With a relative velocity $v = 5 \times 10^2$ m/s we obtain $\langle\tau\rangle \approx 3 \times 10^{-11}$ s.

The frequency

$$\nu_{ik} = |E_i(R) - E_k(R)|\,/h$$

of the radiation emitted or absorbed on the transition $E_i \to E_k$ during a collision depends on the distance $R$ at which the transition takes place (we assume here that the duration of the radiative transition is short compared with the collision time, which means that in Fig. 7.23a, the transition can be drawn as a vertical line). In a gas mixture, the distances $R$ show random fluctuations with a distribution around a mean value $R_m$, which depends on the pressure and temperature of the gas. The frequencies of absorbed or emitted radiation show a corresponding distribution around a most probable value $\nu_{ik}(R_m)$, which may be shifted against the center frequency $\nu_0$ of the unperturbed atom A. The shift $\Delta\nu = \nu_0 - \nu_{ik}(R_m)$ depends on how differently the two energy levels $E_i$ and $E_k$ are shifted at a distance $R_m$(A,B) where the transition probability for emission or absorption has a maximum.

The intensity profile $I(\omega)$ of the collision broadened emission line can be described by

$$I(\omega) \propto \int A_{ik}(R) P_{coll}(R)\; [E_i(R) - E_k(R)]\; \mathrm{d}R\,, \tag{7.75}$$

where $A_{ik}(R)$ is the spontaneous transition probability, which depends on $R$ because the electronic wave functions of the collision pair (AB) depend on $R$. The probability $P_{coll}(R)$ that the distance $R$ lies between $R$ and $R + \mathrm{d}R$ depends on the interaction potential, the density and the temperature of the gas. It can be derived in the following way:

The number of particles B in a spherical shell with radius $R$ around atom A is

$$N_B(R)\,\mathrm{d}R = n_0 4\pi R^2\,\mathrm{d}R \mathrm{e}^{-E_{pot}(R)/kT}\,, \tag{7.76}$$

where $n_0$ is the average density of atoms B. The Boltzmann factor $\exp(-E_{pot}/kT)$ takes into account that the energy of the collision pair depends on the potential interaction energy $E_{pot}(R)$. The probability $P(R)$ is $(N(R)/n_0)$. Inserting this into (7.75) gives the intensity profile of the absorption line

$$I(\omega)\,\mathrm{d}\omega = C^* \left[R^2 \mathrm{e}^{-E^{(i)}_{pot}(R)/kT} \frac{\mathrm{d}}{\mathrm{d}R}\left(E_i(R) - E_k(R)\right)\right] \mathrm{d}R\,. \tag{7.77}$$

Measuring the line profile as a function of the temperature gives

$$\frac{\mathrm{d}I(\omega, T)}{\mathrm{d}T} = \frac{E^{(i)}_{pot}(R)}{kT^2} I(\omega, T)\,. \tag{7.78}$$

This allows the determination of the ground state potential seperately!

From the temperature dependence of the line profile, one can determine the interaction potential for the two corresponding states separately, while from measurements at constant temperature only the difference $\Delta E(R)$ can be obtained.

Frequently, spherical model potentials are substituted into (7.77), such as the Lenard–Jones potential

$$E_{pot}(R) = \frac{a}{R^{12}} - \frac{b}{R^6}\,. \tag{7.79}$$

The coefficients $a$ and $b$ are adjusted for optimum agreement between theory and experiment.

So far we have only discussed elastic collisions, where the energy level of the atom A is only shifted during the collision, but returns to its initial value after the collision, unless the atom has emitted a photon during the collision. The shift of absorption or emission lines caused by elastic collisions corresponds to an energy shift $\Delta E = \hbar\Delta\omega$ between the excitation energy

$\hbar\omega_0$ of the free atom A* and the energy $\hbar\omega$ of the emitted photon. This energy difference is supplied from the kinetic energy of the collision partners. In the case of positive shifts ($\Delta\omega > 0$), the kinetic energy is smaller after the collision than before.

Besides these elastic collisions, inelastic collisions may also occur in which the excitation energy of atom A is either partly or completely transferred into internal energy of the collision partner B, or into translational energy of both partners A and B. Such inelastic collisions are called ***quenching collisions***, because they decrease the number of excited atoms A in level $E_i$ and therefore quench the fluorescence intensity. The total transition probability $A_i$ for the depopulation of level $E_i$ is the sum of radiative and collision-induced probabilities

$$A_i = A_i^{\text{rad}} + A_i^{\text{coll}} \quad \text{with} \quad A_i^{\text{coll}} = n_{\text{B}}\sigma_i\overline{v}\,. \tag{7.80}$$

Inserting the relations

$$\overline{v} = \sqrt{\frac{8kT}{\pi\mu}}\,, \quad \mu = \frac{M_{\text{A}}M_{\text{B}}}{M_{\text{A}}+M_{\text{B}}}\,,$$

$$\text{and} \quad p_{\text{B}} = n_{\text{B}}kT$$

between the mean relative velocity $v$, the partial pressure $p$ of the particles B with number density $n_{\text{B}}$ and the gas temperature $T$ gives the total transition probability

$$A_i = \frac{1}{\tau_i^{\text{spont}}} + ap_{\text{B}} \quad \text{with} \quad a = 2\sigma_{ik}\sqrt{\frac{2}{\pi\mu kT}}\,, \tag{7.81a}$$

which is identical to (7.50) for the lifetime dependence on pressure.

It is evident from (7.64) that this pressure-dependent transition probability causes a corresponding pressure dependent linewidth $\delta\omega$, because it shortens the effective lifetime of the excited level. It can be written as the sum of two damping terms

$$\delta\omega = \delta\omega_n + \delta\omega_{\text{coll}} = \gamma_n + \gamma_{\text{coll}} = \gamma_n + ap_{\text{B}}\,. \tag{7.81b}$$

The collision-induced additional line broadening $ap_{\text{B}}$ is therefore often called ***pressure broadening***.

In a classical model, which describes the excited atom A by a damped harmonic oscillator, the inelastic collisions decrease the amplitude of the oscillation.

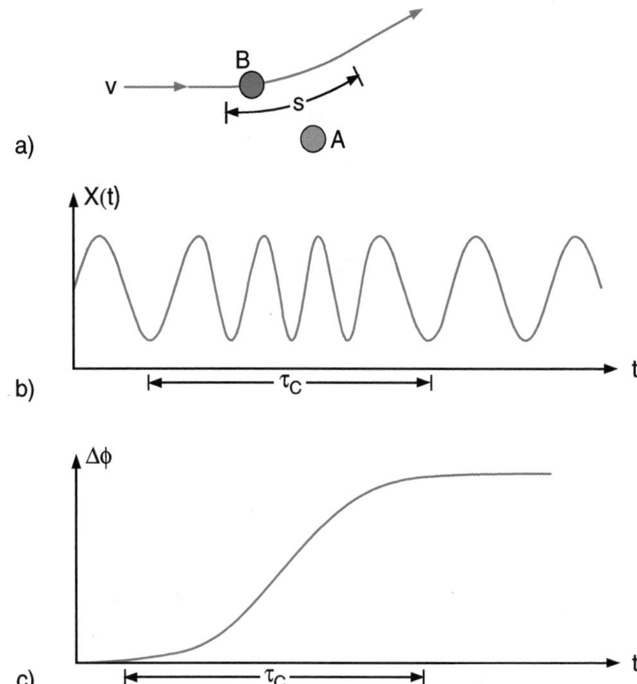


**Fig. 7.24a–c.** Elastic collisions as phase perturbers. (**a**) Classical path of the collision partner B. (**b**) Frequency shift of A during the collision. (**c**) Resulting phase shift

Although the elastic collisions do not change the amplitude, they do change the phase of the oscillation (Fig. 7.24). Because the energy difference between the corresponding states of the atom A changes during the elastic collision time, the frequency $\omega$ of the emitted or absorbed radiation changes by $\Delta\omega(R)$. This change depends on the interaction potential between A and B. Although after the collision the frequency again takes its initial value, the phase-change is (Fig. 7.24)

$$\Delta\varphi = \int_0^\infty \Delta\omega(t)\,\mathrm{d}t\,. \tag{7.82}$$

Elastic collisions are therefore called *phase-changing collisions*. Since the frequency differences $\Delta\omega$ differ for collisions with different distances $R(\text{AB})$, the ensemble of atoms A has suffered random phase changes. The Fourier transform of the radiation gives a Lorentzian line profile, which is broadened and shows a shift of its line center (Fig. 7.23).

The line profile caused by elastic and inelastic collisions is obtained after a somewhat lengthy calculation [7.8] as

**Table 7.3.** Line broadening $\delta\nu_B$ and line shift $\Delta\nu_S$ (in MHz/Pa) for some transitions in alkali atoms colliding with noble gas atoms

| Atomic Transition | Collision partner | | | | | |
|---|---|---|---|---|---|---|
| | He | | Ar | | Xe | |
| | $\delta\nu_b$ | $\Delta\nu_s$ | $\delta\nu_b$ | $\Delta\nu_s$ | $\delta\nu_b$ | $\Delta\nu_s$ |
| Na: $3S_{1/2} \leftrightarrow 3P_{1/2}$ $\lambda = 589.6$ nm | 0.07 | 0.0 | 0.1 | −0.05 | 0.13 | −0.07 |
| K: $4S_{1/2} \leftrightarrow 4P_{1/2}$ $\lambda = 769.9$ | 0.06 | 0.02 | 0.1 | −0.09 | 0.12 | −0.07 |
| Cs: $6S_{1/2} \leftrightarrow 6P_{1/2}$ $\lambda = 894.4$ | 0.08 | 0.05 | 0.08 | −0.07 | 0.09 | −0.06 |

$$I(\omega) = \frac{\left(\frac{\gamma_n+\gamma_{\text{inel}}}{2} + N\overline{v}\sigma_b\right)^2}{(\omega - \omega_0 - N\overline{v}\sigma_s)^2 + \left(\frac{\gamma_n+\gamma_{\text{inel}}}{2} + N\overline{v}\sigma_b\right)^2} . \tag{7.83a}$$

The cross sections

$$\sigma_b = 2\pi \int_0^\infty (1 - \cos\varphi(R))R\,\mathrm{d}R \tag{7.83b}$$

$$\sigma_s = 2\pi \int_0^\infty \sin\varphi(R))R\,\mathrm{d}R \tag{7.83c}$$

are a measure for the line broadening ($\delta v_b$) and the line shift ($\Delta v_s$) by elastic collisions (Table 7.3).

> Both elastic and inelastic collisions result in the broadening of spectral lines. Elastic collisions cause, in addition, a shift of the line center.

The classical models of pressure broadening and shifts can be extended to a more general quantum mechanical treatment [7.9, 10]. This, however, exceeds the scope of this textbook.

## 7.5 X-Rays

In the year 1895, in the German city of Würzburg, *Wilhelm Conrad Röntgen* (1845–1923) (Fig. 7.25) discovered, while experimenting with gas discharge tubes developed by *Phillip Lenard*, that radiation was emitted from these tubes that could penetrate materials such as glass, wood or human tissue. Since he did not know much about the nature of this radiation, he called it X-ray radiation.

**Fig. 7.25.** *Wilhelm Conrad Röntgen* (From E. Bragge; Die Nobelpreisträger, Heinz-Moos-Verlag, München 1964)

In the following years, the importance of these X-rays for material inspection and in particular for medical diagnostics, soon became evident and in 1901 Röntgen was the first physicist to receive the Nobel Prize. In the following 100 years the applications of X-rays have vastly increased [7.11], ranging from material sciences, sterilization of food, numerous medical applications culminating in the development of X-ray tomographs and discoveries of new phenomena in the universe using X-ray astronomy [7.12, 13, 14].

The basic principle of an X-ray tube is shown in Fig. 7.26. Electrons are emitted from a heated cathode, accelerated by a voltage $U$ and impinge onto an anode. In the anode material (for example tungsten) the electrons are stopped and part of their energy $eU$ is converted into radiation with short wavelengths. These X-rays leave the tube through a window and can be collimated by proper apertures.

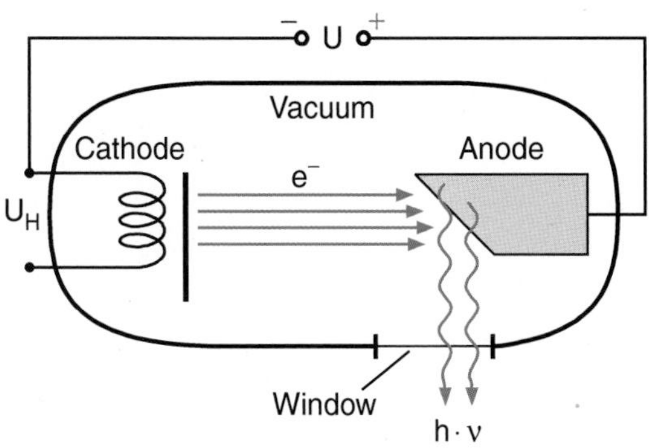


**Fig. 7.26.** Schematic illustration of an X-ray tube

X-rays are produced by two different effects:

1. By decelerating energetic electrons (ranging from keV to MeV) *bremsstrahlung* is produced with a continuous spectral intensity distribution $I(\lambda)$, which depends on the energy of the electrons.
2. The energetic electrons can excite inner shell electrons on transitions to higher non occupied states of in the atoms of the anode. The excited states $E_i$ emit X-rays as spectral lines on transitions $E_i \rightarrow E_k$ with wavelengths $\lambda_{ik}$, characteristic for the anode material. These X-rays are therefore called characteristic X-ray radiation.

We will now discuss both effects in more detail.

### 7.5.1 Bremsstrahlung

Energetic electrons passing through a material with high nuclear charge numbers $Z$, are deflected in the Coulomb field of the positive nuclear charge (Fig. 7.27a). Since, according to electrodynamics, every accelerated or decelerated charge emits radiation with a radiation power proportional to the square of the acceleration, these electrons emit a broad radiation continuum with an intensity distribution depending on the initial electron energy (Fig. 7.28). The high energy limit of this continuum is reached when the total energy $E_{kin} = eU$ of the incident electrons is converted into radiation. This gives the condition

$$h\nu \leq h\nu_{max} = eU \Rightarrow \lambda \geq \lambda_{min} = \frac{hc}{eU} . \tag{7.84}$$

Inserting the quantitative values for $h$, $c$ and $e$ into (7.84) yields the more readily calculable form

$$\boxed{\lambda_{min} = 1234.5(U[\mathrm{V}])^{-1}\,\mathrm{nm}} . \tag{7.85}$$

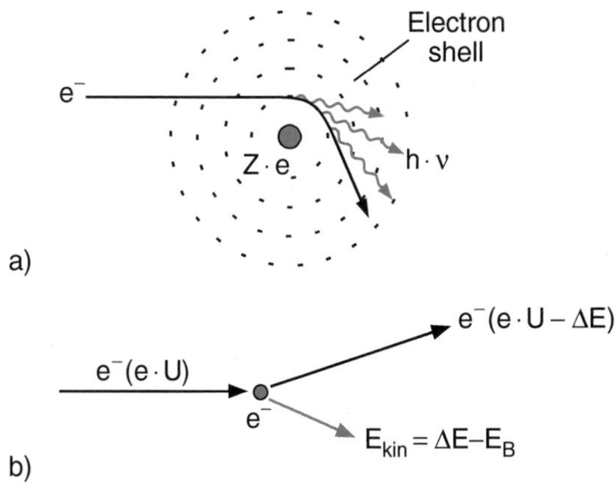


**Fig. 7.27a,b.** The origin of the continuous X-ray radiation (bremsstrahlung). (**a**) Deflection of an incident electron in the Coulomb field of the nucleus of target atoms. (**b**) Inelastic collision of an incident electron with an electron in the shell of a target atom

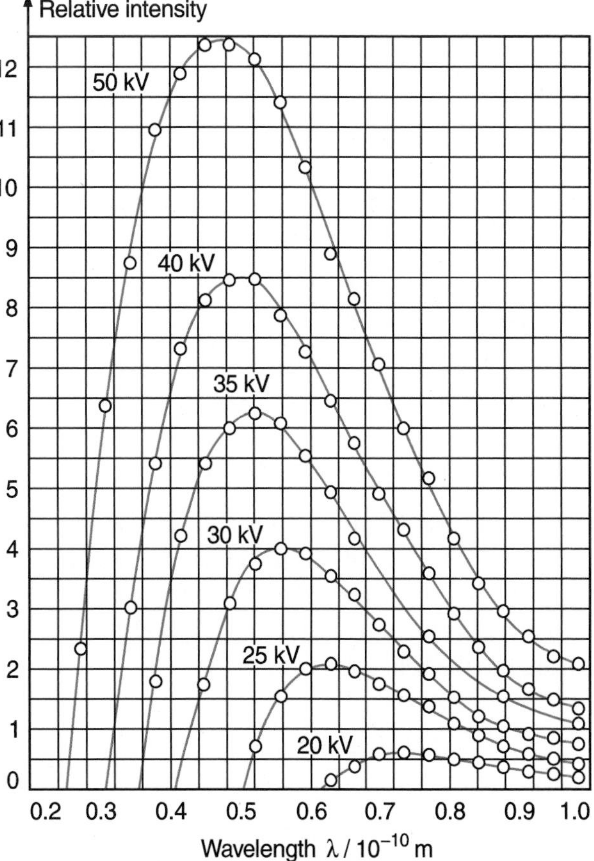


**Fig. 7.28.** Spectral distribution of the bremsstrahlung in a tungsten target for different voltages $U$

**EXAMPLE**

$U = 10\,\text{kV} \Rightarrow \lambda_{min} = 0.12\,\text{nm}$, or $\nu_{max} = 2.5 \times 10^{18} s^{-1}$.
$U = 50\,\text{kV} \Rightarrow \lambda_{min} = 0.024\,\text{nm}$, or $\nu_{max} = 1.25 \times 10^{19} \text{s}^{-1}$.

Of course, the incident electrons can also collide with the electrons in the atomic shells of the anode material (Fig. 7.27b). The outer electrons with a small binding energy are kicked away by collisions with the high energy incident electrons, they collide further with electrons of other atoms and finally convert their energy into heat. Collisions with more tightly bound inner electrons lead to excitations into higher, but still bound, atomic states. This excitation energy is transferred to the characteristic X-ray fluorescence emitted by the excited atoms.

Both contributions represent electromagnetic radiation as was first proved by the English physicist *Charles Glover Barkla* (1877–1944, Nobel Prize 1917), who measured the polarization of the X-radiation. Further convincing proofs were given by the German physicist *Max von Laue* and his assistants *W. Friedrich* and *P. Knipping* and later by the English physicists *William Henry Bragg* (1862–1942) (Fig. 7.29) and his son *William Lawrence Bragg* (1890–1971) who together received the Nobel Prize in 1915. They measured interference and diffraction patterns when single crystals were illuminated by X-rays, which demonstrated that X-rays were electromagnetic waves.

**Fig. 7.29.** *William Henry Bragg* (From E. Bragge; Die Nobelpreisträger der Physik, Heinz Moos-Verlag, München 1964)

### 7.5.2 Characteristic X-Ray-Radiation

The spectral lines of the characteristic X-ray radiation appear only if the energy of the electrons, incident on the anode, is sufficiently high to excite atomic inner shell electrons into higher unoccupied levels (Fig. 7.30) according to the scheme

$$\text{e}^-(E_{\text{kin}}) + \text{A}(E_k) \Rightarrow \text{A}^*(E_i) + \text{e}^-\left(E'_{\text{kin}}\right) \quad (7.86\text{a})$$
$$\text{with} \quad E_{\text{kin}} - E'_{\text{kin}} = E_i - E_k$$
$$\text{A}^*(E_i) \Rightarrow \text{A}(E_k) + h\nu_{ik} \quad (7.86\text{b})$$
$$\text{with} \quad h\nu_{ik} = E_i - E_k\,.$$

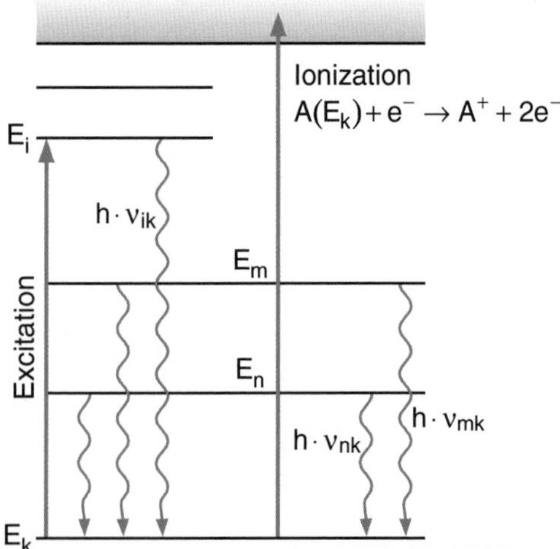


**Fig. 7.30.** Level scheme for the explanation of the characteristic X-ray emission from the anode atoms

EXAMPLE

$E_k(\mathrm{Cu}(1s)) = -8978\,\mathrm{eV}$ (binding energy of an electron in the $1s$ level of the K-shell); $E_i(\mathrm{Cu}(6p)) = -4\,\mathrm{eV}$. The incident electrons can therefore only excite electrons in the K-shell into the level $E_i$, which emits the Cu-K-radiation, if their kinetic energy is above 8974 eV.

The characteristic X-ray radiation appears as sharp lines superimposed on the continuous spectral background of the bremsstrahlung (Fig. 7.31). With increasing voltage $U$ between the cathode and anode, the ratio $\eta = P_{\mathrm{char}}/P_{\mathrm{cont}}$ of the emitted powers of characteristic to continuous radiation increases (Fig. 7.31b). However, even at a voltage $U = 250\,\mathrm{kV}$ it is only about 0.1 for a tungsten anode.

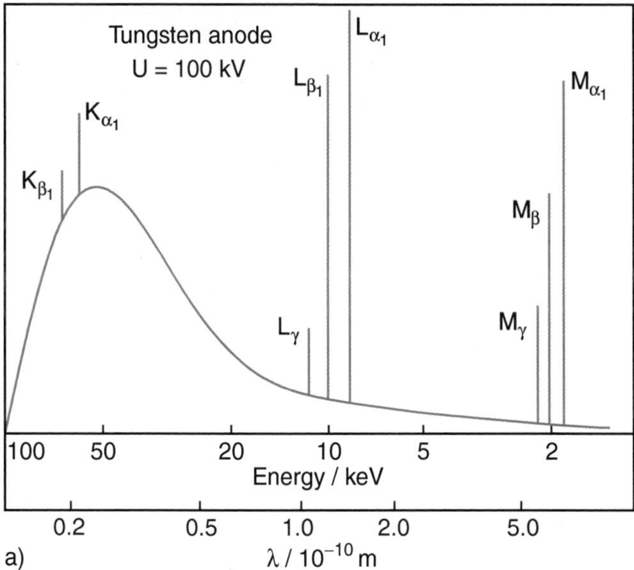


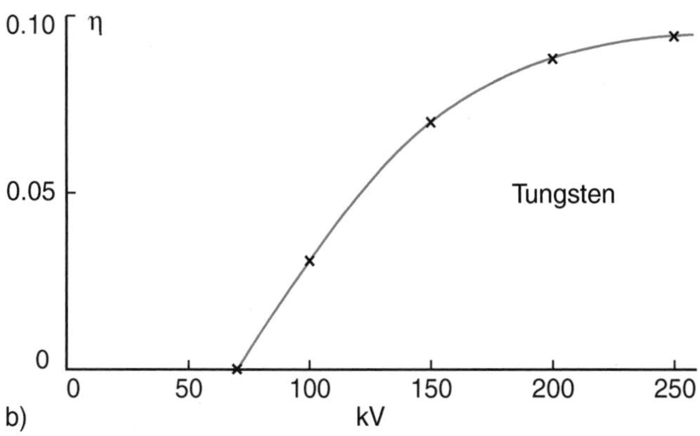


**Fig. 7.31.** (**a**) Continuous X-ray radiation, superimposed by the characteristic lines of tungsten. (**b**) Ratio $\eta = P_{\mathrm{char}}/P_{\mathrm{cont}}$ of emitted characteristic and continuous X-ray radiation power from a tungsten anode as a function of applied voltage $U$

### 7.5.3 Scattering and Absorption of X-Rays

When a parallel X-ray beam passes through a sample in $x$ direction (Fig. 7.32) the transmitted radiation power $P(x)$ decreases after the pathlength d$x$ *by*

$$\mathrm{d}P = -\mu P\,\mathrm{d}x\,. \tag{7.87a}$$

The constant factor $\mu$, which depends on the material of the sample is the ***attenuation coefficient***. Integration over a sample thickness $x$ gives

$$P(x) = P_0 \mathrm{e}^{-\mu x}\,. \tag{7.87b}$$

The attenuation has two reasons: scattering and absorption, which may be both simultaneously effective. The attenuation coefficient can therefore be written as the sum

$$\mu = \mu_s + \alpha$$

of ***scattering coefficient*** $\mu_s$ and ***absorption coefficient*** $\alpha$.

The scattering is produced by the atomic electrons of the sample. In a classical model, each electron can be regarded as a harmonic oscillator induced to forced oscillations under the influence of the incident electromagnetic wave with frequency $\omega$ (Hertzian oscillating dipole). Assume that the electric field vector of the incident wave points in the $y$ direction. Then the oscillating dipole is described by $y = y_0 \sin \omega t$. The amplitude $A$ of the radiation emitted by the oscillator is proportional to the acceleration

$$\ddot{y} = -\omega^2 y_0 \sin \omega t = -\omega^2 y\,. \tag{7.87c}$$

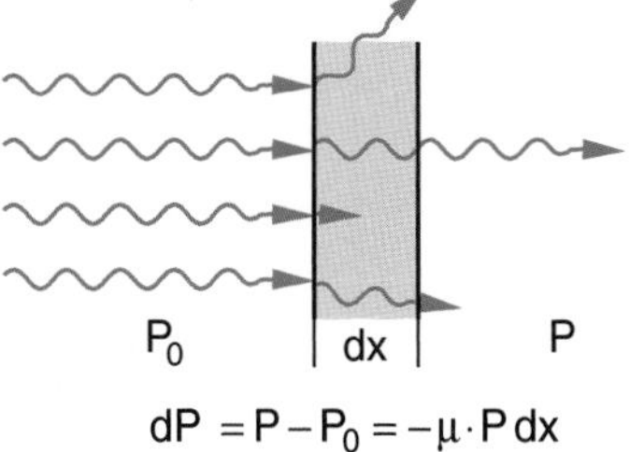


**Fig. 7.32.** Absorption and scattering of X-rays in matter

The emitted power, which is proportional to the absolute square of the amplitude, scales therefore with $\omega^4$! In electrodynamics it is shown that the power emitted by an oscillating dipole is

$$P_s(\vartheta) = \frac{e^2 y_0^2 \omega^4}{32\pi^2 \varepsilon_0 c^3} \sin^2\vartheta \,, \tag{7.87d}$$

where $\vartheta$ is the angle between the dipole axis (in our example the $y$ direction) and the direction of observation. The total scattered power emitted by $N$ scatterers in a solid sample where all atoms stay at their positions, depends on the ratio of wavelength $\lambda$ to the diameter $d$ of the scattering region. If $d \ll \lambda$, the phases of the waves scattered by the different atoms differ only by an amount $\Delta\varphi \ll 2\pi$. All waves interfere coherently and the total amplitude of the scattered wave is proportional to $N$, which implies that the scattered power is proportional to $N^2$! (coherent scattering). If the diameter $d$ is comparable or larger than $\lambda$, constructive and destructive interference occurs and for a non-periodical random arrangement of the scatterers, the total scattered power only scales with $N$ (incoherent scattering).

The scattering can be described by a scattering cross section $\sigma_s$, which defines the area around a scattering atom through which an incident photon has to pass in order to be scattered. If the number density of scatterers is $n$ the scattering coefficient $\mu_s$ is related to the scattering cross section $\sigma_s$ by

$$\mu_s = n\sigma_s, \tag{7.87e}$$

with a thickness $t$ of the scattering sample and an incident power $P_0$ the scattered power is

$$P_s = P_0\left(1 - e^{-\mu_s t}\right) \approx P_0 \cdot \mu_s \cdot t \text{ for } \mu_s \cdot t \ll 1 \,. \tag{7.87f}$$

> The scattering cross section scales with $\omega^4$ or $\lambda^{-4}$. It is therefore much larger in the X-ray region than in the visible range.

## EXAMPLE

When visible radiation with $\lambda = 500\,\text{nm}$ passes through clear water it is attenuated by scattering to $1/e$ of its initial intensity only after a path length of 1 km, while the intensity of X-rays with $\lambda = 0.1\,\text{nm}$ is already reduced to $1/e$ after a 5 mm path length!

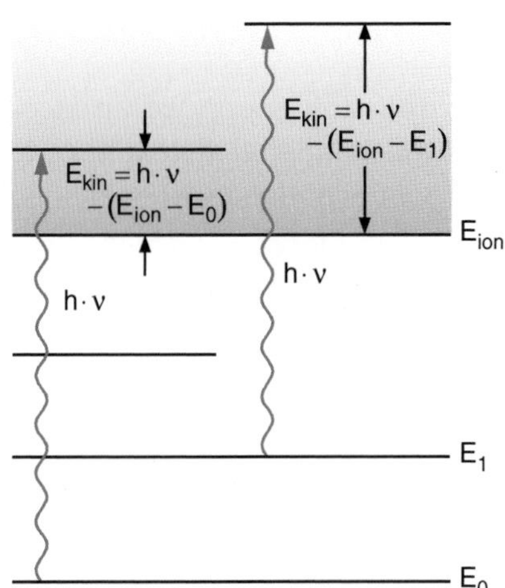


**Fig. 7.33.** The photoeffect

Besides elastic scattering where the wavelength of the scattered radiation is the same as that of the incident radiation, also inelastic scattering can occur (Compton effect, see Sect. 3.1.7). Here the scattered photon with $\nu' < \nu$ may be either scattered again or it may be absorbed. This leads to a complete absorption of the incident power, if the thickness of the sample is sufficiently large.

The absorption of X-rays strongly depends on the absorbing material. It is caused by three different effects.

***Photoeffect***: The X-ray quantum $h\nu$ is absorbed by an atom in the sample in the state $E_k$. This leads for $h\nu > |E_k|$ to photoionization of an inner shell electron (Fig. 7.33) and can be written as

$$h\nu + \mathrm{A}(E_k) \rightarrow \mathrm{A}^+(E_\text{ion}) + e^-(E_\text{kin}) \,. \tag{7.88}$$

Energy conservation demands the relation

$$E_\text{kin}(e^-) = h\nu - (E_\text{ion} - E_k) \,. \tag{7.89}$$

***Compton effect***: The X-ray quantum $h\nu$ "collides" with a "nearly free" electron, which has a binding energy $E_b \ll h\nu$ (see Sect. 3.1.7), transfers only part of its energy onto the electron, and is inelastically scattered according to the scheme:

$$h\nu + e^- \rightarrow h\nu' + e^-(E_\text{kin}) \,. \tag{7.90}$$

With $h(\nu - \nu') = E_\text{kin}(e^-) + E_\text{bind}$ (Fig. 7.34a). The inelastically scattered photon $h\nu'$ can be absorbed by other atoms of the sample (photoeffect).

***Pair formation***: For sufficiently high energies $h\nu > 1\,\text{MeV}$ the X-ray quantum can produce in the material an electron positron pair (Fig. 7.34b)

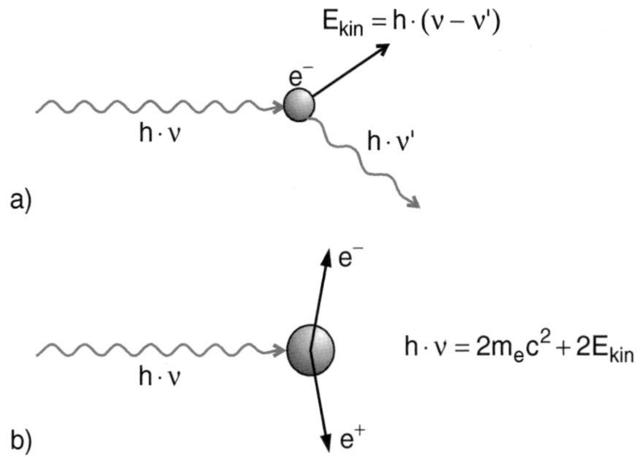


**Fig. 7.34.** (**a**) The Compton effect. (**b**) Pair formation

$$h\nu \to e^- + e^+ + 2E_{kin} \tag{7.91}$$

with $h\nu = 2m_ec^2 + E_{kin}(e^-) + E_{kin}(e^+)$. Each of the two particles must have the same kinetic energy because they have equal masses. This follows immediately from the conservation of momentum.

The relative share of the three processes contributing to the absorption of X-rays strongly depends on the energy $h\nu$. In Fig. 7.35 the contributions of the three processes are plotted as a function of the photon energy. This illustrates that for lead the photoeffect is the dominant absorption process for energies $h\nu <$ 500 keV, but decreases sharply with increasing photon energy. The total absorption cross section has a minimum at photon energies around 3 MeV.

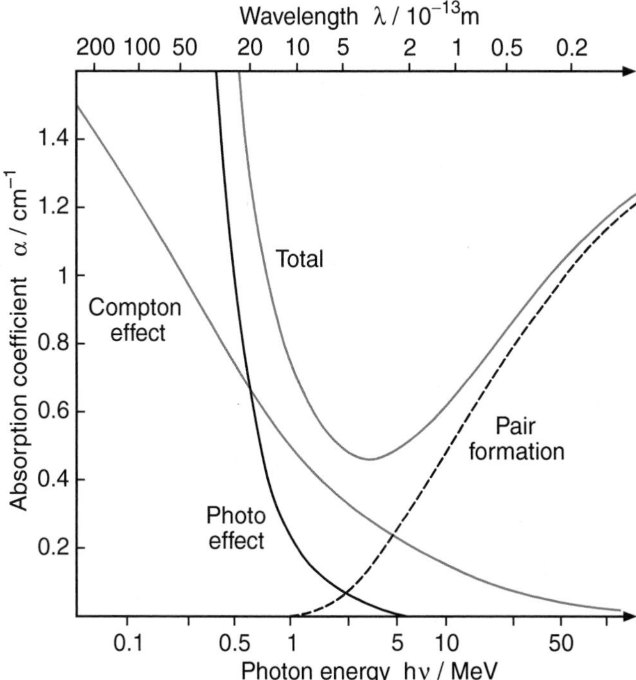


**Fig. 7.35.** Contributions of the photoeffect, the Compton effect and pair formation to the absorption coefficient of X-rays in lead and their dependence on the photon energy

The absorption coefficient

$$\alpha = n\sigma_a \tag{7.92}$$

is the product of particle density n and absorption cross section $\sigma_a$. Often the attenuation of incident X-rays is related to the mass of the absorbing material rather than to its thickness. With a mass density $\rho$ the product $\rho x$ gives the mass per unit area that is traversed by the photon along the pathlength $x$. The equation

$$e^{-\alpha x} = e^{-(\alpha/\rho)\rho x} = e^{-\kappa_a \cdot \rho \cdot x} \tag{7.93}$$

relates the absorption coefficient $\alpha[\mathrm{m}^{-1}]$ with the mass-absorption coefficient

$$\kappa_a = \frac{\alpha}{\rho}\,; \quad [\kappa_a] = 1\,\frac{\mathrm{m}^2}{\mathrm{kg}}\,.$$

The ratio $x_e = 1/(\kappa_a \cdot \rho)$ gives the pathlength $x_e$ after which the intensity of the X-rays has decreased to $1/e$, while $1/\kappa_a$ gives that mass per $\mathrm{cm}^2$ area perpendicular to the X-ray direction that decreases the incident intensity to $1/e$. The mass attenuation coefficient $\kappa_a$ depends on the material and on the photon energy $h\nu$ (Table 7.4).

## EXAMPLE

Lead has a density $\rho = 11.3 \times 10^3\,\mathrm{kg/m^3}$. For X-rays with $\lambda = 0.1\,\mathrm{nm}$ ($\approx 12\,\mathrm{keV}$) the mass-absorption coefficient is $\kappa_a = 7.5\,\mathrm{m^2/kg}$. One therefore needs a mass area density of $0.61\,\mathrm{kg/m^2}$ in order to attenuate the incident X-rays down to $1\% = e^{-4.6}$. This corresponds to a thickness $t = 54\,\mu\mathrm{m}$ of a lead sheet. For X-rays with $\lambda = 0.01\,\mathrm{nm}$ ($\approx 120\,\mathrm{keV}$) is $\kappa_a = 0.5\,\mathrm{m^2/kg}$. Now a thickness of $t = 0.8\,\mathrm{mm}$ is necessary for the attenuation down to 1%.

**Table 7.4.** Mass absorption coefficient $\kappa/(\mathrm{m^2/kg})$ of different absorbing materials for X-rays with photon energies $h\nu$/keV and wavelengths $\lambda$/pm

| $h\cdot\nu$ | $\lambda$ | Air | $H_2O$ | Al | Cu | W | Pb |
|---|---|---|---|---|---|---|---|
| 5 | 246 | 2 | 2.0 | 25 | 24 | 70 | 100 |
| 10 | 123 | 0.5 | 0.52 | 2.6 | 22.4 | 9.53 | 13.7 |
| 50 | 25 | 0.02 | 0.92 | 0.04 | 0.26 | 0.6 | 0.8 |
| 100 | 12 | 0.015 | 0.017 | 0.02 | 0.05 | 0.4 | 0.6 |

The mass-absorption coefficient of a sample with $n$ atoms per $m^3$ can be written as the ratio

$$\kappa_a = \frac{\alpha}{\rho} = \frac{n\sigma_a}{\rho} = \frac{\sigma_a}{m_a} \tag{7.94}$$

of absorption cross section $\sigma_a$ and atomic mass $m_a = \rho/n$, where the absorption cross section

$$\sigma_a = CZ^4\lambda^3 \tag{7.95}$$

strongly depends on the number $Z$ of electrons of each atom and the wavelength $\lambda$.

Experiments show that the absorption cross section $\sigma_a$ is proportional to the fourth power of the nuclear charge $Z$ and the third power of the wavelength $\lambda$.

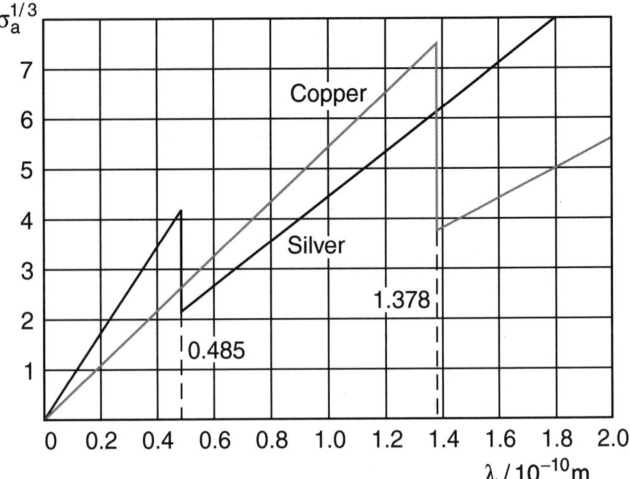


**Fig. 7.36.** Wavelength dependence of absorption coefficient $(\sigma_a)^{1/3}$ showing the absorption K-edges of copper and silver

The constant $C$ depends on the absorbing material, e.g., the packing density of the absorbing atoms and the number of electrons per atom. For absorbing molecular samples the absorption cross sections of the molecules is the sum of the atomic cross sections $\sigma_a$

$$\sigma_{am} = \sum \sigma_{ai} \, .$$

Lead ($Z = 82$) attenuates X-rays, because of the $Z^4$-dependence, about 1580 times more than an equal path length in aluminum ($Z = 13$), and still 100 times more than iron ($Z = 26$). The mass absorption coefficients, however, are only proportional to $Z^3$, because the atomic masses scale with $Z$. It is:

$$n \cdot Z \approx \frac{1}{2} n \cdot A \propto n \cdot m_a$$
$$\Rightarrow \kappa_a = \sigma_a / m_a \propto \sigma_a / Z \, .$$

For equal masses per irradiation area lead attenuates about 30 times more than iron.

Plotting the cubic root of the absorption cross section $\sigma_a^{1/3}$ as a function of $\lambda$ over a large wavelength range (Fig. 7.36) one finds a straight line as predicted by the relation (7.95). However, at certain wavelengths $\lambda_k$, characteristic for the absorbing atoms, the absorption coefficient jumps suddenly and then follows a straight line again. This shows that below these wavelength $\lambda_k$, called the absorption edges, a new absorption channel is opened. The explanation of this behavior is given by Fig. 7.37, which shows schematically the energy level diagram for absorbing transitions of an

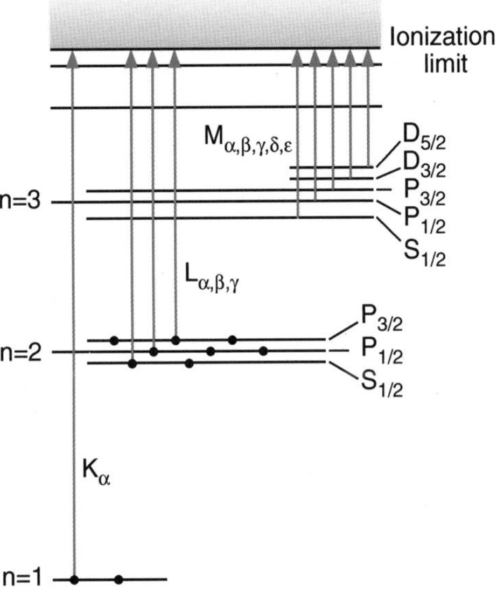


**Fig. 7.37.** Energy level diagram for the explanation of the absorption edges

atom with ionization energy $E_I$. For photon energies $h\nu_k < E_I - E_k$ electrons in atomic shells with principal quantum numbers $n > k$ can be ionized. These electrons can all contribute to the absorption cross section. When the photon energy exceeds the value $h\nu_k$ the electrons in shell $n = k$ can be additionally ionized and therefore add to the absorption cross section, which means that a new absorption channel is opened. According to the new electron shell accessible to absorption of the X-ray quanta, the absorption edges are called K-, L-, or M-edges. Equation (7.95) can be adapted to the real wavelength dependence of the absorption cross section by fitting the constants $C$ for each section between two absorption edges. This yields different values of $C$ for the different sections.

Plotting the square root $\bar{\nu}_k^{1/2}$ of the reciprocal wavelength $\bar{\nu}_k = 1/\lambda_k$ against the nuclear charge number $Z$ (Fig. 7.38) yields the approximate relation

$$\bar{\nu}_k = K_n(Z-S)^2 \Rightarrow \nu_k = cK_n(Z-S)^2 \,, \quad (7.96)$$

where $K_n$ is a constant depending on the principal quantum number $n$, and $S$ is the shielding factor of the nuclear charge $Z \cdot e$ (see Sect. 6.1.1). These relations, discovered empirically by *Henry G.J. Moseley* (1887–1915), can be immediately explained by the term energies of the two corresponding levels with $n = k$ and $n = i$ of the absorbing transition

$$h\nu_{ik} = (Z-S)^2 Ryhc\left(\frac{1}{n_k^2} - \frac{1}{n_i^2}\right) \quad (7.97a)$$

$$\Rightarrow \bar{\nu}_{ik} = (Z-S)^2 Ry\left(\frac{1}{n_k^2} - \frac{1}{n_i^2}\right) \,. \quad (7.97b)$$

Here $Z_{\text{eff}}e = (Z-S)e$ is the effective nuclear charge, which is the real charge $Ze$, partly shielded by the inner electrons, described by the shielding factor $S$ (see Sect. 6.1). If the upper level is above the ionization energy ($n_i = \infty$) (7.97) simplifies to

$$\bar{\nu}_k = \frac{Ry}{n_k^2}(Z-S)^2 \quad (7.97c)$$

and the K-edge with $n_k = 1$ appears at the wavenumber

$$\bar{\nu}(k) = Ry(Z - S(n_k))^2 \,. \quad (7.97d)$$

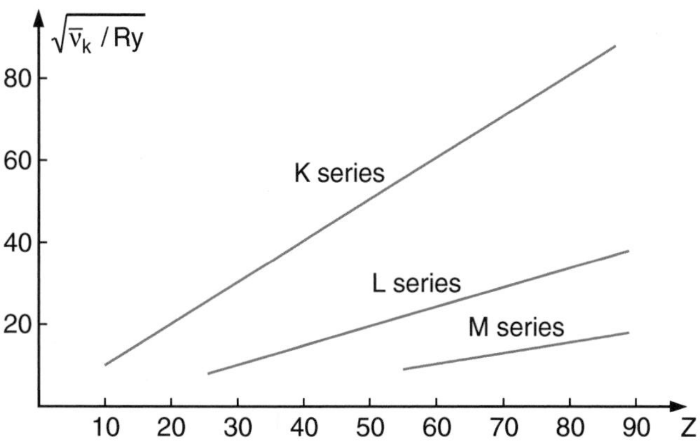


**Fig. 7.38.** Moseley diagram of the $Z$-dependence of the K-absorption edges

Measuring the wavelengths $\lambda_n$ of the different edges gives the shielding factor $S(n)$ for the corresponding electron shells with principal quantum number $n$.

## EXAMPLE

For lead ($Z = 82$) the wavelength at the K-edge is $\lambda_k = 14.8$ pm, which yields the effective nuclear charge number $Z_{\text{eff}} = 80.4$, corresponding to a shielding constant $S = 1.61$. The large shielding factor $S > 1$ shows that besides the remaining $1s$ electron, which gives the main contribution to the shielding, electrons from higher shells also contribute, since their wave functions $\psi(r)$ have a non-negligible value for $r < \langle r(1s)\rangle$.

Moseley used his measurements of K absorption edges to determine the nuclear charge numbers of many elements, where he assumed the shielding factor to be $S = 1$.

Measuring the spectral intervals around the absorption edges with higher resolution reveals a substructure. The edges consist of several closely spaced peaks (Fig. 7.39). The reason for this is the following.

For levels $n$ with an angular momentum $J > 0$, the energy levels $E_n$ split due to two effects. Firstly, levels with the same principal quantum number $n$, but different orbital angular momentum quantum numbers $L$ are degenerate only in the Coulomb potential but have

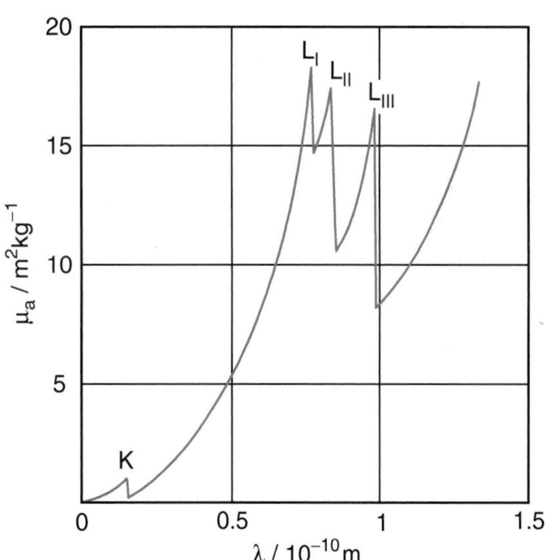


**Fig. 7.39.** Fine structure of the L-absorption edge in the X-ray absorption spectrum of lead

different energies in the real potential of many-electron atoms. Secondly, atoms with an electron spin $S \neq 0$ split, due to spin-orbit interaction (see Sect. 5.5) into fine-structure components with the same principal quantum number $n$, but different values of $J = L + S$ (see Fig. 7.37). Therefore, the absorption edges also show these splittings.

The K-shell with $n = 1$ shows no splitting because the angular momentum quantum numbers of the absorbing electron have to be $l = 0$ and $s = 0$. The levels with $n = 2$ in the L-shell have three sublevels ($l = 0$; $l = 1$, $j = 1/2$; and $l = 1$, $j = 3/2$). Since the fine-structure splittings increase with $Z^4$ they can reach values of several keV for heavy elements.

### 7.5.4 X-ray Fluorescence

If an inner-shell electron is excited from a level $E_k$ into a higher unoccupied level $E_i$ by electron impact or by absorption of X-rays, a vacancy is produced in this inner shell. One of the electrons from higher levels $E_n > E_k$ can fall down into the vacancy while a fluorescence photon $h\nu_{nk} = E_n - E_k$ is emitted, if the transition is allowed (Fig. 7.31).

For a definite excitation energy $E_n$, one therefore generally observes in the fluorescence spectrum many lines with frequencies $\nu_{ik}$ corresponding to all allowed transitions from levels $E_i > E_k$ into the vacancy in level $E_k$ (Fig. 7.40). The whole spectrum can be described by (7.97), which is similar to the Rydberg formula (3.91).

Measurements of the wavelengths $\lambda_{ik} = c/\nu_{ik}$ of these lines allows the determination of the energies of levels in inner shells, which are more difficult to calculate because of strong electron correlation and of the shielding factors $S$, which in turn gives information on the spatial distribution of the wave functions of the electrons involved (see Sect. 6.4).

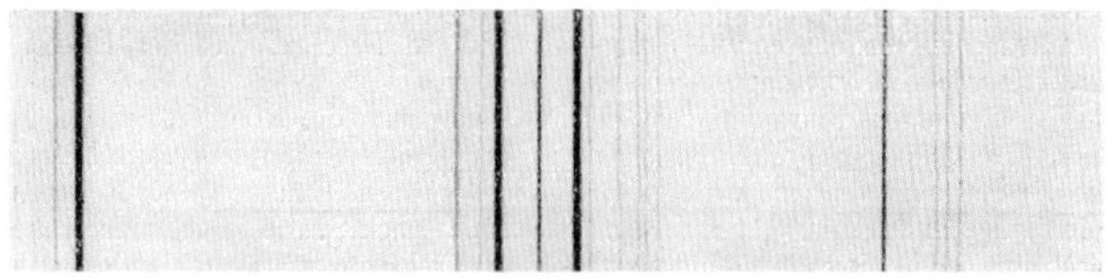

**Fig. 7.40.** The L-fluorescence series of the characteristic X-ray emission of tungsten (from Finkelnburg: Einführung in die Atomphysik, Springer, Berlin, Heidelberg 1967)

### 7.5.5 Measurements of X-Ray Wavelengths

Since the wavelength of X-rays is much smaller than that of visible light, new techniques had to be developed for their measurement. One of these methods uses optical gratings where the X-rays are incident under very small angles against the grating surface (Fig. 7.41). If $\vartheta$ is the angle between the direction of the incident parallel X-ray beam and the plane of the grating, the effective grating constant (i.e., the effective groove separation) is the projection

$$d_{\text{eff}} = d \sin\vartheta \approx d \cdot \vartheta \tag{7.98}$$

of the groove distance $d$ onto the incidence direction. For accurate measurements of the wavelength $\lambda$ the effective grating constant $d_{\text{eff}}$ has to be of the same order of magnitude than $\lambda$.

EXAMPLE

A grating with 1200 grooves per mm has a grating constant $d = 0.83\,\mu\text{m}$. For $\vartheta = 10' \Rightarrow \sin\vartheta = 3\times 10^{-3} \Rightarrow d_{\text{eff}} = 2.5\,\text{nm}$. A wavelength $\lambda = 2.5\,\text{nm}$ corresponds to a photon energy of $h\nu = 0.5\,\text{keV}$. Wavelengths down to 0.2 nm can be measured this way with sufficient accuracy.

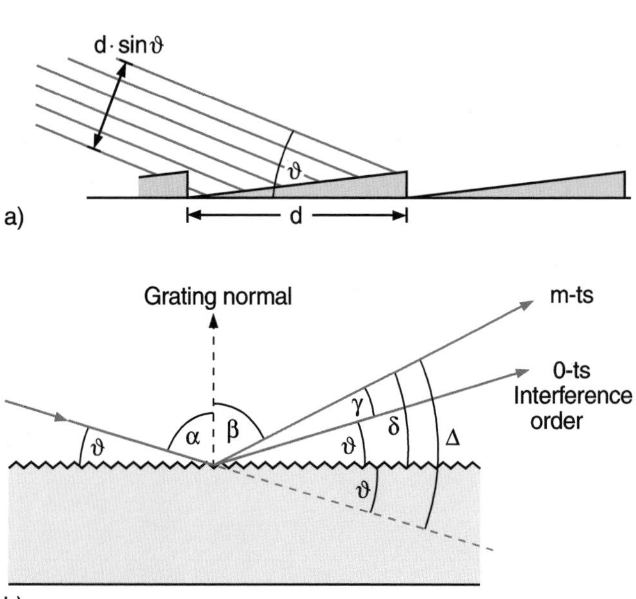


**Fig. 7.41a,b.** Measurement of X-ray wavelengths (**a**) with grazing incidence onto an optical grating with effective grating constant $d \sin\vartheta$ and (**b**) the conditions for the different diffraction orders

For a given angle of incidence $\alpha = 90° - \vartheta$ against the grating normal the diffraction angle $\beta$ is determined by the grating equation

$$d(\sin\alpha - \sin\beta) = m\lambda\ . \tag{7.99}$$

This means that in the direction $\beta$ against the grating normal the path difference between partial beams diffracted by adjacent grooves is $m\lambda$ and therefore a constructive interference occurs in this direction (Fig. 7.42). For very small angles $\vartheta = 90° - \alpha$ and $\gamma = 90° - \beta$ the large angles $\alpha$ and $\beta$ can only be measured with less accuracy than $\vartheta$ and $\gamma$. Inserting $\vartheta$ and $\gamma$ into (7.99) gives

$$d[\cos\vartheta - \cos(\vartheta + \gamma)] = m\lambda\ . \tag{7.100}$$

The $m$th interference order is observed at the angle $\delta = \vartheta + \gamma$ against the grating plane. The total deflection of the diffracted beam against the incident beam is $\Delta = \vartheta + \delta = 2\vartheta + \gamma$. This yields with $\cos\vartheta \approx \cos\gamma \approx 1$ for the wavelength $\lambda$ the relation

$$m\lambda = 2d\sin\frac{\Delta}{2}\sin\frac{\gamma}{2} \approx \frac{d}{2}\Delta\cdot\gamma\ . \tag{7.101}$$

Measuring the total deflection angle $\Delta = 2\vartheta + \gamma$ and the angle $\gamma$ between the 0th diffraction order (regular reflection) and the $m$th order allows the determination of the X-ray wavelength according to (7.101). The grating constant $d$ is calibrated by diffraction of visible light with known wavelength.

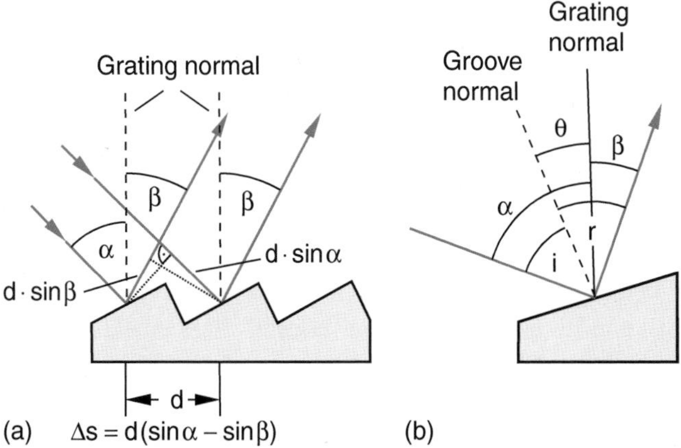


**Fig. 7.42.** (**a**) Illustration of the grating equation (7.99). (**b**) Blaze angle $\Theta$ and direction of maximum reflection

EXAMPLE

$d = 0.83\,\mu\text{m}$, $\lambda = 1\,\text{Å} = 10^{-10}\,\text{m}$, $\vartheta = 10' = 2.8\times10^{-3}\,\text{rad}$. The first interference order ($m = 1$) appears at the angle $\delta = 1.5\times10^{-2}\,\text{rad}$ against the grating plane and under the angle $\Delta = 1.8\times10^{-2}\,\text{rad}$ against the direction of the incident beam.

At such small angles total reflection of the X-rays occurs, because the refractive index $n$ of materials can be smaller than that of the vacuum with $n = 1$. This can be seen as follows.

The real part $n'$ of the complex refractive index $n = n' - i\kappa$ is given by (see Problem 7.15)

$$n^2 = 1 + \sum_i \frac{NZe^2}{\varepsilon_0 m_e(\omega_i^2 - \omega^2)}\ , \tag{7.102}$$

where $N$ is the atom number density, $\omega_i$ are the eigenfrequencies of absorbing transitions of the atoms in the sample and $m_e$ is the mass of the electron. The largest contribution to the sum comes from the transition with the highest frequency $\omega_i$.

If the frequency $\omega$ of the X-rays is higher than all eigenfrequencies (this implies that $\lambda$ is smaller than the K-edge in Fig. 7.36) all terms of the sum in (7.102) become negative and $n^2 < 1$. This means that the sample has a smaller refractive index than the vacuum or the air and the X-rays are totally reflected if their angle of incidence $\alpha$ becomes larger than the critical angle $\alpha_c$ of total reflection defined by

$$\sin\alpha_c = n_{\text{sample}}/n_{\text{air}}\ .$$

EXAMPLE

With $\lambda = 10^{-10}\,\text{m} \Rightarrow \omega \approx 2\times10^{19}\,\text{s}^{-1}$. For copper, the highest eigenfrequency is $\omega_0 = 1\times10^{19}\,\text{s}^{-1}$, which gives $n' = 1 - 1.3\times10^{-5}$. The critical angle for total reflection is then $\sin\alpha_c = sin(90° - \vartheta_c) = n_{\text{sample}}/n_{\text{air}} = 0.999987$, which gives $\vartheta_c = 0.3° = 5\times10^{-3}\,\text{rad}$.

For all angles $\alpha > \alpha_c \Rightarrow \vartheta < \vartheta_c$ total reflection of the incident X-rays is observed.

This means that the total incident intensity is reflected and distributed among the different diffraction orders.

For spectral regions with $n' < 1$ no collecting lenses, based on refraction, are possible. Any X-ray optics, therefore, must use the reflection by collimating mirrors, or Fresnel lenses, based on diffraction and interference.

The most important method for measuring X-ray wavelengths is Bragg diffraction by single crystals (see Sect. 2.4.3).

If a plane electromagnetic wave with wavelength $\lambda$ falls onto a crystal under an angle $\vartheta$ against a crystal plane (Fig. 7.43) all atoms of the crystal can scatter the wave. The partial waves from the different atoms interfere constructively, if the path difference between them is an integer of $\lambda$. This gives the Bragg condition

$$2d_c \sin\vartheta = m\lambda \quad (m = 1, 2, \ldots), \tag{7.103}$$

where $d_c$ is the distance between adjacent parallel crystal planes.

The distance $d_c$ is calibrated by illuminating the crystal with X-rays of known wavelength, measured with the technique, discussed above. When the same crystal is now irradiated with X-rays of unknown wavelengths, they can be determined from measurements of the angle $\vartheta$, where the maximum diffraction occurs. In practice, one measures the angle $2\vartheta$ of the deviation from the incident beam direction.

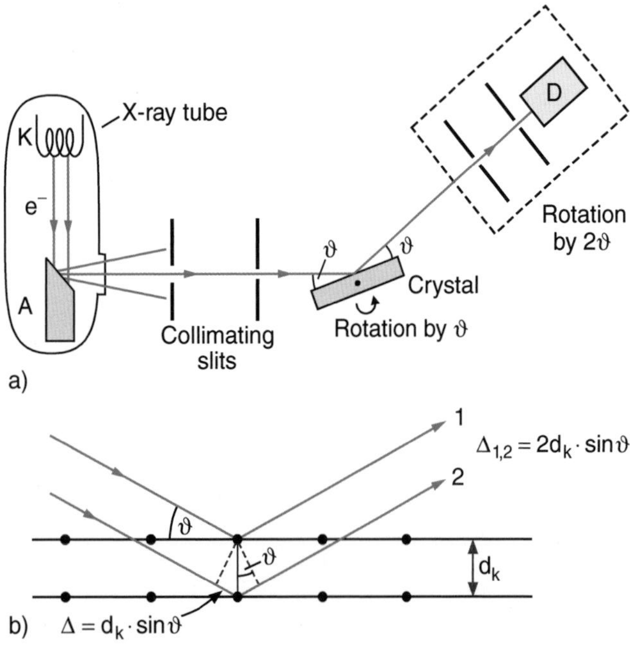


**Fig. 7.43.** (**a**) X-ray spectrometer with rotating crystal. (**b**) Condition for Bragg reflection at parallel crystal planes

## 7.6 Continuous Absorption and Emission Spectra

Transitions between two bound states of atoms or molecules always result in line spectra (see Sect. 3.4) where only discrete frequencies appear in the spectra, which are determined by energy conservation

$$h\nu_{ik} = E_i - E_k \,.$$

If at least one of the two states is not bound (e.g., if it lies above the ionization limit of the atom or above the dissociation energy of the molecule) the emitted or absorbed radiation shows a continuous spectrum.

### EXAMPLES

1. Photoionization of atoms: Here the photon energy $h\nu$ is larger than the binding energy $E_B$ of the atomic electron. The electron can then leave the atom with a kinetic energy
$$E_{kin} = h\nu - E_B \tag{7.104}$$
Varying the frequency $\nu$ changes the kinetic energy accordingly (Fig. 7.44). Every frequency $\nu$ results in an allowed transition and one obtains a continuous absorption spectrum.
2. The inverse process is the radiative recombination, where a free electron with kinetic energy $E_{kin}$ recombines with an ion, which ends up in an excited bound state with binding energy $E_B$, and a photon with energy
$$h\nu = E_{kin} + E_B \tag{7.105}$$
is emitted (Fig. 7.45) This results in a continuous emission spectrum.
3. Bremsstrahlung, discussed in the previous section, is an example for transitions between two unbound states. A free electron with kinetic energy $E_{kin} = eU$ in the X-ray tube is decelerated in the Coulomb field of the nucleus of an atom in the anode and looses the energy $\Delta E = eU - E'_{kin}$, where $E'_{kin}$ is the final energy of the electron after the deceleration.
A second example of bremsstrahlung is the synchrotron radiation emitted by high energy electrons circulating on a curved path in a magnetic field.

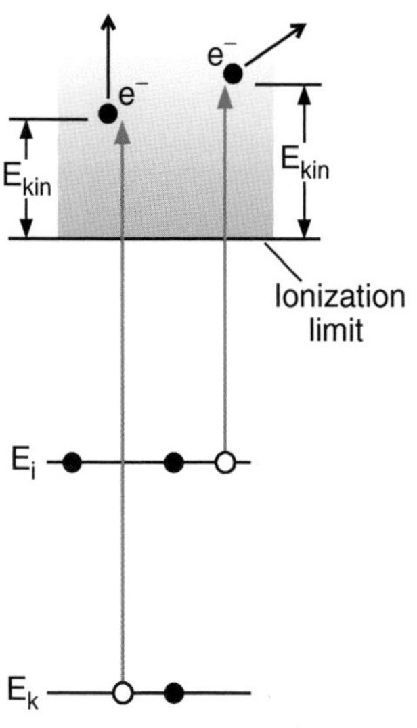


**Fig. 7.44.** Photoionization of an atom in its ground-state $E_k$ or in an excited state $E_i$

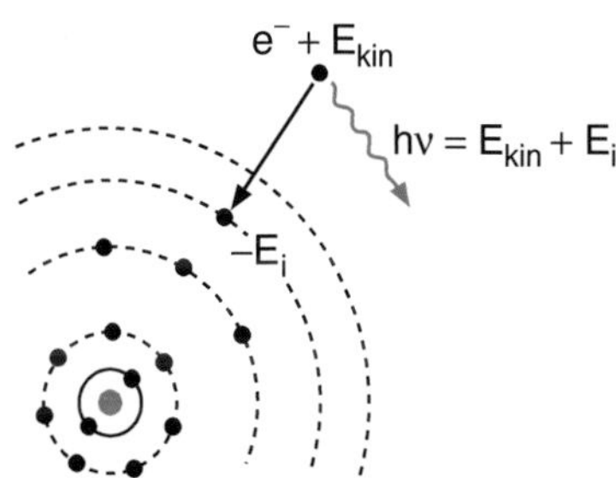


**Fig. 7.45.** Illustration of radiative recombination

We will now discuss such continuous spectra in more detail.

### 7.6.1 Photoionization

Measuring the absorption spectrum of an atom in a bound state $E_k < 0$ with binding energy $E_\mathrm{B}$ one observes with increasing photon energy a series of absorption lines with frequencies (see Sect. 6.6.4)

$$\nu_{ik} = -E_k/h - \frac{c \cdot Ry}{(n_i - \delta_i)^2} , \tag{7.106}$$

which become increasingly dense with increasing principal quantum number $n$. They are caused by transitions from level $E_k$ into bound Rydberg levels. The series converges for $n \to \infty$ towards the ionization limit $h\nu_c = E_\mathrm{B}$ of the atom. For $\nu > \nu_c$ the continuous part of the spectrum appears where the atom is photoionized.

The ions can be collected on the detector with an efficiency of 100% by a small electric field. The number of collected ions is a measure for the number of absorbed photons.

For a density $n_\mathrm{a}$ of absorbing atoms the measured ion rate is

$$\dot{N}_\mathrm{ion} = n_\mathrm{a} \dot{N}_\mathrm{ph} \sigma_\mathrm{PI} V_\mathrm{ion} , \tag{7.107}$$

where $\sigma_\mathrm{PI}$ is the photoionization cross section, $\dot{N}_\mathrm{ph}$ is the incident photon flux per cm$^2$ and $V_\mathrm{ion}$ is the ionization volume. In Fig. 7.46a the experimental arrangement for the measurements of absorption spectra is shown and Fig. 7.46b illustrates schematically the absorption cross section around the transition range from the discrete to the continuous part of the spectrum.

The absorption coefficient passes smoothly from the increasingly dense absorption lines into the continuum. Its value in the continuum is given by the square of the matrix element

$$\boldsymbol{M}_{iE} = \int \psi_i^* \boldsymbol{r} \psi_\mathrm{c}(E)\,\mathrm{d}\tau \tag{7.108}$$

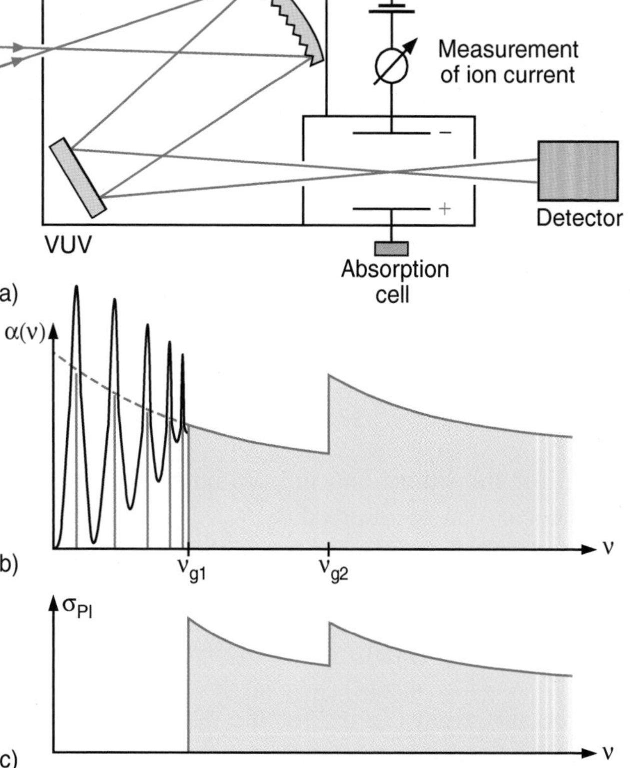


**Fig. 7.46a–c.** Rydberg absorption spectrum with adjacent ionization continuum. (**a**) Experimental setup. (**b**) Frequency dependence of absorption coefficient $\alpha(\mu)$. (**c**) Photoionization cross section $\sigma_\mathrm{PI}(\nu)$

for transitions from a bound state with wave function $\psi_i$ into a continuum state with wave function $\psi_c(E)$ and energy $E$. For $E \to E_{ion}$ the matrix element converges towards the matrix element for bound-bound transitions into Rydberg levels with $n \to \infty$.

The following experimental effect, however, pretends a jump of the absorption coefficient.

Because of the long lifetime $\tau_n$ of Rydberg levels $E_n$, which increases with $n^3$, the natural linewidths of high Rydberg levels $\Delta\nu_n = 1/\tau_n$ becomes very small. The resolvable spectral interval $\Delta\nu_{exp}$ of the spectrograph in Fig. 7.46a is generally much larger than $\Delta\nu_n$. The measured transmitted intensity for small absorptions is for a monochromatic incident radiation

$$I_t(\nu) = I_0 e^{-\alpha(\nu)L} \Rightarrow \Delta I(\nu) = I_0 - I_t \approx \alpha(\nu) L I_0, \tag{7.109}$$

while for radiation with a spectral continuum the absorbed intensity is

$$\begin{aligned}\Delta I_{eff} &= \frac{1}{\Delta\nu_{exp}} \int \Delta I(\nu)\,d\nu \\ &= \frac{L}{\Delta\nu_{exp}} \int\limits_{\nu_0-\Delta\nu_n/2}^{\nu_0+\Delta\nu_n/2} I_0 \alpha(\nu)\,d\nu\,. \end{aligned} \tag{7.110}$$

Since for continuous radiation the intensity $I_0$ barely depends on $\nu$ within a limited spectral interval, we can drag $I_0$ out of the integral in (7.110) and obtain for transitions into Rydberg levels where the absorption coefficient has only noticeable values within the linewidth $\Delta\nu_n$

$$\Delta I_{eff} \approx L I_0 \alpha(\nu_0) \frac{\Delta\nu_n}{\Delta\nu_{exp}}\,,$$

while for transitions into the continuum, all incident frequencies can be absorbed.

Introducing an effective absorption coefficient $\alpha_{eff}$, which reflects the true measured absorption

$$\alpha_{eff} = \Delta I_{eff}/(I_0 L)\,,$$

one obtains from (7.110) for the discrete spectra an effective absorption coefficient

$$\alpha_{eff} = \alpha(\nu_0) \frac{\Delta\nu_n}{\Delta\nu_{exp}}\,, \tag{7.111}$$

which is smaller than the real absorption coefficient $\alpha(\nu_n)$ at the line center by a factor

$$\Delta\nu_n/\Delta\nu_{exp} \ll 1\,.$$

### EXAMPLE

$\Delta\nu_n = 1\,\text{MHz}$, $\Delta\nu_{exp} = 1\,\text{GHz} = 10^3\,\text{MHz} \Rightarrow \alpha_{eff} = 10^{-3}\alpha(\nu_n)$.

Because of the finite spectral resolution of the spectrograph the different lines of the Rydberg spectrum can no longer be resolved for large values of the principal quantum number $n$. In this case, one measures a superposition of several Rydberg transitions and the apparent spectrum seems to be continuous. This shifts the experimentally observed ionization limit to smaller frequencies.

A way out of this deficiency is the measurement of the photo ions instead of the transmitted intensity. Another solution is provided by laser spectroscopy where the natural linewidth can be resolved (see Sect. 10.5).

For transitions into the true continuum is $\Delta\nu_n = \Delta\nu_{exp}$, because all frequencies contribute to the absorption and the above-mentioned problem does not arise.

For doubly excited atoms where two electrons are in excited states (e.g., $ns$, $2p$ for the helium atom, where one electron is in Rydberg levels $ns$ and the other in the $2p$ state), the ionization limit for one of the two excited electrons appears at a higher total energy than for singly excited atoms (Fig. 7.47). The difference is just the excitation energy of the second electron. For our example, the ionization limit is shifted upwards by the excitation energy of the $2p$ electron. For the He atom doubly excited states $(ns, 2p)$ are already for $n = 4$ above the ionization limit of singly excited He atoms $(ns, 1s)$. These doubly excited states can decay by autoionization (see Sect. 6.5), where, due to a correlation between the two electron the electron in the $2p$ state transfers its energy to the Rydberg electron, which can then leave the atom while the $2p$-electron falls back into the $1s$ ground state.

Measuring the absorption spectrum of an atom above its ionization limit, one observes resonances at the energies of doubly excited states that are

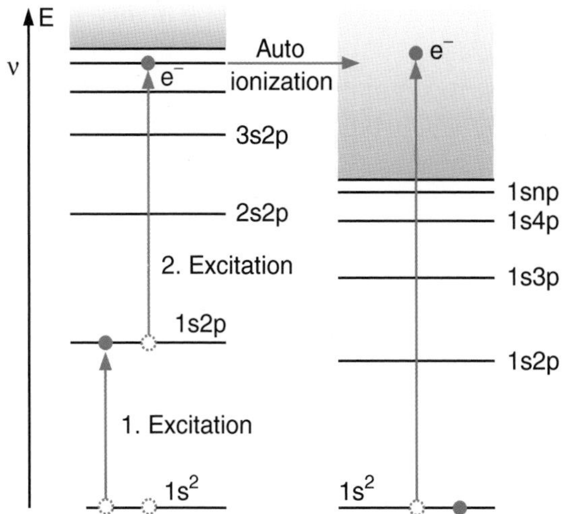


**Fig. 7.47.** Auto-ionization of a double-excited atomic state in competition with single-photon ionization

superimposed on the continuum. They are due to autoionization. The line profiles of these resonances have been studied and explained theoretically by *U. Fano* and are therefore called Fano profiles (Fig. 7.48). They can be understood from the following consideration:

The absorption of a photon reaches two states at the same energy $E$: the doubly excited discrete state with a level width $\Delta E$ and the interval $\Delta E$ around $E$ in the continuum. The wave function of these superimposed states is written for our He example as the linear combination

$$\psi = c_1\psi_k(ns, 2p) + c_2\psi_c(E) . \tag{7.112}$$

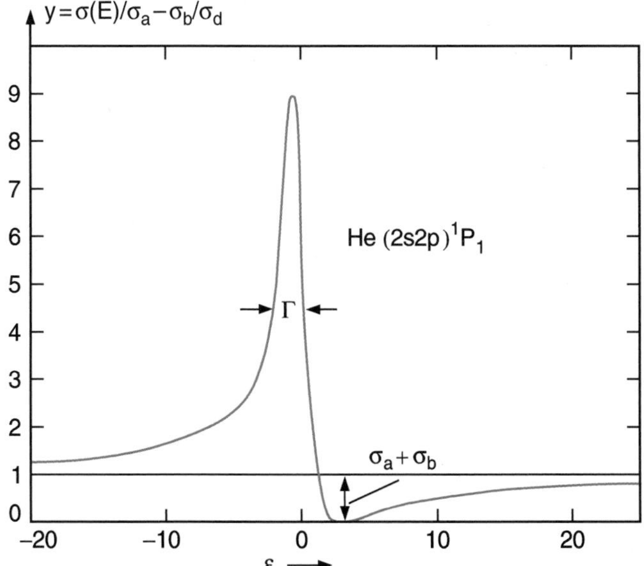


**Fig. 7.48.** Absorption profile of an autoionization line (Fano profile)

The absorption coefficient for the transition from the initial level $E_i$ to the energy $E$ around the autoionization resonance, which is proportional to the square of the matrix element

$$\begin{aligned} \boldsymbol{M}_{iE} &= \int \psi_i^*(E_i)\boldsymbol{r}\left[c_1\psi(ns,2p) + c_2\psi_c(E)\right] \mathrm{d}\tau \\ &= c_1 M_i(ns,2p) + c_2 M_{ic} \end{aligned} \tag{7.113}$$

contains the interference term

$$2c_1c_2M_{ic} \cdot M_i(ns,2p),$$

which depends on the energy difference $\Delta E = E_{\text{cont}} - E(ns,2p)$. While the phase of the continuous wave function depends only weakly on the energy, that of the doubly excited state depends strongly on $\Delta E$, because it changes by $\pi$, when tuning across the resonance profile. If both wave functions are in phase, the absorption cross section becomes maximal, if they have opposite phases the interference is destructively and the absorption becomes minimal.

Fano and Cooper [7.15, 16] have shown that the energy dependence of the absorption cross section around an autoionization resonance can be described by [7.17]

$$\sigma(E) = \sigma_a \frac{(\varepsilon + q)^2}{1 + \varepsilon^2} + \sigma_b , \tag{7.114}$$

where $\varepsilon = \Delta E/(\Gamma/2) = (E - E_r)/(\Gamma/2)$ is the energy difference in units of the halfwidth $\Gamma/2$ of the resonance, $\sigma_a$ is the absorption cross section for the absorption by the doubly excited state (unperturbed by autoionization) and $\sigma_b$ is the background absorption of the direct excitation into the continuum (Fano profile Fig. 7.49).

The Fano parameter $q$ [7.18] stands for the ratio

$$q = -\frac{D_1}{D_2 R_{12}}$$

of the transition amplitude $D_1$ to the perturbed bound state and the product $D_2 R_{12}$ of the transition amplitude $D_2$ to the continuum and the coupling $R_{12}$ between both states. In fact, this product gives the transition amplitude for reaching the continuum state via the bound state. If the transition moments for the two different path ways from the ground state to the continuum state have equal amplitudes but opposite phases, the two contributions cancel each other and the absorption cross section becomes zero [7.17].

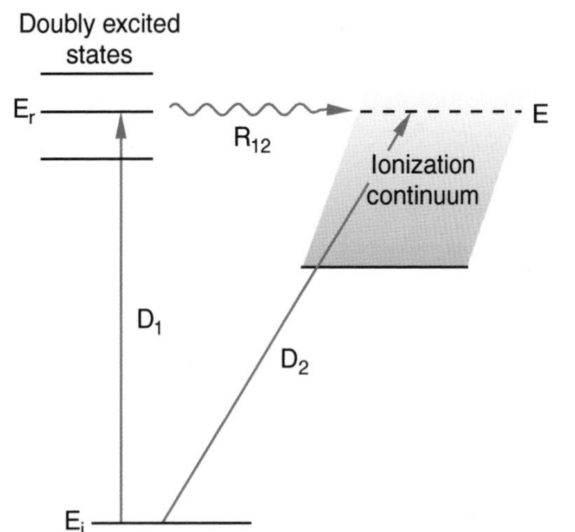

Fig. 7.49. Level scheme for the explanation of Fano profiles

The width $\Gamma = 1/\tau$ (full width at half maximum) is determined by the lifetime $\tau$ of the doubly excited state. Since radiative transitions into lower bound states of the atom have a small probability, the lifetime of autoionizing states is mainly limited by the fast auto ionization process. Typical values for atoms range from $\tau = 10^{-12}$–$10^{-14}$ s.

### 7.6.2 Recombination Radiation

A free electron with velocity $v$ can be captured by an atom or ion into a bound state with binding energy $E_\mathrm{B}$. The energy released at this process can be converted into radiation with a photon energy

$$h\nu = E_\mathrm{kin} - E_i = \frac{m_\mathrm{e}}{2}v^2 - E_i \quad (E_i < 0)\,. \tag{7.115}$$

This process is called ***two-body recombination*** or ***radiative recombination***, contrary to ***three-body recombination***, where the excess energy can be transferred to a third particle, which also takes care of momentum conservation. The three-body recombination, where no radiation is emitted, becomes significant at higher densities of electrons and atoms or ions, because three collision partners have to meet at the same time.

The inverse process of autoionization is the dielectronic recombination. Here a free electron is captured by an atom and the recombination energy is not emitted as radiation but transformed to another bound electron of the atom which is promoted into a higher bound energy level [7.21].

The radiative recombination plays an important role in gas discharges and in other low density plasmas such as in stellar atmospheres. The cross section for electron capture into an atomic state $E_i$ depends on the relative velocity of the two collision partners.

We will consider as an example a low density plasma at local thermal equilibrium with atom or ion density $n_\mathrm{a}$ and electron density $n_\mathrm{e}$. The electron density within the velocity interval between $v$ and $v + \mathrm{d}v$ is $n_\mathrm{e}(v)$, where

$$n_\mathrm{e} = \int n_\mathrm{e}(v)\,\mathrm{d}v\,.$$

The recombination rate is

$$\dot{N}_\mathrm{R} = n_\mathrm{a}\int n_\mathrm{e}(v)\sigma_\mathrm{R}(v)v\,\mathrm{d}v \tag{7.116}$$

and one photon is emitted per recombination event.

In a plasma the velocity distribution of the electrons depends on the electron temperature $T_\mathrm{e}$, which is generally higher than the ion temperature $T_\mathrm{ion}$.

In Fig. 7.50 the dependence of the recombination cross section on the electron temperature is plotted on a double logarithmic scale for two different electron densities. For low densities the recombination rate is low and nearly independent of temperature, while for high electron densities it falls drastically with increasing temperature.

In a plasma at thermal equilibrium with atomic density $n_\mathrm{a}$ the recombination radiation is isotropic. The radiation power emitted into the solid angle $\Delta\Omega = 1$

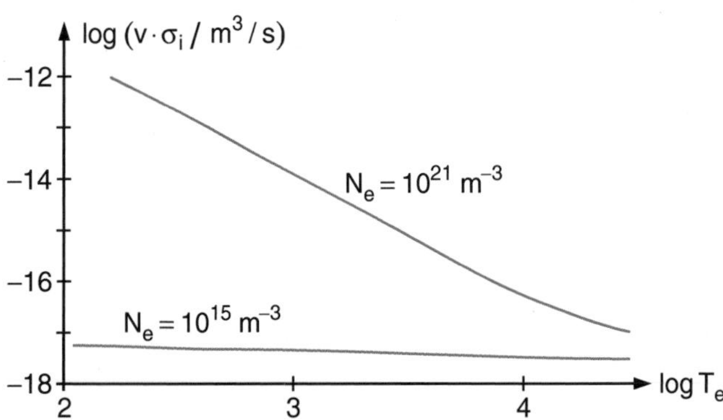

Fig. 7.50. Logarithmic plot of two-body recombination rate $v\sigma_i$ versus the logarithm of the electron temperature $T_\mathrm{e}$ for two different electron densities

Sterad within the frequency interval $d\nu$ is then

$$P(\nu)\,d\nu = \frac{h\nu}{4\pi} n_a n_e(v) v \sigma_R \,d\nu\,. \tag{7.117}$$

From the energy conservation

$$h\nu = (m_e/2)v^2 - E_i \quad (E_i < 0!)\,,$$

the velocity of the recombining electron is

$$v = \left[\frac{2}{m_e}(h\nu + E_i)\right]^{1/2}\,. \tag{7.118}$$

Inserting this into the Maxwellian distribution

$$n_e(v)\,dv = N_e \frac{4v^2}{v_p^3\sqrt{\pi}}\,,e^{-(v/v_p)^2}\,dv \tag{7.119}$$

with the most probable velocity

$$v_p = (2kT/m_e)^{1/2}\,,$$

we obtain with (7.117) the intensity distribution of the recombination radiation as

$$P_\nu\,d\nu = n_a n_e \sigma_R(v) \frac{h^2 v\nu}{m_e \pi^{3/2} v_p^3} e^{-((\nu-\nu_0)/\nu_0)^2}\,d\nu\,, \tag{7.120}$$

with an intensity maximum at the frequency

$$\nu_0 = \left(\frac{1}{2} m_e v_p^2 - E_i\right)/h \tag{7.121}$$

shown schematically in Fig. 7.51 by the black curve for the recombination in a hydrogen plasma

$$H^+ + e^-(E_{kin}) \rightarrow H + h\nu \tag{7.122a}$$

of electrons with protons, which occurs in the atmosphere of hot stars.

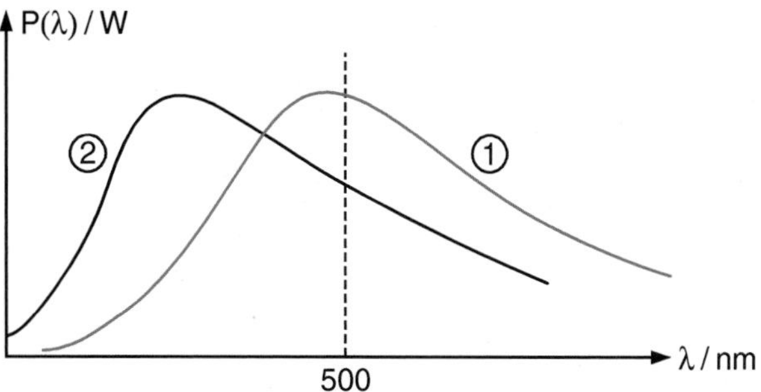


**Fig. 7.51.** Continuous spectrum of radiative recombination in a hydrogen plasma with low electron concentration at $T = 6000\,K$. (1) $= H + e^- \rightarrow H^- + h\nu$. (2) $= H^+ + e^- \rightarrow H + h\nu$

The continuous emission spectrum of our sun, represented by the Planck distribution (3.15) with a temperature of about 5800 K, is an important example for a recombination continuum. The visible and infrared part of it is mainly due to the recombination of neutral hydrogen atoms and electrons, according to the scheme

$$H + e^-(E_{kin}) \rightarrow H^- + h\nu\,, \tag{7.122b}$$

where negative $H^-$ ions are formed. These $H^-$ ions loose their electron by collisions with electrons

$$H^- + e^- \rightarrow H^* + 2e^-\,,$$

which replenishes the supply of neutral H atoms, used again for the recombination process (7.122).

Besides the process (7.122), the recombination of protons and electrons is also present

$$H^+ + e^-(E_{kin}) \rightarrow H^* + h\nu\,. \tag{7.122c}$$

The recombination radiation from this process mainly contributes to the UV part of the continuous spectrum of our sun.

## SUMMARY

- The frequencies $\nu_{ik}$ absorbed or emitted by atoms

$$\nu_{ik} = (E_i - E_\kappa)/h$$

are determined by the energies $E_i$, $E_k$ of the atomic states connected by the radiative transition.
- The relation between the probability $w_\nu B_{ik}$ of absorption by an atom in a radiation field with spectral energy density $w_\nu$ and $w_\nu B_{ki}$ of induced emission is $g_k B_{ki} = g_i B_{ik}$, where $g_i$ and $g_k$ are the statistical weights of the levels, i.e., the number of different wave functions representing this level.
If the total angular momentum quantum number is $J$ the statistical weight is $g = (2J + 1)$.
- The relation between the Einstein coefficients $B_{ik}$ and $A_{ik}$ is $A_{ik} = (8\pi\nu^3/c^3)B_{ik}$.
In a radiation field with one photon per mode the spontaneous radiation probability equals the induced emission probability. In thermal radiation fields at achievable temperatures, the number of photons per mode $n \ll 1$ is very small in the range from near infrared to UV. Here, the spontaneous emission is therefore by far the dominant emission.
- The radiation power absorbed or emitted on a transition $E_i \to E_k$ is proportional to the absolute square of the transition dipole matrix element

$$M_{ik} = e \int \psi_i^* \boldsymbol{r} \psi_k \, \mathrm{d}\tau ,$$

which depends on the wave functions $\psi_i, \psi_k$ of the corresponding atomic states. $M_{ik}$ represents the quantum mechanical average of the classical dipole moment in the two atomic states.
- An electric dipole transition is only allowed, if the selection rules $\Delta L = \pm 1$, $\Delta M_L = 0, \pm 1$, and $\Delta J = 0 \pm 1$ are obeyed, but $J = 0 \to J = 0$ is forbidden. Here $L$ is the orbital angular momentum quantum number $M_L$ its projection and $\boldsymbol{J}$ the quantum number of total angular momentum $\boldsymbol{J} = \boldsymbol{L} + \boldsymbol{S}$.
- For all light atoms (small $Z$) only transitions occur, where the spin quantum number $S$ does not change ($\Delta S = 0$). For atoms with large $Z$ spin-orbit coupling allows weak transitions with $\Delta S \neq 0$.
- Besides electric dipole transitions, higher order transitions, such as electric quadrupole transitions or magnetic dipole transitions, can also occur, but with probabilities that are smaller by several orders of magnitude.
- The mean lifetime $\tau_i = 1/A_i$ of an excited atomic level $\langle i|$ with energy $E_i$ is given by the inverse Einstein coefficient $A_i$ of spontaneous emission on allowed transitions from $\langle i|$ into all other levels $\langle j|$ with energies $E_j < E_i$. Measurements of level lifetimes therefore allow the determination of transition probabilities and matrix elements. They represent a crucial test of the accuracy of calculated wave functions.
- The effective lifetimes are determined by the sum of radiative decay and inelastic collision probabilities. Inelastic collisions shorten the natural lifetime of an atomic level. Elastic collisions perturb the level energies of the emitting atom.
- The linewidths of spectral lines are determined by:
a) The natural linewidth

$$\delta\nu_n = \frac{1}{2\pi}\left(\frac{1}{\tau_i} + \frac{1}{\tau_k}\right)$$

b) The generally much larger Doppler-width

$$\delta\nu_\mathrm{D} = 7.16 \times 10^{-7} \nu_{ik} \sqrt{T/M}$$

$$(M = \text{molar mass})$$

c) Collisions of the emitting or absorbing atom with other atoms or molecules (pressure broadening)
- X-rays are electromagnetic waves with wavelengths in the range of 10 nm to 0.1 nm. The measurement of these wavelengths can be achieved by Bragg reflection or diffraction in single crystals or with optical diffraction gratings under gracing incidence.
- X-rays are produced:
By decelerating electrons with energies in the keV range (continuous emission of bremsstrahlung)
By transitions of atomic electrons from higher energy states in vacant inner shell states (characteristic X-ray emission)

- X-rays are absorbed by:
The photo-effect $A + h\nu \rightarrow A^*$
The Compton effect $h\nu + e^- \rightarrow h\nu' + e^- + E_{kin}$
Pair formation $h\nu \rightarrow e^- + e^+$.
The relative probabilities of these processes depend on the energy $h\nu$ of the X-ray photons.
- Continuous absorption spectra of atoms are observed for the photoionization of atoms. Continuous emission spectra are generated by radiative recombination of free electrons with ions or neutral atoms. Another source of continuous radiation is produced when free electrons are decelerated either in the Coulomb field of atoms (bremsstrahlung) or when high energy electrons (MeV –GeV) are forced, using a magnetic field, onto circular paths (synchrotron radiation).
- The continuous spectrum of our sun in the visible and infrared part is produced by radiative recombination

$$H + e^- \rightarrow H^- + h \cdot \nu ,$$

while the UV part is due to the process

$$H^+ + e^- \rightarrow H + h\nu .$$

## PROBLEMS

1. $10^8$ sodium atoms are excited into the $3^2P_{3/2}$ level ($\tau = 16$ ns) by absorption of light that is linearly polarized in the $x$ direction and propagating into the $z$ direction. The emitted fluorescence follows the angular distribution $I(\vartheta) = I_0 \sin^2 \vartheta$ where $\vartheta$ is the angle against the $x$ direction.

   a) What is the total fluorescence power?
   b) Which fraction of this power is emitted into the solid angle $d\Omega = 0.1$ sterad around $\vartheta = 90°$?

2. a) What is the Doppler width of the Lyman-$\alpha$ line of the H atom at a temperature of $T = 300$ K?
   b) A collimated beam of H atoms (the nozzle diameter is 50 µm, the distance between nozzle and collimating slit is $d = 10$ cm, the width of the slit is $b = 1$ mm) is perpendicularly crossed behind the slit by a parallel beam of a laser tuned across the absorption profile of the Lyman-$\alpha$ line. What is the residual Doppler width?
   c) Compare this width with the natural linewidth ($\tau(2p) = 1.2$ ns).
   d) Is it possible to resolve the hyperfine structure of the $1^2S_{1/2}$ ground state?

3. The spectral width of a line can be limited by the finite interaction time of the atom with the radiation field. What is the minimum interaction time of a calcium atom needed in order to achieve a linewidth of 3 kHz for the transition ${}^1S_0 \rightarrow {}^3P_1$ ($\lambda = 657.46$ nm) with a lifetime $\tau = 0.39$ ms of the upper level? What is the minimum interaction zone of Ca atoms in a collimated beam for an oven temperature of $T = 900$ K?

4. Metastable He atoms in the $2^1S_0$ state in a gas discharge at $T = 1000$ K absorb light on the transition $2^1S_0 \rightarrow 3^1P_1$. The term values of the corresponding levels are 166,272 cm$^{-1}$ and 186,204 cm$^{-1}$, the lifetimes are $\tau(3^1P_1) = 1.4$ ns, $\tau(2^1S_0) = 1$ ms. a) What is the wavelength of the transition? b) What is its natural linewidth? c) What is the Doppler-width?

5. How large is the absorption of a monochromatic wave on the transition in Problem 7.4 relative to the absorption at the line center for an absorption frequency $\nu$ that is 0.1 nm, $0.1\delta\nu_D$, $1\delta\nu_D$ and $10\delta\nu_D$ away from the line center $\nu_0$? Consider the answer for a Gaussian profile and a Lorentzian profile. For which frequency difference is the absorption equal for both profiles?

6. Calculate the velocity and kinetic energy of photoelectrons released from the K shell of molybdenum by the absorption of $K_\alpha$ radiation from silver atoms?

7. What is the recoil energy and recoil velocity of H atoms, initially at rest, when a photon is emitted or absorbed on the transition $n = 2 \leftrightarrow n = 1$? What is the shift of the absorption frequency against the emission frequency? Compare this shift with the natural linewidth and the Doppler-width at 300 K.

▶

8. The quenching cross section for inelastic collisions of excited Na atoms ($3^2P_{1/2}$) with $N_2$ molecules is $\sigma_q = 4 \times 10^{-19}\,m^2$. What is the effective lifetime of the Na ($3^2P_{1/2}$) level with $\tau_{rad} = 16\,ns$ for $N_2$ pressures of 1 mbar, 10 mbar and 100 mbar at a temperature of T = 500 K?
9. Na atoms in a collimated atomic beam with a mean beam velocity of 800 m/s are excited by a perpendicular beam of a tunable laser. What is the minimum collimation ratio: a) In order to resolve the hyperfine structure ($\Delta\nu = 190\,MHz$) of the $3^2P_{1/2}$ level?
10. In order to make the residual Doppler-width smaller than the natural linewidth of the $3S_{1/2} \to 3P_{1/2}$ transition?
11. Compare natural linewidth, Dopplerwidth and collisional broadening of the Lyman-$\alpha$-line ($1s$-$2p$) and the 21 cm line (hyperfine transition $1\,^2S_{1/2}$, $F = 1 \leftrightarrow F = 0$) with $A_{ik}(\alpha) = 10^9\,s^{-1}$ and $A_{ik}(21\,cm) = 2.9 \times 10^{-15}\,s^{-1}$ under the following conditions:

    a) Starlight passes through an interstellar cloud of H-atoms with a density $N = 10^6\,/m^3$ (1 Atom/cm$^3$), temperature $T = 10\,K$, absorption path length $L = 1\,pc = 3 \times 10^{16}\,m$, $\sigma_{coll} = 10^{-19}\,m^2$.

    b) What is the absorption of the two lines with an absorption cross section $\sigma(\text{Lyman}\,\alpha) = 1 \times 10^{-15}\,m^2$, $\sigma(21\,cm) = 3 \times 10^{-26}\,m^2$.

    c) A laser beam with 10 mW power at $\lambda = 3.39\,cm$ and a beam diameter of 1 cm passes through an absorption cell filled with methane $CH_4$ molecules at a pressure of 0.1 mbar and a temperature $T = 300\,K$. What are the ratios of natural linewidth, transit-time broadening, saturation broadening and Doppler-width of the transition $\langle k| \to \langle i|$ with lifetimes $\tau_i = 20\,ms$ and $\tau_k = \infty$?
12. Show, by calculating the integral, that the transition dipole matrix element $\int \psi_{1s}^* \boldsymbol{r} \psi_{2s}\,d\tau$ of the transition $1S \to 2S$ in the H atom is zero. Use the wave function of Table 5.2.
13. What is the value of the transition probability of the transition $1S \to 2P$?
14. What are the transition probability and the natural linewidth of the transition $3s \to 2p$ in the H atom? The lifetimes are $\tau(3s) = 23\,ns$, $\tau(2p) = 2.1\,ns$. Compare the natural linewidth with the Doppler-width of this transition at $T = 300\,K$ and $T = 1{,}000\,K$.
15. Derive (7.102) for small refraction indices ($n - 1 \ll 1$). Is for $n < 1$ the velocity of an electromagnetic wave larger than the velocity $c$ in vacuum? Does this contradict the special theory of relativity?

# 8. Lasers

Laser is an acronym for ***L**ight **A**mplification by **S**timulated **E**mission of **R**adiation* that describes the basic physical principle of its operation. *Gordon*, *Zeiger* and *Townes* [8.1] showed for the first time in 1955 that a microwave could be amplified by $NH_3$ molecules on the inversion transition at $\lambda = 1.26\,\text{cm}$ (Fig. 4.13) if these molecules were prepared in such a way that the upper level of the transition had a larger population than the lower one. With such inverted $NH_3$ molecules inside a microwave cavity, the first "maser" (microwave amplification by stimulated emission of radiation) could be operated.

*Schawlow* and *Townes* published a paper in 1958 with detailed discussions of how the maser principle might be extended into the visible spectral range [8.2]. The first experimental realization of a laser was demonstrated in 1960 by *Maiman*, who built a ruby laser, which was pumped by a helical flashlamp and emitted coherent radiation at $\lambda = 694\,\text{nm}$ [8.3].

Since then, lasers have been developed spanning the whole spectral range from the far infrared down to the vacuum ultraviolet region. They have proved to be valuable tools not only for the solution of many scientific problems but also for numerous technical applications.

In this chapter we will discuss the basic physical principles of lasers, the most important classes of lasers and some interesting novel applications. More detailed discussions can be found in the vast literature on lasers [8.4–7].

## 8.1 Physical Principles

A laser basically consists of three components (Fig. 8.1):

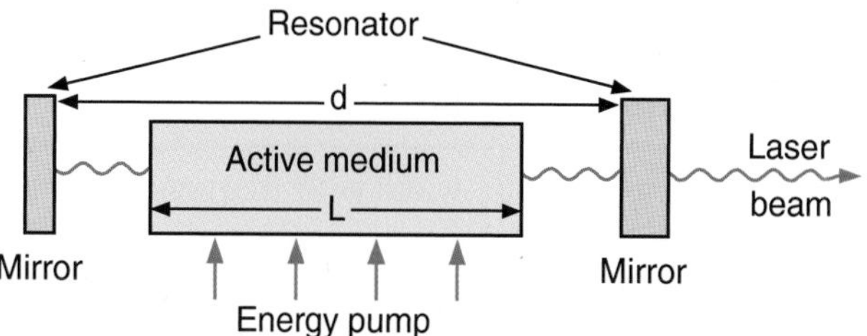


**Fig. 8.1.** Schematic setup of a laser

1. The ***active medium*** where an inverted population $N(E)$ is created by selective energy transfer. This population distribution $N(E)$ deviates strongly from a thermal Boltzmann distribution (Fig. 8.2) in such a way that $N(E_i) > N(E_k)$ for $E_i > E_k$, contrary to a thermal population

$$N(E) \propto \mathrm{e}^{-E/kT} \; .$$

2. The ***energy pump***, (flashlamp, gas discharge, electric current or another laser) that generates the population inversion.
3. The ***optical resonator*** that stores the fluorescence emitted by the active medium in a few modes of the radiation field (see below). In these modes

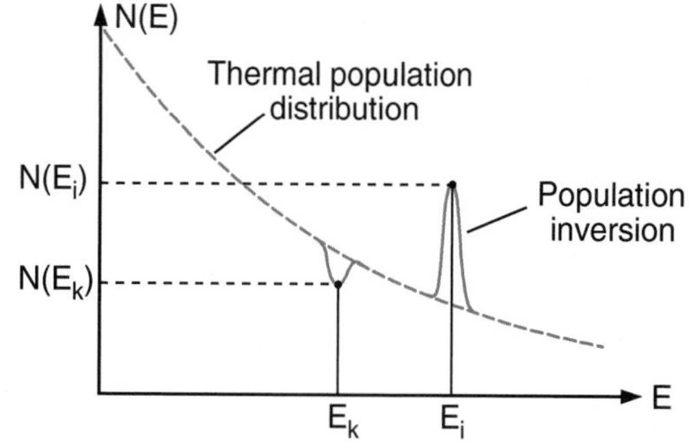


**Fig. 8.2.** Selective population inversion ($N_i > N_k$) in spite of $E_i > E_k$, deviating from a thermal population distribution (*red dashed* curve)

W. Demtröder, *Atoms, Molecules and Photons*, 2nd ed., Graduate Texts in Physics,
DOI 10.1007/978-3-642-10298-1_8, © Springer-Verlag Berlin Heidelberg 2010

the photon number becomes $N_{\text{phot}} \gg 1$. Therefore, in these modes, the induced emission becomes much larger than the spontaneous emission (see Sect. 7.1.1). The optical resonator furthermore reflects the induced emission back into the active medium and allows many paths back and forth through the medium, thus realizing a long amplification path. This converts the light amplifier into a light oscillator if the total amplification exceeds the total losses.

### 8.1.1 Threshold Condition

When an electromagnetic wave with frequency $\nu$ travels in the $z$ direction through a medium (Fig. 8.3) its intensity changes according to Beer's absorption law

$$I(\nu, z) = I(\nu, 0) \cdot \mathrm{e}^{-\alpha(\nu) \cdot z} . \tag{8.1}$$

The frequency-dependent absorption coefficient

$$\alpha(\nu) = [N_k - (g_k/g_i)N_i]\sigma(\nu) \tag{8.2}$$

is determined by the absorption cross section $\sigma(\nu)$ of the transition $N_k \to N_i$, the population densities $N_i$, $N_k$, of the levels with energies $E_i$, $E_k$ with $\Delta E = E_i - E_k = h\nu$, and their statistical weights $g_i$, $g_k$ (the statistical weight of a level with total angular momentum quantum number $J$ is $g = 2J + 1$). For

$$N_i > (g_i/g_k)N_k \Rightarrow \alpha(\nu) < 0 \tag{8.2a}$$

this means that the transmitted wave will be amplified instead of attenuated. Such a deviation (8.2a) from a thermal equilibrium population is called ***inversion*** and the medium where this inversion is realized is called the ***active medium***.

When the active medium with length $L$ is placed between two parallel mirrors (Fig. 8.1) the light wave is reflected back and forth and passes through the active medium many times, where it is amplified each time by the factor

$$G(\nu) = \frac{I(\nu, 2L)}{I(\nu, 0)} = \mathrm{e}^{-2\alpha(\nu) \cdot L} , \tag{8.3}$$

which is larger than 1 for $\alpha(\nu) < 0$.

Unfortunately there are also losses that attenuate the wave. These are reflection, diffraction, absorption, and scattering losses (Fig. 8.4).

***Reflection losses.*** A mirror with reflection coefficient $R$ only reflects the fraction $R < 1$ of the incident intensity. If absorption losses of the mirror can be neglected, the fraction $(1 - R)$ of the incident intensity is transmitted through the mirror.

***Absorption and scattering losses.*** In case of gas lasers, the windows of the tube containing the active medium may absorb and scatter some of the transmitted light. For solid lasers the end surfaces of the laser rod may scatter and reflect some light. Also, the active medium might not have a spatially uniform inversion, leaving locations with $\alpha > 0$. Finally, the mirror surfaces are not perfect. They can scatter light and the reflecting layers can also show small absorptions.

***Diffraction losses.*** Depending on the parameters of the optical resonator (aperture diameter $a$, mirror separation $d$ and radius of curvature $r$ of the mirrors) the wave being reflected back and forth shows an angular spread due to diffraction (see below). This means that only part of the intensity is reflected back into the active medium, which represents a diffraction loss per roundtrip through the resonator.

We will describe the sum of all these losses per roundtrip by the loss factor $\gamma$. The intensity after one roundtrip has decreased (without amplification by the active medium) by the factor $\mathrm{e}^{-\gamma}$:

$$\frac{I(2d)}{I(0)} = \mathrm{e}^{-\gamma} \quad \text{with} \quad \gamma = \gamma_{\text{r}} + \gamma_{\text{sc}} + \gamma_{\text{dif}} . \tag{8.4}$$

Taking into account the amplification by the active medium we obtain the gain factor

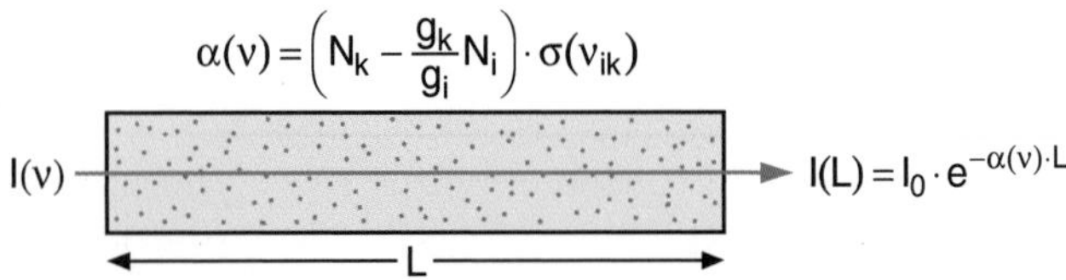


**Fig. 8.3.** Attenuation ($\alpha > 0$) or amplification ($\alpha < 0$) of a light wave passing through a medium

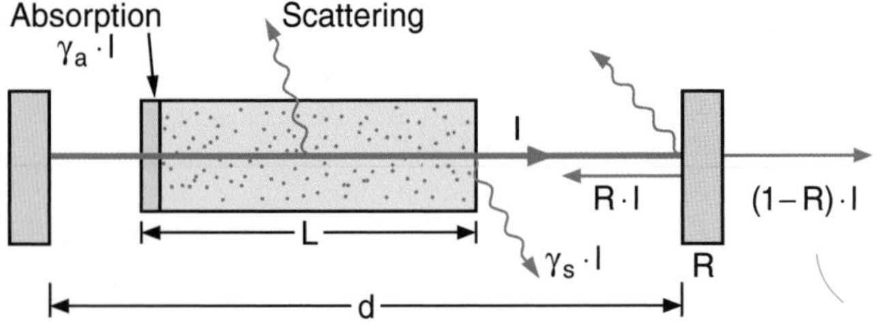


**Fig. 8.4.** Illustration of the losses in a laser resonator

$$G(\nu) = \frac{I(\nu, 2d)}{I(\nu, 0)} = \mathrm{e}^{-(2\alpha(\nu)\cdot L+\gamma)} \,. \tag{8.5}$$

For $G(\nu) > 1$ the amplification overcomes the losses and the light amplifier becomes a light oscillator. According to (8.2) and (8.5) the threshold condition for starting the self-sustained oscillation (i.e., the laser oscillator) is

$$2\alpha(\nu) \cdot L + \gamma \leq 0 \,. \tag{8.5a}$$

Inserting (8.2) this gives

$$2[N_k - (g_k/g_i)N_i] \cdot \sigma(\nu) \cdot L + \gamma \leq 0 \,. \tag{8.5b}$$

The minimum inversion $\Delta N = N_i(g_k/g_i) - N_k$ for lasing must therefore fulfill the threshold condition

$$\boxed{\begin{aligned}\Delta N_{\text{thr}} = N_i(g_k/g_i) - N_k \geq \Delta N_{\text{thr}} \\ = \frac{\gamma(\nu)}{2\sigma(\nu)\cdot L}\end{aligned}} \,. \tag{8.6}$$

If the energy transfer from the pump into the active medium is sufficiently strong to achieve $\Delta N > \Delta N_{\text{threshold}}$ the light will be amplified for each roundtrip, because the amplification exceeds all losses.

**Note:**

The larger the product of absorption cross section $\sigma(\nu)$ times absorption pathlength $L$ the smaller is the threshold inversion $\Delta N_{\text{thr}}$.

The laser oscillation for a continuous laser with time-independent pump power builds up in the following way.

Fluorescence photons, spontaneously emitted by the upper level $E_i$ into the direction of the resonator axis are reflected back into the active medium, where they are amplified, reflected back into the medium by the rear mirror, etc. This results in a photon avalanche with increasing photon number after each roundtrip, if $G(\nu) > 1$ (Fig. 8.5). Part of this radiation power, circulating between the two resonator mirrors is transmitted by one of the mirrors. With increasing photon number the probability of induced emission increases, which decreases the population inversion until it is depleted down to the threshold value. Here, gain and losses are just equal and the laser has reached its stationary state, where the emission is constant. The emitted laser

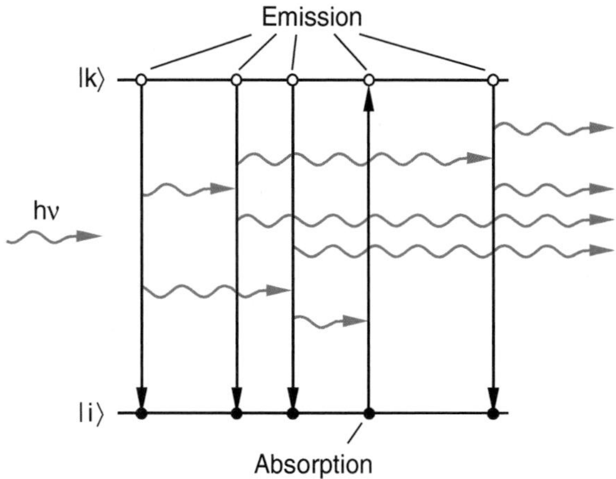


**Fig. 8.5.** Photon avalanche generated by a photon passing through the active medium due to induced emission

power depends on the pump power and the pumping efficiency.

For pulsed lasers the pump power is time-dependent. After a certain pumping time the threshold inversion has been reached. Now laser oscillation starts, which depletes the inversion due to induced emission. The time dependence of the laser output power depends on the relative rates $+\mathrm{d}N_i/\mathrm{d}t$ of pumping and $-\mathrm{d}(N_i - N_k)/\mathrm{d}t$ of inversion depletion by induced emission. For sufficiently strong pumping the laser output power follows the time-dependent pump power and a laser pulse is emitted that is shorter than the pump pulse because it only starts after inversion has been reached and ends when the pump power falls below the threshold value (Fig. 8.6).

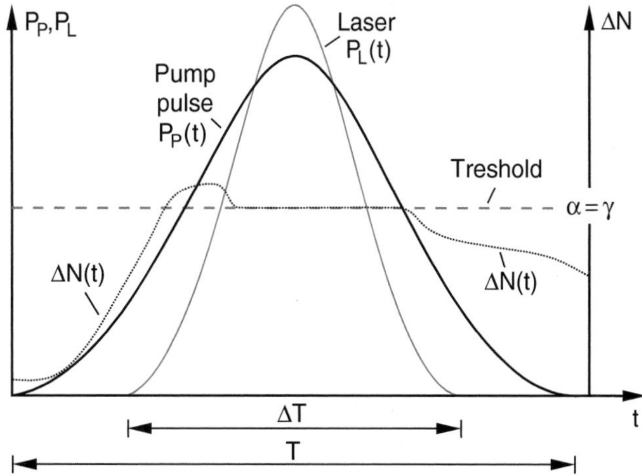


**Fig. 8.6.** Pump-pulse power $P_\mathrm{P}(t)$, laser power $P_\mathrm{L}(t)$, threshold inversion $\alpha = \gamma$ and time-dependent inversion $\Delta N(t)$ for a pulsed laser

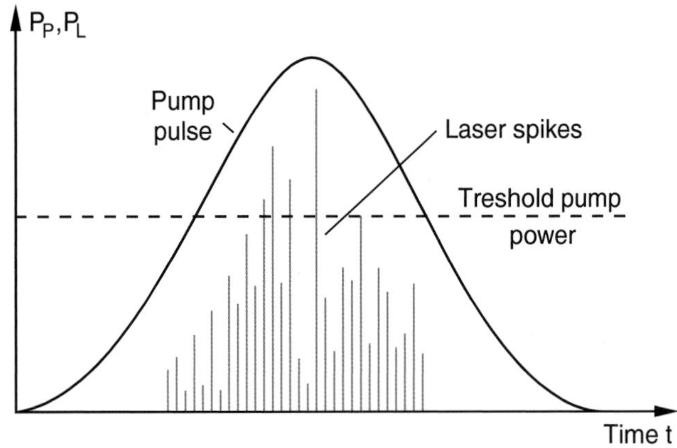


**Fig. 8.7.** Schematic illustration of spikes in the output of a flashlamp-pumped solid-state laser with long relaxation times $\tau_i$, $\tau_k$

In cases of strong depletion by stimulated emission the inversion drops below the threshold already during the pump pulse and the laser emission stops, until the pump has again built up sufficient inversion. Now the laser emission starts again. In such cases (e.g., for the ruby laser) the laser output consists of more or less irregular spikes with short durations, which are emitted while the pump power is above threshold (Fig. 8.7).

The amplification factor is largest for photons with the longest path through the active medium. These are those photons that travel along the resonator axis. Photons emitted into directions inclined to the axis, are not reflected back into the active medium and are therefore less amplified (Fig. 8.8). If their amplification does not reach the threshold value, they cannot contribute to the laser oscillation. Depending on the geometric dimensions of the active medium and the limiting apertures of the optical resonator the laser oscillation is restricted to a small angular divergence around the resonator axis. This results in a laser beam, transmitted through one of the resonator mirrors, which has a small divergence and appears in many cases as nearly parallel light beam with a small diameter.

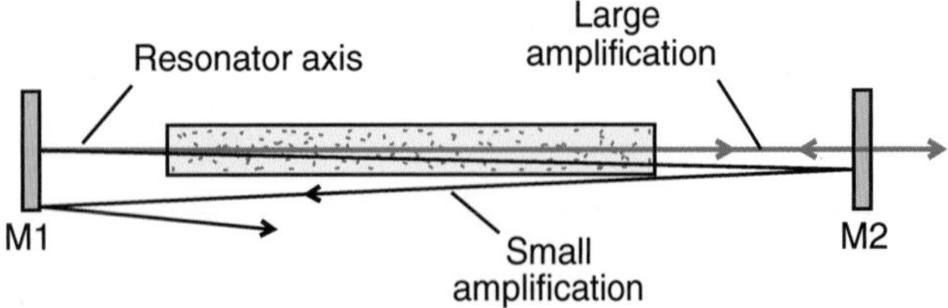


**Fig. 8.8.** The net gain depends on the effective path length through the active medium

### 8.1.2 Generation of Population Inversion

The minimum inversion, required for laser oscillation, can be achieved by a selective pump process, that populates the upper level $E_i$ of the laser transition more strongly than the lower level $E_k$. The pump energy can be transferred either as a pulse (e.g., by flashlamps) or continuously (e.g., by electron impact in a stationary gas discharge). In the first case, laser emission occurs as a pulse, in the second case it occurs continuously (cw = continuous wave operation). We will provide examples of both cases.

The flashlamp-pumped ruby laser historically represents the first demonstration of pulsed laser operation. Its active medium is a cylindrical rod consisting of an $Al_2O_3$ crystal, that is doped with about 1% $Cr^{+++}$ ions. The level scheme of these $Cr^{+++}$ ions is shown in Fig. 8.9. By absorption of light from the flashlamp the ions are pumped from the ground state $E_0$ into the levels $E_1$ and $E_2$, which are strongly broadened by interaction of the excited $Cr^{+++}$-ions with the host crystal. The resulting broad absorption lines overlap with the maximum of the spectral continuum emitted by the flashlamp filled with xenon and can therefore be effectively pumped. The two upper levels transfer part of their excitation energy in a very short time ($10^{-10}$–$10^{-11}$ s) to vibrational energy of the crystal due to a strong interaction with their surroundings. This loss of excitation energy results in fast radiationless transitions into a sharp lower level $E_i$, which is the upper level of the laser transition $E_i \to E_0$ at $\lambda = 694\,\text{nm}$.

In order to achieve population inversion, the number of $Cr^{+++}$ ions in the level $E_i$ must be larger than that in the ground state $E_0$. A direct pumping of

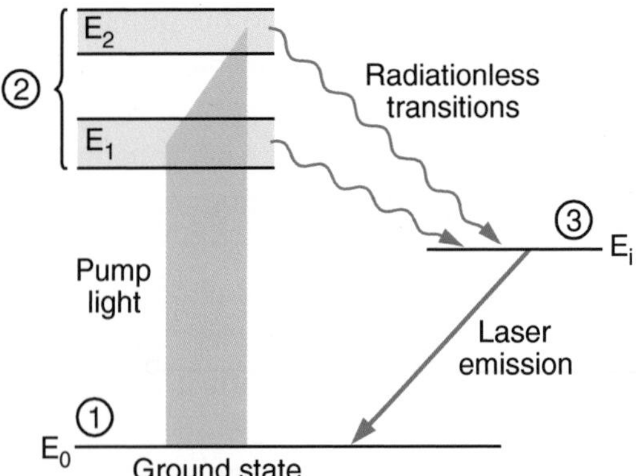


**Fig. 8.9.** Level scheme of the ruby laser

level $E_i$ on a transition $E_0 \to E_i$ could not achieve inversion, because as soon as the populations of both levels become equal, the absorption of the pump light on the laser transition becomes zero and the pump can no longer populate level $E_i$. The intermediate levels $E_1$ and $E_2$ are therefore essential for the realization of laser oscillation. One needs at least three levels, as indicated in Fig. 8.9 by the encircled numbers (where $E_1$ and $E_2$ have been combined into a single level). Such a level scheme for laser operation is called a three level system. The ruby laser is therefore a ***three level laser***.

**Note:**

Under special conditions it is also possible to achieve inversion for a short time in a two-level system, if the pumping time is short compared to all relaxation times of the system and even shorter than the Rabi oscillation time $T_R = \pi \cdot h/(\boldsymbol{M}_{ik} \cdot \boldsymbol{E}(\nu_{ik}))$, where $\boldsymbol{M}_{ik}$ is the matrix element for the transition $i \to k$ and $\boldsymbol{E}$ is the electric field vector of the pump wave. These conditions, however, apply only to very few real systems that are specially designed.

There are several possible experimental configurations of the ruby laser using linear or helical flashlamps (Fig. 8.10). While *Maiman* used the helical design for his first laser, nowadays the linear configuration is preferred. Here the cylindrical ruby rod and the linear flashlamp are placed along the two focal lines of a cylindrical reflector with elliptical cross section (Fig. 8.10a). The light emitted by the flashlamp is focused into the ruby rod from all sides, due to the imaging characteristics of the pump light reflector with elliptical cross section. The mirror surface is coated with dielectric layers with maximum reflection at those wavelengths preferentially absorbed by the $Cr^{+++}$ ions on the transitions $E_0 \to E_1, E_2$. The parallel end faces of the ruby rod are polished and one end face is coated with a highly reflecting layer, the other endface for the laser output with a partially transmitting layer. The flashlamp is fired by discharging a high voltage capacitor through the lamp. While the pump pulse lasts for approximately 1–3 ms, the laser output is a pulse of typically 0.2–0.4 ms, generally consisting of many short (≈ 1 μs) spikes.

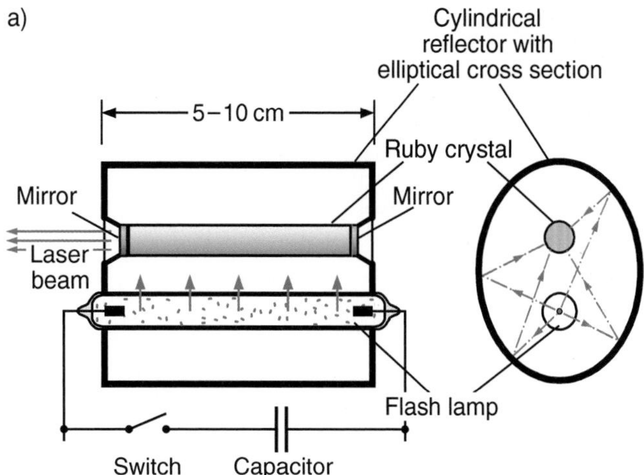


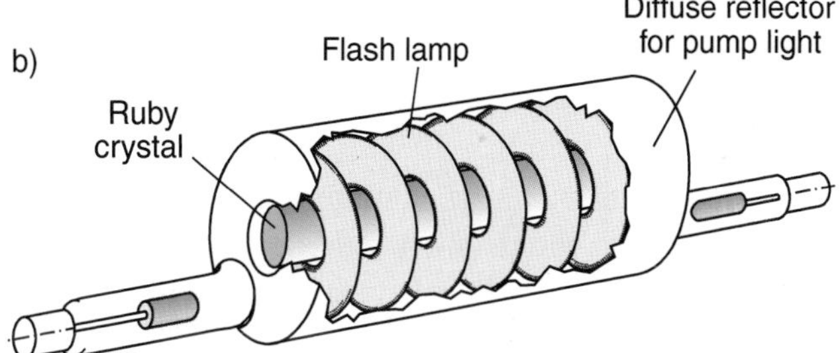


**Fig. 8.10a,b.** Two possible configurations for a pulsed ruby laser. (**a**) Linear flashlamp with cylindrical reflector cavity with elliptical cross section (**b**) Helical flashlamp, originally used by *Maiman* for his first ruby laser

Our second example is the He-Ne-laser, which represents the most commonly used cw gas laser. This laser is based on a four-level system and the pumping is achieved by inelastic collisions of electrons with helium and neon atoms in a stationary gas discharge in a glass tube. Its principle design is shown in Fig. 8.11. A gas discharge is initiated by a high voltage between a cylindrical anode and an aluminum cathode surrounding a glass or quartz capillary (1–4 mm diameter). The power supply for a He-Ne laser has a typical output of 5–10 mA at a voltage of 1 kV. The whole tube is filled with a mixture of about 88% He and 12% Ne at a total pressure of 1–5 mbar. In this discharge (in particular in the narrow capillary, where the current density is high), He and Ne atoms are excited into many high lying energy levels. Most of these levels have a short lifetime and decay by spontaneous emission. In the helium atom there are two metastable states with long lifetimes (see Sect. 6.2). These are the $2\,^3S_1 (\tau = 20\,\text{ms})$ state and the $2\,^1S_0$ state $(\tau \geqslant 600\,\text{s})$, which cannot decay by allowed dipole transitions into lower states. They are populated

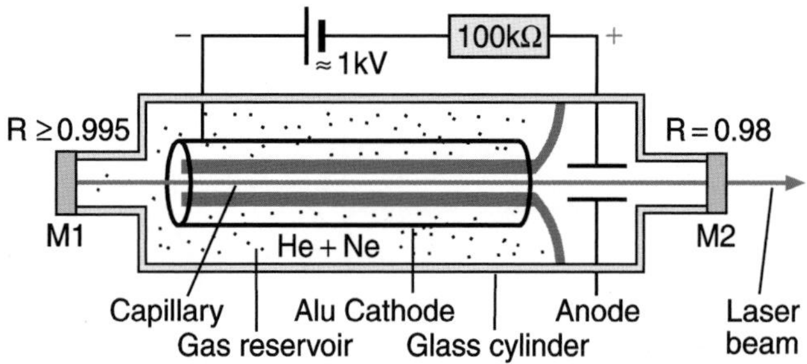


**Fig. 8.11.** Design of a He-Ne laser

not only by electron impact but also by cascading spontaneous emissions from higher levels (Fig. 8.12). In the discharge, therefore, a high concentration of He-atoms in these states is built up.

The electron configuration of excited states in neon is $1s^2\,2s^2\,2p^5\,n'l'$, with $n' = 3, 4, 5, \ldots$. In "Paschen-notation" (L-S-coupling) the sublevels of each configuration are numbered with decreasing energy. For example the $2p^5\,3p$ configuration has ten sublevels ${}^{2S+1}L_J$, which are (with decreasing energy) ${}^1S_0, {}^3P_1, {}^3P_0, {}^3P_2, {}^1P_1, {}^1D_2, {}^3D_1, {}^3D_2, {}^3D_3$ and ${}^3S_1$ numbered by $1, 2, \ldots 10$.

Both metastable He states are in close energy resonance with excited neon levels. By collisions between

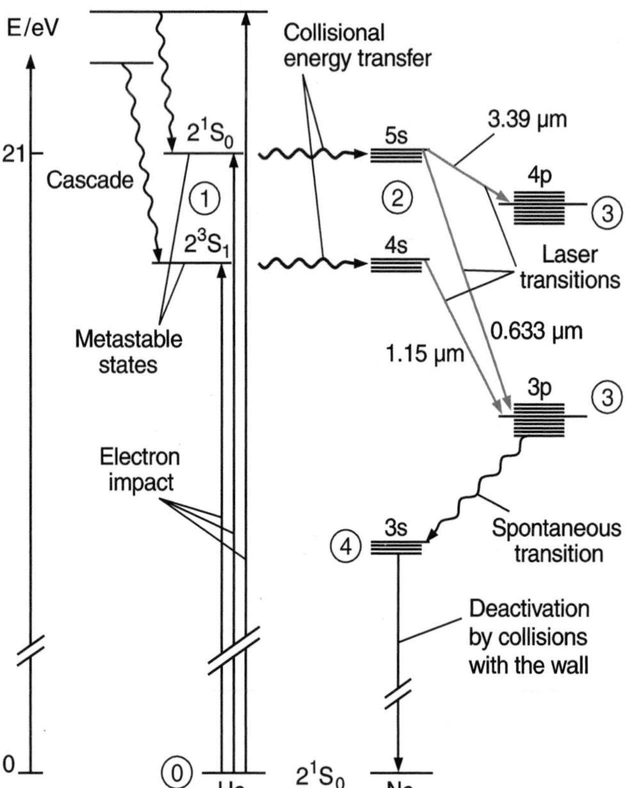


**Fig. 8.12.** Level scheme of the He-Ne laser with three possible laser transitions

excited He atoms and ground state Ne atoms this energy resonance results in large cross sections for the collisional transfer of the excitation energy from the He to the Ne atoms (near resonance collisions of the second kind) according to the scheme:

$$\begin{aligned} &\mathrm{He}^*(2\,{}^1S_0) + \mathrm{Ne}(2\,{}^1S_0) \rightarrow \mathrm{He}(1\,{}^1S_0) + \mathrm{Ne}^*(5s) \\ &\mathrm{He}^*(2\,{}^3S_1) + \mathrm{Ne}(1^1\,S_0) \rightarrow \mathrm{He}(1\,{}^1S_0) + \mathrm{Ne}^*(4s) \end{aligned} \tag{8.7}$$

This energy transfer results in a selective excitation of the neon levels $4s$ and $5s$, which achieve a higher population than the lower levels $4p$ and $5p$, resulting in a population inversion on the transitions $5s \rightarrow 5p$ ($\lambda = 3.39\,\mu$m), $4s \rightarrow 4p$ ($\lambda = 1.15\,\mu$m) and $5s \rightarrow 4p$ ($\lambda = 633$ nm). Such a system, where the laser transition occurs between two excited states, and four levels are involved (the He ground state, a metastable He state and the two Ne levels) is called a ***four-level system***. In Fig. 8.12 the sublevel numbers, explained above, of upper and lower laser levels is indicated on the right side.

Since the population of the lower laser levels is very small, only a small percentage ($\approx 10^{-5}$) of all He atoms needs to be excited into the metastable states, contrary to the three level system of the ruby laser where more than 50% of all $\mathrm{Cr}^{+++}$-ions had to be pumped into the upper laser level. Therefore only about $10^{-6}$ of all neon atoms occupy the upper levels of the laser transitions.

The level scheme shows that laser oscillation is possible for several transitions with different wavelengths. However, only those transitions can reach laser threshold, for which the gain exceeds the losses. The losses can be selected by a proper choice of the resonator mirrors. If the reflection of these mirrors is high for one wavelength but low for the others, laser oscillation can only occur at this favored wavelength.

The gain on the transition $5s \rightarrow 4p$ ($\lambda = 633$ nm) reaches only a few percent for a length of 20 cm in the active medium. Therefore the losses have to be correspondingly low and laser operation could only be achieved after high reflecting dielectric mirrors had been designed with reflectivities of 99.99% for one mirror and 98% for the transmitting mirror. The gain can be increased by using the isotope ${}^3$He instead of ${}^4$He, because here the energy resonance between the metastable He levels and the excited Ne levels is even closer than in ${}^4$He and therefore the cross section for energy transfer from He to Ne is larger.

### 8.1.3 The Frequency Spectrum of Induced Emission

Both the gain $-\alpha(\nu) \cdot L$ and the losses $\gamma(\nu)$ depend on the frequency $\nu$ of the light wave. When the pump process starts, the laser reaches the threshold first for those frequencies where the threshold inversion $\Delta N_{\text{thr}}$ is minimum. The frequency dependence of the gain depends on the active medium. For gaseous media (He-Ne-laser, $Ar^+$-laser) the spectral lines are Doppler-broadened, showing a Gaussian line profile with a width of several GHz ($10^9$ Hz). For solid state lasers or liquid lasers the line width is mainly determined by interaction of the laser atoms, ions or molecules with their surroundings. It is generally much broader than in gases.

The loss factor $\gamma$ mainly depends on the characteristics of the optical resonator. It has minima at the resonance frequencies of the resonator. Therefore lasing starts at those resonator resonances that lie within the spectral gain profile of the active medium. If the spacing between resonator modes is smaller than the spectral width of the gain profile, the laser oscillates on several wavelengths simultaneously. This deteriorates the coherence properties of the laser emission. If oscillation on a single wavelength is required, additional wavelength selecting elements have to be introduced. This can be realized either by a special resonator design or by prisms or optical gratings inside or outside the resonator.

Since the laser resonator plays a central role for the spectral characteristics of laser emission, we will first discuss optical resonators.

## 8.2 Optical Resonators

In Sect. 3.1.2 it was shown that inside a closed resonator a radiation field can exist with an energy density $w_\nu(\nu)$ that is equally distributed over all resonances or modes of the cavity. In the optical spectral range where the wavelength $\lambda$ is small compared to the dimensions of the cavity, the number of modes within the frequency interval $\mathrm{d}\nu$ is (see (3.10b))

$$n(\nu)\,\mathrm{d}\nu = 8\pi(\nu^2/c^3)\,\mathrm{d}\nu\ .$$

For $\nu = 5 \times 10^{14}\,\mathrm{s}^{-1}$ ($\lambda = 600\,\mathrm{nm}$) the number of modes within a Doppler broadened spectral line ($\Delta\nu = 10^9\,\mathrm{s}^{-1}$) is $n(\nu)\Delta\nu = 2.5 \times 10^{14}\,\mathrm{m}^{-3}$. This implies that the spontaneous emission from excited atoms inside a closed cavity is distributed over many modes, which means that the average photon number per mode is very small. In such a closed cavity the induced emission, started by spontaneous photon avalanches, is spread out over many modes. Because the total power emitted by spontaneous and induced emission has to be supplied by the pump energy, one needs an exceedingly high pump power in order to achieve laser oscillation on all these modes. The laser emission would then be distributed over many directions into the solid angle $4\pi$ and the directionality of laser emission would be lost.

Closed cavities, which are used for the realization of masers in the microwave region, where $\lambda$ is comparable with the cavity dimensions, are not suitable for optical lasers.

### 8.2.1 The Quality Factor of Resonators

Assume that the $k$th resonator mode contains the radiation energy $W_k(t)$. If no energy is fed from external sources into this mode, its stored energy will decrease as

$$\frac{\mathrm{d}W_k}{\mathrm{d}t} = -\beta_k \cdot W_k\ , \tag{8.8}$$

which yields the time-dependent stored energy

$$W_k(t) = W_k(0) \cdot \mathrm{e}^{-\beta_k t} \tag{8.9}$$

with the loss factor $\beta_k$. After the time $\tau = 1/\beta_k$ the energy stored in the $k$th mode has decayed to $1/e$ of its initial value at $t = 0$. This time can be regarded as the mean lifetime of a photon stored in this resonator mode. We define the quality factor $Q_k$ of the $k$th resonator mode as $2\pi$ times the ratio of the energy, stored in this mode to the energy loss per oscillation period $T = 1/\nu$ of the radiation with frequency $\nu$:

$$Q_k = -\frac{2\pi\nu \cdot W_k}{\mathrm{d}W_k/\mathrm{d}t}\ . \tag{8.10}$$

Inserting (8.8) and (8.9) gives the relation between the loss factor $\beta_k$ and the quality factor $Q_k$:

$$Q_k = 2\pi\nu/\beta_k\ . \tag{8.11}$$

Since the round trip time in a resonator with mirror separation $d$ is $T = 2d/c$ the loss factor $\gamma$ per roundtrip is then

$$\gamma_k = \beta_k \cdot (2d/c)\,. \tag{8.12}$$

Even if at $t = 0$ the radiation energy, supplied by spontaneous emission, is the same for all modes, those modes with a high $Q$-factor will store this energy for a longer time while those with a low $Q$-factor loose their energy after a short time.

### 8.2.2 Open Optical Resonators

In order to concentrate the induced emission onto a few modes, the $Q$-factor of the resonator must be large for these modes (i.e., the losses must be small), while it should be sufficiently small for all other modes, so that for a given pump power the threshold for laser oscillation is not reached for these modes.

Open resonators, consisting of a suitable arrangement of optical mirrors can fulfill this condition. We will illustrate this by the example of two plane-parallel mirrors $M_1$ and $M_2$ with reflectivity $R_1$ and $R_2$ and with diameters $2a$, which are separated by the distance $d$ (Fig. 8.4). This represents, in fact, a Fabry-Perot interferometer (FPI) used in spectroscopy as a spectral filter with high resolution. There is, however, an essential difference from conventional FPI, where the mirror separation $d$ is small compared to the diameter $2a$ of the mirrors. For the laser resonator the situation is the opposite: here $d \gg 2a$. This makes a large difference with respect to diffraction, which is negligible in a conventional FPI, but essential in a laser resonator.

We will first regard the *reflection losses*.

A light wave reflected back and forth between the mirrors suffers reflection losses and its intensity decreases per roundtrip according to

$$I(2d) = I_0 R_1 R_2 = I_0 \cdot \mathrm{e}^{-\gamma_r}\,. \tag{8.13}$$

The reflection loss factor $\gamma_r$ is defined as

$$\gamma_r = -\ln(R_1 R_2)\,. \tag{8.14}$$

Since the transit time for one roundtrip is $T = 2d/c$, the mean lifetime $\tau$ of a photon stored in the resonator and traveling along the resonator axis is

$$\tau = \frac{2d}{c \cdot \ln(R_1 R_2)} \tag{8.15}$$

if no other losses were present.

**EXAMPLE**

$R_1 = 1$, $R_2 = 0.98$, $d = 0.5\,\mathrm{m} \Rightarrow \gamma_r = 0.02$ and $\tau = 1.5 \times 10^{-7}\,\mathrm{s}$.

We will now discuss the *diffraction losses* of open resonators.

Because of the finite diameter $2a \ll d$ of the mirrors diffraction losses are generally not negligible. This is illustrated by Fig. 8.13c. A plane wave traveling from below onto the mirror $M_1$ is no longer

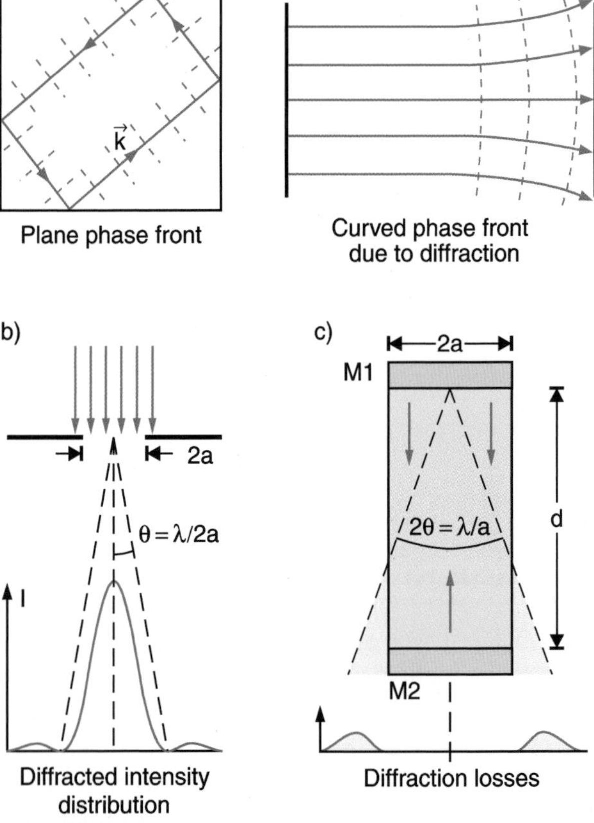


**Fig. 8.13.** (**a**) Plane waves as stationary field solutions in a cubic closed resonator compared with curved wave fronts in an open resonator with diffraction losses. (**b**) Diffraction pattern of a plane wave behind a circular aperture with diameter $2a$, compared in (**c**) to a similar pattern after reflection by a plane mirror of size $2a$

reflected as a plane wave but becomes divergent because of diffraction. This is completely analogous to a plane wave passing through an aperture with diameter $2a$ (Fig. 8.13b). Here the transmitted wave shows an intensity profile

$$I(\Theta) = I_0 \left(\frac{2J_1(x)}{x}\right)^2 \quad \text{with} \quad x = \frac{2\pi a}{\lambda}\sin\Theta \tag{8.16}$$

with a central maximum and higher diffraction orders (see textbooks on optics). The central diffraction maximum has an angular width between the first two nodes of the Bessel function $J_1(x)$ on both sides of the maximum at $x = 0$, which gives

$$\sin\Theta = 1.2\lambda/(2a) \Rightarrow \Theta \approx \lambda/(1.7a)\,. \tag{8.17a}$$

Light with larger diffraction angles does not hit the mirror $M_2$ in Fig. 8.13c and is therefore lost. If the total light power included in the 0th diffraction order, should be reflected by $M_2$ the diffraction angle $\Theta$ has to obey the relation

$$\tan\Theta \cdot d \approx \Theta \cdot d \le a\,.$$

Inserting (8.17a) yields

$$1.7\frac{a^2}{\lambda \cdot d} \ge 1\,. \tag{8.17b}$$

The ratio

$$N_F = a^2/(\lambda d) \tag{8.18}$$

is called the Fresnel number of the resonator. It gives the number of Fresnel zones on the surface of $M_1$, which can be seen from the center A of $M_2$ (Fig. 8.14).

A more detailed calculation shows [8.8, 9] that for $N_F \gg 1$ the diffraction loss factor is $\gamma_d \approx 1/N$. This means that in a resonator with Fresnel number $N_F$ the light power drops after one roundtrip by a factor $\exp(-1/N)$ if only diffraction losses were present. When the light wave makes $m$ roundtrips, the Fresnel number should be $N_F > m \cdot \gamma_R$ if the diffraction losses are to be smaller than the reflection losses.

**EXAMPLE**

For a FPI with $a = 2$ cm and $d = 1$ cm, typically for spectroscopic applications, the Fresnel number for $\lambda = 500$ nm is $N_F = 8 \times 10^4$. The diffraction loss factor is $\gamma_d = 1.2 \times 10^{-5}$ and diffraction losses are therefore negligible. The phase fronts of a wave inside the FPI are planes and the mirror surfaces are nodal planes of the standing wave. These dimensions are, however, not suitable for a laser resonator.

The resonator of a gas laser with plane mirrors (diameter $2a = 0.2$ cm and a separation of $d = 50$ cm) has for $\lambda = 500$ nm a Fresnel number $N_F = 4$. The diffraction losses per roundtrip amount already to 25% and a He-Ne-laser with such a resonator would not reach threshold.

### 8.2.3 Modes of Open Resonators

While the modes of closed cavities can be described as a superposition of plane waves (see Sect. 3.1.2) with amplitudes and phases that are constant on planes perpendicular to the wave vector $\boldsymbol{k}$, in open resonators both quantities are changing across these planes because the diffraction causes a curvature of the wave fronts. Possible modes of open resonators are therefore not plane waves!

The amplitude and phase distribution $A(x, y)$ and $\varphi(x, y)$ of modes in an open resonator with the resonator axis in $z$-direction can be determined in the following way.

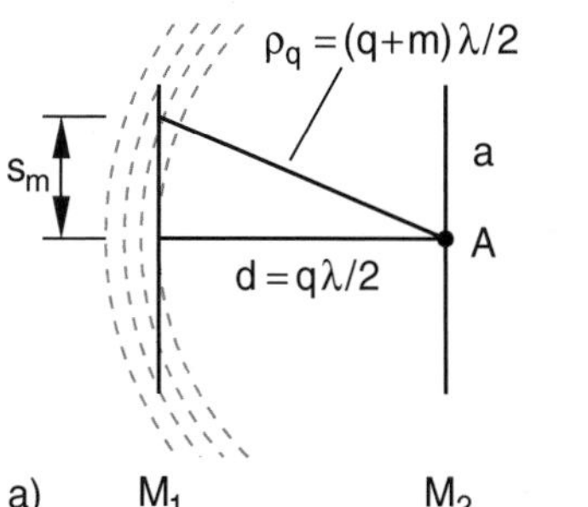


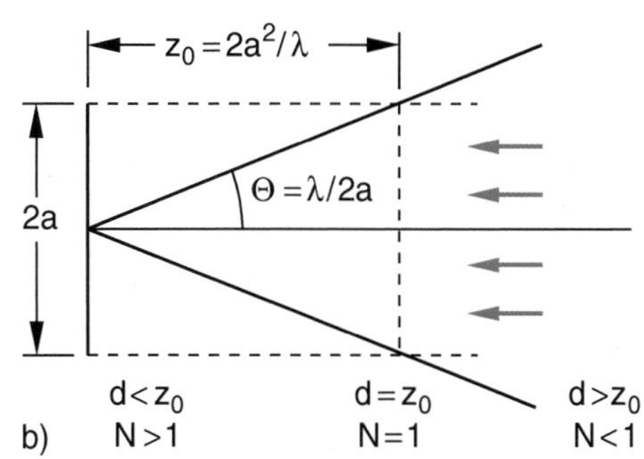


**Fig. 8.14.** (**a**) Fresnel zones on mirror $M_1$, as seen from the center A of the other mirror $M_2$. (**b**) The three regions of $d/a$ with the Fresnel number $N > 1$, $N = 1$, and $N < 1$

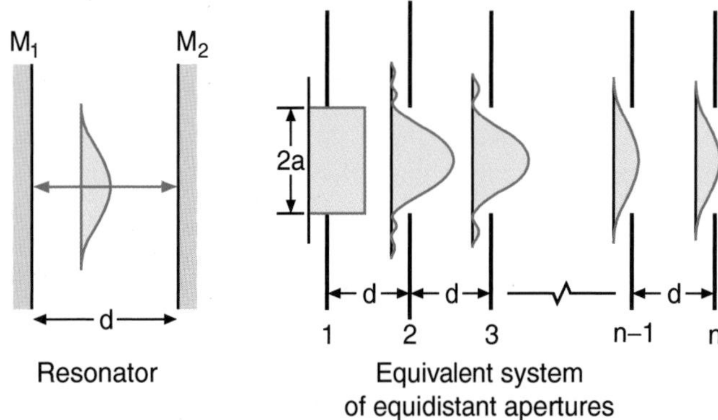


**Fig. 8.15.** The diffraction of a wave traveling back and forth between two mirrors $M_1$ and $M_2$ is equivalent to that of a wave passing through a series of equidistant apertures

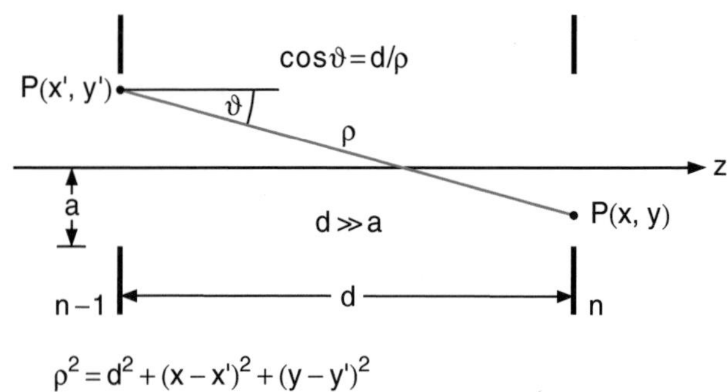


**Fig. 8.16.** Illustration of Eq. (8.20)

The light wave being reflected back and forth between the two resonator mirrors corresponds to, regarding the diffraction effects, a wave passing in $z$-direction through a series of equidistant apertures with the same size as the mirrors (Fig. 8.15). This is shown in optics by Babinet's theorem. When a plane wave passes through the first aperture in the plane $z = 0$ the amplitude distribution $A(x, y)$ will change due to diffraction. The amplitude will at first decrease more at the edges than in the center, until the diffraction losses are equal for all values of $x$ and $y$. We assume that after having passed the $n$th aperture, the diffracted wave will have reached a stationary state, where the relative amplitude distribution $A(x, y)$ will no longer change, although the absolute total amplitude may still decrease. This implies the relation:

$$A_n(x, y) = C \cdot A_{n-1}(x, y) \tag{8.19}$$

where the constant $C$ with $|C| < 1$ does not depend on $x$ and $y$.

The amplitude distribution $A_n(x, y)$ across the $n$th aperture can be calculated from the distribution $A_{n-1}(x', y')$ across the foregoing aperture, using Kirchhoff's diffraction theory. The light emitted by every point $(x', y')$ contributes to the amplitude $A(x, y)$ in the $n$th aperture. From Fig. 8.16 we obtain the relation

$$A_n(x, y) = -\frac{i}{2\lambda} \int\limits_{x'} \int\limits_{y'} A_{n-1}(x', y') \frac{1}{\varrho} \mathrm{e}^{-ik\varrho} \times (1 + \cos\vartheta) \, \mathrm{d}x' \, \mathrm{d}y' \,. \tag{8.20}$$

Inserting (8.19) gives an integral equation for the amplitude $A(x, y)$, which can be generally solved only numerical, except for special cases where analytical solutions are possible [8.10]. The constant factor $C$ in (8.19) is found to be

$$C = (1 - \gamma_\mathrm{d})^{1/2} \cdot \mathrm{e}^{i\varphi} \tag{8.21}$$

where $\gamma_\mathrm{B}$ is the diffraction factor, and $\varphi$ is the phase shift, caused by the curvature of the wave fronts, due to diffraction.

Some solutions of the integral equation (8.20) are illustrated in Fig. 8.17. They correspond to stationary solutions as standing waves between the two resonator mirrors and are called transverse electromagnetic ***(TEM)*** modes of the open resonator. They are labeled by three indices, which give the number of nodes of the standing wave in the $x$-, $y$-, and $z$-directions (Fig. 8.18). The $\mathrm{TEM}_{0,0,q}$ modes with no nodes in $x$- and $y$-direction are called ***fundamental modes***. Their $\boldsymbol{k}$-vector points into the $\pm z$-direction and they have q nodes along the $z$-axis where $q = 2d/\lambda$. Their electric field amplitude distribution $E(x, y)$ shows a Gaussian profile.

Generally, mirrors of circular size are used and the active medium also has a circular cross section. Because of this cylindrical symmetry, cylinder coordinates $(r, \varphi, z)$ are better suited for the description of the

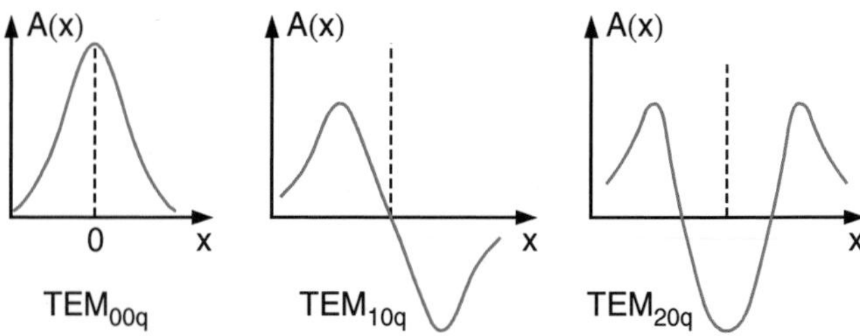


**Fig. 8.17.** One-dimensional electric field distribution in the $x$ direction for some resonator modes

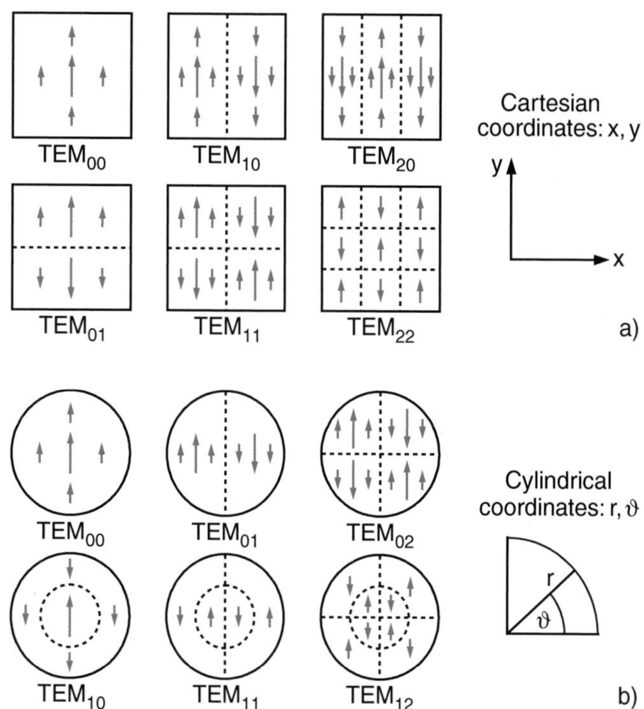


**Fig. 8.18a,b.** Schematic representation of electric field distribution in the $xy$-plane inside the resonator (**a**) In Cartesian coordinates (**b**) In cylindric coordinates

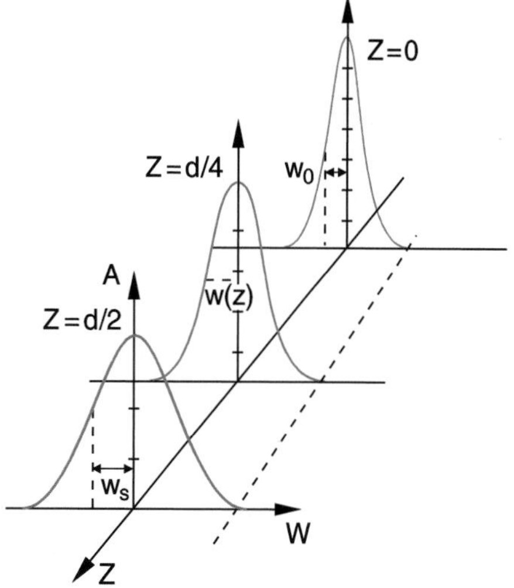


**Fig. 8.19.** Radial intensity profiles $I(r,z)$ and beam waists $w_S(z)$ in a confocal resonator

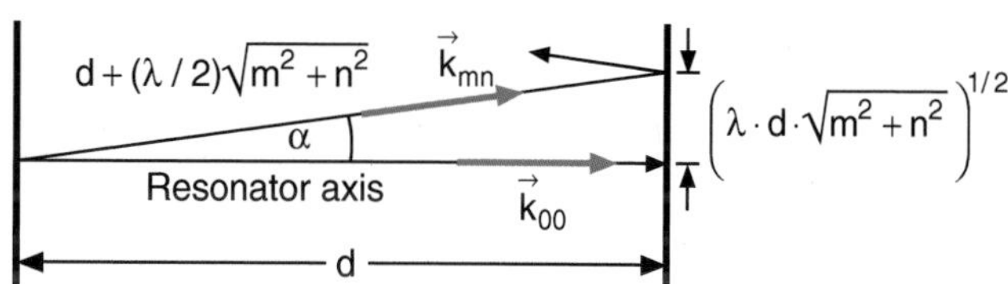


**Fig. 8.20.** Direction of the wave vector $\boldsymbol{k}_{mn}$ of a transverse mode $\mathrm{TEM}_{m,n,q}$ against the resonator axis

amplitude distribution of the modes. The fundamental modes are then described by the radial field amplitude distribution

$$E(r,\varphi,z) = E_0 \mathrm{e}^{-(r/w)^2} ,$$

where $w$ is the beam waist, i.e. for $r = w$ the amplitude has decreased to $E_0/\mathrm{e}$. Because the intensity is related to the electric field amplitude by

$$I = c\varepsilon_0 E^2$$

we obtain the intensity distribution of the fundamental modes

$$I(r,z) = I_0 \mathrm{e}^{-2(r/w(z))^2} \qquad (8.22)$$

where the beam waist $r = w(z)$, for which the intensity has dropped to $I(w) = I(0)/\mathrm{e}^2$ can depend on the $z$-coordinate (Fig. 8.19).

The higher transverse modes $\mathrm{TEM}_{n,m,q}$ with $n, m > 0$ correspond to standing waves with $\boldsymbol{k}$-vectors that are inclined by a small angle $\alpha$ against the resonator axis (Fig. 8.20). For these modes the path length between the two mirrors is

$$s = d + (\lambda/2)(m^2 + n^2)^{1/2} . \qquad (8.23a)$$

The inclination angle is

$$\tan\alpha = \left[(\lambda/d)\cdot(m^2+n^2)^{1/2}\right]^{1/2} . \qquad (8.23b)$$

## EXAMPLE

$d = 50\,\mathrm{cm}$, $\lambda = 500\,\mathrm{nm}$, $m = n = 1 \Rightarrow \tan\alpha = 1.2 \times 10^{-3} \Rightarrow \alpha = 0.07° = 4.2'$.

Resonators with plane mirrors are often not the best choice for two reasons. They have large diffraction losses and they are very critical regarding alignment. A tilt by an angle $\varepsilon$ changes the direction of the reflected beam by $2\varepsilon$, and the reflected beam might not pass back through the active medium (Fig. 8.21a).

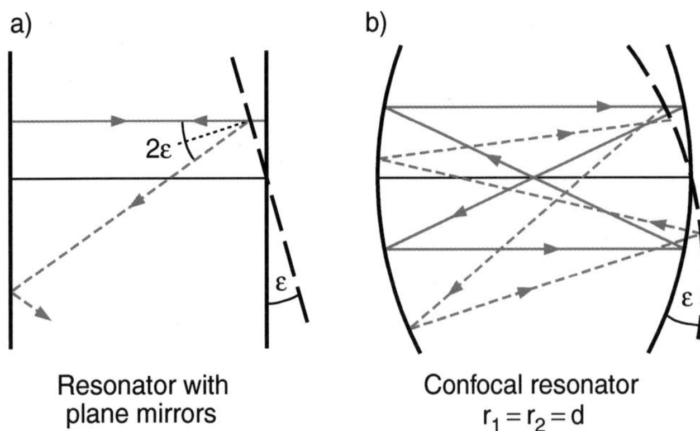


**Fig. 8.21a,b.** Different sensitivities against misalignment for resonators with plane mirrors (**a**) compared to confocal resonators with curved mirrors (**b**)

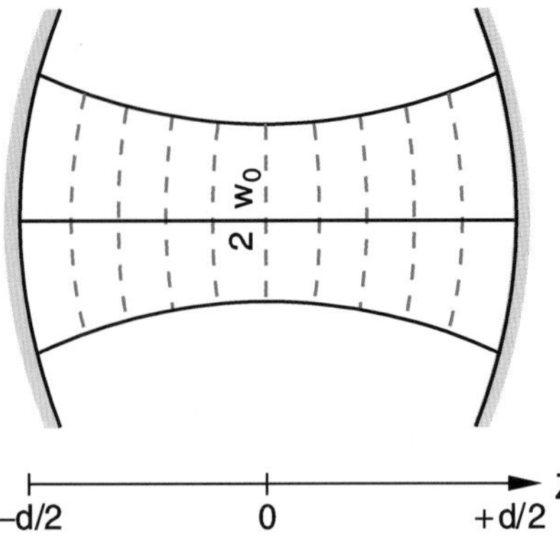


**Fig. 8.22.** Phase fronts at different locations $z$ in a confocal resonator with the mirrors at $z = \pm d/2$

## EXAMPLE

$d = 1$ m and $a = 2$ mm. If the laser beam should pass 50 times through the active medium, the deviation from the correct alignment of the mirrors should not be larger than
$\varepsilon = 2 \times 10^{-3}/50 = 4 \times 10^{-5}\,\text{rad} = \left(2.4 \times 10^{-3}\right)^\circ = 8.5''$.

Spherical mirrors are less critical with respect to alignment, as is shown in Fig. 8.21b for the example of a confocal resonator, where the mirrors with radius of curvature $r$ are separated by the distance $d = r$. The focal points of both mirrors coincide.

Resonators with spherical mirrors have lower diffraction losses, because they refocus the divergent diffracted beam and therefore decrease the beam spot size on the mirrors, if their radius of curvature $r$ and their distance $d$ is chosen properly.

In Fig. 8.19 the beam profile for the fundamental modes are shown for a confocal resonator with two spherical mirrors with equal radii of curvature $r$ and mirror separation $d = r$. The smallest spot size appears in the middle of the resonator at $z = 0$ when the mirrors are at $z = \pm d/2$. In Fig. 8.22 the phase fronts of the fundamental mode in a confocal resonator are illustrated. At $z = 0$ in the middle of the resonator they are plane, at the mirrors they coincide with the mirror surfaces.

### 8.2.4 Diffraction Losses of Open Resonators

The diffraction losses of a standing wave inside a resonator depend on the radial intensity distribution $I(r)$. The larger the intensity at the edges of the mirrors or of limiting apertures inside the resonator, the larger are the diffraction losses. This implies, that the fundamental modes $\text{TEM}_{00q}$ have the lowest diffraction losses while the higher transverse modes with $n, m > 0$ suffer larger losses. In Fig. 8.23 the diffraction losses for the fundamental and for some transverse modes are plotted as a function of the Fresnel number $N_\text{F}$ for resonators with plane mirrors and for confocal resonators with curved mirrors. This illustrates that for confocal resonators diffraction losses are much lower. In fact, a He-Ne-laser can only operate with curved mirrors, because otherwise the diffraction losses would be too high for the small gain achievable in a discharge with only 10–15 cm length.

These diffraction losses offer the possibility to eliminate higher transverse modes and to achieve laser oscillation solely on fundamental modes. The resonator configuration has to be chosen in such a way, that the transverse modes suffer sufficiently high losses, to prevent them from reaching the oscillation threshold.

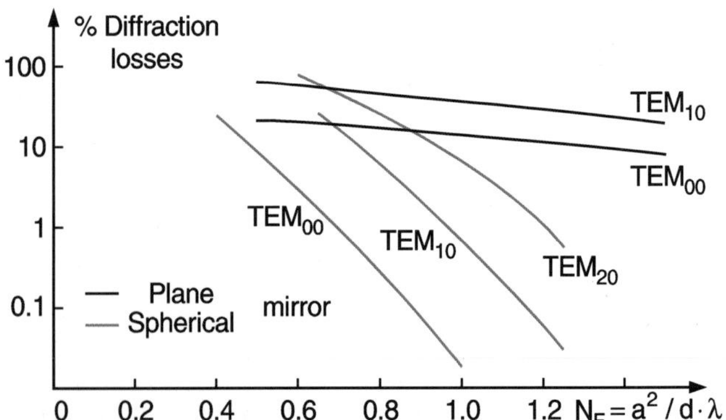


**Fig. 8.23.** Diffraction losses of some modes in resonators with plane and with curved confocal mirrors, as a function of the Fresnel number $F$

**EXAMPLE**

When the gain of the active medium per roundtrip is 10% ($G(\nu) = 1.1$ in (8.3)), the Fresnel number of a confocal resonator has to be $N_F < 0.8$, according to Fig. 8.23, in order to prevent all transverse modes from oscillation. For a wavelength $\lambda = 600\,\text{nm}$ and a mirror separation $d = 50\,\text{cm}$ the limiting aperture must have a diameter of $2a = 2(N_F\lambda d)^{1/2} < 10^{-4}\,\text{m} \approx 1\,\text{mm}$.

### 8.2.5 The Frequency Spectrum of Optical Resonators

For the fundamental modes with $m = n = 0$ a standing wave can build up in a resonator with plane mirrors if an integer multiple of the half-wavelength fits between the mirrors:

$$d = q\lambda/2 \Rightarrow \nu_r = qc/(2d)\,. \tag{8.24a}$$

The resonance frequencies $\nu_r$ of neighboring fundamental modes are separated by

$$\delta\nu_r = \nu_r(q) - \nu_r(q-1) = c/(2d)\,. \tag{8.24b}$$

The spacing $\delta\nu_r$ is called the ***free spectral range*** of the resonantor.

For the transverse modes $\text{TEM}_{nmq}$ the resonance frequencies are obtained from the solutions of the integral equation (8.20), which can be solved analytically for the confocal resonator [8.10]. One obtains

$$\nu_r = \frac{c}{2d}\left(q + \tfrac{1}{2}(m+n+1)\right)\,, \tag{8.24c}$$

which converts to (8.24a) for $m = n = 0$, if $q$ is replaced by $q^* = q + \frac{1}{2}$. When $m + n$ is an even integer, the eigenfrequencies of the transverse modes are just in the mid between two fundamental (also called longitudinal) modes, if $m + n$ is odd, they coincide with a frequency of a fundamental mode.

Standing $\text{TEM}_{nmq}$ waves with these eigenfrequencies have minimum losses. They are stored inside the resonator for a much longer time then waves with nonresonant frequencies. The total losses can be described by the sum

$$\gamma = \gamma_r + \gamma_{sc} + \gamma_{diffr}$$

of the loss factors for reflection losses, scattering and diffraction losses, where $\gamma_{diffr}$ sharply increases with $m$ and $n$.

The threshold condition

$$-2\alpha(\nu)L - \gamma(\nu) > 0$$

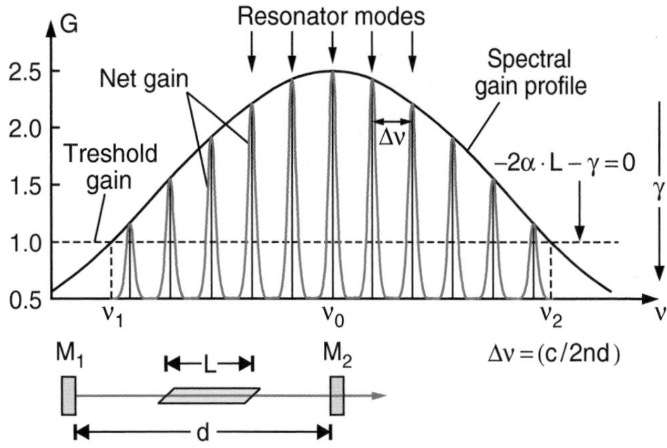


**Fig. 8.24.** Net gain $G(\nu)$ for resonator modes within the gain profile of the active medium. The vertical black lines give the frequencies of a multimode laser oscillating only on fundamental modes $\text{TEM}_{0,0,q}$

is only fulfilled for those resonance frequencies which lie within the spectral gain profile of the amplifying transition of the active medium (Fig. 8.24). The laser emission consists of all these frequencies and the total bandwidth of the laser emission depends on the width of the gain profile above the threshold line $-2\alpha L = \gamma$ in Fig. 8.24.

**EXAMPLES**

1. ***He-Ne Laser***: $d = 50\,\text{cm} \Rightarrow \delta\nu_r = c/2d = 300\,\text{MHz}$. Within the gainprofile with $\Delta\nu_D = 1.5\,\text{GHz}$ are five longitudinal modes. If the discharge tube diameter is $2a < 1\,\text{mm}$, the diffraction losses are too high for transverse modes and the laser oscillates solely on these five fundamental modes.
2. ***Ruby laser***: $d = 10\,\text{cm} \Rightarrow \delta\nu_r = 1.5\,\text{GHz}$. With $2a = 6\,\text{mm}$ the diffraction losses are smaller than the high gain even for higher transverse modes. The width of the gain profile is about 30 GHz. This means that besides about 20 fundamental modes many transverse modes are present in the emission of the ruby laser.

## 8.3 Single Mode Lasers

In order to achieve laser oscillation on a single fundamental mode several measures can be taken.

The simplest one is the shortening of the resonator length $d$ below a value where the mode spacing

$\delta\nu_r = c/(2d)$ becomes larger than one-half of the spectral width of the gain profile at the threshold line. This, however, generally reduces the gain for gas lasers and only small output powers can be achieved. For solid state lasers with a large gain per centimeter, this might be a solution, but the spectral gain profile of these lasers is generally very broad and even short cavities still might result in multimode operation.

The better, and most commonly used method for achieving single mode operation is the insertion of additional frequency selective optical elements into the laser resonator. Such an element can be, for instance, a tiltable plane parallel glass plate with reflecting surfaces on both sides (Fig. 8.25a), which represents a Fabry-Perot etalon. As shown in textbooks on optics, the transmission of this etalon with reflectivity $R$ on both sides is given by

$$T = \frac{1}{1 + F \cdot \sin^2(\delta/2)} \tag{8.25}$$

with

$$F = \frac{4R}{(1-R)^2} \, .$$

The phase shift $\delta = 2\pi\,\Delta s/\lambda$ between two adjacent interfering partial beams with angles of incidence $\alpha$ against the normal to the plate surfaces (Fig. 8.26) is determined by the optical path difference

$$\Delta s = 2t\sqrt{n^2 - \sin^2\alpha} \, . \tag{8.26}$$

From (8.25) it follows that $T = 1$ for $\delta = 2m \cdot \pi$. This is fulfilled for all wavelengths $\lambda_m = \Delta s/m$ i.e. for all frequencies $\nu_m = c/\lambda_m = (c/\Delta s) \cdot m$, $m = 1, 2, 3, \ldots$.

Adjusting the tilting angle $\alpha$ correctly, one of the frequencies $\nu_m$ can coincide with a resonator eigenfrequency inside the gain profile (Fig. 8.25c). Only for this frequency are the total losses small, for all other resonator eigenfrequencies the transmission of the etalon is small and if the reflectivity $R$ of the etalon is sufficiently high the total losses for these frequencies are larger than the gain and they do not reach oscillation threshold (Fig. 8.25d). The laser then oscillates on

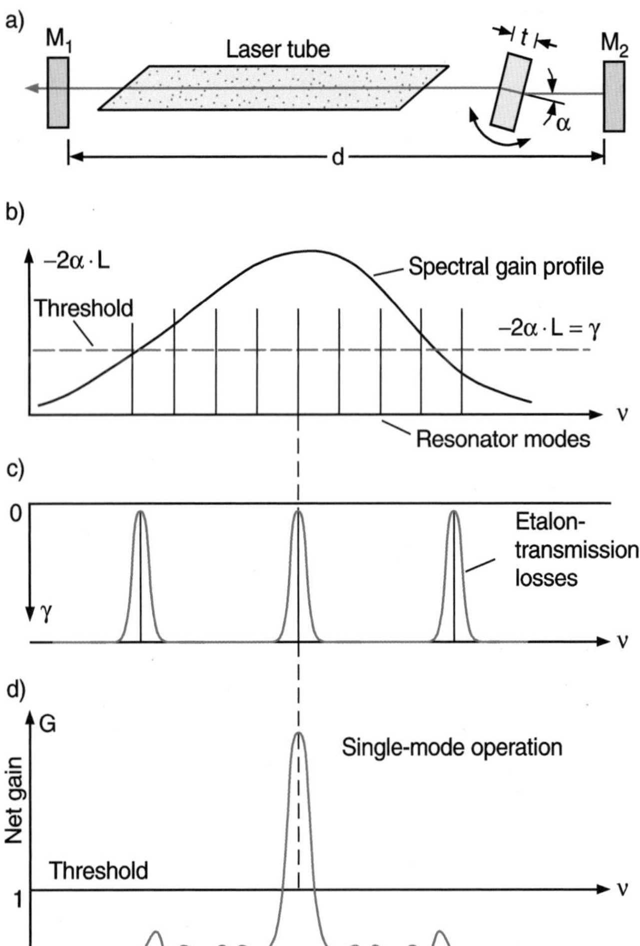


**Fig. 8.25a–d.** Selection of a single resonator mode (**a**) Experimental setup (**b**) Resonator modes within the gain profile of the active medium (**c**) Transmission $T(\nu)$ of the etalon (**d**) Net gain of the laser with the etalon inside the resonator

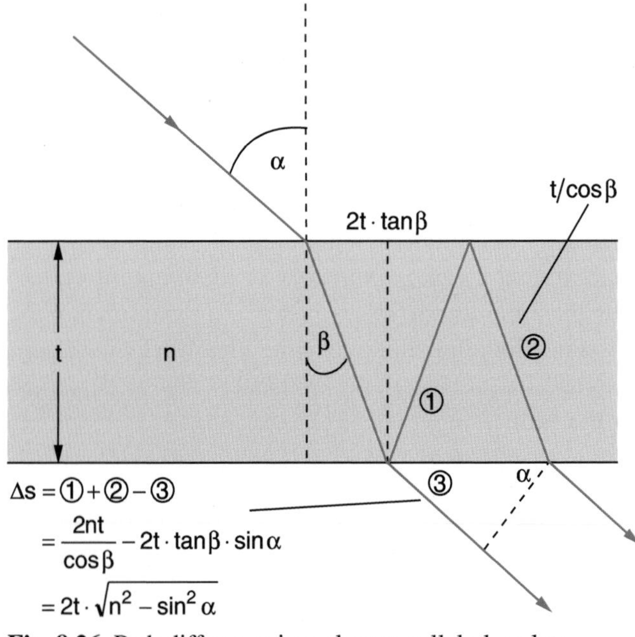


**Fig. 8.26.** Path difference in a plane parallel glas plate

a single fundamental mode if the transverse modes are eliminated by high diffraction losses.

The mean line width of such a single mode laser is mainly determined by technical fluctuations of the optical resonator length $n \cdot d$, where $n$ is the refractive index between the resonator mirrors. Since the laser frequency is given by the eigenfrequency of the resonator

$$\nu_L = \nu_r = q \cdot c/(2nd)$$

fluctuations $\Delta n$ of the refractive index or $\Delta d$ of the resonator length result in corrsponding fluctuations of the laser frequency

$$\frac{-\Delta\nu_L}{\nu_L} = \frac{\Delta n}{n} + \frac{\Delta d}{d} . \tag{8.27}$$

## EXAMPLES

1. If the mirror separation $d = 50\,\text{cm}$ changes by 1 nm, this results in a relative frequency change $\Delta\nu/\nu = 2 \times 10^{-9}$. At a laser frequency of $\nu = 5 \times 10^{14}\,\text{s}^{-1}$ we obtain $\Delta\nu_L = 1\,\text{MHz}$!
2. If the air pressure between the mirrors changes by 1 mbar, this results in a change $\Delta n/n = 2.5 \times 10^{-7}$ of the refractive index, which means a frequency shift of 125 MHz at $\nu = 5 \times 10^{14}\,\text{s}^{-1}$

Such technical fluctuations can be partly compensated, if one of the resonator mirrors is mounted on a piezocrystal (Fig. 8.27). This consists of a material, that changes its length under an external voltage applied to its end faces. If part of the laser output is sent through a very stable Fabry-Perot (Fig. 8.28), the transmitted intensity changes when the laser frequency changes. A photodiode $D_2$ behind the FPI gives an output voltage that reflects this intensity change. The output is compared with a reference voltage provided by the photo diode $D_1$ and the difference is amplified and applied to the piezocrystal, which changes the resonator length and brings the laser frequency back to its wanted value. Such a feedback control system can stabilize the laser frequency within about 1 Hz! New very sophisticated devices can even reach a stability of $1\,\text{mHz} = 10^{-3}\,\text{Hz}$.

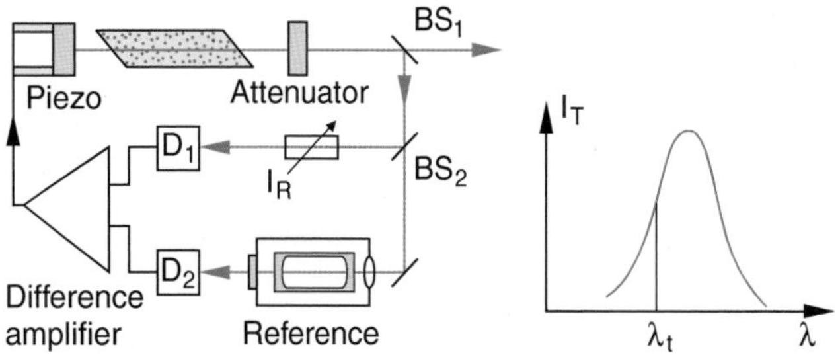


**Fig. 8.28.** Laser wavelength stabilization onto the slope of the transmission $T(\lambda)$ of a stable reference FPI

If all of these technical contributions to the frequency fluctuations are eliminated, there is still a principle physical limitation to the line width of the laser due to the following effect.

The laser emission starts with avalanches of photons induced by spontaneous emission. Since the spontaneous photons are randomly emitted, the amplitudes and phases of these avalanches are random. The total laser output consists of a superposition of such avalanches.

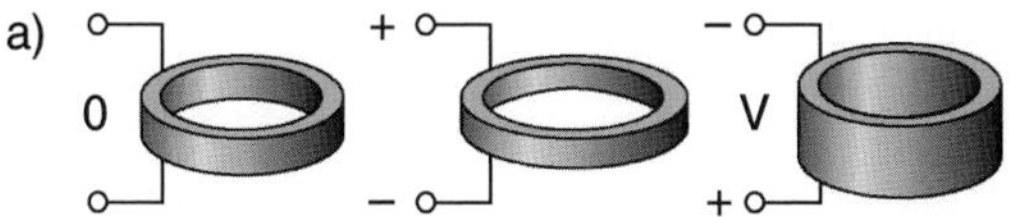


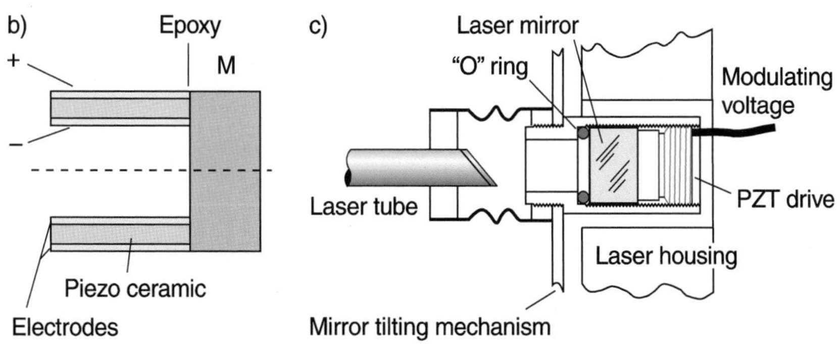


**Fig. 8.27.** (**a**) Piezocylinders and their (exaggerated) change of length with applied voltage (**b**) Laser mirror epoxide on a piezocylinder (**c**) Mirror plus piezomount on a single-mode tunable argon laser

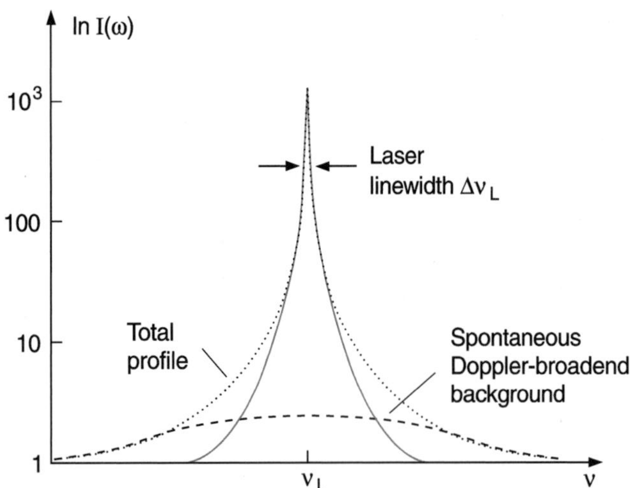


**Fig. 8.29.** Spectral profile of laser emission for an idealized laser without technical perturbations, plotted on a logarithmic scale

This results in amplitude- and phase fluctuations of the laser wave. The amplitude fluctuations are compensated by an inherent feedback mechanism of the active medium: A positive peak in the amplitude reduces the inversion and thus the amplification, while a negative deviation from the average amplitude increases the amplification. Such a feedback mechanism does not work for phase fluctuations, which lead to a finite line width (Sect. 7.4.3). A quantitative derivation gives the famous Schawlow–Townes formula [8.2] for the lower limit of the line width of a single mode laser:

$$\Delta\nu_{\mathrm{L}} = \frac{\pi h \nu_{\mathrm{L}}}{P_{\mathrm{L}}} \cdot \Delta\nu_{\mathrm{r}}^2 \,. \tag{8.28}$$

Here $\Delta\nu_{\mathrm{r}}$ is the width of a resonator resonance for an empty resonator, and $P_{\mathrm{L}}$ is the output power of the single mode laser. In Fig. 8.29 the resulting laser profile is plotted on a logarithmic scale, together with the Doppler-broadened background of the spontaneous emission.

This theoretical limit, which gives for $P_{\mathrm{L}} = 1\,\mathrm{W}$ and $\Delta\nu_N = 1\,\mathrm{MHz}$ a line width of $10^{-6}\,\mathrm{Hz}$ has never been realized in a practical experiment due to the technical perturbations mentioned above. With normal expenditure, a line width of about 100 kHz–1 MHz can be achieved.

**Note:**

A laser width a line width $\Delta\nu_{\mathrm{L}} = 1\,\mathrm{MHz}$ has a coherence length of $\Delta s_{\mathrm{c}} = c/\Delta\nu_{\mathrm{L}} = 300\,\mathrm{m}$! However, for a multimode argon laser with a bandwidth of 5 GHz the coherence length is only $\Delta\nu_{\mathrm{c}} = 6\,\mathrm{cm}$, which is comparable to that of a normal discharge lamp, where a single emission line has been selected.

## 8.4 Different Types of Lasers

The different experimental realizations of lasers can be divided into three main groups according to their active medium:

- Solid-state lasers
- Liquid lasers
- Gas lasers

Each of these types can be operated in a pulsed mode or continuously (cw operation). Depending on the kind of energy transfer from the pump into the active medium we distinguish between optically pumped lasers (e.g., the ruby laser and other solid-state lasers, such as the neodymium laser or the titanium-sapphire laser, and the liquid-dye laser), and electrically pumped lasers (the semiconductor laser and most gas lasers pumped by an electric discharge).

Many types of lasers emit on fixed frequencies, corresponding to discrete transitions in atoms or molecules. Their wavelengths can be changed only slightly within a narrow gain profile of the atomic or molecular transition. We will call them "fixed-frequency lasers."

For spectroscopic applications "tunable lasers" are of particular importance, where the laser wavelength can be tuned over a broader spectral range. These lasers have a broad gain profile and the laser wavelength can be selected within this range by wavelength-selecting optical elements (prism, optical grating or interferometer) inside the laser resonator. Tuning the transmission peak of these elements allows a continuous tuning of the laser wavelength over the whole gain profile. Such single-mode tunable lasers represent an intense, narrowband coherent wavelength-tunable light source, which has proved to be of invaluable advantage for numerous spectroscopic problems.

A completely different concept of tunable lasers uses high energy relativistic electrons from an accelerator as active medium. These electrons are forced onto oscillatory paths in a periodically changing magnetic

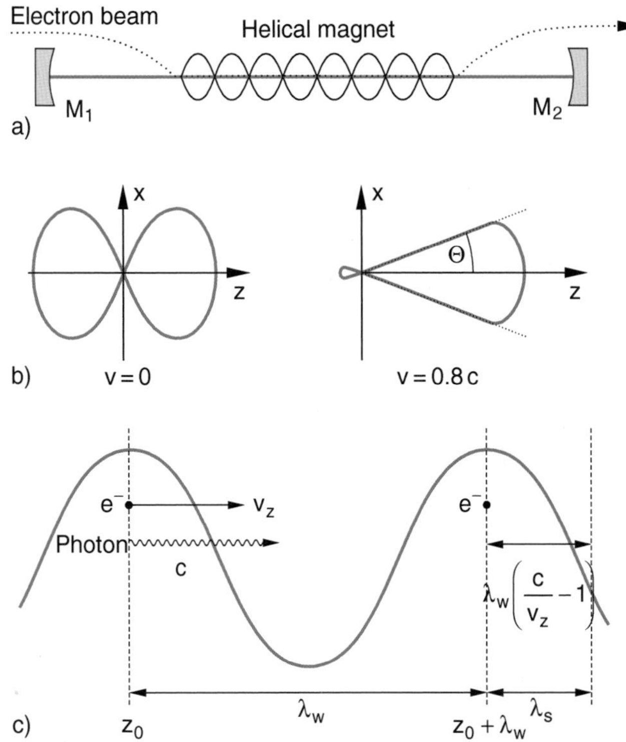


**Fig. 8.30.** (**a**) Schematic arrangement of a free-electron laser (**b**) Radiation of a dipole at rest ($\nu = 0$) and a moving dipole with $\nu \simeq c$ (**c**) Phase-matching condition

**Table 8.1.** Examples of solid-state lasers that can be operated in a pulsed and a cw mode

| Laser type | Active Atom or Ion | Host crystal | Laser-wave-length ($\mu$m) |
|---|---|---|---|
| Ruby laser | $Cr^{+++}$ | $Al_2O_3$ (Saphir) | 0.6943 |
| Neodynium-Glass-laser | $Nd^{+++}$ | Glass | 1.06 |
| Neodynium-YAG-Laser | $Nd^{+++}$ | $Y_3Al_5O_{12}$, $CaF_2$, $CaF_3$ | 1.06 0.9–1.1 |
| Titanium-Sapphire | $Ti^{+++}$ | $Al_2O_3$ | 0.65–1.1 |
| Alexandrit | $Cr^{+++}$ | $BeAl_2O_4$ | 0.7–0.83 |
| Cobalt-laser | $Co^{++}$ | $MgF_2$ | 1.5–2.1 |
| Holmium-laser | $Ho^{+++}$ | YAG | 2.06 |
| Erbium-laser | $Er^{+++}$ | YAG | 2.9 |
| Colour-center laser | vacancies of alkali ions | alkali-halogenid-crystal | 0.8–3.5 depending on the crystal |

field, where the electrons emit radiation. With a properly chosen period length of the alternating magnetic field, the contributions of the radiation from the different segments of the periodic structure superimpose in phase and add up to an intense wave in the forward direction of the average electron path (Fig. 8.30). The wavelength of the coherent emission depends on the energy of the electrons and can extend from the far infrared into the far ultraviolet. With high energy accelerators even the X-ray region can be reached. Such lasers are called ***free-electron lasers*** because their active medium consists of free electrons [8.11].

In the following sections we will discuss the most commonly used laser types.

### 8.4.1 Solid-state Lasers

The active medium of solid-state lasers are cylindrical rods of glass or single crystals, which are doped with special atoms, ions or molecules that can be optically pumped into excited states. The doping concentration varies between 0.1% to about 3%, depending on the kind of host material. In Table 8.1 some examples of solid-state lasers are compiled with their characteristic data.

All these solid-state lasers are optically pumped. Often pulsed flashlamps are used as pump sources, which results in a pulsed laser output. Although ruby lasers were the first lasers, these are being replaced more and more by neodymium lasers, which consist of a glass rod doped with $Nd^{+++}$ ions emitting laser radiation at $\lambda = 1.06\,\mu$m. The advantage of the Nd lasers is based on the fact that it represents a four-level system (Fig. 8.31), which needs less inversion and therefore less pump power than the three-level ruby laser. Its infrared emission can be converted by optical frequency doubling (see Sect. 8.5) into the visible range.

The laser threshold can be further lowered by replacing the glass in the Nd-glass laser by a crystal of yttrium-aluminum-garnet (YAG), which has a higher heat conductivity and can therefore more effectively transfer the excess energy $N(h\nu_p - h\nu_L)$ (produced as heat in the rod when $N$ photons are emitted) to the cooling system.

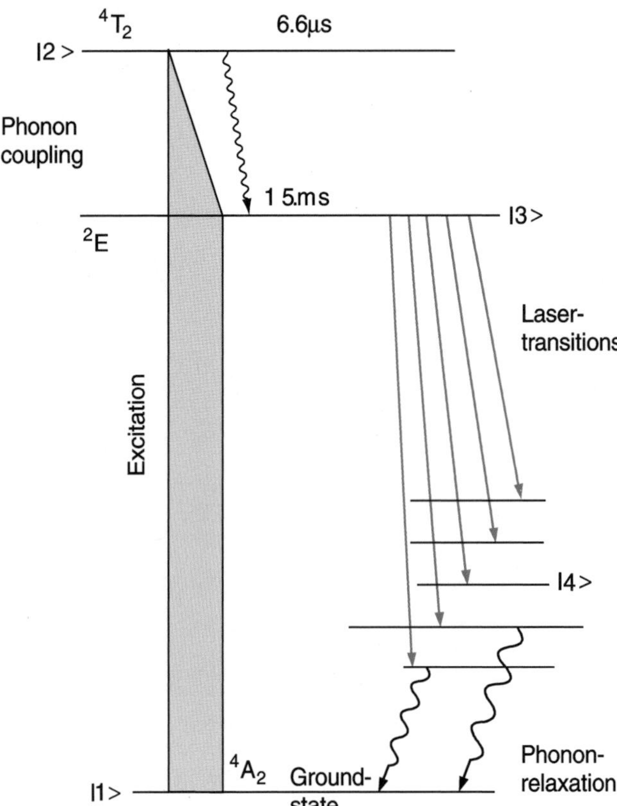


**Fig. 8.31.** Level scheme of the Nd:glas laser

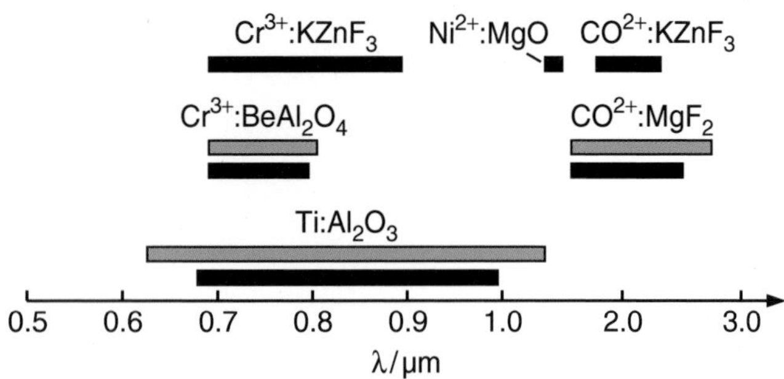


**Fig. 8.33.** Tuning ranges of some solid-state lasers (cw operation: *black*, pulsed operation *blue*)

The pulse durations of these solid-state lasers range from microseconds to milliseconds and the output pulse energies from 1 mJ to about 1 J, which gives peak powers from the kW range to many MW.

In order to achieve higher output powers, the output of the laser oscillator is sent through an optical amplifier, consisting of one or several optically pumped rods, where inversion is achieved (Fig. 8.32). These laser amplifiers have a similar setup as the laser oscillator, but without the mirrors, to prevent self-starting laser oscillation in these stages. The oscillator and amplifier are separated by an optical isolator in order to prevent feedback into the oscillator.

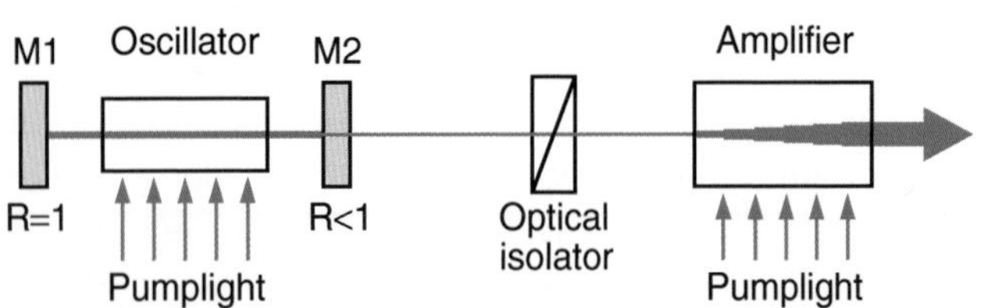


**Fig. 8.32.** Amplification of the laser output power by an optical amplifier

All lasers compiled in Table 8.1 can also be pumped by continuous pump sources, e.g., with continuous lasers. They then emit cw radiation with a wavelength λ that can be tuned within the gain profile of the active medium. Some of these media have a very broad gain profile, such as the Ti:$Al_2O_3$ (titanium-sapphire) laser (Fig. 8.33). The reason for this broad tuning range is as follows. The optically pumped excited states relax in a very short time into a lower level, due to interactions with the vibrating atoms of the host crystal. This level represents the upper laser level. The optical transitions from this level (Fig. 8.34) can terminate on many "vibronic levels" within a low-lying electronic state, corresponding to vibrations of the host crystal $Al_2O_3$ (phonons). These phonons relax very fast into lower levels, thus repopulating the initial state from which the pump process starts.

Another important class of tunable solid-state lasers are color-center lasers, which consist of alkali-halide crystals (e.g., NaCl or KBr), which are transparent in

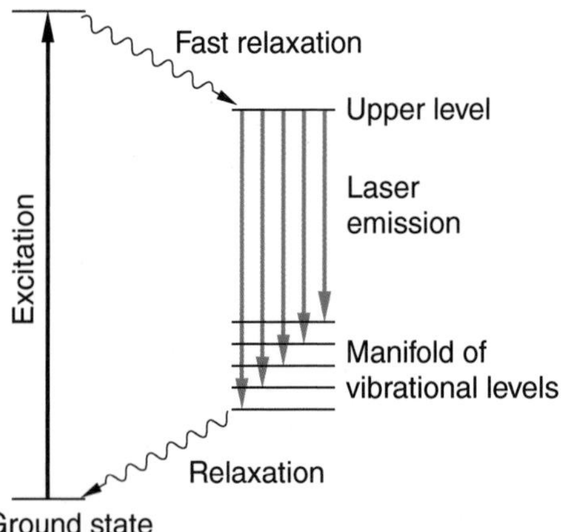


**Fig. 8.34.** Level scheme of vibronic solid-state lasers

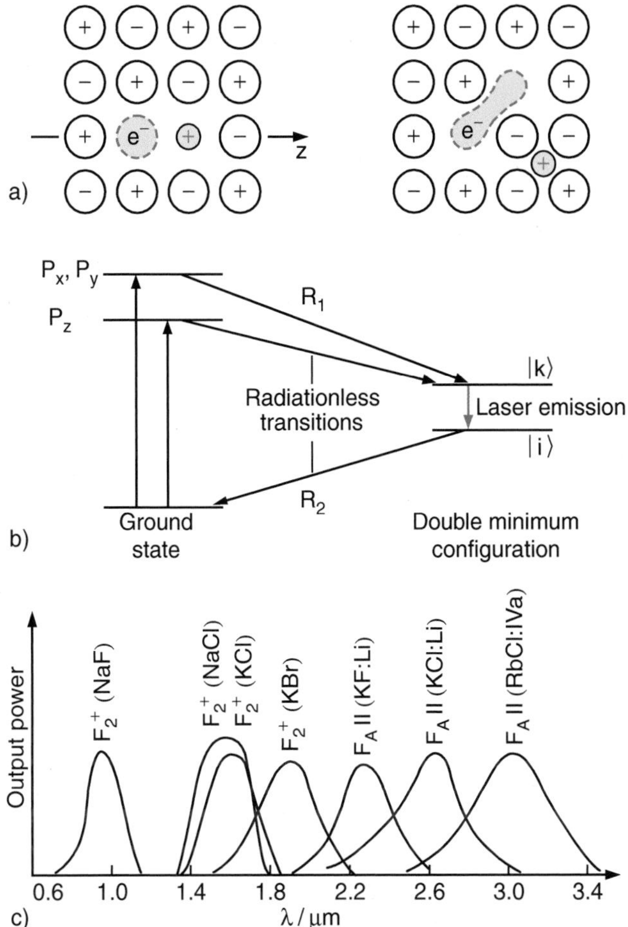


**Fig. 8.35a–c.** Color-center lasers. (**a**) Schematic illustration of defects in the ground state and in the relaxed upper state (**b**) Level scheme (**c**) Tuning ranges with different crystals

the visible. If defects (a missing negative halide ion) are produced in such a crystal by X-ray irradiation, the vacancy spot acts as a potential well for the remaining electron (Fig. 8.35a). The energy levels of this electron can be excited by absorption of visible photons, thus making the crystal appear colored. Therefore these vacancies are called color centers.

When the electron is excited, the forces on the surrounding ions change. This changes their arrangements around the color center and the energy of the initially excited states $|1\rangle$ decreases to level $|k\rangle$. (Fig. 8.35b), which acts as an upper laser level. Similarly to vibronic lasers, the laser emission terminates on many vibronic levels $|i\rangle$ which relax into the initial level $|0\rangle$.

In Fig. 8.35c the tuning ranges of different color center lasers are illustrated.

### 8.4.2 Semiconductor Lasers

The active medium of semiconductor lasers (often called diode lasers) is a *p*-*n* semiconductor diode. An electric current is sent in the forward direction through the diode, which transports electrons from the *n*-into the *p*-section and holes from the *p*- into the *n*-section. At the *n*-*p*-junction the electrons and holes can recombine (i.e., the electrons fall from an energetically higher state in the conduction band into a lower hole state in the valence band) and may emit their recombination energy in the form of electromagnetic radiation (Fig. 8.36). The emitted radiation can be amplified when passing along the *p*-*n*-junction (stimulated recombination). Since the electron density is very high, the amplification is correspondingly large and a path length through the active medium of less than 1 mm is sufficient to reach laser threshold.

The uncoated polished or cleaved end faces of the semiconductor crystal can serve as resonator mirrors. The refractive index of semiconductor materials

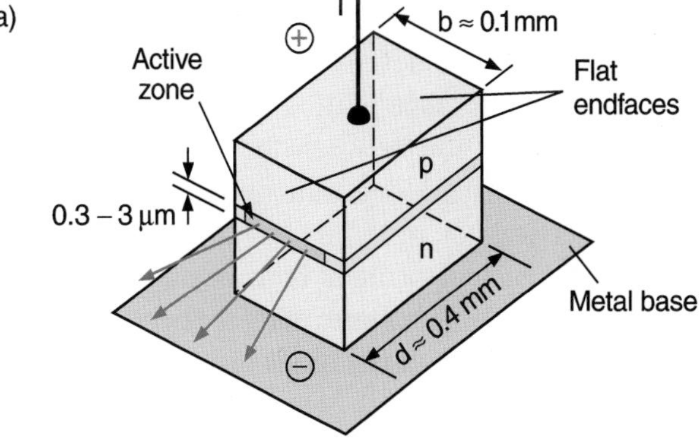


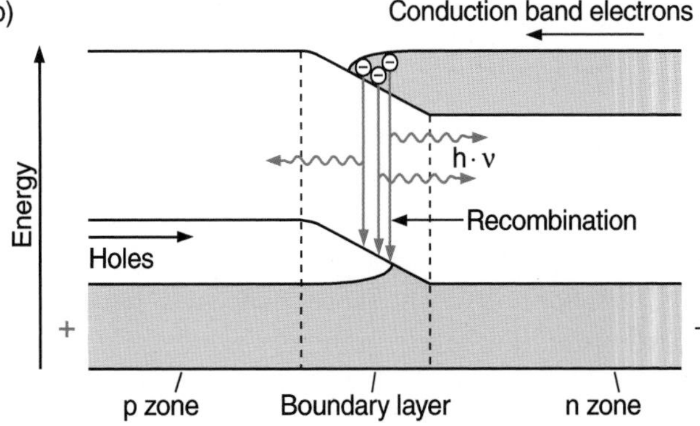


**Fig. 8.36a,b.** Simplified principle of a semiconductor laser. (**a**) Structure of the laser diode (**b**) Level scheme with valence and conduction band and radiative recombination of electrons with holes

is very large. For example for the GaAs (gallium-arsenide) laser emitting at $\lambda = 850\,\mathrm{nm}$ is $n = 3.5$. The reflectivity for vertical incidence is

$$R = \left(\frac{n-1}{n+1}\right)^2 \approx 0.30\,. \tag{8.29}$$

Because of the high gain, this reflection is sufficient to surpass the laser threshold in spite of reflection losses of 0.7 per one-half roundtrip.

Typical output powers of cw semiconductor lasers are 10–50 mW, when they are pumped by an electric current of 100–300 mA. Special arrays of many simultaneously pumped diodes deliver output powers of more than 100 W! The plug-in efficiency of radiation output power to electric input power, defined as the ratio

$$\eta = P_{\mathrm{L}}^{\mathrm{out}}/P_{\mathrm{el}}^{\mathrm{in}} \approx 0.25\,,$$

reaches 25–30%, which is the highest efficiency of all lasers developed so far.

Diode lasers are more and more used for pumping other solid state lasers. Using different semiconductor materials, wide tuning ranges for the diode laser wavelengths can be achieved.

The wavelength range of diode laser emission covers the spectral region between the blue (400nm) to the mid infrared (5μm), depending on the composition of the semiconductor material. Because of this large wavelength span and of their high energy efficiency they replace more and more other optical pumping sources. Meanwhile compact diode laser-pumped solid state laser devices are available, which deliver output powers in the multiwatt range.

### 8.4.3 Dye Lasers

The most important representatives of liquid lasers are dye lasers with various designs, which can be operated in a pulsed as well as in a cw mode. The active media are large dye molecules dissolved in a liquid (e.g., ethylene glycol). These molecules have many vibration-rotation levels in the electronic ground state (singlet $S_0$) and in excited states ($S_i$ or triplet states $T_i$). The energy level scheme is schematically depicted in Fig. 8.37. The strong interaction of the dye molecules with the liquid solvent results in a broadening of the transitions, which is larger than the average spacings

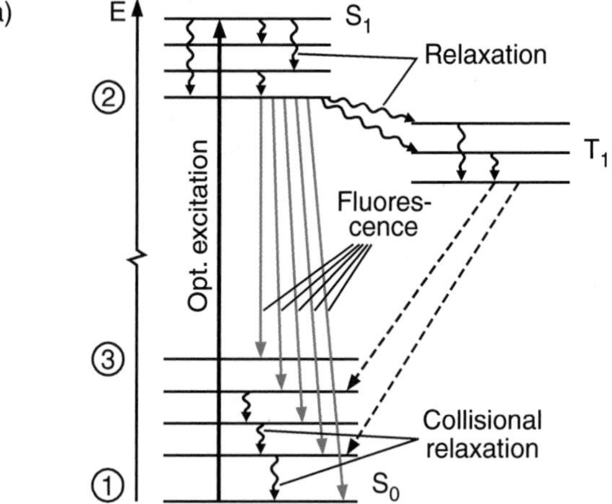


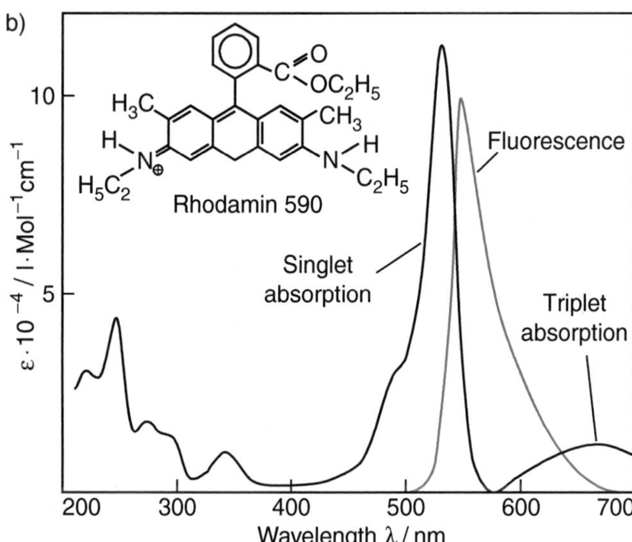


**Fig. 8.37a,b.** Dye laser. (**a**) Level scheme (**b**) Structure of dye molecule rhodamin 590 and absorption fluorescence spectrum

between the different rotational-vibrational transitions. Instead of many discrete lines broad absorption and emission bands appear (Fig. 8.37b).

The pump source (a flashlamp or a pulsed or cw laser) excites the dye molecules from the ground state $S_0$ into many vibration-rotation-levels of the $S_1$ state. Due to the strong interaction with the solvent, the excited molecules relax within a short time ($10^{-10}$–$10^{-12}$ s) into the lowest levels $|2\rangle$ of the $S_1$ state, from where they emit fluorescence on radiative transitions into many vibration-rotation-levels $|3\rangle$ of the $S_0$ state. Since these levels with energies $E > kT$ are not thermally populated at room temperature, population inversion can be reached between these levels $|2\rangle$ and the levels $|3\rangle$, if the former are sufficiently populated by optical pumping. The active medium of the dye laser is therefore a four-level system.

Since the absorption starts from the lower levels $|1\rangle$ and reaches higher levels in $S_1$ while the emission starts from the lowest levels $|2\rangle$ in $S_1$ and terminates on the higher levels $|3\rangle$ in $S_0$, the emission spectrum is red-shifted towards longer wavelengths against the absorption spectrum (Fig. 8.37b).

The dye laser can oscillate on those transitions where the threshold is reached. From the broad emission line profile a specific wavelength can be selected by wavelength-selecting elements inside the laser resonator. Tuning the transmission peaks of these elements results in a corresponding tuning of the laser wavelength. In Fig. 8.38 the tuning ranges for different dyes are shown. This figure illustrates that with different dyes the whole spectral range from 1 μm down to about 400 nm can be covered.

In Fig. 8.39 the experimental design of a flashlamp-pumped dye laser is shown. It is similar to that of the ruby laser in Fig. 8.10, but the solid rod is replaced by a glas tube through which the dye solution is pumped, producing a steady flow of dye molecules through the region pumped by the flashlamp. Because of the broad gain profile a prism is placed inside the resonator in order to select the wanted wavelength. Only that wavelength $\lambda$ can oscillate, for which the laser beam hits the end mirror $M_2$ vertically. All other wavelengths are reflected back under an angle inclined against the resonator axis and do not reach the gain medium again. Wavelength tuning is accomplished by tilting the mirror $M_2$.

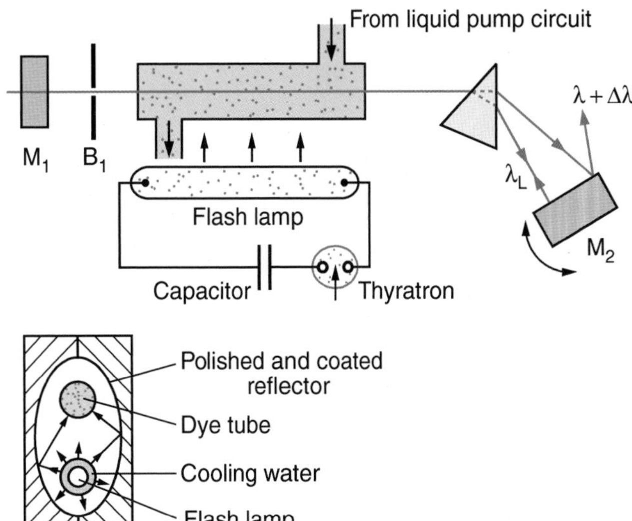


**Fig. 8.39.** Flashlamp-pumped dye laser

In Fig. 8.40 the arrangement is shown for a dye laser, pumped by another pulsed laser (e.g., a nitrogen-laser or an excimer laser (see below)). The pump beam is focused by a cylindrical lens into the dye cell, forming a line focus where inversion is achieved. The narrow dye laser beam is enlarged by telescope optics and falls onto an optical Littrow grating, where the first order diffraction is reflected back into the incident direction. This can be realized with a grating with groove distance $d$, if the incidence angle $\alpha$ is equal to the diffraction angle $\beta$, which gives the grating equation

$$m \cdot \lambda = d(\sin\alpha + \sin\beta) = 2d \cdot \sin\alpha \quad (8.30)$$

with $m = 1$.

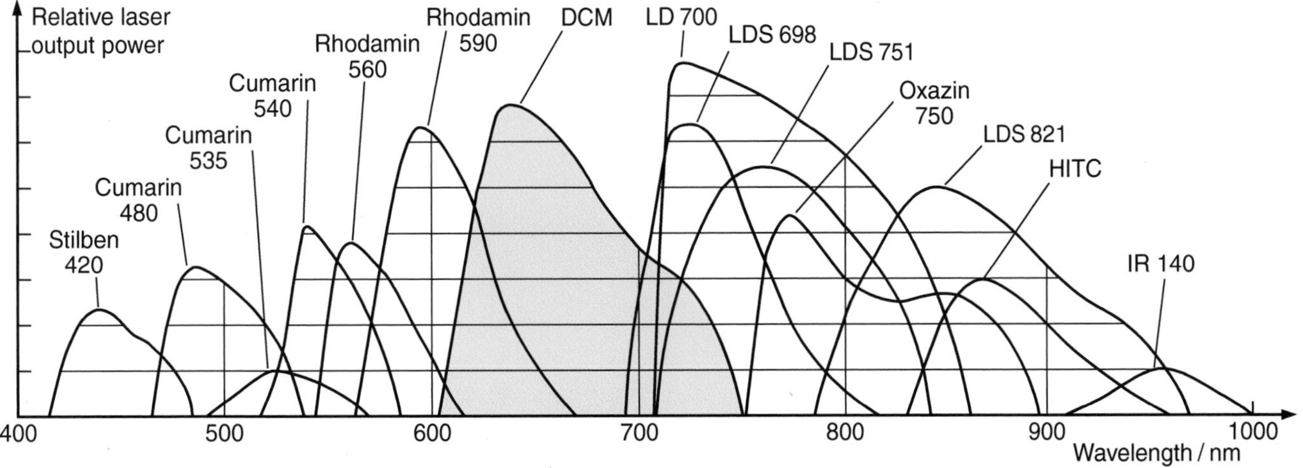


**Fig. 8.38.** Tuning ranges of different dyes as active medium

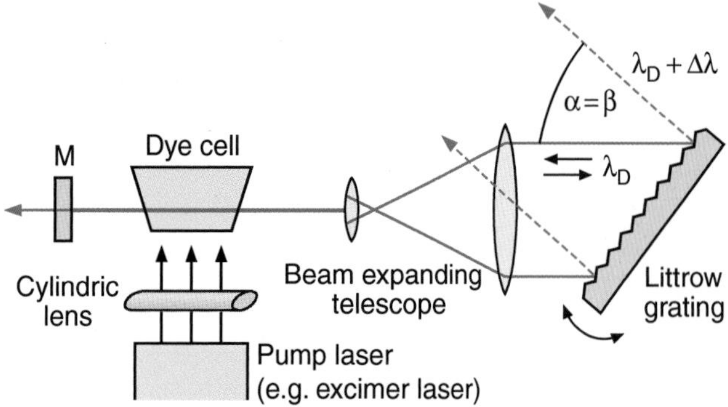


**Fig. 8.40.** Excimer-laser-pumped dye laser with Littrow grating as wavelength-selective reflector

The spectral resolution of the grating

$$\lambda/\Delta\lambda = m \cdot N \tag{8.31}$$

is proportional to the number $N$ of illuminated grooves and the diffraction order $m$. Therefore it is necessary to enlarge the dye laser beam to cover a large number of grooves $N$. Tilting the grating results in a continuous tuning of the laser wavelength.

Since the spectral width $\Delta\nu$ of pulsed lasers with pulse duration $\Delta t$ is principally limited by the Fourier limitation $\Delta\nu = 1/\Delta t$, cw lasers are demanded for really high resolution in the MHz range. In Fig. 8.41 a commercial version of such a single mode cw dye laser is shown.

The active medium is a thin ($\approx 0.5$ mm) liquid jet of the dye solution, which is pumped by an argon laser beam, focused by a spherical mirror into the dye jet. Differently from the previously discussed resonators, four mirrors form a ring-resonator, where no standing laser wave is produced but a wave running only in one direction. This has the advantage that no nodes are present as in a standing wave and the whole inversion of the active medium can contribute to the laser amplification. In order to avoid laser waves in both directions the losses for one direction must be higher than for the other direction. This can be achieved with an optical diode (unidirectional device), consisting of a birefringent crystal and a Faraday polarization rotator, which turns the birefringent rotation back to the input polarization for the wave incident in one direction, but increases the rotation for the other direction. Waves with the wrong polarization suffer large losses at the many Brewster surfaces in the resonator and therefore do not reach the threshold.

Wavelength selection is achieved with a birefringent filter and two Fabry–Perot-etalons with different thicknesses $t$. If the transmission peaks of all these elements are tuned to the same wavelength $\lambda$, the laser will oscillate at this wavelength as a single mode laser. The laser wavelength can be continuously tuned, if all elements and the resonator length are tuned synchronously. This can be realized with special feedback control systems. For details of these devices see [8.12].

### 8.4.4 Gas Lasers

Nearly all gas lasers use gas discharges as active medium. Besides the He-Ne laser already discussed in Sect. 8.1.2 the most important gas lasers are compiled in Table 8.2. Here, we will only briefly discuss the physical principles of their operation.

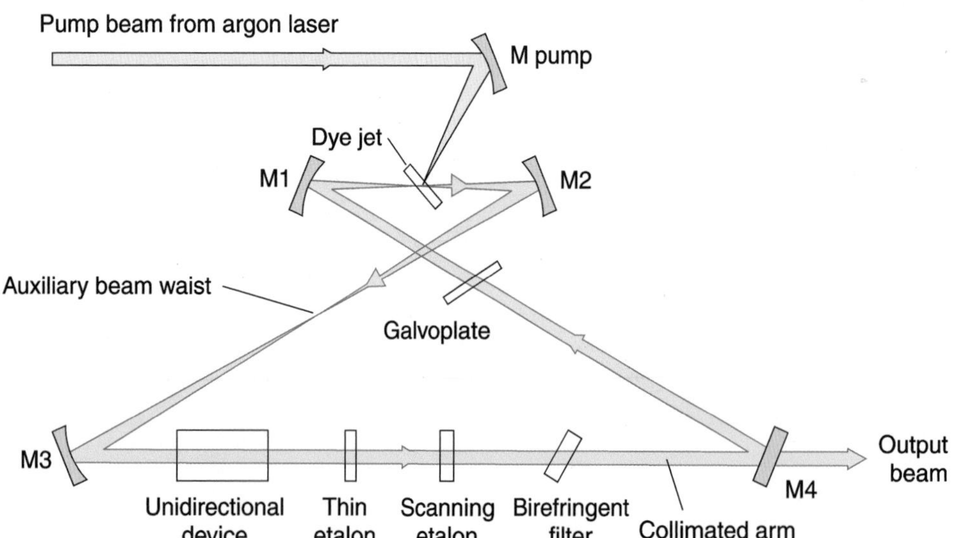


**Fig. 8.41.** Commercial version of a single-mode cw ring dye laser (Spectra-Physics)

**Table 8.2.** Characteristic data of some important types of gas lasers

| Laser type | Laser wavelengths | Output power |
|---|---|---|
| He-Ne-Laser | about 10 transitions with $\lambda = 0.54$–$3.39\,\mu$m | 0.1–100 mW |
| Argon laser | about 20 transitions with $\lambda = 0.35$–$0.53\,\mu$m | 1 W–1 kW |
| $CO_2$-He-$N_2$-laser | about 200 transitions with $\lambda = 9.5$–$10.3\,\mu$m | cw: 1 W – 10 kW<br>pulsed: $\leq$ 1 MW |
| CO-laser | about 300 transitions $\lambda = 4.5$–$6\,\mu$m | cw: several watts |
| Excimer-laser | XeCl: 308 nm<br>KrF: 248 nm<br>ArF: 193 nm<br>$H_2$: 150 $\mu$m | Pulse energies<br>1–400 mJ/pulse<br><br>repet. rate: $\leq$ 200 Hz |
| Chemical lasers | HF, DF: 2–3 $\mu$m and 10–20 $\mu$m | several kW |
| Far infrared lasers pumped by $CO_2$-lasers | several hundred transitions with $\lambda = 50$–$350\,\mu$m | pulsed: mW–W |

The ***argon laser*** oscillates on transitions between different excited levels of argon ions $Ar^+$. It therefore needs a high current discharge (5–50 A with 70–700 A/cm$^2$ current density), where the degree of ionization is high. The excitation of the upper laser levels occurs in two steps:

$$Ar + e^- \rightarrow Ar^+ + 2e^- \tag{8.32a}$$

$$Ar^+ + e^- \rightarrow Ar^{+*}(4p, 4s) + e^- \,. \tag{8.32b}$$

In a capillary of ceramic (length $\approx$ 1 m, diameter $\approx$ 3 mm) current densities of more than 700 A/cm$^2$ are reached at a total discharge current of 50 A. The gas discharge is confined by a longitudinal magnetic field in order to prevent the ions to reach the wall of the capillary where they could damage it by sputtering. The ceramic tube is cooled by a water flow between the tube and an outer cylinder on which the electric wires for the magnetic field are wound. A heated helical cathode supplies the large electron current necessary to maintain the discharge (Fig. 8.42). An elegant technical solution for the transfer of heat ($\approx$ 20–30 KW) to the cooling water is shown in Fig. 8.42b. The discharge runs through small holes (3 mm diameter) in tungsten discs, which are heated by the dissipated power to temperatures up to 1,000 K. These hot discs transfer their energy through copper discs and by radiation to the wall of a ceramic tube with about 40 mm diameter, which is again cooled by water. The larger surface of this tube facilitates the heat transfer.

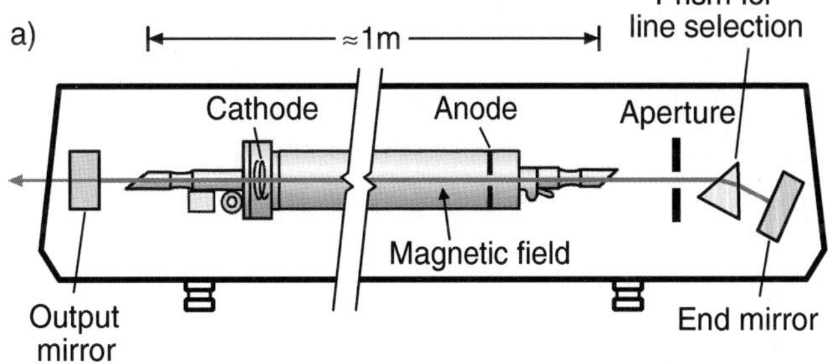


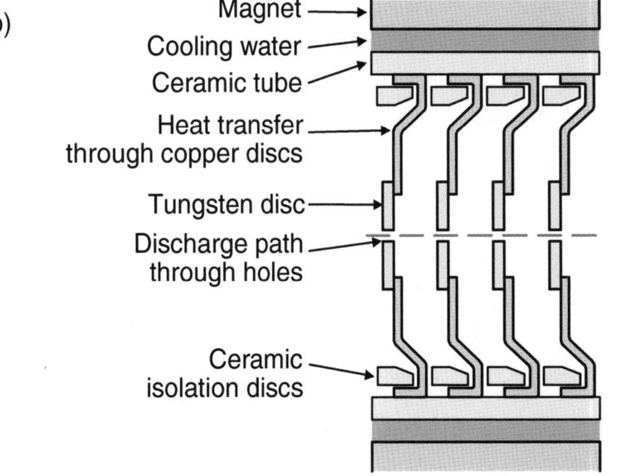


**Fig. 8.42a,b.** Argon ion laser. (**a**) Experimental setup (**b**) Details of the discharge path through holes drilled into tungsten discs and the heat transfer to the water-cooled envelope

Since several upper levels in the $Ar^+$ ions are excited, the laser can reach oscillation threshold for several transitions and therefore oscillates on several wavelengths. A specific wavelength can be selected by a prism inside the resonator (Fig. 8.43). By tilting the mirror $M_2$ the desired wavelength can be chosen.

The efficiency of the argon laser

$$\eta = P_L/P_{electr} \approx 0.1\%$$

is very low. In order to produce 1 W laser output power, more than 10 kW electrical input power are necessary. Most of the power (99.9%), put into the discharge, is converted into heat transferred to the walls of the discharge and has to be taken away by the cooling water.

The ***$CO_2$ laser*** has the highest efficiency of all gas lasers ($\approx 10-20\%$) and for cw operation the highest output power. The active medium is a gas discharge in a mixture of He, $N_2$ and $CO_2$. By electron impact in the discharge excited vibrational levels in the electronic ground states of $N_2$ and $CO_2$ are populated (Fig. 8.44). The vibrational levels $v = 1$ in the $N_2$ molecule and $(\nu_1, \nu_2, \nu_3) = (00^01)$ in the $CO_2$ molecule (see Sect. 10.4) are near-resonant and energy transfer from the $N_2$ molecule to the $CO_2$ molecule becomes very efficient. This populates the $(00^01)$ level in $CO_2$ preferentially, creates inversion between the $(00^01)$ and the $(02^00)$ levels, and allows laser oscillations on many

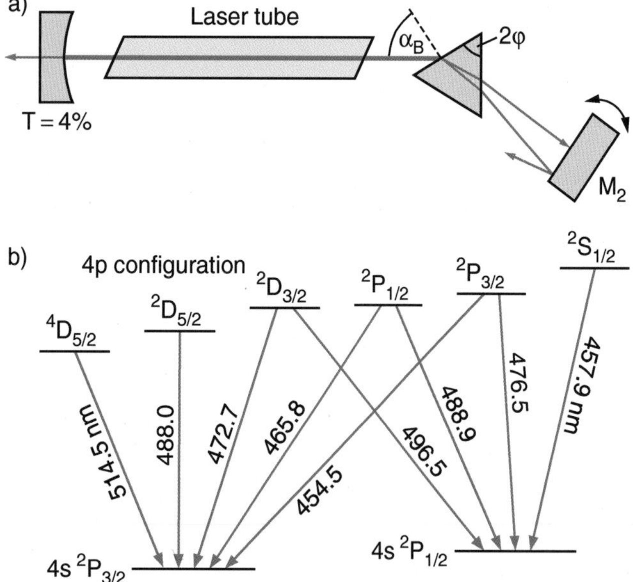


**Fig. 8.43.** (**a**) Laser line selection on a specific transition in argon. (**b**) Level scheme

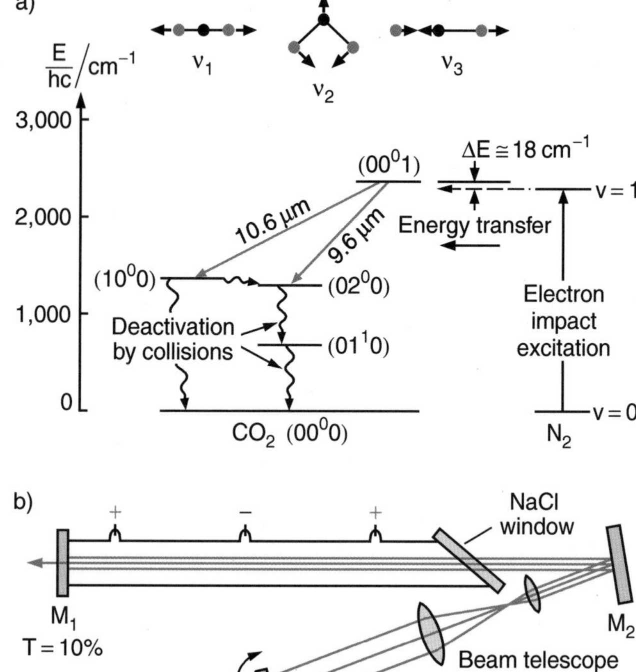


**Fig. 8.44.** (**a**) Level scheme and the three normal vibrational modes of the $CO_2$ molecule. (**b**) $CO_2$ laser with a Littrow-grating for line selection

rotational transitions between these two vibrational states in the wavelength range 9.6–10.6 μm. A single line can be selected by a Littrow-grating, forming one of the resonator end mirrors.

**Note:**

The linear $CO_2$ molecule has three normal modes of vibration, labeled $\nu_1$, $\nu_2$ and $\nu_3$, and depicted in the upper part of Fig. 8.44a (see also Sect. 10.3). The vibrational state of the molecule is described by the number of vibrational quanta in these modes. A state with 1 quantum in $\nu_1$, 2 quanta in $\nu_2$ and 0 quanta in $\nu_3$ is labeled as (120). The bending vibrational mode is twofold degenerate and can have a vibrational angular momentum along the $CO_2$ axis. The number of quanta $nh$ of this vibrational angular momentum is stated as an upper index to the vibrational $\nu_2$ quanta. The upper laser level $(00^01)$ has zero vibrational angular momentum and 1 vibrational quantum in the $\nu_3$ mode.

A powerful gas laser in the UV is the ***excimer laser***, where specific diatomic molecules, called excimers, form the active medium. These excimers

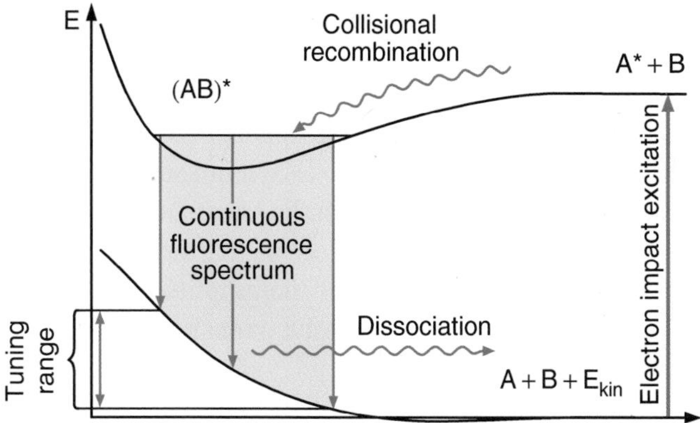


**Fig. 8.45.** Potential curves and fluorescence transitions of an excimer

(excited dimers), are stable in an electronically excited state but unstable in their ground state (Fig. 8.45). Examples are the noble gas halides, such as XeCL, KrCL or ArF. If the stable upper state AB* of the excimer AB is populated, (e.g., by electron excitation of the atom A and recombination $A^* + B \to (AB^*)$, inversion is automatically produced because the lower state is always completely emptied by fast dissociation on a time scale of $10^{-13}$ s, if it is populated by fluorescence from the upper level.

Excimers are therefore ideal candidates for an active laser medium. They have the additional advantage that the emission from the bound upper level terminates on a repulsive potential curve on the dissociative ground state and therefore forms a broad emission continuum. This results in a broad gain profile and the wavelength of the excimer laser can be tuned over a relatively large range.

## 8.5 Nonlinear Optics

The optical frequency of lasers can be doubled in nonlinear optical crystals, thus considerably extending the wavelength range where coherent radiation can be generated. In this section we will briefly discuss the physical principles of optical frequency doubling or mixing under the heading nonlinear optics.

When an optical wave passes through a crystal, it induces the atomic electrons to forced oscillations. For sufficiently small electric field amplitudes $\boldsymbol{E}$ of the wave the elongations of the oscillating electrons are small and the restoring forces are proportional to the elongation (linear range). The induced dipole moments $\boldsymbol{p} = \alpha \cdot \boldsymbol{E}$ are proportional to the field amplitude and the components $P_i$ of the dielectric polarization of the medium induced by the light wave

$$P_i = \varepsilon_0 \sum_j \chi_{ij} E_j \quad (i, j = x, y, z) \tag{8.33}$$

are linearly dependent on $E$, where $\chi_{ij}$ are the components of the tensor $\chi$ of the electric susceptibility. This is the realm of linear optics.

### EXAMPLE

The field amplitude of the sunlight reaching the earth at $\lambda = 500\,\text{nm}$ within a bandwidth of 1 nm is about $E \approx 3\,\text{V/m}$. On the other side the electric field from the Coulomb force, binding the electron to the nucleus is, for a binding energy of 10 eV, about

$$E_\text{B} = -\frac{10\,\text{V}}{10^{-10}\,\text{m}} = 10^{11}\,\text{V/m}\,. \tag{8.34}$$

Therefore, the elongation of the electrons induced by the sunlight (for example, for the Rayleigh scattering) is very small compared with its mean distance from the nucleus and the restoring force within this small elongation is linear to a good approximation.

For much larger light intensities, as can be realized with focused beams of lasers, the nonlinear range of electron elongations can be readily reached. Instead of (8.33) the dielectric polarization has to be written as the expansion

$$P_i = \varepsilon_0 \Bigg[ \sum_j \chi_{ij}^{(1)} E_j + \sum_j \sum_k \chi_{ijk}^{(2)} E_j E_k + \sum_j \sum_k \sum_l \chi_{ijkl}^{(3)} E_j E_k E_l + \ldots \Bigg] \tag{8.35}$$

where $\chi^{(n)}$ is the $n$th order susceptibility, which is represented by a tensor of rank $(n+1)$. The quantities $\chi^{(n)}$ decrease rapidly with increasing $n$. However, for sufficiently high field amplitudes $E$ the higher order terms in (8.35) can be no longer neglected. They form the basis of nonlinear optical phenomena.

When a monochromatic light wave

$$E = E_0 \cos(\omega t - kz) \tag{8.36}$$

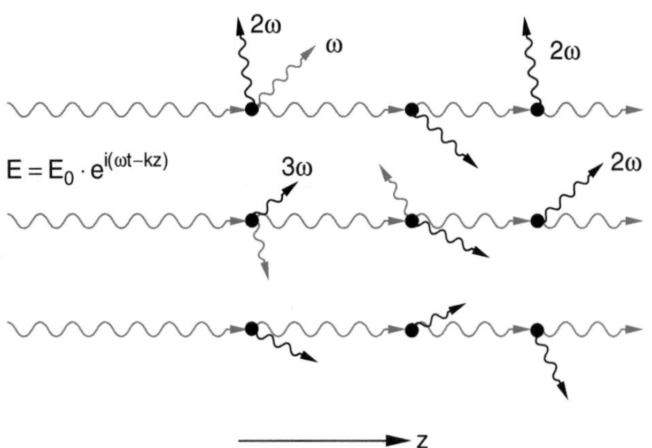


**Fig. 8.46.** Schematic illustration of the generation of optical harmonies under the influence of a strong electromagnetic wave

passes through the medium, the frequency spectrum of the induced polarization $P$ also contains (because of the higher powers $n$ of the field amplitudes $E^n$), besides the fundamental frequency $\omega$, higher harmonics $m\omega$ ($m = 2, 3, 4 \ldots$). This implies: The induced oscillating dipoles emit radiation not only on the frequency $\omega$ (Rayleigh scattering), but also on higher harmonics (Fig. 8.46). The amplitudes $A(m\omega)$ of these emitted waves depends on the magnitude of the coefficients $\chi^{(n)}$ and in a nonlinear way on the amplitude $E_0$ of the incident light wave.

### 8.5.1 Optical Frequency Doubling

If the light wave (8.36) with the electric field vector $\vec{E}_0 = \{E_{0x}, 0, 0\}$ in the $x$-direction field passes through an isotropic medium we obtain from (8.35), for the location $z = 0$, the $x$-component of the dielectric polarization

$$P_x = \varepsilon\left(\chi_{xx}^{(1)} E_{0x} \cos\omega t + \chi_{xxx}^{(2)} E_{0x}^2 \cos^2\omega t + \ldots\right) \tag{8.37}$$

when we neglect all higher order terms $\chi^{(n)}$ with $n > 2$. Similar equations are obtained for the $y$- and $z$-components. Using the relation $\cos^2 x = \frac{1}{2}(1 + \cos 2x)$ we can write (8.37) as

$$P_x = \varepsilon_0\left(\tfrac{1}{2}\chi^{(2)} E_{0x}^2 + \chi^{(1)} E_{0x} \cos\omega t + \tfrac{1}{2}\chi^{(2)} E_{0x}^2 \cos 2\omega t\right) . \tag{8.38}$$

The dielectric polarization contains a constant term $\frac{1}{2}\varepsilon_0\chi^{(2)} E_{0x}^2$, a linear term with frequency $\omega$ and the nonlinear term with $2\omega$. This means that each of the atoms hit by the incident wave radiates a scattered wave that contains the frequency $\omega$ (Rayleigh scattering) and a second harmonic wave with the frequency $2\omega$.

The amplitude of the second harmonic wave is proportional to the square of the amplitude of the incident wave. This means that the intensity $I(2\omega)$ is also proportional to $I^2(\omega)$.

The microscopic second harmonic waves, emitted by the different atoms, can only add up to a macroscopic wave if they are all in phase for all location in the crystal. Since the phase velocity generally depends on the frequency (dispersion), special crystals have to be used in order to match the velocities of the fundamental and the second harmonic wave.

### 8.5.2 Phase Matching

When a plane wave (8.36) passes through the crystal, it generates in each plane $z = z_0$ dipoles with oscillation phases that depend on the phase of the inducing fundamental wave at $z = z_0$. In a neighboring plane, $z = z_0 + \Delta z$, the same phase difference exists between the incident wave and the induced dipoles.

The waves at frequency $\omega$, radiated by the atoms in the plane $z = z_0$ reach the next plane $z = z_0 + \Delta z$ after the same time interval as the incident wave. They therefore superimpose the microscopic waves emitted from atoms in that plane in phase and add up to twice their individual amplitude.

This is, however, not true for the second harmonic waves, because their phase velocity $v_{\mathrm{ph}}(2\omega) = c/n(2\omega)$ differs from that of the incident wave $v_{\mathrm{ph}}(\omega) = c/n(\omega)$ if the refractive index $n(2\omega) \neq n(\omega)$, which is generally the case. The second harmonic wave generated by atoms in the plane $z = z^0$ therefore reaches the plane $z = z_0 + \Delta z$ with another time delay than the incident wave and a phase difference arises between the microscopic second harmonic waves generated in the two planes. After a distance

$$\Delta z = (\lambda/2)/[n(\omega) - n(2\omega)] \tag{8.39}$$

the second harmonic wave generated in the plane $z = z_0$ arrives at the plane $z + \Delta z$ with the opposite phase as the second harmonic waves generated in this

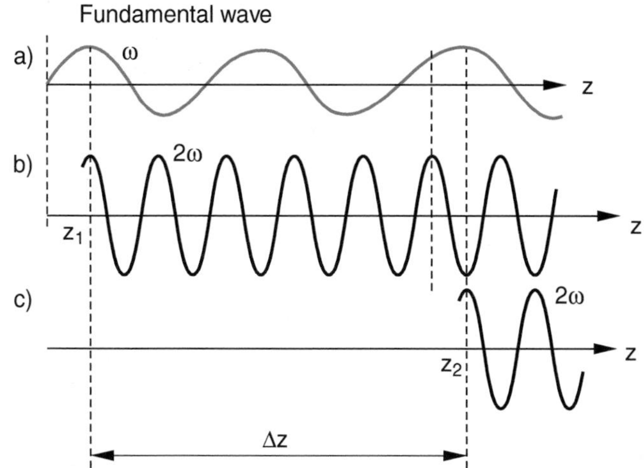


**Fig. 8.47.** Phase shift of $\pi$ between the two harmonic waves with $2\omega$, generated at a point $z_1$ and a point $z_2 = z_1 + \Delta z$

plane and therefore the two contributions interferes destructively (Fig. 8.47).

In summary: In isotropic homogeneous media the second harmonic waves generated in the different planes do not superimpose in phase. Summed over the whole crystal all phase differences between 0 and $2\pi$ occur and the total wave remains very small due to destructive interference of the different microscopic contributions.

A solution to this dilemma is provided by uniaxial birefringent crystals, where the incident light wave is split into an ordinary wave for which the refractive index $n = n_0$ does not depend on the direction, and an extraordinary wave where $n = n_e(\Theta)$ depends on the angle $\Theta$ between the optical axis of the crystal and the propagation direction (Fig. 8.48). At a certain angle $\Theta_P$, called the phase matching angle, the ordinary refractive index $n_0(\omega)$ for the fundamental wave at frequency $\omega$ equals the extraordinary index $n_e(2\omega)$ for the second harmonic wave. In this direction, phase matching is possible for a selected frequency $\omega$. The condition for phase matching can be written as

$$n_e(2\omega) = n_0(\omega) \Rightarrow v_{ph}(\omega) = v_{ph}(2\omega)$$
$$\Rightarrow \boldsymbol{k}(2\omega) = 2\boldsymbol{k}(\omega)\,. \qquad (8.40)$$

If the angle $\Theta$ is changed, phase matching is achieved for another frequency $\omega$, i.e., another wavelength $\lambda$. Therefore the phase matched wavelength can be tuned

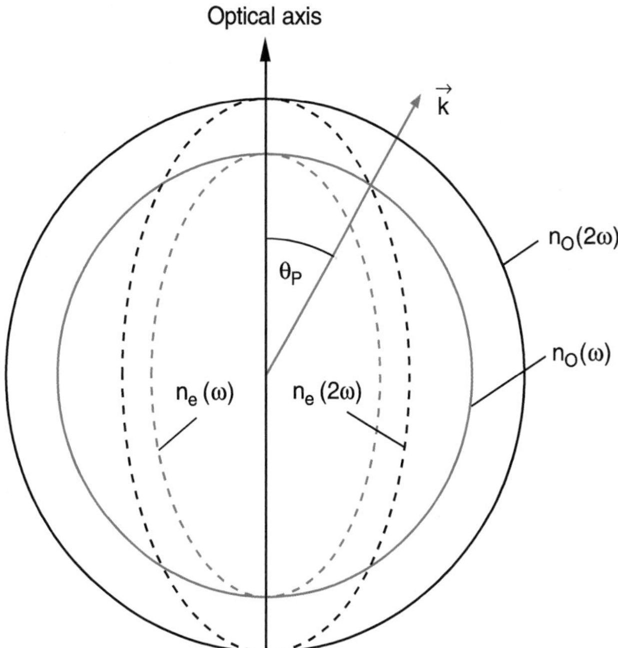


**Fig. 8.48.** Phase matching between fundamental wave with frequency $\omega$ and second harmonics with $2\omega$ in birefringent optical crystals

by tilting the crystal. All microscopic secondary waves at $2\omega$ emitted into the direction $\Theta$ from the dipoles induced by the incident fundamental wave are in phase with the fundamental wave along the whole path through the crystal. Now a macroscopic second harmonic wave can build up, traveling in the same direction as the fundamental wave.

For instance, the red ruby laser emission at $\lambda = 690$ nm is partly converted into UV light at $\lambda = 345$ nm in a properly phase-matched KDP (potassium-dihydrogen phosphate) crystal. With sufficiently large nonlinear coefficients $\chi^{(2)}$ of the doubling crystal and with pulsed incident lasers of high peak powers conversion efficiencies up to $\eta = P(2\omega)/P(\omega) = 40\%$ can be achieved.

With cw lasers the output power is much less and therefore the conversion efficiency $\eta = \chi^{(2)} I(\omega)$ is smaller. One can either focus the laser beam onto the crystal to increase $I(\omega)$ at a given power $P(\omega)$, or the doubling crystal is placed inside an enhancement resonator with highly reflecting mirrors (Fig. 8.49), where the power of the fundamental wave is enhanced by a factor up to 100. With this technique a UV power of more than 50 mW can be achieved for an input power of 500 mW.

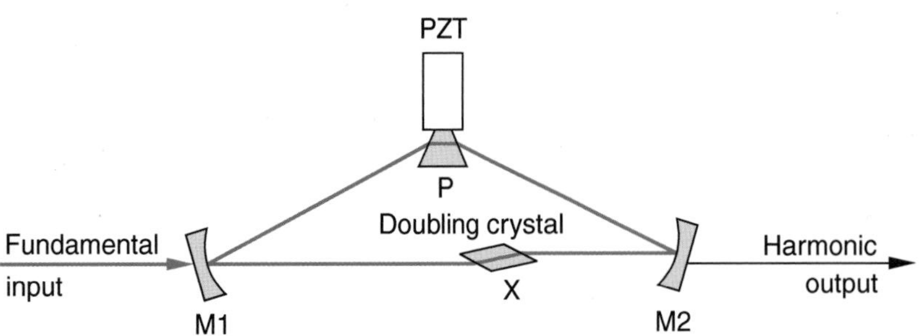


**Fig. 8.49.** Low-loss ring resonator with astigmatic compensation and wide tuning range for optical frequency doubling

### 8.5.3 Optical Frequency Mixing

When two light waves

$$E_1 = E_{01}\hat{e}_x \cos(\omega_1 t - \boldsymbol{k}_1 \boldsymbol{r})$$
$$E_2 = E_{02}\hat{e}_x \cos(\omega_2 t - \boldsymbol{k}_2 \boldsymbol{r})$$

are superimposed in a nonlinear optical medium, the total electric field amplitude $E = E_1 + E_2$ induces a polarization with a nonlinear contribution

$$\begin{aligned} P^{(2)}(\omega) &= \varepsilon_0 \chi^{(2)} \left[E_{01}^2 \cos^2 \omega_1 t + E_{02}^2 \cos^2 \omega_2 t \right. \\ &\quad + 2E_{01}E_{02} \cos\omega_1 t \cdot \cos\omega_2 t] \\ &= \tfrac{1}{2}\varepsilon_0 \chi^{(2)} \left[\left(E_{01}^2 + E_{02}^2\right)\right. \\ &\quad + \left(E_{01}^2 \cos 2\omega_1 t + E_{02}^2 \cos 2\omega_2 t\right) \\ &\quad + 2E_{01}E_{02}(\cos(\omega_1 + \omega_2)t \\ &\quad + \cos(\omega_1 - \omega_2)t)] \;. \end{aligned} \tag{8.41}$$

Besides the second harmonics at frequencies $2\omega_1$ and $2\omega_2$ also waves are generated with the sum frequency $\omega_1 + \omega_2$ and the difference frequency $\omega_1 - \omega_2$, if the phase matching condition can be properly chosen for each of these different contributions. For example, the phase-matching condition for the macroscopic generation of the sum frequency $\omega_3 = \omega_1 + \omega_2$ is

$$\boldsymbol{k}(\omega_1 + \omega_2) = \boldsymbol{k}_1(\omega_1) + \boldsymbol{k}_2(\omega_2) \Rightarrow \tag{8.42a}$$

$$n_2\omega_3 = n_1\omega_1 + n_2\omega_2 \quad \text{with} \quad n_i = n(\omega_i)\;. \tag{8.42b}$$

This condition is generally more readily fulfilled than that for the second harmonic generation, because the directions of the two incident waves (and therefore their wave vectors) can be freely chosen within certain limits, which imposes less restrictions to the selection of the nonlinear crystal.

The possibility of optical frequency mixing has greatly increased the spectral ranges covered by intense coherent light sources. With difference-frequency generation, using two visible lasers, the mid-infrared region can be covered, while sum-frequency generation gives access to the UV range down to $\lambda = 200$ nm. The spectral limitations are given by the spectral regions where the absorption of the nonlinear crystal becomes large. For more information see [8.13–14].

## 8.6 Generation of Short Laser Pulses

The investigation of fast processes induced by the absorption of photons, demand a high time resolution of the detection technique. Examples for such fast processes are the decay of excited states with a short lifetime, the dissociation of molecules or the rearrangement of molecular structure after excitation into higher energy states. This latter process plays an important role in the visual process, where the primarily excited rhodopsin molecules in the retina of the eye undergo many energy transfer process before the excitation energy is transferred into an electrical signal reaching our brain. Such processes could only be studied in detail after the development of ultrashort laser pulses with pulse widths down to about five femtoseconds ($1\,\text{fs} = 10^{-15}$ s).

In this section we will briefly discuss some experimental techniques for the generation of short laser pulses.

### 8.6.1 Q-Switched Lasers

The inversion threshold for obtaining laser oscillation depends on the total losses (see Sect. 8.1.1), which can be expressed by the quality factor ($Q$-factor) of the laser resonator.

The $Q$-value of the $k$th resonator mode is defined as

$$Q_k = -\frac{2\pi\nu W_k}{\mathrm{d}W_k/\mathrm{d}t} = +\frac{\omega}{\gamma_k} T_\mathrm{R} \tag{8.43}$$

and can be expressed by the total loss factor $\gamma_k$ of this mode and the roundtrip time $T_\mathrm{R} = 2d/c$.

The $Q$-switching technique uses the following trick: During the pump process the $Q$-value of the laser resonator is kept so low (i.e., the losses are so high) that the laser threshold is not reached in spite of the growing inversion. At a selected time $t = t_\mathrm{s}$, $Q$ is suddenly switched to a maximum value (Fig. 8.50). This prevents laser oscillation for $t < t_\mathrm{s}$ and allows the population inversion to reach a large value, because it is not depleted by induced emission. When the losses are suddenly switched to a minimum value at $t = t_\mathrm{s}$ the inversion is way above threshold and the amplification of the spontaneous emission starting the photon avalanche is accordingly high. This leads to a fast rising "giant pulse", which depletes the inversion within a short time and therefore terminates itself.

In Fig. 8.51 two possible experimental realizations of $Q$-switched lasers are shown. A fast spinning resonator mirror spoils the $Q$-value for all times, except for the short time span where the mirror surface is perpendicular to the resonator axis. The light from a light-emitting diode LED is reflected by the backside of the spinning mirror onto a photodetector. Its output signal triggers the discharge of the flashlamp, pumping the laser. An electronic delay of the trigger signal can select the time delay between trigger time and vertical position of the spinning mirror.

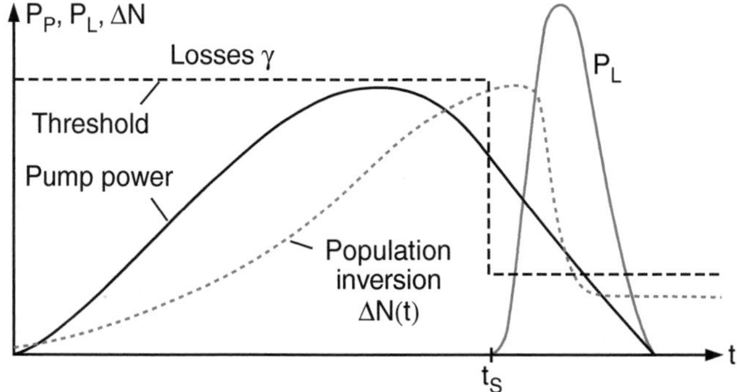


**Fig. 8.50.** Pump power $P_\mathrm{P}(t)$, laser output power $P_\mathrm{L}(t)$, and cavity losses $\gamma(t)$ for a $Q$-switched laser

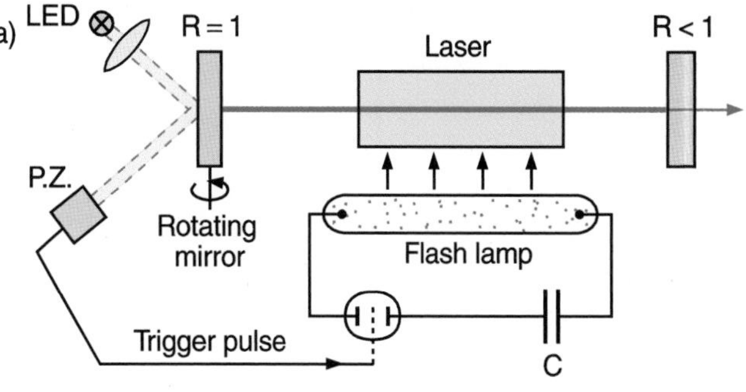


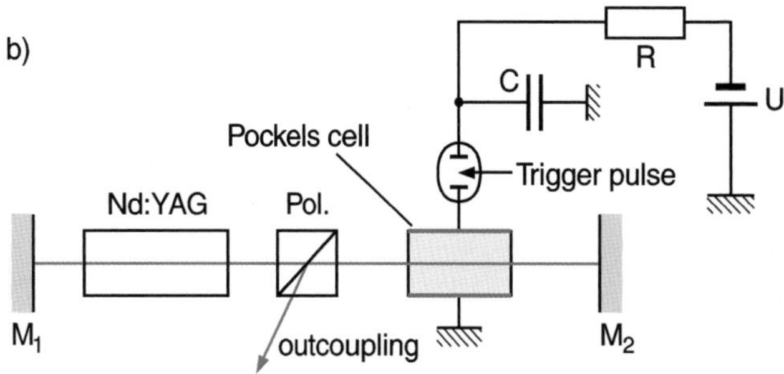


**Fig. 8.51a,b.** Possible realizations of cavity $Q$-switching (**a**) With a rotating resonator mirror (**b**) With an electro-optic switch (Pockels cell) inside the laser resonator

The optimum time delay depends on the duration of the pump pulse and on the lifetime $\tau$ of the upper laser level. The time delay must be smaller than the lifetime $\tau$, because otherwise one looses too much of the upper state population necessary for the amplification of the giant pulse.

Another more commonly employed technique uses a Pockels cell inside the laser resonator for $Q$-switching. A Pockels cell consists of a birefringent crystal that changes its birefringence with an applied electric field. If the crystal is biased in such a way that it rotates the plane of polarization by 45° for one transit, the light transmitting the crystal a second time after reflection by mirror $M_2$ has its plane of polarization turned by 90°. A polarization beam splitter then reflects the beam out of the laser resonator (Fig. 8.51b). At the $Q$-switching time $t_\mathrm{s}$, a high voltage pulse is suddenly applied to the crystal that changes the birefringence and brings the rotation angle per transit to 90° and for the reflected beam to 180°. The beam is now transmitted by the polarization beam splitter and reaches the laser rod.

These techniques generate giant laser pulses with durations of a few ns and peak powers of $10^5$–$10^9$ W, depending on the laser type.

### 8.6.2 Mode-Locking of Lasers

Much shorter pulses can be achieved with the mode-locking technique, which is based on the following principle.

If a light wave with frequency $\omega_0 = 2\pi\nu_0$ and intensity

$$I = I_0 \cos^2 \omega_0 t$$

passes through an optical modulator with modulation frequency $f = \Omega/2\pi$ and transmission

$$T = T_0(1 - a \cdot \sin^2(\Omega t/2)) \tag{8.44}$$

the transmitted intensity is

$$I_t = I \cdot T = I_0 T_0 [1 - a \cdot \sin^2(\Omega/2)t] \cdot \cos^2 \omega_0 t. \tag{8.45}$$

The degree of modulation $a < 1$ depends on the voltage applied to the modulation cell. The Fourier analysis of such a modulated light wave gives a frequency spectrum that consists of the carrier frequency $\nu_0$ and sidebands at frequencies $\nu_0 \pm n \cdot f (n = 1, 2, 3 \ldots)$

Inserting the modulator inside the laser resonator (Fig. 8.52) and choosing the modulation frequency $f = \Omega/2\pi$ to be equal to the frequency separation

$$\Delta\nu = c/2d = f$$

of the longitudinal resonator modes, makes all sidebands resonant with resonator modes. This means that the sidebands can participate in laser oscillation as long as their frequencies lie within the gain profile of the active medium. This leads to a coupling of all resonator modes within the gain profile because the phases of the sidebands are coupled to that of the carrier by the phase of the modulation.

The amplitude of the $k$th mode becomes

$$A_k(t) = T A_{k0} \cos \omega_k t \tag{8.46}$$

$$= T_0 A_{k0} \left[1 - a \sin^2(\Omega/2)t\right] \cos \omega_k t \,.$$

This can be written as

$$A_k(t) = T_0 A_{k0} \left[\left(1 - \frac{a}{2}\right) \cos \omega_k t \right. \tag{8.47}$$

$$\left. + \frac{a}{4}\left[\cos(\omega_k + \Omega)t + \cos(\omega_k - \Omega)t\right]\right] .$$

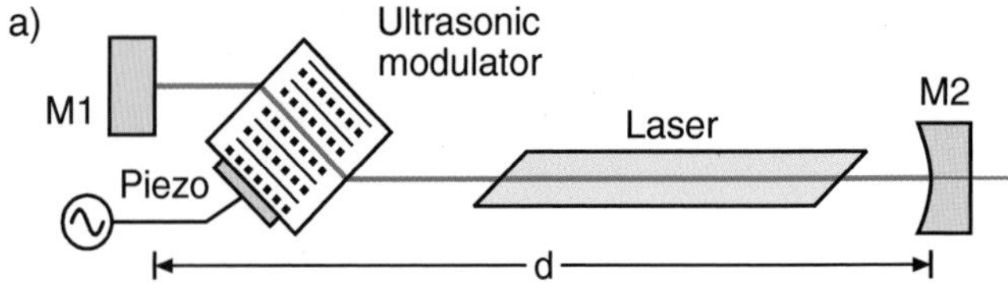


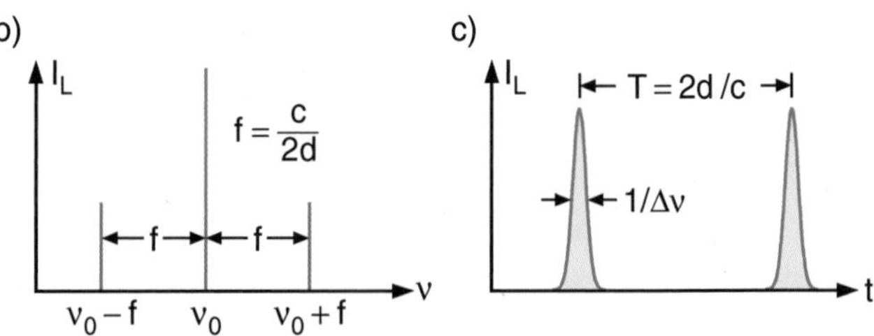


**Fig. 8.52a–c.** Mode-locking of lasers. (**a**) Experimental set-up with an ultrasonic modulator (**b**) Laser frequency $\nu_0$ and the two neighboring side-bands (**c**) Laser output pulses with width $\Delta t \approx 1/\Delta\nu$ and repetition frequency $f = 1/T = c/2d$

The total amplitude of $N = 2m + 1$ coupled modes is then

$$A(t) = \sum_{k=-m}^{+m} A_k(t) = \frac{a}{2} \cdot T_0 A_{k0} \sum_{k=-m}^{+m} \cos(\omega_0 + k \cdot \Omega)t$$

For equal amplitudes $A_k = A_0$ the total time-dependent intensity becomes

$$I(t) \propto A_0^2 \frac{\sin^2\left(\frac{1}{2}N\Omega t\right)}{\sin^2\left(\frac{1}{2}\Omega t\right)} \cos^2 \omega_0 t \,. \tag{8.48}$$

For cw lasers the amplitude $A_0$ is constant in time and (8.48) represents an equidistant sequence of pulses (Fig. 8.53) with a pulse separation

$$T = \frac{2d}{c} = \frac{1}{\Delta\nu} = \Omega/2\pi \,, \tag{8.49}$$

which equals the roundtrip time through the laser resonator. The pulse width

$$\Delta\tau = \frac{2\pi}{(2m+1)/\Omega} = \frac{2\pi}{N\Omega} = \frac{1}{\delta\nu} \tag{8.50}$$

is determined by the number $N = \delta\nu/\Delta\nu = \frac{2d}{c}\delta\nu$ of phase locked modes within the gain profile with spectral width $\delta\nu$ and is therefore inversely proportional to the spectral bandwidth $\delta\nu$ of the gain profile above threshold.

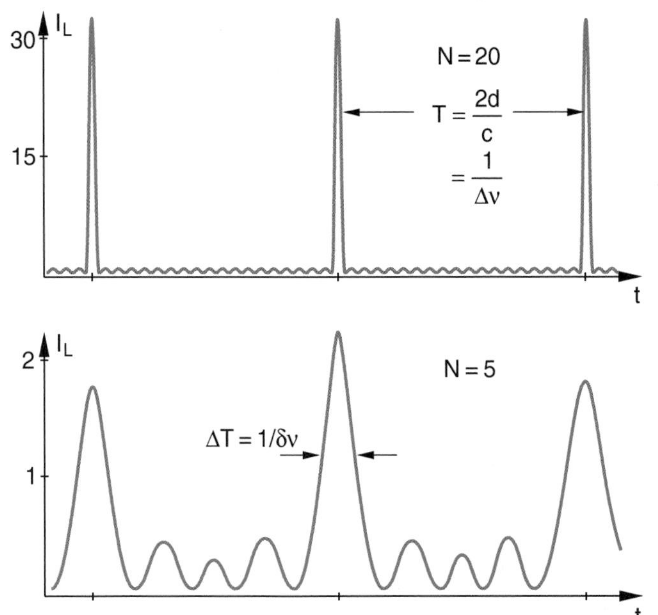


**Fig. 8.53.** Mode-locked pulses, where $N$ modes have been locked. Note the different ordinate scales

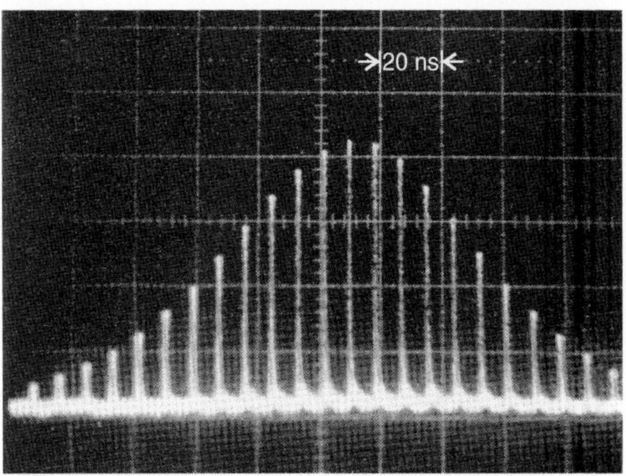


**Fig. 8.54.** Periodic pulse sequence from a pulsed mode-locked Nd:glass laser (W. Rudolf, F.B. Physik, Univ. Kaiserslautern)

The peak power of the pulses is proportional to $N^2$. The pulse energy is proportional to $N^2 \Delta T \propto N$. In between two succesive main pulses $(N-2)$ small maxima appear, which decrease in intensity as $N$ increases (Fig. 8.53).

Contrary to a normal multimode laser that can oscillate simultaneously on many modes with, however, random phases, the mode-locked laser oscillates on many phase-coupled modes, because the modulator enforces a definite phase relation between the oscillating modes.

### EXAMPLES

1. The gain profile of the He-Ne laser has a width of about $\delta\nu = 2\,\text{GHz}$. Mode-locking therefore achieves pulses with a minimum duration of $\Delta\tau = 500\,\text{ps}$.
2. The argon laser has a larger spectral width $\delta\nu = 6\,\text{GHz}$ of its gain profile and allows mode-locked pulses with widths down to $\Delta\tau = 170\,\text{ps}$.
3. The dye laser has a very large spectral bandwidth of about $\delta\nu = 3 \times 10^{13}\,\text{s}^{-1}$. Therefore, pulses down to $\Delta\tau = 3 \times 10^{-14}\,\text{s}$ should be possible. The experimental realization only reaches $\Delta\tau = 3 \times 10^{-12}\,\text{s} = 3\,\text{ps}$. This corresponds to the transit time $\Delta t = \Delta x/c$ of the light through the modulator with length $\Delta x$.

Not only cw lasers, but also pulsed lasers, can be mode-locked. The pulse amplitude is no longer constant but follows the time profile of the gain. In Fig. 8.54 the pulse sequence within the envelope of a single pulse of a mode-locked neodymium-glass laser is shown for illustration.

The shortest laser pulses, obtained so far are generated by a nonlinear effect, called Kerr lens mode-locking. Its basic principle is illustrated in Fig. 8.55.

For sufficiently high intensities, the refractive index is affected by the nonlinear interaction of the light wave with the medium. It can be written as a sum

$$n(\omega, I) = n_0(\omega) + n_2(\omega) \cdot I \tag{8.51}$$

where $n_0(\omega)$ is the normal refractive index and $n_2(\omega) \ll n_0(\omega)$ is the nonlinear part of $n$. The intensity-dependent change of the refractive index is caused by the nonlinear polarization of the atomic electron shells induced by the electric field of the optical wave and is therefore called the optical Kerr effect.

When a laser beam with a Gaussian radial intensity profile $I(r)$ passes through a medium, the refractive index shows a radial gradient with a maximum value of $n$ at the central axis at $r = 0$. The medium then acts like a lens and leads to a focusing of the incident laser beam, where the focal length depends on the laser intensity.

When a laser pulse with the time profile $I(t)$ passes through the medium, the central part of the pulse around its maximum generates the largest gradient of $n(r)$ and therefore the shortest focal length $f_{\min}$

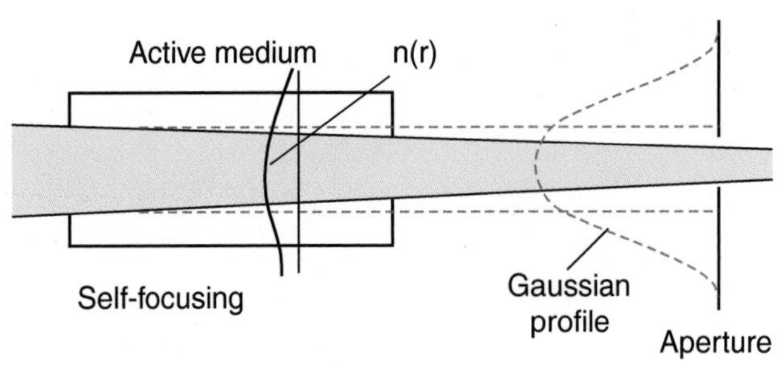


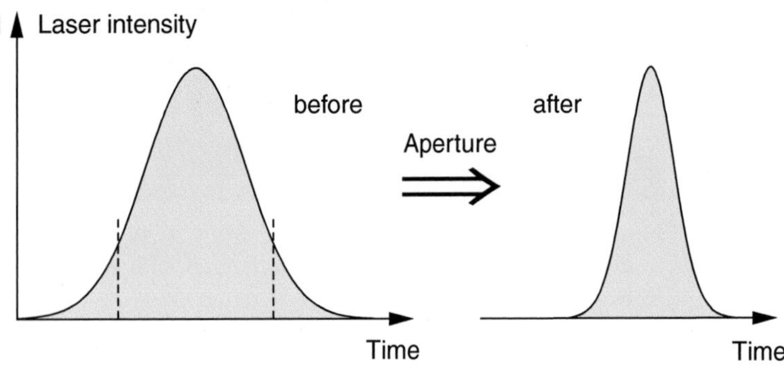


**Fig. 8.55.** Kerr-lens mode-locking

of the Kerr lens. If an aperture is placed at a distance $f_{min}$ behind the Kerr lens, only that part of the pulse $I(t)$ around its maximum at $t = t_0$ is fully transmitted through the aperture. All other parts before and after the maximum produce a longer focal length and therefore have a larger spot size at the aperture and only the central part of the radial beam profile is transmitted through the aperture. These parts of the pulse therefore suffer larger losses and are attenuated. This happens for every roundtrip inside the resonator and leads to a shortening of the pulse duration.

## EXAMPLE

For sapphire $Al_2O_3$ $n_2 = 3 \times 10^{-16}\,\mathrm{cm^2/W}$. For the intensity $I = 10^{14}\,\mathrm{W/cm^2}$ the refractive index changes by $\Delta n = 3 \times 10^{-2}\,n_0$ with $n_0 = 1.76$. For a laser pulse with a wavelength $\lambda = 1\,\mu\mathrm{m}$ this leads to an additional phase shift of the optical phase by $\Delta\varphi = (2\pi/\lambda)\Delta n = 2\pi \times 300 \cdot 1.76$ after a pathlength of 1 cm through the Kerr lens material, which results in a radius of curvature $R = 4\,\mathrm{cm}$ of the wavefront of the light wave. For a Gaussian beam profile with peak intensity $I(0) = 10^{14}\,\mathrm{W/cm^2}$ which would be a plane wave without the Kerr lens, the focal length of the Kerr lens is then $f = 4\,\mathrm{cm}$.

This Kerr lens mode-locking has been successfully applied to the generation of ultrashort light pulses from a Ti:sapphire laser, which has a very broad gain profile and is therefore well suited to allow such short pulses. In Fig. 8.56 a possible experimental realization is shown. The Kerr medium is the Ti:sapphire crystal, which acts simultaneously as active laser medium, and the limiting aperture is placed in front of mirror $M_4$. The Kerr lens changes the focal length and therefore the imaging characteristics of the laser resonator in such a way that for the maximum of the laser pulse the focus lies in the center of the aperture.

With such a device, pulses down to 4 fs have been achieved. For these short pulses the spectral width is

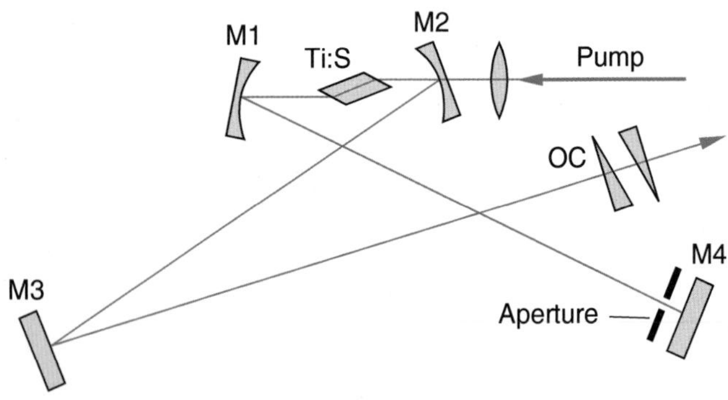


**Fig. 8.56.** Experimental setup for a Kerr-lens mode-locked Ti:sapphire laser (OC = optical compensator)

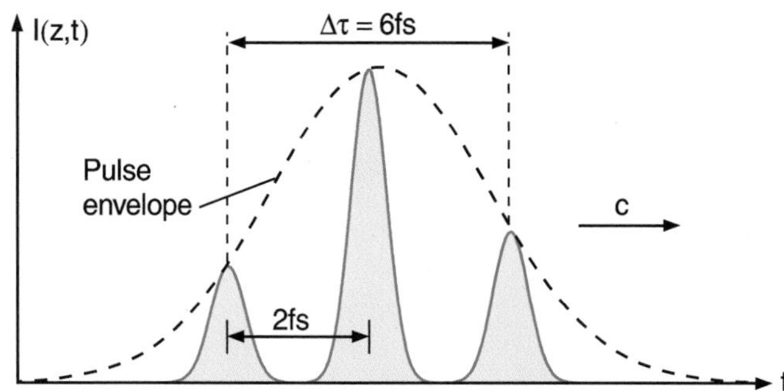


**Fig. 8.57.** Schematic representation of an ultrashort light pulse containing only three optical cycles of the light intensity. The envelope has a half-width of $\Delta\tau = 6\,\text{fs}$ and a spatial extension $\Delta z = c \times \Delta\tau \approx 2\,\mu\text{m}$

very large and any dispersion effects in the laser resonator must be carefully compensated for. For instance, the dielectric mirrors with many reflecting layers generally have a wavelength- dependent phase shift, which would lead to a broadening of the pulse. Therefore special dispersion-compensated mirrors have been designed that avoid this problem [8.15].

For an optical wave at $\lambda = 600\,\text{nm}$ ($\nu = 5 \times 10^{14}\,\text{s}^{-1}$), the optical cycle time is $T_{\text{opt}} = 1/\nu = 2\,\text{fs}$. A light pulse of 6 fs half-width therefore contains only three optical cycles (Fig. 8.57).

### 8.6.3 Optical Pulse Compression

When a short optical pulse is sent through an optical fiber with a core diameter of $5\,\mu\text{m}$, the intensity becomes so high that the refractive index

$$n(\omega) = n_0(\omega) + n_2 \times I(t)$$

is changed by the nonlinear interaction of the medium with the laser pulse. It becomes time-dependent. The nonlinear term $n_2$ can be positive as well as negative, depending on the material and the laser wavelength.

A short pulse of duration $\Delta T$ can be described by the wave packet

$$I(t) = \int\limits_{-\Delta\omega/2}^{+\Delta\omega/2} I(\omega)\mathrm{e}^{i(\omega t - kz)}\,\mathrm{d}\omega\,. \tag{8.52}$$

This represents a superposition of many frequency components within the frequency interval $\delta\nu = 1/\Delta\tau = \Delta\omega/2\pi$ where $I(\omega)$ gives the envelope of the spectral profile.

The linear part $n_0(\omega)$ of the refractive index causes, for normal dispersion ($\mathrm{d}n_0/\mathrm{d}\lambda$), a larger phase velocity for the red components in the pulse than for the blue components. The red components will therefore be at the leading edge and the blue components at the trailing edge of the pulse. This results in spatial- and time-broadening of the pulse.

The nonlinear part $n_2 I(t)$ causes a frequency shift dependent on the intensity. This can be seen as follows. The phase of the wave $E = E_0 \cos(\omega_0 t - kz)$

$$\begin{aligned}\varphi &= \omega_0 t - kz = \omega_0 t - \omega n z/c \\ &= \omega_0 \cdot (t - n_0 z/c) - A \cdot I(t)\,; \quad A = n_2 \omega z/c\end{aligned} \tag{8.53}$$

depends on $I(t)$. Since the frequency

$$\omega = \mathrm{d}\varphi/\mathrm{d}t = \omega_0 - A \cdot \mathrm{d}I/\mathrm{d}t \tag{8.54}$$

is the time derivative of the phase $\varphi$, it is evident from (8.53) that with $A > 0$ the frequency at the leading edge of the pulse ($\mathrm{d}I/\mathrm{d}t > 0$) is decreased and at the trailing edge ($\mathrm{d}I/\mathrm{d}t < 0$) is increased. This phenomenon is called a chirp of the optical pulse, where the optical frequency changes from small to high frequencies over the pulse profile $I(t)$ (Fig. 8.58).

In summary, when passing through an optical medium the optical pulse $I(t)$ becomes broader, caused by the dispersion $n_0(\omega)$, and its spectral profile $I(\omega)$ becomes broader due to the chirp induced by the nonlinear part $n_2 \cdot I(t)$ of the refractive index.

When such a spectrally broadened pulse is sent through a pair of parallel optical gratings (Fig. 8.59), the red components of the pulse are diffracted into another angle $\beta$ than the blue ones. From Fig. 8.59 one can infer the path difference $S$ between the plane phase front of the incident wave at point A and the phase plane at point B as

$$S = S_1 + S_2 = \frac{D}{\cos\beta} + \frac{D\sin\gamma}{\cos\beta}\,, \tag{8.55}$$

where $D$ is the distance between the two parallel gratings. From the grating equation

$$d(\sin\alpha - \sin\beta) = \lambda \tag{8.56}$$

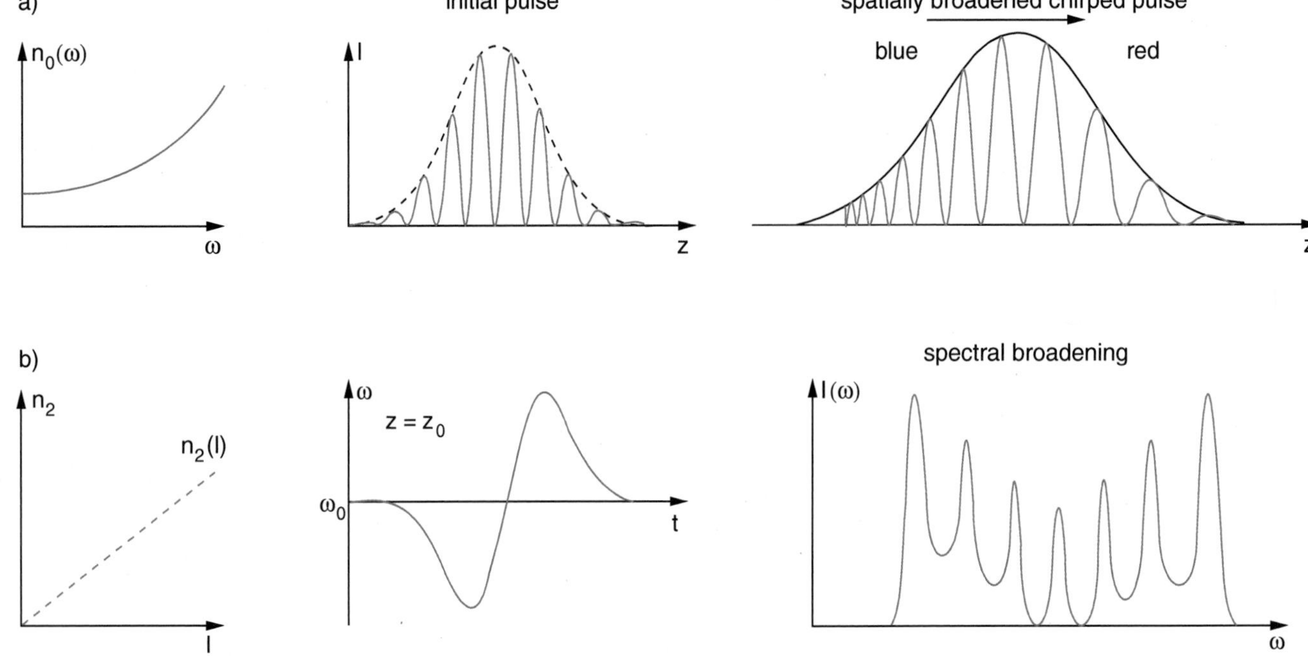


**Fig. 8.58a,b.** Spatial and spectral broadening of a pulse in a medium with normal linear (**a**) and nonlinear (**b**) refractive index

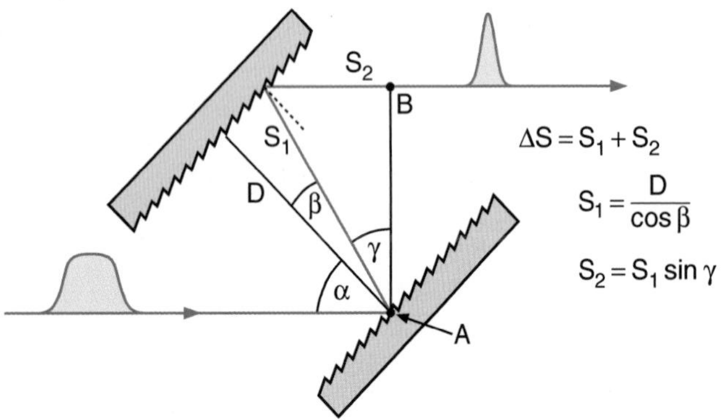


**Fig. 8.59.** Optical pulse compression by a grating pair

for a grating with groove separation $d$ we obtain, after some calculations,

$$\frac{\mathrm{d}S}{\mathrm{d}\lambda} = \frac{\mathrm{d}S}{\mathrm{d}\beta} \cdot \frac{\mathrm{d}\beta}{\mathrm{d}\lambda} = \frac{-D \cdot \lambda}{d^2[1 - \lambda/d - \sin^2\alpha]^{3/2}} . \tag{8.57}$$

This shows that the optical path length through the grating pair increases with increasing wavelength. Choosing the grating separation $D$ sufficiently large, the broadening of the pulse due to the linear dispersion in the optical fiber can be overcompensated for by the grating pair and leads to a shortening of the duration $\Delta T$ of the pulse $I(t)$. The experimental arrangement for the compression of optical pulses after they pass through the fiber is shown in Fig. 8.60.

### 8.6.4 Measurements of Ultrashort Optical Pulses

Since the time resolution even for fast optical detectors is limited to about 100 ps (except for the streak camera, which reaches 1 ps) the measurement of such short pulses can no longer be performed with conventional devices, but demands new ideas. One method is based on optical interferometry (Fig. 8.61). The laser beam is split into two parts that are recombined after having traveled along two different paths with slightly different path lengths. The superposition of the two parts with variable time delay $\tau$ and intensities $I_1(t) = |A_1(t)|^2$ and $I_2(t+\tau) = |A_2(t+\tau)|^2$ gives the total intensity

$$\begin{aligned} I(\tau) &= |A_1(t) + A_2(t+\tau)|^2 \\ &= I_1(t) + I_2(t) + 2A_1(t) \cdot A_2(t+\tau) , \end{aligned} \tag{8.58}$$

which depends on the relative phase between the two optical waves, i.e., on the time delay $\tau$. Although the detector cannot follow the fast optical waves, it measures the time dependent interference pattern $I(\tau)$, if the change of the time delay $\tau$ is sufficiently slow. If the spectral width of the short pulse is large, it contains a superposition of many monochromatic carrier waves with a nearly continuous frequency spectrum. In this case there will be no clear interference pattern and the

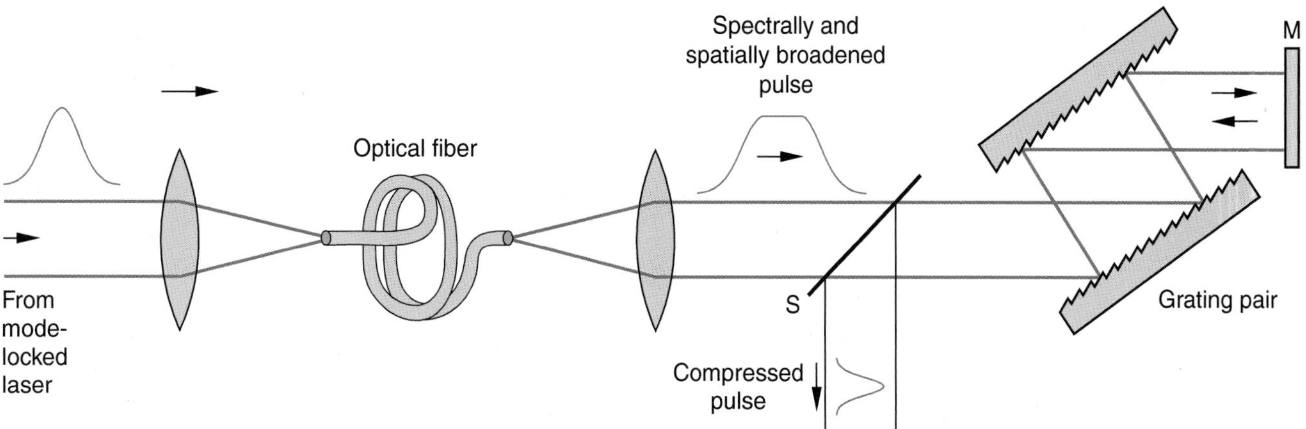


**Fig. 8.60.** Experimental arrangement for the generation of femtosecond pulses by self-phase modulation with subsequent pulse compression by a grating pair [8.16]

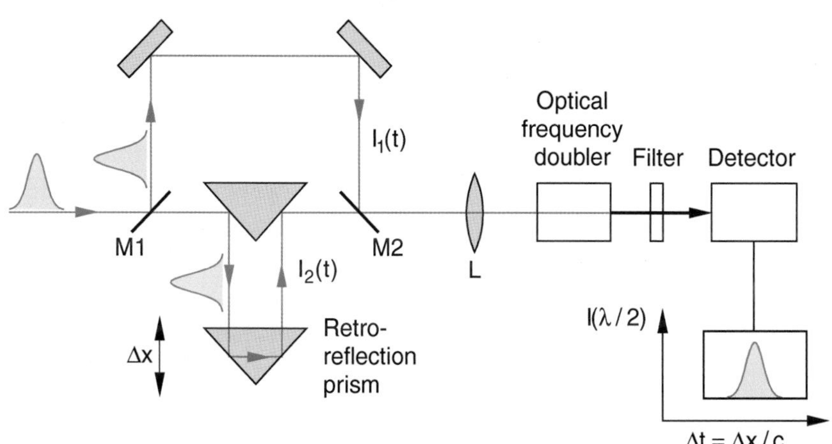


**Fig. 8.61.** Optical interferometry with translation-retroreflecting prism and second harmonic generation for measuring the width of ultrashort pulses

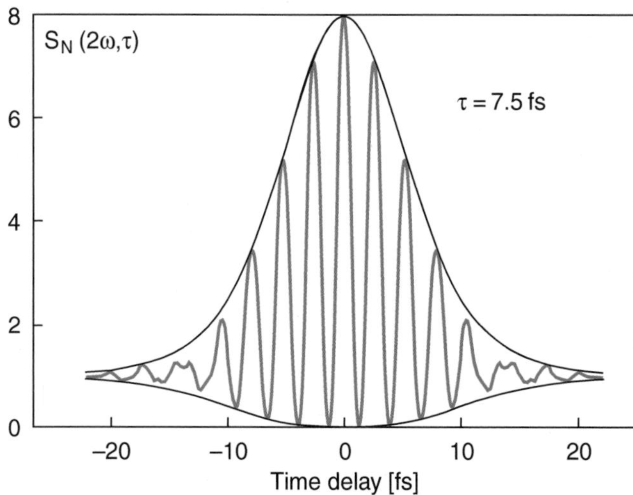


**Fig. 8.62.** Measured femtosecond pulse with only five optical periods of $T = 2.5$ fs within the full half-width of the envelope

detector would measure the sum of the two intensities $I_1 + I_2$, independent on their separation. Here the frequency-doubling of the fundamental wavelength in a nonlinear crystal is a good solution. The intensity

$$\overline{I}(2\omega) \propto |I_1(t) + I_2(t+\tau)|^2 = \overline{I}_1^2 + \overline{I}_2^2 + \overline{2I_1(t)\cdot I_2(t+\tau)}$$

of the second harmonics does depend on the time delay $\tau$. Even if the time constant of the detector is long compared to the pulse width and the detector measures the time average of the pulses, it still gives the true pulse profile $I(t)$.

In Fig. 8.62 an actual experimental result is shown for a pulse with 7.5 fs duration (half-width at half-maximum), which shows the optical cycles with 2.5 fs period, monitored with a detector with a time constant of about 1 ns.

Some applications of these ultrashort pulses are discussed in Chap. 12.

## SUMMARY


- Laser stands for "**L**ight **A**mplification by **S**timulated **E**mission of **R**adiation".
- A laser consists essentially of three components:
  The *energy pump*, which produces inversion in a medium by selective energy transfer into the medium.
  The *active medium* with a population inversion for selected transitions where an electromagnetic wave passing through the active medium is amplified instead of attenuated.
  The *optical resonator*, which stores the radiation power emitted by the active medium in a few resonator modes. In these modes, the number of photons should be large. This ensures that in these modes the induced emission is much stronger than the spontaneous emission.
- Laser oscillation starts at a threshold power delivered by the pump into the active medium, which depends on the critical inversion and the total losses of the lasing modes. At threshold the losses are just compensated by the gain of the active medium.
- The oscillation frequencies of the laser emission are limited by the spectral range where the active medium has sufficient gain. Within the gain profile of the active medium the lasing frequencies are determined by the eigenresonances of the optical resonator.
- The divergence of the emitted laser beam depends on the number of transverse modes participating in laser oscillation. If only fundamental modes contribute to laser emission, the laser beam profile is Gaussian and its divergence is only limited by diffraction effects.
- Single mode lasers, oscillating on a single fundamental mode, can be realized by additional mode selecting elements inside the laser resonator.
- A synchronous tuning of all frequency-selecting elements allows the realization of a single mode laser with a single wavelength tunable across the spectral gain profile of the active medium.
- The active medium can be a solid, a liquid or a gas. Broad gain profiles are provided by semiconductor materials, by dye solutions, by doped crystals with color centers and by vibronic solid state lasers consisting of an insulator, doped by metal ions.
- For some types of lasers, threshold inversion can only be achieved with pulsed pumps (e.g., pulsed Nd:glass lasers or excimer lasers), while most lasers can be operated in a continuous wave mode (cw lasers) as well as in a pulsed mode.
- The time profile of the laser output is limited by the duration of the pump power above threshold.
- By fast switching of the resonator quality factor, short laser pulses in the nanosecond range can be realized ($Q$-switched lasers).
- Coupling of many lasing resonator modes (mode locking) results in even shorter pulses down to about 1 picosecond.
- By pulse compression in optical fibers or by nonlinear gain manipulation inside the laser cavity (Kerr lens mode locking) femtosecond laser pulses have been obtained.


## PROBLEMS

1. a) What is the population ratio $N_i/N_k$ for atoms in a gas for thermal equilibrium at $T = 300\,\text{K}$, if the wavelength of the transition $E_i \to E_k$ is $\lambda = 500\,\text{nm}$ and the angular momentum quantum numbers are $J_i = 1$ and $J_k = 0$?
   b) What is relative absorption of a monochromatic light wave per cm path length through a gas, if the transition probability $A_{ik} = 1 \times 10^8\,\text{s}^{-1}$, the gas pressure $p = 1\,\text{mbar}$ and $10^{-6}$ of all atoms are in the lower state $E_k$ of the transition?
   c) What is the threshold inversion $N_k - N_i$, if the total losses per roundtrip of 10% should be compensated for by the gain over a path length of 20 cm in the active medium?
2. a) Calculate the Doppler-width of the neon line at $\lambda = 633\,\text{nm}$ in a gas discharge with a temperature of $T = 600\,\text{K}$.
   b) How many resonator modes $\text{TEM}_{0,0,q}$ for a resonator length of 1 m can oscillate, if the laser

threshold is at 50% of its maximum value at the line center?

3. An argon laser with a resonator length $d = 1$ m, oscillating at a wavelength $\lambda = 488$ nm can be forced to oscillate on a single mode by inserting a Fabry-Perot etalon inside its resonator.

   a) What is the thickness $t$ of the solid fused quartz etalon with a refractive index $n = 1.5$, if only one etalon transmission maximum should lie within the Doppler broadened gain profile of the argon transition at a discharge plasma temperature of $T = 5000K$?
   b) What is the reflectivity $R$ of the two coated planes of the etalon, if the transmission $T$ of the etalon for the neighboring laser resonator modes should decrease to $T = 1/3$ of that for the selected mode with $T = 1$ at the maximum of the etalon transmission?

4. Assume that the two end mirrors of a laser resonator are connected by invar steel rods with a length $d = 1$ m and a thermal expansion coefficient $\alpha = 12 \times 10^{-6}\,\mathrm{K}^{-1}$.

   a) How much does the laser frequency $\nu$ shift for a temperature change $\Delta T = 1$ K?
   b) If the laser wave inside the resonator passes through 40 cm air at atmospheric pressure, what is the frequency shift for a pressure change of 10 mbar? c) Is the dependence of the cavity's geometric length on the pressure change significant? Give an estimation of this change, using Hooke's law, for the invar rods (the elastic modulus of invar is $E = 10^7\,\mathrm{N/m}^2$, the diameter of the rods is 1 cm).

5. The nearly parallel beam of a laser with wavelength $\lambda = 10\,\mu$m and output power of $P = 10$ W has a beam diameter of $d = 3$ cm. It is focused by a lens with $f = 20$ cm.

   a) How large is the beam waist $w_0$ in the focal plane?
   b) The intensity distribution in this plane is

$$I(r) = I_0 \times \exp[-(r/w_0)^2]\,.$$

   What is the value of $I_0$?
   c) Assume that 10% of the laser power can be used for evaporating material from a steel sheet with thickness $t = 1$ mm placed in the focal plane. How long will it take for the laser beam to produce a hole through the steel sheet, if the evaporation heat is $6 \times 10^6$ J/Kg?

6. A short Fourier-limited laser pulse ($\Delta t = 10$ fs) passes through a medium with refractive index $n = 1.5$ and a dispersion of $\mathrm{d}n/\mathrm{d}\lambda = 4.4 \times 10^4\,\mathrm{m}^{-1}$.

   a) What is the minimum spectral width of the pulse?
   b) After which path length has the width $\Delta t$ of the pulse doubled due to the linear dispersion of the medium?
   c) How large must the intensity be in order to compensate for the pulse spread caused by the linear dispersion if the nonlinear refractive index is $n_2 = 10^{-10}\,\mathrm{cm}^2/\mathrm{W}$?

7. a) What is the quality factor $Q$ of a laser cavity with mirror separation $d = 1$ m, mirror reflectivities $R_1 = R_2 = 0.99$ at a frequency $\nu = 5 \times 10^{14}\,\mathrm{s}^{-1}$, if all other losses (apart from reflection losses) are 2% per roundtrip?
   b) After how much time does the energy stored in the cavity reduce to $1/e$, if at time $t = 0$ the amplification by the active medium suddenly drops to one?
   c) What are the separations $\Delta\nu$ and the half-widths $\delta\nu$ of the longitudinal cavity resonances?

8. Assume the laser oscillation in a cavity mode with $\nu = 4.53 \times 10^{15}\,\mathrm{s}^{-1}$ starts with one photon in this mode. How long does it take until the laser output power in this mode has reached 1 mW, for a net gain $g$ per roundtrip of 5%, a resonator length of $d = 1$ m and mirror transmissions $R_1 = 0$ and $R_2 = 0.02$ if

   a) the net gain $g = -\alpha_0$ is independent of the intensity?
   b) the gain saturation is essential and $\alpha = \alpha_0 + a \times P$ with $a = 0.4\,\mathrm{W}^{-1}\,\mathrm{m}^{-1}$ or $a = 0.55\,\mathrm{W}^{-1}\,\mathrm{m}^{-1}$?

# 9. Diatomic Molecules

In the next two chapters we will discuss the basic physics of molecules. In particular, the following questions shall be addressed and answered:

- Why can neutral atoms combine to form stable molecules, i.e., what is the nature of molecular binding forces?
- What is the internal energy structure of molecules, which not only depends on the distribution of electrons in atoms, but also on the location and motion of atomic nuclei?
- How can chemical reactions, and therefore biological processes, be explained on a molecular level?

First we will discuss diatomic molecules, because their treatment is simpler than that of polyatomic molecules. From the didactical point of view, they provide a good understanding of many problems, which are relevant not only to diatomics but also to polyatomic molecules. In particular, the interaction between two neutral atoms and its dependence on their distance $R$ from each other will give us better insight into the nature of chemical bonds. Also, the model of molecular orbitals, which give an intuitive picture of the spatial electron distribution and play an important role in chemistry, can be explained more readily for diatomic molecules.

Similarly to the situation for atoms, transitions between different molecular levels can be induced by absorption or emission of radiation. Since the energy levels of molecules are not only determined by electronic excitation but also by vibrations of the nuclei or the rotation of the whole molecule around an axis through its center of mass, the spectra of molecules are much more complex than those of atoms, but on the other hand they also give more detailed information about the internal structure and dynamics of molecules. We will discuss the spectra of diatomic molecules in Sect. 9.6, while Chap. 10 deals with the structure, the spectra and dynamics of polyatomic molecules.

We will start with the simplified model of rigid molecules, where the nuclei are clamped at a given position. For diatomic molecules this means that the internuclear distance is fixed and we neglect their vibration. For any arbitrary, but constant, value of $R$ the energy of the molecule is calculated. This yields the potential energy curves $E(R)$ for the different electronic states of the molecule. Later on we will then extend our model to include molecular vibration and rotation.

Computer programs for calculating the potential curves $E(R)$ have developed rapidly and reach meanwhile, at least for diatomic molecules, the same accuracy as the experimental values.

## 9.1 The $H_2^+$ Molecular Ion

The simplest of all molecules is the $H_2^+$ ion, consisting of two identical nuclei (protons) and one electron (Fig. 9.1). The interaction potential $E_{pot}$ between the three particles is

$$E_{pot} = -\frac{e^2}{4\pi\varepsilon_0}\left(\frac{1}{r_A}+\frac{1}{r_B}-\frac{1}{R}\right) . \tag{9.1}$$

When we chose the center of mass of the two protons as the origin of our coordinate system (because of its small mass the electron does not noticeably change the center of mass) we obtain from Fig. 9.1 the relations

$$\boldsymbol{r} = \boldsymbol{R}_A + \boldsymbol{r}_A = \boldsymbol{R}_B + \boldsymbol{r}_B \Rightarrow \boldsymbol{r} = \frac{1}{2}(\boldsymbol{r}_A + \boldsymbol{r}_B)$$

because $\boldsymbol{R}_A = -\boldsymbol{R}_B$. With $\boldsymbol{R} = \boldsymbol{r}_A - \boldsymbol{r}_B$ we can replace $\boldsymbol{r}_A$ and $\boldsymbol{r}_B$ by:

$$\boldsymbol{r}_A = \boldsymbol{r} + \frac{1}{2}\boldsymbol{R} ; \quad \boldsymbol{r}_B = \boldsymbol{r} - \frac{1}{2}\boldsymbol{R} . \tag{9.2}$$

W. Demtröder, *Atoms, Molecules and Photons*, 2nd ed., Graduate Texts in Physics,
DOI 10.1007/978-3-642-10298-1_9, 

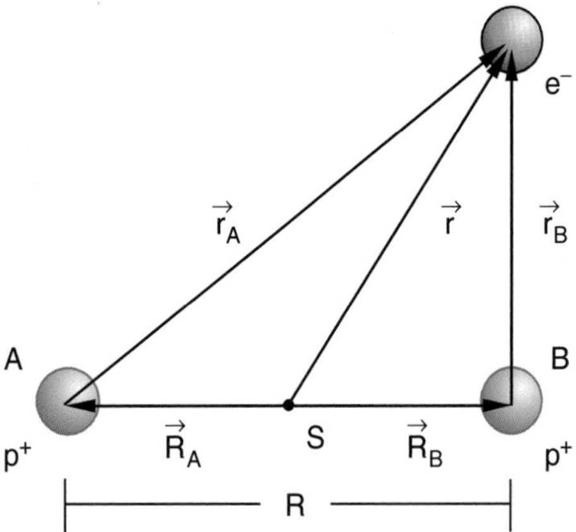


**Fig. 9.1.** $H_2^+$ molecular ion

The Schrödinger equation for this three body problem is:

$$\hat{H}\Phi = E\Phi \tag{9.3a}$$

with

$$\begin{aligned}\hat{H} = &-\frac{\hbar^2}{2M}\left(\Delta_A(\boldsymbol{R}_A) + \Delta_B(\boldsymbol{R}_B)\right) \\ &-\frac{\hbar^2}{2m}\Delta_e(\boldsymbol{r}) + E_{pot}(\boldsymbol{r}, R)\end{aligned} \tag{9.3b}$$

where the first two terms represent the kinetic energy of the nuclei, and the third term that of the electron. The nuclei and the electron move in the potential $E_{pot}(\boldsymbol{r}, R)$.

### 9.1.1 The Exact Solution for the Rigid $H_2^+$ Molecule

Differing from the situation for the H atom, the Schrödinger equation (9.3) cannot be solved analytically. One must therefore use approximations. Because of the much larger mass of the nuclei ($M/m \approx 1836$) their velocities $v$ and their kinetic energy $E_{kin} = p^2/2M$ is much smaller than that of the electron. In a first approximation we can neglect it. In this approximation the two nuclei are clamped at a given distance $R$ (rigid nuclear frame). The nuclear distance $R = |\boldsymbol{R}|$ can be regarded as a freely chosen parameter. This means that (9.3) is solved for all relevant values of $R$. This gives the wave function and the energy $\left(E_{kin} + E_{pot}^{pl}\right)$ of the electron plus the repulsive potential energy between the two nuclei for each chosen value of $R$, i.e., as a function of the parameter $R$.

In this approximation of the rigid $H_2^+$ the Schrödinger equation (9.3) with the potential (9.1) becomes:

$$\begin{aligned}&\left[-\frac{\hbar^2}{2m}\Delta_e(\boldsymbol{r}) - \frac{e^2}{4\pi\varepsilon_0}\left(\frac{1}{r_A} + \frac{1}{r_B} - \frac{1}{R}\right)\right] \\ &\times\Phi(\boldsymbol{r}_A, \boldsymbol{r}_B, R) = E(R)\Phi(\boldsymbol{r}_A, \boldsymbol{r}_B, R)\end{aligned} \tag{9.4}$$

where $\boldsymbol{r}_A$ and $\boldsymbol{r}_B$ depend on the coordinates both of the electron and of the two nuclei. The internuclear distance $R$ must be treated as a fixed parameter, not as variable.

Equation (9.4) can be solved analytically in a similar way as (5.1) for the H atom. Since the potential energy is no longer spherically symmetric but has cylindrical symmetry, it is convenient to use elliptical coordinates:

$$\begin{aligned}&\mu = \frac{r_A + r_B}{R}\;; \quad \nu = \frac{r_A - r_B}{R}\;; \\ &\varphi = \arctan\left(\frac{y}{x}\right)\end{aligned} \tag{9.5}$$

where the location of the two nuclei are the focal points of the ellipsoid with cylindric symmerty and the $z$-axis as symmetry axis is chosen to coincide with the line between the two nuclei (Fig. 9.2). In elliptical coordinates the wave function, which is dependant on three coordinates, is separable into the product

$$\Phi(\boldsymbol{r}_A, \boldsymbol{r}_B, R) = M(\mu) \times N(\nu) \times \phi(\varphi) \tag{9.6}$$

of three functions, where each function depends on only one coordinate. The first factor $M(\mu)$ is constant for $\mu =$ const., i.e. for $r_A + r_B =$ constant. The value of the constant depends on the quantum number associated with $M(\mu)$. This is fulfilled on the surface of an ellipsoid with rotational symmetry around the $z$-axis (Fig. 9.2b). The second factor $N(\nu)$ is constant for $\nu =$ const., for instance for $r_A = r_B$. This is the symmetry plane $z = 0$, perpendicular to the $z$-axis. The third factor is constant for $\varphi =$ const. $\Rightarrow y = c \cdot x$, which gives planes through the $z$-axis.

Similar to the separation of the hydrogen atomic wave function, we obtain three equations for the functions $M(\mu)$, $N(\nu)$ and $\phi(\varphi)$, which are analytically solvable. Since the solutions have to be unambiguous and normalized, this imposes restrictions on the wave functions and leads to definite wave functions and discrete energy eigenvalues $E_n(R)$, which, however, depend on the internuclear separation $R$ (Fig. 9.3). The

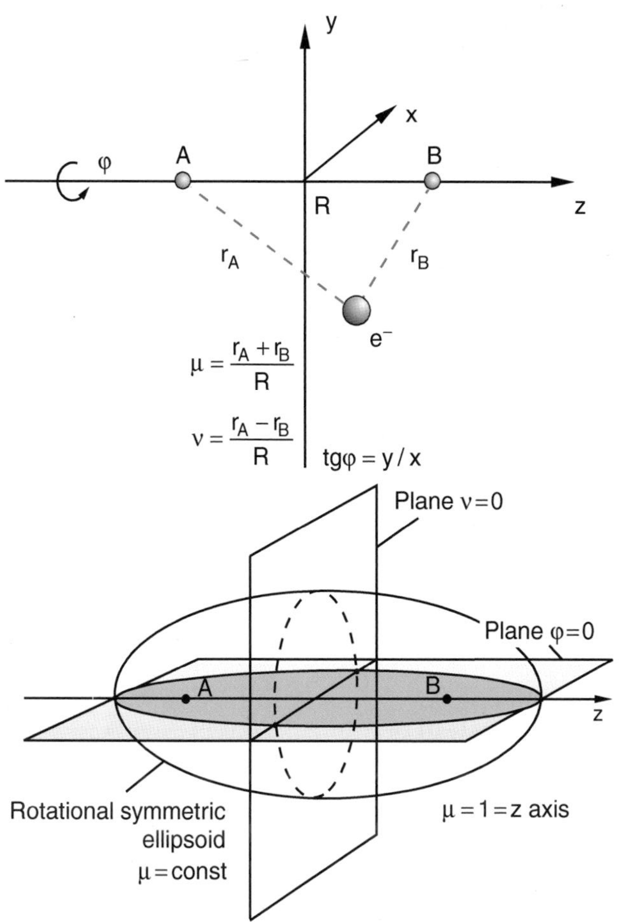


**Fig. 9.2.** Elliptical coordinates and their physical interpretation

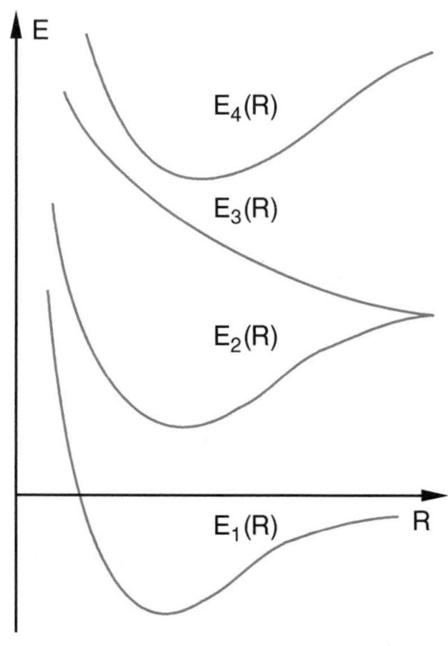


**Fig. 9.3.** Schematic representation of potential curves $E(R)$ corresponding to bound states and repulsive unstable states

surfaces, where $\Phi = 0$ are called nodal surfaces. They separate spatial volumes with positive and negative values of $\Phi$.

> The spatial part $\Phi(\boldsymbol{r}, R)$ of the wave function is called a molecular orbital. Its absolute square $-e \times |\Phi(\boldsymbol{r}, R)|^2$ gives the spatial distribution of the electron charge, for any given value of the nuclear distance $R$.

The nodal surfaces $M(\mu) = 0$ are ellipsoids, $N(\nu) = 0$ are planes perpendicular to the $z$-axis, $\phi(\varphi) = 0$ planes through the $z$-axis (Fig. 9.2).

The functions $E(R)$ can monotonously decrease with increasing $R$ (repulsive unstable energy states) or they can show minima at a certain value of $R$ (stable energy states).

These curves

$$E(R) = \langle E_{\text{kin}}(\mathrm{e}^-)\rangle + \frac{e^2}{4\pi\varepsilon_0}\left(\frac{1}{R} - \left\langle \frac{1}{r_\mathrm{A}} + \frac{1}{r_\mathrm{B}} \right\rangle\right) \tag{9.7}$$

are called potential energy curves, although they also include, besides the potential energy, the time-averaged kinetic energy of the electron. They describe the potential in which the nuclei can vibrate, if we go beyond the approximation of the clamped nuclei.

Since the potential is not spherically symmetric, the angular momentum $\boldsymbol{l}$ of the electron is no longer constant in time. In the classical model the vector $\boldsymbol{l}$ precesses around the internuclear axis. Its absolute value $|\boldsymbol{l}|$ depends generally on the internuclear separation $R$. In quantum mechanics its projection onto the $z$-axis has for a stationary state a well-defined eigen value

$$\langle l_z \rangle = m \cdot \hbar\ , \tag{9.8a}$$

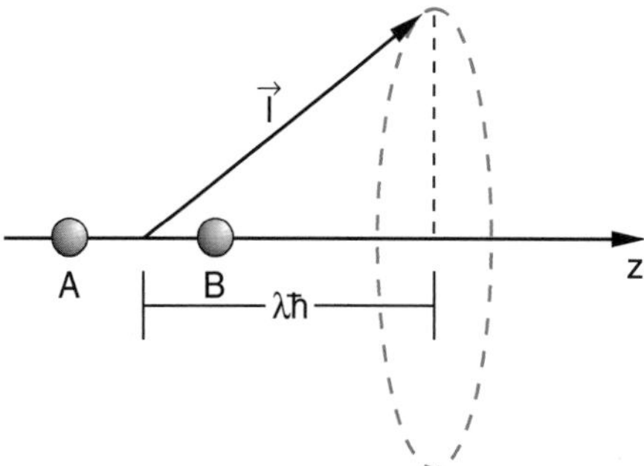


**Fig. 9.4.** The precessing orbital angular momentum $\boldsymbol{l}$ of the electron and its constant projection $l_z = \lambda\hbar$

which is determined by the integer $m = 0, \pm 1, \pm 2, \ldots \pm l$ (Fig. 9.4) and is, for a given potential curve $E(R)$ independent of $R$. The reason for this is that the operator

$$\hat{l}_z = (\hbar/i)\partial/\partial\varphi \tag{9.8b}$$

depends solely on $\varphi$ and not on $R$. If it is applied to the function (9.6) it acts only on the last factor $\phi(\varphi)$ and yields the eigenvalue $m \cdot \hbar$. This is completely analogous to the case of the hydrogen atom in an external magnetic field, where the spherical symmetry is broken and only cylindrical symmetry is left and the vector $\boldsymbol{l}$ precesses around the field axis.

The difference is, however, that in the axial electric field of a nonrotating diatomic molecule the energy of a level does not depend on the direction of the field. This means, that levels with $\langle l_z \rangle = \pm m \cdot \hbar$ have the same energy. Therefore the molecular levels are described by the quantum number $\lambda = |m|$. Instead of (9.8a) we write

$$\langle l_z \rangle = \lambda\hbar\ . \tag{9.8c}$$

Analogous to the nomenclature in atoms, electrons in diatomic molecules are called $\sigma$-electrons for $\lambda = 0$; $\pi$-electrons for $\lambda = 1$ and $\delta$-electrons for $\lambda = 2$ etc. The Latin letters used in atoms are just replaced by Greek letters for molecules.

Due to the precession of the electron around the internuclear axis for $\lambda > 0$ a magnetic field in $z$-direction is produced. If we take into account the electron spin $\boldsymbol{s}$, the magnetic moment $\mu_s$ can have two different orientations in this field, similar to the situation for atoms in the Stern–Gerlach experiment (see Sect. 5.5). The electron spin precesses around the magnetic field direction (which is the $z$-direction) and only its $z$-component

$$\langle s_z \rangle = m_s \cdot \hbar = \pm\tfrac{1}{2}\hbar \tag{9.8d}$$

has definite eigenvalues (Fig. 9.5).

The energy state $E(R)$ of the electron in the clamped nuclei model for a given nuclear distance $R$ is completely defined by the principal quantum number $n$, the quantum number $\lambda$ of the $z$-projection of $\boldsymbol{l}$ and $m_s$ of $\boldsymbol{s}$. This state $E(n, \lambda, m_s)$ can therefore be characterized by the three quantum numbers $n, \lambda, m_s$.

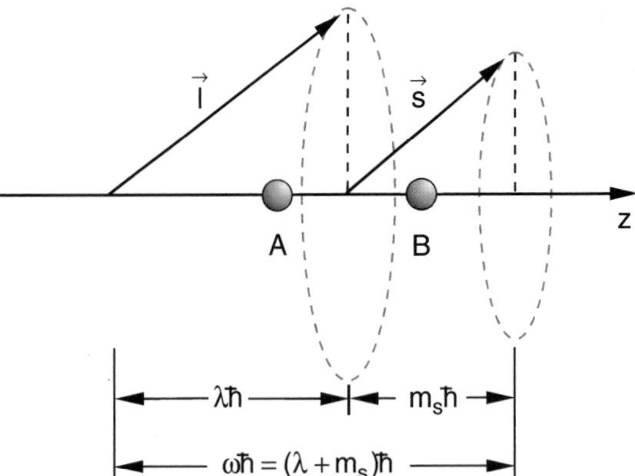


**Fig. 9.5.** The projections $\langle l_z \rangle = \lambda\hbar$ of the orbital angular momentum $\boldsymbol{l}$ and $\langle s_z \rangle = m_s\hbar$ of the electron spin $\boldsymbol{s}$ adds up to the total projection $\omega\hbar = (\lambda + m_s)\hbar$

**Note:**

In the literature the spin projection quantum number $m_s$ is often labeled as $\sigma$. In order to avoid confusion with the state with $\lambda = 0$, which is also called a $\sigma$-state, we will use $m_s$ instead of $\sigma$ in this book.

In Fig. 9.6 the spatial distribution of some molecular orbitals are shown. They are characterized by the three quantum numbers $n$, $l$ and $\lambda$ as $nl\lambda$, where $n$ is a number that labels the electronic states according to increasing energy (analogous to the principal quantum number $n$ in atoms), $l$ is the quantum number of the orbital angular momentum $\boldsymbol{l}$ which is, however, only defined for large $R$, and $\lambda$ the projection quantum number. As in atoms the states with different $l$-values for $R \to \infty$ are labeled with Latin letters ($s$ for $l = 0$, $p$ for $l = 1 \ldots$). A molecular orbital with $n = 1$, $l = 0$, $\lambda = 0$ is named $1s\sigma$ and with $n = 2$, $l = 1$, $\lambda = 1$ is a $2p\pi$ orbital (see Table 9.1).

**Table 9.1.** Quantum numbers and term nomenclature of a molecular electron with principal quantum number $n$, angular momentum quantum number $l$, projection quantum number $\lambda = |m_l|$

| Quantum numbers $n$ | $l$ | $\lambda$ | Term nomenclature |
|---|---|---|---|
| 1 | 0 | 0 | $1s\sigma$ |
| 2 | 0 | 0 | $2s\sigma$ |
| 2 | 1 | 0 | $2p\sigma$ |
| 2 | 1 | 1 | $2p\pi$ |
| 3 | 2 | 1 | $3d\pi$ |
| 3 | 2 | 2 | $3d\delta$ |

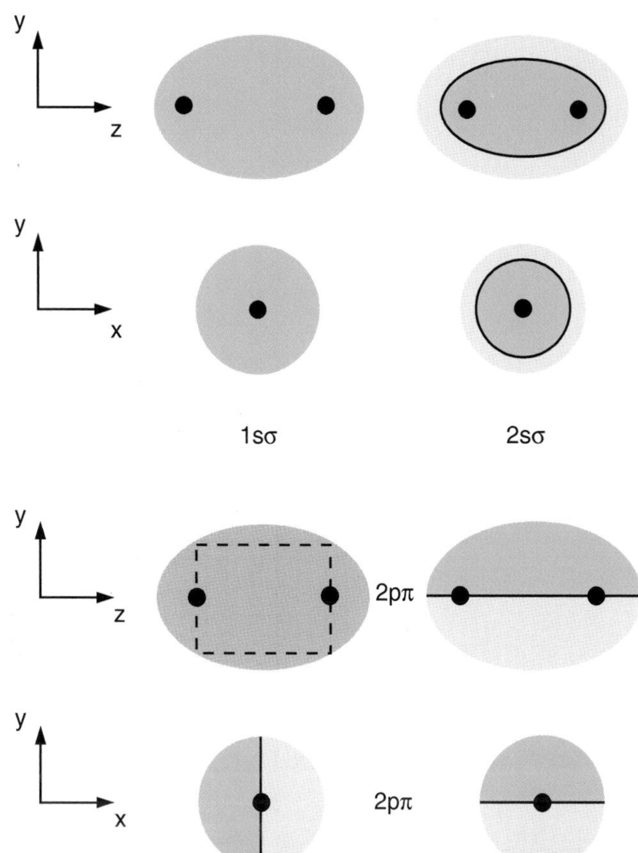


**Fig. 9.6.** Some examples of cross sections of molecular orbitals of the $H_2^+$ ion. In the *blue* regions $\Phi > 0$, in the *grey* ones $\Phi < 0$. The *solid* lines give the location of $\Phi = 0$. The dashed rectangle indicates, that the $y - z$ plane is a nodal plane

### 9.1.2 Molecular Orbitals and LCAO Approximations

Although the rigid $H_2^+$ molecule can be treated rigorously, it is instructive to apply some important approximation methods to the description of this simple molecule and to compare the results with the exact solutions. We can then learn more about the advantages and deficiencies of these methods when they are applied to larger molecules, that cannot be treated exactly.

One of these methods is the LCAO approximation, where the molecular wave function is composed of linear combinations of atomic orbitals of the atoms that form the molecule. The coefficients of the atomic orbitals in this linear combination are optimized in such a way, that the energy, calculated with this molecular wave function becomes a minimum. This optimization procedure is based on the fact that the correct wave function always yields lower energies than the approximate functions [9.1].

Since the absolute square $|\Phi(x, y, z)|^2$ of the wave function $\Phi$ represents the spatial probability density distribution of the electron, which corresponds to the classical orbitals of the electron in the molecule, the molecular wave function is called a molecular orbital.

The $H_2^+$ molecular ion can be composed of an H atom and an $H^+$-ion (= proton) (Fig. 9.7). For the lowest energy state of $H_2^+$, the H atom is in its $1s$ ground state. The atomic orbital of the electron in the H atom is then (see Table 5.2):

$$\phi_A(r_A) = \frac{1}{\sqrt{\pi a_0^3}} e^{-r_A/a_0} \,. \tag{9.9}$$

The electron can be found around either nucleus A or B. Both possibilities lead to the $H_2^+$ ion when the two nuclei are brought to the proper distance $R = R_e$. Since we cannot distinguish between the two possibilities we have to take both into account. We therefore choose the molecular wave function as the linear combination

$$\Phi(\boldsymbol{r}, R) = c_1\phi_A(\boldsymbol{r}_A) + c_2\phi_B(\boldsymbol{r}_B) \tag{9.10}$$

where $\boldsymbol{r}_A = \boldsymbol{r} + \boldsymbol{R}/2$ and $\boldsymbol{r}_B = \boldsymbol{r} - \boldsymbol{R}/2$ can be substituted by the nuclear separation $R$ and the distance $r = |\boldsymbol{r}|$ of the electron from the center of mass (Fig. 9.1).

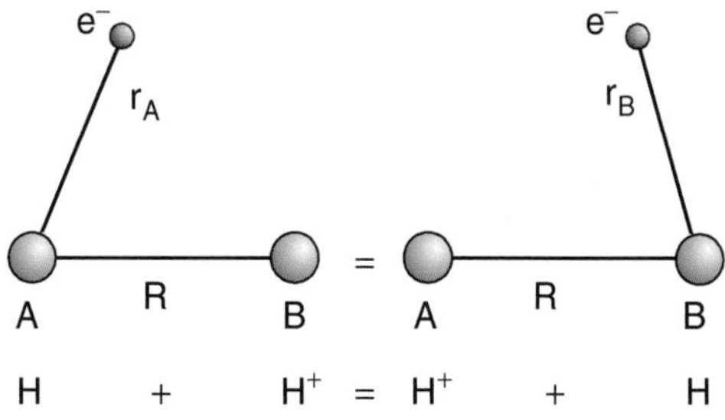


**Fig. 9.7.** Equivalence of the two configurations $H_A + H_B^+$ and $H_A^+ + H_B$

The wave function should be normalized for arbitrary values of $R$. This demands

$$\int |\Phi|^2 \mathrm{d}\boldsymbol{r} = c_1^2 \int |\phi_A(\boldsymbol{r}_A)|^2 \mathrm{d}\boldsymbol{r}_A + c_2^2 \int |\phi_B(\boldsymbol{r}_B)|^2 \mathrm{d}\boldsymbol{r}_B + 2c_1 c_2 \int \phi_A \phi_B \mathrm{d}^3 r \stackrel{!}{=} 1 \quad (9.11)$$

where the integration is performed over the coordinates of the electron.

Since the atomic orbitals are already normalized, the first two integrals have the value 1. From (9.11) we obtain the condition for the coefficients of the normalized molecular wave function:

$$c_1^2 + c_2^2 + 2c_1 c_2 S_{AB} = 1 \quad (9.12)$$

where the integral

$$S_{AB} = \int \phi_A(\boldsymbol{r}_A)\phi_B(\boldsymbol{r}_B)\mathrm{d}\tau \quad (9.13)$$

depends on the spatial overlap of the two atomic wave functions and is therefore called an ***overlap integral***. Its value depends on the internuclear separation $R$, because the integration occurs over the electron coordinates $\boldsymbol{r} = \{x, y, z\}$ which depend on $R$ according to (9.2).

Because of symmetry arguments, we have the condition $|c_1|^2 = |c_2|^2 = |c|^2$. Furthermore, the molecular wave function has to be either symmetric or antisymmetric with respect to the exchange of the two atomic orbitals (which is equivalent to the exchange of the electron between nucleus A and B). This demands $c_1 = \pm c_2$ and yields the normalized molecular wave functions (Fig. 9.8):

$$\Phi_S = \frac{1}{\sqrt{2+2S_{AB}}}(\phi_A + \phi_B) \quad \text{and} \quad (9.14a)$$

$$\Phi_a = \frac{1}{\sqrt{2-2S_{AB}}}(\phi_A - \phi_B)\,. \quad (9.14b)$$

The expectation value for the energy is

$$\langle E \rangle = \int \Phi \hat{H} \Phi \mathrm{d}\tau \quad (9.15)$$

where $\hat{H}$ is the Hamiltonian in the Schrödinger equation (9.4)

$$\hat{H}\Phi = E\Phi$$

of the rigid molecule.

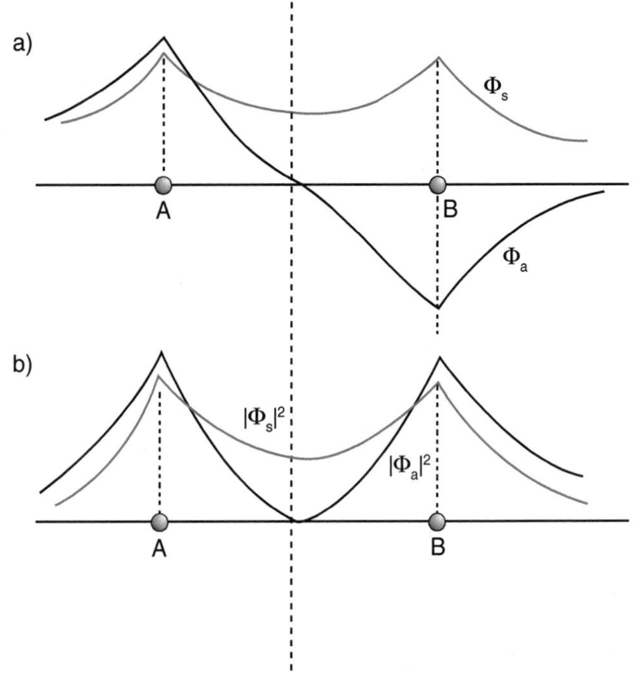


**Fig. 9.8a,b.** Cut through the surfaces with cylindrical symmetry (**a**) $\Phi_S$ and $\Phi_a$ (**b**) $|\Phi_S|^2$ and $|\Phi_a|^2$

Inserting (9.15) with (9.14) into (9.16) gives two energies

$$E_S(R) = \frac{H_{AA} + H_{AB}}{1 + S_{AB}} \quad \text{and}$$
$$E_a(R) = \frac{H_{AA} - H_{AB}}{1 - S_{AB}} \quad (9.16)$$

which depend on the nuclear separation $R$, because the integrals

$$H_{AA} = \int \phi_A \hat{H} \phi_A \,\mathrm{d}\tau_{el}$$
$$H_{BB} = \int \phi_B \hat{H} \phi_B \,\mathrm{d}\tau_{el}$$
$$H_{AB} = \int \phi_A \hat{H} \phi_B \,\mathrm{d}\tau_{el} \quad \text{and}$$
$$S_{AB} = \int \phi_A \phi_B \,\mathrm{d}\tau_{el} \quad (9.17)$$

depend on $R$.

**Note:**

The variables $r_A$ and $r_B$ in the atomic orbitals (9.9) gives the distance between the electron and the nucleus A or B, respectively. The integrals (9.17), therefore,

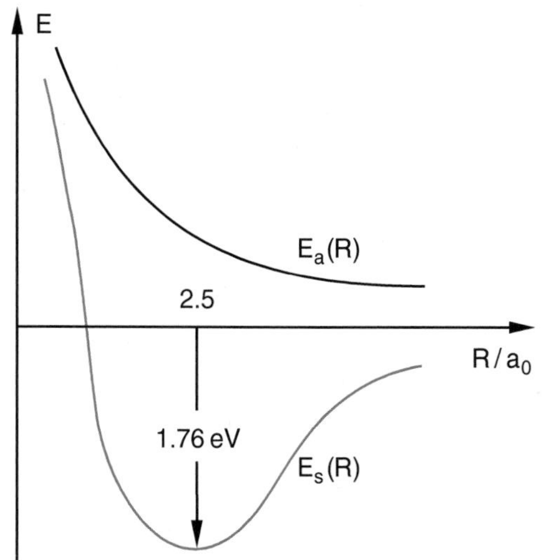


**Fig. 9.9.** Potential curves $E_s(R)$ for the symmetric charge distribution $|\Phi_s|^2$ and $E_a(R)$ for the antisymmetric distribution $|\Phi_a|^2$ in the LCAO approximation

represent two-center integrals, since the atomic orbitals are expressed in coordinates with two different centers. When evaluating the integrals over the electron coordinates $d\tau_{el} = r^2 \sin\vartheta \, dr \, d\vartheta \, d\varphi$, the variables $r_A$ and $r_B$ in the atomic orbitals have to be transformed into a coordinate system with a common origin (see Problem 9.1).

The curves $E_s(R)$ and $E_a(R)$ are shown in Fig. 9.9. They are called potential curves. While $E_s(R)$ has a minimum and corresponds to a bound state, $E_a(R)$ is a monotonic function, falling with increasing $R$. It represents a repulsive potential and corresponds to an unstable molecular state.

It is interesting to investigate the reasons for the binding of $H^+ + H$ in the ground state of $H_2^+$ represented by the wavefunction $\Phi_s$. The energy $E(R)$ is the sum of the kinetic energy of the electron and its potential energy in the attractive force field of the two positively charged nuclei. In addition, the repulsive energy between the two nuclei has to be considered. In Fig. 9.10, both contributions $E_{kin}(R)$ and $E_{pot}(R)$ are plotted together with the total energy $E(R)$ in units of the ionization energy $E_I(H) = 13.6\,\text{eV} = 0.5$ a.u. of the hydrogen atom. The comparison between the results of the simple LCAO approximation and those of the exact calculation gives the following insight:

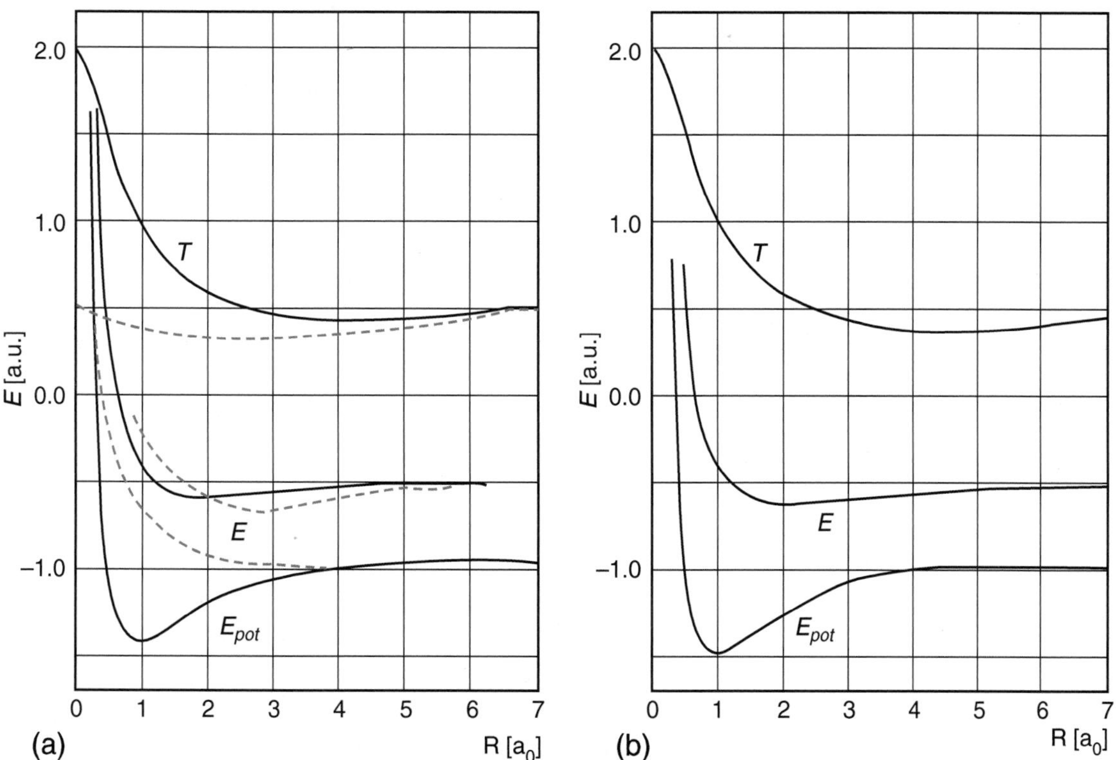


**Fig. 9.10.** Kinetic energy T, potential energy $E_{pot}$ and total energy $E$ of the $H_2^+$ ion: (**a**) LCAO approximation with $\eta = 1$ (*dashed curves*) and with optimized parameter $\eta(R)$ (*solid curves*). (**b**) Exact calculation. a.u = atomic units (Hartree) 1 Hartree = $2 \times E_{ion}(H) = 27.2$ev [9.2]

The LCAO approximation in its simplest form is not in good agreement with the correct solutions. This can be seen, for instance, from the fact that this approximation gives for the limit $R \to 0$:

$$\lim_{R\to 0} E(R) \to -3E_A$$

(without the nuclear repulsion),

as can be seen from the calculation of the integrals (9.17). Since for $R \to 0$ a nucleus with a charge $Ze$ with $Z = 2$ is formed, the energy should be $-Z^2 E_A = -4E_A$.

Two different effects contribute to the binding:

a) The lowering of the potential energy at the equilibrium distance $R_e$, where the total energy $E(R)$ has its minimum. The electron charge distribution, with its maximum in the middle between the two nuclei (Fig. 9.8b), pulls the two nuclei towards each other, due to the Coulomb force between electrons and protons. It acts like a glue that keeps the nuclei together. This is the largest contribution to the binding.

b) The molecular wave function $\Phi_s$ has a larger spatial extension than the atomic 1$s$ orbitals. The spatial uncertainty for the electron is increased and its momentum uncertainty is therefore smaller than for the H-atom, which means that its mean kinetic energy $\langle E_{kin}\rangle = \langle p^2\rangle/2m$ is decreased.

Both effects contribute to the binding in the $H_2^+$ ion. However, their relative magnitude is not correctly described by the LCAO approximation, as can be seen from Fig. 9.10, where the different contributions to the total energy are plotted in units of the ionization energy of the H-atom. In the LCAO-approximation the main contribution to the binding energy comes from the lowering of the kinetic energy, while the exact solution shows that the binding is mainly due to the decrease of the potential energy.

### 9.1.3 Improvements to the LCAO ansatz

The simple LCAO approximation can be essentially improved if instead of the unperturbed atomic orbitals (9.9) modified functions

$$\phi_A = C(1 + \lambda z)e^{-\eta(R)r_A/a_0} \tag{9.18}$$

are used, where the two parameters $\lambda$ and $\eta(R)$ are optimized for each nuclear separation $R$ to bring the energy $E(R)$ to a minimum.

The parameter $\lambda$ in the function (9.18) takes into account, that the charge distribution of the electron is no longer spherically symmetric in the force field of the two protons, but shows a deformation in the $z$-direction, which reduces the spherical symmetry to cylindrical symmetry. Furthermore, the radial distributions $|\Phi(r_A)|^2$ and $|\Phi(r_B)|^2$ will depend on the nuclear separation, which is described by the parameter $\eta(R)$. For $\eta > 1$ the orbitals $\Phi_A$ and $\Phi_B$ are contracted, which causes a lowering of the energy because the electron charge becomes more concentrated between the two nuclei.

The change of the kinetic energy due to the contraction of the atomic orbitals is due to two opposite effects. The contraction of the orbitals gives the electron less space in the $x$- and $y$-directions. This leads to an increase of the mean kinetic energy

$$\left\langle E_{kin}^{\perp}\right\rangle = \tfrac{1}{2}\left\langle\left(v_x^2 + v_y^2\right)\right\rangle$$

due to the uncertainty relation. However, the molecular wave function becomes more expanded in the $z$-direction, because the contraction parameter depends on $z$. This decreases the kinetic energy

$$\left\langle E_{kin\parallel}\right\rangle = \tfrac{1}{2}m\,\langle v_z\rangle^2\ . \tag{9.19}$$

A detailed calculation of both effects shows that a small net decrease of the kinetic energy is obtained for the improved LCAO approximation, which is, however, small compared to the decrease of the potential energy due to the contraction of the electron distribution between the nuclei.

The physical reasons for the molecular binding can be summarized as follows:

When the molecule is formed, the atomic orbitals are deformed in such a way, that the decrease of $E_{pot}(R_e)$ at the equilibrium separation $R_e$ and the decrease of $E_{kin}(R_e)$ overcompensates the energy necessary for the contraction and deformation of the atomic orbitals. This gives an energy minimum at $R = R_e$ below the energy of the separated fragments $H + H^+$.

In Table 9.2 the different contributions to the energy of $H_2^+$ are compiled in units of Hartree = twice the ionization energy of the H-atom (13.6 eV) for the different approximations and for the exact solution.

The binding energy

$$E_B(H_2^+) = E(H, 1s) - E(H_2^+, 1\sigma_g, R = R_e)$$

of the $H_2^+$ ion can be described as the difference between the electronic energy of the atomic $1s$ state and of the molecular $H_2^+$ state at the equilibrium nuclear distance $R = R_e$. This binding energy amounts to only about 1/5 of the binding energy of the electron in the H atom.

**Note:**

The binding energy $E_B$ is here defined as the energy necessary to dissociate the molecule. It is *positive*! One should not be mixed up with the absolute energy at the potential minimum, which is negative!

The different steps for improving the LCAO approximation yield the results for the binding energy and the equilibrium distance $R_e$ at the minimum energy (in units of the Bohr radius $a_0$) compiled in Table 9.3. In Fig. 9.11 the corresponding potential curves illustrate the progress in approaching the exact curves with an increasing number of optimized parameters in the LCAO functions. The best calculated potential curves have been obtained by James and Coolidge [9.3] who used linear combination of 50 basis functions.

**Table 9.2.** Mean kinetic and potential energy of the electron in the $H_2^+$ molecule, given in units of the ionization energy $E_{ion} = 13.6$ eV of the H atom for the different approximations

| Method | $\frac{1}{2}m\overline{v_\perp^2}$ | $\frac{1}{2}m\overline{v_z^2}$ | $\overline{E}_{kin}^{el}$ | $\overline{E}_{pot}$ | $E$ |
|---|---|---|---|---|---|
| LCAO $H_2^+$ $(R = R_e)$ | 0.60 | 0.18 | 0.78 | −1.9 | −1.12 |
| LCAO with $\eta = 1.25$ | 0.92 | 0.28 | 1.20 | −2.4 | −1.2 |
| Exact calculation | 0.95 | 0.30 | 1.25 | −2.5 | −1.25 |
| $H + H^+$ $(R = \infty)$ | 0.67 | 0.33 | 1.0 | −2.0 | −1.0 |

**Table 9.3.** Comparison of binding energy $E_B$ and equilibrium distance $R_e/a_0$ of $H_2^+$ for the different approximations

| Wavefunction $\Phi_s$ | $E_B$/eV | $R_e/a_0$ |
|---|---|---|
| simple LCAO | +1.76 | 2.5 |
| LCAO with optimized $\eta(R)$ but $\lambda = 0$ | +2.25 | 2.0 |
| Inclusion of polarization $(\eta \neq 0, \lambda \neq 0)$ | +2.65 | 2.0 |
| Exact calculation | +2.79 | 2.0 |

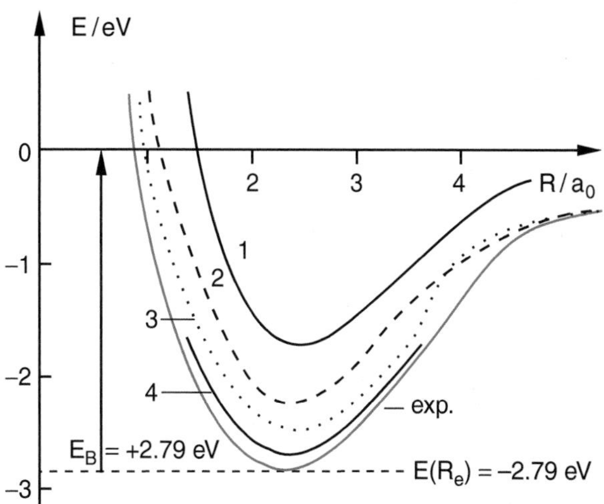


**Fig. 9.11.** Comparison of potential curves $E(R)$ for $H_2^+$ for the different approximations and experimental values (*blue curve*). 1 = simple LCAO, 2 = LCAO with optimized parameter $\eta(R)$, 3 = inclusion of polarization effects by the parameter $\lambda$, 4 = calculations by James and Coolidge

The foregoing discussions have shown that even for the simplest rigid molecule $H_2{}^+$ more refined approximations have to be used in order to reach the accuracy of the exact calculation. This is even truer for more complex molecules. We will at first present the approximate treatment of the second simplest molecule $H_2$ with two electrons.

## 9.2 The $H_2$ Molecule

The $H_2$ molecule consists of two protons as nuclei and two electrons (Fig. 9.12). We therefore have to take into account the interaction between the two electrons. This has the consequence (similar to the situation in the He-atom), that we can no longer separate the wave function into a product of one-dimensional functions, as we could for the $H_2^+$ ion and no exact analytical solution is possible, even for the approximation of the clamped nuclei. We therefore have to use approximation procedures. Furthermore, because two electrons with the

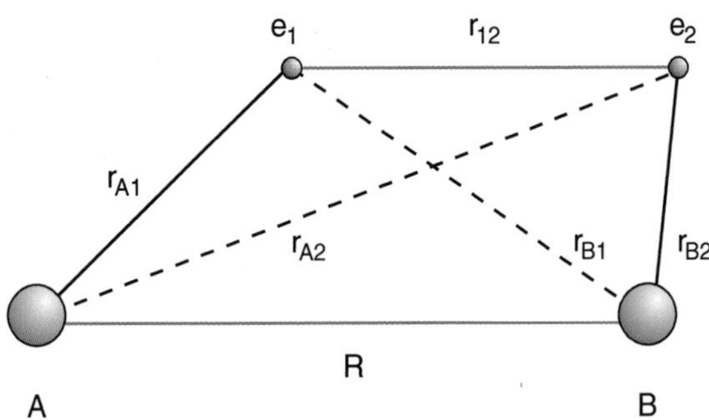


**Fig. 9.12.** $H_2$ molecule

same spatial wave functions are nondistinguishable the Pauli principle has to be obeyed. This means that the total wave function has to be antisymmetric with respect to an exchange of the two electrons.

We will here introduce the two most commonly used approximations: The molecular orbital (MO) approximation and the valence method of Heitler and London, because they provide good insight into the approximate calculation of electronic states of molecules and their physical motivation. Although the two methods seem at first to be quite different, they prove to be equivalent in their improved versions.

### 9.2.1 Molecular Orbital Approximation

The ground state of the $H_2$ molecule dissociates for $R \to \infty$ into two H atoms in the $1s$ state. We therefore choose, similarly to the $H_2^+$ ion, the normalized symmetric linear combination (9.14a)

$$\Phi_s = \frac{1}{\sqrt{2+2S_{AB}}}(\phi_A + \phi_B) \tag{9.20a}$$

of the two $1s$ atomic wave functions $\phi_A$ and $\phi_B$, where the electrons are labeled 1 and 2 and the nuclei A and B. If both electrons are in the molecular ground state the approximate total wave function

$$\Phi(\boldsymbol{r}_1, \boldsymbol{r}_2) = \Phi_s(\boldsymbol{r}_1) \cdot \Phi_s(\boldsymbol{r}_2) \tag{9.20b}$$

can be written as the product of the two molecular orbitals (9.20a). In this approximation the change of the spatial distribution of the molecular orbitals due to the mutual interaction between the two electrons has been neglected.

The ansatz (9.20b) is symmetric with respect to an exchange of the two electrons. Since the Pauli principle demands an antisymmetric total wave function the spin function must be antisymmetric. In this approximation the total wave function, including the spin function, is then

$$\Phi(\boldsymbol{r}_1, \boldsymbol{r}_2, \boldsymbol{s}_1, \boldsymbol{s}_2) = \Phi_s(\boldsymbol{r}_1) \cdot \Phi_s(\boldsymbol{r}_2) \times \frac{1}{2}\sqrt{2} \left[\chi^+(1)\chi^-(2) - \chi^+(2)\chi^-(1)\right] . \tag{9.21a}$$

We introduce the abbreviations

$$\chi^+ = \alpha\,, \quad \chi^- = \beta\,, \quad \phi_A(i) = a(i)\,, \quad \phi_B(i) = b(i)\,,$$

where $\alpha(1)$ means: $m_s(1) = +\frac{1}{2}$, $\beta(1)$ means $m_s(1) = -\frac{1}{2}$.

This gives the shorthand notation of the spatial part (9.20b) as:

$$\Phi(\boldsymbol{r}_1, \boldsymbol{r}_2) = \frac{1}{2+2S_{AB}}[a(1)+b(1)] \times [a(2)+b(2)] \tag{9.20c}$$

while the total wave function (9.21a) including the electron spin can be written as the Slater determinant

$$\Phi(\boldsymbol{r}_1, \boldsymbol{r}_2, \boldsymbol{s}_1, \boldsymbol{s}_2) = \begin{vmatrix} \Phi_s(1)\alpha(1) & \Phi_s(2)\alpha(2) \\ \Phi_s(1)\beta(1) & \Phi_s(2)\beta(2) \end{vmatrix} \tag{9.21b}$$

as can be immediately proved by evaluating the determinant.

The total energy of the two electrons in the rigid $H_2$ molecule is

$$E = \Sigma E_{kin}(i) + E_{pot}\,.$$

The Hamiltonian for the rigid $H_2$ molecule with clamped nuclei is

$$\hat{H} = -\frac{\hbar^2}{2m}(\nabla_1^2 + \nabla_2^2) + \frac{e^2}{4\pi\varepsilon_0}\left(-\frac{1}{r_{A1}} - \frac{1}{r_{B1}} - \frac{1}{r_{A2}} - \frac{1}{r_{B2}} + \frac{1}{r_{12}} + \frac{1}{R}\right). \tag{9.22}$$

With the abrevation

$$\hat{H}_i = -\frac{\hbar^2}{2m}\nabla_i^2 + \frac{e^2}{4\pi\varepsilon_0}\left(-\frac{1}{r_{Ai}} - \frac{1}{r_{Bi}} + \frac{1}{R}\right) \tag{9.23}$$

We can separate this expression into three parts:

$$\begin{aligned}\hat{H} &= \hat{H}_1 + \hat{H}_2 + \frac{e^2}{4\pi\varepsilon_0}\left(\frac{1}{r_{12}} - \frac{1}{R}\right) \\ &= 2\hat{H}(H_2^+) + \Delta E\,.\end{aligned} \tag{9.24}$$

The first two parts both represent the energy of the $H_2^+$ ion, which has only one electron $i = 1, 2$. They have already been discussed in the previous section. The third term in (9.23) describes the Coulomb-repulsion between the two electrons and the two protons. This last term for the nuclear repulsion is subtracted here, because it is already included twice in $\hat{H}_1$ and $\hat{H}_2$.

The energy $E(R)$ of the ground state is given in this approximation as twice the energy of the $H_2^+$ ion plus the energy of electron repulsion minus nuclear repulsion. The corresponding potential curve $E(R)$ has a minimum at $R = R_e \approx 0.7$Å (Fig. 9.13). Here the reference energy $E = 0$ is the energy of the fully separated particles e + e + p + p. The detailed calculation shows that the two contributions of electron and nuclear repulsion in the third term of (9.24) cancel for $R = 1$Å. At the minimum position: $R = R_e(H_2)$ they give only a small negative contribution to the total energy. In this simple molecular orbital approximation the binding energy of the $H_2$ molecule is slightly smaller than twice the binding energy of the $H_2^+$ ion in the LCAO approximation. The binding energy of $H_2$ becomes

$$E_B(H_2) = 2E_B\left(H_2^+\right) + \Delta E = -2.7\,\text{eV}\,,$$

compared with the experimental value $E_B = -4.7$ eV.

A serious drawback of this approximation is the wrong behavior of $E(R)$ for $R \to \infty$. As can be seen from a detailled calculation one obtains $E(\infty) = (22/16)E_B(H)$ instead of the correct value $2E_{ion}(H)$ of the ionization energy of the separated H-atoms. This approximation therefore has to be improved.

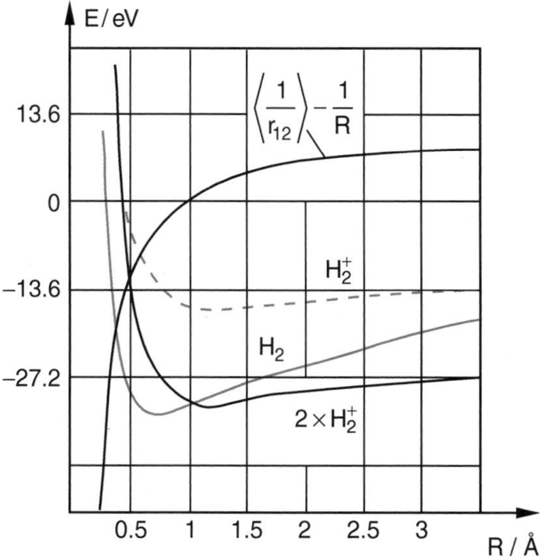


**Fig. 9.13.** Comparison of the potential curves of $H_2^+$ and $H_2$ and the third term in (9.24) in the MO-LCAO approximation (with the kind permission of Prof. Kutzelnigg [9.2])

**Note:**

In the MO approximation the electron repulsion is included in the Hamiltonian, but not in the molecular orbital ansatz for the wave function!

Before we look for improvements of our approximate wave function (9.21), we will discuss another approximation method: the valence bond approximation, developed by *Walter Heitler* (1904–1981) and *Fritz London* (1900–1954) which is therefore called the Heitler–London approximation [9.4].

### 9.2.2 The Heitler–London Method

This method also starts with a molecular orbital ansatz. The ground state molecular orbital with the lowest energy can be populated, according to the Pauli principle, by two electrons with opposite spins. The corresponding wave function

$$\Phi_1 = c_1\phi_A(1) \cdot \phi_B(2) \tag{9.25a}$$

gives the probability amplitude that electron 1 is at nucleus A and can therefore be described by its atomic orbital $\phi_A(1)$, while electron 2 is at nucleus B and is described by the atomic wave function $\phi_B(2)$. However, since the two electrons are undistinguishable,

$$\Phi_2 = c_2\phi_A(2) \cdot \phi_B(1) \tag{9.25b}$$

must be also a wave function, leading to the same charge distribution.

According to the Pauli principle the spatial part of the wave function must be either symmetric or antisymmetric with respect to an electron exchange. Therefore the total spatial part of the normalized wave function can be written as the linear combination of (9.25a) and (9.25b) with $c_1 = c = \pm c_2$

$$\begin{aligned}\Phi_{s,a} &= \Phi_1 \pm \Phi_2 \\ &= c[\phi_A(1)\phi_B(2) \pm \phi_A(2)\phi_B(1)]\,.\end{aligned} \tag{9.26a}$$

Since the atomic wave functions $\phi_A$ and $\phi_B$ are already normalized, we obtain, analogous to the consideration for (9.14), the coefficient

$$c = \left[2\left(1 \pm S_{AB}^2\right)\right]^{-1/2}$$

and the wave function

$$\Phi_{s,a}^{H,L} = \frac{1}{\sqrt{2(1 \pm S_{AB}^2)}} \times [a(1)b(2) \pm a(2)b(1)] \; . \tag{9.26b}$$

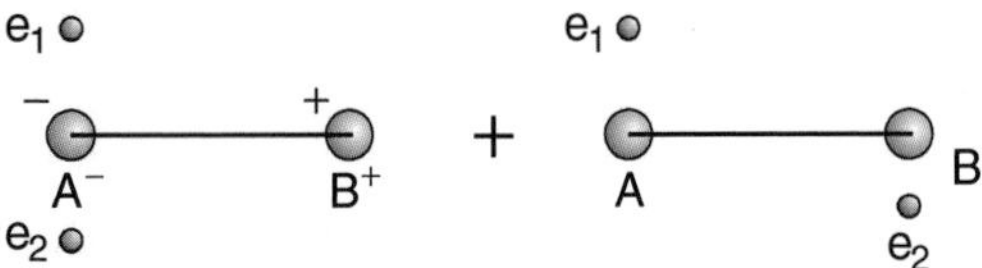


**Fig. 9.14.** The two configurations included in the MO approximation with equal weights

### 9.2.3 Comparison Between the Two Approximations

The difference between the molecular orbital approximation and the Heitler–London ansatz is the following. In the former method a MO ansatz was made for *one* electron, which can be found in the volume covered by $\Phi_A$ as well as by $\phi_B$. The total wave function is therefore the linear combination (9.14a) of $\phi_A$ and $\phi_B$. For the occupation with two electrons the product ansatz (9.21) for the two *molecular* orbitals is used which neglects the mutual repulsion between the two electrons. In the Heitler–London method both electrons are regarded from the beginning and the molecular orbital is written as the product of the two *atomic* orbitals. The final wave function is the linear combination of the two products.

More physical insight into this difference between the wave functions (9.21) and (9.26) can be gained if the multiplication of the two brackets in (9.20c) is explicitly performed. This gives:

$$\Phi_s^{MO} = \frac{1}{2 + 2S_{AB}} \times [a(1)a(2) + b(1)b(2) + a(1)b(2) + a(2)b(1)] \; . \tag{9.27}$$

By comparing with (9.26b) this explicit expression of the MO ansatz illustrates that in the Heitler–London ansatz (9.26b) the first two terms are missing. They describe the situation where both electrons are simultaneously at nucleus A or nucleus B, respectively, i.e., where the ion $H^+H^-$ is formed. In the molecular orbital ansatz this configuration has the same weight as the much more likely configurations $a(1)b(2)$ or $a(2)b(1)$ (Fig. 9.14). It is therefore overestimated, while the ionic form is completely neglected in the Heitler–London ansatz.

In order to compare the molecular binding energy obtained from the two methods we rearrange the Hamiltonian (9.22) into

$$\begin{aligned}\hat{H} &= \left(\frac{-\hbar^2}{2m}\nabla_1^2 - \frac{e^2}{4\pi\varepsilon_0 r_{A1}}\right) + \left(\frac{-\hbar^2}{2m}\nabla_2^2 - \frac{e^2}{4\pi\varepsilon_0 r_{B2}}\right) \\ &\quad - \frac{e^2}{4\pi\varepsilon_0}\left(\frac{1}{r_{A2}} + \frac{1}{r_{B1}} - \frac{1}{r_{12}} - \frac{1}{R}\right) \\ &= \hat{H}_A + \hat{H}_B - \hat{H}_{AB} \; .\end{aligned} \tag{9.28a}$$

Here the division into the different contributions differs from those in (9.24), because $\hat{H}$ is written as the sum of the atomic Hamiltonians $\hat{H}_A$ and $\hat{H}_B$ of the two H-atoms and a third term $H_{AB}$, which contains the attraction of electron 2 by nucleus A and electron 1 by B and the mutual repulsion between the electrons and between the nuclei.

Inserting this Hamiltonian into the Schrödinger equation

$$H\Phi = E\Phi$$

of the rigid $H_2$ molecule yields the total energy

$$E = E(A) + E(B) + \Delta E_B \tag{9.28b}$$

where the first two terms give the energies of the two separated atoms A and B and the third term the difference $E(H_2) - 2E(H)$. A stable molecule is formed only if the energy $\Delta E_B$ of the third term is negative, which means that the total energy of the molecular state is lower than the sum of the ground state energies of the separated atoms. For $\Delta E_B > 0$ the potential curve becomes repulsive. For $R \to \infty$ the binding energy $\Delta E_B(R) \to 0$.

Inserting the symmetric Heitler–London wave function (9.26) into the Hamiltonian (9.28a) the energy

$$E(R) = \int \Phi_s^*(\boldsymbol{r}, R)\, \hat{H} \Phi_s(\boldsymbol{r}, R)\, d\tau \tag{9.29a}$$

can be calculated and gives the binding energy

$$\Delta E(R = R_e) = -3.14\,\text{eV}\,, \tag{9.29b}$$

which is already closer to the experimental value $E_B = -4.7\,\text{eV}$ than the result obtained with the simple MO-LCAO approximation, where the ionic configuration was overemphasized.

### 9.2.4 Improvements to the Approximations

The real share of the ionic configuration $a(1)a(2) + b(1)b(2)$ in the wave function (9.27) can be better quantified if a variable parameter $\lambda$ with $0 < \lambda < 1$ is introduced, changing the wave function (9.27) into

$$\begin{aligned}\Phi_s(\boldsymbol{r}_1, \boldsymbol{r}_1) = c_3[&a(1)b(2) + a(2)b(1) \\ &+ \lambda[a(1)a(2) + b(1)b(2)]]\end{aligned} \tag{9.30}$$

with the normalization constant $c_3$. If $\lambda$ is varied in such a way, that for each internuclear distance $R$ the total energy becomes minimum, a binding energy of $E_B = -4.02\,\text{eV}$ is obtained, a remarkable improvement compared with the simple MO-LCAO approximation.

Of course, the ionic contribution can be also introduced with the weight factor $\lambda$ in an improved Heitler–London version. This gives, instead of (9.26), the same wave function (9.30) as for the improved MO approximation, which means that at this state of improvement, both methods are identical.

The results for the potential curve and the binding energy are still away from the experimental values, which implies that further improvements are necessary. Up to now we have used linear combinations of undistorted atomic 1s wave functions with spherical symmetry. However, when the two atoms approach each other, the electron charge distribution will be distorted due to the interaction between the two electrons and between an electron and the nucleus of the other atom. It will no longer be spherically symmetric but is deformed and also contracted. In order to take this deformation into account, we compose the molecular orbital (9.14a) not only of two $1s$ atomic orbitals, but also of many atomic orbitals including nonspherical ones, and write:

$$\Phi(\boldsymbol{r}_i) = \sum_{k=1}^{N} c_k \phi_k\,. \tag{9.31}$$

In this sum, all atomic orbitals are included that can best reproduce the contraction and deformation of the $1s$ atomic orbitals. The molecular orbital for two electrons can now be written either as the product

$$\Phi(\boldsymbol{r}_1, \boldsymbol{r}_2) = \phi(\boldsymbol{r}_1)\phi(\boldsymbol{r}_2) \tag{9.32a}$$

(MO-LCAO approximation), or as the sum of products of atomic orbitals

$$\Phi(\boldsymbol{r}_1, \boldsymbol{r}_2) = \sum_{i,k} c_{ik} \Phi_i(1)\Phi_k(2) \tag{9.32b}$$

(Heitler–London approximation).

In both cases the coefficients $c_i$, which give the relative contributions of the atomic orbitals to the total molecular wavefunction, are optimized in such a way that the total energy $E(R)$ becomes a minimum for each selected internuclear distance $R$. This can be achieved by solving the equation

$$\frac{\partial}{\partial c_i}\left[\int \Phi^* \hat{H} \Phi \,\mathrm{d}\tau\right] = 0 \tag{9.33}$$

which gives for $N$ coefficients $c_i$ also $N$ equations. Here the energy $E(c_i)$ is regarded as a function of the coefficients $c_i$, since the nuclear separation $R$ is fixed to a selected constant value.

Calculations including 13 functions $\phi_i$ gave a binding energy $E_B(R_e) = -4.69\,\text{eV}$ at the minimum of the potential curve, which already approaches the experimental value very closely.

The most accurate calculations to date were performed by *Kolos et al.* [9.5, 6], who used 50 functions in the expansion (9.31) resulting in a binding energy of $E_B = -4.7467\,\text{eV}$, which agrees with the experimental result within the error bars of the experimental value $E_B^{\text{exp}} = -4.7470 \pm 0.0003\,\text{eV}$.

The results of the different steps in the approximation calculations are illustrated in Fig. 9.15.

The discussion in this section has illustrated that even for the simplest neutral molecule $H_2$, accurate calculations are tedious and demand large computers. The improvements of the simple approximations were guided by physical reasoning and also the selection of proper basis functions in the expansion (9.31) using physical intuition can reduce the number of functions considerably.

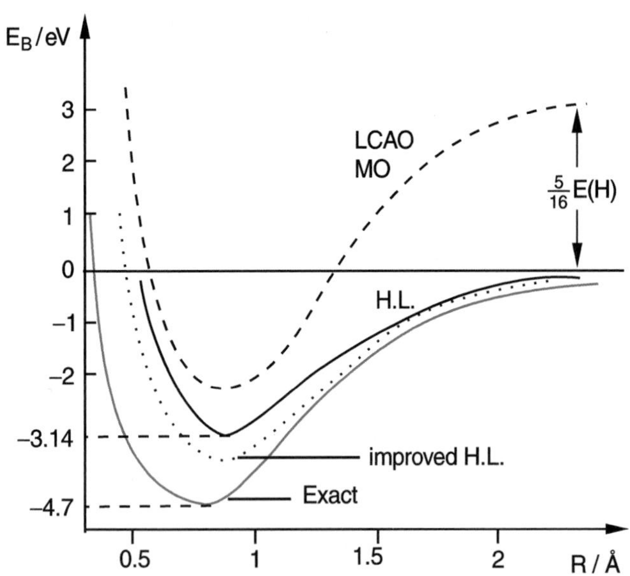


**Fig. 9.15.** Potential curves of the $H_2$ ground state obtained from different approximations. $E_B = 0$ is the energy of the separated ground state H atoms

## 9.3 Electronic States of Diatomic Molecules

Up to now we have only discussed the electronic ground state of the two molecules $H_2^+$ and $H_2$ and have obtained the potential energy curve $E(R)$ by approximate calculation methods. A similar concept is pursued for larger diatomic molecules with nuclear charges $Z_A e$, $Z_B e$ and $N = (Z_A + Z_B)$ electrons. The molecular wave function

$$\Phi(r_1, \dots r_N, R) = \Sigma c_n \phi_n \qquad (9.34)$$

is always written as a linear combination of many basis functions $\phi_n$, which might be atomic orbitals with coordinates centered at the corresponding nucleus A or B, but can also be other functions, such as Gaussian functions or Slater orbitals. They are chosen either by physical intuition to optimize $\Phi$ with as few basis functions as possible, or by computational reasoning aimed at minimizing the computational expenditure.

All of these computations are based on the model of rigid molecules. Their results give the potential energies $E_k(R)$ for the different electronic states and their minimum values $E_k(R_{ek})$, which are arranged in a sequence according to increasing energy.

The different electronic states can be assigned by several criteria:

a) By their energetic order
b) By the symmetry properties of their molecular wave function
c) By the angular momenta of the atomic states forming the molecular state and by the projection of these angular momenta onto the internuclear axis
d) By the vector sum of their electron spins
e) By the atomic states into which they dissociate for $R \to \infty$

Each of these criteria on its own does not allow the unambiguous characterization of a molecular state. For instance there are several different electronic states with the same projection $\Lambda\hbar$ of the orbital angular momentum but different energies or different spins. For a given configuration of the two atoms forming a molecule for decreasing internuclear distance $R$, many different molecular states are possible.

### 9.3.1 The Energetic Order of Electronic States

The different electronic states $E_i(R)$ of diatomics can be characterized by the index $i$, which stands for the various quantum numbers (principal quantum number, angular momentum and spin quantum numbers). While for atoms the principal quantum number $n$ gives the energetic order of the electronic states, this is not generally true for molecules. Only for higher Rydberg states that dissociate into a ground state atom A and a Rydberg atom B, does the energy monotonically increase with increasing principal quantum number $n$ (Fig. 9.16). For lower states the energy difference between states with different angular momenta might be larger than between states with different principal quantum numbers. In this case it is no longer meaningful to use the principal quantum number for the characterization of electronic states.

The historical assignment of molecular electronic states comes from spectroscopic experiments. The ground state is always named the X-state, the next highest state, accessible from the ground state by absorption of photons, is called the A-state, the next one the B-state, etc. If the ground state is a singlet state, only singlet states combine by absorption of radiation with the ground state. The triplet states are named by small Latin letters, a, b, c according to their energetic order. Unfortunately, some new electronic states have been discovered only after the assignment of the known

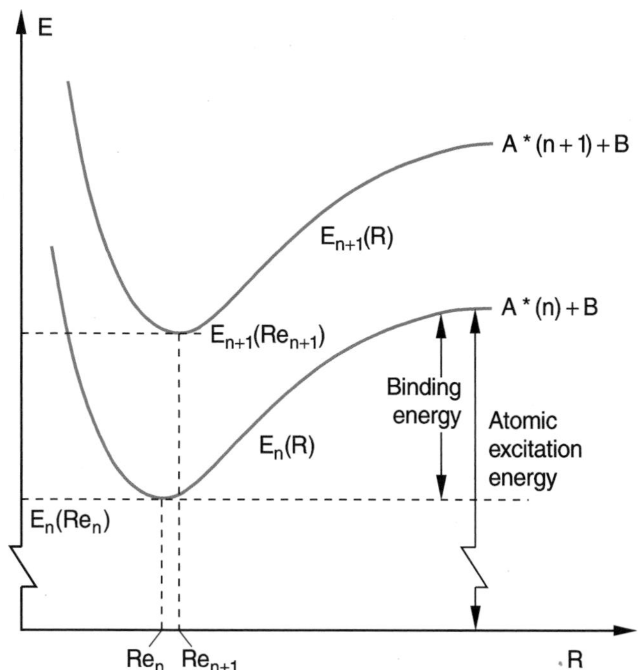


**Fig. 9.16.** Two potential curves of Rydberg states in a diatomic molecule

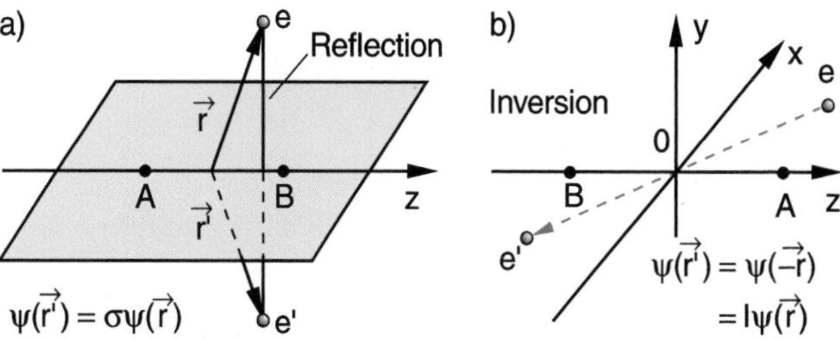


**Fig. 9.17.** (**a**) Reflection of all electron coordinates at a plane through the two nuclei. (**b**) Inversion of all electron coordinates at the origin

states. They are then labeled as A′, B′, ... states. This nomenclature brings some confusion into the scheme, but is still in use.

### 9.3.2 Symmetry Properties of Electronic States

Geometric operations, such as reflections of the molecular electron cloud on a plane through the nuclei or at the origin (chosen as the center of charge) and rotations of the molecule around an axis through the origin, are called symmetry operations if the nuclear frame and the charge distribution are not changed by these operations. This means that $|\Phi|^2$ is invariant against symmetry operations.

We will describe the reflection at a plane through the nuclei by the reflection operator $\sigma$. Since a double reflection brings the molecule back into its original state (Fig. 9.17a) we obtain for the wave function $\Phi$ the relation

$$\sigma(\sigma\Phi) = \sigma^2\Phi = +\Phi \Rightarrow \sigma\Phi = \pm\Phi\,. \qquad (9.35a)$$

We define the symmetry of a wave function $\Phi$ as positive, if

$$\sigma\Phi^+ = +\Phi^+\,. \qquad (9.35b)$$

The wave function has negative symmetry if

$$\sigma\Phi^- = -\Phi^-\,. \qquad (9.35c)$$

The symmetry of a molecular state equals the symmetry of its wave function. It is defined by the symmetries of the two atomic states that form the molecular state.

For homonuclear diatomic molecules with $Z_A = Z_B$ the inversion of all coordinates at the origin is another symmetry operation which defines the ***parity*** of a molecular state (Fig. 9.17b). We can write this as

$$I^2\Phi(\vec{r}) = \Phi(\vec{r})$$
$$I(\Phi(\vec{r}) = \Phi(-\vec{r}) = \pm\Phi(\vec{r}) \qquad (9.36)$$

We can distinguish between two classes of wave functions. They are historically named by the German words *gerade* (for even) and *ungerade* (for odd). The gerade functions have even parity, i.e., $I\,\Phi_g(\vec{r}) = +\Phi_g(-\vec{r}) = \Phi_g(\vec{r})$, while for *ungerade* functions $I\,\Phi_u(\vec{r}) = \Phi_u(-\vec{r}) = -\Phi_u(\vec{r})$ the parity is odd.

**Note** this parity property is only present for diatomics with equal nuclear charges.

### 9.3.3 Electronic Angular Momenta

Each electron in a molecule has an orbital angular momentum $\boldsymbol{l}_i$ and a spin $\boldsymbol{s}_i$. Because the potential, in which the electron moves has no spherical symmetry, the orbital angular momentum is not constant in time but precesses in a nonrotating molecule around the symmetry axis of the molecule. For states with wave functions of cylindrical symmetry, this is the space-fixed internuclear axis, which we choose as the $z$-axis. We characterize the projection

$$l_z = \lambda\hbar \qquad (9.37)$$

by the projection quantum number $\lambda = |m_l|$.

We will at first discuss molecules with only one electron, such as $H_2^+$, or with one electron outside the closed shell, such as $Li_2^+$ (see Sect. 9.1.1).

The precessing electron produces a magnetic field in the $z$-direction. For molecules with small nuclear charge numbers $Z$ the interaction between the magnetic moments of $\boldsymbol{l}_i$ and $\boldsymbol{s}_i$ ($l$-$s$-coupling) is weaker than the coupling of $\boldsymbol{s}$ to this magnetic field in the $z$-direction. In this case both $l_i$ and $s_i$ precess independently around the internuclear axis. The projection of $s$ is called $s_z = \sigma\hbar$, where $\sigma$ is the spin projection quantum number (Fig. 9.18). The molecular electronic state is then characterized by the quantum numbers $\lambda$, $\sigma$ and $s$. The sum of the two projections is named

$$\omega\hbar = \lambda\hbar + \sigma\hbar$$

For molecules with more than one electron the coupling order of the different angular momenta depends on the strength of the different interactions. Similar to the situation in atoms (see Sect. 6.5) different coupling schemes can be realized.

If the interaction between the different orbital angular momenta $l_i$ is larger than the coupling between $l_i$ and $s_i$ (weak spin orbit coupling), than the total orbital momentum

$$\boldsymbol{L} = \sum \boldsymbol{l}_i \tag{9.38a}$$

and the total electron spin

$$\boldsymbol{S} = \sum \boldsymbol{s}_i \tag{9.38b}$$

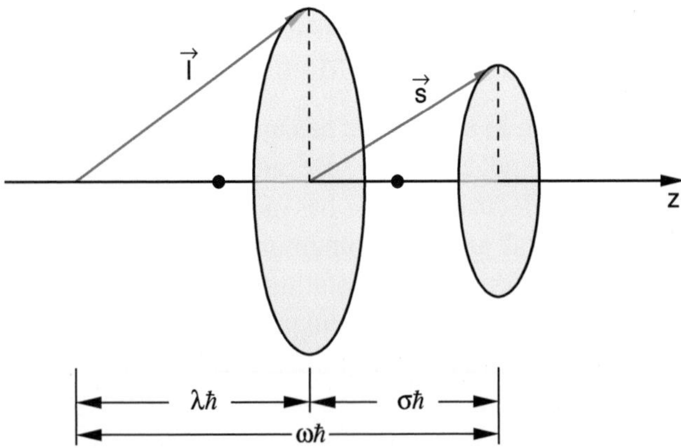


**Fig. 9.18.** Separate precession of $\boldsymbol{l}$ and $\boldsymbol{s}$ around the $z$-axis for weak spin-orbit coupling

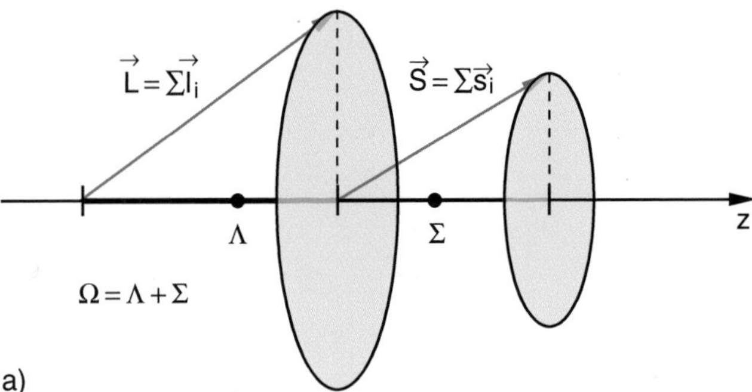


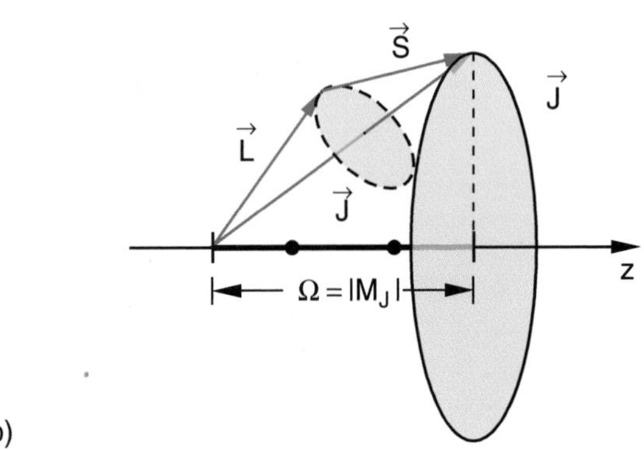


**Fig. 9.19.** (**a**) Separate precession of $\boldsymbol{L} = \sum \boldsymbol{l}_i$ and $\boldsymbol{S} = \sum \boldsymbol{s}_i$ around the internuclear axis $z$ for small spin-orbit coupling. (**b**) Strong spin-orbit coupling with $\boldsymbol{L} + \boldsymbol{S} = \boldsymbol{J}$ and $\boldsymbol{J}$ precesses around the $z$-axis with $J_z = \Omega\hbar$

precess independently around the internuclear axis, forming the constant projections

$$L_z = \Lambda \cdot \hbar \quad \text{and} \quad S_z = \Sigma \cdot \hbar \tag{9.38c}$$

(Fig. 9.19a).

Electronic states with $\Lambda = 0$ are called $\Sigma$-states, those with $\Lambda = 1$ are $\Pi$-states, with $\Lambda = 2$ are $\Delta$-states, etc.

**Note:**

Unfortunately the Greek letter $\Sigma$ stands for states with $\Lambda = 0$, but also for the projection number $|M_S|$.

For large spin-orbit coupling, $\boldsymbol{l}$ and $\boldsymbol{s}$ form the resultant angular momentum

$$\boldsymbol{j} = \boldsymbol{l} + \boldsymbol{s}\,, \tag{9.39a}$$

for each electron and the total electronic angular momentum is

$$\boldsymbol{J} = \Sigma \boldsymbol{j}_i$$

which precesses around the internuclear axis with the projection

$$J_z = \Omega \cdot \hbar \tag{9.39b}$$

(Fig. 9.19b). In these cases the quantum numbers $\lambda, \sigma, \Lambda$ and $\Sigma$ are not well defined, but only $\Omega$.

If the two atomic states into which the molecular state dissociates for $R \to \infty$, have orbital angular momenta $L_A$ and $L_B$ with projections $(M_L)_A$ and $(M_L)_B$ onto the internuclear axis, the resultant projection $\Lambda$ of the molecular state can take the values

$$\Lambda = |M_{LA} + M_{LB}| \; . \tag{9.40}$$

Since $(M_L)_A$ can take the $(2L_A + 1)$ values $-L_A \leqslant M_{LA} \leqslant +L_A$ and $(M_L)_B$ can also take $2L_B + 1$ values, the total possible number of molecular states with different values of $\Lambda$, composed of two given atomic states, can be quite large. Some examples are given in Table 9.4.

EXAMPLE

The combination of an atomic $P$ state with $L = 1$ and a $D$-state with $L = 2$ results in 15 molecular states with $\Lambda = 0, 1, 2, 3$. For $(M_L)_A = -(M_L)_B$ there are three $\Sigma$-states with $\Lambda = 0$. These form the combinations

$$|-(M_L)_A + (M_L)_B| = 0 \; ; \; |+(M_L)_A - (M_L)_B| = 0$$

and

$$(M_L)_A = (M_L)_B = 0 \; .$$

There are six combinations that lead to $\Lambda = 1$, four combinations for $\Lambda = 2$ and two for $\Lambda = 3$.

**Table 9.4.** Quantum numbers $\Lambda$ of molecular states formed by the combination of $P\,(L = 1)$ and $D\,(L = 2)$ atomic states. The number in paranthesis gives the number of possible states

| $(M_L)_A$ | $(M_L)_B$ | $\Lambda$ | Molecular Term |
|---|---|---|---|
| 0 | 0 | 0 | $\Sigma(1)$ |
| 0 | $\pm 1$ | 1 | $\Pi(2)$ |
| $\pm 1$ | 0 | 1 | $\Pi(2)$ |
| $\pm 1$ | $\mp 1$ | 0 | $\Sigma^+, \Sigma^-$ |
| $\pm 1$ | $\mp 2$ | 1 | $\Pi(2)$ |
| 0 | $\pm 2$ | 2 | $\Delta(2)$ |
| $\pm 1$ | $\pm 1$ | 2 | $\Delta(2)$ |
| $\pm 1$ | $\pm 2$ | 3 | $\Phi(2)$ |

### 9.3.4 Electron Spins, Multiplicity and Fine Structure Splittings

Similar to the coupling of orbital angular momenta the spins $S_A$ and $S_B$ of the atomic states A and B define the spin

$$\boldsymbol{S} = \boldsymbol{S}_A + \boldsymbol{S}_B \tag{9.41}$$

of the molecular state formed by combination of the two atomic states as the vector sum of the atomic spins. As for atoms, the multiplicity of a molecular state is given by the number $(2S + 1)$ of possible projections $\Sigma$ of $S$ onto the internuclear axis.

For example, the two H atoms in their $1^2S_{1/2}$ ground state can form singlet molecular states of the $H_2$ molecule with a symmetric spatial wave function and opposite spins or triplet states with an antisymmetric spatial part and parallel spins of the two electrons. In both cases the total wave function always has to be antisymmetric with respect to the exchange of two electrons.

**Note:**

This symmetry is defined with respect to the exchange of any two of the electrons and should not be confused with the geometrical symmetry property with respect to reflection at a plane through the nuclei. While the exchange symmetry is labeled "s" for symmetric and "a" for antisymmetric, the reflection symmetry is labeled as "+" for symmetric and "−" for antisymmetric.

The interaction energy of the electron spin with the molecular magnetic field B produced by the precessing orbital angular momentum is proportional to the scalar product

$$W \propto -\boldsymbol{S} \cdot \boldsymbol{B} \; . \tag{9.42a}$$

Since B is proportional to the projection $\Lambda\hbar$, the spin-orbit interaction energy can be written as

$$W = A\Lambda\Sigma \; , \tag{9.42b}$$

where the constant $A$, which depends on the specific molecular state, is called the ***molecular fine structure constant***. Each molecular electronic state composed of atomic states with angular momenta $L_A$ and $L_B$ and with $\Lambda \neq 0$ and $S \neq 0$ splits into $(2S + 1)$ fine

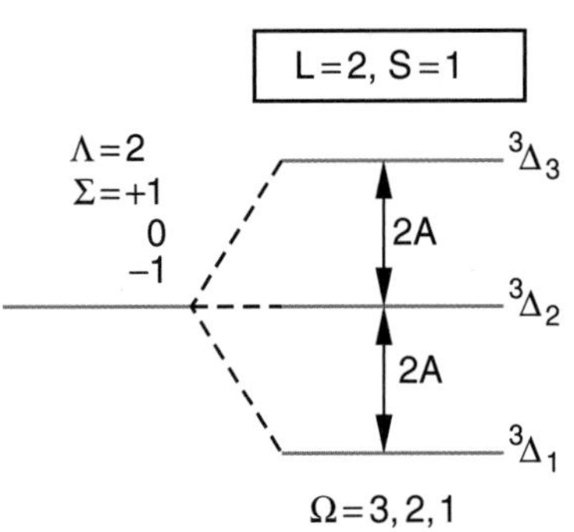


**Fig. 9.20.** Fine structure splitting of a molecular state with $\Lambda = 2$ and $\Sigma = 1$ into the $^{2S+1}\Lambda_\Omega$ equidistant components

structure components, which differ in their spin projection quantum number $\Sigma$. They are labeled by the total projection quantum number

$$\Omega = \Lambda + \Sigma = |(M_L)_\mathrm{A} + (M_L)_\mathrm{B})| + \Sigma$$

(Fig. 9.20). The term values of these fine structure components are

$$T(\Lambda, \Sigma) = T_0 + A\Lambda\Sigma \,. \tag{9.43}$$

**Note:**

Contrary to the situation in atoms the fine structure components are equidistant and are separated by $\Lambda \cdot A$.

For $\Lambda < \Sigma$ fine structure components that differ only in the sign of $\Omega$ are degenerate in the nonrotating molecule.

The complete labeling of a fine structure component is

$$\boxed{^{2S+1}\Lambda_\Omega}$$

(see Fig. 9.20).

### 9.3.5 Electron Configurations and Molecular Ground States

In order to calculate the different electronic states of diatomic molecules with $N$ electrons, the molecular orbitals are determined by using wave functions of the type (9.34). These orbitals are arranged according to increasing energy. For the molecular ground state, each of these orbitals is occupied with two electrons with opposite spins until all electrons are put into the lowest $N/2$ (for even $N$) or $(N+1)/2$ (for odd $N$) orbitals.

For excited electronic states one of the electrons is brought from an occupied into a higher unoccupied orbital.

The molecular orbitals are characterized by the index $n = 1, 2, 3$, labeling the energetic order, the orbital angular momentum quantum number $l$ of the electron, and its projection quantum number $\lambda$.

The calculations give the following energetic order of the orbitals $\psi(n, l, \lambda)$ with the nomenclature shown in Table 9.1:

$$1s\sigma, 2s\sigma, 2p\sigma, 2p\pi, 3s\sigma, 3p\sigma, 3p\pi, 3d\sigma,$$
$$3d\pi, 3d\delta, 4s\sigma, \ldots \,,$$

where the Latin letter indicates the atomic orbital, which gives the major contribution to the molecular orbital.

For homonuclear diatomics each of these orbitals can be realized with gerade as well as with ungerade symmetry. For these molecules there are therefore twice as many orbitals as for heteronuclear molecules.

For the characterization of a molecular state all occupied molecular orbitals are written with an exponent 1 or 2, indicating the number of electrons in this orbital. For example the ground state of the lithium molecule $\mathrm{Li}_2$ with six electrons, formed by two Li-atoms in their $^2S_{1/2}$ ground states is

$$\mathrm{Li}_2\ 1^1\Sigma_\mathrm{g}^+ : (1s\sigma_\mathrm{g})^2(1s\sigma_\mathrm{u})^2(2s\sigma_\mathrm{g})^2 \,.$$

This composition of a molecular state by the orbitals of the different electrons is called the ***electron configuration***.

Since the lowest fully occupied atomic $1s$ orbitals do not contribute much to the molecular binding (they are localized in the atomic $K$-shell around their atomic nucleus) they are often abbreviated by $K$, and the above electron configuration is written as

$$\mathrm{Li}_2\ 1^1\Sigma_\mathrm{g}^+(KK(2s\sigma_\mathrm{g})^2) \,.$$

In Table 9.5 the electron configurations, the internuclear distances $R_\mathrm{e}$ at the minimum energy and the binding energies for the ground states of some homonuclear molecules are listed together with the spin orientations, and in Table 9.6 the same quantities for some heteronuclear molecules are compiled.

**Table 9.5.** Electron configuration in the ground states of some homonuclear diatomic molecules

| | Configuration | State | $R_e$/nm | $E_B$/eV |
|---|---|---|---|---|
| $H_2$ | $(1s\sigma_g)^2$ ↑↓ | $^1\Sigma_g^+$ | 0.074 | 4.476 |
| $He_2^+$ | $(1s\sigma_g)^2\,(1s\sigma_u)$ ↑ | $^2\Sigma_u^+$ | 0.108 | 2.6 |
| $He_2$ | $(1s\sigma_g)^2\,(1s\sigma_u)^2$ ↑↓ | $^1\Sigma_g^+$ | – | 0 |
| $Li_2$ | $K\,K\,(2s\sigma_g)^2$ ↑↓ | $^1\Sigma_g^+$ | 0.267 | 1.03 |
| $B_2$ | $K\,K\,(2s\sigma_g)^2\,(2s\sigma_u)^2$ $(2p\pi_u)^2$ ↑↑ | $^3\Sigma_g^-$ | 0.159 | 3.6 |
| $C_2$: | $KK(2s\sigma_g)^2(2s\sigma_u)^2(2p\pi_u)^4$ | $^1\Sigma_g^+$ | 0.124 | |
| | $KK(2s\sigma_g)^2(2s\sigma_u)^2(2p\pi_u)^3(2p\sigma_g)^1$ ↑↑ | $^3\Pi_u$ | 0.131 | 3.6 |
| $N_2$ | $K\,K\,(2s\sigma_g)^2\,(2s\sigma_u)^2$ $(2p\pi_u)^4\,(2p\sigma_g)^2$ ↑↓ | $^1\Sigma_g^+$ | 0.110 | 7.37 |
| $O_2$ | $K\,K\,(2s\sigma_g)^2\,(2s\sigma_u)^2$ $(2p\sigma_g)^2\,(2p\pi_u)^4\,(2p\pi_g)^2$ ↑↑ | $^3\Sigma_g^-$ | 0.121 | 5.08 |

**Table 9.6.** Electron configurations, equilibrium distances $R_e$ and binding energies of the ground states of some heteronuclear diatomic molecules

| Molecule | Configuration | State | $R_e$/nm | $E_B$/eV |
|---|---|---|---|---|
| LiH | $(1\sigma)^2(2\sigma)^2$ | $^1\Sigma^+$ | 0.160 | 2.52 |
| CH | $(1\sigma)^2(2\sigma)^2$ $(3\sigma)^2 1\pi$ | $^2\Pi$ | 0.112 | 3.65 |
| HF | $(1\sigma)^2(2\sigma)^2$ $(3\sigma)^2 1\pi$ | $^1\Sigma^+$ | 0.092 | 6.11 |
| CO | $K\,K\,(3\sigma)^2(4\sigma)^2$ $(1\pi)^4(5\sigma)^2$ | $^1\Sigma^+$ | 0.128 | 11.09 |
| NO | $K\,K\,(3\sigma)^2(4\sigma)^2$ $(1\pi)^4(2\pi)$ | $^2\Pi$ | 0.115 | 6.50 |

As we have discussed for the $H_2^+$ and $H_2$ molecules the $\sigma_g$ orbitals lead to a negative binding energy, i.e., to bound molecules, while the $\sigma_u$ orbitals result in repulsive potential curves and therefore to unstable molecules. The $\sigma_g$ orbitals are therefore called *bonding orbitals*, and the $\sigma_u$ orbitals called *antibonding*.

In the $H_2$ molecule both electrons are in the bonding $1\sigma_g$ orbital. Without the repulsion between the two electrons the binding energy $E_B(R_e)$ of $H_2$ should be twice as large as for $H_2^+$. The positive repulsion energy decreases the binding energy slightly. The equilibrium distance $R_e$ is slightly smaller than for $H_2^+$, because the larger negative charge of the two electrons between the two nuclei decreases the nuclear repulsion and shifts $R_e$ towards a lower value than for $H_2^+$.

For $He_2^+$ with three electrons, the third electron has to occupy the next higher orbital $1\sigma_u$, which is antibonding. Therefore, the binding energy of $He_2^+$ is smaller than that of $H_2$ and its equilibrium distance $R_e$ is larger.

For the neutral $He_2$ molecule the binding contribution of the two electrons in the $1\sigma_g$ orbital are nearly cancelled by the antibonding energy of the $1\sigma_u$ orbital and the binding energy of $He_2$ is with $E_B = 0.00001\,\text{eV}$ extremely small. For temperatures $T > 1\,\text{K}$ the $He_2$ molecule in its ground state is unstable [9.7].

The $Li_2$-molecule has, in addition to the $He_2$ configuration, two electrons in the bonding $2s\sigma_g$ orbital which contribute to the binding energy. The ground state of $Li_2$ is therefore stable and has a binding energy of $E_B = 1\,\text{eV}$.

The electron configuration in the ground state of the $B_2$ molecule is (Fig. 9.21)

$$B_2(KK(2s\sigma_g)^2(2s\sigma_u)^2(2p\pi_u)^2\,.$$

This configuration can lead to the three molecular states $^3\Sigma_g^-$, $^1\Delta_g$ and $^1\Sigma_g^+$ where the $^3\Sigma_g^-$ state has the lowest energy. Similarly to the situation in atoms (see Sect. 6.2) this is due to Hund's rule that the state with the highest multiplicity has the lowest energy.

The $^1\Delta_g$ state is the first excited state where the two electrons are still in the $2p\pi_u$ state but have antiparallel spins.

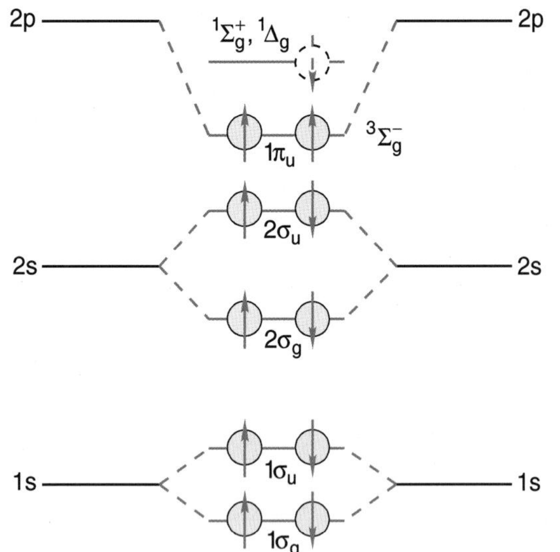


**Fig. 9.21.** Electron configurations for the ground state $^3\Sigma_g^-$ and excited singlet states of the Boron molecule $B_2$

For heteronuclear molecules the distinction between gerade and ungerade states is dropped and there are only half as many orbitals than in homonuclear molecules.

### 9.3.6 Excited Molecular States

If one of the electrons in the occupied orbitals of the ground state is excited into an orbital with higher energy, excited electronic states are formed with energies $E_n(R)$, that depend on the internuclear distance $R$. Within the model of the rigid molecule, we obtain for each of these states a potential energy curve $E(R)$, which converges for $R \to \infty$ towards the dissociation into the separated atoms A + B* or A* + B, i.e., into a ground state atom and an excited atom. In Fig. 9.22 some of these potential curves are shown for the molecular ion $H_2^+$. The principle quantum number is counted for each symmetry type separately, i.e. the lowest $\sigma_u$ state is numbered as $1\sigma_u$, the lowest $\pi_u$ state also as $1\pi_u$, etc. The figure illustrates that for this system all excited states are unstable and dissociate into the fragments $H^+ + H^*$. The reason for this is that the wave functions for the excited states are spatially spread out and therefore the remaining electron charge distribution between the two protons is too small to compensate for the Coulomb repulsion between the two protons.

For molecules with more electrons there are generally many bound excited states. For illustration, the

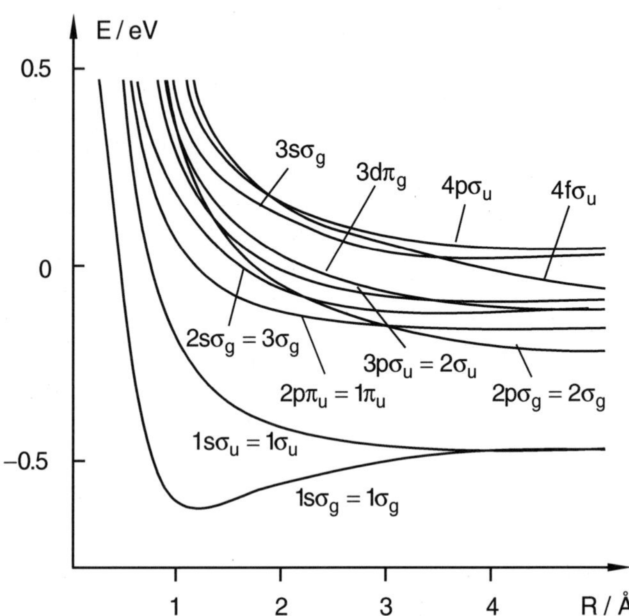


**Fig. 9.22.** Potential curves for the 10 lowest energy states of the $H_2^+$ molecular ion

potential curves of the $Li_2$ are shown in Fig. 9.23. Besides the singly excited states, doubly excited states can also be found where two electrons are excited into higher orbitals. These states dissociate for $R \to \infty$ into two excited Li-atoms. For all these states the core electrons in the $1s$ atomic shells are barely affected by the excitation of one or two electrons from the outer $2s\sigma$ orbital into higher unoccupied orbitals. They do not contribute to the molecular binding.

If one electron is excited into high energy states closely below the ionization limit, molecular Rydberg states are formed. Since this electron is far away from the core, it barely affects the molecular binding. The potential curves of these Rydberg states are therefore similar to the potential curve of the molecular ion.

For the formation of molecules from two atoms with closed inner shells the electrons in these inner shells barely contribute to the molecular binding energy. They stay concentrated around "their" nucleus when the molecule is formed, while the electrons from the atomic valence shell are rearranged into molecular orbitals covering the volume around the two nuclei. The molecular binding energy is therefore mainly caused by these valence electrons. Their spatial rearrangement leads to a minimization of the total energy at a certain distance $R_e$ between the nuclei, which is the reason for the binding of the two atoms into a stable molecule.

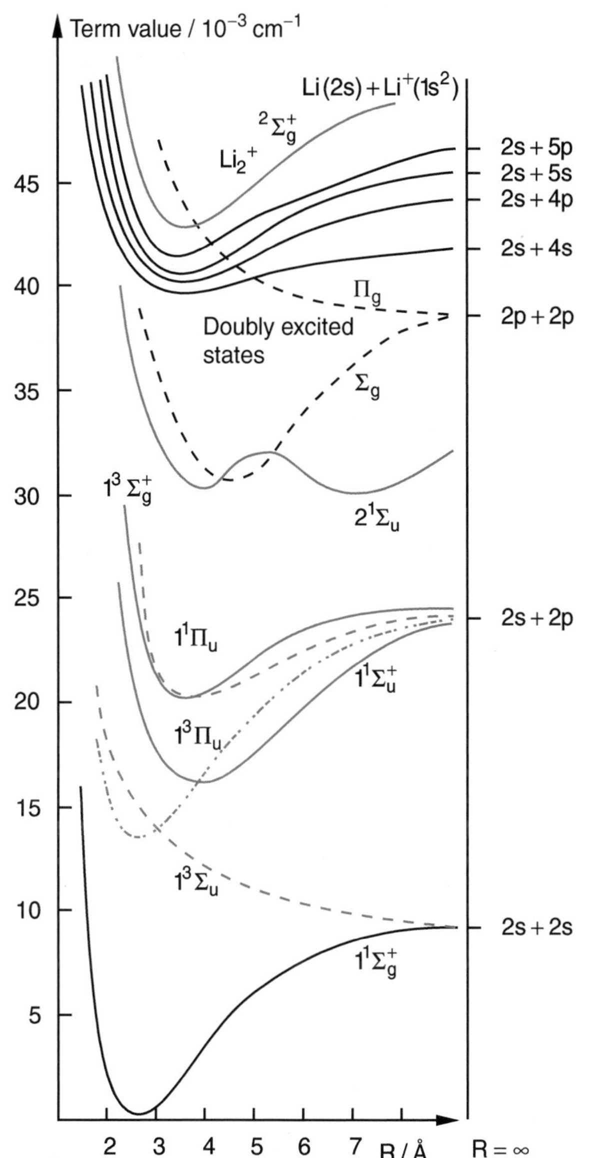


**Fig. 9.23.** Some potential curves of the $Li_2$ molecule up to the ionization limit, including some doubly excited states (*black dashed curves*)

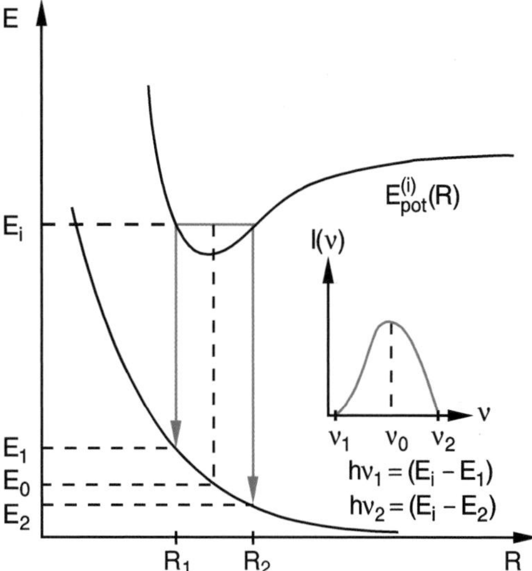


**Fig. 9.24.** Two potential curves of an excimer showing the continuous emission spectrum

## 9.3.7 Excimers

We have seen in Sect. 9.3.5 that two helium atoms in their ground states can not form a stable molecule, because the electrons are so tightly bound to their nucleus, that their wave functions can not form a spatially extended binding orbital. The situation is different for excited He* atoms, where the binding energy of the excited electron is only about $\frac{1}{4}$ of that in the ground state. The electron can then be more readily removed from its nucleus and brought into a binding orbital between the two atoms when another atom approaches the He*.

Indeed $He_2^*$ dimers have a potential curve $E(R)$ with a minimum (Fig. 9.24). This is also true for other noble gases that form, for example, $Ar_2^*$ or $Xe_2^*$ dimers or mixed excited molecules such as Ar*Kr, Ar*Xe or Kr*Xe.

Such molecules, with binding potential curves in excited electronic states but repulsive curves in their ground state, are called ***excimers*** (excited dimers).

Interesting examples of excimers are combinations of noble gas atoms with halogen atoms that have one missing electron in their otherwise filled valence shell. If such a halogen atom approaches an excited noble gas atom, the excited electron of the noble gas atom can gain energy if it fills the hole in the valence shell of the halogen atom forming a molecular ion. This leads to a potential curve $E(R)$ with a minimum.

The fluorescence emitted by these excimers from levels in the stable energy state $E_i$ into the unstable ground state has a broad continuous spectrum, because it terminates on the repulsive potential of the ground state. The photon energy $h\nu = E_i - E_g(R)$ depends on the internuclear distance at which the emission takes place and covers the range $\Delta E(R_1) - \Delta E(R_2)$ when the excited molecule vibrates back and forth between $R_1$ and $R_2$ (Fig. 9.24).

Such excimers are ideal candidates for active laser media. If the upper state can be populated, for instance by collisions $A^* + B$ between excited noble gas atoms $A^*$ and halogen atoms B, inversion is automatically achieved because the lower state is depopulated by dissociation within less than $10^{-12}$ s. A possible excitation scheme is for example

$$\begin{aligned} &Xe + e^- \rightarrow Xe^* + e^- \\ &Xe^* + Cl_2 \rightarrow XeCl^* + Cl \\ &Xe^* + HCl \rightarrow XeCl^* + H\,. \end{aligned} \qquad (9.44)$$

### 9.3.8 Correlation Diagrams

The molecular states can be correlated to the corresponding states of the separated atoms ($R \rightarrow \infty$) and also to the states of a fictive united atom for $R \rightarrow 0$ where all electrons form a shell around a nucleus with charge $(Z_A + Z_B)e$. Since the symmetry and the parity of the electronic states of the system AB must be independent of the distance $R$, only selected states of the separated atoms can combine with molecular states of given symmetry and parity and with such states of the united atom. This means, for example, that gerade states of the united atom correlate with gerade states of the molecule and ungerade with ungerade.

On the left side of Fig. 9.25 the energy levels of the united atom are shown and on the right side the energies of the separated atoms, where for singly excited molecular states one of the atoms is always in its ground state.

Such correlation diagrams, which combine the corresponding states as a function of $R$ by the curves

$$E_n(R) = \int \Phi_n^*(\boldsymbol{r}, R)\, \hat{H} \Phi_n(\boldsymbol{r}, R)\, d\tau \qquad (9.45)$$

are very useful for the correct energy ordering of the molecular states, because the energetic order of the atomic states is not necessarily the same as that of the corresponding molecular states. The strength of the spin-orbit interaction, for example, generally varies with $R$. This has the consequence that the coupling order of the angular momenta $l_i\, s_i$, $L = \Sigma l_i$ and $S = \Sigma s_i$ and the splittings of the fine structure components can change from the united atom ($R = 0$) to the separated atoms ($R = \infty$) (Fig. 9.26). Furthermore, the molecule can not only dissociate into neutral atoms, but also into the ion pair $A^+ + B^-$.

Here a ***noncrossing rule***, postulated by *Eugene Wigner*, is very helpful. It states that curves $E(R)$ of the same symmetry or parity never cross each other. This rule can be proved by quantum mechanical considerations. According to this rule the curve for the

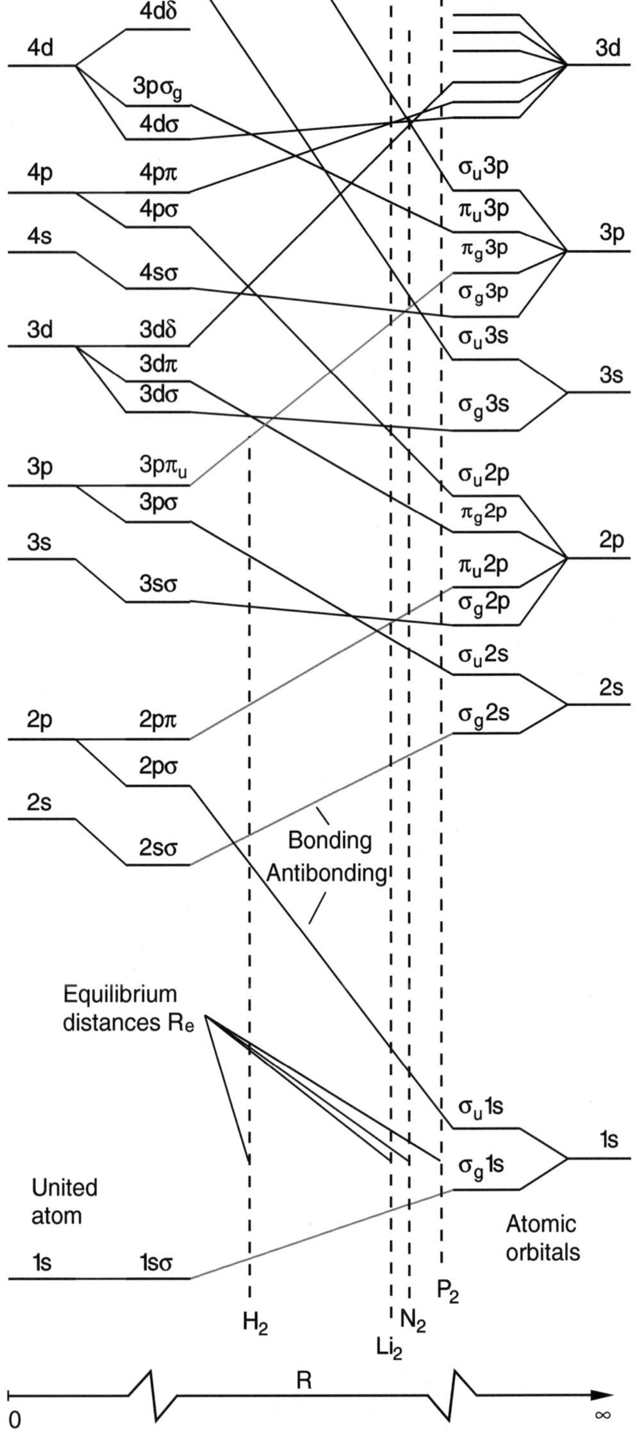


**Fig. 9.25.** Correlation diagram of a homonuclear diatomic molecule. On the right side one of the atoms is always in the $1s$ state, to obtain singly excited molecular states

$1s\sigma_u$ orbital can cross that of the $2s\sigma_g$ orbital but not of the $2s\sigma_u$ orbital.

If the internuclear distances $R_e$ at the minimum of the potential curve $E(R)$ for the different molecular

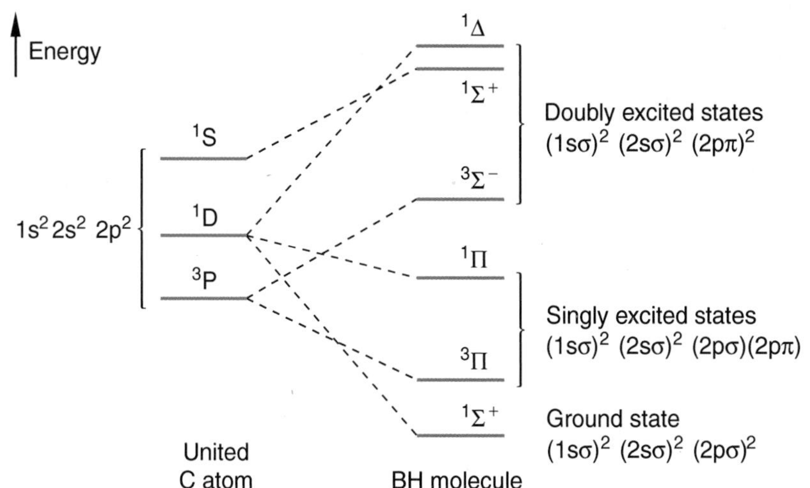


**Fig. 9.26.** Correlation diagram between the united carbon atom and the BH molecule, which both have six electrons

states are known, the correlation diagram gives the correct energetic order of these states.

## 9.4 The Physical Reasons for Molecular Binding

In this section we will answer the question: "Why can two neutral atoms combine to form a stable molecule?". We will see that depending on the specific atoms and on the internuclear distance $R$ several reasons exist for the molecular binding (Fig. 9.27).

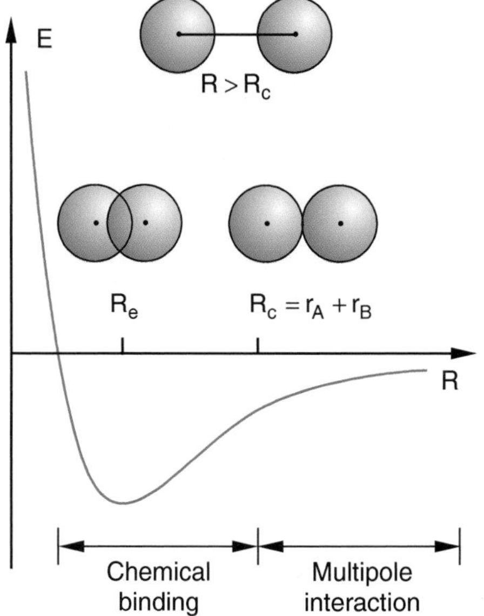


**Fig. 9.27.** Chemical binding with overlap of atomic orbitals is important for $R < R_c$. For $R > R_c$ multipole interaction dominates

### 9.4.1 The Chemical Bond

During the discussion of the $H_2$ molecule, we learned that essentially two effects contribute to the binding energy of the molecule at the equilibrium distance $R_e$.

The first is the spatial rearrangement of the charge distribution of atomic valence electrons. The electron density becomes larger between the two nuclei. This results in an electrostatic attraction between the positive cores of the two atoms (for $H_2$ these are the two protons) and the negative electron charge distribution between them. This effect is emphasized in the valence bond model used in chemistry. In the chemically bound molecule both atoms share one or more valence electrons in a common molecular orbital. This is also described in the LCAO approximation where the molecular orbital is represented by a linear combination of atomic orbitals.

The second reason is of quantum mechanically nature and cannot be explained by a classical model. The molecular orbital has a larger spatial extension then the atomic orbitals. This increases the spatial uncertainty for the electrons and therefore decreases their average momentum $\langle |p| \rangle$ and their kinetic energy $\langle E_{\text{kin}} \rangle = \langle p^2 \rangle / 2m$, according to Heisenberg's uncertainty relation. The combination of both effects leads for stable molecular states to a minimum in the potential curve $E(R)$, since the potential energy $E(R)$ contains the average kinetic energy of the electrons (see Sect. 9.1). This second contribution to the molecular binding is called the exchange interaction, because the two electrons in the atomic orbitals of the LCAO can be

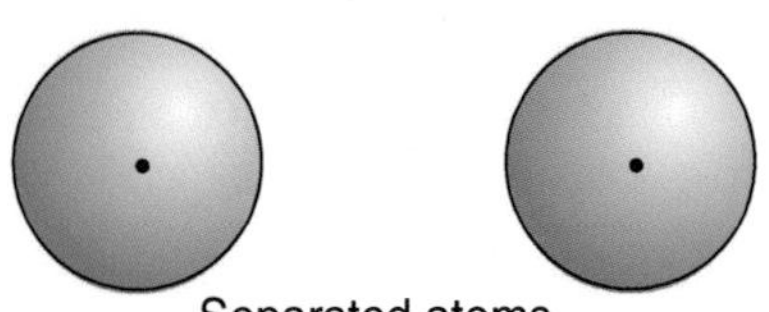


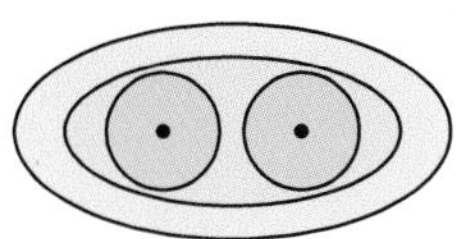


**Fig. 9.28.** Valence bond as increased electron charge between the two nuclei

exchanged since they cannot be distinguished in their common molecular orbital.

Both effects are important at internuclear distances $R < \langle r_A \rangle + \langle r_B \rangle$ that are smaller than the sum of the mean atomic radii $\langle r_A \rangle$ and $\langle r_B \rangle$, which give the extension of the electron clouds in the separated atoms. For distances smaller than this sum, the orbitals of the two atoms can overlap forming molecular orbitals and sharing electrons (Fig. 9.28). Molecular bonds that are formed due to this effect are called ***covalent*** or ***homopolar***.

> One can also describe the chemical binding by energy conservation. If the energy increase necessary to deform the atomic orbitals when the two atoms approach each other is smaller than the decrease of the total energy (potential energy and mean kinetic energy of the electrons) in the rearranged molecular charge distribution, a stable molecule is formed. The nuclear distance $R_e$ and the electron charge distribution always arrange themselves in such a way that the total energy becomes a minimum.

### 9.4.2 Multipole Interaction

For larger internuclear distances $R > \langle r_A \rangle + \langle r_B \rangle$, where the electron clouds of the two atoms no longer overlap, the chemical binding based on the two effects discussed above looses its importance. Nevertheless stable molecules are possible with such large internuclear distances $R$, although their binding energy is smaller. The correct treatment of the effects responsible for these interactions at large distances is based on quantum mechanics. However, good physical insight is already provided by the classical model, which starts from the multipole expansion of an arbitrary charge distribution $\rho(r)$ for an observer at a point $P$ at a distance $R$ from the center of the charge distribution, which is large compared to the extension of $\rho(r)$ (Fig. 9.29). We will discuss this model shortly.

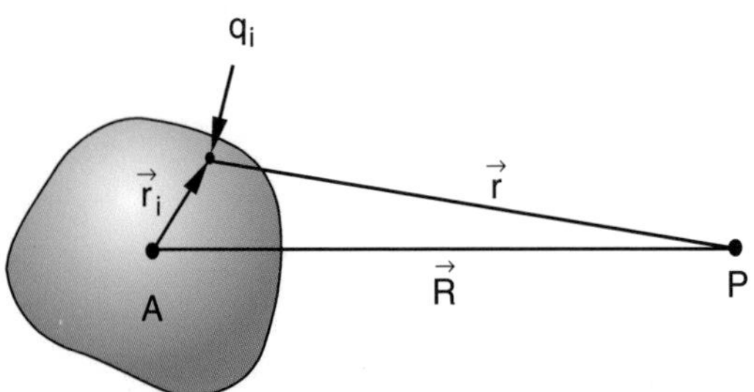


**Fig. 9.29.** Illustration of multipole expansion in (9.46) and (9.47)

The potential $\phi(R)$ at the point $P(R)$ generated by a distribution of charges $q_i(r_i)$ at the locations $r_i$ is

$$\phi(R) = \frac{1}{4\pi\varepsilon_0} \sum \frac{q_i(\boldsymbol{r}_i)}{|\boldsymbol{R} - \boldsymbol{r}_i|} . \tag{9.46}$$

If we choose the origin of our coordinate system to coincide with the nucleus of atom A, the positive charge $q(r_i = 0) = +Z_A \cdot e$ is the nuclear charge of atom A and $q(r_i) = -e$ gives the charge of the $ith$ electron in the electron shell. For $R \gg r$ we can expand (9.46) into the Taylor series and obtain for the potential at a point $P(X, Y, Z)$ with $\boldsymbol{R} = \{X, Y, Z\}$ and $\boldsymbol{r} = \{x, y, z\}$

$$\begin{aligned} \phi(P) = \frac{1}{4\pi\varepsilon_0 R} \Bigg\{ & \sum_i q_i + \frac{1}{R}\left[\frac{X}{R}\sum q_i x_i \right. \\ & \left. + \frac{Y}{R}\sum q_i y_i + \frac{Z}{R}\sum q_i z_i\right] \\ & + \frac{1}{2R^2}\left[\left(\frac{3X^2}{R^2} - 1\right)\sum q_i x_i^2 \right. \\ & + \left(\frac{3Y^2}{R^2} - 1\right)\sum q_i y_i^2 \\ & \left. + \left(\frac{3Z^2}{R^2} - 1\right)\sum q_i z_i^2\right] \\ & + \frac{1}{R^3}[\ldots] + \ldots \Bigg\} \\ = {} & \phi_M\left(\sum q_i\right) + \phi_D\left(\sum \boldsymbol{p}_i\right) \\ & + \Phi_{QM}\left(\sum \widetilde{QM}_i\right) + \ldots \end{aligned} \tag{9.47}$$

The first term $\phi_M$ represents the monopole contribution, which is zero for neutral atoms where $\sum q_i = 0$. For ions it gives the main contribution.

The second term $\phi_D$ describes the potential of an electric dipole with a dipole moment $\boldsymbol{p} = \sum q_i \cdot r_i$, which is the vector sum of the dipole moments $\boldsymbol{p}_i = e_i \boldsymbol{r}_i$ formed by the different electrons and the nucleus at $\boldsymbol{r} = 0$.

The third term $\phi_{QM}$ gives the contribution of the quadrupole moment to the potential, the next terms the higher moments, such as the octopole or hexadecapole which are here neglected.

If another atom B with total charge $q_B$, electric dipole moment $p_B$ and quadrupole moment $QM_B$ is placed at the position $P(R)$, the potential energy of the interaction between A and B can be written as the sum

$$E_{pot}(A,B) = E_{pot}(q_B) + E_{pot}(\boldsymbol{p}_B) + E_{pot}(\widetilde{QM}_B) + \ldots \tag{9.48}$$

where

$$\begin{aligned} E_{pot}(q_B) &= q_B \cdot \phi(p)\,, \\ E_{pot}(\boldsymbol{p}_B) &= +\boldsymbol{p}_B \cdot \textbf{grad}\,\phi(p)\,, \\ E_{pot}(QM_B) &= \widetilde{QM}_B \cdot \textbf{grad}\,\boldsymbol{E}_A\,. \end{aligned} \tag{9.49}$$

The vector gradient $\textbf{grad}\,\boldsymbol{E}_A$ of the electric field $E_A$, produced by atom A is the tensor

$$\textbf{grad}\,\boldsymbol{E} = \left\{ \frac{\partial \boldsymbol{E}}{\partial \boldsymbol{x}}, \frac{\partial \boldsymbol{E}}{\partial \boldsymbol{y}}, \frac{\partial \boldsymbol{E}}{\partial z} \right\}. \tag{9.50}$$

From the expression (9.49) we obtain the following contributions to the interaction energy between A and B.

(a) Two ions with charge $q_A$ and $q_B$ have long-range interactions

$$E_{pot}(q_A, q_B, R) = \frac{1}{4\pi\varepsilon_0}\frac{q_A \cdot q_B}{R} \tag{9.51}$$

which decrease only with $1/R$.

(b) An ion with charge $q_A$ and a neutral atom or molecule with a permanent dipole moment $p_B$, pointing in a direction with an angle $\vartheta$ against the $z$-axis through A and B (Fig. 9.30) experience the interaction energy

$$E_{pot}(\boldsymbol{q}_A, \boldsymbol{p}_B, R) = \frac{-1}{4\pi\varepsilon_0}\frac{q_A p_B \cos\vartheta}{R^2}\,. \tag{9.52}$$

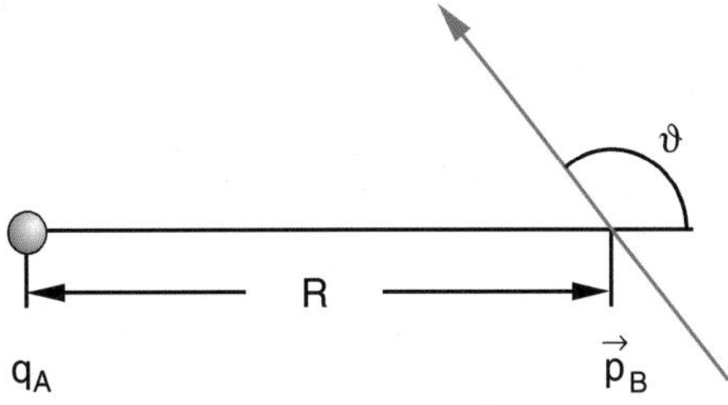


**Fig. 9.30.** Interaction between a charge $q_A$ and an electric dipole moment $\boldsymbol{p}_B$

> The interaction potential between an ion and a neutral atom with permanent dipole moment $\boldsymbol{p}$ is proportional to $1/R^2$ and is zero in the direction perpendicular to the dipole axis.

(c) Two permanent dipoles $p_A$ and $p_B$ with angles $\vartheta_A$ and $\vartheta_B$ against the $z$-axis and angles $\varphi_A$ and $\varphi_B$ against the $x$-axis (Fig. 9.31) have the interaction energy

$$E_{pot}(\boldsymbol{p}_A, \boldsymbol{p}_B, R) = -\boldsymbol{p}_A \cdot \boldsymbol{E}(\boldsymbol{p}_B) = -\boldsymbol{p}_B \cdot \boldsymbol{E}(\boldsymbol{p}_A) \tag{9.53}$$

where

$$\boldsymbol{E}(\boldsymbol{p}_B) = \frac{1}{4\pi\varepsilon_0 R^3}(3p_B \cdot \hat{R} \cdot \cos\vartheta_p - \boldsymbol{p}_B) \tag{9.54}$$

is the electric field at the point A generated by the dipole $p_B$ (Fig. 9.32). The interaction energy then becomes

$$\begin{aligned} &E_{pot}(\boldsymbol{p}_A, \boldsymbol{p}_B, R) \\ &= -\frac{1}{4\pi\varepsilon_0 R^3}[3p_A p_B \cos\vartheta_A \cos\vartheta_B - \boldsymbol{p}_A \cdot \boldsymbol{p}_B] \\ &= -\frac{p_A \cdot p_B}{4\pi\varepsilon_0 R^3}[2\cos\vartheta_A \cos\vartheta_B \\ &\quad - \sin\vartheta_A \sin\vartheta_B \cdot \cos(\varphi_A - \varphi_B)]\,. \end{aligned} \tag{9.55}$$

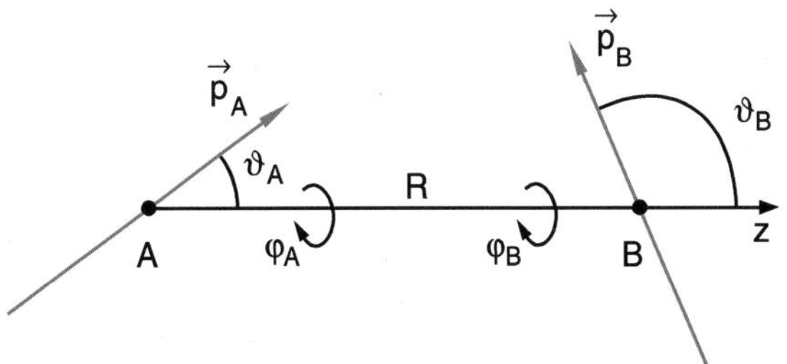


**Fig. 9.31.** Interaction between two dipoles

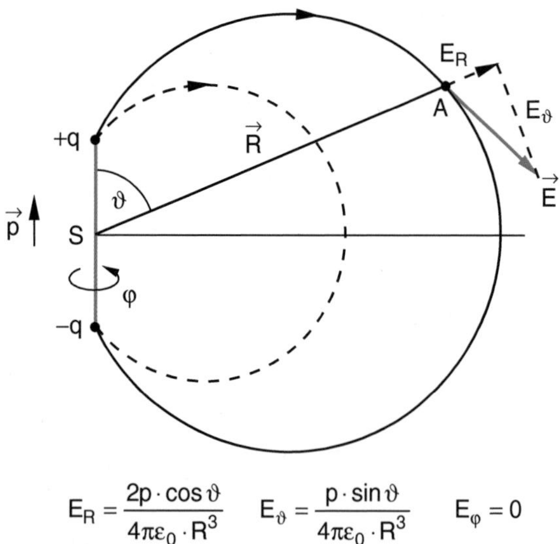


$$E_R = \frac{2p \cdot \cos\vartheta}{4\pi\varepsilon_0 \cdot R^3} \qquad E_\vartheta = \frac{p \cdot \sin\vartheta}{4\pi\varepsilon_0 \cdot R^3} \qquad E_\varphi = 0$$

**Fig. 9.32.** Electric field of a dipole. It has axial symmetry around the dipole axis

> The interaction energy between two permanent electric dipoles is proportional to the product of the two dipole moments and depends on their relative orientation. It decreases as $1/R^3$ with increasing distance much faster than the $1/R$ Coulomb interaction between two charges.

In Fig. 9.33 the interaction energies between two electric dipoles are illustrated for some specific relative orientations of the two dipoles. Positive potential energies mean repulsion between the two dipoles, negative ones attraction. The largest attraction (lowest energy) is obtained when the two dipoles both point into the z-direction (Fig. 9.33d). The largest repulsion (highest potential energy) have two dipoles pointing into opposite $\pm z$ directions (Fig. 9.33e).

### 9.4.3 Induced Dipole Moments and van der Waals Potential

If a neutral atom without permanent dipole moment is placed in an electric field, the opposite forces on the negative electrons and the positive nucleus shift the electron charge distribution into the opposite direction than the nucleus (Fig. 9.34). The centers of the positive and the negative charges no longer coincide as in a neutral atom without permanent dipole moment, and a dipole moment

$$\boldsymbol{p}_A^{\text{ind}} = \alpha_A \boldsymbol{E} \tag{9.56a}$$

is induced by the electric field , which is proportional to the field (Fig. 9.35). The constant $\alpha_A$ is the electric polarizability of atom A and is a measure of the restoring

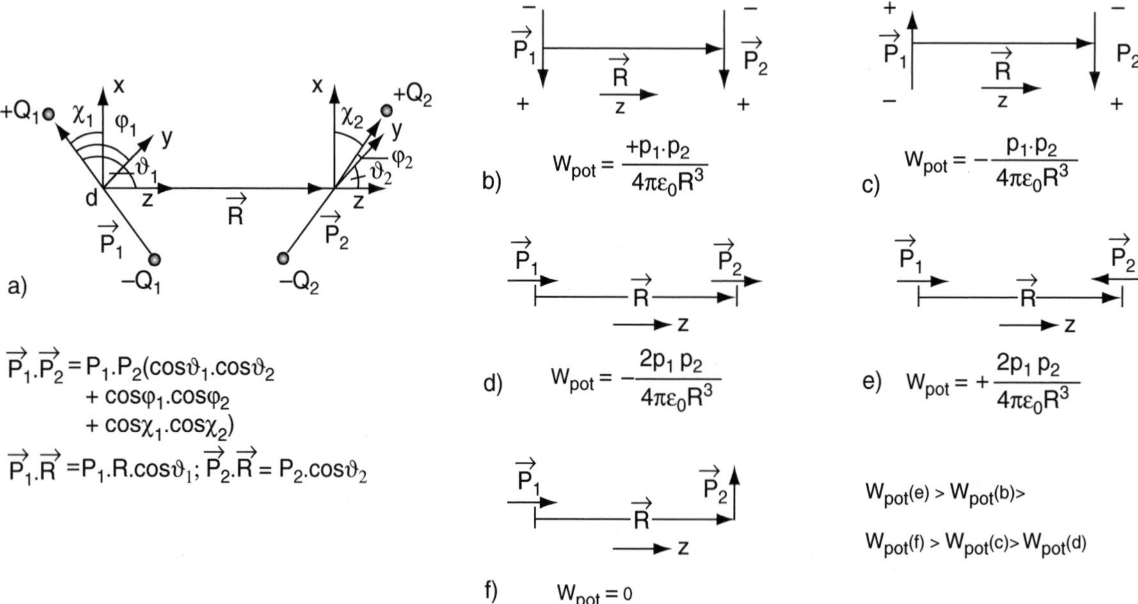


**Fig. 9.33.** Potential energy of two electric dipoles with some specific relative orientations

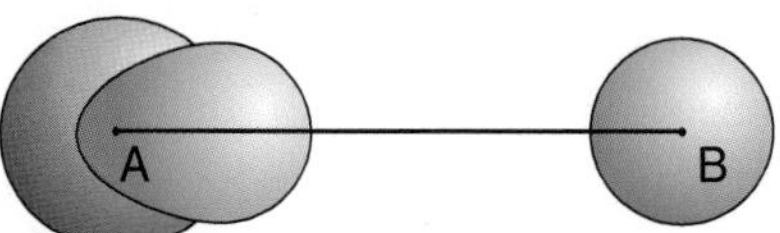


**Fig. 9.34.** Deformation and shift of electron charge distribution of atom A by the interaction with atom B

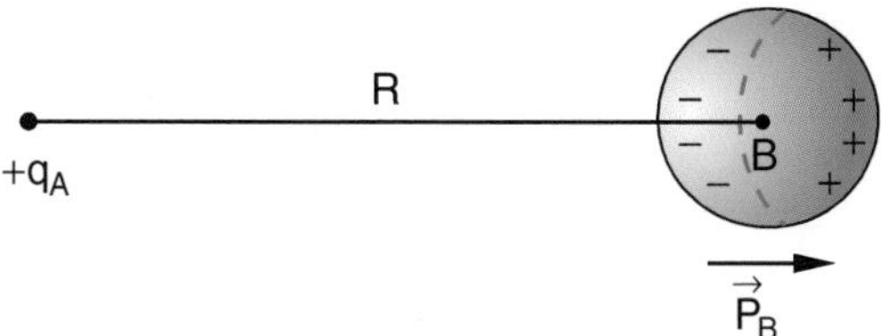


**Fig. 9.35.** The charge $q_A$ produces an induced dipole moment $\boldsymbol{p}_B$

forces in the atom against the displacement and deformation of the electron shell. If the electric field is produced by an ion with charge $q_B$, the induced dipole moment of A becomes

$$\boldsymbol{p}_A^{\text{ind}} = \frac{\alpha_A q_B}{4\pi\varepsilon_0 R^2}\hat{R} \tag{9.56b}$$

where $\hat{R}$ is the unit vector pointing into the direction from B to A.

The potential energy of a neutral atom without permanent dipole moment in an electric field $E$ is

$$E_{\text{pot}} = -\boldsymbol{p}_A^{\text{ind}} \cdot \boldsymbol{E} = -(\alpha_A \boldsymbol{E}) \cdot \boldsymbol{E} . \tag{9.57}$$

If the electric field is produced by an ion with charge $q_B$, the potential energy becomes

$$E_{\text{pot}} = -\frac{\alpha_A q_B^2}{(4\pi\varepsilon_0)^2 R^4} \tag{9.58}$$

If the field is generated by an atom B with permanent dipole moment $\boldsymbol{p}_B$ orientated at an angle $\vartheta_B$ against $\vec{R}$, we obtain from (9.54) and (9.57) the potential energy

$$E_{\text{pot}} = -\frac{\alpha_A p_B^2}{(4\pi\varepsilon_0 R^3)^2}(3\cos^2\vartheta_B + 1) . \tag{9.59}$$

In molecular physics the interaction between two neutral atoms is of particular importance. For a charge distribution on the electron shell that is spherically symmetric on the time average (e.g., for $1s$ electrons in the H atom) the time averaged dipole moment has to be zero. However, there still exists a momentary dipole moment (Fig. 9.36) that produces according to (9.54) a momentary electric field

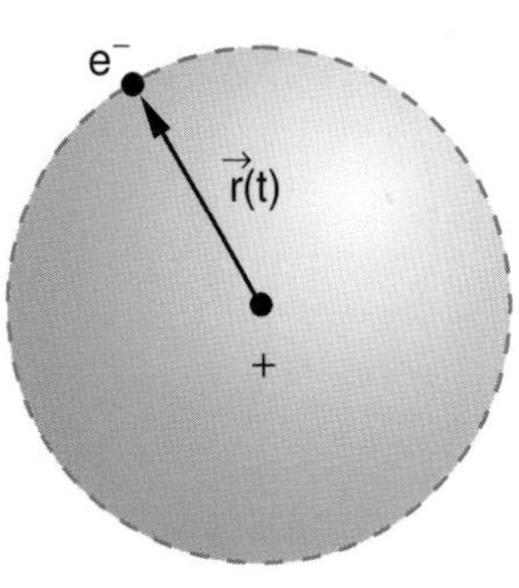


**Fig. 9.36.** Momentary dipole moment of an atom with spherically symmetric charge distribution

$$\boldsymbol{E}_A = \frac{1}{4\pi\varepsilon_0 R^2}(3p_A\hat{R}\cos\vartheta_A - \boldsymbol{p}_A) , \tag{9.60}$$

which is statistically pointing in all directions and has a time average of zero. However, if we place another atom B in the vicinity of A, this field induces a dipole moment in atom B

$$\boldsymbol{p}_B^{\text{ind}} = +\alpha_B \boldsymbol{E}_A , \tag{9.61}$$

which in turn generates an electric field at atom A inducing a dipole moment in A. Now the time average of $\boldsymbol{p}$ or $\boldsymbol{E}$ is no longer zero, because the interaction energy between the two induced dipoles depend on their relative orientation and the positions with minimum energy have a larger probability than those with higher energies. This leads to an attraction between A and B which is called a ***van der Waals interaction*** and is an interaction between two induced dipoles. We will now treat this more quantitatively.

According to (9.55) and Fig. 9.33 the interaction energy between the two dipoles has a maximum negative value, when the two dipoles are parallel ($\vartheta_A = \vartheta_B = 0°$ or $180°$) (Fig. 9.33d) with their dipole moments pointing into the $\pm z$-direction. It has a maximum positive value, when they are antiparallel ($\vartheta_A = 0°$, $\vartheta_B = 180°$), pointing into the opposite $\pm z$-directions (Fig. 9.33e). In the case of induced dipole-dipole interactions both dipole moments are directed along the axis through the two nuclei, which we choose as $z$-axis. This means that $\vartheta_A = \vartheta_B = 180°$ and $\boldsymbol{p} \parallel \boldsymbol{R}$ (Fig. 9.37). From (9.60) we obtain

$$\boldsymbol{E}_A = \frac{2p_A}{4\pi\varepsilon_0 R^3}\hat{R} , \quad \boldsymbol{E}_B = \frac{2p_B}{4\pi\varepsilon_0 R^3}\hat{R} . \tag{9.62}$$

The potential energy of the interaction between the two induced dipole moments $p_A^{\text{ind}}$ and $p_B^{\text{ind}}$ is then

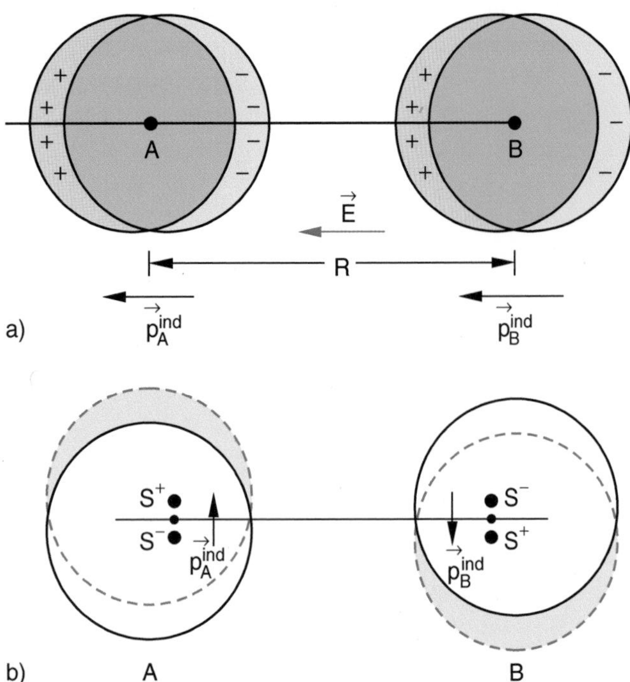


**Fig. 9.37a,b.** Possible orientations of two attraction-induced dipoles with (**a**) parallel (**b**) antiparallel orientation

$$E_{\text{pot}}(R) = -\boldsymbol{p}_{\text{B}}^{\text{ind}} \cdot \boldsymbol{E}_{\text{A}} = -\boldsymbol{p}_{\text{A}}^{\text{ind}} \cdot \boldsymbol{E}_{\text{B}} \,. \tag{9.63}$$

With $\boldsymbol{p}_{\text{A}} = \alpha_{\text{A}} \cdot \boldsymbol{E}_{\text{B}}$ and $\boldsymbol{p}_{\text{B}} = \alpha_{\text{B}} \cdot \boldsymbol{E}_{\text{A}}$ we get from (9.62)

$$E_{\text{pot}}(R) \propto -\boldsymbol{p}_{\text{A}}^{\text{ind}} \cdot \boldsymbol{p}_{\text{B}}^{\text{ind}} = -\alpha_{\text{A}} \cdot \alpha_{\text{B}} \cdot |E|^2 \,, \tag{9.64}$$

which can be written as

$$\boxed{E_{\text{pot}}(R) = -C_1 \frac{\alpha_{\text{A}}\alpha_{\text{B}}}{R^6} = -\frac{C_6}{R^6}} \,, \tag{9.65}$$

with $C_1 = \frac{1}{(4\pi\varepsilon_0)^2}$ and $C_6 = \frac{\alpha_{\text{A}} \cdot \alpha_{\text{B}}}{(4\pi\varepsilon_0)^2}$.

This is the ***van der Waals interaction potential*** between two neutral atoms with the polarizabilities $\alpha_{\text{A}}$ and $\alpha_{\text{B}}$. The constant $C_6$, which is proportional to the product $\alpha_{\text{A}} \cdot \alpha_{\text{B}}$ of the atomic polarizabilities, is called the van der Waals constant.

> The interaction potential between two neutral atoms scales for large separations $R$ as $R^{-6}$. The attraction is much weaker than between charged particles.

**Note** the interaction is attractive (because of the negative sign) and decreases as $1/R^6$ with increasing distance $R$. It is therefore a short range interaction compared with the Coulomb-interaction that is $\propto 1/R$, but has a longer range than the interaction of the chemical bond, which falls off exponentially with increasing $R$. The interaction energy is also negative if the two induced dipoles are orientated antiparallel but both perpendicular to the $z$-axis through their centers (Fig. 9.37b).

The quantum mechanical treatment of the van der Waals interaction is based on the calculations of the atomic charge distributions, perturbed by the mutual interaction between A and B. Since only this perturbation leads to an attraction between the two atoms one needs a perturbation calculation of second order [9.8, 9], which is beyond the scope of this book.

If higher order terms in the multipole expansion are taken into account, the interaction energy between two atoms includes terms with $1/R^8, 1/R^{10}, 1/R^{12}, \ldots$ for the induced quadrupole or octupole interactions. For homonuclear molecules only even powers $n$ of $1/R^n$ can appear for symmetry reasons.

> The multipole interaction between two neutral atoms is only important for internuclear distances $R > \langle r_{\text{A}} \rangle + \langle r_{\text{B}} \rangle$. For smaller values of $R$ the overlap of the electron shells of A and B has to be taken into account, which results in the above-mentioned exchange interaction and the electrostatic interaction due to the increased electron density between the two nuclei.

The total range of $R$-values can, however, be covered by the empirical Lenard–Jones potential

$$E_{\text{pot}}^{\text{LJ}}(R) = \frac{a}{R^{12}} - \frac{b}{R^6} \tag{9.66}$$

where $a$ and $b$ are two parameters that depend on the two atoms A and B and which are adapted to fit best the potential curve obtained either experimentally or by accurate and extensive calculations (Fig. 9.38).

From (9.66) it follows that $E_{\text{pot}}(R) = 0$ for $R = R_0 = (a/b)^{1/6}$. The minimum of $E_{\text{pot}}(R)$ is obtained for $\mathrm{d}E/\mathrm{d}R = 0$. This gives the distance

$$R_{\text{e}} = (2a/b)^{1/6} = 2^{1/6} R_0 \tag{9.67}$$

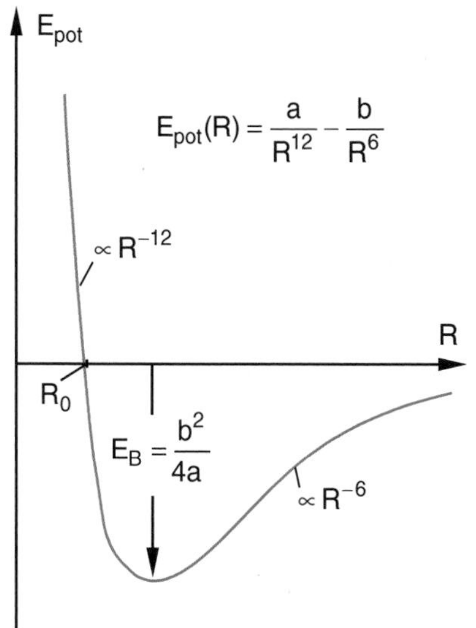


**Fig. 9.38.** Lenard–Jones potential

for the minimum. The binding energy at $R_e$ is then

$$E_B = -E_{pot}(R_e) = b^2/4a \; . \tag{9.68}$$

### 9.4.4 General Expansion of the Interaction Potential

The potential energy $E_{pot}(R)$ of a diatomic molecule can be expanded for $|R - R_e|/R_e < 1$ into a Taylor series around the equilibrium distance $R_e$ of the potential minimum:

$$E_{pot}(R) = \sum_{n=0}^{\infty} \frac{1}{n!} \left( \frac{d[}{\partial} ]^n E_{pot} R^n \right)_{R_e} (R - R_e)^n \; . \tag{9.69a}$$

Because $(\partial^0 E/\partial R^0)_{R_e} = E_{pot}(R_e)$ and $(\partial E/\partial R)_{R_e} = 0$, this gives

$$E_{pot}(R) = E_{pot}(R_e) + \frac{1}{2} \left( \frac{\partial^2 E_{pot}}{\partial R^2} \right)_{R_e} \times (R - R_e)^2 + \ldots \; . \tag{9.69b}$$

In molecular physics the minimum of the ground state potential is generally chosen as $E_{pot}(R_e) = 0$. Instead of the negative energy $E(R_e)$ (which is used if the zero point is chosen as the energy of the separated ground state atoms) the positive energy $E_B = -E(R_e)$ is now used, which gives the binding energy. $E_B$ is slightly larger than the dissociation energy $E_D$, which is the energy necessary to dissociate the molecule from its lowest energy level to the separated atoms at $R = \infty$.

For $|R - R_e|/R_e \ll 1$ the higher order terms with $n > 2$ can be neglected and we obtain a parabolic potential in the vicinity of the potential minimum.

> The potential energy of a diatomic molecule can be approximated in the vicinity of the potential minimum by a parabolic potential (Fig. 9.39).

### 9.4.5 The Morse Potential

In 1929 *P.M. Morse* had already proposed an empirical potential form

$$E_{pot}(R) = E_B \left[ 1 - e^{-a(R-R_e)} \right]^2 \; , \tag{9.70}$$

which represents the attractive part of the potential with a much better approximation to the experimental values than the parabolic potential. This potential converges for $R \to \infty$ correctly towards the dissociation energy $E_D$, while the parabolic potential goes to infinity for $R \to \infty$ (Fig. 9.39). The repulsive part of the Morse potential for $R < R_e$ deviates more severely from the experimental data. (Fig. 9.39) We obtain from (9.70)

$$\lim_{R \to 0} E_{pot}(R) = E_B \left[ 1 - e^{+aR_e} \right]^2 \tag{9.71}$$

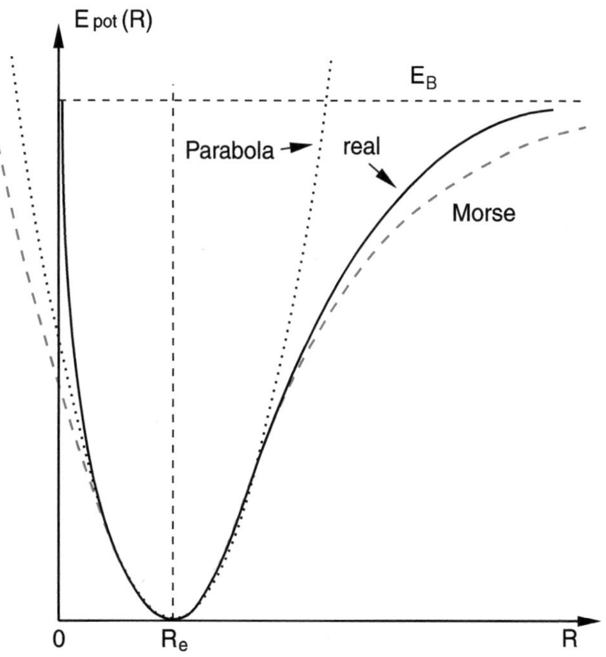


**Fig. 9.39.** Comparison of parabolic and Morse potentials with the real (experimental) potential

while the experimental potential must converge towards the energy levels of the united atom (see Fig. 9.25).

The Morse potential has the great advantage that the Schrödinger equation of two atoms vibrating in this potential can be solved exactly (see Sect. 9.6).

### 9.4.6 Different Binding Types

Depending on the specific atoms of a diatomic molecule and on the internuclear distance, we distinguish between the following different binding types.

> Covalent binding, caused by the exchange of electrons in overlapping atomic orbitals. The redistribution of the electron charge distribution leads to an increase in negative charge between the two positive nuclear charges. It is only important for internuclear distances $R < r_A + r_B$.

Ionic binding between positive and negative ions occurs if the electron exchange between two atoms A and B leads to an increased electron density at atom A and a decreased density at B (Fig. 9.40). Typical examples of ionic binding are combinations of atoms in the first column of the periodic table (alkali atoms with only one electron in the valence shell) with atoms in the seventh column (halogen atoms with one missing electron in the valence shell). Examples are

$$\mathrm{H} + \mathrm{Cl} \rightarrow \mathrm{H}^+\mathrm{Cl}^- \quad \text{and} \quad \mathrm{Na} + \mathrm{I} = \mathrm{Na}^+\mathrm{I}^- . \tag{9.72}$$

> The interaction energy of the ionic bond falls off with $1/R$ and is therefore a long-range interaction.

> The van der Waals interaction between two neutral polarized atoms with a large equilibrium distance $R_e$ is weak and falls off with $1/R^6$. It is therefore a short-range interaction compared to ionic interactions.

For molecules with more than two atoms, another interaction can play an important role, in particular for biological molecules. This is hydrogen bonding, where two atoms are attracted by a $H^+$ ion (i.e., a proton), between the two atoms (Fig. 9.41a). The proton polarizes the two atoms resulting in an attractive interaction with a binding energy that lies between the weak van der Waals energy and the strong valence bond.

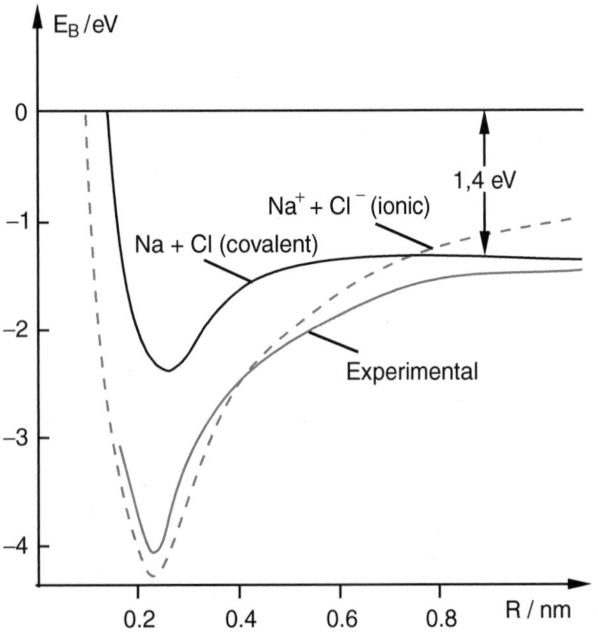

**Fig. 9.40.** Different causes for the binding energy of NaCl, dependent on the internuclear distance $R$

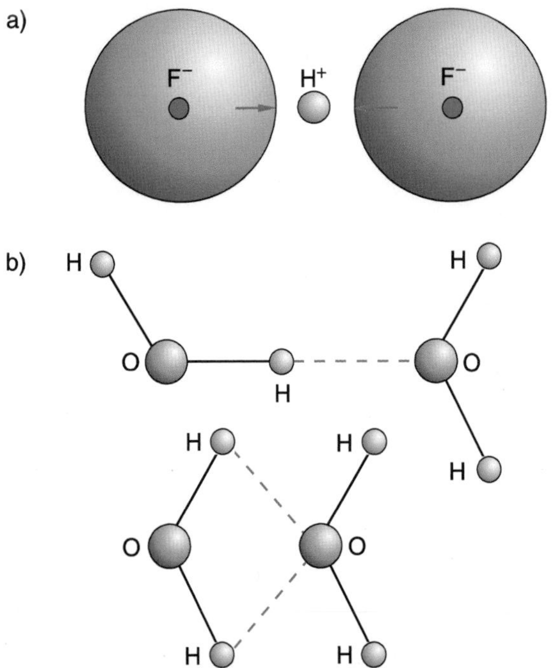


**Fig. 9.41a,b.** Binding by a hydrogen bridge: (**a**) $(HF_2)^-$ (**b**) Two conformations of the water dimer $(H_2O)_2$

Examples of hydrogen bonds are water dimers $(H_2O)_2$ (Fig. 9.41b), which can exist in different conformations. The structure of large biomolecules such as DNA is essentially based on hydrogen bonding.

## 9.5 Rotation and Vibration of Diatomic Molecules

Up to now we have discussed the electronic states of rigid molecules, where the nuclei are clamped to a fixed position. In this section we will improve our model of molecules and include the rotation and vibration of diatomic molecules. This means, that we have to take into account the kinetic energy of the nuclei in the Schrödinger equation $\hat{H}\psi = E\psi$, which has been omitted in the foregoing sections. We then obtain the Hamiltonian

$$\begin{aligned}\hat{H} &= -\frac{\hbar^2}{2}\sum_{k=1}^{2}\frac{1}{M_\mathrm{k}}\nabla_\mathrm{k}^2 - \frac{\hbar^2}{2m_\mathrm{e}}\sum_{i=1}^{N}\nabla_i^2 \\ &\quad + \frac{e^2}{4\pi\varepsilon_0}\left[\frac{Z_1Z_2}{R} + \sum_{i,j}\frac{1}{r_{i,j}} - \sum_i\left(\frac{1}{r_{i1}} + \frac{1}{r_{i2}}\right)\right] \\ &= \hat{E}_\mathrm{kin}^\mathrm{nucl} + E_\mathrm{kin}^\mathrm{el} + E_\mathrm{pot}^0 = \hat{T}_\mathrm{k} + \hat{H}_0 \end{aligned} \tag{9.73}$$

where the first term represents the kinetic energy of the nuclei, the second term that of the electrons, the third represents the potential energy of nuclear repulsion, the fourth that of the electron repulsion and the last term the attraction between the electrons and the nuclei.

### 9.5.1 The Born-Oppenheimer Approximation

Because of their much larger mass, the nuclei in a molecule move much slower than electrons. This implies that the electrons can nearly immediately adjust their positions to the new nuclear configuration when the nuclei move. Although the electronic wave functions $\psi(r, R)$ depend parametrically on the internuclear distance $R$ they are barely affected by the velocity of the moving nuclei. The kinetic energy of the nuclear motion $E_\mathrm{kin} = \frac{1}{2}Mv^2$ is small compared to that of the electrons. We therefore write the total Hamiltonian $H$ in (9.73) as the sum

$$H = H_0 + T_\mathrm{k}$$

of the Hamiltonian $H_0$ of the rigid molecule and $T_\mathrm{k}$ of the kinetic energy of the nuclei. Since the latter is small compared to the total energy of rigid molecule we can regard $T_\mathrm{k}$ as a small perturbation of $H$. In this case the total wave function

$$\psi(\boldsymbol{r}_i, \boldsymbol{R}_\mathrm{k}) = \chi(\boldsymbol{R}_\mathrm{k}) \cdot \Phi(\boldsymbol{r}_i, \boldsymbol{R}_\mathrm{k}) \tag{9.74}$$

can be written as the product of the molecular wave function $\chi(\boldsymbol{R}_\mathrm{k})$ (which depends on the positions $\boldsymbol{R}_\mathrm{k}$ of the nuclei), and the electronic wave function $\Phi(\boldsymbol{r}_i, \boldsymbol{R}_\mathrm{k})$ of the rigid molecule at arbitrary but fixed nuclear positions $\boldsymbol{R}_\mathrm{k}$, where the electron coordinates $\boldsymbol{r}_i$ are the variables and $\boldsymbol{R}_\mathrm{k}$ can be regarded as a fixed parameter. This implies that nuclear motion and electronic motion are independent and the coupling between both is neglected. The total energy $E$ is the sum of the energy $E_n^\mathrm{el}(R)$ of the rigid molecule in the $n$th electronic state, which is represented by the potential curve in Fig. 9.42 and the kinetic energy $(E_\mathrm{vib} + E_\mathrm{rot})$ of the nuclei.

**Note** the total energy is independent of $R$!

This approximation where the interaction between nuclear und electron motion is neglected, was first introduced 1927 by Max Born and Robert Oppenheimer and is therefore called *Born-Oppenheimer approximation*. Because of the neglected interaction the total wave function can be written as the product of the electronic wave function, which describes the molecule

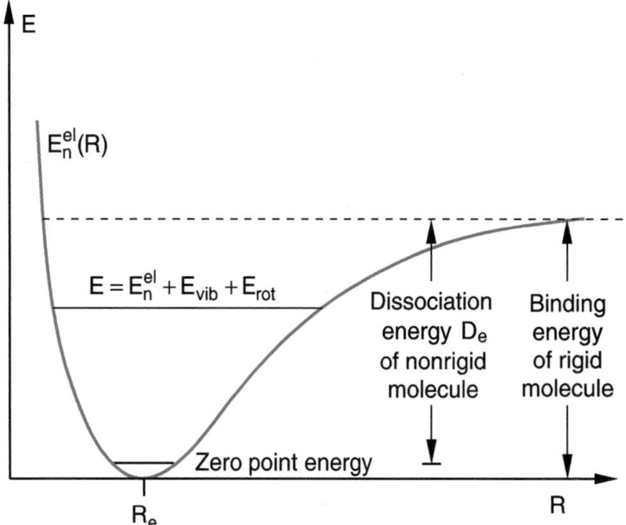


**Fig. 9.42.** Energy $E_n^\mathrm{el}(R)$ of the rigid molecule and total energy $E$ of the nonrigid vibrating and rotating molecule

with clamped nuclei and the nuclear wave function which describes the vibration of the nuclei and the rotation of the whole molecule. Its physical meaning is, that the vibration of the nuclei does not change the electronic state.

This approximation is valid for the electronic ground states of most molecules. For higher electronic states one often observes a breakdown of the BO-approximation, where the vibration can induce a coupling between different electronic states which means that the vibrating nuclei do no longer move in a well defined potential of a single electronic state but changes the electronic configuration during the vibration [9.10].

Inserting this product (9.74) into the Schrödinger equation (9.73) gives the two equations (see Problem 9.4)

$$\hat{H}_0 \Phi_n^{\text{el}}(\boldsymbol{r}, \boldsymbol{R}_\text{k}) = E_n^{(0)} \cdot \Phi_n^{\text{el}}(\boldsymbol{r}, \boldsymbol{R}_\text{k}) \tag{9.75a}$$

$$\left(\hat{T}_\text{k} + E_n^{(0)}\right) \chi_{n,m}(\boldsymbol{R}) = E_{n,m} \chi_{n,m}(\boldsymbol{R}) \,. \tag{9.75b}$$

The first equation describes the electronic wave function $\Phi$ of the rigid molecule in the electronic state $(n, L, \Lambda)$ and $E_n^{(0)}$ is the total electronic energy of this state without the kinetic energy $T_\text{k}$ of the nuclei.

The second equation contains the kinetic energy $T_\text{k}$ of the nuclei and determines the motion of the nuclei in the potential

$$E_n^{(o)} = \left\langle E_{\text{kin}}^{\text{el}} \right\rangle + E_{\text{pot}}(\boldsymbol{r}_i, \boldsymbol{R}_\text{k}) \,, \tag{9.76}$$

which consists of the time average of the kinetic energy of the electrons and the total potential energy of electrons and nuclei. The total energy

$$E_{n,\text{k}} = E_{\text{kin}}^{\text{nuc}} + E_n^{(o)} \tag{9.77}$$

of the nonrigid molecule is the sum of the kinetic energy of the nuclei and the total energy of the rigid molecule. Equation (9.75b) is explicitly written as

$$\left[\left(\frac{-\hbar^2}{2M_\text{A}} \Delta_\text{A} + \frac{-\hbar^2}{2M_\text{B}} \Delta_\text{B}\right) + E_{\text{pot}}^{(n)}(\boldsymbol{R}_\text{k})\right] \chi_{n,m}(\boldsymbol{R}_\text{k}) = E_{n,m} \chi_{n,m}(\boldsymbol{R}_\text{k}) \,. \tag{9.78}$$

In the center of the mass system this translates to

$$\left[\frac{-\hbar^2}{2M} \Delta + E_{\text{pot}}^{(n)}(R)\right] \chi_{n,m}(\vec{R}) = E_{n,m} \chi_{n,m}(\vec{R}) \tag{9.79}$$

where $M = M_\text{A} M_\text{B}/(M_\text{A} + M_\text{B})$ is the reduced mass of the two nuclei and the index $m$ gives the $m$th quantum state of the nuclear movement (vibrational-rotational state).

The important result of this equation is:

The potential energy for the nuclear motion in the electronic state $(n, L, \Lambda)$ depends only on the nuclear distance $R$, not on the angles $\vartheta$ and $\varphi$, i.e., *it is independent of the orientation of the molecule in space.* It is spherically symmetric. The wave functions $\chi = \chi(R, \vartheta, \varphi)$, however, may still depend on all three variables $R, \vartheta$, and $\varphi$.

Because of the spherically symmetric potential equation (9.79) is mathematically equivalent to the Schrödinger equation of the hydrogen atom. The difference lies only in the different radial dependence of the potential. Analogous to the treatment in Sect. 4.3.2 we can separate the wave functions $\chi(R, \vartheta, \varphi)$ into a radial part depending solely on $R$ and an angular part, depending solely on the angles $\vartheta$ and $\varphi$. We therefore try the product ansatz

$$\chi(R, \vartheta, \varphi) = S(R) \cdot Y(\vartheta, \varphi) \,. \tag{9.80}$$

The radial function $S(R)$ depends on the radial form of the potential, while the spherical surface harmonics $Y(\vartheta, \varphi)$ are solutions for *all* spherically symmetric potentials, independent of their radial form.

Inserting the product (9.80) into (9.79) gives, as has been already shown in Sect. 4.3.2, the following equation for the radial function $S(R)$:

$$\frac{1}{\mathrm{d}R^2} \frac{\mathrm{d}}{R} \left(R^2 \frac{\mathrm{d}S}{\mathrm{d}R}\right) + \frac{2M}{\hbar^2} \left[E - E_{\text{pot}}(R) - \frac{J(J+1)\hbar^2}{2MR^2}\right] S = 0 \,. \tag{9.81a}$$

For the spherical surface harmonics $Y(\vartheta, \varphi)$ we obtain the Eq. (4.88), already treated in Sect. 4.4.2:

$$\frac{1}{\sin\vartheta} \frac{\partial}{\partial\vartheta} \left(\sin\vartheta \frac{\partial Y}{\partial\vartheta}\right) + \frac{1}{\sin^2\vartheta} \frac{\partial^2 Y}{\partial\varphi^2} + J(J+1)Y = 0 \,. \tag{9.81b}$$

While the first Eq. (9.81a) describes the vibration of the diatomic molecule, (9.81b) determines its rotation. We will at first discuss eq. (9.81b).

### 9.5.2 The Rigid Rotor

A diatomic molecule with the atomic masses $M_A$ and $M_B$ can rotate around any axis through the center of mass perpendicular to the internuclear axis with the angular velocity $\omega$ (Fig. 9.43). Its rotational energy is then

$$E_{rot} = \tfrac{1}{2} I \omega^2 = J^2/(2I) \,. \tag{9.82}$$

Here $I = M_A R_A^2 + M_B R_B^2 = MR^2$ $(M = M_A M_B/(M_A + M_B))$ is the moment of inertia of the molecule with respect to the rotational axis and $|J| = I\omega$ is its rotational angular momentum. For a rotation around the internuclear axis the contribution of the nuclei to the moment of inertia is vanishingly small, because they lie on this axis. The contribution of the electron shell is also small because of the small electron mass. Therefore the rotational energy for a rotation around this axis is very large and cannot be excited at thermal energies. The electronic contribution has only to be taken into account for electronic states with $\Lambda > 0$ and is then added to the electronic energy (see Sect. 9.5.4). Since the square of the angular momentum

$$|J|^2 = J(J+1)\hbar^2$$

can take only discrete values that are determined by the rotational quantum number $J$, the rotational energies of a molecule in its equilibrium position with an internuclear distance $R_e$ are represented by a series of discrete values

$$E_{rot} = \frac{J(J+1)\hbar^2}{2MR_e^2} \,. \tag{9.83}$$

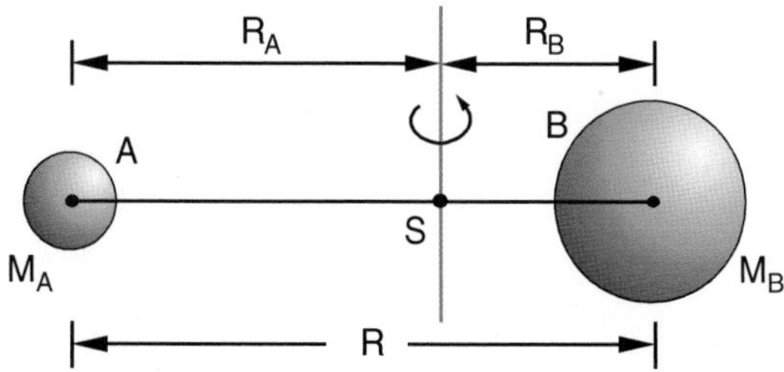


**Fig. 9.43.** Diatomic molecule as a rigid rotor

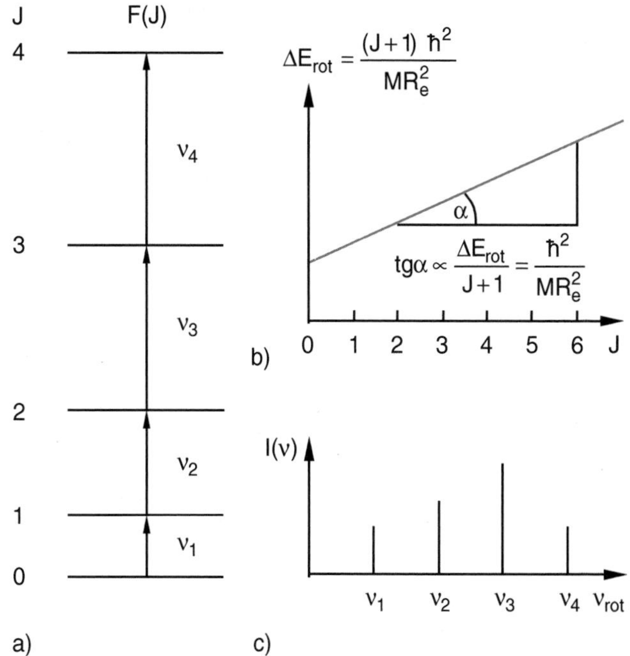


**Fig. 9.44.** (**a**) Energy levels of the rigid rotor (**b**) Separations $\Delta E_{rot} = E_{rot}(J+1) - E_{rot}(J)$ (**c**) Schematic rotational spectrum

The energy separation between the rotational levels $J$ and $J+1$

$$\Delta E_{rot} = E_{rot}(J+1) - E_{rot}(J) = \frac{(J+1)\hbar^2}{MR_e^2} \tag{9.84}$$

increases linearly with $J$ (Fig. 9.44).

This result can also be directly obtained from (9.80). For a fixed nuclear distance $R$ the first term in (9.81a) is zero. Therefore the second term must also be zero, because the sum of the two terms is zero. The kinetic energy of a rigid rotor, which does not vibrate, is $E_{kin} = E_{rot} = E - E_{pot}$, where $E$ is the total energy. The bracket of the second term in (9.80) then becomes for $R = R_e$ equal to (9.83).

In the spectroscopic literature, the rotational term values $F(J) = E(J)/hc$ are used instead of the energies. Instead of (9.83) we write

$$F_{rot}(J) = \frac{J(J+1)\hbar^2}{2hcMR_e^2} = B_e J(J+1) \tag{9.85}$$

with the rotational constant

$$B_e = \frac{\hbar}{4\pi c M R_e^2} , \tag{9.86}$$

which is determined by the reduced mass $M$ and the equilibrium nuclear distance $R_e$. For historical reasons one writes $B_e$ in units of $cm^{-1}$ instead of $m^{-1}$.

## EXAMPLES

1. The $H_2$ molecule has a reduced mass $M = 0.5 M_H = 8.35 \times 10^{-28}$ kg, and the equilibrium distance $R_e = 0.742 \times 10^{-10}\,m \Rightarrow I = 4.60 \times 10^{-48}\,kg\,m^2$. The rotational energies are

$$E_{rot}(J) = 1.2 \times 10^{-21}\, J(J+1)\,\text{Joule} \\ = 7J(J+1)\,\text{meV} .$$

The rotational constant is $B_e = 60.80\,cm^{-1}$.

2. For the HCl molecule the figures are $M = 0.97\,\text{AMU} = 1.61 \times 10^{-27}\,kg$, $R_e = 1.27 \times 10^{-10}\,m \Rightarrow E_{rot} = 2.1 \times 10^{-22}\, J(J+1)\,\text{Joule} = 1.21\,J(J+1)$ meV, $B_e = 10.59\,cm^{-1}$.

In Table 9.7 the equilibrium distances $R_e$ and the rotational constants are listed for some diatomic molecules. The figures show that the rotational energies are within the range of

$$E_{rot} = (10^{-6}\text{–}10^{-2})\, J(J+1)\,\text{eV} .$$

For a rotational angular momentum $J$ the classical relation between $J$, circular frequency $\omega = 2\pi\nu$ and moment of inertia $I$ is $\vec{J} = \vec{\omega} \cdot I$. This gives with $|\vec{J}| = \sqrt{J(J+1)}\hbar$ and $|\vec{\omega}| = 2\pi\nu$ for the rotational period $T_{rot} = 1/\nu$

$$T_{rot} = \frac{2\pi I/\hbar}{\sqrt{J(J+1)}} . \tag{9.87}$$

Depending on the rotational constant $B_e$ the rotational periods range from $T_{rot} = 10^{-14}$ s to $10^{-10}$ s. For $B_e = 1\,cm^{-1}$ one obtains $T_{rot} = 1.6 \times 10^{-11}/\sqrt{J(J+1)}$ s. If an electro-magnetic wave falls onto a sample of molecules it can be absorbed on rotational transitions $J \to J+1$ resulting in absorption lines with frequencies

$$\nu_{rot}(J) = [E(J+1)] - E(J)]/h \tag{9.88a}$$

or, in wavenumber units $cm^{-1}$,

$$\overline{\nu}_{rot}(J) = 2B_e(J+1) . \tag{9.88b}$$

The rotational transitions between levels $J$ and $J+1$ fall into the spectral range with frequencies $\nu = 10^9$–$10^{13}$ Hz, i.e., in the Gigahertz–Terahertz range with wavelengths between $\lambda = 10^{-5}$–$10^{-1}$ m. This spectral region is called the *microwave range*.

In Sect. 9.6.2 we will see, that only molecules with a permanent electric dipole moment can absorb or emit radiation on rotational transitions (except for very weak quadrupole transitions). Therefore homonuclear diatomic molecules show no rotational absorption or emission spectra!

**Table 9.7.** Equilibrium distances and rotational and vibrational constants in units of $cm^{-1}$ for some diatomic molecules

| Molecule | $R_e/pm$ | $B_e$ | $D_e$ | $\alpha_e$ | $\omega_e$ | $\omega_e x_e$ |
|---|---|---|---|---|---|---|
| $H_2$ | 74.16 | 60.8 | $1.6 \times 10^{-2}$ | 3.06 | 4401 | 121.3 |
| $Li_2$ | 267.3 | 0.673 | $9.9 \times 10^{-6}$ | 0.007 | 351.4 | 2.6 |
| $N_2$ | 109.4 | 2.01 | $5.8 \times 10^{-6}$ | 0.017 | 2359.0 | 14.3 |
| $O_2$ | 120.7 | 1.45 | $4.8 \times 10^{-6}$ | 0.016 | 1580.0 | 12.0 |
| $I_2$ | 266.6 | 0.037 | $4.2 \times 10^{-9}$ | 0.0001 | 214 | 0.61 |
| $H^{35}Cl$ | 127.4 | 10.59 | $5.3 \times 10^{-4}$ | 0.31 | 2990 | 52.8 |
| $D^{35}Cl$ | 127.4 | 5.45 | $1.4 \times 10^{-4}$ | 0.11 | 2145 | 27.2 |
| ICl | 232.1 | 0.114 | $4.0 \times 10^{-8}$ | 0.0005 | 384 | 1.50 |
| CO | 112.8 | 1.931 | $6 \times 10^{-6}$ | 0.017 | 2170 | 13.29 |
| NO | 115.1 | 1.705 | $0.5 \times 10^{-6}$ | 0.017 | 1904 | 14.08 |

### 9.5.3 Centrifugal Distortion

A real molecule is not rigid. When it rotates, the centrifugal force acts on the atoms and the internuclear distance widens to a value $R$ where this force $\vec{F}_c = M\omega^2\vec{R}$ is compensated by the restoring force $\vec{F}_r = -(\mathrm{d}E_{pot}/\mathrm{d}R)\hat{R}$ holding the two atoms together, which depends on the slope of the potential energy function $E_{pot}(R)$ (Fig. 9.45).

In the vicinity of the equilibrium distance $R_e$ the potential can be approximated by a parabolic function (see Sect. 9.4.4). This leads to a linear restoring force

$$\boldsymbol{F}_r = -k(R - R_e)\hat{R}\,. \tag{9.89}$$

From the relation centripetal force = restoring force: $M\omega^2 R = k(R - R_e)$ we obtain with $J^2 = I^2\omega^2 = M^2R^4\omega^2$

$$\begin{aligned} M\omega^2 R &= \frac{J(J+1)\hbar^2}{MR^3} \stackrel{!}{=} k(R - R_e) \\ \Rightarrow R &= R_e + \frac{J(J+1)\hbar^2}{MkR^3}\,, \end{aligned} \tag{9.90}$$

which means that the internuclear distance $R$ is widened by the molecular rotation.

If we approximate $R$ on the right side of (9.90) by $R_e$ and $k$ we obtain

$$R = R_e\left(1 + \frac{J(J+1)\hbar^2}{MkR_e^4}\right) = R_e(1+x) \tag{9.91}$$

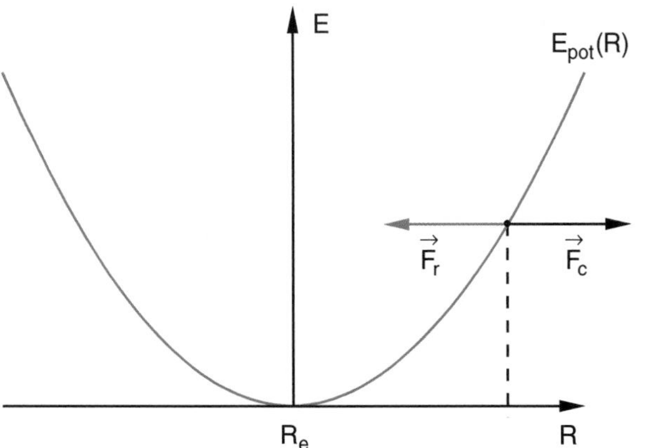


**Fig. 9.45.** Compensation of centrifugal and restoring force in the nonrigid rotating molecule

with $x \ll 1$. This allows us to expand $1/R^2$ into the power series

$$\begin{aligned} \frac{1}{R^2} = \frac{1}{R_e^2}\Bigg[1 &- \frac{2J(J+1)\hbar^2}{MkR_e^4} \\ &+ \frac{3J^2(J+1)^2\hbar^4}{M^2k^2R_e^8} \mp \ldots\Bigg] \end{aligned} \tag{9.92}$$

For $R \neq R_e$ the rotational energy in eq. (9.83) becomes

$$\begin{aligned} E_{rot} &= \frac{J\cdot(J+1)\hbar^2}{2MR^2} = \frac{J(J+1)\hbar^2}{2MR_e^2} - \frac{J^2(J+1)^2\hbar^4}{2M^2kR_e^6} \\ &\quad + \frac{3J^3(J+1)^3\hbar^6}{2M^3k^2R_e^{10}} \pm \ldots\,. \end{aligned} \tag{9.93}$$

> For a given value of the rotational quantum number $J$ the centrifugal widening makes the moment of inertia larger and therefore the rotational energy smaller. This effect overcompensates for the increase in potential energy.

Using the term-values instead of the energies, (9.93) becomes

$$\begin{aligned} F_{rot}(J) = B_e J(J+1) &- D_e J^2(J+1)^2 \\ &+ H_e J^3(J+1)^3 - \ldots \end{aligned} \tag{9.94}$$

with the rotational constants

$$\begin{aligned} B_e &= \frac{\hbar}{4\pi c M R_e^2}\,, \quad D_e = \frac{\hbar^3}{4\pi c k M^2 R_e^6}\,, \\ H_e &= \frac{3\hbar^5}{4\pi c k^2 M^3 R_e^{10}}\,. \end{aligned} \tag{9.95}$$

The spectroscopic accuracy is nowadays sufficiently high to measure even the higher order constant $H_e$. When fitting spectroscopic data by (9.95) this constant, therefore, has to be taken into account.

### 9.5.4 The Influence of the Electron Motion

Up to now we have neglected the influence of the electron motion on the rotation of molecules. In the axial symmetric electrostatic field of the two nuclei in the nonrotating molecule, the electrons precess around the space-fixed molecular $z$-axis. The angular momentum $\boldsymbol{L}(R) = \Sigma \boldsymbol{l}_i(R)$ of the electron shell, which depends on the separation $R$ of the nuclei, has, however, a constant projection

$$\langle L_z \rangle = \Lambda \hbar \tag{9.96a}$$

independent of $R$. For molecular states with electron spin $\boldsymbol{S} \neq 0$ in atoms with weak spin-orbit coupling the spin $\boldsymbol{S}$ precesses independently around the $z$-axis with a projection

$$\langle S_z \rangle = M_s \hbar \, . \tag{9.96b}$$

Both projections add to the total value

$$\Omega \hbar = (\Lambda + M_s)\hbar \, . \tag{9.96c}$$

In the case of strong spin-orbit coupling $\boldsymbol{L}$ and $\boldsymbol{S}$ couple to $\boldsymbol{J}^{\text{el}} = \boldsymbol{L} + \boldsymbol{S}$ with the projection

$$\left\langle J_z^{\text{el}} \right\rangle = \Omega \hbar$$

(see Sect. 9.3.3).

The total angular momentum $\boldsymbol{J}$ of the rotating molecule is now composed of the angular momentum $\boldsymbol{N}$ of the molecular rotation and the projection $\Lambda\hbar$ or $\Omega\hbar$. For $\Omega \neq 0$ the total angular momentum $\boldsymbol{J}$ of the molecule is no longer perpendicular to the $z$-axis (Fig. 9.46).

> Since the total angular momentum of a free molecule without external fields is constant in time, the molecule rotates around the space-fixed direction of $J$ and for $\Lambda \neq 0$ the rotational axis is no longer perpendicular to the molecular $z$-axis.

In a simple model, the whole electron shell can be regarded as a rigid charge distribution that rotates around the $z$-axis. The rotating molecule can then be described as a symmetric top with two different moments of inertia: 1.) The moment $I_1$ of the electron shell rotating around the $z$-axis and 2.) the moment $I_2$ of the molecule (nuclei and electrons) rotating around an axis perpendicular to the $z$-axis. Because the electron masses are very small compared with the nuclear masses, it follows that $I_1 \ll I_2$.

The rotational energy of this symmetric top is

$$E_{\text{rot}} = \frac{J_x^2}{2I_x} + \frac{J_y^2}{2I_y} + \frac{J_z^2}{2I_z} \tag{9.97}$$

with $I_x = I_y = I_2 \neq I_z = I_1$ .

From Fig. 9.46 the following relations can be obtained:

$$\begin{aligned} J_z^2 &= \Omega^2\hbar^2 \\ J_x^2 + J_y^2 &= J^2 - J_z^2 \\ &= \left[J(J+1) - \Omega^2\right]\hbar^2 = N^2\hbar^2 \, . \end{aligned} \tag{9.98}$$

Inserting this into (9.97) gives the term values $F(J) = E_{\text{rot}}(J)/hc$ of the rotational levels

$$\boxed{F(J,\Omega) = B_e\left[J(J+1) - \Omega^2\right] + A\Omega^2} \tag{9.99}$$

with the two rotational constants

$$A = \frac{\hbar}{4\pi c I_1} \gg B_e = \frac{\hbar}{4\pi c I_2} \, . \tag{9.100}$$

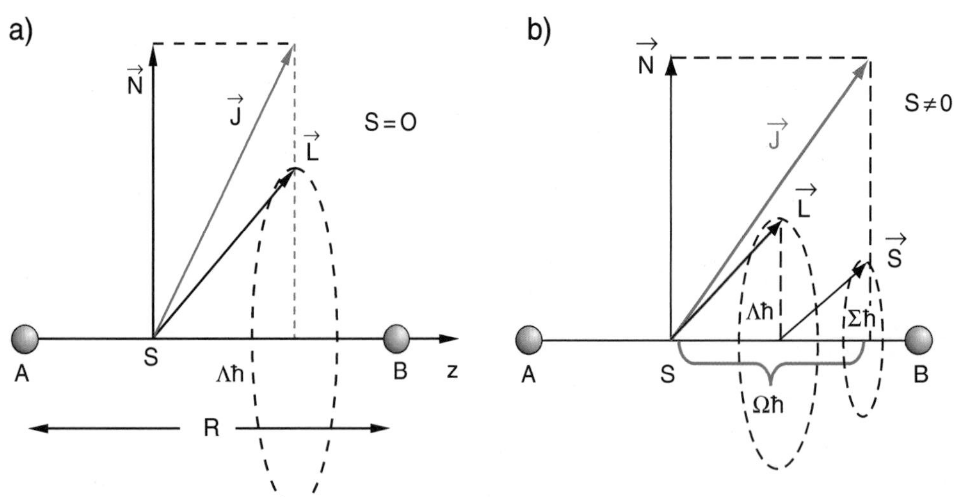


**Fig. 9.46.** Angular momenta of the rotating molecule including the electronic contribution

The term $A\Omega^2$, which does not depend on $J$, is generally added to the electronic energy $T_e$ of the molecular state, since it is constant for all rotational levels of a given electronic state with quantum number $\Omega$. It is therefore also not influenced by the centrifugal distortion.

The ground states of the majority of diatomic molecules are $^1\Sigma$-states with $\Lambda = \Omega = 0$. For these cases (9.99) is identical to (9.85).

### 9.5.5 Vibrations of Diatomic Molecules

For a nonrotating molecule, the rotational quantum number $J$ in (9.81a) is zero. Around the minimum the potential energy $E_{pot} \sim 1/2 \cdot k_r(R - R_e)^2$ can be approximated by a parabola. If we choose the minimum of the potential energy at $R_e = 0$ it becomes

$$E_{pot} \sim 1/2 k_r \cdot R^2 = 1/2 \omega^2 M R^2 \text{ with } \omega = \sqrt{k_r/M}.$$

Inserting this into equ. (9.81a) and replacing the function S by $\psi_{vib} = R \cdot S(R)$ we obtain

$$\frac{d_2\,\psi_{vib}}{dR^2} + \frac{2M}{\hbar^2}\left(E - \frac{1}{2}\omega^2 M\right)\psi_{vib} = 0.$$

The solutions $\psi_{vib}(R)$ of this equation are the vibrational wave functions. They can be written as the product of an exponential function and a Hermite polynom $H_v(R)$ [9.11]

$$\psi_{vib}(R) = \exp[-(M\omega/\hbar)R^2] \cdot H_v(R) \qquad (9.101)$$

where v is the vibrational quantum number (see Sect. 4.2.5 and Table 4.1). For a parabolic potential, the vibrating molecule is a harmonic oscillator, which has been already treated in Sect. 4.2.5. The result obtained there was the quantization of the energy levels.

The energy levels of the harmonic oscillator

$$E(v) = \left(v + \tfrac{1}{2}\right)\hbar\omega \qquad (9.102)$$

depend on the integer vibrational quantum number $v = 0, 1, 2, \ldots$.

They are equally spaced by $\Delta E = \hbar\omega$. The frequency $\omega = \sqrt{k_r/M}$ depends on the constant $k_r = (d^2 E_{pot}/dR^2)_{R_e}$ in the parabolic potential and on the reduced mass $M$ of the molecule. The lowest vibrational level is not $E = 0$ but $E = \frac{1}{2}\hbar\omega$. Some of these vibrational eigenfunctions of the harmonic oscillator are compiled in Table 4.1 and are illustrated in Fig. 4.20.

Although the real potential of a diatomic molecule can be well approximated by a parabolic potential in the vicinity of the potential minimum at $R = R_e$, it deviates more and more for larger $|R - R_e|$ (see Fig. 9.39). This figure also illustrates that the Morse potential is a much better approximation. Inserting the Morse potential

$$E_{pot}(R) = E_B\left[1 - e^{-a(R-R_e)}\right]^2 \qquad (9.103)$$

into the radial part (9.81a) of the Schrödinger equation allows its exact analytical solution (see Problem 9.5). The energy eigenvalues are now:

$$E_{vib}(v) = \hbar\omega_0\left(v + \frac{1}{2}\right) - \frac{\hbar^2\omega_0^2}{4E_D}\left(v + \frac{1}{2}\right)^2 \qquad (9.104)$$

with energy separations

$$\begin{aligned}\Delta E(v) &= E_{vib}(v+1) - E_{vib}(v) \qquad (9.105a)\\ &= \hbar\omega\left[1 - \frac{\hbar\omega}{2E_B}(v+1)\right],\end{aligned}$$

where $E_B$ is the binding energy of the rigid molecule. The vibrational levels are no longer equidistant but separations decrease with increasing vibrational quantum number $v$, in accordance with experimental observations.

The term-values $T_v = E_v/hc$ are

$$T_{vib}(v) = \omega_e\left(v + \tfrac{1}{2}\right) - \omega_e x_e\left(v + \tfrac{1}{2}\right)^2 \qquad (9.105b)$$

with the vibrational constants

$$\omega_e = \frac{\omega_0}{2\pi c}, \quad \omega_e x_e = \frac{\hbar\omega_0^2}{8\pi c E_B} = \omega_e^2\frac{hc}{4E_B}. \qquad (9.105c)$$

The vibrational frequency

$$\omega_0 = a\sqrt{2E_B/M} \qquad (9.106)$$

corresponds to that of a classical oscillator with the restoring force constant $k_r = 2a^2 E_B$. From measurements of $k_r$ (for instance from the centrifugal distortion of rotational levels) and the binding energy $E_B$ the constant $a$ in the Morse potential can be determined.

With the more general expansion of the potential

$$E_{pot}(R) = \sum_n \frac{1}{n!}\left(\frac{\partial^n E_{pot}}{\partial R^n}\right)_{R_e}(R - R_e)^n \qquad (9.107)$$

the Schrödinger equation can only be solved numerically. We will, however, see in Sect. 9.5.7 that the real potential can be very accurately determined from the measured term values of the rotational and vibrational levels.

**Note:**

- The distance between vibrational levels decreases with increasing $v$, but stays finite up to the dissociation energy. This means that only a finite number of vibrational levels fit into the potential well of a bound molecular state. This is in contrast to the infinite number of electronic states in an atom such as the H atom. Here the distance between Rydberg levels converges with $n \to \infty$ towards zero (see (3.88)). This different behavior stems from the different radial dependence of the potentials in the two cases.
- One has to distinguish between the experimentally determined dissociation energy $E_D^{exp}$, where the molecule is dissociated from its lowest vibration level, and the binding energy $E_B$ of the potential well, which is measured from the minimum of the potential (Fig. 9.42). The difference is

$$E_D^{exp} = E_B - \tfrac{1}{2}\hbar\omega \, .$$

### 9.5.6 Interaction Between Rotation and Vibration

Up to now we have looked at the rotation of a nonvibrating molecule and the vibration of a nonrotating molecule. Of course a real molecule can simultaneously rotate and vibrate. Since the vibrational frequency is higher than the rotational frequency by one to two orders of magnitude, the molecule undergoes many vibrations (typically 5–100) during one rotational period (Fig. 9.47). This means that the nuclear distance changes periodically during one full rotation.

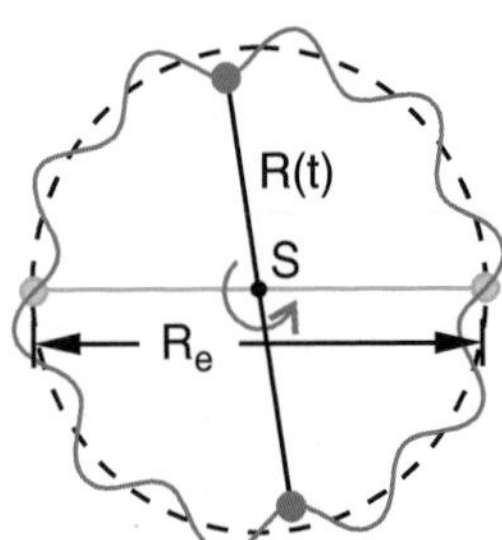


**Fig. 9.47.** Vibrating rotor

**EXAMPLES**

1. For the $H_2$ molecule, $\omega_e = 4401 \text{cm}^{-1} \Rightarrow T_{vib} = 1(c \cdot \omega_e) = 7.7 \times 10^{-15}\text{s}$, while with $B_e = 60.8\,\text{cm}^{-1} \Rightarrow T_{rot} = \left(2c\, B_e \cdot \sqrt{J(J+1)}\right)^{-1} = 2.7 \times 10^{-13}/\sqrt{J(J+1)}\,\text{s}$.
2. For the $Na_2$ molecule, $\omega_e = 150\,\text{cm}^{-1} \Rightarrow T_{vib} = 2.2 \times 10^{-13}\,\text{s}$ and $B_e = 0.15\,\text{cm}^{-1} \Rightarrow T_{rot} = 1.1 \times 10^{-10}/\sqrt{J(J+1)}\,\text{s}$.

Since the total angular momentum $J = I \cdot \omega$ of a freely rotating molecule has to be constant in time, but the moment of inertia $I$ periodically changes, the rotational frequency $\omega$ has to change accordingly with a period $T_{vib}$. Therefore the rotational energy

$$E_{rot} = \frac{J(J+1)\hbar^2}{2mR^2}$$

also varies periodically with a period $T_{vib}$.

> Because the total energy $E = E_{rot} + E_{vib} + E_{pot}$ has to be constant, there is a periodic exchange of rotational, vibrational and potential energy in the vibrating rotor.

In a classical picture this energy flow between potential energy and kinetic energy of vibration and rotation can be illustrated by Fig. 9.48.

The rotational energy obtained from the time-independent Schrödinger equation (9.81a), is the time average over a vibrational period. This time average can be calculated as follows:

The probability to find the nuclei within the interval $\text{d}R$ around the distance $R$ is

$$P(R)\text{d}R = |\psi_{vib}(R)|^2 \text{d}R \, .$$

The quantum mechanical expectation values of $R$ and $1/R^2$ are then

$$\langle R \rangle = \int \psi_{vib}^* R \psi_{vib} \text{d}R \text{ , and} \qquad (9.108)$$

$$\left\langle 1/R^2 \right\rangle = \int \psi_{vib}^* \frac{1}{R^2} \psi_{vib} \text{d}R \, .$$

This gives the mean rotational energy, averaged over one vibrational period

$$\langle E_{rot}(v) \rangle = \frac{J(J+1)\hbar^2}{2M} \int \psi_{vib}^*(v) \frac{1}{R^2} \psi_{vib}(v)\,\text{d}R \, . \qquad (9.109)$$

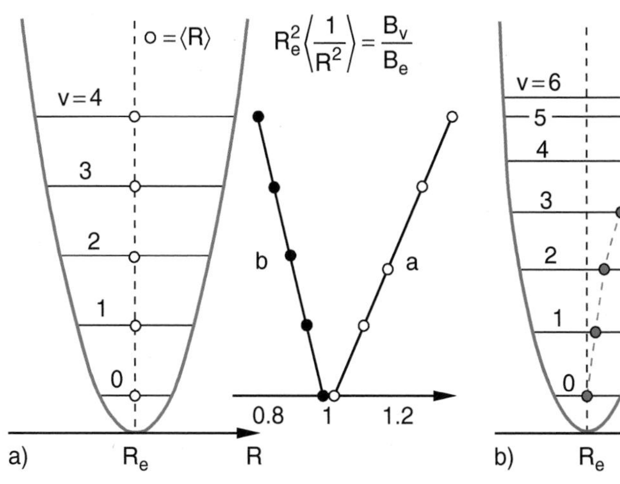


**Fig. 9.48.** Mean internuclear distance $\langle R\rangle$ and rotational constant $B_v \propto \left\langle 1/R^2\right\rangle$ for the harmonic (**a**) and anharmonic (**b**) potential

**Note:**

Even for a harmonic potential the expectation value of $1/R^2$ depends on the vibrational quantum number $v$. It increases with $v$ although $\langle R\rangle$ is independent of $v$ and always equal to $R_e$ (Fig. 9.48). Therefore, the rotational constant $B_v$ of the rotating harmonic oscillator also depends on $v$. For the more realistic anharmonic potential, both $\langle R\rangle$ as well as $\left\langle 1/R^2\right\rangle$ change with $v$. While $\langle R\rangle$ increases $\left\langle 1/R^2\right\rangle$ decreases with increasing $v$.

In order to express the rotational term values by a rotational constant similar to (9.86) or (9.95) we introduce, instead of $B_e$, the rotational constant

$$B_v = \frac{\hbar}{4\pi cM}\int \psi^*_{\text{vib}}(v,R)\frac{1}{R^2}\psi_{\text{vib}}(v,R)\mathrm{d}R \tag{9.110a}$$

averaged over the vibrational motion. It depends on the vibrational quantum number $v$.

For a Morse potential we then obtain

$$B_v = B_e - \alpha_e\left(v+\tfrac{1}{2}\right) \tag{9.111a}$$

where $\alpha_e \ll B_e$. In a similar way an average centrifugal constant

$$D_v = \frac{\hbar^3}{4\pi ckM^2}\int \psi^*_{\text{vib}}\frac{1}{R^6}\psi_{\text{vib}}\mathrm{d}R \tag{9.110b}$$

can be defined, which is related to $D_e$ by

$$D_v = D_e + \beta_e\left(v+\tfrac{1}{2}\right) \quad \text{with} \quad \beta_e \ll D_e\,. \tag{9.111b}$$

For a general potential, higher order constants have to be introduced and one writes

$$B_v = \beta_e - \alpha_e\left(v+\tfrac{1}{2}\right) + \gamma_e\left(v+\tfrac{1}{2}\right)^2 + \ldots \tag{9.112a}$$

$$D_v = D_e + \beta_e\left(v+\tfrac{1}{2}\right) + \delta_e\left(v+\tfrac{1}{2}\right)^2 + \ldots\,. \tag{9.112b}$$

The term value of a rotational-vibrational level can then be expressed as the power series

$$\begin{aligned} T(v,J) = T_e &+ \left[\omega_e\left(v+\tfrac{1}{2}\right) - \omega_e x_e\left(v+\tfrac{1}{2}\right)^2\right.\\ &\left. + \omega_e y_e\left(v+\tfrac{1}{2}\right)^3 + \omega_e z_e\left(v+\tfrac{1}{2}\right)^4 + \ldots\right]\\ &+ \left[B_v J(J+1) - D_v J^2(J+1)^2\right.\\ &\left. + H_v J^3(J+1)^3 \mp \ldots\right]\,. \end{aligned} \tag{9.113a}$$

For a Morse potential this series is reduced to

$$\begin{aligned} T^{\text{Morse}}(v,J) = T_e + \omega_e\left(v+\tfrac{1}{2}\right)\\ - \omega_e x_e\left(v+\tfrac{1}{2}\right)^2 + B_v J(J+1)\\ - D_v J^2(J+1)^2 \end{aligned} \tag{9.113b}$$

where only five constants describe the energies of all levels $(v, J)$ up to energies where the Morse potential is still a good approximation.

### 9.5.7 The Dunham Expansion

In order to also reproduce the rotational-vibrational term values $T(v, J)$ of a rotating molecule for a more general potential (9.107)

$$E_{pot}(R) = \sum_n a_n (R - R_e)^n , \tag{9.114}$$

with

$$a_n = \frac{1}{n!}\left(\frac{\partial^n E_{pot}}{\partial R^n}\right)_{R_e} .$$

Dunham introduced the expansion

$$T(v, J) = \sum_i \sum_k Y_{ik}\left(v + \tfrac{1}{2}\right)^i \left[J \cdot (J+1) - \Lambda^2\right]^k \tag{9.115}$$

where the Dunham coefficients $Y_{ik}$ are fit parameters chosen such that the term values $T(v, J)$ best reproduce the measured term values of rotational levels in vibrational states of the molecule.

With (9.115) the energies of all vibrational-rotational levels of a molecule can be described by a set of molecular constants. These constants are related to the coefficients in the expansion (9.113a) by the relations

$$\begin{aligned} &Y_{10} \approx \omega_e , \quad Y_{20} \approx -\omega_e x_e , \; Y_{30} \approx \omega_e y_e \\ &Y_{01} \approx B_e , \quad Y_{02} \approx D_e , \quad\;\; Y_{03} \approx H_e \\ &Y_{11} \approx -\alpha_e , \; Y_{12} \approx \beta_e , \quad\;\; Y_{21} \approx \gamma_e \end{aligned} \tag{9.116}$$

and also to the coefficients $a_n$ in the general potential expansion (9.114) [9.11].

### 9.5.8 Rotational Barrier

The effective potential for a rotating molecule (see (9.80))

$$E_{pot}^{eff}(R) = E_{pot}^{(v)}(R) + \frac{J(J+1)\hbar^2}{2MR^2} \tag{9.117}$$

includes, besides the potential $E_{pot}(R)$ of the nonrotating molecule, a centrifugal term that depends on the rotational quantum number $J$ and falls off with $R$ as $1/R^2$. For a bound electronic state this leads to a maximum of $E_{pot}^{eff}(R)$ at a distance $R_m$, which can be obtained by setting the first derivative of (9.117) to zero. This distance

$$R_m = \left[\frac{J(J+1)\hbar^2}{M(dE_{pot}/dR)}\right]^{1/3} \tag{9.118}$$

depends on the rotational quantum number $J$ and on the slope of the rotationless potential. (Fig. 9.49)

The minimum of the potential is shifted by the rotation of the molecule from $R_e$ to larger distances and the dissociation energy becomes smaller.

Energy levels $E(v, J)$ above the dissociation energy $E_D$ can be still stable, if they are below the maximum of the centrifugal potential barrier. However, due to the tunnel effect (Sect. 4.2.3) molecules in these levels can dissociate by tunneling through the barrier (Fig. 9.50). This effect is called predissociation by tunneling. The tunnel probability depends exponentially on the width of the barrier and on the energy gap between the maximum of the barrier and the level energy. (see Sect. 4.2.3)

The predissociation rate can be determined by measuring the width $\delta E = h/\tau$ of levels with a lifetime $\tau$. If the predissociation rate is large compared to the radiative decay of a level, the lifetime $\tau$ is mainly determined by predissociation. Measuring $\tau(v)$ for all levels

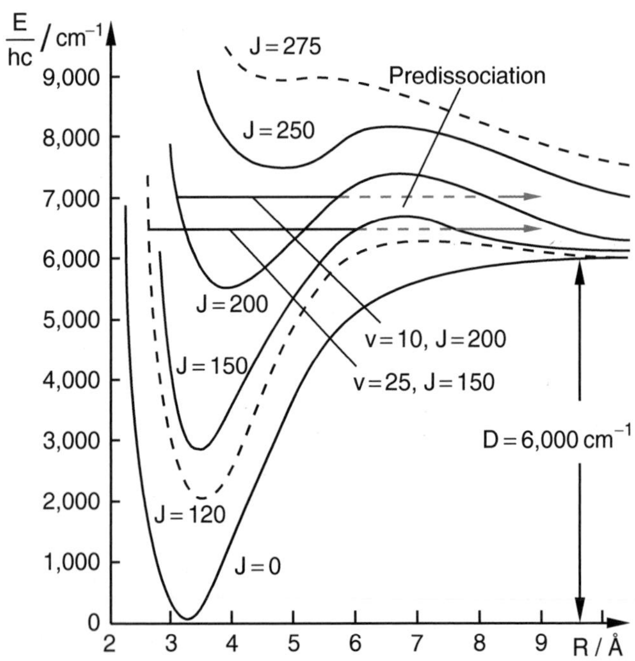


**Fig. 9.49.** Effective potential curves of the rotating $Na_2$ molecule for different rotational quantum number $J$

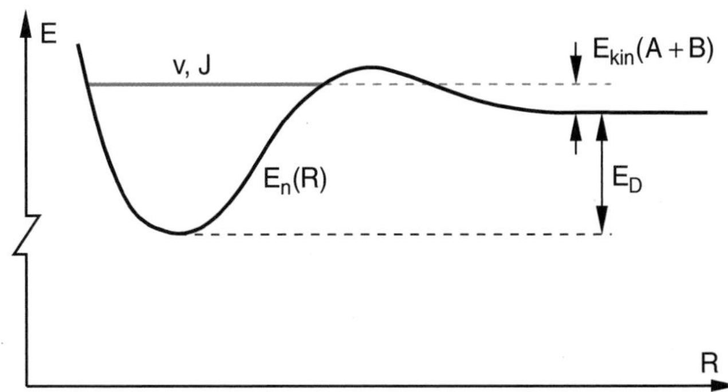


**Fig. 9.50.** Predissociation of a molecule through the rotational barrier

above the dissociation limit gives information on the form and heights of the potential barrier.

The dissociating fragments have a kinetic energy

$$E_{\text{kin}} = E(v, J) - E_{\text{D}}(J = 0) ,$$

which is shared by the two fragments according to their masses.

## 9.6 Spectra of Diatomic Molecules

When a molecule undergoes a transition

$$E_i(n_i, \Lambda_i, v_i, J_i) \leftrightarrow E_k(n_k, \Lambda_k, v_k, J_k)$$

between two molecular states $|i\rangle$ and $|k\rangle$, electromagnetic radiation can be absorbed or emitted with a frequency $\nu = \Delta E/h$. Whether this transition really occurs depends on its transition probability, which is proportional to the absolute square of the dipole matrix element $M_{ik}$ (see Sect. 7.1). The relative intensities of spectral lines can therefore be determined if the matrix elements of the transitions can be calculated. Because of the larger variety of molecular states, with energies depending on the electronic, the rotational and vibrational structure of the molecule, the matrix elements of molecules are more complicated than those of atoms. In this section we will discuss their structure and the molecular spectra derived from them.

For spontaneous emission (fluorescence spectra) the emission probability of a transition $|i\rangle \rightarrow |k\rangle$ is given by the Einstein coefficient $A_{ik}$. According to (7.17) $A_{ik}$ is related to the transition dipole matrix element $M_{ik}$ by

$$A_{ik} = \frac{2}{3} \frac{\omega_{ik}^3}{\varepsilon_0 c^3 h} |\boldsymbol{M}_{ik}|^2 . \tag{9.119a}$$

For the absorption or stimulated emission of radiation the transition probability $P_{ik} = B_{ik} w(\nu_{ik})$ is proportional to the spectral energy density $w(\nu)$ of the radiation field. In Sect. 7.1.3 it was shown that $P_{ik}$ is given by

$$P_{ik} = \frac{\pi}{h^2} E_0^2 \Big| \int \psi_k^* \boldsymbol{\varepsilon} \cdot \boldsymbol{p} \psi_i \, \mathrm{d}\tau \Big|^2 \tag{9.119b}$$

where $\boldsymbol{\varepsilon} = \boldsymbol{E}/|\boldsymbol{E}|$ is the unit vector in the direction of the electric field $\boldsymbol{E}$ of the electromagnetic wave, incident onto the molecules. The transition probability therefore depends on the scalar product $\boldsymbol{\varepsilon} \cdot \boldsymbol{p}$ of electric field vector and electric transition dipole $\boldsymbol{p} = e \cdot \boldsymbol{r}$ of the molecule.

### 9.6.1 Transition Matrix Elements

The dipole matrix element for a transition between two molecular states of a diatomic molecule with wave functions $\psi_i$ and $\psi_k$ is

$$\boldsymbol{M}_{ik} = \iint \psi_i^* \boldsymbol{p} \psi_k \mathrm{d}\tau_{\text{el}} \, \mathrm{d}\tau_{\text{N}} . \tag{9.119c}$$

The integration extends over all $3(Z_A + Z_B)$ electronic coordinates and over the six nuclear coordinates. Often only one of the electrons is involved in the transition. In this case the integration over $\mathrm{d}\tau_{\text{el}}$ only needs to be performed over the coordinates of this electron. The vector $\boldsymbol{p}$ is the dipole operator, which depends on the coordinates of the electrons, involved in the transition and on the nuclear coordinates. In Fig. 9.51 it can be seen that

$$\boldsymbol{p} = -e \sum_i \boldsymbol{r}_i + e(Z_A \boldsymbol{R}_A + Z_B \boldsymbol{R}_B) = \boldsymbol{p}_{\text{el}} + \boldsymbol{p}_{\text{N}} \tag{9.120}$$

where $\boldsymbol{p}_{\text{el}}$ is the contribution of the electrons and $\boldsymbol{p}_{\text{N}}$ that of the nuclei.

**Note** that for homonuclear molecules $Z_A = Z_B$ but $\boldsymbol{R}_A = -\boldsymbol{R}_B$. Therefore $\boldsymbol{p}_{\text{N}} = 0$!

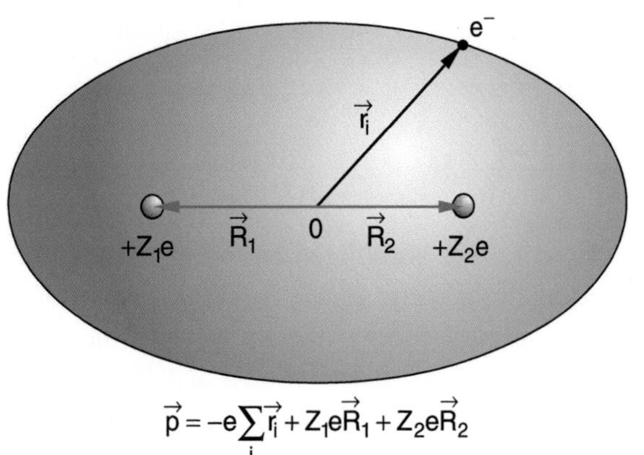


**Fig. 9.51.** Illustration of nuclear and electronic contributions to the molecular dipole moment

Within the adiabatic approximation we can separate the total wave function $\psi(r, R)$ into the product

$$\psi(\boldsymbol{r}, \boldsymbol{R}) = \Phi(\boldsymbol{r}, \boldsymbol{R}) \times \chi_{\mathrm{N}}(\boldsymbol{R}) \tag{9.121}$$

of electronic wave function $\Phi(\boldsymbol{r}, R)$ of the rigid molecule at a fixed nuclear distance $R$ and the nuclear wave function $\chi(\boldsymbol{R})$; which only depends on the nuclear coordinates. Inserting (9.120, 9.121) into (9.119) the matrix elements is written as

$$\boldsymbol{M}_{ik} = \iint \Phi_i^* \chi_{\mathrm{N},i}^* (\boldsymbol{p}_{\mathrm{el}} + \boldsymbol{p}_{\mathrm{N}}) \Phi_k \chi_{\mathrm{N},k}^* \, \mathrm{d}\tau_{\mathrm{el}} \, \mathrm{d}\tau_{\mathrm{N}} \,. \tag{9.122a}$$

Rearranging the different terms gives

$$\begin{aligned} \boldsymbol{M}_{ik} = & \int \chi_i^* \left[ \int \Phi_i^* \boldsymbol{p}_{\mathrm{el}} \Phi_k \, \mathrm{d}\tau_{\mathrm{el}} \right] \chi_k \, \mathrm{d}\tau_{\mathrm{N}} \\ & + \int \chi_1^* \boldsymbol{p}_{\mathrm{N}} \left[ \int \Phi_i^* \Phi_k \, \mathrm{d}\tau_{\mathrm{el}} \right] \chi_k \, \mathrm{d}\tau_{\mathrm{N}} \,. \end{aligned} \tag{9.122b}$$

We distinguish between two different cases (Fig. 9.52):

- The two levels $|i\rangle$ and $|k\rangle$ belong to the same electronic state ($\Phi_i = \Phi_k$). This means that the dipole transition occurs between two vibrational-rotational levels in the same electronic state $\Phi_i$. In this case the first term in the sum (9.122b) is zero because the integrand $\boldsymbol{r}|\Phi_i|^2$ is an ungerade function of the electron coordinates $\boldsymbol{r} = \{x, y, z\}$. The integration from $-\infty$ to $+\infty$ therefore gives zero.

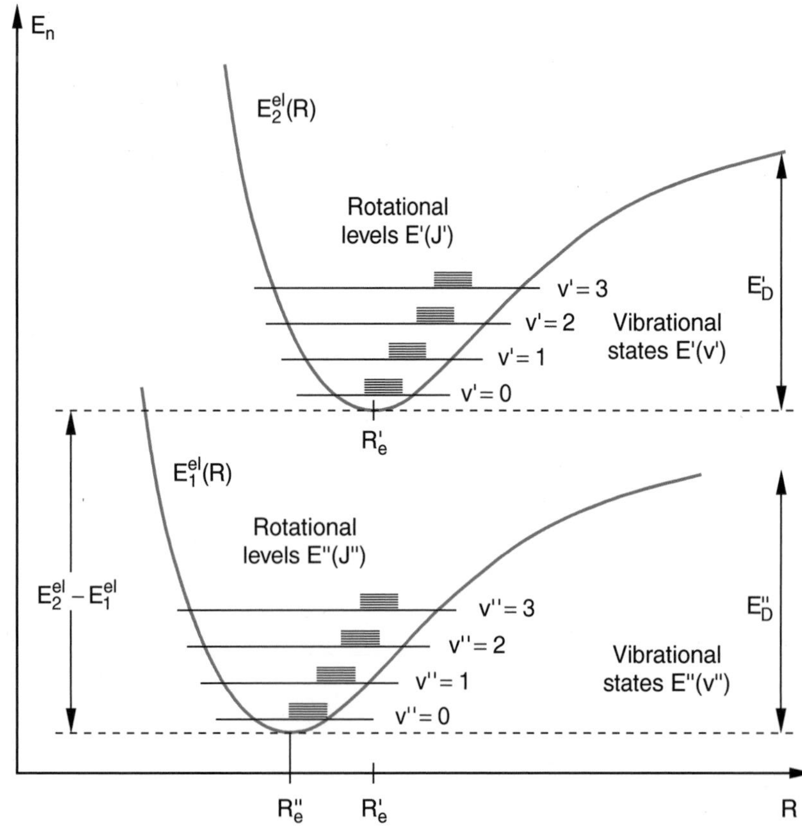


**Fig. 9.52.** Rotational and vibrational levels in two different electronic states of a diatomic molecule

Since the electronic wave functions $\Phi$ are orthonormal, i.e.,

$$\int \Phi_i^* \phi_k \, d\tau_{el} = \delta_{ik} \tag{9.123}$$

the integral over the electronic coordinates in the second term in the sum (9.122b) is equal to one. The matrix element then becomes

$$\boxed{M_{ik} = \int \chi_{i,N} \boldsymbol{p}_N \chi_{k,N} \, d\tau_N} \; . \tag{9.124}$$

The integrand solely depends on the nuclear coordinates, not on the electronic coordinates!

- Transitions between levels in two different electronic states. In this case the integral over the electronic coordinates in the second term of (9.122b) is zero because the $\Phi_i$, $\Phi_k$ are orthonormal. The second term is therefore zero and the matrix element becomes

$$\begin{aligned} M_{ik} &= \int \chi_i^* \left[ \int \Phi_i^* \boldsymbol{p}_{el} \Phi_k \, d\tau_{el} \right] \chi_k \, d\tau_N \\ &= \int \chi_i^* M_{ik}^{el}(R) \chi_k \, d\tau_N \; . \end{aligned} \tag{9.125}$$

We will now discuss both cases separately.

### 9.6.2 Vibrational-Rotational Transitions

All allowed transitions $(v_i, J_i) \leftrightarrow (v_k, J_k)$ between two rotational-vibrational levels in the same electronic state form for $v_i \neq v_k$ the vibrational-rotational spectrum of the molecule in the infrared spectral region between $\lambda = 2 - 20\,\mu\text{m}$. For $v_i = v_k$ we have pure rotational transitions between rotational levels within the same vibrational state, which form the rotational spectrum in the microwave region with wavelengths in the range 0.1–10cm.

The dipole matrix element for these transitions is according to (9.120) and (9.124)

$$M_{ik}^{\text{rot vib}} = e \int \chi_i^* (Z_A \boldsymbol{R}_A + Z_B \boldsymbol{R}_B) \chi_k \, d\tau_N \; . \tag{9.126}$$

For homonuclear diatomic molecules with $Z_A = Z_B$ and $M_A = M_B$ is $R_A = -R_B$ and therefore the integrand is zero $\Rightarrow M_{ik}^{\text{rot vib}} = 0$.

Homonuclear diatomic molecules have no dipole-allowed vibrational-rotational spectra. This means they do not absorb or emit radiation on transitions within the same electronic state. They may have very weak quadrupole transitions.

**Note:**

The molecules $N_2$ and $O_2$, which represent the major constituents of our atmosphere, cannot absorb the infrared radiation emitted by the earth. Other molecules, such as $CO_2$, $H_2O$, $NH_3$ and $CH_4$ do have an electric dipole moment and absorb infrared radiation on their numerous vibrational-rotational transitions. Although they are present in our atmosphere only in small concentrations they can seriously perturb the delicate energy balance between absorbed incident solar radiation and the energy radiated back into space by the earth. If their concentration is increased by only small amounts this can increase the temperature of the atmosphere at the earth's surface (greenhouse effect).

The structure of the vibration-rotation-spectrum and the pure rotation spectrum of heteronuclear molecules can be determined as follows.

Since the interaction potential between the two atoms is spherically symmetric, we choose spherical coordinates for the description of the nuclear wave function $\chi_N(R, \vartheta, \varphi)$.

With $R = |\boldsymbol{R}_A - \boldsymbol{R}_B|$ and $R_A/R_B = M_B/M_A$ (Figs. 9.43 and 9.51) and $\hat{p} = \boldsymbol{p}/|p|$ the dipole moment can be written as

$$\begin{aligned} \boldsymbol{p}_N &= \hat{\boldsymbol{p}} \cdot |p_N| = e \frac{M_B \cdot Z_A - M_A \cdot Z_B}{M_A + M_B} \cdot R \cdot \hat{\mathbf{p}} \\ &= p_o \cdot \hat{p} \text{ with } p_o = e \cdot R \cdot \frac{M_B Z_a - M_a Z_B}{M_A + M_B} \end{aligned} \tag{9.127}$$

The vector $\boldsymbol{p}_N$ of the electric dipole moment is defined in the coordinate system (x,y,z) of the molecular frame while the electric field vector of the electromagnetic wave is defined in the laboratory frame $(\boldsymbol{X};\boldsymbol{Y};\boldsymbol{Z})$.

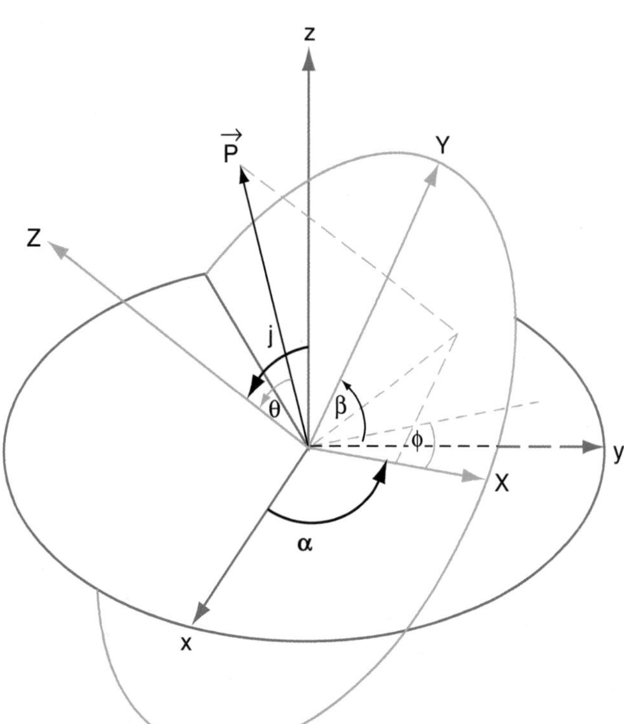

**Fig. 9.53.** Transformation from the molecular frame (x,y,z) into the space-fixed frame (X;Y;Z).

The transformation from the (x,y,z = system into the (***X***;***Y***;***Z*** = system can be achieved through three successive rotations around the axis (x,y,z) by angles α,ß,γ called the *Euler angles* (Fig. 9.53). While the orientation of ***p*** against the axis x,y,z can be described by the angles $\vartheta$ and $\varphi$ (Fig. 9.54), its orientation against the axis ***X***;***Y***;***Z*** is described by the angles $\theta$ and $\Phi$. Since for diatomic molecules $\boldsymbol{p}_N$ is directed along the molecular axis, which is the z-axis, the angle $\vartheta$ is zero and the

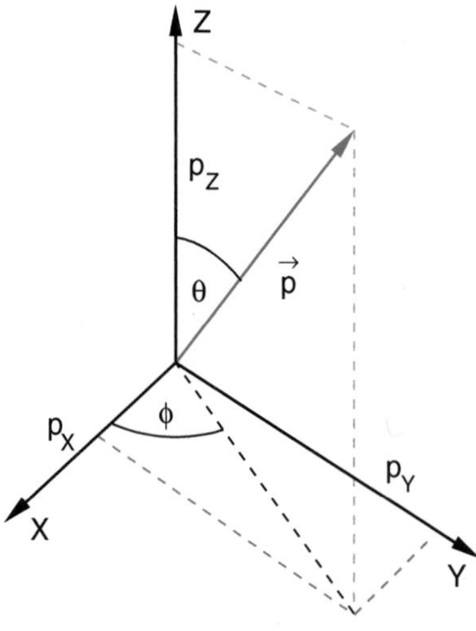

**Fig. 9.54.** Orientation of molecular dipole moment ***p*** in a space-fixed coordinate system

transformation is easy. The components of the dipole moment in the space fixed coordinate system are

$$\begin{aligned} p_X &= p_0 \cdot \sin\mathrm{d}\,\Theta \cos\Phi \\ p_Y &= p_0 \cdot \sin\Theta \sin\Phi \\ p_Z &= p_0 \cdot \cos\Theta\,. \end{aligned}$$

**Note:**

For polyatomic molecules the nuclear dipole moment $\boldsymbol{p}_\mathrm{N}$ can have any direction $(\vartheta,\varphi)$ in the molecular frame. The transformation into the space-fixed coordinate system now has to use the transformation matrix containing the Euler angles (Fig. 9.53).

If the interaction between rotation and vibration is sufficiently weak we can write $\chi_\mathrm{N}$ as the product

$$\chi_\mathrm{N}(R,\vartheta,\varphi) = \psi_\mathrm{vib}(R) Y_J^M(\vartheta,\varphi) \tag{9.128}$$

of the vibrational wave function $\psi_\mathrm{vib} = R \cdot S(R)$ in (9.102) and the rotational wave function $Y_J^M(\vartheta,\varphi)$ for a rotational level with angular momentum $J$ and its projection $M \cdot \hbar$ onto the quantization axis, which is a preferential direction in the laboratory coordinate system. For absorbing transition the quantization axis is, for instance, the direction of the incident electromagnetic wave, or the direction of its ***E***-vector.

Since the vibrational wavefunction $\psi_\mathrm{vib}$ depends solely on the internuclear distance $R$ while the spherical harmonics depends solely on the angles $\Theta$ and $\Phi$ in the space fixed system, we can write the transition dipole matrix element with $\mathrm{d}\tau_\mathrm{N} = R^2\mathrm{dR}\sin\vartheta\,\mathrm{d}\vartheta\,\mathrm{d}\varphi$ as a product of two integrals

$$\begin{aligned} \mathbf{M}_\mathrm{ik} &= \int \psi^*_\mathrm{vib,i} \mathrm{p}_0 \psi_{vib,k} \mathrm{d}R \\ &\quad\times \int Y_{Ji}{}^{\mathrm{Mi}} \hat{\boldsymbol{p}} \mathrm{Y}_\mathrm{Jk}{}^{\mathrm{Mk}} \sin\Theta\vartheta\,\mathrm{d}\Theta\,\mathrm{d}\Phi \end{aligned} \tag{9.129}$$

where we have used the abbreviation $\boldsymbol{p} = \mathrm{p}_0\hat{\boldsymbol{p}}$ and the fact, that the unit vector $\hat{\boldsymbol{p}} = \boldsymbol{p}_\mathrm{N}/|\boldsymbol{p}_\mathrm{N}| = \boldsymbol{p}_\mathrm{N}/\boldsymbol{p}_0$ is directed along the internuclear axis and has an orientation that depends solely on $\Theta$ and $\Phi$.

The transition probability per sec for an absorbing transition between the levels $|\mathrm{i}>$ and $|\mathrm{k}>$ in the same electronic state is according to (7.22)

$$\begin{aligned} &\mathrm{dP_{ik}/dt} = (\pi/\mathrm{h}^2)\cdot \mathrm{E_0}^2 |\textstyle\int \chi_\mathrm{i}^* \boldsymbol{\varepsilon}\cdot \mathbf{p}_\mathrm{N}\, \chi_\mathrm{k} \mathrm{d}\tau|^2 \\ &= (\pi\mathrm{e}^2/\mathrm{h}^2)\mathrm{E_0}^2\mathrm{p}_0^2 |\textstyle\int \chi_\mathrm{i}^* \boldsymbol{\varepsilon}\cdot \hat{\boldsymbol{p}} \chi_\mathrm{k} \mathrm{d}\tau|^2 \\ &= \mathrm{C}|\textstyle\int \chi_\mathrm{i}^* \boldsymbol{\varepsilon}\cdot \hat{\boldsymbol{p}} \boldsymbol{\chi}_\mathbf{k} \mathrm{d}\tau|^2 \text{ with } C = (\pi\mathrm{e}^2/h^2)E_0{}^2)\mathrm{p}_0^2 \end{aligned}$$

$$= C|\int \psi^*_{vib,i}\psi_{vib,k} dR|^2 \times |\int Y_{Ji}{}^{Mi} \boldsymbol{\varepsilon} \cdot \hat{\boldsymbol{p}} Y^{Mk}_{Jk} \sin\Theta d\Theta d\Phi|^2 = C|M^{vib}_{ik}|^2 \cdot |M_{ik}{}^{rot}|^2. \tag{9.130a}$$

The three components of the second integral in the transition matrix element (9.129) are

$$\begin{aligned}
&< J, M|p_X|J', M' > \\
&\quad = p_0 \int\int Y^*{}_{JM} \sin\theta \cdot \cos\Phi Y_{J'M'} \sin\theta d\theta d\Phi \\
&< J, M|p_Y|J'M' > \\
&\quad = p_0 \int\int Y^*{}_{JM} \sin\theta \cdot \sin\Phi Y_{J'M'} \sin\theta d\theta d\Phi \\
&< J, M|p_Z|J'M' > \\
&\quad = p_0 \int\int Y^*{}_{JM} \cos\theta Y_{J'M'} \cdot \sin\theta d\theta d\Phi
\end{aligned} \tag{9.130b}$$

The first factor in (9.130a) describes the vibrational transition $v_i \leftrightarrow v_k$. If the harmonic oscillator functions are used for the vibrational functions $\psi_{vib}(R)$ the calculations of the integral shows that the integral is zero, unless

$$\Delta v = v_i - v_k = 0 \quad \text{or} \quad \pm 1. \tag{9.131}$$

The + sign stands for absorbing, the minus sign for emitting transitions. Transitions with $\Delta v = 0$ are pure rotational transitions within the same vibrational level.

> This selection rule means that for the harmonic oscillator only transitions between neighboring vibrational levels are allowed.

For anharmonic potentials, such as the Morse potential, higher order transitions with $\Delta v = \pm 2, \pm 3, \ldots$ are also observed. Such *overtone-transitions* are, however, much weaker than the *fundamental transitions* with $\Delta v = \pm 1$.

The second integral in (9.129) describes the rotational transitions. It depends on the orientation of the molecular dipole moment $\boldsymbol{p}$ in space.

The amplitude of the radiation emitted into the direction $\boldsymbol{k}$ in space is proportional to the product $|\vec{k} \times \vec{p}|$. For absorbing transitions it is proportional to the scalar product $\boldsymbol{E} \cdot \boldsymbol{p}$ of electric field amplitude and molecular dipole moment $\boldsymbol{p}$. The intensity is proportional to the square of this amplitude.

With the orientation angles $\Theta$ and $\phi$ of $\hat{\boldsymbol{p}} = \boldsymbol{p}/|\boldsymbol{p}|$ against the space-fixed axis $X; Y; Z$ we obtain the relation (Fig. 9.54)

$$\hat{\boldsymbol{\varepsilon}} \cdot \boldsymbol{p} = p_0(\varepsilon_x \sin\Theta \cos\phi + \varepsilon_y \sin\Theta \sin\phi + \varepsilon_z \cos\Theta) \tag{9.132a}$$

where $\varepsilon_i$ is the $i$th component of $\hat{\boldsymbol{\varepsilon}} = \boldsymbol{E}/|\boldsymbol{E}|$ against the space fixed axis $i = X, Y, Z$. The angles can be expressed by the spherical surface harmonics $Y^M_J$:

$$\sqrt{\frac{4\pi}{3}} Y^0_1 = \cos\Theta\,; \quad \sqrt{\frac{8\pi}{3}} Y^{\pm 1}_1 = \mp \sin\Theta \cdot e^{\pm i\phi} \tag{9.132b}$$

which gives

$$\hat{\boldsymbol{\varepsilon}} \cdot \boldsymbol{p} = p\sqrt{\frac{4\pi}{3}} \left( \varepsilon_z Y^0_1 + \frac{-\varepsilon_x + i\varepsilon_y}{\sqrt{2}} Y^1_1 + \frac{\varepsilon_x + i\varepsilon_y}{\sqrt{2}} Y^{-1}_1 \right). \tag{9.132c}$$

Inserting this into the second integral in (9.129) and extracting the components of the space fixed unit vector $\boldsymbol{\varepsilon}$ out of the integral gives for the angular part of the transition probability integrals of the form

$$\int Y^{M_i}_{J_i} Y^{\Delta M}_1 Y^{M_k}_{J_k} \, d\Omega \quad \text{with} \quad \Delta M = 0, \pm 1$$

with the result that these integrals are always zero, except for $\Delta J = J_i - J_k = \pm 1$.

This selection rule is readily understandable, because the absorbed or emitted photon has the spin $s = \pm 1h$ and the total angular momentum of the system photon + molecule has to be conserved.

For the projection quantum number $M$ the selection rules are analogue to that for atoms:

$\Delta M = 0$ for linear polarization of the radiation and $\Delta M = \pm 1$ for circular polarization.

**Note:**

The angle $\vartheta$ is measured against the molecular axis in the molecular coordinate system, while $\Theta$ and $\phi$ are the angles between the molecular dipole moment and the space fixed quantization axis (see above).

In order to save indices in spectroscopic literature the upper state $(v_k, J_k)$ is always labeled with a prime as $(v', J')$, whereas the lower state $(v_i, J_i)$ is labeled with a double prime as $(v'', J'')$. Transitions with

$$\Delta J = J' - J'' = +1$$

are called ***R-transitions***, those with

$$\Delta J = J' - J'' = -1$$

are ***P-transitions***.

All allowed rotational transitions appear in the spectrum as absorption- or emission lines (Fig. 9.55). All rotational lines of a vibrational transition form a ***vibrational band***. Its rotational structure is given by the wavenumbers of all rotational lines

$$\begin{aligned}&\overline{\nu}(v', J \leftrightarrow v'', J'') \\ &= \overline{\nu}_0 + B'_v J'(J'+1) - D'_v J'^2(J'+1)^2 \\ &\quad - \left[B''_v J''(J''+1) - D''_v J''^2(J''+1)^2\right]\end{aligned} \tag{9.133}$$

where $\nu_0$ is the ***band origin***. It gives the position of a fictious $Q$-line with $J' = J'' = 0$. Since this line does not exist in rotational-vibrational spectras of diatomic molecules, there is a missing line at $\nu = \nu_0$ (Fig. 9.55).

Since the rotational constant $B_v = B_e - \alpha_e\left(v + \frac{1}{2}\right)$ generally decreases with increasing $v$ ($\alpha_e > 0$ for most molecules) it follows that for molecules with $B'_e \approx B''_e$ on transitions with $v' > v''$ $B'_v < B''_v$. Plotting $\nu(J = J'')$ for $P$- and $R$ transitions as a function of $\nu$ gives the *Fortrat-diagram* shown in Fig. 9.56. The $R$-lines are on the high frequency side of the band origin $\nu_0$ while the $P$ lines are on the low frequency side. In Fig. 9.57 the vibration-rotation spectrum of HCl is shown with the $P$- and $R$-branch. The lines are split into two components, because the absorbing gas was a mixture of the two isotopomers of HCl with the two atomic isotopes $^{35}$Cl and $^{37}$Cl. Since the rotational and vibrational constants depend on the reduced mass $M$ the lines of different isotopomers are shifted against each other.

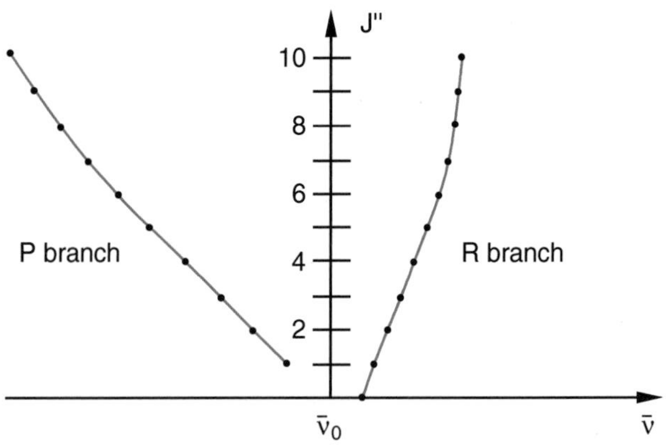


**Fig. 9.56.** Fortrat diagram of the $P$- and $R$-branch of vibrational-rotational transitions

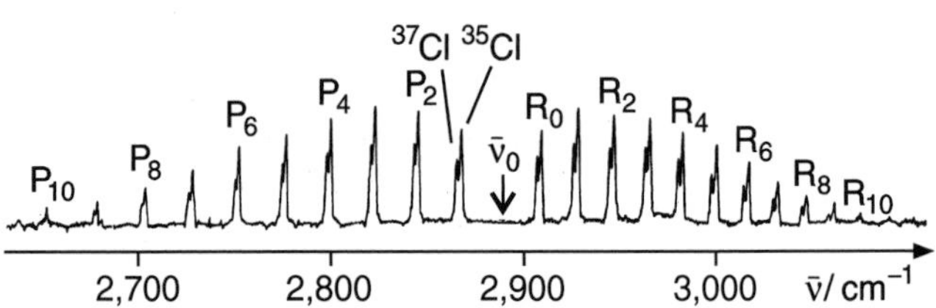


**Fig. 9.57.** Vibrational-rotational absorption of the H$^{35}$Cl and H$^{37}$Cl isotopomers in the infrared region between $\lambda$ = 3.3–3.7 μm

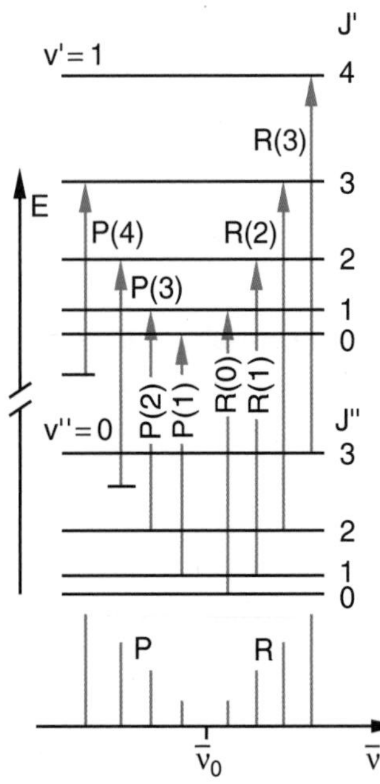


**Fig. 9.55.** $P$ and $R$ rotational transitions between the vibrational levels $v'' = 0$ and $v' = 1$

### 9.6.3 The Structure of Electronic Transitions

We will now evaluate the matrix element (9.125) for electronic transitions. The electronic part $M^{el}_{ik}(R)$ depends on the internuclear distance $R$, because the electronic wave functions $\Phi$ depend parametrically on $R$. In many cases the dependence on $R$ is weak and we can expand $M^{el}_{ik}(R)$ into a Taylor series

$$\boldsymbol{M}^{el}_{ik}(R) = \boldsymbol{M}_{ik}(R_e) + \left(\frac{d\boldsymbol{M}^{el}_{ik}}{R}\right)_{R_e}(R - R_e) + \ldots . \tag{9.134}$$

where the electronic part of the matrix element (9.122) is

$$\boldsymbol{M}_{ik}^{\mathrm{el}}(R) = \int \phi_i^* \vec{p}_{\mathrm{el}} \phi_k d\tau_{\mathrm{el}}$$

In a first approximation only the first term, independent of $R$, is considered, which can be regarded as an average of $M_{ik}(R)$ over the range of $R$-values covered by the vibrating molecule. In this case the constant $M_{ik}(R_{\mathrm{e}})$ can be put before the integral over the nuclear coordinates. Using the normalized nuclear wave functions $\chi_{\mathrm{N}} = \psi_{\mathrm{vib}} \cdot Y(\vartheta, \varphi)$ the matrix element becomes with

$$\boldsymbol{M}_{ik}^{\mathrm{el}} = \int \phi_i^* \vec{p}_{\mathrm{el}} \phi_k d\tau_{\mathrm{el}} \tag{9.135a}$$

$$\boldsymbol{M}_{ik} = \boldsymbol{M}_{ik}^{\mathrm{el}}(R_{\mathrm{e}}) \int \psi_{\mathrm{vib}}^*(v_i) \psi_{\mathrm{vib}}(v_k)\, \mathrm{d}R \cdot \int Y_{J_i}^{M_i} \hat{\boldsymbol{p}}_{\mathrm{el}}(\vec{r}) Y_{J_k}^{M_k} \sin\vartheta\, \mathrm{d}\vartheta\, \mathrm{d}\varphi \tag{9.135b}$$

where $M\hbar$ is the projection of the rotational angular momentum $\boldsymbol{J}$ onto a selected axis (for instance, in the direction of the $\boldsymbol{E}$-vector or the $\boldsymbol{k}$-vector of the incident electromagnetic wave for absorbing transitions, or in the direction from the emitting molecule to the observer for fluorescent transitions). Different from the nuclear part of the molecular dipole moment in Eq. (9.124), which is directed along the internuclear axis, the electronic part of the dipole moment can have any direction in the molecular coordinate system. The transformation to the space fixed system is therefore more complicated and has to use a transformation matrix, which contains the Euler angles.

**Note:**

This approximation of an electric transition dipole moment independent of $R$ is, for many molecules with a strong dependence $M_{ik}^{\mathrm{el}}(R)$, too crude (Fig. 9.58). In such cases also the second term in the expansion (9.134) has to be taken into account.

Since the probability of spontaneous emission is proportional to the square $|M_{ik}|^2$, the intensity of a spectral emission line

$$I(n_i, v_i, J_i \leftrightarrow n_k, v_k, J_k) \propto |M_{ik}^{\mathrm{el}}|^2 \cdot FCF(v_i, v_k) \cdot HL(J_i, J_k) \tag{9.136}$$

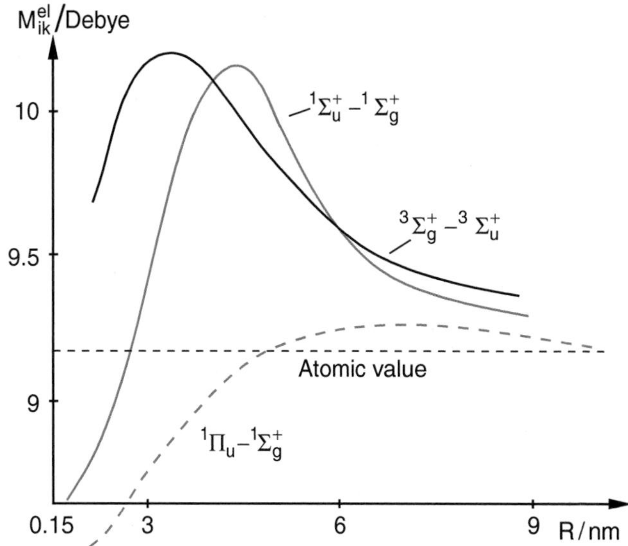


**Fig. 9.58.** Dependence of electronic transition dipole on internuclear distance $R$ for several transitions of the $Na_2$ molecule

is determined by three factors.

The electronic part $|M_{ik}^{\mathrm{el}}|^2$ gives the probability of an electron jump from the electronic state $|i\rangle$ to $|k\rangle$. It depends on the overlap of the electronic wave functions $\Phi_i$ and $\Phi_k$ and their symmetries.

The Franck–Condon factor

$$FCF(v_i, v_k) = \left| \int \psi_{\mathrm{vib}}(v_i) \cdot \psi_{\mathrm{vib}}(v_k)\, \mathrm{d}R \right|^2 \tag{9.137}$$

is determined by the square of the overlap integral of the vibrational wave functions $\psi_{\mathrm{vib}}(v_i)$ and $\psi_{\mathrm{vib}}(v_k)$ in the upper and lower electronic state.

The Hönl–London factor

$$HL(J_i, J_k) = \left| \int Y_{J_i}^{M_i} \hat{\boldsymbol{p}}_{\mathrm{el}} Y_{J_k}^{M_k} \sin\vartheta\, \mathrm{d}\vartheta\, \mathrm{d}\varphi \right|^2 \tag{9.138}$$

depends on the rotational angular momenta and their orientation in space. This factor determines the spatial distribution of the emitted radiation.

An electric dipole transition in fluorescence can only take place if none of these three factors is zero.

The probability of *absorbing transitions* depends according to (9.119b) on the scalar product of the electric field vector $\boldsymbol{E}$ and the dipole moment $\boldsymbol{p}$

$$P_{ik} \propto |\boldsymbol{E} \cdot \boldsymbol{M}_{ik}|^2 .$$

Since only the last factor in (9.136) depends on the orientation of the molecule in space, i.e., the direction of $M_{ik}$ against the electric field vector $\boldsymbol{E}$, only the Hönl–London factor differs for spontaneous emission and absorbing transitions. For the intensity $I = \varepsilon_0 E^2$ of the incident electromagnetic wave we obtain with $\boldsymbol{E} = \boldsymbol{\varepsilon} \cdot |\boldsymbol{E}|$ the transition probability

$$\mathcal{P}_{ik} = \varepsilon_0 E^2 \cdot |M_{ik}^{\mathrm{el}}(R_{\mathrm{e}})|^2 \times \left| \int \psi_{\mathrm{vib}}^{v_i} \cdot \psi_{\mathrm{vib}}^{v_k} \cdot \mathrm{d}R \right|^2 \times \left| \int Y_{J_i}^{M_i} \hat{\boldsymbol{\varepsilon}} \cdot \hat{\boldsymbol{p}} Y_{J_k}^{M_k} \sin\vartheta \, \mathrm{d}\vartheta \, \mathrm{d}\varphi \right|^2 . \quad (9.139)$$

### a) The General Structure of Electronic Transitions

Molecular electronic spectra have structures as shown in (Fig. 9.59).

All allowed transitions $J_i'' \longleftrightarrow J_k'$ between the rotational levels $J_k'$ of a given vibrational level $v'$ in the upper electronic state and $J_i''$ of $v''$ in the lower electronic state form a *band*. In absorption or fluorescence spectra such a band consists of many rotational lines.

Transitions with $\Delta J = 0$ form the **Q-*branch***, those with $\Delta J = J_k' - J_i'' = +1$ the **R-*branch*** and with $\Delta J = -1$ the **P-*branch***. $Q$-branches are only present in transitions where the electronic angular momentum changes by $1\hbar$, (e.g., for $\Sigma \leftrightarrow \Pi$ transitions) in order to compensate for the spin of the absorbed or emitted photon. Electronic transitions with $\Delta\Lambda = 0$ (e.g., between two $\Sigma$-states) have only $P$ and $R$ branches.

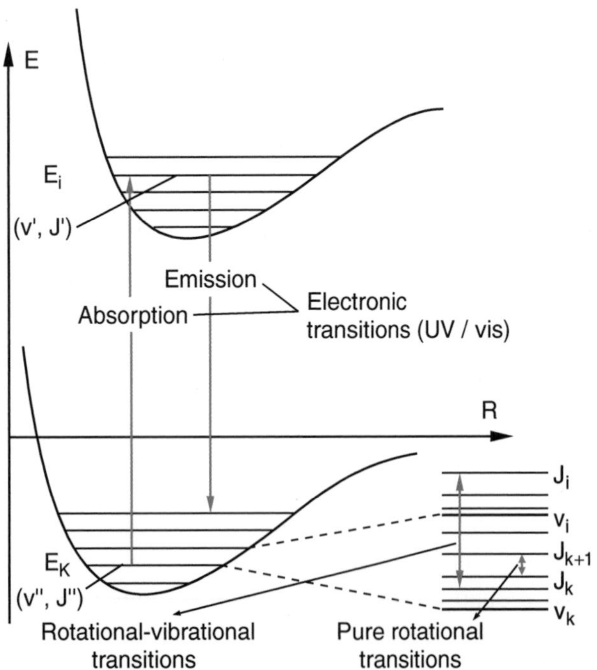


**Fig. 9.59.** Schematic representation of the structure of molecular transitions

The total system of all vibrational bands of this electronic transition is called a *band system*. The total number of lines in such a band system depends not only on the transition probabilities but also on the number of populated levels in the lower or upper electronic state.

The intensities of the lines in the *emission spectrum* are proportional to the population of the emitting upper levels and to the transition probability $A_{ik}$ the photon emission rate is

$$\dot{N}_{\mathrm{phot}} = N_k A_{ik} \quad (9.140\mathrm{a})$$

The intensity is then $I_{ik}^{\mathrm{em}} \propto h \cdot \nu \cdot \dot{N}_{\mathrm{phot}}$. The number of emitting levels depends on the excitation mechanism. Generally, the energy of the upper electronic state is for $T = 300\,\mathrm{K}$ large compared to the thermal energy $kT$. Therefore the thermal population is negligible. Optical pumping with lasers allows the population of single selected levels. In this case the fluorescence spectrum becomes very simple because it is emitted from a single upper level. In gas discharges, many upper levels are excited by electron impact and the number of lines in the emission spectrum becomes very large.

The *absorption spectrum* consists of all allowed transitions from populated lower levels.

Their intensity, as given in Sect. 7.2, is given by

$$I_{ik}^{\mathrm{abs}} = N_i w(\nu) B_{ik} . \quad (9.140\mathrm{b})$$

At thermal equilibrium the population distribution follows a Boltzmann distribution

$$N_i \propto g_i \mathrm{e}^{-E_i/kT} . \quad (9.141)$$

### b) Selection Rules for Electronic Dipole Transitions

The integral (9.135a) becomes zero, if the integrand is an odd function of the electronic coordinates. This gives the selection rules for the symmetry of the electronic wavefunctions $\Phi i$, $\Phi_k$:

$$+ \leftrightarrow - \text{ and } - \leftrightarrow + \text{ but } + \longleftarrow\!\!\not=\!\!\longrightarrow + \text{ and } - \longleftarrow\!\!\not=\!\!\longrightarrow -.$$

For homonuclear diatomics the additional selection rules hold:

$$\mathrm{g} \leftrightarrow \mathrm{u} \text{ and } \mathrm{u} \leftrightarrow \mathrm{g} \text{ but } \mathrm{g} \longleftarrow\!\!\not=\!\!\longrightarrow \mathrm{g} \text{ and } \mathrm{u} \longleftarrow\!\!\not=\!\!\longrightarrow \mathrm{u}$$

From angular momentum conservation the selection rule follows:

$$\Delta\Lambda = 0 \text{ or } \pm 1,$$

If spin-orbit coupling is weak, the following selection rules for the spin quantum number $S$, its projection quantum number $\Sigma$ and the projection number $\Omega = \Lambda + \Sigma$ are valid:

$$\Delta \mathrm{S} = 0; \quad \Delta\Sigma = 0; \quad \Delta\Omega = 0.$$

**EXAMPLES**

The transitions $^1\Sigma \leftrightarrow {}^1\Pi$ or $^1\Sigma \leftrightarrow {}^1\Sigma$ are both allowed, while the transitions $^1\Sigma \longleftarrow\!\!\not=\!\!\longrightarrow {}^3\Pi$ or $^1\Sigma \longleftarrow\!\!\not=\!\!\longrightarrow {}^3\Sigma$ or $^1\Sigma \longleftarrow\!\!\not=\!\!\longrightarrow {}^1\Delta$ are all dipole forbidden.

### c) The Rotational Structure of Electronic Transitions

The wavenumnber of a rotational line in the electronic spectrum of a diatomic molecule corresponding to a transition $(n_i, v_i, J_i) \leftrightarrow (n_k, v_k, J_k)$ is

$$\begin{aligned}\overline{\nu}_{ik} &= (T'_e - T''_e) + \left(T_{\text{vib}}(v') - T_{\text{vib}}(v'')\right) \\ &\quad + \left(T_{\text{rot}}(J') - T_{\text{rot}}(J'')\right)\end{aligned} \tag{9.142}$$

where $T_\text{e}$ gives the minimum of the potential curves $E_\text{pot}(R)$ of the electronic states $|i\rangle$ or $|k\rangle$, $T_\text{vib}$ is the term value of the vibrational state for $J = 0$ and $T_\text{rot}(J)$ the pure rotational term value.

The rotational structure of a vibrational band is then (similarly to the situation for vibrational–rotational transitions within the same electronic state) given by

$$\begin{aligned}\overline{\nu}_{ik} &= \overline{\nu}_0(n_i, n_k, v_i, v_k) + B'_v J'(J'+1) \\ &\quad - D'_v J'^2(J'+1)^2 \\ &\quad - \left[B''_v J''(J''+1) - D''_v J''^2(J''+1)^2\right].\end{aligned} \tag{9.143}$$

In (9.143), the rotational constant $B'_v$ in the upper state can either be larger or smaller than $B''_v$ in the lower electronic state. This depends on the binding energies and the equilibrium distances $R_\text{e}$ in the two states. The ***Fortrat-Diagrams*** shown in Fig. 9.60 has a different structure for each of the two cases.

At those $J$-values where the curve $\nu(J)$ becomes vertical, the density of rotational lines within a given spectral interval has a maximum. The derivative $\mathrm{d}\nu/\mathrm{d}J$ changes its sign. For the case $B''_v > B'_v$ the positions $\nu(J)$ of the rotational lines increase for $R$-lines before the maximum and then decrease again (Fig. 9.60a). The position $\nu_\text{h}$ of this line pileup is called the ***band head***. For $B''_v > B'_v$ the $R$-lines show a band head at the high frequency side of the band, while for $B''_v < B'_v$ the $P$-lines accumulate in a band head at the low frequency side (Fig. 9.60b). The line density may

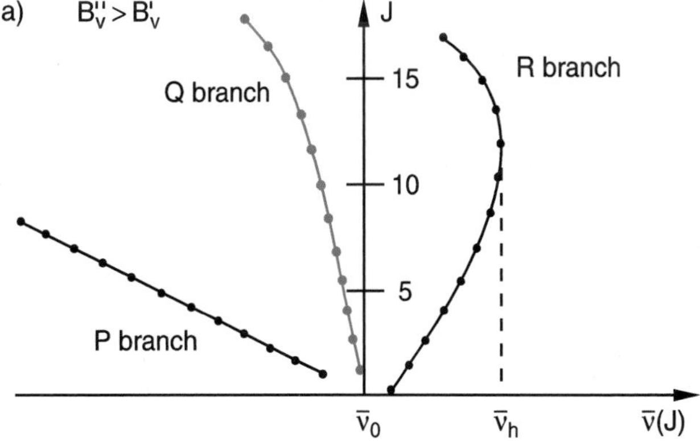


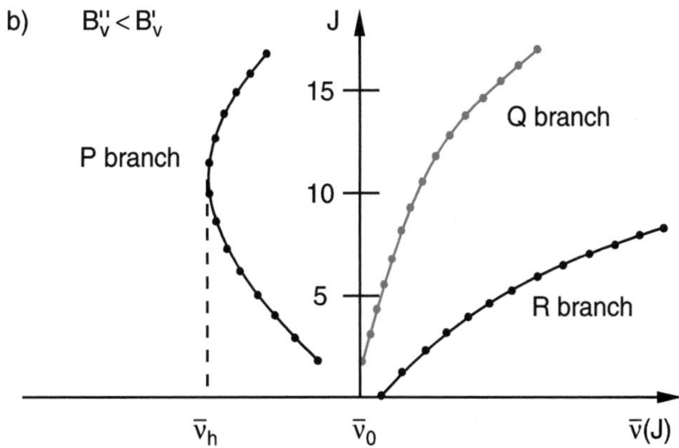


**Fig. 9.60a,b.** Fortrat-diagram for $P$, $Q$ and $R$ branches in electronic transitions: (**a**) $B''_v > B'_v$ (**b**) $B''_v < B'_v$

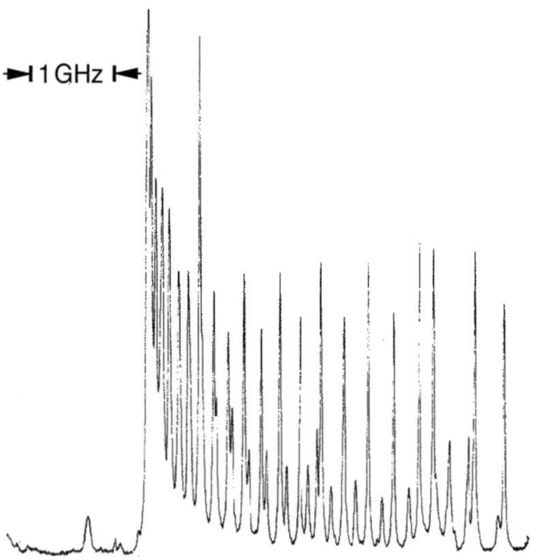


**Fig. 9.61.** Band head of the vibrational band $v' = 9 \leftarrow v'' = 14$ of the electronic transition $C\,^1\Pi_u \leftarrow X\,^1\Sigma_g^+$ of the $Cs_2$ molecule, recorded with sub-Doppler-resolution

become so high, that even with very high spectral resolution the different lines cannot be resolved. This is illustrated by Fig. 9.61, which shows the rotational lines in the electronic transition of the $Cs_2$ molecule around the band head, taken with sub-Doppler resolution.

In molecular electronic spectra taken with photographic detection and medium resolution where only part of the rotational lines are resolved, a sudden jump of the blackening on the photoplate appears at the band head while the line density gradually decreases with increasing distance from the band head. The band appears shadowed (Fig. 9.62) to the opposite frequency side of the band head. For $B_v'' > B_v'$ the band is red-shadowed and the band head is on the blue side of the band, while for $B_v'' < B_v'$ the band is blue-shadowed and the band head appears on the red side.

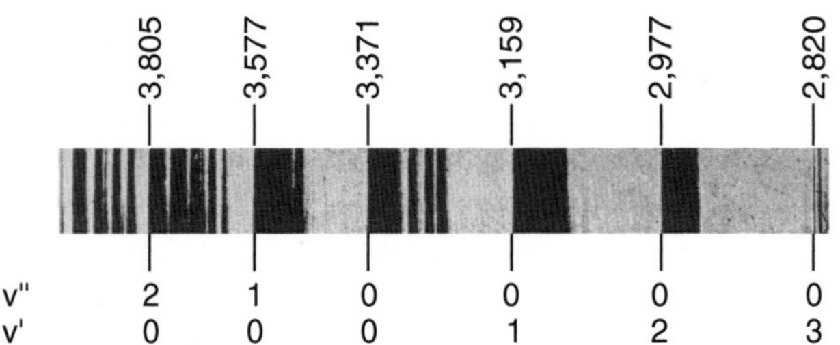


**Fig. 9.62.** Photographic recording of the band structure in the electronic transition $3\Pi_g \leftarrow 3\Pi_u$ of the $N_2$ molecule. The wavelengths of the band heads are given in Å $= 0.1$ nm above the spectrum (with the kind permission of the late Prof. G. Herzberg [G. Herzberg: Molecular Spectra and Molecular Structure Vol. I (van Nostrand, New York, 1964)])

In cases where the electronic transition allows $Q$-lines, their spectral density is higher than that of the $P$- and $R$-lines. For $B_v' = B_v''$ all $Q$-lines $Q(J)$ have the same position. For $B_v' > B_v''$ their positions $\nu(J)$ increase with increasing $J$ (Fig. 9.60a) while for $B_v' < B_v''$ they decrease (Fig. 9.60b).

### d) The Vibrational Structure and the Franck–Condon Principle

The vibrational structure of electronic transitions is governed by the Franck–Condon factor (9.137), which in turn depends on the overlap of the vibrational wave functions in the two electronic states. In a classical model, which gives intuitive insight into electronic transitions, the absorption or emission of a photon occurs within a time interval that is short compared to the vibrational period $T_{vib}$ of the molecule. In a potential diagram (Fig. 9.63) the electronic transitions between the two states can be then represented by vertical arrows. This means, that the internuclear distance $R$ is the same for the starting point and the final point of the transition. Since the momentum $p = h\nu/c$ of the absorbed or emitted photon is very small compared to that

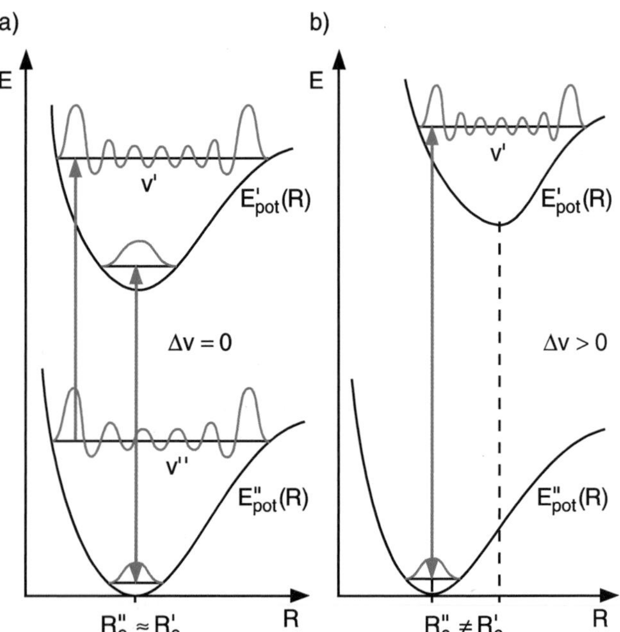


**Fig. 9.63.** Illustration of the Franck-Condon principle for vertical transitions with $\Delta v = 0$ (**a**) and $\Delta v > 0$ in case of potential curves with $R_e'' = R_e'$ and $R_e' > R_e''$

of the vibrating nuclei, the momentum $p$ of the nuclei is conserved during the electronic transition. Also, the kinetic energy $E_{\text{kin}} = p^2/2M$ does not change. From the energy balance

$$\begin{aligned} h\nu &= E'(v') - E''(v'') \\ &= E'_{\text{pot}}(R) + E'_{\text{kin}}(R) - \left[E''_{\text{pot}}(R) + E''_{\text{kin}}(R)\right] \\ &= E'_{\text{pot}}(R^*) - E''_{\text{pot}}(R^*) \end{aligned} \tag{9.144}$$

it follows that the electronic transition takes place at a nuclear distance $R^*$ where the kinetic energies of the vibrating nuclei in the upper and lower state are equal, i.e., $E'_{\text{kin}}(R^*) = E''_{\text{kin}}(R^*)$. This can be graphically illustrated by the difference potential

$$U(R) = E''_{\text{pot}}(R) - E'_{\text{pot}}(R) + E(v') \tag{9.145}$$

introduced by *Mulliken* (Fig. 9.64). The electron jump from one electronic state into the other takes place at such a value $R^*$, where Mulliken's difference potential intersects the horizontal energy line $E = E(v'')$, where

$$U(R^*) = E(v'') .$$

In the quantum mechanical model, the probability for a transition $v' \leftrightarrow v''$ is given by the Franck–Condon factor (9.137). The ratio

$$\mathcal{P}(R)\,\mathrm{d}R = \frac{\psi'_{\text{vib}}(R)\psi''_{\text{vib}}(R)\,\mathrm{d}R}{\int \psi'_{\text{vib}}(R)\psi''_{\text{vib}}(R)\,\mathrm{d}R} \tag{9.146}$$

gives the probability that the transition takes place in the interval $\mathrm{d}R$ around $R$. It has a maximum for $R = R^*$.

If the two potential curves $E'_{\text{pot}}(R)$ and $E''_{\text{pot}}(R)$ have a similar $R$-dependence and equilibrium distances $R'_e \approx R''_e$ the $FCF$ for transitions with $\Delta v = 0$ are maximum and for $\Delta v \neq 0$ they are small (Fig. 9.63a). The larger the shift $\Delta R = R'_e - R''_e$ the larger becomes the difference $\Delta v$ for maximum $FCF$ (Fig. 9.63b).

### 9.6.4 Continuous Spectra

If absorption transitions lead to energies in the upper electronic state above its dissociation energy, unbound states are reached with non-quantized energies. The absorption spectrum then no longer consists of discrete lines but shows a continuous intensity distribution $I(\nu)$. A similar situation arises, if the energy of the upper state is above the ionization energy of the molecule, similarly to atoms (see Sect. 7.6).

In the molecular spectra the ionization continuum is, however, superimposed by many discrete lines

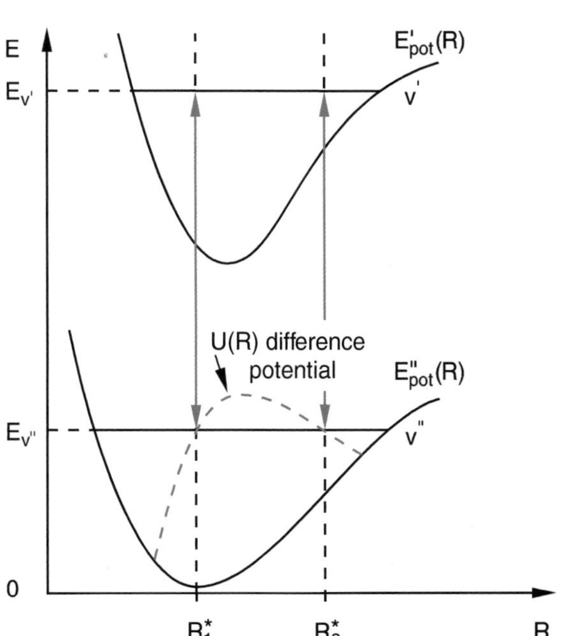


**Fig. 9.64.** Illustration of the Mulliken-difference potential $V(R) = E''_{\text{pot}}(R) - E'_{\text{pot}}(R) + E(v')$

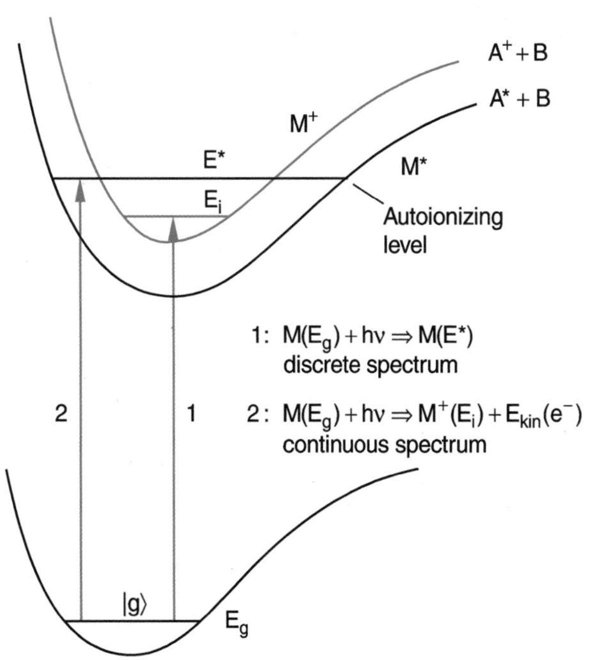


**Fig. 9.65.** Excitation (1) of a bound Rydberg level in the neutral molecule and (2) of a bound level in the molecular ion $M^+$

that correspond to transitions into higher vibrational-rotational levels of bound Rydberg states in the neutral electron. Although the electronic energy of these Rydberg states is still below the ionization limit, the additional vibrational-rotational energy brings the total energy above the ionization energy of the non-vibrating and non-rotating molecule (Fig. 9.65).

Such states can decay by autoionization into a lower state of the molecular ion, where part of the kinetic energy of the vibrating and rotating molecular core is transferred to the Rydberg electron, which then gains sufficient energy to leave the molecule (Fig. 9.66). The situation is similar to that in doubly excited Rydberg atoms where the energy can be transferred from one excited electron to the Rydberg electron (see Sect. 6.6.2). However, while this process in atoms takes place within $10^{-13}$–$10^{-15}$ s, due to the strong electron-electron interaction, in molecules it is generally very slow (between $10^{-6}$–$10^{-10}$ s), because the coupling between the motion of the nuclei and the electron is weak. In fact, within the adiabatic approximation it would be zero! The vibrational or rotational autoionization of molecules represent a breakdown of the Born–Oppenheimer approximation. The decay of these levels by autoionization is slow and the lines appear sharp. In Fig. 9.66 an example of the excitation scheme of autoionizing Rydberg levels is shown. The Rydberg levels are generally excited in a two-step process from the ground state $|g\rangle$ to level $|i\rangle$ by absorption of a photon from a laser and the further excitation $|i\rangle \rightarrow |k\rangle$ by a photon from another laser. The autoionization of the Rydberg level $|k\rangle$ is monitored by observation of the resultant molecular ions. A section of the autoionization spectrum of the $Li_2$-molecule with sharp lines and a weak continuous background, caused by direct photoionization, is shown in Fig. 9.67. The lines have

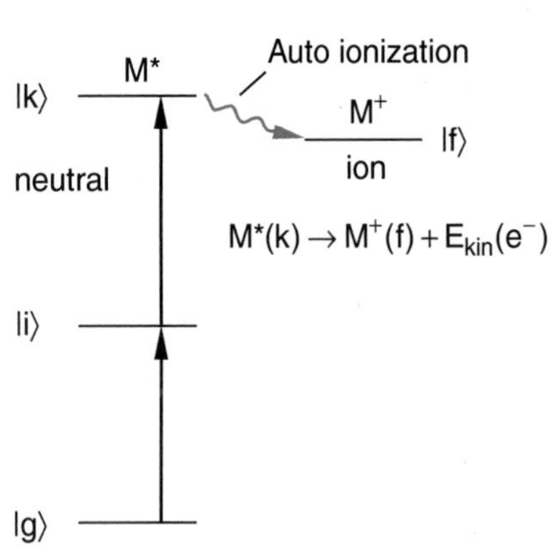


**Fig. 9.66.** Two-step excitation of a molecular Rydberg level $|k\rangle$, which transfers by auto ionization into a lower level $|f\rangle$ of the molecular ion. The difference energy is given to the free electron

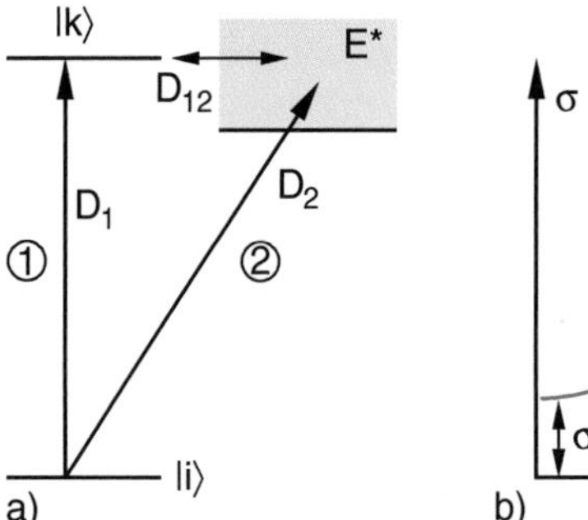


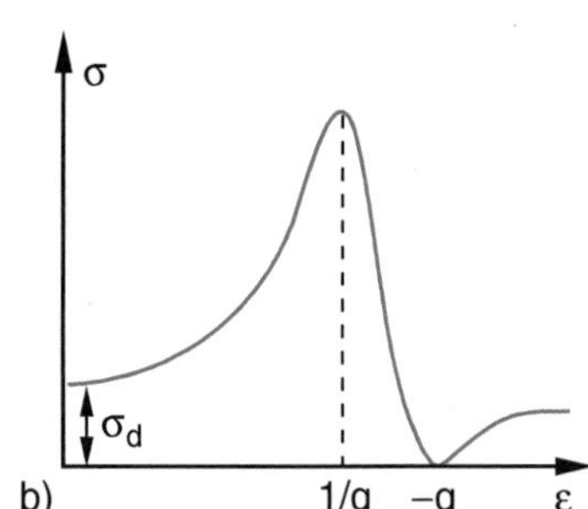


**Fig. 9.68.** (**a**) Interference of two possible excitation pathways to the energy $E^*$ in the ionization continuum (**b**) Resultant Fano-profile with asymmetric line shape. $\sigma_d$ is the absorption crosssection for direct photoionization

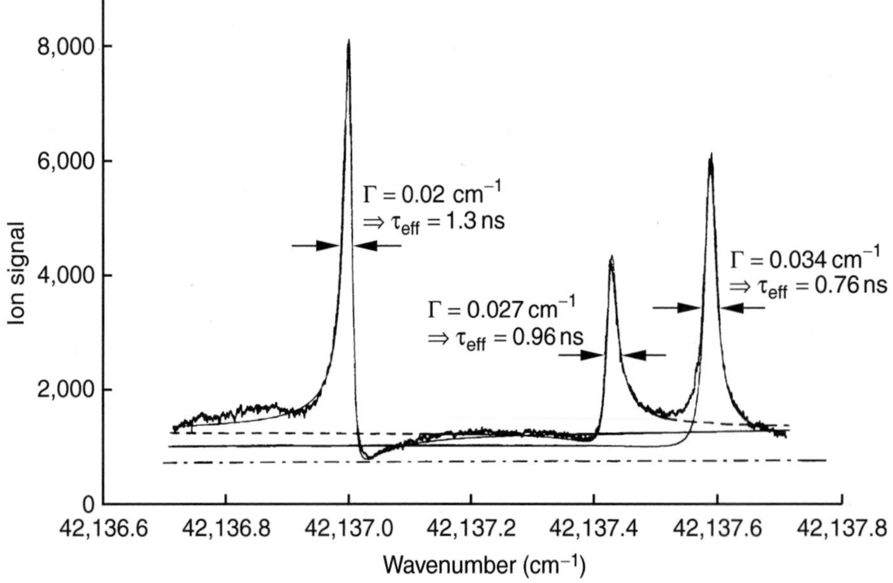


**Fig. 9.67.** Section of the auto ionization spectrum of the $Li_2$ molecule with three lines showing Fano profiles with differend widths.

an asymmetric line profile called a ***Fano-profile*** [9.12]. The reason for this asymmetry is an interference effect between two possible excitation paths to the energy $E^*$ in the ionization continuum, as illustrated in Fig. 9.68:

1. The excitation of the Rydberg level $|k\rangle$ of the neutral molecule from level $|i\rangle$ with the probability amplitude $D_1$ with subsequent autoionization,

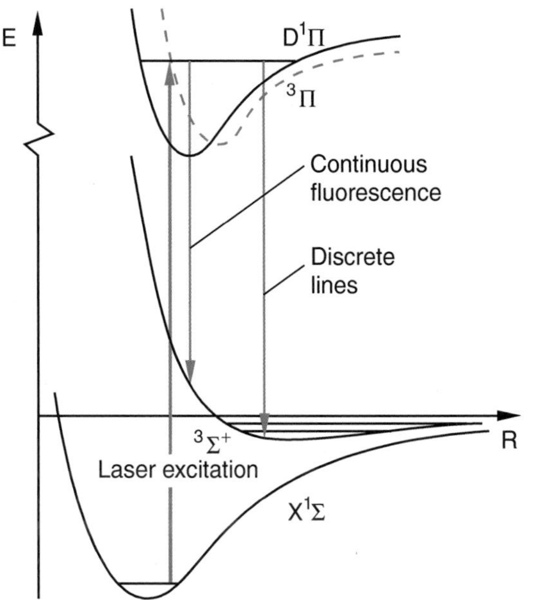


**Fig. 9.69.** Level scheme of the NaK molecule with excitation and discrete and continuous emission spectrum

2. The direct photoionization from level $|i\rangle$ with the probability amplitude $D_2$.

When the frequency of the excitation lasers is tuned, the phase of the transition matrix element does not change much for path 2, but much more for path 1, because the frequency is tuned over the narrow resonance of a discrete transition. The total transition probability

$$\mathcal{P}_{\text{if}} = |D_1 + D_2|^2$$

therefore changes with the frequency of the excitation laser because the interference is on one side of the resonance destructive on the other constructive, resulting in an asymmetric line profile.

Continuous spectra can also appear in emission, if a bound upper level is excited that emits fluorescence into a repulsive lower state. Such a situation is seen in excimers, which have stable excited upper states but an unstable ground state (Fig. 9.69). For illustration, the emission spectrum of an excited state of the NaK alkali molecule is shown in Fig. 9.70. This state is a mixture of a singlet and a triplet state, due to strong spin-orbit coupling. Therefore transitions from this mixed state into lower singlet as well as triplet states becomes allowed. While the emission into the stable singlet ground state $X\,^1\Sigma$ shows discrete lines, the emission

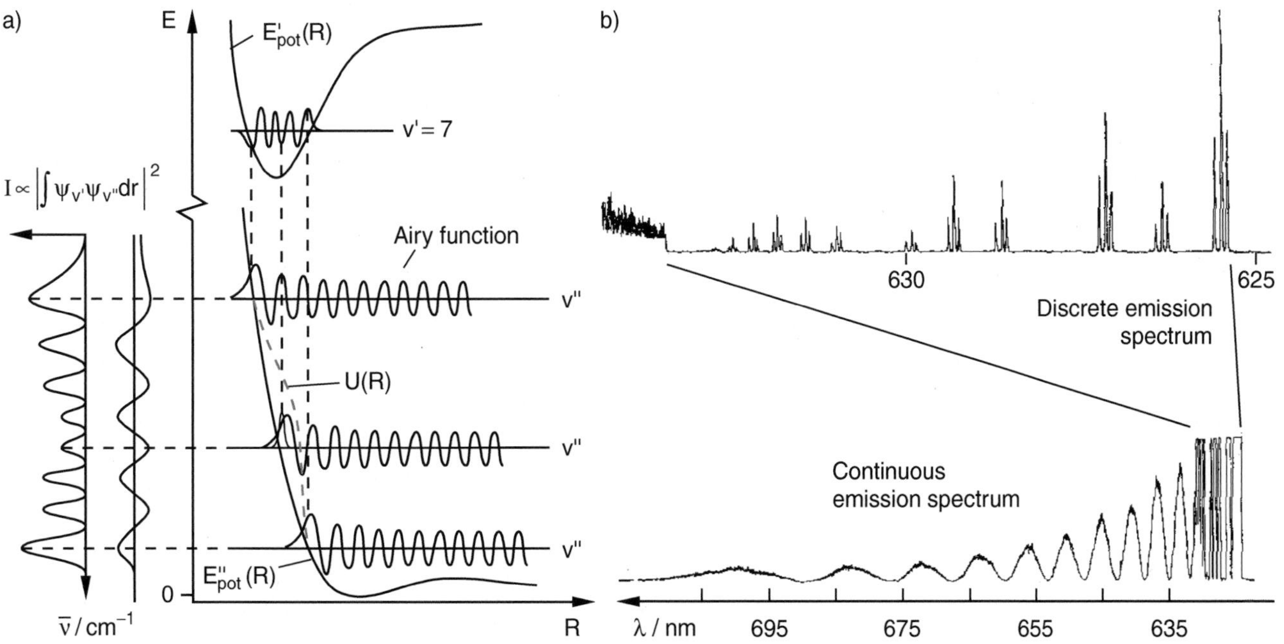


**Fig. 9.70.** (**a**) Vibrational overlap and Franck-Condon factor for the continuous emission (**b**) Measured emission spectrum of the NaK molecule

into the weakly bound lowest triplet state $3\,^3\Sigma$ shows, on the short wavelength side, a section of discrete lines terminating at bound vibrational-rotational levels in the shallow potential well of the $a\,^3\Sigma$ state and, on the long wavelength side, a modulated continuum terminating on energies above the dissociation limit of the $a\,^3\Sigma$ state. The intensity modulation reflects the FCF, i.e., the square of the overlap integral between the vibrational wave function of the bound level in the upper electronic state with the function of the unstable level in the repulsive potential above the dissociation energy of the lower state which can be described by an Airy function. The frequency $\nu = E'(R) - E''(R)$ and the wavelength $\lambda = c/\nu$ of the emission depends on the internuclear distance $R$ because the emission terminates on the Mulliken potential of the repulsive lower state (dashed blue curve in Fig. 9.70). The number $q = v' - 1$ of nodes in the fluorescence spectrum gives the vibrational quantum number $v'$ of the emitting level.

## SUMMARY

- For the simplified model of a rigid diatomic molecule, the electronic wave functions $\psi(r, R)$ and the energy eigenvalues $E(R)$ can be approximately calculated as a function of the internuclear distance $R$. The wave functions are written as a linear combination of atomic orbitals (LCAO approximation) or of other suitable basis functions.
- In a rotating and vibrating molecule the kinetic energy of the nuclei is generally small compared to the total energy of a molecular state. This allows the separation of the total wave function $\psi(r, R) = \chi_N(R)\Phi^{el}(r, R)$ into a product of a nuclear wave function $\chi(R)$ and an electronic function $\Phi^{el}(r, R)$, which depends on the electronic coordinates $r$ and only contains $R$ as a free parameter. This approximation, called the adiabatic or *Born–Oppenheimer approximation*, neglects the coupling between nuclear and electron motion. The potential equals that of the rigid molecule and the vibration and rotation takes place in this potential.
- Within this approximation the total energy of a molecular level can be written as the sum $E = E_{el} + E_{vib} + E_{rot}$ of electronic, vibrational and rotational energy. This sum is independent of the nuclear distance R.
- The electronic state of a diatomic molecule is characterized by its symmetry properties, its total energy $E$ and by the angular momentum and spin quantum numbers. For one-electron systems these are the quantum numbers $\lambda = l_z/h$ and $\sigma = s_z/h$ of the projections $l_z$ of the electronic orbital angular momentum and $s_z$ of the spin $s$ onto the internuclear $z$-axis. For multi-electron systems $L = \Sigma l_i$, $S = \Sigma s_i$, $\Lambda = L_z/h = \Sigma \lambda_i$, and $M_s = \Sigma\sigma_i = S_z/h$. Although the vector $\boldsymbol{L}$ might depend on $R$, the projection $L_z$ does not.
- The potential curves $E_{pot}(R)$ are the sum of mean kinetic energy $\langle E_{kin}\rangle$ of the electrons, their potential energy and the potential energy of the nuclear repulsion. If these potential curves have a minimum at $R = R_e$, the molecular state is stable. The molecule vibrates around the equilibrium distance $R_e$. If $E_{pot}(R)$ has no minimum, but monotonically decreases with increasing $R$ the state is unstable and it dissociates.
- The vibration of a diatomic molecule can be described as the oscillation of one particle with reduced mass $M = M_A M_B/(M_A + M_B)$ in the potential $E_{pot}(R)$. In the vicinity of $R_e$ the potential is nearly parabolic and the vibrations can be well-approximated by a harmonic oscillator. The allowed energy eigenvalues, defined by the vibrational quantum number $v$, are equidistant with a separation $\Delta E = \hbar\omega$. For higher vibrational energies the molecular potential deviates from a harmonic potential. The distances between vibrational levels decrease with increasing energy. A good approximation to the real potential is the Morse-potential, where $\Delta E_{vib}$ decreases linearly with energy. Each bound electronic state has only a finite number of vibrational levels.
- The rotational energy of a diatomic molecule $E_{rot} = J(J+1)h^2/2I$ is characterized by the rotational quantum number $J$ and the moment of

inertia $I = MR^2$. Due to the centrifugal force $F_c$ the distance $R$ increases slightly with $J$ until $F_c$ is compensated by the restoring force $F_r = -\mathrm{d}E_{pot}/\mathrm{d}R$ and the rotational energy becomes smaller than that of a rigid molecule.

- The absorption or emission spectra of a diatomic molecule consists of:
  a) Pure rotational transitions within the same vibrational level in the microwave region
  b) Vibrational-rotational transitions within the same electronic state in the infrared region
  c) Electronic transitions in the visible and UV region
- The intensity of a spectral line is proportional to the product $N \cdot |M_{ik}|^2$ of the population density $N$ in the absorbing or emitting level and the square of the matrix element $M_{ik}$.
- Homonuclear diatomic molecules have neither a pure rotational spectrum nor a vibrational-rotational spectrum. They therefore do not absorb in the microwave and the mid-infrared region, unless transitions between close electronic states fall into this region.
- The electronic spectrum consists of a system of vibrational bands. Each vibrational band includes many rotational lines. Only rotational transitions with $\Delta J = 0;\ \pm 1$ are allowed. The intensity of a rotational transition depends on the Hönl-London factor and those of the different vibrational bands are determined by the Franck-Condon factors, which are equal to the square of the vibrational overlap integral.
- Continuous absorption spectra arise for transitions into energy states above the dissociation energy or above the ionization energy. Continuous emission spectra are observed for transitions from bound upper states into a lower state with a repulsive potential.

## PROBLEMS

1. How large is the Coulomb repulsion of the nuclei in the $H_2^+$ ion and the potential energy of the electron with wave function $\Phi^+(r,R)$ at the equilibrium distance $R_e = 2a_0$? First calculate the overlap integral $S_{AB}(R)$ in (9.13) with the wave function (9.9). What is the mean kinetic energy of the electron, if the binding energy is $E_{pot}(R_e) = -2.65\,\mathrm{eV}$? Compare the results with the corresponding quantities for the H atom.
2. How large is the electronic energy of the $H_2$ molecule (without nuclear repulsion) for $R = R_e$ and for the limiting case $R = 0$ of the united atom?
3. a) Calculate the total electronic energy of the $H_2$ molecule as the sum of the atomic energies of the two H atoms minus the binding energy of $H_2$.
   b) Compare the vibrational and rotational energy of $H_2$ at a temperature $T = 300\,\mathrm{K}$ with the energy of the first excited electronic state of $H_2$.
4. Prove that the two separated equations (9.75) are obtained when the product ansatz (9.74) is inserted into the Schrödinger equation (9.73).
5. Show that the energy eigenvalues (9.104) are obtained when the Morse potential (9.103) is inserted into the Schrödinger equation (9.80).
6. What is the ionization energy of the $H_2$ molecule when the binding energies of $H_2$ and $H_2^+$ are $E_B(H_2) = -4.48\,\mathrm{eV}$ and $E_B(H_2^+) = -2.65\,\mathrm{eV}$ and the ionization energy of the H atom $E_{Io} = 13.6\,\mathrm{eV}$?
7. Calculate the frequencies and wavelengths for the rotational transition $J = 0 \to J = 1$ and $J = 4 \to J = 5$ for the HCl molecule. The internuclear distance is $R_e = 0.12745\,\mathrm{nm}$. What is the frequency shift between the two isotopomers $H^{35}Cl$ and $H^{37}Cl$ for the two transitions? What is the rotational energy for $J = 5$?
8. If the potential of the HCl molecule around $R_e$ is approximated by a parabolic potential $E_{pot} = k(R - R_e)^2$ a vibrational frequency $\nu_0 = 9 \times 10^{13}\,\mathrm{s^{-1}}$ is obtained. What is the restoring force constant $k$? How large is the vibrational amplitude for $v = 1$?

# 10. Polyatomic Molecules

With an increasing number of atoms in a molecule, the complexity of the molecular structure and the many possibilities for different isomeric configurations of these atoms make investigations of polyatomic molecules a challenging task. Because of the larger number of degrees of freedom $f = 3q - 3$ for the internal motions (vibrations and rotations) of the $q$ atoms in the molecule, the dynamics of such a system of $q$ nuclei and $N = \sum Z_i$ electrons plays a more important role than in diatomics. The potential energy can no longer be described by a potential curve as in diatomics but by an $n$-dimensional energy surface in an $(n+1)$-dimensional space of the nuclear coordinates. The dimension $n$ ($n = 3q - 7$ for linear, $n = 3q - 6$ for planar and $n = 3q - 5$ for nonplanar molecules) depends on the number $q \geq 3$ of atoms.

We will start our discussion with molecules composed of three atoms, where many of the features of polyatomic molecules are already present but the overall complexity is smaller than for larger molecules. The extension of the methods developed here to larger molecules is often straight forward. As in the previous chapter we will begin with a model of rigid molecules and discuss the electronic states of these molecules, independent of vibrations and rotations, which will be treated later in separate sections.

## 10.1 Electronic States of Polyatomic Molecules

The molecular orbital model introduced in Sect. 9.3 provides good access to the electronic structure and geometrical arrangement of polyatomic molecules. The molecular orbitals are chosen as linear combinations of atomic orbitals of the atoms forming the molecule. Since the atomic valence electrons give the main contribution to the chemical bond and the inner atomic electrons are not significantly affected by the formation of the molecule, the number of atomic orbitals in the linear combination can be restricted to these valence orbitals. A measure of the strength of a chemical bond is the overlap integral

$$\int_{AB} S_{ab}\, d\tau_{el} = \int \varphi_A^* \cdot \varphi_B\, d\tau_{el} \qquad (10.1)$$

of the atomic orbitals contributing to the binding energy. This will be illustrated by the following examples.

### 10.1.1 The $H_2O$ Molecule

For the formation of the $H_2O$ molecule we have to consider the two $1s$ orbitals of the two H atoms and the four valence orbitals $2s$, $2p_x$, $2p_y$, $2p_z$ of the oxygen atom, since the two $1s$ electrons in the inner shell of the O atom do not contribute much to the binding energy. The electron configuration of the O atom in the valence shell is $2s^2, 2p_x, 2p_y, (2p_z)^2$ (Fig. 10.1a).

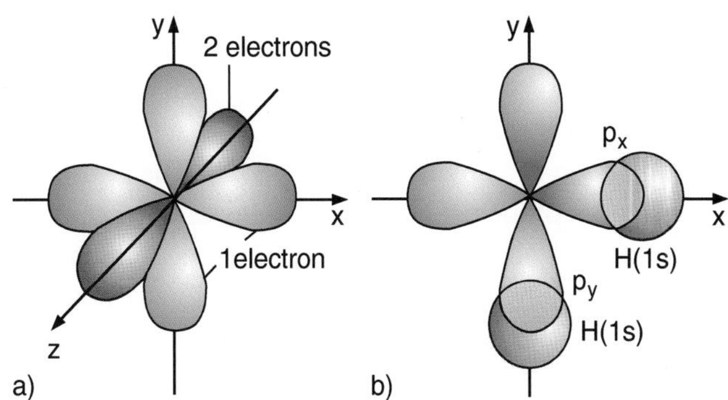


**Fig. 10.1.** (**a**) The atomic orbitals of the valence electrons of the O atom. (**b**) Formation of the $H_2O$ molecule without hybridization

W. Demtröder, *Atoms, Molecules and Photons*, 2nd ed., Graduate Texts in Physics,
DOI 10.1007/978-3-642-10298-1_10, 

As a first approximation only the two unpaired electrons in the $2p_x$ and $2p_y$ atomic orbitals are used to form molecular orbitals with the two $1s$ atomic orbitals of the two H atoms, where each of the two molecular orbitals contains one electron from the H atom and one from the O atom (Fig. 10.1b). This increases the electron density between the O atom and the two H atoms and therefore leads to an attraction between the atoms which implies that the total energy decreases. The overlap integral and the binding energy become large. The two electrons must have opposite spins since they are in the same molecular orbital.

In this approximation the binding molecular orbitals are the linear combinations

$$\Phi_1 = c_1\phi(1s_{\mathrm{H}}) + c_2\phi(2p_x)\,, \tag{10.2a}$$
$$\Phi_2 = c_3\phi(1s_{\mathrm{H}}) + c_4\phi(2p_y)\,. \tag{10.2b}$$

Each of them is occupied with two electrons. According to this model we expect a bent structure of the $H_2O$ with a binding angle of $\alpha = 90°$. The experimental value is $\alpha = 105°$. The reason for this (relatively small) discrepancy is the following:

Due to the interaction between the electrons of the H atoms and those of the O atom, the electron shell of the O atom is slightly deformed (see the discussion in Sect. 9.1.3). Therefore the $2s$ orbital in the O atom is no longer spherically symmetric but is better described by the linear combination (see (Fig. 10.2a))

$$\phi_{\mathrm{h}} = c_{21}\phi(2s) + c_{12}\phi(2p)\,. \tag{10.3}$$

This combination of $s$ and $p$ orbitals leads to a shift of the center of the charge distribution (Fig. 10.2b), which results in a better overlap of the oxygen atomic wave function $\phi_{\mathrm{h}}$ with the $1s$ orbital of the H atom and therefore to an increase of the binding energy for each of the two bonds. Therefore the terms $c_2\phi(2p_x)$ and $c_4\phi(2p_y)$ should be replaced by the orbitals (10.3). The optimum contribution of the p-function in the linear combination (10.3) can be obtained by varying the coefficients $c_{2i}$ until the binding energy becomes maximum and the total energy minimum. The chemical bonds obtained with such "hybrid orbitals" are no longer orthogonal to each other. Taking into account all polarization and exchange effects, the experimental value $\alpha = 105°$ of the binding angle can be indeed verified by a quantum mechanical calculation (Fig. 10.3).

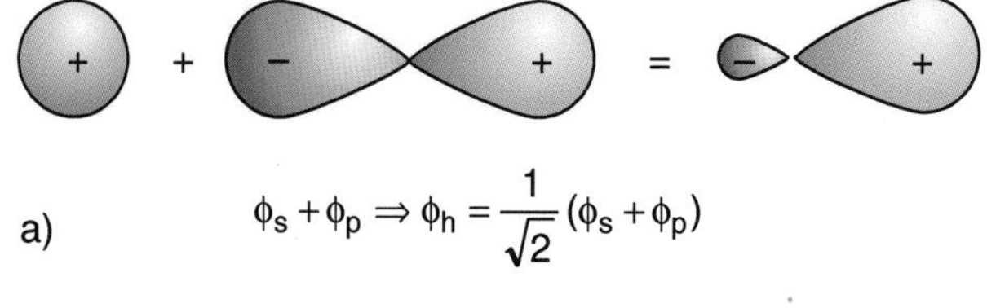


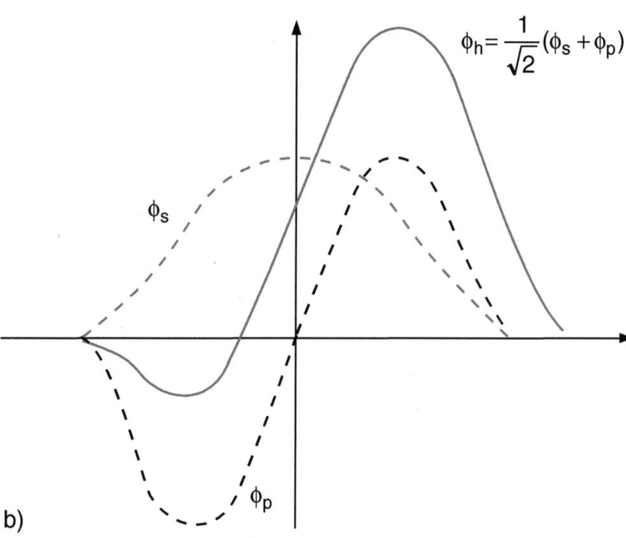


Fig. 10.2. (a) Deformation of a $p_z$ orbital, described by a linear combination of $s$ and $p_z$ orbitals. (b) Shift of the charge distribution by the linear combination of $s$ and $p$ functions

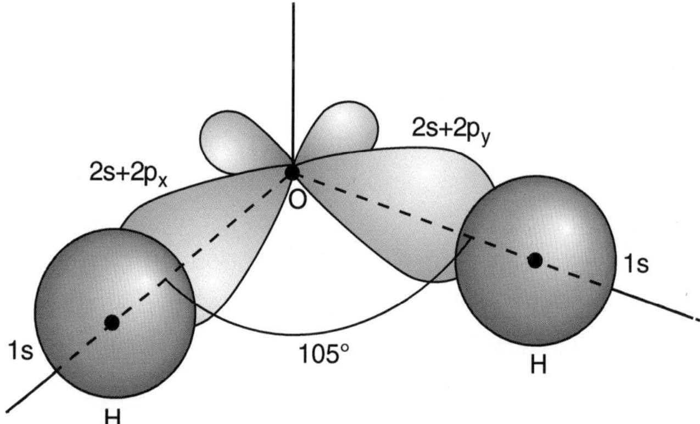


Fig. 10.3. Change of $H_2O$ geometry by hybridization of the atomic orbitals of the O atom

### 10.1.2 Hybridization

The deformation of the atomic orbitals by the interaction between the atoms of a molecule results in a spatial distribution of these orbitals that can no longer be described by a single unperturbed atomic orbital but needs a linear combination of at least two atomic orbitals. This "mixture" of atomic orbitals is called

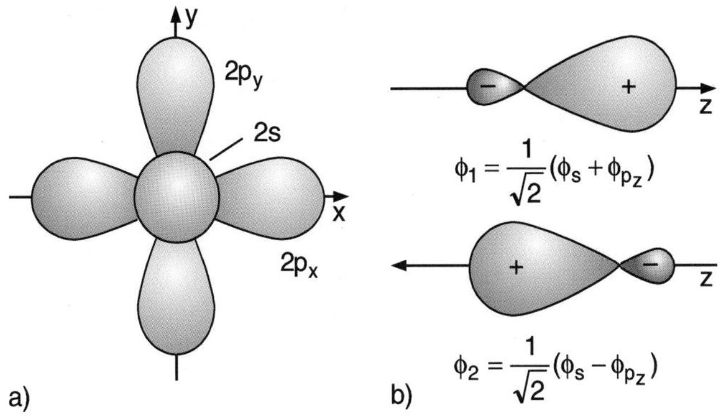


**Fig. 10.4a,b.** *sp* hybridization of atomic orbitals for the valence electrons of the C atom. (**a**) $2p_x$, $2p_y$ and $2s$ atomic orbitals of the O atom (**b**) The hybrid functions

"hybridization". We will illustrate this by the example of the carbon atom and its combination with other atoms.

The electron configuration in the electronic ground state of the C atom is

$$(1s)^2(2s)^2(2p_x)(2p_y)\,.$$

The C atom has two unpaired electrons in the $2p_x$ and $2p_y$ atomic orbitals. The simple LCAO model would predict two directional valence bonds of these orbitals combined with atomic orbitals of atoms attached to the C atom, with a binding angle $\alpha = 90°$ (Fig. 10.4a) between the two valence bonds analogous to the case for $H_2O$ before hybrid orbitals had been introduced.

In many cases it is energetically favorable if also one of the $2s$ electrons contributes to the binding, because the linear combination $2s + 2p$ of the atomic orbitals can result (similarly to the case of the $H_2O$ molecule) in a larger overlap with the atomic orbital of the atom attached to the C atom. If the energy, necessary to bring one $2s$ electron into the hybrid atomic orbital, is overcompensated by the gain in binding energy (i.e., by the decrease of the total energy) then the hybridization is favorable for the molecule by minimizing its total energy.

The combination of $s$ and $p$ atomic orbitals is called $sp$ hybridization, which we will now discuss. For the optimum binding to other atoms each carbon valence orbital should be occupied by one electron. Since the $p_x$ and $p_y$ orbitals are already occupied by one electron each from the C-atom, only the $2p_z$-orbital is free to form the hybrid-orbital. We consider the two atomic hybrid functions

$$\begin{aligned}\phi_1 &= c_1\phi(s) + c_2\phi(p_z)\,,\\ \phi_2 &= c_3\phi(s) - c_4\phi(p_z)\,,\end{aligned} \tag{10.4}$$

which can be composed from the $2s$ and the $2p_z$ atomic orbitals. The two functions have to be normalized and they are orthogonal to each other, which means that

$$\int \phi_i\phi_k\,\mathrm{d}\tau_{\mathrm{el}} = \delta_{ik} \quad \text{with} \quad \delta_{ik} = \begin{cases}1 & \text{for} \quad i = k\\ 0 & \text{for} \quad i \neq k\end{cases}\,. \tag{10.5}$$

Inserting the $sp$-hybrid orbitals (10.4) into the integral yields the conditions

$$c_1 = c_2 = c_3 = c_4 = \frac{1}{\sqrt{2}}\,,$$

for the coefficients $c_i$. The two $sp$-hybrid orbitals then become

$$\begin{aligned}\phi_1 &= \frac{1}{\sqrt{2}}\left[\phi(s) + \phi(p_z)\right]\,, \\ \phi_2 &= \frac{1}{\sqrt{2}}\left[\phi(s) - \phi(p_z)\right]\,.\end{aligned} \tag{10.6}$$

If we assume, that $\phi(s)$ and $\phi(p)$ can be described by the hydrogen $2s$ and $2p_z$ functions (see Table 4.2) the angular part of $\phi$ becomes

$$\phi_{1,2}(\vartheta) = \frac{1}{2\sqrt{2\pi}}\left[1 \pm \sqrt{3}\cos\vartheta\right]\,, \tag{10.7}$$

where $\vartheta$ is the angle against the $z$-axis. This shows that $|\phi_1|^2$ becomes maximum for $\vartheta = 0°$, $|\phi_2|^2$ for $\vartheta = 180°$.

> *sp* hybridization leads to two directional bonds pointing in opposite directions. This results in the formation of linear molecules if no other bonds are present.

**Note:**

We have here assumed that the angular parts of the s- and p-atomic orbitals can be superimposed with equal contributions ($c_1 = c_2$), although their radial parts are different. This is, of course, an approximation, which can be justified, because the atomic orbitals are centred at different atoms and their angular dependence is considered in the overlap region.

This $sp_z$ hybridization gives the C atom two additional free bonds and, including the $p_x$ and $p_y$ orbitals with their unpaired electrons, altogether four bonds are available for bonding to other atoms.

For instance, when two O atoms are attached to the C atom, the largest overlap of atomic orbitals of the C atom with those of the O atoms is achieved by the $sp_z$ hybridization, which yields two strong bonds in opposite directions. The $CO_2$ molecule O=C=O should therefore be linear, which agrees with the experimental results.

Another example is the acetylene molecule $C_2H_2$ where each of the two C atoms has two bonds in opposite directions formed by $sp_z$ hybrid orbitals. One of the bonds is for an H atom and one is for the other C atom giving the linear structure H–C≡C–H. The third bond between the two C atoms is formed by the other $2s$ electrons of the C-atoms. For the linear molecules $CO_2$ or $C_2H_2$ the $2p_x$ and $2p_y$ orbitals are not involved in the bonds along the $\pm z$-axis. The $sp$-hybridization includes only the $p_z$-orbital.

For some molecules including a C atom it is energetically more favorable if one of the $2s$ electrons in the C atom forms a hybrid function with *two p* electrons. For this $sp^2$ hybridization we have to form three atomic hybrid orbitals as linear combinations of $2s$, $2p_x$ and $2p_y$ functions. Analogously to the discussion above on $sp$ hybridization we obtain for the $sp^2$ hybridization, when we take into account the normalization and the orthogonality of the three functions, the three hybrid orbitals

$$\phi_1(sp^2) = \frac{1}{\sqrt{3}}\phi(s) + \sqrt{\frac{2}{3}}\phi(p_x) \tag{10.8}$$

$$\phi_2(sp^2) = \frac{1}{\sqrt{3}}\phi(s) - \frac{1}{\sqrt{6}}\phi(p_x) + \frac{1}{\sqrt{2}}\phi(p_y)$$

$$\phi_3(sp^2) = \frac{1}{\sqrt{3}}\phi(s) - \frac{1}{\sqrt{6}}\phi(p_x) - \frac{1}{\sqrt{2}}\phi(p_y)$$

with angular parts

$$\phi_1(\varphi) = \frac{1}{2\sqrt{\pi}}\left[\frac{1}{\sqrt{3}} + \sqrt{2}\cos\varphi\right] \tag{10.9}$$

$$\phi_2(\varphi) = \frac{1}{2\sqrt{\pi}}\left[\frac{1}{\sqrt{3}} - \frac{1}{\sqrt{2}}\cos\varphi + \sqrt{\frac{3}{2}}\sin\varphi\right]$$

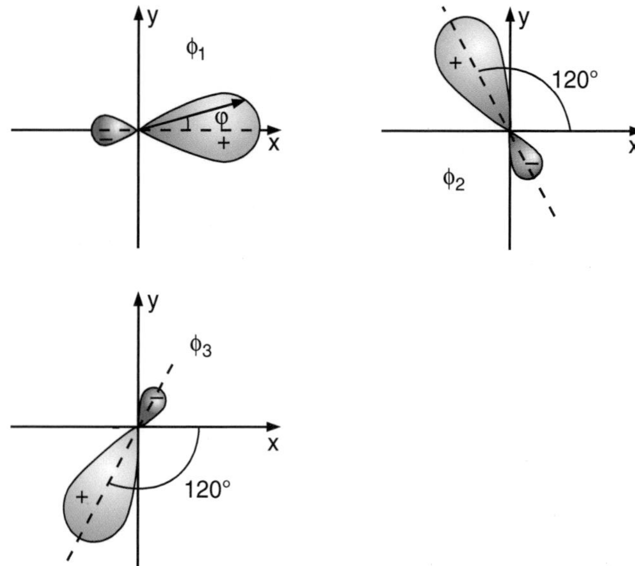


**Fig. 10.5.** $sp^2$ hybridization with the three hybrid orbitals pointing into the directions $\varphi = 0°$, $120°$ and $240°$ from the $x$-axis

$$\phi_3(\varphi) = \frac{1}{2\sqrt{\pi}}\left[\frac{1}{\sqrt{3}} - \frac{1}{\sqrt{2}}\cos\varphi - \sqrt{\frac{3}{2}}\sin\varphi\right].$$

Here $\varphi$ is the angle against the positive $x$-axis (Fig. 10.5). The three functions have their maxima at $\varphi = 0$ for $\phi_1$, $\varphi = 120°$ for $\phi_2$ and $\varphi = 240°$ (or $-120°$) for $\phi_3$.

> The $sp^2$ hybridization results in three bonds in directions $\varphi = 0°$, $120°$ and $240°$ from the $x$-axis, which are all in the $xy$-plane.

Finally we will discuss $sp^3$ hybridization, which explains the structure of the methane molecule $CH_4$. Here we have to combine the $2s$ orbital with all three $p$ orbitals $p_x$, $p_y$ and $p_z$. The four normalized and orthogonal hybrid functions composed of linear combinations of these four atomic orbitals are

$$\phi_1 = \frac{1}{2}\phi(s) + \frac{1}{2}\sqrt{3}\phi(p_z) \tag{10.10}$$

$$\phi_2 = \frac{1}{2}\phi(s) + \sqrt{\frac{2}{3}}\phi(p_x) - \frac{1}{2\sqrt{3}}\phi(p_z)$$

$$\phi_3 = \frac{1}{2}\phi(s) - \frac{1}{\sqrt{6}}\phi(p_x) + \frac{1}{\sqrt{2}}\phi(p_y) - \frac{1}{2\sqrt{3}}\phi(p_z)$$

$$\phi_4 = \frac{1}{2}\phi(s) - \frac{1}{\sqrt{6}}\phi(p_x) - \frac{1}{\sqrt{2}}\phi(p_y) - \frac{1}{2\sqrt{3}}\phi(p_z).$$

Inserting the angular parts of these functions, we obtain for the $sp^3$ hybridization, four hybrid orbitals that point into the corners of a tetrahedron (Fig. 10.6).

Besides the composition of $s$ and $p$ orbitals, hybrid orbitals can also be formed by linear combinations including $d$ orbitals if these orbitals are occupied in the atoms involved in the molecular binding. They represent directional bonds, which can result in different geometrical structures of molecules. For instance, $sp^2d$ hybridization leads to four bonds, all in the same plane, directed toward the corners of a square. The geometric structure depends on the number of valence electrons involved in the hybridization. For example, molecules $AB_4$ (A, B = arbitrary atoms) with eight valence electrons show a tetrahedral geometry while those with ten valence electrons form a bipyramidal structure, and those with 12 valence electrons have a planar quadratic structure. In Table 10.1 some examples are compiled.

These simple considerations show that it is possible to predict the geometric structure of many (but not all!) molecules from the atomic orbitals involved in the bonds between the atoms. The binding molecular orbitals are those that maximize the electron density between adjacent atoms and give a minimum total energy.

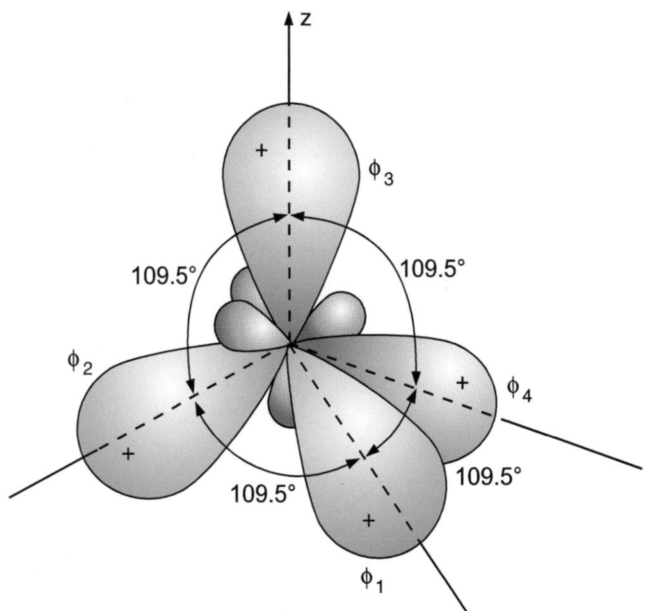


**Fig. 10.6.** The four hybrid functions for the $sp^3$ hybridization

**Table 10.1.** The different types of hybridization with the number of hybrid orbitals and the resultant molecular geometries

| Hybrid type | Number | Geomety | Example |
|---|---|---|---|
| $sp$ | 2 | linear | $C_2H_2$ |
| $sp^2$ | 3 | plane, 120° | $C_2H_4$ |
| $sp^3$ | 4 | tetrahedron | $CH_4$ |
| $sp^2d$ | 4 | planar aquare | $XeF_4$ |
| $sp^3d$ | 5 | triangular bypyramid | $PCl_5$ |
| $sp^3d^2$ | 6 | octohedron | $SF_6$ |

The basic principle of hybridization is the minimization of the total energy and the optimization of the negative binding energy. This is achieved by optimization of the overlap integral between two atoms.

In Fig. 10.7 the value of the overlap integral $S$ of two hybrid atomic orbitals

$$\phi_i = \frac{1}{\sqrt{1+\lambda^2}}\left[\phi(s) + \lambda\phi(p)\right] \quad i = 1,2\,. \quad (10.11)$$

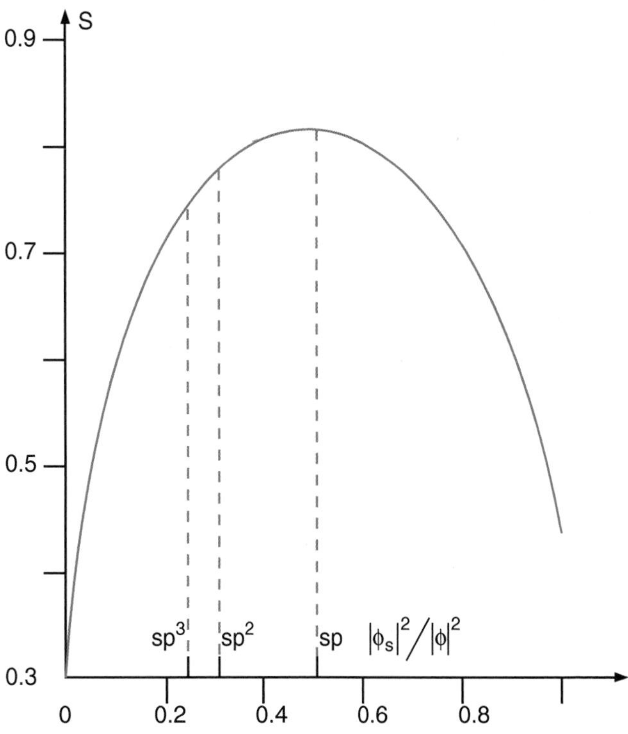


**Fig. 10.7.** Value of the overlap integral $S$ of the hybrid atomic orbitals for the C – C bond as a function of $|\phi(s)|^2/|\phi|^2$ for a C–C-distance of $(4/3)a_0$ [10.14]

forming a C − C bond is plotted as a function of the relative share $|\phi(s)|^2/|\phi|^2$ of the $\phi(s)$ part of the hybrid function. The plot shows that for the $sp$ hybridization the optimum share between $\phi(s)$ and $\phi(p)$ is 50%. The hybridization increases the value of the overlap integral $S$ from $S = 0.3$ without hybridization ($\lambda = 0$) to $S = 0.82$ for $\lambda = 1$. For $sp^2$ hybridization $\lambda = \sqrt{2}$, while for the optimum $sp^3$ hybridization we obtain the optimum value $\lambda = \sqrt{3}$.

The energy necessary to lift the $s$ electron into the hybrid orbital is overcompensated by the increase of the binding energy.

> All molecules in their electronic ground states have a geometry where the total energy has a minimum.

### 10.1.3 The $CO_2$ Molecule

The molecular orbitals of the $CO_2$ molecule are obtained by taking into account the $4 \cdot 3 = 12$ atomic valence orbitals: $2s$, $2p_x$, $2p_y$ and $2p_z$ for each of the three atoms (Fig. 10.8). The normalized and orthogonal linear combinations of these 12 atomic orbitals are 12 molecular orbitals with energies $E_n$. They are ordered with increasing energy and can be occupied by two electrons in each orbital. Since there are only 16 valence electrons for the ground state of $CO_2$ (four from the C atom and 6 from each O atom), only the eight lowest orbitals are filled with electrons. The coefficients in the linear combinations have to be optimized in such a way that the overlap of the electron wavefunctions between two atoms becomes maximum and the total energy of the molecule becomes a minimum.

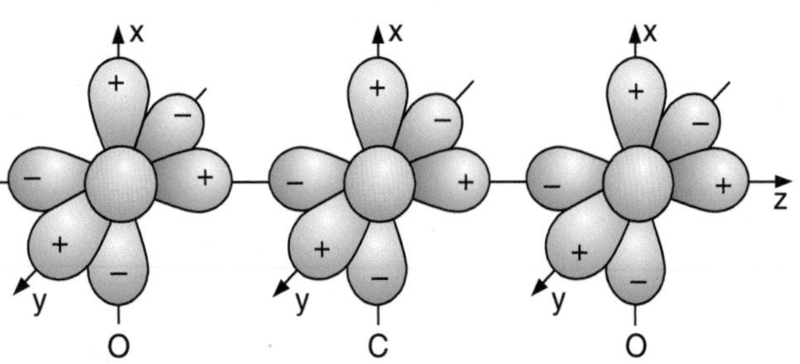


**Fig. 10.8.** The unperturbed atomic orbitals for the valence electrons of O and C atoms

In the C atom the $2s$ orbital is occupied by two electrons, and each of the $2p_x$ and $2p_y$ orbitals have one electron. In the O atom one $2p$ orbital has two electrons and the other two have one electron each. Without hybridization only two $p$ orbitals of the C atom and two $2p$ orbitals for each O atom would be available. A larger overlap is obtained by $sp$ hybridization, where one $2s$ and one $2p$ orbital from each of the three atoms combine to form six atomic hybrid orbitals. From these atomic hybrid orbitals molecular orbitals are formed which allow a better overlap of the atomic orbitals of two neighbouring atoms. Calculating the energies of these molecular hybrid orbitals with optimized coefficients gives the energetic order shown in Fig. 10.9.

The main contribution to the binding energy comes from those molecular orbitals that are formed as linear combinations of the $sp_z$ hybrid orbitals from the O atom and the $s$ or $p_z$ orbitals from the C atom. Since the maxima for this molecular orbital are along the $z$-axis, they favor a linear configuration of the $CO_2$ molecule. Atomic orbitals that do not contribute to the binding (e.g., the orbitals in Fig. 10.8 pointing in the $x$ or $y$ direction) are called *"lone pairs."*

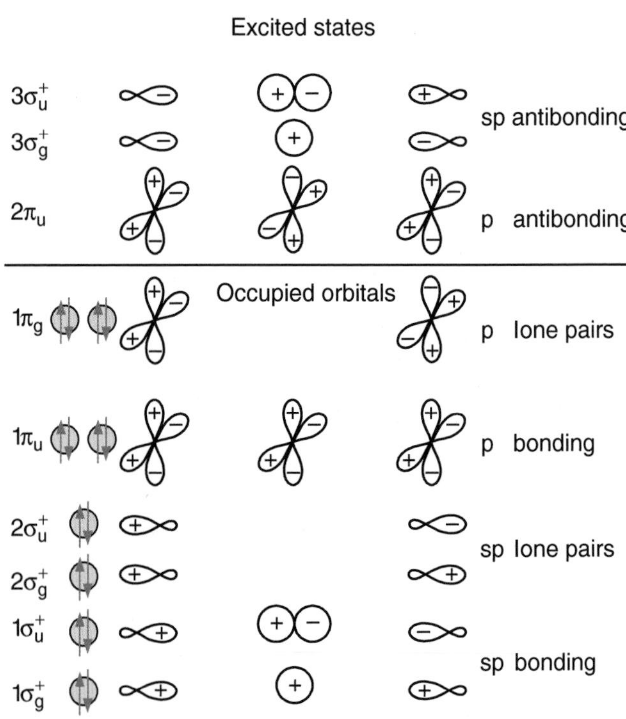


**Fig. 10.9.** Energetic order of the $CO_2$ molecular orbitals formed by $sp$-hybrid atomic orbitals

For linear molecules, the electronic states can be classified according to the projection $\Lambda\hbar = \sum \lambda_i \hbar$ of the electronic angular momentum $L = \sum l_i$, analogously to the nomenclature for diatomics. Since for the four lowest orbitals $\lambda = 0$, they are $\sigma$ orbitals. Their symmetry with respect to an inversion at the centre of charge can be "gerade" or "ungerade." The four lowest orbitals are then $1\sigma_g^+$, $1\sigma_u^+$, $2\sigma_g^+$, $2\sigma_u^+$ (see Sect. 9.3).

### 10.1.4 Walsh Diagrams

The binding angle of triatomic molecules can be estimated from a diagram first proposed by Walsh [10.1]. In this diagram the energy dependence $E(\alpha)$ on the binding angle $\alpha$ is plotted for all occupied molecular orbitals. We will illustrate this for the case of dihydrides $AH_2$ (A stands for an arbitrary atom). In Fig. 10.10 the lowest orbitals of $AH_2$ molecules and their energetic order are shown for the case where A is an atom of the second period in the periodic system. Figure 10.11 illustrates the dependence $E(\alpha)$ of the energy of these orbitals on the binding angle $\alpha$. The lowest $1\sigma_g$ orbital is not included in Fig. 10.11 because it does not contribute to the binding. One can see that the $\sigma_g$ and the $\sigma_u$ orbitals have their minimum energy for $\alpha = 180°$, they therefore favor the linear configuration. The $\pi$ orbitals, on the other hand, have their minimum energy for $\alpha = 90°$. They tend to a bent molecule. If the sum $\sum n_i E_i(\alpha)$ of the energies of all occupied orbitals weighted with the electron occupation number $n = 1$ or 2 is calculated, the real binding angle of the molecule in its ground state is that angle where the sum becomes a minimum. For instance, in the $H_2O$ molecule with eight valence electrons, the four lowest orbitals are occupied with two electrons each. The total energy for the ground state of $H_2O$ has a minimum when $\alpha = 105°$.

If one of the electrons is excited into a higher unoccupied orbital $\phi_k$, the angle $\alpha$ can change depending on the slope of $E_k(\alpha)$. If the energy of the excited orbital decreases with decreasing angle $\alpha$, the binding angle in the excited state becomes smaller than in the ground state. If it increases, $\alpha$ becomes larger. For instance, if an electron is removed from the $1\pi_u$ orbital with the energy dependence $\mathrm{d}E_\pi/\mathrm{d}\alpha > 0$ the binding angle must increase. Indeed the molecular ion $H_2O^+$ has a binding angle of $\alpha = 110°$ compared to 105° for the neutral molecule.

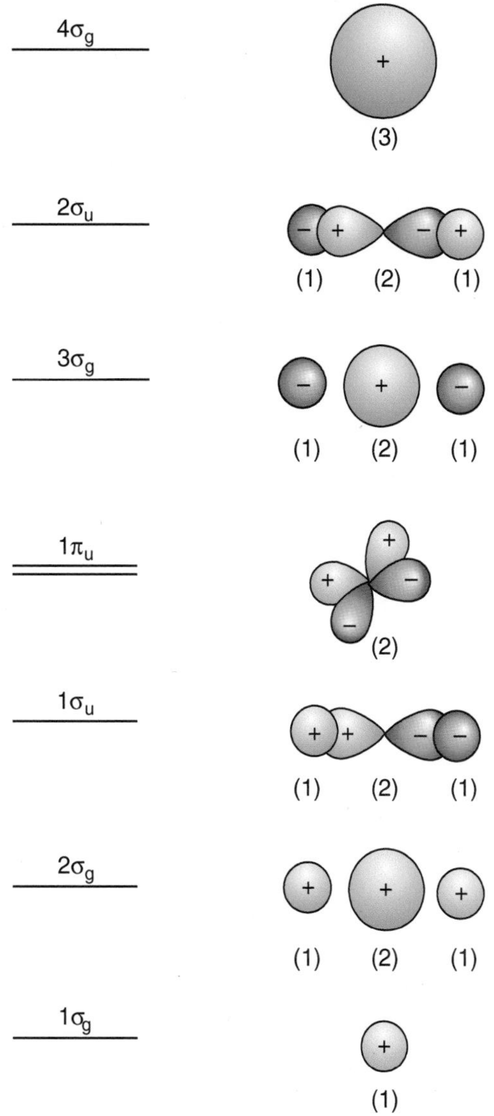


**Fig. 10.10.** Energy levels and orbitals of the $AH_2$ molecule

For the $NH_2$ radical with seven valence electrons the two $\sigma$ orbitals $2\sigma_g$ and $1\sigma_u$ and the lowest $1\pi_u$ orbital are occupied with two electrons each and the upper $1\pi_u$ orbital with an energy $E$ that does not depend on $\alpha$, has only one electron. We therefore expect for $NH_2$ a binding angle $\alpha$ close to that for $H_2O$. The experimental value is $\alpha = 103.4°$, which is indeed close to $\alpha = 105°$ for the $H_2O$ molecule.

Similar Walsh diagrams can be constructed for other triatomic molecules, $XY_2$, where X and Y are arbitrary atoms (Fig. 10.12). Examples are $CO_2$, $NO_2$, $SO_2$, $CS_2$ and $CF_2$. If the occupied molecular orbitals and their dependence $E(\alpha)$ are known, such diagrams can immediately show the molecular geometry, for example, that $CO_2$ is linear in its ground state but $NO_2$ is bent with a binding angle $\alpha = 134°$ [10.2, 3].

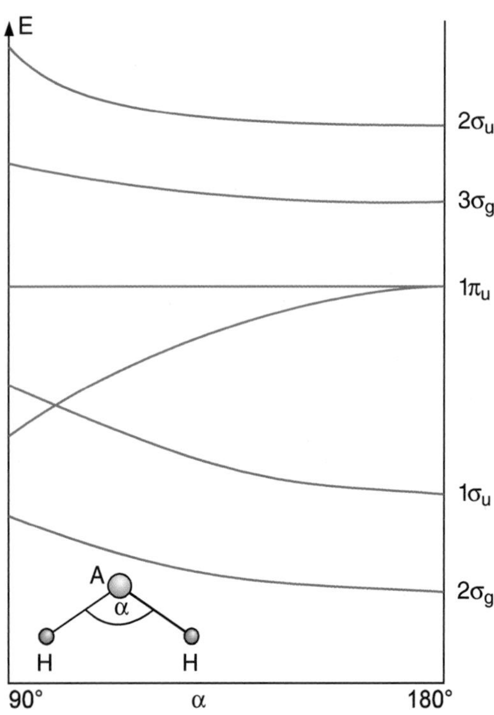


**Fig. 10.11.** Walsh diagram for $AH_2$ molecules

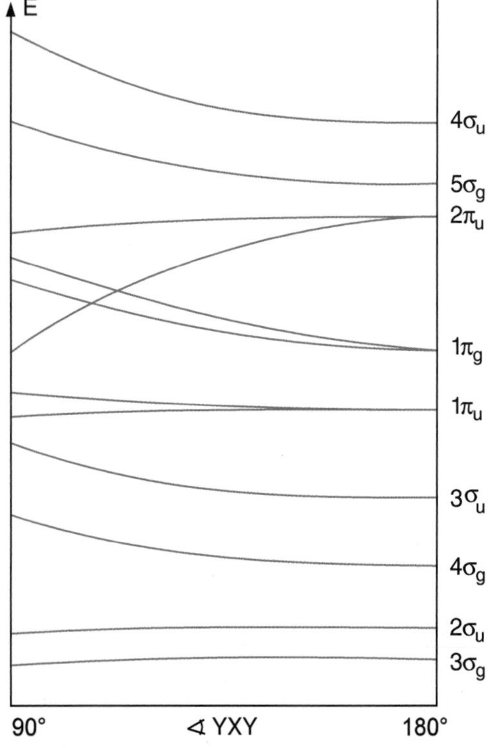


**Fig. 10.12.** Walsh diagram for $XY_2$ molecules

## 10.2 Molecules with more than Three Atoms

While all triatomic molecules have a planar structure, the geometry of molecules with more atoms can be linear, planar or nonplanar, i.e., three-dimensional. This gives more possibilities for ***isomers*** (these are molecules with the same atoms but different arrangements of these atoms and therefore with different geometric structure). The geometry depends on the electronic structure of the molecule and it can be different for excited electronic states from that of the ground state.

For nonlinear molecules there is no electronic angular momentum, because there is no symmetry axis around which the electrons could precess. Therefore the electronic states of nonlinear molecules can not be characterized by a quantum number $\Lambda$. They are specified by possible symmetry operations that bring the nuclear frame from an initial position into identical positions. It turns out that all such symmetry operations form the elements of a symmetry group characteristic of the specific molecule. All molecules can be classified according to their symmetry group and there is a big advantage in doing so. The mathematical tools of group theory can be applied to the description of molecular structure and transitions between molecular states. This allows the decision to be made whether a transition is allowed or forbidden without actually calculating the matrix elements.

The full group theoretical treatment is, however, beyond the scope of this textbook and the reader is referred to the literature [10.4, 5, 6].

In the following sections we will illustrate the basic features of electronic states of polyatomic molecules and their treatment within the framework of molecular orbitals and we will illustrate this by some examples.

### 10.2.1 The $NH_3$ Molecule

The N atom has three unpaired electrons in the three $p_x$, $p_y$ and $p_z$ orbitals (see Fig. 10.13). We therefore expect three directed bonds with the $1s$ orbitals of the three H atoms, which should be perpendicular to each other. The overlap between the $1s$ orbitals of the H atoms and the atomic orbitals of the N atom becomes larger when $sp$-hybrid atomic orbitals are formed instead of the pure $p$ orbitals of the N

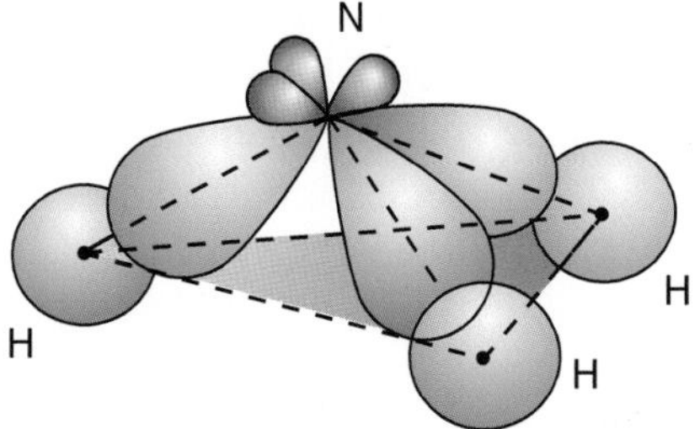


**Fig. 10.13.** *sp*-hybrid orbitals of the valence electrons of the N atom, optimizing the bonding energy of the $NH_3$ molecule

atom. The hybridization enlarges the angles between the bonds from 90° to 107.3°. The structure of the nonplanar $NH_3$ molecule is that of a pyramid with an equilateral triangle as basis plane and three triangles as side planes (Fig. 4.14). The asymmetric charge distribution generates an electric dipole moment $p = 5 \times 10^{-30}\,\mathrm{C\,m} = 1.5\,\mathrm{Debye}$ directed along the pyramid axis from the N atom towards the center of the basis triangle of the three H atoms.

The potential energy $E(h)$ as a function of the heights $h$ of the N atom above the plane of the three H atoms has a maximum for $h = 0$ (Fig. 10.14) and two minima for $h = \pm h_0$, because the N atom can be above or below the plane of the H atoms. The two mirror image configurations are not distinguishable. When the molecule vibrates (see Sect. 10.3) the N atom can tunnel through the potential barrier, where the tunnel frequency is small compared with the vibrational frequency. This is illustrated by a classical model in Fig. 10.15.

In order to obtain the energy eigenvalues of the vibrating and tunneling molecule the two nondistinguishable configurations with the N atom above and below the plane of the H atoms have to be taken into account. The vibrational wave function $\psi_{\mathrm{vib}}$ therefore

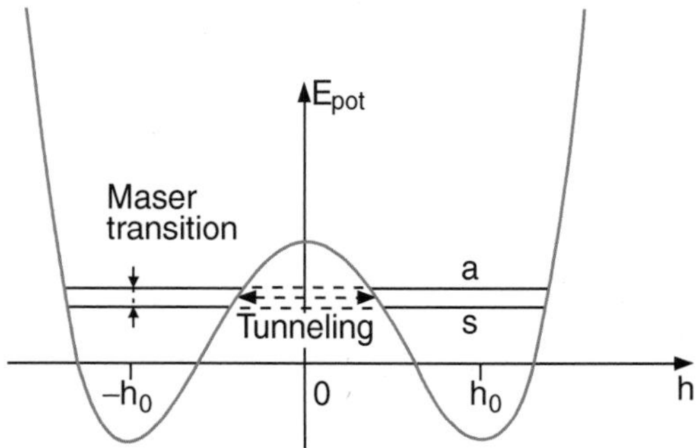


**Fig. 10.14.** Double minimum potential $E_{\mathrm{pot}}(h)$ of $NH_3$ with tunneling splitting of vibrational levels

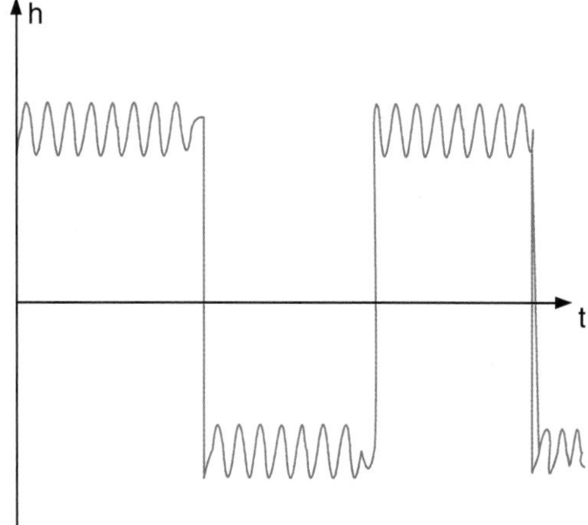


**Fig. 10.15.** Classical model of the vibration of the N atom against the plane $h = 0$ of the three H atoms

has to be a linear combination of the two possibilities $\psi_1$ and $\psi_2$. We can form a normalized symmetric and an antisymmetric combination

$$\psi_{\mathrm{s}} = N(\psi_1 + \psi_2)\,, \quad \psi_{\mathrm{a}} = N(\psi_1 - \psi_2) \qquad (10.12)$$

of harmonic oscillator functions, which lead to vibrational sublevels with slightly different energies. The normalization factor $N$ is chosen such, that

$$\int |\psi_{\mathrm{s}}|^2\,\mathrm{d}\tau_{\mathrm{vib}} = \int |\psi_{\mathrm{a}}|^2\,\mathrm{d}\tau_{\mathrm{vib}} = 1\,.$$

The splitting increases with increasing vibrational energy but is still small compared to the vibrational energy.

This inversion splitting of the vibrational levels of $NH_3$ has become famous, since the first maser, developed in 1955 by *C.H. Townes* and *A.L. Schawlow*, was based on a transition between two inversion levels of $NH_3$ in the microwave region.

**Note:**

In the quantum mechanical formalism of the time-independent Schrödinger equation the wave functions (10.12) of the two states with well defined energies are spread out over both sides of the potential in Fig. 10.14. The classical model in Fig. 10.15 describes the dynamical process of the tunnelling N-atom, where the tunnel frequency is smaller than the vibration frequency. The classical model has one deficiency: It gives the impression that one can know at which time the

N-atom is on one side of the plane of the three H-atoms. This is, however, not possible according to quantum mechanics.

### 10.2.2 Formaldehyde and Other $H_2AB$ Molecules

The formaldehyde molecule $H_2C = O$ has a planar geometry in its electronic ground state where all four atoms are in the $yz$-plane (Fig. 10.16a). In its first excited state, however, it shows a nonplanar pyramidal structure, where the two H atoms are above and below the $yz$-plane (Fig. 10.16b). The molecule has 12 valence electrons, which occupy the six lowest molecular orbitals, where the four lowest are $\sigma$ type orbitals while the next two are $\pi$ type orbitals (Fig. 10.16c). One of these $\pi$ orbitals contributes to the bonding between the C and the O atom, the other $\pi$ orbital is nonbonding. If an electron is excited from the nonbonding orbital into a higher bonding orbital the bonding energy of this excited state is larger than in the ground state. The potential energy in the first excited state has, similarly to $NH_3$, two minima separated by a potential barrier. The two H atoms can tunnel through this barrier with a tunnel frequency that is higher than in $NH_3$, because of the smaller masses of the H atoms compared to that of the N atom in $NH_3$.

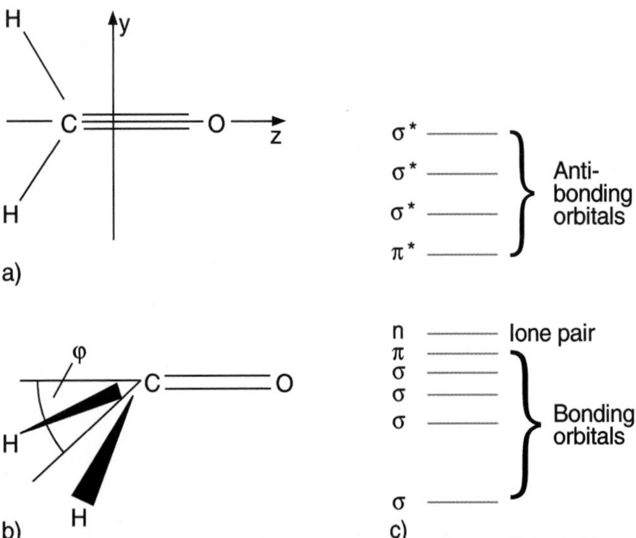


**Fig. 10.16a–c.** Geometry of the formaldehyde molecule. (**a**) Planar ground state geometry. (**b**) Nonplanar geometry of the first excited state. (**c**) Energetic order of the molecular orbitals for $H_2AB$ molecules

A similar molecular orbital scheme as in Fig. 10.16c applies to all other $H_2AB$ molecules, although there are not many stable molecules of this type. The $H_2O_2$ or $H_2C_2$ molecules are examples of $H_2AB$ molecules with $A = B$.

### 10.2.3 Aromatic Molecules and π-Electron Systems

In the foregoing sections we have discussed localized bonds in molecules. Here the spatial distribution of the charge of valence electrons contributing to the bonding is restricted to a narrow volume between the two bonding atoms.

There is an important class of molecules, called conjugated or aromatic molecules, where delocalized electrons play an essential role. One example is the butadiene molecule $C_4H_6$ (Fig. 10.17), where alternating single and double bonds between the C atoms occur. The electrical conductivity of these molecules along the direction of the carbon chain is much larger than in molecules with localized bonds. This hints already to easily moving delocalized electrons. It turns out that these delocalized electrons belong to overlapping atomic $p$ orbitals, which result in molecular $\pi$ orbitals. The spatial distribution of the electron charge $e|\Phi|^2$ is determined by the molecular orbital $\Phi$. In Fig. 10.18 the wave functions $\Phi$ for the four lowest molecular orbitals are schematically shown. The bent chain of the four C atoms is here approximated as a straight line. This figure illustrates again that the energy increases with the number of nodes of the wave function. The larger the delocalization of the wave function (i.e. the smaller the number of nodes) the lower is the energy.

We will now discuss this using the benzene molecule $C_6H_6$ as an example (Fig. 10.19).

Spectroscopic experiments have shown that benzene has a planar structure in its electronic ground state.

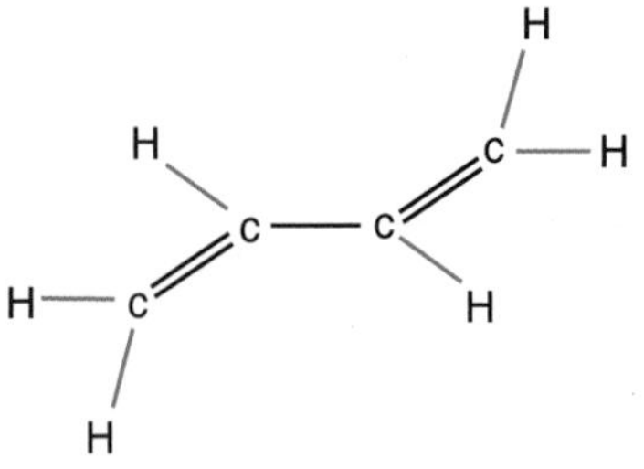

**Fig. 10.17.** Butadiene molecule

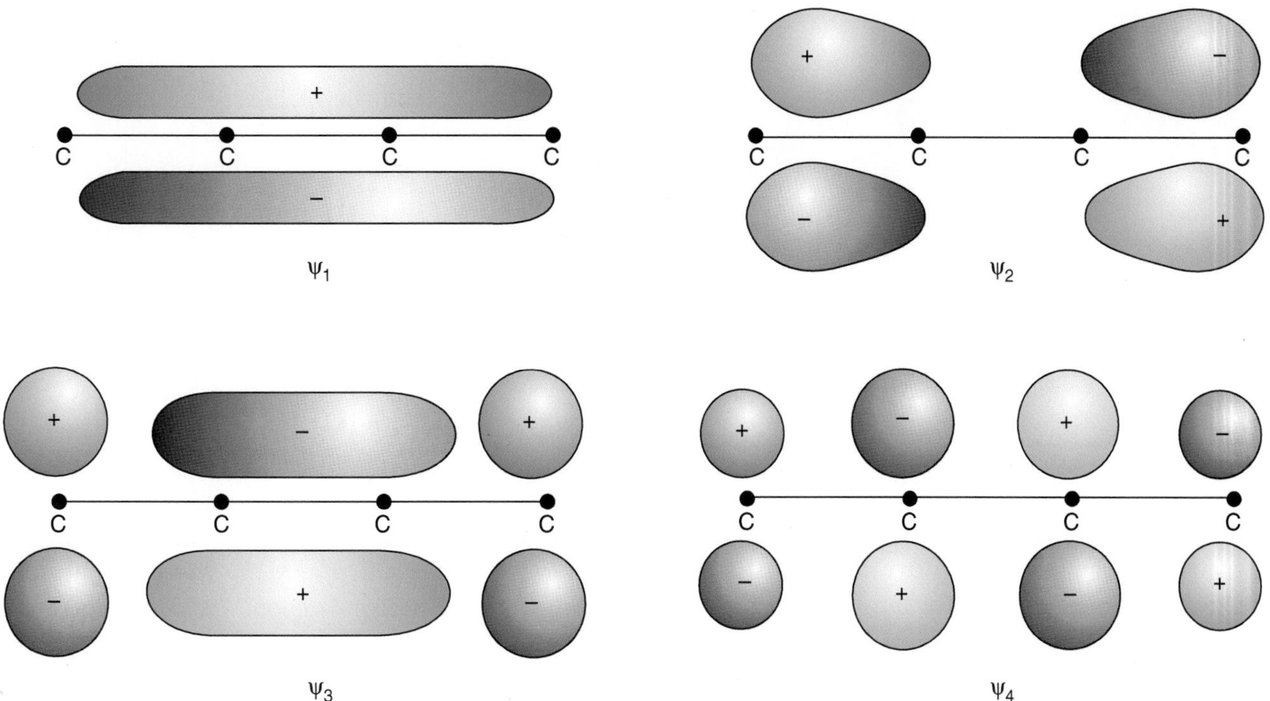


**Fig. 10.18.** Schematic illustration of the $\pi$ orbitals for the four lowest energy states of butadiene

The six C atoms form a hexagon with bond angles $\alpha = 120°$ between the C – C bonds. As has been discussed in Sect. 10.1.2 this bond angle suggests an $sp^2$ hybridization of the atomic orbitals of the C atoms. We can therefore conclude that localized $\sigma$-type bonding between neighboring C atoms and between C and H atoms are present, where one valence electron of the C atom is involved in each of these bonds. This takes three electrons from each C atom and one electron from each H atom (Fig. 10.19a).

There is still one valence electron per C atom left over in the $p_z$ orbital, altogether six electrons that are not involved in the $sp^2$ hybridization. These $p_z$ orbitals are directed perpendicular to the $xy$-plane of the molecule (Fig. 10.19b). They can be used for another bonding type that is not localized. This can be understood as follows. There are two nondistinguishable possibilities for how two electrons from neighboring $p_z$ orbitals with opposite spins can contribute to the bonding, which are illustrated in Fig. 10.19c,d. For all six $p_z$ electrons this gives six different combinations. The total electronic wave function for these $p_z$ electrons is a linear combination

$$\Phi = \sum c_i \phi_i \tag{10.13}$$

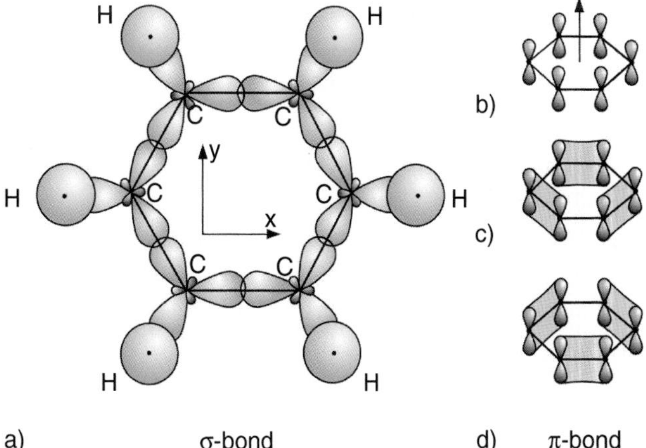


**Fig. 10.19a–d.** The benzene molecule $C_6H_6$. (**a**) $\sigma$-type molecular orbitals formed with $sp^2$-hybrid atomic orbitals from the C atoms (**b**) $\pi$-type orbitals (**c**), (**d**) two nondistinguishable $\pi$-orbital configurations

of the $p_z$ orbitals of the six C atoms, forming a molecular $\pi$ orbital.

This linear combination spreads the wave function over all C atoms in the hexagon. We therefore call it "delocalized." The delocalized electrons are uniformly spread out around the ring of the C atoms and they contribute to the stability of the planar structure.

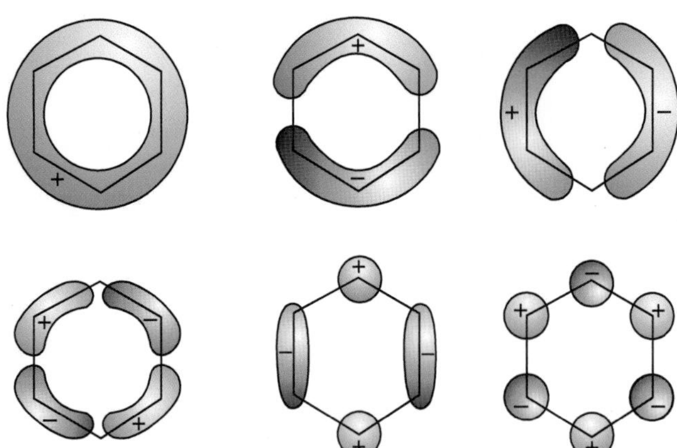


**Fig. 10.20.** Delocalized $\pi$ orbitals of the six lowest energy states of the benzene molecule $C_6H_6$

The spatial distribution of the wave functions for the $\pi$-electron system are shown for the six lowest states of $C_6H_6$ in Fig. 10.20.

In a simple model, electrons that have equal probabilities to be found anywhere along an interval $L$ (in our case $L$ is the circumference of the hexagon) can be treated like electrons in a potential box (see Sect. 4.2.4). They have discrete energy levels

$$E_n = \frac{n^2 h^2}{2m_e L^2}\,; \quad n = 1, 2, 3, \ldots\,, \tag{10.14}$$

which are defined by the condition $L = n\lambda_{DB} = n \cdot h/p$ and $E_{kin} = p^2/2m$.

For our benzene molecule example the C − C distance is 140 pm $\Rightarrow L = 6 \times 140\,\text{pm} = 840\,\text{pm}$.

Transitions between energy levels $n \to n+1$ have the energy

$$h\nu = \Delta E = \frac{h^2(2n+1)}{2m_e L^2}\,. \tag{10.15}$$

Inserting the numerical values for $n = 1$ gives $\Delta E = 1 \times 10^{-18}\,\text{J} \Rightarrow \nu = 1.5 \times 10^{15}\,\text{s}^{-1} \Rightarrow \lambda = 200\,\text{nm}$.

Excitation of the $\pi$ electrons of benzene therefore results in absorption bands in the UV region around $\lambda = 200\,\text{nm}$.

The results of this crude model agree fairly well with the observed absorption wavelength of $\lambda = 220\,\text{nm}$. The difference stems from the fact that the interaction between the electrons has been neglected in this model.

The bonding in aromatic molecules is based on two effects:
Localized bonding between neighboring C atoms and the C and H atoms, which are brought about by $sp^2$-hybrid atomic orbitals of the C atoms, and delocalized $\pi$ orbitals, which extend along all C atoms (for benzene, over all six C atoms in the ring).

## 10.3 Rotation of Polyatomic Molecules

While for diatomic and linear polyatomic molecules rotation is only possible around an axis through the center of mass ***perpendicular to the molecular symmetry axis***, for nonlinear polyatomic molecules there are more possibilities. A free three-dimensional rotor can rotate around any axis $\omega$ through the centre of mass $S$ (Fig. 10.21). The moment of inertia depends on the orientation of the rotation axis in the molecular coordinate system $(x, y, z)$, which is generally chosen in such a way that the main symmetry axis of the molecule coincides with the $z$-axis.

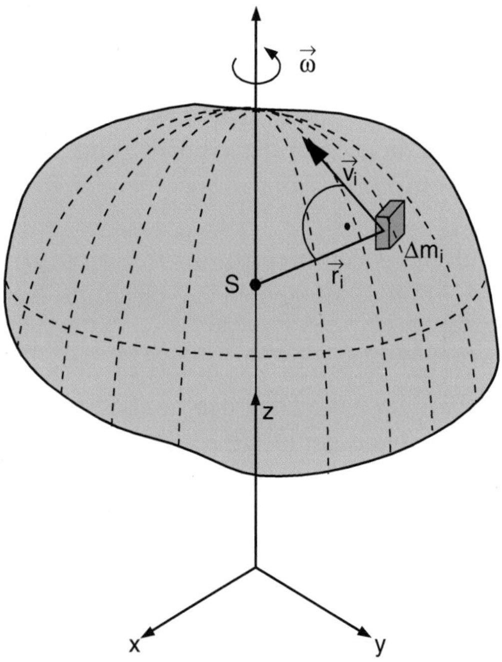


**Fig. 10.21.** Rotation of a rigid body around an arbitrary axis through the center of mass

Although in principle rotations around arbitrary axes through the center of mass are possible, not all of these rotations are stable. The rotation axis can change its orientation against the axes of the molecular coordinate system and of the laboratory coordinate system. However, the total angular momentum $J$ of a free molecule without external torque is always constant. The molecule therefore rotates around a momentary rotational axis $\boldsymbol{\omega}$ that nutates around the space fixed direction of the angular momentum (Fig. 10.22).

**Note:**

The mutation is sometimes called "torque-free precession", while the real precession only occurs when an external torque acts on the molecule.

The moment of inertia $I$ with respect to the rotation axis depends on the orientation of this axis in the molecular coordinate system. It can be described by a tensor. This can be seen for the example of a rigid rotating body as follows (Fig. 10.21).

The angular momentum $\boldsymbol{L}_i$ of a mass element $\Delta m_i$ with a distance $r_i$ from the rotation axis is

$$\boldsymbol{L}_i = \Delta m_i(\boldsymbol{r}_i \times \boldsymbol{v}_i) = \Delta m_i(\boldsymbol{r}_i \times (\boldsymbol{\omega}_i \times \boldsymbol{r}_i)) . \tag{10.16a}$$

This can be rearranged to

$$\boldsymbol{L}_i = \Delta m_i \left[ \left( r_i^2 \boldsymbol{\omega} \right) - (\boldsymbol{r}_i \cdot \boldsymbol{\omega})\boldsymbol{r}_i \right] . \tag{10.16b}$$

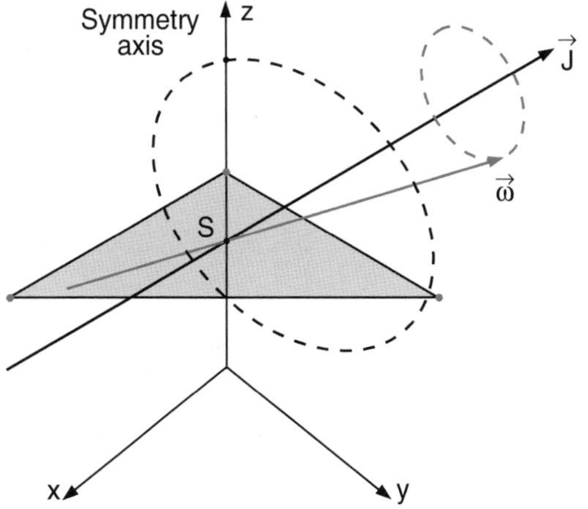


**Fig. 10.22.** The momentary rotation axis $\boldsymbol{\omega}$ and the molecular symmetry axis nutate around the space fixed angular momentum $\boldsymbol{J}$

The angular momentum of the whole rotating body is then

$$\boldsymbol{L} = \int \left[ r^2 \boldsymbol{\omega} - (\boldsymbol{r} \cdot \boldsymbol{\omega})\boldsymbol{r} \right] \mathrm{d}m . \tag{10.17a}$$

This vector equation corresponds to the three equations for the components

$$\begin{aligned} L_x &= I_{xx}\omega_x + I_{xy}\omega_y + I_{xz}\omega_z , \\ L_y &= I_{yx}\omega_x + I_{yy}\omega_y + I_{yz}\omega_z , \\ L_z &= I_{zx}\omega_x + I_{zy}\omega_y + I_{zz}\omega_z , \end{aligned} \tag{10.17b}$$

where the coefficients $I_{ik}$ are abbreviations for the expressions

$$\begin{aligned} I_{xx} &= \int \left( r^2 - x^2 \right) \mathrm{d}m , \quad I_{yy} = \int \left( r^2 - y^2 \right) \mathrm{d}m, \\ I_{zz} &= \int \left( r^2 - z^2 \right) \mathrm{d}m , \\ I_{xy} &= I_{yx} = -\int xy \, \mathrm{d}m , \quad I_{yz} = I_{zy} = -\int yz \, \mathrm{d}m, \\ I_{xz} &= I_{zx} = \int xz \, \mathrm{d}m \end{aligned} \tag{10.18}$$

as can be seen by inserting $I_{ik}$ into (10.17a) and using the relations $r^2 = x^2 + y^2 + z^2$ and $\boldsymbol{r} \cdot \boldsymbol{\omega} = x\omega_x + y\omega_y + z\omega_z$. The components $I_{ik}$ can be arranged in a matrix and (10.17b) can then be written as

$$\begin{pmatrix} L_x \\ L_y \\ L_z \end{pmatrix} = \begin{pmatrix} I_{xx} & I_{xy} & I_{xz} \\ I_{yx} & I_{yy} & I_{yz} \\ I_{zx} & I_{zy} & I_{zz} \end{pmatrix} \cdot \begin{pmatrix} \omega_x \\ \omega_y \\ \omega_z \end{pmatrix} \tag{10.19a}$$

or in the abbreviated form as

$$\boldsymbol{L} = \tilde{I} \cdot \boldsymbol{\omega} .$$

Mathematically $\tilde{I}$ is a tensor. The scalar product of this tensor with the vector $\boldsymbol{\omega}$ gives the vector $\boldsymbol{L}$.

Plotting $1/\sqrt{I}$ as a function of the direction against the coordinate axes gives an ellipsoid with its center at the origin (Fig. 10.23).

When choosing the axis of the molecular coordinate system in such a way that two of the axes coincide with the smallest and the largest moment of inertia, the inertia tensor can be transformed into a diagonal form. The three terms $I_a$, $I_b$, and $I_c$ in the diagonal are called the **principal moments of inertia** around the principal axes $a$, $b$ and $c$. If two of these moments are equal, the system is called a ***symmetric rotor***, if

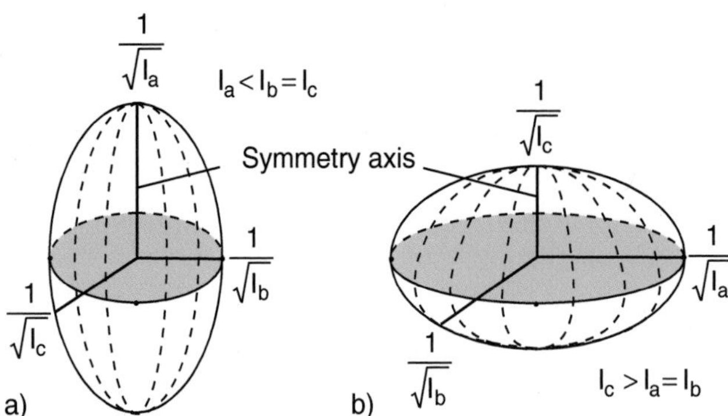


**Fig. 10.23a,b.** Moment of inertia ellipsoid for a symmetric top. (**a**) Prolate top, (**b**) oblate top

all three are different, we have an ***asymmetric rotor***. All rigid bodies with a rotational symmetry axis are symmetric rotors. For $I_a < I_b = I_c$ the moment of inertia for a rotation around the symmetry axis is smaller than the other two. We call this a ***prolate symmetric top*** (Fig. 10.23a). For $I_c > I_a = I_b$ we have an ***oblate symmetric top*** (Fig. 10.23b).

Note that the body can rotate around any arbitrary axis $\omega$ through the center of mass (Fig. 10.24).

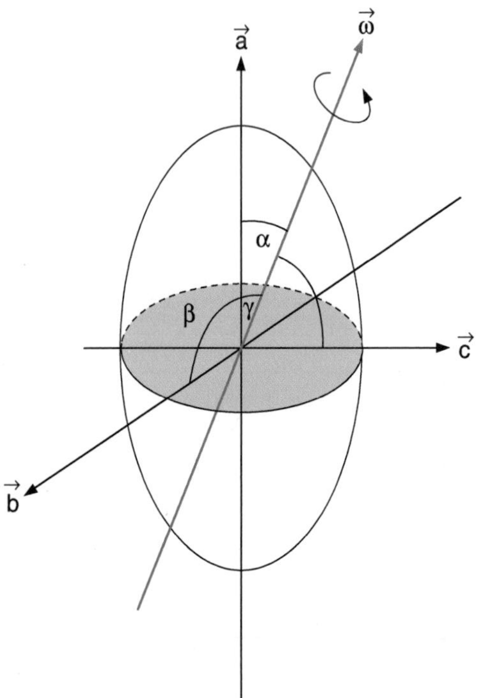


**Fig. 10.24.** The three principal axes of inertia for a rigid rotating body and the momentary rotation axis $\boldsymbol{\omega}$

In Fig. 10.25 the principal axes are shown for the rotation of the methyl iodide molecule $ICH_3$, which is a symmetric rotor. The moments around the $b$- and $c$-axis are equal but different from that around the $a$-axis, which is the symmetry axis. Both the momentary rotation axis and the molecular symmetry axis precess around the space fixed direction of the rotational angular momentum $J$ (Fig. 10.25b).

Any rotation around an arbitrary rotation axis through the centre of mass can always be composed of rotations around the principal axes of the molecule.

If the angular momentum $J$ is written as

$$\boldsymbol{J} = \{J_a, J_b, J_c\} = \{\omega_a I_a, \omega_b I_b, \omega_c I_c\}\,. \tag{10.20}$$

The rotational energy $E_{\text{rot}} = \frac{1}{2} I \omega^2$ is

$$\begin{aligned} E_{\text{rot}} &= \frac{1}{2}\left(\omega_a^2 I_a + \omega_b^2 I_b + \omega_c^2 I_c\right) \\ &= \frac{J_a^2}{2I_a} + \frac{J_b^2}{2I_b} + \frac{J_c^2}{2I_c}\,. \end{aligned} \tag{10.21}$$

Since the principal axes are precessing around the space fixed rotational angular momentum $J$, the components $J_a$, $J_b$ and $J_c$ are generally not constant in time although the vector $\boldsymbol{J}$ and the rotational energy $E_{\text{rot}}$ are both constant.

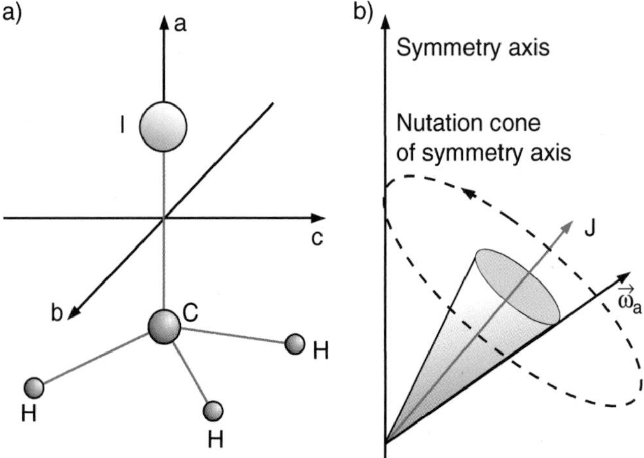


**Fig. 10.25.** (**a**) Principle axes of inertia for the methyl iodide molecule $ICH_3$ which is a prolate symmetric top with the $a$-axis as the symmetry axis. (**b**) Momentary rotation axis $\boldsymbol{\omega}_a$ and symmetry axis of a torque-free symmetric top both mutating around the space-fixed direction of the angular momentum $\boldsymbol{J}$

### 10.3.1 Rotation of Symmetric Top Molecules

The description of the rotation of polyatomic molecules is greatly simplified for symmetric top molecules where two of the principal moments of inertia are equal. The symmetry axis of the molecule precesses around the space fixed rotational momentum $J$, unless $J$ and $\omega$ coincide with the symmetry axis (Fig. 10.25b).

We assume the $a$-axis as the symmetry axis of the prolate symmetric top. Then $I_a \neq I_b = I_c$ and (10.21) can be simplified to

$$E_{\text{rot}} = \frac{J_a^2}{2I_a} + \frac{J_b^2 + J_c^2}{2I_b} = \frac{J^2 - J_a^2}{2I_b} + \frac{J_a^2}{2I_a} \tag{10.22}$$

$$= \frac{J^2}{2I_b} + J_a^2\left(\frac{1}{2I_a} - \frac{1}{2I_b}\right).$$

In the quantum mechanical description the angular momentum and one of its components can be simultaneously measured. We choose the symmetry axis as the quantization axis and as the $z$-axis of the molecular coordinate system. We then obtain the eigenvalues

$$\left\langle \hat{J}^2 \right\rangle = J(J+1)\hbar^2\,; \quad \left\langle \hat{J}_a \right\rangle = K\hbar \tag{10.23}$$

where $K\hbar$ is the projection of $\boldsymbol{J}$ onto the symmetry axis of the symmetric top. The projection quantum number $K$ can take all $(2J+1)$ values $-J \leqslant K \leqslant +J$. The rotational energy of a symmetric top molecule is then

$$\boxed{E_{\text{rot}} = \frac{J(J+1)\hbar^2}{2I_b} + K^2\hbar^2\left(\frac{1}{2I_a} - \frac{1}{2I_b}\right)}\,. \tag{10.24}$$

We introduce, similarly to the situation for diatomic molecules in Sect. 9.5.2, the rotational constants

$$A = \frac{\hbar}{4\pi c I_a}\,, \quad B = \frac{\hbar}{4\pi c I_b} = C = \frac{\hbar}{4\pi c I_c} \tag{10.25}$$

and obtain for the rotational term values $F_{\text{rot}} = E_{\text{rot}}/hc$ of the prolate top with $A > B = C$:

$$\boxed{F_{\text{rot}} = BJ(J+1) + (A-B)K^2}\,. \tag{10.26a}$$

Different from the case of diatomic molecules we obtain not a single rotational term ladder but a multitude of ladders, one for each $K$ value. Since $J$ cannot be smaller than $K$, each rotational term ladder starts with a $J$ value $J_{\text{min}} = K$.

**Note:**

1. The rotational energies depend on $K^2$. Therefore each rotational level is twofold degenerate except for $K = 0$, because terms with $+K$ have the same energy as those with $-K$.
2. In the literature the principal axis are labelled in such a way, that $A \geq B \geq C$. In this nomenclature a prolate symmetric top is described by $A > B = C$ with the $a$-axis as symmetry axis, while for the oblate top $A = B > C$ the $c$-axis is chosen as symmetry axis (Fig. 10.23). The rotational term values of an oblate symmetric top are:

$$\boxed{F_{\text{rot}} = B \cdot J(J+1) + (C-B)K^2}\,. \tag{10.26b}$$

The energetic order of the rotational term values depends on the relative magnitude of the rotational constants $A$ and $B$. For $A > B = C$ (***prolate symmetric top***) the energies for a given value of $J$ increase with increasing $K$ ((10.26a) and Fig. 10.26a). For $C < B = A$ (***oblate symmetric top***) they decrease ((10.26b) and Fig. 10.26b).

Examples for symmetric top molecules are all linear molecules, such as $CO_2$, NCO or HCN, but also a large group of planar molecules with a symmetry axis perpendicular to the molecular plane, such as $Na_3$, and nonplanar molecules with a symmetry axis such as methyl iodide (Fig. 10.21a).

The intensity of a rotational line in absorption is proportional to the power

$$\Delta P(J_1, K_1 \to J_2, K_2) = [N(J_1, K_1) - N_2(J_2, K_2)] \times B_{12} w_\nu(\nu_{12}) h\nu_{12} \tag{10.27}$$

absorbed in the transition $(J_1, K_1) \to (J_2, K_2)$ in a radiation field with spectral energy density $w_\nu(\nu_{12})$.

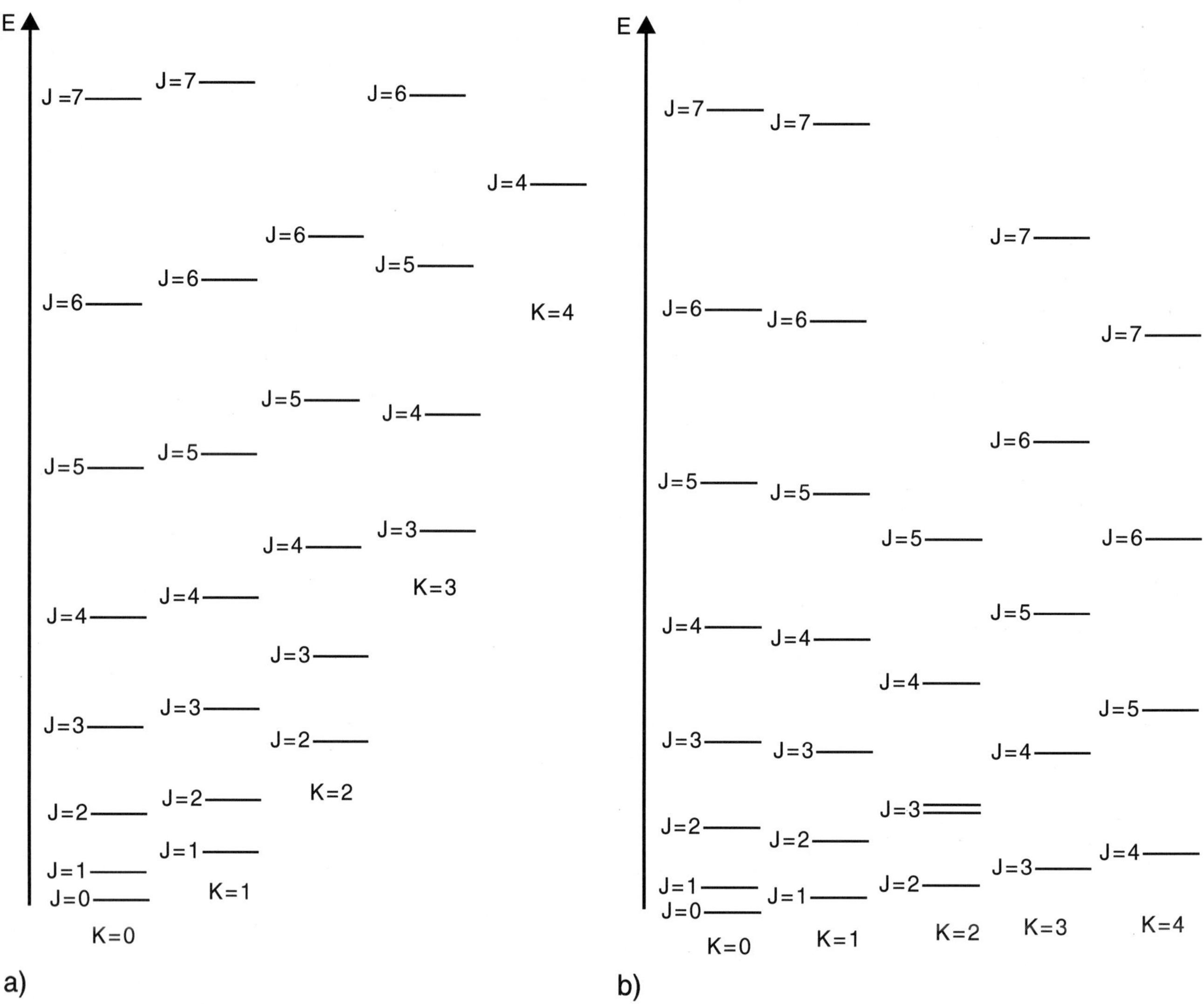


**Fig. 10.26a,b.** Rotational energy levels $E(J,K)$ (**a**) for a prolate, (**b**) for an oblate symmetric top

The population density $N(J,K)$ of a rotational levels at thermal equilibrium is given by the Boltzmann distribution

$$N(J,K) = \frac{N}{Z} g_k (2J+1) \mathrm{e}^{-E_{\mathrm{rot}}/kT} , \qquad (10.28)$$

where $N = \sum N(J,K)$ is the total population in all rotational levels and

$$Z = \sum (2J+1) \times \mathrm{e}^{-E_{\mathrm{rot}}/kT}$$

is the partition function, which acts as a normalization factor such that $\sum N(J,K) = N$, and $g_k = 2 - \delta_{k0}$ i.e. $g_k = 1$ for $K = 0$ and $g_k = 2$ for $K \neq 0$.

At room temperature the energy difference $\Delta E$ between two rotational levels is generally small compared with $kT$. We can therefore write (10.27) with $h\nu = \Delta E$ as

$$\Delta P = B_{12} w_\nu h\nu \cdot N_1 \left(1 - \mathrm{e}^{\Delta E/kT}\right) \approx B_{12} w_\nu N_1 \frac{(\Delta E)^2}{kT} . \qquad (10.29)$$

This shows that the intensity of a rotational transition is proportional to the population density $N_1$ of the absorbing level and to the ratio $(\Delta E)^2/kT$.

The wavenumber of a rotational transition is

$$\overline{\nu}_{\mathrm{rot}} = F(J_2, K_2) - F(J_1, K_1) = 2B(J+1) \quad \text{for } \Delta J = \pm 1 , \ \Delta K = 0 \qquad (10.30)$$

$$\overline{\nu}_{\text{rot}} = 2B(J+1) \pm (A-B)(2K+1)$$
$$\text{for } \Delta K = \pm 1 \, .$$

### 10.3.2 Asymmetric Rotor Molecules

For most polyatomic molecules, all three rotational constants are different. There is no symmetry axis for the definition of the quantization axis and therefore the projection $K\hbar$ is not defined. The theoretical treatment of such molecules is more complicated and exceeds the level of this textbook.

Often, two of the principal moments of inertia are not very different and the molecule can be regarded as a "near symmetric top." In such cases the asymmetric top molecule can be characterized by its two limiting cases $I_a < I_b = I_c \Rightarrow A > B = C$ (prolate symmetric top) and $I_c > I_a = I_b \Rightarrow C < A = B$ (oblate symmetric top). The projection of the rotational angular momentum $J$ onto the symmetry $a$-axis of the prolate top is labeled $K_a$ and that onto the $c$-axis of the oblate top by $K_c$. The rotational level of the asymmetric top is characterized by the two numbers $K_a$ and $K_c$ as $J_{K_a,K_c}$. Note, however, that $K_a$ and $K_c$ are not real quantum numbers, since the corresponding components of the angular momentum vector both do not commute with the rotational Hamiltonian. For each rotational quantum number $J$, several sublevels $(K_a, K_c)$ exist, where all combinations of $K_a$ and $K_c$ are allowed for which $K_a + K_c \leqslant J$ or $\leqslant J+1$, depending on the symmetry properties of the rotational level.

## 10.4 Vibrations of Polyatomic Molecules

In a molecule with $N$ atoms each atom has three degrees of freedom for its motions. The molecule therefore has $3N$ degrees of freedom. The motion of the center of mass in the three-dimensional space has three degrees of freedom. Nonlinear molecules have another three rotational degrees of freedom for the rotation around three independent axes. Therefore $3N-6$ degrees of freedom are left for the molecular vibrations (for linear molecules $3N-5$, because there are only rotations around two axes perpendicular to the molecular axis). All vibrations of the molecule can therefore be composed of $3N-6$ (or $3N-5$ for linear molecules) vibrational modes.

### 10.4.1 Normal Vibrations

If the nuclei of a molecule are removed from their equilibrium positions, where the potential energy has a minimum, restoring forces try to bring the nuclei back to their positions of minimum energy. As long as the displacements are sufficiently small, the potential is harmonic and the restoring forces are linear functions of the displacements, resulting in harmonic oscillations of the nuclei. For sufficiently small vibrational amplitudes all possible vibrations of a molecule can be composed as linear combinations of so-called "normal vibrations." These are vibrations where all nuclei move at the same time through their equilibrium position. They therefore move either in phase or with opposite phases.

All vibrations of a molecule at rest must have a total momentum $P = 0$. Otherwise the center of mass of the molecule would move and we would have a translation. Also, the total angular momentum $J$ of the vibrating, nonrotating molecule must be zero. This leaves for a nonlinear molecule with $N$ atoms $(N-6)$ degrees of freedom for vibrations, for a linear molecule $N-5$, because have are only rotations around 2 axis perpendicular to the molecular axis possible. The $3N-6$ normal vibrations of a nonlinear molecule can be found by looking for all simultaneous displacements of the nuclei (in phase or with opposite phases), that fulfill the two conditions $P = 0$ and $J = 0$.

In Fig. 10.27 the normal vibrations of a nonlinear and a linear triatomic molecule are illustrated. The nonlinear molecule has $3N-6 = 9-6 = 3$ normal vibrations, the linear molecule $3N-5 = 4$, where two of the vibrations (the bending vibrations $\nu_2$ in the $x$ and in the $y$ directions, i.e., in the plane of the figure and perpendicular to it) have the same energy because of the cylindrical symmetry of the potential.

### 10.4.2 Quantitative Treatment

We will now discuss the normal vibrations of a molecule more quantitatively.

The $3N$ coordinates of the $N$ nuclei in the molecular frame are $x_i$, $y_i$, $z_i$ $(i = 1, 2, \dots N)$ and their

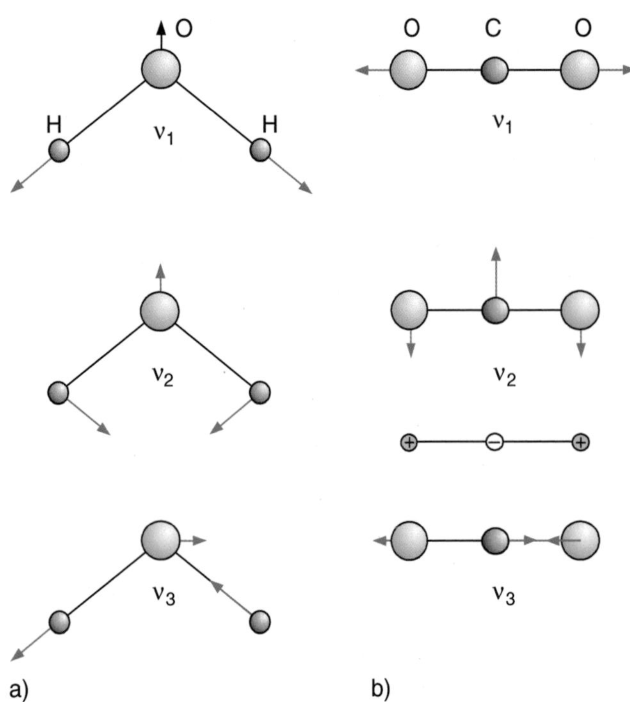


**Fig. 10.27a,b.** Normal vibrations for two triatomic molecules. (**a**) Nonlinear molecule $H_2O$, (**b**) linear molecule $CO_2$

equilibrium positions are $x_{i0}$, $y_{i0}$, $z_{i0}$. We define the displacements during a vibration as

$$\begin{aligned} \xi_1 &= x_1 - x_{10}, \quad \xi_2 = y_1 - y_{10}\,, \\ \xi_3 &= z_1 - z_{10}, \quad \xi_4 = x_2 - x_{20} \ldots \xi_{3N} = z_N - z_{N0}, \end{aligned} \tag{10.31}$$

with indices running from 1 to $3N$ (Fig. 10.28).

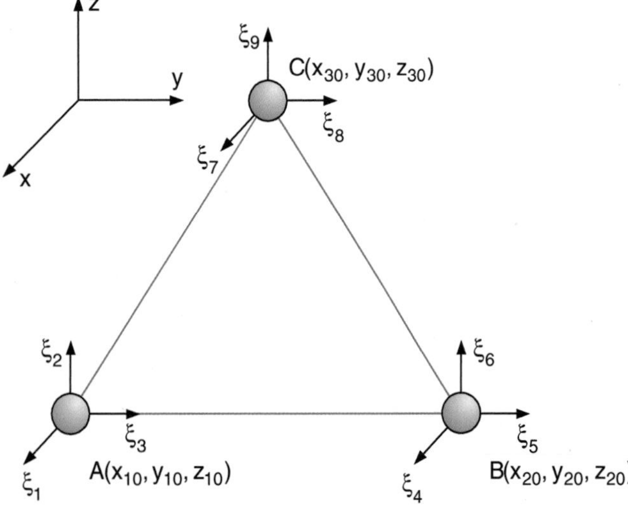


**Fig. 10.28.** Coordinates $(x, y, z)$ of equilibrium positions and displacements $\xi_i$ of the nuclei A, B, C in a vibrating triatomic molecule

The potential $E_{\text{pot}}(\xi_1 \ldots \xi_{3N})$ of the vibrating nuclei depends on the displacements $\xi_i$. For sufficiently small values of $\xi$ we can expand the potential into a Taylor series

$$\begin{aligned} E_{\text{pot}} &= E_0 + \sum_1 \left(\frac{\partial E_{\text{pot}}}{\partial \xi_i}\right)_0 \xi_i \\ &\quad + \frac{1}{2}\sum_{i,j}\left(\frac{\partial^2 E_{\text{pot}}}{\partial \xi_i \partial \xi_j}\right)_0 \xi_i \xi_j + \ldots \end{aligned} \tag{10.32}$$

where we neglect all higher order terms beyond the quadratic term.

If we choose the energy of the minimum of the potential at $\xi = 0$ as $E_{\text{pot}}(\xi = 0) = 0$ and take into account that for the minimum $\partial E_{\text{pot}}/\partial \xi_i)_0 = 0$ we obtain

$$E_{\text{pot}}\,(\xi_1 \ldots \xi_2) = \frac{1}{2}\sum_{i,j} b_{ij}\xi_i\xi_j \tag{10.33}$$

$$\text{with} \quad b_{ij} = \left(\frac{\partial^2 E_{\text{pot}}}{\partial \xi_i\, \partial \xi_j}\right)_0 .$$

The components of the restoring force are

$$F_i = -\partial E_{\text{pot}}/\partial \xi_i\,. \tag{10.34}$$

The equations of motion for the vibrating nuclei are therefore

$$F_i = m_1 \frac{\mathrm{d}^2 \xi_i}{\mathrm{d}t^2}\,. \tag{10.35}$$

With the introduction of mass-weighted coordinates

$$q_i = \sqrt{m_k}\xi_i\,,$$

where $m_k$ is the mass of the nucleus involved in the oscillation $q_i$, we obtain from (10.33–10.35) a system of $3N$ homogeneous differential equations

$$\ddot{q}_i + \sum_{j=1}^{3N} b_{ij} q_i = 0\,. \tag{10.36}$$

This is a coupled system of equations because the coefficients $b_{ij}$ couple the motion of the $i$th nucleus with that of the $j$th nucleus. It describes the motions of $N$ coupled harmonic oscillators with the time dependent mass-weighted displacements

$$q_i = d_i\,\cos(\omega_i t + \varphi_i) \tag{10.37}$$

from their equilibrium positions. These oscillators have the oscillation amplitudes $d_i$, the frequencies $\omega_i$ and the phases $\varphi_i$. Inserting (10.37) into (10.36) gives the relationship between the frequencies $\omega_i$ and the potential parameters $b_{ij}$. Generally the restoring force for the displacement $q_i$ will be affected by the other displacements $q_k$ because the nondiagonal potential parameters $b_{ij}$ cause couplings between the different oscillations. Only for certain initial conditions can all nuclei oscillate with the same frequency and the same phase. Such selected vibrations of the molecule are called ***normal vibrations***. They can be obtained in the following way.

Equation (10.36) can be written as the vector equation

$$\ddot{\boldsymbol{q}} + \tilde{B} \cdot \boldsymbol{q} = 0 \,, \tag{10.38}$$

where $\tilde{B} = (b_{ij})$ is the matrix with the components $b_{ij}$ and $\boldsymbol{q} = \{q_1, q_2, \dots q_{3N}\}$ is the mass-weighted displacement vector of all $N$ nuclei. If $\tilde{B}$ is a diagonal matrix

$$\tilde{B} = \lambda \cdot \tilde{E}$$

$$\text{with } \tilde{E} = \begin{pmatrix} 1\;0\;0\dots & \\ 0\;1 & \\ \vdots \quad 1 & \\ 0 & 1 \end{pmatrix}$$

where $\tilde{E}$ is the diagonal unity matrix, the system of equations (10.38) would become $3N$ *uncoupled* equations for the displacements $q_i$ with the solutions

$$q_i = A_i \cos \omega t \,, \quad i = 1, 2, \dots 3N \,, \tag{10.39}$$

where all nuclei move with the same frequency $\omega_i = \sqrt{\lambda_i}$ and pass through their equilibrium positions $q_i = 0$ at the same time. In order to meet this situation we have to find a coordinate system in which the matrix $B$ becomes diagonal.

This leads to the equation

$$\tilde{B} \cdot \boldsymbol{q} = \lambda \tilde{E} \cdot \boldsymbol{q} \Rightarrow \left(\tilde{B} - \lambda \tilde{E}\right) \boldsymbol{q} = 0 \,, \tag{10.40}$$

which has nontrivial solutions only, if the determinant

$$\det |\tilde{B} - \lambda \tilde{E}| = 0 \tag{10.41}$$

is zero. For each solution $\lambda_i$ of (10.41) one obtains from (10.40) a set of $3N$ values $q_{ki}(t)$ $(k = 1, 2, \dots 3N)$, which represent the displacements of all $N$ nuclei as a function of time. For a given value of $k$ all $q_{ki}$ can be regarded as the components of a vector

$$Q_k = A_k \cos(\omega_k t + \varphi_k) \text{ with } \omega_k = \sqrt{\lambda_k} \,, \tag{10.42}$$

which describes the simultaneous mass weighted displacements of all $N$ nuclei for the $k$th vibrational mode. The vector $\boldsymbol{Q}_k$ is called the ***normal coordinate*** of the $k$th normal vibration with frequency $\omega_k = \sqrt{\lambda_k}$.

With these normal coordinates the coupled equations (10.38) become uncoupled:

$$\ddot{Q}_k + \omega_k^2 Q_k = 0 \quad (k = 1, 2, 3, \dots 3N) \,. \tag{10.43}$$

The physical reason for this uncoupling is the fact that for each normal coordinate all nuclei oscillate in phase with the same frequency, i.e., the whole molecule oscillates on the frequency of one of the normal vibrations. The mathematical reason is that the potential and kinetic energy written in these normal coordinates become pure quadratic forms without nondiagonal coupling terms

$$E_{\text{kim}} = \frac{1}{2} \sum_k \dot{Q}_k^2 = 0 \,; \quad E_{\text{pot}} = \frac{1}{2} \sum_k \lambda_k Q_k^2 \tag{10.44}$$

if in the expansion (10.32) of the potential energy higher order terms are neglected.

The solutions of (10.43) are the normal vibrations (10.42).

In the quantum mechanical representation, the different normal vibrations of the molecule can be described like vibrations of a linear harmonic oscillator with the energy

$$E(v) = \hbar\omega \left(v_i + \tfrac{1}{2}\right) \,, \tag{10.45}$$

where $v_i$ is the number of vibrational quanta in this normal vibration.

Within the approximation of a harmonic potential any arbitrary vibration of a polyatomic molecule can be composed as a linear combination of normal vibrations with the total energy

$$E_{\text{vib}} = \sum_k \hbar\omega_k \left(v_k + \frac{d_k}{2}\right) \,, \tag{10.46}$$

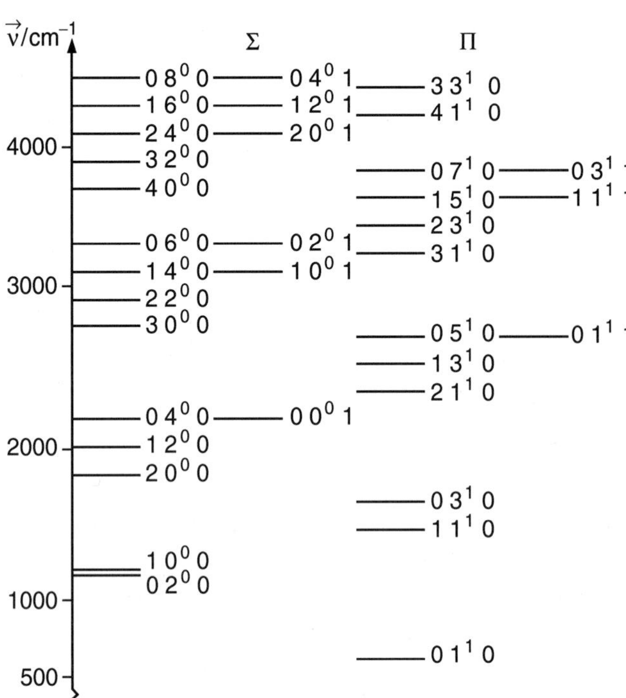


**Fig. 10.29.** The vibrational energy levels of the $CO_2$ molecule as linear combinations of the normal vibrations $(\nu_1, \nu_2, \nu_3)$ with $v_i$ vibrational quanta. The $\Sigma$ levels have the vibrational angular momentum $l = 0$, the $\Pi$-levels have $l = 1$

where $d_k$ gives the degree of degeneracy of the $k$th normal vibration. For doubly degenerate vibrations $d = 2$.

In Fig. 10.29 the vibrational energy levels of the $CO_2$ molecule and their composition of the three normal vibrations are illustrated. The nomenclature is $(\nu_1, \nu_2^l, \nu_3)$, where $\nu_1$ is the symmetric stretch vibration, $\nu_2$ is the bending vibration and $\nu_3$ is the asymmetric stretch vibration (see Fig. 10.27). The assignment (1 0 2) means that the vibration is composed of one vibrational quantum of $\nu_1$, no quantum of the bending vibration $\nu_2$ and two quanta of the asymmetric stretch vibration $\nu_3$.

The degenerate bending vibrations have a peculiarity. They can be composed of two linear bending vibrations in the $xz$- and $yz$-plane, which both have the same vibrational energy. If these two vibrations are superimposed with a phase difference of $\pi/2$, the nuclei perform circular motions around the $z$-axis (Fig. 10.30). This results in an angular momentum $l\hbar$ in the $z$ direction, which is called the ***vibrational angular momentum***. The quantum number $l$ is written as an exponent of the number $v_2$ of vibrational quanta in the bending vibration $\nu_2$. The nomenclature $(1\,3^1\,0)$ in Fig. 10.29, for instance, means $v_1 = 1$, $v_2 = 3$, $l = 1$ and $v_3 = 0$.

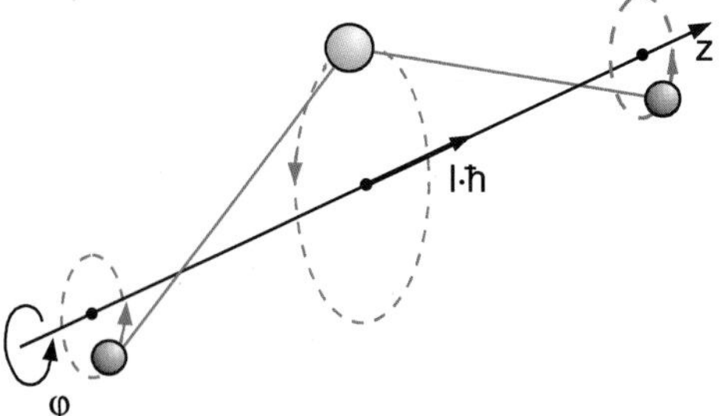


**Fig. 10.30.** Vibrational angular momentum of two degenerate bending vibrations $\nu_2$ with a phase difference $\pi/2$

In the system of normal coordinates the nuclei perform harmonic oscillations in a parabolic potential, where all nuclei vibrate with the same frequency and with either the same or the opposite phase.
For sufficiently small vibrational amplitudes any arbitrary vibration of the molecule can be composed of normal vibrations.

For higher vibrational energies the approximation of the harmonic potential is no longer valid and higher order terms in the potential expansion (10.32) have to be taken into account. Now, the total vibrational energy is no longer the sum of the energies of the normal vibrations, but coupling terms arise, and instead of (10.46) we obtain

$$\begin{aligned} &E_{\text{vib}}(v_1, v_2 \ldots v_{3N-6}) \qquad (10.47)\\ &= \sum \hbar\omega_k \left(v_k + \frac{d_k}{2}\right) \\ &\quad + \sum_{i,k} x_{ik} \left(v_i + \frac{d_k}{2}\right)\left(v_k + \frac{d_k}{2}\right) + \ldots . \end{aligned}$$

This implies that the displacement of a nucleus during its vibration affects the restoring force for the other nuclei, and therefore their vibrational energy.

### 10.4.3 Couplings Between Vibrations and Rotations

Similar to the situation in diatomic molecules the vibrations of polyatomic molecules change the moments of inertia and therefore the rotational constants. Furthermore the rotations lead to a centrifugal stretching of the molecule, which changes the mean distances between the atoms. Like in diatomics the dependence of the rotational constants on the vibrational quantum numbers can be written in the approximation of a parabolic potential:

$$\begin{aligned} A_{v_k} &= A_0 - \sum \alpha_k^{(A)} \left( v_k + \frac{d_k}{2} \right), \\ B_{v_k} &= B_0 - \sum \alpha_k^{(B)} \left( v_k + \frac{d_k}{2} \right), \\ C_{v_k} &= C_0 - \sum \alpha_k^{(C)} \left( v_k + \frac{d_k}{2} \right). \end{aligned} \tag{10.48}$$

For polyatomic molecules, however, a new coupling between rotation and vibration can occur, which is caused by Coriolis forces in the rotating and vibrating molecule and can mix different normal vibrations. This is illustrated in Fig. 10.31, where the molecule rotates around an axis $\omega$ in the $y$ direction perpendicular to the drawing $xz$-plane. For the symmetric stretch vibration $\nu_1$ with velocities $\vec{v}_i$ of the nuclei the Coriolis forces try to decrease or increase the rotational energy, depending on the phase of the vibration. Here we therefore have a rotational-vibrational coupling. The Coriolis force can, however, also couple two different vibrations. For example, if the nuclei perform a bending vibration $\nu_2$ their velocities are directed in the $\pm x$ direction. Therefore a Coriolis force

$$\boldsymbol{F} = 2m(\boldsymbol{v} \times \boldsymbol{\omega}) \tag{10.49}$$

acts in the $z$ direction on the nucleus with mass $m$ and velocity $v$. This force moves nuclei 1 and 3 into the $-z$ direction, nucleus 2 into the $+z$ direction, and therefore induces the asymmetric stretch vibration $\nu_3$ of the molecule. The rotation of the molecule therefore couples via the Coriolis force two normal vibrations. This coupling also affects the frequency of the bending vibration because the total restoring force including the Coriolis force has changed.

For the other component of the bending vibrations, where the velocities of the nuclei are parallel to the rotational axis, the Coriolis force is zero. This component is therefore not affected and its frequency is not changed.

We see that the molecular rotation lifts the degeneracy of the two bending vibrational modes. The degenerate level in the nonrotating molecule splits into two components in the rotating molecule. For more information on vibrations of polyatomic molecules see [10.4, 5].

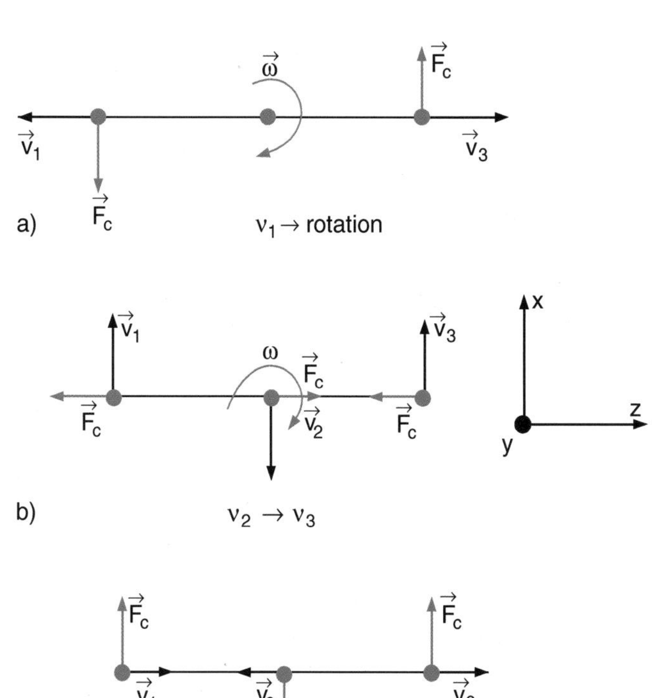


**Fig. 10.31.** (**a**) Coupling of rotation and normal vibration $\nu_1$ by Coriolis forces. (**b**), (**c**) Coupling between $\nu_2$ and $\nu_3$ vibrations by Coriolis forces in the rotating molecule

## 10.5 Spectra of Polyatomic Molecules

Because of the larger manifold of molecular levels the spectra of polyatomic molecules are more complex than those of diatomics. The larger possibilities of couplings between vibrations of the nuclei and motions of the electrons often cause a breakdown of the Born–Oppenheimer approximation and the separation of the total wave function into an electronic part and a nuclear part is in many cases not possible.

However, the general selection rule still holds that only those transitions between molecular levels $|i\rangle$ and $|k\rangle$ are allowed, for which at least one of the components of the transition dipole matrix element

$$M_{ik} = \int \psi_i^* p \psi_k \, d\tau_{el} \, d\tau_N \tag{10.50}$$

is different from zero. This means that the integrand has to be totally symmetric with respect to all symmetry operations of the molecule in question.

Besides these electric dipole transitions, much weaker magnetic dipole transitions or electric quadrupole transitions are possible, just like in diatomic molecules.

The wavenumber of a transition between the level $|i\rangle = \left(n'', v_1'', v_2'' \dots v_{3N-6}'', J'', K''\right)$ in the lower electronic state and level $|k\rangle = \left(n', v_1', v_2' \dots v_{3N-6}', J', K'\right)$ in the upper electronic state (where $n$ stands for all electronic quantum numbers), is given by the term difference

$$\begin{aligned}\overline{\nu} = &\left(T_e' - T_e''\right) + G'\left(v_1', v_2' \dots v_{3N-6}'\right) \\ &- G''\left(v_1'', v_2'' \dots v_{3N-6}''\right) \\ &+ F'(J', K') - F''(J'', K'') .\end{aligned} \tag{10.51}$$

where $T_e$ is the electronic term value at the minimum of the potential energy, $G$ the vibrational and $F$ the rotational term values. All transitions between given vibrational levels in the upper and lower state form a vibrational band consisting of all allowed rotational lines in this band. All different bands of an electronic transition make up a band system. The pure electronic transition frequency $\nu_e = T_e' - T_e''$ is called the origin of the band system.

As for diatomic molecules we can again distinguish between pure rotational spectra (in the microwave range), vibrational-rotational transitions (in the mid-infrared) and electronic transitions, ranging from the near infrared to the far ultraviolet region.

The total dipole moment of the molecule can be written as the sum of an electronic and a nuclear part

$$p = p_e + p_N . \tag{10.52}$$

Inserting this into the matrix element (10.50) and assuming that $M_{ik}^{el}$ is independent of the nuclear coordinates we obtain within the adiabatic approximation

$$\psi = \Phi^{el} \cdot \psi^{vib} \cdot \psi^{rot} = \Phi^{el} \cdot \chi_N \tag{10.53}$$

for the transition dipole matrix element the sum of two terms (see Sect. 9.6)

$$\begin{aligned}M_{ik} = &\int \Phi_i^{el} p_{el} \Phi_k^{el} \, d\tau_{el} \times \int \chi_{N_i} \chi_{N_k} \, d\tau_N \\ &+ \int \Phi_i^{el} \Phi_k^{el} \, d\tau_{el} \times \int \chi_{N_i} p_N \chi_{N_k} \, d\tau_N .\end{aligned} \tag{10.54}$$

For transitions within the same electronic state, the first summand is zero, because $|\Phi^{el}|^2$ is totally symmetric and the integrand in the first integral has the symmetry of the vector $p_{el}$. The integral over all electronic coordinates therefore vanishes. We need only to consider the second line in (10.54), which can be written, because $\int \Phi_i^{el} \Phi_k^{el} \, d\tau_{el} = \delta_{ik}$, as

$$(M_{ik})_{vib\,rot} = \int (\psi_{vib}\psi_{rot})_i \, p_N (\psi_{vib}\psi_{rot})_k \, d\tau_N . \tag{10.55}$$

For electronic transitions between two different electronic states the first line in (10.54) becomes

$$M_{ik} = M_{ik}^{el} \times (\text{FCF}) \times \text{H} , \tag{10.56}$$

where FCF is the Franck-Condon factor and H the Hönl–London factor.

The situation is quite similar to that for diatomic molecules (see Sect. 9.6). The difference lies in the larger manifold of vibrational and rotational levels for polyatomic molecules.

We will first discuss transitions between levels within the same electronic state.

### 10.5.1 Vibrational Transitions within the Same Electronic State

These transitions represent the infrared spectrum of the molecule. Generally the dipole moment of a molecule depends on the molecular displacements during a vibration. We can write

$$p_N(Q) = p_N(0) + \sum_{n=1}^{3N-6} \left(\frac{\partial p_N}{\partial Q_n}\right)_0 Q_n , \tag{10.57}$$

where the normal coordinate $Q_n$ represents the displacement vector of the nuclei for the $n$th normal vibration. Inserting this into (10.55) gives the transition dipole matrix elements

$$(\boldsymbol{M}_{ik})_{\text{vib}} = \boldsymbol{p}_N(0)\int \psi_i^{\text{vib}}\psi_k^{\text{vib}}\,\mathrm{d}\tau_{\text{vib}} + \sum_{n=1}^{3N-6}\left(\frac{\partial \boldsymbol{p}_N}{\partial Q_i}\right)_0 \int \psi_i^{\text{vib}} Q_n \psi_k^{\text{vib}}\,\mathrm{d}\tau_{\text{vib}}\,. \tag{10.58a}$$

For transitions between different vibrational levels the first integral vanishes because the vibrational wave functions $\psi_{\text{vib}}$ are orthonormal. We therefore obtain the matrix elements for vibrational transitions within the same electronic state

$$(\boldsymbol{M}_{ik})_{\text{vib}} = \sum_{n=1}^{3N_6}\left(\frac{\partial \boldsymbol{p}_N}{\partial Q_n}\right)_0 \int \psi_i^{\text{vib}} Q_n \psi_k^{\text{vib}}\,\mathrm{d}\tau_{\text{vib}}\,. \tag{10.58b}$$

Vibrational transitions of polyatomic molecules within the same electronic state do not depend on a possible permanent dipole moment, but on the change of the dipole moment with the vibrational displacement of the nuclei for the normal vibration.

In Fig. 10.32 the dependence of the dipole moment on the normal coordinate $Q$ is shown for the three normal vibrations of the $CO_2$ molecule. While for the symmetric stretch vibration $\nu_1$ the dipole moment does not depend on $Q$, it changes with $Q$ for the other two normal vibrations. Therefore, infrared absorption of $CO_2$ is only possible for the $\nu_2$ and $\nu_3$ normal vibrations. They are called ***infrared-active***, while $\nu_1$ is infrared-inactive.

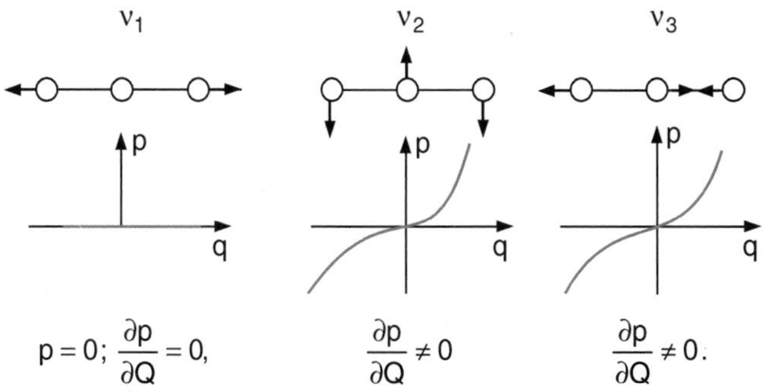


**Fig. 10.32.** Dependence of the electric dipole moment $p$ on the vibrational displacement from the equilibrium position for the three normal vibrations of $CO_2$

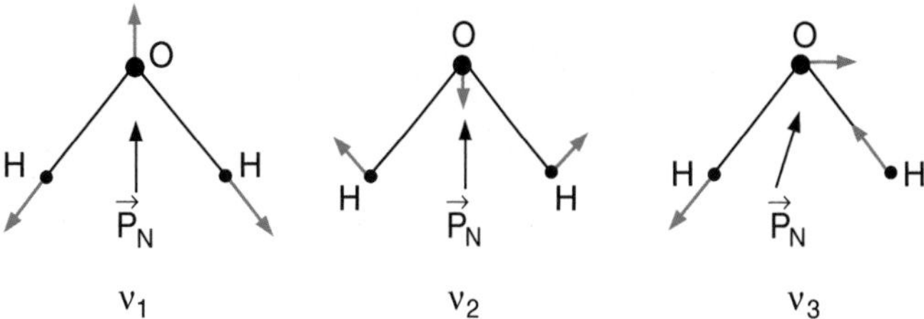


**Fig. 10.33.** Dependence of nuclear dipole moment on the nuclear displacements for the three normal vibrations of $H_2O$

The dipole moment vector can not only change its amount with $Q$ but also its direction with respect to the molecular axes. This is illustrated in Fig. 10.33 for the $H_2O$ molecule. Here the dipole moment changes its amount for the $\nu_1$ and $\nu_2$ vibrations and its direction for the $\nu_3$ vibration. Therefore all three normal vibrations are infrared-active.

The transitions into the first excited vibrational levels (e.g., the transition (0 0 0) → (0 1 0) for $CO_2$), are called ***fundamental bands***. There are also transitions possible into higher vibrational levels (e.g., (0 0 0) → (0 0 3) or (0 3 0)). They are called "***overtone transitions,***" with transition probabilities that are smaller by several orders of magnitude than for the fundamental bands.

It is also possible to excite several vibrational modes simultaneously (Fig. 10.34), e.g., on the ***combination transition*** (0 0 0) → (1 1 3). If the energy of these higher vibrational states is above 1.6 eV the absorption wavelengths are below $\lambda = 700\,\text{nm}$ and therefore fall into the visible range.

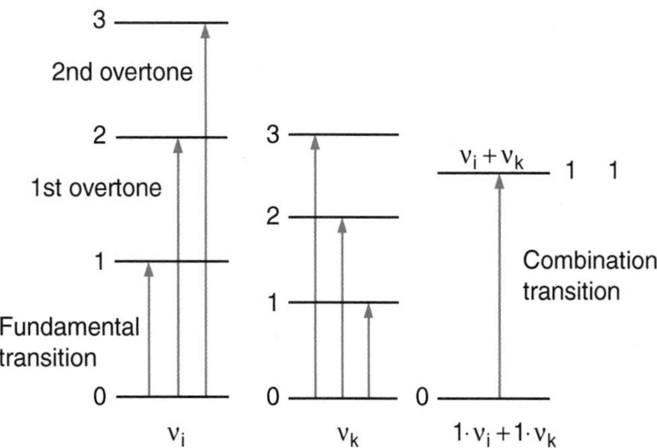


**Fig. 10.34.** Fundamental, overtone, and combination transitions

Infrared absorption of molecules only occurs on those vibrational transitions where the electric dipole moment changes. While the fundamental transitions fall into the infrared region, overtone transitions can contribute to the visible absorption of molecules.

### 10.5.2 Rotational Structure of Vibrational Bands

Because of the larger manifold of rotational levels the rotational structure of polyatomic molecules is more complex than that of diatomics.

The greatest similarity to diatomic spectra can be found for linear polyatomic molecules, which can only rotate around an axis perpendicular to the molecular axis. They are described like diatomics by a single rotational constant $B$ and the rotational structure of the spectra is quite analogous to that of diatomics. They can be regarded as special symmetric tops with the top axis along the molecular axis, where the projection quantum number $K$ is always zero, because the rotational angular momentum is perpendicular to the top axis, if we neglect the electronic orbital momentum (see Sect. 9.5.4).

True symmetric top molecules with $K \geqslant 0$ have rotational levels $(J, K)$ that depend on two quantum numbers $J$ and $K$ (Section 10.3.1).

The selection rules for rotational transitions within a vibrational band depend on the direction of the transition dipole moment $M_{ik}$ against the symmetric top axis $z$. If $\boldsymbol{M_{ik}} \parallel z$ only transitions with $\Delta K = 0$ are allowed (***parallel bands***) and the selection rules are:

$$\begin{aligned} &\Delta K = 0\,, \quad \Delta J = \pm 1 \quad &&\text{for } K = 0\,, \\ &\Delta K = 0 \quad \Delta J = \pm 1, 0 \quad &&\text{for } K \neq 0\,. \end{aligned} \tag{10.59}$$

When the transition moment is perpendicular to the top axis we speak of ***perpendicular bands*** with the selection rules

$$\Delta K = \pm 1\,, \quad \Delta J = 0, \pm 1\,. \tag{10.60}$$

For molecules with inversion symmetry a further symmetry selection rule holds for transitions between two rotational levels in different vibrational states:

$$+ \leftrightarrow -\,, \; + \not\leftrightarrow +\,, \; - \not\leftrightarrow -\,. \tag{10.61}$$

The intensity

$$I(J, K) \propto \text{FCF} \times H_{JK} \times N \times g_{JK} \times \mathrm{e}^{-E(J,K)/kT} \tag{10.62}$$

of a rotational line depends on the Franck–Condon factor FCF, the Hönl–London–factor $H_{JK}$ and the population density $N$ of the sample molecules at temperature $T$. The population density of the absorbing level $|J, K\rangle$ is $N_{JK} = N \cdot g_{JK} \cdot \mathrm{e}^{-E(J,K)/kT}$ with the statistical weight factor $g_{JK}$. The Hönl–London factor, which gives the relative intensities of rotational lines within a vibrational band, is slightly different for $P$, $Q$ and $R$ lines (see Sect. 9.6.3). For parallel bands with $\Delta K = 0$ one obtains:

$\Delta J = -1$ ($P$-branch)

$$H_{KJ} = \frac{J^2 - K^2}{J(2J+1)} \tag{10.63a}$$

$\Delta J = 0$ ($Q$-branch)

$$H_{KJ} = \frac{K^2}{J(J+1)} \tag{10.63b}$$

$\Delta J = +1$ ($R$-branch)

$$H_{KJ} = \frac{(J+1)^2 - K^2}{(J+1)(2J+1)}\,. \tag{10.63c}$$

**Note** there are no $Q$ transitions for $K = 0$.

For perpendicular bands with $\Delta K = \pm 1$ the Hönl–London factors are

$\Delta J = -1$

$$H_{KJ} = \frac{(J-1 \mp K)(J \mp K)}{J(2J+1)} \tag{10.64a}$$

$\Delta J = 0$

$$H_{KJ} = \frac{(J+1 \pm K)(J \mp K)}{J(J+1)} \tag{10.64b}$$

$\Delta J = +1$

$$H_{KJ} = \frac{(J+2 \pm K)(J+1 \pm K)}{(J+1)(2J+1)}\,. \tag{10.64c}$$

The spectra of asymmetric tops where each rotational level $(J, K_a, K_c)$ is characterized by the rotational quantum number $J$ and two indices $K_a$ and $K_c$ follow selection rules that are mainly based on the symmetry properties of the rotational levels [10.5, 6, 7, 8].

### 10.5.3 Electronic Transitions

A single vibrational band of transitions between a given vibrational level in the lower and in the upper electronic states consists of many sub-bands: For each $K$ value a rotational $P$, $Q$ and $R$ branch is possible. The rotational lines with different $K$ values may overlap, resulting in a spectrum with a complex structure (Fig. 10.35).

The intensity of the different vibrational bands depends, as for diatomic molecules, on the overlap of the vibrational wave functions in the upper and lower state. However, the vibrational levels can have different symmetries, different from the case of diatomics. Due to symmetry selection rules only transitions between those vibrational levels are allowed for which the integrand in (10.50) is totally symmetric. Since the ground state (electronic and vibrational ground state) is generally totally symmetric, the total symmetry of the upper level, determined by the symmetries of the electronic state, the vibrational and rotational level must be the same as that of one of the three components of the dipole moment.

Only those vibrational bands are allowed for which the product $\Phi_{\text{el}} \times \psi_{\text{vib}}$ in the upper state has the same symmetry as one of the components of the dipole moment.

If the coupling between the electronic wave function and the vibrational wave function cannot be neglected, the distinction between the vibrational and electronic part is no longer possible. The energy levels characterized by the product $\Phi_{el} \times \psi_{\text{vib}}$ are called ***vibronic levels*** (vibrational-electronic levels) [10.9].

This coupling makes electronic transitions allowed, even if the electronic part of the matrix element is zero, as long as the product function fulfills the condition that it has the same symmetry as one of the components of the dipole moment. Such vibronically allowed electronic transition can be found in many molecules.

The relative intensities of the rotational transition follow the same formulas (10.60) as for transitions between rotational-vibrational levels within the same electronic state. The difference is, however, that

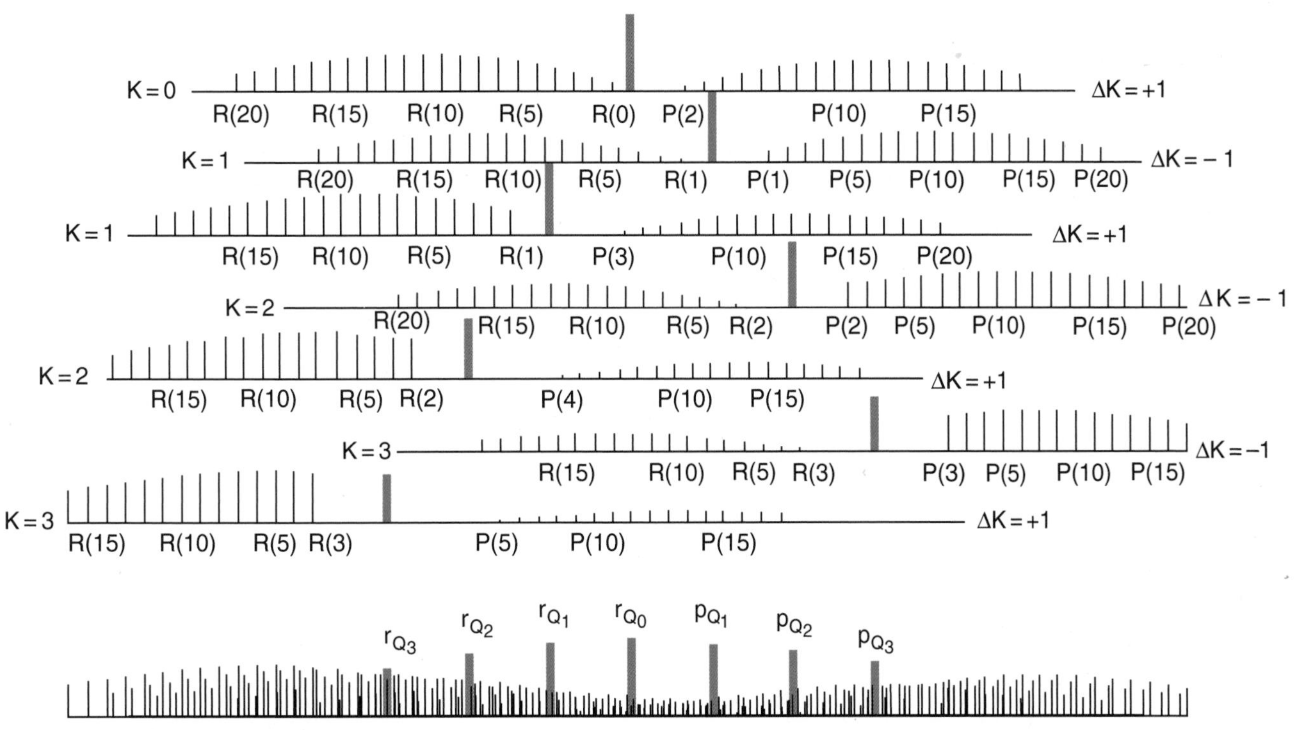


**Fig. 10.35.** Calculated spectrum of a perpendicular band of a symmetric top with $A' \approx A''$ and $B' \approx B''$. The rotational lines of the $Q$ branches are not resolved (with kind permission of the late Prof. Herzberg [10.9])

now the electronic part $p_e$ of the dipole moment instead of the nuclear part $p_N$ is responsible for the transitions.

The symmetry selection rules for the different molecules can be derived with the help of group theory, which will not be discussed here (see [10.6, 7, 8]).

## 10.6 Clusters

Clusters are aggregates of $N$ atoms or molecules in a transition regime between molecules and solid microparticles or liquid droplets. The number $N$ ranges from $N = 3$–$100$ for small clusters to $N = 10^6$ for large clusters. Their properties are intermediate between those of isolated monomers (atoms or molecules) and those of the bulk solid state. One difference to molecules is the fact, that they are not found in appreciable numbers in an equilibrium vapour. Another difference is the delocalization of their valence electrons like in solid conductors. However, a difference to solids is the much larger importance of surface effects and the question arises, how many metal cluster atoms are necessary to reach the metallic properties of a solid metal? They can be weakly bound (e.g., the van der Waals clusters $He_N$ or $Ar_N$, (Fig. 10.36), but can be also strongly bound by valence forces, e.g., silicon clusters $Si_N$ or carbon clusters $C_N$ (Fig. 10.37). The

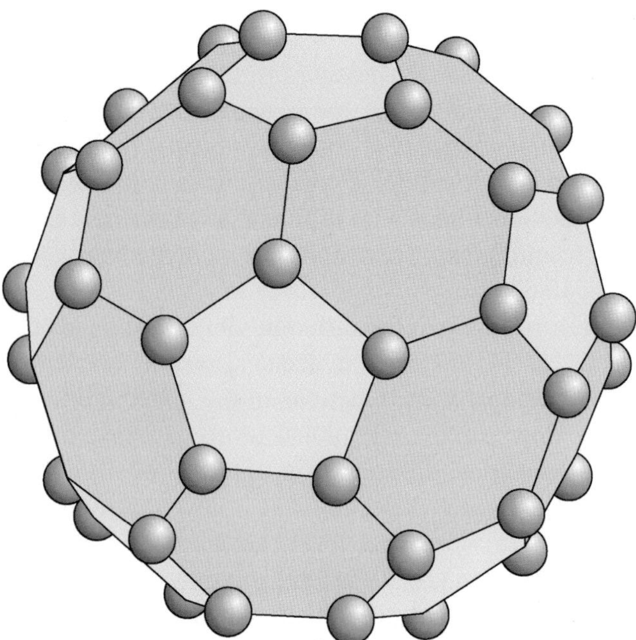

**Fig. 10.37.** Carbon cluster $C_{60}$, called "fullerene", or "buckey-ball"

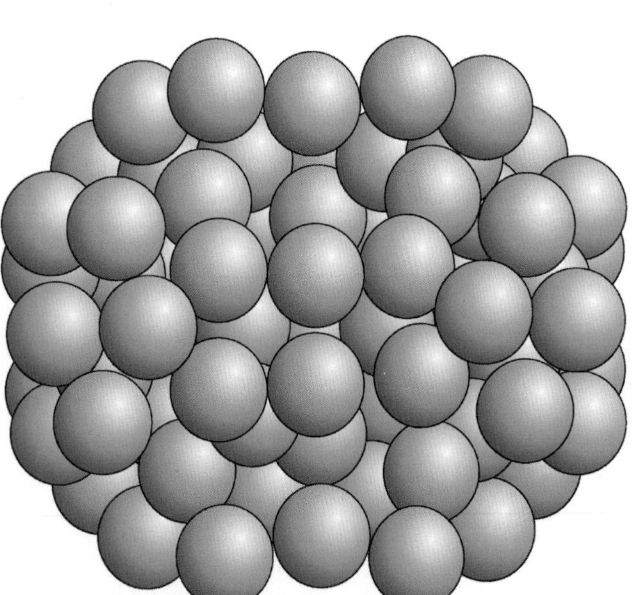

**Fig. 10.36.** Noble gas cluster $Ar_{105}$

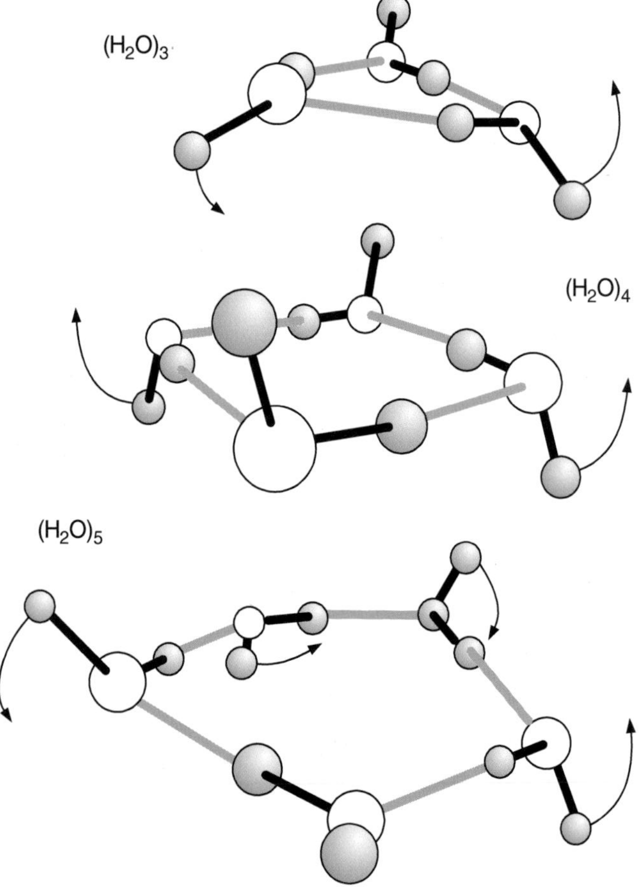


**Fig. 10.38.** Small water clusters

large clusters with $10^6$ atoms have a spherical geometry, a diameter of about 40 nm and are therefore still small compared to dust particles or particles in cigarette smoke. The ratio of surface atoms to bulk atoms is much larger than in solid microparticles. Therefore surface effects play on important role.

Clusters represent the transition range from molecules to solid particles or liquid drops.

The study of their properties can give much information on the processes of condensation and evaporation, or more generally on the physics of phase transitions on a molecular level. Therefore the studies of clusters have been greatly intensified in recent years and nowadays many laboratories are investigating their properties with different experimental techniques (see Chap. 11).

We can classify clusters by their constituents in atomic clusters (e.g., $C_N$) or molecular clusters, such as $(H_2O)_N$. Another classification scheme is based on the binding strength. There are van der Waals clusters (noble gas clusters), clusters with hydrogen bonding (for example, $(H_2O)_N$ (Fig. 10.38)), valence bond clusters such as $Si_N$ or $C_N$ and clusters with ionic bonds such as $(NaCl)_N$. The form of the clusters can be nearly spherical, cubic or rhombohedral. In Fig. 10.39 the calculated structures of $(NaCl)_N$ clusters are shown for increasing cluster size.

The structure of stable clusters is determined by the total binding energy: Those clusters with a given number $N$ are most stable which have the maximum number of bonds between nearest neighbours.

Another classification uses the number $N$ of atoms or molecules in the cluster:

$N = 3$–12: Microclusters; all atoms are surface atoms
$N = 13$–100: Small clusters. Quantum effects govern the characteristics of these clusters.
$N = 100$–$10^6$ : large clusters. The ratio of surface to bulk atoms decreases with $N$.
$N > 10^6$: Microparticles or liquid droplets. Their properties approach those of solids or liquids.

For small clusters, most of the atoms are at the surface. With an increasing number of atoms the relative share of the surface atoms decreases (Table 10.2), which implies that the influence of surface effects becomes smaller with increasing cluster size.

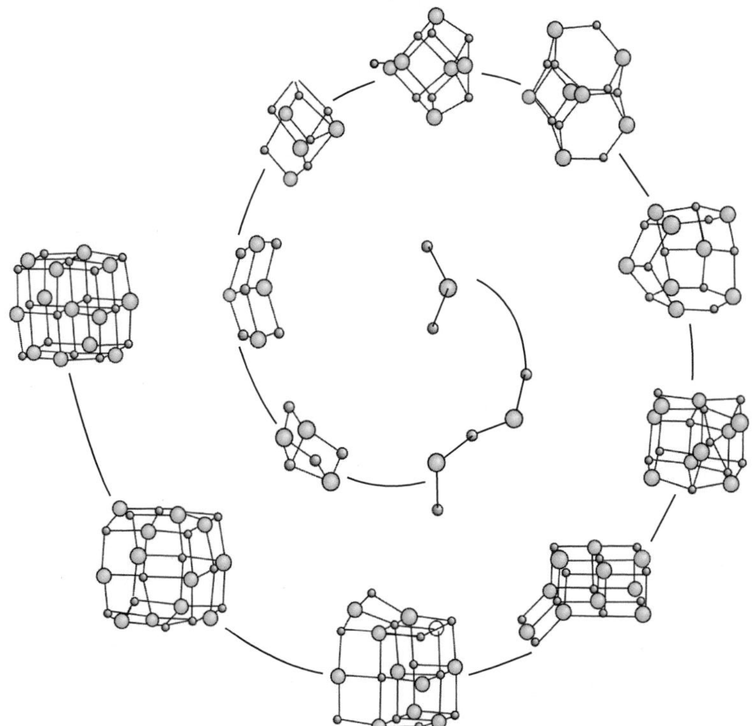

**Fig. 10.39.** Calculated structures of $(NaCl)_N$ clusters with increasing number $N$ of NaCl molecules [10.10]

**Table 10.2.** Mean radius of a spherical cluster and ratio $N_s/N$ of surface atoms as a function of total number $N$ of cluster atoms, assuming spherical atoms with an atomic radius $r_{at} = 0.22$ nm

| $N$ | $R$/nm | $N_s/N$ |
|---|---|---|
| 10 | | 1 |
| $10^2$ | 1.03 | 0.8 |
| $10^3$ | 2.2 | 0.4 |
| $10^4$ | 4.8 | 0.23 |
| $10^5$ | 10.0 | 0.08 |
| $10^6$ | 21.5 | 0.02 |
| $10^{10}$ | 480 | 0.002 |

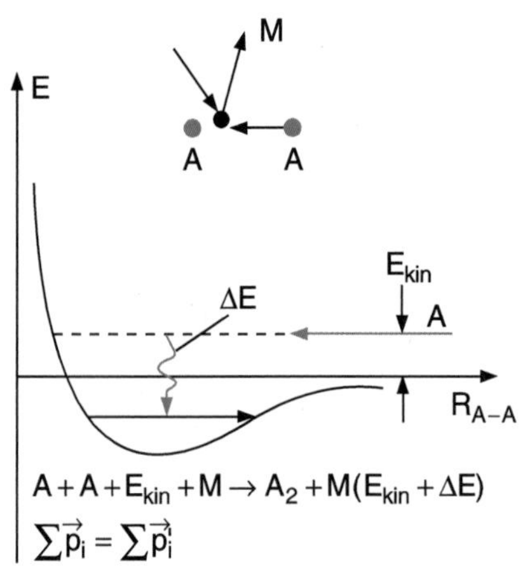


**Fig. 10.41.** Recombination of two atoms with the kinetic energy $E_{kin}$ of their relative motion by collision with a third partner M that takes away the energy $\Delta E > E_{kin}$

### 10.6.1 Production of Clusters

Clusters can be produced in many ways. The most commonly used techniques are the condensation of atoms or molecules in a supersaturated vapor, or the formation of clusters during the adiabatic expansion of a supersonic jet from a vapor reservoir at high pressure, where the vapor is mixed with a noble gas, through a small nozzle into a vacuum chamber (Fig. 10.40). During the adiabatic expansion the atoms cool down because their relative kinetic energy is transferred into directional flow energy. This means that all particles move with nearly the same velocity and the relative velocity becomes very small. If two atoms approach each other during the expansion of the jet, they can recombine (Fig. 10.41) if their small energy $E_{kin}$ of relative motion can be transferred to a third collision partner $M$, which can be either a noble gas atom or the wall of the nozzle. This collision process forms a diatomic molecule, which can again collide with another atom to form a triatomic molecule, etc. If the number of possible cluster-forming collisions during the expansion is sufficiently large, this can result in large cluster sizes.

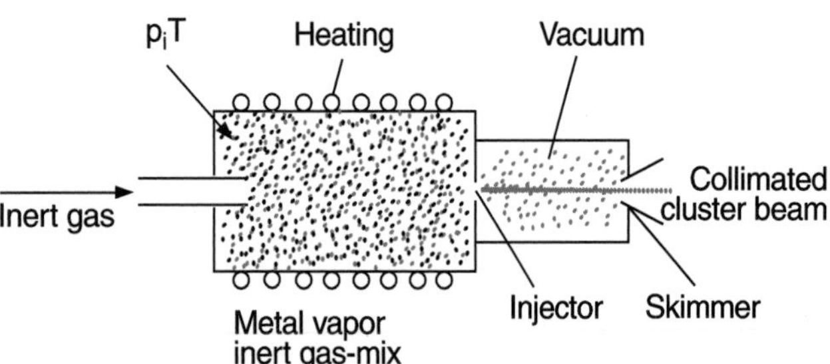


**Fig. 10.40.** Generation of metal clusters by adiabatic expansion of a mixture of noble gas and metal vapor

### 10.6.2 Physical Properties of Clusters

The cluster size, i.e., the number $N$ can be inferred from mass spectrometric measurements. The neutral clusters are injected into the ion source of a mass spectrometer (Sect. 2.7) and are ionized either by photoionization or by electron impact. The mass selected cluster ions are detected and their abundance is measured as a function of $N$. This mass distribution of cluster ions, illustrated in Fig. 10.42 for $Na_N^+$ cluster ions, gives the wanted distribution of the neutral cluster only if the ionization does not result in fragmentation. Ionization with photons of selected energy $h\nu$, which is only slightly higher than the ionization energy of the selected cluster, fulfills this condition fairly well.

From Fig. 10.42 it is obvious that the abundance of Na clusters with certain $N$ values (magic numbers $N = 8, 20, 40$) is particularly high. This favors the assumption that these cluster sizes are more stable than the others and has lead to a shell-structure model of clusters where the atoms in a cluster arrange themselves in shells around the inner atoms in order to form a densely packed structure. This model seems to be correct for noble gas clusters. However, for metal clusters, it turns out that the electron structure of the atomic valence electrons is responsible for the stability of the clusters with magic atom numbers. In a "jellium model" of metal clusters, where the valence electrons can freely move

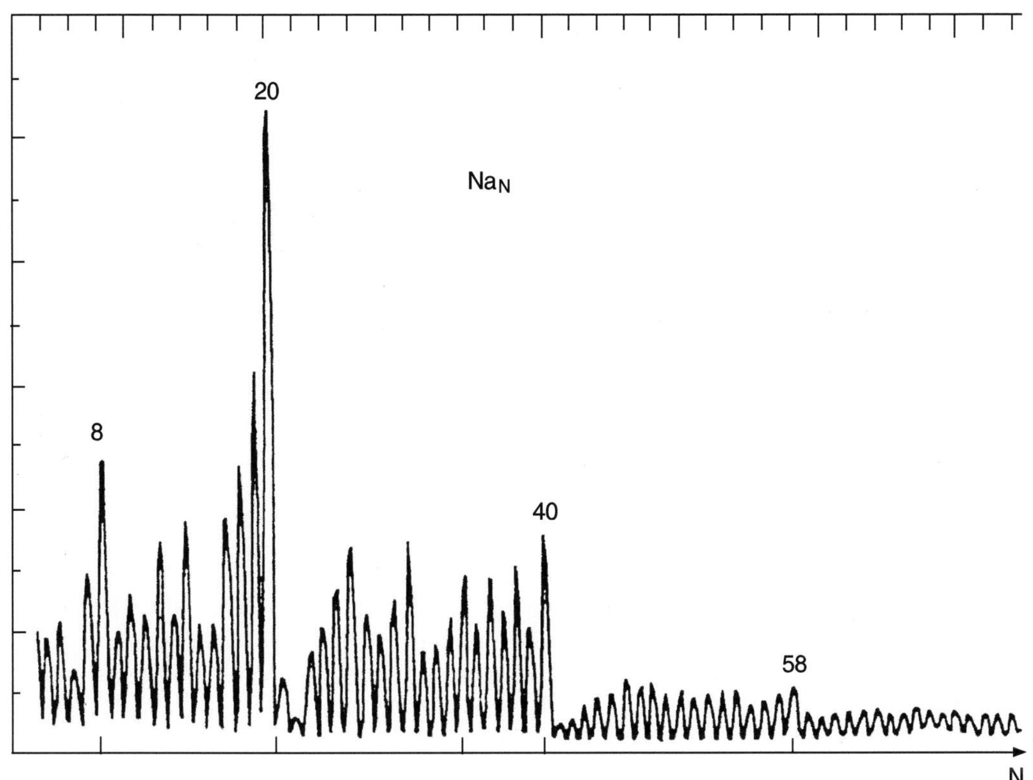


**Fig. 10.42.** Mass spectrum of $Na_N^+$ clusters, showing the magic numbers $N = 8, 20, 40, 58$

within the cluster, the electrons can be treated as particles in a three-dimensional box with the same size as the cluster. Different from the one-dimensional box in Sect. 4.2.4 the levels are not only characterized by the quantum number $n$, but also by the angular momentum quantum number $L$. These levels are now filled according to the Pauli principle. The highest stability is reached for completely filled levels. The experimentally found magic numbers are indeed identical with the number of valence electrons up to a closed shell, i.e., a completely filled level $(n, L)$ in the three-dimensional potential box, which corroborates the jellium model.

The ionization energy $E_{\mathrm{ion}}(N)$ depends on the number $N$ of atoms but does not show a smooth convergence towards the work function of the bulk metal for $N \to \infty$. It has maxima for certain values of $N$, again reflecting the shell structure of the electron cloud according to the jellium model. The same behavior is found for the binding energy of clusters (Fig. 10.43).

A very interesting question is the dependence of the melting temperature of a cluster on the cluster size. At this temperature the cluster switches from having a solid-like behavior with a geometrically well-defined structure to a liquid microdroplet with spherical size [10.11]. Since the atoms at the surface of the cluster have a different energy than those inside the cluster, the fraction $N_S/N$ of surface atoms will certainly influence the melting temperature.

How can the temperature of a cluster be determined? One method used for large cold helium clusters is the following. The He clusters are sent through a gas of molecules, where some clusters pick up a molecule, which can either be attached to the cluster surface or enclosed in the middle of the cluster where it adapts to the temperature of the cluster. The spectroscopy of the rotational structure of the molecule allows the determination of the temperature from the intensity distribution of the rotational lines [10.12].

It turns out that for some clusters the melting temperature differs from the freezing temperature, i.e., it makes a difference whether the phase transition is approached from higher temperatures or from lower ones. The study of this effect will improve our understanding, on a microscopic scale, of superheating or supercooling phenomena, observed for bulk materials such as glass or water.

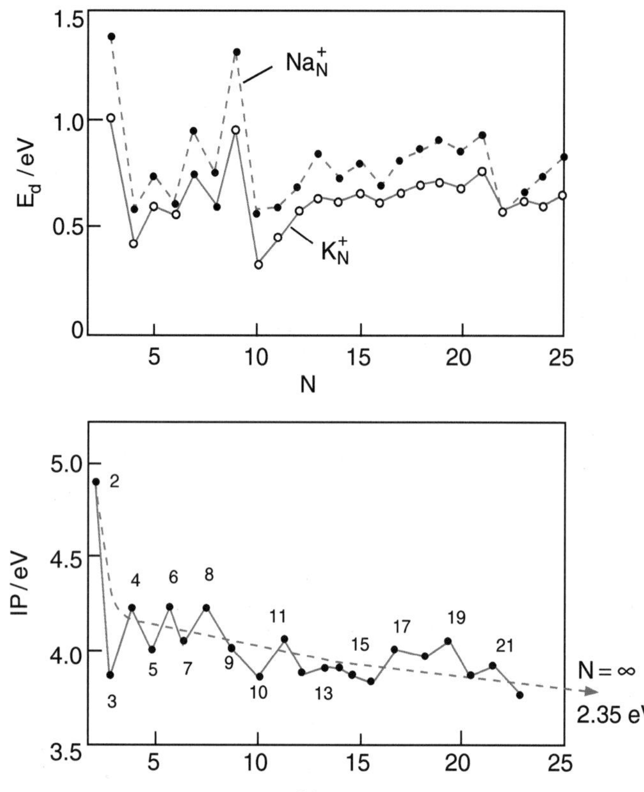


**Fig. 10.43.** (**a**) Dissociation energy of sodium and potassium clusters. (**b**) Ionization potentials IP($N$) for sodium clusters $(\mathrm{Na})_N$

Another interesting phenomena is connected with the photographic process in the emulsion of a photoplate exposed to light, where chemical reactions of silver halide are responsible for the blackening of the photoplate. Detailed studies have shown that silver clusters $\mathrm{Ag}_N$ with $N \geqslant 4$ start the nucleation for the formation of silver halide grains.

The optical properties of metal clusters have received much attention because the clusters form an intermediate range between molecules with transitions between discrete electronic states and bulk metal with broad transitions between the valence and the conduction band. The question of whether broad resonances in the optical range found for alkali clusters are due to surface plasmons or are just unresolved absorption spectra between a dense manifold of discrete states is still unsettled.

The colorful church windows produced in the middle age gain their brilliant colors from gold clusters imbedded in the glass.

## 10.7 Chemical Reactions

Chemical reactions are based on collisions between atoms or molecules. In order to understand the dynamics and the energy balance of such a reaction one has to know the interaction potentials between the reacting molecules and the bonding energies of reaction partners and reaction products. Therefore spectroscopy is a very valuable tool for studying chemical reactions on a molecular level.

We consider the chemical reaction

$$m \cdot \mathrm{A} + n \cdot \mathrm{B} \rightarrow \mathrm{A}_m\mathrm{B}_n\,, \tag{10.65}$$

where $m$ atoms or molecules A react with $n$ partners B to form the product $\mathrm{A}_m\mathrm{B}_n$. This reaction may proceed directly or via intermediate steps. Sometimes catalysts are necessary to start the reaction. Not all molecules A or their partners B in the gas container where the reaction takes place may participate in the reaction.

With the concentrations $n_\mathrm{A}$ and $n_\mathrm{B}$ of the reaction partners (number of particles per volume) the reaction rate (number of reactions per volume and time) of the reaction (10.65) is

$$R = k_\mathrm{R} \cdot n_\mathrm{A}^m \cdot n_\mathrm{B}^n\,. \tag{10.66}$$

The factor $k_\mathrm{R}$ is the ***reaction constant*** or ***rate constant***. The sum $m+n$ of the exponents $m$ and $n$ is called the ***order of the reaction***, because it gives the number of reaction partners necessary for the formation of one molecule of the reaction product.

### 10.7.1 First Order Reactions

Assume that only one species of molecules A is present, which can decay into products X when energy is transferred to A (for instance by photon absorption or by increasing the temperature of the gas of particles A, i.e., by increasing the relative kinetic energy of the colliding species A). The reaction rate

$$R = -\mathrm{d}n_\mathrm{A}/\mathrm{d}t = k_\mathrm{R}^{(1)} \cdot n_\mathrm{A} \tag{10.67}$$

is proportional to the concentration $n_\mathrm{A}$. This implies that (10.67) represents a first order reaction. The first order reaction constant $k^{(1)}$ has the unit $[k^{(1)}] = 1\,\mathrm{s}^{-1}$. Integration of (10.67) gives (Fig. 10.44)

$$n_\mathrm{A}(t) = n_\mathrm{A}(0) \cdot \mathrm{e}^{-k_\mathrm{R}^{(1)}t} = n_\mathrm{A}(0)\mathrm{e}^{-t/\tau}\,. \tag{10.68}$$

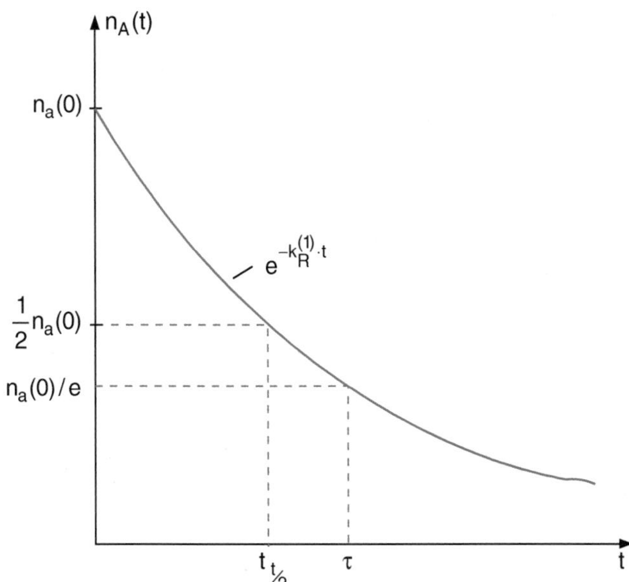


**Fig. 10.44.** Decay $n_a(t)$ for a first-order reaction

The mean lifetime $\tau = 1/k_R^{(1)}$, after which the concentration $n_A$ has decayed to $1/e$ of its initial value $n_A(0)$ due to the reaction (10.67), is independent of the concentration $n_A$.

Examples are the thermal decomposition of molecules with a low dissociation energy, or the photo-induced dissociation of molecules, such as $I_2 + h\nu \rightarrow I + I$.

### 10.7.2 Second Order Reactions

Often two equal species A (atoms or molecules) can recombine to a molecule $A_2$. This can, however, only happen if the excess energy, due to the kinetic energy of their relative motion, is taken away by a third collision partner (Fig. 10.41), which can be another molecule M (Fig. 10.45) or the wall of the reaction chamber. Otherwise the molecule $A_2$ can not be stabilized.

The reaction rate is now

$$R = -\frac{dn_A}{dt} = k_R^{(2)} n_A^2 . \tag{10.69}$$

This is a second order reaction, because the reaction rate is proportional to the square of the concentration $n_A$. The dimension of the second order reaction rate constant is $\left[k_R^{(2)}\right] = \mathrm{m^3\,s^{-1}}$.

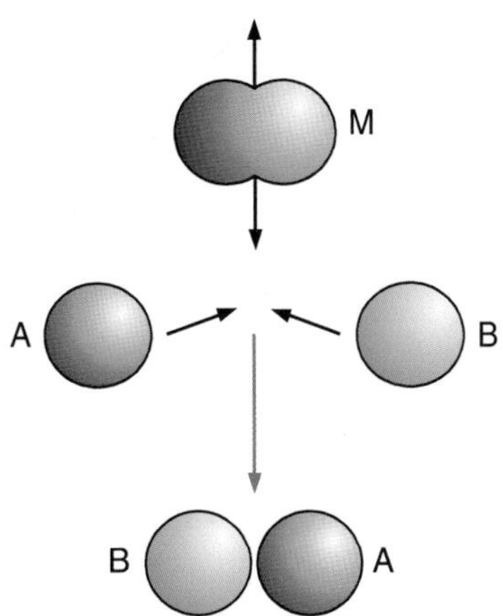


**Fig. 10.45.** Schematic representation of the recombination collision $A + B + M \rightarrow (AB) + M + E_{kin}$

Integration of (10.69) yields

$$\int \frac{dn_A}{n_A^2} = -\int k_R^{(2)}\, dt$$

$$\Rightarrow n_A(t) = \frac{n_A(0)}{1 + k_R^{(2)} n_A(0) t} . \tag{10.70}$$

The concentration $n_A(t)$ decays as a hyperbolic function. After a time

$$t = \tau_{1/2}^{(2)} = \frac{1}{k_R^{(2)} n_A(0)} \tag{10.71}$$

the concentration has decayed to one half of its initial value $n_A(0)$.

**Note** that for a second order reaction the decay time of the concentration depends on the initial concentration!

Another type of second order reactions is the reaction

$$A + B \rightarrow AB$$

(Fig. 10.45), where two different species A and B are present in the reaction chamber. Their reaction rate is given by the equation

$$\frac{dn_A}{dt} = \frac{dn_B}{dt} = -k_R^{(2)} n_A n_B . \tag{10.72}$$

For equal initial concentrations $n_A(0) = n_B(0)$ we obtain equations analogous to (10.70) for $n_A(t)$ and $n_B(t)$.

For $n_A(0) \neq n_B(0)$ we can use the substitutions

$$n_A(t) = n_A(0) - x(t)$$
$$n_B(t) = n_B(0) - x(t) .$$

Inserting this into (10.72), integration gives the ratio

$$\frac{n_A(t)}{n_B(t)} = \frac{n_A(0)}{n_B(0)} e^{-k_R^{(2)}[n_B(0)-n_A(0)]t} . \tag{10.73}$$

For $n_A(0) = n_B(0)$ the ratio does not change with time, while for $n_A(0) < n_B(0)$ it decays exponentially with time, which means that $n_A(t)/n_A(0)$ decays faster than $n_B(t)/n_B(0)$.

### 10.7.3 Exothermic and Endothermic Reactions

When a reaction

$$AB + CD \rightarrow AC + BD$$

proceeds during a collision between the reaction partners, chemical bonds have to be broken (in our example the bonds A – B and C – D) and new bonds are formed.

This can only happen if the electron clouds of the reaction partners overlap during the collision in order to rearrange the electron distribution for the formation of new bonds. The collision partners have to approach each other sufficiently close. This rearrangement of the electrons needs energy. Therefore, energy must generally be fed into the colliding system during the approach of the two reaction partners. This energy may be supplied by the kinetic energy of the collision partners or by photo excitation. When the reaction products are formed, energy is released, because the energy of the molecules AC and BD is smaller than the energy of the free atoms A, B, C and D by the amount of the bonding energies of AC and BD.

The reaction can be formally divided into two steps by introducing the concept of the collision complex. At first the two collision partners AB and CD form with a rate constant $k_1$ the collision complex K = (ABCD)*, which then can decompose with the rate constant $k_2$ into the reaction products AC and BD (Fig. 10.46).

$$AB + CD \rightarrow K^* \rightarrow AC + BD . \tag{10.74}$$

The collision complex cannot only decompose into the end products AC + BD, but can also decay back into the intial reactants AB + CD. If the reaction probability

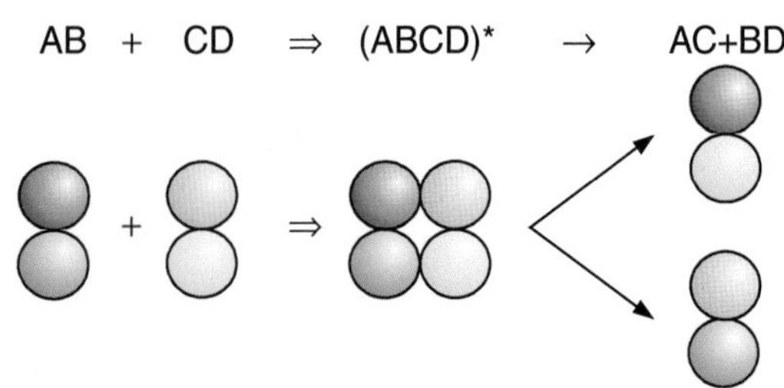


**Fig. 10.46.** The concept of collision complex in the reaction AB + CD → AC + BD

for the former is $\kappa$, it is $(1-\kappa)$ for the latter. The total reaction rate for the reaction (10.74) is therefore

$$k_+ = \kappa \times k_1 \times k_2 \tag{10.75}$$

and for the backward reaction AC + BD → K* → AB + CD it is

$$k_- = k_3 \times (1-\kappa) \times k_4 . \tag{10.76}$$

where $k_3$ is the rate constant for the reaction AC + BD → (ABCD)* and $k_4$ for the decay (ABCD)* → AB + CD. Both reactions can be illustrated by a schematic reaction diagram (Fig. 10.47), where the potential energy during the reaction is plotted as a function of the reaction coordinate, which symbolizes the progress of the reaction over time. During the collision the distance between the reaction partners at first decreases and after the formation of the collision complex it increases again. The reaction coordinate is therefore not a geometrical coordinate, but rather represents a time sequence.

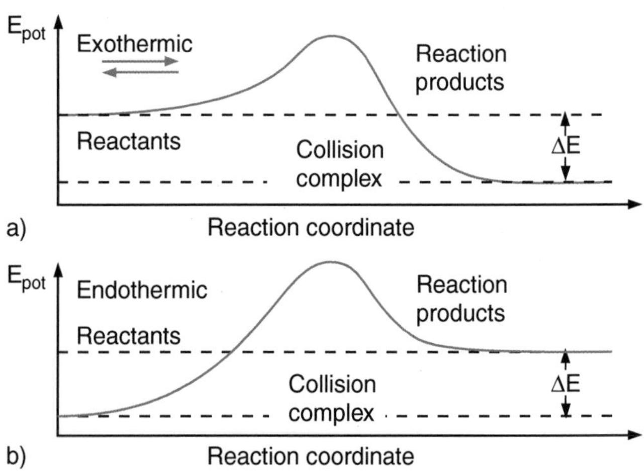


**Fig. 10.47a,b.** Potential energy as a function of the reaction coordinate for an (**a**) exothermic and (**b**) endothermic reaction

If the energy necessary to form the collision complex is smaller than the energy released by its decay into the reaction products the reaction is called ***exothermic***. Exothermic reactions release more energy to the surroundings than they need for their initiation. For ***endothermic*** reactions the energy necessary for the formation of the reaction complex is larger than the energy released at its decay. Endothermic reactions therefore need net energy fed into the system.

For the two examples shown in Fig. 10.47 a potential barrier has to be overcome during the reaction. It gives the energy necessary to form the collision complex. The energy difference $\Delta E$ is released to the surroundings for exothermic reactions and has to be fed into the system for endothermic reactions.

For exothermic reactions, the energy excess $\Delta E$ is transferred into kinetic energy of the reaction products. This leads to a temperature increase of the gas in the reaction chamber. The reaction products can be also in excited states and can release their excitation energy by emission of photons or by transferring this energy during further collisions into kinetic energy.

The inverse reaction of an exothermic reaction must be endothermic:

$$\mathrm{AB} + \mathrm{CD} \underset{\text{endothermic}}{\overset{\text{exothermic}}{\rightleftarrows}} \mathrm{AC} + \mathrm{BD}\,. \tag{10.77}$$

This implies that the inverse reaction is always possible, if the reaction partners get sufficient energy to overcome the reaction barrier.

In a gas at thermal equilibrium, both directions in Eq. (10.77) are possible. The exothermic reaction has, however, a higher probability than the inverse endothermic reaction. Under stationary conditions in a gas of reacting molecules there will be an equilibrium concentration of reactants and reaction products, which depends on the temperature of the gas, on the energy difference $\Delta E$ and on the statistical weights of the particles on the two sides of (10.77).

The equilibrium of forward and backward reaction is governed by the equilibrium constant

$$K = \mathrm{k}_+/k_- = [AB] \cdot [CD]/[AC] \cdot [BD]$$

which depends on the concentrations of the reaction partners AB and CD and the reaction products AC and BD.

The goal of the kinetic theory of chemical reactions is the determination of absolute rate constants for both directions of reactions like (10.77), based on the properties of reactants and reaction products, such as bonding energies, interaction potentials and number of degrees of freedom, that can be activated at a given temperature. This allows one to calculate the temperature-dependent ratio of reactants to reaction products under thermal equilibrium at a temperature $T$.

### 10.7.4 Determination of Absolute Reaction Rates

We will discuss the determination of absolute reaction rates for two examples.

The most thoroughly measured and calculated reaction is the exchange reaction for collisions of hydrogen atoms H with molecules $H_2$, according to the scheme

$$\mathrm{H}_a + \mathrm{H}_b\mathrm{H}_c\ (v = 0, 1) \rightarrow \mathrm{H}_a\mathrm{H}_b\ (v = 0, 1) + \mathrm{H}_c\,. \tag{10.78a}$$

In order to be able to distinguish between the atoms $\mathrm{H}_a$, $\mathrm{H}_b$ and $\mathrm{H}_c$ in the actual experiment, the isotope deuterium D is used, and the reaction is then modified to

$$\mathrm{D} + \mathrm{H}_2\ (v = 0, 1) \rightarrow \mathrm{HD}\ (v = 0, 1) + \mathrm{H}\,, \tag{10.78b}$$

where the reaction products HD can be readily distinguished by their different mass from the $H_2$ molecule. Since the vibrational energy of the $H_2$ and HD molecules significantly differs for the two isotopomers, the influence of vibrational excitation of one reactant on the reaction rate and the probability of internal energy of the reaction product can be accurately determined. The relevant energy diagram is shown in Fig. 10.48. The reaction barrier for the reaction (10.78b) has a height of 0.12 eV. The vibrational energy of $H_2$ with $E_{\mathrm{vib}}(v = 1) = 0.5\,\mathrm{eV}$ is much larger than the barrier. This implies that the reaction probability for $v = 1$ should be much larger than for $v = 0$ [10.13].

The experiments yield reaction coefficients

$$k_{\mathrm{R}}^{(2)}(\mathrm{H}_2\ (v = 1)) \geq 10^{-5}\,\mathrm{m}^3/\mathrm{s}$$

while

$$k_{\mathrm{R}}^{(2)}(\mathrm{H}_2\ (v = 0)) \approx 2.5 \times 10^{-10}\,\mathrm{m}^3/\mathrm{s}$$

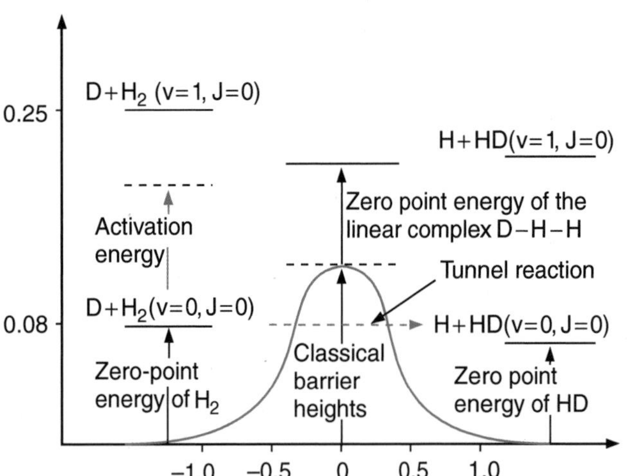


**Fig. 10.48.** Reaction barrier for the exchange reaction $H_2 + D \rightarrow HD + H$

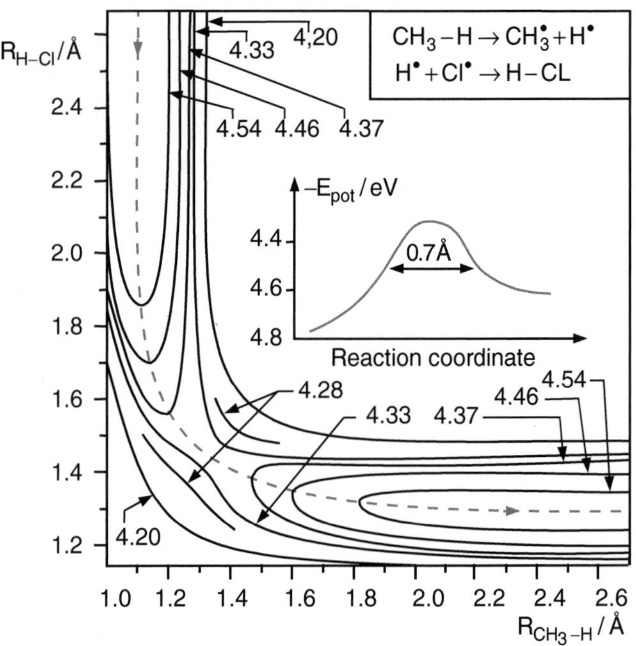


**Fig. 10.49.** Energy contour lines of the potential surface $E_{pot}(R_{C-H}, R_{CH_3-H})$ for the reaction $CH_4Cl + Cl \rightarrow CH_3 + HCl$

is about $4 \times 10^4$ times smaller. This result agrees with quantum mechanical calculations.

The second example, taken from [10.14], is the reaction

$$CH_4 + Cl^* \rightarrow CH_3 + HCl\,. \tag{10.79a}$$

When the chlorine atom Cl approaches the $CH_4$ molecule, one C − H-bond has to be broken in order to initiate the reaction (10.79a). We will assume that the other three C − H bonds will not be significantly affected. In the model of the reaction complex the reaction is written as

$$\begin{aligned} CH_4 + Cl^* &\rightarrow [CH_4Cl]^* = \left[CH_3 - H^\bullet + Cl^\bullet\right] \\ &\rightarrow CH_3 + HCl\,, \end{aligned} \tag{10.79b}$$

where the star means excited species and the dots on the H and Cl atoms indicates that the atoms act like radicals with a free chemical bond.

In order to determine the time-dependent potential during the reaction we have to calculate the energy of the complex as a function of the distances between the three particles $CH_3$, H and Cl. This can be achieved with the help of ab initio quantum chemical calculations, discussed in Sect. 9.2. The result of such calculations is plotted in Fig. 10.49 in a two-dimensional diagram of equipotential energy curves $E_{pot}(R_{CH_3,H}, R_{H,Cl}) = \text{constant}$, which depend on the distances $R_{CH_3,H}$ between $CH_3$ and H and $R_{H,Cl}$ between H and Cl. The ordinate of this diagram gives the distance between the nuclei of the reaction product HCl and the abscissa the distance between the H atom and the radical $CH_3$. The reaction path is represented by the blue dashed curve. The reactants approach each other from the upper left part of the diagram where the Cl atom is still far away from the $CH_4$ molecule. The reaction barrier is reached in the lower left part and the reaction products leave to the lower right part with a kinetic energy that equals the energy difference $\Delta E$ of the potential energy at the maximum and at infinite distance $R$ between the reaction products.

## 10.8 Molecular Dynamics and Wave Packets

Up to now we have only discussed stationary states of molecules, where the energies of molecular levels is determined by the stationary wave functions obtained from the time-independent Schrödinger equation.

Often the situation arises that energy is fed into the molecule at time $t = 0$ (for instance, by absorption of a photon or by electron impact or by a collision with another atom or molecule). The question now is, how and how fast the energy is redistributed within the molecule or is given away by emission of photons or by inelastic collisions. The time dependent fate of the molecule after its excitation is of great importance for all chemical reactions (see foregoing section). Molecular dynamics is the essential feature of life. Stationary molecules cannot induce any processes and the study of the stationary energy states is mainly of interest because it is an essential help for the understanding of molecular dynamics (see Sect. 12.2).

The description of time-dependent processes in molecules demands time-dependent wave functions as solutions of the time-dependent Schrödinger equation. We have already discussed in Sect. 3.3 that moving localized particles have to be described by wave packets instead of stationary wave functions. This wave packets best describe the classical movement of particles because the classical velocity is represented by the group velocity of the wave packet.

The wave packet description also builds a bridge between the quantum mechanical and the classical model for the vibration and rotation of molecules. We will here discuss this for the example of diatomic molecules, although it is also applicable to polyatomic molecules.

If the motion of the vibrating atoms in a molecule shall be described with sufficient spatial resolution, the still resolvable time interval $\Delta t$ must be small compared to the vibrational period $T_{\text{vib}} = 1/\nu_{\text{vib}}$ of molecules vibrating with the frequency $\nu_{\text{vib}}$. Because of the uncertainty relation

$$\Delta E \cdot \Delta t \geqslant h \tag{10.80}$$

we obtain for the minimum still resolvable energy interval

$$\Delta E \gg h \cdot \nu_{\text{vib}} = \hbar\omega_{\text{vib}} . \tag{10.81}$$

This implies that single vibrational levels cannot be resolved (as in stationary spectroscopy with sufficient spectral resolution), but only a superposition of neighboring vibrational levels within the energy interval $\Delta E$ which can be described by the wave packet

$$\psi(x,t) = \sum_n \phi_n(x) \mathrm{e}^{-\mathrm{i}[(E_n/\hbar)t - k_n x]} . \tag{10.82}$$

The wave packet moves within the potential of the vibrating diatomic molecule (Fig. 10.50) with the group velocity

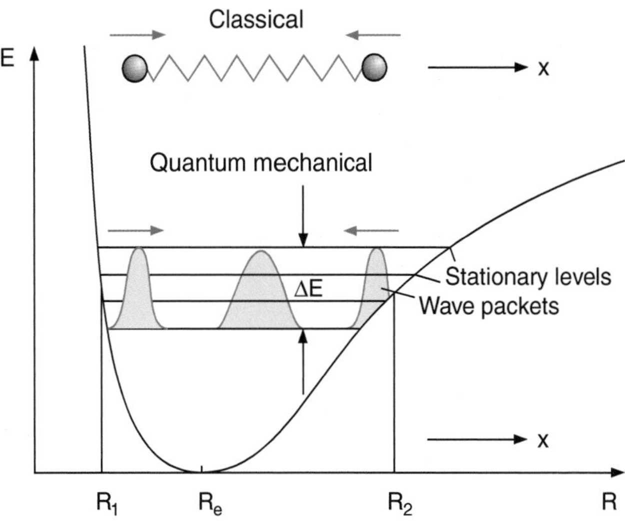


**Fig. 10.50.** Wave packet representation of a vibrating diatomic molecule

$$v_{\text{G}} = \frac{\partial\omega}{\partial k} = \frac{1}{\hbar}\frac{\partial E}{\partial k} \tag{10.83}$$

between the two turning points $R_1$ and $R_2$. The form of the wave packet periodically changes during its motion within the potential, because its kinetic energy $E_{\text{kin}} = E - E_{\text{pot}}(R)$ changes periodically and therefore the phase velocities

$$\begin{aligned} v_{\text{ph}}(\omega) &= \frac{\omega}{k} = \frac{1}{\hbar}\frac{E_{\text{kin}}}{k} \\ &= \frac{1}{\hbar k}\big(E - E_{\text{pot}}(R)\big) \end{aligned} \tag{10.84}$$

of the partial waves $\phi_n$ also change. At the turning points $R_1$ and $R_2$ of the potential is $E = E_{\text{pot}}$ and the kinetic energy becomes zero.

This motion of the wave packet can be experimentally monitored by pump and probe experiments using femtosecond laser pulses (see Sect. 12.2). It can be also calculated if the potential is known, and with modern computer programs it can be visualized on a computer screen where the motion can be viewed in a slow motion version, where the real velocity is slowed down by a factor of $10^{14}$ (also see Table 1 in the Appendix, where the computer simulation of the motion of a wave packet in a one-dimensional potential box is shown in an $x$-$t$ two-dimensional diagram).

Since the potential surfaces of polyatomic molecules can nowadays be calculated with sufficient accuracy, the motion of the atoms in vibrating polyatomic molecules can be visualized and gives a more vivid impression of the dynamics of vibrating molecules.

The time-dependent description is in particular important for electronic processes, such as dissociation of electronically excited molecules, or the rearrangement of the electron cloud following excitation of the molecule which can result in structural changes (isomerization) or in the transfer of electronic into vibration energy. Such processes play an important role in biological molecules. One example is the visual process, where rhodopsin molecules in the retina of the eye are excited by absorption of photons and pass through several structural changes within femto- to pico-seconds. The details of these processes could be only investigated after the application of ultra short laser pulses and new spectroscopic techniques such as time-resolved Raman spectroscopy and pump-and probe spectroscopy.

## SUMMARY


- For polyatomic molecules with $q$ atoms the potential energy $E_{\text{pot}}(R)$ as a function of the nuclear geometry is represented by an $(n-1)$-dimensional surface in an $(n-1)$-dimensional space. For planar molecules is $n = 3q - 6$, for non-planar molecules is $n = 3q - 5$. The nuclei vibrate on this surface and the restoring force for a displacement of the nuclei from the equilibrium position at the minimum of the surface is given by the slope grad $E_{\text{pot}}$ of the surface.
- The localized bonding between two atoms of the molecule is described by the spatial overlap of atomic valence orbitals. Often the hybridization (linear combination of atomic orbitals involved in the chemical bond) increases the bonding energy. The kind of hybridization determines the geometrical direction of the localized bonds: $sp$ hybridization leads to a linear geometry, $sp^2$ leads to three bonds with angles of 120° between the bond directions.
- For aromatic molecules, delocalized $\pi$ orbitals are essential for the bonding and geometrical structure of the molecules. The energy levels of $\pi$ orbitals depend on the spatial extension of the orbitals and can be approximately calculated as for a particle in a potential box.
- The rotation of nonlinear molecules is possible around any axis through the center of mass. It can be described by the space fixed rotational angular momentum $J$ and the three principal moments of inertia. The rotation axis is generally not constant in time but changes its direction in the molecular frame unless it points into the direction of one of the principal axes.
- For a symmetric top, two of the principal moments of inertia are equal. The rotational energy of a symmetric top is determined by the two differing moments of inertia, by the angular momentum $J$ and its projection $K\hbar$ onto the top axis.
- Any vibration of a molecule can be described as a superposition of some of the $(3q-6)$ normal vibrations ($3q-5$ for linear molecules), as long as the vibrational amplitudes are sufficiently small to stay within the parabolic part of the potential. The total vibrational energy is the sum of all vibrational quanta in the normal vibrations involved.
- For higher vibrational energies the increasing anharmonicity of the potential leads to nonlinear couplings between the different normal vibrations and the total vibrational energy is no longer the sum of the energies of the normal vibrations but contains coupling terms.
- The spectra of polyatomic molecules consist of pure rotational transitions in the microwave region, vibrational-rotational transitions within the same electronic state in the mid-infrared region and electronic transitions in the range from the near infrared to the ultraviolet region.
- Infrared vibrational transitions are allowed if the nuclear dipole moment $\boldsymbol{p}$ depends on the nuclear displacements. Normal vibrations $Q$ where $\mathrm{d}\boldsymbol{p}/\mathrm{d}Q \neq 0$ are called infrared active, those with $\mathrm{d}\boldsymbol{p}/\mathrm{d}Q = 0$ are infrared inactive. The absorption probability for infrared inactive transitions vanishes.

- The rotational structure of vibrational and of electronic transitions of linear molecules is similar to that of diatomic molecules. For symmetric tops it consists of $K$ subbands. For any given $K$ value when $K \neq 0$ $P$, $Q$ and $R$ transitions are possible, for $K = 0$ no $Q$-lines appear.
- Electronic transitions are allowed, if the integrand of the transition dipole moment is symmetric for all symmetry operations of the molecule. Often the wave functions cannot be separated into an electronic and a nuclear part. The symmetry selection rule is, however, still valid for the total wave function.
- Clusters are aggregates of many like (homonuclear) or different (heteronuclear) atoms or molecules. They represent a transition regime between molecules and solid microparticles or liquid microdrops. They can be classified, according to their size, into small clusters ($N =$ 3–100), medium-sized clusters ($N = 100$–1,000) and large clusters $N > 10^3$. They can also be classified by the type of atoms into noble gas clusters, metal clusters, covalent atom clusters or molecular clusters.
  With respect to the strength of their bonding we distinguish between van der Waals clusters, ionic clusters, covalent clusters and metal clusters with delocalized electrons.
- Chemical reactions are based on collisions between atoms or molecules where the atomic composition of the collision partners changes.
  The order of a chemical reaction is determined by the number of reactants. The velocity of a reaction is described by rate constants, which depend on the potential energy as a function of the reaction coordinate, and on the kinetic and internal energy of the reactants.
- Many reactions show a potential barrier, which has to be overcome by the reactants in order for the reaction to happen. Reactions where the final kinetic energy of all reaction products is larger than the initial energy of the reactants are called exothermic. If it is smaller the reactions are endothermic. The energy released in exothermic reactions is converted into translational energy and leads to a temperature increase of the reaction product in the reaction chamber. For endothermic reactions, energy has to be fed into the system in order to realize the reaction.
- For the description of molecular dynamics wave packets, instead of stationary wave functions, have to be used. The energy resolution $\Delta E$ is restricted by the wanted time resolution $\Delta t$ according to the uncertainty relation

$$\Delta E \Delta t \geqslant h\ .$$

## PROBLEMS

1. The nuclear geometry of the $Na_3$ molecule is an acute angled triangle with the apex angle $\alpha = 80°$ and the side-length $s = 0.324$ nm. What are the three principal moments of inertia and the three rotational constants $A$, $B$ and $C$. Prove that $1/A + 1/B = 1/C$. Is this true for any planar molecule?
2. The rotational constants of the $NO_2$ molecule are $A = 8.00\,\mathrm{cm}^{-1}$, $B = 0.434\,\mathrm{cm}^{-1}$ and $C = 0.410\,\mathrm{cm}^{-1}$. Determine the geometry of the molecule.
3. Determine the normal vibrations of the linear acetylene molecule $C_2H_2$ by drawing the arrows for the atomic displacements. Which of these vibrations are infrared active?
4. The frequencies of the normal vibrations of the linear $CO_2$ molecule are $\nu_1 = 1388 cm^{-1}$, $\nu_{21} = 667\,\mathrm{cm}^{-1}$ and $\nu_3 = 2349 cm^{-1}$. Calculate the restoring forces between the C and the O atoms.
5. With the Walsh diagrams as a guide, determine the electronic configurations and molecular orbitals for the ground state and the first three excited states of the $BH_2$ molecule.
6. The vibrational constants in the excited A state of the $Na_2$ molecule are $\omega_e = 159\,\mathrm{cm}^{-1}$, $\omega_2 x_e = 1.2\,\mathrm{cm}^{-1}$. A short laser pulse of duration 130 fs excites vibrational levels around $v' = 15$. How many levels are excited? What is the period $T$ for a full oscillation of the wave packet?

# 11. Experimental Techniques in Atomic and Molecular Physics

The goals of all experimental investigations in atomic and molecular physics are:

1. To gain information about the structure of atoms and molecules and their mutual interactions.
2. To determine the bonding and ionization energies and to investigate electric and magnetic moments and their influence on the interaction energy
3. To acquire more details about time dependent processes in atoms and molecules, i.e., about the molecular dynamics, which govern all atomic and molecular processes, such as chemical reactions and the interactions of photons with matter. They are the basis for all biological processes and therefore for life on earth.

In order to reach these goals, a large variety of different experimental techniques have been developed. They can be classified into three categories:

a) Spectroscopic techniques
b) Measurements of collision cross sections
c) Investigations of macroscopic phenomena.

**Spectroscopic Techniques.** Here the absorption or emission of electromagnetic radiation by free atoms or molecules is observed. From measurements of the wavelengths of spectral lines the energy levels can be determined. The intensities of spectral lines give information about transition probabilities, which in turn depend on the wave functions of the energy levels and on their symmetries. The knowledge of wavelengths and intensities allows one to set up a model of the structure of the molecule and its potential energy surface. Measurements of fine and hyperfine splittings can determine magnetic moments and their interactions and Stark splittings in external electric fields give information on electric dipole moments.

Couplings between different electronic states or between vibrational and electronic states cause deviations of the line positions from their expected values, which are called ***perturbations***. The investigations of such perturbations give information on the type and strength of these couplings.

Measurements of collision-induced line broadenings and energy transfer by inelastic collisions are ways to learn about the interaction potential between the collision partners.

Essential for the power of such spectroscopic measurements are the achievable spectral resolution and the sensitivity.

In cases where spectroscopy can be applied this technique represents the most powerful and most accurate of the three categories.

**Measurements of Collision Cross Sections.** Collisions can be also investigated without spectroscopic techniques by observing the scattered particles after elastic or inelastic collisions have occurred. Here the integral or differential cross sections are measured which, similarly to the measurements of collisional line broadening, give information on the interaction potential between the collision partners. Measurements of inelastic collisions allow one to follow the energy transfer between the partners from which the internal structure of the partners can often be inferred. Reactive collisions represent the primary processes of chemical reactions.

**Investigations of Macroscopic Phenomena.** The transport characteristics of molecular gases, such as diffusion, heat conduction or viscosity, also give information about the interactions between molecules or atoms. Here the properties of single molecules are not directly studied, but are inferred from the average effects of a large number of molecules ($N > 10^{20}$). In

W. Demtröder, *Atoms, Molecules and Photons*, 2nd ed., Graduate Texts in Physics,
DOI 10.1007/978-3-642-10298-1_11, © Springer-Verlag Berlin Heidelberg 2010

the same category belong measurements of relationships between thermodynamic properties (pressure $p$, volume $V$ and temperature $T$) of real gases, such as those given by the van der Waals equation or the virial expansion of the general gas equation (see Sect. 2.2). Although such macroscopic measurements are easier and less expensive, the information about molecular properties are not as detailed as in the former measurement category because the results are averages over the velocity distribution and over the random orientations based on a large number of molecules. Details like the deviation of the molecular shape from spherical symmetry can therefore only be indirectly inferred.

The information obtained from the different experimental techniques often supplement each other. Spectroscopic techniques mainly give information on bound states of molecules or atoms not far from the equilibrium geometry. Collision experiments, on the other hand, probe the long-range part of the interaction potential.

During recent years, the introduction of laser spectroscopy has allowed several novel experimental techniques that combine scattering experiments with spectroscopic preparation of the collision partners with internal state selection and spectroscopic detection of the scattered particles. This combination gives much more detailed information than could be obtained from one of the two methods alone.

Of particular importance for the study of molecular dynamics was the development of ultrashort laser pulses that allow the experimenter to follow in real-time fast relaxation processes, the dissociation of molecules and the formation of new bonds in chemical reactions.

In this chapter the most important experimental techniques, including new developments, are presented. This should give the reader more detailed information on the way our knowledge about molecular structure and dynamics is obtained from experimental results [11.1, 2].

## 11.1 Basic Principles of Spectroscopic Techniques

An important criterion of a spectroscopic techniques is its sensitivity, which is defined as the minimum number of photons absorbed during a transition that can be still detected. This is also a measure of the minimum number of atoms or molecules that can be still monitored. The ultimate sensitivity is reached when single atoms or molecules can be detected.

The sensitivity of absorption spectroscopy can also be defined by the minimum absorption coefficient $\alpha(\nu)$ that can still be measured.

When an electromagnetic wave with frequency $\nu$ and intensity $I_0$ passes through an absorbing sample, the transmitted intensity is (Fig. 11.1)

$$I_t(\nu) = I_0 e^{-\alpha(\nu)x} . \quad (11.1)$$

The spectral absorption coefficient $\alpha(\nu_{ik})$ of a transition $E_k \rightarrow E_i$ is determined by the absorption cross section $\sigma_{ik}$, the difference $N_k - N_i$ of the populations densities and the ratio $g_k/g_i$ of the statistical weights:

$$\alpha(\nu_{ki}) = [N_k - (g_k/g_i)N_i]\sigma(\nu_{ki}) . \quad (11.2)$$

The statistical weights $g = (2J + 1)$ give the number of possible orientations of the total angular momentum $J$ of the levels. The absorption coefficient $\sigma_{ik}$ is related to the Einstein coefficient $B_{ik}$ by

$$B_{ki} = \frac{c}{h\nu_{ki}} \int_{\nu=0}^{\infty} \sigma(\nu)\,d\nu , \quad (11.3)$$

where the integration extends over all frequencies, although it is mainly only the interval $\Delta\nu$ within the absorption line profile that contributes to the absorption.

For small absorptions the exponential function in (11.1) can be expanded into

$$I_t(\nu) \approx I_0(1 - \alpha(\nu)x) .$$

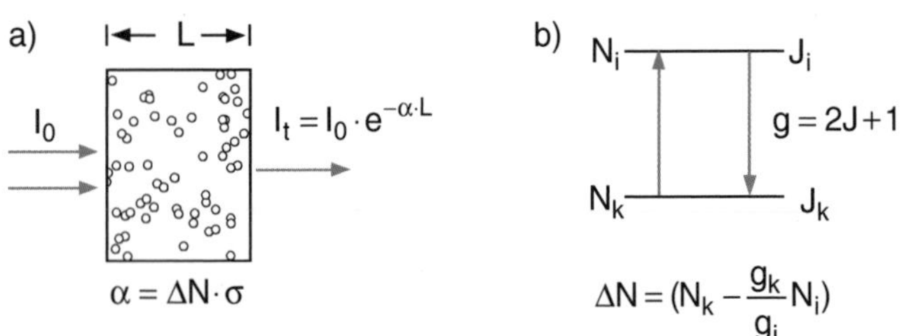


**Fig. 11.1.** Absorption of a monochromatic wave by a sample of molecules with level densities $N_k$, $N_i$

We then obtain from (11.2) for an absorption length $L$

$$\begin{aligned}\frac{\Delta I}{I_0} = \frac{I_0 - I_t}{I_0} &\approx \alpha(\nu)L \qquad (11.4)\\ &= [N_k - (g_k/g_i)N_i]\,\sigma(\nu_{ik})L\\ &= \Delta N_{ik} \cdot \sigma(\nu_{ik})L\,.\end{aligned}$$

The minimum detectable intensity change $\Delta I$ mainly depends on possible fluctuations of the incident intensity $I_0$ and on other noise sources such as detector noise. We summarize all noise contributions to a total noise $\Delta I_{\text{noise}}$. Absorptions $\Delta I = I_0\alpha L < \Delta I_{\text{noise}}$ need special electronic devices to still be detectable. For the intensity change $\Delta I$ the detector gives an output signal $S \propto \Delta I$, while the noise contributes the amount $\delta S \propto \Delta I_{\text{noise}}$. A measure for the sensitivity is the achievable signal-to-noise ratio $S/\delta S$. From the condition $S \geq \delta S \Rightarrow \Delta I \geq \Delta I_{\text{noise}}$ we obtain with (11.2)–(11.4) the minimum detectable number density of absorbing molecules as

$$\boxed{\Delta N \geq \frac{1}{\sigma(\nu_{ki})L(S/\delta S)}}\,. \qquad (11.5)$$

This shows that for a high sensitivity the path length $L$ should be as long as possible, the signal-to-noise ratio as large as possible and transitions with a large absorption cross section should be selected. We will later see how this demand can be met experimentally.

For all spectroscopic techniques the ***spectral resolving power*** $R$ plays an essential role. It is defined as

$$R = \lambda/\Delta\lambda_{\min}\,, \qquad (11.6)$$

where $\Delta\lambda_{\min}$ is the minimal spectral interval that can be still resolved, i.e., two spectral lines separated by $\Delta\lambda > \Delta\lambda_{\min}$ can be recognized as two separated lines. This minimal interval $\Delta\lambda_{\min}$ depends on the profile of the spectral lines. A criterion postulated by Lord *Rayleigh* is the following (Fig. 11.2). Two spectral lines can be regarded as resolved, when the two partly overlapping line profiles have a dip between the two maxima with intensity $I_{\max}$, which does not exceed $(8/\pi^2)I_{\max} \approx 0.8I_{\max}$.

The spectral resolving power depends on the spectroscopic instrument used for the separation of different wavelengths of absorbing or emitting transitions. These can either be spectrometers or interferometers. We will briefly discuss both types of instruments.

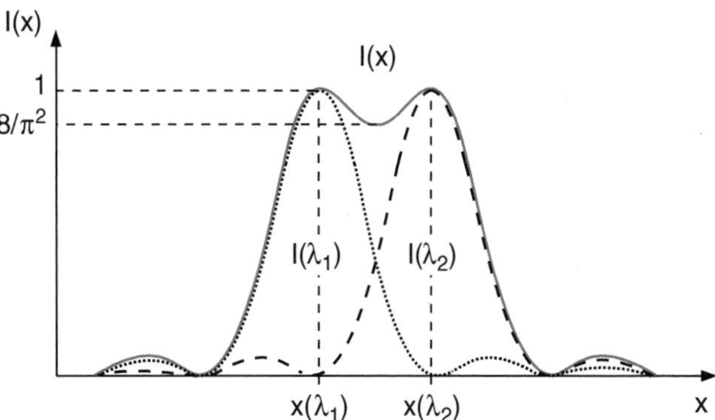


**Fig. 11.2.** Rayleigh definition of the resolution of two spectral lines

## 11.2 Spectroscopic Instruments

We will here present two different kinds of spectroscopic instruments:

- Wavelength selectors (monochromators, spectrographs and interferometers)
- Radiation detectors (thermopiles, photodiodes, photomultipliers and CCD devices)

### 11.2.1 Spectrometers

All spectrometers are instruments that allow the spatial dispersion of different wavelengths. This dispersion can be reached by refracting prisms (prism spectrometer, Fig. 11.3) or by diffraction (grating spectrometers, Fig. 11.4). The incident radiation is focused onto the entrance slit $S_1$, which is imaged by two lenses $L_1$ and $L_2$ or by two spherical mirrors $M_1$ and $M_2$ onto the detection plane. The dispersing element in the parallel part of the radiation bundle causes a wavelength-dependent change $\Delta\Theta = (\mathrm{d}\Theta/\mathrm{d}\lambda)\Delta\lambda$ of the propagation direction and thus a shift $\Delta x$ of the position of the slit image in the observation plane (Fig. 11.5). The

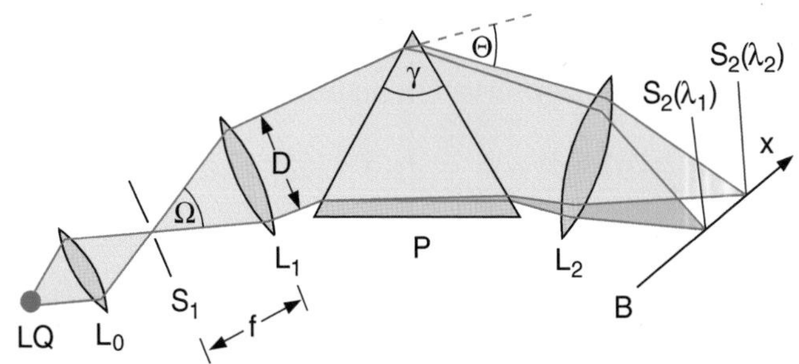


**Fig. 11.3.** Prism spectrograph

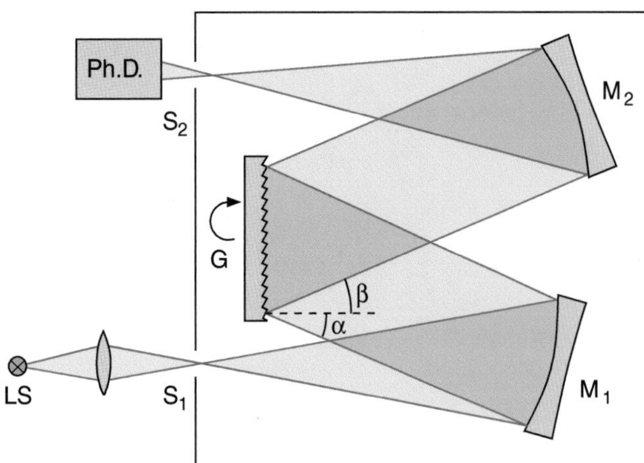


**Fig. 11.4.** Grating monochromator (Ph.D. = photodetector, G = optical grating)

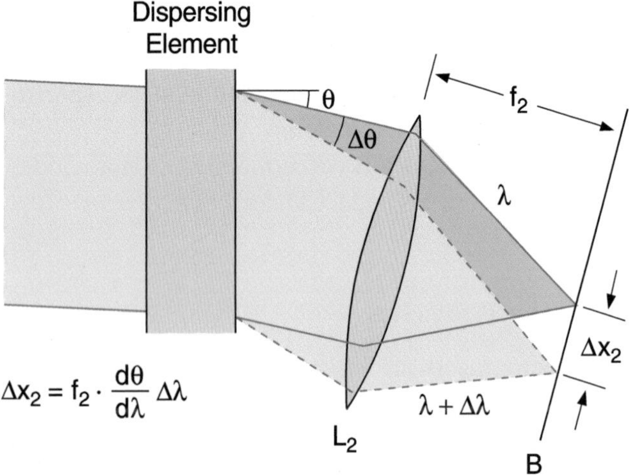


**Fig. 11.5.** Angular dispersion $\mathrm{d}\theta/\mathrm{d}\lambda$ and linear dispersion $\mathrm{d}x/\mathrm{d}\lambda$ of a dispersing element

angular dispersion $\mathrm{d}\Theta/\mathrm{d}\lambda$ depends on the dispersion of the prism material or on the separation of the grooves in a diffraction grating.

With a focal length $f_2$ of $L_2$ or $M_2$ the linear dispersion of the spectrometer is

$$\mathrm{d}x/\mathrm{d}\lambda = f_2 \mathrm{d}\Theta/\mathrm{d}\lambda \,. \tag{11.7}$$

Two spectral lines with wavelengths $\lambda_1$ and $\lambda_2$ are separated in the observation plane by

$$\Delta x_2 = (\mathrm{d}x/\mathrm{d}\lambda)\Delta\lambda = f_2(\mathrm{d}\Theta/\mathrm{d}\lambda)\Delta\lambda \,, \tag{11.8}$$

where $\Delta\lambda = \lambda_1 - \lambda_2$.

When an exit slit with variable width $\delta x_2$ is mounted in the observation plane, a spectral interval

$$\Delta\lambda = \delta x_2/(\mathrm{d}x/\mathrm{d}\lambda) \tag{11.9}$$

is transmitted. The central wavelength of this interval can be tuned by turning the grating in Fig. 11.4 or by tilting the observation arm behind $L_2$ in Fig. 11.3. Instead of the exit slit $S_2$ modern spectrometers often use a CCD array with 500–2000 pixels, a pixel width of about $\delta x = 10$–$20\,\mu\mathrm{m}$ and a total width $D = 10$–$40\,\mathrm{mm}$. With this device a spectral interval of $\Delta\lambda = D/(\mathrm{d}x/\mathrm{d}\lambda)$ can be covered simultaneously with a spectral resolution $\delta\lambda = \delta x/(\mathrm{d}x/\mathrm{d}\lambda)$.

The width $\delta x_2$ of the slit image is related to the width $\delta x_1$ of the entrance slit by

$$\delta x_2 = (f_2/f_1)\delta x_1 \,.$$

The minimum resolvable spectral interval

$$\Delta\lambda_{\mathrm{min}} = \delta x_2/(f_2 \mathrm{d}\Theta/\mathrm{d}\lambda) = \delta x_1/(f_1 \mathrm{d}\Theta/\mathrm{d}\lambda) \tag{11.10}$$

is generally limited by the width $\delta x_1$ of the entrance slit or by the width of one pixel of the CCD array. With decreasing width $\delta x_1$ the spectral resolution $\lambda/\Delta\lambda$ increases. However, this is only true down to a minimum slit width $\delta x_{\mathrm{min}}$, where diffraction starts to play the dominant role. One has to distinguish between two different effects of diffraction:

1. The diffraction by the limiting aperture $a$ in the spectrometer (Fig. 11.6) given by the diameter of the lenses or mirrors or by the grating or prism. This diffraction leads to a broadening of the image of the entrance slit. Instead of the geometrical image, a diffraction pattern is observed with a central maximum, which has a full width $\delta x_2 \approx 2f_2\lambda/a$ between the first two minima at both sides of the central maximum. The spectral resolution is not further increased when the entrance slit width is decreased below $\delta x_{\mathrm{min}} = 2(f_1/f_2)\delta x_2 = 2f_1\lambda/a$. The minimum resolvable spectral interval is then
$$\delta\lambda \geq 2(\lambda/a)/(\mathrm{d}\Theta/\mathrm{d}\lambda) \,. \tag{11.11}$$
2. The much larger diffraction by the narrow entrance slit $\delta x_1 \ll a$. This, however, does *not* influence the spectral resolution, but leads to a decrease of the transmitted radiation power if the diffraction angle $\Delta\alpha = \lambda/\delta x_1$ becomes larger than the acceptance angle $\alpha = a/f_1$ of the instrument (Fig. 11.7).

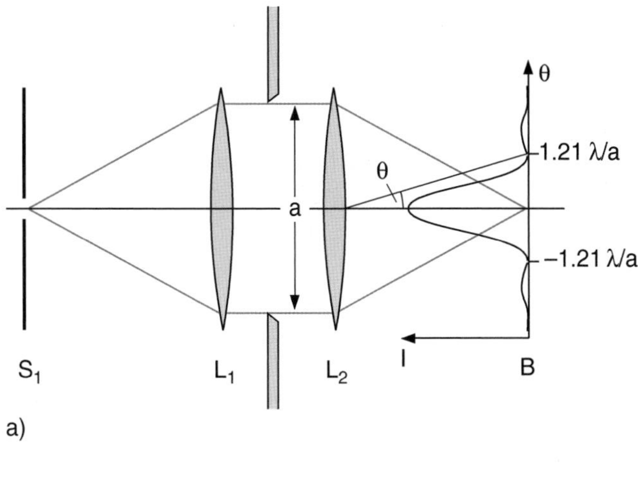


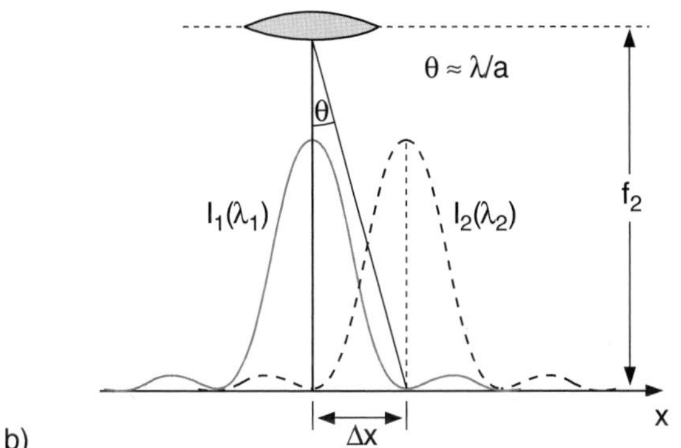


**Fig. 11.6.** (**a**) Broadening of line profiles by diffraction at the limiting aperture $a$ of a spectrometer (**b**) Diffraction-limited resolution

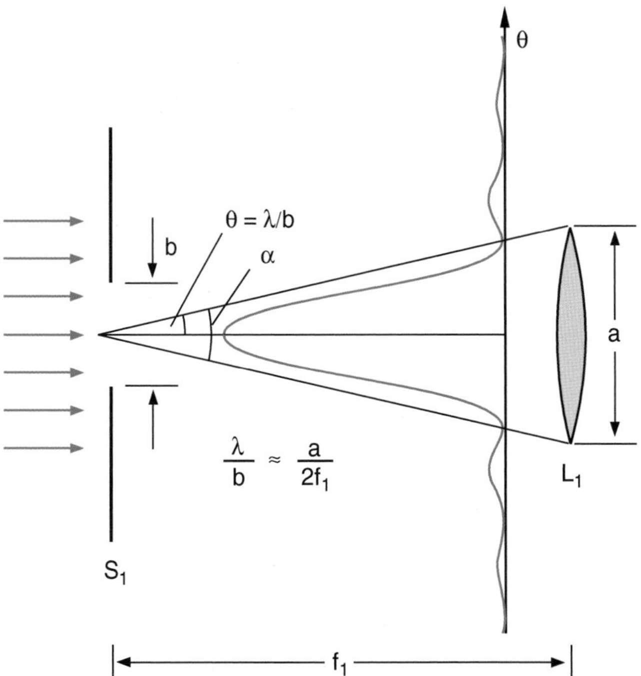


**Fig. 11.7.** The diffraction at the entrance slit results in increasing intensity losses if $2(\lambda/b) > a/f_1$

## EXAMPLE

For a spectrometer with $f_1 = f_2 = 1$ m and a limiting aperture with diameter $a = 10$ cm, the full width of the central diffraction maximum is $\delta x_2 = 20\lambda$. For a wavelength of $\lambda = 500$ nm this would be, without diffraction, the geometrical image of an entrance slit with 12 μm slit width. The minimum resolvable spectral interval is then, with the condition $(\mathrm{d}x/\mathrm{d}\lambda) \cdot \Delta\lambda > \delta x_{\min}$ according to (11.11), $\Delta\lambda \geq (10^{-6}/10^{-1})/(\mathrm{d}\Theta/\mathrm{d}\lambda)$. For an angular dispersion $\mathrm{d}\Theta/\mathrm{d}\lambda = 10^{-4}\,\mathrm{nm}^{-1}$ this gives the principal lower limit for the resolvable spectral interval $\Delta\lambda \geq 10^{-1}$ nm. The resolving power is then $\lambda/\Delta\lambda = 5 \times 10^3$.

### a) Prism Spectrometer

If a parallel light beam falls under an angle $\alpha$ onto a prism (Fig. 11.8) it is refracted twice and the deflection angle $\Theta$ depends on the index of refraction $n$ on the angle of incidence $\alpha$ and on the apex angle $\varepsilon$ of the prism. The minimum deflection is observed when the light beam passes the prism parallel to the base $g$ (symmetrical case with $\alpha_1 = \alpha_2$). From geometrical optics one can derive for this situation the relation

$$\sin\left(\frac{\Theta+\varepsilon}{2}\right) = n\sin(\varepsilon/2)\,. \tag{11.12}$$

because: $\theta = 2(\alpha - \beta)$ and $2\beta = \varepsilon \Rightarrow \alpha = \frac{\Theta+\varepsilon}{2}$
$\sin\alpha = n \cdot \sin\beta = n \cdot \sin(\varepsilon/2) = \sin\left(\frac{\Theta+\varepsilon}{2}\right)$
The angular dispersion $\mathrm{d}\Theta/\mathrm{d}\lambda$ can be written as

$$\begin{aligned}\frac{\mathrm{d}\Theta}{\mathrm{d}\lambda} &= \left(\frac{\mathrm{d}n}{\mathrm{d}\Theta}\right)^{-1} \cdot \frac{\mathrm{d}n}{\mathrm{d}\lambda} \\ &= \frac{2\sin(\varepsilon/2)}{\sqrt{1-n^2\sin^2(\varepsilon/2)}}\frac{\mathrm{d}n}{\mathrm{d}\lambda}\,.\end{aligned} \tag{11.13a}$$

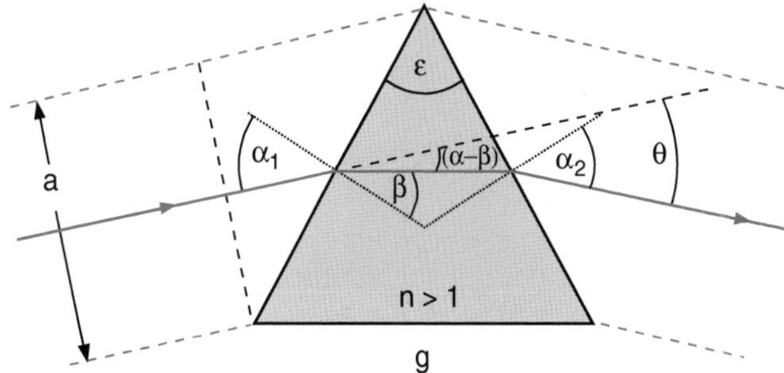


**Fig. 11.8.** Refraction of light by a prism at minimum deviation where $\alpha_1 = \alpha_2 = \alpha$ and $\theta = 2\alpha - \varepsilon$

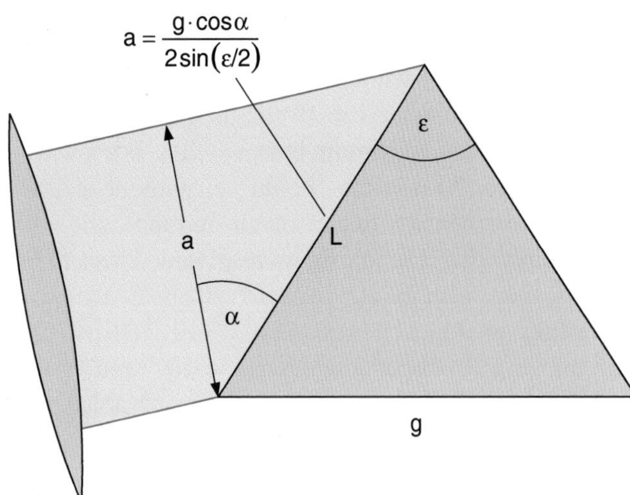


**Fig. 11.9.** Limiting aperture $a$ in a prism spectrograph, determined by the size of the prism

This shows that the angular dispersion *does not depend on the size of the prism but on the apex angle* $\varepsilon$!

For $\varepsilon = 60°$ (11.13a) becomes

$$\frac{d\theta}{d\lambda} = \frac{dn/d\lambda}{\sqrt{1-(n/2)^2}} . \tag{11.13b}$$

However, if the size of the prism is the limiting aperture in the spectrometer, it is important for diffraction effects (Fig. 11.9). With the maximum beam diameter $a = g\cos\alpha_1$, the diffraction-limited resolving power is, according to (11.11) for $\varepsilon = 60°$ and symmetrical arrangement,

$$\frac{\lambda}{\Delta\lambda} = g\frac{dn}{d\lambda} . \tag{11.14}$$

> The principal limit for the resolving power is given by the base length $g$ of the prism and the dispersion $dn/d\lambda$ of the prism material.

**EXAMPLE**

A glass prism with $g = 5\,\text{cm}$ has for $\lambda = 400\,\text{nm}$ the values $n = 1.53$ and $dn/d\lambda = 1\times10^{-4}\,\text{nm}^{-1} = 10^5\,\text{m}^{-1}$. This gives a limit $\lambda/\Delta\lambda = 5\times10^{-2}\times10^5 = 5\times10^3$.

## b) Grating Spectrometers

An optical grating consists of many narrow grooves (0.5–10 μm width) cut into a glass or fused quartz plate, which are separated by a distance $d$ (Fig. 11.10a).

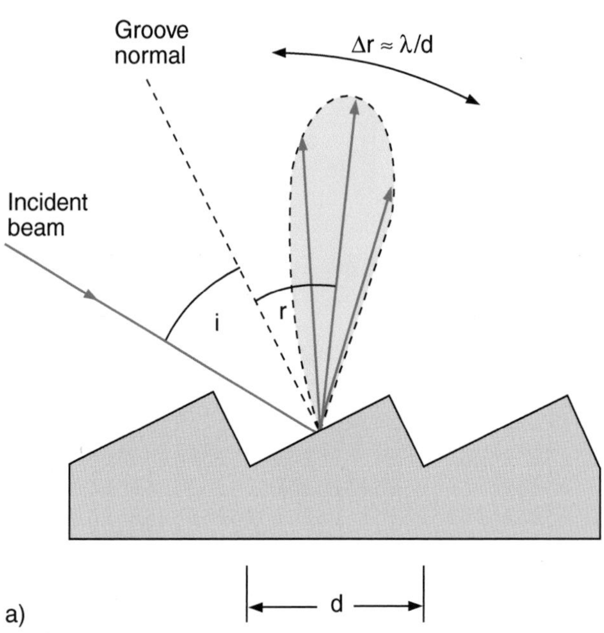


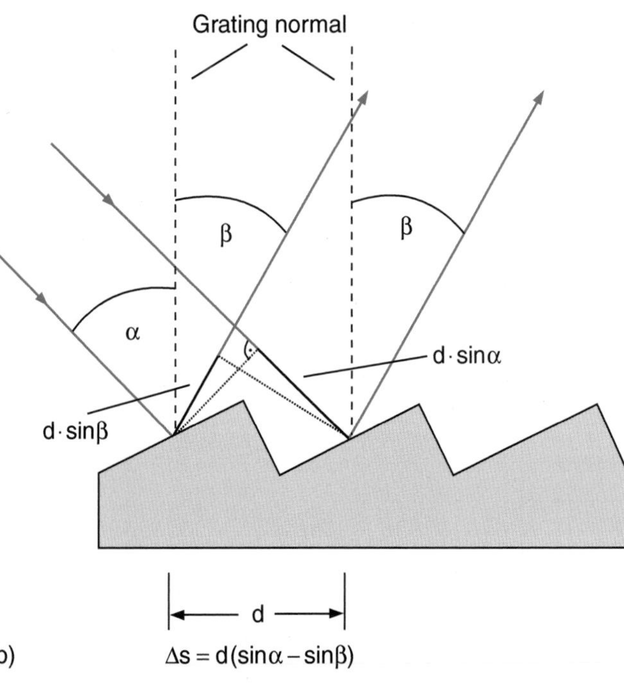


**Fig. 11.10.** (**a**) Reflection of incident light from a single groove into the diffraction angle $\Delta r = \lambda/d$ around the specular reflection angle $r = i$. (**b**) Illustration of the grating Equation (11.15)

The grooves are covered with a reflecting layer. Due to the narrow width of the grooves the reflected light has a diffraction-limited angular distribution around the geometrical reflection angle $r$, which equals the incident angle $i$.

The basic design of a grating spectrometer in shown in Fig. 11.4. The parallel radiation striking the grating with an angle $\alpha$ against the normal to the grating plane is reflected by each of the grooves into an angular cone defined by diffraction. The interference between the different rays is constructive if the phase difference between neighboring partial waves is $\Delta\varphi = m \cdot 2\pi$. According to Fig. 11.10b the path difference between neighboring partial waves is

$$\Delta s = d(\sin\alpha - \sin\beta) \tag{11.15a}$$

for $\Delta\varphi = m2\pi \Rightarrow \Delta s = m\lambda$.

We therefore obtain, as a condition for constructive interference of all reflected partial waves, the grating equation

$$\boxed{d(\sin\alpha \pm \sin\beta) = m\lambda}\ , \tag{11.15b}$$

where the positive sign must be used if the reflected beam is on the same side of the grating normal as the incident beam and the negative sign if they are on opposite sides.

## EXAMPLE

For $d = 1.5\,\mu\text{m}$, $\lambda = 500\,\text{nm}$, $m = 1$ and $\alpha = 30°$ the reflected light is observed at $\beta = 10.6°$.

The angular intensity distribution of the reflected light can be obtained when all interfering partial waves reflected by the different grooves are added with their appropriate phases. The phase difference between neighboring beams is

$$\Delta\varphi = \frac{2\pi}{\lambda} \cdot d(\sin\alpha \pm \sin\beta)\ . \tag{11.16}$$

If we choose the phase of the partial wave reflected from the first groove as $\phi_0 = 0$ than the superposition of all partial amplitudes gives the total amplitude

$$A_\mathrm{R} = \sqrt{R} \sum_{m=0}^{N-1} A_\mathrm{g} \mathrm{e}^{\mathrm{i}m\phi}\ , \tag{11.17}$$

where $R(\beta)$ is the reflectivity of the grating surface and $A_\mathrm{g}$ is the amplitude of the wave incident on one of the grooves.

The total intensity $I = \varepsilon_0 c A_\mathrm{R} A_\mathrm{R}^*$ is then

$$I_\mathrm{R} = I_0 R \frac{\sin^2(N\phi/2)}{\sin^2(\phi/2)}\ . \tag{11.18}$$

This intensity distribution $I(\phi)$ is plotted in Fig. 11.11 for two different numbers of grooves $N$. It consists of sharp principal maxima for $\phi = m \cdot 2\pi$ and $N-2$ small side maxima between the two principal maxima with amplitudes that decrease with increasing $N$.

Optical gratings have about $N = 10^6$–$10^7$ grooves. This implies that for a given angle $\alpha$ of incidence and a given wavelength $\lambda$, the diffracted light is only reflected onto a very narrow angular interval $\Delta\beta$ around $\beta_m$.

The intensity profile of the principal maxima around $\beta_m$ can be obtained as follows.

With $\beta = \beta_m + \varepsilon$ ($\varepsilon \ll \beta$) we can write

$$\sin(\beta_m + \varepsilon) \approx \sin\beta_m + \varepsilon \cdot \cos\beta_m\ .$$

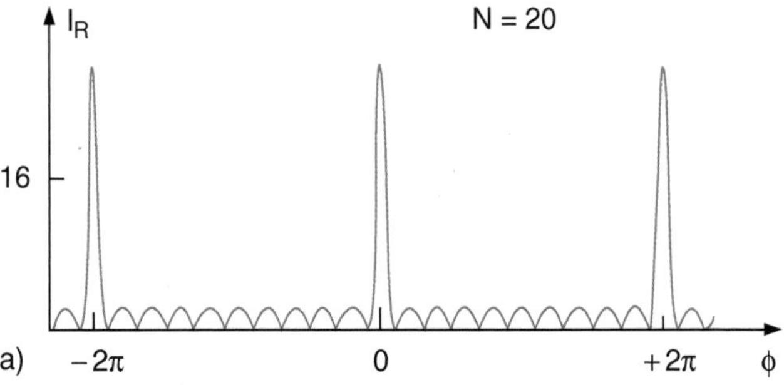


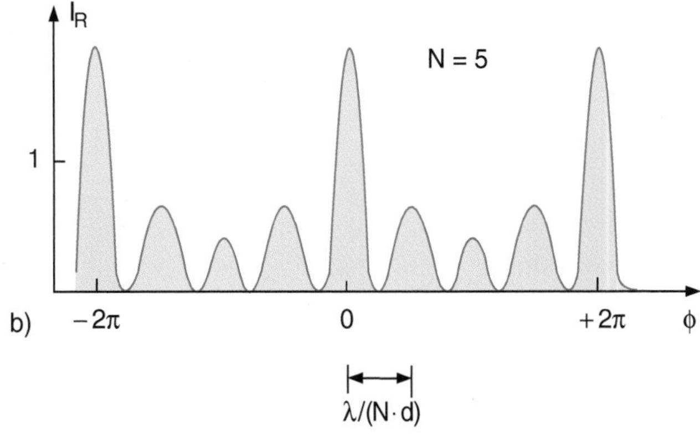


**Fig. 11.11a,b.** Intensity distribution of light reflected by $N$ grooves as a function of the phase difference $\phi = (2\pi/\lambda) \times d(\sin\alpha \pm \sin\beta)$ between neighboring partial waves. Note the different scales in (**a**) and (**b**)

Inserting (11.16) into (11.18) we obtain

$$\begin{aligned} I_\mathrm{R} &= I_0 R \frac{\sin^2(N\delta/2)}{\sin^2(\delta/2)} \\ &\approx R \cdot I_0 N^2 \frac{\sin^2(N\delta/2)}{(N\delta/2)^2} \end{aligned} \tag{11.19}$$

with $\delta = (2\pi d/\lambda)\varepsilon \cos\beta_m \ll 1$. This is the same intensity distribution as that of an illuminated slit with width $D = N \cdot d$.

The first minima on both sides of the principal maximum occur for

$$N\delta = 2\pi \Rightarrow \varepsilon_{1,2} = \pm\lambda/(Nd\cos\beta_m\ . \tag{11.20}$$

The distribution with the maximum at $\beta_m$, therefore, has a full width between the first two minima $I = 0$ on both sides of $\beta_m$

$$\Delta\beta = \frac{\lambda}{Nd\cos\beta_m}\ . \tag{11.21}$$

The angular dispersion for a given angle $\alpha$ is obtained by taking the derivative of (11.15b).

This gives with $\mathrm{d}\beta/\mathrm{d}\lambda = (\mathrm{d}\lambda/\mathrm{d}\beta)^{-1}$

$$\frac{\mathrm{d}\beta}{\mathrm{d}\lambda} = \frac{m}{d\cos\beta} = \frac{\sin\alpha \pm \sin\beta}{\lambda\cos\beta}\ , \tag{11.22}$$

which illustrates that the angular dispersion is determined solely by the angles $\alpha$ and $\beta$ and not by the number of grooves.

The resolving power can be immediately derived from (11.22) and the width $\Delta\beta$ in (11.21) of the principal diffraction maximum. Since two spectral lines with wavelengths $\lambda$ and $\lambda + \Delta\lambda$ are just resolved if the maximum of $I(\lambda)$ falls into the first minimum of $I(\lambda + \Delta\lambda)$ (Fig. 11.2), this gives the condition

$$\frac{\mathrm{d}\beta}{\mathrm{d}\lambda} \cdot \Delta\lambda \geq \frac{\lambda}{Nd\cos\beta} \tag{11.23a}$$

or

$$\frac{\sin\alpha \pm \sin\beta}{\lambda\cos\beta}\Delta\lambda \geq \frac{\lambda}{Nd\cos\beta}\ . \tag{11.23b}$$

This reduces with (11.15) to

$$\boxed{\frac{\lambda}{\Delta\lambda} \geq N \cdot m}\ . \tag{11.24}$$

The theoretical resolving power of a diffraction grating is the product of the interference order $m$ and the total number $N$ of illuminated grooves.

Since

$$Nd(\sin\alpha \pm \sin\beta) = \Delta s_m$$

is the maximum path difference between the partial waves reflected from the first and the $n$th groove, we can also write Eq. (11.24) as

$$\boxed{\frac{\lambda}{\Delta\lambda} = \frac{\Delta s_m}{\lambda}}\ . \tag{11.25}$$

The spectral resolution of any spectrometer is equal to the maximum path difference between interfering beams in the instrument, measured in units of the wavelength $\lambda$.

## EXAMPLE

A grating with a ruled area of $10 \times 10\,\mathrm{cm}^2$ and $10^3$ grooves per mm allows in second order ($m = 2$) a spectral resolving power of $R = 2 \times 10^5$ if all grooves are illuminated. Two spectral lines at $\lambda = 500\,\mathrm{nm}$, which are separated by $\Delta\lambda = 2.5 \times 10^{-3}\,\mathrm{nm}$, should be resolvable. However, because of diffraction at the limiting aperture $a$ (given by the size of the grating) the practically achievable resolution is only about $1 \times 10^5$.

The linear dispersion $\mathrm{d}x/\mathrm{d}\lambda$ is for $\alpha = \beta = 30°$ and $f_1 = f_2 = 1\,\mathrm{m} : \mathrm{d}x/\mathrm{d}\lambda = 2\,\mathrm{mm/nm}$. For an entrance slit width $d = 50\,\mu\mathrm{m}$, a spectral resolution $\Delta\lambda = 0.025\,\mathrm{nm}$ can be achieved.

Modern grating spectrometers are often developed without the two mirrors in Fig. 11.4. A concave grating images the entrance slit onto the detector plane, where a CCD array monitors a large spectral range simultaneously (Fig. 11.12). With an entrance slit width $d = 10\,\mu\mathrm{m}$ matched to the width of one pixel (size $10 \times 40\,\mu\mathrm{m}^2$) in the CCD array (size $40\,\mu\mathrm{m} \times 30\,\mathrm{mm}$) and a linear dispersion of $\mathrm{d}x/\mathrm{d}\lambda = 0.2\,\mathrm{mm/nm}$, a spectral resolution of $\Delta\lambda = 0.05\,\mathrm{nm}$ is reached with a simultaneous detection of a spectral range of 200 nm. For more detailled information see [11.3, 4]

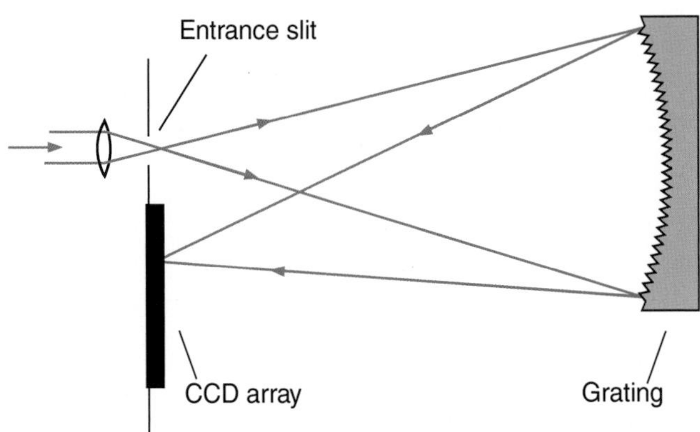


**Fig. 11.12.** Spectrometer with concave grating and CCD detector array

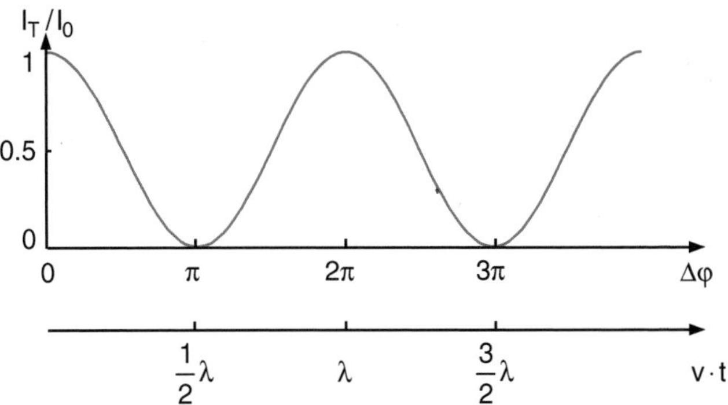


**Fig. 11.14.** Transmitted intensity for a two-beam interferometer as a function of the phase difference between the two interfering beams

### 11.2.2 Interferometers

Interferometers are based on the interference of two or more coherent partial waves. The principle of two-beam interference is illustrated in Fig. 11.13. The wave, emitted from a source S is split by a beam splitter $BS_1$ into two partial beams that travel different path lengths $s_1$ and $s_2$ before they are again superimposed at beam splitter $BS_2$. The total amplitude $E = E_1 + E_2$ at P depends on the phase difference

$$\Delta\varphi = (2\pi/\lambda)(s_2 - s_1) \tag{11.26}$$

and therefore on the path difference $\Delta s = s_2 - s_1$ and on the wavelength $\lambda$.

The total transmitted intensity is

$$I_t = c\varepsilon_0 |E_1 + E_2|^2 \,. \tag{11.27a}$$

With $E_1 = E_0 \cos(\omega t + \varphi_1)$ and $E_2 = E_0 \cos(\omega t + \varphi_2)$ we obtain for the transmitted intensity, averaged over a period of the optical wave (because detectors can not follow the optical cycle and monitor the time average $\langle I(t)\rangle$ (Fig. 11.14), which is for $E_1 = E_2 = E$:

$$\begin{aligned}\langle I(t)\rangle = I(\Delta s) &= c\varepsilon_0 E^2(1+\cos(\Delta\varphi))\\ &= 2I_0 \cos^2(\Delta\varphi/2)\end{aligned} \tag{11.27b}$$

with $\Delta\varphi = \varphi_2 - \varphi_1$.

From the measured period $\Delta\varphi = 2\pi \Rightarrow \Delta s = \lambda$ between two maxima of $I(\Delta s)$ the wavelength $\lambda$ can be determined [11.5, 6].

#### a) Michelson Interferometer

A possible experimental realization of the two-beam interference is the Michelson interferometer (Fig. 11.15), where the incoming parallel beam is split by a beam splitter into two partial beams, which are reflected by the mirrors $M_1$ and $M_2$ and are again superimposed at BS before they reach the detector D

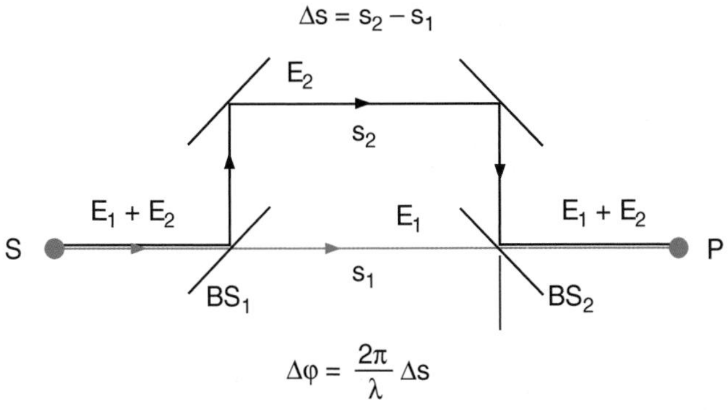


**Fig. 11.13.** Principle of two-beam interference

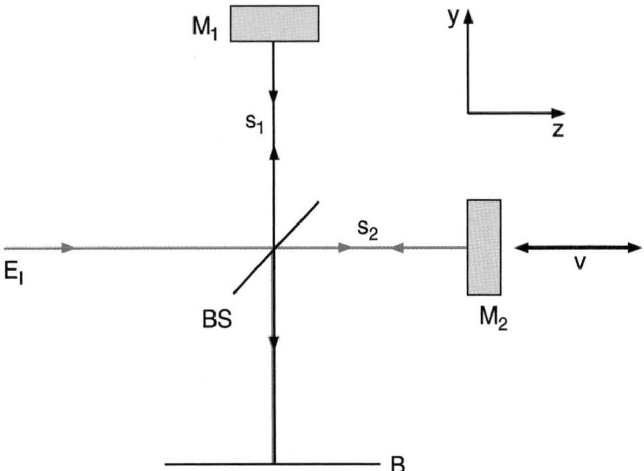


**Fig. 11.15.** Michelson interferometer

in the observation plane B. If $M_2$ is translated by $\Delta x$, the path difference between the two interfering beams changes by $\Delta s = 2\Delta x$.

If the maxima for the transmitted intensity are counted for a measured translation $L = m\lambda$ of mirror $M_2$ the wavelength $\lambda$ can be determined.

### EXAMPLE

For $L = 10\,\text{cm}$ and $\lambda = 500\,\text{nm}$ the number of counts $m$ is $m = \Delta s/\lambda = 2L/\lambda = 2 \times 10^{-1}/(5 \times 10^{-7}) = 4 \times 10^5$. If $L$ can be measured within $1\,\mu\text{m}$ the relative error of the wavelength measurement is $\Delta\lambda/\lambda = 10^{-5}$. Generally the length $L$ is measured interferometrically with a known wavelength $\lambda_R$ and an accuracy of 10 nm, which gives a relative uncertainty of $10^{-8}$ for the wavelength determination.

## b) Fabry–Perot Interferometer

If many partial waves interfere we speak of multiple-beam interference. We will illustrate this by interference at a plane-parallel glass plate (Fig. 11.16).

If the plane wave

$$E = A_0 e^{i(\omega t - kr)}$$

falls onto such a glass plate with an angle $\alpha$ against the surface normal the wave is split into a reflected part with amplitude $\sqrt{R} \cdot A_0$ and a transmitted part with amplitude $A_0\sqrt{1-R}$. The transmitted part is partly reflected at the lower surface and partly transmitted. We obtain for the amplitudes of the different partial waves in Fig. 11.16 the following relations:

$$\begin{aligned}
|A_1| &= \sqrt{R}|A_0|\,, \qquad |B_1| = \sqrt{1-R}|A_0|\,,\\
|C_1| &= \sqrt{1-R}|A_0|\,, \quad |D_1| = (1-R)|A_0|\,,\\
|A_2| &= \sqrt{1-R}|C_1| = (1-R)\sqrt{R}|A_0|\,,\\
|B_2| &= \sqrt{R}|C_1| = R \cdot \sqrt{(1-R)}|A_0|\,,\\
|A_3| &= \sqrt{1-R}|C_2| = R^{3/2}(1-R)|A_0| \ldots, \text{etc.}
\end{aligned} \tag{11.28}$$

This gives the general equation for the amplitudes $A_i$ of the reflected waves

$$|A_{i+1}| = R|A_i| \quad \text{for} \quad i \geq 2 \tag{11.29a}$$

and for the transmitted amplitudes

$$|D_{i+1}| = R|D_i| \quad \text{for} \quad i \geq 1\,. \tag{11.29b}$$

The path difference between two reflected waves $i$ and $i+1$ is, according to Fig. 11.17,

$$\begin{aligned}
\Delta s &= n(\overline{AB} + \overline{BC}) - \overline{AD}\\
&= \frac{2nd}{\cos\beta} - 2d\tan\beta\sin\alpha
\end{aligned}$$

with $\sin\alpha/\sin\beta = n$ this gives

$$\Delta s = 2d\sqrt{n^2 - \sin^2\alpha}\,, \tag{11.30}$$

which results in a phase difference

$$\Delta\varphi = 2\pi\,\Delta s/\lambda + \delta\varphi\,,$$

where $\delta\varphi$ accounts for possible phase shifts at the reflection which depend on the polarization of the wave.

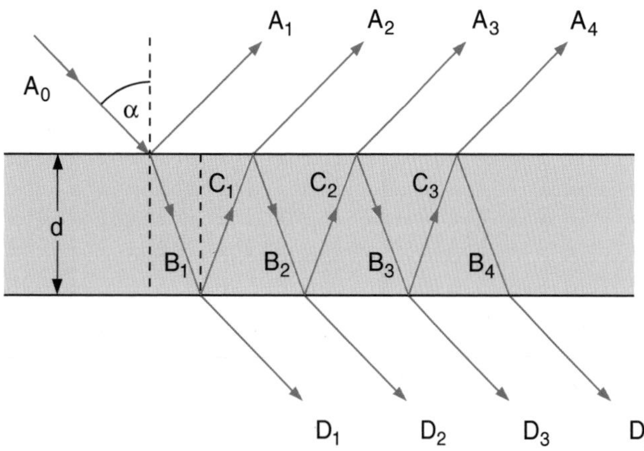


**Fig. 11.16.** Multiple beam interference at a plane-parallel glass plate, covered with reflecting surfaces

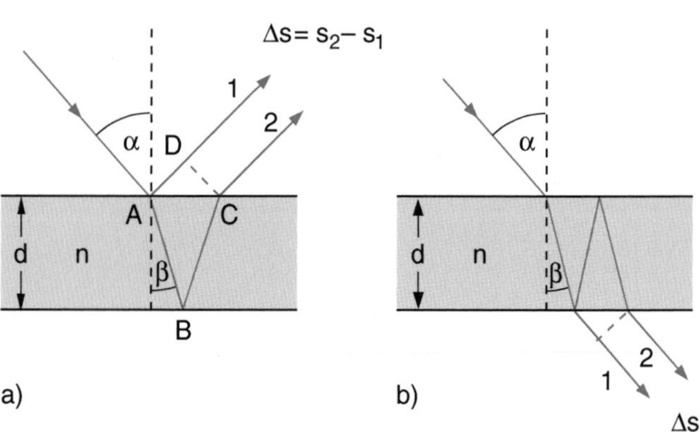


**Fig. 11.17a,b.** Path difference between two beams (**a**) reflected by (**b**) transmitted through a plane-parallel glass plate

The total amplitude of the reflected wave is

$$
\begin{aligned}
A &= \sum_{m=1}^{p} A_m \mathrm{e}^{\mathrm{i}(m-1)\Delta\varphi} \qquad (11.31)\\
&= \pm A_0\sqrt{R}\\
&\quad \cdot \left[1-(1-R)\mathrm{e}^{\mathrm{i}\Delta\varphi} - R(1-R)\mathrm{e}^{2i\Delta\varphi} + \ldots\right]\\
&= \pm A_0\sqrt{R}\left[1-(1-R)\mathrm{e}^{\mathrm{i}\Delta\varphi}\sum_{m=0}^{p-2} R^m \mathrm{e}^{\mathrm{i}m\Delta\varphi}\right].
\end{aligned}
$$

In the case of highly reflecting surfaces the number $p$ of reflections is large. For $p \to \infty$ the sum of the geometric series is

$$\sum_{m=0}^{\infty} R^m \mathrm{e}^{\mathrm{i}m\Delta\varphi} = \frac{1}{1-R\mathrm{e}^{\mathrm{i}\Delta\varphi}}$$

and the total reflected amplitude becomes

$$A = \pm\sqrt{R}A_0 \frac{1-\mathrm{e}^{\mathrm{i}\Delta\varphi}}{1-R\mathrm{e}^{\mathrm{i}\Delta\varphi}}\,. \qquad (11.32)$$

The intensity of the reflected wave is therefore

$$I_\mathrm{R} = c\varepsilon_0|A|^2 = I_0 R\frac{2-2\cos\Delta\varphi}{1+R^2-2R\cos\Delta\varphi}\,, \qquad (11.33a)$$

which can be written with $1-\cos x = 2\sin^2(x/2)$ as

$$I_\mathrm{R} = I_0 \frac{4R\sin^2(\Delta\varphi/2)}{(1-R)^2+4R\sin^2(\Delta\varphi/2)}\,. \qquad (11.33b)$$

In a similar way, one finds for the intensity of the transmitted wave

$$I_\mathrm{t} = I_0 \frac{(1-R)^2}{(1-R)^2+4R\sin^2(\Delta/2)}\,. \qquad (11.34)$$

These are called *Airy formulas*. With the abbreviation

$$F = 4R/(1-R)^2 \qquad (11.35)$$

we can write the Airy formulas as

$$
\boxed{
\begin{aligned}
I_\mathrm{R} &= I_0 \frac{F\sin^2(\Delta\varphi/2)}{1+F\sin^2(\Delta\varphi/2)}\\
I_\mathrm{t} &\approx I_0 \frac{1}{1+F\sin^2(\Delta\varphi/2)}
\end{aligned}
}\,. \qquad (11.36)
$$

The reflected and the transmitted intensities depend on the phase differences $\Delta\varphi = 2\pi\,\Delta s/\lambda$ and therefore on the wavelength $\lambda$. For $\Delta\varphi = 2m\pi \Rightarrow \Delta s = m\lambda$ the transmitted intensity $I_t(2\pi) = I_0$ becomes maximum, i.e., the transmission of the interferometer becomes 100% and the reflected intensity is zero. For $\Delta\varphi = (2m+1)\pi$ the transmitted intensity has minima

$$I_t^{\min} = \left(\frac{1-R}{1+R}\right)^2 I_0\,. \qquad (11.37)$$

In Fig. 11.18 the transmittance $T = I_\mathrm{t}/I_0$ is plotted for different values of the reflection coefficient $R$ as a function of $\Delta\varphi$. For a given thickness $d$ of the plane-parallel plate and a given angle $\alpha$ of incidence $\Delta s$ is fixed and $\Delta\varphi$ solely depends on $\lambda$. The abscissa in Fig. 11.18 can then also be scaled in units of frequency $\nu = c/\lambda$. The frequency separation of two neighboring maxima with $\Delta\varphi_1 = 2m\pi$ and $\Delta\varphi_2 = 2(m+1)\pi$ is the free spectral range

$$\delta\nu = \frac{c}{\Delta s} = \frac{c}{2d\sqrt{n^2-\sin^2\alpha}}\,. \qquad (11.38a)$$

For vertical incidence ($\alpha = 0$) this reduces to

$$\boxed{\delta\nu(\alpha=0) = \frac{c}{2nd}}\,. \qquad (11.38b)$$

The multiple-beam interference is used in the ***Fabry–Perot interferometer*** (FPI) for accurate wavelength measurements. The technical realization uses either a plane-parallel plate of fused quartz covered with reflecting layers on both sides (Fig. 11.19a) or two plates with one reflecting and one antireflecting layer that are aligned in such a way that between the reflecting surfaces a plane parallel air layer is formed

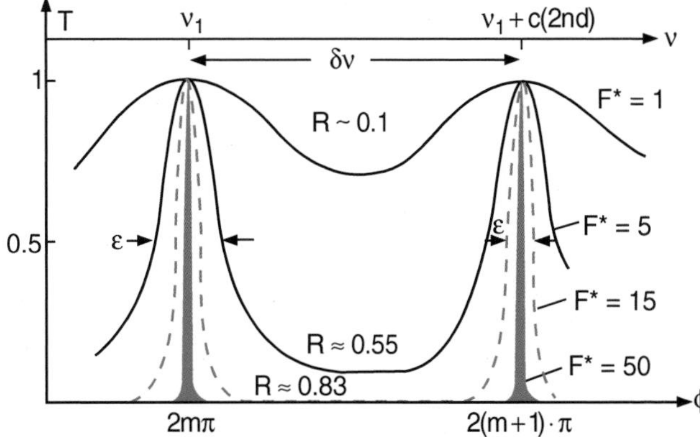


**Fig. 11.18.** Transmittance of an absorption-free multiple-beam interferometer as a function of the phase difference $\phi$ for different values of the reflectivity $R$ and the finesse $F^*$

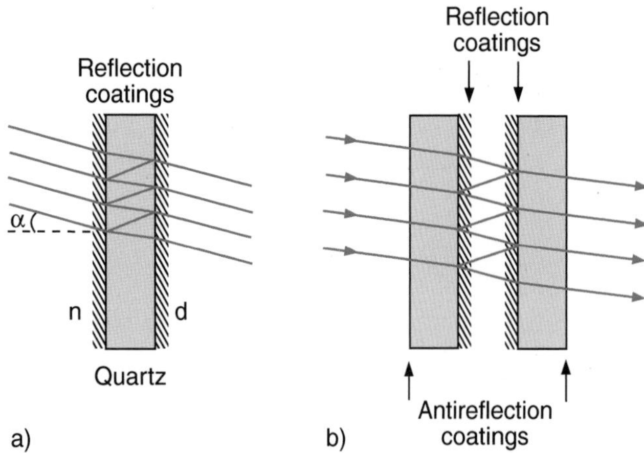


**Fig. 11.19a,b.** Two possible realizations of a Fabry–Perot interferometer: (**a**) solid etalon, (**b**) air-spaced plane-parallel reflecting surfaces

(Fig. 11.19b). The advantage of the first version is that no further alignment is necessary, but the fabrication of such a plate with a parallelism of better than $\lambda/100$ over the whole surface is expensive. In the second version the plate separation and therefore the free spectral range can be chosen arbitrarily, but the two plates have to be aligned very carefully.

If the FPI is illuminated with divergent monochromatic light, only those angles $\alpha$ are transmitted for which $\Delta s = 2m\pi$. Behind the FPI, a system of bright rings is observed on a dark background (Fig. 11.20). The diameters of these rings are a measure of the wavelength $\lambda$, and are described as follows.

The angles $\beta_m$ of the transmitted rays inside the plate with refractive index $n$ are given according to (11.30) by the condition

$$m\lambda = 2nd\cos\beta_m\,, \tag{11.39}$$

where $m$ is the integral order of the interference. If the inner ring has the order $m_0$ the $p$th ring has the order $m = m_0 - p$ (note that with increasing $\beta$ the order $m$ decreases). For small angles $\beta$ we can approximate $\cos\beta \approx 1 - \beta^2/2$ and obtain from (11.39):

$$\begin{aligned}(m_0 - p)\lambda &= 2nd\left(1 - \beta_p^2/2\right) \\ &= 2nd\left[1 - \frac{1}{2}\left(\frac{\alpha_p}{n}\right)^2\right].\end{aligned} \tag{11.40}$$

When the interference pattern is imaged by the lens $L_2$ with focal length $f_2$, the ring diameter $D_p$ of the $p$th ring is $D_p = 2f_2\alpha_p$. Inserting this into (11.40) gives

$$D_p^2 = \frac{8n^2f^2}{2nd}[2nd - (m_0 - p)\lambda]\,. \tag{11.41}$$

For the smallest ring with $p = 0$, (11.40) becomes

$$m_0\lambda = 2nd\left(1 - \frac{\alpha_0^2}{2n^2}\right), \tag{11.42}$$

which can be also written as

$$(m_0 + \varepsilon)\lambda = 2nd\,.$$

The excess

$$\varepsilon = \frac{d\alpha_0^2}{n\lambda} \tag{11.43}$$

with $\varepsilon < 1$ is the fractional interference order, which depends on the wavelength $\lambda$ and the separation $d$ of the two plane-parallel surfaces. For $\varepsilon = 0$ the inner ring diameter becomes $D_0 = 0$ and the optical path length $nd$ equals an integer multiple $m_0$ of the wavelength $\lambda$. With the excess $\varepsilon$, Eq. (11.41) becomes

$$D_p^2 = \frac{8n^2f^2}{(m_0 + \varepsilon)}(p + \varepsilon)\,. \tag{11.44}$$

A linear fit of the measured values $D_p^2$ versus the ring number $p$ yields a straight line, which intersects the ordinate axis at $p + \varepsilon = 0$ and allows the determination of $\varepsilon$ and therefore of $\lambda$. If $\lambda$ changes by one free spectral range the integer $m_0$ changes by one, which cannot directly be noticed from the ring system.

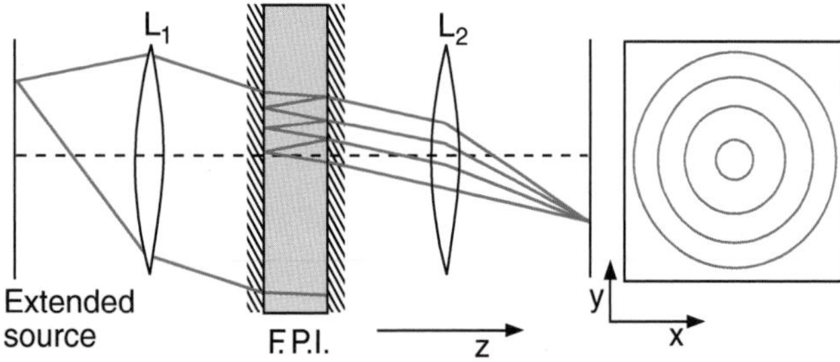


**Fig. 11.20.** Light from an extended monochromatic source forms a ring system after transmission through a Fabry–Perot interferometer

> **Note** that with a Fabry–Perot interferometer the wavelength can only be determined modulo a free spectral range $\delta\lambda$, because two wavelengths $\lambda$ and $\lambda + \delta\lambda$ have the same ring system. For the unambiguous determination of $\lambda$ the wavelength has to be known beforehand within at least one free spectral range.

The dispersion of the FPI (i.e. the dependence of the ring diameter $D$ on the wavelength $\lambda$) is obtained from (11.39) as

$$\frac{\mathrm{d}D}{\mathrm{d}\lambda} = f\frac{\mathrm{d}\beta}{\mathrm{d}\lambda} = \frac{f}{\lambda_m \sin\beta} . \tag{11.45}$$

The half-width $\Delta\nu$ of the transmission peaks can be calculated from (11.36) with $F = 4R/(1-R)^2$ as

$$\Delta\nu = \frac{c(1-R)}{2nd\pi\sqrt{R}} = \frac{2\delta\nu}{\pi\sqrt{F}} . \tag{11.46}$$

Introducing the finesse $F^*$ of the FPI

$$\boxed{F^* = \frac{\pi\sqrt{R}}{1-R} = \frac{\pi}{2}\sqrt{F}} , \tag{11.47}$$

which is a measure for the number of interfering beams in the FPI, the half-width $\Delta\nu$ can be expressed by the free spectral range $\delta\nu$

$$\Delta\nu = \frac{1}{F^*}\delta\nu . \tag{11.48a}$$

This gives the spectral resolution

$$\boxed{\frac{\Delta\nu}{\nu} = \frac{1}{F^*}\frac{\delta\nu}{\nu}} . \tag{11.48b}$$

**EXAMPLE**

Assume an FPI as a plane-parallel glass plate with $n = 1.5$, $d = 1\,\mathrm{cm}$ and reflectivities $R = 0.98$ for both surfaces. We then obtain $F = 9800$ and a finesse $F^* = 155.5$. The free spectral range is $\delta\nu = 10^{10}\,\mathrm{s}^{-1}$ and the half-width of the transmission peaks is $\Delta\nu = \delta\nu/F^* = 64\,\mathrm{MHz}$. The spectral resolution for a wavelength $\lambda = 500\,\mathrm{nm} \Rightarrow \nu = 6\times10^{14}\,\mathrm{s}^{-1}$ is $\nu/\Delta\nu = 9\times10^6$. According to (11.25) it can be also expressed as the maximum path difference $\Delta s = F^*\cdot 2nd = 155.5\cdot 3 = 466\,\mathrm{cm}$ in units of the wavelength $\lambda = 500\,\mathrm{nm}$. This gives $\lambda/\Delta\lambda = \Delta s/\lambda = 9\times10^6$, which must, of course, agree with the value for $\nu/\Delta\nu$.

Comparing the spectral resolution of the prism spectrograph, grating spectrometer and Fabry–Perot interferometer shows that the prism instrument has the lowest and the FPI the highest spectral resolution. However, there are also drawbacks for the different instruments.

While the wavelength determination with a prism spectrometer is not as accurate but unambiguous, the grating spectrometer has to know the interference order $m$ in Eq. (11.15). A spectral line with wavelength $\lambda_1$ appears for $m = 1$ at the same position in the detector plane as a line with $\lambda_2 = \lambda_1/2$, measured in second order with $m = 2$. Since $m$ can only take small numbers (usually $m = 1$ or 2) a broadband spectral filter in front of the spectrometer can solve this ambiguity.

For a Fabry–Perot interferometer the interference order is much higher (for our example above $m = 6\times10^6$ and it is much harder to determine the actual integer value of $m$. One has to measure the wavelength within one free spectral range beforehand using other instruments in order to determine the absolute value of $m$ [11.7].

### 11.2.3 Detectors

The sensitive detection of electromagnetic radiation is of fundamental importance for spectroscopic measurements. Therefore, the optimum selection of a detector can be essential for the success of the experiment. The choice of a detector depends on the wavelength to be monitored. We will here present only a few types of detectors with their spectral response and their time behavior. The following features are important for choosing the best detector:

- The spectral response, which determines the wavelength range the detector can be used for.
- The absolute sensitivity $S(\lambda) = V_\mathrm{S}/P$, defined as the ratio of the output signal $V_\mathrm{S}$ and the incident radiation power $P$. If the output signal is a voltage (e.g., for thermo-elements or for photovoltaic devices) the sensitivity $S(\lambda)$ is expressed in units of volts per watt. For output currents (e.g., for photomultipliers) $S(\lambda)$ is given in amperes per watt. With the sensitive area $A$ of the detector, the power received by the detector can be written as the product $P = A\cdot I$, where $I$ is the incident intensity.
- The noise figure of the detector, expressed by the *noise equivalent input power* NEP, means an incident radiation power that generates the same output signal as the detector noise itself. Although the signal-to-noise ratio $S/N$ is in most cases limited

by the noise of the incident radiation, the detector noise might further reduce it [11.8, 9].

### a) Thermal Detectors

Thermal detectors measure the input power of the incident radiation independent of the wavelength $\lambda$. They can therefore be used within the whole spectral range from microwaves to UV radiation. The absorbed radiation power results in an increase of the detector temperature, which depends on the heat capacity $H$ and the heat conductivity $G$ for heat transfer from the detector to the surroundings. All characteristic features of the detector that depend on its temperature can be used for monitoring the incident radiation power. In most cases the temperature-dependent electrical conductivity is utilized as a monitor [11.10].

If the fraction $\beta$ of the incident radiation power $P$ is absorbed by the detector, its temperature increase is determined from the energy balance

$$\beta P = H\frac{\mathrm{d}T}{t} + G(T - T_\mathrm{S})\,, \tag{11.49}$$

where $T_\mathrm{S}$ is the temperature of the surroundings.

If the time-independent incident power $P_0$ is switched on at time $t = 0$, the integration of (11.49) gives

$$T = T_\mathrm{S} + \frac{\beta P_0}{G}\left(1 - \mathrm{e}^{-(G/H)t}\right)\,. \tag{11.50}$$

The temperature rises from the initial value $T_\mathrm{S}$ at $t = 0$ to the temperature $T = T_\mathrm{S} + \Delta T$ for $t = \infty$ (Fig. 11.21). The temperature rise

$$\Delta T = \frac{\beta P_0}{G} \tag{11.51}$$

is inversely proportional to the thermal losses $G$ and does not depend on the heat capacity $H$.

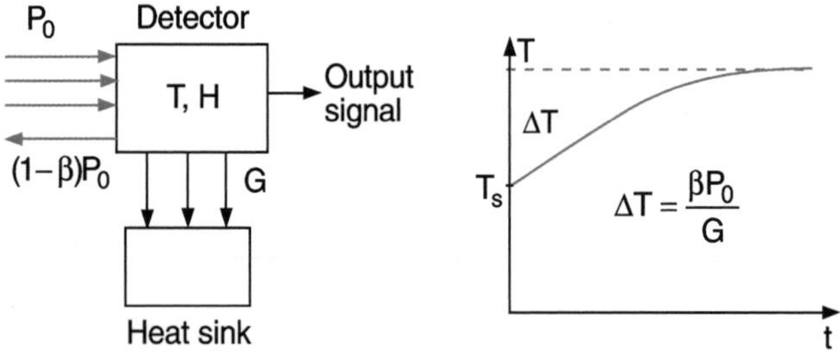


**Fig. 11.21.** Thermal detector and its temperature $T$ when at $t = 0$ a constant power $\beta P_0$ is absorbed by the detector

Small values of $G$ make the detector sensitive, but also slow.

With time-dependent radiation power

$$P = P_0(1 + a\cos\Omega t) \quad |a| \le 1$$

one obtains from (11.49) a detector temperature

$$T(\Omega) = T_\mathrm{S} + \Delta T(1 + \cos(\Omega t + \phi) \tag{11.52}$$

that depends on the modulation frequency $\Omega$ and which has a phase lag $\phi$ against the phase of the radiation determined by

$$\tan\phi = \Omega H/G\,. \tag{11.53}$$

The modulation amplitude

$$\Delta T = \frac{a\beta P_0 G}{\sqrt{G^2 + \Omega^2 H^2}} \tag{11.54}$$

decreases with $\Omega$. At the frequency $\Omega_\mathrm{g} = G/H$ the amplitude $\Delta T$ has dropped by a factor of $\sqrt{2}$ from its DC value for $\Omega = 0$.

We define as the time constant of the detector the value

$$\tau = 1/\Omega_\mathrm{g} = H/G\,. \tag{11.55}$$

A fast and sensitive detector should have a small heat capacity $H$ and a small heat conductivity $G$.

In Fig. 11.22 some thermal detectors are shown. The first one is a thermopile that consists of many thermo-elements in a series, where one end is in thermal contact with the backside of a thin metal sheet and the other is connected to a heat sink kept at constant temperature. If the radiation is incident on the blackened surface of the sheet, its temperature rises and the temperature difference to the heat sink is monitored by the thermo-elements. The output signal equals the sum of the thermovoltages of the different thermocouples.

The second example in Fig. 11.22b is a bolometer, where the voltage $U = RI$ across the bolometer with electrical resistance $R$ is measured when a constant current $I$ is sent through it.

The temperature change $\Delta T$ of the bolometer induced by the absorbed incident radiation causes a resistance change

$$\Delta R = (\mathrm{d}R/\mathrm{d}T)\Delta T \tag{11.56}$$

and therefore an output signal

$$\Delta U = I\Delta R\,.$$

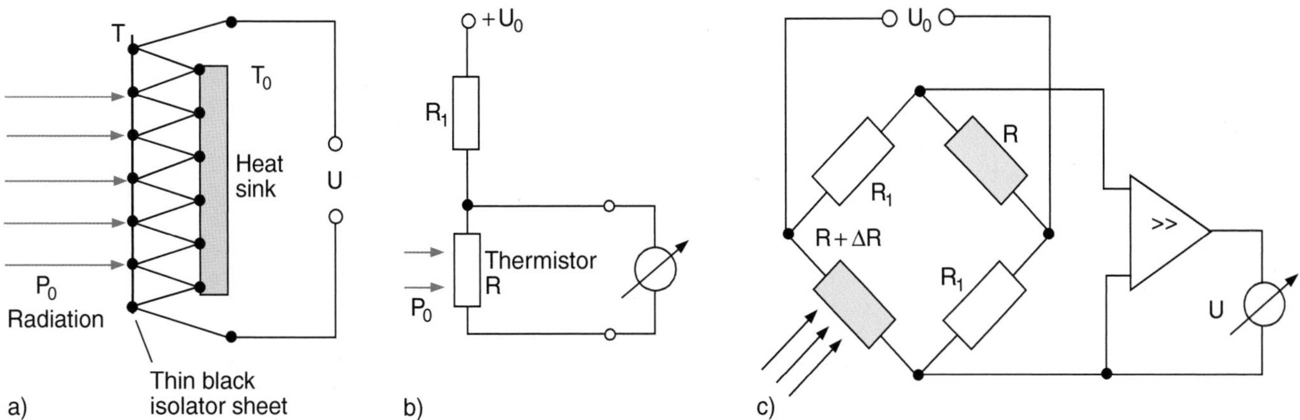


**Fig. 11.22a–c.** Thermal detectors and their circuits. (**a**) Thermopile, consisting of many thermoelements in series. (**b**) Thermistor with electrical resistance $R$. The change of electrical resistance $\Delta R(T)$ with temperature results in a voltage change $\Delta U$. (**c**) Balanced bridge circuit for sensitive detection of $\Delta R$

Inserting (11.51) gives the relation between the absorbed radiation energy $\beta \cdot P_0$ and the output signal $\Delta U$:

$$\Delta U = I(\mathrm{d}R/\mathrm{d}T)\beta P_0/G\,. \qquad (11.57)$$

The sensitivity of the bolometer depends on the derivative $\mathrm{d}R/\mathrm{d}T$, which becomes very large for semiconductor materials at low temperatures, where the electrical resistance $R$ decreases very steeply with increasing $T$. Therefore, sensitive bolometers are generally made of doped semiconductors and are used at temperatures below 4 K.

## EXAMPLE

At $T = 1.5\,\mathrm{K}$ the electric resistance of a doped silicon semiconductor bolometer is $R = 10\,\mathrm{k\Omega}$. With a current of $100\,\mathrm{\mu A}$ the voltage across the bolometer is $U = 1\,\mathrm{V}$. The derivative $\mathrm{d}R/\mathrm{d}T$ at $T = 1.5\,\mathrm{K}$ is about $10^{-4}\,\mathrm{V/K}$. With $\beta = 0.9$ and $G = 10^{-15}\,\mathrm{W/K}$ a radiation power of $P_0 = 10^{-12}\,\mathrm{W}$ generates an output signal $\Delta U = 10^{-5}\,\mathrm{V} = 10\,\mathrm{\mu V}$. This is readily detectable because the detector noise at these low temperatures is below this value [11.11].

A particularly large derivative $\mathrm{d}R/\mathrm{d}T$ is obtained at the critical temperature of superconductivity, where the system jumps within a very narrow interval $\Delta T$ from the normal to the superconducting state. If the device can be maintained at this temperature by feedback control, the incident radiation power can be inferred from the feedback control signal.

### b) Photodiodes

Photodiodes are doped semiconductor radiation detectors that can be used either as photoconductive or as photovoltaic devices. When the semiconductor diode is irradiated, electrons are excited from the valence into the conduction band (Fig. 11.23). This produces free charge carriers and increases the conductivity of the diode. If an external voltage is applied in the forward direction, the electrical resistance $R$ of the diode decreases and with it the voltage drop across the diode. The output signal is

$$\Delta U = I\Delta R = aP_0\,. \qquad (11.58)$$

These photoconductive devices require an external voltage $U_0$.

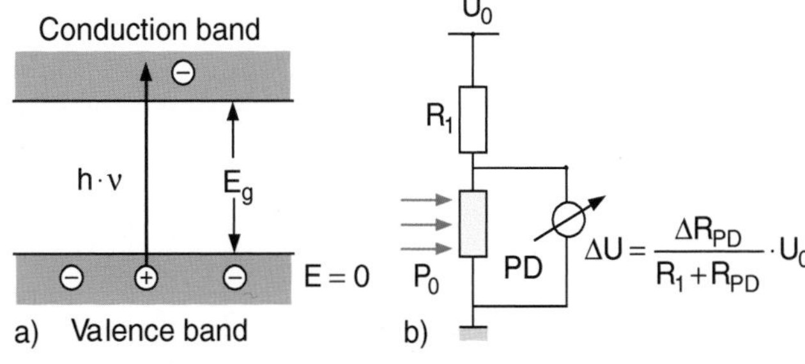


**Fig. 11.23.** (**a**) Excitation of electrons from the valence into the conduction band by photoabsorption. (**b**) Measuring the resistance change $\Delta R(P_0)$ as a voltage change $\Delta U$

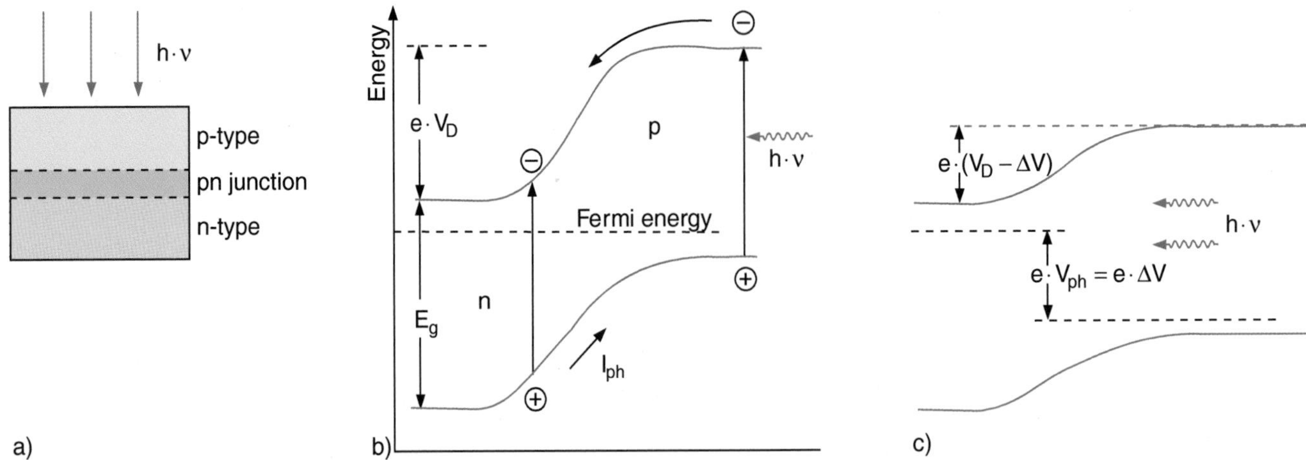


**Fig. 11.24a–c.** Photovoltaic semiconductor detector. (**a**) Schematic structure, (**b**) band energies around the *p*-*n* junction without illumination, (**c**) with illumination

The photovoltaic semiconductor detectors on the other hand generate the output voltage without an external supply. They are based on the following principle, which is illustrated in Fig. 11.24.

In the non-illuminated p-n junction of the diode, the diffusion of the electrons from the n-region into the p-region and of the holes into the opposite direction produces a charge separation and therefore a diffusion voltage $V_D$ across the junction. This diffusion voltage cannot be detected by an external voltmeter because it is just compensated by the opposite contact potentials between the connections to the voltmeter and the n- or p-part at both end faces of the diode.

If the p-n junction is illuminated, electron-hole pairs are created by photon absorption. The electrons are driven by the diffusion voltage into the n-region and the holes into the p-region. This leads to a decrease $\Delta V_D$ of the diffusion voltage, which appears as the photovoltage

$$V_{photo} = \Delta V_D \tag{11.59a}$$

across the open electrodes of the diode.

If these electrodes are connected through an amperemeter, the photocurrent

$$i_{photo} = -\eta e \phi A \tag{11.59b}$$

is measured. Here $\eta$ is the quantum efficiency of the detector, $\phi = I_0/h\nu$ the incident photon flux density (number of photons per cm$^2$ per s) and $A$ is the sensitive area of the detector.

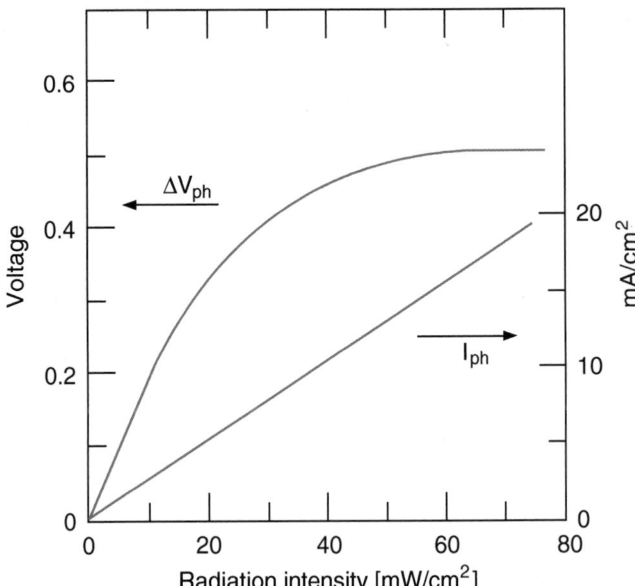


**Fig. 11.25.** Output voltage $\Delta V_{Ph}$ of photodiode with open ends and photocurrent $I_{Ph}$ of a photodiode with shortened output

The illuminated p-n photodetector can be used either as a photocurrent generator or as a voltage source, depending on the external resistor $R$ between the electrodes. The output voltage can, however, never exceed the diffusion voltage $V_D$ (Fig. 11.25).

### c) Photomultipliers

Photomultipliers are very sensitive detectors for low-level radiation from the near-infrared to the ultraviolet region. They consist of a photocathode and a number of amplifying dynodes in an evacuated glass cylinder (Fig. 2.42). The spectral response depends on the materials used for the photocathode. Some spectral sensitivity curves of typical photocathodes are shown in Fig. 11.26.

The photoelectrons, generated under illumination of the photocathode with an efficiency $\eta = n_e/n_{ph}$, are accelerated and focused by a voltage of around 100 V between the cathode and the first dynode onto the first dynode with a copper beryllium surface. Here they produce, by electron impact, secondary electrons, which are again accelerated to the second dynode. If each electron impinging onto a dynode generates $q > 1$ secondary electrons, one photoelectron produces at the $N$th dynode an electron avalanche of $q^N$ electrons, which are imaged onto the anode. Here the electron avalanche pulse, induced by a single photoelectron, charges the output capacitance $C$ up to a voltage $V_{out} = q^N e/C$.

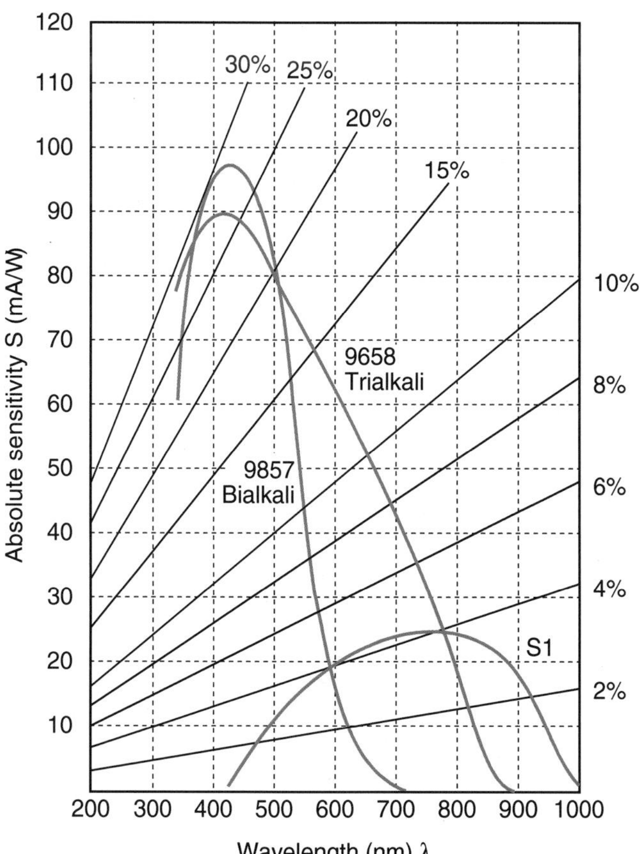


**Fig. 11.26.** Spectral response curves $S(\lambda)$ of several photocathodes and quantum efficiency lines $\eta = n_e/h \cdot \nu = \eta_e \lambda/(h \cdot c)$

In the case of continuous illumination one obtains a continuous photocurrent. For $n$ photons/s impinging on the photocathode with a quantum efficiency $\eta$, the photocurrent emitted by the cathode is $n_{phot}\eta e$ and the anode current is $i_a = n_{phot}\eta e q^N$. Across the resistor $R$ between the anode and ground, a dc voltage $V = Ri_a$ is measured [11.12].

## EXAMPLE

With a quantum efficiency $\eta = 0.2$ of the photocathode and the amplification factor $q = 4$ the photocurrent at the anode of a photomultiplier with $N = 10$ dynodes is $i_a = n_{phot} \cdot 0.2 \cdot 1.6 \times 10^{-19} \cdot 4^{10}$ A. With illuminating radiation $\lambda = 500$ nm and $P_0 = 10^{-15}$ W $\Rightarrow$ $n_{phot} = 6500/s$ we obtain a photocurrent emitted from the cathode of $i_c = 6500 \cdot \eta \cdot e/s = 2.1 \times 10^{-16}$ A and at the anode of $i_a = 2.1 \times 10^{-10}$ A.

A single photoelectron generates an electron pulse with the charge $Q = e \cdot 4^{10} = 1.7 \times 10^{-13}$ C. With a capacitance of 10 pF at the photomultiplier output, this gives a voltage pulse of $U = Q/C = 17$ mV.

## 11.3 Microwave Spectroscopy

Transitions between molecular rotational levels in the same vibrational state fall into the microwave range. Accurate measurement of rotational absorption lines allows the determination of the moments of inertia and therefore of the geometry of the molecule. Also, transitions between atomic or molecular hyperfine levels or Stark splittings can be induced by microwaves, which gives information on nuclear spins and electric or magnetic moments of molecules.

Microwave spectroscopy is the most accurate method for the determination of molecular structure in the electronic ground state or for measuring electric and magnetic dipole moments.

For polyatomic molecules, the microwave spectrum generally consists of many lines and it is not always straight forward to assign all lines unambiguously to the correct transitions. Here, the substitution of one or more atoms in the molecule by their isotopes is helpful and the comparison of the microwave spectra of different isotopomers generally allows the unambiguous determination of the rotational constants and the molecular structure [11.13].

A possible experimental arrangement for microwave absorption spectroscopy is shown in Fig. 11.27. Microwaves are generated by microwave oscillators (klystron, carcinotron or gun oscillator), are focused by a Teflon lens transparent to microwaves, and are sent through a long microwave guide that contains the absorbing gas at the temperature $T$.

At thermal equilibrium the population density of molecular levels follows a Maxwell–Boltzmann distribution

$$\frac{N_i}{N_k} = \frac{g_i}{g_k} e^{-\Delta E/kT} . \tag{11.60}$$

Inserting this into (11.2) gives the absorption coefficient

$$\alpha(\nu_{ki}) = N_k \left(1 - e^{-\Delta E/kT}\right) \sigma(\nu_{ki}) . \tag{11.61}$$

At room temperature $T = 300$ K the thermal energy $kT$ is much larger than the photon energy $h\nu = \Delta E$.

For example, for $\nu = 3$ GHz, $h\nu = 2 \times 10^{-24}$ J, while for $T = 300$ K, $kT = 4 \times 10^{-21}$ J, i.e., 2000 times larger. Therefore, $h\nu/kT \ll 1$ and we can expand the exponential to obtain

$$\begin{aligned} \alpha(\nu_{ki}) &\approx N_k(\Delta E/kT)\sigma(\nu_{ki}) \\ &= N_k(h\nu/kT)\sigma(\nu_{ki}) . \end{aligned} \tag{11.62}$$

Although the density of molecules in the absorbing level might be large, the difference between the lower and upper state population is small and therefore the absorption rate is only slightly larger than the stimulated emission rate. The net absorption can be expressed by an effective density $N_{\text{eff}}$ of absorbing molecules, which is smaller than the real number density $N_k$ by a factor $h\nu/kT$, i.e., $N_{\text{eff}} = N_k(h\nu/kT)$.

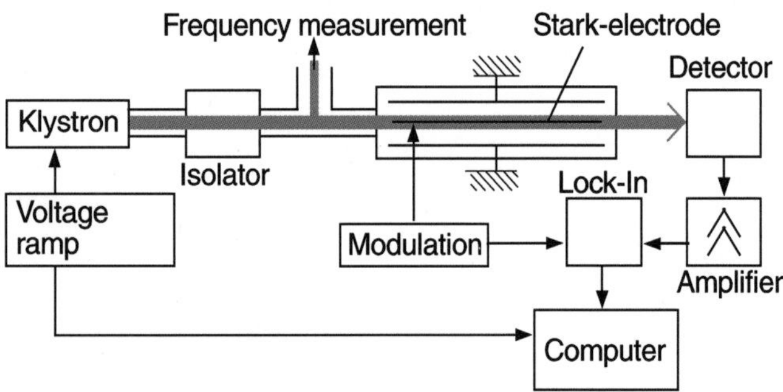


**Fig. 11.27.** Possible experimental arrangement for microwave spectroscopy in a Stark-cell where the voltage of the central electrode against the outer electrodes at ground potential is modulated

## EXAMPLE

At a gas pressure of 10 mbar the total molecular density $N = \sum N_i$ at $T = 300$ K is about $N = 3 \times 10^{23}/\text{m}^3$. The population distributes over many rotational-vibrational levels. We assume that 1% of all molecules are in the absorbing level with energy $E_k$ and that the ratio $h\nu/kT$ is about $10^{-3}$. The effective net absorption coefficient then becomes $\alpha = 3 \times 10^{18}\sigma(\nu_{ik})\text{m}^2/\text{m}^3$. With the absorption cross section $\sigma = 10^{-23}\,\text{m}^2$ the relative absorption for 1m absorption path length becomes $\Delta I/I_0 = 3 \times 10^{-5}$. In order to detect this small absorption the fluctuations of the microwave intensity $I_0$ should not be larger than this value.

The sensitivity can be greatly enhanced by frequency modulation of the microwave. The frequency of the incident wave with intensity

$$I(t) = I_0 \cos^2[2\pi \nu_{\text{m}}(1 + a\cos(2\pi f t)t] \tag{11.63}$$

oscillates around the central frequency $\nu_{\text{m}}$. When $\nu_{\text{m}}$ is tuned across an absorption line, the absorption coefficient $\alpha(\nu)$ and therefore the detected signal are modulated at the frequency $f$ (Fig. 11.28). The transmitted intensity can be expanded into the Taylor series

$$I_{\text{t}}(\nu) = I_{\text{t}}(\nu_{\text{m}}) + \sum \frac{a^n}{n!} \left(\frac{\text{d}^n I_{\text{t}}}{\text{d}\nu^n}\right)_{\nu_{\text{m}}} \nu_{\text{m}}^n \cos^n(2\pi f t) . \tag{11.64}$$

If the phase-sensitive detector (lock-in amplifier) only transmits that part of the signal that is modulated at frequency $f$, the detected signal

$$I_{\text{t}}(\nu_{\text{m}}) - I_{\text{t}}(\nu) = \Delta I_{\text{t}} = -a \cdot \nu_{\text{m}} \cdot \left(\frac{\text{d}I_{\text{t}}}{\text{d}\nu}\right)_{\nu_{\text{m}}} \cos(2\pi f t) \tag{11.65}$$

is proportional to the first derivative $\text{d}I_{\text{t}}/\text{d}\nu$ of the transmitted intensity and to the modulation index $a$.

The fluctuations of the incident microwave intensity $I_0$ are generally distributed over a broad frequency range, depending on the kind of noise, influencing $I_0$. The phase sensitive detector suppresses most of the

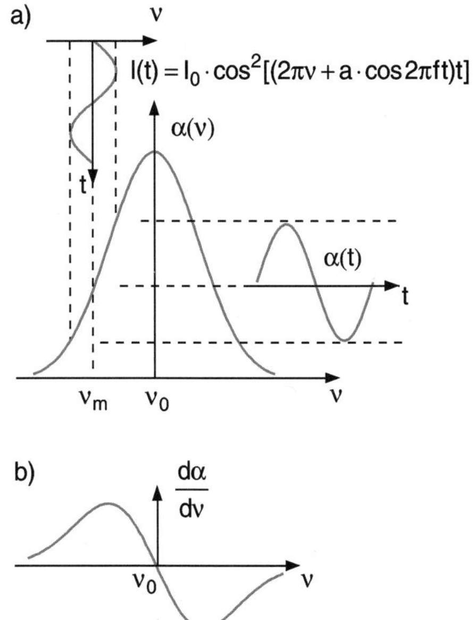


Fig. 11.28. (a) Absorption $\alpha(t)$ for frequency-modulated radiation. (b) Derivative $\mathrm{d}\alpha/\mathrm{d}\nu$ of the absorption coefficient

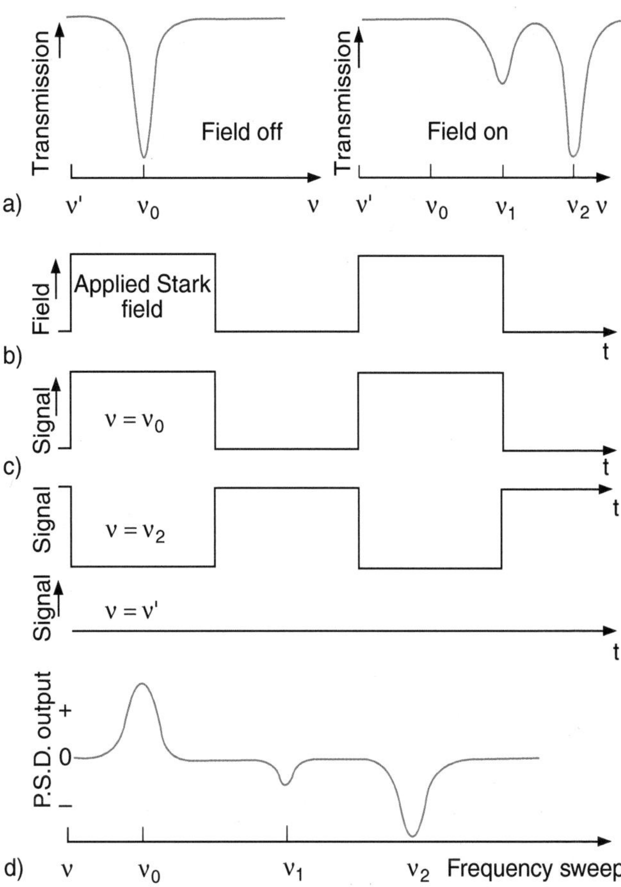


Fig. 11.29. (a) Absorption line without and with external electric field. (b) Square wave modulation of Stark field. (c) Sign of the signal for $\nu_0$, $\nu_2$ und a frequency $\nu'$ on the lower side of the absorption line. (d) Resultant spectrum when the microwave frequency $\nu$ is swept over the absorption lines

noise because it only transmits the contributions within a small frequency interval $\delta\nu$ around the frequency $f$. The width $\delta\nu = 1/\tau$ depends on the time constant $\tau$ of the detector. This can also be illustrated in the following way. The detector always measures the difference $\Delta I_t(f) = (I_0 - I_t)|_f$ at the frequency $f$. If $I_0$ fluctuates so too does $I_t$. Because generally $(I_0 - I_t) \ll I_0$, the fluctuations are mostly cancelled by measuring the difference $\Delta I$.

Instead of modulating the microwave frequency, the absorption frequency can be modulated, which has the same effect on the reduction of noise. This absorption modulation can be achieved by an external ac electric field, which causes a Stark shift of the molecular levels (if the molecule possesses an electric dipole moment) The shift depends on the rotational quantum number $J$ and is therefore different for the two levels $\langle J|$ and $\langle J+1|$ of a rotational transition $J \to J+1$. The frequency $\nu_0$ of the transition is shifted and split into the different Stark components.

The electric field is generated between two long parallel electrodes in the absorption cell (Fig. 11.27), which are separated by the distance $d$ and are charged to the voltages $+U_0$ and $-U_0$. With $U_0 = 10\,\mathrm{kV}$ and $d = 4\,\mathrm{mm}$ the electric field is $E = 25\,\mathrm{kV/cm}$. Another technical alternative is to use only one charged thin metal sheet in the middle of the absorption cell, where the electric field between this electrode and the grounded walls of the cell has opposite directions in the upper and lower half of the cell (Fig. 11.29).

To illustrate the high spectral resolution, a section of the microwave absorption spectrum of the CO molecule is shown in Fig. 11.30. As was outlined in Sect. 9.5, the frequencies of the rotational lines increases linearly with the rotational quantum number $J$ and the lines are therefore equidistant. The frequency difference depends on the rotational constant and differs for the two isotopomers $^{12}$CO und $^{13}$CO.

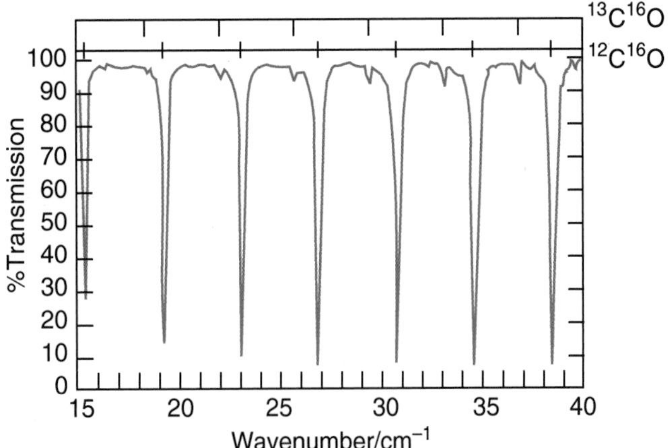


**Fig. 11.30.** Microwave absorption spectrum of $^{12}$CO (*intense lines*) and the rare isotopomer $^{13}$CO (*weak lines*) showing some absorption lines of pure rotational transitions $J \to J+1$ [11.14]

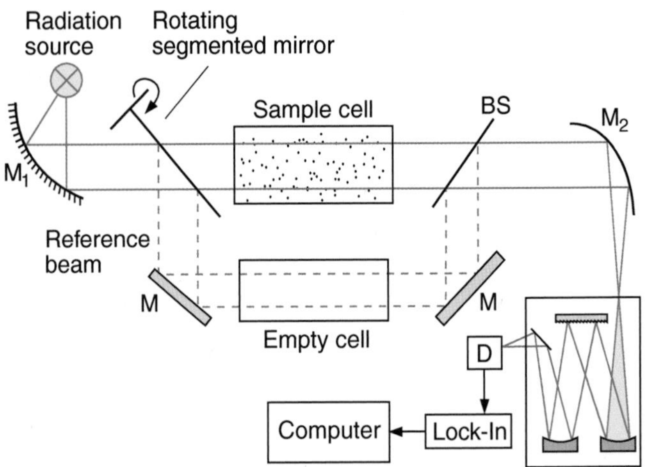


**Fig. 11.31.** Classical infrared spectrometer with reference, rotating segmented mirror and lock-in detection

## 11.4 Infrared Spectroscopy

The vibrational-rotational transitions of most molecules, and also many electronic transitions between excited electronic states of atoms and molecules, fall into the infrared spectral region. Since the vibrational bands are characteristic for each molecule, this spectral range is also called the *fingerprint region* of molecular spectroscopy [11.14, 15].

Several different techniques have been developed for accurate and sensitive infrared spectroscopy. They can also be used in the visible and UV region.

### 11.4.1 Infrared Spectrometers

In Fig. 11.31 the basic experimental setup is shown for classical infrared absorption spectroscopy. The radiation source is a high-pressure mercury lamp or a hot tungsten pin. Both sources emit a spectral continuum, where the hot tungsten pin (globar) with a temperature $T$ emits the continuous blackbody thermal radiation with a maximum at $\lambda_{max} \propto 1/T$ (see Sect. 3.1). For $T = 1{,}000$ K the maximum of the intensity distribution is at $\lambda_{max} = 3\,\mu$m. The radiation is collimated by the mirror $M_1$ and is sent as a parallel beam through the absorption cell with length $L$, before it is focused by $M_2$ onto the entrance slit of a spectrometer, which disperses the spectral continuum and transmits the desired wavelengths. A detector behind the spectrometer monitors the transmitted intensity $I_t(\nu)$. The spectral resolution is limited by the resolving power of the spectrometer.

To increase the sensitivity, a difference method is used: A rotating segmented mirror sends the parallel beam alternately through the sample cell and through an identical reference cell, which is, however, empty. The difference

$$\Delta I(\nu) = I_R - I_t(\nu) = \alpha(\nu) \cdot L \cdot I_0 \qquad (11.66a)$$

between the reference intensity $I_R$ and the intensity $I_t$ of the signal beam transmitted through the absorption cell is measured.

Dividing the measured signal $\Delta I(\nu)$ by the incident intensity $I_0$ gives

$$\Delta I / I_0 = \alpha(\nu) \cdot L\,, \qquad (11.66b)$$

which shows that for the detection of the relative absorption by measuring the ratio $\Delta I/I_0$, the fluctuations of the incident intensity $I_0$ are eliminated.

### 11.4.2 Fourier Transform Spectroscopy

Fourier transform spectroscopy has many advantages compared to conventional absorption spectroscopy. It is based on interferometry with a Michelson interferometer (Sect. 11.2.2), where the path length $L_2$ of one of the arms is continuously and uniformly changed by mirror $M_2$ moving with constant speed $v$ (Fig. 11.15). Assume the radiation source emits a monochromatic wave with amplitude

$$E(\omega) = A_0 \cos \omega_0 t$$

and intensity

$$I(\omega) = c\varepsilon_0 E^2 = c\varepsilon_0 A_0^2 \cos^2 \omega_0 t = I_0 \cos^2 \omega_0 t \,. \qquad (11.67)$$

The two interfering partial beams have the amplitudes

$$A_i = \sqrt{RT} A_0 \,,$$

where $R$ and $T$ are the reflectivity and transmittance of the beam splitter BS. Because of the different optical path lengths $s_1$ and $s_2$, the interference intensity at the detector plane

$$\begin{aligned} I_t &= c\varepsilon_0 RT A_0^2 \qquad (11.68)\\ &\quad \times [\cos(\omega_0 t + ks_1) + \cos(\omega_o t + ks_2)]^2 \\ &= RT I_0 \big[ \cos^2(\omega_0 t + ks_1) + \cos^2(\omega_0 t + ks_2) \\ &\quad + \cos(2\omega_0 t + k(s_1 + s_2)) + \cos(k(s_1 - s_2)) \big] \end{aligned}$$

depends on the path difference $\Delta s = s_1 - s_2$.

The detector can not follow the fast optical oscillations with frequency $\omega_0$. The detector signal is therefore proportional to the time average $\langle I(t) \rangle$. With $\langle \cos \omega_0 t \rangle = 0$ and $\langle \cos^2 \omega_0 t \rangle = \frac{1}{2}$ we obtain from (11.68), with $s_2 = s_1 + vt$ and $k = \omega_0 / c$,

$$S(t) \propto \langle I(t) \rangle = RT I_0 \left[ 1 + \cos\left( \omega_0 \frac{v}{c} t \right) \right] . \qquad (11.69)$$

Instead of the frequency $\omega_0$, the detector measures the much lower frequency $\Omega = \omega_0 v / c$ for the transmitted intensity averaged over the time constant $\tau$ of the detector.

> In the Michelson interferometer, with one mirror uniformly moving with velocity $v$, the optical frequency $\omega_0$ of the radiation source it transformed to the much smaller frequency $\omega_0 v / c$ of the output signal.

## EXAMPLE

$v = 3\,\text{cm/s}$, $\omega_0 = 10^{14}/\text{s} \Rightarrow (v/c)\omega_0 = 10^4/\text{s}$

In Fig. 11.32a the signal $\langle I(t) \rangle$ is plotted for monochromatic incident radiation as a function of the phase difference $\delta = \omega_0 (v/c) t$ depending on the product $v \cdot t$. The maxima occur if the path difference $\Delta s = vt$ becomes an integer multiple of the wavelength $\lambda$.

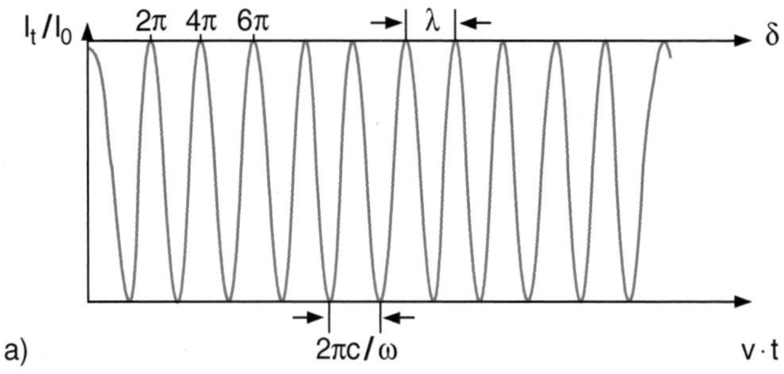


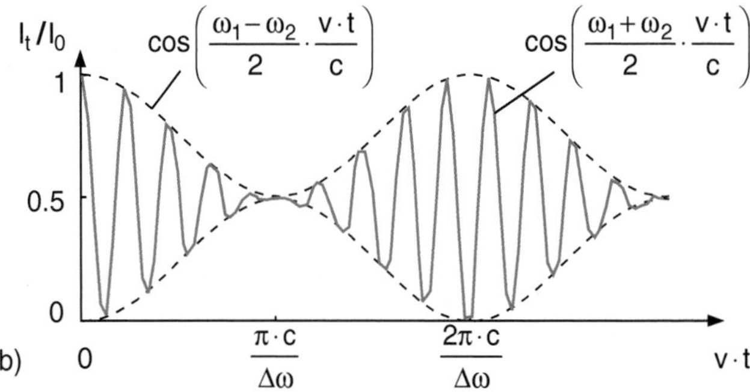


**Fig. 11.32a,b.** Normalized intensity $I_t(t)/I_0$ transmitted through a Michelson interferometer with moving mirror as a function of $v \cdot t$. (**a**) for monochromatic incident intensity. (**b**) for a superposition $I_{01}(\omega_1) + I_{02}(\omega_2)$

Mathematically, the spectrum $I(\omega)$ of the radiation source can be obtained through a Fourier transformation of the measured signal $S(t) \propto \langle I(t) \rangle$, because (11.69) can be written in the form

$$I(\omega) = \lim_{\tau \to \infty} \int_{t=0}^{\tau} S(t) \cos\left( \omega \frac{v}{c} t \right) \mathrm{d}t \qquad (11.70)$$

as can be proved by inserting (11.69) into the integral.

If the radiation source emits two frequencies $\omega_1$ and $\omega_2$, the two partial beams with frequency $\omega_1$ and the two beams with $\omega_2$ interfere with each other independently. The interference between $\omega_1$ and $\omega_2$ averages to zero, because the phase difference between two beams with different frequencies periodically changes between 0 and $2\pi$. If the detector time constant $\tau$ is larger than $2\pi/(\omega_1 - \omega_2)$ it cannot follow these rapid changes. The measured intensity averaged over a time period $\tau$ is then

$$\langle I_t(t) \rangle = \langle I_1 \rangle + \langle I_2 \rangle \,.$$

The interferogram is simply the superposition of the interferograms of two monochromatic waves with frequencies $\omega_1$ and $\omega_2$. This is illustrated in Fig. 11.32b) for the case of equal amplitudes of the two waves. Inserting $I_0 = I_1 \cos(\omega_1 t) + I_2 \cos(\omega_2 t)$ into (11.69) gives

$$S(t) \propto \langle I_T(t) \rangle \tag{11.71a}$$
$$= RT\overline{I}_0 \left[ 1 + \cos\left(\omega_1 \frac{v}{c} t\right) + \cos\left(\omega_2 \frac{v}{c} t\right) \right] .$$

With the frequencies

$$\omega_1 = \frac{\omega_1 + \omega_2}{2} + \frac{\omega_1 - \omega_2}{2} \tag{11.72a}$$
$$\omega_2 = \frac{\omega_1 + \omega_2}{2} - \frac{\omega_1 - \omega_2}{2} \tag{11.72b}$$

this converts to

$$S(t) \propto RT\overline{I}_0 \left[ 1 + 2\cos\left( \frac{\omega_1 - \omega_2}{2} \frac{v}{c} t \right) \times \cos\left( \frac{\omega_1 + \omega_2}{2} \frac{v}{c} t \right) \right] . \tag{11.71b}$$

From Fig. 11.32b we can obtain a vivid illustration of the spectral resolving power of Fourier spectroscopy. In order to find the frequencies $\omega_1$ and $\omega_2$ from the measured signal $S(t)$, the path $s_2(t) = vt$ traveled by the moving mirror during the measurement has to be sufficiently large to include at least one beat period $T = (2\pi c/v)/(\omega_1 - \omega_2)$. The minimum frequency interval $\delta\omega = (\omega_1 - \omega_2)$ is related to the minimum time $\Delta t = \Delta s/v$ for the measurement by

$$\frac{v}{t}\delta\omega \geq \frac{2\pi}{\Delta t} \Rightarrow \delta\omega \geq \frac{2\pi c}{\Delta s} . \tag{11.73a}$$

This can be stated in other words. The minimum path difference $\Delta s = v\Delta t$ between the two interfering partial waves, measured in units of the wavelength $\lambda$ gives the resolving power $\omega/\delta\omega$ of the interferometer, because

$$\frac{\omega}{\delta\omega} = \frac{2\pi\nu}{\delta\omega} = \frac{2\pi c \Delta s}{\lambda \cdot 2\pi c} = \frac{\Delta s}{\lambda} . \tag{11.73b}$$

If the radiation source emits radiation on many frequencies, the detector signal $S(t)$ becomes more complicated [11.16]. In any case, however, the relation

$$S(t) = a \int\limits_0^\infty \overline{I}_0(\omega) \left[ 1 + \cos\left( \omega \frac{v}{c} t \right) \right] \mathrm{d}\omega \tag{11.74a}$$

holds and the Fourier transform of the signal $S(t)$, measured during the time interval $\Delta t = \tau$, yields the spectral intensity spectrum

$$\overline{I}(\omega) = \lim_{\tau \to \infty} \frac{b}{\tau} \int\limits_0^\tau S(t) \cos\left( \omega \frac{v}{c} t \right) \mathrm{d}t \tag{11.74b}$$

of the radiation source [11.16].

Mathematically, the Fourier transform requires that the limits of the integral extend from 0 to $\infty$. Experimentally only a finite maximum value $\Delta s_{max}$ can be realized, depending on the construction of the interferometer. This deficiency can be taken into account by introducing a gate function $G(t)$, which is multiplied with the signal $S(t)$. This can be, for instance, a rectangular function

$$G(t) = \begin{cases} 1 \text{ for } 0 \leq t \leq t_{max} \\ 0 \text{ for } t > t_{max} \end{cases} \quad \text{with } t_{max} = \Delta s_{max}/v .$$

The spectrum of the source is now expressed by the Fourier transform

$$\overline{I}(\omega) = \int\limits_0^\infty S(t) G(t) \cos\left( \omega \frac{v}{c} t \right) \mathrm{d}t . \tag{11.75}$$

The Fourier transform of a rectangular function $f(x)$ is (analogously to the diffraction-limited intensity, transmitted through a rectangular slit) given by the function $(\sin x/x)^2$, which has a principal maximum for $x = 0$ and many small side maxima for $x = (2m + 1)\pi/2$. Therefore, the Fourier transform (11.76) gives, for each line in the spectrum, a diffraction-like structure. This is perturbing for a dense spectrum, because the side maxima of a strong line might overlap with the principal maxima of adjacent weak lines. Therefore a Gaussian gate function that does not produce these diffraction-like features is used instead of the rectangular function. The optimum choice for such a gate function allows one to reproduce the source spectrum without severe perturbations (the use of such an optimized gate function is called *apodization*) [11.17].

For measuring absorption spectra, a radiation source with continuous emission spectrum is used and

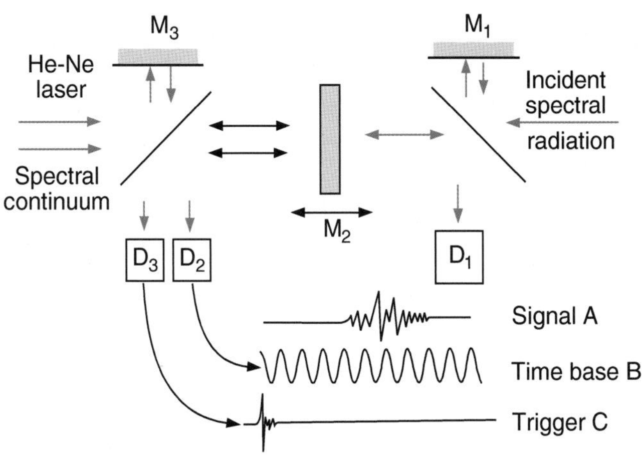


**Fig. 11.33.** Schematic design of a Fourier spectrometer

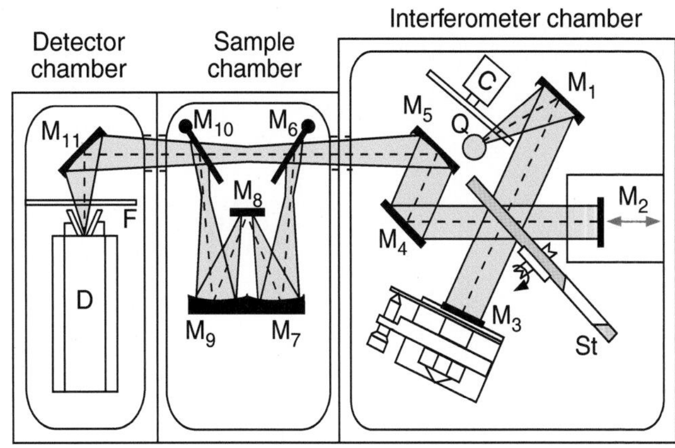


**Fig. 11.34.** Fourier spectrometer (Polytec FIR 30)

the radiation passes through the absorption cell before it reaches the detector. In order to enhance the sensitivity, a special mirror configuration lets the radiation pass through the absorption cell several times, thus increasing the absorption path length $L$. The basic operation of a Fourier interferometer is explained in Fig. 11.33. The incident radiation to be measured comes from the right. The signal S(t) is measured by the detector $D_1$, while the mirror $M_2$ is uniformly moving. The realized increase in path difference for the moving mirror is from a few centimeters up to some meters depending on the construction of the instrument.

Since the path difference $\Delta s$ has to be measured very accurately it is determined by optical interferometry. The parallel beam of a He-Ne laser is sent into the left part of the instrument, which forms a second Michelson with the same moving mirror $M_2$. The detector $D_2$ measures the interferogram of the monochromatic laser radiation, which gives the pattern shown in Fig. 11.32a and serves as an accurate time base, because the wavelength of the He-Ne laser is known very precisely. In order to fix the time $t = 0$, where both arms of the interferometer have equal path lengths and therefore $\Delta s = 0$, a broad spectral continuum is sent to the interferometer and its interferogram, which gives approximately a delta-function at $t = 0$ is detected by $D_3$. A commercially developed Fourier interferometer is depicted in Fig. 11.34. For illustration of the attainable sensitivity and spectral resolution the Fourier spectrum of very weak overtone band in acetylene $C_2H_2$ with rotational resolution is shown in Fig. 11.35.

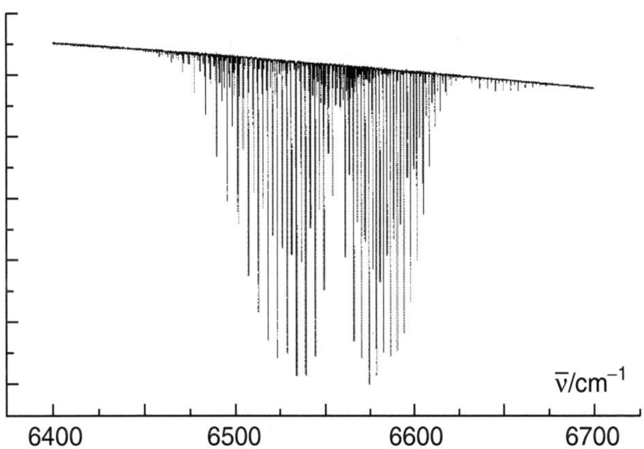


**Fig. 11.35.** Section of the Fourier spectrum of a weak overtone band (10100) ← (00000) of acetylene $C_2H_2$. (Platz T., PhD thesis, Kaiserslautern)

> The big advantages of Fourier spectroscopy are the attainable large spectral resolving power and the better signal-to-noise ratio at measuring times that are much less than in conventional spectroscopy with spectrometers.

The reason for this is that all frequency components of the spectrum are measured simultaneously, while, for instance, in microwave spectroscopy the radiation source has to be tuned in the course of time over all spectral intervals. If the total measured spectral interval $\Delta\omega$ is divided into $N$ sections $\delta\omega$ ($\Delta\omega = N\delta\omega$), where $\delta\omega$ is the smallest resolvable spectral interval, the signal $S$ for a given measuring time is $N$ times larger than

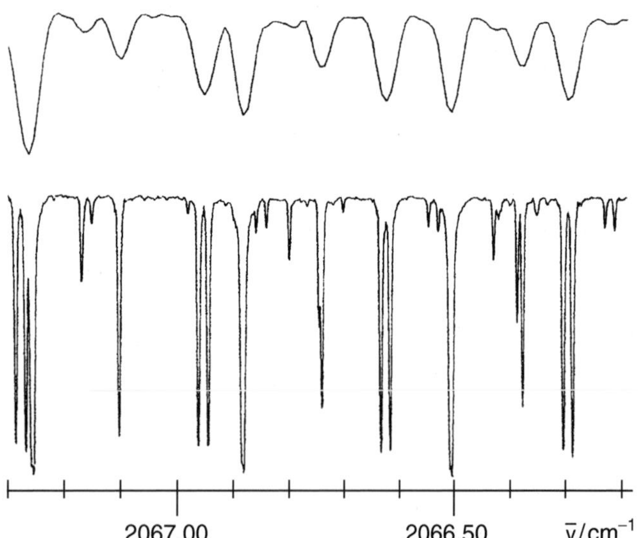


**Fig. 11.36.** Section of the OCS absorption spectrum. Comparison between classical infrared spectroscopy (*upper trace*) and Fourier spectroscopy with a 1-m maximum path difference

in conventional spectroscopy. The achievable signal-to-noise ratio is then $\sqrt{N}$ times larger. The improvement in spectral resolution and S/N-ratio is illustrated by the OCS-spectra in Fig. 11.36.

### EXAMPLE

For a measured spectral interval $\Delta\overline{\nu} = 1{,}000\,\mathrm{cm}^{-1}$ $\Rightarrow \Delta\omega = 2\pi \times 3 \times 10^{12}\,\mathrm{s}^{-1}$. If the smallest resolvable interval is $\delta\overline{\nu} = 0.1\,\mathrm{cm}^{-1} \Rightarrow N = 10^4$. In order to achieve the same signal-to-noise ratio as in conventional spectroscopy one needs with Fourier spectroscopy only 1% of the measuring time that is necessary when a monochromator with spectral resolution $\delta\overline{\nu}$ is tuned over the whole spectral interval $\Delta\nu$.

## 11.5 Laser Spectroscopy

The introduction of lasers to spectroscopy has enlarged the possibilities of spectroscopic investigations enormously regarding spectral resolution and sensitivity, which have been enhanced by several orders of magnitude.

We will briefly discuss in this section some techniques that exemplify the high sensitivity attainable and the special techniques with exceedingly high spectral resolution where the Doppler width of the absorption lines can be overcome and "Doppler-free" spectral resolution is possible.

Of particular interest is the investigation of ultrafast phenomena in atoms and molecules, which can nowadays be studied with ultrashort laser pulses with a time resolution of about 1 fs($10^{-15}$ s).

For a more detailed representation the reader is referred to specialist literature [11.18–20].

### 11.5.1 Laser-Absorption Spectroscopy

Absorption spectroscopy with monochromatic wavelength-tunable lasers resembles, in many aspects, microwave spectroscopy (see Sect. 11.2). The advantage of lasers is, however, their large tuning range, which exceeds that of clystrons by far, and furthermore their possible application in many spectral regions from the far infrared to the extreme ultraviolet [11.21].

A typical experimental setup is shown in Fig. 11.37. The output beam of a tunable laser is split by a 50% beam splitter BS2 into a reference beam and a signal beam that passes through the absorption cell. Another beam splitter BS1 directs a small fraction of the laser beam into a long Fabry–Perot interferometer with mirror separation $d$, which gives equidistant frequency markers separated by $\Delta\nu = c/2d$.

The advantages of this arrangement compared with conventional spectroscopy without lasers can be summarized as follows:

- There is no need for a spectrometer with wavelength dispersion because the tunable laser is already monochromatic and the absorption lines

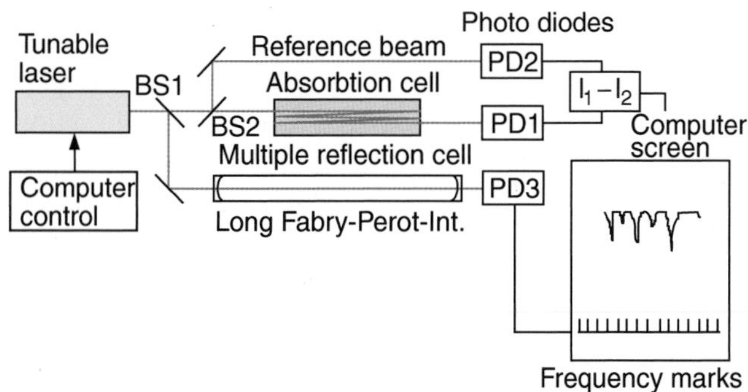


**Fig. 11.37.** Schematic experimental setup for laser absorption spectroscopy with reference beam and frequency marker FPI

appear as dips in the transmitted laser intensity when the laser is scanned through the spectrum.

- The spectral resolution is not limited by any instrument, but solely by the width of the absorption lines, which is generally given by their Doppler width (see Sect. 7.4.2). With Doppler-free techniques even this limitation can be overcome.
- Because of the good collimation of the parallel laser beam a long path length $L$ through the absorption cell can be realized, for instance, by multipath arrangements with spherical mirrors (see below). This enhances the sensitivity (since the absorption $\Delta I = I_0 \alpha L$ is proportional to the absorption length $L$) and allows the detection of weak transitions or of tiny densities of absorbing molecules.
- If the spectral width $\delta\nu_L$ of the laser radiation is smaller than the absorption line width $\Delta\nu_a$ (which is the case for single-mode lasers) the measured difference $\Delta I = I_0 - I_T = I_0 \alpha L$ is larger than in the case $\delta\nu > \Delta\nu_a$ where $\Delta I = I_0 \alpha L(\Delta\nu_a/\delta\nu)$.

The latter advantage is often erroneously neglected. Its importance shall be illustrated by the following example.

In conventional spectroscopy the grating spectrometer may have a resolution of $\delta\nu = 0.5\,\mathrm{cm}^{-1}$, while the Doppler width of the absorption lines in the visible is typical about $\Delta\nu_a = 0.03\,\mathrm{cm}^{-1}$. This gives for the same absorption $\alpha L$ at the center of an absorption line an intensity change $\Delta I$ that is 16 times smaller resulting in a sensitivity that is also 16 times smaller than for absorption spectroscopy with a laser that has a bandwidth $\Delta\nu_L \ll \Delta\nu_a$.

The sensitivity can be further enhanced by modulation of the laser frequency, similarly to the situation in microwave spectroscopy. Here we will, however, consider the case where the frequency modulation is larger than the line width of the absorbing transitions.

When the laser beam from a tunable diode laser passes through an electro-optic modulator (this is a crystal that changes its refractive index proportional to the voltage applied to electrodes on opposite sides of the crystal) the optical path length periodically changes and with it the phase of the optical wave. This phase modulation results in a frequency modulation because the frequency $\omega = \mathrm{d}\varphi/\mathrm{d}t$ is the derivative of the phase $\varphi$. The modulation of the optical carrier frequency $\omega$ creates sidebands at frequencies $\omega \pm \Omega$, where $\Omega$ is the modulation frequency. Differently from pure frequency modulation, the phases of the two sidebands are opposite (Fig. 11.38). If the transmitted intensity (this is the sum of the intensities of carrier and sidebands) is detected by a phase-sensitive detector (lock-in) tuned to the modulation frequency $\Omega$, the measured signal is zero if there is no absorption, because the detector measures the two differences $I_c - I_+$ and $I_c - I_-$ of the intensities $I_c$ of the carrier and $I_\pm$ of the two sidebands, which have opposite signs and therefore cancel each other. If, however, one of the sidebands coincides with an absorption line, this sideband is attenuated and the balance is lost. Fluctuations of the incident intensity appears in the carrier and the sidebands and therefore cancel for the difference measurement. The schematic arrangement for absorption spectroscopy with a tunable diode laser is shown in Fig. 11.39.

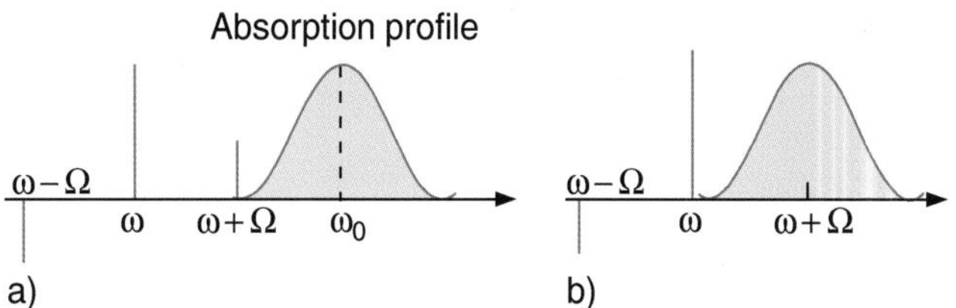


**Fig. 11.38a,b.** Principle of phase-modulated absorption spectroscopy

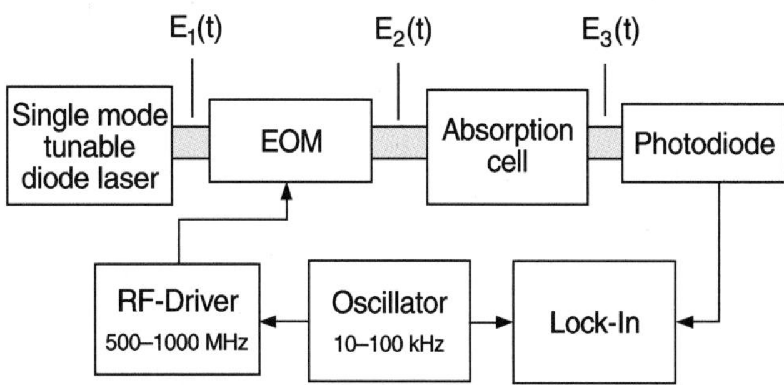


**Fig. 11.39.** Absorption spectroscopy with a tunable single mode diode laser with phase-modulation

The enhanced sensitivity is illustrated in Fig. 11.40, which compares the absorption measurement of a weak overtone absorption line of the $H_2O$ molecule with a non-modulated diode laser and the signal obtained with the modulation technique. The signal-to-noise ratio is enhanced by two orders of magnitude.

### 11.5.2 Optoacoustic Spectroscopy

An absorption cell with volume $V$ may contain $N$ absorbing molecules, which can be excited into the

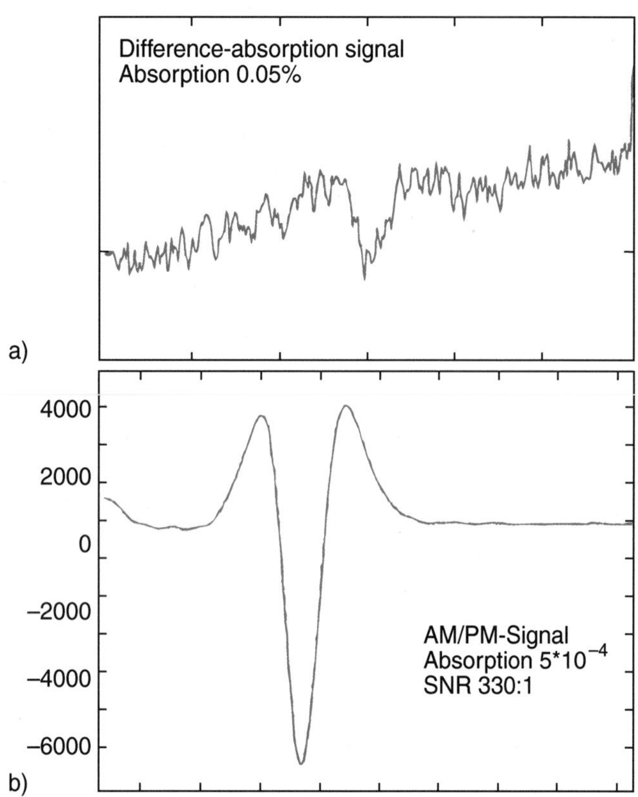


**Fig. 11.40a,b.** Rotational absorption line of a vibrational overtone transition $(1,2,1) \leftarrow (0,0,0)$ for the water molecule $H_2O$ (**a**) measured without modulation (**b**) measured with phase modulation of the laser beam

energy level $E_i = E_k + h\nu$ by absorption of a photon $h\nu$ (Fig. 11.41). If the excited molecule collides with other atoms or molecules in the cell, it can transfer its excitation energy into translational energy. If $N_1$ molecules are excited, the kinetic energy $E_{kin} = \frac{3}{2}NkT$ of the species, increases by

$$\Delta E_{kin} = N_1 \cdot h \cdot \nu = \frac{3}{2}N_1 k \Delta T$$

The temperature rise $\Delta T$ can be obtained from $N_1 \cdot h\nu = \frac{3}{2}Nk\Delta T \Rightarrow$

$$\Delta T = \frac{N_1 h\nu}{\frac{3}{2}Nk} . \tag{11.76}$$

For a closed cell, the total number of molecules $N$ is constant. A temperature rise therefore increases the gas pressure $p = nkT$ with $n = N/V$ by $\Delta p = nk\Delta T$.

> The energy of the absorbed photons is converted into kinetic energy resulting in a pressure increase.

So far we have neglected the radiative decay of the excited level. If the emitted photons are not absorbed within the cell, their energy is not converted into a pressure increase. The transferred energy is now

$$\Delta E = N_1 h\nu \frac{1}{1 + \tau_{coll}/\tau_{rad}} .$$

The conversion becomes more efficient as the ratio of radiative lifetime to the collisional deactivation time of the excited level $(\tau_{rad}/\tau_{coll})$ increases.

If the exciting laser beam is chopped at a frequency $f$, which is smaller than the inverse energy transfer

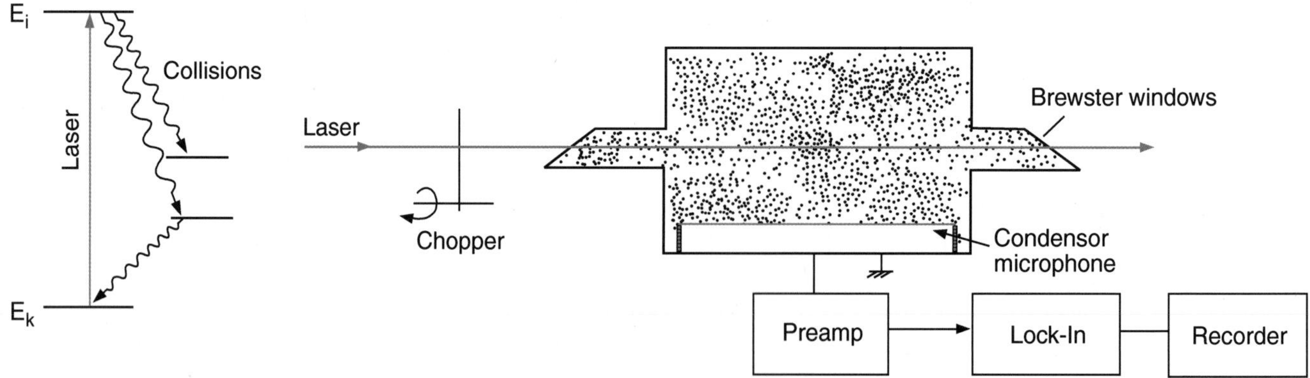


**Fig. 11.41a,b.** Photoacoustic spectroscopy: (**a**) level scheme, (**b**) schematic experimental arrangement

time $1/T_{tr} = (1/\tau_{coll} + 1/\tau_{rad})$ the pressure in the cell is modulated at the frequency $f$. A sensitive microphone in the side wall of the cell detects these pressure changes, which represent acoustic waves. When the chopping frequency is chosen to coincide with one of the acoustical eigenresonances of the cell, standing acoustic waves are generated with amplitudes that are dependent on the acoustic quality factor $Q$ of the cell and are much higher than for the nonresonant case.

The technique is quite sensitive, because the acoustic resonator acts as an amplifier and stores the energy converted from the excited molecules into standing acoustic waves for a time $\tau \propto Q$. When the laser wavelength is tuned across a molecular absorption spectrum one obtains for each absorption line an acoustic signal and altogether an optoacoustic spectrum. As an illustration Fig. 11.42 shows the optoacoustic spectrum of a very weak overtone band in acetylene, with the different rotational lines. Since the absorbed optical photon energy is converted into acoustic energy the method is named *optoacoustic* or *photoacoustic spectroscopy* [11.22, 23].

The sensitivity can be further enhanced when the acoustic cell is placed inside an optical multiple pass cell (Fig. 11.43). The laser beams inside this cell with spherical mirrors form a hyperboloid with cylindrical symmetry and excite acoustic standing waves on those locations in the acoustic cell where the standing acoustic waves have maximum amplitude. This optimizes the conversion efficiency for the energy transfer into the acoustic mode.

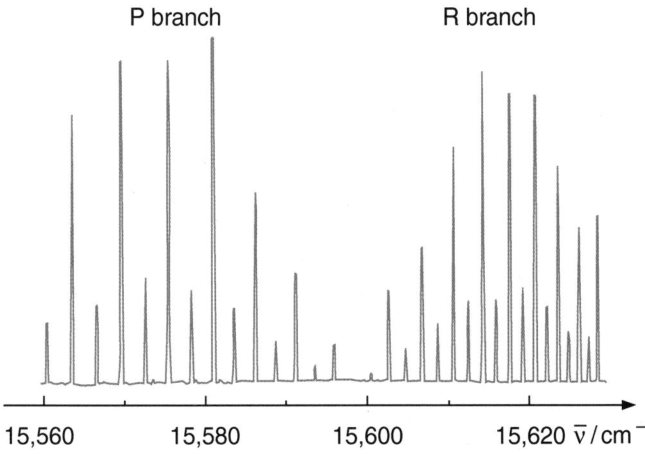


**Fig. 11.42.** Optoacoustic spectrum of the overtone band of acetylene around $\bar{\nu} = 15,600\,\text{cm}^{-1}$

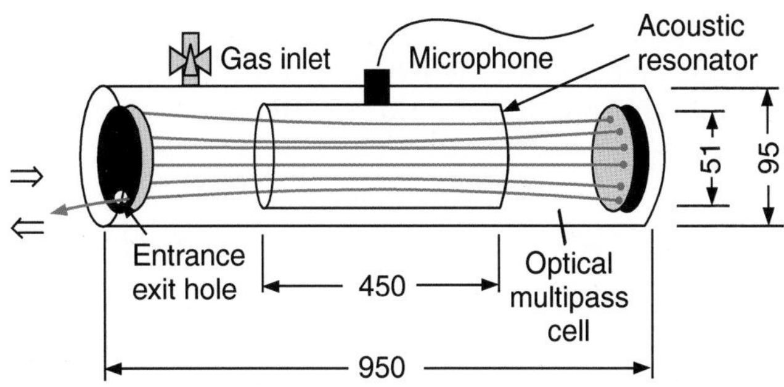


**Fig. 11.43.** Acoustic resonance cell inside an optical multipass cell. All measures are in mm

### 11.5.3 Optogalvanic Spectroscopy

For the spectroscopy of high-lying atomic, molecular or ionic levels, gas discharges are convenient because many of these levels are populated by electron impact, in particular, metastable levels with long lifetimes. The discharge impedance depends on the number of electrically charged carriers, i.e., electrons and ions, which are created by electron impact ionization from excited levels of the neutral species. If the population of these excited levels is changed, for instance by absorption of photons, the electron density will change and therefore the impedance of the discharge. If the discharge is fed through a ballast resistor R by a power supply with constant voltage, the impedance change causes a corresponding change of the discharge current and therefore a voltage change across resistor R. This forms the basis for optogalvanic spectroscopy.

The experimental arrangement is shown in Fig. 11.44. The discharge tube (which might be a hollow cathode discharge or a tube used for the He-Ne laser or the argon ion laser) is irradiated by the chopped output beam of a tunable laser (dye laser or semiconductor laser). When the laser wavelength

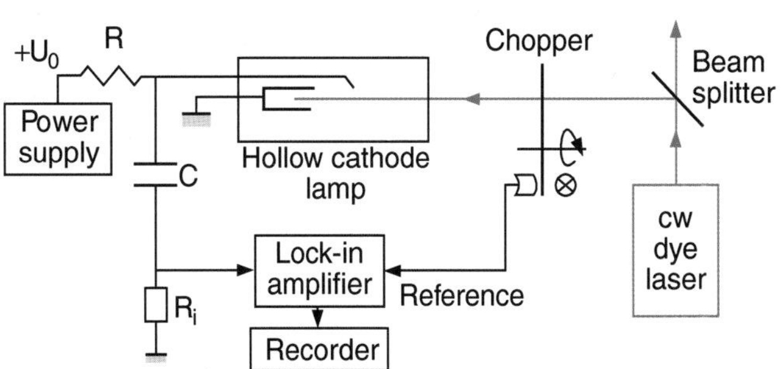


**Fig. 11.44.** Experimental arrangement of optogalvanic spectroscopy in a hollow cathode discharge

coincides with an absorption line of species in the discharge, this causes a modulation of the level densities $N_i$, $N_k$ and therefore of the discharge current at the chopping frequency and a corresponding voltage modulation $\Delta U$ across the ballast resistor R. This modulated voltage is coupled through a capacitance C into a lock-in amplifier and recorded by a computer. If the laser wavelength is tuned to a transition $E_i \to E_k$ the population of the two levels changes due to laser-induced transitions by

$$\Delta N_i = N_{i0} - N_{i\mathrm{L}} = -\Delta N_k \; .$$

The corresponding voltage change is then

$$\begin{aligned} \Delta U &= R \Delta I \\ &= a[\Delta N_i PI(E_i) - \Delta N_k PI(E_k)] \,, \end{aligned} \tag{11.77}$$

where $PI(E_i)$ is the ionization probability of level $E_i$.

The optogalvanic signals might be positive or negative, depending on the difference of ionization probabilities of the levels $E_i$ and $E_k$.

Most gas discharge tubes are filled with noble gases. However, if the noble gas is mixed with other volatile components their spectra can be also measured by this technique. Even molecules can be inserted into the discharge. Since some molecules might be fragmented by electron impact, the spectra of the mother molecules and their fragments overlap and the assignment might be difficult.

In a hollow cathode discharge, the ions impinge on the inner walls of the hollow cathode and release atoms and ions of the wall material by sputtering processes. With higher discharge currents the spectra of these sputtered materials become more and more prominent. This is illustrated in Fig. 11.45, where the optogalvanic spectrum of aluminum, copper and iron obtained in a hollow cathode discharge is shown.

Investigations of intensity and time behavior of optogalvanic signals give information on radiative decay constants of highly excited levels and on collision cross sections for electron impact excitation and ionization. These parameters are important for understanding plasma characteristics [11.24].

Since in the gas discharge many intermediate levels are populated, it is possible to reach by one-photon absorption very high atomic or molecular levels. One example is the investigation of atomic Rydberg levels by optogalvanic spectroscopy [11.25, 26].

### 11.5.4 Cavity-Ringdown Spectroscopy

A very sensitive technique for detecting small absorptions is cavity-ringdown spectroscopy, which is based on measurements of the decay time of optical resonators filled with an absorbing gas [11.28]. Its basic principle can be understood as follows.

When a short laser pulse with power $P_0$ is sent through the optical resonator of length $L$ with two highly reflecting mirrors (reflectivities $R_1 = R_2 = R$) and

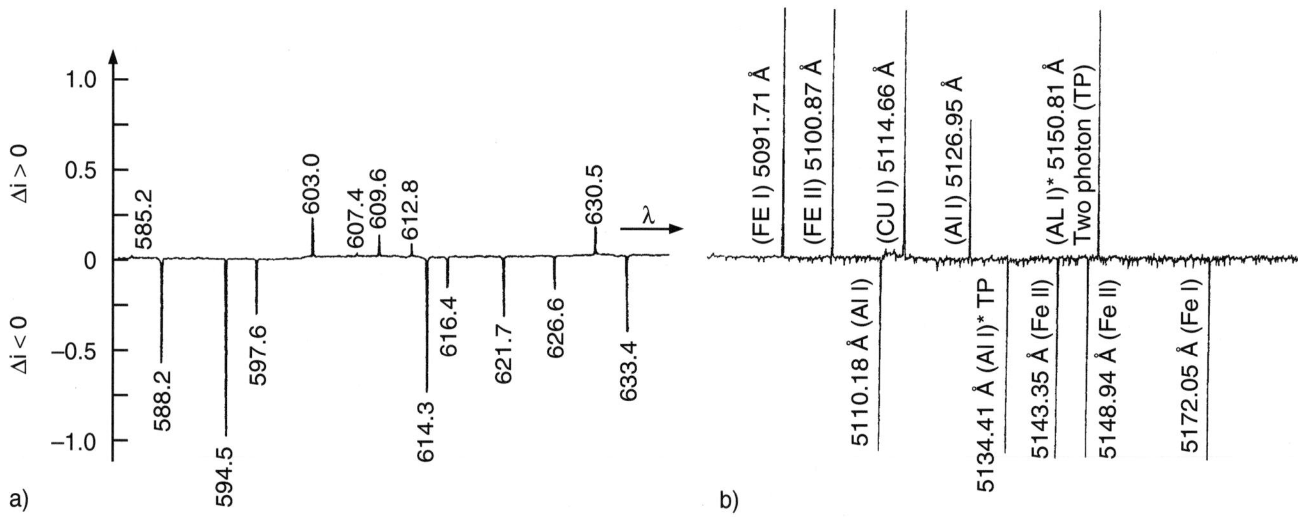


**Fig. 11.45a,b.** Optogalvanic spectrum (**a**) of a neon discharge (1 mA, $p = 1$ mbar), generated with a broadband cw dye laser [11.24], (**b**) of Al, Cu, and Fe vapor sputtered in a hollow cathode that was illuminated with a pulsed dye laser [11.27]

total absorption losses $A$ (scattering, diffraction and mirror absorption, but excluding the absorption by the gas) the transmission

$$T = 1 - R - A \ll 1$$

is very small. The pulse will be reflected back and forth between the two mirrors (Fig. 11.46), while for each roundtrip a small fraction will be transmitted through the end mirror and can reach the detector. The detector receives a series of pulses separated by the roundtrip time $T = 2L/c$ with decreasing power from pulse to pulse.

The power of the first transmitted pulse is

$$P_1 = T^2 \mathrm{e}^{-\alpha L} P_0 , \tag{11.78a}$$

where $\alpha$ is the absorption coefficient of the gas in the resonator. After each roundtrip the pulse power decreases by an additional factor $R^2 \cdot \exp(-2\alpha L)$. After $m$ roundtrips the power has decreased to

$$\begin{aligned} P_m &= \left(R\mathrm{e}^{-\alpha L}\right)^{2m} P_1 \\ &= \left[(1 - T - A)\mathrm{e}^{-\alpha L}\right]^{2m} P_1 , \end{aligned} \tag{11.78b}$$

which can be written with $\ln(1 - x) \approx x$ for $x \ll 1$ as

$$P_m = P_1 \mathrm{e}^{2m[\ln R - \alpha L]} \approx P_1 \mathrm{e}^{-2m[T + A + \alpha L]} . \tag{11.78c}$$

If the detector time constant is large compared to the pulse width it just detects the envelope of the pulse amplitudes and records an exponential decay with the decay time

$$\tau_1 = \frac{L/c}{T + A + \alpha L} . \tag{11.79}$$

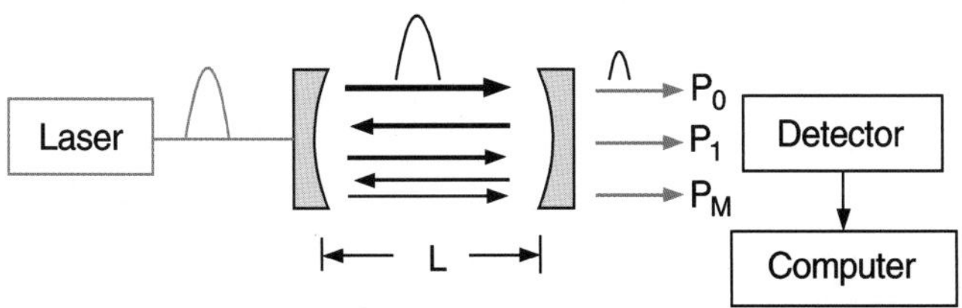


**Fig. 11.46.** Principle of cavity ring-down spectroscopy with pulsed lasers

Without absorbing gas in the resonator ($\alpha = 0$) the decay time will be lengthened to

$$\tau_2 = \frac{L/c}{T + A} . \tag{11.80}$$

From the difference

$$\frac{1}{\tau_1} - \frac{1}{\tau_2} = \alpha \cdot c \tag{11.81}$$

the absorption coefficient $\alpha$ can be determined as a function of the laser wavelength $\lambda$. The minimum detectable absorption is limited by the reflectivity $R$, the unavoidable losses $A$ of the resonator and by the accuracy of measuring the decay times $\tau_1$ and $\tau_2$. This accuracy is limited by the attainable signal-to-noise ratio.

The physical reason for the large sensitivity is the long effective absorption path, which is

$$L_{\mathrm{eff}} = L/(1 - R + A) . \tag{11.82}$$

## EXAMPLE

For $L = 1\,\mathrm{m}$, $R = 0.999$, $A = 0.001 \Rightarrow L_{\mathrm{eff}} = 500\,\mathrm{m}$.

The experimental setup is shown in Fig. 11.47. The incoming laser beam has to be mode-matched to the fundamental $\mathrm{TEM}_{00q}$ resonator mode. Otherwise transverse modes are excited, which have much higher diffraction losses (see Sect. 8.2). A careful alignment of the system is required to obtain optimum results. When the laser wavelength is tuned across the absorption spectrum the maxima of the absorption coefficient $\alpha$ at the center of the absorption lines lead to decay time minima and a computer program converts these measured minima into the absorption coefficient $\alpha(\lambda)$. In

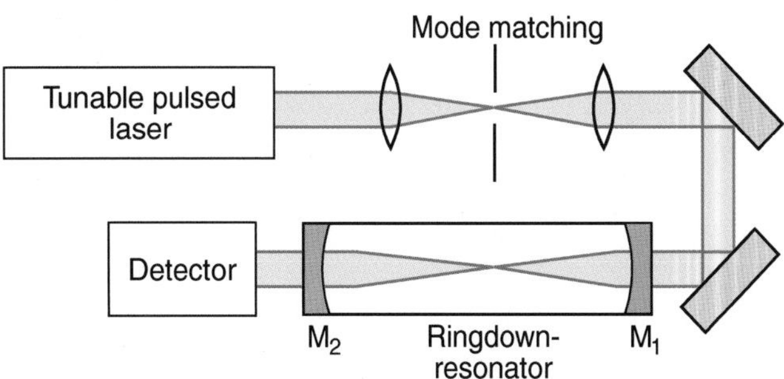


**Fig. 11.47.** Experimental setup with mode-matching optics

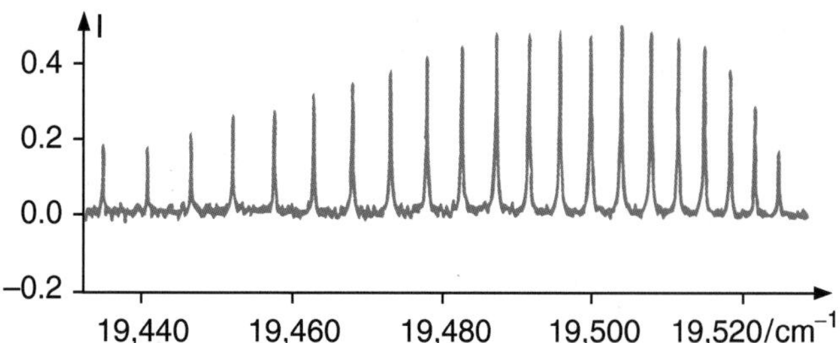


**Fig. 11.48.** Section of the rotational lines on the overtone band $(2,0,5) \leftarrow (0,0,0)$ of the HCN molecule, measured with CRDS [11.29]

Fig. 11.48 the rotational spectrum of the overtone transition in the HCN molecule, measured by Romanini and Lehman [11.29] is shown, illustrating the good signal-to-noise ratio.

The following conditions should be met to realize the high sensitivity and spectral resolution:

1. Due to the spectral bandwidth of the laser pulse many fundamental resonator modes within the bandwidth $\delta\omega_R$ can be excited. In order to resolve the absorption lines with spectral width $\delta\omega_a$ the laser bandwidth $\delta\omega_L$ should be smaller than the absorption line width.
2. The relaxation time of the resonator must be longer than that of the excited molecules. This demands high reflection coatings of the cavity mirrors ($R > 0.999$) and careful alignment.

**EXAMPLE**

With a resonator length $L = 0.5\,$m, a mirror reflectivity $R = 0.998$, resonator losses per roundtrip $A = 0.001$ and mirror transmission $T = 0.0001$ per mirror, the decay time of the empty resonator is $\tau_2 = 0.5/(3 \times 10^8 \times 0.002) = 8.33 \times 10^{-7}\,$s. With an absorption coefficient $\alpha = 10^{-6}/\text{cm} = 10^{-4}/\text{m}$ the absorption per path is $\alpha L = 5 \times 10^{-5}$ and the decay time $\tau_2$ of the filled resonator becomes $\tau_2 = 0.5/(3 \times 10^8 \times 0.00205) = 8.13 \times 10^{-7}\,$s. The relative change of $\tau$ is only $\Delta\tau/\tau = 0.024 = 2.4\%$. With a laser pulse duration $\Delta t_L = 10\,$ns the laser bandwidth is $\delta\omega_L = 10^8\,\text{s}^{-1} \Rightarrow \delta\nu_L = 16\,$MHz, which is small compared to the Doppler width of about 1 GHz of the absorption lines.

With a reflectivity $R = 0.9995$ and transmission $T = 0.00005$ and resonator losses $A = 0.0001$ The decay time $\tau_2$ becomes $\tau_2 = 1 \cdot 10^{-5}\,$s and $\tau_1 = 8.3 \cdot 10^{-6}\,$s. Now the difference becomes $1.7 \cdot 10^{-6}$s which is 17% of the decay time $\tau_2$. This demonstrates the importance of low loss resonators and high reflectivity mirrors.

### 11.5.5 Laser-Induced Fluorescence Spectroscopy

So far we have mainly discussed different techniques of absorption spectroscopy. Laser-induced fluorescence spectroscopy is based on selective excitation of one or a few levels in the upper state of atoms or molecules. The fluorescence spectrum, emitted by this excited level, is dispersed by a spectrometer and the dispersed fluorescence lines are recorded by a photomultiplier or a CCD array (Fig. 11.49). If the fluorescence is emitted from a single upper rotational-vibrational level of a molecule the allowed transitions represent a relatively simple spectrum compared to an emission spectrum from a gas discharge, where many upper levels are populated and the total fluorescence is the superposition of all fluorescence series emitted by the different upper levels.

The wavenumbers of the fluorescence lines are equal to the differences between the term value of the upper level and those of the terminating lower levels. If the upper level is the same for all lines the separations of the fluorescence lines just give the energy differences between the levels in the lower state (Fig. 11.50). If the Franck–Condon factors for transitions into high vibrational levels of the electronic ground state are sufficiently large (this depends on the excited upper vibrational level), the vibrational levels in the electronic

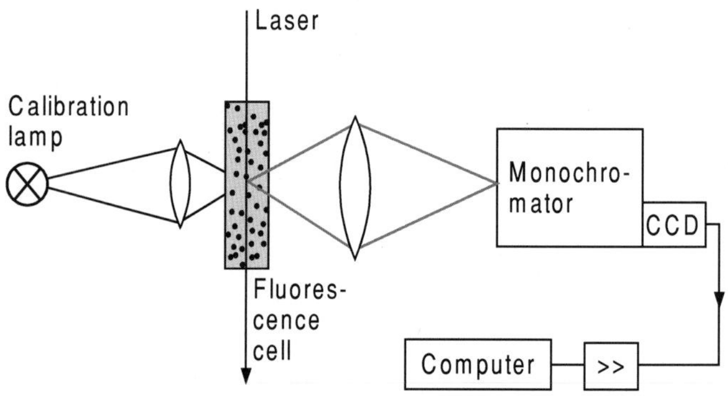


**Fig. 11.49.** Schematic setup for measuring the dispersed laser-induced fluorescence spectrum. The calibration lamp (thorium hollow cathode) provides spectral lines for wavelength calibration

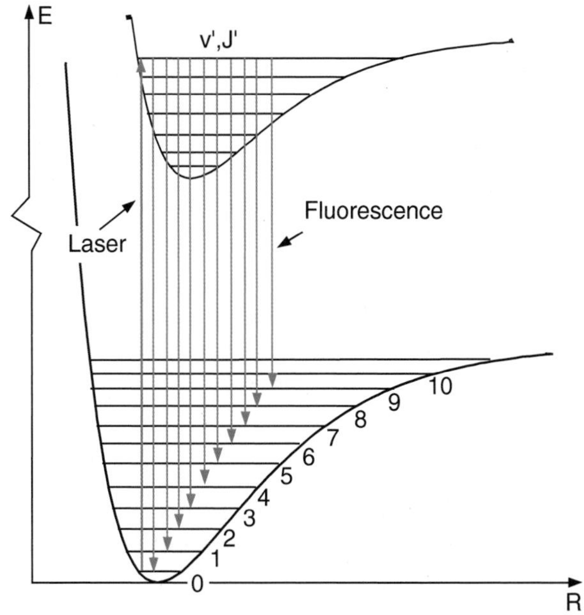


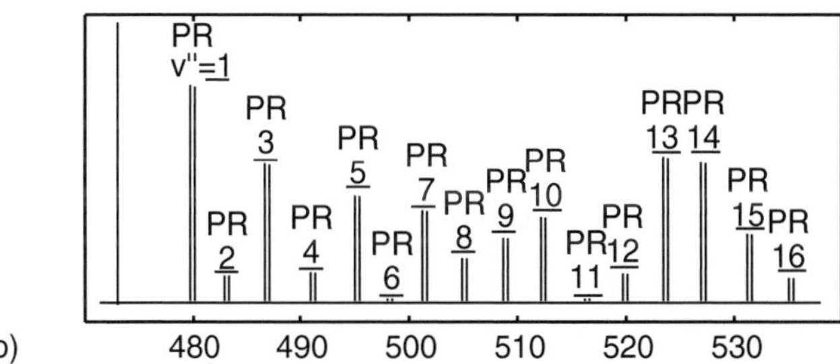


**Fig. 11.50a,b.** Laser-induced fluorescence. (**a**) Level scheme. (**b**) Fluorescence spectrum of the $Na_2$ molecule, emitted from the selectively excited level $B^1\Pi_u$ ($v' = 6$, $J' = 27$)

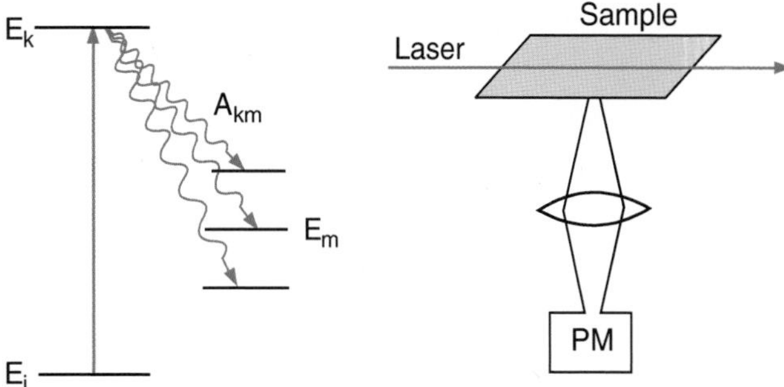


**Fig. 11.51.** Level scheme and experimental arrangement for fluorescence excitation spectroscopy

ground state can be measured up to the dissociation energy. This allows a very accurate determination of the potential curve [11.30].

In another technique the total fluorescence intensity (without dispersion by a spectrometer) is measured as a function of the wavelength of the exciting laser (Fig. 11.51). In this case the fluorescence is just used for monitoring the absorption, because each absorbed photon creates a fluorescence photon if other deactivation processes of the excited level (for instance collision-induced radiationless transitions) can be neglected. This ***excitation spectroscopy*** is a very sensitive version of absorption spectroscopy, as can be seen from the following estimation:

Assume that $\dot{n}_a$ photons per s are absorbed. The fluorescence photons, imaged onto the cathode of a photomultiplier, create $\dot{n}_{pe}$ photoelectrons per second. With the quantum efficiency of the photocathode $\eta_{ph} = \dot{n}_{pe}/\dot{n}_{fl}$, the rate of absorbed laser photons $\dot{n}_a$, the fluorescence quantum yield $\eta_{fl} = \dot{n}_{fl}/\dot{n}_a$, the geometrical collection efficiency $\delta$ of the fluorescence photons onto the photocathode, the measured rate of photoelectrons is given by the signal

$$S = \dot{n}_{pe} = \dot{n}_a \eta_{fl} \eta_{ph} \delta\ . \qquad (11.83)$$

The absorption rate

$$\dot{n}_a = N_k \sigma_{ki} n_L \Delta x \qquad (11.84)$$

depends on the number density $N_k$ of molecules in the absorbing level $\langle k|$, the absorption cross section $\sigma_{ki}$, the number $n_L$ of laser photons incident per s and cm$^2$ onto the sample and the absorption path length $\Delta x$.

## EXAMPLE

With a quantum efficiency $\eta_{ph} = 0.2$ of the multiplier photocathode, a collection efficiency of $\delta = 0.1$, which demands a collection optics for the fluorescence that covers a solid angle of $0.4\pi$, a fluorescence yield $\eta_k = 1$ for the excited level $\langle k|$ and a laser photon flux of $N_L = 3 \times 10^{18}$/s that corresponds to a laser power of 1 W at $\lambda = 500$ nm, an absorption rate $n_a = 10^4$/s (which means a relative absorption of the laser of $\Delta I_L/I_L = 3 \times 10^{-15}$) gives a photoelectron rate of $n_{pe} = 200$/s. If the dark current of the photomultiplier is $n_{pe}(0) = 50$, this gives a signal to background ratio of 4, which demonstrates that relative absorptions of less than $10^{-15}$ can still be detected. This represents a much higher sensitivity than any direct absorption measurement!

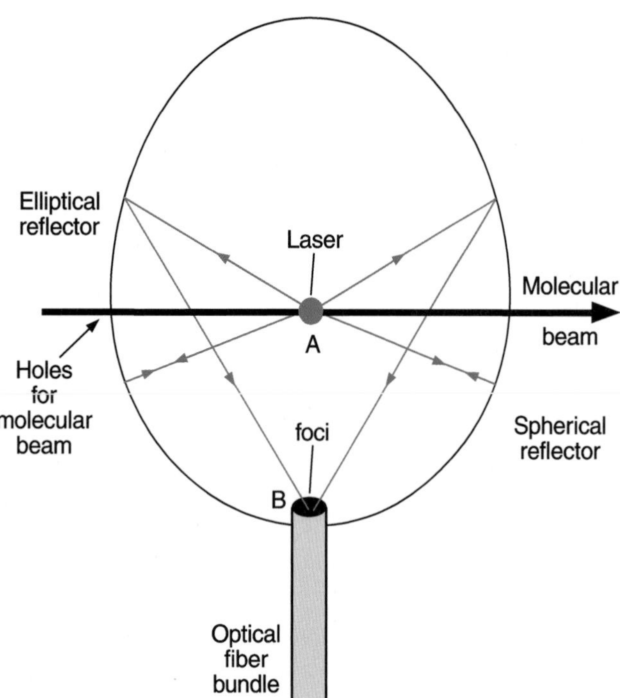


**Fig. 11.52.** Efficient collecting system for laser-induced fluorescence, consisting of an elliptical mirror in the upper half and a spherical mirror in the lower half of the plane

The collection efficiency for fluorescence photons can be enlarged by an elliptical mirror arrangement (Fig. 11.52). If the intersection A of laser and molecular beams is placed in one focal point of the ellipsoid, the fluorescence is imaged into the second focal point B where an optical fiber bundle transmits it either directly to the photomultiplier or to the entrance slit of the spectrometer.

### 11.5.6 Ionization Spectroscopy

The most sensitive detection technique is ionization spectroscopy. Here a laser is tuned through the spectrum of interest and excites for each absorbing transition a selected upper level. Differently from fluorescence-excitation spectroscopy, here not the fluorescence is monitored, but the molecule in the excited level is ionized by a second laser (Fig. 11.53). If this laser is sufficiently intense, the ionizing transition can be saturated, which means that every excited molecule is ionized before it can emit a fluorescence photon or can be deactivated by collisions. The ions are collected by electric fields, accelerated and imaged onto the cathode of an open ion multiplier. This is a device that

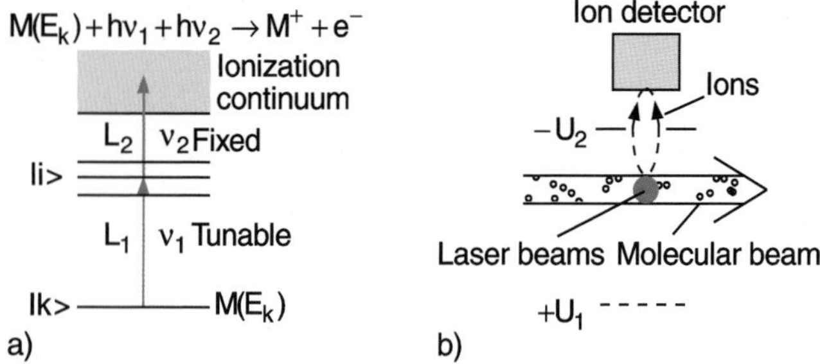


**Fig. 11.53a,b.** Ionization spectroscopy. (**a**) Level scheme for resonant twp-photon ionization. (**b**) Experimental setup for photo-ionization in a molecular beam

is quite similar to a photomultiplier. Only the photocathode is replaced by a metal cathode, where the ions, impinging with energies of a few keV, produce electrons by ion bombardment. These electrons are further accelerated and multiplied like in a photomultiplier.

With an optimum ion collecting system design, each ion produced through photoionization, can be imaged onto the ion multiplier and produces a voltage pulse at the multiplier output that can be counted.

If the first laser with wavelength $\lambda_1$ is tuned through the spectral region of interest and the second laser with a fixed wavelength $\lambda_2$ ionizes all excited molecules the measured ion rate $N_{\text{ion}}(\lambda_1)$ essentially gives the absorption spectrum $\alpha(\lambda_1)$ of transitions from the absorbing levels in the electronic ground state to the levels excited by laser L1. The sensitivity of this method is, however, higher by several orders of magnitude than in absorption spectroscopy, where the attenuation of the transmitted laser intensity is monitored.

In favorable cases, single atoms or molecules can be detected [11.31].

The number of ions produced per second is determined similarly to (11.84) by

$$\dot{N}_{\text{ion}} = N_i \sigma_{iI} \dot{N}_{\text{L2}} , \tag{11.85}$$

where $N_i$ is the stationary population of the excited level $\langle i|$ and $\dot{N}_{\text{L2}}$ is the photon flux of laser L2 incident on the molecules in level $\langle i|$.

The stationary density $N_i$ of molecules in level $\langle i|$ is determined from the balance between the excitation and depopulation processes. If $R_i$ is the total relaxation probability of the excited level $\langle i|$ (besides the ionization probability $P_{iI} = \sigma_{iI}\dot{N}_{L2}$), we obtain

$$\frac{dN_i}{t} = N_k\sigma_{ki}\dot{N}_{L1} - N_i(R_i + \sigma_{iI}\dot{N}_{L2}) = 0$$

$$\Rightarrow N_i = N_k\frac{\sigma_{ki}\cdot\dot{N}_{L1}}{R_i + \sigma_{iI}\dot{N}_{L2}} . \qquad (11.86)$$

For a stationary population $N_k$ of molecules in the absorbing level $\langle k|$ the refilling rate must be equal to the excitation rate which equals the rate $\dot{n}_a$ of absorbed photons. The refilling rate is supplied either by relaxation processes of other levels into level $\langle k|$ or by the flow of molecules into the excitation volume.

This gives for the measured ion rate (11.85)

$$\dot{N}_{ion} = N_k\frac{\sigma_{ki}\dot{N}_{L1}}{1 + R_i/(\sigma_{iI}\dot{N}_{L2})} . \qquad (11.87a)$$

For $(\sigma_{iI}\dot{N}_{L2}) \gg R_i$ the ion rate becomes

$$\dot{N}_{ion} = N_k\sigma_{ki}\dot{N}_{L1} = \dot{n}_a \qquad (11.87b)$$

and the measured signal

$$S = \dot{N}_{ion}\delta\eta \qquad (11.87c)$$

becomes for $\delta = \eta = 1$ equal to the rate $\dot{n}_a$ of absorbed photons on the transition $k \to i$. Here $\delta$ is the collection efficiency of the ions onto the multiplier cathode and $\eta$ the detection efficiency of the multiplier for one impinging ion.

With proper ion collecting optics, $\delta = 1$ can be reached. For ions with kinetic energies of several keV the multiplier efficiency is $\eta = 1$. If the second laser is sufficiently intense, $P_{iI} \gg R_i$. Then (11.87c) gives for $\delta = \eta = 1 : S_I = \dot{n}_a$. This means that every photon of the first laser, absorbed on the transition $k \to i$ is converted into a signal pulse.

> Under favorable conditions every absorbed photon can result in a signal count. With sufficiently large laser intensities this implies that single atoms or molecules can be detected.

## EXAMPLE

Assuming an absorption cross section $\sigma_{ki} = 10^{-14}\,\text{cm}^2$, a stationary density $N = 10^9\,\text{cm}^{-3}$ (corresponding to a gas pressure of $10^{-7}$ mbar). If 0.1% of all molecules are in the absorbing level $\langle k|$ the density of absorbing molecules is $N_k = 10^6\,\text{cm}^{-3}$. The absorption coefficient is $\alpha = N_k\sigma_{ki} = 10^{-8}/\text{cm}$. With a path length $\Delta x = 1\,\text{cm}$, $10^{-8}$ of all incident photons are absorbed. When the number of photons per s of the first laser is $\dot{N}_{L1} = 3\times10^{16}$ (10 mW), we obtain $n_a = 3\times10^8\,\text{s}^{-1}$. Under stationary conditions this absorption rate must be compensated for by an equal refilling rate. In a gas at room temperature the mean molecular velocity is about $500\,\text{m/s} = 5\times10^4\,\text{cm/s}$. At a density of $10^6\,\text{cm}^{-3}$ absorbing molecules, the refilling rate could even compensate a depopulation rate of $10^{10}\,\text{s}^{-1}$ in a volume of $1\,\text{cm}^3$.

### 11.5.7 Laser Spectroscopy in Molecular Beams

In many cases, spectral resolution is limited by the Doppler width of the absorption lines. This limitation often prevents the recognition of finer details in the spectrum, such as hyperfine splittings, Zeeman splittings in weak external fields or the rotational structure in molecular spectra. *Doppler-free* techniques that can overcome the limitations set by the Doppler width are then demanded for getting the full information from atomic or molecular spectra. We will discuss in the following sections some of these Doppler-free techniques.

An elegant way to reduce the Doppler width is the reduction of the velocity distribution of atoms or molecules in collimated molecular beams (Fig. 11.54).

Molecules effusing from a reservoir R through a small hole A into a vacuum chamber have to pass through a narrow slit B with width $b$ at a distance $d$

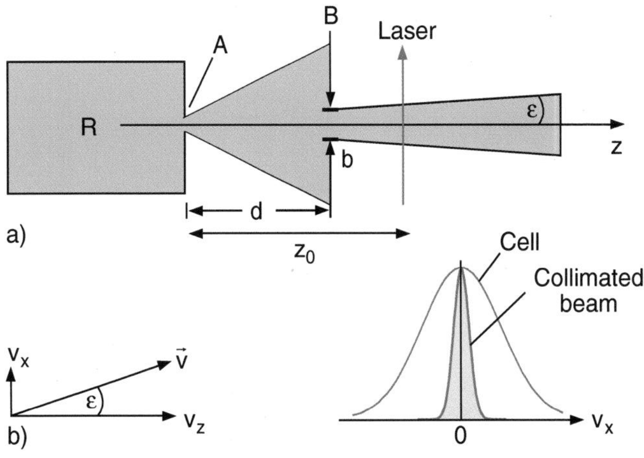


**Fig. 11.54.** Laser spectroscopy in a collimated molecular beam

from A downstream in the molecular beam in order to reach the interaction region with the laser beam. If we choose the molecular beam axis as the $z$-axis and place the slit in the $y$ direction, the $v_x$ component of the molecules passing through the slit is reduced by the geometrical factor

$$v_x \le (b/2d)\overline{v} = v_z \tan\varepsilon\ , \qquad (11.88)$$

where $\overline{v} = (8kT/\pi m)^{1/2}$ is the mean velocity of particles with mass $m$, when the temperature of the reservoir is $T$, $v_z$ is the velocity component parallel to the beam axis and $\varepsilon$ with $\tan\varepsilon = b/2d$ is the collimation angle.

If the beam of a tunable single mode laser crosses the molecular beam in $x$ direction (i.e., perpendicular to the molecular beam axis) behind the slit, only molecules within the narrow interval $v_x \le (b/2d)\overline{v}$ contribute to the absorption, which means that the width $\Delta\nu$ of the absorption line is reduced compared to its Doppler width in a gas at thermal equilibrium (Fig. 11.54).

## EXAMPLE

With $b = 1\,\text{mm}$, $d = 100\,\text{mm} \Rightarrow \tan\varepsilon = 5\times 10^{-3}$. Instead of a typical Doppler width of $\Delta\nu_\text{D} = 1\,\text{GHz}$ in the optical range, the line width is now reduced by a factor of 200 to $\Delta\nu_\text{D}^\text{red} = 5\,\text{MHz}$. This is already in the range of the natural line width (see Sect. 7.4.1).

The quantitative treatment starts with the density of molecules in the beam with velocities $v$ within the interval $\text{d}v$

$$n(v,r,\theta) = C\frac{\cos\theta}{r^2}nv^2\text{e}^{-(v/v_\text{p})^2}\text{d}v\ , \qquad (11.89)$$

where the normalization factor $C = (4/\sqrt{\pi})/v_\text{p}^3$ assures that the total density $n$ of the molecules is

$$n = \int n(v)\,\text{d}v$$

and $v_\text{p} = (2kT/m)^{1/2}$ is the most probable velocity. The angle $\theta$ is measured against the $z$-axis. The spectral profile of the absorption coefficient is

$$\alpha(\omega,x) = \int\limits_{-v_z\tan\varepsilon}^{+v_z\tan\varepsilon} n(v_x,x)\sigma(\omega,v_x)\,\text{d}v_x\ . \qquad (11.90)$$

where $n(v_x,x)\text{d}v_x$ is the density of molecules in the intervall $\text{d}v_x$ at the location $(x,z_0)$. With $v_x = (x/r)v \Rightarrow \text{d}v_x = (x/r)\text{d}v$ and $\cos\Theta = z/r$ (Fig. 11.55) we derive from (11.89) for the molecular density

$$n(v_x,x)\text{d}v_x = C\cdot n\frac{z}{x^3}v_x^2\text{e}^{-[(rv_x/xv_\text{p})^2]}\text{d}v_x\ . \qquad (11.91)$$

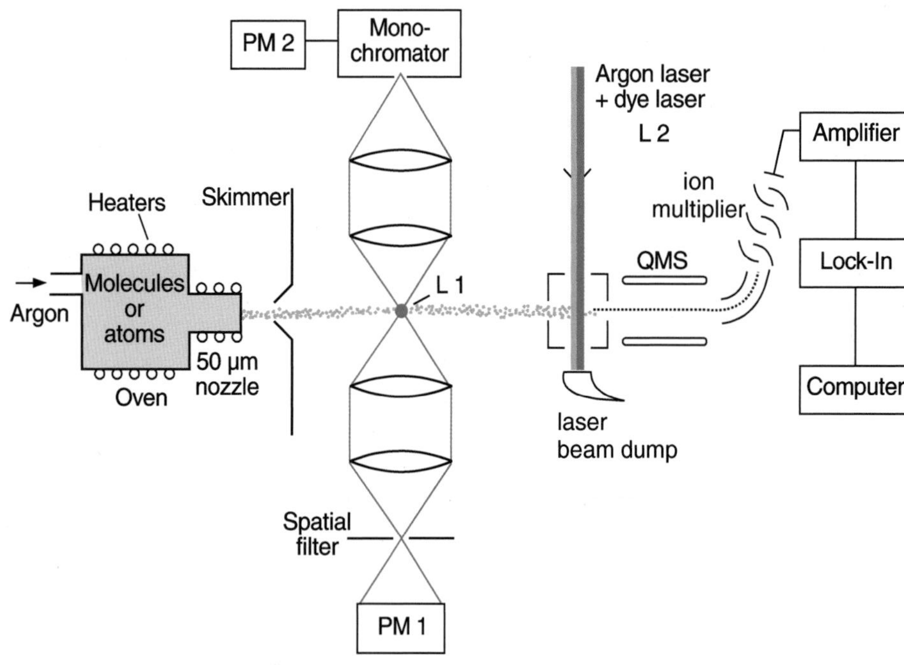


**Fig. 11.55.** Sub-Doppler laser spectroscopy in a collimated molecular beam, where the total fluorescence is measured by PM 1 and the dispersed fluorescence spectrum by PM 2. The two photon-induced ions are mass selected by the quadrupole mass spectrometer QMS and are detected by an ion-multiplier

The absorption cross section $\sigma(\omega, v_x)$ is a Lorentzian profile, Doppler-shifted by $k \cdot v_x$ (see Sect. 7.4.2)

$$\sigma(\omega, v_x) = \sigma_0 \frac{(\gamma/2)^2}{(\omega - \omega_0 - kv_x)^2 + (\gamma/2)^2} . \tag{11.92}$$

Inserting (11.91) and (11.92) into (11.90) gives, with $\mathrm{d}v_x = \mathrm{d}\omega/k$, the spectral profile of the absorption coefficient

$$\alpha(\omega) = a \int \frac{e^{-\left[\frac{c(\omega-\omega_0')}{v_\mathrm{p} \sin \varepsilon \omega_0'}\right]^2}}{\left(\omega - \omega_0'\right)^2 + (\gamma/2)^2} \mathrm{d}\omega_0' \tag{11.93}$$

with $\omega_0' = \omega_0(1 + v_x/c)$.

This is the convolution of a Lorentzian and a Gaussian line profile, which is called a ***Voigt profile*** (Fig. 7.22). The Gaussian profile has a width

$$\Delta\omega_\mathrm{D}^\mathrm{red} \cdot \sin \varepsilon ,$$

reduced by a factor $\sin \varepsilon$ against the Doppler width $\Delta\omega_\mathrm{D}$ in a gas at thermal equilibrium.

High resolution absorption spectra can be measured by monitoring the total fluorescence $I_\mathrm{fl}(\lambda_\mathrm{L})$ as a function of the laser wavelength (excitation spectroscopy) or by resonant two-photon ionization, where the ions are extracted by an electric field from the intersection volume of molecular and laser beams onto the ion multiplier. In Fig. 11.55, PM1 measures the total fluorescence, while PM2 behind a monochromator monitors the dispersed fluorescence at a fixed laser wavelength. If the two-photon ionization occurs in the ion source of a mass spectrometer the ions can be mass selected. This is very helpful for the assignment if many different masses (e.g., isotopes or different cluster sizes in a cluster beam) are present.

As an illustration of the attainable spectral resolution Fig. 11.56 shows the hyperfine structure of a rotational transition of the $Na_2$ molecule in the electronic transition from the $X^1\Sigma_\mathrm{g}$ ground state into the mixed $A^1\Sigma_\mathrm{u} - a^3\Pi_\mathrm{u}$ upper state, where the two upper states are mixed by spin-orbit coupling. For symmetry reasons only components with nuclear spin quantum number $I = 1$ and 3 are possible in a rotational level with even rotational quantum number $J' = 24$.

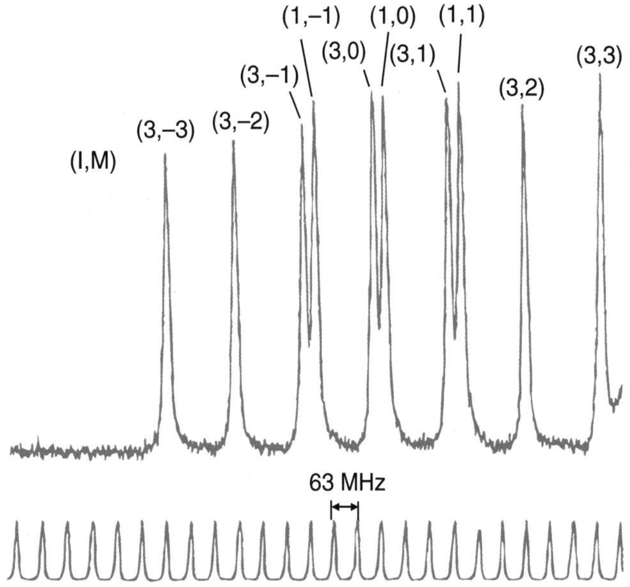


**Fig. 11.56.** The hyperfine components of the rotational line R(23) in the system $a^3\Pi_0 \leftarrow X^1\Sigma_\mathrm{g}$ of the $Na_2$ molecule, measured in a collimated molecular beam. The components are labelled by the nuclear spin quantum number $I$ and the spin projection quantum number $M$ [11.32]

### 11.5.8 Nonlinear Laser Spectroscopy

Several spectroscopic techniques that achieve sub-Doppler resolution are based on the nonlinear interaction of molecules with the laser radiation. When a plane electromagnetic wave passes in $x$ direction through an absorbing gas, we saw in Sect. 11.1 that the attenuation $\mathrm{d}I$ of the intensity is related to the absorption coefficient $\alpha$ by

$$\mathrm{d}I = -\alpha I \mathrm{d}x . \tag{11.94}$$

The absorption coefficient

$$\alpha(\omega) = [N_k - (g_k/g_i)N_i]\sigma(\omega) = \Delta N \cdot \sigma(\omega) \tag{11.95}$$

is determined by the population difference $\Delta N$ and the absorption cross section $\sigma$. This gives for (11.94) the relation

$$\mathrm{d}I = -\Delta N \cdot \sigma(\omega) \cdot I \cdot \mathrm{d}x . \tag{11.96}$$

For sufficiently small intensities $I$ the population densities $N_i$ and $N_k$ are not much affected by absorption transitions, because relaxation processes can refill the population $N_k$ of the absorbing level (Fig. 11.57a). In this case the absorption coefficient $\alpha$ is independent of

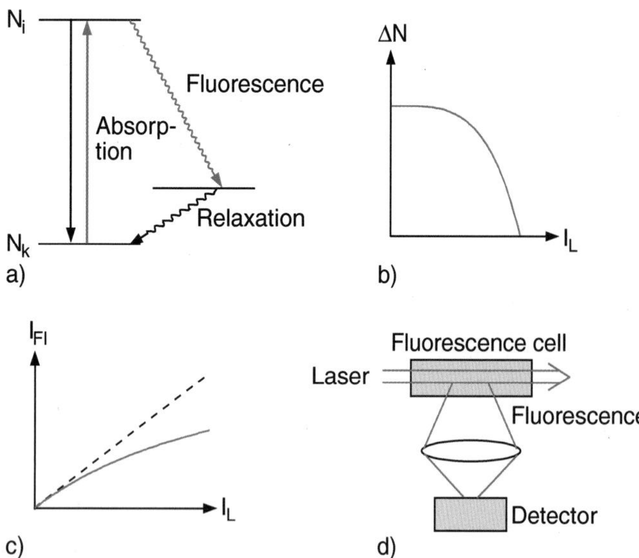


**Fig. 11.57a–d.** Saturation of a molecular transition. (**a**) Level scheme (**b**) Population difference $\Delta N$. Intensity of laser-induced fluorescence (**c**) as a function of laser intensity (**d**) measuring $I_{\mathrm{Fl}}(I_{\mathrm{L}})$

the intensity $I$ and integration of (11.94) yields *Beer's* law of linear absorption

$$I = I_0 \mathrm{e}^{-\alpha x} = I_0 \mathrm{e}^{-\Delta N \sigma x} . \tag{11.97}$$

For larger intensities the absorption rate can exceed the relaxation rates refilling the absorbing level $\langle k|$. This means that the population $\mathrm{N}_k$ decreases with increasing intensity $I$ and therefore the absorption coefficient also decreases. Instead of (11.96) we have to write

$$\mathrm{d}I = -\Delta N(I) \cdot I \cdot \sigma \cdot \mathrm{d}x . \tag{11.98}$$

The change $\mathrm{d}I$ of the intensity and therefore the absorbed power depends in a nonlinear way on the incident intensity. We can write the intensity dependent population of the absorbing level $\langle k|$ as a power series

$$N_k = N_{k0} + \frac{\mathrm{d}N_k}{\mathrm{d}I} I + \frac{1}{2}\frac{\mathrm{d}^2 N_k}{\mathrm{d}I^2} I^2 + \ldots \tag{11.99a}$$

and a corresponding relation for the upper level $\langle i|$, where $\mathrm{d}N_k/\mathrm{d}I < 0$ and $\mathrm{d}N_i/\mathrm{d}I > 0$.

For the population difference we obtain (Fig. 11.57b):

$$\Delta N(I) = \Delta N_0 + \frac{\mathrm{d}(\Delta N)}{\mathrm{d}I} I + \frac{1}{2}\frac{\mathrm{d}^2(\Delta N)}{\mathrm{d}I^2} I^2 + \ldots . \tag{11.99b}$$

Inserting this into (11.98) gives

$$\mathrm{d}I = -\left[\Delta N_0 \sigma I + \frac{\mathrm{d}}{\mathrm{d}I}(\Delta N) I^2 \sigma + \ldots\right] \mathrm{d}x . \tag{11.100}$$

The first term describes the linear absorption, the second term is quadratically dependent on $I$ and diminishes the absorption because $\mathrm{d}(\Delta N)/\mathrm{d}I < 0$.

The nonlinear absorption can be demonstrated by measuring the intensity of the laser-induced fluorescence $I_{\mathrm{fl}}(I_{\mathrm{L}})$ as a function of the laser intensity (Fig. 11.57c). One can see that at first $I_{\mathrm{fl}}$ increases linearly with $I_{\mathrm{L}}$, but for higher laser intensities the increase is less than linear, because the absorption coefficient decreases and therefore the relative absorption of the laser intensity $\mathrm{d}I_{\mathrm{L}}/I_{\mathrm{L}}$. For large laser intensities the fluorescence intensity approaches a constant value (saturation), which is limited by the rate of the relaxation processes refilling the absorbing level $\langle k|$.

This saturation of the absorption can be used for Doppler-free spectroscopy, as will be outlined in the next section.

### 11.5.9 Saturation Spectroscopy

We consider a sample of atoms or molecules in the gas phase with a Doppler-broadened absorption profile around the center frequency $\omega_0$. When a monochromatic laser beam with frequency $\omega$ passes in the $x$ direction through the sample, only those molecules that are Doppler-shifted into resonance with the laser frequency can absorb the laser photons. Since the Doppler shift is $\Delta\omega = kv_x$, these molecules must have velocity components $v_x$ that satisfy the relation

$$\omega = \omega_0(1 + kv_x) . \tag{11.101}$$

Because of saturation, the population $N_k$ in the absorbing level of these molecules decreases and $N_i$ increases accordingly. In the velocity distribution $N_k(v_x)$ of all molecules, a narrow dip is burnt while a corresponding peak appears in the distribution $N_i(v_x)$ of molecules in the upper level (Fig. 11.58a).

If the laser beam is reflected by a mirror back into the sample, its $\boldsymbol{k}$-vector is reversed and molecules with opposite velocity components interact with the reflected beam (Fig. 11.58b). Now for $\omega_{\mathrm{L}} \neq \omega_0$ two

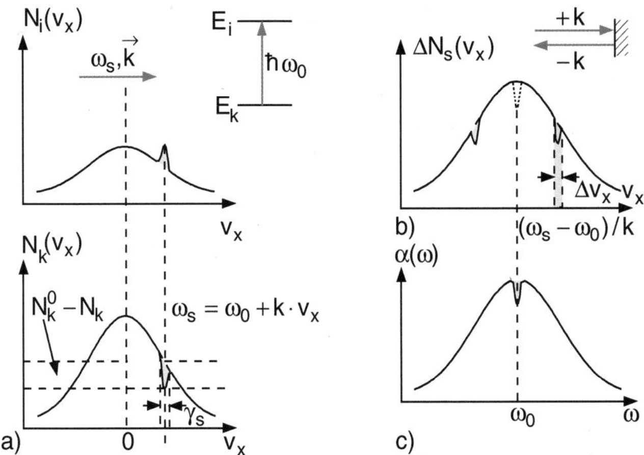


**Fig. 11.58.** (**a**) Velocity-selective saturation. (**b**) Dips in the population difference $\Delta N(v_x)$ for $\Omega_L \neq \Omega_0$ (**c**) Lamb-dip at $\Omega = \Omega_0$ of the Doppler-broadened profile of molecules in a standing laser wave

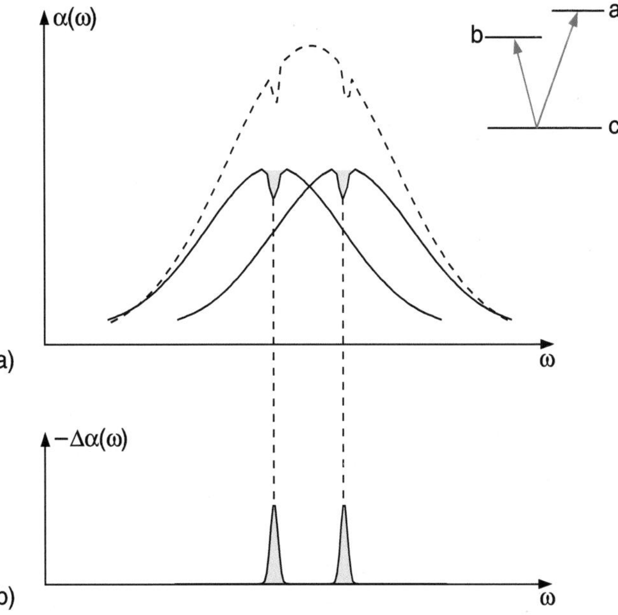


**Fig. 11.59.** Overlapping Doppler profiles of two close transitions with a clear separation of their Lamb dips

holes are burnt into the velocity distribution of the inversion $\Delta N(v_x)$, i.e., two different velocity classes interact with the two laser beams.

When the laser frequency $\omega_L$ is tuned over the Doppler-broadened absorption profile the two holes move towards each other when $\omega_L$ approaches the center frequency $\omega_0$ of the molecular transition. For $\omega_L = \omega_0$, only one velocity class in the interval $\mathrm{d}v_x$ around $v_x = 0$ interacts with both laser beams. The molecules experience the double intensity and therefore the population inversion $\Delta N$ will be more decreased for $\omega_L = \omega_0$ than for $\omega_L \neq \omega_0$. The absorption coefficient $\alpha(\omega)$ with a Doppler-broadened profile, therefore has a dip around the center frequency $\omega_0$ (Fig. 11.58c). This dip, caused by saturation of the population, is called a ***Lamb dip*** after *Willis Lamb* (Nobel Prize, 1955), who explained this saturation effect quantitatively.

The width of the Lamb dip equals the natural line width of the molecular transition, broadened by saturation and collisions. It is about two orders of magnitude smaller than the Doppler width.

This narrow Lamb dip is used for Doppler-free spectroscopy (called saturation or Lamb dip spectroscopy). Assume two transitions from a common level $\langle c|$ into two slightly split levels $\langle a|$ and $\langle b|$. If the splitting is smaller than the Doppler width, the two Doppler-broadened spectral lines cannot be resolved. However, the two Lamb dips of these transitions are well separated (Fig. 11.59).

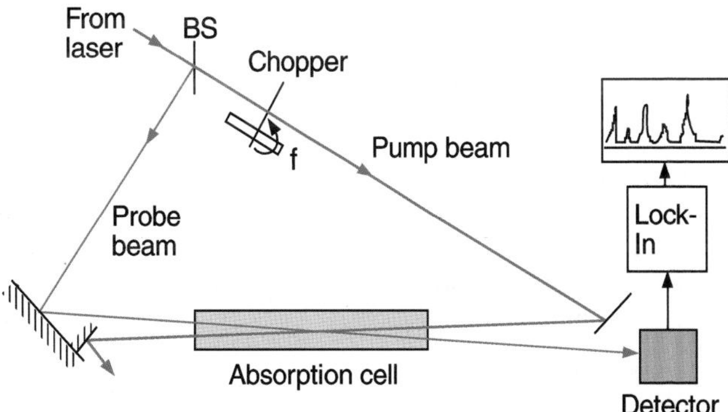


**Fig. 11.60.** Possible experimental setup for saturation spectroscopy

A possible experimental arrangement is shown in Fig. 11.60). The beam of a wavelength-tunable laser is split by beam splitter BS into a strong pump beam and a weak probe beam that pass into opposite directions through the sample cell. The detector measures the transmitted probe beam intensity as a function of the laser frequency $\omega_L$. Each time the laser frequency coincides with the center frequency of a molecular transition, a Lamb peak appears in the transmitted intensity, because the absorption exhibits a dip at this frequency.

If the absorbed laser power is measured instead of the transmitted intensity (for instance by monitoring

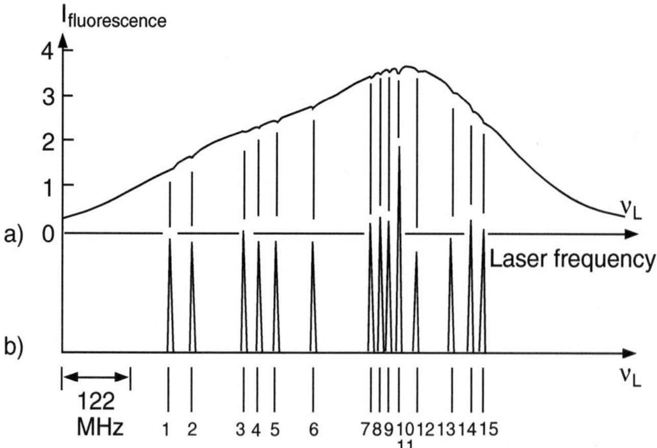


**Fig. 11.61a,b.** Saturation spectrum of the hyperfine components of a rotational transitions of the $I_2$ molecule. (**a**) Doppler-broadened overlapping absorption profiles with small Lamb dips. (**b**) Elimination of the Doppler-broadened background by chopping the pump beam

the laser-induced fluorescence) the Lamb dips can be immediately seen. This is demonstrated in Fig. 11.61a, which shows the unresolved Doppler-broadened profiles of 15 hyperfine components of a rotational transition in the visible absorption spectrum of the iodine molecule $I_2$ with well-resolved Lamb dips.

The Doppler-broadened background can be eliminated when the pump beam is periodically chopped. Now a lock-in detector measures the difference of the transmitted probe intensity with the pump beam on and off. The result is a Doppler-free spectrum (Fig. 11.61b) with a better signal-to-noise ratio.

Even with low-power lasers, saturation spectroscopy can be performed when the sample is placed inside the laser resonator where the intensity is much higher than outside the resonator (Fig. 11.62). The standing wave inside the resonator can be composed of a back and forth traveling wave. Therefore the conditions for producing a Lamb dip are automatically fulfilled. Either the laser-induced fluorescence from the sample molecules can be monitored, which exhibits the Lamb dips, or the laser intensity transmitted through one of the resonator mirrors is measured, which shows corresponding narrow peaks in the laser output power at the center frequencies of the absorption lines of the sample molecules. The laser wavelength can be tuned by changing continuously the length of the laser resonator. If the resonator length is additionally modulated while the laser wavelength is scanned, the first derivative of the Lamb dips or peaks is measured with a lock-in, tuned to the modulation frequency. In Fig. 11.63 such a derivative spectrum of the rotational transition $B^3\Pi_u\ (v' = 58, J' = 99) \leftarrow X^1\Sigma_g^+\ (v'' = 1, J'' = 98)$ in the $I_2$ molecule is shown with its 21 hyperfine components.

The attainable high spectral resolution is demonstrated in Fig. 11.64, which shows two hyperfine

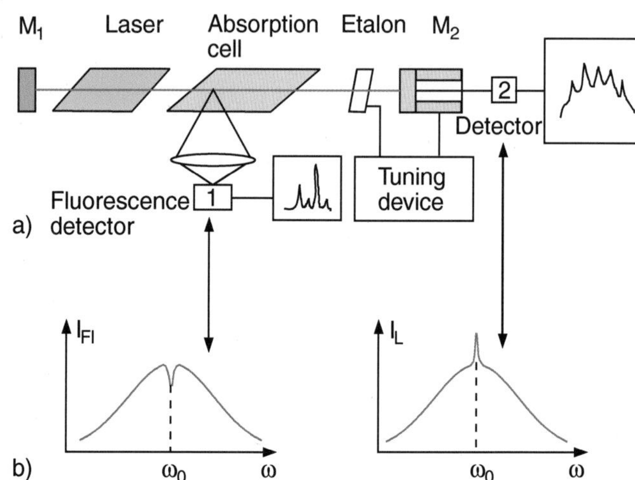


**Fig. 11.62.** Saturation spectroscopy inside the laser resonator

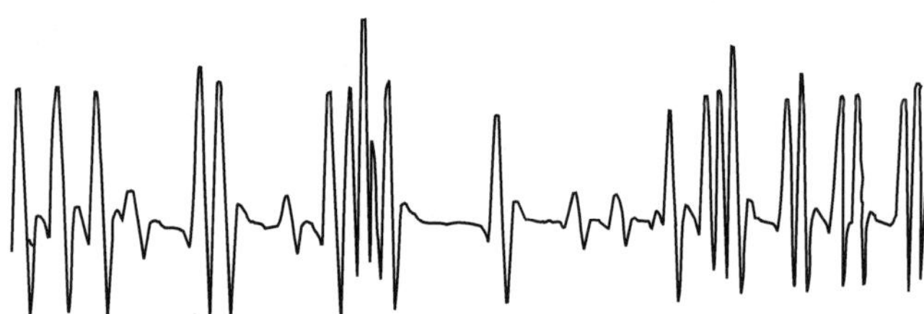

**Fig. 11.63.** Derivative saturation spectrum of the iodine transition $B^3\Pi_u\ (v'=58,\ J'=99) \leftarrow X^1\Sigma_g\ (v''=1,\ J''=98)$ with 21 hfs components

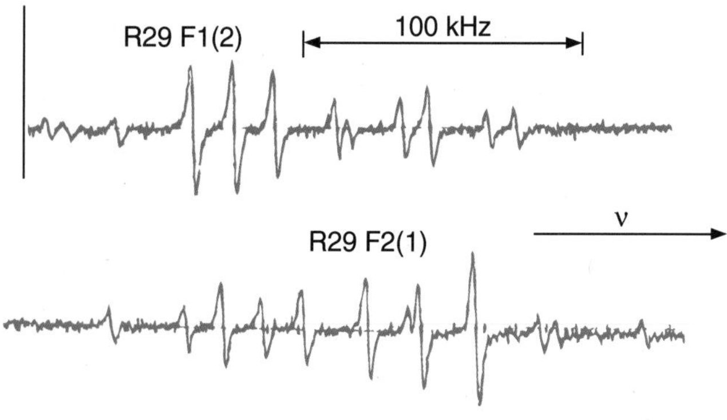


**Fig. 11.64.** Hyperfine and super hyperfine structures of a rotational-vibrational transition in $SF_6$ [11.33]

components F1 and F2 in the spectrum of $SF_6$, where each of the two components is further split into "super-hyperfine components," due to Coriolis effects and spin-rotation couplings in the rotating molecule. The spectral resolution is here better than 3 kHz [11.33]!

### 11.5.10 Doppler-Free Two-Photon Spectroscopy

For sufficiently strong laser intensities, an atom or molecule can simultaneously absorb two photons $\hbar\omega_1$ and $\hbar\omega_2$ from two lasers, or two photons $\hbar\omega_1$ from the same laser. Depending on the relative orientations of the two-photon spins, optical transitions with $\Delta L = 0$ or $\Delta L = \pm 2$ are induced. Two-photon transitions are weaker by several orders of magnitude than allowed one-photon transitions. Therefore, lasers with sufficiently high intensities are demanded for their observation. The probability of two-photon absorption is greatly enhanced if a molecular level $E_\text{m}$ is close to the energy $E_k + \hbar\omega_1$ or $E_k + \hbar\omega_2$ (Fig. 11.65).

For a two-photon transition from the initial lower level $\langle k|$ to the final level $\langle f|$ in a molecule at rest, the energy conservation demands

$$E_f - E_k = \hbar(\omega_1 + \omega_2)\,. \qquad (11.102)$$

When the molecule moves with velocity $v$, the frequency $\omega$ of the light wave is shifted in the molecular frame to $\omega' = \omega - \boldsymbol{k}\cdot\boldsymbol{v}$. The resonance condition (11.102) now becomes

$$E_f - E_k = \hbar(\omega_1 + \omega_2) - \hbar\boldsymbol{v}(\boldsymbol{k}_1 + \boldsymbol{k}_2)\,. \quad (11.103)$$

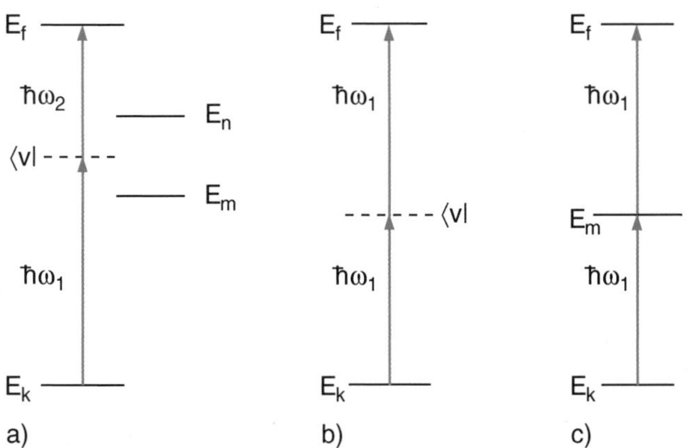


**Fig. 11.65a–c.** Two-photon absorption (**a**) with different photon energies, (**b**) with $\omega_2 = \omega_1$. (**c**) Resonant two photon transitions with a real intermediate level $E_m$

If the two photons come from two beams of the same laser traveling in opposite directions, we have the situation that $\omega_1 = \omega_2$ and $\boldsymbol{k}_1 = -\boldsymbol{k}_2$. This has the consequence that the last term in (11.103) containing the velocity of the molecule becomes zero.

> The two-photon absorption becomes, in this case, independent of the molecular velocity, meaning that all molecules within the velocity distribution contribute to the two-photon absorption, which is now Doppler-free.

The experimental arrangement is shown in Fig. 11.66. The sample is irradiated by the focussed beam of a tunable dye laser and by the focussed beam reflected by the spherical mirror M. The two-photon transition is monitored by the laser-induced fluorescence emitted from the upper level $E_f$ into intermediate levels $E_\text{m}$ that are connected to $\langle f|$ by allowed one-photon transitions. A Faraday-rotator acts as an optical diode and prevents the reflected laser beam from passing back into the laser, because this could cause instabilities to the laser.

When the laser frequency $\omega$ is tuned over the two-photon resonance, the signal consists of a narrow peak (produced by two photons with opposite $\boldsymbol{k}$ vectors) and a Doppler-broadened background, produced by two photons from the same beam with parallel $\boldsymbol{k}$ vectors (Fig. 11.67). The Doppler width is twice as large as that of a one-photon transition with frequency $\omega$. The probability that the two photons come from opposite

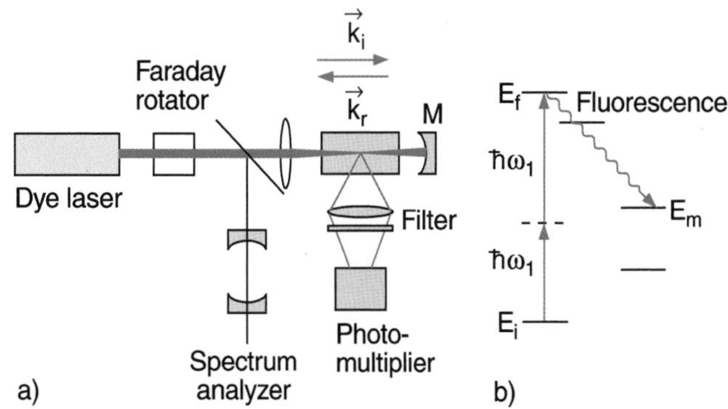


**Fig. 11.66.** Experimental arrangement for Doppler-free two-photon absorption with fluorescence detection

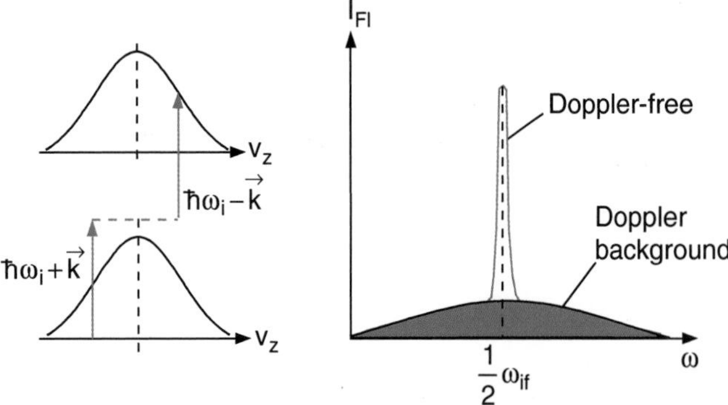


**Fig. 11.67.** Doppler-free line profile of two-photon absorption with Doppler-broadened background, which has here been exaggerated

beams is twice as large as that for two photons from the same beam. This implies that twice as many molecules contribute to the narrow peak than to the broad background. The narrow peak is therefore $2 \times \Delta\omega_D/\Delta\omega_n$ times as high as the background. Since the Doppler width $\Delta\omega_D$ is about two orders of magnitude larger than the natural line width $\Delta\omega_n$ the background can be neglected.

In Fig. 11.68 the Doppler-free spectrum of lead atoms on the transition $7p3P_0 \leftarrow 6p^2 3P_0$ at $\lambda =$ 450.4 nm illustrates the high spectral resolution allowing the measurement of the isotope shifts [11.34].

More extensive representation of two-photon spectroscopy can be found in [11.35].

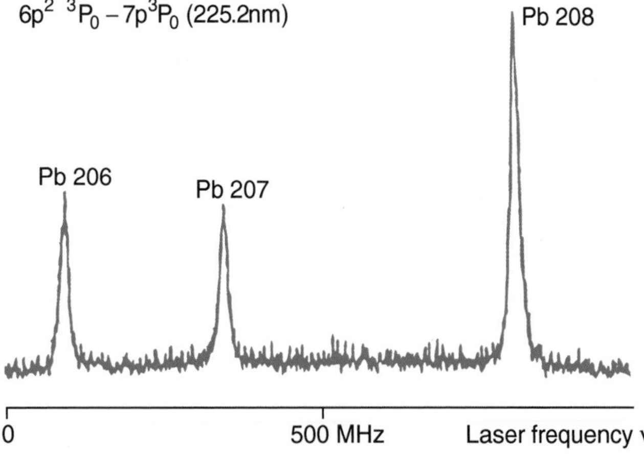


**Fig. 11.68.** Measurement of the isotope shift of the stable lead isotopes measured with Doppler-free two-photon spectroscopy around $\lambda_{exc} = 450$ nm and monitored via fluorescence [11.34]

# 11.6 Raman Spectroscopy

Raman spectroscopy has proved to be a powerful tool for investigating the vibrational structure of molecules in their electronic ground state. We will see that this method is often complementary to infrared spectroscopy. It has seen a great renaissance since lasers have been used as intense light sources, which have vastly increased its possibilities.

## 11.6.1 Basic Principles

Raman spectroscopy is based on the inelastic scattering of light by molecules and was first discovered by *Chandrasekhara Raman* in 1928 (Nobel Prize 1930). In the context of photons we can describe it as follows. A photon $\hbar\omega_0$ collides with a molecule in level $E_k$ (Fig. 11.69a), transfers part of its energy to the internal energy of the molecule, which is excited to the final level $E_f$, and the scattered photon $\hbar\omega_s$ has less energy, i.e., its frequency has decreased:

$$\hbar\omega_0 + M(E_k) \to M^*(E_f) + \hbar\omega_s \qquad (11.104)$$

$$\text{with} \quad \hbar(\omega_0 - \omega_s) = E_f - E_i > 0\,.$$

The energy difference $\Delta E = E_f - E_i$ may appear as vibrational, rotational or electronic energy of the molecule (Fig. 11.69b). The inelastic scattered radiation is called ***Stokes radiation***.

If the photon $\hbar\omega_0$ is scattered by an excited molecule, super-elastic scattering may occur, where the excitation energy is transferred to the scattered photon,

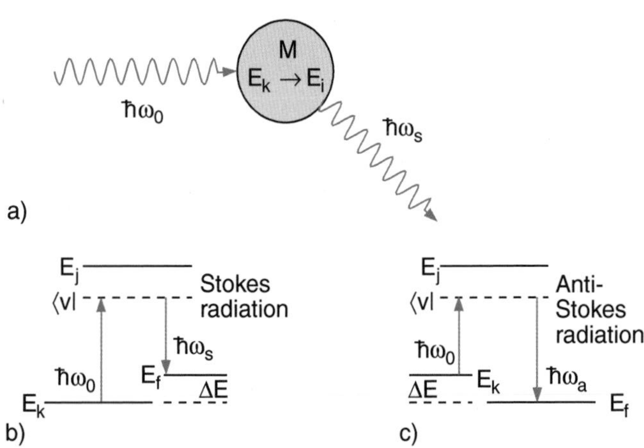


**Fig. 11.69a–c.** Raman effect. (**a**) Raman effect as inelastic photon scattering. Level scheme for the generation of (**b**) Stokes radiation and (**c**) of anti-Stokes radiation

which now has a higher energy than the incident photon (Fig. 11.69c). This super-elastically scattered radiation is called ***anti-Stokes radiation***.

In the energy level scheme of Fig. 11.69 the intermediate state $E_v = E_k + \hbar\omega_o$ during the scattering process is often formally described as a virtual state, which, however, is not a real stationary eigenstate of the molecule. If this virtual state coincides with one of the molecular eigenstates one speaks of the "***resonance Raman effect***".

In the spectrum of the scattered light, different frequencies may appear: the elastically scattered radiation, called Rayleigh-scattering, the inelastic Stokes contribution of the Raman scattering and the super-elastic anti-Stokes lines (Fig. 11.70).

The classical description of the Raman effect starts with the electric dipole moment

$$\boldsymbol{p}(E) = \boldsymbol{p}_0 + \tilde{\alpha}\boldsymbol{E}\,, \tag{11.105}$$

written as the sum of an eventual permanent moment $\boldsymbol{p}_0$ and a field-dependent contribution $\boldsymbol{p}_{\text{ind}} = \tilde{\alpha}\boldsymbol{E}$, where $\tilde{\alpha}$ is the electric polarizability, which is a tensor of rank two with components $\alpha_{ij}$.

The dipole moment and the polarizability can both depend on the nuclear displacements $q_n$ of the vibrating molecule. For small displacements from the equilibrium position we can expand the two quantities into the Taylor series

$$\boldsymbol{p}(q) = \boldsymbol{p}(0) + \sum_{n=1}^{Q}\left(\frac{\partial \boldsymbol{p}}{\partial q}\right)_0 q_n + \ldots \tag{11.106a}$$

$$\alpha_{ij}(q) = \alpha_{ij}(0) + \sum_{n=1}^{Q}\left(\frac{\partial \alpha_{ij}}{\partial q_n}\right)_0 q_n\,, \tag{11.106b}$$

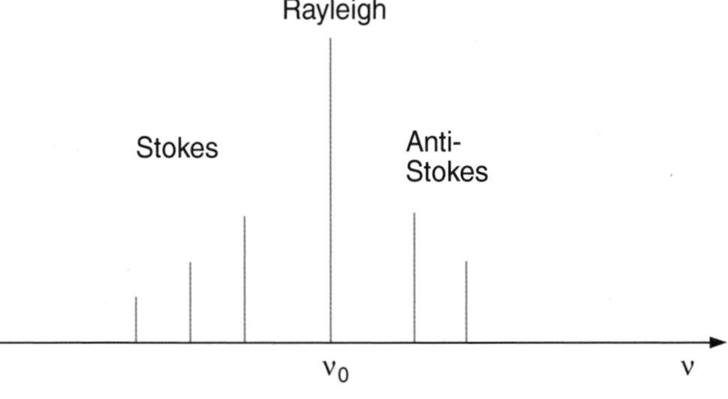


**Fig. 11.70.** Schematic spectrum of elastic Rayleigh scattering and Raman scattering with Stokes and anti-Stokes lines

where $Q = 3N - 6$ (or $3N - 5$ for linear molecules) is the number of normal vibrational modes and $\boldsymbol{p}(0)$ and $\alpha(0)$ represent dipole moment and polarizability at the equilibrium position $q = 0$.

If we describe the $n$th normal vibration with

$$q_n(t) = q_{n0}\cos(\omega_n t) \tag{11.107a}$$

and the electric field amplitude as

$$\boldsymbol{E}(t) = \boldsymbol{E}_0\cos\omega t\,. \tag{11.107b}$$

We can insert (11.106–11.107) into (11.105) and obtain for the oscillating dipole moment

$$\begin{aligned}\boldsymbol{p}(t) = \boldsymbol{p}_0 &+ \sum_{n=1}^{Q}\left(\frac{\partial \boldsymbol{p}}{\partial q_n}\right)_0 q_{n0}\cos(\omega_n t) \\ &+ \alpha_{ij}(0)E_0\cos\omega t \\ &+ \frac{1}{2}\boldsymbol{E}_0\sum_{n=1}^{Q}\left(\frac{\partial \alpha_{ij}}{\partial q_n}\right)_0 q_{n0}[\cos(\omega+\omega_n)t \\ &+ \cos(\omega-\omega_n)t]\,.\end{aligned} \tag{11.108}$$

The first term represents the permanent dipole moment of the molecule. The second term is responsible for the infrared spectrum, where the intensities of the transitions depend on the derivatives $\partial p/\partial q_n$ (see Sect. 9.5.1). The third term is responsible for the elastic Rayleigh scattering, while the last term describes the Raman scattering, where the intensities of the Raman lines depend on the derivatives $\partial\alpha/\partial q_n$ of the polarizability. In Fig. 11.71 the dependences $\partial p/\partial q$ and $\partial\alpha/\partial q$ are shown for the three normal vibrations of the linear $CO_2$ molecule. The change of the dipole moment with the vibrational displacements is only nonzero for the bending vibration $\nu_2$ and for the asymmetric stretch vibration $\nu_3$. These normal modes are therefore called *infrared active.* On the other hand the polarizability $\alpha$ changes for the symmetric stretch vibration $\nu_1$, which is called *Raman active*. This shows that infrared and Raman spectroscopy supplement each other.

There are many molecules with vibrational modes that are infrared as well as Raman active. However, there are always vibrations that are only either infrared or Raman active.

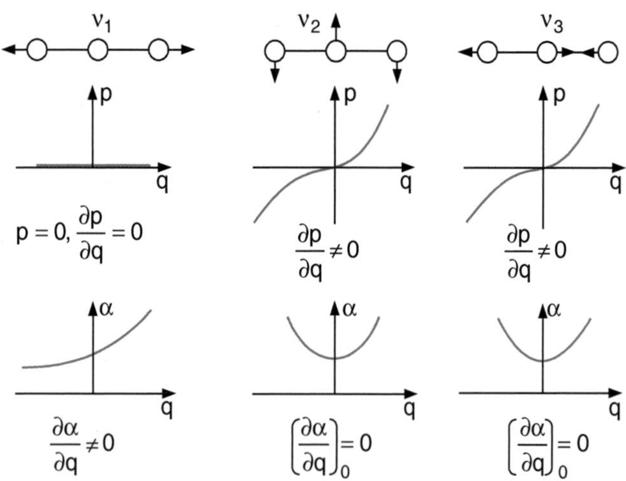


**Fig. 11.71.** Dependence $\frac{\partial p}{\partial q}$ of dipole moment and $\frac{\partial \alpha}{\partial q}$ of polarizability on the normal vibrations of the $CO_2$ molecule

### 11.6.2 Coherent Anti-Stokes Raman Spectroscopy

Even with strong pump beams the intensity of spontaneously scattered Raman light is often very weak. Here a new technique is useful, which is based on the coherent amplification of molecular vibrations by interaction with two incident laser waves. The frequencies $\omega_1$ and $\omega_2$ of the two lasers are chosen in such a way that $\omega_1 - \omega_2$ equals the vibrational frequency $\omega_{\text{vib}}$ of a Raman active normal vibration of the molecule. This technique is called coherent anti-Stokes Raman spectroscopy (CARS). Its principle is illustrated in Fig. 11.72. Instead of the spontaneously scattered Stokes radiation induced by the pump wave with $\omega_1$, the downward transition from the virtual level $\langle v_1|$ to the excited vibrational level $E_f$ is now caused by stimulated emission induced by the second laser with $\omega_2 = \omega_1 - E_f/\hbar$. This produces a much larger population in level $E_f$ than could be achieved with spontaneous Stokes radiation. Starting from level $E_f$ a second photon $\hbar\omega_1$ from laser $L_1$ induces the anti-Stokes radiation with frequency $\omega_a = 2\omega_1 - \omega_2$ with $\omega_2 = \omega_s$. The anti-Stokes wave is emitted as a coherent wave if the three waves are in phase. This can be achieved if the phase-matching condition

$$\boldsymbol{k}_a = 2\boldsymbol{k}_{L1} - \boldsymbol{k}_s \tag{11.109}$$

with $\vec{k}_s = \vec{k}_{L2}$ is fulfilled (Fig. 11.72b). This condition represents the conservation of momentum for the four participating photons. In gases, where the dispersion is small, all vectors have to be parallel, while in liquids or solids, where dispersion cannot be neglected, the phase-matching condition requires the angle $\alpha$ between the pump beam and the anti-Stokes wave. The two incident laser beams are also not collinear, but have different directions with an angle $\beta$ between them.

A possible experimental arrangement for CARS is shown in Fig. 11.73. The sample cell is placed inside the resonator of a single-mode argon ion laser, which delivers the pump beam with $\omega_1$. The output beam of a tunable dye laser is coupled collinearly to the beam from $L_1$ into the sample cell by the prism. The CARS

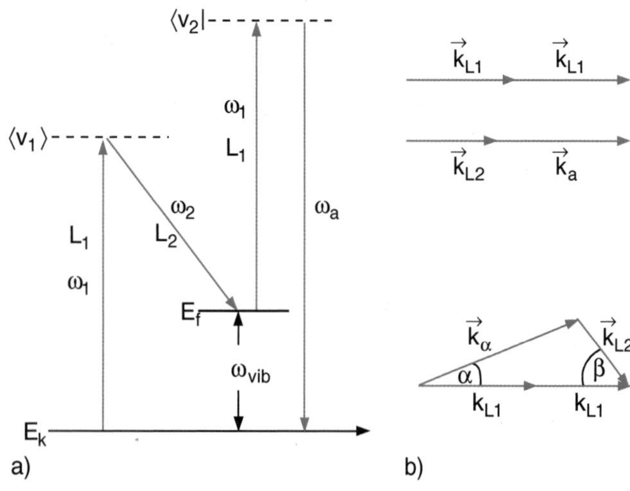


**Fig. 11.72a,b.** CARS spectroscopy. (**a**) Level scheme. (**b**) Wave vector conditions for phase matching in gases (*above*) and for liquids or solids with dispersion (*below*)

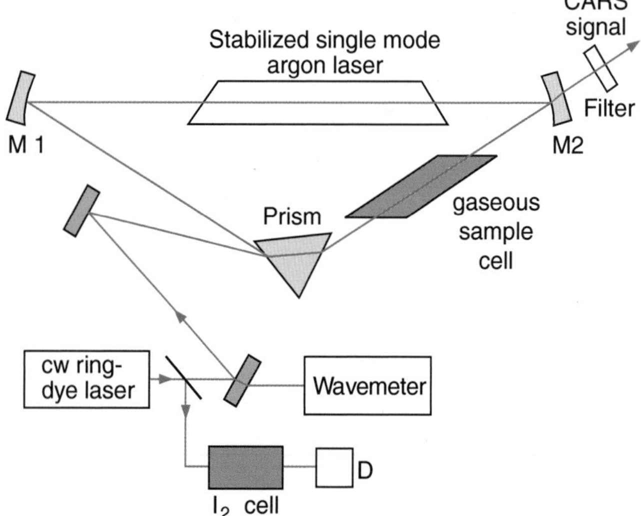


**Fig. 11.73.** Schematic arrangement of a cw CARS spectrometer with intracavity excitation of the sample [11.36]

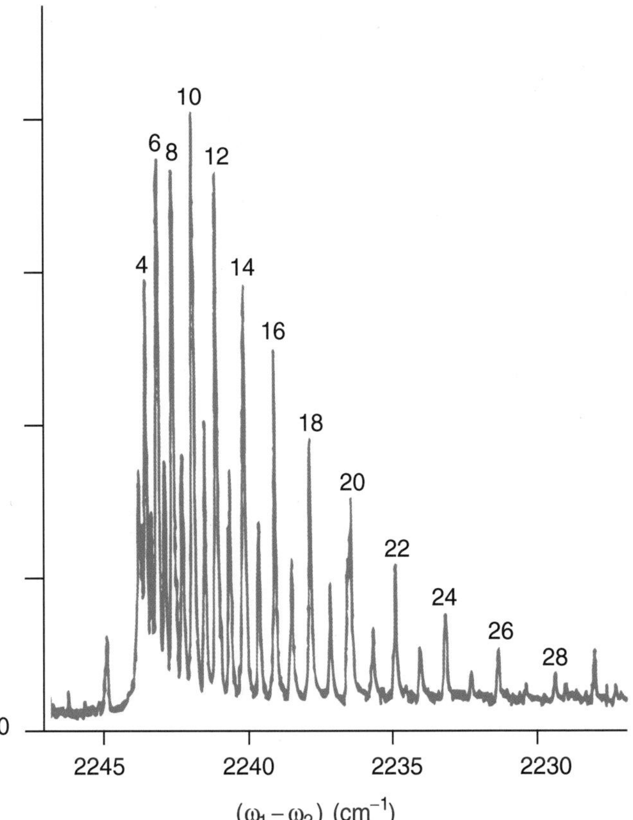


**Fig. 11.74.** Collinear beam CARS spectrum of $N_2$ ($\vartheta = 3$) in glow discharge at 5 mbar pressure. The discharge current is 80 mA. $Q$-lines are labeled by their $J$ numbers [11.37]

signal is emitted through the resonator mirror $M_2$ and a spectral filter suppresses all other frequencies.

CARS spectroscopy is finding more and more applications in molecular spectroscopy, in environmental sciences and also for the solution of technical problems. First of all it gives information about the vibrational-rotational structure of molecules for vibrational modes that are not accessible to infrared spectroscopy (Fig. 11.74). Secondly the population distribution $N(v, J)$ of vibrational-rotational levels can be inferred from the intensities of the CARS signals. This allows the determination of the temperature of a gas with high spatial resolution. One example is the spatial temperature distribution in flames, which is an indicator of the efficiency of chemical reactions in the flame. Its knowledge can serve for the optimization of the burning process and the minimization of the emission of pollutant gases.

The advantages of CARS can be summarized as follows:

- The intensity of the anti-Stokes signal is by far larger than in spontaneous Raman spectroscopy.
- The spectral resolution is only limited by the bandwidth of the two incident lasers. If these are single mode lasers the resolution is at least two orders of magnitude higher the for the Doppler-limited spontaneous Raman spectroscopy.
- Since the magnitude of the CARS signal is proportional to the product of the intensities of the two incident lasers the spatial resolution can be optimized by focusing the two laser beams into the sample.
- The anti-Stokes wave forms a highly collimated beam and the detector can therefore be far away from the interaction region where the sample is placed. This diminishes spontaneous background, which is particularly important if spectroscopy in hot gases (e.g., furnaces) is performed, where the strong continuous black body radiation overlaps the signal. The intensity of this background radiation decreases with the square of the distance from the sample contrary to the collimated CARS beam.

The disadvantage of CARS is the necessarily large experimental expenditure with two lasers, which must be single mode, if the spectral resolution should be in the sub-Doppler region.

With pulsed lasers time dependent processes in molecules and their influence on the change of level populations can be studied by CARS. One example is the vibrational distribution and its change with time in biological molecules after excitation by photons, where the pathway of energy transfer and isomerization of the initially excited species can be followed with pico- to femtosecond resolution. The investigation of these processes is, for instance, important for the understanding of photosynthesis or of the visual process in our eyes and of the different steps of energy transfer from the photoexcitation of the antenna molecules in retina cells to the electrical pulse transmitted to the brain [11.38].

## 11.7 Spectroscopy with Synchrotron Radiation

Most processes induced by visible light involve excitation of valence electrons of atoms or molecules. In order to excite inner shell electrons into free places in outer shells, ultraviolet or even X-ray photons are necessary. This inner shell excitation of molecules often results in fragmentation of the molecules. The investigation of the different fragments and their energies as a function of the excitation energy gives valuable information on the coupling between inner shell electrons (which generally do not contribute to the molecular binding) and electrons in binding orbitals.

In atomic physics the calculations of electron energies in inner shells require relativistic corrections, and correlation effects are much more dominant than in valence shells.

For a long time, intense radiation sources were missing in this spectral range. With the development of synchrotrons and storage rings this deficiency has been removed and the number of experiments in atomic and molecular physics using synchrotron radiation sources has vastly increased.

The basic physics of synchrotron radiation can be explained by relativistic electrodynamics. Every accelerated charged particle emits radiation (see Sect. 7.5.1). In a synchrotron, electrons are accelerated to energies of several GeV ($10^9$ eV) and circulate with velocities close to the velocity of light ($v \approx c$) in an evacuated ring stabilized by a magnetic field. The intensity and the spectral distribution of the synchrotron radiation

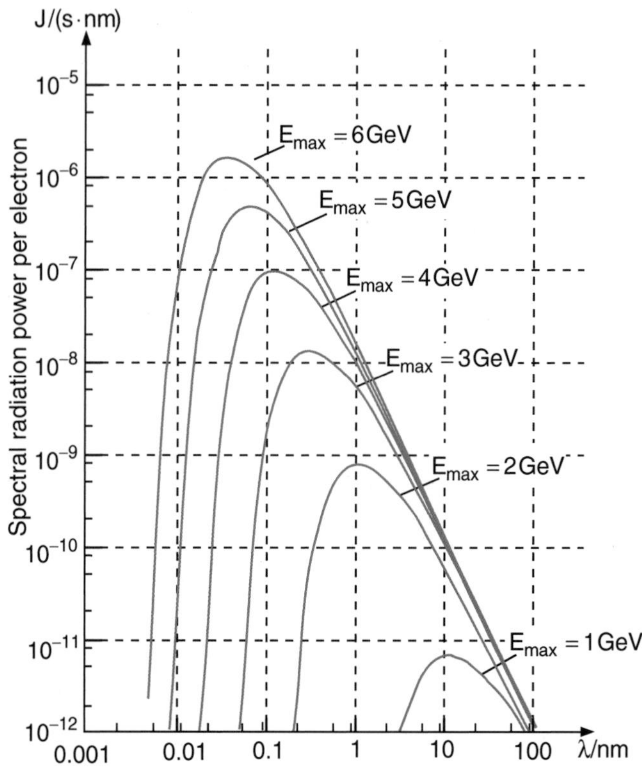


**Fig. 11.75.** Spectral distribution of synchrotron radiation per electron for different electron energies in a ring with $R = 31.7$ m [11.39]

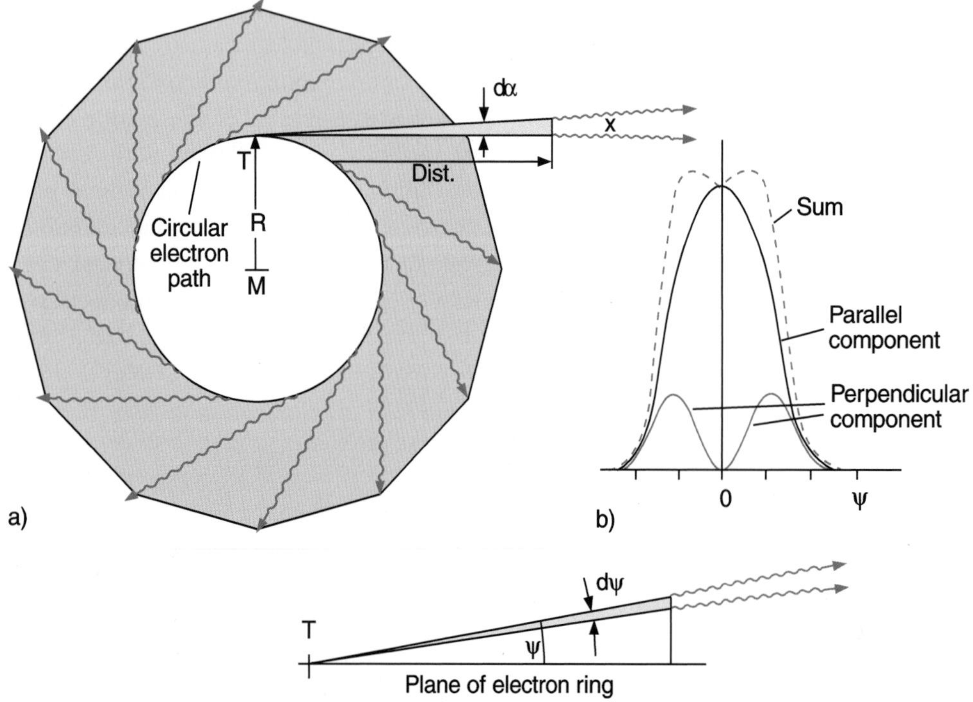


**Fig. 11.76a,b.** Schematic illustration of emission and polarization characteristics of synchrotron radiation. (**a**) View from above the plane of the electron path. (**b**) Intensities of the polarization components parallel and perpendicular to this plane as a function of the emission angle $\psi$ against this plane

depend on the number of circulating electrons, their energy and the radius of curvature of their path. In Fig. 11.75 the spectral distributions for the synchrotron radiation of a single electron are given for different energies of the electron and for a radius $R = 31.7\,\text{m}$ (which is the radius of the storage ring DORIS in Hamburg). This demonstrates that, e.g., for electron energies of 4 GeV the maximum of the spectral distribution lies at $\lambda = 0.1\,\text{nm}$, i.e., in the X-ray region, but extends from 0.02–100 nm. The synchrotron radiation covers the whole region from the extreme UV into the near UV and even up to the visible range.

Typical currents in storage rings are about 1A, which corresponds to $6 \times 10^{18}$ electrons passing through a cross section of the ring per second.

At these highly relativistic energies the synchrotron radiation is emitted from any point along the ring into a small solid angle around the tangent to the ring (Fig. 11.76). It is concentrated in the plane of the ring and is mainly polarized parallel to this plane, while a weaker component emitted under a small angle $\psi$ against the plane, is polarized perpendicular to the plane.

For most experiments the broad spectral continuum of the synchrotron radiation has to be dispersed by a monochromator. The narrow beam of the synchrotron radiation is focused by a torodial mirror onto the entrance lit of the monochromator (Fig. 11.77). A spherical grating (Rowland grating) acts as a wavelength disperser and focusses the desired wavelength of the diffracted radiation onto the exit slit. Turning the grating tunes the wavelength that is imaged onto the exit slit. Most experiments (e.g., absorption or excitation spectroscopy) are performed with this monochromized radiation behind the primary monochromator.

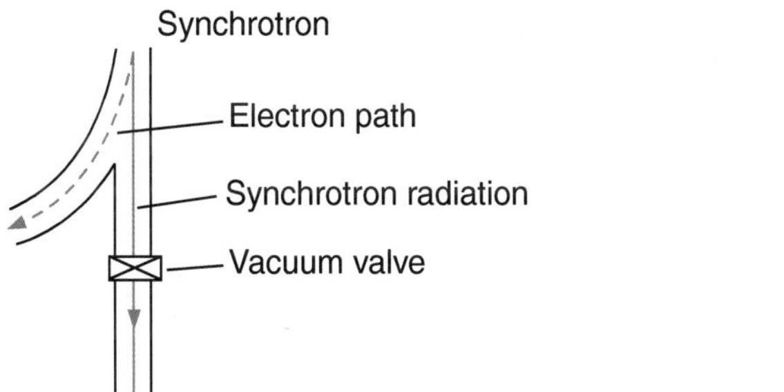

**Fig. 11.77.** Experimental setup for spectroscopy with monochromatic synchrotron radiation with tunable wavelength for absorption and fluorescence spectroscopy

## 11.8 Electron Spectroscopy

Detailed investigations of collisions between electrons and atoms provide important information about the energy levels of atoms and the basic mechanisms in gas discharges or in stellar atmospheres where such collisions play a major role. In fact, one of the first experiments to prove the quantum-theoretical model of discrete energy eigenstates of atoms was the Franck–Hertz experiment (Sect. 3.4.4). Studies of collisions between electrons and molecules serve to measure the electron distribution in molecules and to prove orbital models of theoretical chemistry. In particular, exchange effects and the correlation between electrons can be tested when spin-polarized electrons are used as collision partners. Furthermore polarization effects in the electron shell of atoms and molecules due to the interaction between the incident electron and the atomic electrons can be determined by such experiments.

### 11.8.1 Experiments on Electron Scattering

When an electron collides with an atom, it can be elastically scattered or it can loose energy by inelastic collisions or through electron impact ionization.

In elastic collisions only the direction of the electron momentum $p$ is changed, its amount $|p|$ remains constant. The momentum transfer depends on the interaction potential between electron and atom (see

Sect. 2.8). If the impact parameter $b$ is larger than the radius of the atomic electron shell, the interaction potential is that between an induced dipole moment (due to the polarization of the electron cloud) and a negative charge (Fig. 11.78) and is therefore proportional to $R^{-4}$ (Sect. 9.4). For smaller impact parameters, the interaction potential is more complicated and depends on the radial electron distribution in the atom.

For inelastic collisions, atoms or molecules are excited into higher energy levels. This excitation can be determined by measuring the energy loss of the colliding electron or by detecting the fluorescence emitted from the excited levels (Franck–Hertz experiment). For sufficiently high energies of the incident electrons, single or even double ionization can occur:

$$\begin{aligned} e^- + A &\rightarrow A^+ + 2e^- \,, \\ e^- + A &\rightarrow A^{++} + 3e^- \,. \end{aligned} \tag{11.110}$$

The most detailed information can be obtained when the highest possible energy resolution is achieved. This demands, that all incident electrons have the same energy.

Since electrons emitted from a hot cathode have a thermal velocity distribution they have to be monochromized. This can be performed by a cylindrical electrostatic energy selector (Sect. 2.6.3).

All electrons passing through a slit behind the energy selector (Fig. 11.79), where the two condensor plates have radii of curvatures $R_1$ and $R_2$ have a kinetic energy (see Sect. 2.6.3 and problem 2.15)

$$E^0_{\text{kin}} = \frac{1}{2} eU/\ln(R_1/R_2) \,, \tag{11.111}$$

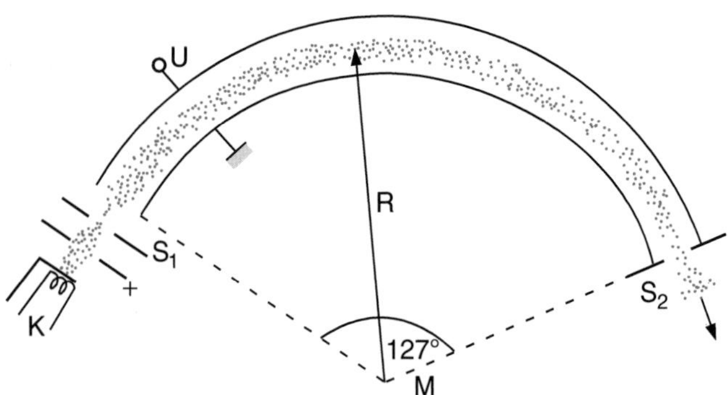


**Fig. 11.79.** Monochromizing electrons, emitted from a cathode K, by a 127° electrostatic cylinder condensator

which can be varied with the voltage $U$ between the two electrodes of the cylindrical selector.

The experimental arrangement for electron scattering experiments is illustrated in Fig. 11.80. The fluorescence, emitted from the excited atoms, which cross the electron beam in an atomic beam perpendicular to the scattering plane, is monitored by a photomultiplier. The monochromized incident electrons, scattered by an angle $\vartheta_1$, enter the entrance slit of a second energy selector and are detected by $D_1$. Measuring the rate of electrons with a defined energy loss

$$\Delta E = E^0_{\text{kin}} - E_1 \,,$$

and scattered by the angle $\vartheta_1$ as a function of the kinetic energy of the incident electrons, yields the differential cross section for inelastic collisions that have lead to

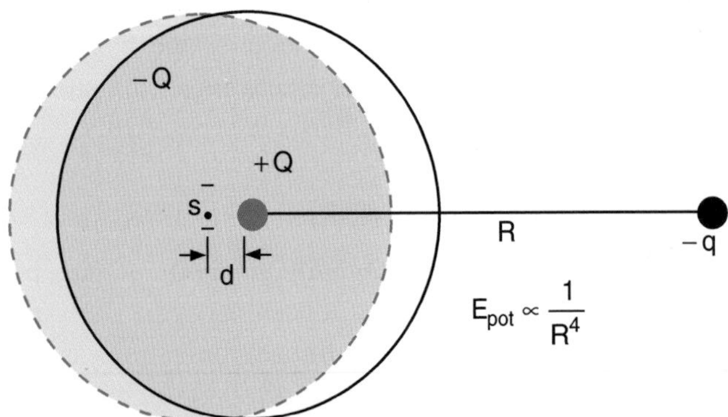


**Fig. 11.78.** Shift of negative electron cloud by the electric field of a negative charge $-q$, generating the induced dipole moment $\boldsymbol{p}^{\text{ind}} = (Q \cdot d)\hat{R}$

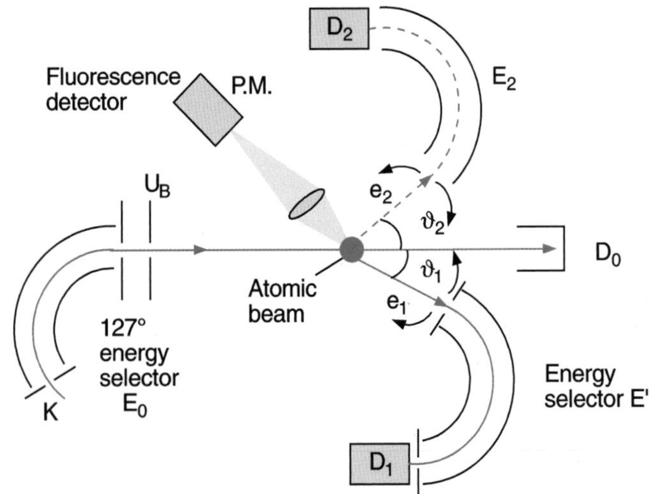


**Fig. 11.80.** Apparatus for measuring inelastic and ionizing collisions between electrons and atoms with coincidence detection of scattered and ionized electrons

the excitation of a defined energy level in the atom. Varying the voltage $U_1$ at the second energy analyzer allows one to measure elastic collisions $\left(E_1 = E^0_{\text{kin}}\right)$ as well as inelastic collisions with any energy loss $\Delta E$.

For ionizing collisions

$$e^- + A \rightarrow A^+ + 2e^- ,$$

a third energy selector is installed that monitors the second electron scattered by the angle $\vartheta_2$ and its kinetic energy $E^{(2)}_{\text{kin}}$. The two electrons are measured in coincidence. This means that both detectors give an output signal only if the two signals appear at the same time, i.e., in a time interval $\Delta t$ around a time $t_0$. The energy $E_2$ of the second electron is determined by the ionization energy $E_{\text{ion}}$ of the atom and by the energies $E_0$ of the incident electron and $E_1$ of the other scattered electron. Energy conservation demands that

$$E_0 = E_{\text{ion}} + E_1 + E_2 . \qquad (11.112)$$

Such coincidence experiments with energy selection and angular resolution allow the determination of a threefold differential cross section $\mathrm{d}^3\sigma/(\mathrm{d}E_1\mathrm{d}\Omega_1\mathrm{d}\Omega_2)$. They are experimentally very difficult but give, on the other hand, the most detailed information on the ionization process, because they show at which impact parameter the ionization probability becomes maximum and what influence the electrostatic interaction between the two electrons after the collision has on the angular distribution of the two electrons. For illustration, the measured angular distributions $N_1(E_1, \vartheta_1)$ and $N_2(E_2, \vartheta_2)$ of the two electrons for the electron impact ionization of helium atoms is shown in Fig. 11.81. The two electrons are detected in nearly opposite directions (partly due to the post-collision Coulomb interaction). The angular width of the distributions give information on the momentum distribution of the atomic electron before the collision.

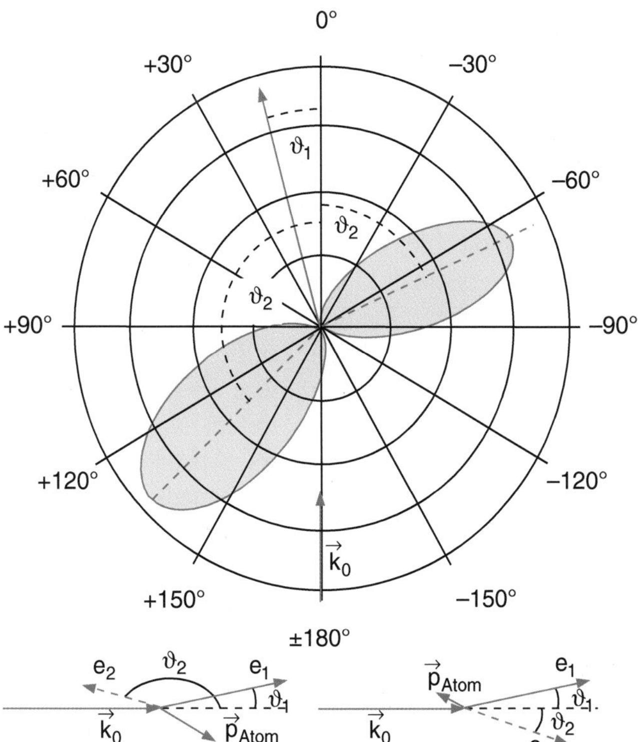


**Fig. 11.81.** Angular distribution $N(\vartheta_2)$ of electrons in the process $e^- + \text{He} \rightarrow \text{He}^+ + e_1^- + e_2^-$ with $E_0 = 100\,\text{eV}$, $E_1 = 70\text{eV}$, $E_2 = 5.4\text{eV}$, $\vartheta_1 = 15°$ [11.40]

### 11.8.2 Photoelectron Spectroscopy

When atoms or molecules are irradiated with monochromatic light of frequency $\nu$, a photoelectron can be emitted if the photon energy $h\nu$ is larger than the ionization energy

$$h\nu + \text{M} \rightarrow \text{M}^+ + e^-(E_{\text{kin}}) \qquad (11.113)$$

with kinetic energy

$$E_{\text{kin}} = h\nu - (E_{\text{ion}} + E(\text{M}^{+*})) . \qquad (11.114)$$

The ion $\text{M}^+$ can be in its ground state or in bound excited states $\text{M}^{+*}(E_i)$ (Fig. 11.82). Measuring the kinetic energy of the photoelectron produced by the known photon energy $h\nu$ allows the determination of discrete energy levels of the ion $\text{M}^+$. The rate of photoelectrons with kinetic energy $E_{\text{kin}}$ gives information on the probability of exciting these levels and on its photon energy dependence.

As an example of photoelectron spectroscopy of solid metal surfaces, a schematic experimental setup is shown in Fig. 11.83. The photoelectrons, emitted under the angle $\vartheta$ against the surface normal, are detected after their energy analysis by a cylindrical electrostatic energy analyzer. Such measurements give the energy distribution of the electrons in the metal before the photon releases them.

In Fig. 11.84 the photoelectron spectrum obtained after photoionization of $CS_2$ molecules by the He

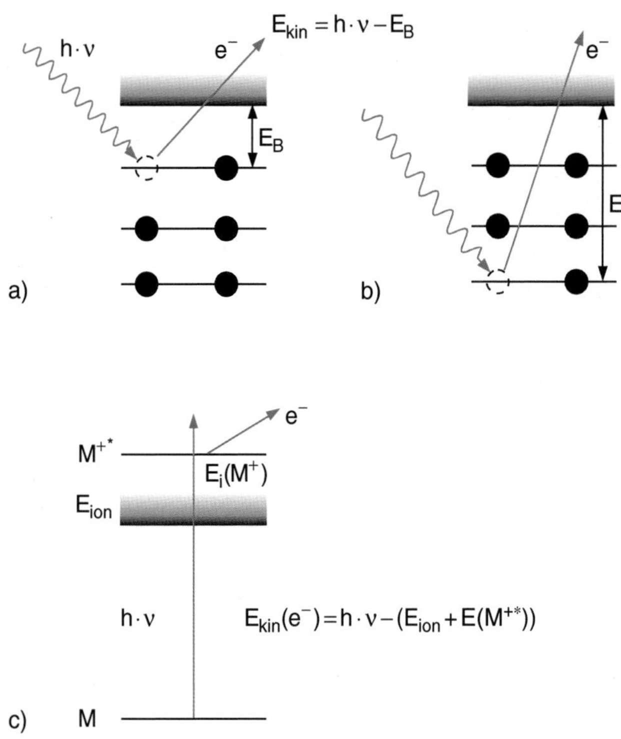


**Fig. 11.82a–c.** Photoionisation (**a**) of an electron in the valence shell of an atom (**b**) of an inner shell electron with X-rays (**c**) of a molecule, where the molecular ion remains in an excited state

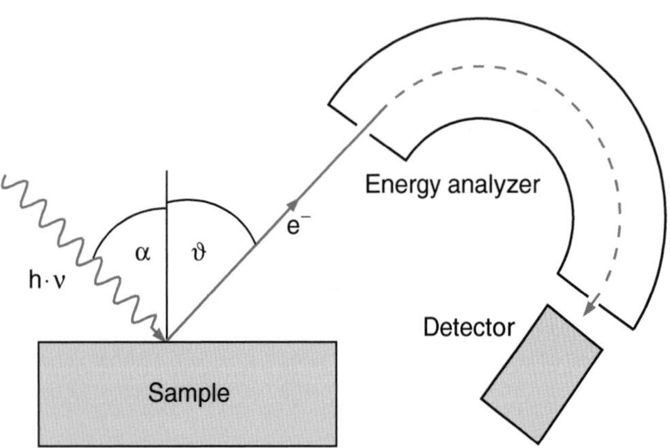


**Fig. 11.83.** Experimental arrangement for photoelectron spectroscopy of solids

resonance line at $\lambda = 50.8\,\text{nm}$ is shown. It illustrates the excitation of a progression of the bending vibration $\nu_2$ in the $CS_2^+$ ion. From the energy loss $\Delta E = h\nu - E(v_2)$ of the electrons, the vibrational energies in the molecular ion can be inferred. The energy distribution $N_{pe}(\Delta E)$ also gives information on the transition

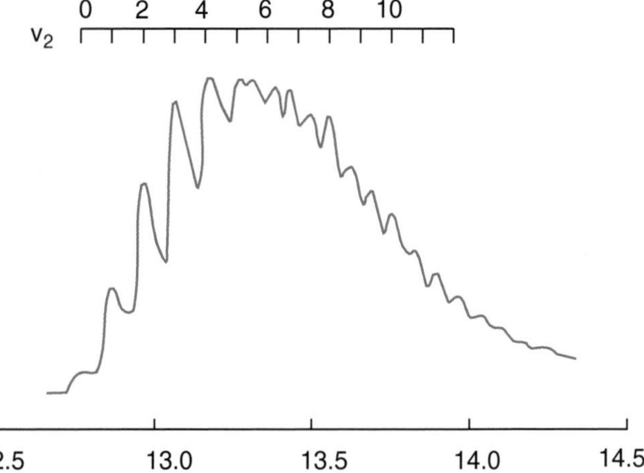


**Fig. 11.84.** Photo electron spectrum obtained after the photoionization of $CS_2$ molecules by photons from a helium resonance lamp at $\lambda = 50.8\,\text{nm}$ [11.41]

probability for populating the different vibrational levels $|v_2\rangle$ in the process.

$$CS_2 + h\nu \rightarrow CS_2^+(v_2) + e^-(E_{kin}) .$$

For electron spectroscopy of atoms or molecules in the gas phase, collimated molecular beams that are crossed perpendicularly by the photon beam are generally used. The photon energy from a helium discharge lamp that emits a spectral continuum is selected using a concave grating. In Fig. 11.85 the arrangement is shown for measuring threshold photoelectrons of atoms and molecules in a collimated beam. These electrons, with very small kinetic energies are produced by photons with energy $h\nu$, just at the ionization threshold.

Higher energy resolution can be achieved if lasers are used for photoionization. Since cw lasers in the far UV for one-photon ionization are not available (unless frequency doubling of UV-lasers is used), a two-step excitation is utilized, where an intermediate atomic or molecular level is excited by the first laser and then ionization is achieved by a second laser. In Fig. 11.86 a modern apparatus for such photoelectron experiments with two-step excitation is shown [11.42]. The level scheme explains the situation for potassium atoms. The slow photoelectrons can attach to other molecules with a large electron affinity in the crossing volume of atomic and molecular beams and the resulting negative ions $M^-$ are selected through a quadrupole mass spectrometer and are imaged onto the detector.

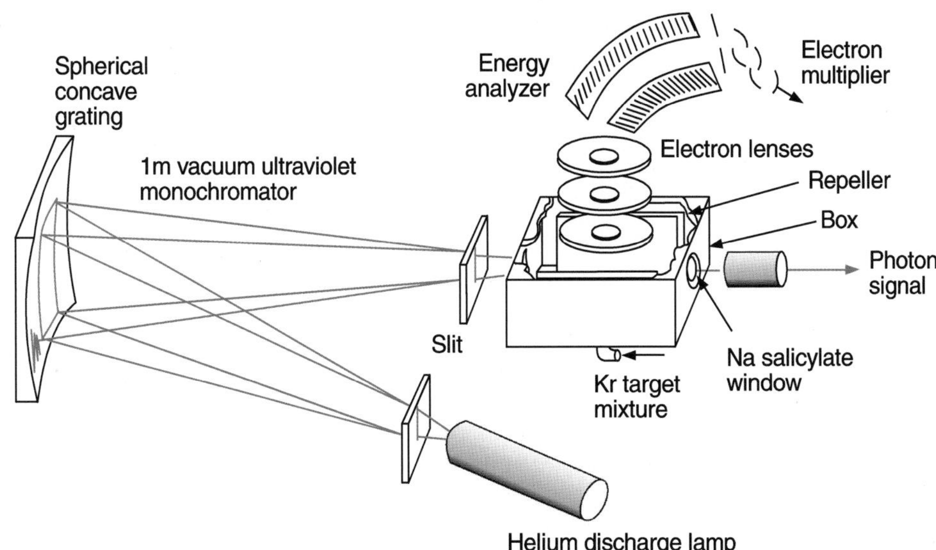


**Fig. 11.85.** Photoelectron spectroscopy in a molecular beam with monochromized UV radiation and energy analyzer for the electrons. The atomic beam is vertical. The nozzle is at the lower part of the vacuum chamber

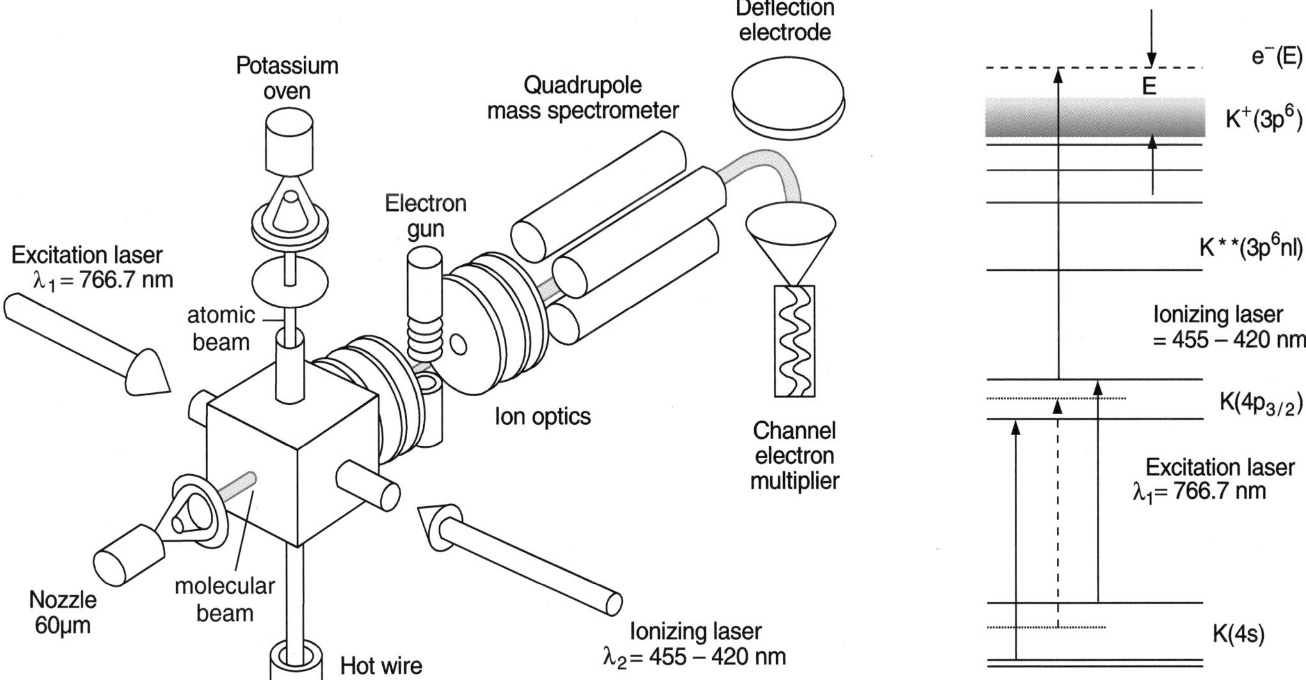


**Fig. 11.86.** Stepwise ionization of K atoms in an atomic potassium beam by two lasers and mass selective detection of the anions $M^-$ produced by detachment of the photoelectron to molecules M in a crossed beam arrangement [11.42]

### 11.8.3 ZEKE Spectroscopy

Over the last years, a modification of photoelectron spectroscopy has been developed, where a tunable laser excites levels in the molecular ion, but only photoelectrons with nearly zero kinetic energy are selected for detection [11.43]. If the photoexcitation is performed in a collimated molecular beam the slow photoelectrons can be collected by a weak electric field and imaged onto the electron detector. This field is only switched on with a delay time $\Delta t$ after the pulsed excitation at $t = 0$. Therefore the levels of the molecular ion are excited under field-free conditions and do not suffer a Stark shift. The fast photoelectrons have already left

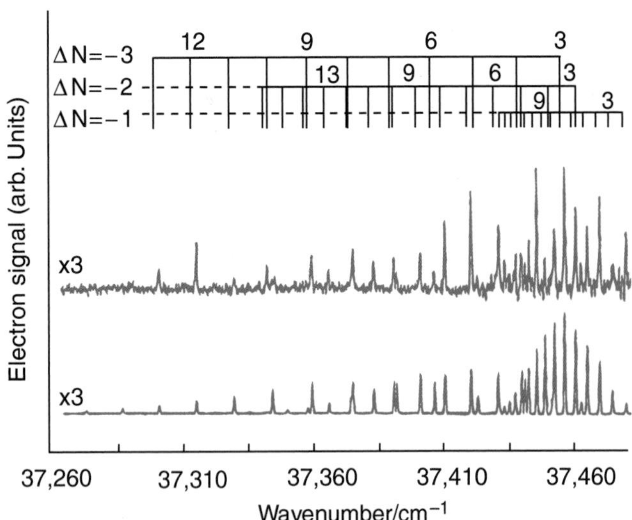


**Fig. 11.87.** ZEKE spectrum of $ND_4$ with rotational resolution. *Upper trace*: experimental spectrum. *Lower trace*: simulation [11.44] (Merkt, J.C.P. 106, 1 1997)

the excitation region during the delay time $\Delta t$, but the slow electrons have not moved far and can be collected with high efficiency. The energy resolution is mainly limited by the spectral bandwidth of the pulsed laser.

In Fig. 11.87 the ZEKE spectrum of $ND_4$ is shown as a function of the wavenumber of the exciting laser. The peaks correspond to photoexcitations starting from different rotational levels in the ground state of the neutral $ND_4$ and reaching the level $v' = 0\ J' = 0$ in the $ND_4^+$ ion electronic ground state.

## 11.9 Measurements of Magnetic and Electric Moments in Atoms and Molecules

Many molecules posses a magnetic dipole moment $\boldsymbol{p}_\mathrm{m}$ due to the orbital angular momenta or the spins of their electrons or due to nuclear spins. In an external magnetic field $\boldsymbol{B}$ they experience a torque

$$\boldsymbol{D} = \boldsymbol{p}_\mathrm{m} \times \boldsymbol{B} \tag{11.115}$$

which tries to orient the molecules in such a way, that $\boldsymbol{p}_\mathrm{m}$ is parallel to $\boldsymbol{B}$, because then the potential energy

$$W_\mathrm{pot} = -\boldsymbol{p}_\mathrm{m} \cdot \boldsymbol{B} \tag{11.116}$$

has a minimum. The mean macroscopic magnetization of a sample volume with N molecules is

$$<\mathbf{M}> = \Sigma \mathbf{p_m} = \mathrm{N} \cdot <\cdot \mathrm{p_m}> = \mathrm{N} \cdot \eta \cdot \mathrm{p_m} \tag{11.117}$$

where the average magnetic moment $<\mathrm{p_m}> = \eta \cdot \bar{\mathrm{p}}_\mathrm{m}$ is smaller than $\mathrm{p_m} = |\mathbf{p_m}|$ because of the following reason:

At temperatures $T > 0$ the orientation of the molecular magnetic moments is partly destroyed by the thermal motion of the molecules which try to establish thermal equilibrium, where all orientation are equally probable. The mean magnetisation of N molecules in the gas phase is proportional to the ratio

$$\langle M_\mathrm{m} \rangle \propto \frac{\langle W_\mathrm{pot} \rangle}{\frac{3}{2}kT} = \frac{\vec{p}_\mathrm{m} \cdot \vec{B}}{\frac{3}{2}kT} \tag{11.118}$$

of mean magnetic to mean kinetic energy of the molecules.

For a rotating molecule the situation is more complex. In the molecular frame the magnetic dipole moment has a well defined direction determined by the total electronic angular momentum. For the free rotating molecule the vector $\boldsymbol{p}_\mathrm{m}$, precesses around the space fixed rotational axis, defined by the total angular momentum $J$. Only the average moment $\overline{p}_\mathrm{m} = \boldsymbol{p}_\mathrm{m} \cdot \boldsymbol{J}$ which equals the projection of $\boldsymbol{p}_\mathrm{m}$ onto $\boldsymbol{J}$ is observed (Fig. 11.88). In an external magnetic field the angular momentum $\vec{J}$ and with if the projection $\bar{p}_\mathrm{m}$ precesses around the field axis and therefore the average magnetic potential energy of the molecule is given by

$$\langle W_\mathrm{pot} \rangle = -\frac{(\boldsymbol{p}_\mathrm{m} \cdot \boldsymbol{J}) \cdot (\boldsymbol{J} \cdot \boldsymbol{B})}{\boldsymbol{J}^2} . \tag{11.119}$$

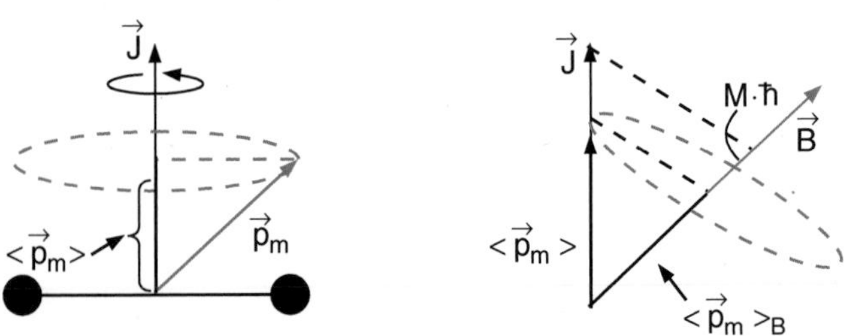


**Fig. 11.88.** Average magnetic dipole moment in a rotating molecule and its projection onto the direction of an external magnetic field

If the centres of the nuclear charges and the electron charges do not coincide, the molecule possesses an electric dipole moment $\boldsymbol{p}_{el}$. Examples are the molecules HCl. $H_2O$ of NaCl, which have either ionic binding or a strong electric polarisability, which leads to a charge transfer and a separation of the two centres of charge.

In an external homogeneous electric field $\boldsymbol{E}$ a torque

$$\boldsymbol{D} = \boldsymbol{p}_{el} \times \boldsymbol{E} \qquad (11.120)$$

acts on these molecules resulting in a potential energy

$$W_{pot} = -\boldsymbol{p}_{el} \cdot \boldsymbol{E} \,. \qquad (11.121)$$

In inhomogeneous magnetic or electric fields additional forces

$$\boldsymbol{F}_m = \boldsymbol{p}_m \cdot \textbf{grad}\, \boldsymbol{B} \quad \text{and} \quad \boldsymbol{F}_{el} = \boldsymbol{p}_{el} \cdot \textbf{grad}\, \boldsymbol{E} \qquad (11.122)$$

are present, which try to pull the molecules into the direction of either strong or weak fields, depending on the Zeeman- or Stark levels $(J, M)$ of the molecule.

These magnetic or electric moments of molecules have great importance for scientific and technical applications. Examples are the orientation of molecules in liquid crystals which influences the optical transmission or reflection of these devices, the nuclear magnetic resonance spectroscopy and its application in biology and medicine, or the adiabatic demagnetisation used for reaching extremely low temperatures. It is therefore of great interest to study these moments in more detail.

### 11.9.1 The Rabi-Method of Radio-Frequency Spectroscopy

*Isidor Isaac Rabi* (1898–1988) developed a molecular beam method for the precision measurements of magnetic or electric moments or of hyperfine splittings in atoms or molecules [11.45]. He received for this ingenious technique the Nobel prize in 1944. The principle of his technique is represented in Fig. 11.89. The atoms or molecules effuse from the reservoir R through a small hole H into the vacuum chamber. After they

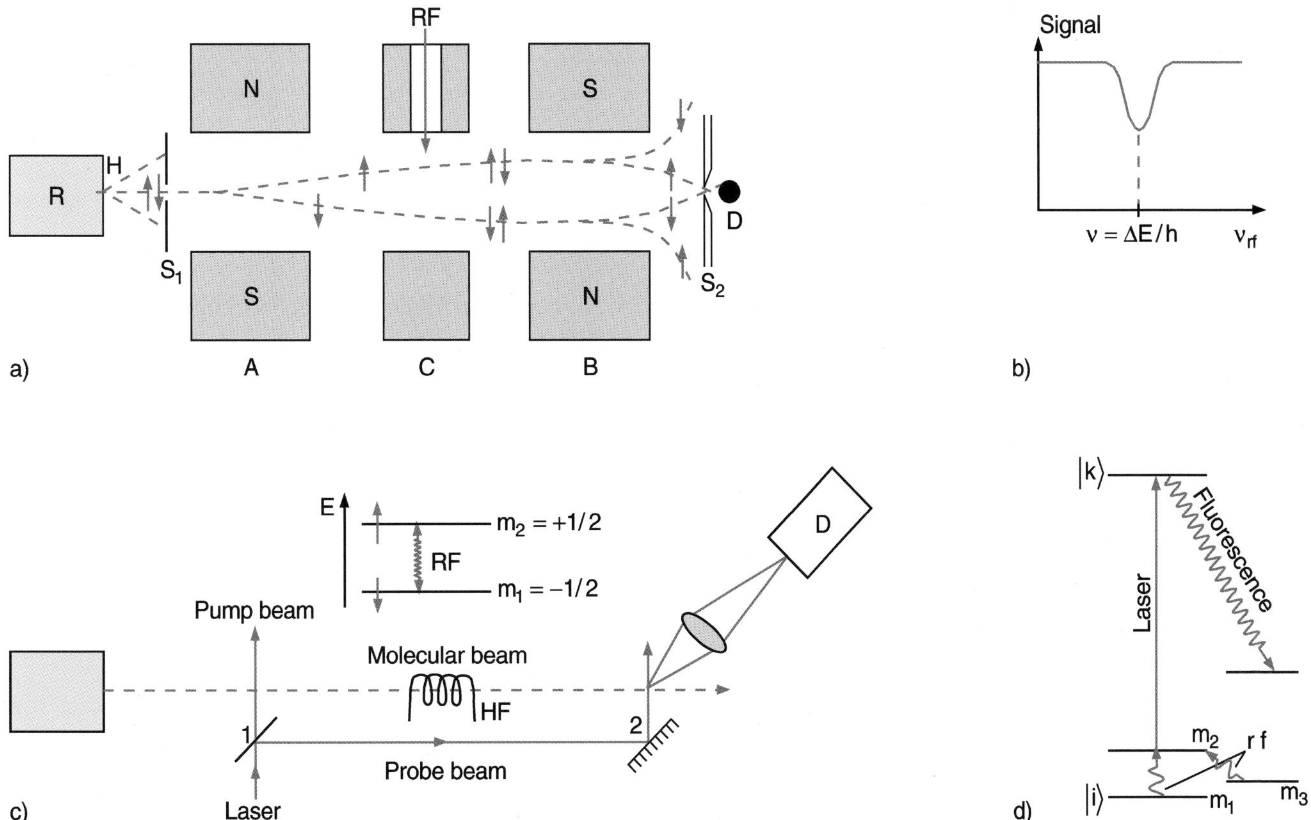


**Fig. 11.89a–d.** Rabi method. (**a**) Experimental setup. (**b**) Detector signal as a function of the radio frequency $\nu_{rf}$. (**c**) Modern laser version of Rabi method. (**d**) Level scheme for optical pumping and radio frequency transitions

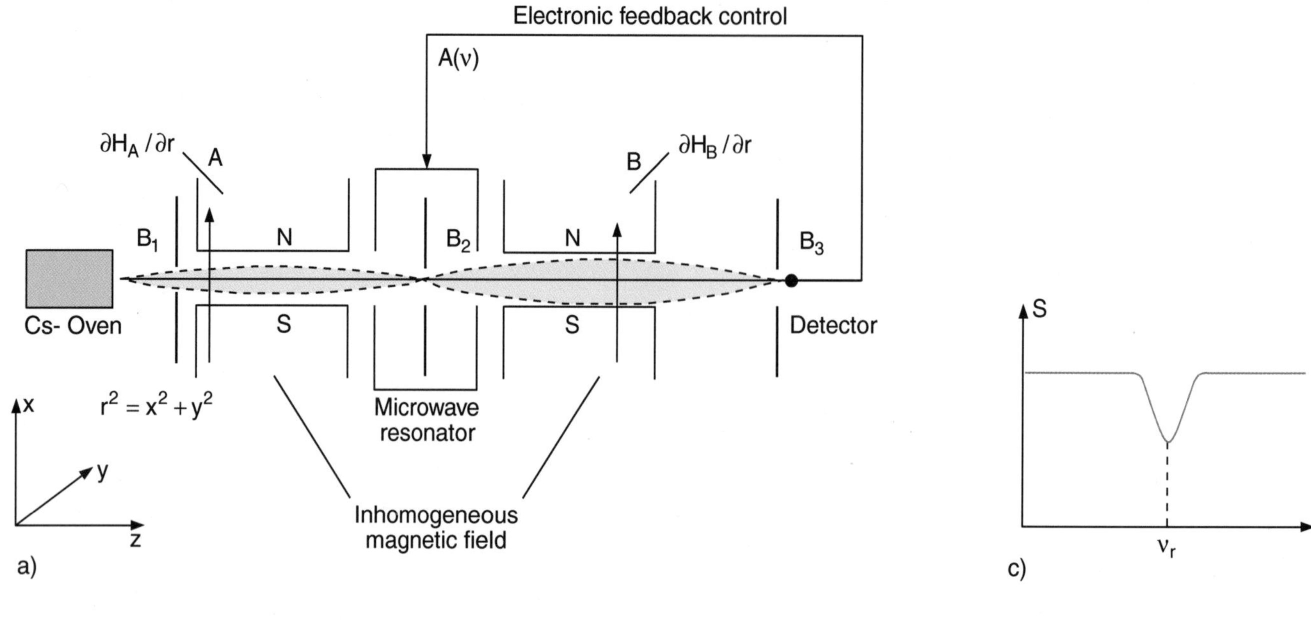


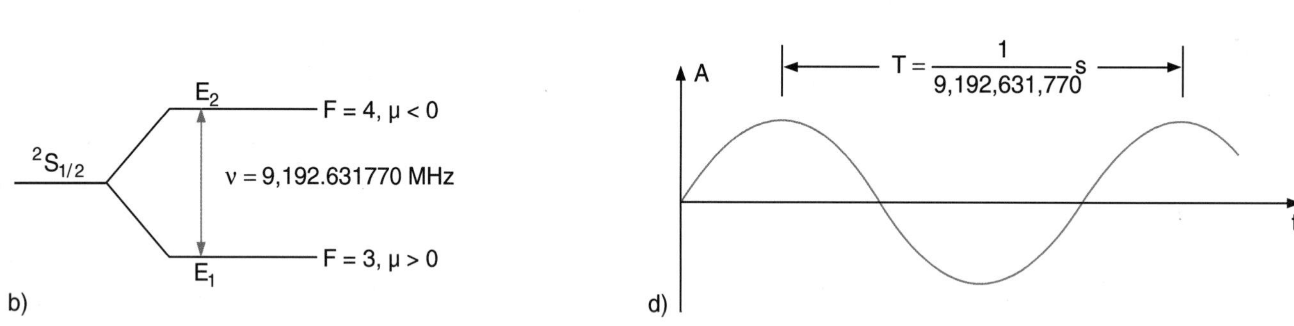


**Fig. 11.90a–d.** Cesium atomic clock. (**a**) Experimental setup. (**b**) Hyperfine splitting of the Cs($^2S_{1/2}$) ground state. (**c**) Detector signal as a function of the microwave frequency. (**d**) Stabilized microwave

have been collimated by the slit $S_1$ they pass through an inhomogeneous magnetic field $\boldsymbol{A}$ where they are deflected by the force $\boldsymbol{F} = \boldsymbol{p}_\text{m} \cdot \textbf{grad}\, \boldsymbol{B}$. In the following homogeneous field $C$ they do not experience a force but the projection $(p_\text{m})_z$ of their magnetic moment is quantized, i.e. it can only take discrete values $m \cdot g_j \cdot \mu_B \cdot B$ where $\mu_\text{B}$ is the Bohr-magneton (see Sect. 5.2). The Zeeman energies of the molecules in this field are

$$E_\text{M} = E_0 + m \cdot g_j |p_\text{m}| \cdot |B| / \sqrt{J(J+1)}$$
$$-m \le J \le +m. \qquad (11.123)$$

In the following inhomogeneous magnetic field $B$, which is opposite to the field in $A$ the molecules experience the opposite force as in $A$ and are deflected back onto the detector behind a slit $S_2$.

Now a radio-frequency is fed through a hole into the region of the homogeneous field $C$. If its frequency $\nu$ matches the Zeeman splitting $\Delta\nu = g_J |p_\text{m}| \cdot |B|/h$, it induces transitions between the Zeeman-components which changes their population. This results in a different deflection in the inhomogeneous field $B$ because the deflecting force $F$ depends on the magnitude and the direction of the magnetic moment which are different for the different $m$-levels. The molecules cannot pass through the slit $S_2$ and the signal decreases (Fig. 11.89b).

A modern version of the Rabi-method replaces the two inhomogeneous fields $A$ and $B$ by two laser beams (Fig. 11.89c), which cross the molecular beam perpendicular [11.46]. The laser frequency is tuned to the wanted transition $\langle k| \to \langle i|$. For sufficiently high laser intensities the transition can be readily saturated, i.e. the population of the absorbing level decreases (only a few milliwatts, focussed into the molecular beam, are sufficient). This saturation causes a decrease of the absorption of the second laser beam from the same laser and the laser-induced fluorescence in the crossing point 2 decreases. If the molecules are irradiated between the two crossing points by a radio-or microwave, which induces transitions from other levels into the depleted level, its population increases again and the signal in point 2 becomes larger. Measuring the signal as a function of the radio-frequency allows the determination of levels splittings between hyperfine-levels or rotational levels with a very high accuracy. The linewidth of these signals is mainly limited by the interaction time of the molecules with the rf-field, ie. by their transit time through the rf-field region.

The Rabi-Method is the basis of the cesium atomic clock (Fig. 11.90). The cesium atoms emerge from an oven through a narrow hole into the vacuum. The inhomogeneous magnetic fields A and B are realized by sixpole magnets which acts as magnetic lenses. The field A focuses the divergent atomic beam into the centre of a microwave resonator located between A and B and the field B refocuses the beam onto the detector. The microwave resonator is tuned to a transition between the hyperfine components $F = 3$ and $F = 4$ in the $^2S_{1/2}$ state of the cesium atom $^{133}$Cs at a frequency of $\nu = 9.192\,\mathrm{GHz}$. Since the magnetic moment is different for the two hfs-levels the deflection in the $B$-field changes when a transition took place in the microwave cavity. A feedback signal delivered from the detector keeps the microwave frequency always at the centre of the hyperfine-transition. Both, the classical and its laser version are used and are compared with regard to the optimum frequency stability which reaches a relative standard deviation of $\Delta\nu/\nu = 10^{-13}$. Up to now this cesium atomic clock provides the primary frequency and time standard. It might be, however, soon replaced by more stable laser devices.

## 11.9.2 Stark-Spectroscopy

The levels of atoms or molecules with a permanent electric dipole moment split in an electric field into $(2J + 1)$ Stark-components, where $J$ is the total angular momentum quantum number of the level.

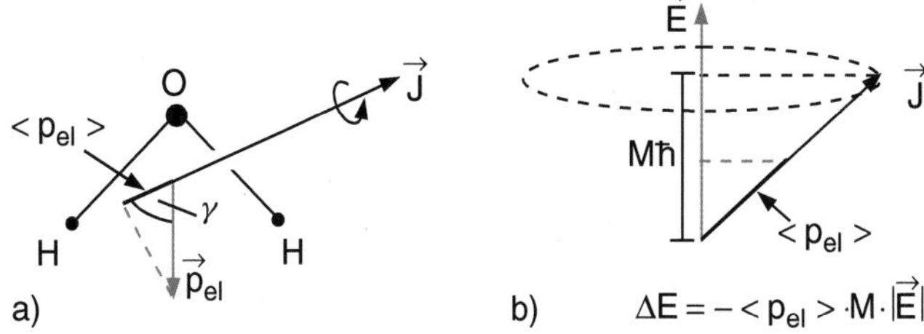


**Fig. 11.91.** (**a**) Average electric dipole moment in the rotating $H_2O$ molecule. (**b**) Projection $M\hbar$ of angular momentum $\boldsymbol{J}$ on the electric field direction and Stark shift $\Delta E$

Without external field the total angular momentum $\boldsymbol{J}$ has a constant magnitude and a constant orientation. If the direction of $\boldsymbol{p}_{\mathrm{el}}$ does not coincide with the direction of $\boldsymbol{J}$, the electric dipole moment $\boldsymbol{p}_{\mathrm{el}}$ precesses in the rotating molecule around $\boldsymbol{J}$ and the time-averaged component is, according to Fig. 11.91

$$\langle \boldsymbol{p}_{\mathrm{el}} \rangle = |\boldsymbol{p}_{\mathrm{el}}| \cdot \cos\gamma = |\boldsymbol{p}_{\mathrm{el}}| \cdot \frac{K}{\sqrt{J(J+1)}}\,, \tag{11.124}$$

where $K \cdot \hbar$ is the projection of $J$ onto the symmetry axis of the molecule which for symmetric top molecules coincides with the direction of $\boldsymbol{p}_{\mathrm{el}}$ (Fig. 11.91a). In an external electric field $\boldsymbol{E}$ the dipole moment $\boldsymbol{p}_{\mathrm{el}}$ precesses around the field direction and therefore also $\boldsymbol{J}$ precesses with a constant projection $M \cdot \hbar$ onto the field direction.

The energy shift of a level

$$\Delta E = -\langle \boldsymbol{p}_{\mathrm{el}} \rangle \cdot \boldsymbol{E} = |\boldsymbol{p}_{\mathrm{el}}| \cdot |\boldsymbol{E}| \cdot \frac{K \cdot M}{J(J+1)} \tag{11.125}$$

is proportional to the electric field $E$ (***linear Stark-effect***).

For a diatomic molecule the electric dipole moment points into the direction of the internuclear axis. Without electronic angular momentum the rotational angular momentum $J$ is perpendicular to this axis, which means that $K = 0$ and therefore $\langle p_{\mathrm{el}} \rangle = 0$.

For diatomic molecules in $^1\Sigma$ states there is no linear Stark-shift, even if they posses a permanent electric dipole moment.

There is, however, a second order effect. When the electric dipole rotates in the electric field, its potential energy is periodically changing and therefore the rotation is no longer uniform. The rotation is slower for positions where the potential energy is maximum. Therefore the average energy differs from that of the field-free case. There is a Stark-shift of second order which is proportional to the square $E^2$ of the electric field.

Even without a permanent electric dipole moment the electric field can polarise the charge distribution in the molecule which results for a non-rotating molecule in an induced electric moment

$$\boldsymbol{p}_{\text{el}}^{\text{ind}} = \tilde{\alpha} \cdot \boldsymbol{E} \,, \tag{11.126}$$

where $\tilde{\alpha}$ is the tensor of the molecular polarizability, which is a measure of the displacements of the charges from their equilibrium positions and generally depends on the direction against the molecular frame. Therefore $\boldsymbol{p}_{\text{el}}^{\text{ind}}$ generally does not point into the same direction as the electric field, but forms the angle $\beta$ against $\boldsymbol{E}$ and processes around the field direction.

Now the energy shift is

$$\Delta E = -\boldsymbol{p}_{\text{el}}^{\text{ind}} \cdot \boldsymbol{E} = -(\tilde{\alpha} \cdot \boldsymbol{E}) \cdot \boldsymbol{E} = \sum_{i,j} \alpha_{ij} E_i E_j \tag{11.127}$$

which means that the Stark-shift is proportional to the square of the electric field $E$ (quadratic Stark-effect).

In Fig. 11.92 the quadratic Stark-shift is illustrated by the example of the excited levels $3\,^2P_{3/2}$, $3\,^2P_{1/2}$ and the ground state level $3\,^2S_{1/2}$ in the sodium atom. The components with a projection $\pm M$ have the same energy, because for the quadratic Stark-effect the energy does not depend on the sign of $M$ but only on the amount $|M|$. Therefore only the level $P_{3/2}$ splits into two Stark components, while the other two levels $3^2\,S_{1/2}$ and $3^2\,P_{1/2}$ with $J = 1/2$ are only shifted but do not split.

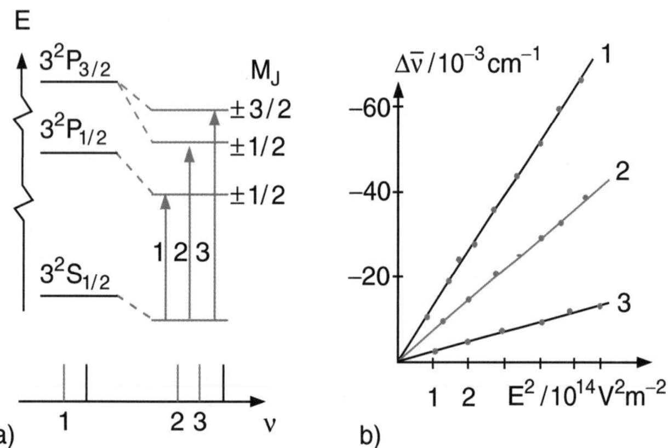


**Fig. 11.92.** Quadratic Stark effect of the three Stark components of the sodium resonance lines $^2S_{1/2} \to {}^2P_{1/2,3/2}$

Measurements of the line-shifts as a function of $E^2$ give a straight line (Fig. 11.92b), which proves the quadratic Stark-effect. Since the Stark-shift is smaller in the $^2S_{1/2}$ state as in the $P$-states. The Stark-components of the transitions are therefore shifted towards smaller frequencies.

From the measurement of the Stark shifts the polarisability $\alpha$ can be obtained according to (11.127). As was outlined in Sect. 9.4.3 $\alpha$ is responsible for the van-der-Waals binding of molecules at large internuclear distances and influences many macroscopic properties of matter, such as the melting temperature of van-der Waals-solids or the viscosity of noble gases.

The experimental arrangement can be that of the Rabi-method where the magnetic fields are replaced by electric fields [11.47]. For Doppler-limited spectroscopy also absorption cells can be used with electrodes on the side-walls for producing the electric field.

Instead of tuning the laser frequency across the transitions between Stark-components the laser frequency can be kept constant and the Stark field is tuned until the the laser frequency is in resonance between an optical transition between the Stark-components in the lower and upper level [11.48].

## 11.10 Investigations of Atomic and Molecular Collisions

Detailed studies of elastic, inelastic and reactive collisions between atoms and molecules have made an important contribution to our understanding of the structure of atoms and molecules, the interaction potential between the collision partners and of the microscopic process of chemical reactions. One example is the development of the Rutherford model of the atom (Sect. 2.8), which was derived from the results of experiments on scattering α particles by gold atoms. These experiments proved for the first time that most of the atomic mass is concentrated in the very small volume of the atomic nucleus. Another example is the study of inelastic collisions of electrons with atoms in the Franck–Hertz experiment (Sect. 3.4.4), which proved the discrete level structure of atoms and confirmed Bohr's atomic model.

Because of their importance for basic physics and in applications, an essential part of experimental efforts in atomic and molecular physics has been devoted to the investigation of collision processes. This can be seen from the fact that there are several important biannual international conference series on atomic and molecular collisions [11.49].

### 11.10.1 Elastic Scattering

When a parallel beam of particles A with particle flux $\dot{N}_A$ (number of particles per $cm^2$ per second) passes into the $x$ direction through a volume $V$ filled with particles B of density $n_B$ (Fig. 11.93) the flux $\dot{N}_A$ has decreased after a path length $x = d$ through the volume filled with particles B to

$$\dot{N}_A(x = d) = \dot{N}_A(x = 0)e^{-n_B \cdot \sigma_{int} \cdot d} \tag{11.128}$$

(see Sect. 2.8.1). The quantity $\sigma_{int}$ is the integral elastic collision cross section. The ratio of $\sigma_{int}$ to the cross section $q$ of the incident parallel beam describes the probability that a particle A is scattered by one particle B out of the beam, i.e., that it is deflected by an angle $\vartheta > \vartheta_{min}$ and cannot reach the detector behind the aperture S. This illustrates that the measured integral scattering cross section not only depends on the interaction potential, but also on the angular resolution $\vartheta_{min}$ of the apparatus. For $\vartheta_{min} = 0$, the classical integral cross section $\sigma_{int}$ would become infinite because collisions with zero scattering angles, i.e., infinite impact parameters $b$ would contribute to it.

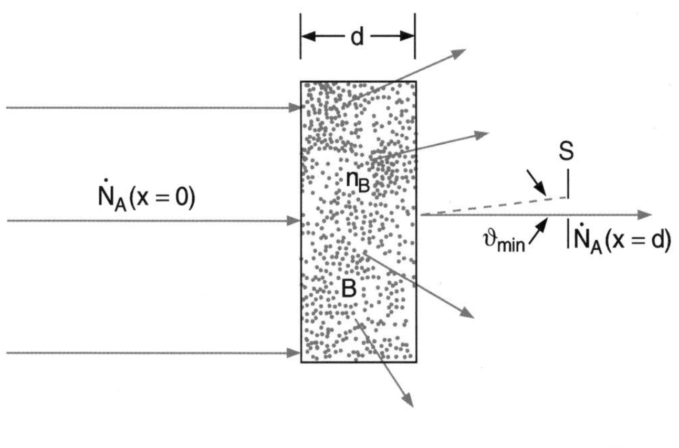


**Fig. 11.93.** Attenuation of a parallel beam of particles A after passing through a scattering volume with particle density $n_B$

In the quantum mechanical model $\vartheta_{min}$ cannot become zero and therefore the maximum impact parameter $b_{max}(\vartheta_{min})$ remains finite. This can be seen as follows.

In the center-of-mass system the two colliding particles can be replaced by one particle with the reduced mass $\mu$ moving in the spherically symmetric potential that equals the interaction potential between the two particles (Sect. 2.8.2). For elastic collisions the momentum does not change its amount but only its direction. For small deflection angles we have the relation (Fig. 11.94)

$$\vec{\vartheta} \approx \tan\vartheta = \frac{\Delta p}{p} = \frac{b\Delta p}{bp} = \frac{|\Delta \boldsymbol{L}|}{|\boldsymbol{L}|}, \tag{11.129}$$

where $|\boldsymbol{L}| = b\mu v = n\hbar$ is the orbital angular momentum of the particle with mass $\mu$, velocity $v$ and impact parameter $b$, which has to be an integer multiple of $\hbar$. Therefore the minimum value of $\Delta L$ is $\Delta L_{min} = \hbar$. This gives the minimum deflection angle

$$\vartheta_{min} = \frac{\hbar}{b_{max}\mu v} = \frac{\lambda_{dB}}{2\pi b_{max}}, \tag{11.130}$$

where $\lambda_{DB} = h/(\mu v)$ is the de Broglie wavelength. With $\vartheta_{min} > 0$ a maximum impact parameter $b_{max}$ is determined and the integral cross section $\sigma_{int} = \pi b_{max}^2$ remains finite.

> The smallest scattering angle is therefore determined by the ratio of the de Broglie wavelength $\lambda_{DB}$ to the maximum impact parameter $b_{max}$.

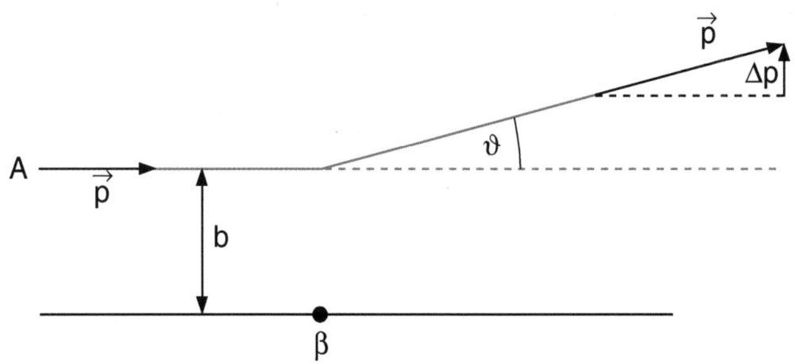


**Fig. 11.94.** For small deflection angles $\vartheta$ and large impact parameters $b$ is $\tan\vartheta = \Delta p/p$

In Equation (2.109) a relation between scattering angle $\vartheta$ and potential interaction energy $E_{pot}(R)$ was given. For large impact parameters the path of the particle is only slightly bent and we can approximate the closest distance $r_{min}$ by $r_{min} = b$. Under these conditions the interaction energy $E_{pot}$ is very small compared to the initial kinetic energy $E_{kin}(\infty) = E_0$. When we expand for $E_{pot}/E_0 \ll 1$ the square root under the integral in (2.142) we obtain the result

$$\vartheta_{min}(b_{max}) = \frac{\lambda_{DB}}{b_{max}} \propto \frac{E_{pot}(b_{max})}{E_0} . \tag{11.131}$$

For an attractive potential $E_{pot} = -C_n/r^n$ this gives

$$\frac{h}{\mu v b_{max}} \propto \frac{C_n}{b_{max}^n \mu v^2} . \tag{11.132}$$

Measuring the velocity-dependent integral cross section

$$\sigma_{int} = \pi b_{max}^2 \propto \left(\frac{C_n}{hv}\right)^{2/(n-1)} \tag{11.133}$$

one can therefore derive the constant $C_n$ and the exponent $n$ of the interaction potential.

The measurements can be done with the apparatus shown in Fig. 11.95, where two beams of particles A and B cross each other perpendicularly. The scattering volume $V$ is defined by the overlap region of the two beams. The minimum scattering angle of particle A is determined by the distance $d$ between scattering volume and aperture $B_2$ and the slit width $b_2$ of $B_2$ in front of the detector. The velocity $v$ of the relative movement

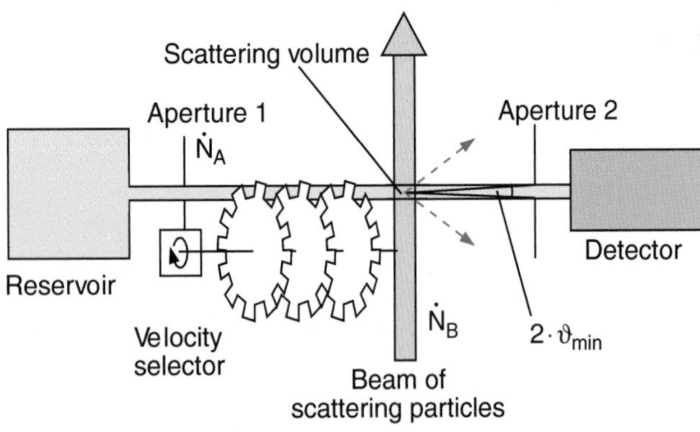


**Fig. 11.95.** Measurement of integral cross section with velocity-selected molecules

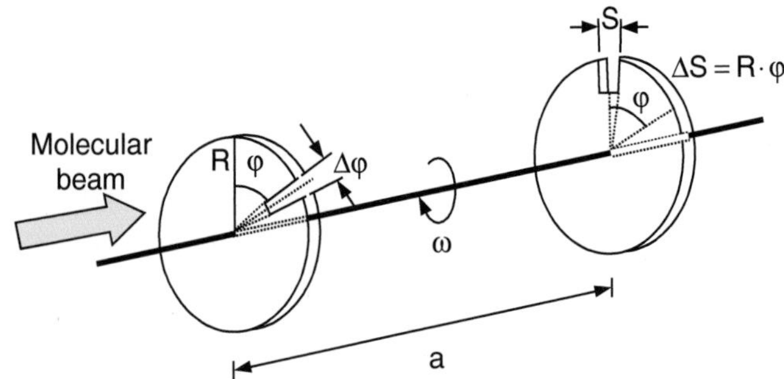


**Fig. 11.96.** Principle of a mechanical velocity selector

can be varied with the rotational speed of the velocity selector, which consists of a series of cog-wheels mounted on the same axis. The principle of the velocity selector can be illustrated by Fig. 11.96, which shows the simplified version consisting of two discs with radius $R$ and slits of width $s$ in a distance $a$. If the angular displacement of the two slits is $\varphi$ and the rotational angular speed of the discs is $\omega$, only particles are transmitted through both slits of the rotating discs for which the flight time $T = a/v$ equals the time $R\varphi/(R\omega) = \varphi/\omega$. The velocity of these particles is then

$$v = a \cdot \omega/\varphi . \tag{11.134a}$$

The width $\Delta v$ of the transmitted velocity interval is given by the angular width $\Delta\varphi$ of the two slits

$$\Delta v = v(\Delta\varphi/\varphi) . \tag{11.134b}$$

More information about the interaction potential at smaller distances is obtained from measurements of differential cross sections as was already outlined for the Rutherford scattering in Sect. 2.8.6.

Similarly to Fig. 11.95, two crossed molecular beams are used (Fig. 11.97). The difference is that now only molecules scattered into the solid angle $\Delta\Omega$ around the deflection angle $\vartheta$ are received by the detector, while in measurements of the integral cross section the attenuation of the incident beam is determined, which is a measure for the total number of scattered particles integrated over all scattering angles from $\vartheta_{min}$ to $\vartheta = 180°$.

There is a problem for nonmonotonic potentials, such as the Lenard–Jones potential (9.67), where different impact parameters $b_1$ and $b_2$ can lead to the same

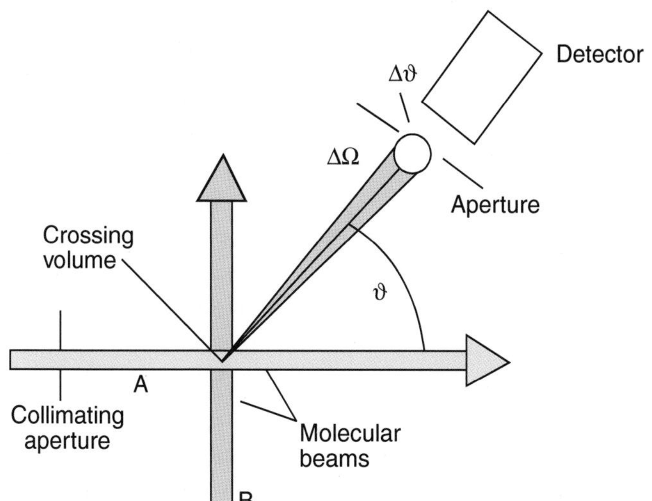


**Fig. 11.97.** Measurement of differential cross section for collisions in crossed molecular beams

scattering angle $\vartheta$. This is illustrated in Fig. 11.98. At the internuclear distances $r_1$ and $r_2$ the slopes of the potential have equal magnitude but opposite sign. The forces $F = -$ **grad** $E_{\text{pot}}$ are therefore equal in magnitude, but $F_1$ is repulsive and $F_2$ is attractive. Two particles A with impact parameters $b_1$ and $b_2$ approaching on opposite sides of the line $b = 0$ are scattered into the same direction $\vartheta$.

In such cases it can not be decided which of the two impact parameters has contributed to the measured signal and therefore the superposition of the probability amplitudes has to be considered (see Sect. 3.5.2). This leads to interference effects because the two possibilities correspond to two different paths of the particles through the potential and their de Broglie wavelengths are therefore shifted in a different way, which results in different phase shifts $\varphi_n$.

The intensity of the de Broglie wave scattered by the deflection angle $\vartheta$ is

$$I(\vartheta) \propto |A_1(b_1, \varphi_1) + A_2(b_2, \varphi_2)|^2 , \qquad (11.135)$$

where the $A_i$ are the amplitudes of the scattered waves.

The intensity $I(\vartheta)$ shows interference maxima and minima, which are produced in a similar way as the refraction and reflection of light by raindrops resulting in the observation of the rainbow.

Therefore the interference effects in the differential cross section are called *rainbow scattering*.

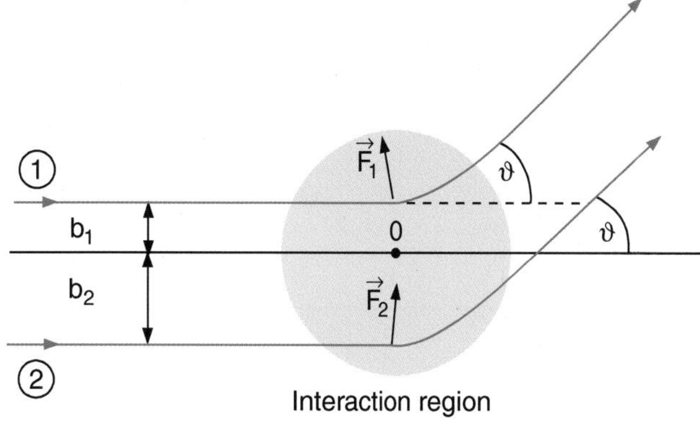


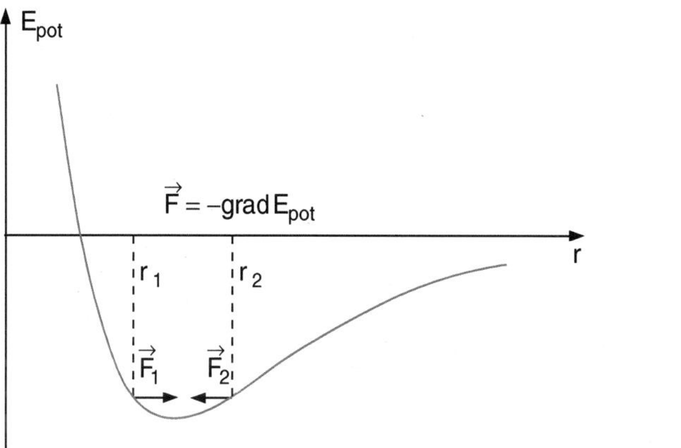


**Fig. 11.98.** Interference effects for scattering in a nonmonotonic potential where different impact parameters $b$ can lead to identical deflection angles $\vartheta$

In Fig. 11.99 the deflection function $\vartheta(b)$ for a non-monotonic potential is illustrated, showing three impact parameters $b_1$, $b_2$, and $b_3$ leading to the same values of $|\vartheta|$. Since the scattering is symmetric around the axis $b = 0$, the differential cross sections $\sigma(\vartheta) = \sigma(-\vartheta)$ must be equal.

According to (2.147) the differential cross section is proportional to $\mathrm{d}b/\mathrm{d}\vartheta = (\mathrm{d}\vartheta/\mathrm{d}b)^{-1}$, i.e., to the inverse slope of the deflection function $\vartheta(b)$. In the classical model $\mathrm{d}\sigma/\mathrm{d}\Omega$ should become infinite for $b = b_r$ and $b = 0$. In the quantum mechanical treatment, the impact parameters are only defined with an uncertainty that equals the de Broglie wavelength. Therefore there are only maxima instead of infinite poles in the scattering cross section (Fig. 11.99c).

The measurements of the positions and heights of these rainbow maxima and their dependence on the energy of the incident particles (i.e., on their de Broglie

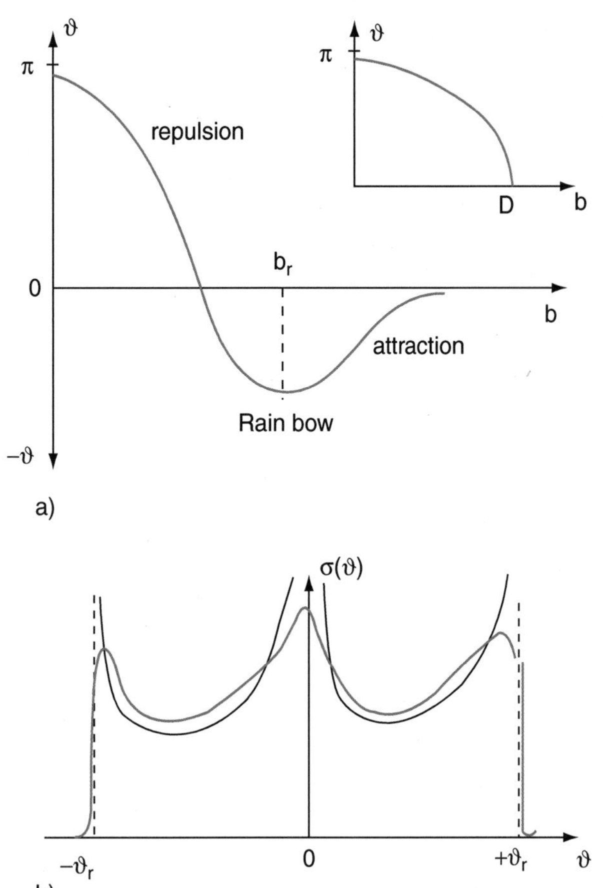


**Fig. 11.99a–b.** (a) Deflection function $\vartheta(b)$ for a potential with minimum. The insert shows $\vartheta(b)$ for hard spheres with diameter $D$. (b) Cross section for elastic collisions according to the classical model (black curve) and the quantum mechanical model (blue curve)

wavelength) gives very detailed information of the radial dependence of the interaction potential, because the phase differences are very sensitive against changes of $E_{\text{pot}}(r)$ [11.50].

### 11.10.2 Inelastic Scattering

During the inelastic collision of a particle A with a particle B part of the kinetic energy of A is transferred into internal energy of B. If B is an atom, this is electronic excitation energy, if B is a molecule, the internal energy may be rotational, vibrational or electronic energy

$$A(E_{\text{kin}}) + B(E_{\text{i}}) \rightarrow A + B(E_{\text{f}}) + \Delta E_{\text{kin}} \quad (11.136)$$

with $\Delta E_{\text{kin}} = E_{\text{f}}(\text{B}) - E_{\text{i}}(\text{B})$

The collision induces in the target particle B a transition from the initial internal energy $E_{\text{i}}$ to a final state with energy $E_{\text{f}}$, where we have assumed that the internal state of A has not changed.

The probability for this inelastic process depends on the interaction potential between A and B, on the impact parameter $b$ and on the relative kinetic energy of the collision partners. If all of these parameters should be included in the experimental investigation, one has to know the initial kinetic energy, the initial and final internal energies $E_i$ and $E_{\text{f}}$ of particle B and the impact parameter, i.e., the scattering angle $\vartheta$. Such a scattering experiment can be called an ideal experiment, because is allows the determination of all wanted details [11.52]. A possible experimental arrangement for such a detailed experiment, which combines the method of laser spectroscopy with molecular beam techniques [11.53] is shown in Fig. 11.100.

The molecules A pass before the collision region through the beam of laser $\text{L}_1$ (pump laser), which induces a transition $|i\rangle \rightarrow |k\rangle$ in the molecule. Already with moderate laser powers the transition can be saturated and the initial level $|i\rangle$ can be nearly completely depopulated. The molecules are scattered in the crossing volume with a beam of particles B. Those molecules A, deflected by the angle $\vartheta$ which is defined by an aperture in front of the detector, are excited by a second laser $\text{L}_2$ on the transition $|f\rangle \rightarrow |j\rangle$. The

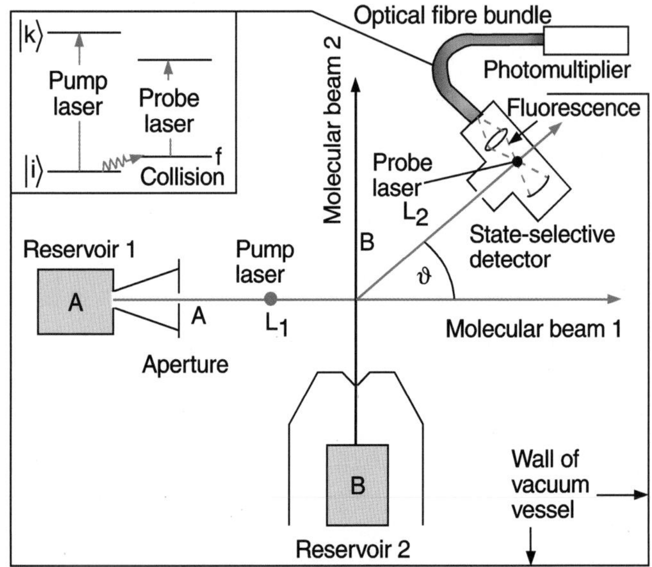


**Fig. 11.100.** Measuring state selective differential cross sections by laser excitation before and after the collision [11.46]

laser-induced fluorescence is collected with carefully designed optics and imaged by a fiber bundle onto a photomultiplier. The photomultiplier output signal $S$ is a measure for the population $N_f$ of the scattered particles.

If the pump laser $L_1$ is periodically chopped the population $N_i$ is modulated at the chopping frequency. The probe laser $L_2$ now measures the difference $\Delta S = S(L_1 \text{ on}) - S(L_1 \text{ off})$. The difference of the signals gives the rate of those molecules that are deflected by the angle $\vartheta$ and have been transferred during the collision from level $|i\rangle$ to level $|f\rangle$.

Measuring the dependence $\Delta S(\vartheta)$ gives direct information on the dependence of the inelastic cross section on the impact parameter and tells us which part of the potential has the largest influence on inducing the transition. They teach us, for instance, which impact parameters mainly contribute to rotational or vibrational energy transfer, or how the initial internal energy of one collision partner influences the probability of a collision-induced transition [11.51–53].

### 11.10.3 Reactive Scattering

In Sect. 10.7 chemical reactions were described by rate equations, which give the macroscopic picture of these reactions averaged over a large number of reaction partners. On the molecular scale, these reactions are the result of many individual molecular collisions, e.g.,

$$\begin{aligned} &AB + C \rightarrow AC + B \\ &AB + CD \rightarrow AC + BD\,. \end{aligned} \tag{11.137}$$

The probability of such reactive collisions depends on the relative kinetic energy, on the internal energy of the reactants (for instance the vibrational energy of AB) and on the interaction potential. Since the interaction potential is generally not spherically symmetric, steric effects may play an essential role, i.e., the reaction probability may strongly depend on the relative orientation of the reacting molecules.

Often a reaction takes place only if the kinetic energy exceeds a threshold (reaction barrier), which in turn can depend on the internal energy of the reactants.

For the experimental investigation of reactive collisions the rate of reaction products B or AC, scattered by the angle $\vartheta$, are measured as a function of the initial relative kinetic energy of the reactants. This gives the differential reaction cross section $d\sigma/d\Omega$. The integration over all scattering angles yields the integral cross section $\sigma_{\text{total}}(v)$ as a function of the relative velocity of the collision partners. The velocity-dependent rate reaction is related to the integral cross section by

$$\begin{aligned} R(AB + C \rightarrow AC + B) &= k_R^{(2)} n_{AB} \cdot n_C \\ &= v \cdot \sigma_{\text{tot}}(v) \cdot n_{AB} \cdot n_C \end{aligned}$$

where the second order rate constant $k_R^{(2)}(v)$ is

$$k_R^{(2)}(v) = v \cdot \sigma_{\text{total}}(v)\,.$$

The average reaction constants

$$\begin{aligned} \langle k_R(v)\rangle &= \frac{1}{\langle v\rangle} \int_{v=0}^{\infty} k_R(v)\,dv \\ &= \frac{1}{\langle v\rangle} \int p(v) v \cdot \sigma(v)\,dv \end{aligned} \tag{11.138}$$

where $p(v)$ is the probability that the relative velocity of the reaction partners is within the interval $dv$.

These reaction constants introduced in Sect. 10.7 for reactions in gas cells are mean values, averaged over the thermal distribution of the relative velocity of the reaction partners. They can be calculated from the measured cross sections.

The reaction constant $k(i \rightarrow f)$ for the reaction from the state $|i\rangle$ of the reaction partners to the final state $|f\rangle$ of the reaction product

$$k(i \rightarrow f) = \int p(\boldsymbol{v}) \cdot \sigma_{i\rightarrow f}(v) v \cdot d\boldsymbol{v}\, d\Omega \tag{11.139}$$

may not only depend on the absolute value of the velocity but also on the direction of $\boldsymbol{v}$.

For spherical symmetric potentials $\sigma_{i\rightarrow f}$ does not depend on the direction of $v$. Integration of the solid angle $d\Omega$ yields with

$$\begin{aligned} &d\boldsymbol{v} = v^2 dv d\Omega \\ &k_{i\rightarrow f}(v) = 4\pi v^3 p(v) \sigma_{i\rightarrow f} dv \end{aligned} \tag{11.140}$$

Inserting for $p(v)$ the Maxwell-Boltzmann distribution

$$p(v) = (\mu/2\pi k_B T)^{3/2} \exp(-\mu v^2/2k_B T) \tag{11.141}$$

with the reduced mass $\mu = m_{AB} \cdot m_C/(m_{AB} + m_C)$ of the reaction partners and the translational energy

$E_{tr} = ½\mu v^2$, we obtain for the reaction constant $k_{i\to f}(T)$ as a function of the temperature $T$

$$k_{i\to f}(T) = 8k_B T/\pi\mu) \int E_{tr} \exp(-E_{tr}/k_B T)\sigma_{i\to f} (E_{tr})(1/k_B T)^2 dE_{tr} \quad (11.142)$$

**EXAMPLE**

For $\sigma_{i\to f}$ $\int$ $= \pi d_{AB-C}(1-E_{tr}/E_0)$ for $E_{tr} \geq E_0$; $=0$ for $E_{tr} < E_0$

the integration of (11.142) gives

$$k_{i\to f}(T) = (8k_B T/\pi\mu^2)^{1/2}\, \pi d^2_{AB-C} \cdot \exp(-E_0/k_B T)$$

The reaction constant grows exponentially with rising temperature $T$. The experimental technique used for the investigation of reactive collisions depends on the kind of reaction partners. In early experiments, reactions involving alkali atoms were mostly studied because they can be readily and efficiently detected by ***Langmuir–Taylor detectors***. These are heated tungsten wires that have an electron work function which is higher than the ionization energy of the alkali atoms. If an alkali atom hits the wire it becomes ionized because the electron gains more energy when it is captured in the conduction band than is necessary to remove it from the alkali atom. The alkali ions evaporate from the hot wire and are accelerated by an electric field onto a Faraday cup or an electron multiplier. Meanwhile the detection techniques have been improved, using laser-induced fluorescence for state-specific detection of atoms or molecules. Therefore reactive scattering experiments in crossed beams have been extended to a larger variety of different collision parameters [11.54]

## 11.11 Time-Resolved Measurements of Atoms and Molecules

While free atoms or molecules are stable in their ground states, they decay in higher energy states either by emission of fluorescence (radiative transitions) or by collision-induced deactivation processes (radiationless transitions). If collisions transfer the atom into its ground state, the fluorescence intensity is diminished and the collisions are called *quenching collisions*.

In molecules, the excitation energy can be transferred even in the absence of collisions to other excited levels (for instance, high-lying vibrational levels in the electronic ground state) if these levels are coupled to the initially-excited level. Such an internal energy transfer can result in isomerization or fragmentation of the molecule.

All these time-dependent processes fit into the area of molecular dynamics. For experimental investigations, time-resolving techniques are demanded, where the necessary time resolution ranges from milliseconds down to femtoseconds.

We will present here some techniques that have already been used for quite some time. The latest experimental developments of producing and measuring ultrafast laser pulses and their applications to molecular dynamics will be left to the next chapter.

### 11.11.1 Lifetime Measurements

If atoms or molecules are excited by a short pulse (light pulse or electron pulse) into an energetically higher state $E_i$, the population $N_i(t)$ decays due to emission of radiation or to inelastic collisions. In Sect. 7.3 we saw that the time-dependent population is given by

$$N_i(t) = N_i(0) \cdot e^{-t/\tau_{eff}} . \quad (11.143)$$

The effective lifetime $\tau_{eff}$ is determined by

$$\frac{1}{\tau_{eff}} = \frac{1}{\tau_{spont}} + n_B \overline{v}_{AB} \sigma_i^{inel} = A_i + R_i \quad (11.144)$$

with the spontaneous transition probability $A_i = \sum A_{ik}$ written as the sum over the probabilities $A_{ik}$ of transitions from level $\langle i|$ into all accessible lower levels $\langle k|$ plus the collision-induced depopulation probability

$$R_i = n_B \overline{v}_{AB} \sigma_i^{inel} . \quad (11.145)$$

Since the total fluorescence rate $R_{fl} = N_i A_i$ is proportional to the population $N_i$, its time dependence follows the same equation as (11.143), and the measurement of the time-dependent fluorescence yields the effective lifetime $\tau_{eff}$ of level $\langle i|$. Its dependence on the density $n_B$ of collision partners allows the determination of the inelastic total cross section. The extrapolation to $n_B = 0$ gives the spontaneous lifetime $\tau_{spont}$ of level $\langle i|$.

The short light pulses for the excitation of level $\langle i|$ in the spectral range from 400–2000 nm are delivered by pulsed lasers, where the UV range with $\lambda < 400\,\text{nm}$ is generally covered by optical frequency doubling of visible lasers. In the spectral range below 200 nm, pulses of synchrotron radiation from electron storage rings are available with pulse widths of a few picoseconds.

The experimental technique for measuring lifetimes depends on the desired time resolution and the wanted repetition rate of the excitation pulses.

For a time resolution $\Delta t > 10^{-9}\,\text{s}$ and a pulse repetition rate $f < 10^3\,\text{s}^{-1}$ the apparatus required is shown schematically in Fig. 11.101. A pulsed laser excites the molecules in a cell and the laser-induced fluorescence is imaged onto the photomultiplier. The output pulse of the photomultiplier is amplified and delivered to a fast oscilloscope, where it can be directly viewed.

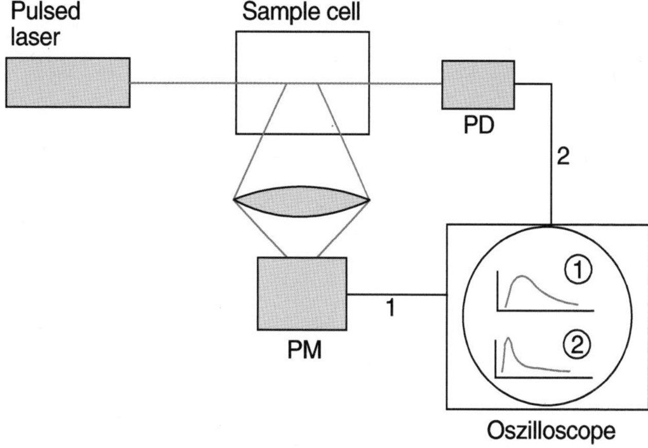


**Fig. 11.101.** Excitation of levels with a pulsed laser (pulse profile monitored by PD 2) and measurement of the decay curve of the laser-induced fluorescence with a photomultiplier PM

Another technique uses a gated integrator. Here, several time intervals $t_n + \Delta t$ with $\Delta t \ll \tau$ are selected by controllable electronic gates (Fig. 11.102). The output signal of the $n$-th gate is proportional to the intensity of the fluorescence

$$I(t, \Delta t) = \frac{1}{\Delta t} \int\limits_{t_n}^{t_n+\Delta t} I(t)\,\mathrm{d}t \qquad (11.146)$$

integrated over the gate width $\Delta t$. For an exponential decay of the fluorescence, the ratio

$$\frac{I(t_1, \Delta t)}{I(t_2, \Delta t)} = e^{-(t_2-t_1)/\tau} \qquad (11.147)$$

$$\Rightarrow \ln \frac{I(t_1, \Delta t)}{I(t_2, \Delta t)} = -\frac{1}{\tau}(t_2 - t_1)$$

allows the determination of the lifetime $\tau$.

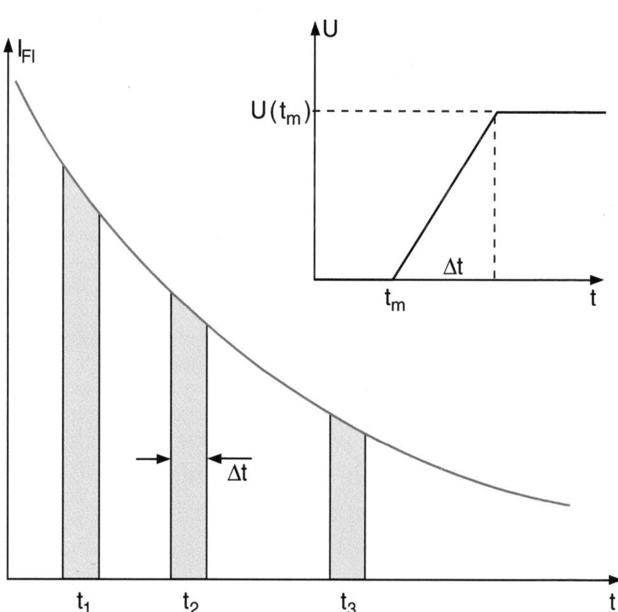


**Fig. 11.102.** Principle of lifetime. Measurements with a gated integrator and a sample and hold circuit

A more accurate technique (Fig. 11.103a) uses cw radiation for excitation, modulated at the frequency $\Omega$. The incident intensity is then

$$I_\text{L}(t) = \tfrac{1}{2} I_0(1 + a \sin^2 \Omega t) \cos^2 \omega_{ik} t\ . \qquad (11.148)$$

The time-dependent population of the excited level is obtained from

$$\frac{\mathrm{d}N_i}{\mathrm{d}t} = +N_k B_{ki} w_\text{L} - N_i(A_i + R_i + B_{ik} w_\text{L})\ , \qquad (11.149)$$

where the depopulation by induced emission has been included. The energy density $w_\text{L}$ of the radiation is related to its intensity by $I_\text{L} = c \cdot w_\text{L}(t)$.

Inserting (11.143) and (11.148) into (11.149) gives the fluorescence power

$$P_\text{Fl} = N_i A_i h\nu \qquad (11.150)$$

$$= b\left[1 + \frac{a \sin(\Omega t + \varphi)}{\left[1 + (\Omega \tau_\text{eff})^2\right]^{1/2}}\right] \cos^2 \omega_{ik} t\ ,$$

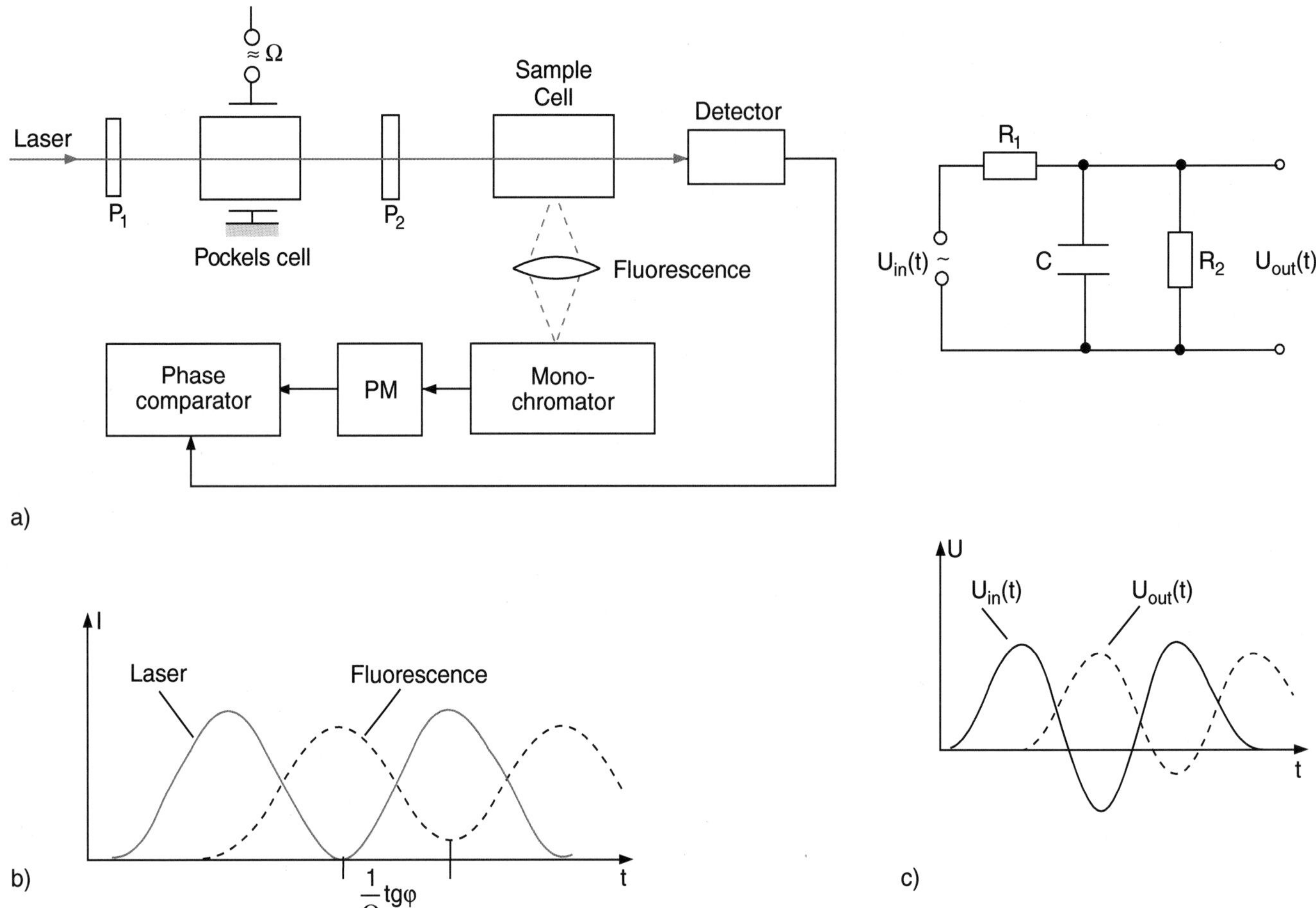


**Fig. 11.103a–c.** Phase-shift method for measuring lifetimes of excited atomic or molecular levels. (**a**) Experimental arrangement and (**b**) equivalent electric network

where the constant $b \propto N_0 \sigma_{0i} I_L V$ depends on the density $N_0$ of molecules in the absorbing level $\langle 0|$, the absorption cross section $\sigma_{0i}$, the laser intensity $I_L$ and the excitation volume $V$, seen by the fluorescence detector.

Equation (11.150) shows that the emitted fluorescence is modulated at the same frequency $\Omega$ as the exciting laser radiation but has a phase shift $\varphi$ (Fig. 11.103b) that is related to the lifetime $\tau$ by

$$\tan \varphi = \Omega \tau \, . \tag{11.151}$$

The mathematically equivalent problem is represented by the electronic circuit in Fig. 11.103c, where an ac input voltage $U_{in}(t) = U_0 \cdot \cos \Omega t$ generates an output voltage $U_{out}(t) = U_1 \cdot \cos(\Omega t - \varphi)$. The output voltage has the same mathematical form as the fluorescence intensity if one substitutes the mean lifetime $\tau$ by the time constant $RC$ with $R = R_1 R_2/(R_1 + R_2)$ and the laser intensity $I(t)$ by the input current $I_{in}(t) = (U_{in} - U_{out})/R_1$.

With a cw mode-locked laser, pulse widths of a few picoseconds can be achieved with repetition rates of several MHz. The energy of each pulse is small, in particular, if optical frequency doubling has to be used in order to obtain wavelengths in the UV region. Therefore only a few molecules are excited per pulse and the detection probability for a fluorescence photon per excitation pulse is small compared to one. In such a case the single photon counting technique, shown in Fig. 11.104, is preferable. The photomultiplier detects single fluorescence photons, which produce electric output pulses that are amplified and fed into a time-to-amplitude converter. This device generates a fast voltage ramp $U(t) = U_0(t - t_0)$ that is started by the excitation pulse at time $t_0$ and stopped by the fluorescence pulse at time $t = t_1$. The output voltage $U(t_1) = U_0(t_1 - t_0)$ is formed into a pulse with

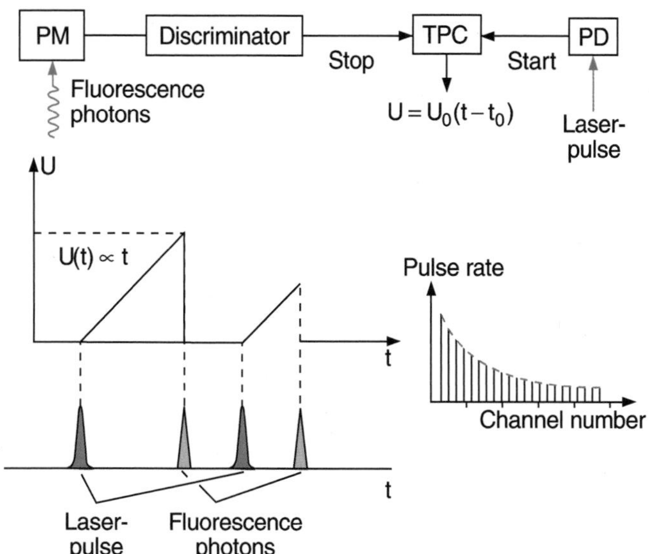


**Fig. 11.104.** Schematic diagram for lifetime measurements using single photon counting with a time-to-pulse height converter and a multichannel analyzer

amplitude $U(t_1)$, which is stored in a multichannel analyzer where the different pulses are sorted into different channels according to their amplitude. Each channel corresponds to a definite time difference $\Delta t = t_1 - t_0$. The number of pulses stored in channel $n$ is proportional to the fluorescence intensity at time $t_n$, because the probability $P(t)\Delta t$ that a fluorescence photon from a single atom is emitted during the time interval from $t$ to $t + \Delta t$,

$$P(t)\Delta t = I(t)\Delta t / \int_0^\infty I(t)\,\mathrm{d}t \tag{11.152}$$

$$= N_{\mathrm{ph}}(t)\Delta t / \left[\int_0^\infty N_{\mathrm{ph}}(t)\,\mathrm{d}t\right]$$

is proportional to the number of fluorescence photons emitted in the same time interval by many atoms that have all been excited at the same time. The amplitude distribution of the pulses in the multichannel analyzer

$$N(U)\Delta U = a\mathrm{e}^{-t/\tau_{\mathrm{eff}}} \tag{11.153}$$

directly gives the decay curve of the excited molecules.

The technique is called ***single photon counting with delayed coincidence***, because for each exciting laser pulse the corresponding fluorescence photon, delayed by the time $t$, is detected coincidentally [11.55].

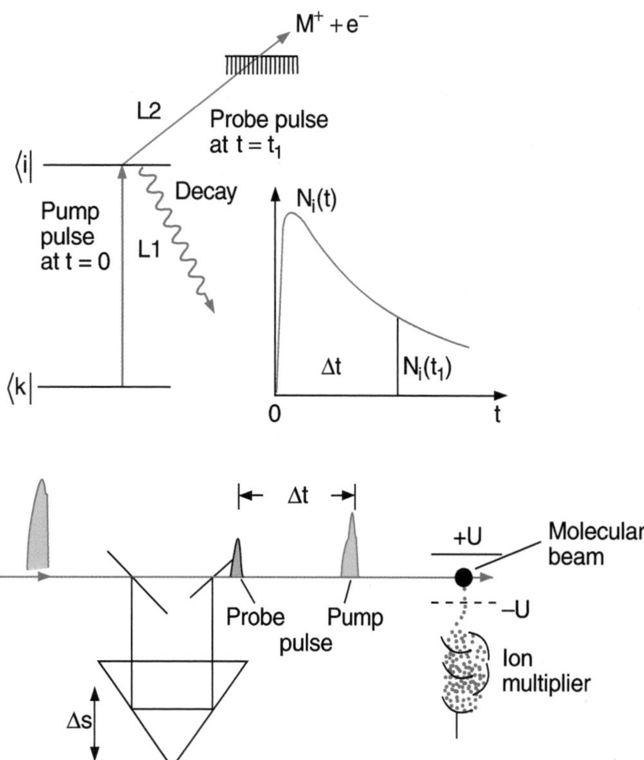


**Fig. 11.105.** Pump-and-probe technique

For very high time resolution in the pico- and femto-second range, electronic detectors are not fast enough. Here a pump-probe techniques can be used. A short pump pulse at time $t = 0$ excites the sample into a definite level $|i\rangle$, which decays over the course of time. A second weak probe pulse with variable delay $\Delta t$ against the pump pulse interrogates the population of the decaying level (Fig. 11.105). This can be realized in different ways. If the probe pulse can ionize the molecules in the excited level $|i\rangle$, the number of detected ions as a function of the delay time $\Delta t$ is directly proportional to the number $N_i(\Delta t)$. Another way of monitoring the time-dependent population $N_i(t)$ uses the total fluorescence emitted from a level $|j\rangle$, excited by the probe laser on the transition $|i\rangle \to |j\rangle$. Its intensity $I_{\mathrm{fl}}(\Delta t)$ measured as a function of the delay time $\Delta t$ is again proportional to the population $N_i$. In many cases the stimulated emission induced by the probe pulse on a downward transition from $|i\rangle$ to lover levels $|m\rangle$, which is proportional to $(N_i(\Delta t) - N_m)$ can be utilized as a monitor. If $N_m = 0$, the intensity of stimulated emission is again proportional to $N_i(\Delta t)$. With these pump and probe techniques, a time resolution in the femtosecond range can be achieved, even when the detector has a much lower time constant.

### 11.11.2 Fast Relaxation Processes in Atoms and Molecules

The time resolved nonradiative relaxation of a level caused by collisions can be used for the determination of collision cross sections and reaction rates. Here the pump and probe techniques can be also applied to levels in the electronic ground state, where no fluorescence is emitted and therefore some of the other techniques do not work. In Fig. 11.106 the principle is shown for measuring the relaxation of vibrational levels in the electronic ground state of molecules. A short pulse from the pump laser tuned to a transition $|k\rangle \to |i\rangle$ depletes the lower level $|k\rangle$ of molecules in a gas cell. Collisions with other atoms or molecules refill it again, because they try to reestablish thermal equilibrium. The fluorescence intensity $I_{\text{fl}}(\Delta t)$, induced by the pulse of the probe laser on a transition $|k\rangle \to |j\rangle$ is measured as a function of the delay time between pump and probe pulse. This intensity is proportional to the population $N_i(\Delta t)$. It is a measure for the total inelastic collision rates for energy transfer from all other levels into level $|i\rangle$.

The individual contributions of collision rates for population transfer from a specific level $|m\rangle$ into level $|k\rangle$ can also be measured with a slight modification. Now the probe laser is tuned to a transition $|m\rangle \to |j\rangle$. The fluorescence $I_{\text{fl}}$ gives the time-dependent population of level $|m\rangle$. The ratio of $\Delta N_m/\Delta N_k$ yields the relative contribution of the specific collision-induced transition $|m\rangle \to |k\rangle$ to the total refilling rate of level $|k\rangle$.

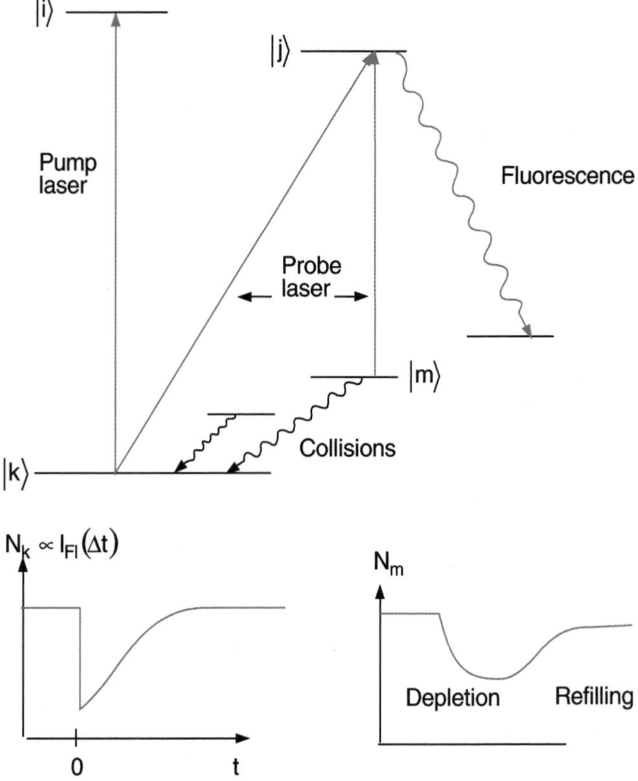


**Fig. 11.106.** Level scheme and time-dependent level populations when a short pump laser pulses depletes level $|k\rangle$, which is refilled by collisional transfer from other levels

The pump-probe technique with ultrashort pulses from femtosecond lasers has an increasing number of applications. Some of them will be discussed in the next chapter.

## SUMMARY

- There are three principal methods for the investigation of the structure and dynamics of atoms, molecules and their intermolecular interactions. These are:
  - Spectroscopic techniques
  - Scattering experiments
  - Investigation of macroscopic phenomena, such as transport properties of gases, and measurements of the virial coefficients in the equation of state for real gases
- In spectroscopic experiments the sources of information are wavelengths, intensities, polarization properties and spectral profiles of absorption and emission lines, which allow the determination of energy term values, transition probabilities and interaction potentials.
- Spectra of rotational transitions and of fine or hyperfine splittings can be measured with microwave spectroscopy.
- Molecular vibrations are mainly detected using infrared and Raman spectroscopy, where the two methods supplement each other. Transitions where the electric dipole moment changes are infrared active, those where the polarizability changes are Raman active.
- The spectral resolution of classical spectroscopy is generally limited by the resolving power of spectrographs. In Fourier spectroscopy it is limited by the maximum path difference between the two interfering beams. Laser spectroscopic techniques allow the resolution of the real line profiles of atomic or molecular transitions.
- Sub-Doppler spectroscopy is possible in collimated molecular beams or by nonlinear techniques that are based on the velocity selective saturation of molecular transitions.
- Magnetic and electric moments can be determined with radio frequency spectroscopy in molecular beams in combination with the deflection of molecules in inhomogeneous fields (Rabi method)
- Photoelectron spectroscopy measures energy and angular distribution of photoelectrons produced by photoionization of atoms or molecules. It gives information on the energies and wave functions of inner-shell electrons and their influence on molecular structure and binding energies.
- Elastic, inelastic and ionizing collisions of electrons with atoms or molecules allow the determination of excitation and ionization cross sections as a function of the electron energy and give information on correlation effects between bound electrons in atoms and molecules.
- Measurements of elastic and inelastic collision between atoms and molecules allow the determination of interaction potentials. The most detailed information is obtained from differential cross sections, from which in some cases, one can also determine potentials with nonspherical symmetric.
- In the ideal scattering experiment, differential cross sections are measured for selected inelastic transitions where the initial and the final state of one collision partner is determined together with the scattering angle. This opens the possibility of studying chemical reactions in detail on the molecular level.
- The investigation of collision processes by laser spectroscopic techniques allows the determination of integral collision cross sections in cells and of differential cross section in crossed molecular beams. The preparation of selected states before the collision and the analysis of states after the collision yields cross sections for individual collision-induced state-to-state transitions.
- Time-resolved measurements can follow up dynamical processes in atoms and molecules, such as radiative decay, collision-induced transitions or isomerization and dissociation of molecules.

  Experimental techniques for lifetime measurements include excitation with intensity-modulated light, single-photon counting with delayed coincidence or fast pump-and-probe techniques. The time resolution is currently a few femtoseconds and reaches in special cases the sub femtosecond regime.

## PROBLEMS

1. A microwave passes through HCl gas at a pressure of $p = 1$ mbar. What is the relative absorption for a pathlength of 1 m on the transition $J = 1 \to J = 2$ in the vibrational ground state, when the absorption cross section is $\sigma_{ki} = 10^{-18}\,\mathrm{m}^2$
   a) at the temperature $T = 100$ K,
   b) for $T = 300$ K?

   Use the data of Table 9.7.
2. Show that $I(\omega)$ in (11.70) is the Fourier transform of the signal $S(t)$.
3. An optical diffraction grating with 1200 grooves per mm is illuminated with the yellow sodium line with incidence angle $\alpha = 30°$. What is the distance between the two fine structure components at $\lambda_1 = 588.9$ nm and $\lambda_2 = 589.5$ nm in the detector plane of the spectrograph with a focal length $f = 1$ m of the imaging mirror?
4. a) An absorption cell filled with hydrogen gas $H_2$ is illuminated by the output beam of the argon ion laser at $\lambda = 488$ nm. Where is the Stokes line for $\Delta v = 1$ and $\Delta J = 0$?
   b) What is the minimum resolving power of the spectrograph, if the rotational Raman line $J'' = 0 \to J' = 1$ have to be separated from the Rayleigh line?
5. A laser beam with 100 mW power passes through a cell filled with a gas with absorption coefficient $\alpha = 10^{-6}\,\mathrm{cm}^{-1}$. How many fluorescence photons with $h \cdot \nu = 2.48$ eV are emitted per cm of path length of the laser beam if each absorbed laser photon produces a fluorescence photon?
   What is the output current of a photodetector, if all fluorescence photons in a solid angle of 0.2 sterad are collected, the photocathode has a quantum efficiency of 20% and the current amplification of the detector is $10^6$?
6. The collimation angle of a sodium atomic beam is $\varepsilon = 2°$. What is the residual Doppler width when a laser beam perpendicular to the atomic beam excites the atoms into the $3P$ state?
   What is the maximum value of $\varepsilon$, when the hyperfine structure of the levels below should be resolved.
   a) The $3^2P_{1/2}$ level ($\Delta\nu_{\mathrm{HFS}} = 190$ MHz)
   b) The $3^2P_{3/2}$ level ($\Delta\nu_{\mathrm{HFS}} = 16$ MHz, 34 MHz, 59 MHz)
7. a) How large should be the gradient of a 20-cm long magnetic field in order to deflect sodium atoms in the $3^2S_{1/2}$ state by an angle of 3° when the atoms fly with a velocity $v = 600$ m/s in the $z$-direction through a Rabi molecular beam apparatus?
   b) How many photons must every atom absorb from a laser beam perpendicular to the atomic beam, in order to achieve the same deflection by photon recoil?
8. The quenching cross section of the deactivation of excited sodium atoms $\mathrm{Na}(3P_{1/2})$ for collisions with $N_2$-molecules is $\sigma_q = 10^{-15}\,\mathrm{cm}^2$. At which pressure in a cell with a Na-$N_2$-mixture ($n_{\mathrm{Na}} \ll n_{N_2}$) has the effective lifetime $\tau_{\mathrm{eff}}$ decreased to $0.5\,\tau_{\mathrm{spont}}$ ($\tau_{\mathrm{spont}} = 16$ ns) at a temperature 400 K?
9. Explain, how the energy transfer cross section for the collision-induced transition $E_m \to E_i$ can be obtained from the measurement of the stationary populations $N_m$ and $N_i$?

# 12. Modern Developments in Atomic and Molecular Physics

Over the last few years, several very interesting new developments in atomic and molecular physics were initiated that have considerably widened our understanding of the interaction of light with matter and opened new possibilities for many applications. In this chapter we will briefly discuss some of the experiments that have pushed forward these developments.

## 12.1 Optical Cooling and Trapping of Atoms

For many investigations of atoms, their thermal velocity limits the spectral resolution. Although the Doppler-free techniques presented in the previous chapter can overcome the Doppler-limit, atoms with velocity $v$ only spend a limited time $\Delta t = d/v$ inside the interaction region with a laser beam of diameter $d$ and therefore the "time-of-flight" broadening $\Delta\nu = 1/\Delta t$ of spectral lines still presents a principal limit to the resolution.

In particular, for precision measurements it is desirable to keep atoms inside the observation region as long as possible, free from interactions with other particles or with the wall of the vacuum chamber. In order to reach this goal, one has to reduce the atom velocity and trap the atoms for a sufficiently long time in a small and well-defined volume away from any walls of the vacuum vessel.

This has become possible by new techniques of optical cooling and trapping, which allows one to reduce the temperature of an atomic gas down to below 1 μ K, without condensation taking place.

### 12.1.1 Photon Recoil

Assume an atom with mass $m$ is moving with the velocity $v$ into the $z$ direction. When a photon $h\nu$ with momentum $\hbar\boldsymbol{k}$ traveling into the opposite $-z$ direction is absorbed by the atom, the total momentum $p = m\boldsymbol{v} + \hbar\boldsymbol{k}$ must be conserved (Fig. 12.1). After photon absorption, the atom has a lower velocity

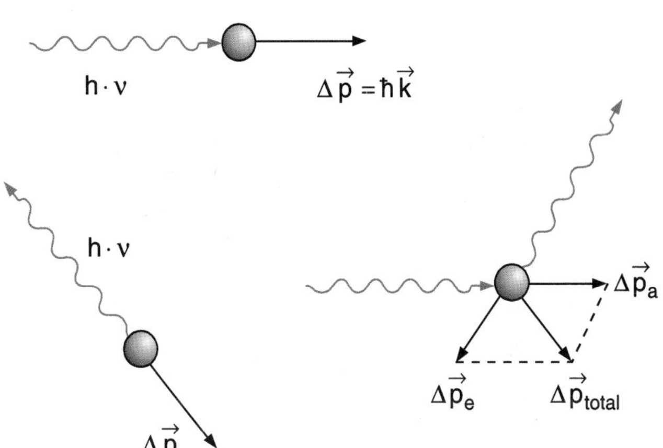


**Fig. 12.1a–c.** Momentum transfer by recoil effect by (**a**) absorption of a photon, (**b**) emission of a photon, and (**c**) total transfer for one absorption-emission cycle

$$\boldsymbol{v}' = \boldsymbol{v} - \hbar\boldsymbol{k}/m\ .$$

The velocity change $|\Delta\boldsymbol{v}| = |\boldsymbol{v}' - \boldsymbol{v}| = |\hbar\boldsymbol{k}/m| = h\nu/(mc)$ is, however, very small.

**EXAMPLE**

Sodium atoms with $m = 23\,\text{AMU} = 3.8\times10^{-26}\,\text{kg}$ absorb, on their resonance transition $3S \to 3P_{1/2}$, light with wavelength $\lambda = 589\,\text{nm}$. The photon energy is $h\nu = 2\,\text{eV} \Rightarrow \Delta v = h\nu/(mc) = 3\,\text{cm/s}$. If the initial velocity is $v = 600\,\text{m/s}$ the deceleration down to $v = 0$ requires the absorption of $2\times10^4$ photons.

The excited atom emits a fluorescence photon, which also causes a recoil and therefore a transfer of

W. Demtröder, *Atoms, Molecules and Photons*, 2nd ed., Graduate Texts in Physics,
DOI 10.1007/978-3-642-10298-1_12, 

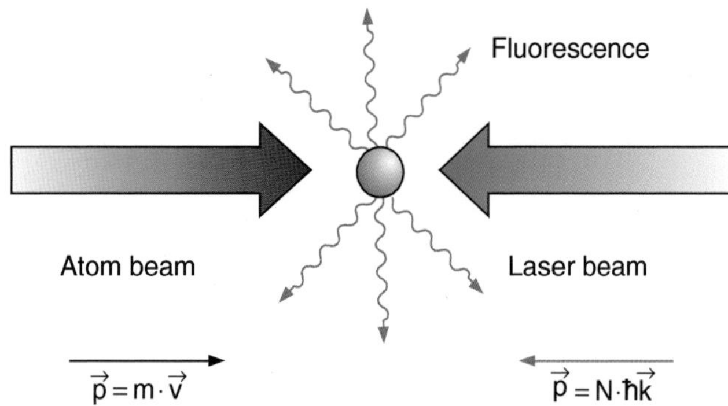


**Fig. 12.2.** For many absorption-emission cycles, the average recoil of the fluorescence photons is zero and the net momentum transfer for $N$ absorbed photons is $\Delta \boldsymbol{p} = N\hbar \boldsymbol{k}$

momentum to the atom. However, the direction of the emitted fluorescence photons is randomly distributed over the whole solid angle. If the atom undergoes many absorption-emission cycles during its interaction time with a laser beam, the time average over the momentum transfer by the emission becomes zero (Fig. 12.2).

How many absorption-emission cycles can be realized?

If we choose a true two-level system, where the absorption takes place on the transition $\langle 1| \to \langle 2|$ and the fluorescence from the excited level $\langle 2|$ terminates only on the initial level $\langle 1|$ we have a closed system, where the atom is never transferred to levels other than $\langle 1|$ or $\langle 2|$. After excitation of level $\langle 2|$ with a mean lifetime $\tau_2$ by a laser photon at time $t$ the fluorescence photon is emitted with an average delay time $\tau_2$. This means that the initial level $\langle 1|$ is only populated again at the time $t + \tau_2$. The next laser photon can therefore excite the same atom no earlier than at time $t + \tau_2$. The fluorescence rate is $N_2 A_2 = N_2/\tau_2$. Since the population $N_2$ cannot be larger than $N_1$ (otherwise, stimulated emission will bring the atoms back to level $\langle 1|$) the absorption rate cannot be larger than the fluorescence rate and the maximum rate of absorption emission cycles is $R_a = 1/(2\tau_2)$ [12.1].

## EXAMPLES

a) For sodium atoms, the lifetime of the upper $3^2P_{1/2}$ level is $\tau_2 = 16\,\text{ns}$. The maximum number of absorbed photons per second is then $N_{ph} = 6.3 \times 10^7\,/\text{s}$. With an initial velocity of $v_0 = 600\,\text{m/s}$ it takes $2 \times 10^4$ absorbed photons in order to bring the atom to rest with $v = 0$. This takes only $3 \times 10^{-4}\,\text{s}$. The deceleration $a = -\text{d}v/\text{d}t = -2 \times 10^6\,\text{m/s}^2$ is about 200,000 times larger than the earth's acceleration $g = 9.8\,\text{m/s}^2$!

b) For magnesium atoms, with $m = 24\,\text{AMU}$, which absorb on the resonance singlet transition at $\lambda = 285.2\,\text{nm}$, the deceleration is even larger because of the higher photon energy $h\nu = 3.7\,\text{eV}$ and the shorter lifetime $\tau = 2\,\text{ns}$ of the upper level. The result is that $\Delta v = -6\,\text{cm/s}$ per absorbed photon and the required number of absorbed photons to completely stop the atoms is $1.3 \times 10^4$. This gives a minimum stopping time of $3 \times 10^{-5}\,\text{s}$. During the deceleration time, the atoms travel along a path length of about 1 cm, $\left(s = -\frac{1}{2}at^2\right)$ if their velocity was initially $v_0 = 600\,\text{m/s}$.

How can such a true two-level system be realized?

In Fig. 12.3 the level scheme of the sodium transition $3S_{1/2} \to 3P_{3/2}$ is shown with the hyperfine components characterized by the quantum number $F$ of the total angular momentum $\boldsymbol{F} = \boldsymbol{s} + \boldsymbol{L} + \boldsymbol{I}$ of electron spin $\boldsymbol{s}$, orbital angular momentum $\boldsymbol{L}$ and nuclear spin $\boldsymbol{I}$. If the excitation laser is tuned to the transition $F'' = 2 \to F' = 3$, the only allowed fluorescence

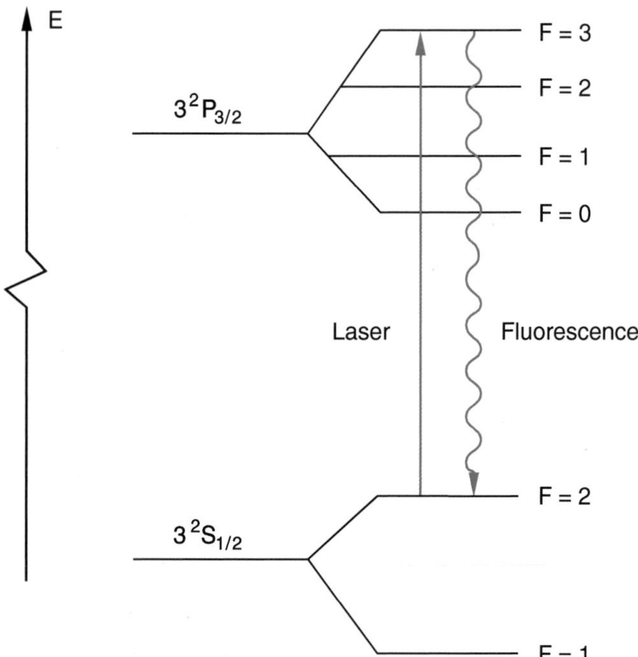


**Fig. 12.3.** Realization of a true two-level system

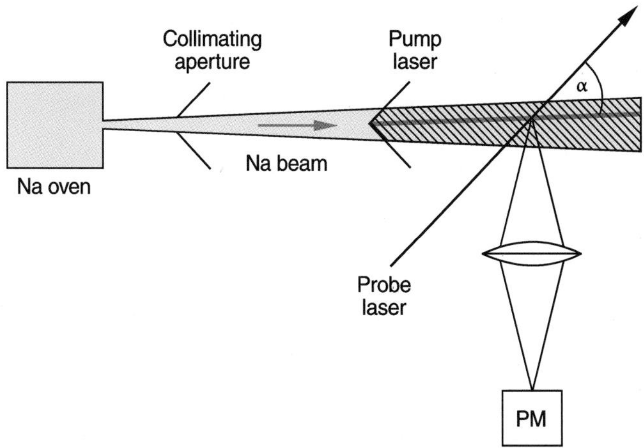


**Fig. 12.4.** Simplified experimental realization for the deceleration in a collimated beam by photon recoil

transition is $F' = 3 \rightarrow F'' = 2$ because $\Delta F = 0, \pm 1$. With this selective excitation a true two-level system is realized.

The experimental arrangement for decelerating atoms in a collimated atomic beam is shown in Fig. 12.4. The pump-laser beam is directed along the molecular beam axis opposite to the velocity of the atoms. The final velocity of the atoms is monitored by a weak probe laser crossing the molecular beam with an angle $\alpha$ against the beam axis. The absorption of the probe laser is Doppler-shifted against the resonance frequency $\omega_0$ by the amount $\Delta\omega = \boldsymbol{v} \cdot \boldsymbol{k} = vk \cos\alpha \Rightarrow \Delta\nu = (v/\lambda)\cos\alpha$.

During the deceleration process the following problem arises. Because of the Doppler-shift the frequency $\omega$ of the pump laser has to be tuned to $\omega = \omega_0 - kv = \omega_0 - kv_z$, which changes in time while the velocity decreases. Two solutions have been found: Either the laser frequency or the absorption frequency of the atoms must be tuned such that they are synchronized with the changing Doppler-shift. When using diode lasers, the laser frequency can be shifted with the electric current through the diode. The absorption frequency of the atoms can be shifted by Zeeman tuning in a magnetic field $B(z)$, which decreases with increasing $z$ in such a way that the Zeeman transition always stays in resonance with the fixed laser frequency (Fig. 12.5).

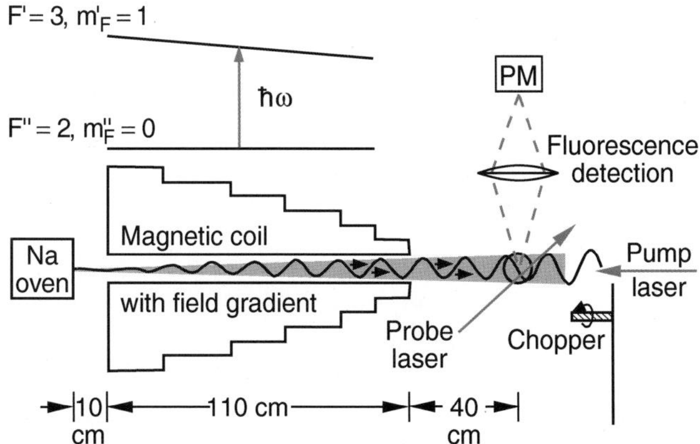


**Fig. 12.5.** Laser cooling of atoms in a collimated beam with a fixed laser frequency and Zeeman tuning of the atomic absorption frequency

### 12.1.2 Optical Cooling of Atoms

In the previous section, "one-dimensional cooling" was discussed, where only one velocity component of the atoms was decelerated. The final goal is the deceleration of all three velocity components resulting in a real three-dimensional cooling and a corresponding reduction of the gas temperature.

When atoms in a gas cell are irradiated from six laser beams pointing into the $\pm x, \pm y, \pm z$ directions (Fig. 12.6) the atoms suffer a recoil if they absorb photons from the six beams. The absorption probability depends on the detuning from the resonance frequency $\omega_0$. An atom with a velociy $\boldsymbol{v}$ can absorb laser photons

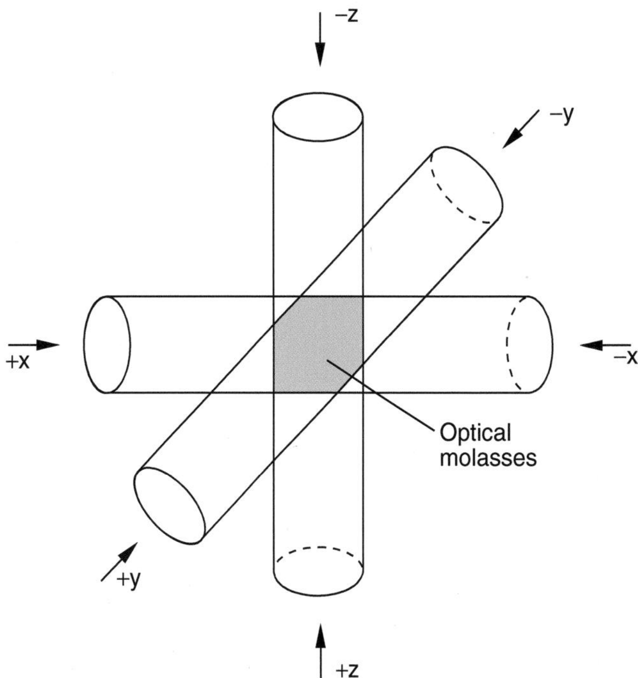


**Fig. 12.6.** Optical molasses with six pair-wise counterpropagating laser beams

in the frequency range $\omega = \omega_0 + \boldsymbol{k} \cdot \boldsymbol{v} \pm \gamma$, where $\gamma$ is the homogeneous line width (natural line width + pressure broadening + saturation broadening of the absorbing transition) with the probability (see Sect. 7.4 and eq. 11.92)

$$P(\omega) = \frac{P_0(\gamma/2)^2}{(\omega_\mathrm{L} - \omega_0 - \boldsymbol{k} \cdot \boldsymbol{v})^2 + (\gamma/2)^2} . \tag{12.1}$$

Due to the thermal velocity distribution of the atoms the total absorption profile is a Doppler-broadened Gaussian profile. The absorption frequencies of the two laser beams pointing into the opposite direction have opposite Doppler shifts (Fig. 12.7). The frequency dependence of the absorption rates for photons from two opposite laser beams is then

$$R^{\pm}(v) = \frac{R_0}{1 + \left(\frac{\omega_\mathrm{L} - \omega_0 \pm \boldsymbol{k}\boldsymbol{v}}{\gamma/2}\right)^2} . \tag{12.2}$$

The total net recoil force component $F_i$ $(i = x, y, z)$ is

$$F_i = -\left[R^+(v_i) - R^-(v_i)\right] \hbar \boldsymbol{k} . \tag{12.3}$$

Inserting (12.2) into (12.3) gives with $\delta = \omega_\mathrm{L} - \omega_0$

$$F_i = + \frac{16 R_0 \delta k v \hbar k}{\gamma^2 \left[1 + \frac{8}{\gamma^2}\left(\delta^2 + (kv)^2\right) + \left(\frac{\delta^2 - (kv)^2}{\gamma^2/4}\right)^2\right]} . \tag{12.4a}$$

For $kv \ll \omega_\mathrm{L} - \omega_0 = \delta$ the net force becomes

$$F_i = +a v_i \quad \text{with} \quad a \approx \frac{16 \delta \hbar k^2 R_0}{\gamma^2 \left[1 + (2\delta/\gamma)^2\right]^2} , \tag{12.4b}$$

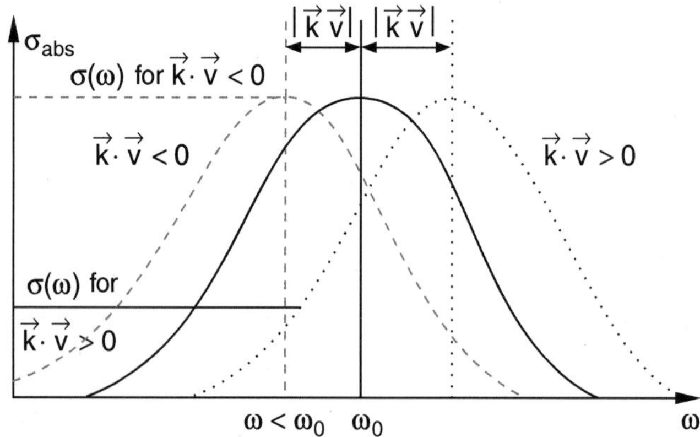


**Fig. 12.7.** For $\omega < \omega_0$, the absorption probability is larger for $\boldsymbol{k} \cdot \boldsymbol{v} < 0$ than for $\boldsymbol{k} \cdot \boldsymbol{v} > 0$

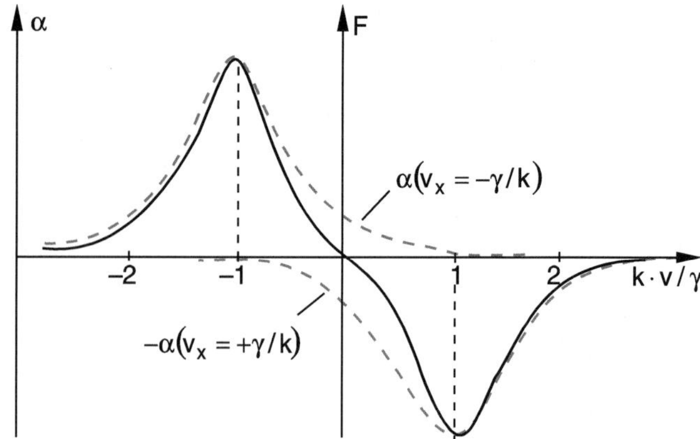


**Fig. 12.8.** Frictional force $F(\mathrm{v})$ (solid curve) and absorption coefficient $\alpha(\mathrm{v})$ (dashed curve) in an optical molasses for a red detuning $\delta = -\gamma$. The *dashed curve* shows the absorption profiles by atoms moving with $v_x = \pm\gamma/k$ for a single laser beam propagating in the $x$ direction

which is proportional to the atomic velocity. Its dependence on the detuning $\delta$ is illustrated in Fig. 12.8.

When the laser frequency $\omega_\mathrm{L}$ is slightly smaller than the resonance frequency $\omega_0$ $(\delta < 0)$ atoms moving towards a laser beam have a larger probability to absorb a photon than those atoms that move in the direction of the beam. Therefore in this case the atoms are pushed towards the overlap region of the six laser beams [12.2].

An atom with mass $M$, moving inside the overlap region of the six laser beams experience a force with the three components $F_i(v_i) = -a v_i$ $(i = x, y, z)$, that damps its velocity. From the relation

$$\mathrm{d}v/\mathrm{d}t = F/M \Rightarrow \mathrm{d}v/v = -\frac{a}{M}\mathrm{d}t$$

we obtain the time-dependent velocity components

$$v_i = v_{i0} \mathrm{e}^{-(a/M)t} . \tag{12.5}$$

The velocity of an atom that experiences photon recoil in the overlap region of six laser beams in $\pm x, \pm y$ and $\pm z$ directions, decreases exponentially with the damping time $t_\mathrm{D} = M/a$.

The ensemble of atoms under the influence of these frictional forces is called ***optical molasses***.

EXAMPLE

For Na atoms one obtains, for a detuning $\delta = 2\gamma$ and with the data of the example in the previous section, the constant $a = 1 \times 10^{-20}$ Ns/m and a damping time $t_D = 2.3\,\mu$ s. The atoms move in this overlap region like particles in a viscous molasses.

What are the lowest temperatures that can be reached by optical cooling?

The smaller the velocity of the atoms become, the smaller is their Doppler-shift and the closer the laser frequency $\omega_L$ must approach the center frequency $\omega_0$ of the atomic transition in order to stay in resonance. If the velocity $v$ of the atoms has decreased to $v < \gamma/k$, the detuning $\delta = \omega_L - \omega_0$ has to be smaller than the homogeneous width $\gamma$ of the atomic transition. The cooling efficiency decreases with decreasing $\delta$ and for $\omega_L = \omega_0$ the net cooling force becomes zero.

Because of the recoil effect during the absorption or emission of photons each atom performs a random motion comparable to the Brownian motion (see Sect. 2.3.1). Although the time average $\langle v \rangle$ of the atomic velocities is zero, the mean square value $\langle v^2 \rangle$ increases by the random recoil forces, which means that the temperature increases. This increase has to be compensated by the cooling force which only acts for $\delta < 0$. The minimum temperature $T_{min}$ is reached when even for $\delta < \gamma$ the energy $\Delta E < \hbar\gamma$ taken away from the atom by optical cooling can just compensate for the thermal energy $k_B T$, added to the atom by the random recoil momentum. This gives for the minimum temperature the "*Doppler limit*"

$$T_{min} = \hbar\gamma / k_B \,. \qquad (12.6)$$

EXAMPLE

For a natural linewidth $\gamma = 10\,\text{MHz}$ the minimum temperature is $T_{min} = 72\,\mu$ K.

Experiments have shown, however, that even lower temperatures below 1 µ K can be reached. Therefore another cooling effect has to be present, which is related to the shift of atomic energy levels in the field of the standing laser waves. The optical transition probability depends on the electric field strength in the standing laser wave and leads to the effect, that on the average the absorbed photon has a smaller energy than the emitted fluorescence photon, thus taking away energy from the atom (Sisyphus cooling) [12.3, 4].

This cooling technique results in a lower limit for the attainable temperature of atoms with mass M, which is called **the recoil limit**

$$T_{recoil} = h^2/(2M \cdot \lambda^2 \cdot k_B).$$

EXAMPLE

For the Na-atom with M = 23AMU and $\lambda = 590$nm one obtains

$$T_{recoil} = 1.2 \cdot 10^{-6}\text{K} = 1.2\mu\,\text{K}$$

### 12.1.3 Optical Trapping of Atoms

Although optical cooling decreases the volume of an atomic cloud in the velocity space, it does not concentrate them in real space. This can be achieved with a device called a magneto-optical trap (MOT), illustrated in Fig. 12.9. The MOT consists of a pair of anti-Helmholtz coils through which a current is sent into opposite directions. With the $z$-axis as the symmetry axis, the magnetic field produced by the coils in the vicinity of the trap center at $z = 0$ is

$$B(z) = bz \,. \qquad (12.7)$$

The magnetic field is zero at the center $z = 0$ and increases linearly with increasing distance from the center. The field lines are indicated as blue curves in Fig. 12.9a.

In the magnetic field the atomic energy levels experience a splitting into Zeeman components with the energies

$$E_i = -\boldsymbol{p}_m \cdot \boldsymbol{B} = g_F \mu_B m_F |\boldsymbol{B}| \,. \qquad (12.8)$$

Where $\mu_B$ is the Bohr-magneton, $F$ is the quantum number of the total angular momentum (including nuclear spin) and $m_F$ the magnetic quantum number $-F \le m_F \le F$.

In Fig. 12.10b the $z$-dependence of the Zeeman energy is plotted for the two levels with $F = 0$ and $F = 1$.

The six laser beams passing through the MOT are circularly polarized (Fig. 12.10a). They induce transitions with $\Delta m_F = \pm 1$. If the laser frequency $\omega_L$ is

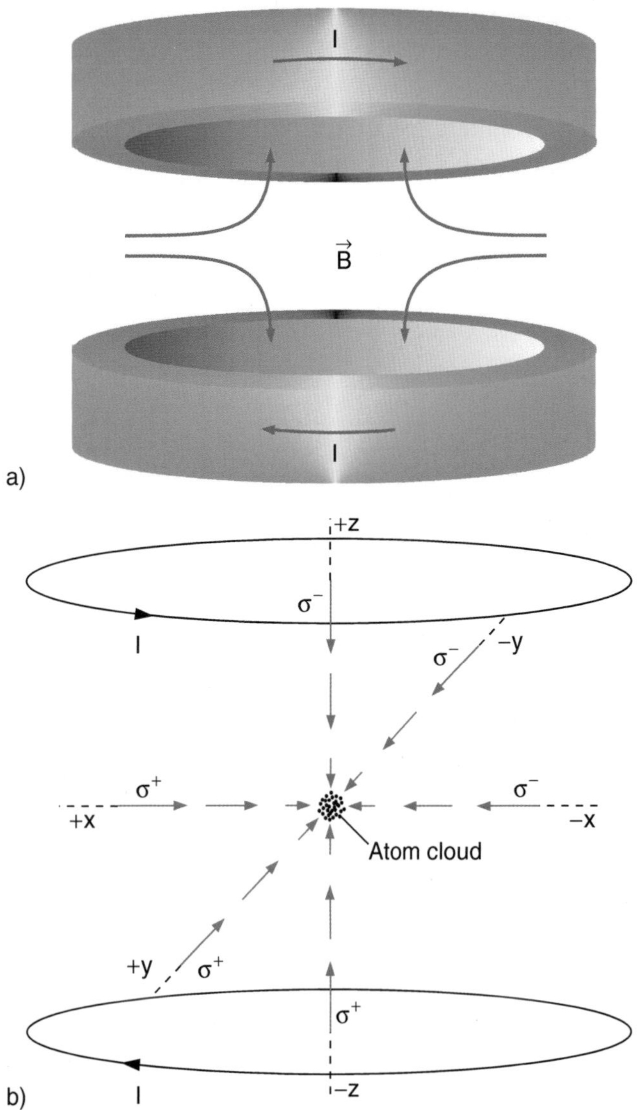


**Fig. 12.9.** Principal arrangement of the magneto-optical trap MOT

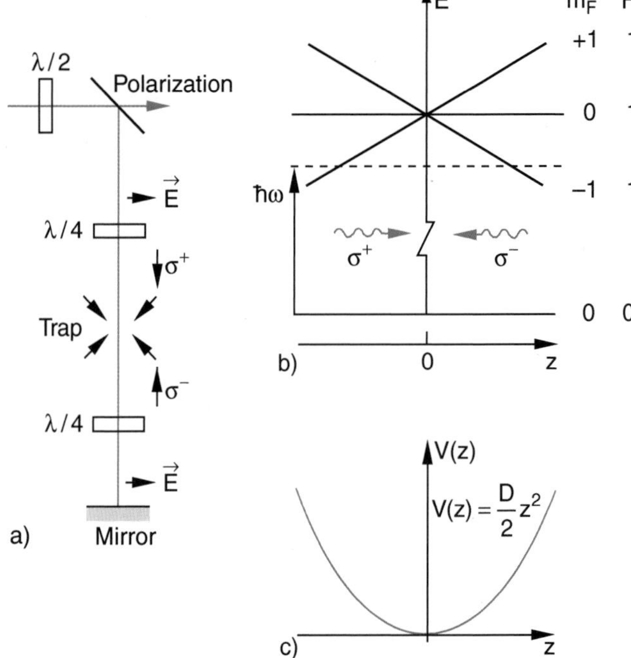


**Fig. 12.10.** (**a**) Generation of two counter propagation $\sigma^+$ and $\sigma^-$ polarized laser beams. (**b**) Zeeman levels in the MOT. (**c**) Trap potential

below the resonance frequency $\omega_0$, atoms in the region $z>0$ absorb preferentially $\sigma^-$-light inducing $\Delta m_F = -1$ transitions, while for $z<0$ mainly $\sigma^+$-light is absorbed. For $z=0$ the absorption rates for $\sigma^+$ and for $\sigma^-$ light are equal. Since for $\omega<\omega_0$ atoms moving towards the laser beam propagation suffer a larger average recoil than those moving into the direction of the beam propagation (see Sect. 12.1.2) this pushes all atoms moving outwards back towards the center. The gas of cold molecules is therefore compressed into a dense cloud around the center of the MOT.

The $z$-dependent force acting on the atoms is (similar to the discussion in the previous section)

$$F_z(z) = R_{\sigma^+}(z)\hbar \boldsymbol{k}_{\sigma^+} + R_{\sigma^-}(z)\hbar \boldsymbol{k}_{\sigma^-}\,, \tag{12.9}$$

determined by the difference of the transferred momentum per second (note, that $\boldsymbol{k}_{\sigma^+}$ and $\boldsymbol{k}_{\sigma^-}$ are antiparallel). For a Lorentzian absorption profile with halfwidth $\gamma$ the absorption rates become

$$R_{\sigma^\pm} = \frac{R_0}{1+\left[\frac{\omega_L-\omega_0\pm p_m bz/\hbar}{\gamma/2}\right]^2}\,. \tag{12.10}$$

In the vicinity of $z=0$ we can expand the bracket in (12.10) and obtain for the restoring force acting on the atoms

$$F_z = -Dz \quad \text{with} \quad D = R_0 p_m b \frac{16k\delta}{\gamma^2(1+4\delta^2/\gamma^2)^2}\,. \tag{12.11}$$

This is a force that linearly increases with $z$. Because $Fz = -\partial E_{pot}/\partial z$ we can attribute to the MOT a harmonic potential

$$E_{pot}(z) = \frac{1}{2}Dz^2\,, \tag{12.12}$$

which stabilizes the atoms around $z=0$.

Including the velocity-dependent force, discussed in the previous section, the total force acting on the atoms is

$$F_z = -Dz - av_z \tag{12.13}$$

resulting in a damped oscillation of the atoms with mass $M$ around the trap center with a frequency

$$\Omega = \sqrt{D/M} \tag{12.14}$$

and the damping constant

$$\beta = a/M\,. \tag{12.15}$$

## EXAMPLES

For Rubidium atoms with $M = 1.4 \times 10^{-25}\,\mathrm{kg}$ the absorption wavelength is $\lambda = 785\,\mathrm{nm} \Rightarrow k = 8 \times 10^6\,\mathrm{m}^{-1}$. With a laser detuning $\delta = \gamma$ and an absorption rate $R_0 = \gamma/2$ one obtains $a = 4 \times 10^{-21}\,\mathrm{Ns/m}$. With a magnetic field gradient $b = 0.1\,\mathrm{T/m}$ and the magnetic moment $p_\mathrm{m} \approx \mu_\mathrm{B}$ the constant $D = 2.37 \times 10^{-18}$ can be calculated. This gives the oscillation frequency $\omega = 4{,}100\,\mathrm{s}^{-1}$ and the damping constant $\beta = 1.2 \times 10^{-2}$. The atoms relax with a time constant of 12 ms after about 50 oscillation periods against the trap center.

---

Up to now we have only regarded the movement of the atoms in the $z$ direction.

For the $x$ and $y$ directions, similar considerations hold. The magnetic field of the anti-Helmholtz coils is a quadrupole field. From Maxwell's equation div$B = 0$ follow the relations

$$\frac{\partial B_x}{\partial x} = \frac{\partial B_y}{\partial y} = -\frac{1}{2}\frac{\partial B_z}{\partial z}\,. \tag{12.16}$$

The restoring forces in the $x$ and $y$ directions are therefore one half of that in the $z$ direction.

**Note:**

Besides these restoring forces due to the recoil by photon absorption, there is, of course, also the magnetic force

$$\boldsymbol{F}_\mathrm{m} = -\boldsymbol{p}_\mathrm{m} \cdot \mathbf{grad}\,\boldsymbol{B}\,. \tag{12.17}$$

Inserting the relevant figures for the experimentally realized field gradients, it turns out that this force is much smaller then the recoil force.

For more detailled information on magneto-optical traps see [12.5, 6].

### 12.1.4 Bose–Einstein Condensation

If the density $n$ of cold atoms becomes so large, that the mean distance $d$ between atoms is smaller than the de Broglie wavelength

$$d < \lambda_\mathrm{DB} = \frac{h}{mv} \Rightarrow n > \frac{1}{\lambda_\mathrm{dB}^3} \tag{12.18a}$$

the different atoms can no longer be distinguished, because each atom can be only localized within its de Broglie wavelength. More detailed calculations show that above a critical density

$$n_\mathrm{c} = \frac{2.612}{\lambda_\mathrm{DB}^3} \tag{12.18b}$$

the properties of the gas drastically change. Since the average kinetic energy is

$$\frac{1}{2}m\left\langle v^2\right\rangle = \frac{3}{2}k_\mathrm{B}T$$

we can express the de Broglie wavelength by

$$\lambda_\mathrm{DB} = \frac{h}{\sqrt{3mk_\mathrm{B}T}} \tag{12.19}$$

and obtain for the critical density the expression

$$n_\mathrm{c} = 13.57(mk_\mathrm{B}T)^{3/2}/h^3\,, \tag{12.20}$$

which depends on the temperature $T$ of the atomic gas. Below a critical temperature $T_\mathrm{c}$ a phase transition takes place for a gas of bosonic atoms from a classical gas to a degenerate gas of indistinguishable atoms. Bosonic atoms have an integer total spin quantum number and each energy level can be occupied by many atoms, different from Fermonic atoms with a spin of 1/2, where each energy level can be occupied by at most two particles. At a sufficiently low temperature, all bosonic atoms occupy the lowest possible energy state. Such an atomic ensemble represents a degenerate gas, which is called a **Bose–Einstein condensate (BEC)** after the Indian mathematician *Satyendra Nath Bose*, who in 1922 developed the concept of indistinguishable photons and gave a mathematical proof of Planck's radiation formula. He sent his ideas to *Albert Einstein* who generalized Bose's model to include not only photons but all particles with integer spin quantum numbers. They predicted the behavior of Bosons for transitions from a classical to a degenerate gas. Both authors published the results in 1924 [12.7].

## EXAMPLE

The Na atom has an electron spin of $1/2$ and a nuclear spin of $3/2$. The total spin quantum number in the $^2S_{1/2}$ ground state is $F = 1$ or $F = 2$. The sodium atom is therefore a bosonic atom. The Li atom has the two isotopes $^6$Li and $^7$Li with nuclear spins $I = 1$ and $I = 3/2$. Together with the electron spin $s = 1/2$ the total spin quantum number for $^6$Li is odd (it is a Fermion) and for $^7$Li it is even (it is a boson).

Inserting the relevant figures into equation (12.20), it turns out that the temperature reached by optical cooling is still not low enough to reach the critical temperature for BEC. Therefore, additional cooling techniques have to be applied. One of these methods is based on evaporation cooling. It is in principle the same technique one uses for cooling a hot cup of coffee by blowing over the liquid surface. This removes the fastest molecules in the vapor phase above the surface and decreases the average kinetic energy, i.e., the temperature of the remaining molecules. The basic principle applied to the molecules in the MOT is the following [12.8].

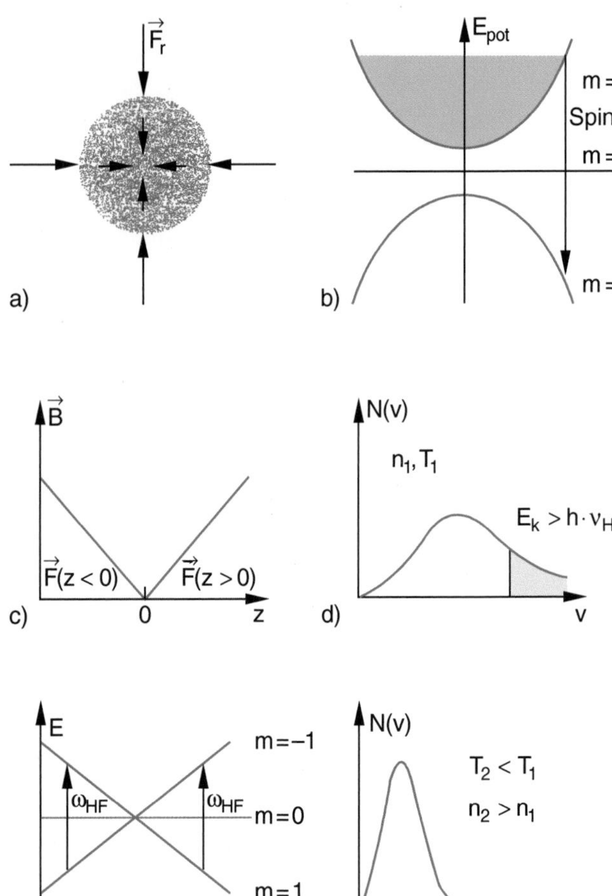


**Fig. 12.11a–f.** Evaporation cooling in a magnetic trap. (**a**) Atomic cloud in the MOT. (**b**) Potential of the pure magnetic trap. (**c**) Forces in the magnetic trap. (**d**) Velocity distribution of atoms. The *shadowed area* is released from the trap. (**e**) Radiofrequency induced spin flips. (**f**) Velocity distribution of cooled atoms

At first the cold atoms are transferred from a MOT to a pure magnetic trap that has no cooling lasers, by pushing them with a laser beam into the wanted direction. In this trap that is formed by an inhomogeneous magnetic field, they are kept not by the recoil forces of lasers like in the MOT, but by the magnetic force $F = -p_m$ **grad** $B$ due to their magnetic moment $p_m$, (Fig. 12.11a,b). Although this force is much smaller than the recoil force, the atoms, which are now already very cold, can be trapped since their kinetic energy is small. Here they have the potential energy $E_{pot} = -\boldsymbol{p}_m \cdot \boldsymbol{B}$. Due to the distribution of their thermal energy they fill the trap potential up to the energy $E_{max}$.

Now the trap is irradiated by a radio-frequency that induces flips of the electron spin $s$, when its frequency $\nu_{rf} = \Delta E/h$ matches the energy difference $\Delta E = E_\uparrow - E_\downarrow = g_s\mu_B \cdot B = 2\mu_B B$ (see Sect. 5.6).

The atoms that have suffered a spin flip are automatically pushed out of the trap because their potential is now repulsive (Fig. 12.11b). Choosing the frequency in such a way that only the hottest atoms can make a spin flip, results in a decrease of the mean energy of the trapped atoms. During the evaporation of atoms, the temperature drops below the critical temperature. The radial density profile of the atoms becomes very narrow because all atoms condense into the lowest potential energy state of the magnetic trap increasing the atom density at the center. This can be monitored by sending a widened weak laser beam through the condensate, which is partly absorbed (Fig. 12.12). With a CCD-array, the transmitted intensity is detected, which is a measure of the spatial dependence of the absorption and thus of the density of atoms (Fig. 12.12a–d).

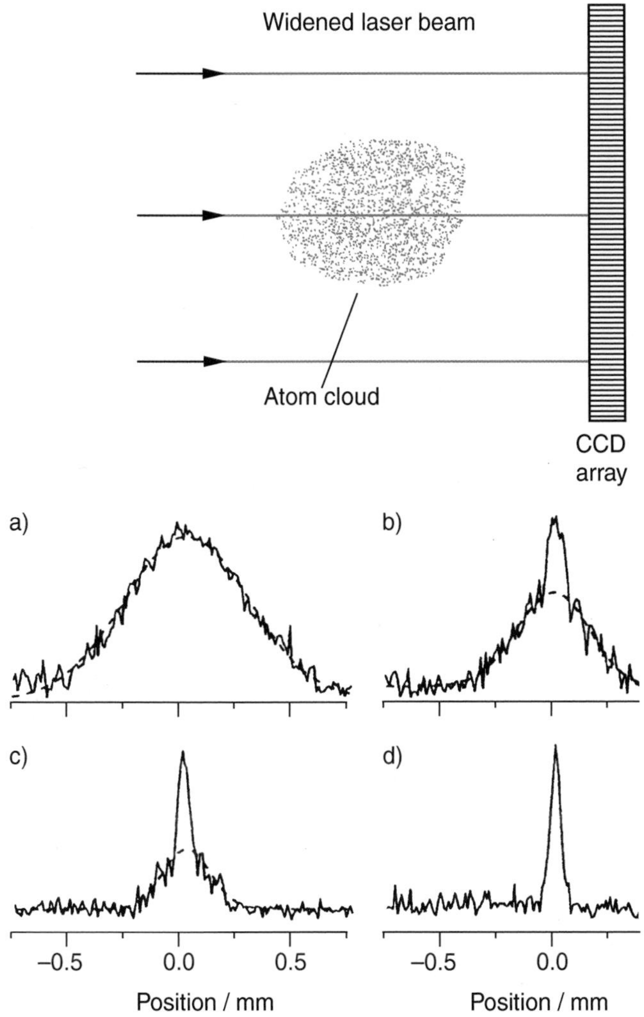


**Fig. 12.12a–d.** Monitoring BEC by the spatial absorption profile of the transmitted probe laser. Density profiles of atomic cloud with $T = 1.2\,\mu\mathrm{K}$ (**a**), 310 nK (**b**), 170 nK (**c**), and below the critical temperature of BEC (**d**) [12.8]

By this technique the first experimental proof of Bose–Einstein condensation was reported in 1996 by *E. Cornell* and *C. Wieman* in Boulder, Colorado [12.8] and independently by *W. Ketterle* and his group at MIT in Boston [12.9, 10]. Both groups received the Nobel Prize in Physics in 2001 for this pioneering work. Meanwhile BEC of multi-component atoms have been realized, where Fermions are cooled by collisions with cold Bosons (sympathetic cooling) [12.11]. Recently, even Bose–Einstein condensation for molecules has been achieved [12.12].

The BEC behaves in many aspects like a supraliquid with zero viscosity. Spectroscopy of atoms in a Bose–Einstein condensate gives a lot of information on collective effects of many atoms, which are all in the same state [12.13]. One of the fascinating effects is the release of a coherent beam of atoms out of the BEC. Such a coherent beam represents a large flux of atoms that all have nearly the same energy. Because of its resemblance to a coherent beam of photons, which is a laser beam, it is called an atom laser. Two beams from the same BEC can interfere with each other if they are spatially overlapping [12.14].

These atom lasers can find applications in atom interferometry [12.15] and in surface science for inspecting surfaces with a spatial resolution in the subnanometer range. Recent investigations of the characteristics of such condensates for different atoms concentrate on the dynamics in the BEC, such as collision processes, formation of molecules, generation of turbulences such as vortices with quantized angular momenta, frictional effects and factors influencing the lifetime of a BEC. Of interest are the differences and similarities between BEC and superfluid liquids and electromagnetic induced optical transparency.

### 12.1.5 Molecular Spectroscopy in a MOT

Since the relative velocity of cold atoms in a MOT is very small, they stay for a longer time together when they approach each other. During this time the atom pair can be excited by a laser into a stable molecular state (Fig. 12.13). By a second laser stimulated emission can be induced terminating on levels of the electronic ground state below the dissociation threshold. In this way molecules can be formed by photo-association. If both lasers are narrow band single mode lasers spectroscopy of levels close to the dissociation energy of the diatomic molecule can be performed with ultrahigh resolution.

For these high-lying levels, the mean distance between the two atoms can become very large and several small effects are significant for the small binding energy, such as magnetic interactions between electron and nuclear spins, or retardation effects due to the finite time delay at which a change of the electron position at one atom influences the second atom. Such effects can even cause a second shallow minimum in the potential curve of the diatomic molecule at internuclear distances of about 10–30 nm besides the deep minimum at the equilibrium distance $R_e \approx$

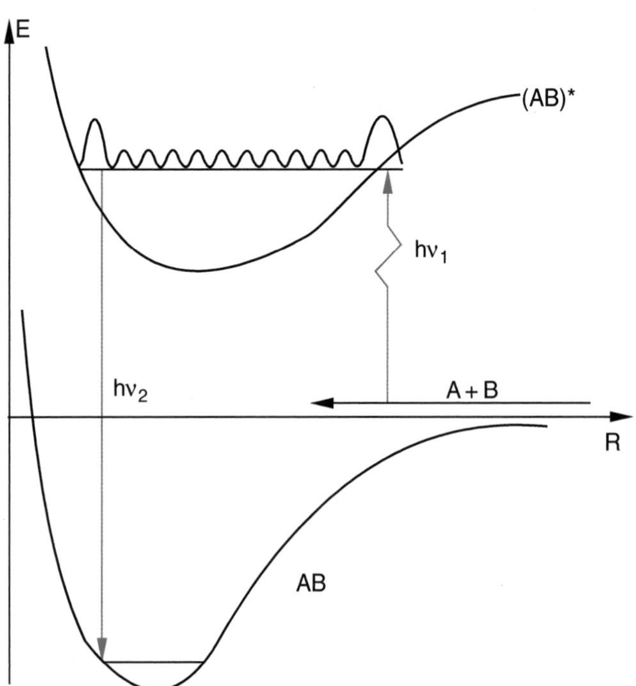


**Fig. 12.13.** Formation of cloud molecules by photoassociation of a pair A + B

0.2–0.3 nm [12.16]. A very ingenious method to produce Bose-Einstein condensation of molecules from atoms in a BEC is based on the energy difference of atoms with electron spin and of molecules in a magnetic field [12.18].

In Fig. 12.14 the potential energy of an atomic collision pair $Cs + Cs$ and that of the molecule $Cs_2$ is plotted as a function of the magnetic filed strength B. The field dependence $E_{pot}(B)$ is stronger for the collision pair with each atom in the hyperfine level with the quantum numbers $F = 3$, $M_F = 3$. $L = 0$ and the magnetic moment $\mu = 1.5\mu_B$ than for the molecule in the 4g hyperfine level with $F = 4$ and $\mu = 0.93\mu_B$. At a critical field strength $B_c$ the lines cross each other. This implies that at this magnetic field strength the energy of the atom pair is the same as that of the molecule in the specific level (see insert in Fig. 12.14) This resonant condition is called Feshbach resonance [12.17]. At magnetic fields $B < B_c$ the molecule has a lower energy than the atom pair and is therefore stable. When the magnetic field is slowly (adiabatically) scanned from higher fields over the Feshbach resonance to lower fields, stable molecule can be formed. This can be monitored by the corresponding decrease of the atomic density in the BEC. When the field B is again increased beyond $B_c$ the molecules become unstable and the atomic density increases again.

Since the dimer $Cs_2$ has an integer total spin ($S = 0$ and $I = 3$) it is a Boson and can form a Bose-Einstein condensate [12.18]. These dimers represent huge molecules with internuclear separation of about 100nm! Their spectroscopy gives information about interatomic forces at large distances and allows determining the very last vibration level directly below the dissociation energy.

The study of molecule formation at these extremely low temperatures gives information on the velocity dependence of reactive collisions and on the dependence of collision cross sections on the relative spin orientation.

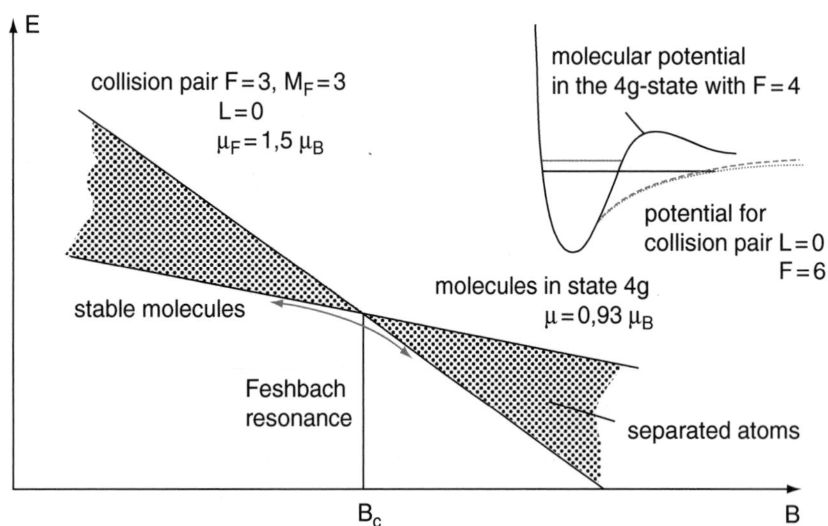


**Fig. 12.14.** Potential energy of the collision pair $Cs(F = 3,\ M_F = 3) + Cs(F = 3,\ M_F = 3)$ and the molecule $Cs_2$ as a function of an external magnetic field [12.17]

## 12.2 Time-resolved Spectroscopy in the Femtosecond Range

With the development of femtosecond lasers, a whole new area of research has begun. Now molecular vibrations or the formation and dissociation of molecules can be observed in real-time. The old dream of photochemists to be able to control chemical reactions by light has, at least for some favorable examples, become true.

In this section we will discuss some exciting examples regarding these new developments in the application of femtosecond lasers.

### 12.2.1 Time-resolved Molecular Vibrations

Stationary spectroscopy measures time averages of molecular states. For example, the average internuclear distance of diatomic molecules in the vibrational level $\langle v|$ obtained from measurements of the rotational constant $B_v$, is the time average over many vibrational periods (see Sect. 9.5).

With femtosecond optical pulses the motion of the vibrating nuclei can be made "visible" in a similar way as fast periodic macroscopic motions can be visualized by a stroboscopic technique where the body is made visible by pulsed illumination with a repetition frequency that is adapted to that of the motion. We will illustrate the technique using the $Na_2$ molecule as an example (Fig. 12.15).

A femtosecond laser pulse (pump-pulse) with time duration $\Delta t$ excites the molecules from their ground state into a superposition of vibrational levels $v'$ within the energy interval $\Delta E = h/\Delta t$ in an electronically excited state. The different vibrational levels are excited coherently and form a wave packet $|\psi(R,t)|^2$, which moves in the $(R,t)$ diagram of Fig. 12.16 periodically between the minimum value $R_{min}$ and the maximum value $R_{max}$ in the potential of the upper state (which is, for our example, the $D^1\Sigma_u$ state of $Na_2$). If the excited molecule is now further excited by a second laser pulse it can be ionized. The ionization depends on the internuclear distance $R$ at which the second pulse finds the molecule. For $R = R_{max}$ the probe laser pulse can reach the ionization limit for the dissociation

$$Na_2{}^* + h\nu \rightarrow Na(3s) + Na^+ ,$$

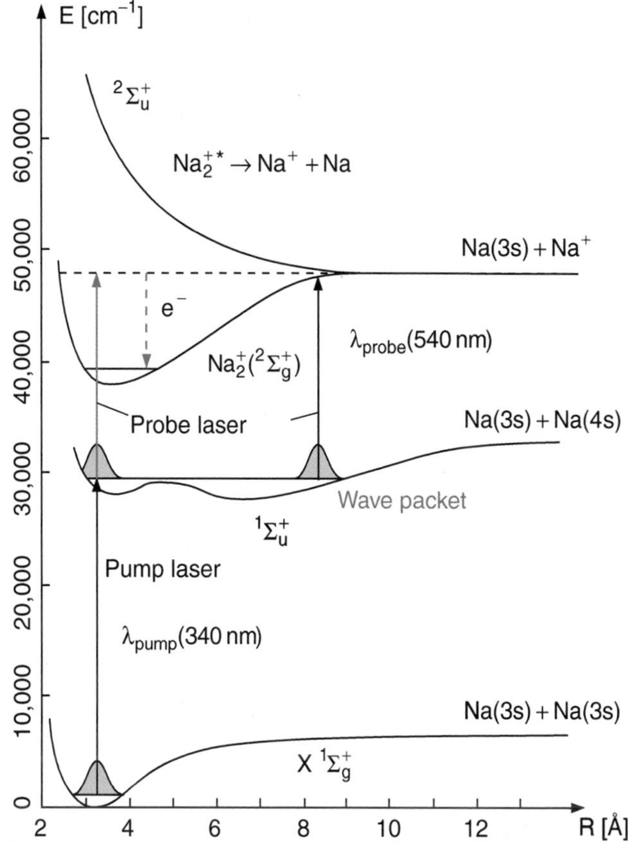


**Fig. 12.15.** Level scheme for the pump and probe technique applied to the $Na_2$ molecule [12.19]

whereas for $R = R_{min}$ it cannot reach the repulsive potential curve of the $Na_2^+(^1\Sigma_u)\Sigma_u$ potential but ends in a virtual level of the system $Na_2^+ + e^-$, which immediately converts into a real bound level $(v_1^+ J^+)$ in the $^2\Sigma_g^+$ state of the molecular ion $Na_2^+$ because

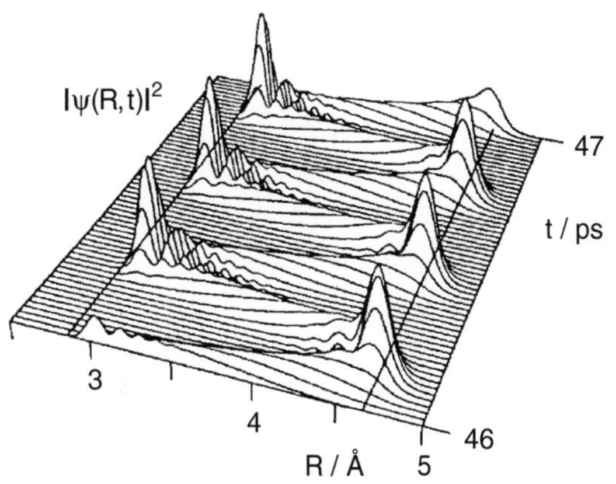


**Fig. 12.16.** Motion of wave packet in a bound potential

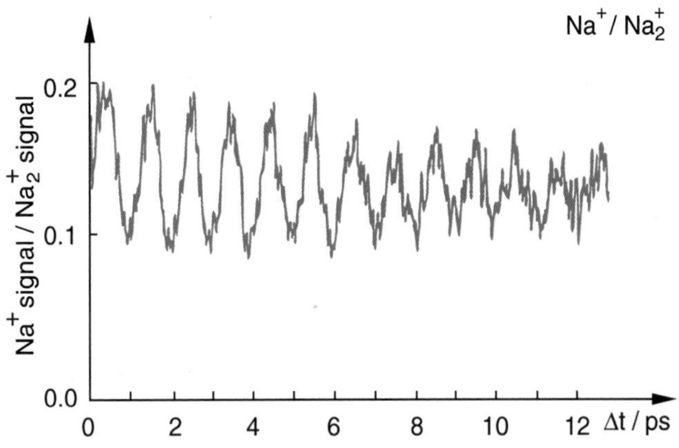


**Fig. 12.17.** Ratio $S\left(\mathrm{Na}^+\right)/S\left(\mathrm{Na}_2^+\right)$ as a function of delay time $\Delta t$ between pump and probe pulse [12.20]

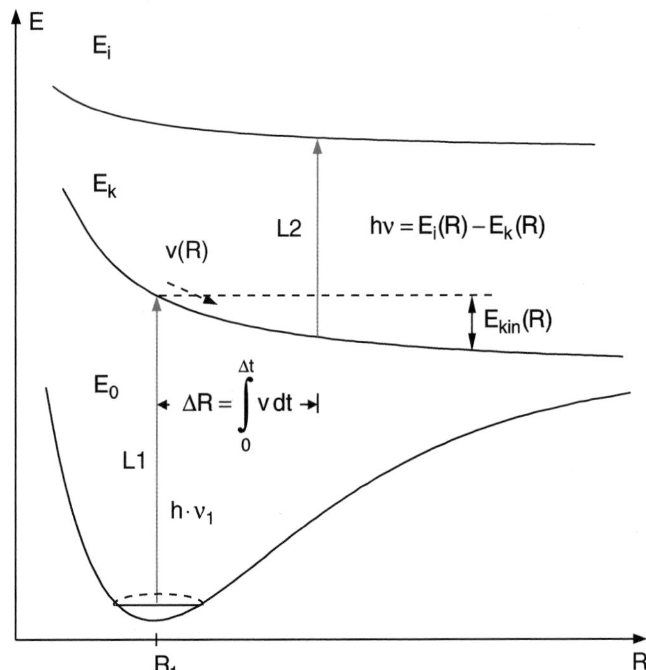


**Fig. 12.18.** Level scheme for probing molecular photodissociation on a femtosecond scale

the photo-electron carries away the excess energy as kinetic energy.

If the ratio $N_1(\mathrm{Na}^+)/N_2\left(\mathrm{Na}_2^+\right)$ is measured as a function of the delay time between the pump- and probe-pulse, the oscillatory curve of Fig. 12.17 is obtained, from which the vibrational period in the intermediate level of the $^1\Sigma_\mathrm{u}$ state can be inferred.

### 12.2.2 Femtosecond Transition State Dynamics

With femtosecond laser pulses, short-lived transient states of molecular systems can be studied "in real-time". One example is the photo-dissociation of a diatomic molecule, which has a bound ground state $|0\rangle$, but a repulsive potential in the excited state $|k\rangle$ (Fig. 12.18). If the pump laser pulse reaches the excited state potential at an internuclear distance $R_1$ the fragments of the dissociated molecule have a kinetic energy $E_\mathrm{kin}(R) = E_\mathrm{pot}(R_1) - E_\mathrm{pot}(R = \infty)$. Which depends on the slope of the potential $E_\mathrm{pot}(R)$ in the upper state as a function of $R$. If the probe pulse has a delay $\Delta t$, it excites the molecule further into the state $|i\rangle$ at a separation $R = R_1 + \Delta R$ with $\Delta R = \int_0^{\Delta t} v\,\mathrm{d}t$. The excitation into the state $|i\rangle$ can be detected by the fluorescence emitted from $|i\rangle$. The maximum fluorescence is reached when the separation $R$ corresponding to the time delay $\Delta t$ has a value for which $\Delta E(R) = E_i(R) - E_k(R) = h\nu_2$.

Varying the time delay $\Delta t$ one can map out the difference $\Delta E(R)$ of the two potential curves.

**EXAMPLE**

The iodine atoms in the dissociating $\mathrm{I}_2$ molecule have with $E_\mathrm{kin} = 15\,\mathrm{meV}$ a velocity $v = 100\,\mathrm{m/s}$. After a time delay of $\Delta t = 10\,\mathrm{ps}$ their separation $R$ has increased by 1 nm. In order to map out the potential curve with a resolution of 0.1 nm, a time resolution of 1 ps is demanded.

A second example is the photodissociation of the NaI molecule, which shows at the internuclear distance $R_\mathrm{c}$ an avoided crossing of the ground state potential, dissociating into $\mathrm{Na}^+ + \mathrm{I}^-$ and the excited state potential, which converges towards the lower energy of the neutral products $\mathrm{Na} + \mathrm{I}$ (Fig. 12.19a). If the molecule is excited at the distance $R_1$ into the repulsive part of the excited state potential, the wave packet moves "downhill" towards larger separations $R$. At the avoided crossing of the two potential curves part of the wave packet is transmitted and forms neutral Na and I atoms. The other part is reflected back towards small $R$ values, until it is again reflected at the inner part of the repulsive potential. It reaches the crossing point again where it is partly transmitted and partly reflected, etc.

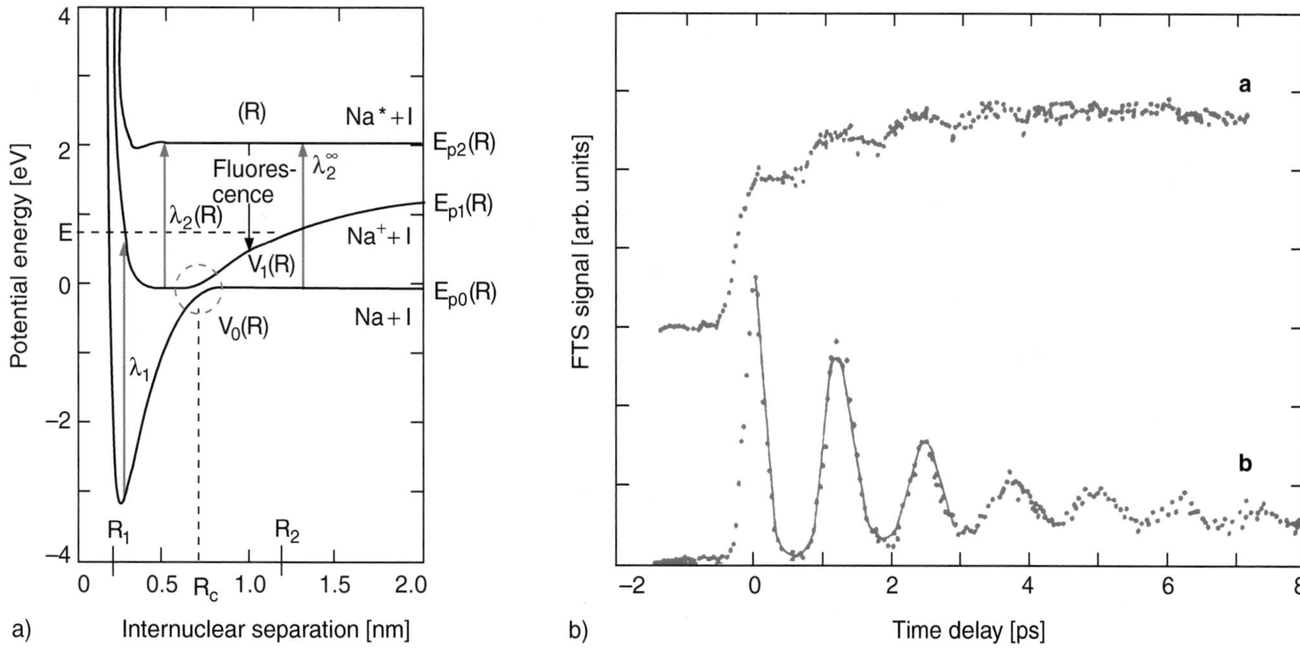


**Fig. 12.19.** (**a**) Potential diagram of NaI with the pump transition at $\lambda_1$ and the tunable probe pulse at $\lambda_2(R)$. (**b**) Fluorescence intensity $I_{\mathrm{Fl}}(\Delta t)$ as a function of the delay time $\Delta t$ between pump and probe pulses with $\lambda_2$ tuned to the atomic Na* transition (*curve a*) and $\lambda_2$ tuned to $\lambda_2(R)$ with $R < R_c$ (*curve b*) [12.21]

When the wavelength $\lambda_2$ of the probe laser is tuned to the atomic Na transition, the laser-induced atomic resonance fluorescence is a measure for the number of Na atoms formed by the molecular excitation (curve (a) in Fig. 12.19b). If the wavelength $\lambda_2$ is tuned to the energy difference $\Delta E(R < R_c)$ the probe laser maps the oscillatory movement of the reflected part of the wave packet (curve (b)).

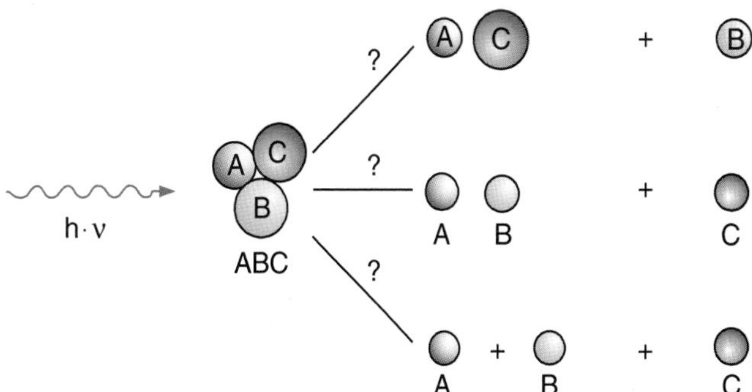


**Fig. 12.20.** Possible reaction channels for photo-excitation of a triatomic molecule ABC

### 12.2.3 Coherent Control

If a polyatomic molecule is excited into a dissociating state, the dissociation products depend on the preparation of the excited state by the excitation pulse (Fig. 12.20). For example, when a molecule ABC is excited, there are three different decay channels.

$$\mathrm{ABC} + h\nu \rightarrow (\mathrm{ABC})^* \rightarrow \begin{cases} \nearrow \mathrm{AB} + \mathrm{C} \\ \mathrm{ABC} + h\nu \\ \searrow \mathrm{A} + \mathrm{BC} \end{cases} \qquad (12.21)$$

Which of the three channels is actually realized, depends on the spatial distribution and on the phase of the wave function $\psi(r, t)$ in the excited state of ABC, which in turn can be influenced by the wavelengths, the time profiles and the phase distribution of the different wavelength-components in the excitation pulse.

The coherent control technique tries to optimize a wanted reaction channel and to minimize all unwanted ones by a proper choice of these pulse parameters. This is performed by a learning algorithm and works as follows [12.22].

The pump pulse is sent onto an optical grating, that diffracts the different wavelengths of the pulse components into different directions (Fig. 12.21). A collimating lens forms a parallel beam where the different wavelengths are spatially shifted against each

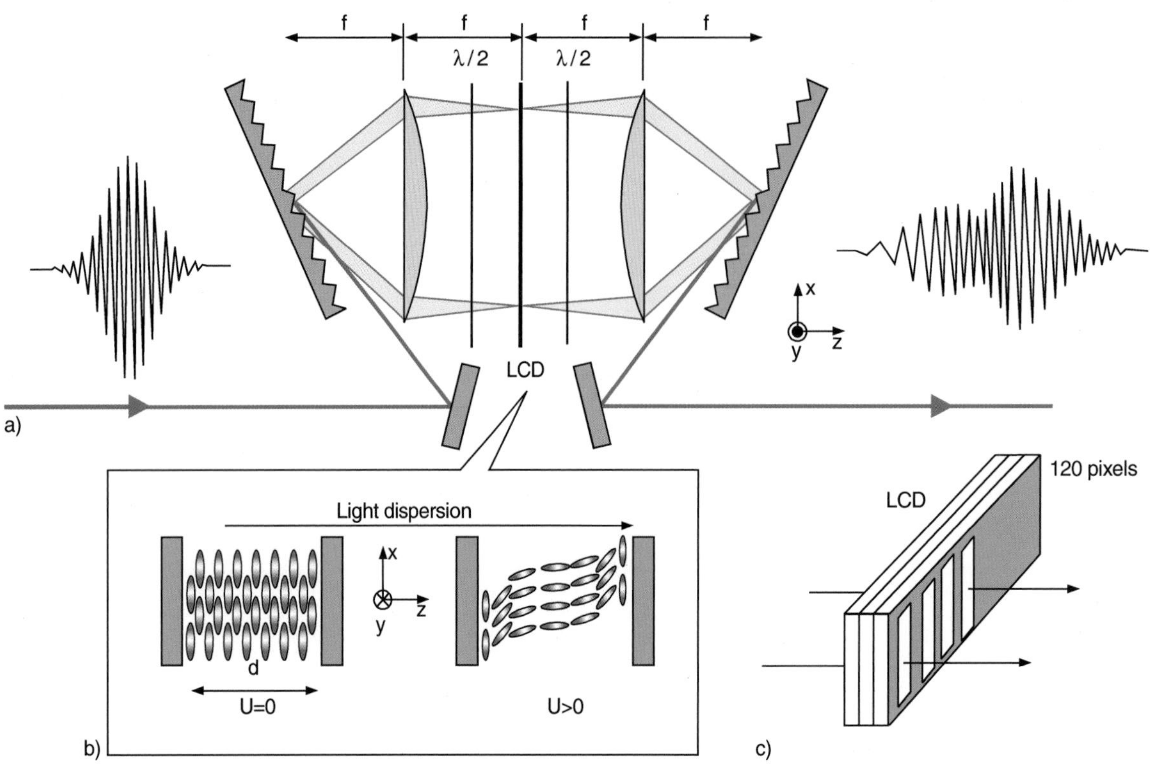


**Fig. 12.21.** (**a**) Optical arrangement for optimizing femtosecond laser pulses. (**b**) Orientation of the molecules in the liquid of a single pixel. (**c**) Schematic illustration of the whole mask with 120 pixels where only 4 are shown. [12.24]

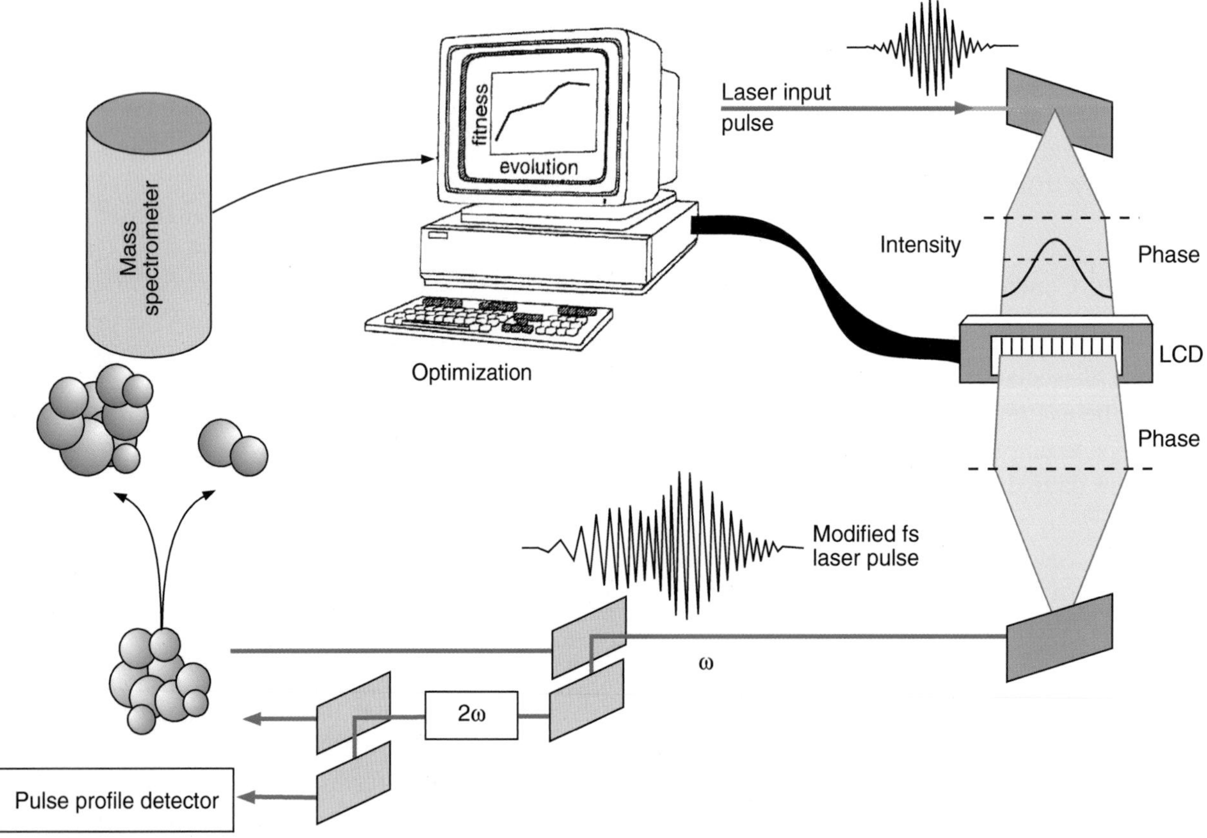


**Fig. 12.22.** Flow chart of coherent control of molecular reactions [12.24]

other due to the dispersion by the grating. The parallel beam passes through a liquid crystal array consisting of many pixels. When an electric voltage $U$ is applied to these pixels the molecules in the liquid crystal between the electrodes are partly oriented and the index of refraction $n$ changes, where $\Delta n \propto U$. This causes a phase shift of the transmitted optical wave. Since the voltage for each pixel can be chosen separately the phase of each wavelength-component passing through one of the pixels can be controlled separately.

A second lens focuses the parallel beam onto a second grating, where the different wavelengths are again superimposed and leave the system as a parallel output beam. Because of the different phase shifts for the different wavelengths the intensity distribution $I(t)$ has changed. The time profile of the output pulse and the intensity distribution $I(\lambda)$ of the different wavelength components can be varied within certain ranges by changing the voltages across the different pixels of the liquid crystal.

In order to optimize a specific reaction the output pulse is sent into the reaction chamber and the wanted products are measured. The signal is fed into a computer that controls the different voltages of the pixels. These voltages are now changed and the computer checks whether the wanted signal has increased or decreased. With a special learning algorithm (e.g., survival of the fittest) the output laser pulse is optimized in such a way, that it produces the maximum wanted signal of the investigated reaction (Fig. 12.22). For favorable cases the selectivity of the reaction could be enhanced by one order of magnitude through this optimized coherent control system.

For more details see [12.23, 24, 25, 26].

## 12.3 Optical Metrology with New Techniques

For many scientific and technical applications, accurate measurements of wavelengths, frequencies and time are essential. One example is the precise determination of fundamental physical constants, such as the fine structure constant, the Rydberg-constant, the gravitational constant and their possible changes with time. Such changes must be very small, since they could not be detected up to now. On the other hand they would severely influence cosmological models and might lead to a revision of fundamental physical principles.

A second example is the synchronization of the detection systems in radio telescopes at different locations on earth, which demands very accurate and stable frequencies in order to perform radio interferometry with high angular resolution of radio signals from distant galaxies.

On the technical side, the accuracy of the global positioning system (GPS) is limited by the frequency stability of the signals superimposed from different satellites.

It is therefore worthwhile to look for the optimum method for generating ultrastable frequencies.

### 12.3.1 Frequency Comb

The frequency spectrum of a cw mode-locked laser, emitting a regular train of short pulses, consists of a comb of equally spaced frequency components that represent the fundamental modes of the laser resonator (see Sect. 8.5.2). The total spectral range $\Delta\nu = 1/\Delta t$ of the frequency comb depends on the width $\Delta t$ of the pulses. With femtosecond pulses ($\Delta t = 50\,\mathrm{fs}$) the spectral range extends over $\Delta\nu = 2\times10^{13}\,\mathrm{s}^{-1} = 20\,\mathrm{THz}$.

This spectral range can be greatly increased by focusing the laser beam into a special optical fiber, where the spectrum is considerably broadened by self-phase modulation. The index of refraction depends on the intensity of the laser pulses and a temporal phase change causes a frequency shift, since $\delta\nu = (1/2\pi)\cdot \mathrm{d}\varphi/\mathrm{d}t$ (see Sect. 8.5.3). With sufficiently intense laser pulses, a total spectral width of the frequency comb over more than one decade (e.g., from $\lambda = 1064\,\mathrm{nm}$ to $500\,\mathrm{nm}$) has been achieved, which corresponds to a frequency interval of 320 THz.

**EXAMPLE**

With a resonator length $d = 50\,\mathrm{cm}$ the spacing between the modes is $\Delta\nu_\mathrm{m} = 300\,\mathrm{MHz}$. With a total spectral width of 320 THz the frequency comb includes $1.07\times10^6$ modes.

Measurements have proved that the spectral components of the comb are precisely equidistant, even in the far wings of the comb. This opens the possibility to compare different frequency standards with very high

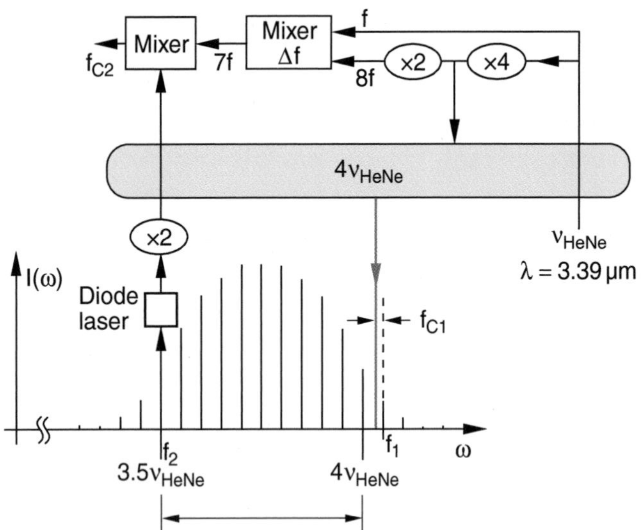


**Fig. 12.23.** Absolute frequency measurement of the He-Ne laser at $\lambda = 3.39\,\mu$m with the optical frequency comb [12.27]

precision and to measure absolute optical frequencies. The methods works as follows.

Presently, the cesium clock (Fig. 11.90) still represents the international frequency standard. The mode spacing $\Delta\nu_m$ of the frequncy comb is locked to the frequency $\nu_{Cs}$ in such a way that $\nu_{Cs} = m\Delta\nu_m$ with $m \approx 30$. This locking can be realized by tuning the length $d$ of the laser resonator ($\Delta\nu_m = c/2d$), until the repetition rate of the mode-locked pulses equals a subharmonic of the cesium frequency.

The frequency difference $N\Delta\nu_m$ between the $n$-th and the $(n+N)$-th mode of the comb is now precisely known.

In order to measure the absolute frequency of a stabilized laser, the 4th harmonic of a He-Ne laser at $\lambda = 3.39\,\mu$m is compared with the frequency $f_1$ of the closest mode in the frequency comb (Fig. 12.23). For a mode separation of 300 MHz the beat frequency $f_{c1} = 4\nu_{\text{He-Ne}} - f_1$ is smaller than 150 MHz and can be therefore counted with a fast electronic counter.

Now the frequency $\nu_{\text{He-Ne}}$ is mixed in a nonlinear optical crystal with the doubled frequency $2 \cdot 4\nu_{\text{He-Ne}}$ and the difference frequency $\Delta\nu = 8\nu_{\text{He-Ne}} - 1\nu_{\text{He-Ne}} = 7\nu_{\text{He-Ne}}$ is compared with the doubled frequency $2\nu_D$ of a tunable diode laser stabilized on

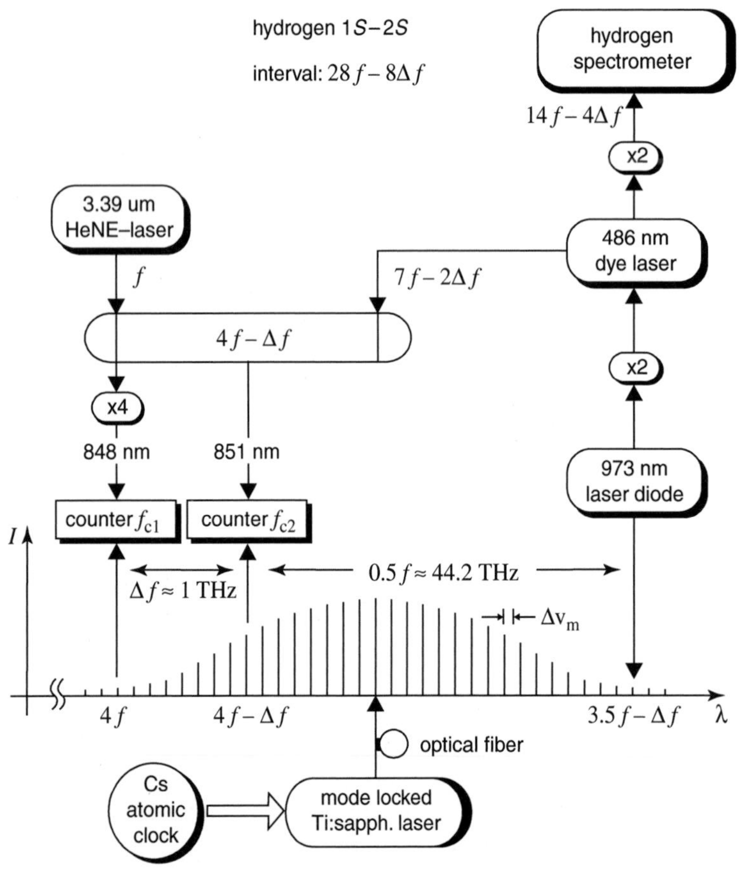


**Fig. 12.24.** Novel frequency chain for the measurement of the hydrogen $1S-2S$ interval. The frequency comb of a mode-locked laser is used to measure two large optical frequency differences [12.27]

a mode $f_2$ of the frequency comb, which is 584498 modes away from the mode $f_1$ onto which $4\nu_{\text{He-Ne}} - f_{c1}$ is stabilized. The difference frequency $7\nu_{\text{He-Ne}} - 2\nu_{\text{D}} = f_{c2}$ is counted.

We have now the equations:

$$f_1 - f_2 = 584498\,\Delta\nu_{\text{m}} + f_{c1}$$

$$\begin{aligned} 4f_{\text{He-Ne}} &= f_1 - f_{c1} \\ 7f_{\text{He-Ne}} &= 2f_2 - f_{c2} \\ \Rightarrow f_{\text{He-Ne}} &= 8f_{\text{He-Ne}} - 7_{\text{He-Ne}} \\ &= 2 \times 584,498\,\Delta\nu_{\text{m}} + f_{c2} \end{aligned}$$

which relates the cesium frequency $\nu_{\text{Cs}}$ to the frequency $\nu_{\text{He-Ne}}$ of the He-Ne laser since the matching frequency $f_{c1}$ is counted and known within 1 Hz [12.27].

In a similar way, the absolute frequency of the $1S \to 2S$ transition in the hydrogen atom can be determined with the help of the frequency comb (Fig. 12.24). In order to realize this, the frequency-doubled output of a dye-laser with $\lambda = 486$ nm is stabilized onto the two-photon transition $1S$–$2S$ of atomic hydrogen at $\lambda = 121.5$ nm, which has a natural line width of only 1.3 Hz (!), because the lifetime of the metastable $2S$ level is about 0.13 s. Therefore the frequency of the laser can be stabilized onto this narrow transition with a stability of a few Hertz. The direct comparison of this optical frequency with the microwave frequency of the cesium standard can be performed with an accuracy of $\Delta\nu/\nu \leq 5 \times 10^{-16}$ [12.28, 29].

### 12.3.2 Atomic Clocks with Trapped Ions

Very accurate and stable frequencies can be generated if the laser frequency is stabilized onto a narrow transition of cold ions that are kept for a long time in an ion trap [12.30]. This ion trap can be either a Penning trap (Fig. 12.8) or an rf quadrupole trap (Fig. 12.25), called a Paul trap after its inventor *Wolfgang Paul* (Nobelprize 1989).

Here, a single ion can be trapped and optically cooled in the trap. After it has reached its lowest temperature, it oscillates with very small amplitudes around the trap center. The ion is excited from its ground state by two different lasers into two upper states. This is illustrated in Fig. 12.25 for the mercury ion $\text{Hg}^+$ with the ground state $5d^{10}6s(^2S_{1/2})$, which can be excited by the pump laser with $\lambda = 194$ nm into the short-lived upper level $5d^{10}6p(^2P_{1/2})$ with $\tau = 2.3$ ns.

The fluorescence, emitted from this level, serves as a monitor for the excitation. With cw excitation the fluorescence photon rate is about $2.2 \times 10^8$ /s.

A second laser, called the clock laser, is tuned to the dipole forbidden, but quadrupole-allowed transition from the $^2S_{1/2}$ ground state to the $^2D_{5/2}$ level with a lifetime of $\tau = 100$ ms. This transition is very narrow ($\Delta\nu = 1.5$ Hz!). If the clock laser can be stabilized onto the center of this narrow transition, a very high frequency stability ($\nu/\Delta\nu \geq 10^{16}$) can be achieved. Since the cold ion is kept within a small volume

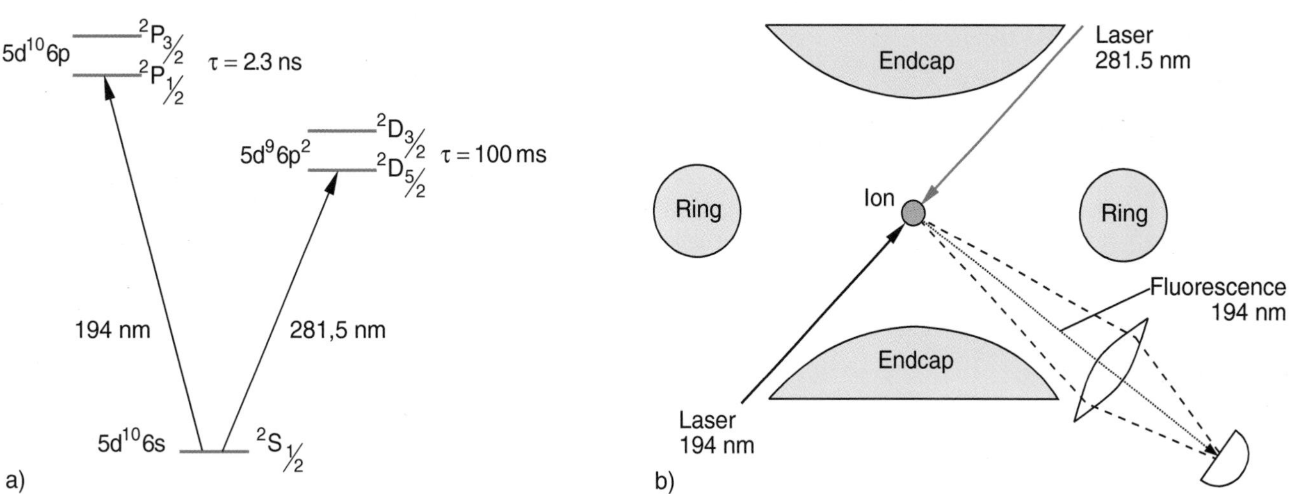


**Fig. 12.25.** (**a**) Level scheme of the $\text{Hg}^+$ ion with pump laser and clock laser transitions. (**b**) Paul trap with single ion and detection of the fluorescence

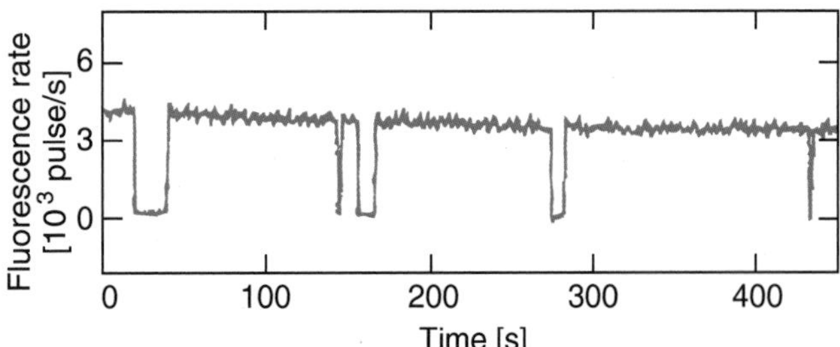


**Fig. 12.26.** Experimental demonstration of quantum jumps of a single ion [12.31]

around the center of the trap, where the electric field is nearly zero, possible Stark-shifts are very small and do not limit the absolute precision of the frequency measurement.

With this system it is possible to observe single quantum jumps. If only a single ion is present in the trap, the ground state is empty for about 100 ms when a clock transition has taken place before it is refilled by the fluorescence photon from the $^2D_{5/2}$ level. During this time, the pump laser cannot excite the UV fluorescence and the signal is interrupted (Fig. 12.26). The rate of these dark periods is a measure for the frequency of the clock transition. The longer the dark periods, the closer is the clock laser frequency to the center of the clock-transition [12.31].

When more than one ion is present in the trap, the Coulomb repulsion between the ions pushes the ions away from the center and their positions in the trap are determined by the minimum total potential energy of the trap potential plus the Coulomb repulsion. Below a critical temperature $T_c$, the ions arrange themselves in a geometrical structure like in a solid crystal, however with distances between the ions, that are several orders of magnitude larger than in the solid. Such an arrangement is called a Wigner crystal (Fig. 12.27).

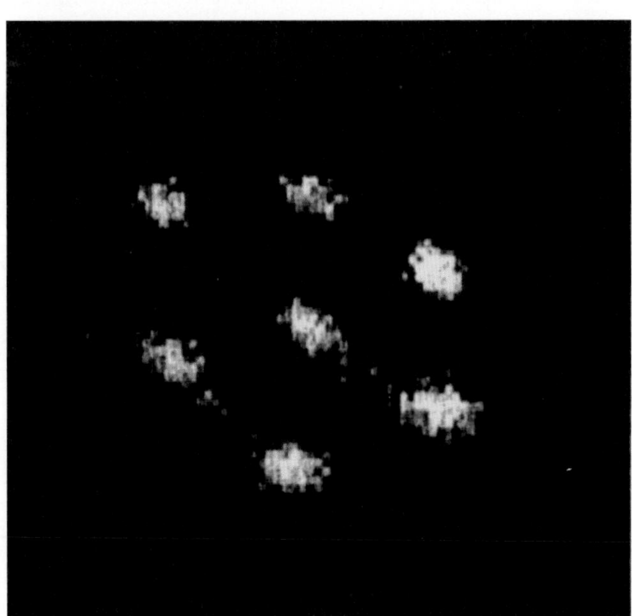

**Fig. 12.27.** Photograph of a Wigner crystal of seven trapped ions, taken with a microscope and an image intensifier. The distance between the ions is about 20 μm [12.32]

Such crystals, where the forces are well known, are good candidates to study properties of solid crystals, where the superposition of several effects often masks the pure interaction between close neighboring atoms [12.32, 33].

## 12.4 Squeezing

The electric field of a monochromatic light wave can be represented by

$$E(t) = E_0(t)[\cos(\omega t + \boldsymbol{k}\cdot\boldsymbol{r} + \varphi(t))] \tag{12.22a}$$

or by

$$E(t) = E_1(t)\cos(\omega t + \boldsymbol{k}\cdot\boldsymbol{r}) + E_2(t)\sin(\omega t + \boldsymbol{k}\cdot\boldsymbol{r}) \tag{12.22b}$$

with $\tan\varphi = -E_2/E_1$. The electric field amplitudes $E_1$ and $E_2$ are called the quadrature amplitudes.

Even when all technical noise has been eliminated there are still small fluctuations of the amplitude and the phase of the wave, because the wave consists of $N$ photons per second that are emitted in a random way. Therefore the photon number $N$ fluctuates within the range $N \pm \sqrt{N}$ and also the time sequence of the photons (which corresponds to the phase of the wave) shows similar fluctuations. These fluctuations are caused by the quantum structure of light and they represent principle limits, set by the uncertainty relation, which generally cannot be overcome.

### EXAMPLE

Assume an optical detector with the quantum efficiency $\eta < 1$ is illuminated by $N$ photons per sec. This leads to a minimum relative fluctuation $\Delta S/S$ of the signal $S$, which for a bandwidth $\Delta f$ is given by

$$\frac{\Delta S}{S} = \frac{\sqrt{N\eta\Delta f}}{N\eta} = \sqrt{\frac{\Delta f}{\eta N}}\,. \tag{12.23}$$

For a radiation power of 1 mW at $\lambda = 600\,\text{nm} \rightarrow N = 3 \times 10^{15}\,\text{s}^{-1}$. With a bandwidth $\Delta f = 100\,\text{Hz}$ (time constant of 10 ms) and a quantum efficiency of 20% we obtain a minimum fluctuation $\Delta S/S = 4 \times 10^{-7}$. If the output signal is a voltage of 1 V it fluctuates with $\Delta U = 0.4\,\mu\,\text{V}$.

---

In Fig. 12.28 the fluctuations are illustrated in two different ways. In the $E(t)$ diagram the fluctuation ranges of amplitude and phase are indicated by the dotted curves. In the $E_1$-$E_2$ diagram the uncertainty range, due to the fluctuations $\delta E_1$ and $\delta E_2$, are marked by the area of the blue circle. It can be also characterized in a polar vector diagram by the fluctuations $\delta\varphi$ of the direction of the vector $E_0$ and $\delta E_0$ of its length.

Assume, the field amplitude in (12.22) represents the electric field of a single mode laser. If it is normalized in such a way, that

$$\left\langle E^2\right\rangle = \left\langle E_1^2\right\rangle + \left\langle E_2^2\right\rangle = \frac{\hbar\omega}{2\varepsilon_0 V}\,, \tag{12.24}$$

where $V$ is the volume of the mode, $\varepsilon_0$ the dielectric constant, the minimum uncertainty product can then be written as [12.30]

$$\Delta E_1 \Delta E_2 \geq 1\,. \tag{12.25}$$

A coherent state of the single mode laser (also called a Glauber state), can be written as a linear superposition of states with photon occupation numbers $N_k$. This describes a Poisson distribution for the probability of finding $N_k$ photons, with the maximum at the mean value $< N >$. For such a coherent state the uncertainties have minimum values and are symmetric, i.e.,

$$\Delta E_1 = \Delta E_2 = 1\,. \tag{12.26}$$

Such a state has the minimum possible uncertainty.

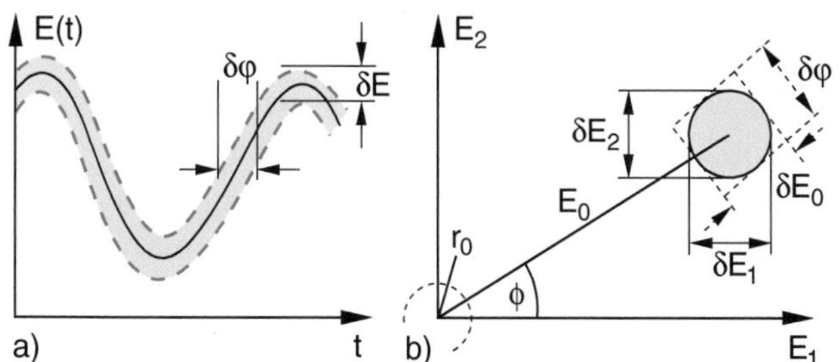


**Fig. 12.28a,b.** Amplitude and phase uncertainties of a laser wave shown in an amplitude–time diagram (**a**) and in a polar phase diagram (**b**)

The question is now whether it is possible to reduce the amplitude noise below this limit. Since the uncertainty relation cannot be violated, the phase fluctuations then have to become larger. For certain experiments this would still increase the signal-to-noise ratio. For other experiments it might be desirable to reduce the phase fluctuations at the cost of higher amplitude fluctuations.

In the diagram of Fig. 12.28b this decrease of the fluctuations of one component at the cost of the other means a squeezing of the uncertainty circle into an ellipse. The techniques used to change the relations between amplitude-and phase-fluctuations in favor of the desired experimental results are therefore called *squeezing* [12.34].

Coherent light shows phase-independent noise. This can be experimentally demonstrated by the two-beam Mach–Zehnder interferometer shown in Fig. 12.29.

The monochromatic laser beam is split by the beam splitter $BS_1$ into two partial beams with amplitudes $E_1$ and $E_2$, which are superimposed again by $BS_2$. One of the beams passes through an optical wedge, where the optical path-length can be continuously varied, which causes a variable phase shift $\Delta\varphi$ between the two partial waves. The two detectors $PD_1$ and $PD_2$ monitor the mean intensities

$$\left\langle I_{1,2}\right\rangle = \frac{1}{2}c\varepsilon_0\left[\left\langle E_1^2\right\rangle + \left\langle E_2^2\right\rangle \pm 2E_1E_2\cos\Delta\varphi\right] \tag{12.27}$$

averaged over many cycles of the optical field with frequency $\omega$. For equal amplitudes $E_1 = E_2$ the detected intensities are

$$\begin{aligned}&\langle I_1\rangle = \langle I_0\rangle\cos^2(\Delta\varphi/2)\,,\langle I_2\rangle = \langle I_0\rangle\sin^2(\Delta\varphi/2)\,,\\&\Rightarrow \langle I_1\rangle + \langle I_2\rangle = \langle I_0\rangle\,.\end{aligned} \tag{12.28}$$

The two detectors in Fig. 12.29 measure the projection of $E$ in Fig. 12.28b onto the axes $E_1$ and $E_2$, respectively. A variation of the phase $\varphi$ in one arm of the interferometer corresponds to a rotation of the vector $E_0$. In Fig. 12.29b the intensity $\langle I(\Delta\varphi)\rangle$ is plotted as a function of the phase-difference $\Delta\varphi$ between the two beams. By a proper choice of $\Delta\varphi$ for the arrangement in Fig. 12.29, one can measure the fluctuations $\langle\delta E_1\rangle$ and $\langle\delta E_2\rangle$ separately. If the frequency spectra of the intensities $I_1(f)$ and $I_2(f)$ are measured by a spectrum analyzer, it turns out that at higher frequencies,

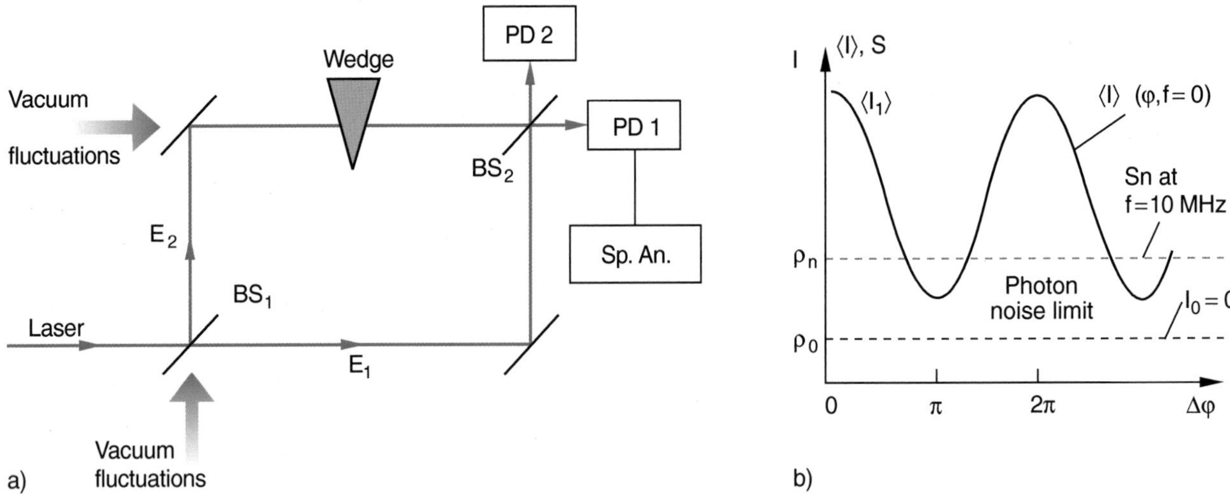


**Fig. 12.29.** (**a**) Mach–Zehnder interferometer with variable phase delay $\varphi$ realized by an optical wedge. (**b**) Detected mean intensity $\langle I \rangle$ measured at $f = 0$, and phase-independent photon noise power density, measured at $f = 10\,\text{MHz}$ with and without input intensity $I_0$

where the technical noise is negligible, the noise power spectral density $\rho_n(f)$ becomes independent of the frequency $f$. The noise power densitiy $\rho_n$ measured at a frequency $f = 10\,\text{MHz}$ is independent of $\Delta\varphi$. This can be understood as follows.

The intensity fluctuations are due to the random emission of photons, which are uncorrelated in the two interfering beams. Although the mean intensities $\langle I_1 \rangle$ and $\langle I_2 \rangle$ depend on $\Delta\varphi$, the fluctuations do not! The detected noise power density $\rho_n \sim \sqrt{N} \sim \sqrt{I_0}$ shows the same noise level for the minimum of $I(\Delta\varphi)$ as for the maximum.

If the incident laser beam in Fig. 12.29a is blocked, the noise level does not go to zero, but has a lower limit $\rho_0$ that is attributed to the zero-point fluctuations of the vacuum field, which is also present in a darkroom. This noise adds to the detector noise. The interferometer has two inputs: the coherent light field and the vacuum field, which is also present without the light field. Because the fluctuations of these two inputs are not correlated, their noise power densities add. The maximum achievable signal to noise ratio is then

$$\frac{S}{\rho_n} \propto \frac{N}{\sqrt{N} + \rho_0} . \tag{12.29}$$

In the phase diagram in Fig. 12.30, the circular uncertainty area with radius $\sqrt{\rho_0}$ around the origin $E_1 = E_2 = 0$ corresponds to this vacuum noise power density $\rho_0$.

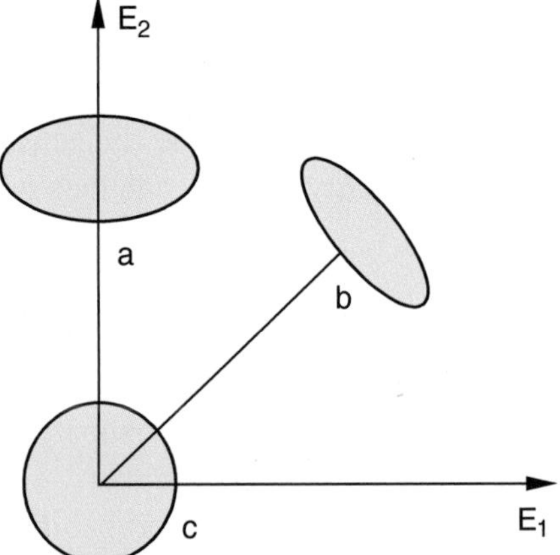


**Fig. 12.30.** Uncertainty areas for different squeezing conditions. *a* $\langle E_1 \rangle = 0$, $\Delta E_2$ is squeezed but $\Delta\varphi$ is larger than in the nonsqueezed case. *b* General case of squeezing of $\Delta E$ at the expense of increasing $\Delta\varphi$. *c* Uncertainty area of zero-point fluctuations with $\langle E_1 \rangle = \langle E_2 \rangle = 0$

The preparation of squeezed states tries to minimize the uncertainty of one of the two quantities $\delta E$ or $\delta\varphi$ at the expense of the increased uncertainty of the other. The uncertainty area is then squeezed into an ellipse. Although the area of this ellipse is slightly larger than the circular area of the unsqueezed state, one may still gain in the signal-to-noise ratio, if the minimized quantity determines the noise level of the detected signal.

This can be achieved by starting with a normal coherent state of the incoming wave and introducing a correlation between the fluctuations of $E_1$ and those of $E_2$, which means between amplitude and phase fluctuations. How can this be achieved?

The experimental realization of squeezing is based on the nonlinear interaction of an optical wave in a nonlinear medium. One example is illustrated in Fig. 12.31, which shows a Mach–Zehnder interferometer similar to that in Fig. 12.29. However, one of the light beams with amplitude $E_1(t)$ passes through a medium with a nonlinear index of refraction

$$n(E) = n_0 + n_2 E^2 \tag{12.30}$$

that depends on the intensity of the light wave. Assume that the incident wave with amplitude

$$E_{\mathrm{in}}(t) = E_{\mathrm{in}1} + \delta_{\mathrm{a}} E(t) + \delta_{\mathrm{ph}} E(t) \tag{12.31}$$

has amplitude fluctuations described by $\delta_{\mathrm{a}} E(t)$ and phase fluctuations represented by $\delta_{\mathrm{ph}} E(t)$. If the medium has negligible absorption the amplitude of the outgoing wave is the same as that of the incident wave. The phase of the wave, however, is affected by the index of refraction. We obtain, after a pathlength $L$ through the medium,

$$\begin{aligned}\varphi_{\mathrm{out}} &= \varphi_{\mathrm{in}} + 2\pi n L/\lambda \\ &= \varphi_{\mathrm{in}} + \frac{2\pi L}{\lambda}\left[n_0 + n_2(E_1 + \delta E_{\mathrm{a}1})^2\right] .\end{aligned} \tag{12.32}$$

Inserting this into (12.22a) we obtain

$$E_1^{\mathrm{out}} = E_1^{\mathrm{in}} \quad \text{and} \quad \delta E_{\mathrm{a}1}^{\mathrm{out}} = \delta E_{\mathrm{a}1}^{\mathrm{in}}(t) , \tag{12.33}$$

which means that the amplitude and its fluctuation have not changed for $E_1$. For the phase fluctuations the situation is different. With $\delta E_{\mathrm{ph}} = E_1 \delta\varphi$ one obtains

$$\begin{aligned}\delta E_{\mathrm{ph}}^{\mathrm{out}} &= \delta E_{\mathrm{ph}}^{\mathrm{in}} + \frac{2\pi n_0 n_2 L}{\lambda}\left(E_1^{\mathrm{in}}\right)^2 \delta E_{\mathrm{a}}^{\mathrm{in}} \\ &= \delta E_{\mathrm{ph}}^{\mathrm{in}} + 2 r_{\mathrm{Kerr}} \delta E_{\mathrm{a}}^{\mathrm{in}}(t) ,\end{aligned} \tag{12.34}$$

with the optical Kerr parameter

$$r_{\mathrm{Kerr}} = \frac{2\pi n_0 n_2 L \left(E_1^{\mathrm{in}}\right)^2}{\lambda} .$$

The nonlinear medium couples the amplitude- and phase fluctuations. The amplitude $E_1$ changes the index of refraction, which in turn affects the phase. The

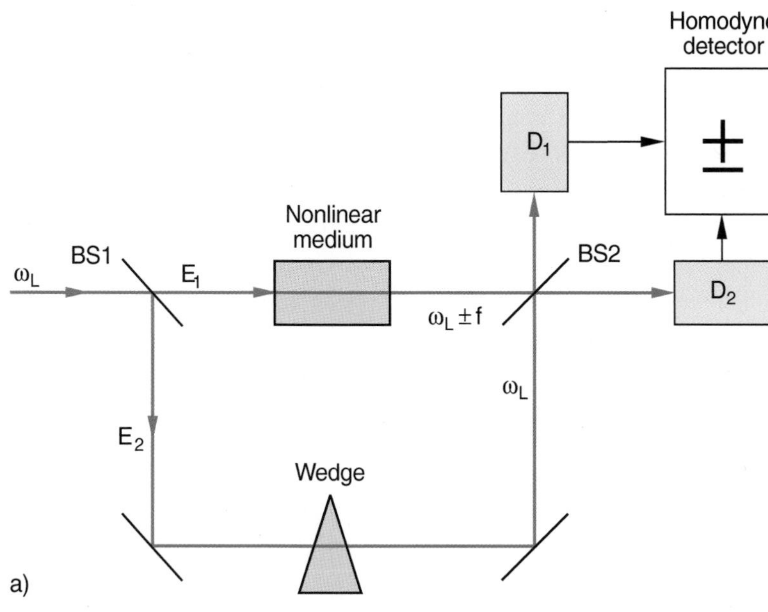


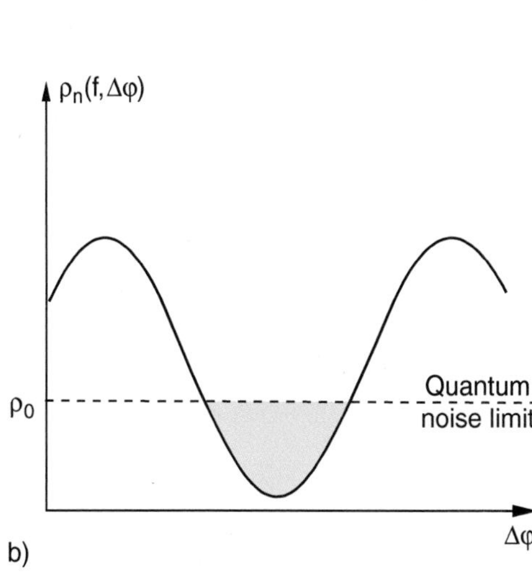


**Fig. 12.31a,b.** Schematic diagram of squeezing experiments with a nonlinear medium in a Mach–Zehnder interferometer. (**a**) Experimental arrangement, (**b**) Homodyne signal $\Delta S = S_1 - S_2 \propto \rho_n(f, \Delta\varphi)$ as a function of the phase difference $\Delta\varphi$ at a frequency $f = 10\,\mathrm{MHz}$

total noise has increased because the nonlinear medium adds fluctuations to the phase. In the quadrature diagram, based on (12.22b), each point in the upper part of Fig. 12.32 represents certain instantaneous fluctuations $\delta E_a$, $\delta E_{ph}$. If we look at the distribution of the fluctuations we obtain the upper diagrams in Fig. 12.32, where each dot represents the instantaneous value $\delta E_a^{in}(t)$, $\delta E_{ph}^{in}(t)$ on the left side and $\delta E_a^{out}(t)$, $\delta E_{ph}^{out}(t)$ on the right side. The lower graphs show the corresponding histograms of the density of points. One can see that the distribution of the amplitude fluctuations are narrower for the outgoing wave than for the incident wave, while for the phase distribution, the opposite is true.

This shows, that the arrangement of Fig. 12.31 reduces the amplitude fluctuations at the expense of increasing phase fluctuations.

Squeezing can be detected by the interferometer setup shown in Fig. 12.31, where the nonlinear medium is placed in one arm of the interferometer. The system acts as heterodyne detector where the local oscillator is the strong partial wave in the lower arm of the interferometer, which interferes with the much weaker squeezed light in the upper arm. The two detectors measure the two output signals $E_1^{out}$ and $E_2^{out}$. The difference $\Delta S = S_1 - S_2$ of the corresponding detector

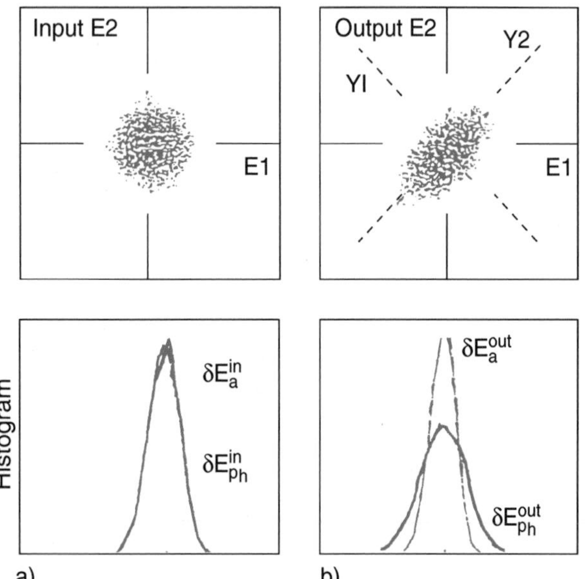


**Fig. 12.32.** Effect of a medium with nonlinear refractive index on the fluctuation of a light wave [12.32]

signals is fed into a spectrum analyzer, which is set to a fixed frequency $f$. This gives a signal that represents the fluctuations of the output signals at the frequency $f$, that is proportional to the noise power density $\rho_n(f)$. Now the phase of the local oscillator is continuously changed by the optical wedge and the signal $\Delta S(\Delta\varphi)$ is measured as a function of the phase difference between the two waves. The result is illustrated in Fig. 12.31b and shows that for certain phase differences the fluctuations decrease below the noise level of two beams with uncorrelated fluctuations, which would be independent of $\Delta\varphi$.

Note the difference from Fig. 12.29, where the output intensity $\langle I(\varphi)\rangle$ is plotted, while here the fluctuations $\delta I \sim \rho_n(\Delta\varphi)$ are shown.

An interesting possible application of squeezing might be the detection of gravitational waves by optical interferometry [12.35]. Such waves travel with the velocity of light through space and cause a contraction or expansion of space, depending on the amplitude and phase of the wave. As predicted by Einstein, gravitational waves should be emitted from sources where large masses are accelerated. Examples are supernovae explosions of stars or close binary stars, which are two stars bound by mutual gravitation that move around their common center of mass. If the two components have a small distance and therefore experience a large acceleration, they can emit sufficiently intense gravitational waves that might be strong enough to be detected on earth. The amplitude of the gravitational wave depends on its frequency. In Fig. 12.33, calculated amplitudes are shown for some astronomical sources. Estimations show that a supernova explosion in the Virgo cluster would generate gravitational waves that cause a relative space contraction on earth as small as $10^{-21}$! This means that a length of 1 m is changed by $10^{-21}$ m, which is $10^{11}$ times smaller than the diameter of the hydrogen atom.

Up to now the sensitivity of gravitational wave detectors has not been high enough to detect such small effects, but new, large optical interferometers are under construction that might reach the necessary sensitivity for detecting gravitational waves from supernovae explosions in nearby galaxies [12.36].

The basic arrangement of these optical gravitational wave antennas shown in Fig. 12.34 consists of a Michelson-type interferometer with two long arms that are orthogonal to each other. The side lengths of

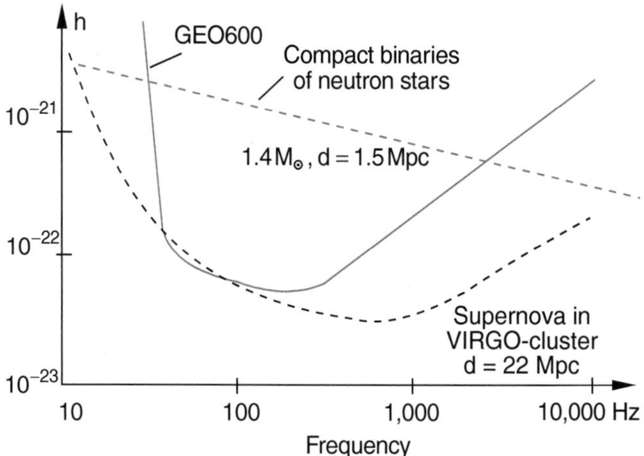


**Fig. 12.33.** Calculated amplitudes $h(f)$ of gravitational waves, emitted from two different sources, compared with the detection sensitivity of GEO 600, a gravitational wave detector close to Hannover

these devices range from 60 m to several kilometers. With multi-pass arrangements using high reflectivity mirrors the path length can be extended by a factor of about 100. The beam of a well-stabilized laser (frequency stability better than 1 Hz!) with the mean power $P_0 = \frac{1}{2}c\varepsilon_0 A_0^2$ that delivers $N = P_0/h\nu$ photons per second is split by the beam splitter with reflectivity $R$ and transmittance $(1 - R)$ into two partial beams with amplitudes $A_1 = A_0\sqrt{R}$ and $A_2 = A_0\sqrt{1-R}$ which travel back and forth through the two orthogonal long arms before they are again superimposed and interfere with each other. Because each of the two beams has been once reflected and once transmitted by the beamsplitter BS the amplitudes of the two interfering beams are both $A_0\sqrt{R}\cdot(1-R)$ and their pha se difference is $\Delta\varphi$. The detector $D_1$ measures the

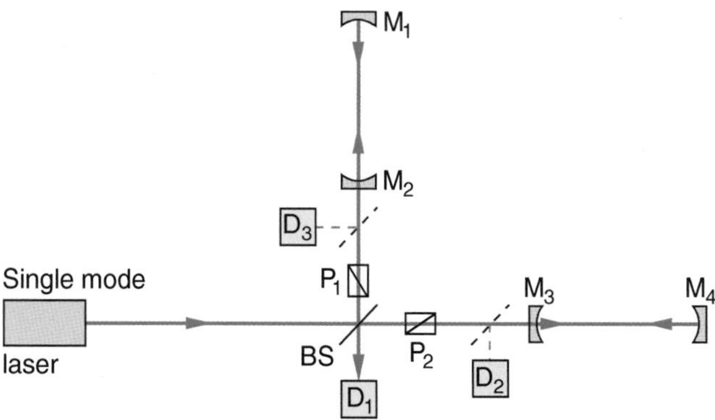


**Fig. 12.34.** Improved Michelson-interferometer as gravitational wave detector

interference intensity at the exits of the interferometer. The detected mean power, averaged over the oscillation period of the light

$$\bar{P} = \frac{1}{2}c\varepsilon_0 A_0^2 R(1-R)[1+\cos\Delta\varphi]^2 \qquad (12.35)$$

depends on the reflectivity $R$ of the beam splitter and on the phase difference $\Delta\varphi$ between the two interfering waves. The two polarizers $P_1$ and $P_2$ take care that the two interfering waves are strictly linearly polarized and the splitting ratio of the beam splitter, which depends on the polarization state of the waves, remains constant.

If a gravitational wave arrives at the detector, the arms of the interferometer are differently contracted or expanded depending on the direction and on the phase of the gravitational wave. Assume the gravitational wave causes a change $\Delta L$ of one of the two long arms resulting in a phase change $\Delta\varphi_g$. Then the detector $D_1$ with quantum efficiency $\eta$ delivers for $R = 0.5$, $A_3 = A_4 = A_0/2$ the signal

$$S \propto \frac{N}{4}\eta\left[1+\cos(\Delta\varphi_0+\Delta\varphi_g+\delta\varphi_n)\right]^2 , \qquad (12.36)$$

where $\Delta\varphi_0$ is the phase difference without gravitational wave and $\delta\varphi_n$ is the phase fluctuation due to noise. The photon noise of the two interfering beams with $N/2$ photons each is proportional to $\sqrt{N/2}$. The detectors $D_2$ and $D_3$ measure the light powers and the noise of the partial waves in the two arms before they interfere. Part of the noise is due to random fluctuations of the photon number $N$. All three detectors measure these fluctuations. However, the difference signal $\Delta S = S_1 - (S_2 - S_3)$ does not depend on technical fluctuations of the incident light power but is sensitive to phase fluctuations [12.36].

It turns out that the highest sensitivity can be achieved if the power at the exit of the interferometer is minimum, i.e., the phase difference should be $\Delta\varphi_0 = \pi$. Then the average difference signal $\Delta S$ becomes for $\delta\varphi_n \ll \Delta\varphi_g$

$$\overline{\Delta S} \sim (N/4)\left[1-\cos(\Delta\varphi_g)\right] = \frac{N}{2}\sin^2\left(\Delta\varphi_g/2\right) . \qquad (12.37)$$

It would be zero without a gravitational wave. The noise power of the two beams at the detector adds quadratically. It is mainly due to the fluctuations in photon numbers $S_n \sim \sqrt{N/2+N/2} = \sqrt{N}$, and the phase fluctuations $\delta\varphi_n$, because the other noise sources cancel for the difference $S_2 - S_3$.

The signal-to-noise ratio becomes with $\Delta\varphi_g = 2\pi\Delta L/\lambda = (2\pi h/\lambda)L$ with $h = \Delta L/L$

$$\frac{\Delta S}{\delta S} = \frac{N\sin^2(\Delta\varphi_g/2)}{\sqrt{N + N\sin^2(\delta\varphi_n/2)}} = \frac{\sqrt{N}\sin^2(\pi hL/\lambda)}{1+\sqrt{N}\sin^2(\delta\varphi_n/2)} . \tag{12.38}$$

Since $h$ and $\delta\varphi_n$ are both very small, this can be written as

$$\frac{\Delta S}{\delta S} = \frac{\sqrt{N}(\pi hL/\lambda)^2}{1+\sqrt{N}(\delta\varphi_n/2)^2} . \tag{12.39}$$

This is larger than unity for

$$h^2 > \frac{1+\sqrt{N}(\delta\varphi_n/2)^2}{\sqrt{N}(L/\lambda)} . \tag{12.40}$$

The minimum detectable length change $h = \Delta L/L$, induced by the gravitational wave is therefore limited by the length $L$ of the interferometer arms, the maximum available photon flux $N$ and the phase fluctuations $\delta\varphi_n$. Recent improvements have allowed a sensitivity of $h = 2\times 10^{-22}/\sqrt{Hz}$ for the GEO 600 interferometer.

If a beam of squeezed light is coupled into the second input port of the interferometer the phase noise $\delta\varphi_n$ can be decreased (see above), which decreases the noise level and increases the sensitivity.

There are definite plans to launch a huge interferometer, called LISA (large interferometer space antenna) into space. It will consist of three satellites at the corner of an equilateral triangle with distances of $5\times 10^6$ km between the satellites (Fig. 12.35). The center of the triangle moves 20° behind the earth on the earth's orbit around the sun. The plane of the triangle is inclined by 60° against the ecliptic plane of the earth's path. It can be shown that for this configuration the three satellites sit in stable local potential minima (Lagrange points), caused by the superposition of the gravitational forces from sun and earth [12.37].

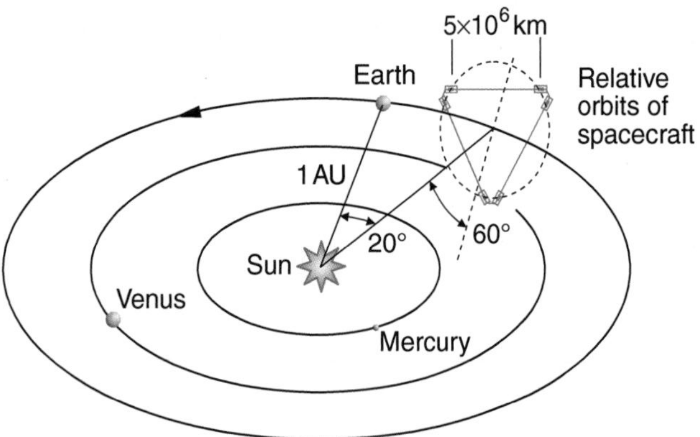


**Fig. 12.35.** LISA concept. Six spacecraft in a triangle, with a pair at each vertex. Only four are required for the basic interferometry

The side arms of this huge interferometer have to be stabilized within a micron, which is certainly a challenge for the mechanical stability and the electronic stabilization system [12.38].

## 12.5 New Trends in Quantum Optics

The experimental progress achieved in the development of lasers with better frequency stability and of high sensitivity detection techniques has allowed the experimental realization of ideas that could only be illustrated in former times with "Gedanken experiments" and are related to fundamental principles of quantum mechanics. Examples are the "which way" experiments, the "quantum non-demolishing" experiments, and "Schrödinger's cat", which we will briefly discuss in this section.

### 12.5.1 Which Way Experiments

Assume a light beam hits a 50% beam splitter (Fig. 12.36a). The two detectors A and B then both detect 50% of the incident intensity. If the intensity is dropped to such a low value that only one photon is ever on the way from the source to the detectors, then either A or B is hit by the photon, because the photon cannot split. Averaged over many measuring results the number of counts will be, however, equal for A and B, thus reproducing the classical result.

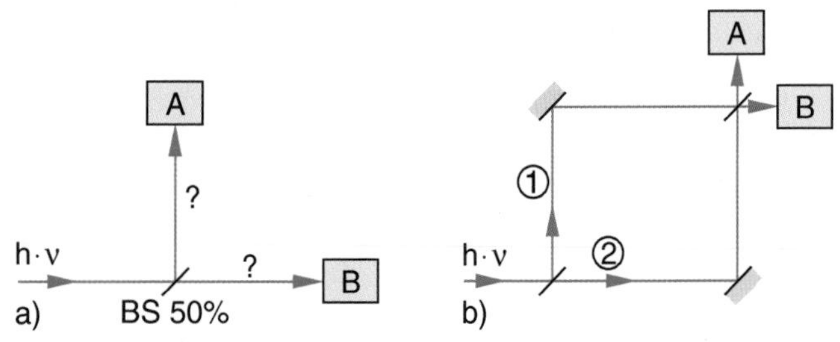


**Fig. 12.36.** (**a**) The incoming photon hits either detector A or detector B. (**b**) Depending on the path difference between 1 and 2 only detector A receives photons or only the detector B

Now we supplement our device to the Mach–Zehnder interferometer in Fig. 12.36b. The photon can either travel along path 1 or path 2. One would again expect that each of the two detectors A and B is hit on the average by 50% of all incident photons. But the experiment tells us that for a proper adjustment of the two path lengths 1 and 2, only detector A receives photons but not B. For another adjustment the opposite is true. If we, however, block one of the paths 1 or 2, both detectors again receive 50% of all incident photons.

The photon apparently has information about both paths 1 and 2.

In Sect. 3.5 we discussed Young's double-slit experiment with electrons and photons. The astonishing result was that interference structures could be observed even if only one photon or one electron at a time was passing through one of the slits. The essential point was that we did not know through which of the two slits the photon or electron had passed. When, in a Gedanken experiment, the electron was illuminated behind the slits by a photon, it could be decided through which of the two slits the electron had passed. This knowledge about "which way" the particle has taken destroyed the interference pattern. Although the argument that the momentum, transferred to the electron by absorption of the photon, would smear out the interference structure, is convincing, this is not the real reason for the disappearance of the interference.

This was experimentally confirmed by Rempe and his group [12.39] in a "which way experiment" based on an atom interferometer, which is explained in Fig. 12.37. An incoming collimated beam A of cold rubidium atoms from a MOT (see Sect. 12.1) passes through two separated standing light waves. The detuning $\Delta\omega = \omega_L - \omega_0$ of the light frequency from the atomic resonance $\omega_0$ is so large that spontaneous emission is negligible. When entering the standing light field with intensity $I = I_0 \cos^2(k_L z)$, the atoms experience a periodic potential $U(z) = U_0 \cos^2(k_L z)$ with $U_0 \sim I_0/\Delta\omega$ and some of the atoms are Bragg reflected. This splits the incoming atomic beam into two beams B and C: the Bragg reflected beam B and the transmitted beam C. The angle between the two beams corresponds to the relative transferred momentum $\Delta p/p$.

When entering the second standing light field both beams B and C are again split into a diffracted and a transmitted beam. Far away from the two light fields,

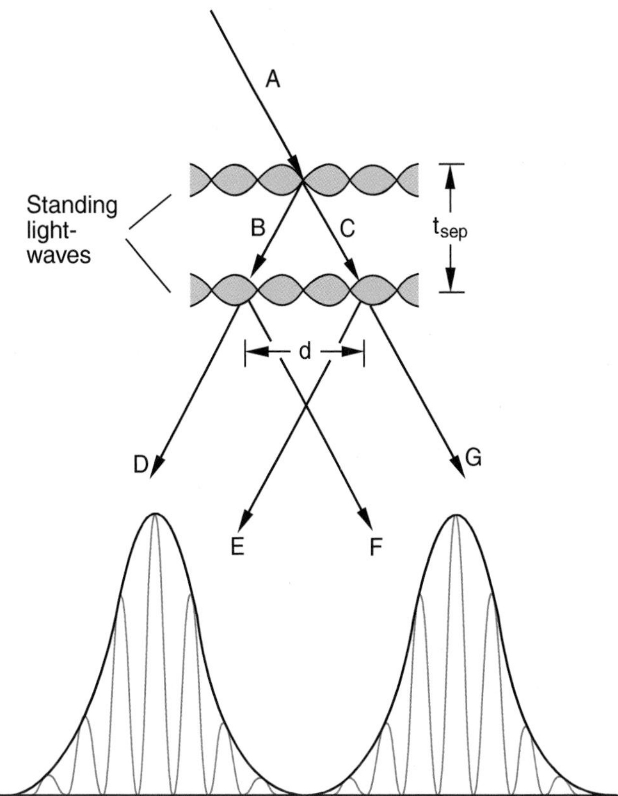

**Fig. 12.37.** Scheme of the atom interferometer. The incoming atomic beam A is split into two beams. Beam C is transmitted and beam B is Bragg-reflected from a standing light wave. The beams are not exactly vertical, because a Bragg condition must be fulfilled. After free propagation for a time $t_{sep}$ the beams are displaced by a distance $d$. Then the beams are split again with a second standing light wave. In the far field, a spatial interference pattern is observed

the interference between beam D and E and between F and G is observed.

Now a microwave with frequency $\omega_{mw} \approx 3\,\text{GHz}$ is added to the device, which induces transitions between the hyperfine components $F = 2$ and $F = 3$ in the $^2S_{1/2}$ state of $^{85}$Rb (Fig. 12.38). The laser frequency is tuned halfway between the $|2\rangle \rightarrow |e\rangle$ and the $|3\rangle \rightarrow |e\rangle$ transitions. The detuning then has equal magnitudes but opposite signs.

The wave function of the atoms experience a phase shift upon Bragg reflection, that depends on the internal state of the atom. For atoms in state $|2\rangle$ the detuning is negative, the atom sees a negative light shift potential and suffers a phase jump of $\pi$ upon reflection (which is equivalent to the reflection of a light wave at an optically thinner medium), while atoms in state $|3\rangle$ see a positive light shift potential and do not suffer a phase shift.

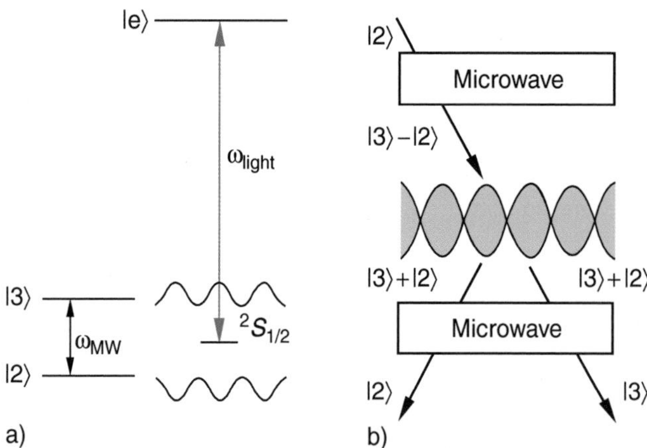


**Fig. 12.38a,b.** Storage of which way information. (**a**) Simplified level scheme of $^{85}$Rb. The excited state ($5^2P_{3/2}$) is labelled $|e\rangle$. The ground state ($5^2S_{1/2}$) is split into two hyperfine sates with total angular momentum $F = 2$ and $F = 3$, which are labeled $|2\rangle$ and $|3\rangle$, respectively. The standing light wave with angular frequency $\omega_{\text{light}}$ induces a light shift for both ground states which is drawn as a function of position. (**b**) The beam splitter produces a phase shift that depends on the internal and external degree of freedom. A Ramsey scheme, consisting of two microwave $\pi/2$ pulses, converts this phase shift into a population difference (see text)

This phase shift can be converted into a population difference when the microwave represents a $\pi/2$-pulse converting the internal state of the atom into a superposition $(|3\rangle + |2\rangle)/\sqrt{2}$. After having passed through the first light field, the internal state of the Bragg-reflected atoms is changed into $(|3\rangle - |2\rangle)/\sqrt{2}$, while the transmitted atoms are still in the superposition state $(|3\rangle + |2\rangle)/\sqrt{2}$. Now the atoms are again exposed to a $\pi/2$-pulse of the second microwave field, which brings the transmitted atoms back to state $|3\rangle$ and the reflected atoms to state $|2\rangle$. We can now distinguish between the reflected and the transmitted beam, because the two beams contain atoms in different states.

The question is now, whether the "which way information" destroys the interference pattern behind the second standing light field. The experiment proves that indeed the interference fringes disappear.

The mere fact that the information is stored and *could* be read out always destroys the interference, even if this information is not really read out! Also when it is read out, there is no interference observed.

The novelty of this result is, that there is a negligible recoil by the absorption of the microwave. Therefore the smearing out of the interference structure by the momentum transfer in the Gedanken experiment for the modified Young's two-slit interference is not the real reason for the disappearance of the interference structure, but the possible knowledge about the path the atoms have taken instead.

The somehow strange state that describes the situation where a particle can go one way, as well as the other, is called an ***entangled state***. Such a state does not appear in classical physics and it is a typical quantum mechanical phenomenon. Its properties lead to some surprising results, which have been the subject of great concern. After a controversial discussion between Einstein and Bohr about the consequences, Einstein, together with *Boris Podolsky* and *Nathan Rosen* wrote a seminal paper that intended to illustrate, by a paradox situation called the Einstein–Podolsky–Rosen paradox, that quantum mechanics is incomplete. We will explain it with the following example [12.40].

### 12.5.2 The Einstein–Podolski–Rosen Paradox

We consider the emission of two photons $h\nu_1$ and $h\nu_2$ into opposite directions by an atom on a cascade transition from the upper state $^1S$ via an intermediate state $^1P$ to the ground state $^1S$ (Fig. 12.39). Since the angular momentum of the atom before and after the emission is $L = 0$, the two photons have to have opposite spin orientations, i.e., one has $\sigma^+$ the other $\sigma^-$ polarization. Unless $\nu_1 \neq \nu_2$ and we can measure the frequencies, we do not know, which of the two photons is $\sigma^+$ and which is $\sigma^-$ polarized.

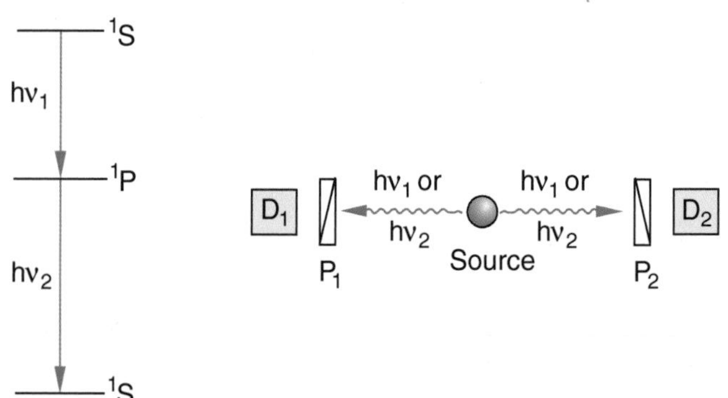


**Fig. 12.39.** Schematic illustration of the Einstein–Podolsky–Rosen paradox

If, however, the detector $D_1$ measures the polarization state of the photon emitted to the left side, we can tell which polarization state of the photon emitted to the right side is measured by $D_2$. Even if the two detectors are separated by a large distance, the measurement of $D_2$ is determined by the result of $D_1$. For a long time the controversial dispute was about the question, how the information obtained by the measurement at $D_1$ is communicated to $D_2$. The situation is even more astonishing, when the polarizer in front of detector $D_1$ is switched to the other polarization state after the photon has been emitted by the atom but before it reaches $D_1$. Even now the detector $D_2$ measures the opposite state of polarization as the new state of $D_1$. Can information between the detectors be transmitted with a speed greater than the speed of light?

Einstein believed that quantum theory could not be complete, that there might be "hidden parameters" that had not yet been detected [12.40]. Theoretical proofs given by *J.S. Bell* [12.41] and experiments performed by *A. Aspect* [12.42], have shown, however, that there are no hidden parameters. The paradox can be explained, if quantum theory is regarded as a nonlocal theory. This means that the description of microparticles is not bound to a specific location. The two photons represent an entangled state, where the physical parameters of one photon are determined by those of the other.

The paradox can also be explained in the following way. Two particles with spin up or down are emitted by a source into two opposite directions. We know that the total spin of the two particles must be zero, because the emitting source has zero angular momentum. The wave function of the system can be described as

$$|\psi\rangle = \frac{1}{\sqrt{2}}(|\uparrow\rangle|\downarrow\rangle + |\downarrow\rangle|\uparrow\rangle) \,. \tag{12.41}$$

After the particles are far apart and do not interact with each other, we measure the spin of the particle emitted to the right side. Then we know its spin and we can predict the spin of the other particle. This means that its spin is determined by the knowledge of the spin of the other particle.

The situation of this paradox is similar to that of Young's two-slit experiment and can be explained in the same way [12.40, 43].

### 12.5.3 Schrödinger's Cat

*Erwin Schrödinger* used an illustrative example to show that quantum mechanics at this time still had some unanswered questions about the validity of entangled states.

Assume a cat is put in a black box where two switches can be activated by the cat. One switch does nothing, the other opens a bottle with poisonous gas that kills the cat immediately. There is no information transferred from the inside to the outside of the box. The observer from the outside only knows that one of the switches has been opened by the cat, but he does not know which of the two. He therefore does not know whether the cat is dead or alive. From a quantum mechanical point of view the cat is therefore in an entangled state (12.46) and is simultaneously dead and alive. Our common sense tells us that this is nonsense.

The question is, how far the concept of entangled states can be transferred to the macroscopic world. We remember that entangled states can only be formed from a coherent superposition of states. The coherence is destroyed by relaxation processes. The larger the number of atoms, the higher the probability of relaxation. Therefore, systems with many atoms, which have been coherently prepared at time $t = 0$ loose this coherence within a time interval, which decreases rapidly with increasing number of atoms. *Haroche* and coworkers have shown that even for a system of three atoms, the coherence generated by simultaneous absorption of a pulse laser lasted for less than a few microseconds [12.44]. Therefore entangled states for a cat with more than $10^{25}$ atoms cannot exist.

### 12.5.4 Entanglement and Quantum Bits

The which way experiments discussed in the previous section require the realization of a coherent superposition

$$|\psi\rangle = (|\psi_a\rangle + |\psi_b\rangle)/\sqrt{2} \tag{12.42}$$

of quantum states, where $\psi_a$ describes the quantum state of the system when only path "a" is open and path "b" is closed and $\psi_b$ the state when path "b" is open and "a" is closed. This coherent superposition plays the possibly essential role in quantum computers [12.45]. While a classical computer operates with bits (this is a system that allows two possible values 0 and 1), a quantum computer is based on qubits, which are represented by a two-level system that can be in the levels $|0\rangle$ and $|1\rangle$. Unlike classical computers, the quantum systems have the possibility of coherence and superposition of states.

The general state of the system can be described by the qubit

$$|Q\rangle = \alpha|0\rangle + \beta|1\rangle \quad \text{with} \quad |\alpha|^2 + |\beta|^2 = 1 , \tag{12.43}$$

which is a coherent superposition of the two states $|0\rangle$ and $|1\rangle$. If we measure the qubit, we will find that it has the value $|0\rangle$ with the probability $|\alpha|^2$ and the value $|1\rangle$ with the probability $|\beta|^2$. This seems to bear an uncertainty, but the important point is that for a coherent superposition one can always find a basis in which the value of the qubit $|Q\rangle$ is well defined. This can be illustrated by a simple example.

If we consider the specific state

$$|Q\rangle = (|0\rangle + |1\rangle)/\sqrt{2} \tag{12.44}$$

the qubit will be with 50% probability in state $|0\rangle$ and with 50% in state $|1\rangle$. With the so-called Hadamar transformation

$$H(0) \to \frac{1}{\sqrt{2}}(|0\rangle + |1\rangle) \ , H(1) \to \frac{1}{\sqrt{2}}(|0\rangle - |1\rangle) \tag{12.45}$$

the qubit $|Q\rangle$ of (12.44) is transformed into the well defined state $|0\rangle$ because $H(Q) = |0\rangle$.

**Note:**

This Hadamar transformation is only possible for a coherent superposition of states. For an incoherent mixture, such a transformation would not be possible because in this case there will be a pure addition of probabilities $\alpha^2$ and $\beta^2$ (not of amplitudes $\alpha$ and $\beta$) without any interference.

The Hadamar transformation can be physically realized using a 50% beam splitter (Fig. 12.40). We will denote a particle above the beam splitter to be in the state $|0\rangle$ and below with $|1\rangle$.

If $\alpha$ is the probability amplitude that the particle is incident from above and $\beta$ that for the particle incident from below, the incident state can be characterized by

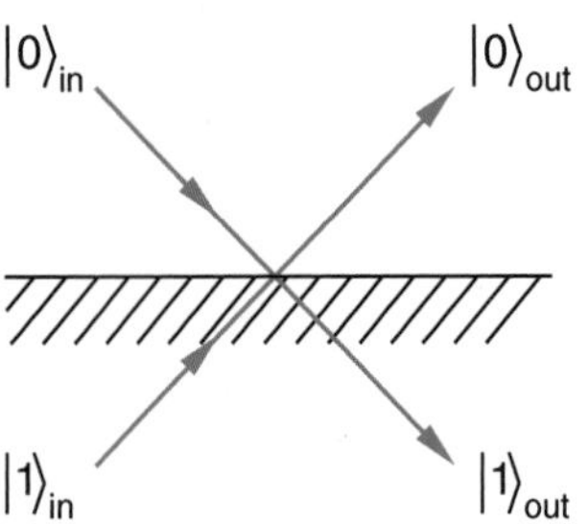


**Fig. 12.40.** 50% beam splitter as optical example for a Hadamar transformation

the superposition (the qubit)

$$|Q\rangle_{\text{in}} = \alpha|0\rangle_{\text{in}} + \beta|1\rangle_{\text{in}} . \tag{12.46}$$

With a 50% beamsplitter the reflected and the transmitted amplitudes are $A_{\text{out}} = A_{\text{in}}/\sqrt{2}$. If the amplitude of the particle wave incident from below suffers a phase jump of $\pi$ upon reflection, its sign is reversed. Now $(\alpha + \beta)$ is the probability amplitude for finding the particle in the upper outgoing direction and $(\alpha - \beta)$ for the lower direction. The output state is therefore

$$\begin{aligned} |Q\rangle_{\text{out}} &= \frac{1}{\sqrt{2}}[(\alpha + \beta)|0\rangle + (\alpha - \beta)|1\rangle] \\ &= H|Q\rangle_{\text{in}} . \end{aligned} \tag{12.47}$$

This is exactly the result of the Hadamar transformation (12.45).

The Mach–Zehnder interferometer in Fig. 12.41 corresponds to a sequence of two Hadamar transformations

$$|Q\rangle_{\text{out}} = HH|Q\rangle_{\text{in}} . \tag{12.48}$$

When the Hadamar transformations (12.45) are applied to the general Qubit (12.42) one recognizes that

$$HH|Q\rangle_{\text{in}} = |Q\rangle_{\text{in}} . \tag{12.49}$$

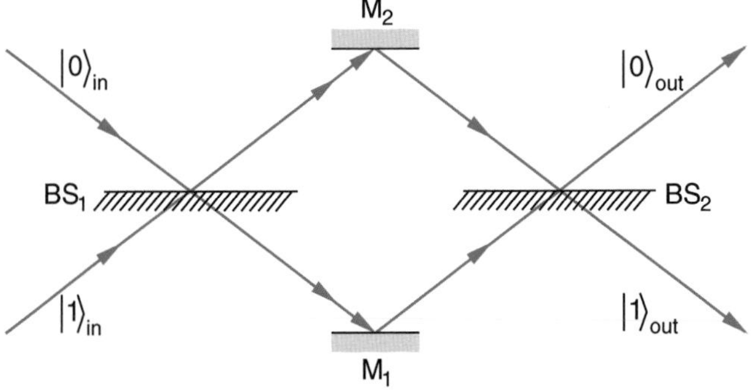


**Fig. 12.41.** Mach–Zehnder interferometer as optical realization of a sequence of two Hadamar transformations

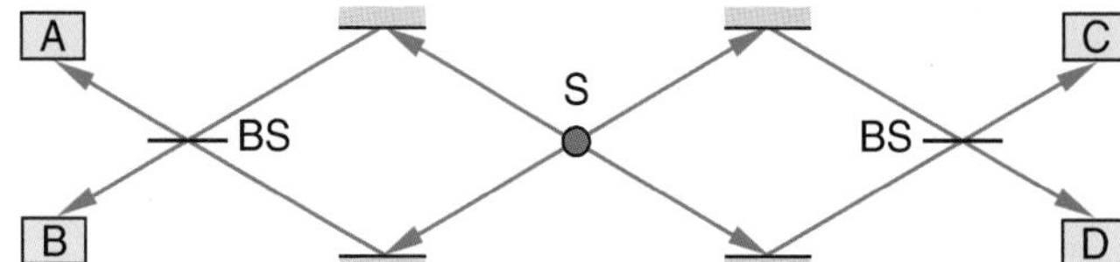


**Fig. 12.42.** Emission of two photon pairs in an entangled state by the sequence $S$

> This means that a Mach–Zehnder interferometer has an output state that is identical to the input state.

Assume a source S emits a pair of photons with opposite momenta in such a way that one photon is emitted to the right and the other in the opposite direction to the left (Fig. 12.42). If the photon to the left is emitted into the upper half space the photon to the right has to be emitted into the lower half space, because both photons of a pair should be antiparallel, otherwise the total momentum would not be zero. In our description by qubits, the left photon, which we call photon 1 is in the state $|0\rangle_1$ if it is in the upper beam and in the state $|1\rangle_1$ if it is in the lower beam. Photon 2, emitted to the right, is in the corresponding states $|0\rangle_2$ or $|1\rangle_2$. Although we do not know which of the two photons is in the upper or in the lower beam, we do know that photon 1 is in the upper beam if photon 2 is in the lower one and vice versa. We can describe this situation by a two-particle superposition state

$$|Q\rangle_e = \frac{1}{\sqrt{2}}(|0\rangle_1|1\rangle_2 + |1\rangle_1|0\rangle_2) , \qquad (12.50)$$

which is an ***entangled state***, as discussed above. It is represented by a superposition of two qubits as product states and has interesting properties. Neither of the two qubits has a definite value. One cannot tell whether qubit 1 is in the state "o" or in "1". The same is true for qubit 2. However, as soon as the value of one of the two qubits has been determined by a measurement, the other one has to have the opposite value.

The concept of entangled states plays a key role for quantum computers and quantum communication and cryptography. For entangled states the information received by the right detector in Fig. 12.42 could be predicted if a measurement at the left detector was performed, even if the two detectors were far apart.

For more details see [12.45, 46, 47].

### 12.5.5 Quantum Gates

Any computer needs gates for performing mathematical operations. These can be, for example,"not-gates", "and-gates", "or-gates" or "nor gates". Their basic principle is illustrated in Fig. 12.44. The "not-gate" can be realized by a simple inverter: There is only one input channel and one output. If the input is 1, the output is 0 and vice versa. The other gates consist of devices with two input channels and one output channel. The input signals can be 1 or 0 realized by different voltage steps. The input voltages can trigger a flip-flop circuit to produce the output voltage. For the "or-gate" the output is 1, if at least one of the input channels has the input 1. For the "and-gate" the output is 1 only if both input channels are 1, otherwise it is 0. The "nor-gate" represents the inversion of the "or-gate", i.e. the output is 1 only if both input channels are 0.

These gates in a classical computer correspond to quantum gates in quantum computers. One example is the Hadamar gate

$$H = \frac{1}{\sqrt{2}}\begin{pmatrix} 1 & 1 \\ 1 & -1 \end{pmatrix}$$

which transfers the q-bit $|0\rangle$ at the input to the entangled state $(1/\sqrt{2})(|0\rangle + |1\rangle)$ at the output and $|1\rangle \rightarrow (1/\sqrt{2})(|0\rangle - |1\rangle)$.

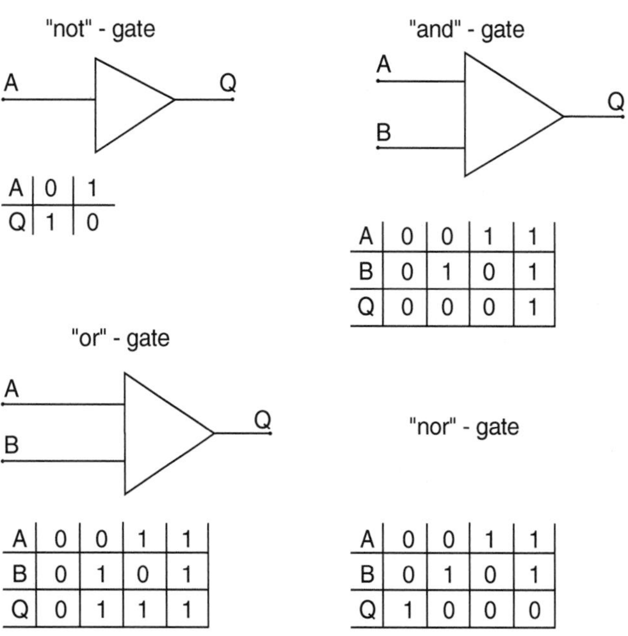


| A | 0 | 1 |
|---|---|---|
| Q | 1 | 0 |

| A | 0 | 0 | 1 | 1 |
|---|---|---|---|---|
| B | 0 | 1 | 0 | 1 |
| Q | 0 | 0 | 0 | 1 |

| A | 0 | 0 | 1 | 1 |
|---|---|---|---|---|
| B | 0 | 1 | 0 | 1 |
| Q | 0 | 1 | 1 | 1 |

| A | 0 | 0 | 1 | 1 |
|---|---|---|---|---|
| B | 0 | 1 | 0 | 1 |
| Q | 1 | 0 | 0 | 0 |

**Fig. 12.43.** Different computer gates

Another example is the Pauli X-gate

$$X = \begin{pmatrix} 0 & 1 \\ 1 & 0 \end{pmatrix}$$

which corresponds to the classical NOT gate and transfers the input q-bit $|0\rangle \to |1\rangle$ and $|1\rangle \to |0\rangle$

or the Pauli Z-gate

$$Z = \begin{pmatrix} 1 & 0 \\ 0 & -1 \end{pmatrix}$$

which performs the transformations $|0\rangle \to |0\rangle$ and $|1\rangle \to -|1\rangle$.

If we would like to compute an arbitrary function by reversible gates, we need another gate. One possibility is the **Toffoli gate**, proposed in 1980 by Toffoli.

This gate has a 3-bit input and output. If the first two bits are set, it flips the third bit as the following table shows:

| INPUT | | | OUTPUT | | |
|---|---|---|---|---|---|
| 0 | 0 | 0 | 0 | 0 | 0 |
| 0 | 0 | 1 | 0 | 0 | 1 |
| 0 | 1 | 0 | 0 | 1 | 0 |
| 0 | 1 | 1 | 0 | 1 | 1 |
| 1 | 0 | 0 | 1 | 0 | 0 |
| 1 | 0 | 1 | 1 | 0 | 1 |
| 1 | 1 | 0 | 1 | 1 | 1 |
| 1 | 1 | 1 | 1 | 1 | 0 |

The Toffoli gate is universal; this means that for any *boolean function* $f(x_1, x_2, \ldots, x_m)$, there is a circuit consisting of Toffoli gates which takes $x_1, x_2, \ldots, x_m$ and some extra bits set to 0 or 1 and outputs $x_1, x_2, \ldots, x_m, f(x_1, x_2, \ldots, x_m)$, and some extra bits (called garbage). Essentially, this means that one can use Toffoli gates to build systems that will perform any desired boolean function computation in a reversible manner [Wikipedia Toffoli gate].

The question is now how quantum gates can be realized experimentally in such a way, that they can be useful for quantum computers. There are several proposals and we will here only mention one of them, namely a linear array of ions, trapped in a linear Paul trap (Fig. 12.44a). Such ions can interact with radiation from two lasers tuned to specific wavelengths of a three level scheme (Fig. 12.44b). The ground state $|0\rangle$ of the ions is connected to an excited level $|1\rangle$ by a weak transition, (e.g. a dipole forbidden transition), and to another excited level $|2\rangle$ by a strong transition which is used for cooling the ion and for detection of a quantum jump from $|0\rangle$ to $|1\rangle$. The two levels $|0\rangle$ and $|1\rangle$ form a pair of entangled states and realize the qubit $|Q\rangle = \alpha|0\rangle + \beta|1\rangle$ in (12.46). If a $\pi/2$-pulse is applied to the transition $|0\rangle \to |1\rangle$ the two levels are equally populated, while a $\pi$-pulse inverts the population. The state of the qubit can be measured by the presence (the ground state is occupied i.e. $\beta = 0$) or the absence (the ground state is empty, i.e. $\alpha = 0$) of fluorescence induced by the strong transition $|0\rangle \to |1\rangle$. How can such a qubit be transferred to another ion?

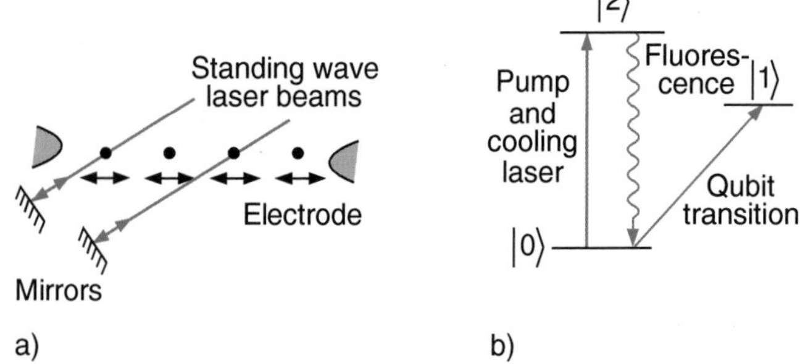


**Fig. 12.44.** Realization of qubits and quantum gates by interaction of laser radiation with ions in a linear trap

The Coulomb interaction between neighbouring ions in the linear trap is by far the strongest interaction. It differs for the different states $|0\rangle$ or $|1\rangle$ of the ion. The laser-cooled ions perform small oscillations around their equilibrium positions. Since they are all trapped in the harmonic trap potential, there are certain common normal oscillations of all ions (see Sect. 10.4, where normal vibrations of linear molecules are discussed). Since the different ions are coupled by their Coulomb interaction the excitation of one ion may change the interaction with the neighbouring ions and therefore the common normal oscillation mode of all ions in the trap. If this change can be detected, the information has been transferred from one qubit of an excited ion to the centre of mass motion of the ions. From here it can be further transferred to the internal state of another ion. This would represent a coherent interaction between two qubits realizing the construction of a quantum-gate.

If the laser beams are inclined against the axis of the linear trap, the movement of the ion would cause a definite Doppler-shift of its absorption frequency. The normal oscillation modes of the ions cause sidebands in the absorption spectrum.The frequencies of these sidebands differ for the different normal modes If the laser wavelength is tuned to such a sideband, it will only interact with ions, oscillating in this specific mode.

As is shown in [12.48] a two qubit logic quantum gate can be implemented with this device. In principle, every quantum algorithm can be realised by using only single- and two-qubit gates. This approach, however, is unfavourable for more complex tasks and would rapidly reach the limits of current implementations. Hence, physicists worldwide strive for the efficient realisation of multi-qubit gates. The scientists in Innsbruck [12.48, 49] succeeded in implementing a three-qubit gate acting on three trapped calcium ions representing the qubits. The target qubit of the so-called Toffoli gate will only be switched when both control qubits are set to "1", while in all other cases the target qubit will not be changed.

This novel gate does not only augment the set of available quantum gates, but also raises the achievable efficiency. Thomas Monz, junior scientist at the experiment, explains: "A Toffoli gate based on the conventional approach would require a sequence of six controlled switch operations. In comparison, our approach is three times faster while operating at a reduced error rate." Applications of the Toffoli gate lie within quantum error correction or quantum mechanical prime factorization. Thus, it represents an important component of a future quantum computer.

## SUMMARY

- Optical cooling of atoms means the reduction of their velocity by momentum transfer from the absorbed photon to the counter-propagating atom. Atoms in a gas are cooled in the three-dimensional interaction region of six laser beams in the $\pm x$, $\pm y$, $\pm z$ directions, if the laser frequency is tuned to the red side of the atomic absorption frequency.
- Atoms can be trapped in a magneto-optical trap (MOT) which consists of an anti-Helmholtz pair of coils, producing a cylindrical symmetric inhomogeneous magnetic field which causes a Zeeman splitting of the absorption line that increase with increasing distance from the trap centre. The trapping is caused by atomic recoil, due to photon absorption which depends on the distance from the trap centre.
- The atomic ensemble inside the trap is called optical molasses. The atoms in this molasses perform damped harmonic oscillations around the trap centre.
- The lower limit $T_{\min} = h\gamma/k_B$ of the temperature, achieved with optical cooling is set by the random recoil of spontaneously emitted photons. It depends on the homogeneous linewidth $\gamma$ of the optical transition. Lower temperatures can be reached by evaporation cooling.
- If the density of the cooled and trapped atoms becomes so high, that the mean distance between atoms becomes smaller than their deBroglie wavelength, a phase transition to a new state of matter occurs, called Bose–Einstein-condensation (BEC).
- Cold molecules can be produced by association of two atoms with small relative velocity, achieved by absorption of a photon by this atom pair and subsequent stimulated emission into a stable molecular state. Another technique uses resonances with molecular bound states in a variable magnetic field (Feshbach resonances).
- With femtosecond laser pulses online observations of dissociation of molecules are possible, using a pump-and probe technique.
- Coherent control means controlled shaping of amplitude $E(t)$ and phases $\varphi(t)$ of ultrashort pulses by a thin mask of many liquid crystal pixels, with refractive indices that are controlled by external voltages. If these pulses are absorbed by molecules they can produce a wanted coherent superposition of wave-functions in the excited molecular state and can suppress unwanted combinations. This favours wanted chemical reactions.
- The output of a mode-locked laser consists of many equidistant frequencies, called a frequency comb. The separation of adjacent frequencies in this comb equals the mode separation of the laser ▶

cavity. The frequency comb can be used for ultra-precise direct frequency-comparison between the caesium clock in the microwave region with optical frequencies and improves the measurements of absolute optical frequencies by several orders of magnitude.

- New and more precise atomic clocks can be realized with single ions, trapped in a Penning trap. The very narrow forbidden transition in such ions is used as the clock transition.
- Squeezing means the reduction of amplitude noise or phase noise below the uncertainty limit. If the amplitude noise is pushed below this limit the phase noise increases accordingly and vice versa. It can be experimentally realized by Mach–Zehnder interferometers where a nonli near absorption medium is placed in one of the interferometer arms.
- Which way-experiments investigate the loss of coherence in the superposition of two partial beams of a light source, if the way of the photons through the device is known.
- Entangled states represent a coherent superposition of atomic states. Such states are specific for quantum mechanics. They do not exist in classical physics. If one of the components of an entangled state is measured, the other is known, even if the pair of states is separated by long distances. This is proved by measurements of the Einstein–Rosen–Podolsky paradoxon, which can be only explained if it is assumed that quantum mechanics is a nonlocal theory.
- Entangled states play an essential role in teletransportation of information and in quantum computers, where they are used as $q$-bits instead of bits in the classical computer.

## PROBLEMS

1. Sodium atoms with a velocity $v_0 = 700\,\mathrm{m/s}$ in a collimated atomic beam are decelerated by absorption of laser photons on the $D_2$ line until they come to rest. If the optimum absorption-emission-cycle is assumed, what are the deceleration path length, the deceleration time and the absolute value of the deceleration?
2. What is the damping time of Na-atoms in an optical molasses, when the laser frequency is red-shifted by 10 MHz from the central absorption frequency of the $D_2$-line, if the absorption rate at line centre is $R_0 = 10^6\,\mathrm{s}^{-1}$?
3. Calculate the restoring force acting on Na-atoms in a MOT under the conditions of Problem 12.2 if the magnetic field gradient is $b = 0.01\,\mathrm{T/m}$. What is the harmonic potential of the trap and at which distance from the trap centre has the density of atoms decreased to half of the density $N(r = 0)$, if the temperature of the atoms is $T = 1\,\mu\,\mathrm{K}$. What are the oscillation frequency and damping constant of the atomic movements in the trap?
4. Calculate the mean distance between Na-atoms in a gas with density $10^{13}\,\mathrm{cm}^{-3}$. At which temperature becomes the deBroglie wavelength equal to the mean distance? What is the volume of the BEC at the critical temperature in a magnetic trap with a field gradient of $b = 10^{-3}\,\mathrm{T/m}$?
5. How many vibrational levels around $v = 20$ are coherently excited in the $B$-State of the $Cs_2$ molecule by a laser pulse with Gaussian time profile of half width $\Delta t = 20\,\mathrm{fs}$, if the vibrational constants are $\omega_e = 30\,\mathrm{cm}^{-1}$ and $\omega_e x_e = 0.04\,\mathrm{cm}^{-1}$? What is the time period of the oscillating wave packet? What is the distance between the two turning points of the wave packet movement, if a Morse potential is assumed? Which fraction of this distance is the extension of the wave packet?
6. Estimate, how stable the length of each of the two arms in a Michelson-type gravitational wave detector has to be, if the output intensity of the two interfering laser beams should not change by more than $10^{-8}\,I_0$, apart from fluctuations of $I_0$. Why can a length stability of much better than one atomic diameter be realized?

# Chronological Table for the Development of Atomic and Molecular Physics

≈ 440 BC *Empedocles* assumes, that the whole world is composed of 4 basic elements: Fire, Water, Air and Soil.

≈ 400 BC *Leucippos* and his disciple *Democritus* claim, that the world consists of small indivisible particles, called atoms, which are stable and nondestructable.

≈ 360 BC *Plato* attributes four regular regular geometric structures composed of triangles and squares (Platonic bodies) to the four elements and postulates that these structures and their interchange represent the real building blocks of the world.

≈ 340 BC *Aristotle* contradicts the atomic theory and assumes that the mater is continuous and does not consist of particles.

300 BC *Epicurus* revives the atomic model and assumes that the atoms have weight and spatial extension.

200 BC –1600 AC *no real progress in atomic physics.*

1661 *Robert Boyle* fights in his book: "*The Sceptical Chemist*" for the atomic model, which states, that all matter consists of atoms which differ in size and form for the different elements. He defines the terms "*chemical element*" and "*chemical compound*".

1738 *Daniel Bernoulli* assumes that heat can be explained as the movement of small particles. He may be regarded as the father of the kinetic gas theory.

1808 *John Dalton* supports in his book "*A New System of Chemical Philosophy*" the atomic hypothesis by describing his experiments on careful weighing the masses of reactants and reaction products of a chemical reaction. The results of these experiments lead to the "law of constant proportions": Each chemical element consists of equal atoms which form with simple number ratios molecules as building blocks of chemical compounds.

1811 *Amedeo Avogadro* derives from the laws of *Gay-Lussac* ($\Delta p/p = \Delta T/T$) and *Boyle-Marriot* ($p \cdot V =$constant for $T =$constant) the conclusion that all gases contain under equal conditions (pressure and temperature) the same number of particles per volume.

1820 John Herapath publishes his conception of a Kinetic theory of gases.

1857 *Rudolf J.E. Clausius* develops further the kinetic gas theory, founded by J. Herapath and *D. Bernoulli*.

1860 *Gustav Robert Kirchhoff* and *Robert Bunsen* create the foundations of spectral analysis of the chemical elements.

1865 *Joseph Lohschmidt* calculates the absolute number of molecules contained in $1\,\text{cm}^3$ of a gas under normal conditions ($p = 1$ atm, $T = 300\,\text{K}$).

1869 *Lothar Meyer* and *D.I. Mendelejew* establish (independent of each other) the *Periodic System of the Elements*.

1869 *Johann Wilhelm Hittorf* discovers the *cathode rays* in gas discharges.

1870 *James Clark Maxwell* gives the mathematical foundations to the kinetic gas theory. He defines the atoms as "absolute and unchangeable building blocks of matter".

1884 *Ludwig Boltzmann* develops from statistical grounds the distribution function for the energy of a system of free atoms in a constant volume. Together with the Austrian physicist *Josef Stefan* he derives the *Stefan–Boltzmann radiation law*.

W. Demtröder, *Atoms, Molecules and Photons*, 2nd ed., Graduate Texts in Physics,
DOI 10.1007/978-3-642-10298-1, 

1885 *Johann Jakob Balmer* finds the Balmer-formula for the spectral lines of the hydrogen atom.

1886 *Eugen Goldstein* discovers the "*Kanal-Strahlen*" (anode rays).

1886 *Heinrich Hertz* detects experimentally the electromagnetic waves predicted by Maxwell's theory and discovers 1887 the photo-electric effect and performs first experiments on the absorption of cathode rays.

1888 *Phillip Lenard* further investigates the absorption of the cathode rays.

1895 *Wilhelm Conrad Röntgen* discovers, while working on the properties of cathode rays a new kind of radiation which he called X-rays (first Nobel Prize in Physics 1901).

1896 *Henry Becquerel* first discovers radioactivity (Nobel prize 1903).

1898 *Marie Curie* separates different radioactive elements (Polonium and Radium) from minerals (Nobel prizes for Physics 1903 and Chemistry 1911).

1900 *Max Planck* presents his new theory of black body radiation, introducing the energy quanta $h \cdot \nu$ of the radiation field. This is nowadays regarded as the birth year of quantum physics (Nobel prize 1918).

1905 *Albert Einstein* develops his theory of Brownian motion. He explains the photoelectric effect using Planck's light quantum hypothesis (Nobel prize 1921).

1906 Charles Glover Barkla discovers the characteristic X-rays of the elements (Nobel prize 1917).

1909 *Robert Millikan* measures the elementary charge $e$ with his oil-droplet experiment (Nobel prize 1923).

1911 *Ernest Rutherford* and his coworkers investigate the scattering of $\alpha$-particles by gold nuclei and postulates his atomic model. This can be regarded as the foundation of modern atomic physics (Nobel prize for Chemistry 1908).

1912 *Max von Laue* (Nobelprize 1914) and his coworkers demonstrate, that X-rays represent electro-magnetic waves by observing the diffraction of X-rays by crystals.
Shortly later *William Henry Bragg* (Nobelprize 1915) confirms this result and furthermore shows that both the characteristic and the continuum radiation could be polarized.

1913 *Niels Bohr* (Nobelprize 1922) develops his new atomic model, based on the Rutherford model and the quantum hypothesis of Planck.

1913 *Henry Moseley* finds periodic regularities for the absorption frequencies of X-rays by different atoms and is able to determine the nuclear charge number $Z$ of the atoms from his measurements of absorption edges.
*James Franck* and *Gustav L. Hertz* investigate the inelastic collisions of electrons with atoms (Franck–Hertz experiment). Nobel prize 1925.

1919 *Arnold Sommerfeld* comprises all known facts and models of atoms in his famous textbook: "*Atombau und Spektrallinien*" and refines the atomic model of Bohr.

1921 *Otto Stern* and *Walter Gerlach* investigate the deflection of atoms in an inhomogeneous magnetic field and demonstrate the quantization of the component of the atomic angular momentum.

1923 *Arthur Holly Compton* (Nobel prize 1927) explains the inelastic scattering of X-rays by electrons (Compton effect) using the model of light quanta.

1924 *Louis de Broglie* (Nobel prize 1929) introduces the concept of matter waves.

1925 *S.A. Goudsmit* and *G.E. Uhlenbeck* explain the anomalous Zeeman effect by introducing the electron spin, postulated theoretical already 1924 by *W. Pauli*.

1925 *W. Pauli* (Nobel prize 1945) introduces the exclusion principle (Pauli-principle) which states that every existing atomic state occupied by more than on electron must be described by a wavefunction (product of spatial part and spin function) which is antisymmetric with respect to the exchange of two electrons.

1925 *Erwin Schrödinger* (Nobel prize 1933) extends the ideas of deBroglie about matter waves to a general wave-mechanics which is based on a special wave equation, called the Schrödinger equation.

1927 *Wolfgang Pauli* gives a mathematical description of the electron spin in form of quadratic "spin-matrices" with two rows and two columns. *Werner Heisenberg* (Nobel prize

1932) develops together with *Max Born* (Nobel prize 1954) and *Pascual Jordan* the mathematical concept of quantum mechanics, represented by matrices. He derives the uncertainty relations.

1928 *J.C. Davisson* (Nobel prize 1937) and *L.H. Germer* prove experimentally the wave nature of electrons by observing the diffraction pattern of electrons passing through thin crystalline foils.
*Paul Dirac* (Nobel prize 1933) develops a relativistic theory of Quantum Mechanics.
*Chandrasekhara Venkata Raman* (Nobel prize 1930) discovers the inelastic scattering of light by molecules (Raman-effect).

1932 *E. Ruska* (Nobel prize 1986) constructs the first electron microscope.

1936 *I. Rabi* (Nobel prize 1944) demonstrates a new techniques of radiofrequency spectroscopy in molecular beams for the precise measurement of magnet moments.

1944 *G.Th. Seaborg* (Nobelprize for Chemistry 1951) identifies the first tran-uranium elements.

1947 *Polykarp Kusch* (Nobel prize 1955): Measurement of the magnetic moment of the electron.
*Willis Lamb* (Nobel prize 1955): Measurement of the energy difference (Lamb-shift) between the $2S_{1/2}$ and $2P_{1/2}$ levels in the hydrogen atom.

1947 *John Bardeen* develops together with *W.H. Brattain* and *W. Shockley* the transistor (Nobel prize 1956).

1948 *Felix Bloch* and *Edward Mills Purcell* (Nobel prize 1952) demonstrate the nuclear magnetic resonance technique NMR.

1948 *J. Schwinger*, *R.P. Feynman* and *S. Tomonaga* (Nobel prize 1965) Theoretical formulation of quantum field theory (quantum Electrodynamics).

1950 *A. Kastler* (Nobel prize 1966) and *J. Brossel* demonstrate the experimental technique of optical pumping using incoherent light sources before the invention of the laser.

1953 *F.H. Crick* and *J.D. Watson* prove experimentally by X-ray diffraction the double helix structure of DNA (Nobel prize for medicine 1963).

1954 *N.G. Basow*, *A.M. Prochorov* and *Ch. Townes* (Nobel prize 1964) develop the theoretical foundations of the maser principle, based on Kastler's idea of optical pumping. First experimental verification of the $NH_3$-maser by *J.P. Gordon*, *H.J. Zeiger* and *Ch. Townes*.

1957 Explanation of supra-conductivity by *John Bardeen*, *Leon Cooper* and *J. Robert Schrieffer* (BCS theory) Nobel prize 1972.

1958 *Rudolf Mößbauer*: Recoil-free emission and absorption of $\gamma$-quants by atomic nuclei (Mößbauer effect) (Nobel prize 1972).

1959 *Arthur Schawlow* (Nobel prize 1995) and Charles Townes give detailed description for the extension of the maser principle to the optical range.

1960 First experimental realization of an optical maser (ruby laser) by *Th. Maiman*.

1961 The first He-Ne-laser is constructed by *W.R. Bennet* and *A. Javan*, based on detailed investigations of atomic collision processes in gas discharges.

1966 The dyelaser is developed indepently by *F.P. Schäfer* and *P.A. Sorokin*.

1971 *G. Herzberg* receives the Nobel prize in Chemistry for his centennial work on molecular spectroscopy.

1980 First proposals for optical cooling of atoms by photon recoil by *Th.W. Hänsch*, *A. Schawlow* and *V. Letokhov*.

1982 Development of tunnel microscopy by *G. Binning* and *H. Rohrer*, where single atoms on surfaces can be observed (Nobel prize 1986).

1986 Discovery of high temperature supra conductivity by *J. Bednarz* and *K.A. Müller* (Nobel prize 1987).

1988 Nobel prize to *H. Michel*, *J. Deisenhofer* and *R. Huber* for the elucidation of the primary process in the photosynthesis of green plants using femtosecond laser spectroscopy.

1989 Nobel prize to *Norman Ramsey*, *H. Dehmelt* and *Wolfgang Paul* for the experimental storage and trapping of neutrons, ions and electrons in electromagnetic traps.

1991 Pulse-Fourier-transform NMR spectroscopy: Nobel prize to *Richard Ernst*.

1992 Manipulation of single atoms on surfaces using the atomic force microscope.

1994 Optical cooling of free atoms in the gas phase; observation of optical molasses.

1995 Realization of magneto-optical traps; cooling of atoms below 1 $\mu$K by Sysiphos cooling, developed by *C. Cohen-Tannoudji* and coworkers.

First realization of Bose–Einstein Condensation (BEC) by *C. Wieman*, *E. Cornell* and indepently by *W. Ketterle*, using the combination of optical cooling and evaporation cooling to reach temperatures below 100 nK (Nobel prize 2001).

1997 Nobel prize to St. Chu C. Cohen-Tannoudji and W. Phillips for their developments of experimental techniques to cool and trap atoms with laser light.

1998 First demonstration of a continuous coherent beam of cold atoms from a BEC (atom laser) by *Th.W. Hänsch* and coworkers.

2000 Development of optical frequency comb.

2001 Production of very cold molecules by recombination of atoms in a BEC.

2005 First observation of Bose Einstein Condensation of molecules.

2006 Nobel prize to J. Hall and Th. Hänsch for the development of optical precision spectroscopy based on the optical frequency comb, to R. Glauber for his contributions to the quantum theory of optical coherence.

# Solutions to the Exercises

## Chapter 2

1. a) The mean distance is

$$\bar{d} = \sqrt[3]{\frac{1}{2.6 \times 10^{25}}}\,\text{m}$$
$$= 3 \times 10^{-9}\,\text{m} \approx 15\,\text{atom diameters}\,.$$

b) The filling factor $\eta$ is:

$$\eta = \frac{4}{3}\pi R^3 n = \frac{4}{3}\pi\, 10^{-30} \times 2.6 \times 10^{25}$$
$$= 1.1 \times 10^{-4} = 0.01\%\,.$$

c) The mean free path length is

$$\Lambda = \frac{1}{n\sigma\sqrt{2}}$$
$$\sigma = \pi(2r)^2 = 4\pi R^2 = 1.3 \times 10^{-19}\,\text{m}^2$$
$$n = 2.6 \times 10^{25}\,\text{m}^{-3}$$
$$\Rightarrow \Lambda = \frac{1}{\sqrt{2} \times 3.3 \times 10^6}\,\text{m} = 2.2 \times 10^{-7}\,\text{m}$$
$$= 220\,\text{nm}\,.$$

2. The mass density is:

$$\varrho_\text{m} = (0.78 \times 28 + 0.21 \times 32 + 0.01 \times 40) \times n\text{AMU}$$
$$1\,\text{AMU} = 1.66 \times 10^{-27}\,\text{kg}, \quad n = 2.6 \times 10^{25}/\text{m}^3$$
$$\Rightarrow \varrho_\text{m} = (21.8 + 6.72 + 0.4) \times 2.6 \times 10^{25} \times 1.66 \times 10^{-27}\,\text{kg/m}^3$$
$$= 1.25\,\text{kg/m}^3$$

3. a) $1\,\text{g}\,^{12}\text{C} = \frac{1}{12}\text{mol}$

$$\Rightarrow N = 6 \times 10^{23}/12 = 5 \times 10^{22}\,.$$

b) $1\,\text{cm}^3\,\text{He} \mathrel{\hat{=}} \frac{10^{-3}}{22.4} V_\text{M}$

$$\Rightarrow N = \frac{6 \times 10^{23} \times 10^{-3}}{22.4} = 2.7 \times 10^{19}\,.$$

c) $1\,\text{kg}\,N_2 \mathrel{\hat{=}} \frac{6 \times 10^{23}}{28} \times 10^3$ molecules

$$\Rightarrow N = 4.3 \times 10^{25}\,\text{atoms}\,.$$

d) $10\,\text{dm}^3\,H_2$ at $10^6$ Pa $\mathrel{\hat{=}}$ $100\,\text{dm}^3$ at $10^5$ Pa = 1 atm

$$\Rightarrow \nu = \frac{100}{22.4} \approx 4.5\,\text{moles}$$

$\Rightarrow N = 4.5 \times 6 \times 10^{23} = 2.7 \times 10^{24}$ molecules = $5.4 \times 10^{24}$ H-atoms.

4. $p = nkT;\, n = 1\,\text{cm}^{-3} = 10^6\,\text{m}^{-3}$

$$\Rightarrow p = 10^6 \times 1.38 \times 10^{-23} \times 10\,[\text{Pa}]$$
$$= 1.38 \times 10^{-16}\,\text{Pa} \approx 1.38 \times 10^{-18}\,\text{mbar}\,.$$

Such low pressures cannot be obtained in laboratories. Because of outgassing of the walls of a vacuum chamber and backstreaming of gas through the vacuum pump the lowest achievable pressure in the lab is around $10^{-10}$ Pa $\approx 10^{-12}$ mbar.

5. $100\,°\text{N} \mathrel{\hat{=}} 273\,\text{K} \Rightarrow 1\,°\text{N} \mathrel{\hat{=}} 2.73\,\text{K}$. The mean energy per atom and degree of freedom for a fixed temperature must be independent on the chosen temperature scale. The new Boltzmann constant $k_\text{N}$ is then obtained from:

$$\Rightarrow \frac{1}{2}k_\text{B}T_\text{K} = \frac{1}{2}k_\text{N}T_\text{N}$$
$$\Rightarrow k_\text{N} = \frac{1}{2.73}k_\text{B} = 5.1 \times 10^{-24}\,\text{J/°N}\,.$$

The boiling point of water, measured in the new temperature scale is:

$$T_S = \left(100 + \frac{100}{2.73}\right)\,°\text{N} = 136.6\,°\text{N}\,.$$

6. The sound velocity $v_{\text{Ph}}$ in a gas at pressure $p$ and density $\varrho$ is

$$v_{\text{Ph}} = \sqrt{\kappa p/\varrho} \quad \text{with} \quad \kappa = C_p/C_V = \frac{f+2}{f}\,,$$

where $C_p$ is the molar specific heat at constant pressure, $C_V$ at constant volume and $f$ is the number of degrees of freedom.

$$\Rightarrow\ v_{\text{Ph}}^2 = \kappa p/\varrho\,.$$

From the general gas equation for a mole volume $V_M$

$$pV_M = RT \ \Rightarrow\ R = \frac{pV_M}{T} = \frac{p}{\varrho}\frac{M}{T}\,,$$

$(M = \text{mole mass})$

we obtain:

$$v_{\text{Ph}}^2 = \kappa RT/M \Rightarrow R = \frac{M \cdot v_{\text{Ph}}^2}{\kappa \cdot T}$$

For radial acoustic resonances the acoustic wavelength is

$$n\lambda = r_0 \ \Rightarrow\ v_{\text{Ph}} = \nu_n\lambda = (\nu_n/n)r_0$$

The general gas constant $R$ is then obtained as

$$R = \frac{v_{\text{Ph}}^2 M}{\kappa T} = \frac{\nu_n^2 r_0^2 M}{n^2 \kappa T}$$

For argon $\kappa = (f+2)/f = 5/3$, $M = 40$ g/mole. Measuring the frequencies $\nu_n$ for different values of the integers $n$ $(n = 1,2,3,\ldots)$ yields the gas constant $R$, because $M, r_0, \kappa, T$ are known.

7. The effective mass of the collodial particle is:

$$m^* = m - \frac{4}{3}\pi r^3 \varrho_{\text{liquid}} = \frac{4}{3}\pi r^3(\varrho_{\text{part}} - \varrho_{\text{liquid}}) = 7.74 \times 10^{-18}\,\text{kg},$$

$$m = 4.0 \times 10^{-17}\,\text{kg}\,.$$

The vertical distribution is

$$n(z) = n_0 e^{-m^* gz/kT}$$

$$\Rightarrow \frac{n(h_1)}{n(h_2)} = e^{-(m^* g/kT)(h_1 - h_2)}$$

$$\Rightarrow k = \frac{m^* g \Delta h}{T \ln(n_1/n_2)} = \frac{7.7 \times 10^{-18} \times 9.81 \times 6 \times 10^{-5}}{290 \ln(49/14)} = 1.25 \times 10^{-23}\,\text{J/K}\,.$$

The best value accepted today is

$$k = 1.38 \times 10^{-23}\,\text{J/K}\,.$$

$$N_A = R/k = \frac{8.3}{1.25 \times 10^{-23}\,\text{mol}} \approx 6.02 \times 10^{23}/\text{mol}$$

$$M = N_A m = 6.02 \times 10^{23} \times 4.76 \times 10^{-14}\,\text{g/mol} = 3 \times 10^{10}\,\text{g/mol}\,.$$

When a colloid molecule has an average mass number of $10^4$ AMU the nanoparticle consists of about $3 \times 10^6$ molecules.

8. a) If the first diffraction order is at an angle $\beta_1 = 87°$, the incidence angle $\alpha$ can be obtained from the grating equation (see Fig. S.1):

$$d(\sin\alpha - \sin\beta_1) = \lambda \Rightarrow$$

$$\sin\alpha = \frac{\lambda}{d} + \sin\beta_1 = \frac{5 \times 10^{-10}}{0.83 \times 10^{-6}} + 0.99863 = 0.99923$$

$$\Rightarrow\ \alpha = 87.75°\,.$$

The second diffraction order appears at:

$$\sin\beta_2 = \sin\alpha - \frac{2\lambda}{d} = 0.99803 \ \Rightarrow\ \beta_2 = 86.40°\,.$$

The difference is $\Delta\beta = 0.6°$. For $\alpha = 88.94°$ we would obtain: $\Delta\beta = 0.75°$.

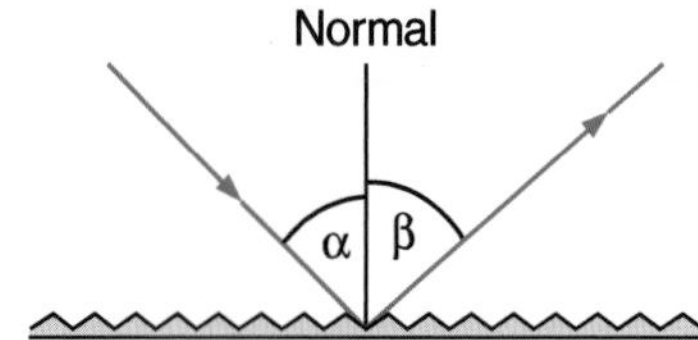


**Fig. S.1.**

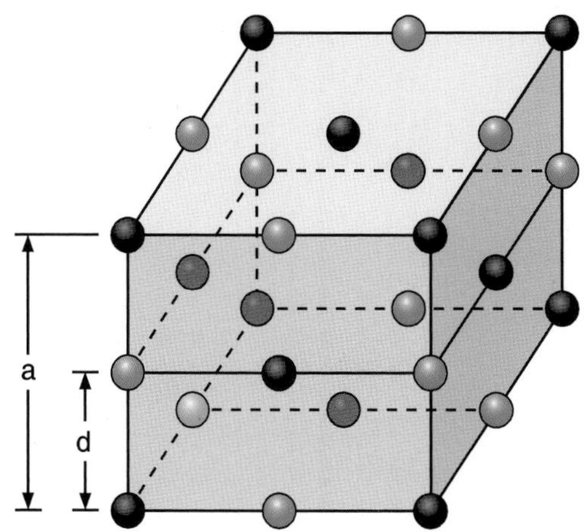


**Fig. S.2.**

b) The Bragg condition is

$$2d \sin\alpha = \lambda$$

$$\Rightarrow d = \frac{\lambda}{2\sin\alpha} = \frac{2 \times 10^{-10}}{2 \times 0.358}\,\text{m} = 2.79 \times 10^{-10}\,\text{m}\,.$$

This is half the side length $a$ of the elementary cell of the cubic crystal (Fig. S.2).

$$\Rightarrow a = 0.56\,\text{nm}\,.$$

NaCl has a face-centered cubic (fcc) crystal structure. Each elementary cell is occupied by four NaCl molecules. The molecular mass of NaCl is $(23 + 35) = 58$ AMU. The number of molecules per m$^3$ is:

$$N = \frac{4}{5.6^3 \times 10^{-30}}\,\text{m}^3 = 2.5 \times 10^{28}\,\text{m}^{-3}\,,$$

The mass of one molecule NaCl is:

$$m_{\text{NaCl}} = \frac{\varrho}{N} = \frac{2.1 \times 10^3}{2.28 \times 10^{28}}\,\text{kg} = 9.2 \times 10^{-26}\,\text{kg}\,.$$

In 1 mole of NaCl (= 58 g) are $N_A$ molecules.

$$\Rightarrow N_A = \frac{5.8 \times 10^{-2}}{9.2 \times 10^{-26}}\,\text{mole}^{-1} = 6.3 \times 10^{23}\,\text{mole}^{-1}\,.$$

c) From the Bragg condition

$$2d \sin\vartheta = m\lambda$$

we obtain for $m = 1$ the side length $a = 2d$ of the elementary cell of the fcc crystal as

$$a = \frac{\lambda}{\sin\vartheta} = 6.6 \times 10^{-10}\,\text{m}\,.$$

The radius $r_0$ of the spheres is (according to Sect. 2.4.3)

$$r_0 = \frac{1}{4}\sqrt{2}a = 2.33 \times 10^{-10}\,\text{m}$$

$$\Rightarrow V_{\text{sph}} = \frac{4}{3}\pi r_0^3 = 5.3 \times 10^{-29}\,\text{m}^3\,.$$

The filling factor is

$$\eta = \frac{4 \times \frac{4}{3}\pi r_0^3}{a^3} = \frac{16\pi \times 2\sqrt{2}}{3.64} \approx 0.78\,.$$

9. The van der Waals equation for 1 mole is:

$$\left(p + \frac{a}{V_M^2}\right)(V_M - b) = RT$$

($V_M$ = mole volume).

$$\Rightarrow pV_M - pb + \frac{a}{V_M} - \frac{ab}{V_M^2} = RT\,.$$

$$\Rightarrow pV_M\left(1 - \frac{b}{V_M} + \frac{a/p}{V_M^2} - \frac{ab/p}{V_M^3}\right) = RT\,.$$

This can be written as

$$pV_M(1 - x) = RT \quad \text{with} \quad (x \ll 1)\,.$$

With

$$\frac{1}{1-x} \approx 1 + x$$

$$\Rightarrow pV_M = RT\left(1 + \frac{b}{V_M} - \frac{a/p}{V_M^2} + \frac{ab/p}{V_M^3}\right)\,.$$

Comparison with the virial equation (vires = forces)

$$pV_M = RT\left(1 + \frac{B(T)}{V_M} + \frac{C(T)}{V_M^2}\right)$$

gives the coefficients:
$B(T) = b = 4$ times the eigenvolume of all molecules in $V_M$, and

$$C(T) = -\frac{a}{p}\,,$$

The ratio $p_i = a/V_M^2$ is called the "internal pressure" = pressure caused by the mutual attraction of the molecules. The term $C(T)/V_M^2$ gives the ratio $p_i/p$ of "internal pressure" to external pressure.

10. a) If a parallel beam of atoms A per m$^2$ per s hits atoms B at rest, the scattering cross section is

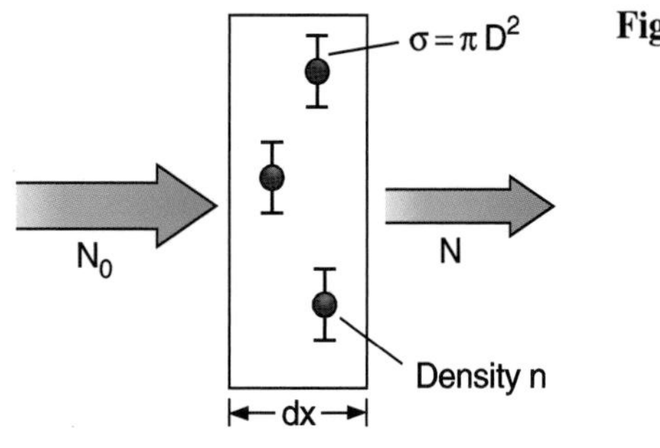


Fig. S.3.

(Fig. S.3) $\sigma = \pi(r_1 + r_2)^2$. For equal atoms (A = B) $r_1 = r_2 \Rightarrow \sigma = 4\pi r^2 = \pi D^2$ with $D = 2r$.
The number of particles scattered out of the beam along the path d$x$ through a gas of atoms B with density $n_\mathrm{B}$ is

$$\mathrm{d}N = -N n_\mathrm{B} \sigma \,\mathrm{d}x$$

$$\Rightarrow N(x) = N_0 \mathrm{e}^{-n_\mathrm{B}\sigma x} .$$

The mean free path length for which $N(x) = N_0/\mathrm{e}$ is

$$\Lambda = \frac{1}{n_\mathrm{B}\sigma} .$$

b) In a gas with thermal equilibrium the particles have an isotropic Maxwellian velocity distribution. The mean time between two collisions is:

$$\tau = \frac{1}{n\sigma|\boldsymbol{v}_r|}$$

where the relative velocity is

$$\boldsymbol{v}_r = \boldsymbol{v}_1 - \boldsymbol{v}_2$$

$$\Rightarrow \boldsymbol{v}_r^2 = \boldsymbol{v}_1^2 + \boldsymbol{v}_2^2 - 2\boldsymbol{v}_1 \cdot \boldsymbol{v}_2$$

$$\Rightarrow \left\langle v_r^2 \right\rangle = \left\langle v_1^2 \right\rangle + \left\langle v_2^2 \right\rangle, \text{ because } \langle \boldsymbol{v}_1 \cdot \boldsymbol{v}_2 \rangle = 0$$

$$\text{for A = B is} \left\langle v_1^2 \right\rangle = \left\langle v_2^2 \right\rangle = \left\langle v^2 \right\rangle$$

$$\Rightarrow \left\langle v_r^2 \right\rangle = 2\left\langle v^2 \right\rangle .$$

$$\Rightarrow \tau = \frac{1}{\sqrt{2} n\sigma\sqrt{\langle v^2 \rangle}} ,$$

$$\Lambda = \tau\sqrt{\langle v^2 \rangle} = \frac{1}{\sqrt{2} n\sigma} .$$

11. a) For a longitudinal magnetic field $\boldsymbol{B}$ with a length $L = 4f$, the time of flight between the two apertures in Fig. 2.67 is, according to (2.99b),

$$T = \frac{2\pi m}{eB} \quad \text{with} \quad T = L/v_z . \tag{1}$$

With an acceleration voltage $U$ the velocity $v_z$ is from

$$eU = \frac{m}{2} v_z^2 \Rightarrow v_z = \sqrt{2eU/m} . \tag{2}$$

Inserting in (1) gives

$$L = \frac{2\pi m}{eB} v_z \Rightarrow L^2 = \frac{4\pi^2 m^2}{e^2 B^2} \frac{2eU}{m}$$

$$\frac{e}{m} = \frac{8\pi^2}{L^2} \frac{U}{B^2} ,$$

$$\Rightarrow \frac{\delta(e/m)}{e/m} \leq \left|\frac{2\delta L}{L}\right| + \left|\frac{2\delta B}{B}\right| + \left|\frac{\delta U}{U}\right|$$

$$= 4 \times 10^{-3} + 2 \times 10^{-4} + 1 \times 10^{-4}$$

$$= 4.3 \times 10^{-3} .$$

In order to set $L = 4f$ one has to determine the focal length $f$. This is achieved by shifting the aperture $A_2$ until maximum transmission is reached.
Assume the maximum deviation of the electrons from the axis is $a = 5$ mm;

with $L = 100$ mm we obtain

$$\Rightarrow \sin\alpha \approx \frac{5}{25} = 0.2 \,\mathrm{rad}$$

(Fig. S.4). If the position of the focus is shifted by $\Delta L$ from its optimum value, the radius of the convergent electron beam is enlarged by $\Delta L \tan\alpha$ from $r_0$ to $r_0 + \Delta L \tan\alpha$. If the current $I$ of electrons flows through the aparture with radius $r_0 = 0.5$ mm (can be measured within $\Delta I/I = 10^{-3}$) the shift can be seen, if the area of the

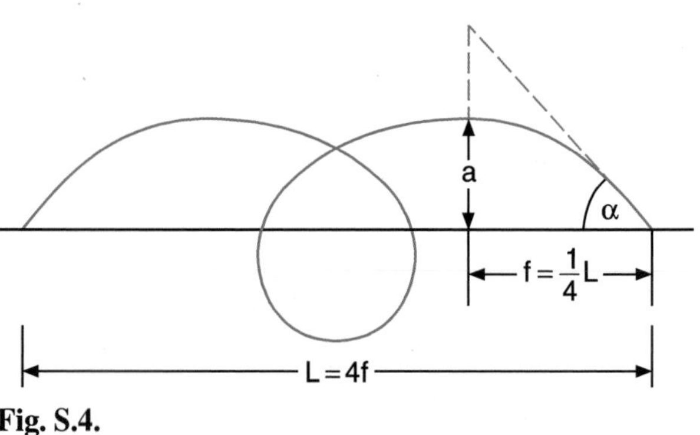


Fig. S.4.

aparture has increased to $\pi r_0^2\left(1+10^{-3}\right)$.

$$\pi\left[(r_0+\Delta r)^2-r_0^2\right]\le 10^{-3}\pi r_0^2$$

$$\Rightarrow \Delta r \lesssim 10^{-3} r_0 = 2.5\times 10^{-4}\,\text{mm},$$

$$\Delta r = \Delta L \tan\alpha$$

$$\Rightarrow \Delta L \le 2.5\times 10^{-4}\,\text{mm}/0.2$$

$$= 1.25\times 10^{-3}\,\text{mm},$$

$$\Rightarrow \Delta L/L = 1.25\times 10^{-5}\,.$$

The geometrical uncertainty of $\Delta L$ from the measurement of $L$ is therefore much larger than that of the uncertainty of measuring the transmitted current. The maximum relative error is therefore not affected by the uncertainty of optimising the transmitted current but rather by the mechanical accuracy of length measurement.

b) The maximum deflection from the straight path is $\delta x < 10^{-3} b$, because the current $I \propto b$ can be measured within $\Delta I/I = 10^{-3}$. The deviation from the $z$-axis is

$$x = \frac{1}{2} a t^2 \quad \text{with} \quad a = \frac{1}{m}\left(eE_x - evB_y\right) = \frac{1}{m} F_x\,.$$

With $t = L/v$ and $v^2 = 2eU/m$ we obtain

$$x = \frac{1}{2m} F_x \frac{L^2}{2eU}$$

$$\Rightarrow \delta x = \frac{\partial x}{\partial F_x}\delta F_x + \frac{\partial x}{\partial L}\delta L + \frac{\partial x}{\partial U}\delta U$$

$$\Rightarrow \left|\frac{\delta x}{x}\right| = \left|\frac{\delta F_x}{F_x}\right| + 2\left|\frac{\delta L}{L}\right| + \left|\frac{\delta U}{U}\right|\,.$$

With $\delta L/L = 1.25\times 10^{-5}$, $\delta F_x/F_x = \delta E_x/E_x + \delta B_y/B_y = 2\times 10^{-4}$, $\delta U/U = 10^{-4}$ we get

$$\frac{\delta x}{x} = 3.3\times 10^{-4}$$

$$\Rightarrow \delta x = 3.3\times 10^{-5}\,\text{mm}$$

for $x = b = 0.1$ mm. The uncertainty $\delta x < 10^{-3} b = 1\times 10^{-4}$ mm due to the uncertainty of measuring the current $I$ is larger here than in problem a) because of the uncertainties of measuring $E$, $B$ and $U$. The ratio $e/m = E^2/(2UB^2)$ can then be measured within

$$\frac{\delta(e/m)}{e/m} = 2\left|\frac{\delta E}{E}\right| + \left|\frac{\delta U}{U}\right| + 2\left|\frac{\delta B}{B}\right|$$

$$\le \left|\frac{\delta x}{x}\right| < 10^{-3}\,.$$

12. a) The vertical force is gravitation. If the flight direction is the $x$-direction the flight time is:

$$t_f = \frac{L}{v} = \frac{2}{300}\,\text{s} = 6.7\times 10^{-3}\,\text{s}\,.$$

The vertical deflection is

$$\Delta_1 z = \frac{1}{2} g t^2 = \frac{1}{2}\cdot 9.81\cdot 6.7^2\cdot 10^{-6}\,\text{m}$$

$$= 2.2\times 10^{-4}\,\text{m} = 0.22\,\text{mm}\,.$$

The divergence of the beam is:

$$\Delta\vartheta = \frac{b_1 + d_0}{2\cdot L} = \frac{4\times 10^{-6}}{2} = 2\times 10^{-5}\,.$$

The width of the beam at a distance $d_2 = 200$ cm downstream of $S_1$ is:

$$\Delta_2 z = d_2\cdot\Delta\vartheta = 4\times 10^{-3}\,\text{m}\,.$$

The deflection by gravity therefore changes the beam intensity, transmitted through a slit $S_2$ by the fraction $2.2\times 10^{-4}/4\times 10^{-3} = 5.5\times 10^{-2} = 5.5\%$.

b) The deflection of atoms with mass $M$ and charge $\Delta q$ is:

$$\Delta z = \frac{1}{2}\frac{E\cdot\Delta q}{M} t^2$$

$$= \frac{1}{2}\cdot 5\times 10^6\cdot\Delta q\cdot 6.7^2\times 10^{-6}\,\text{m}$$

$$= 1.1\times 10^2\cdot\frac{\Delta q}{M}\,[\text{m}]\,.$$

Assume $M = 4\,\text{AMU} = 6.7\times 10^{-27}$ kg (He-atoms) and a sensitivity of $\Delta z_{\min} = 10\,\mu$m than the minimum value of $\Delta q$ is $\Delta q_{\min} = 6\times 10^{-35}\,\text{C} = 3.8\times 10^{-16}\,e$.

c) The change $\Delta z$ of the deflection is $2\Delta z = 2.2\times 10^2\,\Delta q/M$ [m]. The relative change of the intensity, transmitted through $S_2$ is, according to a)

$$\frac{\Delta I}{I} = \frac{2.2\times 10^2\Delta q/M}{4\times 10^{-3}}\,.$$

If a relative change of $10^{-4}$ can be still measured, than the minimum charge difference

$$\Delta q = \frac{10^{-4}\times 4\times 10^{-3}}{2.2\times 10^2}\cdot M$$

can be measured. Inserting $M = 6.7\times 10^{-27}$ kg gives

$$\Delta q = 1.2\times 10^{-35}\,\text{C}\,.$$

13. From $mv^2/R = evB$ and $f_0 = R/\sin\varphi$ one obtains

$$B = \frac{mv}{eR} = \frac{mv}{ef_0 \sin\varphi} = \frac{1}{ef_0 \sin\varphi}\sqrt{2meU}\,.$$

With $eU = 10^3\,\text{eV} = 1.6 \times 10^{-16}\,\text{J}$, $m = 40\,\text{AMU} = 40 \times 1.66 \times 10^{-27}\,\text{kg}$, $\sin\varphi = \sin 60° = \frac{1}{2}\sqrt{3}$, $f_0 = 0.8\,\text{m} \Rightarrow B = 4.2 \times 10^{-2}$ Tesla.

14. According to (2.100) the focal length $f$ is:

$$f = \frac{4\sqrt{\phi_0}}{\int\limits_0^{z_0} \frac{2a\,\mathrm{d}z}{\sqrt{\phi_0 + az^2}}} = \frac{2\sqrt{\phi_0/a}}{\int\limits_{z=0}^{z_0} \frac{\mathrm{d}z}{\sqrt{(\phi_0^2/a) + z^2}}}$$

$$= \frac{2\sqrt{\phi_0/a}}{\ln\left(z + \sqrt{(\phi_0^2/a) + z^2}\right)\Big|_0^{z_0}}$$

$$= \frac{2\sqrt{\phi_0/a}}{\ln\left(\frac{z_0 + \sqrt{\phi_0^2/a + z_0^2}}{\sqrt{\phi_0^2/a}}\right)}\,.$$

15. a) The potential of the cylindric condenser can be obtained from the Laplace equation $\Delta\phi = 0$, which is written in cylindric coordinates $(r, \varphi, z)$ as:

$$\frac{1}{r}\frac{\partial}{\partial r}\left(r \cdot \frac{\partial \phi}{\partial r}\right) = 0\,,$$

with the solution

$$\phi = c_1 \ln r + c_2\,.$$

The cylinder surfaces $r = R_1$ and $r = R_2$ are at the fixed potentials $\phi_1$ and $\phi_2$. This gives:

$$c_2 = \phi_1 - c_1 \ln R_1$$

$$c_1 = \frac{\phi_2 - \phi_1}{\ln(R_2/R_1)}$$

$$\Rightarrow \phi(r) = \phi_1 + \frac{\phi_2 - \phi_1}{\ln(R_2/R_1)} \cdot \ln(r/R_1)\,.$$

The electric field $E(r)$ is:

$$E(r) = -\frac{\partial \phi}{\partial r} = \frac{\phi_1 - \phi_2}{\ln(R_2/R_1)} \cdot \frac{1}{r}\,.$$

The optimum path of the ions through the center $R_0 = (R_1 + R_2)/2$ of the cylindric sector field is obtained from:

$$\frac{mv_0^2}{R_0} = e \cdot E(R_0) = \frac{2e}{R_1 + R_2} \frac{\phi_1 - \phi_2}{\ln(R_2/R_1)}\,.$$

The optimum voltage is then:

$$U = \phi_2 - \phi_1 = \frac{R_1 + R_2}{2R_0}\frac{m}{e}v_0^2 \ln\left(\frac{R_2}{R_1}\right)$$

$$= \frac{mv_0^2}{e} \ln\left(\frac{R_2}{R_1}\right)$$

$$= \frac{2eV_0}{e} \ln(R_2/R_1)\,.$$

b) Assume an ion enters the cylindrical field at $\varphi = 0$ and $r = R_0 = (R_1 + R_2)/2$ with the velocity $v_0$. Assume it deviates at time $t$ from its optimum path $r = R_0$ by $\delta r$. The equation of motion $F = m \cdot a$ than becomes for the radial motion:

$$m\delta\ddot{r} = m \cdot \frac{v^2}{r} + e \cdot E(r + \delta r)\,. \quad (3)$$

Expansion of $E$ into a Taylor series yields:

$$E(R_0 + \delta r) = E(R_0) + \left(\frac{\mathrm{d}E}{\mathrm{d}r}\right)_{R_0} \delta r + \ldots. \quad (4)$$

From a) we obtain:

$$\frac{\mathrm{d}E}{\mathrm{d}r} = \frac{U}{\ln(R_2/R_1)} \cdot \frac{1}{r^2}\,. \quad (5)$$

Inserting this into (4) and (3) gives:

$$\delta\ddot{r} - \frac{v_0^2}{r^3}R_0^2 + \frac{v_0^2}{R_0}\left(1 - \frac{\delta r}{R_0}\right) = 0$$

$$\frac{1}{r^3} = \frac{1}{R_0^3(1 + \delta r/R_0)^3} \approx \frac{1}{R_0^3} - \frac{3}{R_0^4}\delta r + \ldots$$

$$\Rightarrow \delta\ddot{r} - \frac{v_0^2}{R_0}\left(1 - 3\frac{\delta r}{R_0} - 1 + \frac{\delta r}{R_0}\right) = 0$$

$$\Rightarrow \delta\ddot{r} + \frac{2v_0^2}{R_0^2}\delta r = 0\,.$$

With $\omega_0 = v_0/R_0$ this becomes:

$$\delta\ddot{r} + 2\omega_0^2 \delta r = 0$$

$$\Rightarrow \delta r = R_0\left(\sin\left(\sqrt{2}\omega_0 \cdot t\right)\right)\,.$$

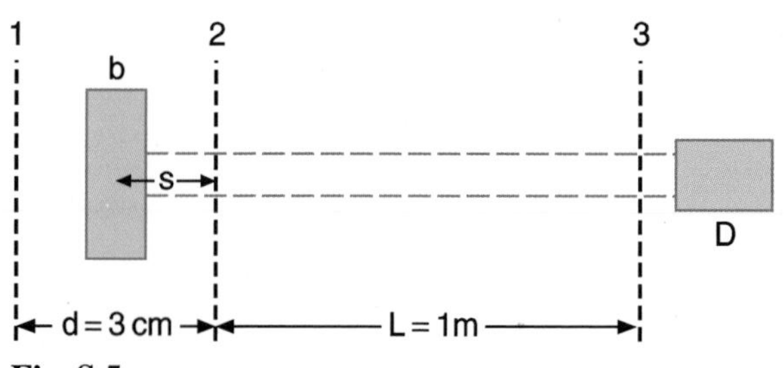


**Fig. S.5.**

After the time $t = \pi/\left(\sqrt{2}\omega_0\right) \Rightarrow \varphi = \pi/\sqrt{2} = 127°$. The deviation $\delta r$ becomes zero. The cylindrical condenser with an angle $\varphi = 127°$ therefore acts as focussing device.

16. An ion produced at the location $x$ travels a distance $s = \frac{1}{2}at_1^2$ with $a = eE/m$ in the electric field $E = U/d$ to the grid 2 (Fig. S.5).

$$\Rightarrow t_1 = \sqrt{\frac{2ms}{eE}}$$

$$v_1 = (eE/m)t_1 = \sqrt{\frac{2eEs}{m}}.$$

The drift time in the field-free region between grid 2 and 3, where the ion moves with constant velocity, is:

$$t_2 = L/v_1 = L\sqrt{\frac{m}{2eEs}}.$$

The total flight time is:

$$T = t_1 + t_2 = \sqrt{\frac{m}{eE}}\frac{2s + L}{\sqrt{2s}}.$$

The time difference $\Delta T$ for ions of equal mass, produced at $s_1 = (d+b)/2$ and $s_2 = (d-b)/2$ (i.e., at the opposite edges of the ionization volume) is

$$\Delta T_1 = \sqrt{\frac{m}{eE}}\left(\frac{2s_1 + L}{\sqrt{2s_1}} - \frac{2s_2 + L}{\sqrt{2s_2}}\right)$$
$$= \sqrt{\frac{m}{eE}}\left(\frac{d + b + L}{\sqrt{d+b}} - \frac{d - b + L}{\sqrt{d-b}}\right).$$

For $m = 100\,\text{AMU} = 1.66 \times 10^{-25}$ kg, $b = 2$ mm, $d = 30$ mm, $L = 1$ m one obtains:

$$\Delta T_1 = 1.018 \times 10^{-5}(5.769 - 6.143)\text{s}$$
$$= -3.811\,\mu\text{s}.$$

The ion with $s = (d+b)/2$ has a *shorter* flight time than an ion starting from $s = (d-b)/2$, because it is accelerated longer and has a larger velocity.

Two ions, with masses $m_1$ and $m_2$, both starting from the middle of the ionization volume ($s = d/2$) have the flight time difference:

$$\Delta T_2 = \frac{d + L}{\sqrt{edE}}\left(\sqrt{m_1} - \sqrt{m_2}\right).$$

Inserting the numerical values for $m_1 = 110$ AMU, $m_2 = 100$ AMU gives:

$$\Delta T_2 = 1.49 \times 10^8 \times 2 \times 10^{-14}\,\text{s}$$
$$\approx 3\,\mu\text{s}.$$

Two masses can be separated if their flight time difference $\Delta T_2$ is at least $\Delta T_1$ or larger. The mass resolution of our example is only $\Delta m/m \approx 10$.

It can be greatly increased by the McLaren arrangement with two different accelerating electric fields.

b) The increase of mass resolution in the reflectron can be seen as follows: Assume the lengths of the two arms of the reflectron to be $L_1 = L_2 = L$. Two ions, generated at two different locations have velocities $v_1$ and $v_2$ with $v_i = (2eE \cdot s_i/m)^{1/2}$. Their flight time is: $T_i = 2L/v_i$ without penetrating into the reflection field $E_\text{r}$. Here they penetrate a distance $d_\text{r}$ determined by the energy balance:

$$\frac{m}{2}v_i^2 = e \cdot E_\text{r} \cdot d_\text{r} \Rightarrow d_\text{r} = mv_i^2/(2eE_\text{r}).$$

The deceleration time is obtained from

$$d_\text{r} = \frac{1}{2}e\frac{E_\text{r}}{m}t^2.$$

Their time for penetration and reflection is therefore

$$t = 2 \cdot (2d_\text{r} \cdot m/eE_\text{r})^{1/2}.$$

Inserting $s_i = (d \pm b)/2$ we obtain for the total flight time of an ion through the reflectron:

$$T_i = \frac{2L}{v_i} + 2\frac{mv_i}{eE_\text{r}}.$$

Inserting $s_i = (d \pm b)/2 \Rightarrow v_i = [(2eE/m)(d \pm b)/2]^{1/2}$.

We can calculate the maximum time difference for ions generated at $s_i = (d \pm b)/2$ as:

$$\Delta T = \frac{2L}{\sqrt{eE/m}}\left(\frac{1}{\sqrt{b-d}} - \frac{1}{\sqrt{b+d}}\right)$$
$$+ \frac{2m}{eE_\text{r}}\sqrt{\frac{eE}{m}}\left[\sqrt{d-b} - \sqrt{d+b}\right].$$

With $\text{d}\Delta T/\text{d}b = 0$ one obtains the optimum retarding field $E_\text{r}$ as

$$E_\text{r} = \frac{E}{L}\frac{d - b - (d^2 - b^2)^{1/2}}{(d^2 - b^2)^{1/2} - (d-b)^2(d^2 - b^2)^{-1/2}}.$$

c) The width of the ion beam at the exit of the sector field is (see Fig. 2.74a)

$$b_2 = b_1 + \frac{2m}{eB}\Delta v$$

$$\frac{mv^2}{R} = qvB \;\Rightarrow\; R = \frac{mv}{qB}$$

$$\frac{m}{2}v^2 = qU \;\Rightarrow\; R = \frac{1}{B}\sqrt{\frac{2Um}{q}}$$

$$\Rightarrow\; m = \frac{R^2B^2q}{2U}\,.$$

In order to seperate two masses $m_1$ and $m_2$, the condition

$$2(R_1 - R_2) \geq b_2$$

has to be met. The mass resolution is then

$$\frac{m}{\Delta m} = \frac{R^2B^2q}{2U}\,\frac{2U}{B^2q\left(R_1^2 - R_2^2\right)} = \frac{R^2}{R_1^2 - R_2^2} \approx \frac{R_\mathrm{m}}{2(R_1 - R_2)}$$

with

$$R_\mathrm{m} = \frac{1}{2}(R_1 + R_2)\,.$$

Since $2(R_1 - R_2) > b_2$

$$\Rightarrow\; \frac{m}{\Delta m} \leq \frac{R_\mathrm{m}}{b_2}\,.$$

Assuming $R_\mathrm{m} = 0.3\,\mathrm{m}$, $b_2 = 1\,\mathrm{mm}$

$$\Rightarrow\; \frac{\Delta m}{m} \leq 300\,.$$

This is better than our simple TOF spectrometer, but worse than the reflectron.

17. The torque $\boldsymbol{D}$ on a particle with mass $m$ is:

$$\boldsymbol{D} = \boldsymbol{r} \times \boldsymbol{F}\,,$$

where $\boldsymbol{F}$ is the force acting on the particle at a distance $\boldsymbol{r}$ from the center.
For centro-symmetric force fields is

$$\hat{F} = f(r)\cdot\hat{r}$$

$$\Rightarrow\; \boldsymbol{D} = f(r)\cdot\boldsymbol{r}\times\hat{r} = 0\,.$$

The torque is the time derivative of the angular momentum $\boldsymbol{L} = \boldsymbol{r}\times\boldsymbol{p} = m\cdot\boldsymbol{r}\times\boldsymbol{v}$

$$\frac{\mathrm{d}\boldsymbol{L}}{\mathrm{d}t} = \boldsymbol{D}\,.$$

Since $\boldsymbol{D} = 0 \Rightarrow L = \text{const.}$

18. a) According to (2.163a) the impact parameter is

$$b = \frac{qZe}{4\pi\varepsilon_0\mu v_0^2}\cot(\vartheta/2)\,,\quad \text{with}$$

$$q = 2e\,,\quad Z = 79\,,$$

$$\varepsilon_0 = 8.85\times 10^{-12}\,\mathrm{A\,s/V\,m}\,,$$

$$\cot 45° = 1\,,\quad \frac{\mu}{2}v_0^2 = 5\,\mathrm{MeV} = 8\times 10^{-13}\,\mathrm{J}$$

$$\Rightarrow\; b = 2.27\times 10^{-14}\,\mathrm{m} = 22.7\,\mathrm{Fermi}\,.$$

b) For the backwards-scattered particles ($\vartheta = 180°$) we obtain the minimum distance $r_\mathrm{min}$ at the turning point from the energy balance

$$\frac{\mu}{2}v_0^2 = \frac{qZe}{4\pi\varepsilon_0 r_\mathrm{min}}$$

$$\Rightarrow\; r_\mathrm{min} = \frac{qZe}{2\pi\varepsilon_0\mu v_0^2} = 4.54\times 10^{-14}\,\mathrm{m}\,.$$

c) For $\vartheta = 90°$ the impact parameter is $b_0 = 2.27\times 10^{-14}\,\mathrm{m}$. All particles with $b < b_0$ are scattered into the angular range $90° < \vartheta \leq 180°$.
In order to estimate the maximum value of $b$ (i.e., the smallest deflection angle $\vartheta$) we assume $b_\mathrm{max} = \overline{d}/2$, where $\overline{d}$ is the average distance between two gold atoms in the scattering gold foil. In this case the $\alpha$-particle passes between two gold atoms and experiences a net deflection force of zero.
The number density of gold atoms is

$$n_V = \frac{\varrho N_\mathrm{A}}{M} = 6\times 10^{22}/\mathrm{cm}^3$$

with the mass density $\varrho = 19.3\,\mathrm{g/cm^3}$; the Avogadro number $N_\mathrm{A} = 6\times 10^{23}/\mathrm{mole}$ and the molar mass $M = 197\,\mathrm{g/mole}$.
The number density $n_\mathrm{F}$ per $\mathrm{cm}^2$ of gold atoms in the foil with thickness $t = 5\times 10^{-6}\,\mathrm{m}$ is $n_\mathrm{F} = n_V t = 3\times 10^{19}/\mathrm{cm}^2$. With $b_\mathrm{max} = \frac{1}{2}\overline{d} = \frac{1}{2}/\sqrt{n_\mathrm{F}} = 9.1\times 10^{-11}\,\mathrm{cm} = 9.1\times 10^{-13}\,\mathrm{m}$ the scattering cross section is

$$\sigma = \pi b_\mathrm{max}^2 \approx 2.6\times 10^{-20}\,\mathrm{cm}^2\,.$$

The fraction of atoms scattered into the angular range $\vartheta \geq 90°$ is

$$\frac{N(\vartheta \geq 90°)}{N(\vartheta \leq 180°)} = \frac{\pi b_0^2}{\pi b_{\max}^2} = \left(\frac{2.27 \times 10^{-14}}{9.1 \times 10^{-13}}\right)^2 = 6 \times 10^{-4}.$$

d) $b(\vartheta = 45°) = a \cot 22.5° = 2.71\,a$ with $a = qZe/(4\pi\varepsilon_0 \mu v_0^2) \Rightarrow a = 2.27 \times 10^{-14}\,\text{m}$.

$$\frac{N(45° \leq \vartheta \leq 90°)}{N(\vartheta \leq 180°)} = \frac{\pi\left[(2.41)^2 a^2 - a^2\right]}{\pi b_{\max}^2} = \frac{4.8a^2}{b_{\max}^2} = \frac{4.8 \times 2.27^2 \times 10^{-28}}{8 \times 10^{-21}} = 3.1 \times 10^{-7}.$$

19. The rate $N(\vartheta)$ of particles scattered into the angular range $\vartheta_1 \leq \vartheta \leq \vartheta_2$ is for Rutherford scattering:

$$N(\vartheta)\Delta\vartheta \propto \int_{\vartheta_1}^{\vartheta_2} \frac{\sin\vartheta \,\mathrm{d}\vartheta}{\sin^4(\vartheta/2)} = \int_{\vartheta_1}^{\vartheta_2} \frac{2\cos(\vartheta/2)}{\sin^3(\vartheta/2)}\,\mathrm{d}\vartheta = \left[-\frac{2}{\sin^2(\vartheta/2)}\right]_{\vartheta_1}^{\vartheta_2}.$$

This allows the calculation of the ratio

$$\frac{N(1° \pm 0.5°)}{N(5° \pm 0.5°)} = \frac{46{,}689}{214.4} = 218.$$

For the Thomson model we obtain for a medium scattering angle $\overline{\vartheta} = 2 \times 10^{-4}\,\text{rad}$ and an average number $m$ of scattering events in the gold foil, according to the numerical value given in Problem 12.15c:

$$m = n_F \sigma = 3 \times 10^{19} \times 3 \times 10^{-16} \approx 10^4$$
$$\Rightarrow \langle\vartheta\rangle = \sqrt{m}\,\overline{\vartheta} = 2 \times 10^{-4} \times 10^2\,\text{rad} \approx 1.2° = 2 \times 10^{-2}\,\text{rad} \approx 1.2°,$$

$$N(\vartheta)\Delta\vartheta \propto \int_{\vartheta_1}^{\vartheta_2} \sin\vartheta\, \mathrm{e}^{-(\vartheta/\langle\vartheta\rangle)^2}\,\mathrm{d}\vartheta \approx \int \vartheta \mathrm{e}^{-(\vartheta/\langle\vartheta\rangle)^2}\,\mathrm{d}\vartheta = \left[\frac{\langle\vartheta\rangle^2}{2}\mathrm{e}^{-(\vartheta/\langle\vartheta\rangle)^2}\right]_{\vartheta_1}^{\vartheta_2}$$

$$\Rightarrow \frac{N(1° \pm 0.5°)}{N(5° \pm 0.5°)} = \frac{\mathrm{e}^{-0.17} - \mathrm{e}^{-1.56}}{\mathrm{e}^{-14} - \mathrm{e}^{-21}} \approx 7.5 \times 10^5.$$

This shows that the scattering rate decreases much stronger with increasing $\vartheta$ than for the Rutherford scattering.

20. a)

$$\frac{\mu v_0^2}{2} = \frac{Ze^2}{4\pi\varepsilon_0 r_{\min}},$$
$$r_{\min} = 5 \times 10^{-15}\,\text{m}, \quad Z = 29,$$
$$\mu = \frac{1 \times 63}{64} = 0.98\,\text{AMU}$$
$$\Rightarrow \frac{\mu}{2}v_0^2 = \frac{29 \times 1.6^2 \times 10^{-38}}{4\pi \times 8.85 \times 10^{-12} \times 5 \times 10^{-15}}\,\text{J} = 1.33 \times 10^{-12}\,\text{J}$$
$$\Rightarrow \frac{m}{2}v_0^2 = 1.36 \times 10^{-12}\,\text{J} = 8.5\,\text{MeV}.$$

b) For $\vartheta < 180°$ is

$$r_{\min} = b\left[1 - \frac{E_{\text{pot}}(r_{\min})}{\frac{\mu}{2}v_0^2}\right]^{-1/2}.$$

With $r_{\min} = 5 \times 10^{-15}\,\text{m} \Rightarrow b = 1.775 \times 10^{-15}\,\text{m}$

$$\Rightarrow \cot(\vartheta/2) = b/a \quad \text{with} \quad a = \frac{Ze^2}{4\pi\varepsilon_0 \mu v_0^2}$$
$$\Rightarrow \vartheta \geq 113.4°.$$

## Chapter 3

1. With the de Broglie relation

$$\lambda_{\text{dB}} = \frac{h}{p}$$
$$\Rightarrow v = \frac{h}{m\lambda} = \frac{6.63 \times 10^{-34}}{10^{-10} \times 1.67 \times 10^{-27}}\,\frac{\text{m}}{\text{s}} = 3.97 \times 10^3\,\text{m/s}.$$

Thermal neutrons have at $T = 300\,\text{K}$ a mean velocity $\overline{v} = 2.2 \times 10^3\,\text{m/s}$ and a kinetic energy of $E_{\text{kin}}^{\text{thermal}} = 40\,\text{meV}$. Our neutron is slightly super thermal. Its kinetic energy is

$$E_{\text{kin}} = \frac{m}{2} v^2 = 1.31 \times 10^{-20}\,\text{J} = 82\,\text{meV}\,.$$

2. The average energy per mode is

$$\langle E \rangle = \langle n \rangle \cdot h\nu\,,$$

where $\langle n \rangle$ is the average number of photons in this mode. If $P_n$ is the probability, that a mode contains $n$ photons, it follows for thermal equilibrium

$$P_n = \frac{\mathrm{e}^{-n \cdot h\nu / k_B T}}{\sum_n \mathrm{e}^{-nh\nu/k_B T}} \Rightarrow \sum_{n=0}^{\infty} P_n = 1\,.$$

$$\Rightarrow \frac{P_{n+1}}{P_n} = \mathrm{e}^{-h\nu/k_B T} = x\,.$$

The geometrical series $\sum_{n=0}^{\infty} x^n$ has for $x < 1$ the value

$$\sum_n x^n = \frac{1}{1-x}\,.$$

With the relation:

$$\sum_{n=0}^{\infty} n \cdot x^n = x \cdot \frac{\mathrm{d}}{\mathrm{d}x}\left(\sum x^n\right) = \frac{x}{(1-x)^2}$$

we obtain

$$\langle n \rangle = \sum_{n=0}^{\infty} n \cdot P(n) = \sum n \cdot \frac{x^n}{\sum x^n} = \frac{x}{1-x}$$

$$= \frac{1}{\mathrm{e}^{h\nu/k_B T} - 1}\,.$$

The mean energy per mode is then

$$\langle E \rangle = \frac{h\nu}{\mathrm{e}^{h\nu/k_B T} - 1}\,.$$

3. Differentiation of $S_\nu^*$ in (3.16) gives:

$$\frac{\partial S_\nu^*}{\partial \nu} = \frac{6h\nu^2}{c^2} \frac{1}{\mathrm{e}^{h\nu/k_B T} - 1} - \frac{2h\nu^3}{c^2} \frac{\mathrm{e}^{h\nu/k_B T} \cdot h/k_B T}{(\mathrm{e}^{h\nu/k_B T} - 1)^2}$$

$$= 0$$

$$\Rightarrow 3 - \frac{h\nu}{k_B T} \mathrm{e}^{h\nu/k_B T} \left(\mathrm{e}^{h\nu/k_B T} - 1\right)^{-1} = 0\,.$$

With $x = h\nu/k_B T$ this gives

$$3 = \frac{x}{1 - \mathrm{e}^{-x}} \Rightarrow x = 3\left(1 - \mathrm{e}^{-x}\right)\,.$$

This equation can be only solved numerically. The solution is:

$$x_m = 2.8215$$

$$\Rightarrow \nu_m = 2.8215\,kT/h = 5.873 \times 10^{10}\,\text{s}^{-1} \cdot T\,[\text{K}]\,.$$

With $\lambda = c/\nu \Rightarrow \mathrm{d}\lambda = -(c/\nu^2)\,\mathrm{d}\nu$ we obtain instead of (3.16) for $S_\lambda^*$ the expression

$$S_\lambda^* = \frac{2\pi hc^2}{\lambda^5} \frac{1}{\mathrm{e}^{hc/(\lambda kT)} - 1}\,.$$

With $\mathrm{d}S_\lambda^*/\mathrm{d}\lambda = 0$ we obtain in a similar way

$$\lambda_m = \frac{2.88 \times 10^{-3}\,[\text{m}]}{T\,[\text{K}]}\,.$$

Note, that $\mathrm{d}\lambda/\mathrm{d}\nu$ decreases with increasing frequency $\nu$. The distribution $S_\lambda^*$, which gives the radiation flux per constant interval $\mathrm{d}\lambda$, therefore differs from $S_\nu^*$ which is given for constant interval $\mathrm{d}\nu$. The maximum of $S_\nu^*$ at $\nu_m$ is *not* at $\nu_m = c/\lambda_m$!

4. a) Energy conservation demands

$$h\nu = \Delta E_{\text{kin}}^{\text{el}} \tag{1}$$

$$= m_0 c^2 \left[\frac{1}{\sqrt{1 - v_2^2/c^2}} - \frac{1}{\sqrt{1 - v_1^2/c^2}}\right]$$

The conservation of momentum requires:

$$\hbar \boldsymbol{k} = \frac{m_0 \boldsymbol{v}_2}{\sqrt{1 - v_2^2/c^2}} - \frac{m_0 \boldsymbol{v}_1}{\sqrt{1 - v_1^2/c^2}} \tag{2}$$

$$\hbar^2 k^2 = \frac{h^2\nu^2}{c^2}$$

$$= \frac{m_0^2 v_1^2}{1 - v_1^2/c^2} + \frac{m_0^2 v_2^2}{1 - v_2^2/c^2} - \frac{2m_0^2 \boldsymbol{v}_1 \cdot \boldsymbol{v}_2}{\sqrt{\left(1 - v_1^2/c^2\right)\left(1 - v_2^2/c^2\right)}}\,. \tag{3}$$

Taking the square of (1) gives:

$$h^2\nu^2 = m_0^2 c^4 \left[\frac{1}{1 - v_1^2/c^2} + \frac{1}{1 - v_2^2/c^2} - \frac{2}{\sqrt{\left(1 - v_1^2/c^2\right)\left(1 - v_2^2/c^2\right)}}\right]\,. \tag{4}$$

A comparison of (3) and (4) gives, after rearranging the terms:

$$\left(c^2 - v_1^2\right)\left(c^2 - v_2^2\right) = \left(c^2 - \boldsymbol{v}_1 \cdot \boldsymbol{v}_2\right)^2$$
$$\Rightarrow (\boldsymbol{v}_1 - \boldsymbol{v}_2) = 0 \Rightarrow \boldsymbol{v}_1 = \boldsymbol{v}_2$$
$$\Rightarrow \nu = 0!$$

This means that photoabsorption by a free electron is not possible. The absorption can only take place in the presence of an atom, which can compensate the photon recoil. In the Compton effect the scattered photon has the momentum $\hbar \boldsymbol{k}_s \neq \hbar \boldsymbol{k}$ and the energy $h\nu_s < h\nu$.

b) The momentum of the photon is:

$$|\boldsymbol{p}_{\text{phot}}| = \frac{h\nu}{c}.$$

For $h\nu = 0.1\,\text{eV} = 1.6 \times 10^{-20}\,\text{J}$ ($\lambda = 12\,\mu\text{m}$)

$$\Rightarrow p_{\text{phot}} = \frac{1.6 \times 10^{-20}}{3 \times 10^8}\frac{\text{J s}}{\text{m}} = 5.3 \times 10^{-29}\,\text{N s}$$

For $h\nu = 2\,\text{eV}$ ($\lambda = 600\,\text{nm}$)

$$\Rightarrow p_{\text{phot}} = 1.07 \times 10^{-27}\,\text{N s}$$

For $h\nu = 2\,\text{MeV}$

$$\Rightarrow p_{\text{phot}} = 1.07 \times 10^{-21}\,\text{N s}$$

The recoil velocity of a hydrogen atom with the above momenta would be

$$v_1 = \frac{p}{m} = 3.2 \times 10^{-2}\,\text{m/s} \text{ for } h\nu = 0.1\,\text{eV},$$
$$v_2 = \frac{p}{m} = 6.4 \times 10^{-1}\,\text{m/s} \text{ for } h\nu = 2\,\text{eV},$$
$$v_3 = \frac{p}{m} = 6.4 \times 10^{5}\,\text{m/s} \text{ for } h\nu = 2\,\text{MeV}.$$

In the first case the atom would not be pushed out of resonance for the Lyman α-line, for the last case it would be completely Doppler-shifted out of resonance.

5. The first diffraction minimum appears at the diffraction angle $\alpha$ with

$$\sin\alpha = \frac{\lambda}{b} = \frac{h}{bp} = \frac{h}{b\sqrt{2mE_{\text{kin}}}}.$$

The full width between the two minima at both sides of the central maximum is

$$B = 2D\sin\alpha = \frac{2Dh}{b\sqrt{2mE_{\text{kin}}}} > b,$$

which should be larger than the slit width $b$. This gives the condition

$$b < \left(\frac{2Dh}{\sqrt{2mE_{\text{kin}}}}\right)^{1/2}.$$

For $D = 1\,\text{m}$ and $E_{\text{kin}} = 1\,\text{keV} = 1.6 \times 10^{-16}\,\text{J}$ we obtain:

$$b_{\max} = \left[\frac{2 \times 6.6 \times 10^{-34}}{\sqrt{2 \times 9.11 \times 10^{-31} \times 1.6 \times 10^{-16}}}\right]^{1/2}\text{m}$$
$$= 8.81 \times 10^{-6}\,\text{m} = 8.81\,\mu\text{m}.$$

6. The radii of the Bohr orbitals are

$$r_n = \frac{n^2}{Z}a_0.$$

a) For $n = 1$, $Z = 1 \Rightarrow r_1 = a_0 = 5.29 \times 10^{-11}\,\text{m}$.

b) For $n = 1$, $Z = 79 \Rightarrow r_1 = 6.70 \times 10^{-13}\,\text{m}$.

The velocities of the electron are

$$v = \frac{h}{2\pi m_e r_n} = \frac{Z\hbar}{m_e a_0 n^2}.$$

a) $Z = 1, n = 1$:

$$\Rightarrow v_1 = 2.19 \times 10^6\,\text{m/s} = 7.3 \times 10^{-3}\,c = \frac{c}{137}.$$

b) $Z = 79, n = 1$:

$$v_1 = 1.73 \times 10^8\,\text{m/s} = 0.577\,c.$$

In case b) the relativistic effects become very large and have to be taken into account. We can calculate the relativistic velocity:

$$E_{\text{kin}} = (m - m_0)c^2 = m_0c^2\left(\frac{1}{\sqrt{1 - v^2/c^2}} - 1\right)$$
$$= -E_n = Z^2\frac{Ry^*}{n^2}.$$

For $n = 1$ we obtain:

$$v = c\sqrt{1 - \left(\frac{m_0c^2}{m_0c^2 + E_1}\right)^2},$$

with $m_0c^2 = 0.5\,\text{MeV}$ we obtain:

$$E = \frac{79^2 \times 13.5}{1}\,\text{eV} = 0.084\,\text{MeV}$$
$$\Rightarrow v = c\sqrt{1 - \left(\frac{0.5}{0.584}\right)^2}$$
$$= 0.517\,c.$$

The relative error of the nonrelativistic calculation is

$$\frac{\Delta v}{v}=\frac{0.06}{0.517}c=0.116\,c=11.6\%\,.$$

c) The relativistic mass increase is

$$\begin{aligned}\Delta m=m-m_0&=m_0\left(\frac{1}{\sqrt{1-v^2/c^2}}-1\right)\\&=m_0\left(\frac{1}{\sqrt{1-0.517^2}}-1\right)\\&=0.17\,m_0\,.\end{aligned}$$

The relativistic energy correction is (see Sect. 5.4)

$$\Delta E_{\rm r}(n=1,Z=1)=9\times10^{-4}\,\text{eV}\,.$$

For $Z=79$ it is

$$\Delta E_{\rm r}(n=1,Z=79)=5.6\,\text{eV}\,.$$

7. After the mean life time $\tau$ the number of neutrons have decayed to $1/\text{e}$ of the initial value and after the time $\tau\ln 2$ to $1/2$ of the initial value. During this time they travel a distance $x=v\tau\ln 2$.
The velocity of the neutrons is

$$\begin{aligned}v&=\frac{h}{m\lambda}\Rightarrow x=\frac{h\tau\ln 2}{m\lambda}\\&=\frac{6.62\times10^{-34}\times900\times0.69}{1.67\times10^{-27}\times10^{-9}}=2.4\times10^5\,\text{m}\,.\end{aligned}$$

The decay time of the neutrons could be measured by trapping them in a magnetic quadrupole trap with the geometry of a circle. With a radius $r=1$ m, they travel $(2.4\times10^5/2\pi)=4\times10^4$ times around the circle before they decay, if no other losses are present.

8. The wavelength of the Lyman α-line can be obtained from the relation

$$h\nu=\frac{hc}{\lambda}=Ry^*\left(1-\frac{1}{4}\right)$$

$$\Rightarrow\lambda=\frac{4}{3\,Ry}\quad\text{with}\quad Ry=Ry^*/hc\,.$$

a)

$$Ry\left({}^3\text{H}\right)=Ry_\infty\cdot\frac{\mu}{m_{\rm e}}\quad\text{with}\quad\mu=\frac{m_{\rm e}m_{\rm N}}{m_{\rm e}+m_{\rm N}}\,,$$

where $m_{\rm N}$ is the mass of the nucleus.

$$\begin{aligned}\Rightarrow Ry\left({}^3\text{H}\right)&=Ry_\infty\frac{1}{1+m_{\rm e}/m_{\rm N}}\\&\approx Ry_\infty\frac{1}{1+\frac{1}{3\cdot1836}}\\&=0.999818\,Ry_\infty\\&=1.0971738\times10^7\,\text{m}^{-1}\,.\end{aligned}$$

The wavelength of Lyman $\alpha$ $n=2\to n=1$ is then:

$$\lambda=\frac{4}{3Ry}=1.215\times10^{-7}\,\text{m}=121.5\,\text{nm}\,.$$

b) For positronium ($\text{e}^+\text{e}^-$)

$$\mu=m_{\rm e}/2\Rightarrow Ry(\text{e}^-\text{e}^+)=\frac{1}{2}Ry_\infty$$

$$\Rightarrow\lambda=243.0\,\text{nm}\,.$$

9. At room temperature ($T=300$ K) only the ground state is populated. Therefore all absorbing transitions start from the ground state with $n=1$. The photon energies are then

$$h\nu_n=a\left(1-\frac{1}{n^2}\right)$$

$$h\nu_{n+1}=a\left(1-\frac{1}{(n+1)^2}\right)$$

with $\lambda=c/\nu$ we obtain

$$\frac{\lambda_1}{\lambda_2}=\frac{\nu_{n+1}}{\nu_n}=\frac{1-1/(n+1)^2}{1-1/n^2}\,.$$

With $\lambda_1=97.5$ nm, $\lambda_2=102.8$ nm $\Rightarrow\lambda_1/\lambda_2=0.948$. For $n=2\Rightarrow\lambda_1/\lambda_2=0.843$, for $n=3\Rightarrow\lambda_1/\lambda_2=0.948$.
The two lines therefore belong to $n=3$ and $n=4$.
The constant $a$ can be determined from

$$\nu_n=\frac{c}{\lambda_n}=\frac{a}{h}\left(1-\frac{1}{n^2}\right)$$

with $\lambda_3=102.8$ nm we obtain

$$\begin{aligned}a&=\frac{hc}{\lambda_3}\frac{1}{1-1/3^2}\\&=\frac{hc}{\lambda_3}\cdot\frac{9}{8}=2.177\times10^{-18}\,\text{J}=Ry^*\,.\end{aligned}$$

The lines therefore belong to transitions in the hydrogen atom with $Z=1$, $n=3$ and $n=4$.

10. Since the resolving power of the spectrograph is assumed to be

$$\left|\frac{\lambda}{\Delta\lambda}\right| = \left|\frac{\nu}{\Delta\nu}\right| = 5\times 10^5$$

the difference $\Delta\nu$ of two adjacent lines in the Balmer spectrum has to be $\Delta\nu \geq \nu/(5\times 10^5)$. The frequencies of the Balmer series are

$$\nu_n = \frac{Ry^*}{h}\left(\frac{1}{2^2}-\frac{1}{n^2}\right) \quad n \geq 3\,.$$

The ratio $\nu/\Delta\nu$ is then

$$\frac{\nu}{\Delta\nu} = \frac{\nu_n}{\nu_{n+1}-\nu_n} = \frac{\frac{1}{4}-\frac{1}{n^2}}{\frac{1}{n^2}-\frac{1}{(n+1)^2}} \leq 5\times 10^5$$

$$\Rightarrow \frac{n^2-4}{4-4\left(\frac{n}{n+1}\right)^2} \leq 5\times 10^5$$

$$\Rightarrow n \leq 158\,.$$

Another way of solving this problem is as follows:

$$\nu(n) = \frac{Ry^*}{h}\left(\frac{1}{4}-\frac{1}{n^2}\right)\,.$$

For large $n$ we can regard $\nu(n)$ as a continuous function of $n$ and obtain by differentiating:

$$\frac{\mathrm{d}\nu}{\mathrm{d}n} = \frac{2Ry^*}{h}\frac{1}{n^3}$$

$$\Rightarrow \Delta\nu \approx \frac{2Ry^*}{h}\frac{1}{n^3}\Delta n$$

with $\Delta n = 1$ we get:

$$\frac{\nu}{\Delta\nu} = \frac{1}{2}\left(\frac{1}{4}-\frac{1}{n^2}\right)n^3\,.$$

Since $n^2 \gg 4$:

$$\frac{n^3}{8} \leq 5\times 10^5 \Rightarrow n^3 \leq 4\times 10^6$$

$$\Rightarrow n \leq 158\,.$$

11. For the uncertainty $\Delta r = a$ the kinetic energy of the electron is

$$E_{\text{kin}} \geq \frac{\hbar^2}{2ma^2}\,.$$

Its potential energy at a distance $a$ from the nucleus is

$$E_{\text{pot}} = -\frac{2e^2}{4\pi\varepsilon_0 a}\,.$$

The total energy is then

$$E \geq \frac{\hbar^2}{2ma^2} - \frac{2e^2}{4\pi\varepsilon a}\,.$$

From the condition $\mathrm{d}E/\mathrm{d}a = 0$ for the minimum energy we obtain

$$a_{\min} = \frac{2\pi\varepsilon_0\hbar^2}{me^2} = \frac{a_0}{2}$$

$$\Rightarrow E_{\text{pot}}(a_{\min}) = -\frac{4e^2}{4\pi\varepsilon_0 a_0}$$

$$= -4E_{\text{pot}}(\text{H}, n=1)$$

$$= -108\,\text{eV}\,,$$

$$E_{\text{kin}} = -\frac{1}{2}E_{\text{pot}} = +54\,\text{eV}\,.$$

## Chapter 4

1. Inserting the ansatz $\psi(\boldsymbol{r},t) = g(t)\cdot f(\boldsymbol{r})$ into the time-dependent Schrödinger equation (4.7b) one obtains, after division by $f(\boldsymbol{r})\cdot g(t)$,

$$\mathrm{i}\hbar\frac{1}{g(t)}\frac{\partial g(t)}{\partial t} = -\frac{\hbar^2}{2m}\frac{1}{f(\boldsymbol{r})}\Delta f(\boldsymbol{r}) = C\,.$$

Since the left side of this equation depends solely on $t$, the right side solely on $\boldsymbol{r}$, both sides have to be constant, which we name $C$. The right side gives the time-independent Schrödinger equation (4.6) for $C = E - E_{\text{pot}}$. Then the left side becomes

$$\frac{\partial g(t)}{\partial t} = \frac{E-E_{\text{pot}}}{\mathrm{i}\hbar}g(t)$$

$$\Rightarrow g(t) = g_0\mathrm{e}^{-\mathrm{i}E_{\text{kin}}/\hbar\cdot t}\,.$$

For a free particle $E_{\text{pot}} = 0 \Rightarrow E_{\text{kin}} = E$. The function $g(t)$ then represents the phase factor

$$g(t) = g_0\mathrm{e}^{-\mathrm{i}(E/\hbar)t} = g_0\mathrm{e}^{-\mathrm{i}\omega t}$$

with $E = \hbar\omega$.

2. The reflectivity $R = 1-T$ can be derived from (4.26a) when we insert:

$$\frac{E}{E_0} = \frac{0.4}{0.5} = 0.8\,,$$

$$\alpha = \frac{1}{\hbar}\sqrt{2m(E_0-E)}$$

$$= \frac{2\sqrt{1.67\times10^{-27}\times0.1\times1.6\times10^{-22}}}{1.05\times10^{-34}}\,\mathrm{m}^{-1}$$
$$= 2.2\times10^{9}\,\mathrm{m}^{-1}$$

with $a = 1\times10^{-9}\,\mathrm{m} \Rightarrow \alpha\cdot a = 2.2$.

$$\Rightarrow T = \frac{0.2}{0.2+0.3125\times\sinh^2(2.20)} = 0.126\,,$$

i.e., 12.6% of all particles are transmitted, 87.4% are reflected.

3. For the negative potential step the same derivation can be used as for the positive step. The reflection coefficient is

$$R = \frac{|B|^2}{|A|^2} = \left|\frac{k-k'}{k+k'}\right|^2 .$$

We abbreviate:

$$k = \frac{1}{\hbar}\sqrt{2mE}\,;\quad k' = \frac{1}{\hbar}\sqrt{2m(E-E_0)}$$

with $E_0 < 0$ and obtain

$$R = \frac{E-E_0/2-\sqrt{E(E-E_0)}}{E-E_0/2+\sqrt{E(E-E_0)}}\,.$$

For $E_0 = 0 \Rightarrow R = 0$, for $E_0 \to -\infty \Rightarrow R = 1$. For $E_0 = -E$ the kintetic energy of the particle becomes for $x > 0$ $2E_{\mathrm{kin}}(x<0)$. The reflection coefficient is then

$$R = \frac{3-2\sqrt{2}}{3+2\sqrt{2}} = 0.029\,.$$

For $E = -\frac{1}{2}E_0$ we obtain:

$$R = \frac{2-\sqrt{3}}{2+\sqrt{3}} = 0.072\,.$$

This shows that with increasing step heights $R$ increases. This is completely analogous to the reflection of optical waves at a boundary between two media with refractive indices $n_1$ and $n_2$, where

$$R = \left|\frac{n_1-n_2}{n_1+n_2}\right|^2 = \frac{\Delta n^2}{(2n)^2}$$

with $n = \frac{1}{2}(n_1+n_2)$.

4. With

$$\psi_1 = A\mathrm{e}^{\mathrm{i}k_1x} + B\mathrm{e}^{-\mathrm{i}k_1x}\,,$$
$$\psi_2 = C\mathrm{e}^{\mathrm{i}k_2x} + D\mathrm{e}^{-\mathrm{i}k_2x}\,,$$
$$\psi_3 = A'\mathrm{e}^{\mathrm{i}k_1x}$$
$$k_1 = \left(2mE/\hbar^2\right)^{1/2},$$
$$k_2 = \left(2m(E-E_0)/\hbar^2\right)^{1/2} = \mathrm{i}\alpha$$

we obtain from the boundary condition (4.25) the relations:

$$A+B = C+D\,,$$
$$C\mathrm{e}^{\mathrm{i}k_2a} + D\mathrm{e}^{-\mathrm{i}k_2a} = A'\mathrm{e}^{\mathrm{i}k_1a}\,,$$
$$k_1(A-B) = k_2(C-D)\,,$$
$$k_2\left(C\mathrm{e}^{\mathrm{i}k_2a} - D\mathrm{e}^{-\mathrm{i}k_2a}\right) = k_1A'\mathrm{e}^{-\mathrm{i}k_1a}\,.$$

This set of equations yields the results:

$$A = \left[\cos k_2a - \mathrm{i}\frac{k_1^2+k_2^2}{2k_1k_2}\sin k_2a\right]\mathrm{e}^{\mathrm{i}k_1a}A'\,,$$
$$B = \mathrm{i}\frac{k_2^2-k_1^2}{2k_1k_2}\sin k_2a\mathrm{e}^{\mathrm{i}k_1a}A'\,.$$

The reflection coefficient with $\cos^2 x = 1-\sin^2 x$

$$R = \frac{|B|^2}{|A|^2} = \frac{\left(k_1^2-k_2^2\right)^2\sin^2 k_2a}{4k_1^2k_2^2+\left(k_1^2-k_2^2\right)^2\sin^2 k_2a}$$

and the transmission coefficient is:

$$T = \frac{|A'|^2}{|A|^2} = \frac{4k_1^2k_2^2}{4k_1^2k_2^2+\left(k_1^2-k_2^2\right)^2\sin^2 k_2a}\,.$$

One can readily prove that $R+T=1$. With $k_1^2 = (2mE/\hbar^2)$ and $k_2^2 = 2m(E-E_0)/\hbar^2$ the transmission becomes

$$T = \frac{4E(E-E_0)}{4E(E-E_0)+E_0^2\sin^2\left[\frac{a}{\hbar}\sqrt{2m(E-E_0)}\right]}\,. \qquad (1)$$

Dividing the nominator and denominator by $(4EE_0)$ and using the relation $\sin(\mathrm{i}x) = \mathrm{i}\sinh x$ we obtain for $E > E_0$ the result (4.26a). For $E > E_0$ the transmission becomes $T = 1$ for

$$\frac{a}{\hbar}\sqrt{2m(E-E_0)} = n\pi$$
$$\Rightarrow \lambda = \frac{h}{\sqrt{2m(E-E_0)}} = \frac{2a}{n}\,,\qquad n = 1,2,3\,.$$

For a potential well with depth $E_0$ the potential energy is $E_{\text{pot}} < 0$ if we choose $E_{\text{pot}} = 0$ outside the well. In (4.26a) one has to change the sign of $E_0$. With the numerical data of Problem 4.2 ($E = 0.4\,\text{eV}$, $E_0 = -0.5\,\text{meV}$, $a = 1\,\text{nm}$) we then obtain from (4.26a)

$$T = \frac{1+0.8}{1+0.8+0.31\sin^2\left(a\sqrt{2m \times 0.9\,\text{meV}}/\hbar\right)} = \frac{1.8}{1.8+0.31\sin^2(6.46)} = 0.994\,,$$

where we have used $\sin ix = i\sinh x$.

5. For a potential well with infinitely high walls the bound energy levels for a well with depth $(-E_0)$ are

$$E_n = \frac{\hbar^2}{2m}\frac{\pi^2}{a^2}n^2 \le E_0\,.$$

Inserting the numerical values $a = 0.7\,\text{nm}$, $E_0 = 10\,\text{eV}$ gives:

$$E_n = 1.1 \times 10^{-49}\frac{n^2}{m}\,\text{J} \le 1.6 \times 10^{-18}\,\text{J}\,.$$

a) Electrons with mass $m = 9.1 \times 10^{-31}$ kg:

$$E_n = 1.2 \times 10^{-19} n^2\,\text{J}$$

$$\Rightarrow n^2 \le \frac{1.6 \times 10^{-18}}{1.2 \times 10^{-19}} = 12.9 \Rightarrow n \le 3\,.$$

There are only three bound levels in the well.
b) Protons with mass $m = 1.67 \times 10^{-27}$ kg:

$$\Rightarrow E_n = 5.59 \times 10^{-23}\,\text{J} \times n^2$$
$$\Rightarrow n^2 \le 2.4 \times 10^4$$
$$\Rightarrow n \le 155\,.$$

c) With the exact solution for the potential well with finite wall heights the wave functions are no longer zero for $x = 0$ and $x = a$, but they penetrate a little bit into the wall regions (Fig. S.6). As was discussed in Sects. 4.2.2 and 4.2.4 we now have the wave functions

$$\psi_{\text{I}} = A_1 e^{\alpha x} \quad \text{for} \quad x \le 0$$

with

$$\alpha = \frac{1}{\hbar}\sqrt{2m(E_0 - E)}\,;$$

$$\psi_{\text{II}} = A_2 \sin(kx + \varphi) \quad \text{for} \quad 0 \le x \le a\,,$$

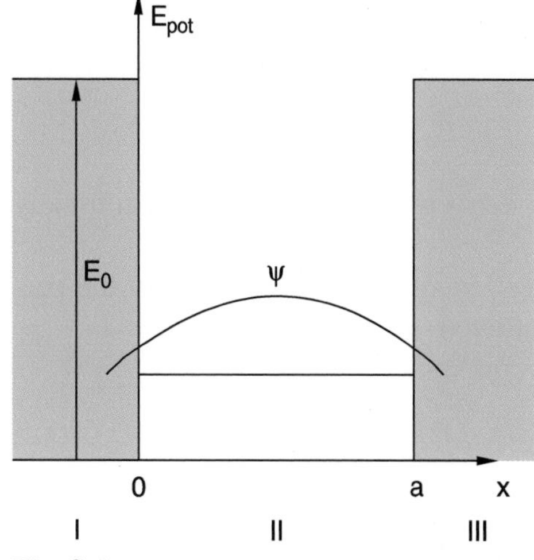

**Fig. S.6.**

where the phase $\varphi$ determines the amplitude of $\psi_{\text{II}}$ for $x = 0$ and $x = a$.

$$\psi_{\text{III}} = A_3 e^{-\alpha x} \quad \text{for} \quad x \ge a\,.$$

From the boundary conditions

$$\psi_{\text{III}}(0) = \psi_{\text{II}}(0)\,; \quad \psi_{\text{II}}(a) = \psi_{\text{III}}(a) \tag{1}$$

we obtain:

$$A_1 = A_2 \sin\varphi\,, \tag{2}$$
$$A_3 = A_2 \sin(ka + \varphi)e^{-\alpha a}\,. \tag{3}$$

From the continuity of the derivatives one obtains:

$$\psi'_{\text{I}}(0) = \psi'_{\text{II}}(0) \Rightarrow \alpha A_1 = kA_2 \cos\varphi\,.$$

This gives with (2):

$$\alpha = k\cot\varphi \Rightarrow \varphi = \arctan(k/\alpha)\,. \tag{4}$$

From the condition

$$\left.\frac{d(\ln\psi_{\text{II}})}{dx}\right|_{x=a} = \left.\frac{d(\ln\psi_{\text{III}})}{dx}\right|_{x=a}$$

one obtains:

$$\varphi = -ka - \text{arccot}(k/\alpha) + n\pi\,. \tag{5}$$

The comparison of (4) and (5) gives for $k$ the condition:

$$ka = n\pi - 2\text{arccot}(k/\alpha)$$

and the energy levels:

$$E_n = \frac{\hbar^2 k_n^2}{2m} = \frac{\hbar^2}{2ma^2}[n\pi - 2\,\text{arccotg}(k/\alpha)]^2\,.$$

For $E_0 = \infty \Rightarrow \alpha = \infty$ and the arccotg becomes zero. This gives the results

$$k_n = n\pi/a \Rightarrow E_n = \frac{\hbar^2\pi^2}{2ma^2}n^2 ,$$

which was derived from the well with infinitely high walls.

6. At the lowest energy (zero-point energy) the particle is restricted to the interval $\Delta x = x_2 - x_1$ between two points

$$x_{1,2} = \pm(2E_{\text{pot}}/D)^{1/2} ,$$

which are the intersections of the energy level

$$E(v = 0) = \frac{1}{2}\hbar\sqrt{D/m} .$$

With the parabolic potential

$$E_{\text{pot}} = \frac{1}{2}Dx^2$$

we obtain:

$$\Delta x = 2\left(\hbar\sqrt{D/m}\Big/ D\right)^{1/2} = 2\left(\hbar\Big/\sqrt{D\cdot m}\right)^{1/2} .$$

7. The $x$-component of $\hat{L}$ is

$$\hat{L}_x = -\mathrm{i}\hbar\left(y\frac{\partial}{\partial z} - z\frac{\partial}{\partial y}\right)$$

$$\frac{\partial}{\partial z} = \frac{\partial r}{\partial z}\frac{\partial}{\partial r} + \frac{\partial \vartheta}{\partial z}\frac{\partial}{\partial \vartheta} + \frac{\partial \varphi}{\partial z}\frac{\partial}{\partial \varphi}$$

$$\Rightarrow \hat{L}_x = -\mathrm{i}\hbar\left[\left(y\frac{\partial r}{\partial z} - z\frac{\partial r}{\partial y}\right)\frac{\partial}{\partial r} + \left(y\frac{\partial \vartheta}{\partial z} - z\frac{\partial \vartheta}{\partial y}\right)\frac{\partial}{\partial \vartheta} + \left(y\frac{\partial \varphi}{\partial z} - z\frac{\partial \varphi}{\partial y}\right)\frac{\partial}{\partial \varphi}\right] . \qquad (1)$$

with

$$r = \sqrt{x^2 + y^2 + z^2}$$

$$\Rightarrow \frac{\partial r}{\partial z} = \frac{z}{r}, \quad \frac{\partial r}{\partial y} = \frac{y}{r} .$$

$$\vartheta = \arccos\left(\frac{z}{\sqrt{x^2 + y^2 + z^2}}\right)$$

$$\Rightarrow \frac{\partial \vartheta}{\partial z} = \frac{(z^2/r^2) - 1}{\sqrt{x^2 + y^2}}$$

$$\frac{\partial \vartheta}{\partial y} = \frac{zy/r^2}{\sqrt{x^2 + y^2}}$$

$$\varphi = \arctan\frac{y}{x} \Rightarrow \frac{\partial \varphi}{\partial y} = \frac{x}{x^2 + y^2}, \quad \frac{\partial \varphi}{\partial z} = 0 .$$

Inserting these relations into (1) gives:

$$\hat{L}_x = -\mathrm{i}\hbar\left[0\frac{\partial}{\partial r} - \frac{y}{\sqrt{x^2 + y^2}}\frac{\partial}{\partial \vartheta} - \frac{zx}{x^2 + y^2}\frac{\partial}{\partial \varphi}\right] = +\mathrm{i}\hbar\left[\sin\varphi\frac{\partial}{\partial \vartheta} + \cot\vartheta\cos\varphi\frac{\partial}{\partial \varphi}\right] .$$

The components $\hat{L}_y$ and $\hat{L}_z$ can be obtained in an analogous way. In order to obtain $\hat{L}^2$ we use the relation:

$$\hat{L}^2 = \hat{L}_x^2 + \hat{L}_y^2 + \hat{L}_z^2$$

and we have to calculate $\hat{L}_x^2$, $\hat{L}_y^2$ and $\hat{L}_z^2$:

$$\hat{L}_x^2 = -\hbar^2\left(\sin\varphi\frac{\partial}{\partial \vartheta} + \cot\vartheta\cos\varphi\frac{\partial}{\partial \varphi}\right) \times \left(\sin\varphi\frac{\partial}{\partial \vartheta} + \cot\vartheta\cos\varphi\frac{\partial}{\partial \varphi}\right) .$$

The differential operators $\partial/\partial\vartheta$ and $\partial/\partial\varphi$ act on all functions, after multiplication of the two brackets, stand behind these operators,
This yields the four terms:

$$\sin\varphi\frac{\partial}{\partial \vartheta}\sin\varphi\frac{\partial}{\partial \vartheta} = \sin^2\varphi\frac{\partial^2}{\partial \vartheta^2} ;$$

$$\sin\varphi\frac{\partial}{\partial \vartheta}\left(\cot\vartheta\cos\varphi\frac{\partial}{\partial \varphi}\right) = \sin\varphi\cos\varphi\left(-\frac{1}{\sin^2\vartheta}\frac{\partial}{\partial \varphi} + \cot\vartheta\frac{\partial}{\partial \vartheta}\frac{\partial}{\partial \varphi}\right) ;$$

$$\cot\vartheta\cos\varphi\frac{\partial}{\partial \varphi}\left(\sin\varphi\frac{\partial}{\partial \vartheta}\right) = \cot\vartheta\left(\cos^2\varphi\frac{\partial}{\partial \vartheta} + \cos\varphi\sin\varphi\frac{\partial}{\partial \vartheta}\frac{\partial}{\partial \varphi}\right) ;$$

$$\cot\vartheta\cos\varphi\frac{\partial}{\partial \varphi}\left(\cot\vartheta\cos\varphi\frac{\partial}{\partial \varphi}\right) = \cot^2\vartheta\left(-\cos\varphi\sin\varphi\frac{\partial}{\partial \varphi} + \cos^2\varphi\frac{\partial^2}{\partial \varphi^2}\right) .$$

Similar terms are obtained for $\hat{L}_y^2$ and $\hat{L}_z^2$. The addition finally gives (4.111) when the relation

$$\cot\vartheta\frac{\partial}{\partial\vartheta}+\frac{\partial^2}{\partial\vartheta^2}=\frac{1}{\sin\vartheta}\frac{\partial}{\partial\vartheta}\left(\sin\vartheta\frac{\partial}{\partial\vartheta}\right)$$

is used.

8. The wave function within the range $x<0$ and $x>a$ is for a penetration depth $\delta x$

$$\psi(\delta x)=C\mathrm{e}^{(\mathrm{i}/\hbar)\sqrt{2m(E_0-E)}\delta x}\,.$$

The probability of finding a particle in this range is proportional to $|\psi(\delta x)|^2$. It decreases to $1/\mathrm{e}$ for

$$\delta x=\frac{\hbar}{\sqrt{2m(E_0-E)}}\,.$$

Example: $m=9.1\times10^{-31}$ kg (electron mass), $E=\frac{1}{2}E_0$, $E_0=1\,\mathrm{eV}=1.6\times10^{-19}\,\mathrm{J}$

$$\Rightarrow\ \delta x=\frac{1.06\times10^{-34}}{\sqrt{1.82\times10^{-30}\times0.8\times10^{-19}}}\,\mathrm{m}$$
$$=0.28\,\mathrm{nm}\,.$$

9. For $E=\frac{1}{2}E_0$ we obtain from (4.26a):

$$T=\frac{0.5}{0.5+0.5\sinh^2 2\pi}$$
$$=1.4\times10^{-5}\,.$$

With the approximation (4.26b) we obtain for $E=0.5\,E_0$:

$$T=4\mathrm{e}^{-4\pi}=1.395\times10^{-5}\,,$$

which is practical identically to the value of the correct calculation.
For $E=\frac{1}{3}E_0$:

$$T=\frac{2/3}{2/3+3/4\sinh^2\left(2\pi\sqrt{2}\right)}$$
$$=5.1\times10^{-8}\,.$$

Maximum transmission $T=1$ is reached for $E>E_0$ if $\sin^2(\mathrm{i}a\alpha)=0$

$$\Rightarrow\ a\sqrt{2m(E-E_0)}=n\pi\hbar=\frac{n}{2}h$$
$$\Rightarrow\ a=\frac{n}{2}\lambda\,,$$

where

$$\lambda=\frac{h}{\sqrt{2m(E-E_0)}}$$

is the de Broglie wavelength of the particle within the range of the barrier.
b) $E=0.8\,\mathrm{eV}$, $E_0=1\,\mathrm{eV}\Rightarrow E/E_0=0.8$, $m=9.1\times10^{-31}$ kg.

$$\alpha=\sqrt{2m(E-E_0)}/\hbar$$
$$=\frac{\sqrt{2\times9.1\times10^{-31}(0.2\times1.6\times10^{-19})}}{1.06\times10^{-34}}\,\mathrm{m}^{-1}$$
$$=2.28\times10^{9}\,\mathrm{m}^{-1}$$
$$a=10^{-9}\,\mathrm{m}\ \Rightarrow\ \sinh^2(\alpha a)=24.4$$
$$\Rightarrow\ T=\frac{-0.2}{0.2-0.28\times24.4}\approx0.03\,.$$

For $E=1.2\,\mathrm{eV}$

$$\Rightarrow\ T=\frac{-0.2}{-0.2-0.208\sinh^2 2.28}=0.625\,.$$

10. The energy levels in the two-dimensional quadratic potential well are, according to (4.66):

$$E\left(n_x,n_y\right)=\frac{\hbar^2\pi^2}{2ma^2}\left(n_x^2+n_y^2\right)\le E_{\max}\,. \tag{2}$$

Inserting the numerical values

$$m=9.1\times10^{-31}\,\mathrm{kg}\,,$$
$$a=10^{-8}\,\mathrm{m}\,,$$
$$E_{\max}=1\,\mathrm{eV}$$

gives the conditions

$$\left(n_x^2+n_y^2\right)\le2.66\times10^2$$
$$\Rightarrow\ n_x,n_y\le16\quad\text{and}\quad n_x^2+n_y^2\le266\,.$$

All possible levels can be visualized as points in a two-dimensional space with the axis $n_x$ and $n_y$. They fill a quarter of a circular area $(\pi/4)\left(n_x^2+n_y^2\right)$ because $n_x,n_y\ge0$. There are therefore approximately $(\pi/4)\times266=208$ energy levels that obey these conditions. Some of them are degenerate. These are levels with the same value of $n_x^2+n_y^2$. Examples are: $n_x=n_y=5$ and $n_x=1$, $n_y=7$ and $n_x=7$, $n_y=1$.

## Chapter 5

1. The expectation value of $r$ is defined as

$$\langle r\rangle=\int\psi^*r\psi\,\mathrm{d}\tau$$

with $d\tau = r^2 \sin\vartheta \, dr \, d\vartheta \, d\varphi$. In the $1s$ state of the H atom

$$\psi = \frac{1}{\sqrt{\pi} a_0^{3/2}} e^{-r/a_0}$$

$$\Rightarrow \langle r \rangle = \frac{1}{\pi a_0^3} 4\pi \int_0^\infty e^{-2r/a_0} r^3 \, dr$$

$$= \frac{4}{a_0^3} \frac{3!}{(2/a_0)^4} = \frac{3}{2} a_0 \, .$$

The expectation value of $r$ is therefore larger than the Bohr radius $a_0$! The expectation value of $1/r$ is

$$\left\langle \frac{1}{r} \right\rangle = \frac{1}{\pi a_0^3} 4\pi \int_0^\infty e^{-2r/a_0} r \, dr$$

$$= \frac{4}{a_0^3} \frac{a_0^2}{4} = \frac{1}{a_0} \, .$$

For the $2s$ state the wave function is

$$\psi(2s) = \frac{1}{4\sqrt{2\pi} a_0^{3/2}} \left(2 - \frac{r}{a_0}\right) e^{-r/2a_0}$$

$$\Rightarrow \langle r \rangle = \frac{4\pi}{16 \times 2\pi a_0^3} \int_0^\infty \left(2 - \frac{r}{a_0}\right)^2 e^{-r/a_0} r^3 \, dr$$

$$= \frac{1}{8a_0^3} \int_0^\infty \left[ 4r^3 e^{-r/a_0} - \frac{4r^4}{a_0} e^{-r/a_0} + \frac{r^5}{a_0^2} e^{-r/a_0} \right] dr$$

$$= \frac{1}{8a_0^3} \left[ 24a_0^4 - 96a_0^4 + 120a_0^4 \right] = 6a_0 \, .$$

A similar calculation for $\langle 1/r \rangle$ yields

$$\left\langle \frac{1}{r} \right\rangle = \frac{1}{4a_0} \, .$$

2. The excitation energy of $E_e = 13.3\,\text{eV}$ can populate the upper levels $E_n$ with energies

$$E_n = IP - \frac{Ry^*}{n^2} \le E_e$$

$$\Rightarrow n^2 \le \frac{Ry^*}{IP - E_a} = \frac{13.6}{13.6 - 13.3} = 45.3$$

$$\Rightarrow n \le 6 \, ,$$

where $IP = 13.6\,\text{eV}$ is the ionization potential. Therefore all lines appear in the emission that start from levels $n \le 6$. These are for $n = 6$

$$\begin{aligned} 6s &\to 5p\,, 4p\,, 3p\,, 2p\,, \\ 6p &\to 5s\,, 4s\,, 3s\,, 2s\,, 1s\,, \\ 6d &\to 5p\,, 4p\,, 3p\,, 2p\,, \\ 6f &\to 5d\,, 4d\,, 3d\,, \\ 6g &\to 5f\,, 4f\,, \end{aligned}$$

and similar expressions for $n = 5, 4, 3, 2$.
Since all terms with equal quantum number $j = l + s$ are degenerate, many of these lines are coincident in energy and are not separated.

3. a) For the ground state $1s$ of the H atom $r = a_0$ according to Bohr's model. (Note, however, that $\langle r \rangle = \frac{3}{2} a_0$, see Problem 5.1.)
The excitation energy of $12.09\,\text{eV}$ reaches levels with energies

$$E_n = IP - \frac{Ry^*}{n^2} = 12.09\,\text{eV}$$

$$\Rightarrow n^2 = \frac{13.599}{13.599 - 12.09} = 9 \Rightarrow n = 3 \, .$$

Since $r \propto n^2$ the Bohr radius becomes $r(n = 3) = 9a_0$.
b) For the excitation energy $E_e = 13.387\,\text{eV}$ we obtain in a similar way

$$n^2 = \frac{13.599}{0.212} = 64 \Rightarrow n = 8 \, .$$

$r(n = 8) = 64a_0$.

4. In the classical model,

$$\boldsymbol{\mu}_e = -e/(2m_e)\boldsymbol{l} \Rightarrow \frac{\boldsymbol{\mu}_e}{\boldsymbol{l}} = -\frac{e}{2m_e}$$

is constant and independent of the principal quantum number $n$.
In the quantum-mechanical description the expectation values:

$$\langle \mu_z \rangle = -m_l \frac{e\hbar}{2m_e} \, , \quad -l \le m_l \le +l \, ,$$

$$\left\langle \mu_e^2 \right\rangle = l(l+1) \frac{e^2\hbar^2}{4m_e^2} \, .$$

Although the number of possible components $\langle \mu_z \rangle$ depend on $l$, the values are still independent of $n$.

5. a) The velocity of the electron on the lowest Bohr orbit is $v_1 = c/137$. Its relative mass is then

$$m(v) = \frac{m_0}{\sqrt{1 - v^2/c^2}} \approx m_0 \left(1 + \frac{1}{2}\frac{v^2}{c^2}\right)$$
$$= m_0 \left(1 + 2.66 \times 10^{-5}\right) .$$

The mass increase $\Delta m_1 = m - m_0 = 2.66 \times 10^{-5} m_0$. For $n = 2$, because of $v \propto 1/n$, $v(2s) = 3.65 \times 10^{-3} c$

$$\Rightarrow m(v) = m_0 \left(1 + 6.6 \times 10^{-6}\right)$$
$$\Rightarrow \Delta m_2 = 6.6 \times 10^{-6} m_0 .$$

The difference is:

$$\delta m = \Delta m_1 - \Delta m_2$$
$$= 2.0 \times 10^{-5} m_0 = 1.8 \times 10^{-35}\,\text{kg} .$$

b) The energy difference is:

$$\Delta E = E(2s) - E(1s) = 10\,\text{eV} = 1.6 \times 10^{-18}\,\text{J} .$$

The potential energy $E_{\text{pot}} = -2E_{\text{kin}}$. Since

$$E = E_{\text{pot}} + E_{\text{kin}} = \frac{1}{2} E_{\text{pot}} \Rightarrow E_{\text{pot}} = 2 \cdot E .$$

The difference in potential energy corresponds to a mass difference

$$\Delta m = m_1 - m_2 = 2\Delta E/c^2$$
$$= -3.6 \times 10^{-35}\,\text{kg} .$$

Both effects are opposite in sign. Because the relative velocity effect $m_v(1s) > m_v(2s)$ is only 1/2 of the potential energy effect, the mass of the atom is larger in the $2s$ state than in the $1s$ state.

6. a) The angular momentum of a spherical body, rotating around an axis through its mass center is

$$|s| = I\omega = \frac{2}{5} m_e r^2 \omega = \sqrt{3/4}\hbar .$$

The velocity at the equator is

$$v_{\text{equator}} = r\omega = \frac{5}{2}\sqrt{0.75}\hbar/(m_e r) .$$

For $r = 1.4 \times 10^{-15}$ m, $m_e = 9.1 \times 10^{-31}$ kg we obtain

$$v = 1.8 \times 10^{11}\,\text{m/s} \gg c\,!$$

For $r = 10^{-18}$ m $\Rightarrow$ $v = 2.5 \times 10^{14}$ m/s. This is a contradiction to the special theory of relativity and shows that the model of a charged sphere for the electron is not correct. The definition of the electron spin as the angular momentum of a sphere with mass $m_e$ is wrong!

b) The rotational energy is:

$$E_{\text{rot}} = \frac{1}{2} I\omega^2 = \frac{1}{5} m_e r^2 \omega^2 = \frac{1}{5} m_e v_{\text{eq}}^2 .$$

For $r = 1.4 \times 10^{-15}$ m $\Rightarrow$ $E_{\text{rot}} = 6 \times 10^{-9}$ J, while the mass energy $E_0 = m_e c^2 = 8 \times 10^{-14}$ J is much smaller. This shows again that the mechanical model of the electron can not be correct.

7. The Zeeman splitting of the $2^2S_{1/2}$-state is, according to (5.7),

$$\Delta E_s = g_j \mu_B B ,$$

with $g_j = g_s \approx 2$. For the $3^2P_{1/2}$ state it is (see Fig. S.7):

$$\Delta E_p = g_j \mu_B B$$

with

$$g_j = 1 + \frac{\frac{1}{2}\frac{3}{2} + \frac{1}{2}\frac{3}{2} - 1 \times 2}{2\frac{1}{2}\frac{3}{2}} = \frac{2}{3} .$$

The four Zeeman lines (Fig. S.7) appear as two pairs, where the smaller distance is

$$\Delta\nu_1 = \frac{2}{3}\mu_B B/h$$

and the larger distance is

$$\Delta\nu_2 = \frac{4}{3}\mu_B B/h .$$

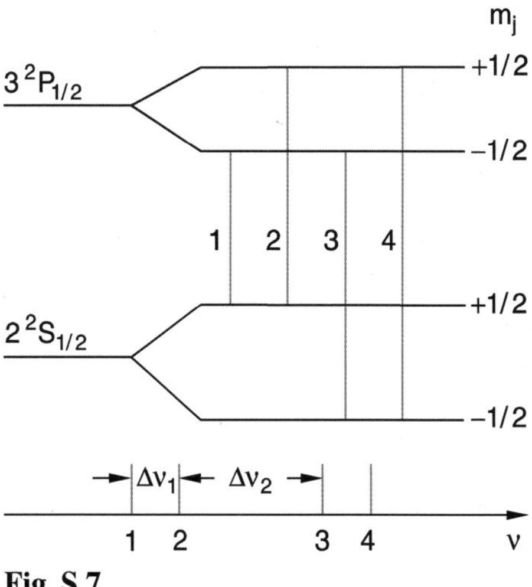


**Fig. S.7.**

For $B = 1\,$Tesla, $\mu_B = 9.27 \times 10^{-24}\,$J/T we obtain

$$\Delta\nu_1 = 9.3 \times 10^9\,\mathrm{s}^{-1}\,,$$
$$\Delta\nu_2 = 1.86 \times 10^{10}\,\mathrm{s}^{-1}\,.$$

a) In order to resolve these components with a central frequency $\nu = 4.5 \times 10^{14}\,\mathrm{s}^{-1}$ the spectral resolving power has to be

$$\left|\frac{\nu}{\Delta\nu}\right| = \left|\frac{\lambda}{\Delta\lambda}\right| = \frac{4.5 \times 10^{14}}{9.3 \times 10^9} = 4.8 \times 10^4\,.$$

The resolving power of a grating spectrograph is

$$\left|\frac{\lambda}{\Delta\lambda}\right| \le mN\,,$$

where $m$ is the interference order and $N$ the number of illuminated grooves. For $m = 2$ we then obtain:

$$N \ge \frac{4.8 \times 10^4}{2} = 24{,}000\,.$$

b) The Fabry–Perot interferometer with mirror separation $d = 1\,$cm has a free spectral range

$$\delta\nu = \frac{c}{2d} = \frac{3 \times 10^8}{2 \times 10^{-2}}\,\mathrm{s}^{-1} = 1.5 \times 10^{10}\,\mathrm{s}^{-1}\,.$$

With a finesse

$$F^* = \frac{\pi\sqrt{R}}{1-R}$$

the minimum separation of two resolvable lines is:

$$\Delta\nu = \delta\nu/F^*\,.$$

With $R = 0.95 \Rightarrow F^* = 61$ and

$$\Delta\nu = \frac{1.5 \times 10^{10}}{61} \approx 2.5 \times 10^8\,\mathrm{s}^{-1}\,.$$

In order to resolve all four Zeeman lines, the magnetic field has to be at least $B \ge 0.026\,$T.

8. The potential energy of a magnetic dipole in a magnetic field is:

$$E = -\boldsymbol{\mu} \cdot \boldsymbol{B}\,.$$

The magnetic moment of the proton is

$$\mu_p = 2.79\mu_K\,.$$

The separation of the two hyperfine components is $\Delta E = 5.58\mu_K \cdot \boldsymbol{B}$. The line with $\lambda = 21\,$cm, corresponding to the transition between the HFS components has a frequency

$$\nu = \frac{\Delta E}{h} = \frac{c}{\lambda}$$

with $\Delta E = 9.46 \times 10^{-25}\,$J. With $\mu_K = 5.05 \times 10^{-27}\,$J/T the magnetic field is

$$B = \frac{\Delta E}{5.58\mu_K} = \frac{9.46 \times 10^{-25}}{5.58 \times 5.05 \times 10^{-27}}\,\mathrm{T}$$
$$= 33.5\,\mathrm{T}\,.$$

The magnetic field produced by the electron in the 1$s$ state of hydrogen at the location of the proton is therefore with 33.5 T much larger than fields obtained in the lab.

9. The frequencies of the transitions are given by

$$h\nu = Ry^*\left(\frac{1}{1} - \frac{1}{4}\right) = \frac{3}{4}Ry^*\,.$$

The Rydberg constant

$$Ry^* = \frac{e^4}{8\varepsilon_0^2 h^2}\mu \quad \text{with} \quad \mu = \frac{m_n m_e}{m_n + m_e}$$

depends on the mass $m_n$ of the nucleus. For the H atom

$$\mu = m_e\frac{1}{1 + \frac{1}{1836}} = m_e \cdot 0.999456\,.$$

For the D = ${}^2_1$H isotope

$$\mu = m_e\frac{1}{1 + \frac{1}{3672}} = m_e \cdot 0.999728\,.$$

For the isotope T = ${}^3_1$H it is

$$\mu = m_e \cdot 0.999818\,.$$

The wavenumber and frequency of the Lyman $\alpha$-lines are then:

$$\overline{\nu}\left({}^1_1\mathrm{H}\right) = 82{,}258.2\,\mathrm{cm}^{-1}$$
$$\Rightarrow \nu = 2.466039 \times 10^{15}\,\mathrm{s}^{-1}\,,$$
$$\overline{\nu}\left({}^2_1\mathrm{D}\right) = 1.00027\overline{\nu}\left({}^1_1\mathrm{H}\right) = 82{,}280.6\,\mathrm{cm}^{-1}\,.$$

The difference is

$$\Delta\overline{\nu}_1 = \overline{\nu}\left({}^2_1\mathrm{D}\right) - \overline{\nu}\left({}^1_1\mathrm{H}\right)$$
$$= 22.4\,\mathrm{cm}^{-1}$$
$$\overline{\nu}\left({}^3_1\mathrm{T}\right) = 1.00036\overline{\nu}\left({}^1_1\mathrm{H}\right) = 82{,}288.0\,\mathrm{cm}^{-1}$$
$$\Delta\overline{\nu}_2 = \overline{\nu}\left({}^3_1\mathrm{T}\right) - \overline{\nu}\left({}^1_1\mathrm{H}\right)$$
$$= 29.8\,\mathrm{cm}^{-1}\,.$$

The hyperfine-splittings are:

a) ${}_1^1$H: $E_{\text{HFS}} = \pm\boldsymbol{\mu}\cdot\boldsymbol{B} \Rightarrow \Delta E = 2|\boldsymbol{\mu}|\cdot|\boldsymbol{B}|$ with $B = 35\,\text{T}$ (see Problem 5.8) and $\mu = 2.79\mu_K$ $\Rightarrow \Delta E = 5.58\,\mu_K\cdot\boldsymbol{B} = 9.43\times10^{-25}\,\text{J} \hat{=} 5.9\times 10^{-6}\,5.9\times 10\text{eV}$.

b) ${}_1^2$D: The internal magnetic field is caused by the electron and therefore the same for all three isotopes. The two hyperfine components have the energy

$$E_{\text{HFS}} = \frac{A}{2}\cdot[F(F+1) - J(J+1) - I(I+1)]\ .$$

With $F = 3/2$, $J = 1/2$ we obtain

$$E_{\text{HFS}} = \frac{A}{2}\ .$$

With $F = 1/2 \Rightarrow F_{\text{HFS}} = -A$. The splitting is then: $\Delta F = \frac{3}{2}A$ with

$$A = g_I\mu_K\cdot B_J/\sqrt{J(J+1)}$$
$$= 2g_I\mu_K B_J/\sqrt{3}\ .$$

c) ${}_1^3$T: Here is $I = 3/2$; $\Rightarrow F = 2$ and $F = 1$

$$E(F=2) = +\frac{3}{4}A\ ;\quad E(F=1) = -\frac{5}{4}A$$
$$\Rightarrow \Delta E = 2A = 4g_I\mu_K B_J/\sqrt{3}\ .$$

## Chapter 6

1. The potential experienced by the second electron in the He atom is (see Fig. 6.8)

$$\phi(r_2) = -\frac{Ze}{4\pi\varepsilon_0 r_2} + \frac{e}{4\pi\varepsilon_0}\int\frac{|\psi_{1s}(r_1)|^2}{r_{12}}\,\mathrm{d}\tau.$$
$$\mathrm{d}\tau_1 = r_1^2\sin\vartheta_1\sin\varphi\,\mathrm{d}r_1\mathrm{d}\vartheta_1\mathrm{d}\varphi_1 \qquad (1)$$
$$r_{12}^2 = r_1^2 + r_2^2 - 2r_1r_2\cos\vartheta$$
$$\Rightarrow r_{12}\,\mathrm{d}r_{12} = r_1r_2\sin\vartheta\,\mathrm{d}\vartheta$$

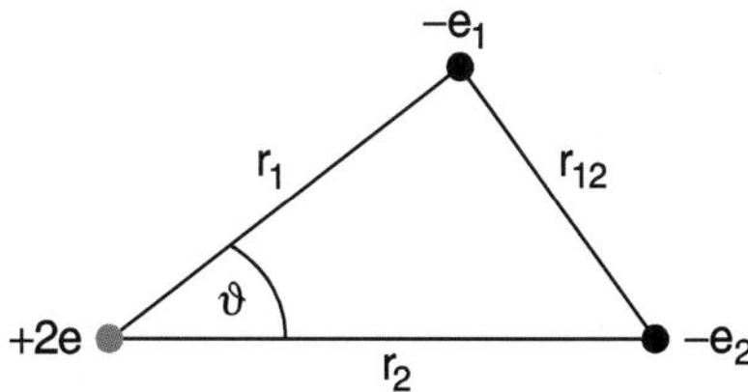


**Fig. S.8.**

$$\psi_{1s} = \frac{Z^{3/2}}{\sqrt{\pi}a_0^{3/2}}\mathrm{e}^{-Zr_1/a_0}\ .$$

Inserting these relations with $Z = 2$, into the integrand we obtain for the integral

$$I = \int\frac{|\psi_{1s}|^2}{r_{12}}r_1^2\sin\vartheta\,\mathrm{d}r_1\,\mathrm{d}\vartheta\,\mathrm{d}\varphi$$
$$= \frac{Z^3\pi}{\pi a_0^3}\left[\int\limits_{r_1=0}^{r_2}\frac{\mathrm{e}^{-Z2r_1/a_0}r_1^2}{r_1r_2}\mathrm{d}r_1\int\limits_{r_{12}=r_2-r_1}^{r_2+r_1}\mathrm{d}r_{12}\right.$$
$$\left.+\int\limits_{r_1=r_2}^{\infty}\frac{\mathrm{e}^{-Z2r_1/a_0}r_1}{r_2}\mathrm{d}r_1\int\limits_{r_{12}=r_1-r_2}^{r_1+r_2}\mathrm{d}r_{12}\right]. \qquad (2)$$

For $\vartheta = 0$ we have (Fig. 5.8)

$$r_{12} = \begin{cases} r_2 - r_1 & \text{for } r_1 < r_2\,, \\ r_1 - r_2 & \text{for } r_1 > r_2\,. \end{cases}$$

For $\vartheta = \pi \Rightarrow r_{12} = r_1 + r_2$. The integration of the first term in (2) gives

$$I_1 = \left(-\frac{r_2a_0}{Z} - \frac{a_0^2}{Z^2} - \frac{a_0^3}{2r_2Z^3}\right)\mathrm{e}^{-2Zr_2/a_0}$$
$$+\frac{a_0^3}{2r_2Z^3}$$

and for the second term:

$$I_2 = \left(\frac{a_0r_2}{Z} + \frac{a_0^2}{2Z^2}\right)\mathrm{e}^{-2Zr_2/a_0}\ .$$

This gives, with $r = r_2$, the total potential felt by the second electron:

$$\phi(r) = -\frac{(Z-1)e}{4\pi\varepsilon_0 r} - \frac{e}{4\pi\varepsilon_0}\left(\frac{Z}{a_0} + \frac{1}{r}\right)\mathrm{e}^{-2Zr/a_0}$$

with $Z = 2$ for the He atom.

2. The charge density of the two $1s$-electrons is approximately

$$\eta = -2e\cdot\psi^2(1s) = -\frac{2e}{b^3\cdot\pi}\mathrm{e}^{-2R/b}$$

with $b = a_0/Z_{\text{eff}}$ .

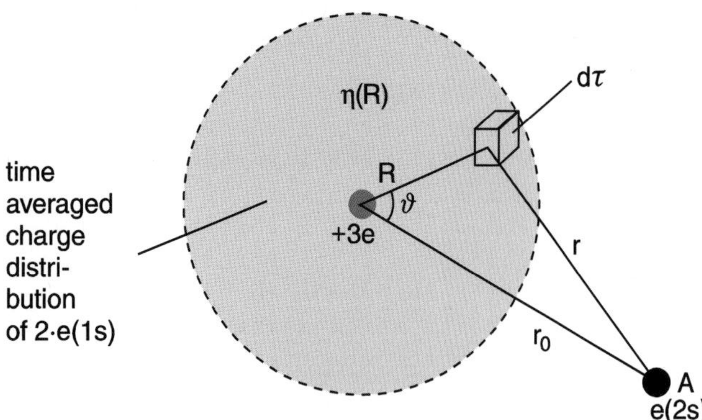


**Fig. S.9.**

The potential, experienced by the $2s$ electrons in the field of the nucleus and the two $1s$-electrons is (Fig. S.9):

$$\phi = \frac{-3e}{4\pi\varepsilon_0 r_0} + \frac{1}{4\pi\varepsilon_0}\int \frac{\eta}{r}\,\mathrm{d}\tau \;.$$

The integral can be solved as follows:

$$\int \frac{\eta}{r}\,\mathrm{d}\tau$$

$$= -\frac{2e}{b^3\pi}\int\limits_{R=0}^{\infty}\int\limits_{\vartheta=0}^{\pi}\int\limits_{\varphi=0}^{2\pi}\frac{\mathrm{e}^{-2R/b}}{r}R^2\sin\vartheta\,\mathrm{d}R\,\mathrm{d}\vartheta\,\mathrm{d}\varphi$$

With $r^2 = R^2 + r_0^2 - 2r_0R\cos\vartheta \Rightarrow r\mathrm{d}r = r_0R\cdot\sin\vartheta\,\mathrm{d}\vartheta$ the integral then becomes:

$$\int \frac{\eta}{r}\,\mathrm{d}\tau = -\frac{4e}{b^3}\int\limits_{R=0}^{\infty}\int\limits_{r}\frac{R}{r_0}\mathrm{e}^{-2R/b}\,\mathrm{d}R\,\mathrm{d}r \;.$$

Is the point A (location of $2s$-electron) outside the charge distribution of the $1s$-electrons ($r_0 > R$) the integration over $r$ extends from $r_0 - R$ to $r_0 + R$. For $r_0 < R$ ($2s$-electron inside the core) it extends from $R - r_0$ to $R + r_0$. This gives:

$$\int \frac{\eta}{r}\,\mathrm{d}\tau = -\frac{4e}{b^3}\left[\int\limits_{R=0}^{r_0}\int\limits_{r=r_0-R}^{r_0+R}\frac{R}{r_0}\mathrm{e}^{-2R/b}\,\mathrm{d}R\,\mathrm{d}r + \int\limits_{R=r_0}^{\infty}\int\limits_{r=R-r_0}^{R+r_0}\frac{R}{r_0}\mathrm{e}^{-2R/b}\,\mathrm{d}R\,\mathrm{d}r\right]$$

$$= -\frac{4e}{b^3}\left[\int\limits_{R=0}^{r_0}\frac{2R^2}{r_0}\mathrm{e}^{-2R/b}\,\mathrm{d}R + \int\limits_{R=r_0}^{\infty}2R\cdot\mathrm{e}^{-2R/b}\,\mathrm{d}R\right].$$

The integrals can be solved analytically

$$\int \frac{\eta}{r}\,\mathrm{d}\tau = -\frac{2e}{r_0}\left[1 + \mathrm{e}^{-2r_0/b}\left(1 + \frac{r_0}{b}\right)\right].$$

The potential then becomes with $b = a_0/Z_{\mathrm{eff}}$ and when we rename $r_0 \to r$:

$$\phi(r) = \frac{-e}{4\pi\varepsilon_0 r}\left[1 + 2\mathrm{e}^{-2Z_{\mathrm{eff}}\cdot r/a_0}\left(1 - \frac{-rZ_{\mathrm{eff}}}{a_0}\right)\right].$$

3. The mean distance between the atoms is $\overline{d} = n^{-1/3}$. The de Broglie-wavelength $\lambda_{\mathrm{dB}}$ becomes larger than $\overline{d}$ for

$$\frac{h}{m\overline{v}} > \overline{d} \Rightarrow \overline{v} < \frac{h}{mn^{-1/3}} \,.$$

The mean velocity is

$$\overline{v} = \sqrt{\frac{8kT}{\pi m}} \,.$$

This gives the condition for the temperature $T$:

$$T < \frac{\pi h^2 n^{2/3}}{8k\cdot m} \,.$$

Example: $n = 10^{12}/\mathrm{cm}^3 = 10^{18}/\mathrm{m}^3$ and $m = 23\,\mathrm{AMU} = 23\times1.66\times10^{-27}\,\mathrm{kg}$

$$\Rightarrow T < 3.3\times10^{-7}\,\mathrm{K} = 330\,\mathrm{nK}\,.$$

Below this temperature the particles are no longer distinguishable because their location can only be defined within a volume $\lambda_{\mathrm{dB}}^3$. The atoms form a Bose–Einstein condensate of identical particles.

4. The potential energy of the two electrons is (Fig. S.10)

$$E_{\mathrm{pot}} = -\frac{4e^2}{4\pi\varepsilon_0 r_1} + \frac{e^2}{4\pi\varepsilon_0(2r_1)} = -\frac{7}{8}\frac{e^2}{\pi\varepsilon_0 r_1} = -\frac{7}{4}\frac{e^2}{\pi\varepsilon_0 a_0}\,,$$

with $r_1 = a_0/2$ .

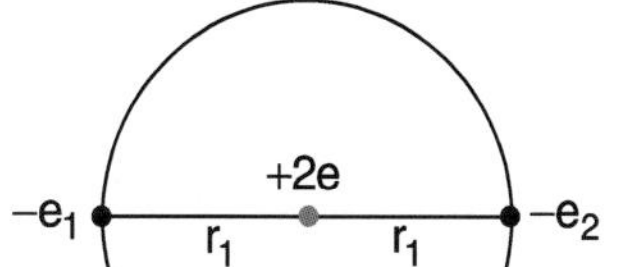


**Fig. S.10.**

The kinetic energy of the two electrons is:

$$E_{\text{kin}} = \frac{2m_e v^2}{2} = m_e v^2 .$$

With

$$v = \frac{Zh}{2\pi m a_0} = \frac{h}{\pi m a_0}$$

we obtain:

$$E_{\text{kin}} = \frac{h^2}{\pi^2 m a_0^2} .$$

The total energy of the system is then:

$$E = -\frac{7}{4}\frac{e^2}{\pi\varepsilon_0 a_0} + \frac{h^2}{\pi^2 m a_0^2} .$$

Inserting the numerical values yields

$$E = -1.30 \times 10^{-17}\,\text{J} = -82\,\text{eV} .$$

The experimental result is $E = -78.9\,\text{eV}$. This shows that our simple model approaches the real situation quite well. The small difference comes partly from the fact that we have neglected the relativistic mass increase.

5. The expectation value of the potential energy $e \cdot \phi(r)$ of an electron in the $2s$ state is

$$\langle E_{\text{pot}} \rangle = e \int \psi_2^* \phi(r) \psi_2 \, \mathrm{d}\tau_2 .$$

Assuming a spherically symmetric wave function $\psi_2(2s)$ as in the H atom (because the electron moves in a potential that is essentially a Coulomb potential with the effective charge $Z_{\text{eff}} = 1$) we can write:

$$|\psi_2|^2 = \frac{1}{32\pi a_0^3}\left(2 - \frac{r}{a_0}\right)^2 \mathrm{e}^{-r/a_0}$$

$$\Rightarrow E_{\text{pot}} = -\frac{e^2}{4\pi\varepsilon_0 \times 32\pi a_0^3}$$

$$\times \left[\int \left(\frac{1}{r} + \left(\frac{1}{r} + \frac{Z}{a_0}\right)\mathrm{e}^{-2Zr/a_0}\right) \times \left(2 - \frac{r}{a_0}\right)^2 \mathrm{e}^{-r/a_0}\,\mathrm{d}\tau\right]$$

$$\int = \int_{r=0}^{\infty} \left[\left(\frac{4}{r} - \frac{4}{a_0} + \frac{r}{a_0^2}\right)\mathrm{e}^{-r/a_0} + \left(\frac{4}{r} + \frac{4(Z-1)}{a_0} - \frac{(4Z-1)r}{a_0^2} + \frac{Zr^2}{a_0^3}\right) \times \mathrm{e}^{-(2Z+1)r/a_0}\right] r^2\,\mathrm{d}r \int_{\vartheta=0}^{\pi} \sin\vartheta\,\mathrm{d}\vartheta \int_0^{2\pi} \mathrm{d}\varphi$$

$$= 4\pi \int_{r=0}^{\infty} \left[\left(4r - \frac{4r^2}{a_0} + \frac{r^3}{a_0^2}\right)\mathrm{e}^{-r/a_0} + \left(4r + \frac{4r^2}{a_0} - \frac{7r^3}{a_0^2} + \frac{2r^4}{a_0^3}\right)\mathrm{e}^{-5r/a_0}\right]\mathrm{d}r$$

$$= 4\pi\left[4a_0^2 - 8a_0^2 + 6a_0^2 + \frac{4}{25}a_0^2 + \frac{8}{125}a_0^2 - \frac{42}{625}a_0^2 + \frac{48}{3125}a_0^2\right]$$

$$= 4\pi\frac{6788}{3125}a_0^2$$

$$\Rightarrow E_{\text{pot}}(2s) = -\frac{e^2}{4\pi\varepsilon_0 a_0}0.272 = -7.39\,\text{eV} .$$

A similar calculation for the $3s$ state gives:

$$E_{\text{pot}}(3s) = -3.19\,\text{eV}$$
$$\Rightarrow \Delta E_{\text{pot}} = 4.2\,\text{eV}$$

because $E_{\text{kin}} \approx -\frac{1}{2}E_{\text{pot}}$ we obtain

$$\Delta E = \Delta E_{\text{pot}} + \Delta E_{\text{kin}} \approx \frac{1}{2}\Delta E_{\text{pot}} \approx 2.1\,\text{eV} .$$

This agrees fairly well with the experimental value $\Delta E_{\text{exp}} = 2.3\,\text{eV}$, obtained from the difference of the wavenumbers $\overline{\nu}(3s \leftrightarrow 1s) - \overline{\nu}(2s \leftrightarrow 1s)$ of the two-photon allowed transition.

6. The largest mean distance between the electrons is realized if the total spin of all electrons has

its maximum allowed value. Since for this case the spin function is symmetric, the spatial part of the wave function is antisymmetric with respect to exchanging two electrons. This means that the wave function $\psi(\boldsymbol{r}_1, \boldsymbol{r}_2)$ has nodes for $\boldsymbol{r}_1 = \boldsymbol{r}_2$. Therefore the potential energy of the mutual repulsion is minimized, which gives a minimum for the total energy.

7. With the screening constant $S$, the potential energy of a Rydberg electron can be written as

$$E_{\text{pot}} = -\frac{(Z-S)e^2}{4\pi\varepsilon_0 r} .$$

For the total energy (note that $E_{\text{kin}} = -\frac{1}{2}E_{\text{pot}}$) we obtain

$$E = -\frac{(Z-S)e^2}{8\pi\varepsilon_0 r} = E_{\text{kin}} + E_{\text{pot}} .$$

According to the Rydberg formula (6.32) the energy can be also expressed by

$$E = -\frac{-(Z-S)^2 Ry^*}{n^2} = -\frac{Ry^*}{(n-\delta)^2} .$$

The comparison yields the relation

$$S = Z - \frac{n}{n-\delta}$$

between the screening constant $S$ and the quantum defect $\delta$. For $\delta \to 0$ we obtain $S = Z - 1$, which means a Coulomb potential with $Z_{\text{eff}} = Z - S = 1$.

8. For the myonic atom the reduced mass is

$$\mu = \frac{m_\mu \cdot m_{\text{N}}}{m_\mu + m_{\text{N}}} .$$

With $m_\mu = 206.76\, m_{\text{e}}$, $m_{\text{N}} = 140 \times 1836\, m_{\text{e}}$ we obtain $\mu = 206.6\, m_{\text{e}}$.

$$\Rightarrow Ry^*_\mu = 206.6 \cdot Ry^{*\infty}$$

$$\Rightarrow E_n = -\frac{206.6 Ry^{*\infty} Z^2}{n^2}$$

The energy of the photon on a tansition $n = 2 \to n = 1$ is

$$h\nu = 206.6 Ry^{*\infty} Z^2 \left(\frac{1}{1} - \frac{1}{4}\right) = \frac{3}{4} 60^2 \times 206.6 \times 13.6\,\text{eV} = 7.59 \times 10^6\,\text{eV} = 7.59\,\text{MeV} .$$

The radius $r_n^\mu$ of the myon is

$$r_n^\mu = \frac{n^2}{Z}\frac{a_0}{206.6} .$$

The smallest radius of the electron is $r_1^{\text{el}} = a_0/Z$. The condition $r_1^{\text{el}} = r_n^\mu$ gives:

$$\frac{n^2}{206.6} = 1 \Rightarrow n \approx 14 .$$

The radius $r_{14}^\mu$ of the myon is about the same as the lowest radius $r_1^{\text{el}}$ of the electron orbit.

9. The potential energy of the electron with the wave function $\psi(\boldsymbol{r})$ is (see Problem 6.1)

$$E_{\text{pot}} = +e \int |\psi(\boldsymbol{r})|^2 \phi(r)\, \text{d}\boldsymbol{r} .$$

For the $3s$ electron the probability to find the electron inside the $n = 2$ shell is larger than for the $3p$ electron, because the $3p$-functions has a node at $r = 0$. Therefore the shielding of the nuclear charge is smaller for the $3s$ electron than for the $3p$ electron. The potential energy is lower and therefore its total energy.

10. The potential energy of the second electron in the $\text{H}^-$ ion is

$$E_{\text{pot}}(r_2) = +e\phi(r) ,$$

where the potential $\phi(r)$ has been calculated in Problem 6.1 and where we have to insert $Z = 1$.

$$\Rightarrow \phi(r) = -\frac{e}{4\pi\varepsilon_0}\left(\frac{1}{a_0} + \frac{1}{r}\right) \text{e}^{-2r/a_0} .$$

The wave functions of the second electron can be approximated by the hydrogen wave function

$$\psi(r) = \frac{1}{\sqrt{\pi} a_0^{3/2}} \text{e}^{-r/a_0}$$

when we neglect the repulsion between the two electrons. The potential energy is then

$$E_{\text{pot}}^{\text{min}} = +e \int |\psi(1s)|^2 \phi(r)\, \text{d}\tau = -\frac{4\pi e^2}{4\pi\varepsilon_0 \pi a_0^3} \int_0^\infty \left(\frac{1}{a_0} + \frac{1}{r}\right) \text{e}^{-4r/a_0} r^2\, \text{d}r = -\frac{e^2}{\pi\varepsilon_0 a_0^3}\frac{3}{32} a_0^2 = -\frac{3}{8}\frac{e^2}{4\pi\varepsilon_0 a_0} = -10.2\,\text{eV} .$$

The crude approximation gives a binding energy

$$E_B = E_{pot} + E_{kin} = -5.1\,\mathrm{eV}$$

of the second electron. It is higher than the experimental value of $-2.5\,\mathrm{eV}$, because we have neglected the repulsion between the two electrons.

11. The energy of the state with principal quantum number $n$ is

$$E_n = -Ry^* \frac{Z_{eff}^2}{n^2} \Rightarrow Z_{eff}^2 = -\frac{n^2 E_n}{Ry^*}\,; \quad E_n < 0\,.$$

With $Ry^* = 13.6\,\mathrm{eV}$ one obtains:

$$Z_{eff}^2(n=2) = \frac{4 \times 5.39}{13.6} = 1.58$$
$$\Rightarrow Z_{eff} = 1.26\,.$$

The nuclear charge $Ze$ with $Z = 3$ is screened by the two 1$s$ electrons by 1.74$e$.
For the Rydberg level with $n = 20$ we obtain:

$$Z_{eff}^2(n=20) = \frac{400 \times 0.034}{13.6} \approx 1$$
$$\Rightarrow Z_{eff} = 1\,.$$

For high Rydberg levels the screening of the nuclear charge $Ze$ is nearly complete by the $(Z-1)$ electrons of the atomic core.

12. For all alkali atoms there is a single electron in the valence shell with principal quantum number $n = 2$ (Li), $n = 3$ (Na), $n = 4$ (K), $n = 5$ (Rb), $n = 6$ (Cs). The larger the $n$ is, the better the shielding of the nuclear charge $Ze$ by the $Z-1$ electrons in the core. This implies that $Z_{eff}$ decreases with increasing $n$ and the binding energy of the valence electron (this is the ionization energy of the atom) decreases with increasing $n$.
The experimental technique for the determination of the binding energy is, for instance, the photoionization of the atom.

$$\mathrm{A}(E_n) + h\nu \rightarrow \mathrm{A}^+ + \mathrm{e}^-(E_{kin})\,.$$

One can measure the frequency $\nu_g$, where

$$E_{kin}(\mathrm{e}^-) = 0 \Rightarrow -E_n = h\nu\,.$$

The approximate calculation of the binding energy is based on the Hartree method (see Sect. 6.4.2), which converges rapidly because the charge distribution of the core electrons, which form closed shells is spherically symmetric. The valence electron moves in a spherical potential, which is nearly a Coulomb potential ($\sim 1/r$) for larger $r$, but deviates from it for small $r$.

13. The potential energy of the electron is

$$E_{pot}(x) = -\frac{e^2}{4\pi\varepsilon_0 r} - eE_0 x$$
$$\text{with} \quad x = r\cos\alpha\,.$$

In the $x$-direction $\alpha = 0 \Rightarrow \cos\alpha = 1$. We than can write:

$$\frac{\mathrm{d}E_{pot}}{\mathrm{d}r} = \frac{e^2}{4\pi\varepsilon_0 r^2} - eE_0\,.$$

The maximum of the potential barrier is at $\mathrm{d}E_{pot}/\mathrm{d}r = 0$

$$\Rightarrow r_m = \left(\frac{e}{4\pi\varepsilon_0 E_0}\right)^{1/2}$$

$$\Rightarrow E_{pot}(r_m) = -\sqrt{\frac{e^3 E_0}{\pi\varepsilon_0}}\,.$$

Without the external field $E_{pot}(r) \Rightarrow 0$ for $r \rightarrow \infty$. The lowering of the ionization potential is, therefore,

$$\Delta(IP) = -\sqrt{\frac{e^3 E_0}{\pi\varepsilon_0}}\,.$$

Due to the tunnel effect, the effective decrease of $IP$ is even slightly larger.

## Chapter 7

1. a) The total emitted energy is

$$W_{Fl} = N\left(3^2 P_{3/2}\right) h\nu = 10^8 \times 3.4 \times 10^{-19}\,\mathrm{J}$$
$$= 3.4 \times 10^{-11}\,\mathrm{J}\,.$$

The time dependent fluorescence power can be calculated as

$$P_{Fl} = P_0 \mathrm{e}^{-t/\tau}\,,$$

where excitation at $t = 0$ is assumed.

The emitted energy is related to the power by

$$W_{\text{Fl}} = \int_0^\infty P_{\text{Fl}}\,\mathrm{d}t = P_0 \int_0^\infty \mathrm{e}^{-t/\tau}\,\mathrm{d}t = \tau \cdot P_0$$

$$\Rightarrow P_0 = 3.4 \times 10^{-11}\,\text{J}/1.6 \times 10^{-8}\,\text{s} = 2.1 \times 10^{-3}\,\text{W}\,.$$

b) The angular distribution is

$$W(\vartheta) = W_0 \sin^2\vartheta\,,$$

$$W_{\text{total}} = W_0 \int_{\varphi=0}^{2\pi} \int_{\vartheta=-\pi/2}^{+\pi/2} \sin^2\vartheta\,\mathrm{d}\vartheta\,\mathrm{d}\varphi = 2\pi W_0 \left[\frac{1}{2}\vartheta - \frac{1}{4}\sin 2\vartheta\right]_{-\pi/2}^{+\pi/2} = \pi^2 W_0$$

$$\Rightarrow W_0 = W_{\text{total}}/\pi^2\,.$$

Within the angular range $\Delta\vartheta = 0.1$ Sterad around $\vartheta = 90°$ the energy

$$W(\vartheta = \pi/2 \pm \Delta\vartheta/2, \varphi = 0 \pm \Delta\varphi/2) = \Delta\varphi W_0 \times \int_{\vartheta=\pi/2-\Delta\vartheta/2}^{\pi/2+\Delta\vartheta/2} \sin^2\vartheta\,\mathrm{d}\vartheta = W_0 \cdot \Delta\varphi \left(\frac{1}{2}\Delta\vartheta + \frac{1}{2}\sin\Delta\vartheta\right)$$

is emitted. For $\Delta\vartheta \ll 1$ we have:

$$W = W_0 \cdot \Delta\varphi\Delta\vartheta\,.$$

The solid angle $\Delta\Omega$ is

$$\Delta\Omega = \Delta\vartheta\,\Delta\varphi = 0.1\,\text{Sterad}$$

$$\Rightarrow W = 0.1\,W_0 = \frac{0.1}{\pi^2} W_{\text{total}} \approx 0.01\,W_{\text{total}}\,.$$

This means that 1% of the total energy is emitted in a direction perpendicular to the dipole axis into the solid angle $\Delta\Omega = 0.1$ Sterad.

2. a) The Doppler width is

$$\delta\nu_{\text{D}} = 7.16 \times 10^{-7} \nu_0 \sqrt{T/M}\left[\sqrt{\text{mol}/(\text{g K})}\right].$$

With $T = 300$ K, $M = 1$g/mol, $\nu_0 = 2.47 \times 10^{15}\text{s}^{-1}$ we obtain

$$\delta\nu_{\text{D}} = 3.06 \times 10^{10}\,\text{s}^{-1} = 30.6\,\text{GHz}$$

$$\Rightarrow |\Delta\lambda_{\text{D}}| = \frac{c}{\nu^2}\delta\nu_{\text{D}} = 1.5 \times 10^{-3}\,\text{nm}\,.$$

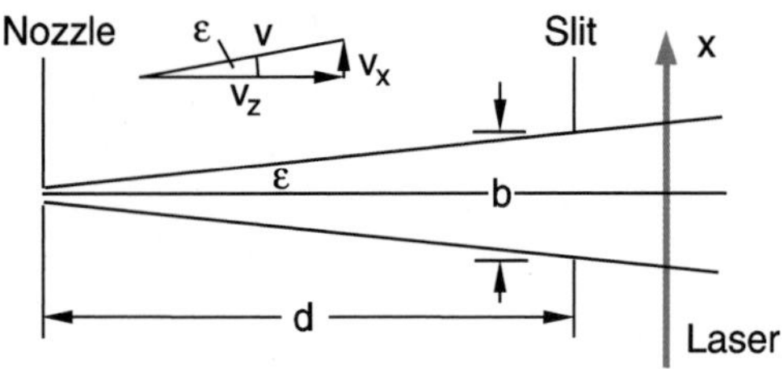


**Fig. S.11.**

b) The collimation ratio of the atomic beam is (Fig. S.11)

$$\varepsilon = \frac{b}{2d} = \frac{1}{200}\,.$$

Since the nozzle diameter is small compared to $b$ we can regard the nozzle as a point-like source of atoms. The transverse velocity distribution $f(v_x)$ is determined by $|v_x| < v\sin\varepsilon$. The reduced Doppler width is then:

$$(\delta\nu_{\text{D}})_{\text{beam}} = \delta\nu_{\text{D}}(v)\sin\varepsilon\,.$$

With $\sin\varepsilon = 5 \times 10^{-3}$ and $\delta\nu_{\text{D}} = 3.06 \times 10^{10}\,\text{s}^{-1}$ we obtain:

$$(\delta\nu_{\text{D}})_{\text{beam}} = 1.5 \times 10^8\,\text{s}^{-1} = 150\,\text{MHz}\,.$$

c) The natural line width is

$$\delta\nu_{\text{n}} = \frac{1}{2\pi\tau} = \frac{10^9}{2\pi\,1.2}\text{s}^{-1} = 132\,\text{MHz}\,.$$

The measured line width is

$$\delta\nu_{\text{exp}} \approx \sqrt{(\delta\nu)^2_{\text{beam}} + (\delta\nu_{\text{n}})^2} \approx 200\,\text{MHz}\,.$$

The hyperfine splitting is

$$\delta\nu_{\text{HFS}}(1s) = 1.4 \times 10^9\,\text{s}^{-1}\,.$$

It can therefore be readily resolved.

3. The natural line width of the Ca transition is

$$\delta\nu_{\text{n}} = \frac{1}{2\pi\tau} = \frac{10^4}{2\pi\,3.9}\text{s}^{-1} = 410\,\text{Hz}$$

The interaction time $\Delta T$ should be longer than:

$$\Delta T \geq \frac{1}{2\pi\Delta\nu} = \frac{1\,\text{s}}{2\pi\,3 \times 10^3} = 53\,\mu\text{s}$$

in order to reach a line width of $\Delta\nu = 3$ kHz. The natural line width is negligible compared with the transit time broadening.

The mean velocity of the Ca atoms is at an oven temperature $T = 900\,\mathrm{K}$

$$\overline{v} = \sqrt{\frac{8kT}{\pi m}} = \sqrt{\frac{8 \times 1.38 \times 10^{-23} \times 9 \times 10^2}{\pi \times 40 \times 1.66 \times 10^{-27}}}\,\frac{\mathrm{m}}{\mathrm{s}}$$
$$= 690\,\mathrm{m/s}\,.$$

The minimum interaction zone is then

$$\Delta s = \overline{v}\,\Delta T = 6.9 \times 10^2 \times 5.3 \times 10^{-5}\,\mathrm{m}$$
$$= 3.7 \times 10^{-2}\,\mathrm{m} = 37\,\mathrm{mm}\,.$$

4. a) The wavelength $\lambda$ of a transition between energy levels with term values $T_i$, $T_k$ is

$$\lambda_{ik} = \frac{1}{T_i - T_k} = \frac{1}{19{,}932}\mathrm{cm} = 501.7\,\mathrm{nm}\,.$$

b) The natural line width is:

$$\delta\nu_\mathrm{n} \le \frac{1}{2\pi\tau_i} + \frac{1}{2\pi\tau_k} = \frac{10^9}{2\pi\,1.4} + \frac{10^3}{2\pi}$$
$$= 1.14 \times 10^8\,\mathrm{s}^{-1} = 114\,\mathrm{MHz}\,.$$

c) The Doppler width is

$$\delta\nu_\mathrm{D} = 7.16 \times 10^{-7}\,\nu_0\sqrt{T/M}[\mathrm{mol/(g\,K)}]$$
$$\nu_0 = \frac{c}{\lambda} = \frac{3 \times 10^8}{5.017 \times 10^{-7}}\mathrm{s}^{-1}$$
$$= 5.98 \times 10^{14}\,\mathrm{s}^{-1}$$
$$T = 10^3\,\mathrm{K}, \quad M = 4\,\mathrm{g/mol}$$
$$\Rightarrow \delta\nu_\mathrm{D} = 6.77 \times 10^9\,\mathrm{s}^{-1} = 6.77\,\mathrm{GHz}\,.$$

5. a) The Lorentzian line profile with half-width $\delta\nu_\mathrm{n}$ is:

$$\alpha(\nu) = \alpha(\nu_0)\frac{(\delta\nu_\mathrm{n}/2)^2}{(\nu - \nu_0)^2 + (\delta\nu_\mathrm{n}/2)^2}\,.$$

A wavelength difference $\Delta\lambda = 0.1\,\mathrm{nm}$ corresponds to a frequency difference

$$\Delta\nu = (\nu_1 - \nu_0) = \frac{c}{\lambda^2}\Delta\lambda = 1.2 \times 10^{11}\,\mathrm{s}^{-1}$$
$$\Rightarrow \alpha(\nu_1) = \alpha(\nu_0)\frac{(0.57 \times 10^8)^2}{(1.2 \times 10^{11})^2 + (0.57 \times 10^8)^2}$$
$$= 2.25 \times 10^{-7}\,\alpha(\nu_0)\,.$$

For a Doppler-broadened absorption profile

$$\alpha(\nu) = \alpha(\nu_0)\mathrm{e}^{-[(\nu_1-\nu_0)^2/(2\delta\nu_\mathrm{D})^2]\ln 2}$$

we obtain with $\delta\nu_\mathrm{D} = 6.77 \times 10^9\,\mathrm{s}^{-1}$ for $\nu_1 - \nu_0 = 0.1\,\delta\nu_\mathrm{D}$:

$$\alpha(\nu_1) = \alpha(\nu_0)\mathrm{e}^{-0.028\ln 2}$$
$$= 0.98\,\alpha(\nu_0)\,.$$

For a Lorentzian profile with $\delta\nu_\mathrm{n} = 1.14 \times 10^8\,\mathrm{s}^{-1}$ we have:

$$\nu_1 - \nu_0 = 0.1\,\delta\nu_\mathrm{D} = 6.77 \times 10^8\,\mathrm{s}^{-1}$$
$$\Rightarrow \alpha(\nu_1) = \alpha(\nu_0)\frac{(5.7 \times 10^7)^2}{(6.77 \times 10^8)^2 + (5.7 \times 10^7)^2}$$
$$= 7 \times 10^{-3}\,\alpha(\nu_0)\,.$$

For $\nu_1 - \nu_0 = \delta\nu_\mathrm{D}$ we obtain for the Gaussian profile:

$$\alpha(\nu_1) = 0.146\,\alpha(\nu_0)$$

and for the Lorentzian profile:

$$\alpha(\nu_1) = 7 \times 10^{-5}\,\alpha(\nu_0)\,.$$

For $\nu_1 - \nu_0 = 10\,\delta\nu_\mathrm{D}$

$$\Rightarrow \alpha(\nu_1) = \alpha(\nu_0)\mathrm{e}^{-278} = 2 \times 10^{-120}\,\alpha(\nu_0)$$

for the Gaussian profile and

$$\alpha(\nu_1) = \alpha(\nu_0)\frac{(5.7 \times 10^7)^2}{(6.77 \times 10^{10})^2}$$
$$= 7 \times 10^{-7}\,\alpha(\nu_0)\,.$$

for the Lorentzian profile.
Here the absorption by the Gaussian profile is negligible compared to the absorption by the Lorentzian profile.
b) The relative absorption at $\nu = \nu_1$ of the two profiles becomes equal for:

$$\frac{(\delta\nu_\mathrm{n}/2)^2}{(\nu_1 - \nu_0)^2 + (\delta\nu_\mathrm{n}/2)^2} = \mathrm{e}^{-[(\nu_1-\nu_0)/0.6\,\delta\nu_\mathrm{D}]^2}$$
$$\Rightarrow \ln\left[(\delta\nu_\mathrm{n}/2)^2 + (\nu_1 - \nu_0)^2\right] - \ln(\delta\nu_\mathrm{n}/2)^2$$
$$= \left[(\nu_1 - \nu_0)/0.6\,\delta\nu_\mathrm{D}\right]^2\,.$$

The numerical solution (which can be obtained with the program "Mathematica") depends on the ratio $x = \delta\nu_\mathrm{n}/\delta\nu_0$. For $x = 0.01 \Rightarrow \nu_1 - \nu_0 = 439.6\,\delta\nu_\mathrm{n}$. For $x = 0.1 \Rightarrow \nu_1 - \nu_0 = 27.8\,\delta\nu_\mathrm{n}$. For $x = 1 \Rightarrow \nu_1 - \nu_0 = 0.895\,\delta\nu_\mathrm{n}$.

6. The frequency of the $K_\alpha$ lines of silver can be estimated for an effective charge $Z_{\text{eff}} = Z - 1$ with $Z = 47$ from the relation:

$$h\nu = Ry^*(Z-1)^2\left(\frac{1}{n_1^2} - \frac{1}{n_2^2}\right).$$

For $n_1 = 1, n_2 = 2$ and $Ry^* = 13.6\,\text{eV}$ we obtain:

$$\begin{aligned} h\nu &= 13.6 \times 46^2 \times \frac{3}{4}\,\text{eV} = 21.6\,\text{keV} \\ &= 3.45 \times 10^{-15}\,\text{J} \\ \Rightarrow \nu &= 5.22 \times 10^{18}\,\text{s}^{-1} \\ \lambda &= \frac{c}{\nu} = 5.75 \times 10^{-11}\,\text{m} = 0.575\,\text{Å}. \end{aligned}$$

The experimental value is $h\nu = 21.9\,\text{keV}$, $\lambda = 0.562\,\text{Å}$. The ionization energy of molybdenum is (see, e.g., the American Handbook of Physics)

$$IP\left({}^{42}\text{Mo}\right) = 20.0\,\text{keV}.$$

The kinetic energy of the photoelectron is then

$$E_{\text{kin}} = h\nu - IP = (21.9 - 20.0)\,\text{keV} = 1.9\,\text{keV}.$$

The velocity is

$$v = \sqrt{2E_{\text{kin}}/m_e} = 2.6 \times 10^7\,\text{m/s} = 8.6 \times 10^{-2}\,c.$$

7. The recoil momentum is

$$\boldsymbol{p} = \hbar\boldsymbol{k} \quad \text{with} \quad |\boldsymbol{k}| = \frac{2\pi}{\lambda}.$$

The recoil energy is

$$E_{\text{kin}} = \frac{p^2}{2m} = \frac{\hbar^2k^2}{2m} = \frac{(h\nu)^2}{2mc^2} = \frac{1}{2}\frac{E_{\text{phot}}^2}{mc^2}.$$

For the transition $n = 2 \to n = 1$ in the H atom is $h\nu = 10.2\,\text{eV}$ and $mc^2(\text{proton}) = 938.8\,\text{MeV}$

$$\Rightarrow E_{\text{kin}} = \frac{1}{2}\frac{10.2^2}{938.8 \times 10^6}\,\text{eV} = 5.5 \times 10^{-8}\,\text{eV}.$$

The velocity of the H atom after emission of the photon is:

$$\begin{aligned} v &= \frac{p}{m} = \frac{\hbar k}{m} = \frac{h\nu}{mc^2}c \\ &= \frac{10.2}{9.38 \times 10^8} \times 3 \times 10^8\,\text{m/s} \\ &= 3.3\,\text{m/s}. \end{aligned}$$

$$\Rightarrow \frac{\Delta\nu_{\text{recoil}}}{\nu} = \frac{v}{c} = 1.09 \times 10^{-8}$$

$$\begin{aligned} \Rightarrow \Delta\nu_{\text{recoil}} &= 1.09 \times 10^{-8} \times 2.47 \times 10^{15}\,\text{s}^{-1} \\ &= 2.7 \times 10^7\,\text{s}^{-1}. \end{aligned}$$

The natural line width of the Lyman α-line is

$$\delta\nu_n = \frac{1}{2\pi\tau(2p)} = 7.4 \times 10^7\,\text{s}^{-1}$$

since $\tau(2p) = 2.1\,\text{ns}$.
The recoil shift is therefore only $\Delta\nu_{\text{recoil}} = 0.35\,\delta\nu_n$. The Doppler width at $T = 300\,\text{K}$ is:

$$\delta\nu_D = 3.06 \times 10^{10}\,\text{s}^{-1} \Rightarrow \Delta\nu_{\text{recoil}} \ll \delta\nu_D.$$

8. The effective lifetime is determined by

$$\frac{1}{\tau_{\text{eff}}} = \frac{1}{\tau_n} + n\sigma\overline{v}_r,$$

where $\overline{v}_r$ is the mean relative velocity. From the relation $p = nkT$ we obtain for $p = 1\,\text{mbar} = 10^2\,\text{Pa}$

$$n = \frac{p}{kT} = \frac{10^2\,\text{Pa}}{1.38 \times 10^{-23} \times 500}\,\text{m}^{-3}.$$

$$\Rightarrow n = 1.45 \times 10^{22}\,\text{m}^{-3}$$

$$\overline{v}_r = \sqrt{\frac{8kT}{\pi m}}$$

with

$$m = \frac{m_1 m_2}{m_1 + m_2} = 0.55m(\text{Na}) = 2.1 \times 10^{-26}\,\text{kg}$$

$$\Rightarrow \overline{v}_r = \sqrt{\frac{8 \times 1.38 \times 10^{-23} \times 500}{2.1\,\pi \times 10^{-26}}}\,\frac{\text{m}}{\text{s}} = 915\,\text{m/s}.$$

This gives

$$\begin{aligned} \frac{1}{\tau_{\text{eff}}} &= \left(\frac{10^9}{16} + 1.45 \times 10^{22} \times 4 \times 10^{-19} \times 961\right)\text{s}^{-1} \\ &= 6.81 \times 10^7\,\text{s}^{-1} \\ &\Rightarrow \tau_{\text{eff}} = 14.7\,\text{ns}. \end{aligned}$$

The effective lifetime is smaller by a factor $\frac{14.7}{16} = 0.919$ than the natural lifetime $\tau = 16\,\text{ns}$. For $p = 10\,\text{mbar}$ the second term becomes

$$n\sigma\overline{v}_r = 5.3 \times 10^7\,\text{s}^{-1}$$

and the effective lifetime shortens to

$$\tau_{\text{eff}} = 8.7\,\text{ns} = 0.54\,\tau_n.$$

For $p = 100\,\text{mbar} \Rightarrow \tau_{\text{eff}} = 1.7\,\text{ns} = 0.11\,\tau_n$.

9. The residual Doppler width in the collimated atomic beam is

a) $(\delta\nu_D)_{\text{res}} = \sin\varepsilon\,\delta\nu_D < 190\,\text{MHz}$.

With $\delta\nu_D = 7.16 \times 10^{-7}\nu_0\sqrt{T/M}\,\text{mol/(g K)} =$

$2 \times 10^9\,\text{s}^{-1}$ with $\nu_0 = 5.09 \times 10^{14}\,\text{s}^{-1}$, $M = 23\,\text{g/mol}$, $T = 695\,\text{K}$, $\overline{v} = 800\,\text{m/s}$

$$\Rightarrow \sin\varepsilon \leq \frac{190}{2000} = 0.095\,.$$

b) If the residual Doppler width should be equal to the natural line width $\delta\nu_n = 10\,\text{MHz} \Rightarrow$

$$\sin\varepsilon = \frac{\delta\nu_n}{\delta\nu_D} = \frac{10^7}{2 \times 10^9} = 5 \times 10^{-3}\,.$$

10. a) From the Einstein coefficient $A_{ik} = 3 \times 10^{-15}\,\text{s}^{-1} = \frac{1}{\tau}$ we obtain the natural line width of the HFS transition with $\lambda = 21\,\text{cm}$

$$\delta\nu_n = \frac{A_{ik}}{2\pi} = 5 \times 10^{-16}\,\text{s}^{-1}\,.$$

The natural line width of the HFS transition is therefore extremely narrow. The Doppler width is

$$\delta\nu_D = 7.16 \times 10^{-7}\frac{3 \times 10^8}{0.21}\sqrt{10}\,\text{s}^{-1} = 3.2\,\text{kHz}\,.$$

The collision broadening is:

$$\delta\nu_{coll} = \frac{n\sigma\overline{v}_r}{2\pi}\,.$$

With $\overline{v}_r = \sqrt{\frac{8kT}{\pi m}} = 460\,\text{m/s}$ with $m = \frac{m_1 m_2}{m_1+m_2}$

$$\Rightarrow \delta\nu_{coll} = 7.3 \times 10^{-20}\,\text{s}^{-1}\,.$$

The collisional broadening is therefore completely negligible. For the Lyman $\alpha$-line we obtain:

$$\delta\nu_n = \frac{A_{ik}}{2\pi} = 1.6 \times 10^8\,\text{s}^{-1}\,,$$
$$\delta\nu_D = 5.6 \times 10^9\,\text{s}^{-1}\,,$$
$$\delta\nu_{coll} = 7.3 \times 10^{-13}\,\text{s}^{-1}\,.$$

The Doppler width is by far the prominent broadening.

b) At a temperature $T = 10\,\text{K}$ the ratio of the populations is

$$\frac{N(F=1)}{N(F=0)} = 3 \cdot \text{e}^{-h\nu/kT} \approx 3 \cdot 0.994\,.$$

The population difference is

$$\Delta N = N(F=0) - \frac{1}{3}N(F=1) = 0.006 \cdot N(F=0)\,.$$

The absorption coefficient is

$$\alpha = \left[N(F=0) - \frac{1}{3}N(F=1)\right] \cdot \sigma_{abs} = 0.006 \cdot N(F=0) \cdot \sigma_{abs}\,.$$

With $\sigma_{abs} = 3 \times 10^{-26}\,\text{m}^2$, $N = 10^6\,\text{m}^{-3}$ and $L = 3 \times 10^{16}\,\text{m}$ we obtain

$$\alpha \cdot l = 5.4 \times 10^{-6}\,.$$

The absorption of the 21 cm line is therefore negligible. For the Lyman $\alpha$-line the situation is different:

$$\alpha = \sigma \cdot N = 1 \times 10^{-15} \cdot 10^6\,\text{m}^{-1} = 10^{-9}\,\text{m}^{-1}\,.$$

After $10^9$ m the intensity has decreased to $(1/\text{e})I_0$. The interstellar cloud with $L = 3 \times 10^{16}\,\text{m}$ is therefore completely opague. Lyman-$\alpha$-light is completely absorbed.

c) The natural line width of the methane transition is

$$\delta\nu_n = \frac{1}{2\pi \times 2 \times 10^{-2}}\,\text{s}^{-1} = 7.96\,\text{s}^{-1}\,.$$

This extremely small line width is used for stabilizing the wavelength $\lambda = 3.39\,\mu\text{m}$ of the He-Ne laser (see Chap. 8).

$$\delta\nu_D = 7.16 \times 10^{-7}\frac{c}{\lambda}\sqrt{T/M(\text{mol/(g K)})}$$

with $\lambda = 3.39\,\mu\text{m}$, $M = 16\,\text{g/mol}$, $T = 300\,\text{K}$

$$\Rightarrow \delta\nu_D = 274\,\text{MHz}\,.$$

The mean velocity of the $CH_4$ molecules is

$$\overline{v} = \sqrt{\frac{8kT}{\pi m}} = 630\,\text{m/s}\,.$$

$\Rightarrow$ Transit time

$$t_{trans} = \frac{0.01\,\text{m}}{630\,\text{m/s}} = 1.6 \times 10^{-5}\,\text{s}$$

$$\Rightarrow \delta\nu_{trans} = \frac{1}{2\pi t_{trans}} \approx 10^4\,\text{s}^{-1}\,.$$

The only way to reduce transit-time broadening is the enlargement of the laser beam diameter or a reduction of the velocity by cooling the gas.

11. According to [7.13] the matrix element for the transition $1s \leftrightarrow 2s$ is

$$\begin{aligned}\boldsymbol{M}_{ik} &= e\int \psi(2s)\boldsymbol{r}\psi(1s)\,\mathrm{d}\tau \\ &= \frac{1}{4\pi\sqrt{2}a_0^3} \\ &\quad\times\int\left(2-\frac{r}{a_0}\right)\mathrm{e}^{-r/(2a_0)}\boldsymbol{r}\mathrm{e}^{-r/a_0}\,\mathrm{d}\tau \\ &= C\int\limits_r\int\limits_\vartheta\int\limits_\varphi\left(2-\frac{r}{a_0}\right)\mathrm{e}^{-3r/(2a_0)}\boldsymbol{r} \\ &\qquad\times r^2\sin\vartheta\,\mathrm{d}r\,\mathrm{d}\vartheta\,\mathrm{d}\varphi\,,\end{aligned}$$

$$\begin{aligned}(M_{ik})_x &= C\int\int\limits_{\vartheta=-\pi/2}^{+\pi/2}\int\limits_{\varphi=0}^{2\pi}\left(2-\frac{r}{a_0}\right)\mathrm{e}^{-3r/(2a_0)} \\ &\qquad\times xr^2\sin\vartheta\,\mathrm{d}r\,\mathrm{d}\vartheta\,\mathrm{d}\varphi\,.\end{aligned}$$

With $x = r\cdot\sin\vartheta\,\cos\varphi$ the integration over $\varphi$ yields

$$\int\limits_{\varphi=0}^{2\pi}\cos\varphi\,\mathrm{d}\varphi = 0\,.$$

A similar result is obtained for $(M_{ik})_y$ with $y = r\,\sin\vartheta\,\sin\varphi$. For the third component $(M_{ik})_z$ we obtain with $z = r\,\cos\vartheta$

$$\int\limits_0^{2\pi}\mathrm{d}\varphi = 2\pi$$

but

$$\int\limits_{\vartheta=-\pi/2}^{+\pi/2}\sin\vartheta\,\cos\vartheta\,\mathrm{d}\vartheta = \frac{1}{2}\sin^2\vartheta\,\Big|_{-\pi/2}^{+\pi/2} = 0\,.$$

12. The transition probability is, according to (7.17),

$$\begin{aligned}&A_{ik}(1s\to 2p) \\ &= \frac{2}{3}\frac{e^2\omega_{ik}^3}{\varepsilon_0 c^3 h}\left|\int\psi_i^*(2p)\boldsymbol{r}\psi_k(1s)\,\mathrm{d}\tau\right|^2\end{aligned}$$

The $2p$ level has three components with $m = 0, \pm 1$ that are degenerate without external magnetic field. There are therefore three degenerate transitions with different polarizations on the $1s \to 2p$ transition. If the quantization axis is chosen as the $z$-axis, the matrix element $M_x + \mathrm{i}M_y$ describes transitions with $\Delta m = +1$, $M_x - \mathrm{i}M_y$ with $\Delta m = -1$ and $M_z$ with $\Delta m = 0$. The total transition probability is proportional to

$$|M_{ik}|^2 = |(M_{ik})_x|^2 + (M_{ik})_y^2 + (M_{ik})_z^2\,,$$

$$|M_x + \mathrm{i}M_y|^2 = M_x^2 + M_y^2\,.$$

Using the wave functions of Table 5.2 we obtain:

$$\begin{aligned}\left(M_x + \mathrm{i}M_y\right) &= \frac{1}{8\pi a_0^4}\int\limits_r\int\limits_\vartheta\int\limits_\varphi \mathrm{e}^{-r/a_0}(x+\mathrm{i}y)r \\ &\quad\times\mathrm{e}^{-r/2a_0}\sin\vartheta\mathrm{e}^{-\mathrm{i}\varphi}r^2\sin\vartheta\,\mathrm{d}\varphi\,\mathrm{d}\vartheta\,\mathrm{d}r\,.\end{aligned}$$

With $x = r\,\sin\vartheta\,\cos\varphi$; $y = r\,\sin\vartheta\,\sin\varphi$ one obtains $x + \mathrm{i}y = r\,\sin\vartheta\,\mathrm{e}^{\mathrm{i}\varphi}$

$$\begin{aligned}\Rightarrow\left(M_x + \mathrm{i}M_y\right) &= \frac{1}{8\pi a_0^4}\int\limits_{r=0}^{\infty}r^4\mathrm{e}^{-3r/(2a_0)}\,\mathrm{d}r \\ &\quad\times\int\limits_{\vartheta=0}^{\pi}\sin^3\vartheta\,\mathrm{d}\vartheta\int\limits_{\varphi=0}^{2\pi}\mathrm{d}\varphi\,.\end{aligned}$$

The first integral has the value $256a_0^5/81$, the second integral is $4/3$ and the third is $2\pi$. This gives:

$$(M_x + \mathrm{i}M_y)^2 = \left(\frac{256}{243}a_0\right)^2$$

$$\Rightarrow A_{ik}(\Delta m = \pm 1) = \frac{2}{3}\frac{e^2\omega_{ik}^3 a_0^2}{\varepsilon_0 c^3 h}\left(\frac{256}{243}\right)^2.$$

With $\omega_{ik} = 2\pi\,2.47\times 10^{15}\,\mathrm{s}^{-1}$ the transition probability becomes:

$A_{ik}(\Delta m = \pm 1) = 1.25\times 10^{10}\,\mathrm{s}^{-1}$.

An analogous calculation for $M_z$ with $z = r\,\cos\vartheta$ gives

$$\begin{aligned}M_z &= \frac{1}{4\pi\sqrt{2}a_0^4}\int\limits_{r=0}^{\infty}r^4\mathrm{e}^{-3r/(2a_0)}\,\mathrm{d}r \\ &\quad\times\int\limits_{\vartheta=0}^{\pi}\cos^2\vartheta\,\sin\vartheta\,\mathrm{d}\vartheta\int\limits_{\varphi=0}^{2\pi}\mathrm{d}\varphi \\ &= \frac{1}{4\pi\sqrt{2}a_0^4}\frac{256}{81}a_0^5\frac{2}{3}2\pi \\ &= \frac{256}{243\sqrt{2}}a_0\end{aligned}$$

$$\Rightarrow A_{ik}(\Delta m = 0) = \frac{1}{2}A_{ik}(\Delta m \pm 1)\,.$$

13. The $3s$ level can only decay into the $2p$ level. Therefore the transition probability for the transition $3s \to 2p$ is

$$A_{ik} = \frac{1}{\tau(3s)} = \frac{10^9}{23}\,\mathrm{s}^{-1} = 4.3\times 10^7\,\mathrm{s}^{-1}\,.$$

The natural line width is

$$\begin{aligned}\delta\nu_\mathrm{n} &= \frac{1}{2\pi}\left(\frac{1}{\tau(3s)}+\frac{1}{\tau(2p)}\right)\\ &= \frac{1}{2\pi}\left(4.3\times 10^7 + 4.76\times 10^8\right)\,\mathrm{s}^{-1}\\ &= 83\,\mathrm{MHz}\,.\end{aligned}$$

The Doppler-width is

$$\delta\nu_\mathrm{D} = 7.16\times 10^{-7}\,\nu_0\sqrt{T/M[\mathrm{mol/(g\,K)}]}\,.$$

With

$$\nu_0 = \frac{1}{h}Ry^*\left(\frac{1}{2^2}-\frac{1}{3^2}\right) = 4.57\times 10^{14}\,\mathrm{s}^{-1}\,,$$

$$M = 1\,\mathrm{g/mol}\,,\quad T = 300\,\mathrm{K}\,,$$

this gives

$$\delta\nu_\mathrm{D} = 5.67\times 10^9\,\mathrm{s}^{-1} = 5.67\,\mathrm{GHz}$$

$$\Rightarrow \frac{\delta\nu_\mathrm{n}}{\delta\nu_\mathrm{D}} = 0.014$$

for $T = 300\,\mathrm{K}$ and 0.008 for $T = 1000\,\mathrm{K}$.

14. The incident wave $E_0\cdot \mathrm{e}^{\mathrm{i}(\omega t-kz)}$ induces the electrons of the medium to forced oscillations. If the amplitude of this oscillation is within the range of linear optics (the restoring force is proportional to the elongation of the atomic electron) we can describe the atomic electron by a harmonic oscillator. Its equation of motion is:

$$m\ddot{x} + b\dot{x} + Dx = -eE_0\mathrm{e}^{\mathrm{i}(\omega t-kz)}\,. \tag{1}$$

With $D/m = \omega_0^2$, $\gamma = b/m$ we can make the ansatz

$$x = x_0\mathrm{e}^{\mathrm{i}\omega t}\,.$$

Inserting this into (1) gives

$$x_0 = -\frac{eE_0/m}{\left(\omega_0^2-\omega^2\right)+\mathrm{i}\gamma\omega}\,. \tag{2}$$

The induced dipole moment $p = p_0\cdot \mathrm{e}^{\mathrm{i}\omega t}$ with $p_0 = e\cdot x_0$

$$\Rightarrow p = \frac{e^2E}{m\left(\omega_0^2-\omega^2+\mathrm{i}\gamma\omega\right)}\,. \tag{3}$$

Since $p = \alpha(\omega)\cdot E$ we obtain for the polarizability

$$\alpha(\omega) = \frac{e^2}{m\left[\left(\omega_0^2-\omega^2\right)+\mathrm{i}\gamma\omega\right]} \tag{4}$$

The wave equation for a wave travelling through a medium is:

$$\Delta E = \mu\mu_0\varepsilon\varepsilon_0\frac{\partial^2 E}{\partial t^2} = \frac{1}{v_\mathrm{ph}^2}\frac{\partial^2 E}{\partial t^2}\,, \tag{5}$$

where

$$v_\mathrm{ph} = \frac{1}{\sqrt{\mu\mu_0\varepsilon\varepsilon_0}} = \frac{c}{\sqrt{\mu\varepsilon}}$$

is the phase velocity of the wave. In dielectric media is the magnetic susceptibility $\mu\approx 1$ and we obtain:

$$v_\mathrm{ph} = \frac{c}{\sqrt{\varepsilon}} = \frac{c}{n} \Rightarrow n = \sqrt{\varepsilon}\,. \tag{6}$$

Inserting the relation: $D = \varepsilon_0 E + P$ into (1) we obtain the equation, equivalent to (5):

$$\Delta E = \frac{1}{c^2}\frac{\partial^2 E}{\partial t^2} + \frac{1}{\varepsilon_0 c^2}\frac{\partial^2 P}{\partial t^2}\,. \tag{7}$$

The dielectric polarization is for a wave with its $\boldsymbol{E}$-vector in $x$-direction ($\boldsymbol{E} = \{E_x,0,0\}$):

$$P_x = N\cdot\alpha\cdot E_x = N\cdot\alpha\cdot E_0\cdot \mathrm{e}^{\mathrm{i}(\omega t-kz)}\,, \tag{8}$$

where $N$ is the number of oscillating dipoles per $\mathrm{m}^3$ and $\alpha$ is the polarizability.
Inserting (8) into (7) gives:

$$-k^2E_x = -\frac{\omega^2}{c^2}E_x - \frac{\omega^2 N\alpha}{\varepsilon_0 c^2}E_x$$

$$\Rightarrow k^2 = \frac{\omega^2}{c^2}(1+N\alpha/\varepsilon_0) \tag{9}$$

with $v_\mathrm{ph} = c/n = \omega/k \Rightarrow n = ck/\omega$, where $n$ is the refractive index.

$$\Rightarrow n^2 = 1 + N\alpha/\varepsilon_0\,. \tag{10}$$

Inserting (4) into (10) yields

$$n^2 = 1 + \frac{Ne^2}{m\left[\omega_0^2-\omega^2+\mathrm{i}\gamma\omega\right]}\,.$$

with $n = n' - \mathrm{i}\kappa \quad\Rightarrow\quad n^2 = n'^2 - \kappa^2 - 2\mathrm{i}n'\kappa$, where $\kappa$ is the absorption coefficient and $n'$ the real part of the complex index of refraction. For small damping ($\gamma\omega \ll \omega^2$) the summation over all atomic resonances gives the result (7.102).

In order to transport a signal, the monochromatic wave has to be modulated. The envelope of this modulated wave represents a wave packet with the group velocity $v_g = \mathrm{d}\omega/\mathrm{d}k$.

## Chapter 8

1. a) The ratio of the populations $N_i$, $N_k$ is:

$$\frac{N_i}{N_k} = \frac{g_i}{g_k}\mathrm{e}^{-h\nu/kT}$$

$$g_i = 2J_i + 1 = 3\,, \quad g_k = 2J_k + 1 = 1$$

$$\Rightarrow \frac{N_i}{N_k} = 3\mathrm{e}^{-(hc/\lambda)/kT} = 3\mathrm{e}^{-96} = 6.6 \times 10^{-42}\,.$$

The thermal population of the upper level $|i\rangle$ is therefore completely negligible.

b) The relative absorption of the incident wave with intensity $I_0$ is

$$A = \frac{I_0 - I_\mathrm{t}}{I_0}\,, \quad \text{with}$$

$$I_\mathrm{t} = I_0\mathrm{e}^{-\alpha L} \approx (1 - \alpha L)I_0 \quad \text{for} \quad \alpha L \ll 1$$

$$\Rightarrow A \approx \alpha L = N_k\sigma_{ki}L\,.$$

The population density $N_k = 10^{-6}N$ can be obtained from the equation $p = NkT$

$$\Rightarrow N_k = 10^{-6}\,p/kT\,.$$

For a pressure $p = 10^2$ Pa we obtain:

$$N_k = \frac{10^{-6} \times 10^2}{1.38 \times 10^{-23} \times 300}\,\mathrm{m}^{-3} = 10^{16}\,\mathrm{m}^{-3} = 10^{10}\,\mathrm{cm}^{-3}\,.$$

The absorption coefficient $\alpha$ is related to the Einstein coefficient $B_{ki}$ as follows: The absorbed power is:

$$\frac{\mathrm{d}W_{ki}}{\mathrm{d}t} = B_{ki}h\nu\, w_\nu(\nu)\,.$$

The spectral energy density $w_\nu(\nu)$ is related to the spectral intensity $I_\nu(\nu)$ by $w_\nu(\nu) = I_\nu(\nu)/c$. The power absorbed by one atom is

$$\frac{1}{N_k}\frac{\mathrm{d}W_{ki}}{\mathrm{d}t} = \frac{1}{N_k}I(\nu)\int_0^\infty \alpha(\nu)\,\mathrm{d}\nu \approx \frac{1}{N_k}I(\nu_0)\alpha_0(\nu_0)\delta\nu\,,$$

where $\delta\nu$ is the full half-width of the absorption profile. With $\alpha_{ki} = N_k\sigma_{ki}$ we obtain:

$$\sigma_{ki} = \frac{h\nu}{c}B_{ki}/\delta\nu = \frac{c^2 A_{ik}}{8\pi\nu^2\delta\nu} = \frac{\lambda^2 A_{ik}}{8\pi\delta\nu}\,.$$

Inserting the Doppler width $\delta\nu_\mathrm{D} \approx 10^9\,\mathrm{s}^{-1}$ we obtain:

$$\sigma_{ki} = 10^{-15}\,\mathrm{m}^2 = 10^{-11}\,\mathrm{cm}^2$$

$$\Rightarrow A = 10^{10} \times 10^{-11} \times 1 = 0.1\,.$$

This means that 10% of the incident light power is absorbed.

c) For the compensation of 10% total losses the condition $-2\alpha L \geq 0.1$ must be met.

$$\Rightarrow ((g_k/g_i)N_i - N_k)\,\sigma_{ki}L \geq 0.05\,.$$

With $g_k = 1$, $g_i = 3$ and $\sigma_{ki} = 10^{-11}\,\mathrm{cm}^2$ this gives:

$$\left(\frac{1}{3}N_i - N_k\right)10^{-11} \times 20 \geq 0.05\,.$$

With $N_k = 10^{10}/\mathrm{cm}^3$ the upper state population density then has to be

$$N_i \geq 3.075 \times 10^{10}/\mathrm{cm}^3\,.$$

For equal statistical weight factors $g_i = g_k$ the minimum population $N_i$ would be

$$N_i \geq 1.025\,N_k = 1.025 \times 10^{10}\,\mathrm{cm}^{-3}\,.$$

This would require a population in the upper state, which is 2.5% higher than in the lower state in order to reach laser threshhold.

2. a) The Doppler width is

$$\delta\nu_\mathrm{D} = 7.16\frac{c}{\lambda}\sqrt{T/M}10^{-7}\,\mathrm{s}^{-1}\,.$$

With $\lambda = 632.8\,\mathrm{nm}$, $T = 600\,\mathrm{K}$, $M = 20\,\mathrm{g/mol}$ this gives:

$$\delta\nu_\mathrm{D} = 1.86 \times 10^9\,\mathrm{s}^{-1} = 1.86\,\mathrm{GHz}\,.$$

b) The mode separation is

$$\delta\nu = \frac{c}{2d} = \frac{3 \times 10^8}{2 \times 1} = 150\,\mathrm{MHz}\,.$$

The number of longitudinal modes within the full half-width of the Doppler profile is then

$$m = \frac{1.86 \times 10^9}{1.5 \times 10^8} = 12\,.$$

3. a) The frequency separation of the transmission maxima of the etalon is

$$\delta\nu_E = \frac{c}{2nt}\,.$$

If this should be larger than the Doppler width

$$\delta\nu_D = 7.16 \times 10^{-7}\frac{c}{\lambda}\sqrt{\frac{5000}{40}} = 5 \times 10^9\,\mathrm{s}^{-1}$$

at $\lambda = 488\,\mathrm{nm}$, the thickness $t$ of the etalon should be

$$t = \frac{c}{2n\delta\nu_E} \leq \frac{c}{2n\delta\nu_D} = \frac{3\times 10^8}{2\times 1.5\times 5\times 10^9}\,\mathrm{m} = 2\times 10^{-2}\,\mathrm{m} = 2\,\mathrm{cm}\,.$$

b) The intensity $I_z$ transmitted through the Fabry–Perot interferometer is:

$$I_t = I_0\frac{1}{1+F\sin^2(\pi\Delta s\nu/c)}$$

with $F = 4R/(1-R)^2$. The finesse

$$F^* = \frac{\pi\sqrt{R}}{1-R} = \frac{\pi\sqrt{F}}{2} = \frac{\Delta\nu_m}{\Delta\nu_E}$$

equals the ratio of frequency separation $\Delta\nu_m$ of transmission maxima to the half-width $\Delta\nu_E$ of the transmission peaks. The transmission of the FPI has decreased from the maximum $I_t = I_0$ to $I_t = \frac{1}{3}I_0$ for

$$F\sin^2(\pi\Delta s\nu/c) = 2 = F\sin^2\left[\frac{\pi\Delta s}{c}(\nu_0 + \Delta\nu_L)\right]$$

$$\Rightarrow \sin\left(\frac{\pi\Delta s}{c}\Delta\nu_L\right) = \sqrt{\frac{2}{F}}\,.$$

If $\nu_0$ is the frequency of the transmission peak, the condition

$$\sin^2(\pi\Delta s\nu_0/c) = 0 \Rightarrow \pi\Delta s\nu_0/c = m \quad \text{(integer)}$$

must be met. For $\lambda = 488\,\mathrm{nm}$, $\Delta s = 2nt = 6\,\mathrm{cm}$ $\Rightarrow m = 122{,}950$. The frequency condition $\Delta\nu_L$ of longitudinal modes is

$$\Delta\nu_L = \frac{c}{2d} = 125\,\mathrm{MHz}\,.$$

With

$$\sin\left(\frac{\pi\Delta s}{c}\Delta\nu_L\right) = \sin\left(\frac{6\pi}{3\times 10^{10}}1.25\times 10^8\right) = 0.078$$

we obtain

$$\sqrt{\frac{2}{F}} = 0.075 \Rightarrow F^* = \frac{\pi\sqrt{F}}{2} = 28.5\,.$$

The finesse $F^*$ must be at least 28.5 in order to select a single longitudinal mode. With $F^* = \pi\sqrt{R}/(1-R)$ we obtain for the reflectivity $R$ of the Fabry–Perot

$$R > 0.89\,.$$

4. a) The frequencies $\nu_m$ of the resonator modes are

$$\nu = \frac{c}{\lambda} = m\cdot\frac{c}{2L} \quad \text{because} \quad L = m\lambda/2\,.$$

With $L = 1\,\mathrm{m}$, $\nu = 5\times 10^{14}s^{-1} \Rightarrow m = 3.33\times 10^6$,

$$\begin{aligned} L &= L_0(1+\alpha T)\,,\\ \Delta L &= L - L_0 = L_0\alpha\Delta T\\ &= 1\times 12\times 10^{-6}\times 1\,\mathrm{m}\\ &= 1.2\times 10^{-5}\,\mathrm{m}\,,\\ \frac{\Delta\nu}{\nu} &= \frac{\Delta L}{L} = 1.2\times 10^{-5}\\ \Rightarrow \Delta\nu &= 1.2\times 10^{-5}\times 5\times 10^{14}\,\mathrm{s}^{-1}\\ &= 6\times 10^9\,\mathrm{s}^{-1}\,. \end{aligned}$$

Since the mode distance is only $\Delta\nu_L = 150\,\mathrm{MHz}$ the laser frequency jumps during the temperature change after a shift of about $\Delta\nu/2 \approx 75\,\mathrm{MHz}$ back to the next mode. Temperature tuning alone does not allow a continuous frequency shift of more than 75 MHz, unless the resonator contains additional frequency selective optical elements.

b) The refractive index $n$ can be written as

$$n = 1 + aN \Rightarrow n - 1 \propto N\,,$$

where $N$ is the number density of air molecules. If the air pressure changes by $\Delta p = 10\,\mathrm{mbar} = 1\%$, the same relative change occurs for $(n-1)$. For air at atmospheric pressure is $n - 1 = 2.7\times 10^{-4}$. The change for $\Delta p = 1\%$ is then $\Delta(n-1) = 2.7\times 10^{-6}$. The optical path length inside the resonator changes then by

$$\Delta(nL) = 2.7\times 10^{-6}\times 0.2\,\mathrm{m} = 5.4\times 10^{-7}\,\mathrm{m}$$

$$\Rightarrow \Delta\nu = \frac{\Delta(nL)}{L}\nu_0 = 2.7\times 10^8\,\mathrm{s}^{-1}\,.$$

Also in this case the laser frequency jumps back to the next resonator mode, if $\Delta\nu > \Delta\nu_m/2$.

5. a) The classical diffraction theory yields for the diffraction of a plane wave with wavelength $\lambda$ by a circular aperture with diameter $d$ the angular divergence of the central diffraction maximum

$$\Delta\alpha = 1.2\frac{\lambda}{d}.$$

For a laser beam with a Gaussian intensity profile with half-width $w_0$ at the focusing lens with focal length $f$ we obtain for the half-width of the local spot:

$$w_0 = f\frac{\lambda}{\pi w_s}.$$

With $w_s = d/2$ we obtain:

$$w_0 = f\frac{\lambda}{\pi d/2} = 42\,\mu\text{m}.$$

This differs from the result for a plane wave by the factor $1.2\pi/2 \approx 1.9$.
b) The intensity is obtained from

$$\begin{aligned} P &= 2\pi \int_0^\infty r \cdot I(r)\,\mathrm{d}r \\ &= 2\pi I_0 \cdot \int_0^\infty r \cdot \mathrm{e}^{-(r/r_0)^2}\,\mathrm{d}r = \pi w_0^2 \cdot I_0 \end{aligned}$$

$$I_0 = \frac{P}{\pi w_0^2} = \frac{10 \times 10^{12}\,\text{W}}{\pi \cdot 42^2\,\text{m}^2} = 1.8 \times 10^9\,\text{W/m}^2.$$

c) Only $10\% = 1\,\text{W}$ can be used for the evaporation. The mass of the evaporated material with thickness $D$ is

$$\begin{aligned} M &= \varrho \cdot \pi w_0^2 D \\ &= 8 \times 10^3 \pi \times 42^2 \times 10^{-12} \times 10^{-3}\,\text{kg} \\ &= 4.4 \times 10^{-11}\,\text{kg}. \end{aligned}$$

The heat of evaporation is

$$W_e = 6 \times 10^6\,\text{J/kg}.$$

The necessary energy for evaporation is

$$W = W_e \cdot M = 2.6 \times 10^{-4}\,\text{J}.$$

The time needed for the evaporation of the mass $M$ is:

$$t = W/P = 2.6 \times 10^{-4}\,\text{s} = 0.26\,\text{ms}.$$

Since a considerable fraction of the absorbed energy is lost by heat conduction into the cold surroundings, one needs, in fact, about ten times as long, i.e. $t = 2.6\,\text{ms}$.

6. a) The minimum spectral width is

$$\Delta\nu \geq 0.5/\Delta T.$$

With $\Delta T = 10^{-14}\,\text{s} \Rightarrow \Delta\nu \geq 0.5 \times 10^{14}\,\text{s}^{-1}$. For a wavelength $\lambda = 600\,\text{nm}$ this corresponds to a spectral width

$$\Delta\lambda = 6 \times 10^{-8}\,\text{m} = 60\,\text{nm}.$$

b) The spatial pulse length is in the beginning:

$$\Delta s_0 = \frac{c\Delta T}{n} = 2 \times 10^{-6}\,\text{m} = 2\,\mu\text{m}.$$

After passing through a dispersive medium with length $L$, the pulse length has broadened. The difference of optical path length for the different wavelengths $\lambda$ within the spectral pulse profile is

$$\begin{aligned} \Delta(nL) &= L\frac{\mathrm{d}n}{\mathrm{d}\lambda}\Delta\lambda = L \cdot 4.5 \times 10^4 \times 6 \times 10^{-8} \\ &= 2.64 \times 10^{-3}\,L. \end{aligned}$$

In order to keep $\Delta s \leq 4\,\mu\text{m}$, the condition

$$\frac{1}{n}\Delta(nL) = (\Delta s - \Delta s_0) \leq 2\,\mu\text{m}$$

has to be fulfilled.

$$\begin{aligned} \Rightarrow L &= \frac{n\,2.0 \times 10^{-6}\,\text{m}}{2.64 \times 10^{-3}} = 1.1 \times 10^{-3}\,\text{m} \\ &= 1.1\,\text{mm}, \end{aligned}$$

for $n = 1.5$. This means that after the passage through a 1.1-mm thick glass plate the pulse length has already doubled from 2 to 4 $\mu$m.
c) Because of the nonlinear, intensity-dependent part of the refractive index the wavelength $\lambda$ is red-hifted during the pulse rise time ($\mathrm{d}I/\mathrm{d}t > 0$). It is blue-hifted at the trailing edge ($\mathrm{d}I/\mathrm{d}t > 0$).
If the material is chosen in such a way that the linear part of $n$ in the surrounding of $\lambda_0$ (central wavelength of the spectral pulse profile) shows anomalous dispersion ($\mathrm{d}n/\mathrm{d}\lambda > 0$), the red part is delayed more than the blue part of the pulse. This compensates the opposite effect of the nonlinear part of $n$. Complete compensation demands

$$\frac{\mathrm{d}}{\mathrm{d}\lambda}(n_0(\lambda) + n_2 I) = 0.$$

7. The resonator quality factor is defined as

$$Q_k = -2\pi\nu \frac{W_k}{\mathrm{d}W_k/\mathrm{d}t},$$

where

$$W_k(t) = W_k(0)\mathrm{e}^{-\gamma_k t}$$

is the energy stored in the $k$th mode of the resonator. If $\gamma_k$ are the losses per round trip, the power after one round trip has decreased to

$$P = P(0)\mathrm{e}^{-\gamma_L}.$$

The loss factor $\gamma_L$ is composed of reflection losses and other losses (diffraction, scattering, absorption). The reflection losses per round trip are

$$\gamma_R = -\ln(R_1 R_2) = 0.02.$$

If the other losses together are also 0.02, we obtain

$$\gamma_L = 0.04.$$

Since the round trip time of a light wave in a resonator with mirror separation $d$ is

$$T = \frac{2d}{c}$$

the losses per sec and are

$$\gamma = \gamma_L/T = \frac{c}{2d}\gamma_L.$$

With $d = 1\,\mathrm{m} \Rightarrow \gamma = 1.5 \times 10^8 \times 0.04\,\mathrm{s}^{-1} = 6 \times 10^6\,\mathrm{s}^{-1}$. The quality factor is then for a frequency $\nu = 5 \times 10^{14}\,\mathrm{s}^{-1}$:

$$Q_k = \frac{2\pi\, 5 \times 10^{14}}{6 \times 10^6} = 5.2 \times 10^8.$$

Per oscillation period of the light the power decreases by the fraction

$$\eta = \frac{2\pi}{5.2 \times 10^8} = 1.2 \times 10^{-8}.$$

It takes $1/\gamma \approx 1.7 \times 10^{-7}\,\mathrm{s}$ until the power in the resonator mode has decreased to $1/\mathrm{e}$ of its initial value.

8. After one round trip the power has increased by the factor

$$\frac{P_1}{P_0} = \mathrm{e}^{-(2\alpha d + \gamma)}.$$

With a net gain $-(2\alpha d + \gamma) = 0.05$ we obtain

$$\frac{P_1}{P_0} = \mathrm{e}^{0.05} = 1.05.$$

The time $T$ for a round trip is

$$T = \frac{2d}{c} = \frac{2}{3}10^{-8}\,\mathrm{s}.$$

The time-dependent power then becomes:

$$P(t) = P_0 \exp\left[-\frac{2\alpha d + \gamma}{2d/c}t\right].$$

With a mirror transmission $T = 0.02$ the output power of 1 mW demands a power of 50 mW inside the resonator. The initial power $P(0)$ is given by one photon, i.e.,

$$P(0) = \frac{h\nu c}{2d} = 4.5 \times 10^{-10}\,\mathrm{W}.$$

a)

$$\frac{P}{P_0} = \mathrm{e}^{0.05/(0.666 \times 10^{-8}\,\mathrm{s})t}$$

$$\Rightarrow t = \frac{1\,\mathrm{s}}{7.5 \times 10^6} \ln \frac{P}{P_0}$$
$$= 1.3 \times 10^{-7} \ln \frac{5 \times 10^{-2}}{4.5 \times 10^{-10}}\,\mathrm{s}$$
$$= 25 \times 10^{-7}\,\mathrm{s} = 2.5\,\mu\mathrm{s}.$$

b) If we take into account saturation effects, the gain depends on the laser power inside the resonator. We obtain:

$$\frac{\mathrm{d}P(t)}{\mathrm{d}t} = \frac{\text{gain per round trip}}{\text{round trip time}} P(t)$$
$$= -\frac{1}{T}\left[(-2\alpha_0 d + \gamma) - 2ad\,P(t)\right] P(t).$$

With $T = 2d/c$, $(-2\alpha_0 d + \gamma) = 0.05$ this gives

$$\frac{\mathrm{d}P}{\mathrm{d}t} = \frac{1}{T}(0.05 - 2adP)P.$$

This is a nonlinear differential equation

$$\dot{y} - Ay + By^2 = 0$$

with

$$A = \frac{0.05}{T}, \quad B = \frac{2ad}{T}.$$

Division by $y^2$ gives:

$$\frac{\dot{y}}{y^2} - \frac{A}{y} + B = 0.$$

Substitution:

$$z(t) = \frac{1}{y(t)} \Rightarrow \dot{z} = -\frac{1}{y^2}\dot{y},$$
$$\Rightarrow \dot{z} + Az - B = 0.$$

The solution of the homogeneous equation ($B=0$) gives:

$$z = Ce^{-At}.$$

The solution of the inhomogeneous equation ($B \neq 0$) is:

$$z = C(t)e^{-At} \Rightarrow \dot{z} = \left(\dot{C} - CA\right)e^{-At}.$$

Inserting this into the inhomogeneous equation yields

$$\dot{C} - CA + AC = Be^{At}$$
$$\Rightarrow C = \frac{B}{A}e^{At} + D$$
$$\Rightarrow z = \frac{B}{A} + De^{-At}$$
$$\Rightarrow y = \frac{1}{B/A + De^{-At}}$$
$$\Rightarrow P(t) = \frac{1}{2ad/0.05 + De^{-0.05t/T}}.$$

For $t = 0 \Rightarrow P = P_0$

$$\Rightarrow D = \frac{1}{P_0} - 40ad$$
$$\Rightarrow P(t) = \frac{P_0}{40adP_0 + (1 - 40adP_0)e^{-0.05t/T}}.$$

With $P(t_1) = 50\,\text{mW} \Rightarrow P(t_1)/P_0 \approx 10^8$ because $P_0 = 4.5 \times 10^{-10}\,\text{W}$. The denominator therefore has to be $10^{-8}$. With $a = 0.4\,\text{W}^{-1}\,\text{m}^{-1}$, $d = 1\,\text{m}$ we obtain

$$40 \times 0.4 \times 4.5 \times 10^{-10} + e^{-0.05t/T} = 10^{-8}$$
$$\Rightarrow e^{-0.05t/T} = 2.8 \times 10^{-9}$$
$$\Rightarrow t = -\frac{T}{0.05}\ln 2.8 \times 10^{-9}$$
$$\Rightarrow t = +\frac{T}{0.05}\ln 3.57 \times 10^{8}$$
$$= 20\,T\,19.7 = 394\,T.$$

With $T = \frac{2}{3}10^{-8}\,\text{s} \Rightarrow 263 \times 10^{-8}\,\text{s} = 2.63\,\mu\text{s}$. With $a = 0.55\,\text{W}^{-1}\,\text{m}^{-1}$ $t$ increases to

$$t = 20\,T \ln 10^{10}\,\text{s} = 30.7\,\mu\text{s}.$$

## Chapter 9

1. The potential energy of the Coulomb repulsion of the two protons is

$$E_{\text{pot}} = \frac{e^2}{4\pi\varepsilon_0 \cdot 2a_0} = 2.3 \times 10^{-18}\,\text{J} = 13.6\,\text{eV}.$$

The potential energy of the electron in the state $|\phi^+\rangle$ is

$$E_{\text{pot}} = -\frac{e^2}{4\pi\varepsilon_0}\int |\phi^+|^2\left(\frac{1}{r_A} + \frac{1}{r_B}\right)d\tau.$$

The wave function is

$$\phi^+ = \frac{\phi_A + \phi_B}{\sqrt{2 + S_{AB}}}$$

Inserting $\phi_A$ and $\phi_B$ from (9.9) gives:

$$|\phi^+|^2 = \frac{1}{2\pi a_0^3}\,\frac{e^{-2r_A/a_0} + e^{-2r_B/a_0} + 2e^{-(r_A+r_B)/a_0}}{1 + S_{AB}}$$

$$E_{\text{pot}} = -\frac{e^2}{8\pi^2\varepsilon_0 a_0^3(1 + S_{AB})}$$
$$\int\left[\frac{e^{-2r_A/a_0} + e^{-2r_B/a_0} + 2e^{-(r_A+r_B)/a_0}}{r_A} + \frac{e^{-2r_A/a_0} + e^{-2r_B/a_0} + 2e^{-(r_A+r_B)/a_0}}{r_B}\right]d\tau.$$

With elliptical coordinates

$$\mu = \frac{r_A + r_B}{R};$$
$$\nu = \frac{r_A - r_B}{R};$$
$$\varphi = \arctan(y/x)$$

where the two protons are sitting in the focal points. With these coordinates and

$$d\tau = \frac{R^3}{8}\left(\mu^2 - \nu^2\right)d\mu\,d\nu\,d\varphi,$$

the integral can be solved analytically. With

$$r_A = \frac{R}{2}(\mu + \nu),$$
$$r_B = \frac{R}{2}(\mu - \nu),$$

the overlap integral $S_{AB}$ becomes:

$$S_{AB} = \frac{1}{\pi a_0^3}\int e^{-(r_A+r_B)/a_0}d\tau$$

$$= \frac{R^3}{8\pi a_0^3}\left[\int\limits_{\mu=1}^{\infty} \mu^2 \cdot \mathrm{e}^{-R\mu/a_0}\,\mathrm{d}\mu \int\limits_{\nu=-1}^{+1} \mathrm{d}\nu \cdot \int\limits_{\varphi=0}^{2\pi} \mathrm{d}\varphi \right.$$

$$\left. - \int\limits_{\mu=1}^{\infty} \mathrm{e}^{-R\mu/a_0}\mathrm{d}\mu \int\limits_{\nu=-1}^{1} \nu^2\,\mathrm{d}\nu \int\limits_{\varphi=0}^{2\pi} \mathrm{d}\varphi \right].$$

The integration yields

$$S_{AB} = \mathrm{e}^{-R/a_0}\left(1 + \frac{R}{a_0} + \frac{R^2}{3a_0^2}\right).$$

This gives for the potential energy

$$E_{pot} = -\frac{C \cdot R^2}{2\pi} \int\limits_{\mu}\int\limits_{\nu}\int\limits_{\varphi} \left[\mathrm{e}^{-\varrho(\mu+\nu)} + \mathrm{e}^{-\varrho(\mu-\nu)} \right.$$

$$\left. + \mathrm{e}^{-\varrho\mu}\right] \mu\,\mathrm{d}\mu\,\mathrm{d}\nu\,\mathrm{d}\varphi$$

with

$$C = \frac{e^2}{8\pi\varepsilon_0 a_0^3 (1 + S_{AB})}; \quad \varrho = R/a_0$$

$$\Rightarrow E_{pot} = -C \cdot R^2 \int\limits_{\mu}\int\limits_{\nu} \mu \left[\mathrm{e}^{-\varrho(\mu+\nu)} + \mathrm{e}^{-\varrho(\mu-\nu)} \right.$$

$$\left. + \mathrm{e}^{-\varrho\mu}\right] \mu\,\mathrm{d}\mu\,\mathrm{d}\nu$$

$$= -C \cdot R^2 \int\limits_{\mu=1}^{\infty} \mu \mathrm{e}^{-\varrho\mu}\,\mathrm{d}\mu$$

$$\times \int\limits_{\nu=-1}^{+1} \left(\mathrm{e}^{-\varrho\nu} + \mathrm{e}^{+\varrho\nu} + 1\right) \mathrm{d}\nu$$

$$= -2CR^2 \left[1 + \frac{1}{\varrho}\left(\mathrm{e}^{\varrho} - \mathrm{e}^{-\varrho}\right)\right]$$

$$\times \int\limits_{1}^{\infty} \mu \cdot \mathrm{e}^{-\varrho\mu}\,\mathrm{d}\mu$$

$$= -2Ca_0^2 \left[1 + \frac{1}{\varrho}\left(\mathrm{e}^{\varrho} - \mathrm{e}^{-\varrho}\right)\right]$$

$$\times (\varrho + 1)\mathrm{e}^{-\varrho}$$

$$= -2Ca_0^2 \left[\left(1 + \frac{1}{\varrho}\right)\left(1 - \mathrm{e}^{-2\varrho}\right) \right.$$

$$\left. + (1 + \varrho)\mathrm{e}^{-\varrho}\right].$$

For $\varrho = 2 \Rightarrow S_{AB} = 0.586 \Rightarrow$

$$2Ca_0^2 = \frac{e^2}{4\pi\varepsilon_0 a_0 \cdot 1.586}$$

$$\Rightarrow E_{pot} = -\frac{27.2\,\mathrm{eV}}{1.586} \cdot 1.879 = -32.2\,\mathrm{eV}.$$

The kinetic energy of the electron can be obtained from the energy relation

$$E_{kin}(\mathrm{e}^-) + E_{pot}(\mathrm{e}^-) + E_{pot}(\text{protons})$$
$$= E(\mathrm{H}) + E_B,$$

where $E_B$ is the binding energy of $H_2^+$ and $E(\mathrm{H})$ is the energy in the groundstate of the H-atom.

$$\Rightarrow E_{kin}(\mathrm{e}^-) = -13.6\,\mathrm{eV} - 2.65\,\mathrm{eV} + 32.2\,\mathrm{eV}$$
$$- 13.6\,\mathrm{eV}$$
$$= 2.35\,\mathrm{eV}$$

This approximate method gives a too small value. The real value is $E_{kin} = 12\,\mathrm{eV}$, compared to $E_{kin}(\mathrm{H}) = 13.6\,\mathrm{eV}$.

2. For $R \to 0$ the $H_2$-molecules converge towards the He-atom $^2_2\mathrm{He}$. Therefore the energy of the electrons has to converge towards the groundstate energy of the He-atoms. (Note that the two missing neutrons do not affect this energy.) This is, according to Sect. 6.1:

$$E(\mathrm{He}) = -78.9\,\mathrm{eV}.$$

This value can be composed of the energy

$$E_1 = -2 \times 4 \times 13.6\,\mathrm{eV} = -108.8\,\mathrm{eV}$$

without repulsion between the two electrons, and the Coulomb energy

$$E_2 = +29.9\,\mathrm{eV}$$

of this repulsion.
The energy of the electron is:

$$E^{el}(H_2, R = R_e) = 2 \cdot E(\mathrm{H}) + E_B(H_2) - \frac{e^2}{4\pi\varepsilon_0 R_e},$$

where $E(\mathrm{H}) = -13.6\,\mathrm{eV}$ is the electronic energy of the H-atom, $E_B(H_2) = -4.7\,\mathrm{eV}$ is the binding energy of the $H_2$-molecule and $e^2/(4\pi\varepsilon_0 R_e) = 0.3 \times 10^{-17}\,\mathrm{J} = 19.4\,\mathrm{eV}$ is the nuclear Coulomb repulsion at a distance $R_e = 0.074\,\mathrm{nm}$ between the two protons. The result is:

$$E^{el}(H_2, R = R_e) = -51.1\,\mathrm{eV}.$$

This energy is composed of negative potential energy, due to electron–proton attraction, positive potential energy of electron–electron repulsion and positive kinetic energy of the two electrons.

3. a) The total energy of the rigid $H_2$-molecule (including Coulomb repulsion between the two nuclei) is, according to (9.2)

$$\begin{aligned} E(H_2) &= 2E(H) + E_B(H_2) \\ &= -27.2\,\text{eV} - 4.7\,\text{eV} \\ &= E^{el}(H_2, R_e) + \frac{e^2}{4\pi\varepsilon_0 R_e} \\ \Rightarrow\ E^{el}(H_2, R_e) &= -19.4\,\text{eV} - 31.9\,\text{eV} \\ &= -51.3\,\text{eV}\,. \end{aligned}$$

In order to separate the $H_2$-molecule into 2 electrons and 2 protons one has to put the energy $E = E^{el}(H_2, R_e) - e^2/(4\pi\varepsilon_0 R_e) = 31.9\,\text{eV}$ into the molecule.

b) At a temperature of 300 K the mean vibrational energy is: $E_{vib} = kT$ (potential + kinetic energy of the vibrating nuclei). The rotational energy (2 degrees of freedom for rotations around two possible axis perpendicular to the internuclear axis) is

$$\begin{aligned} E_{rot} &= 2 \cdot \frac{1}{2} kT = kT \\ &\Rightarrow\ E_{vib} + E_{rot} = 2kT\,. \end{aligned}$$

The relation between $kT$ and $E$ is:

$$1\,\text{eV} \,\hat{=}\, kT \quad \text{for} \quad T = 11{,}604\,\text{K}\,.$$

$\Rightarrow$ at $T = 300\,\text{K}$:

$$2kT \,\hat{=}\, 2 \cdot \frac{300}{11.604}\,\text{eV} = 52\,\text{meV}\,.$$

This is very small compared to the electronic energy difference $E^{el} \approx 10\,\text{eV}$ between the first excited electronic state and the groundstate of $H_2$.

The correct calculation of $E_{vib}$ and $E_{rot}$ has to take into account the quantization of the energy levels. The population of a level $(v, J)$ is:

$$N(v, J) = \frac{(2J+1)e^{-(E_{rot}+E_{vib})/kT}}{\sum_{v,J=0}^{\infty}(2J+1)e^{-(E_{rot}+E_{vib})/kT}}$$

and the vibrational-rotational energy is then

$$E_{vib,\,rot} = \sum_{v,J=0}^{\infty} N(v, J) \cdot E(v, J)$$

$$\text{with}\quad E_{vib} = \left(n + \frac{1}{2}\right) \cdot \hbar\omega_{vib}\,; \qquad E_{rot} = J(J+1)hc \cdot B_{rot}\,.$$

The results do not differ much from the estimation above.

4. In the product

$\psi(r, R) = \chi(R) \cdot \psi_{el}(r, R)$

of nuclear wave functions $\chi(R)$ and electronic wave function $\psi_{el}(r, R)$ the parameter $R$ is not a variable. The wave function $\psi(r, R)$ is a function of $r$, which can be calculated for any arbitrary but fixed value of $R$.

The Schrödinger equation is:

$$\begin{aligned} &-\frac{\hbar^2}{2M}\sum_{k=1}^{2} \Delta_k(\chi \cdot \psi_{el}) \qquad (1) \\ &\quad -\frac{\hbar^2}{2m_e}\Delta_e(\chi \cdot \psi_{el}) + E_{pot} \cdot \chi \cdot \psi_{el} = E\chi\psi_{el}\,. \end{aligned}$$

Multiplying the equation with $\psi_{el}^*$ and integrating over the electron coordinates, gives with

$$\int \psi_{el}^* \psi_{el}\, d\tau = 1$$

the equation:

$$\begin{aligned} &-\frac{\hbar^2}{2M}\sum_{k=1}^{2}\Delta_k\chi - \left(\int \psi_{el}^* \frac{\hbar^2}{2m_e}\Delta_e\psi_{el}\, d\tau_{el}\right) \cdot \chi \\ &\qquad + \chi \cdot \int \psi_{el}^* E_{pot}\psi_{el}\, d\tau = E \cdot \chi \end{aligned}$$

because the operator $\Delta_k$ acts only upon $\chi$ and $\Delta_e$ only upon $\psi_{el}$.

The time-averaged potential energy of the nuclei, averaged over the motion of the electron is

$$\begin{aligned} \langle E_{pot}(R)\rangle &= \left\langle E_{kin}^{el}\right\rangle + \left\langle E_{pot}^{el}\right\rangle + \left\langle \frac{e^2}{4\pi\varepsilon_0 R}\right\rangle \\ &= -\int \psi_{el}^* \frac{\hbar^2}{2m_e}\Delta_e\psi_e\, d\tau_e \\ &\quad + \int \psi_{el}^*\left(E_{pot}^{el} + \frac{e^2}{4\pi\varepsilon_0 R}\right)\psi_{el}\, d\tau_{el} \\ \Rightarrow\ &-\frac{\hbar^2}{2M}\sum_k \Delta_k\chi + \langle E_{pot}(R)\rangle\,\chi = E\chi\,. \end{aligned}$$

The equation for the electrons in the rigid molecule is obtained from (1) with $\sum_k E_{\text{kin}}^{\text{nucl}} = 0$.

5. The Schrödinger equation (9.80) is for the non-rotating molecule ($J = 0$), $M = M_1 - M_2/(M_1 + M_2)$:

$$\frac{1}{R^2}\frac{\mathrm{d}}{\mathrm{d}R}\left(R^2\frac{\mathrm{d}S}{\mathrm{d}R}\right) + \frac{2M}{\hbar^2}\left[E - E_{\text{pot}}(R)\right]S = 0\,. \tag{2}$$

Introducing the function $\chi(R) = R\cdot S(R)$ (2) becomes:

$$\frac{\mathrm{d}^2\chi}{\mathrm{d}R^2} + \frac{2M}{\hbar^2}\left[E - E_{\text{pot}}(R)\right]\chi = 0\,.$$

With the relative elongation $\varrho = (R - R_e)/R_e$ of the vibrating nuclei and the Morse-potential

$$E_{\text{pot}}(R) = E_D\left(1 - e^{-a\varrho}\right)^2$$

we obtain:

$$\frac{\mathrm{d}^2\chi}{\mathrm{d}R^2} + \frac{2M}{\hbar^2}\left[E - E_D\left(1 - e^{-a\varrho}\right)^2\right]\chi = 0\,. \tag{3}$$

With the trial solution:

$$\chi = z^{A\sqrt{1-\varepsilon}}e^{-z/2}\cdot u\,,$$

$$z = 2A\cdot e^{-a\varrho}\,;\quad \varepsilon = \frac{E}{E_D}\,,$$

$$A^2 = \frac{2E_D\cdot MR_e^2}{\hbar^2 a^2}$$

Equation (3) becomes the Laguerre differential equation:

$$\frac{\mathrm{d}^2u}{\mathrm{d}z^2} + \frac{\mathrm{d}u}{\mathrm{d}z}\left(\frac{2A\sqrt{1-\varepsilon}+1}{z} - 1\right) + u\cdot\frac{A - \frac{1}{2} - A\sqrt{1-\varepsilon}}{z} = 0\,.$$

This equation has the eigenvalues (see text books on differential equations)

$$\varepsilon = 1 - \left(1 - \frac{v + 1/2}{A}\right)^2 = \frac{2}{A}\left(v + \frac{1}{2}\right) - \frac{1}{A^2}\left(v + \frac{1}{2}\right)^2\,,$$

where $v = 0, 1, 2$ is the vibrational quantum number. With the vibrational frequency

$$\omega_e = \frac{2E_D}{hc\cdot A}$$

of molecular vibrations around the equilibrium distance $R_e$ we obtain:

$$E(v) = h\cdot c\cdot\omega_e\left(v + \frac{1}{2}\right) - \frac{h^2c^2\omega_e^2}{4E_D}\left(v + \frac{1}{2}\right)^2\,.$$

For the harmonic oscillator the second term is zero. This quadratic term has the consequence, that the vibrational energy levels are no longer equidistant, but the distance $\Delta E = E(v+1) - E(v)$ decreases with increasing $v$.

For

$$hc\omega_e\left(v + \frac{1}{2}\right) = 2E_D$$

the dissociation limit is reached. The maximum possible quantum number $v$ is then:

$$v_{\max} = \frac{2E_D}{hc\omega_e} - \frac{1}{2}\,.$$

6. The ionization energy of $H_2$ is (Fig. S.12):

$$\begin{aligned} E^{\text{ion}}(H_2) &= E_B(H_2) + E^{\text{ion}}(H) - E_B\left(H_2^+\right) \\ &= (4.48 + 13.6 - 2.65)\,\text{eV} \\ &= 15.43\,\text{eV}\,. \end{aligned} \tag{4}$$

The binding energies $E_B(H_2)$ and $E_B\left(H_2^+\right)$ are defined as the energies from the minimum of the potential curve to the dissociation limit. The measured values of the dissociation energy refers to the lowest vibrational level. In this case the energy relation is:

$$\begin{aligned} E^{\text{ion}}(H_2, v = 0) &= E_D(H_2, v = 0) + E^{\text{ion}}(H) \\ &\quad - E_D\left(H_2^+, v = 0\right)\,, \end{aligned} \tag{5}$$

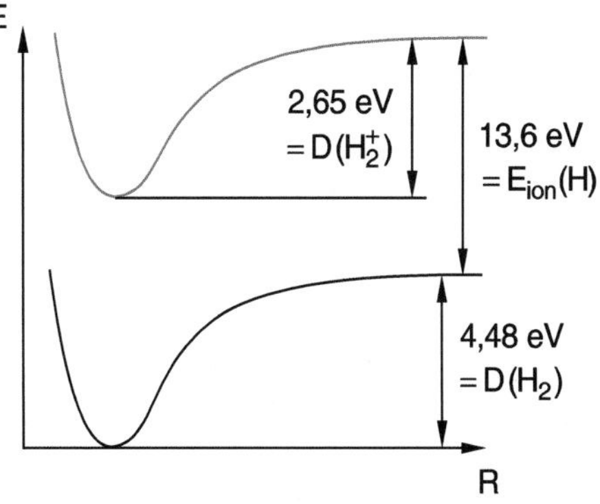


**Fig. S.12.**

which differs from (4) by the difference of the zero-point energies

$$\Delta E_{ZP} = E_{vib}(H_2, v=0) - E_{vib}\left(H_2^+, v^+=0\right) .$$

7. With $R_e = 1.2745\,\text{Å} \Rightarrow$

$$B_e = \frac{\hbar}{4\pi c\mu R_e^2} ,$$

$$\mu = \frac{m_1 \cdot m_2}{m_1 + m_2}$$

$$= \frac{1\cdot 35}{36} = 0.9722\,\text{AMU} \quad \text{for } H^{35}Cl$$

$$= \frac{1\cdot 37}{38} = 0.9737\,\text{AMU} \quad \text{for } H^{37}Cl$$

$$\Rightarrow B_e\left(H^{35}Cl\right) = 10.68\,\text{cm}^{-1} ,$$

$$\Rightarrow B_e\left(H^{37}Cl\right) = 10.659\,\text{cm}^{-1} .$$

For $H^{35}Cl$ we obtain:

$$\nu_{rot}(J=0 \to J=1) = 2cB_e$$
$$= 6\times 10^{10}\cdot 10.68\,\text{s}^{-1} = 64.1\times 10^{10}\,\text{s}^{-1}$$
$$= 641\,\text{GHz} ,$$
$$\nu_{rot}(J=4 \to J=5) = 10cB_e$$
$$= 3204\,\text{GHz} = 3.204\,\text{THz} .$$

For $H^{37}Cl$ we obtain:

$$\nu_{rot}(J=0 \to J=1) = 6\times 10^{10}\cdot 10.66\,\text{s}^{-1}$$
$$= 639.6\,\text{GHz} \Rightarrow \lambda = c/\nu = 4.7\,\text{cm} ,$$
$$\Delta\nu_{rot}\left(H^{35}Cl - H^{37}Cl\right) = 1.4\,\text{GHz}$$
$$\nu_{rot}(J=4 \to J=5) = 3.198\,\text{THz}$$
$$\Rightarrow \lambda = 0.94\,\text{cm} .$$

The rotational energy $E_{rot}(J=5)$ is:

$$E_{rot}/hc = J\cdot(J+1)\cdot B_e = 30\,B_e$$
$$= 320.4\,\text{cm}^{-1} \Rightarrow E_{rot} = 39.7\,\text{meV} \quad \text{for } H^{35}Cl$$
$$= 319.8\,\text{cm}^{-1} \Rightarrow E_{rot} = 39.6\,\text{meV} \quad \text{for } H^{37}Cl$$

8. The vibrational frequency for the model of a harmonic oscillator is:

$$\nu = \frac{1}{2\pi}\sqrt{k/\mu} \Rightarrow k = 4\pi^2\nu^2\cdot\mu .$$

For $H^{35}Cl$ is $\mu = 0.9722\,\text{AMU}$

$$\Rightarrow k = 4\pi^2\cdot 9\times 10^{13}\cdot 0.9722\cdot 1.66\times 10^{-27}\,\text{kg/s}^2$$
$$= 513\,\text{kg/s}^2 .$$

With $E_{pot} = k\cdot(R-R_e)^2$

$$\Rightarrow R - R_e = \sqrt{E_{pot}/k} .$$

The energy of the vibrational level $v = 1$ is

$$E_{vib} = \frac{3}{2}h\cdot\nu .$$

At the turning point is $E_{kin} = 0 \Rightarrow E_{vib} = E_{pot} = \frac{3}{2}h\nu$

$$\Rightarrow k\cdot(R-R_e)^2 = \frac{3}{2}h\cdot\nu$$

$$\Rightarrow R - R_e = \left(\frac{3}{2}\cdot 6.6\times 10^{-34}\cdot 9\times 10^{13}\Big/513\right)^{1/2}$$
$$= 1.32\times 10^{-11}\,\text{m} = 0.132\,\text{Å} .$$

The vibrational amplitude in the level $v = 1$ is therefore only about 10% of the internuclear distance $R_e$:

$$(R - R_e)/R_e \leqq 0.104 .$$

## Chapter 10

1. The three principal rotational axes are perpendicular to each other and intersect in the center of mass S (Fig. S.13), which divides the heights $h$ of the triangle in the ratio 2 : 1. If the sides of the triangle are denoted as $s$ we obtain the three moments of inertia for rotations around the axis $i$:

$$I_a = I_1 = 2m\left(\frac{h}{3}\right)^2 + m\left(\frac{2h}{3}\right)^2$$
$$= \frac{2}{3}mh^2 = \frac{2}{3}ms^2\cos^2(\alpha/2) ,$$

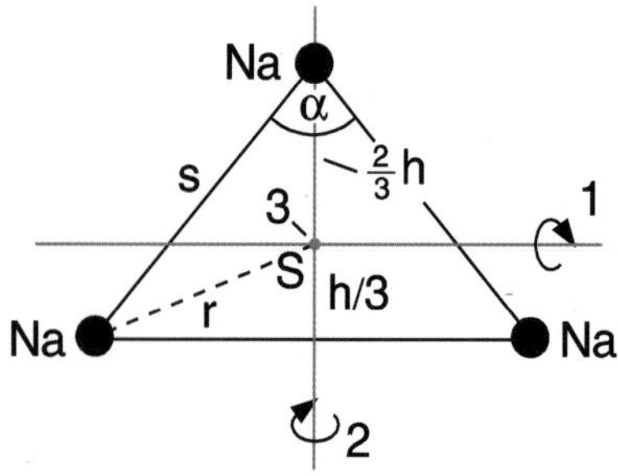


**Fig. S.13.**

where $m$ is the mass of the Na-atom.

$$I_b = I_2 = 2mx^2 = 2ms^2 \sin^2(\alpha/2)$$

$$I_c = I_3 = 2mr^2 + m\left(\frac{2}{3}h\right)^2$$

with $r^2 = x^2 + \left(\frac{h}{3}\right)^2$ we obtain:

$$\begin{aligned} I_3 &= 2mx^2 + \frac{2}{3}mh^2 \\ &= 2ms^2\left(\sin^2(\alpha/2) + \frac{1}{3}\cos^2(\alpha/2)\right) . \end{aligned}$$

For $\alpha = 80°$ this gives:

$$\begin{aligned} &I_1 = 0.39\, m \cdot s^2 ; \quad I_2 = 0.83\, m \cdot s^2 \\ &I_3 = 1.22\, m \cdot s^2 \\ \Rightarrow\ &I_3 = I_1 + I_2 . \end{aligned}$$

The last equation is true for all planar molecules. The $Na_3$-molecule represents an asymmetric rotor with $I_1 \neq I_2 \neq I_3 \neq I_1$.
With $m = 23 \times 1.66 \times 10^{-27}$ kg; $s = 3.24\,\text{Å} = 3.24 \times 10^{-10}$ m

$$\begin{aligned} \Rightarrow\ I_a = I_1 &= 1.56 \times 10^{-45}\,\text{kg}\,\text{m}^2 \\ I_b = I_2 &= 3.32 \times 10^{-45}\,\text{kg}\,\text{m}^2 \\ I_c = I_3 &= 4.85 \times 10^{-45}\,\text{kg}\,\text{m}^2 \end{aligned}$$

The rotational constants are defined as:

$$\begin{aligned} A &= \frac{\hbar}{4\pi c \cdot I_a} = 17.85\,\text{m}^{-1} = 0.1785\,\text{cm}^{-1} \\ B &= \frac{\hbar}{4\pi c \cdot I_b} = 8.388\,\text{m}^{-1} = 0.0839\,\text{cm}^{-1} \\ C &= \frac{\hbar}{4\pi c \cdot I_c} = 5.742\,\text{m}^{-1} = 0.0574\,\text{cm}^{-1} \end{aligned}$$

One can prove, that within the accuracy of rounding up and down the relation

$$\frac{1}{A} + \frac{1}{B} = \frac{1}{C}$$

is valid for the planar $Na_3$-molecule. This is, however, only strictly valid for the non-vibrating molecule.

2. The values of the moments of inertia can be obtained from the rotational constants.

$$\begin{aligned} A &= 800\,\text{m}^{-1} \\ B &= 43.4\,\text{m}^{-1} \\ C &= 41.0\,\text{m}^{-1} \end{aligned}$$

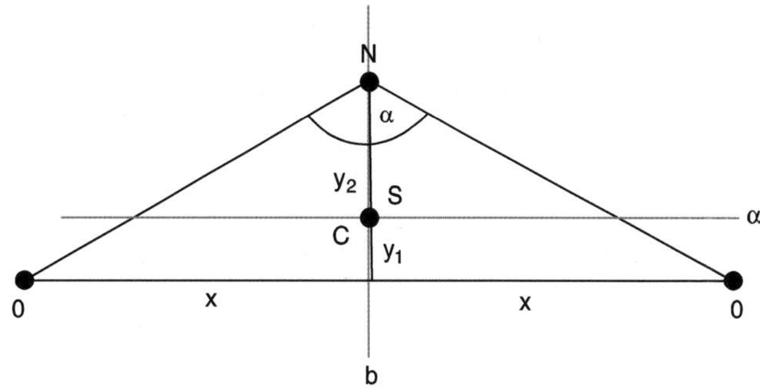


**Fig. S.14.**

$$\begin{aligned} I_a &= \frac{\hbar}{4\pi c \cdot A} = 3.50 \times 10^{-47}\,\text{kg} \cdot \text{m}^2 \\ I_b &= 64.3 \times 10^{-47}\,\text{ks m}^2\ 10^{-47} \\ I_c &= 68.3 \times 10^{-47}\,\text{kg m}^2\ 10^{-47} \end{aligned}$$

From Fig. S.14 we find:

$$\frac{y_1}{y_2} = \frac{m(\text{N})}{2m(\text{O})} = \frac{14}{32} = 0.4375$$

$$\begin{aligned} I_a &= 2m(0)y_1^2 + m(\text{N})y_2^2 \\ &= y_1^2\left[2m(\text{O}) + m(\text{N})\left(\frac{32}{14}\right)^2\right] \\ &= (32 + 73)\,\text{AMU} \cdot y_1^2 \\ &= 105\,\text{AMU}\,y_1^2 = 1.74 \times 10^{-25}\,\text{kg} \cdot y_1^2 \end{aligned}$$

$I_b = 2m(\text{O}) \cdot x^2$ Inserting the numerical value $I_1$

$$\begin{aligned} &= 3.5 \times 10^{-47}\,\text{ks} \cdot \text{m}^2 \text{ yields } y_1 = 1.42 \times 10^{-11}\,\text{m} \\ &\Rightarrow y_2 = y_1/0.438 = 3.246 \times 10^{-11}\,\text{m} \\ &\Rightarrow y = 4.67 \times 10^{-11}\,\text{m} \\ &= 32\,\text{AMU} \cdot x^2 \\ &= 0.53 \times 10^{-25}\,\text{x}^2 \cdot \text{kg} \cdot \text{m}^2 \end{aligned}$$

Inserting the value of $I_b$ gives
$x = 11.10^{-10}$ m
With $\tan(\alpha/2) = x/y \Rightarrow \alpha = 134°$
From $S = \sqrt{x^2 + y^2} \Rightarrow S = 1.19\,\text{Å}$

3. The linear molecule $C_2H_2$ with 4 atoms has $3 \cdot 4 - 5 = 7$ normal vibrations, where the two vibrations $\nu_4$ and $\nu_5$ are degenerate (Fig. S.15). There are therefore 5 different vibrational frequencies $\nu_1, \nu_2, \ldots, \nu_5$. Transitions from the $(0,0,0,0,0)$-vibrational ground state to excited vibrational states are infrared active, if the dipole moment of

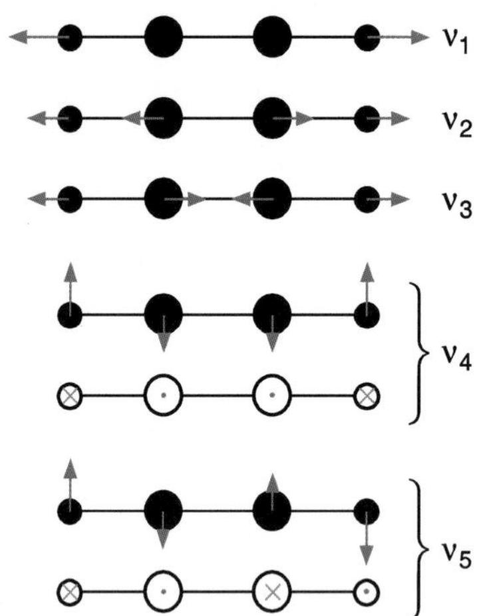


**Fig. S.15.**

the upper state is different from that of the ground state. These are the vibrations $\nu_3$ and $\nu_5$.
The Raman active transitions are: $\nu_1, \nu_2$ and $\nu_4$ (Fig. S.15).

4. In the approximation of harmonic oscillations the frequencies $\nu$ are related to the force constants $k$ by

$$\nu = \frac{1}{2\pi}\sqrt{k/\mu}\,,$$

where $\mu$ is the reduced mass.
For the normal vibration $\nu_1$ the C-atom remains at rest and the two O-atoms oscillate with opposite phases (Fig. S.16). The restoring forces for the two C–O bonds are equal with opposite directions. The reduced mass is

$$\mu = \frac{2m(\mathrm{O}) \cdot m(\mathrm{C})}{2m(\mathrm{O}) + m(\mathrm{C})} = \frac{2 \cdot 16 \cdot 12}{44}\,\mathrm{AMU} = 8.73\,\mathrm{AMU}\,.$$

The restoring force constant $k_1$ is then:

$$\begin{aligned} k_1 &= 4\pi^2\nu_1^2 \cdot \mu = 4\pi^2 c^2 \overline{\nu}_1^2 \mu \\ &= 4\pi^2 \cdot 9 \times 10^{20} \cdot 1388^2 \times 8.73\,\mathrm{AMU} \\ &= 9.96 \times 10^2\,\mathrm{kg/s^2} \approx 1000\,\mathrm{kg\,s^{-2}}\,. \end{aligned}$$

The restoring force $F_r = -k \cdot (R - R_e)$ at the maximum elongation $(R - R_e) = 1.3 \times 10^{-11}\,\mathrm{m}$ is then $F_r = -1.3 \times 10^{-8}\,\mathrm{N}$. For the bending vibration $\nu_2$ the amplitude $\Delta y$ of the C-atom is

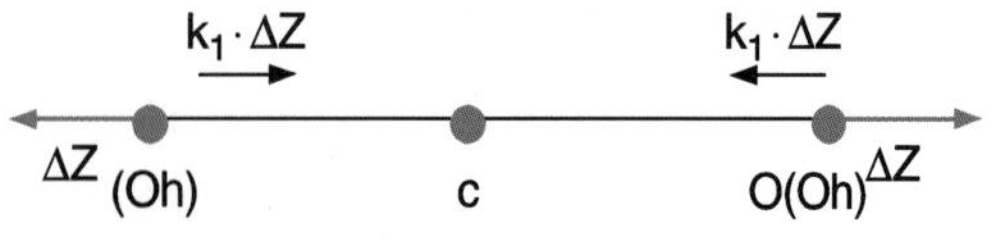

**Fig. S.16.**

$$\Delta y_1(\mathrm{C}) = -\frac{2m(\mathrm{O})}{m(\mathrm{C})} \cdot \Delta y_2(\mathrm{O}) = -\frac{32}{12}\Delta y_2\,.$$

The time dependent distance $d(t)$ between the C-atom and one O-atom of the vibrating molecule is (Fig. S.17):

$$\begin{aligned} d &= \left[d_0^2 + (\Delta y_1 - \Delta y_2)^2\right]^{1/2} \\ &= \left[d_0^2 + (1.375\,\Delta y_1)^2\right]^{1/2} \\ &= d_0\left[1 + 1.9\left(\frac{\Delta y_1}{d_0}\right)^2\right]^{1/2} \\ &\approx d_0\left(1 + 0.95\left(\frac{\Delta y_1}{d_0}\right)^2\right)\,. \end{aligned}$$

The change $(d - d_0)$ during the bending vibration is then

$$d - d_0 = 0.95\Delta y_1^2/d_0\,.$$

The restoring force is

$$F = -k_2(d - d_0) = -k_2 \cdot 0.95\Delta y_1^2/d_0\,.$$

The energy is then

$$E_{\mathrm{vib}} = h \cdot \nu = h \cdot \overline{\nu} \cdot c = 2 \cdot \frac{1}{2} \cdot k_2(d - d_0)^2\,.$$

On the other side is

$$k_2 = 4\pi^2 c^2 \overline{\nu}_2^2 \cdot \mu = 231\,\mathrm{kg\,s^{-2}}\,.$$

This gives a change

$$\begin{aligned} d - d_0 &= (h\overline{\nu} \cdot c/k_2)^{1/2} \\ &= 7.6 \times 10^{-10}\,\mathrm{m} = 0.076\,\text{Å}\,. \end{aligned}$$

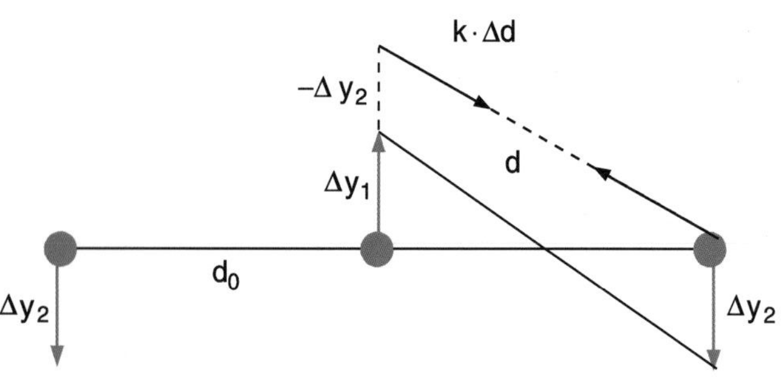


**Fig. S.17.**

This should be compared with the average distance $d_0 \approx 1.2\,\text{Å}$. The vibrational amplitude is therefore only 6% of the C–O-distance.

5. a) The ground state configuration of $BH_2$ is $\ldots(2a_1)^2(1b_2)^2(3a_1)^1 X^2A_1$. According to the Walsh diagram in Fig. 10.11 the energy of the $(2a_1)$-orbital and of the $(1b_2)$-orbital decreases from $\alpha = 90°$ to $\alpha = 180°$, while that of the $(3a_1)$ strongly increases. The total energy has a minimum (from an estimation of the curves in Fig. 10.11) at $\alpha \approx 125°$–$135°$. The correct value is $\alpha = 131°$.
   b) In the first excited state the electron in the $(3a_1)$-orbital is excited into the $(1b_1)$-orbital. The energy of this orbital is independent of $\alpha$ and therefore the excited state leads to a linear configuration ($\alpha = 180°$) which is labelled as $(2\sigma_g)^2(2\sigma_u)^2(1\pi_u)^1 A^2\pi_u$.
6. In the approximation of the Morse-potential the energy levels are

$$
\begin{aligned}
E(v) &= h \cdot c \cdot \left[15.5\omega_e - \omega_e x_e (15.5)^2\right] \\
&= hc\left(2176.2\,\text{cm}^{-1}\right) \\
&= 4.3 \times 10^{-20}\,\text{J} = 0.27\,\text{eV}\,.
\end{aligned}
$$

The energy difference between neighbouring vibrational levels is:

$$
\begin{aligned}
\Delta E &= E(v=15) - E(v=14) \\
&= hc[\omega_e - 30\omega_e x_e] \\
&= hc\left[(159-36)\,\text{cm}^{-1}\right] = hc \cdot \left[123\,\text{cm}^{-1}\right] \\
&= 0.015\,\text{eV}\,.
\end{aligned}
$$

The Fourier-limited frequency width of the 30 fs laser pulse is

$$
\begin{aligned}
\Delta\nu &= \frac{1}{2\pi\,\Delta t} \approx 15.8 \times 10^{12}\,\text{s}^{-1} \\
\Rightarrow \Delta E &= h \cdot \Delta\nu = 11.5 \times 10^{-21}\,\text{J} \\
&= 6.3 \times 10^{-2}\,\text{eV} = 0.063\,\text{eV}\,.
\end{aligned}
$$

Therefore about 4–5 vibrational levels are simultanously excited. For $v' = 15$ the nuclear vibrations can be treated classically.

$$
\begin{aligned}
&\Rightarrow \Delta E = h \cdot \nu_{\text{vib}} = hc \cdot \left[123\,\text{cm}^{-1}\right] \\
&\Rightarrow \nu_{\text{vib}} = 3 \times 10^{10} \cdot 123\,\text{s}^{-1} \\
&\qquad\;\; = 3.7 \times 10^{12}\,\text{s}^{-1} \\
&\Rightarrow T_{\text{vib}} = \frac{1}{\nu_{\text{vib}}} = 2.7 \times 10^{-13}\,\text{s} = 270\,\text{fs}\,.
\end{aligned}
$$

## Chapter 11

1. The net absorption coefficient $\alpha$ is, according to (11.2) given by

$$\alpha = \left[N_k - (g_k/g_i)N_i\right]\sigma_{ki}\,. \qquad (1)$$

In our case is

$$g_k = 2J_k + 1 = 3\,; \quad g_i = 2J_i + 1 = 5\,.$$

The transition wave number is

$$
\begin{aligned}
\overline{\nu} &= B_e[J_i(J_i+1) - J_k(J_k+1)] \\
&= 4B_e = 42.36\,\text{cm}^{-1}\,.
\end{aligned}
$$

The frequency of the absorption line is:

$$\nu = c \cdot \overline{\nu} = 1.27 \times 10^{12}\,\text{s}^{-1}\,.$$

In this case is $h \cdot \nu \ll kT$

$$
\begin{aligned}
&\Rightarrow \frac{N_i}{g_i} = \frac{N_k}{g_k} \cdot \mathrm{e}^{-h\nu/kT} \approx \frac{N_k}{g_k}\left(1 - \frac{h\nu}{kT}\right) \\
&\Rightarrow \alpha \approx N_k \frac{h\nu}{kT}\sigma_{ki}\,. \qquad (2)
\end{aligned}
$$

At a pressure $p = NkT$ the density of molecules is $N = p/kT$. With $p = 1\,\text{mbar} = 10^2\,\text{Pa}$

$$\Rightarrow N = \frac{10^2\,\text{m}^{-3}}{1.38 \times 10^{-23} \cdot T/\text{K}}\,.$$

For $T = 100\,\text{K} \Rightarrow N = 7.25 \times 10^{22}\,\text{m}^{-3}$. The population density of the absorbing level $N_k(J_k)$ is then:

$$N_k(J_k) = \frac{(2J_k+1) \cdot N}{Z} \cdot \mathrm{e}^{-E_{\text{rot}}/kT}\,, \qquad (3)$$

where

$$Z = \sum_n g_n \mathrm{e}^{-E_n/kT}$$

is the partition function, which is a normalization factor, that assures $\sum_n N_n = N$.
For

$$\Delta E_{\text{rot}} = E_{\text{rot}}(J+1) - E_{\text{rot}}(J) \ll kT$$

we can approximate the partition function $Z$ by the integral over a continuous variable $J$:

$$Z \approx \int_0^\infty (2J+1)\mathrm{e}^{-hc\cdot B_\mathrm{e} J(J+1)/kT}\,\mathrm{d}J = \frac{kT}{hcB_\mathrm{e}} .$$

For our example of the HCl-molecule is $J_k = 1$, $B_\mathrm{e} = 10.59\,\mathrm{cm}^{-1}$ and $Z = 6.56 \times 10^{-2} \cdot T/\mathrm{K}$. For $T = 100\,\mathrm{K}$ this becomes $Z = 6.56$ and the ratio

$$\frac{N_k}{N} \approx \frac{3}{6.56} \cdot \mathrm{e}^{-0.3} \approx 0.34 .$$

This means that 34% of all molecules are in the level with $J_k = 1$.
Inserting (3) in (2) gives

$$\alpha = \frac{(2J_k+1)h^2 c\nu B_\mathrm{e}}{(kT)^2}\mathrm{e}^{-E_{\text{rot}}/kT} \cdot \sigma \cdot N .$$

With the numerical values: $N = 7.25 \times 10^{18}\,\mathrm{m}^{-3}$, $B_\mathrm{e} = 1059\,\mathrm{m}^{-1}$, $J_k = 1$, $T = 100\,\mathrm{K}$ one abtains

$$\alpha = 1.9 \times 10^{-3}\,\mathrm{m}^{-1} .$$

For $T = 300\,\mathrm{K}$ the absorption coefficient becomes

$$\alpha = 2.1 \times 10^{-4}\,\mathrm{m}^{-1} .$$

2. The two functions

$$f(t) = \frac{1}{\sqrt{2\pi}} \int_{-\infty}^{+\infty} g(\omega) \cdot \mathrm{e}^{-\mathrm{i}\omega t}\,\mathrm{d}\omega$$

$$g(\omega) = \frac{1}{\sqrt{2\pi}} \int_{-\infty}^{+\infty} f(t) \cdot \mathrm{e}^{+\mathrm{i}\omega t}\,\mathrm{d}t$$

form a Fourier-pair. With the relation

$$\mathrm{e}^{\mathrm{i}\omega t} = \cos\omega t + \mathrm{i}\sin\omega t$$

the real cosinus-Fourier-transform of $g_\mathrm{c}(\omega)$ is

$$f_\mathrm{c}(t) = \sqrt{\frac{2}{\pi}} \int_0^\infty g_\mathrm{c}(\omega)\cos\omega t\,\mathrm{d}\omega$$

$$g_\mathrm{c}(\omega) = \sqrt{\frac{2}{\pi}} \int_0^\infty f_\mathrm{c}(t)\cos\omega t\,\mathrm{d}t .$$

The computer, used to perform these transformations in Fourier-Sepctroscopy subtracts the constant background term in (11.74a), which is independent of the time $t$. Substituting $f_\mathrm{c}(t)$ by $\sqrt{\pi/2}S(t)$ and $g_\mathrm{c}(\omega)$ by $\bar{I}(\omega \cdot v/c)$ one obtains

$$\bar{I}(\omega \cdot v/c) = \int_0^\infty S(t)\cos(\omega \cdot v/c)\,\mathrm{d}t .$$

3. The grating equation is

$$d(\sin\alpha + \sin\beta) = \lambda .$$

For our example is

$$d = \frac{1}{1200}\mathrm{mm} = 0.833\,\mu\mathrm{m},$$

$$\alpha = 30^\circ \Rightarrow \sin\alpha = \frac{1}{2},$$

$$\sin\beta_1 = -\frac{1}{2} + \frac{588.9}{833.3} = +0.2067$$

$$\Rightarrow \beta_1 = 11.93^\circ = 0.2082\,\mathrm{rad},$$

$$\sin\beta_2 = -\frac{1}{2} + \frac{589.5}{833.3} = 0.2074\,\mathrm{rad}$$

$$\Rightarrow \beta_2 = 11.97^\circ = 0.2089\,\mathrm{rad}.$$

The angular difference is

$$\Delta\beta = 0.04^\circ = 7.36 \times 10^{-4}\,\mathrm{rad} .$$

In the focal plane of the imaging mirror with focal length $f = 1\,\mathrm{m}$ is the lateral distance between the two spectral lines:

$$\Delta s = f \cdot \Delta\beta = 1 \cdot 7.36 \times 10^{-4}\,\mathrm{m} = 0.736\,\mathrm{mm}.$$

With a slitwidth $d < 360\,\mu\mathrm{m}$ of the entrance slit the two lines can be separated.

4. a) The shift of the Raman line against the Rayleigh line is for $\Delta v = 1$, $\Delta J = 0$ for the $H_2$-molecule $\Delta\overline{\nu} = 4395\,\mathrm{cm}^{-1}$. The wave number of the argon-ion laser line is

$$\overline{\nu}_{\mathrm{L}} = \frac{10^7}{488} = 20{,}492\,\mathrm{cm}^{-1}$$
$$\Rightarrow\ \overline{\nu}_{\mathrm{R}} = (20{,}492 - 4395)\mathrm{cm}^{-1} = 16{,}097\,\mathrm{cm}^{-1}$$
$$\Rightarrow\ \lambda_{\mathrm{R}} = 490.9\,\mathrm{nm}\,.$$

b) The term value of the rotational level $J = 1$ is with $B_{\mathrm{e}} = 60.8\,\mathrm{cm}^{-1}$

$$T(J = 1) = J \cdot (J + 1)B_{\mathrm{e}} = 121.6\,\mathrm{cm}^{-1}\,.$$

The difference between the Rayleigh and Raman line is

$$\Delta\overline{\nu} = 121.6\,\mathrm{cm}^{-1} \ \Rightarrow\ \Delta\lambda = \lambda^2 \cdot \Delta\overline{\nu} = 2.9\,\mathrm{nm}\,.$$

The spectral resolving power of the spectrometer should be at least

$$R \geq \frac{\lambda}{\Delta\lambda} = \frac{488}{2.9} = 168\,,$$

which can be achieved already with a small grating or prism spectrometer.

5. According to Beer's absorption law the transmitted laser power is

$$P_{\mathrm{t}} = P_0 \cdot \mathrm{e}^{-\alpha x} \approx P_0\,(1 - \alpha x) \quad \text{for} \quad \alpha x \ll 1\,.$$

The power, absorbed per cm path length is

$$\Delta P = \alpha \cdot P_0 = 10^{-6} \cdot 10^{-1}\,\mathrm{W} = 10^{-7}\,\mathrm{W}\,.$$

For a wavelength $\lambda = 500\,\mathrm{nm}$ is

$$h \cdot \nu = 2.48\,\mathrm{eV} = 3.97 \times 10^{-19}\,\mathrm{W\,s}\,.$$

The absorbed power $\Delta P$ corresponds to

$$N = \frac{\Delta P}{h \cdot \nu} = \frac{10^{-7}\,\mathrm{W}}{3.97 \times 10^{-19}\,\mathrm{W\,s}}$$
$$= 2.5 \times 10^{11}\ \text{absorbed photons/s}$$

which generate $2.5 \times 10^{11}$ fluorescence photons. The photo detector receives

$$\frac{\Delta\Omega}{4\pi} \cdot N = \frac{0.2}{4\pi} \cdot 2.5 \times 10^{11} = 4 \times 10^9\,\text{photons/s}\,.$$

With a quantum efficiency $\eta = 0.2$ of the detector this gives $0.2 \cdot 4 \times 10^9 = 8 \times 10^8$ photoelectrons/s. If the amplification of the photomultiplier is $G = 10^6$ the output current is

$$I_{\mathrm{A}} = 8 \times 10^8 \cdot 1.6 \times 10^{-19} \cdot 10^6 = 0.13\,\mathrm{mA}\,.$$

6. With a collimation angle $\varepsilon = 2°$ the residual Doppler width of an absorption line is

$$\Delta\nu_{\mathrm{D}}(\text{beam}) = \sin\varepsilon \cdot \Delta\nu_{\mathrm{D}}(\text{cell})\,.$$

For a temperature $T$ is

$$\Delta\nu_{\mathrm{D}}(\text{cell}) = 7.16 \times 10^{-7}\nu_0\sqrt{T/M}\,.$$

With $T = 500\,\mathrm{K}$, $M = 23\,\mathrm{g/Mol}$, $\nu_0 = 5.09 \times 10^{14}\,\mathrm{s}^{-1}$

$$\Rightarrow\ \Delta\nu_{\mathrm{D}} = 1.7 \times 10^9\,\mathrm{s}^{-1}$$
$$\Rightarrow\ \Delta\nu_{\mathrm{D}}(\text{beam}) = (\sin 2°) \cdot 1.7 \times 10^9\,\mathrm{s}^{-1}$$
$$= 5.9 \times 10^7\,\mathrm{s}^{-1}\,.$$

a) In order to resolve the hyperfine structure of the $3\,^2P_{1/2}$ level, the condition

$$\Delta\nu_{\mathrm{D}}(\text{beam}) = \left(1.7 \times 10^9 \cdot \sin\varepsilon\right)\mathrm{s}^{-1} < 190\,\mathrm{MHz}$$

must be fullfilled. This gives

$$\sin\varepsilon < 0.11 \Rightarrow \varepsilon < 6.4°\,.$$

b)

$$\sin\varepsilon < \frac{16}{1700} = 9.4 \times 10^{-3} \Rightarrow \varepsilon = 0.55°\,.$$

Here one has to take into account, that the natural line width is already $\Delta\nu_{\mathrm{n}} = 10\,\mathrm{MHz}$. The absorption profile is the convolution of Lorentzian and Gaussian profiles and the total linewidth is

$$\Delta\nu \approx \sqrt{\Delta\nu_{\mathrm{n}}^2 + (\Delta\nu_{\mathrm{D}}\sin\varepsilon)^2}$$
$$\Rightarrow\ \Delta\nu_{\mathrm{D}} \cdot \sin\varepsilon \leq \left((\Delta\nu)^2 - \Delta\nu_{\mathrm{n}}^2\right)^{1/2}\,\mathrm{s}^{-1}$$
$$= 10^6 \cdot \sqrt{(16)^2 + (10)^2}\,\mathrm{s}^{-1}$$
$$= 10^6 \cdot \sqrt{156}\,\mathrm{s}^{-1}$$
$$\leq 12.5\,\mathrm{MHz}$$
$$\Rightarrow\ \sin\varepsilon \leq \frac{12.5 \times 10^6}{1700 \times 10^6} = 0.0074$$
$$\Rightarrow\ \varepsilon \leq 0.43°\,.$$

7. a) The transverse force $F_x$ acting on the atoms flying into the $z$-direction is

$$F_x = -|p_m \cdot \mathbf{grad}\,B|\,.$$

The magnetic moment in the $^2S_{1/2}$ state is mainly due to the electron spin, i.e. $p_m = \mu_{\mathrm{B}} = 9.27 \times 10^{-24}\,\mathrm{J/Tesla}$. The deflection angle $\alpha$ of the sodium atoms is

$$\tan\alpha = \frac{v_x}{v_z}\,.$$

The velocity $v_x$ is after a flight time $t = L/v_z$

$$v_x = \mu_B \cdot |\,\mathbf{grad}\;\boldsymbol{B}| \cdot \frac{t}{m} = \frac{\mu_B}{m}|\,\mathbf{grad}\;\boldsymbol{B}| \cdot \frac{L}{v_z}$$

For $\alpha = 3° \Rightarrow \tan\alpha = 0.052$. With $L = 0.2\,\text{m}$, $v_z = 600\,\text{m/s} \Rightarrow v_x = 0.052 \cdot 600\,\text{m/s}$

$$\Rightarrow \mathbf{grad}\;B = \frac{m \cdot v_x \cdot v_z}{\mu_B \cdot L} = \frac{m \cdot v_z^2 \tan\alpha}{\mu_B \cdot L}$$
$$= \frac{23 \cdot 1.66 \times 10^{-27} \cdot 36 \times 10^4 \cdot 0.052}{9.3 \times 10^{-24} \cdot 0.2}$$
$$= 3.8 \times 10^2\,\text{T/m}\,.$$

b) The photon transfers the momentum

$$\Delta p = h \cdot \nu / c\,.$$

The momentum of the sodium atoms in $z$-direction is

$$p_z = m \cdot v_z\,.$$

The deflection angle for the absorption of one photon is

$$\tan\alpha = \frac{\Delta p}{p_z} = \frac{h \cdot \nu}{c \cdot m \cdot v_z}$$
$$\text{with} \quad h \cdot \nu = 2.1\,\text{eV}\,.$$

For $n$ absorbed photons it is:

$$\tan\alpha = n \cdot \frac{h \cdot \nu}{c \cdot m \cdot v_z}\,.$$

In order to reach a deflection of $\alpha = 3°$ one needs

$$n = \frac{c \cdot m \cdot v_z}{h \cdot \nu} \cdot \tan\alpha$$
$$= \frac{3 \times 10^8 \cdot 23 \cdot 1.66 \times 10^{-24} \cdot 600}{2.1 \cdot 1.6 \times 10^{-19}}$$
$$= 1 \times 10^3\ \text{photons}\,,$$

With a laser beam diameter $d = 1\,\text{cm}$ the time of flight of the atoms through the laser beam is

$$t = d/v_z = 1.6 \times 10^{-5}\,\text{s}\,.$$

The minimum absorption rate is then

$$R_{\text{abs}} = n/t = 10^3/\left(1.6 \times 10^{-5}\right) = 6.3 \times 10^7\,\text{s}^{-1}\,,$$

which implies that the mean cycle time absorption-fluorescence should be $\tau_c = 1/R_{\text{abs}} = 16\,\text{ns}$. The spontaneous life time is $\tau_{\text{sp}} = 16\,\text{ns}$. Since the minimum mean cycle time is $\tau_c \approx 2\,\tau_{\text{sp}}$ the maximum rate can be only $R_{\max} = 3.15 \times 10^7\,\text{s}^{-1}$. The laser beam diameter therefore has to be enlarged to $d = 2\,\text{cm}$.

8. The effective life time is

$$\frac{1}{\tau_{\text{eff}}} = \frac{1}{\tau_{\text{spont}}} + n \cdot \sigma \cdot \overline{v}_r\,.$$

If $\tau_{\text{eff}} = \frac{1}{2}\tau_{\text{spont}}$ we obtain

$$n \cdot \sigma \overline{v}_r = \frac{1}{\tau_{\text{spont}}}\,.$$

With $p = n \cdot k \cdot T \Rightarrow$

$$\frac{p \cdot \sigma \cdot \overline{v}_r}{k \cdot T} = \frac{1}{\tau_{\text{spont}}} \Rightarrow p = \frac{k \cdot T}{\sigma \cdot \overline{v}_r \cdot \tau_{\text{spont}}}\,.$$

The mean relative velocity $\overline{v}_r$ is

$$\overline{v}_r = \sqrt{\frac{2 \cdot 8kT}{\pi \cdot m}}$$
$$\text{with } m = \frac{m_1 \cdot m_2}{m_1 + m_2} = \frac{23 \cdot 28}{51} = 12.6\,\text{AMU}$$
$$\Rightarrow p = \frac{\sqrt{\pi \cdot m \cdot kT}}{4\sigma \cdot \tau_{\text{spont}}}$$
$$= \frac{\sqrt{\pi \cdot 12.6 \cdot 1.66 \times 10^{-24} \cdot 1.38 \times 10^{-23} \cdot 400}}{4 \cdot 10^{-19} \cdot 1.6 \times 10^{-8}}\,\text{Pa}$$
$$= 0.93 \times 10^5\,\text{Pa} = 0.93\,\text{bar}\,.$$

9. Assume the level $E_i$ is selectively excited by a cw laser with an excitation rate $R_{\text{exc}}$. If the total deactivation rate is

$$D = N_i\left(A_i + n \cdot \sigma_q \cdot \overline{v}\right)\,,$$

where $n$ is the density of collision partners. We obtain $D = R$ under stationary conditions. This yields the stationary population density

$$N_i = \frac{D}{A_i + n \cdot \sigma_q \cdot \overline{v}} = \frac{R}{A_i + n\sigma_q\overline{v}}\,,$$

where $\sigma_q$ is the total deactivation cross section for nonradiative transitions. The fluorescence power, emitted by level $E_i$ is

$$P_i = N_i \cdot A_i\,.$$

If level $E_m$ is populated by collisional energy transfer from level $E_i$ we obtain:

$$\frac{\mathrm{d}N_m}{\mathrm{d}t} = n_i \cdot n \cdot \sigma_{i\to m} \cdot \overline{v} - N_m \cdot A_m = 0$$
$$\Rightarrow \sigma_{i\to m} = \frac{N_m}{N_i} \cdot \frac{A_m}{n \cdot \overline{v}} = \frac{P_m}{P_i} \cdot \frac{A_i}{n \cdot \overline{v}} .$$

Measuring the relative total fluorescence powers $P_m/P_i$ yields $\sigma_{i\to m}$, if $A_i = 1/\tau_i$ is known from lifetime measurements of level $E_i$.

## Chapter 12

1. The lifetime of the $3\,^2P_{3/2}$ level is $\tau = 16\,\mathrm{ns}$, the optimum absorption-emission cycle period is $2\tau = 32\,\mathrm{ns}$, the absorption rate $R = 1/(2\tau) = 10^9/64 \approx 1.6 \times 10^7\,\mathrm{s}^{-1}$. Each absorbed photon transfers the momentum $\Delta p = h \cdot \nu/c$ onto the Na-atom and decreases its velocity by $\Delta v = \Delta p/m = h \cdot \nu/(m \cdot c) \approx 3\,\mathrm{cm/s}$.
The velocity decreases per s is then $a = 3 \cdot 1.6 \times 10^7 = 4.8 \times 10^7\,\mathrm{cm/s^2} = 4.8 \times 10^5\,\mathrm{m/s^2}$. In order to bring the Na atom to rest, $N = 700/0.03 = 23{,}333$ absorptions are necessary. This takes a time

$$T = \frac{23{,}333}{1.6 \times 10^7} \approx 1.5 \times 10^{-3}\,\mathrm{s} = 1.5\,\mathrm{ms} .$$

The deceleration path length is

$$\begin{aligned}\Delta z &= v_0 T - \frac{1}{2} a T^2 \\ &= 1.5 \times 10^{-3} \\ &\quad \times \left(700 - \frac{1}{2} \cdot 4.8 \times 10^5 \cdot 1.5 \times 10^{-3}\right)\,\mathrm{m} \\ &= 1.5 \times 10^{-3}(700 - 360)\,\mathrm{m} \\ &= 0.51\,\mathrm{m} .\end{aligned}$$

2. The net force on the atoms is

$$\boldsymbol{F} = -a \cdot \boldsymbol{v} \quad \text{with}$$
$$a = \frac{16\delta\hbar k^2}{\gamma^2\left[1 + (2\delta/\gamma)^2\right]^2} .$$

With $\delta = 2\pi \times 10^7\,\mathrm{s}^{-1}$, $k = \frac{2\pi}{\lambda} = 0.94 \times 10^7\,\mathrm{m}^{-1}$, $\gamma = \frac{1}{\tau} \approx 6 \times 10^7\,\mathrm{s}^{-1}$, $R_0 = 10^6$

$$\begin{aligned}\Rightarrow a &= \frac{16 \cdot 2\pi \cdot 10^7 \cdot 1.06 \times 10^{-34} \cdot 0.94^2 \cdot 10^{14} \cdot 10^6}{4\pi^2 \cdot 10^{14}[1 + (4\pi/6)^2]^2} \\ &= \frac{15 \times 10^{-7}}{182 \times 10^{14}} = 8.2 \times 10^{-23}\,\mathrm{N\,s/m} .\end{aligned}$$
$$\Rightarrow a/M = \frac{8.2 \times 10^{-23}}{23 \times 1.66 \times 10^{-24}} = 2.1 \times 10^3\,\mathrm{s}^{-1}$$
$$\Rightarrow \tau_{\mathrm{damp}} = \frac{M}{a} = 0.48 \times 10^{-3}\,\mathrm{s} = 480\,\mu\mathrm{s} .$$

3. The restoring force in $z$-direction is

$$F_z = -D \cdot z ,$$

where the force constant is

$$D = R_0 \cdot p_m \cdot b \cdot \frac{16k\delta}{\gamma^2\left(1 + 4\delta^2/\gamma^2\right)^2} .$$

With

$$\begin{aligned} k &= \frac{2\pi}{\lambda} = 0.94 \times 10^7\,\mathrm{m}^{-1} , \\ \gamma &= \frac{1}{\tau} = 6 \times 10^7\,\mathrm{s}^{-1} , \\ R_0 &= 10^6\,\mathrm{s}^{-1} \quad \text{and} \quad \delta = 2\pi \cdot 10^7\,\mathrm{s}^{-1} , \\ p_m &\approx \mu_\mathrm{B} = 9.28 \times 10^{-24}\,\mathrm{J/T} , \\ b &= 10^{-2}\,\mathrm{T/m}\end{aligned}$$

we obtain

$$\begin{aligned} D &= 10^6 \cdot 9.28 \times 10^{-24} \cdot 10^{-2} \\ &\quad \times \frac{16 \cdot 0.94 \times 10^7 \cdot 2\pi \cdot 10^7}{36 \times 10^{14}\left(1 + 16\pi^2/36\right)^2}\,\mathrm{N/m} \\ &= 3.7 \times 10^{-2}\,\mathrm{N/m} .\end{aligned}$$

The oscillation frequency is:

$$\begin{aligned}\Omega &= \sqrt{D/M} \\ &= \sqrt{3.7 \times 10^{-2}/(23 \cdot 1.66 \times 10^{-24})\,\mathrm{s}^{-1}} \\ &= 3 \times 10^{10}\,\mathrm{s}^{-1} .\end{aligned}$$

The damping constant is

$$\beta = \frac{a}{M} = 2.1 \times 10^3\,\mathrm{s}^{-1}$$
$$\Rightarrow \tau_{\mathrm{damp}} = 480\,\mu\mathrm{s} .$$

The atoms perform $3 \times 10^{10} \cdot 4.8 \times 10^{-4} = 1.4 \times 10^7$ oscillations, before the oscillation amplitude has decreased to $1/e$ of its initial value.

4. The mean distance $\overline{d}$ is related to the density $N$ by

$$\overline{d} = N^{-1/3} \Rightarrow \overline{d} = \sqrt[3]{10^{-13}}\,\text{cm} = 0.46 \times 10^{-4}\,\text{cm} = 4.6 \times 10^{-7}\,\text{m}\,.$$

The de Broglie wavelength is

$$\lambda_{\text{dB}} = \frac{h}{\sqrt{3mk_\text{B}T_\text{c}}} = \overline{d} = 4.6 \times 10^{-7}\,\text{m}$$
$$\Rightarrow T_\text{c} = \frac{h}{3mk_\text{B}\overline{d}^2} = \frac{6.6 \times 10^{-34}}{3\cdot 23 \times 1.66 \times 10^{-27} \cdot 1.38 \times 10^{-23} \cdot 6.7^2 \times 10^{-14}} = 9.3 \times 10^{-7}\,\text{K} = 930\,\text{nK}\,.$$

BEC occurs under these conditions at $T_\text{c} = 930\,\text{nK}$. The atoms in the trap can reach a region, where the magnetic field energy $E_\text{M} = -\boldsymbol{\mu}_\text{m} \cdot \boldsymbol{B}$ equals their kinetic energy $E_\text{kin} = \frac{1}{2}m\overline{v}^2$ with $\overline{v}^2 = \frac{3kT}{m}$.

$$\Rightarrow E_\text{kin} = \frac{3}{2}kT = \frac{3}{2} \cdot 1.38 \times 10^{-23} \cdot 9.3 \times 10^{-7}\,\text{J}\,.$$

The magnetic field is

$$B = b \cdot r \Rightarrow r = \frac{B}{b} = \frac{E_\text{m}}{\mu_\text{m} \cdot b} = \frac{\frac{3}{2}kT}{\mu_\text{m} \cdot b}\,.$$

With $\mu_\text{m}\left(\text{Na}\,^2S_{1/2}\right) \approx \mu_\text{B} = 9.27 \times 10^{-24}\,\text{J}\,\text{T}^{-1}$

$$\Rightarrow r = \frac{1.5 \times 1.38 \times 10^{-23} \cdot 9.3 \times 10^{-7}}{9.27 \times 10^{-24} \cdot 10^{-3}}\,\text{m} = 2.08 \times 10^{-3}\,\text{m} = 2.08\,\text{mm}\,.$$

With a density of $10^{13}\,\text{cm}^{-3}$ there are $N = 10^{13} \cdot \frac{4}{3}\pi r^3 = 3700$ atoms in the trap.

5. The energy of a vibrational level with quantum number $v$ is:

$$E(v) = hc\left[\omega_\text{e}\left(v + \frac{1}{2}\right) - \omega_\text{e}x_\text{e}\left(v + \frac{1}{2}\right)^2\right]\,.$$

The energy width of the femtosecond pulse with Gaussian time profile, for which $\Delta\nu \cdot \Delta T = 0.44$ is

$$\Delta E = h \cdot \Delta\nu = 0.44h/\Delta T\,.$$

With $\Delta T = 2 \times 10^{-14}\,\text{s} \Rightarrow \Delta E = 1.5 \times 10^{-20}\,\text{J} = 0.09\,\text{eV}$. The energy separation of the vibrational levels is

$$\Delta E = \frac{1}{2}(E(v+1) - E(v-1)) = \frac{1}{2}hc\,[2\omega_\text{e} - (4v+2)\omega_\text{e}x_\text{e}]\,.$$

For $v = 20$ this gives with $\omega_\text{e} = 30\,\text{cm}^{-1}$, $\omega_\text{e}x_\text{e} = 0.04\,\text{cm}^{-1}$

$$\Delta E = \frac{1}{2}hc\,[2\omega_\text{e} - 82\omega_\text{e}x_\text{e}] = 5.61 \times 10^{-22}\,\text{J} = 3.5 \times 10^{-3}\,\text{eV}\,.$$

The laser pulse can therefore excite

$$n = \frac{0.09}{0.0035} = 26$$

vibrational levels of the $\text{Cs}_2$-molecule.

6. The intensity at the output of the Michelson interferometer is:

$$I_\text{t} = \frac{I_0}{2} \cdot \cos^2(\Delta\varphi/2)\,,$$

where $\Delta\varphi = 2\pi \cdot \Delta s/\lambda$ is the phase difference between the two interfering beams. If $\Delta s$ changes by $\delta \Rightarrow \Delta\varphi$ changes by $\delta\varphi = 2\pi\delta/\lambda$. The intensity change

$$I_\text{t}\left(\frac{\Delta\varphi + \delta\varphi}{2}\right) - I_\text{t}\left(\frac{\Delta\varphi}{2}\right) \le 10^{-8}\,I_0$$

should be smaller than $10^{-8}I_0$. With $\Delta\varphi = \pi \Rightarrow$

$$\cos\left(\frac{\Delta\varphi + \delta\varphi}{2}\right) = \cos\frac{\Delta\varphi}{2}\cos\frac{\delta\varphi}{2} - \sin\frac{\Delta\varphi}{2}\sin(\delta\varphi/2) = -\sin(\delta\varphi/2) \approx -\delta\varphi/2$$

$$\Rightarrow I_\text{t} = \frac{I_0}{4} \cdot \delta\varphi \Rightarrow \delta\varphi \le 4 \times 10^{-8}\,\text{rad}\,.$$

The phase of a plane wave is sensitive to the position of the plane mirror surface, averaged over the whole surface. Deviations from an ideal plane, caused by the atomic structure, are averaged out.

# References


### Textbooks on Atomic Physics

- H. Haken, A.C. Wolf: The Physics of Atoms and Quanta, 6th ed. (Springer, Berlin, Heidelberg 2000)
- K. Krane: Modern Physics (John Wiley & Sons, New York 1992)
- C. Food: Atomic Physics (Oxford Univ. Press, Oxford 2004)
- D. Büdker: Atomic Physics: An Explanation through Problems and Solutions (Oxford Univ. Press, Oxford 2004)
- H. Friedrich: Theoretical Atomic Physics (Springer, Berlin, Heidelberg 1998)
- Max Born: Atomic Physics, 8th ed. (Dover Bibl., Mineola, N.Y. 1989)
- R.P. Feynman, R.B. Leighton, M. Sounds: The Feynman Lectures on Physics, Vol. 3 (Reading, Mass. 1965); See also: Springer Handbook of Atomic, Molecular and Optical Physics (Springer, Berlin, Heidelberg 2004)
- B. Bederson, H. Walther (eds.) Advances in Atomic, Molecular and Optical Physics Vol. 1–49 (Academic Press, New York 1965–2005)
- I. Estermann (ed.): Methods of Experimental Physics Academic Press, Reading, Mass. 1959–2002

### Textbooks on Molecular Physics

- G. Herzberg: Molecular Spectra and Molecular Structure Vol. 1–4 (van Nostrand, New York 1964–66)
- J.M. Hollas: Modern Spectroscopy, 2nd. ed. (John Wiley & Sons, Leicester 1992)
- J.D. Graybell: Molecular Spectroscopy (McGraw Hill, New York 1988)
- J.L. McHale: Molecular Spectroscopy (Prentice Hall, Upper Saddle River, N.J. 1999)
- J.M. Brown: Molecular Spectroscopy (Oxford Univ. Press, Oxford 1998)
- H. Haken, H.Ch. Wolf: Molecular Physics and Elements of Quantum Chemistry, 2nd ed. (Springer, Berlin, Heidelberg 2004)
- P.W. Atkins: Molecular Quantum Mechanics, 4th ed. (Oxford Univ. Press 2005)

### Chapter 1

1.1. R.P. Wayne: Chemistry of Atmospheres (Clarendon Press, Oxford 1991)

1.2. J.V. Iribarne, H.R. Cho: Atmospheric Physics (Kluwer, Dordrecht 1980)

1.3. J.M. Wallace, P.V. Hobbs: Atmospheric Science Vol. 96 (Academic Press, New York 2006) J. Frederic: Principles of Atmospheric Science (Jones & Bartlett Publ. Sudbury MA, USA 2008)

1.4. D.G. Andrews: An Introduction to Atmospheric Physics (Cambridge Univ. Press, Cambridge 2000)

1.5. T.E. Graedel, Paul J. Crutzen: Atmospheric Change (W.H. Freeman, San Francisco 1993); G. Guyot: Physics of the Environment and Climates (John Wiley & Sons, New York 1998); J.H. Seinfeld, Sp.N. Pandu: Atmospheric Chemistry and Physics (Wiley Interscience, New York 1987)

1.6. G. Seisenberger et al.: Real-Time Single Molecule Imaging of the Infection Pathway of an Adeno-associated Virus. Science **294**, 1929 (2001)

1.7. http://www.google.de/search?hl=de&q=Single+Molecule+Detection&btnG=Google-Suche&meta=&aq=f&oq=

1.8. N. Marme et al. Identification of single point mutations in mycobacterial 16S rRNA sequences by confocal single molecule fluorescence spectroscopy. Nucleic Acids research 2006, Vol. **34**, No13, 90

1.9. A. Unsöld, B. Baschek: The New Cosmos, 5th ed. (Springer 2005)

1.10. St. W. Stahler, F. Palla: The Formation of stars (Willy VCH, Weinheim 2004)

1.11. A. Siegman: Lasers (University Science Books 1986); O. Svelto: Principles of Lasers, 5th ed. (Springer, Berlin, Heidelberg, 2009)

1.12. M. Crawford, N. Holonyak, F.A. Kiss, Jn.: In: Pursuit of the Ultimate Lamp. Scient. Am. Febr. 2001 **284**, 49; E.F. Schubert: Light-Emitting Diodes (Cambridge Univ. Press, Cambridge 2003)

1.13. S. Reineke et al: White Organic light-emittions diodes with fluorescence tube efficiency. Nature *459*, 234 (2009)
1.14. D. Bäuerle: Laser Processing and Chemistry, 3rd ed. (Springer, Berlin, Heidelberg 2000)
1.15. A. Zangwill: Physics at Surfaces (Cambridge Univ. Press, Cambridge 1988)
1.16. S.A. Diddams, T.W. Hänsch et al.: Direct Link between microwave and optical frequencies with a 300 THz femtosecond Pulse, Phys. Rev. Lett. **84**, 5102 (2000); M. Fischer, T.W. Hänsch et al.: New Limits on the Drift of Fundamental Constants from Laboratory Measurements, Phys. Rev. Lett. **92**, 230802 (2004)
1.17. Th. Brody: The Philosophy behind Physics (Springer, Berlin, Heidelberg 1993)
1.18. W. Heisenberg: Physics and Philosophy: (Prometheus Books Amherst, New York 1999); Marc Lange: An Introduction to the Philosophy of Physics (Blackwell Publ., Oxford 2002)
1.19. R.M.J. Cotterill: Biophysics (John Wiley & Sons, New York 2002)
1.20. B. Nolting: Methods in Modern Biophysics (Springer, Berlin, Heidelberg 2003)
1.21. Ph. Nelson: Biological Physics (Freeman, San Francisco 2007)
1.22. Lily E. Kai: Who wrote the book of life? A history of the genetic code (Stanford Univ. Press 2000)

## Chapter 2

2.1. Isaac Asimov: The History of Physics (Walker & Company, New York 1984); B. Pullmann: The Atom in the History of Human Thought (Oxford Univ. Press, Oxford 2002)
2.2. J.D. Bernal: A History of Classical Physics (Barnes & Noble, Basking Ridge, N.J. 1997); J. McDonnell: The Concept of Atoms from Demokritus to John Dalton (Edwin Mellen Press, New York 1992); J. Agassi: The Continious Revolution: A History of Physics from the Greeks to Einstein (McGraw Hill 1968); R.E. Peierls: Atomic History (Springer, Berlin, Heidelberg, 1997)
2.3. R. Purrington: Physics in the Nineteenth Century (Rutgers Univ. Press, Camden, N.J. 1997); C. Cercignani: Boltzmann, the Man who trusted Atoms (Oxford Univ. Press, Oxford 1999)
2.4. H. Krach: Quantum Generations: A History of Physics in the Twentieth Century (Princeton Univ. Press, Princeton 2002)
2.5. E. Segrë: Modern Physicists and their Discoveries, Vol. 1: From Falling Bodies to Radio Waves, Vol. 2: From X-Rays to Quarks (W.H. Freeman, San Francisco 1984)
2.6. J. Mehra, H. Rechenberg: The Historical Development of Quantum Theory (Springer, Berlin, Heidelberg 1982–2000)
2.7. J. Dalton: A New System of Chemical Philosophy (Bickerstaff London 1808) reproduced as facsimile by William Dawson & Sons, London, Science Classics Library (New York 1964) and Cambridge Library Collection, Cambridge 2010
2.8. R.D. Deslattes: The Avogadro-Constant, Ann. Rev. Phys. Chemistry **31**, 435 (1980)
2.9. M.R. Moldower, J.P.M. Trusler, T.J. Edwards, J.B. Mehl, R.S. Davis: Measurement of the Universal Gas Constant using a spherical acoustic resonator, Phys. Rev. Lett. **60**, 249 (1988)
2.10. D. Attwood: Soft X-Rays and Extrem UV-Radiation. Principles and Applications (Cambridge Univ. Press 1999)
2.11. K. Fujii et al: Determination of the Avogadro constant by accurate measurement of the molar volume of a silicon crystal. Metrologia **36**, 455 (1999)
2.12. U. Bonse, M. Hart: An X-Ray-interferometer, Appl. Phys. Lett. **6**, 155 (1965)
http://e1.physik.uni-dortmund.de/xtm/
B. Kramer (ed.): The Art of Measurement (VCH, Weinheim 1988)
2.13. http://e1.physik.uni-dortmund.de/xtm/
2.14. U. Bonse, W. Graef: X-Ray and Neutron Interferometry, in: X-Ray Optics, Topics in Appl. Phys. Vol.**22**, (Springer Berlin, Heidelberg)
2.15. A. Einstein: Investigations on the Theory of Brownian Motion (Dover, New York 1956)
A. Borodin, P. Salmimen: Handbook of Brownian Motion (Birkhäuser, Boston, MA 1996)
2.16. E. Kappler: Die Brown'sche Molekularbewegung, Naturwissenschaften **27**, 649 (1939)
2.17. K.S. Birdi: Scanning Probe Microscope: Applications in Science and Technology (CRC-Press, Boca Raton 2003)
2.18. St. Flegler, J. Heckman, K.L. Klomparens: Scanning and Transmission Electron Microscope (Oxford Univ. Press, Oxford 1995)
2.19. E.W. Müller: Feldemission, Ergebn. exakter Naturwiss. XXVII 290–360 (1953)
2.20. D.B. Williams, C.B. Carter: Transmission Electron Microscopy (Plenum Press, New York 1996)
2.21. R.F. Egerton: Physical Principles of Electron Microscopy: An Introduction to TEM, SEM and AEM (Springer Berlin, Heidelberg 2008)
2.22. D. Chescoe, P.J. Goodhew: The Operation of Transmission and Scanning Electron Microscopy (Oxford Science Publ., Oxford 1990)

2.23. D. Breger: The Art of the Scanning Electron Microscope (Columbia Univ. Press, New York 1995)

2.24. C.B. Gilmore: The Unseen Universe; Photographs from the Scanning Electron Microscope (Schocken Books, New York 1974)

2.25. D.A. Bonnell (ed.): Scanning Tunneling Microscopy and Spectroscopy (VCH, Weinheim 1993)

2.26. J.A. Stroscio, W.J. Kaiser (eds.): Scanning Tunneling Microscopy. In: Methods of Experimental Physics, Vol. 27 (Academic Press, New York 1993)

2.27. C. Bai: Scanning Tunneling Microscopy and its Applications: Springer Series in Surface Science Vol. **32** (2000)

2.28. R. Wiesendanger and H.J. Guntherodt: Theory of STM and related Scanning Probe Methods. Springer Series in Surface Science Vol. **3**, (1998)

2.29. D.M. Eigler, E.K. Schweitzer: Positioning single atoms with a scanning tunneling microscope, Nature **344**, 524 (1990)

2.30. A. Kühnle, G. Meyer, S.W. Hla, K.-H. Rieder: Understanding atom movement during lateral manipulation with the STM tip using a simple simulation method, Surface Science **499**, 15 (2002)

2.31. S.H. Cohen: Atomic Force Microscopy/Scanning Tunneling Microscopy (Plenum Press, New York 1995)

2.32. D.Y. Lee: High Speed and higly accurate Tip-scanning atomic force microscope (VDM Verlag 2008)

2.33. S. Morita: Non-Contact Atomic Force Microscopy (Springer, Berlin, Heidelberg 2002)

2.34. Paul E. West: Introduction to Atomic Force Microscopy: Theory, Practice and Applications http://www.afmuniversity.org/Cover.html

2.35. E.J.Giessibl: Advances in Atomic Force Microscopy. Rev. Mod. Phys. **75**, 949 (2003)

2.36. W. Wien: Handbuch der Experimentalphysik Vol. 14, Chapter: "Kanalstrahlen" (Springer, Berlin 1927)

2.37. Robert A. Millikan: Nobel Lecture (Elsevier Publ. Comp. Amsterdam 1965)

2.38. V.W. Hughes, L. Schulz (eds.): Sources of Atomic Particles. In: Methods of Experimental Physics, Vol. 4: Atomic and Electron Physics (Academic Press, San Diego 1988)

2.39. J.P. Guzowsky and G.M.Hieftje: Gas Sampling Glow Discharge: A versatile Ionization Source for Gas Chromatography Time of Flight Mass Spectrrometry. Analytical. Chem.72, 3812 (2000)

2.40. I.G. Brown: The Physics and Technology of Ion Sources (Wiley,New York, 2$^{nd}$ ed. 2004

2.41. R.A. Lyttleton, H. Bondi: On the Physical Consequence of a general excess of charge, Proc. Roy. Soc. **A252**, 313 (1959)

2.42. G. Gallinaro, M. Marinelli, G. Morpurgo: Electric Neutrality of Matter, Phys. Rev. Lett. **38**, 1255 (1977)

2.43. M. Szilagyi: Electron and ion optics (Plenum Publ. Corporation, New York 1988)

2.44. H. Liebl: Applied Charged Particle Optics (Springer, Berlin, Heidelberg 2007)

2.45. P.W. Hawkes, E. Kasper (eds.): Principles of Electron Optics (Academic Press, New York 1996)

2.46. R.F. Egerton: Energy Loss Spectroscopy in the Electron Microscope (Plenum Press, New York 1996)

2.47. F. Zhang and Z. Zhang (eds): Progress in Transmission Electron Microscopy: Springer Series in Surface Science Vol. **38** and **39** (2001)

2.48. L. Reimer and H. Kohl: Transmission Electron Mircroscopy: Physics of Image Formation. (Springer Berlin, Heidelberg 2008)

2.49. E. De Hoffmann and V. Strobant: Mass Spectrometry: Principles and Applications 3rd. ed. (John Wiley & Sons, New York 2007)

2.50. J.Th. Watson, O.D. Sparkman: Introduction to Mass Spectrometry: Instrumentation, Applications and Strategies for Data Interpretation. (Wiley, 4$^{th}$ ed. New York 2007)

2.51. J.H. Gross: Mass Spectrometry: A Textbook 2nd ed. (Springer, Berlin, Heidelberg 2006)

2.52. J.R. de Laeter: Application of Inorganic Mass Spectrometry (Wiley Interscience, New York 2001)

2.53. J. Mattauch: Massenspektrographie und ihre Anwendungen und Probleme der Atom- und Kernchemie. Ergebnisse der exakten Naturwiss. **19**, 170 (1940)

2.54. W.C. Wiley, I.H. McLaren: Time-of-flight mass spectrometer with improved resolution, Rev. Scient. Instrum. **26**, 1150 (1955)

2.55. E.W. Schlag (ed.): Time of Flight Mass Spectrometry and its Applications (Elsevier, Amsterdam 1994)

2.56. D.M. Lubmann: Lasers and Mass Spectrometry (Oxford Univ. Press, Oxford 1990)

2.57. M.M. Kappes: Experimental studies of gas-phase main group clusters, Chem. Rev. **88**, 369 (1988)

2.58. W. Paul: Elektromagnetische Käfige für geladene und neutrale Teilchen, Phys. Blätter **46**, 227 (1990)
W. Paul: Angew. Chemie Int. Ed. Engl. **29**, 739 (1990)

2.59. http://www.abrf.org/ABRFNews/1996/sep96iontrap.html

2.60. L.S. Brown, G. Gabrielse: Geonium Theory: Physics of a single electron or ion in a Penning trap, Rev. Mod. Phys. **58**, 233 (1986)

2.61. G. Bollen, R.B. Moore, G. Savard, H. Stoltzenberg: The accuracy of heavy ion mass measurement using time of flight ion cyclotron resonance in a Penning trap, J. Appl. Phys. **68**, 4355 (1990)

2.62. J. Chadwick et al. (eds.): Collected papers of Lord Rutherford (Vieweg, Braunschweig 1963)

2.63. E. Rutherford, J. Chadwick: Ellis: Radiation from Radioactive Substances (Cambridge 1930, p. 197)


## Chapter 3


3.1. I. Newton: Philosophiae Naturalis Principia Mathematica (London 1686); and: Optics, a treatise of the reflexion, inflexion and colours of light (London 1704)

3.2. C.S. Huygens: Traité de la lumiére (Paris 1676); (see: The New Encyclopadia Britannica)

3.3. M. Planck: Physikalische Abhandlungen und Vorträge Bd. 1–3 (Vieweg, Braunschweig 1958); M. Planck: Über das Gesetz der Energieverteilung im Normalspektrum, Ann. Phys. **4**, 553 (1901)

3.4. see also: J. Mehra, H. Rechenberg: The Historical Development of Quantum Mechanics (Springer, Berlin, Heidelberg 1982–2000)

3.5. R.A. Millikan: A direct Photoelectric Determination of Planck's $h$. Phys. Rev. **7**, 355 (1916)

3.6. see for instance: J. Mehra: The Solvay Conferences in Physics (D. Riedel, Dordrecht 1975)

3.7. R.V. Pound, G.A. Rebka: Apparent weight of photons, Phys. Rev. Lett. **4**, 337 (1960)

3.8. J.L. Snider: New Measurements of the Solar Gravitational Redshift, Phys. Rev. Lett. **28**, 853 (1972)

3.9. L. de Broglie: Ondes et corpiseiles (Hermann, Paris 1930)

3.10. C.S. Adams, M. Siegel, J. Mlynek: Atom optics, Phys. Reports **240**, 145 (1994)

3.11. P. Meystre: Atom Optics (Springer, Berlin, Heidelberg 2001)

3.12. V. Bonse: Recent Advances in X-ray and Neutron Interferometry, Physica B **151**, 7 (1988)

3.13. P.R. Berman (ed.): Atomic Interferometry (Academic Press, New York 1996)

3.14. S.A. Werner, H. Rauch: Neutron Interferometry (Oxford Univ. Press, Oxford 2000)

3.15. Max Born, R.J. Blin-Stoyle, J.M. Radcliffe: Atomic Physics (Dover Publications Mineola, N.Y., 8th ed. 1989)

3.16. C. Cohen-Tannoudji, B. Diu, F. Laloë: Quantum Mechanics Vol I (John Wiley & Sons, New York)

3.17. L. Rosenfeld (ed.): Nils Bohr's Collected Work (1972–...) Nils Bohr: On the Quantum Theory of Line Spectra. Kgl. Dan. Vid. Selsk. Skr. Nat. Mat. Atd Series **8**, 4, p. 1–118 (1918)

3.18. J. Franck, G. Hertz: Über Zusammenstöße zwischen Gasmolekülen und langsamen Elektronen. Verband der deutschen Physikal. Gesellschaft **15**, 373 (1913); see also: J. Mehra, H. Rechenberg: The Historical Development of Quantum Mechanics (Springer 1982–2000)

3.19. G. F. Hanne: What really happens in the Franck-Hertz experiment with mercury? Am. J. Phys. Vol *56*, 698 (1988)

3.20. J. Levine: Early Gravitational Wave Detection Experiments: Physics in Perspective *6*, 42–75 (Birkhausen, Basel 2004)

3.21. St. Dürr, G. Rempe: Wave-Particle Duality in an Atom Interferometer, Adv. Atomic, Molecular and Optical Physics Vol. **41** (1999)

3.22. H. Paul: Introduction Quantum Optics From Light Quanta to Teleportation (Cambridge Univ. Press, Cambridge 2004)

3.23. M.O. Scully, B.G. Englert, H. Walther: Quantum optical tests of complementary, Nature **351**, 111 (1991)

3.24. W. Schleich: Quantum Optics in Phase Space (Wiley VCH, Weinheim 2001)

3.25. M.O. Scully, M.S. Zubairy: Quantum Optics (Cambridge Univ. Press, Cambridge 1997); Chr. Gerry, P. Knight: Introductory Quantum Optics (Cambridge Univ. Press, Cambridge 2004)


## Chapter 4

**For more detailled representations of quantum mechanics see, for instance:**


4.1. A. Messiah: Quantum Mechanics (Dover Publ., Mineola, N.Y. 2000)

4.2. C. Cohen-Tannoudji: Quantum Mechanics Vol. I + II, Wiley Interscience 1996

4.3. R. Shankar: Quantum Mechanics 2nd ed. (Plenum Press, New York 1994)

4.4. G.L. Squires: Problems in Quantum Mechanics (Cambridge Univ. Press, Cambridge 1995)

4.5. D.J. Griffiths: Introduction to Quantum Mechanics (Prentice Hall, Upper Saddle River N.J. 1994)

4.6. Yung-Kuo Lim et al.: Problems and Solutions on Quantum Mechanics (World Scientific Publ. Singapore 1998)

4.7. R. Gilmore: Elementary Quantum Mechanics in one dimension (John Hopkins Univ. Press, Baltimore 2004)


## Chapter 5


5.1. O. Stern, W. Gerlach: Der experimentelle Nachweis des magnetischen Moments des Silberatoms, Zeitschrift für Physik **8**, 110 (1992) and **9**, 349 (1922)

5.2. J.D. Jackson: Classical Electrodynamics 3rd. edition (Wiley, New York 1998)
5.3. D. Hestenes: Quantum Mechanics from Selfinteraction, Foundations of Physics **15**, 63 (1985)
5.4. W.E. Lamb, R.C. Retherford: Fine Structure of the Hydrogen Atom by a Microwave Method, Phys. Rev. **72**, 241 (1947)
5.5. T.W. Hänsch: High Resolution Spectroscopy of Hydrogen. In: G.F. Bassani, M. Inguscio, T.W. Hänsch (eds.): The Hydrogen Atom (Springer, Berlin, Heidelberg 1989)
5.6. T. Udem et al.: Phase Coherent Measurement of the Hydrogen 1S–2S transition frequency. Phys. Rev. Lett. **79**, 2646 (1997)
5.7. Niels Bohr: The Correspondence Principle (Elsevier Science Publ., Amsterdam 1976)
W. Krajewski: Correspondence Principle and Growth of Science (Springer, Berlin, Heidelberg 1999)
5.8. Cl. Cohen-Tannoudji: Photons and Atoms, Introduction to Quantum Electrodynamic, New Edition (Wiley Interscience, New York 1997)
5.9. W. Greiner: Quantum Electrodynamics 3rd ed. (Springer, Berlin, Heidelberg 2003)
5.10. Richard P. Feynman: Quantum Electrodynamics (Perseus Books Group, Philadelphia PA 1998)


## Chapter 6


6.1. W. Meyer: Kaiserslautern, private communication
E.R. Davidson: Single-Configuration Calculations of Excited States of Helium, J. Chem. Phys. **42**, 4199 (1965)
6.2. J. Emsley: The Elements (Oxford Univ. Press, Oxford 1996)
S. Fraga: Handbook of Atomic Data (Elsevier Science, Amsterdam 1977)
6.3. P.S. Krstic, F. Ownby, D.R. Schultz (eds.): Atomic and Molecular Data and their Applications, Conference Proceedings, Gatlinburg 2002 (Am. Inst. of Physics, New York 2002)
6.4. H. Friedrich: Theoretical Atomic Physics 2nd ed. (Springer, Berlin, Heidelberg 2002)
6.5. T.F. Gallagher: Rydberg Atoms (Cambridge Univ. Press, Cambridge 1994)
6.6. I.C. Percival: Planetory Atoms, Proc. Royal Soc. London A **353**, 289 (1977)
6.7. J. Boulmer, P. Camus, P. Pillet: Double Rydberg spectroscopy of the barium atom, J. Opt. Soc. Am. B **4**, 805 (1987)
6.8. B. Laus et al.: X-ray emission during the muonic cascade in H. Phys. Rev. Lett **80**, 3041 (1998)
6.9. L. Willmann, K. Jungmann: The Muonium Atom as a probe of Physics beyond the Standard Model, In: Lecture Notes in Physics **499**, 49 (1997)
6.10. G. Backenstoß: Antiprotonic Atoms. In: Atomic Phys. **10**, 147 (North Holland Publ. Comp., Amsterdam 1987)
6.11. St. Chu: Laser Spectroscopy of Positronium and Myonium, In: G.F. Bassani, M. Inguscio, T.W. Hänsch (eds.): The Hydrogen Atom, p. 114 (Springer, Berlin, Heidelberg 1989)
6.12. L.M. Simons (ed.): Electromagnetic Cascade and Chemistry of Exotic Atoms. Ettore Majorana Center, Science Series Vol. 52 (Plenum Publ., New York 1991)
6.13. D. Gotta et al.: Nucl. Phys. A **660**, 283 (1999)
6.14. M. Hori et al.: Observation of Cold Long-Lived Antiprotonic Helium Ions. Phys. Rev. Lett. **94**, 063401 (2005)
T. Yamazati et al.: Antiprotonic Helium. Physics Reports **366**, 183 (2002)
6.15. G. Gabrielse et al.: Background-Free Observation of Cold Anti-Hydrogen, Phys. Rev. Lett. **89**, 213401 (2002)
6.16. J.L. Basdevant, J. Dalibart: The Spectrum of Positronium, In: The Quantum Mechanics Solver. (Springer, Berlin, Heidelberg 2000); R.S. Vallery et al.: Phys. Rev. Lett. **90**, 203402 (2003)
6.17. V.W. Hughes: Recent Advances in Myonium, In: G.F. Bassani, M. Inguscio, T.W. Hänsch (eds.): The Hydrogen Atom (Springer 1989)
6.18. F.E. Maas et al.: A measurement of the $1S$-$2S$ transition frequency in muonium. Phys. Lett. A **187**, 247 (1994)


## Chapter 7


7.1. H. Haken, A.C. Wolf: The Physics of Atoms and Quanta 6th ed. (Springer, Berlin, Heidelberg 2000)
7.2. H.S. Friedrich: Theoretical Atomic Physics 2nd ed. (Springer, Berlin, Heidelberg 2004)
7.3. G. Grynberg, B. Cagnac: Doppler-free multiphoton spectroscopy, Report on Progress in Physics **40**, 791 (1977)
7.4. J.R. Lakowicz (ed.): Topics in Fluorescence Spectroscopy Vol 5: Nonlinear and Two-Photon Induced Fluorescence (Plenum Press, New York 1997)
7.5. L.J. Curtis: Atomic Structure and Lifetimes (Cambridge Univ. Press, Cambridge 2003)
7.6. R.E. Imhoff, F.H. Read: Measurements of Lifetimes of atoms and molecules, Reports on Progress in Physics **40**, 1 (1977)

7.7. D.V. O'Connor, D. Phillips: Time Correlated Single Photon Counting (Academic Press, New York 1984)

7.8. K.M. Sando, Shih-I. Chu: Pressure broadening and laser-induced spectral line shapes (Adv. Atom. Mol. Phys. **Vol 25**, Academic Press, New York 1988)

7.9. I.I. Sobelman, L.A. Vainshtein, E.A. Yukov: Excitation of Atoms and Broadening of Spectral Lines (Springer, Berlin, Heidelberg 1995)

7.10. U. Fano, A.R.P. Rai: Atomic Collisions and Spectra (Academic Press, New York 1986)

7.11. J. Als-Nielsen, Des McMorrow: Elements of Modern X-ray Physics (John Wiley & Sons, Chichester 2001)

7.12. J. Graham Brown: X-rays and their applications (Plenum Publ., New York 1966); D. Attwood: Soft X-Rays and Extreme UV Radiation: Principles and Applications (Cambridge Univ. Press 1999)

7.13. A.C. Fabian (ed.): Frontiers of X-ray Astronomy (Cambridge Univ. Press, Cambridge 2004)

7.14. F.D. Seward, Ph.A. Charles: Exploring the X-Ray Universe (Cambridge Univ. Press, Cambridge 1995)

7.15. A. Temkin (ed.): Autoionization: Recent Developments (Plenum Press, New York 1994)

7.16. L. Adhya, K. Rai Dastidar: Deviation from the Fano Profile in resonance enhanced multiphoton ionization. Phys. Rev. A **50**, 3537 (Oct. 1994)

7.17. U. Fano, L. Fano: Atomic and Molecular Physics (Univ. Chicago Press 1959)

7.18. U. Fano, J.W. Cooper: Line profiles in the far-UV absorption spectra of rare gases. Phys. Rev. *137*, A1364 (1965)

7.19. T.K. Fang, T.N. Chang: Determination of profile parameters of a Fano resonance. Phys. Rev. *A57*, 4407 (1998)

7.20. U. Fano: Effects of Configuration Interaction on intensities and phase shifts. Phys. Rev. *124*, 1866 (1961)

7.21. M.J. Seaton: Di-electronic recombination to states with large principal quantum numbers. J. Phys. B. Atom. Mol. Opt. Phys. **31**, L1017 (1998)


## Chapter 8


8.1. J.P. Gordon, H.J. Zeiger, C.H. Townes: Molecular microwave oscillator and new hyperfine structure in the microwave spectrum of $NH_3$. Phys. Rev. **95**, 282 (1954)

8.2. A.L. Schawlow, C.H. Townes: Infrared and optical masers, Phys. Rev. **112**, 1940 (1958)

8.3. T.H. Maiman: Stimulated optical radiation in ruby, Nature **187**, 493 (1960)

8.4. A. Siegmann: Lasers (Oxford Univ. Press, Oxford 1986)

8.5. O. Svelto: Principles of Lasers 5th edn. (Plenum, New York 2009)

8.6. J. Hecht: Understanding Lasers Wiley-IEEE press, 3rd edition (New Jersey 2008)

8.7. Ch.H. Townes: How the Laser Happened (Oxford Univ. Press, Oxford 2001)

8.8. N. Hodgson, H. Weber: Optical Resonators: Fundamentals, Advanced Concepts and Applications (Springer, Berlin, Heidelberg 1997)

8.9. R. Kossowsky et al.: Optical Resonators: Science and Engineering (Kluwer Academic Publ. Dordrecht 1998)

8.10. G.D. Boyd, H. Kogelnik: Generalized confocal resonator theory, Bell Syst. Techn. Journal **41**, 1347 (1962)

8.11. E.L. Saldin, E. Schneidmiller, M. Yunkow: The Physics of Free Electron Lasers. (Springer, Heidelberg, New York 2000)

8.12. W. Demtröder: Laser Spectroscopy 4rd ed. (Springer, Berlin, Heidelberg 2008)

8.13. D.L. Mills: Nonlinear Optics, 2nd ed. (Springer, Berlin, Heidelberg, New York 1998).

8.14. N. Bloembergen: Nonlinear Optics 4th ed. (World Scientific, Singapore 1996)

8.15. R. Szipöcz, A. Köbazi-Kis: Theory and designs of chirped dielectric laser mirrors. Appl. Phys. B65, 115 (1997)

8.16. C.V. Shank, R.L. Fork, R. Yen, R.W. Stolen, W.J. Tomlinson: Compression of femtosecond optical pulses, Appl. Phys. Lett. **40**, 761 (1982)

8.17. H.W. Schröder et al: A high power single mode CW dye ring laser. Appl. Phys. *14*, 377 (1978)


## Chapter 9


9.1. R. McWeen (ed.): Coulson's Valence (Oxford Univ. Press, Oxford 1980)

9.2. W. Kutzelnigg: Einführung in die theoretische Chemie (Verlag Chemie, Weinheim 1994); I.W. Levine: Quantum Chemistry 5th ed. (Prentice Hall, Upper Saddle River, N.J., 1999)

9.3. H.M. James, A.S. Coolidge: The ground state of the hydrogen molecule. J. Chem. Phys. **1**, 825 (1933)

9.4. H. Heitler, F. London: Wechselwirkung neutraler Atome und homöopolare Bindung nach der Quantenmechanik, Zeitschrift für Physik **44**, 455 (1927)

H.B. Grey: Chemical Bonds: An Introduction to Atomic and Molecular Physics (University Science Books, Sausalito, CA 1994)

C.A. Coulson: The Shape and Structure of Molecules (Oxford Univ. Press,1987)

9.5. W. Kolos, C.C.J. Rothaan: Accurate Electronic Wavefunctions for the $H_2$-Molecule, Rev. Mod. Phys. **32**, 219 (1960)
9.6. W. Kolos, L. Wolniewicz: Nonadiabatic Theory for Diatomic Molecules and its Application to the Hydrogen Molecule, Rev. Mod. Phys. **35**, 473 (1963)
9.7. R.E. Grisente et al.: Determination of Bond Length and Binding Energy of the Helium Dimer, Phys. Rev. Lett. **85**, 2284 (2000)
9.8. J.O. Hirschfelder (ed.): Intermolecular Forces (Wiley & Sons, New York 1954); J.O. Hirschfelder, Ch.F. Curtis, R.B. Byrd: Molecular Theory of Gases and Liquids (Wiley & Sons, New York 1954)
9.9. J. Goodishman: Diatomic Interaction Potential Theory Vol. I + II (Academic Press, New York 1973); see Maksic (ed.): The Concept of the Chemical Bond (Springer, Berlin, Heidelberg 1990)
9.10. H. Lefebvre, R.W. Field: Perturbations in the Spectra of Diatomic Molecules, 2nd ed. (Academic Press New York (2000))
9.11. J.L. Dunham: The Energy Levels of a Rotating Vibrator, Phys. Rev. **41**, 721 (1932)
9.12. J.W. Cooper, U. Fano, F. Prats: Classification of two-electron excitation levels of helium, Phys. Rev. Lett. **10**, 518 (1963)


## Chapter 10


10.1. A.D. Walsh: J. Chem. Soc. **1953**, 2260 (1953)
10.2. J.K. Burdett: Chemical Bonds, A Dialog (J. Wiley & Sons, Chichester 1997)
10.3. J.M. Hollas: High Resolution Spectroscopy 2nd ed. (John Wiley & Sons, Chichester 1998)
10.4. E.B. Wilson Jr., J.C. Decius, P.C. Cross: Molecular Vibrations (McGraw Hill, New York 1954)
H.C. Allen, P.C. Cross: Molecular Vib-Rotors (Wiley Interscience, New York 1963)
10.5. G. Herzberg: Molecular Spectra and Molecular Structure, Vol II: Infrared and Raman Spectra (van Nostrand Reinhold, New York 1963)
10.6. D.S. Schonland: Molecular Symmetry and Spectroscopy (Van Nostrand Reinhold Comp., London 1971)
10.7. J.M. Hollas: Symmetry in Molecules (Chapman & Hall, London 1972)
10.8. R.L. Carter: Molecular Symmetry and Group Theory (John Wiley & Sons, New York 1997)
10.9. G. Herzberg: Molecular Spectra and Molecular Structure III: Electronic Spectra and Electronic Structure of Polyatomic Molecules (Van Nostrand Reinhold Comp., New York 1966)
10.10. G. Benedek, T.P. Martin, G. Pacchioni (eds.): Elemental and Molecular Clusters (Springer, Berlin, Heidelber 1988)
10.11. H. Haberland (ed.): Clusters of Atoms and Molecules I+II (Springer, Berlin, Heidelberg 1994)
10.12. K, von Haeften, M. Havenith: He-droplets. A fluid with unusual properties. In: W.F. Schmidt, E. Illenberger (eds.) Electron Excitations in Liqufied Rare Gases (American Scientific Publ. 2005)
10.13. K.D. Rinnen, D.A.V. Kliner, R.N. Zare: The $H + D_2$ reaction. J. Chem. Phys. **91**, 7514 (1989)
10.14. N.H. March, J.F. Mucci: Chemical Physics of Free Molecules (Plenum Press, New York 1992)


## Chapter 11


11.1. A.C. Melissonos, J. Napolitano: Experiments in Modern Physics 2nd ed. (Academic Press, New York 2003)
B. Bederson, H. Walther (eds.): Advances in Atomic, Molecular and Optical Physics Vol 1–49 (Academic Press, New York)
11.2. E. Wolf (ed.): Progress in Optics, Vol. 1–44 (North-Holland Publ., Amsterdam 1961–2005)
11.3. E. Popov, E.G. Loewen: Diffraction Gratings and Applications (Dekker, New York 1997)
11.4. M.D. Perry et al.: High efficiency multilayer dielectric diffraction gratings, Opt. Lett. **20**, 140 (1995)
11.5. W.H. Steel: Interferometry (Cambridge Univ. Press, Cambridge 1967)
11.6. P. Hariharan: Optical Interferometry (Academic Press, New York 1986)
11.7. J.M. Vaughan: The Fabry–Perot Interferometer. (Hilger, Bristol 1989)
11.8. G.H. Rieke: Detection of Light. From the Ultraviolet to the Submillimeter (Cambridge Univ. Press, Cambridge 1994)
11.9. J.J. Keyes (ed.): Optical and Infrared Detection 2nd ed. (Springer, Berlin, Heidelberg 1980)
11.10. E.H. Putley: Thermal Detectors, in [11.9], p. 71
11.11. M. Zen: Cryogenic Bolometers. in: Atomic and Molecular Beam Methods, Vol. I ed. by G. Scales (Oxford Univ. Press, Oxford 1988)
11.12. See for instance the information sheets on photomultipliens, issued by the manufacturers RCA, EMI, Hamamatsu, available on the web
11.13. Ch.H. Townes, A.L. Schalow: Microwave Spectroscopy (Dover Publications, Mineola, N.Y. 1975)
11.14. J.W. Fleming, J. Chamberlain: Infrared Phys. **14**, 277 (1974)

11.15. B.H. Stuart et al.: Modern Infrared Spectroscopy (John Wiley & Sons 1996)
H. Günzler, H.V. Gremlich: IR-Spectroscopy: An Introduction (Wiley VCH, Weinheim 2002)

11.16. P.R. Griffith, J.A. DeHaseth: Fourier Transform Infrared Spectroscopy (Wiley Interscience, 2nd ed., 1986); R.R. Williams: Spectroscopy and the Fourier Transform (John Wiley & Sons, New York 1995)

11.17. B.C. Smith: Fundamentals of Fourier Transform Infrared Spectroscopy (CRC-Press, Boca Raton, Fl. USA); J. Kauppinen, J. Partanen: Fourier Transforms in Spectroscopy (Wiley, New York 2001)

11.18. W. Demtröder: Laser Spectroscopy 3rd ed. (Springer, Berlin, Heidelberg 2003)

11.19. J.C. Lindon, G.E. Trautner, J.L. Holmes: Encyclopedia of Spectroscopy and Spectrometry Vol. I–III (Academic Press, London 2000)

11.20. F. Träger, (ed): Springer Handbook of Lasers and Optics. (Springer, Heidelberg 2007)

11.21. J. Sneddon (ed.): Lasers in Analytical Atomic Spectroscopy (Wiley, New York 1997)

11.22. A. Rosencwaig: Photoacoustic Spectroscopy (Wiley, New York 1980)

11.23. J. Xiu, R. Stroud: Acousto-Optic Devices. Principles, Design and Applications (Wiley, New York 1992)

11.24. B. Barbieri, N. Beverini, A. Sasso: Optogalvanic Spectroscopy, Rev. Mod. Phys. **62**, 603 (1990)

11.25. M.A. Zia,, M.A. Baig: Laser optogalvanic spectroscopy of the even parity Rydberg states of atomic mercury. J.Opt. Soc. Am. B22, 2702 (2005)

11.26. V.N. Ochkin, N.G. Prebrazhensky, N.Y. Shaparev: The Optogalvanic Effect (Chem. Rub. Comp. Cleveland Ohio 1999)

11.27. K. Narayanan, G. Ullas, S.B. Rai: A two step optical double resonance study of a FE–Ne-hollow cathode discharge using optogalvanic detection, Opt. Commun. **184**, 102 (1991)

11.28. P. Zalicki, R.N. Zare: Cavity Ringdown spectroscopy for quantitative absorption measurements, J. Chem. Phys. **102**, 2708 (1995)

11.29. D. Romanini, K.K. Lehmann: Ring-down cavity absorption spectroscopy of the very weak HCN overtone bands with six, seven and eight stretching quanta, J. Chem. Phys. **99**, 6287 (1993)

11.30. G. Höning, M. Cjajkowski, M. Stock, W. Demtröder: High resolution spectroscopy of $Cs_2$. J. Chem. Phys. **71**, 2138 (1979); M. Raab, H. Weickenmeier, W. Demtröder: The dissociation energy of the cesium dimer, Chem. Phys. Lett. **88**, 377 (1982)

11.31. G. Hurst, M.G. Payne: Principles and Applications of Resonance Ionisation Spectroscopy, ed. by D.S. Kliger (Academic Press, New York 1983)

11.32. J.B. Atkinson, J. Becker, W. Demtröder: Hyperfine structure of the 625 nm band in the $a^3\Pi_\mu \leftarrow X\sum_S^1$ transition of $Na_2$, Chem. Phys. Lett. **87**, 128 (1982)

11.33. B. Bobin, C.J. Bordé, C. Bréant: Vibration-rotation molecular constants for the ground state and $\nu_3 = 1$ states of $SF_6$ from saturated absorption spectroscopy, J. Mol. Spectrosc. **121**, 91 (1987)

11.34. A. Timmermann: High resolution two-photon spectroscopy of the $6p^2\,{}^3P_0 - 7p\,{}^3P_0$ transition in stable lead isotopes, Z. Physik A **286**, 93 (1980)

11.35. G. Grynberg, B. Cagnac: Doppler-free multiphoton spectroscopy, Rep. Progress Phys. **40**, 791 (1977)

11.36. H.W. Schrötter, H. Frunder, H. Berger, J.P. Boquitlon, B. Lavorel, G. Millet: High Resolution CARS and Inverse Raman Spectroscopy, in: Adv. Nonlinear Spectroscopy Vol. **3**, 97 (Wiley, New York 1987)

11.37. J.P. Taran: CARS-Spectroscopy and Applications, in: Applied Laser Spectroscopy. ed. by W. Demtröder and M. Inguscio (Plenum Press, New York 1990)

11.38. W. Zinth et al.: Femtosecond Spectroscopy and model calculations for an understanding of the primary reactions in bacterio-rhodopsin, in: Ultrafast Phenomena XII, ed. by T. Elsässer et al. (Springer, Berlin, Heidelberg 2000)

11.39. Ch. Kunz: Synchrotron Radiation. Techniques and Applications (Springer, Berlin, Heidelberg 1979); J.A.R. Samson, D.L. Lederer: Vacuum Ultraviolet Spectroscopy (Academic Press, New York 2000)

11.40. P. Schlemmer, M.K. Srivastava, T. Rösel, H. Ehrhardt: Electron impact ionization of helium at intermediate collision energies, J. Phys. B **24**, 2719 (1991)

11.41. N.H. March, J.F. Mucci: Chemical Physics of Free Molecules (Plenum Press, New York 1992)

11.42. H. Hotop, M.W. Ruf, M. Allan, I.I. Fabrikant: Resonance and Treshold Phenomena in Low-Energy Electron Collisions with Molecules and Clusters, Adv. Atomic, Molecular and Optical Physics **49**, 85 (2003)

11.43. E.W. Schlag: ZEKE-Spectroscopy (Cambridge Univ. Press, Cambridge 1998)

11.44. R. Signorell, F. Merkt, H. Palm: Structure of the Ammonium Radical from a Rotationally Resolved Photoelectron Spectrum, J. Chem. Phys. **106**, 6523 (1997)

11.45. See for instance: N.F. Ramsey: Molecular Beams 2nd edition (Clarendon Press, Oxford 1989)
11.46. K. Bergmann: State Selection via optical methods. in: Atomic and Molecular Beam Methods, ed. by G. Scoles, Vol. 1 (Oxford Univ. Press, Oxford 1988)
11.47. J.C. Zorn, T.C. English: Molecular Beam electric resonance spectroscopy, Adv. Atom. Mol. Physics **9**, 243 (1973)
11.48. K. Uehara, T. Shimizu, K. Shimoda: High resolution Stark Spectroscopy of molecules by infrared and far-infrared masers. IEEE J. Quant. Electr. **4**, 728 (1968)
11.49. Proc. International Conf. on the Physics of Electronic and Atomic Collisions ICPEAC I–XXVI (North Holland Publ., Amsterdam 1959–2010)
11.50. see for instance: HMI-information in English on the web: www.hmi.de/bereiche/info/dualismus/kernregenbogen_en.html
11.51. M.A.D. Fluendy, K.P. Lawley: Chemical Applications of Molecular Beam Scattering (Chapman & Hall, London 1973)
11.52. K. Bergmann: State selection via optical methods. In: Atomic and Molecular Beam Methods, ed. by G. Scoles (Oxford University Press, Oxford 1989)
11.53. K. Bergmann, U. Hefter, J. Witt: State-to-State differential cross sections for rotational transitions in $Na_2$ + He collisions, J. Chem. Phys. **71**, 2726 (1979)
11.54. F. Leomon et al.: Crossed-beam universal detection reactive scattering of radical beams, Mol. Phys. **108**, 1097 (2010)
11.55. P.V. O'Connor, P. Phillips: Time Correlated Single Photon Counting (Academic Press, New York 1984)


## Chapter 12


12.1. T.W Hänsch, A.L. Schawlow: Cooling of gases by laser radiation, Opt. Commun. **13**, 68 (1975)
12.2. D. Sesko, C.G. Fam, C.E. Wieman: Production of a cold atomic vapor using diode-laser cooling, J. Opt. Soc. Am. B **5**, 1225 (1988)
12.3. J. Dalibard, C. Cohen-Tannoudji: Laser cooling below the Doppler-limit by polarization gradients: simple theoretical model. J. Opt. Soc. Am. B **6**, 2023 (1989)
12.4. A. Arimondo, W.D. Phillips, F. Strumia (eds.): Laser Manipulation of Atoms and Ions (North Holland Publ. Amsterdam 1992)
12.5. H.I. Metcalf, P. van der Straaten: Laser Cooling and Trapping (Springer, Berlin, Heidelberg 1999)
12.6. K. Sengstock, W. Ertmer: Laser Manipulation of Atoms, Adv. At. Mol. and Opt. Physics **35**, 1 (1995)
12.7. S.N. Bose: Planck's Gesetz und Lichtquantenhypothese, Z. Physik **26**, 178 (1924); A. Einstein: Sitz. Berichte Preuss. Akademie Berlin **22**, 261 (1924)
12.8. M.H. Andersen, J.R. Ensher, M.R. Matthews, C.E. Wieman and E.A. Cornell: Observation of Bose–Einstein Condensation in a Dilute Atomic Gas, Science **269**, 198 (1995)
12.9. K.B. Davis, M.O. Mewes, M.A. Joffe, M.R. Andrews, W. Ketterle: Evaporative Cooling of Sodium Atoms, Phys. Rev. Lett. **74**, 5202 (1995)
12.10. W. Ketterle, N.J. van Druten: Evaporative cooling of trapped atoms. Adv. Atom.Molec. Optical Physics Vol.37, 181 (1996).
12.11. Ch. Becker: Multicomponent Bose Einstein Condensates Dr. Hut Verlag München 2009
12.12. S. Jochim, M. Bartenstein, R. Grimm: Bose–Einstein Condensation of Molecules, Science Express 13. Nov. 2003/Science 1093280; and: Physics Today, October 2003
12.13. A. Griffin, D.W. Snoke, S. Stringari (eds.): Bose–Einstein Condensation (Cambridge Univ. Press, Cambridge 1995)
12.14. S. Martelucci (ed.): Bose–Einstein Condensates and Atom Laser (Kluwer Academic Publ., New York 2000)
12.15. P. Bermann (ed.): Atom Interferometry (Academic Press, San Diego 1997);
A. Widera, Th.W. Hänsch et al.: Measurements of Atomic Scattering Properties. Phys. Rev. Lett. **92**, 160 406-1 (2004)
12.16. W.C. Stwalley, K.-H. Uang: Pure Long Range Molecules, Phys. Rev. Lett. **41**, 1164 (1978)
12.17. C. Chin et al.: Observation of Feshbach-like resonances in collisions between ultracold molecules. Phys. Rev. Lett. **94**, 123201 (2005)
12.18. M. Mark, T. Kraemer, J. Harbig, C. Chin, H.C. Nägerl, R. Grimm : Efficient creation of molecules from a cesium Bose-Einstein condensate. Europhys. Lett. **69**, 706 (2005)
12.19. T. Baumert, M. Grosser, R. Thalweiler, G. Gerber: Femtosecond time-resolved molecular photoionisation: The $Na_2$-System, Phys. Rev. Lett. **67**, 3753 (1991)
12.20. E. Schreiber: Femtosecond Real Time Spectroscopy of Small Molecules and Clusters (Springer, Berlin, Heidelberg 1998)
12.21. A.H. Zewail: Femtochemistry (World Scientific, Singapore 1994)

12.22. M. Shapiro, P. Brummer: Coherent control of atomic, molecular and electronic processes, Adv. Atom. Mol. and Opt. Phys. **42**, 287 (2000)

12.23. A. Assion, G. Gerber et al.: Control of chemical reactions by feedback-optimized phase shaped femtosecond laser pulses, Science **282**, 119 (1998)

12.24. T. Brixner, N.H. Damrauer, G. Gerber: Femtosecond Quantum Control, Adv. Atom. Mol. and Opt. Phys. **46**, 1–56 (2001)

12.25. A. Rice, M. Zhao: Optical Control of Molecular Dynamics (Wiley, New York 2000)

12.26. D. Zeidler, S. Frey, K.L. Kompa, M. Motzkus: Evolutionary algorithm and their applications to optimal control studies, Phys. Rev. A **64**, 023420 (2001)

12.27. J. Reichert, T.W. Hänsch et al.: Phase coherent vacuum ultraviolet to radiofrequency comparison with a mode-locked laser, Phys. Rev. Lett. **84**, 3232 (2000); S.A. Didamus, T.W. Hänsch: Direct link between microwave and optical frequencies with a 300 THz femtosecond pulse, Phys. Rev. Lett **84**, 5102 (2000); Th. Udem, R. Holzwarth, T.W. Hänsch: Optical Frequency Metrology, Nature **416**, 233 (2002)

12.28. N. Kolchevsky, M. Fischer, S.G. Karshenboim, T.W. Hänsch: High Precision Optical Measurement of the 2S Hyperfine-Interval in Atomic Hydrogen, Phys. Rev. Lett. **92**, 033003 (2004)

12.29. S.G. Karshenboim et al. (eds.): The Hydrogen Atom. Precision Physics of Simple Atomic Systems (Springer, Berlin, Heidelberg 2001)

12.30. S.A. Didamus et al.: An optical clock based on a single trapped $^{199}Hg^{+}$-Ion. Science **293**, 825 (2001)

12.31. T. Sauter, R. Blatt, W. Neuhauser, P.E. Toschek: Quantum jumps in a single ion, Physica Scripta **22**, 128 (1988)

12.32. R. Blümel, W. Walther et al.: Phase transitions of stored laser-cooled ions, Nature **334**, 309 (1988)

12.33. J. Javamainen: Laser Cooling of trapped ion-clusters, J. Opt. Soc. Am. B **5**, 73 (1988)

12.34. H.A. Bachor: A Guide to Experiments in Quantum Optics, 2nd ed. (Wiley VCH, Weinheim 2004)
H. Paul: Introduction to Quantum Optics. From Light Quanta to Quantum Teleportation (Cambridge Univ. Press 2004)

12.35. P.S. Paulson: Fundamentals of Interferometric Gravitational Wave Detectors (World Scientific, Singapore 1994)

12.36. F. Seifert, P. Kwee, M. Heurs, B. Wilke and K. Danzmann : Laser Power Stabilization for second generation gravitational wave detectors. Opt. Lett. **31**, 2000 (2006)

12.37. B. Bhawal: Physics of interferometric gravitational wave detectors. Pramana J. Phys. 63, 645 (1994)

12.38. LISA: www. International Technology Education Association: The Technology Teacher (Febr. 2004)

12.39. St. Dürr, G. Rempe: Wave-Particle Duality in an Atom Interferometer, Adv. Atomic, Molecular and Optical Physics Vol. 41 (1999)

12.40. J.S. Bell, A. Aspect: Speakable and Unspeakable in Quantum Mechanics. Collected Papers on Quantum Philosophy (Cambridge Univ. Press, Cambridge 2004)

12.41. J.S. Bell: On the Einstein–Rosen–Podolsky-Paradox, Physics **1**, 195 (1964); A. Afriat: The Einstein–Podolsky–Rosen-Paradox (Plenum Press, New York 1998)

12.42. A. Aspect, P. Grangier, G. Roger: Experimental Tests of Bell's inequality using timer-varying analyzers. Phys. Rev. Lett. **49**, 1804 (1982)

12.43. A. Afriat, F. Sellen: The Einstein–Rosen–Podolsky paradox in atomic, nuclear and particle physics. (Plenum Press, New York 1999)

12.44. S. Haroche, M. Brune, J.M. Raimond: Schrödinger cats and entanglement experiments in cavity QED. Laser Spectroscopy XIII, ed. by Zhi-jiang Wang, Zhi-ming Zhang and Yu-zhu Wang (World Scientific, Singapore 1998)

12.45. D. Bouwmeester, A. Ekert, A. Zeilinger (eds.): The Physics of Quantum Information, 3rd printing (Springer, Berlin, Heidelberg 2001)

12.46. C.P. Williams, S.H. Clearwater: Explanations in Quantum Computing (Springer, Berlin, Heidelberg 1997)

12.47. John Preskill: Lecture Notes on Quantum Computing, home page Ph219/CS219; D. Deutsch, A. Ekert: Quantum Computation. Physics World, March 1998

12.48. J.I. Cirac, P. Zoller, J.F. Poyatos: Quantum Gates and Quantum Computation with trapped ions. in: The Physics of Quantum Information, ed. by D. Bouwmeester, A. Ekert and A. Zeilinger (Springer, Berlin, Heidelberg 2001)

12.49. T. Monz, K. Kim, W. Hänsel, M. Riebe, A. S. Villar, P. Schindler, M. Chwalla, M. Hennrich, and R. Blatt, "Realization of the quantum Toffoli gate with trapped ions Physical Review Letters 102, 040501 (2009), arXiv:0804.0082

# Subject Index


$a$–$b$-diagram 62
absolute mass 53
absorption
– continuous 280
– Doppler-free two-photon 186
– effective coefficient 282
– losses 290
– minimum coefficient 422
– nonlinear 456
– relative 440
– spectrum 112
absorption coefficient 273
absorption edges 276
abundance
– isotopic 64
actinides 216
Airy formulas 431
$\alpha$-decay 137
amplifier
– optical 306
amplitude
– quadrature 504
angular momentum 152
– conservation 254
– coupling 224
– electronic 341
– operator 152
– quantum number 153
– rotational 359
– vibrational 402
anti-proton 240
anti-Stokes radiation 461
apodization 442
approximation
– adiabatic 357
– Born–Oppenheimer 378
– breakdown of the Born–Oppenheimer 403
– Heitler–London 337
– LCAO 331
– MO 336
– MO-LCAO 339
– self-consistent field 223
Argon laser 311
*Aristoteles* 8
asymptotic solution 162
atom
– alkali 218
– anti-hydrogen 240
– carbon 229
– doubly excited 282
– exotic 237
– helium 201
– images 20
– interferometry 495
– kaonic 239
– laser 495
– multielectron 221
– myonic 238
– pionic 239
– quantum structure 111
– stability 117
– thermal He 101
– tin 228
atomic beam 44
atomic clock 503
atomic force microscope 29
atomic interference pattern 99
atomic mass unit (AMU) 10
atomic orbital 334
– 1$s$ deformation 339
– deformation 334
atomic polarizability 354
atomic spectrum 112
atomic unit 335
atomic weight 10
attenuation coefficient 273
Auger
– effect 234
– process 233
autoionization 233, 282, 378
*Autrecourt, N.* 8
average number of photons per mode 249
*Avogadro, A.* 11
Avogadro constant 12, 14, 17

$B_2$ molecule 345
Balmer's formula 116
Balmer series 172
band 374
– fundamental 405
– head 375
– origin 372
– parallel 406
– perpendicular 406
– system 374
– vibrational 372
beam
– atomic 44
– molecular 453
beam profile for the fundamental modes 300
BEC 493
*Beer* 456
Beer's law 456
bending vibration 402
benzene 392
*Bernoulli, D.* 9
beryllium 211
*Berzelius, J.J.* 10
binding
– covalent 356
– ionic 356
– types 356
binding energy 335, 338, 364
Biot–Savart's law 178
blue-shadowed 376
Bohr magneton 169
Bohr radius 115, 162

Bohr's atomic model 113
bolometer 434
*Boltzmann, L.* 8
Boltzmann distribution 249
bond
– chemical 349, 383
– covalent 350
– homopolar 350
bonding
– hydrogen 356
– localized 394
– localized $\sigma$-type 393
Born's statistical interpretation 123
boron 211
Bose–Einstein condensation 493
boson 205
bound-bound transition 282
boundary conditions 133
Bragg diffraction 280
Bragg reflection 100
bremsstrahlung 271, 280
Brownian motion 20

carbon 211
CARS 462
cathode rays 34
CCD array 424
center of mass 159
centrifugal distortion 361
chaotic movement 120
charge distribution 70
chemical laser 311
clamped nuclei 336
classical particle paths 120
classical wave description 95
*Clausius, R.J.* 8, 9
cloud chamber 24
cluster 408
– mass distribution 410
– melting temperature 411
– size 409
– temperature 411
$CO_2$-laser 312
$CO_2$ molecule 388
coherence length 304
coherent control 499
coincidence experiments 467
CO-laser 311
collision
– broadening 267
– central 68
– charge exchange 40
– cross section 421
– elastic 268
– inelastic 269
– phase-changing 269
– quenching 480
– radius 267
– reactive 421, 479
– time 267
– uenching 269
collision broadened emission line 268
collision complex 414
collision-induced rate 261
colour-center laser 305
Compton effect 91, 274
Compton wavelength 93
concept of nature 124
condition
– phase-matching 462
configuration
– electron 224
– interaction 224
– two-electron 231
contribution
– monopole 351
conversion efficiencies 315
coordinate
– elliptical 328
– inversion 341
– mass-weighted 400
Coriolis force 403
correlation diagrams 348
correspondence principle 194
Coulomb force 34
Coulomb-repulsion 337
coupling
– between vibrations and rotations 402
– fine structure energies 225
– intermediate cases 229
– $j$–$j$ 228
– $L$–$S$ 225
– of angular momenta 224
– schemes 224
– spin-orbit 207, 342
– terms 402
– vector diagram 228
covalent binding 356
cross section
– differential 65, 66, 69, 75
– integral 66, 69
– integral scattering 65
– threefold differential 467
crystal
– birefringent 310
– body-centered cubic 32
– face-centered cubic 32
– liquid 501
– uniaxial birefringent 315
cyclotron frequency 63

*Dalton, J.* 9, 10
damping constant 262, 493
De Broglie wavelength 97
deflection angle 72
degenerate 145, 165
delocalized 393
– $\pi$ orbitals 394
*Democritus* 7, 8
density
– effective 438
– energy 81
– mode 85
– momentum 82
– packing 32
– population 398
– power 83
– probability 106, 120
derivative
– first 438
detector 433
– Langmuir–Taylor 480
– phase-sensitive 445
– thermal 434
– time constant 434
determinant
– Slater 223
device
– photoconductive 435
– photovoltaic 435
diagram
– Slater 230
– Walsh 389
diatomic molecules
– vibrations 363
difference-frequency generation 316
diffraction
– atoms 98
– Bragg 280
– electron 97, 108

– losses 290, 296, 300
– X-ray 15
diffusion 30
– coefficient 14
– process 22
dipole
– approximation 252
– electric 351
– Hertzian 250
dipole moment
– induced 352
– momentary 353
– total 404
Dirac equation 159, 190
discharge
– hollow cathode 448
dispersion
– angular 424
– linear 424
– of FPI 433
– of the wave packet 109
displacement
– mass weighted 401
dissociation energy 364
distribution
– angular 467
– function 21
– Gaussian 73
– radial of atomic electrons 209
– temperature 463
Doppler broadening 264
Dunham coefficients 366
duo-plasmatron 41

eigenfunction 150
– common 155
eigenvalue 150
eigenvolume 29
Einstein coefficient 248
– for absorption 248
Einstein–de Haas effect 177
Einstein–Podolski–Rosen paradox 512
electric field
– perpendicular 46
electric quadrupole moment 258
electron 35
– configuration 224, 344
– delocalized 392
– diffraction 97, 108
– free 37
– independent 203, 221
– mass 41
– mechanical model 197
– motion 361
– multiplier 39
– optics 47
– Rydberg 235
– shell 209
– spin 204
electronic transition 372
electron microscope 25, 52, 102
electron model 195
electron radius
– classical 196
electron shell 168
electron spin 176
element
– alkali 216
– alkaline earth 216
– chemical properties 217
– piezo 27
– trans-uranium 216
elliptical coordinate 328
elliptical mirror arrangement 452
emission
– coefficient 37
– field 37
– induced 248
– secondary electron 38
– spectrum 112
– spontaneous 248
– thermal 37
*Empedocles* 8
energetic order
– of the orbitals 344
energy
– binding 335
– density 81
– dissociation 364
– equipotential curves 416
– fine structure coupling 225
– ionization 117, 212, 411
– mean rotational 364
– of Rydberg levels 236
– rotational 362
– surface 383
– transfer 291
– zero-point 139
energy density
– spectral 249
energy pump 289
*Epicurus* 8
equation
– uncoupled 401
equilibrium distances 360
Erbium laser 305
excess 432
excimer 312, 347
excitation
– inner-shell 233
– one-electron 231
– simultaneous 232
– single electron 232
exotic atom 237
expansion
– adiabatic 410
– Dunham 366
expansion of the universe 44
expectation value 149, 166
– for the location $x$ 149
– of the kinetic energy 150
experiment
– ideal 478

Fabry–Perot-etalon 310
factor
– Hönl–London 404, 406
fadenstrahlrohr 42
Fano
– profile 283
Fano parameter 283
Fano profile 283, 378, 379
Faraday's law 14
Faraday polarization rotator 310
femtosecond 316
Fermi-contact interaction 186
fermion 205
field
– axially symmetric 47
field ionisation 137
finesse 433
fine structure 178
– component 179, 344
– constant 181
– coupling energies 225
– labeling of component 225
– Sommerfeld's constant 180
– splittings 343
fine-structure component 278
fingerprint region 440
flashlamp
– helical 293

floating ball 45
fluctuation
– of the laser frequency 303
– phase 304
– technical 303
fluorescence 260
– laser-induced 450
focal length
– of a magnetic lens 51
– of a magnetic sector field 52
focusing
– double 57
– velocity-independent 55
force
– Coriolis 403
– Coulomb 34
– nuclear 76
– restoring 400
formaldehyde molecule 392
formula
– modified Rydberg 221
– Rydberg 234
– Schawlow–Townes 304
Fortrat-diagram 372, 375
Fourier limitation 310
Fourier transformation 441
Franck–Condon factor 373
Franck–Condon principle 376
Franck–Hertz experiment 118
free-electron laser 305
free spectral range 431
frequency
– cyclotron 63
– modulation 438
– tunnel 391
frequency selective optical elements 302
frequency spectrum
– of optical resonators 301
frequency spectrum of induced emission 295
Fresnel number 297
full-width at half-maximum 261
function
– amplitude 304
– deflection 477
– hyperbolic 413
– partition 398
– spherical harmonic 147
– symmetric spin 204

gain factor 290
gas
– discharges 34
– effusion 44
gas constant 12
gas laser 310
gate function 442
*Gay-Lussac, J.L.* 11
Gedanken-experiment 122
generator
– photocurrent 436
gerade 341
giant pulse 317
globar 440
*Gordon* 289
grating
– concave 428
– equation 427
– Rowland 465
grating equation 279
grating pair 322
gravitational redshift 94
greenhouse effect 369
ground state
– helium 206
– of the H atom 172
gyromagnetic ratio 176

$H_2$ molecule 335
$H_2AB$ molecule 392
$H_2O$ molecule 383
halogens 218
hard spheres 18, 70
Hartree–Fock method 224
Hartree method 222
$He_2$ molecule 345
$He_2^{2+}$ 345
heat
– conduction 30
– specific 13
Heisenberg's uncertainty principle 106
He–Ne laser 293
Hermitian polynomials 141
*Heraclitus* 8
hidden parameter 513
Holmium laser 305
Hönl–London factor 373
*Humboldt, A.v.* 11
hybridization 384
– $sp$ 385
– $sp^2$ 386
– $sp^3$ 386
hydrogen atom 164
hyperfine component 185
hyperfine constant 185
hyperfine structure 184, 455

imaging
– of electron beams 47
impact parameter 67, 77
induced dipole moment 352
induced emission 248
infrared-active 405, 461
inner-shell excitation 233
integral
– overlap 387
intensity 82, 95
– of a rotational line 397
– spectral 82
intensity distribution of the fundamental modes 299
interaction
– configuration 224
– dipole–dipole 187
– exchange 349
– Fermi-contact 186
– multipole 350
– octupole 354
– rotation-vibration 364
– van der Waals 353
– virtual 191
interference
– effects 477
– fractional order 432
– multiple-beam 430
interference pattern 96
interference phenomena 121
interferometer
– Fabry–Perot 431
– Mach–Zehnder 17
– Michelson 429, 440
interferometry
– atom 495
– neutron 100
– X-ray 16
inversion 290
inversion tunneling 138
ion 33
– free 39
– $H_2^+$ 327
– optics 45

– sources 40
– trapped 503
ionic contribution 339
ionization
– electron impact 39
– energy 117
– field 234
– thermal 40
ion multiplier 452
ionosphere 40
isomers 390
isotope 64
isotopic abundance 64

jellium model 410

kanalstrahlen 34
Kerr lens 320
– mode-locking 319
kinetic gas theory 17
Kirchhoff's diffraction theory 298
Kirchhoff's law 83
*Kolos* 339
K-shell 209

*Lamb, Willis* 457
Lamb dip 457
Lamb shift 191
Landé factor 178, 183
lanthanides 216, 218
Laplace operator 130
laser
– argon 311
– atom 495
– chemical 311
– CO 311
– $CO_2$ 312
– color-center 305, 306
– diode 307
– dye 308
– erbium- 305
– excimer 312
– fixed-frequency 304
– flashlamp-pumped dye 309
– free-electron 305
– gas 310
– He–Ne 293
– Holmium 305
– line width 303
– mode-locking 318
– neodymium 305
– neodyniumm-glass- 305
– pulsed 291
– $Q$-switched 316
– ruby flashlamp-pumped 292
– semiconductor 307
– single mode 301
– solid-state 305
– three level 293
– Titan-Sapphire 305, 320
– tunable diode 445
– ultrashort pulses 422
lattice constant 16
LCAO
– improvements 334
learning algorithm 501
Legendre's polynomials 147
Lenard–Jones potential 33, 268, 354, 476
lens
– converging 47
– electro-optic 47
– magnetic 50
leucht-electron 218, 247
*Leucippus* 7
level
– vibronic 306, 407
$Li_2$ molecule 345
lifetime 232
– effective 261
– mean 260, 413
light
– circular polarized 254
line
– kernel 262
– profile 261
– wings 262
linewidth
– natural 262, 263
liquid crystal 501
LISA 510
lithium 210
Littrow-grating 312
localization of the particle 106
lone pairs 388
Lorentzian profile 263
Lorentz transformation 179
L-shell 209
Lyman $\alpha$-line 266

*Mach, Ernst* 7
macroscopic measurements 422
magic numbers 410
magnetic moment
– nuclear 185
magnetization 177
magnetron
– motion 64
– movement 63
*Maiman* 289, 293
maser 391
mass
– absolute 53
– photon 94
– relative atomic 53
mass absorption coefficient 275
mass spectrometer 53
– quadrupole 61
– time-of-flight 58
mass spectrum 58
– $Na_n$ 60
Mathieu's differential equations 62
matrix elements 250, 368
matter wave 102
– function 104
*Maxwell, J.C.* 8, 9
Maxwell–Boltzmann distribution 265
mean free path length 29, 65
mean quadratic deviation 150
measuring process 123
medium
– active 289
mercury lamp 250
method
– ab-initio 222
– Hartree 222
– Hartree–Fock 224
metrology
– optical 501
microscope
– atomic force 29
– electron 25, 52, 102
– scanning electron 26
– transmission electron 25
– tunneling 27
microwave range 360
mode 84
– density 85
– fundamental 298
– of open resonator 297
– phase-coupled 319
– spectral density 85

- TEM 298
- transverse 299
model
- jellium 410
- semiclassical 116
- shell-structure 410
modified Rydberg formula 221
modulation
- frequency 438
- phase 445
modulator
- electro-optic 445
mole 12
- volume 12
molecular beam 453
molecular dynamics 416
molecular polarizability 474
molecule 11
- acetylene 386
- aromatic 392
- asymmetric rotor 399
- $B_2$ 345
- butadiene 392
- $CO_2$ 388
- diatomic 327
- formaldehyde 392
- $H_2AB$ 392
- $H_2$ 335
- $H_2O$ 383
- $He_2$ 345
- $Li_2$ 345
- $NH_3$ 390
- polyatomic 383
- rigid $H_2^+$ 328
- symmetric top 397
moment
- of inertia 394, 395
- principal 395
- quadrupole 351
*Morse, P.M.* 355
Morse potential 355, 365
MOT 491
M-shell 209
multielectron atom 221
multiplet 224
- structure 226
multiplicity 206, 231, 343
multipole interaction 350
multipole transition 257
myonium 241, 242

Na D-line 266
neodynium-glass-laser 305
neutron 65
- interferometry 100
- spectrometer 100
$NH_3$ molecule 390
noise equivalent input power NEP 433
noncrossing rule 348
nonlinear absorption 456
nonlinear optic 313
normalization constant 339
normal vibration 399, 401
nuclear force 76
nuclear $g$-factor 185
nuclear magneton 185

observable 150
octupole interaction 354
oil droplet 35
one-dimensional box
- energy 139
operator 149, 150
- commutable 151
- Hermitian 151
- of the angular momentum 152
- reflection 341
optic
- electron 47
- nonlinear 313
optical amplifier 306
optical fiber bundle 452
optical frequency mixing 316
optical pulse
- compression 321
- ultrashort 322
optical reflection coefficient 134
optical trapping 491
orbital
- antibonding 345
- bonding 345
- hybrid 384
- molecular 329, 331
- $sp$-hybrid 385
orbital angular momentum 161
orbital magnetic moment 169
order
- energetic 340
ortho-helium 209
oscillator
- harmonic 141, 363
- microwave 438
*Ostwald, W.* 7
overlap integral 332
overtone-transitions 371

$P$-transition 372
pair formation 274
para-helium 209
parity 256
*Parmenides* 8
particle
- free 130, 131
- in a box 138
particle aspects of light 95
particle model 81
particle wave 101
partition function 86
Paschen–Back effect 187
path
- elliptical 168
Pauli principle 205
penetration depth 134
periodic system 216
*Perrin, J.B.* 13
perturbations 421
phase grating 99
phase matching 314
- angle 315
phase velocity 102
photo-association 495
photodiode 435
photodissociation 498
photoeffect 37, 274
photoelectric effect 89
photoionization 40, 280, 281
photomultiplier 436
photon 86, 93
- mass 94
- spin 93
photon avalanche 291
photon number per mode 295
photon recoil 191, 487
photosphere 40
$\pi$-meson 239
Planck's constant 86
Planck's formula 249
Planck's radiation law 84, 86
planetary atom 236
planetary model 113

plasma 40
plasma-frequency 71
*Plato* 8
Pockels cell 317
point-like charge 197
pointing vector 82
Poisson distribution 95
polarizability
– atomic 354
– molecular 474
polarization
– beam splitter 317
– dielectric 313
polarized
– circularly 93
– linearly 93
population inversion 292
positronium 241
potential
– barrier 391
– difference 377
– effective 222
– interaction 268
– Lenard–Jones 33, 268, 354, 476
– monotonic 68
– Morse 355, 365
– parabolic 355
– spherically symmetric 145
– van der Waals 352
potential barrier 132
potential curves 333
potential energy
– curves 329
potential step 134
potential well 140
power
– resolving 442
– spectral resolving 423, 442
predissociation 366
pressure broadening 269
principle
– building-up 209
– Pauli 205
probability
– density 106, 120
process
– diffusion 22
– electrolytic 15
– photographic 412
– visual 463
product-ansatz 161
profile
– Fano 283
– Gaussian line 266
– line 269
– Lorentzian 263
– spectral 263
– Voigt 267
property
– optical 412
– symmetry 341
proton 65
*Proust, J.L.* 9
pump and probe experiments 417

$Q$-factor 296
$Q$-line 372
quality factor 295
quantization axis 154, 188
quantum bits 513
quantum defect 221
quantum electrodynamics 124, 159
quantum nondemolishing experiment 124
quantum number 163, 330
– angular momentum 153
– magnetic 153, 254, 255
– projection 342
– spin 256
– spin projection 330
quantum structure of atoms 111
quark 185
qubit 514

$R$-transition 372
radiation
– anti-Stokes 461
– blackbody 82
– cavity 83
– power 264
– recombination 284
– spectral 83
– Stokes 460
– synchrotron 463
– thermal 249
radiation field
– isotropic 253
radiative recombination 280
radical
– $NH_2$ 389
raisin cake model 71
*Raman, Chandrasekhara* 460
Raman active 461
Raman effect
– resonance 461
Ramsauer effect 140
random walk 73
rate constant 412
*Rayleigh* 423
Rayleigh–Jeans radiation law 85
reaction
– absolute rates 415
– chemical 412
– constant 412
– coordinate 414
– endothermic 415
– exchange 415
– exothermic 415
– first order 412
– inverse 415
– order 412
– rate 412
– second order 413
recombination
– radiation 284
– radiative 284
– rate 284
– three-body 284
– two-body 284
red-shadowed 376
reduced mass 360
reflection
– Bragg 100
– loss factor 296
– losses 290, 296
reflection coefficient 134, 136
reflector
– cylindrical 293
reflectron 60
refractive index 319
relative absorption 440
relative atomic mass 53
relative motion 159
– of electron and nucleus 161
relativistic correction 173
relativistic energy shift 173
relativistic mass correction 174
repulsive potential 333
resolving power 442
resonator
– enhancement 315
– open optical 296
– optical 289, 295

restoring force 400
ring-resonator 310
rotation 394
rotational barrier 366
rotational constants 361
rotational energy 362
rotational term values 359
rotation-vibration interaction 364
rotor
– asymmetric 396, 399
– rigid 359
– symmetric 395
– vibrating 364
Rowland arrangement 171
Rutherford
– atomic model 73
– scattering formula 74
Rydberg atom
– doubly excited 378
Rydberg constant 115, 116, 194
Rydberg formula 234
Rydberg state 234, 236

scattering
– classical 66
– electron 465
– inelastic 478
– losses 290
– plane 67
– potential 67
– rainbow 477
– reactive 479
– super-elastic 460
scattering coefficient 273
scattering experiment 70
*Schawlow, A.L.* 289, 391
Schawlow–Townes formula 304
Schrödinger's cat 513
Schrödinger equation 129
– time-dependent 130
secondary emission coefficient 38
second harmonic 314
selection rule 247, 253
– for harmonic oscillator 371
– parity 255
semiclassical model 116
sensitivity 422
– absolute 433
shell-structure model 410
shielding constant 202, 216
short laser pulses 316
signal-to-noise ratio 423
silver halide grains 412
single mode laser 301
size
– cluster 410
Slater determinant 336
Slater orbitals 340
solar constant 87
spatial distribution
– of the electron 166
spatial resolution 109
spectral gain profile 301
spectral intensity 82
spectral radiation 83
spectral resolving power 423, 442
spectral response 433
spectrograph
– parabola 54
spectrometer 423
– grating 426
– infrared 440
– ion-cyclotron-resonance 63
– neutron 100
– prism 425
spectroscopy
– cavity-ringdown 448
– electron 465, 468
– excitation 451
– Fourier transform 440
– infrared 440
– ionization 452
– Lamb dip 457
– laser-absorption 444
– microwave 437
– nonlinear laser 455
– optoacoustic 445
– optogalvanic 447
– photoacoustic 447
– photoelectron 467
– Raman 460
– saturation 456
– time-resolved 497
– two-photon 459
– ZEKE 469
spectrum
– absorption 112
– atomic 112
– continuous 280, 377
– continuous emission 285
– emission 112
– helium 208
– iron 228
– of CO 439
– of our sun 285
– photoelectron 467
spin
– absolute value 176
– electron 204
– magnetic moment 176
spin-orbit coupling 178
– constant 179
splitting
– inversion 391
squeezing 504
stable solutions 62
star
– binary 508
Stark-effect
– linear 473
– quadratic 474
Stark shift 439
Stark-spectroscopy 473
state
– doubly excited atomic 232
– electronic 340, 341
– entangled 512
– excited atomic 231
– excited molecular 346
– metastable 232
– metastable excited 264
– metastable He 294
– Rydberg 234, 236
– singlet 205
– triplet 205
statistical interpretation 105
statistical weight 249, 422
Stefan–Boltzmann's law 88
Stefan–Boltzmann constant 89
stellar atmospheres 267
Stern–Gerlach experiment 175
Stern–Vollmer plot 261
Stokes radiation 460
Stokes' law 36
storage ring DORIS 465
structure
– bipyramidal 387
– rotational 406
subshell 210
sum frequency 316
supra-liquid 495
surface
– equipotential 48

susceptibility
– electric 313
symmetric top 362
– molecules 397
– near 399
– oblate 396, 397
– prolate 396, 397
symmetry
– axis 394
– group 390
– operations 390
– selection rules 407
system
– four-level 294
– $\pi$-electron 392
– two-level 293

technique
– Doppler-free 453
temperature
– absolute 19
term assignment 188
term values 344
theory
– nonlocal 513
theory of the photo effect 90
thermal dissociation
– of $H_2$ 171
thermopile 434
Thomas factor 179
Thomson's atomic model 71
threshold condition 290
Titan-Sapphire laser 320
Titan-Sapphire-laser 305
TOF 58
– broadening 487
– design 60
total dipole moment 404
*Townes, C.H.* 289, 391
transformation
– Hadamar 514
transition
– bound-bound 282
– combination 405
– electronic 372
– forbidden 252
– fundamental 371
– magnetic dipole 257, 259
– matrix elements 367
– multipole 257
– overtone 405
– probability 247, 250
– quadrupole 257
– radiationless 480
– radiative 480
– spontaneous 248
– state dynamics 498
– two-photon 186, 259
– vibrational-rotational 369
transport
– coefficients 29
trap
– magneto-optical 491
tunnel effect 135, 366
two-dimensional box 144

ultrashort optical pulse 322
ultraviolet catastrophe 85
uncertainty relation for energy and time 110
ungerade 341

vacuum ultraviolet 171
valence method 336
van der Waals constant 354
van der Waals equation 29
van der Waals interaction 353
van der Waals potential 352
velocity 235
– group 104
velocity selector 476
vibration
– asymmetric stretch 402
– bending 402
– normal 399, 401
– symmetric stretch 402
vibrational constants 363
vibration period 143
virtual interaction 191
viscosity
– of a gas 30
Voigt profile 267, 455
volume
– atomic 212

wave
– gravitational 508
wave description 81
wave equation 130
wavefunction 163
– normalized 164
– of the harmonic oscillator 142
– parity 341
wave packet 103, 416, 497
– dispersion 109
– spread of the 132
– width 108
wavelength-tunable light source 304
which way experiment 510
Wien's law 88
*Wigner, Eugene* 348
work-function 37, 89

X-ray 270
– characteristic 271
– characteristic fluorescence 272
– diffraction 15
– fluorescence 278
– scattering 273
– total reflection 279
– wavelength 278

YAG 305
Young's double slit experiment 121

Zeeman component 169, 255
Zeeman effect 168
– anomalous 181
Zeeman splitting 170
*Zeiger* 289

Printed by Printforce, the Netherlands